ADVANCED LEVEL
PHYSICS

Sixth Edition

Books by M. Nelkon
Published by Heinemann
ADVANCED LEVEL PRACTICAL PHYSICS (*with J. Ogborn*)
ADVANCED LEVEL PHYSICS EXERCISES AND EXAMPLES
SCHOLARSHIP PHYSICS
SOLUTIONS AND TEST PAPERS (*A-level*)
REVISION NOTES IN PHYSICS
Book II. Optics, Waves, Sound, Heat, Properties of Matter
PRINCIPLES OF ATOMIC PHYSICS AND ELECTRONICS
ELEMENTARY PHYSICS, Book I and II (*with A. F. Abbott*)
MATHEMATICS OF PHYSICS (*with J. H. Avery*)
ELECTRICAL PRINCIPLES (*with H. I. Humphreys*)
ELECTRONICS AND RADIO PRINCIPLES (with *H. I. Humphreys*)

Published by Collins Educational
PRINCIPLES OF PHYSICS (*GCSE standard*)
EXERCISES IN ORDINARY LEVEL PHYSICS

Published by Edward Arnold
ELECTRICITY

Published by Blackie
HEAT

Books by P. Parker
Published by Arnold
ELECTRONICS

Cover Photograph

Close process control demonstrated by the Cobra laser robot cutting an egg-shell without damaging the membrane. (Courtesy of UKAEA, Culham Laboratory).

ADVANCED LEVEL
PHYSICS

Sixth Edition

MICHAEL NELKON
M Sc CPhys F InstP AKC
formerly Head of the Science Department
William Ellis School, London

PHILIP PARKER
M Sc F InstP AMIEE
Late Senior Lecturer in Physics
The City University, London

HEINEMANN
EDUCATIONAL

**Heinemann Educational a division of
Heinemann Educational Books Ltd
Halley Court, Jordan Hill, Oxford OX2 8EJ**

*Oxford London Edinburgh
Madrid Athens Bologna Paris
Melbourne Sydney Auckland
Ibadan Nairobi Harare Gaborone
Singapore Tokyo Portsmouth NH (USA)*

ISBN 0 435 68668 2
© M. Nelkon and Mrs. P. Parker 1958, 1964, 1970, 1977, 1982, 1987
First published as one volume 1958
Reprinted four times
Second edition 1964
Reprinted three times
Third edition (SI) 1970
Reprinted four times
Fourth edition 1977
Reprinted three times
Fifth edition 1982
Reprinted four times
Sixth edition 1987

94 95 96 97 98 17 16 15 14 13 12 11 10 9 8

*Illustrations drawn by
George Hartfield Illustrators, Cheam, Surrey*

*Designed by The Pen and Ink Book Co. Ltd, London
Set in 10/11 Monophoto Times by
Advanced Filmsetters (Glasgow) Ltd
Printed in England by Clays Ltd, St Ives plc*

Preface to Sixth Edition

In this edition the text has been revised to cover the common core of the new syllabuses of the major Examining Boards. The text has also been extended where possible to deal with some options or further topics but specialist books listed by the Boards must be consulted.

Important formulae, laws and main points have been highlighted throughout the text. This should help the student in his or her understanding and in revision. In addition, many worked examples on all branches of the subject have been given to illustrate basic points. The exercises contain a selection of updated questions.

It is hoped that the general treatment of all branches of the subject will continue to assist students who intend to specialise in a science or in medicine or in allied subjects.

Some of the more important points in the text are as follows:

Mechanics The new syllabus has allowed a fuller treatment of fundamental points, which should help the non-mathematical student, and many worked examples are preceded by an analysis of the problem. In dynamics, the emphasis is on vectors and components, on the relation between force and momentum and conservation of momentum, and on energy. In equilibrium of forces, the polygon of forces and couples have been discussed. In gravitational theory, field strength and potential have been applied to the earth-moon system. Rotational dynamics and fluid motion are now in a separate chapter.

Solid Materials has been extended and I am very much indebted to Dr. M. Crimes, head of science, Woodhouse Sixth Form College, London, for his valuable contribution.

Electricity Field strength and potential in the electric field, and their relation, have been compared with corresponding concepts in the gravitational field. Charge-discharge in *C-R* circuits is given prominence. In current electricity, circuits include Kirchhoff's laws and in electromagnetic induction the inductor-resistor circuit is fully discussed.

Optics Geometrical optics has been reduced in accordance with the new syllabuses but lenses and optical instruments have been included. Optical fibres and their applications in telecommunications have been added.

Waves The general properties of waves are fully covered. In diffraction of light, special consideration has been given to the resolving power of optical and radio telescopes.

Heat Gas equations and the kinetic theory are followed by an introduction to thermodynamics, including the Carnot cycle, entropy changes and their statistical interpretation.

Electronics A new chapter on basic analogue and digital circuits has been added to cover the new syllabuses. I am very much indebted to I. Lovat, senior physics master, Malvern College, Worcestershire, for his valuable contribution.

Atomic Physics There is now a more straight-forward account of energy levels and the wave-particle duality is again given prominence.

I am grateful to the following for their kind assistance with this edition: Dr. M. Crimes for new exercise questions; Professor R. S. Ellis, University of Durham, for arranging to provide a photograph showing the use of optical fibres in astronomy and to Dr. P. Gray, Anglo-Australian Observatory, for the photograph and caption; Professor R. S. Shaw, Royal Free Hospital, Hampstead, London, for a photograph illustrating the use of ultrasonics in medical physics. I am also indebted to Dr. M. Jaffar, Quaid-I-Azam University, Pakistan, and other correspondents abroad for helpful comments, and to A. Gee, Queens' College, Cambridge, for assistance.

I also acknowledge with thanks the expert advice and unfailing courtesy of the editor Shirley Cooley, of Stephen Ashton, Richard Gale, Louise Rice and Trevor Hook of the Publishers, and of George Hartfield for illustrations.

I am also very grateful to the following for their considerable assistance with preparation of the previous editions: J. H. Avery, formerly Stockport Grammar School: M. V. Detheridge, William Ellis School, London; S. S. Alexander, formerly Woodhouse School and The Mount School for Girls, Mill Hill, London; Dr. M. Crimes, Woodhouse School, London; Rev. M. D. Phillips, Ampleforth College, Ampleforth; Mrs. J. Pope, formerly Middlesex Polytechnic; C. F. Tolman, Whitgift School, Croydon; D. Deutsch, formerly Clare College, Cambridge; P. Betts, Barstable School, Basildon, Essex; R. D. Harris, Ardingly College, Sussex; R. Croft, The City University; M. P. Preston, Lewes; N. Phillips, Loughborough University; and Dr. L. S. Julien, University of Surrey.

Note to Reprint

In this reprint, I have taken the opportunity to add some basic matter on Energy and Telecommunications in a section on 'Further Topics' at the back and to include some recent A-level questions.

I am indebted to Dr. J. M. R. Graham, Department of Aeronautics, Imperial College of Science and Technology, London, for his expert advice on the Wind Power section, to Dr. D. Lane and Dr. R. J. Taylor, Energy Technology Support Unit, Harwell, for information on Wave Power, and to Dr. M. Crimes, Woodhouse Sixth Form College, London, and M. V. Detheridge, Highgate School, London, for their generous assistance with this section. I am also very grateful to K. Danks, Brunel Technical College, Bristol, for revision of points in the Materials section.

Publisher's Note

Since the first publication of *Advanced Level Physics*, the revisions for reprints and new editions have been undertaken by Mr. Nelkon owing to the death of Mr. Parker.

Contents

Part One: Mechanics, Solid Materials

1 Dynamics   3

Linear motion. Velocity. Acceleration. Vectors, scalars. Equations of motion. Free-fall, *g*. Graphs. Vector addition and subtraction. Components. Projectiles.

Laws of motion. Inertia, mass, weight. Force. Gravitational force. $F = ma$. Linear momentum. Impulse. Action, reaction in systems. Conservation of momentum. Inelastic and elastic collisions.

Work, energy power. Work calculations. Power of engine. Kinetic, and gravitational potential, energy. Gravitational momentum and energy changes. Dimensions.

2 Circular Motion, Gravitation, S.H.M.   48

Circular motion Angular speed. Acceleration in circle. Centripetal forces. Banked track. Conical pendulum. Bicycle motion.

Gravitation Kepler's laws. Newton's law of gravitation. *G* and measurement. Earth-moon system. Weightless. Earth satellites. Parking orbit. Earth density. Mass of Sun. Gravitation potential. Velocity of escape. Satellite potential and kinetic energy.

Simple harmonic motion Formula for acceleration, velocity, displacement. Graphs of kinetic and potential energy. Oscillations in spring-mass system. Potential and kinetic energy exchanges. Springs in series and parallel. Simple pendulum. Oscillation of liquid in U-tube.

3 Forces in Equilibrium, Forces in Fluids   94

Forces in equilibrium Adding forces—vector and parallelogram methods. Resolved components. Couple and moment. Triangle and polygon of forces. Equilibrium of forces. Centre of mass, centre of gravity.

Forces in fluids Pressure formula. Atmospheric pressure. Density. Archimedes' principle and proof. Flotation. Stokes' law. Terminal velocity.

4 Further Topics in Mechanics and Fluids   114

Rotational Dynamics Torque and angular acceleration. Rotational kinetic energy. Work done in rotation. Angular momentum. Conservation of angular momentum and applications. Rolling motion down inclined plane.

Fluid Motion Bernoulli principle. Filter pump, aerofoil lift, carburettor, Venturi meter. Pitot-static tube principle.

5 Elasticity, Molecular Forces, Solid Materials 133

Elasticity Proportional and elastic limits. Hooke's law. Elastic limit. Yield point. Breaking stress. Young modulus measurement. Force in bar. Energy stored. Energy per unit volume.

Molecular Forces Molecular separation. Intermolecular forces and potential. Properties of solids—Hooke's law, thermal expansion. Latent heat of vaporisation. Bonds between atoms and molecules.

Solid Materials Crystalline, amorphous, glassy, polymeric solids. Imperfections in crystals. Dislocations. Elastic and plastic deformation. Ductile and brittle substances. Slip plane. Work hardening. Annealing. Cracks and effect. Toughness and hardness. Composite materials. Polymers. Structure and mechanical properties. Branching and cross-linking. Thermosets and thermoplastics. Hysteresis of rubber. Wood.

Part Two: Electricity

6 Electrostatics 177

Coulomb's law. Permittivity. Field strength (intensity) and field patterns. Point charge, sphere, parallel plates. Gauss's law and applications. Potential and values for conductors. Potential gradient. Relation to field strength. Equipotentials.

7 Capacitors 212

Parallel-plate formula. *C* by vibrating reed and high impedance voltmeter. Dielectric and relative permittivity and polarisation. Capacitors in series and parallel. Connected capacitors. Energy formulae. Charging and discharging in *C-R* circuit. Time-constant.

8 Current Electricity 241

Current Electricity Charge carriers in materials. $I = nAve$. Ohmic and non-ohmic conductors. Circuit laws—series and parallel. Kirchhoff's laws. E.m.f., internal resistance. Terminal p.d. Maximum power in load.

Energy, Power. Heating effect of current. Thermoelectricity.

9 Measurements by Potentiometer and Wheatstone Bridge 280

Potentiometer Principle of comparing and measuring e.m.f. and p.d. Calibration of ammeter and voltmeter. Comparison of resistances. Thermoelectric e.m.f.

Wheatstone bridge Network and relation between resistances. Resistivity. Temperature coefficient of resistance. Thermistor and application.

10 Magnetic Field and Force on Conductor · 301

Magnetic Fields Magnetic flux and flux-density B of solenoid, straight conductor, narrow circular coil.
Force on Conductor $F = BIl$. Torque $= BANI$. Application to moving-coil meter, simple motor. Force on charge, $F = Bev$. Hall voltage. Hall probe applications.

11 Magnetic Fields of Current-Carrying Conductors · 323

Value of B for infinite and finite solenoid. Straight conductor. Narrow circular coil. Helmholtz coils. Fields and force between parallel currents. Ampere definition. Ampere balance. Biot-Savart law for B-values. Ampere's law.

12 Electromagnetic Induction · 342

Flux linkage. Faraday, Lenz laws. $E = Blv$. Dynamo, transformer. Lorenz method for R. Charge and flux linkage. Current in L, R circuit. Self and mutual inductance. Energy in coil.

13 A.C. Circuits · 383

Peak, r.m.s. values in sine and square-wave voltages. Single components, R,C,L—phase. Series circuits—impedance, phase. Resonance in L,C,R circuit. Power in circuits. Parallel circuit. Bridge rectifier.

Part Three: Geometrical Optics, Waves, Wave Optics, Sound Waves

14 Geometrical Optics · 411

Reflection—spherical and paraboloid mirrors. Refraction at plane surface. $n \sin i = $ constant. Total internal reflection, critical angle. Refraction through Prism. Maximum and minimum deviation. Spectrometer. Dispersion by prism.
Optical fibres. Monomode, multimode, optical paths. Maximum incidence. Absorption and dispersion. Conversion of light signal to sound.

15 Lenses, Optical Instruments · 435

Lenses Converging, diverging lenses. Images. Calculations.
Optical Instruments Refractor and reflector telescopes. Magnifying power. Eye-ring. Simple microscope.

16 Further Topics in Optics 456

Compound microscope, lens camera—f-number, depth of field. Spherical and chromatic aberrations.

17 Oscillations and Waves 465

Oscillations, resonance, phase. Ultrasonics. Longitudinal, transverse waves. Wave speed. Progressive and stationary (standing) waves in sound and light. Reflection, refraction, diffraction, interference, polarisation. Electromagnetic wave spectrum and properties. Speeds of matter and electromagnetic waves.

18 Wave Theory of Light, Speed of Light 501

Wave Theory Huygens' principle. Reflection and refraction at surfaces. Refractive index and speed. Critical angle on wave theory.
 Speed of light—Foucault, speed in liquid, Michelson method.

19 Interference of Light Waves 515

Principle of superposition. Coherent sources. Phase. Young's two-slit fringes. Air-wedge fringes. Newton's rings. Blooming of lens. Colours in thin films.

20 Diffraction of Light Waves 539

Diffraction at single slit. Intensity variation. First minimum. Resolving power of telescope objective.
 Multiple slits. Diffraction grating and orders. Wavelength by diffraction grating. Hologram principle.

21 Polarisation of Light Waves 561

Transverse waves. Plane-polarisation by Polaroid. Polarisation by reflection. Brewster angle. Double refraction. Nicol prism. Electric vector and polarisation. Applications—stress-analysis.

22 Characteristics of Sound Waves 572

Pitch, loudness, quality. Intensity of sound. Beats and application. Doppler principle in sound—linear and circular motion. Doppler principle in light. Red shift.

23 Waves in Pipes and Strings 590

Stationary (standing) waves. Nodes and antinodes. Waves in pipes. Displacement, pressure. Fundamental, overtones. End-correction. Speed of sound in pipe.

Waves in strings. Fundamental and overtones. Sonometer. A.c. mains frequency. Longitudinal wave in string.

Part Four: Heat

24 Introduction: Temperature, Heat, Energy 619

Temperature, Heat, Energy. Temperature—types of thermometers and scales. Heat and temperature. Zeroth law. Heat and energy. Conservation of energy.

25 Thermometry 625

Fixed points, triple point. Gas thermometer and standard temperature. Constant volume gas thermometer. Thermoelectric thermometer. Platinum resistance thermometer. Calculations of temperature. Other thermocouples.

26 Heat Capacity, Latent Heat 635

Heat capacity. Specific heat capacity. Measurement by electrical method. Continuous flow method. Advantages. Newton's law of cooling. Heat loss and temperature fall. Cooling correction.

Specific latent heat. Electrical method, method of mixtures.

27 Gas Laws, Thermodynamics, Heat Capacities 651

Boyle's law. Volume, pressure and temperature. Ideal gas law. $pV = nRT$. Connected gas containers. Dalton's law. Unsaturated and saturated vapours.

Thermodynamics. Work done by gas. Internal energy of gas. First law of thermodynamics. Work from graphs. Reversible isothermal and adiabatic changes.

Heat capacities at constant volume and pressure. Enthalpy.

28 Kinetic Theory of Gases 683

Assumptions for ideal gas. Pressure formula. Root-mean-square speed. Temperature and kinetic theory. Boltzmann constant. Graham's law of diffusion. Maxwell distribution of molecular speeds.

29 Transfer of Heat: Conduction and Radiation 698

Conduction and temperature gradient. Thermal conductivity. Electrical analogy. Temperature gradient in good and bad conductors. Measuring conductivity of good and bad conductors. Double glazing.

Radiation and infra-red rays. Use of thermopile for absorption and radiation. Black-body radiation. Energy distribution among wavelengths. Wien law. Stefan law and applications. Sun's temperature.

30 Further Topics in Heat 732

Kinetic Theory. Real Gases. Thermodynamics. Radiation. Kinetic Theory and degrees of freedom. Values of molecular heat capacities and ratios. Mean free path. Viscosity.

Real gases and critical phenomena. Andrews' experiments on p-V curves for carbon dioxide. Critical temperature. Real gas laws. Boyle temperature. Joule-Kelvin effect. Van der Waals equation.

Part Five: Electrons, Electronics, Atomic Physics

31 Electron Motion in Fields, Cathode-Ray Oscilloscope 757

The Electron. Oil-drop experiment, Millikan experiment. Thermionic emission. Properties of electron. _Electron motion in magnetic field_. Circular path, applications. _Electron motion in electric field_. Parabolic path, energy gain. Charge-mass ratio measurements. Electron mass. Helical path of electrons.

Cathode-ray oscilloscope. Voltage supplies. Time-base. Focusing. Controls. Uses for a.c. voltage, frequency, phase, clock.

32 Junction Diode. Transistor and Applications 786

Energy bands in solids. Electron, hole carriers. Intrinsic, extrinsic semi-conductors. Effect of temperature. _P-n junction_, Barrier p.d. Rectification. Bridge rectifier. Zener diode.

Transistor, n-p-n, p-n-p. Current flow. Common-emitter characteristics. Amplifier. Current gain. Phase change. Saturation, cut-off. Transistor switch. Use as amplifier. Logic gate.

33 Analogue and Digital Electronics 809

Analogue electronics. Voltage gain. Non-inverting and inverting amplification of Opamp. Characteristics of Opamp—impedance, saturation, virtual earth. Power supplies. Off-set voltage. Frequency characteristic. Closed loop gain. Summing amplifier. Positive feedback. Square-wave oscillator. Astable multivibrator. Sine wave oscillator. Voltage comparator. Switching circuits. Integrator.

Digital electronics. Logic gates—NOT, AND, NAND, OR, NOR. Use of NAND gates. EOR, ENOR gates. Heating control. Half-Adder. Full-Adder. Bistable. Clocked SR flip-flop. Binary four-bit counter.

34 Photoelectricity, Energy Levels, X-Rays, Wave-Particle Duality 844

Photoelectricity. Demonstration. Intensity. Threshold value. Wave theory defect. Einstein photoelectric theory. Photons. Stopping potential. Experiment for Planck constant. Photocells.

Energy levels of hydrogen. Excitation. Ionisation potential. Frequency values in spectrum. Emission, absorption. Fraunhofer lines. Laser principle. Bohr's theory of hydrogen atom.

X-rays. Nature and properties. Crystal diffraction. Bragg law. Moseley law. X-ray spectrum. Minimum wavelength value. Absorption spectra.

Wave-particle duality. Electron diffraction. De Broglie formula. Momentum and energy. Duality. Compton effect.

35 Radioactivity, Nuclear Energy 879

Geiger-Müller tube. Count rate, voltage characteristic. Quenching. Solid state detector. Counter. Ratemeter. Experiments on alpha particles, beta particles and gamma rays. Nature of particles and rays. Half-life. Decay constant. Measuring long and short half-lives. Carbon dating. Cloud chambers.

Nucleus. Geiger-Marsden experiment. Scattering law. Protons, neutrons. Radioactive disintegration. Nuclear reactions. Mass spectrometer. Isotopes.

Nuclear energy. Einstein mass-energy relation. Mass unit. Binding energy, nuclear forces. Nuclear stability. Energy from isotopes. Nuclear fusion. Chain reaction. Fission. Thermonuclear reaction.

36 Further Topics 918

Nuclear energy. Fossil fuels. Geothermal energy. Wind power. Tidal Power. Wave Power.

Optical fibre telecommunications. LED. Semiconductor laser. Photodiode. Radio telecommunications. Aerials and standing waves. Types of aerials. Amplitude modulation. Sidebands. AM radio receiver.

Answers to Exercises 929

Index 937

Acknowledgements

Thanks are due to the following Examining Boards for their kind permission to reprint past questions. Answers are the sole responsibility of the author, and not of the appropriate examining boards.

London University School Examinations (*L.*)
Oxford and Cambridge Schools Examination Board (*O. & C.*)
Joint Matriculation Board (*JMB.*)
Cambridge Local Examinations Syndicate (*C.*)
Oxford Delegacy of Local Examinations (*O.*)
Associated Examining Board (*AEB.*)
Welsh Joint Education Committee (*W.*)

I am indebted to the following for kindly supplying photographs and permission to reprint them:

The late Lord Blackett and the Science Museum, Fig. 35.18; Head of Physics Department, The City University, London, Figs. 19.14, 20.2, 21.5; Dr. B. H. Crawford, National Physical Laboratory, Fig. 17.20; R. Croft, The City University, Figs, 19.6, 20.4, 20.17; Hilger and Watts Limited, Figs. 19.11, 19.15, 20.9, 20.12; National Chemical Laboratory, Fig. 34.22(i); N. Phillips, Loughborough University, Fig. 20.20; late Sir G. P. Thomson and the Science Museum, Fig. 34.22(ii); late Sir J. J. Thomson, Fig. 35.19; United Kingdom Atomic Energy Authority, Figs. 6.8, 35.24; The Worcester Royal Porcelain Company Limited and Tom Biro, Fig. 29.26.

Part 1
Mechanics. Solid Materials

1

Dynamics

Linear Motion, Vectors, Free-fall, Graphs, Projectiles

Dynamics *is the science of motion. It deals with the velocity, acceleration, force and energy of large objects such as cars and aeroplanes and tiny objects such as the electrons in your television set which produce the pictures. Dynamics also helps in investigations on the motion of athletes or the motion of a ball bowled in cricket or hit in golf.*

Before you play a game like football or tennis, you need to learn the basic skills and how to apply them. In the same way, we start with the main points in dynamics and show how they are applied in velocity, acceleration, free-fall in gravity and motion graphs.

Plate 1A *Steve Cram of England winning the 1500 metres in a new world record time of 3 min 25.67 s at Nice, France. Said Aouita of Morocco is second.*

Associated Press Ltd

Motion in Straight Line, Velocity

If a runner, moving in a straight line, takes 10 s to run 100 m, the average *velocity* in this direction

$$= \frac{\text{distance}}{\text{time}} = \frac{100\,\text{m}}{10\,\text{s}} = 10\,\text{m/s or } 10\,\text{m s}^{-1}$$

The term 'displacement' is given to the distance moved in a constant direction, for example, from L to C in Figure 1.1 (i). So

---

velocity is the rate of change of displacement.

---

or 'change in displacement/time taken'.

Velocity can be expressed in *metre per second* (m s^{-1}) or in *kilometre per hour* (km h^{-1}). By calculation, $36\,\text{km h}^{-1} = 10\,\text{m s}^{-1}$.

If a car moving in a straight line travels equal distances in equal times, no matter how small these distances may be, the car is said to be moving with *uniform* velocity. The velocity of a falling stone increases continuously, and so is a *non-uniform* velocity.

If, at any point of a journey, Δs is the small change in displacement in a small time Δt, the velocity v is given by $v = \Delta s / \Delta t$. In the limit, using calculus notation,

$$v = \frac{ds}{dt}$$

We call ds/dt the *instantaneous velocity* at the time or place concerned. The term 'mean velocity' refers to appreciable or finite times and finite distances.

Vectors and Scalars

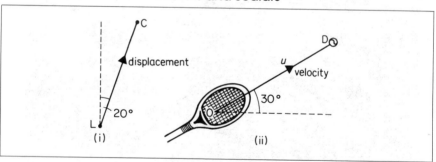

Figure 1.1   *Vectors*

Displacement and *velocity* are examples of a class of quantities called *vectors* which have both magnitude and direction. They may therefore be represented to scale by a line drawn in a particular direction. For example, Cambridge is 80 km from London in a direction 20° E. of N. We can therefore represent the displacement between the cities in magnitude and direction by a straight line LC 4 cm long 20° E. of N., where 1 cm represents 20 km, Figure 1.1 (i). Similarly, we can represent the velocity u of a ball leaving a racket at an angle of 30° to the horizontal by a straight line OD drawn to scale in the direction of the velocity u, the arrow on the line showing the direction, Figure 1.1 (ii).

Unlike vectors, *scalars* are quantities which have magnitude but no direction. A car moving along a winding road or a circular track at 80 km h^{-1} is said to have a *speed* of 80 km h^{-1}. 'Speed' is a quantity which has no direction but only magnitude, like 'mass' or 'density' or 'temperature'. These quantities are examples of scalars.

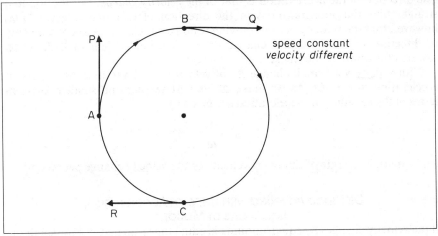

Figure 1.2 *Velocity (vector) and speed (scalar)*

The distinction between speed and velocity can be made clear by considering a car moving round a circular track at say $80 \, \text{km h}^{-1}$, Figure 1.2. At every point on the track the *speed* is the same—it is $80 \, \text{km h}^{-1}$. At every point, however, the *velocity* is different. At A, B or C, for example, the velocity is in the direction of the corresponding tangent AP, BQ or CR if we 'freeze' the motion of the car. So even though they have the same magnitude or size, the three velocities are all different because they point in different directions.

Acceleration in Linear Motion

In a 100 metres race, sprinters aim to increase their velocity to a maximum in the shortest time, so they *accelerate* from starting blocks. This acceleration can have a very high value.

The acceleration, symbol a, of an object is defined as the *rate of change of velocity* or

$$a = \frac{\textbf{velocity change}}{\textbf{time taken}}$$

So if a car accelerates from $15 \, \text{m s}^{-1}$ to $35 \, \text{m s}^{-1}$ in 5 s, then

$$\text{average acceleration, } a = \frac{(35-15) \, \text{m s}^{-1}}{5 \, \text{s}} = 4 \, \text{m s}^{-2}$$

Note carefully that the time unit for acceleration is 's^{-2}' because the time 'second' is repeated twice.

A car moving with constant (uniform) velocity has zero acceleration, from the definition. If a car brakes, its velocity may decrease from $30 \, \text{m s}^{-1}$ to $20 \, \text{m s}^{-1}$ in 5 s. In this case the car has a retardation or deceleration, or, mathematically, a *negative* acceleration. So

$$a = -\frac{30-20}{5} = -2 \, \text{m s}^{-2}$$

Note carefully that

acceleration is a vector.

The direction of the acceleration is that of the *velocity change*. For motion in a straight line, the acceleration is in the direction of that line. We see later, however, that the velocity of a car moving round a circular track keeps changing in direction and the velocity change or acceleration is then in a direction towards the centre of the track.

With a finite very small change Δv of velocity in a finite time Δt, the mean acceleration $a = \Delta v/\Delta t$. As we make Δv and Δt smaller and smaller, then, in terms of the calculus, the acceleration a is given by

$$a = \frac{dv}{dt}$$

where dv/dt is the rate of change of velocity or the velocity change per second.

Distance travelled with Uniform Acceleration, Equations of Motion

If the velocity changes by equal amounts in equal times, no matter how small the time-intervals may be, the acceleration is said to be *uniform*. Suppose that the velocity of a car moving in a straight line with uniform acceleration a increases from a value u to a value v in a time t. Then, from the definition of acceleration,

$$a = \frac{v-u}{t}$$

from which
$$v = u + at . \qquad . \qquad . \qquad . \qquad . \qquad (1)$$

Suppose a train with a velocity u accelerates with a uniform acceleration a for a time t and attains a velocity v. The distance s travelled by the object in the time t is given by

$$s = \text{average velocity} \times t$$
$$= \tfrac{1}{2}(u+v) \times t$$

But
$$v = u + at$$
$$\therefore s = \tfrac{1}{2}(u+u+at)t$$
$$\therefore s = ut + \tfrac{1}{2}at^2 \qquad . \qquad . \qquad . \qquad . \qquad . \qquad (2)$$

Also, $t = (v-u)/a$ from (1), then

$$s = \text{average velocity} \times t = \frac{(v+u)}{2} \times (v-u)/a$$

$$= (v^2 - u^2)/2a$$

Simplifying,
$$v^2 = u^2 + 2as \qquad . \qquad . \qquad . \qquad . \qquad (3)$$

Equations (1), (2), (3) are the equations of motion of an object moving in a straight line with uniform acceleration. When an object undergoes a uniform *deceleration* or *retardation*, for example when brakes are applied to a car, a has a *negative* value.

Examples on Equations of Motion

1   An aeroplane lands on the runway with a velocity of $50\,\mathrm{m\,s^{-1}}$ and decelerates at $10\,\mathrm{m\,s^{-2}}$ to a velocity of $20\,\mathrm{m\,s^{-1}}$. Calculate the distance travelled on the runway.

(*Analysis*   No time is mentioned. So we use $v^2 = u^2 + 2as$.)
    Here $u = 50\,\mathrm{m\,s^{-1}}$, $v = 20\,\mathrm{m\,s^{-1}}$, $a = -10\,\mathrm{m\,s^{-2}}$.

Using the formula $v^2 = u^2 + 2as$ to find s, then

$$20^2 = 50^2 + (2 \times -10 \times s) = 50^2 - 20s$$

So

$$400 = 2500 - 20s$$

$$s = \frac{2500 - 400}{20} = \frac{2100}{20} = 105\,\text{m}$$

2 A car moving with a velocity of $15\,\text{m s}^{-1}$ accelerates uniformly at the rate of $2\,\text{m s}^{-2}$ to reach a velocity of $20\,\text{m s}^{-1}$.

Find   (i) the time taken,   (ii) the distance travelled in this time.

(*Analysis*   (i) We need time t. So we use $v = u + at$.   (ii) We need distance s. Knowing t, we can use $s = ut + \frac{1}{2}at^2$ or, without t, use $v^2 = u^2 + 2as$)

(i) Using

$$v = u + at$$

$$\therefore 20 = 15 + 2t$$

$$\therefore t = \frac{20 - 15}{2} = 2\cdot5\,\text{s}$$

(ii) Using

$$s = ut + \frac{1}{2}at^2$$

$$s = (15 \times 2\cdot5) + \frac{1}{2} \times 2 \times 2\cdot5^2$$

$$= 37\cdot5 + 6\cdot25 = 43\cdot75\,\text{m}$$

Motion Under Gravity, Free-fall

When an object falls to the ground under gravitational pull, experiment shows that the object has a constant or uniform acceleration of about $9\cdot8\,\text{m s}^{-2}$ or $10\,\text{m s}^{-2}$ approximately, while it is falling. The numerical value of this acceleration is usually denoted by the symbol g. Drawn as a vector quantity, g is represented by a straight vertical line with an arrow on the line pointing downwards, Figure 1.3 (i).

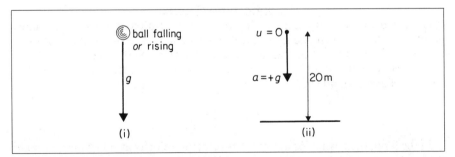

Figure 1.3   *Motion under gravity—free-fall*

Suppose that a ball is dropped from a height of 20 m above the ground, Figure 1.3 (ii). Then the initial velocity $u = 0$, and the acceleration $a = g = 10\,\text{m s}^{-2}$ (approx.). When the ball reaches the ground, $s = 20\,\text{m}$. Substituting in $s = ut + \frac{1}{2}at^2$, then

$$s = 0 + \frac{1}{2}gt^2 = 5t^2$$

$$\therefore 20 = 5t^2 \quad \text{or} \quad t = 2\,\text{s}$$

So the ball takes 2 seconds to reach the ground.

If a cricket-ball is thrown vertically upwards, it slows down owing to the attraction of the earth. The magnitude of the deceleration is $9.8 \, \text{m s}^{-2}$, or g. Mathematically, a deceleration can be regarded as a negative acceleration in the direction along which the object is moving; and so $a = -9.8 \, \text{m s}^{-2}$ in this case.

Examples on Motion under Free-fall (Gravity)

1 A ball is thrown vertically upwards with an initial velocity of $30 \, \text{m s}^{-1}$. Find (i) the time taken to reach its highest point, (ii) the distance then travelled. (Assume $g = 10 \, \text{m s}^{-2}$.)

(*Analysis* (i) We need time t. So we can use $v = u + at$. (ii) We need distance s. So we can use $s = ut + \frac{1}{2}at^2$.)

(i) Here $u = 30 \, \text{m s}^{-1}$, $v = 0$ at highest point, since ball momentarily at rest, $a = -g = -10 \, \text{m s}^{-2}$. From $v = u + at$,

$$0 = 30 + (-10)t \quad \text{or} \quad 10t = 30 \text{ and so } t = 3 \, \text{s}$$

(ii) $$\text{Distance } s = ut + \frac{1}{2}at^2$$

$$= (30 \times 3) + \frac{1}{2} \times (-10) \times 3^2$$

$$= 90 - 45 = 45 \, \text{m}$$

2 A lift is moving down with an acceleration of $3 \, \text{m s}^{-2}$. A ball is released $1.7 \, \text{m}$ above the lift floor. Assuming $g = 9.8 \, \text{m s}^{-2}$, how long will the ball take to hit the floor?

(*Analysis* We need time t. As we have distance s, we can use $s = ut + \frac{1}{2}at^2$.)
Acceleration of ball relative to lift, $a = 9.8 - 3 = 6.8 \, \text{m s}^{-2}$.
Here $u = 0$, $a = 6.8 \, \text{m s}^{-2}$, $s = 1.7 \, \text{m}$. From $s = ut + \frac{1}{2}at^2$

$$1.7 = 0 + \frac{1}{2} \times 6.8 \times t^2 = 3.4 \, t^2$$

So $$t^2 = 1.7/3.4 = 0.5. \text{ Then } t = \sqrt{0.5} = 0.7 \, \text{s}$$

Distance-Time Graphs

When the distance, s of a car moving in a constant direction from some fixed point is plotted against the time t, a *distance-time* (*s-t*) *graph* of the motion is obtained. The velocity of the car at any instant is given by the change in distance per second at that instant. So at E in Figure 1.4, if the change in distance s is Δs and this change is made in a time Δt,

$$\text{velocity at E} = \frac{\Delta s}{\Delta t}$$

In the limit, then, when Δt approaches zero, the velocity at E becomes equal to the *gradient of the tangent to the curve* at E. Using calculus notation, $\Delta s / \Delta t$ then becomes equal to ds/dt (p. 4). So the gradient of the tangent at E is the instantaneous velocity at E.

Velocity at E = gradient of *s-t* graph at E

If the distance-time graph is a straight line CD, the gradient is constant at all points; so the car is moving with a *uniform* velocity, Figure 1.4. If the distance-time graph is a curve CAB, the gradient varies at different points. The car then moves with non-uniform velocity. At the instant corresponding to A the velocity is zero, since the gradient at A of the curve CAB is zero.

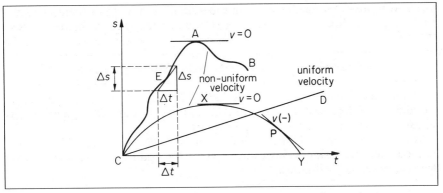

Figure 1.4   *Distance* (s)–*time* (t) *graphs* (*constant direction*)

When a ball is thrown upwards, the graph of the height s reached at any instant t is represented by the parabolic curve CXY in Figure 1.4. The gradient at X is zero, illustrating that the velocity of the ball at its maximum height is zero. From C to X the curve has a positive gradient (velocity upwards). From X to Y, as at P, the gradient is negative (velocity downwards).

Velocity-Time Graphs, Acceleration and Distance

When the velocity of a moving train is plotted against the time, a 'velocity-time (v-t) graph' is obtained. Useful information can be deduced from this graph, as we shall see shortly.

If the velocity is uniform, the velocity-time graph is a straight line parallel to the time-axis, as shown by line (1) in Figure 1.5. If the train increases in velocity steadily from rest, the velocity-time graph is a straight line, line (2), inclined to the time-axis. If the velocity change is not steady, the velocity-time graph is curved. In Figure 1.5, the velocity-time graph OAB represents the velocity of a train starting from rest which reaches a maximum velocity at A, and then comes to rest at the time corresponding to B.

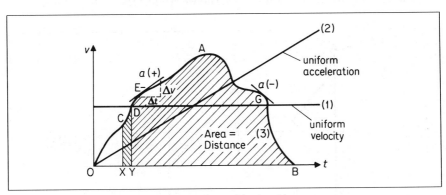

Figure 1.5   *Velocity* (v)–*time* (t) *curves*

Acceleration is the 'rate of change of velocity', that is, the change of velocity per second. The gradient to the curve at any point such as E is given by:

$$\frac{\text{velocity change}}{\text{time}} = \frac{\Delta v}{\Delta t}$$

where Δv represents a small change in v in a small time Δt. In the limit, the ratio $\Delta v / \Delta t$ becomes dv/dt, using calculus notation, or the gradient of the tangent at E. So

---

acceleration at E = gradient of velocity-time graph at E

At the peak point A of the curve OAB the gradient is zero, that is, the acceleration is then zero. From O to A, the gradient at any point such as E is upward or positive, so the train is accelerating.

At any point, such as G, between A, B the gradient to the curve is negative because the graph slopes downwards. Here the train has a *deceleration* or decrease in velocity with time.

Distance Travelled

We can also find the distance travelled from a velocity-time graph. In Figure 1.5, suppose the velocity increases in a very small time-interval XY from a value represented by XC to a value represented by YD. Since the small distance travelled = average velocity × time XY, the distance travelled is represented by the *area* between the curve CD and the time-axis, shown shaded in Figure 1.5. By considering every small time-interval between OB in the same way, it follows that

---

distance = AREA between *v-t* graph and time-axis

This result applies to any velocity-time graph, whatever its shape.

Example on Graphs

A rubber ball is thrown vertically upwards from the ground and falls on a horizontal smooth surface at the ground. The ball then bounces up and down with decreasing velocity.
   (a) Draw the velocity-time graph of its motion.
   (b) From your graph, show how the maximum height of the ball can be found when it is initially thrown up and the distance it falls when it first reaches the ground.

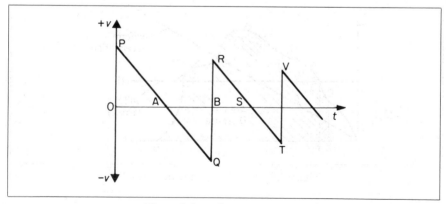

Figure 1.6   *Velocity–time graph*

(a) The graph is shown in Figure 1.6. The ball is thrown up with a velocity which we shall take as a positive velocity. As the ball rises, its velocity decreases to zero at A along the straight line PA.

Having reached its maximum height, the ball now falls. Its downward velocity is a negative one and it falls with the same numerical value of acceleration g. So the gradient of AQ is the same as PA and PAQ is a straight line.

At Q, the ball is about to hit the ground. A moment later the ball rebounds and its velocity is now high and positive. So the rebound velocity is represented by BR, where QBR is very nearly a vertical line. The velocity now varies along RSTV as explained. The lines PQ and RT are parallel because their gradients are equal to the acceleration g. The rebound velocity decreases as the ball continues to bounce, as shown.

(b) The maximum height above the ground is the area of triangle OAP. The height or distance s it falls is the area of triangle AQB, since AQ is the velocity-time graph as the ball falls.

Vector Addition

Two vectors such as a velocity of $3\,\text{m}\,\text{s}^{-1}$ and a velocity of $4\,\text{m}\,\text{s}^{-1}$ can not be added without taking into account their direction.

As an example, suppose a ship is moving with a velocity of $4\,\text{m}\,\text{s}^{-1}$ in a direction OA relative to the sea and a girl runs across the deck with a velocity of $3\,\text{m}\,\text{s}^{-1}$ in a direction OB at an angle of $60°$ to the ship's velocity. In one second the ship moves a distance OA which represents 4 m according to some scale and the girl then moves in the direction OB a distance of 3 m. So the *resultant* velocity of the girl is in some direction OC between OA and OB.

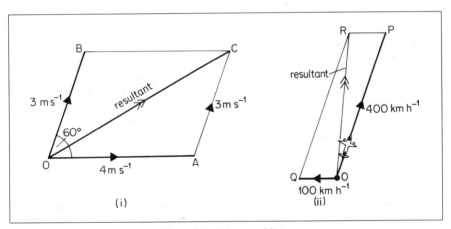

Figure 1.7 *Vector addition*

Two vectors can be added by drawing a *parallelogram* of the vectors. In Figure 1.7 (i), we draw a line OA to represent to scale $4\,\text{m}\,\text{s}^{-1}$ and then draw OB at an angle of $60°$ to OA to represent $3\,\text{m}\,\text{s}^{-1}$ on the same scale. The parallelogram OACB is now drawn. *The diagonal OC through O represents the vector sum or resultant of the two velocities in magnitude and direction.* Drawing or calculation shows that OC is about $6\,\text{m}\,\text{s}^{-1}$ and is $37°$ to OB. This is the velocity of the girl relative to the sea as she runs across the deck.

Another useful way of adding vectors is to draw the ship's velocity OA to scale and then *from A* to draw the girl's velocity AC on the same scale. The line OC now represents the sum or resultant of the two vectors. The result is the same as the parallelogram method but quicker.

An aeroplane travelling at $400 \, km \, h^{-1}$ in a direction OP is blown off-course by a wind of velocity $100 \, km \, h^{-1}$ blowing in a direction OQ. Figure 1.7 (ii). To find the resultant velocity we add the two vectors OP ($400 \, km \, h^{-1}$) and PR ($100 \, km \, h^{-1}$) as shown. The sum is OR in magnitude and direction. Otherwise, the parallelogram method can be used.

Vector Subtraction

The *relative velocity* of two cars is found by *subtracting* the two velocities.

Suppose that a car X is travelling with a velocity v along a road 30° E. of N., and a car Y is travelling with a velocity u along a road due east, Figure 1.8 (i).

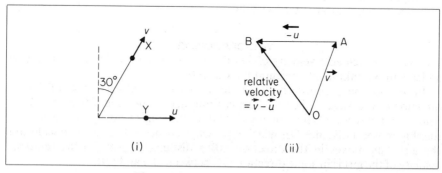

Figure 1.8 *Subtraction of velocities*

Arrows above the velocity letters show they are vectors. So the velocity of X relative to Y = difference in velocities $= \overrightarrow{v} - \overrightarrow{u} = \overrightarrow{v} + (-\overrightarrow{u})$. Suppose OA represents the velocity, v, of X in magnitude and direction, Figure 1.8 (ii). Since Y is travelling due east, a velocity AB numerically equal to u but in the due *west* direction represents the vector $(-\overrightarrow{u})$. The vector sum of OA and AB is OB, which therefore represents in magnitude and direction the velocity of X minus that of Y. By drawing an accurate diagram of the two velocities, OB can be found.

Example on Vector Subtraction

A car is moving round a circular track with a constant speed v of $20 \, m \, s^{-1}$, Figure 1.9 (i). At different times the car is at A, B and C respectively. Find the velocity change
(a) from A to C,
(b) from A to B.

(a) Velocity change from A to C $= \overrightarrow{v_C} - \overrightarrow{v_A} = (+20) - (-20)$

$$= 40 \, m \, s^{-1} \text{ in the direction of C}$$

(b) Velocity change from A to B $= \overrightarrow{v_B} - \overrightarrow{v_A} = \overrightarrow{v_B} + (-\overrightarrow{v_A})$

In Figure 1.9 (ii), PQ represents the vector $\overrightarrow{v_B}$ or $20 \, m \, s^{-1}$ and QR represents $-\overrightarrow{v_A}$ or $20 \, m \, s^{-1}$.

So $\qquad PR = \overrightarrow{v_B} - \overrightarrow{v_A} = \sqrt{20^2 + 20^2} = 28 \, m \, s^{-1}$ (approx.)

and its direction θ relative to v_B is 45°.

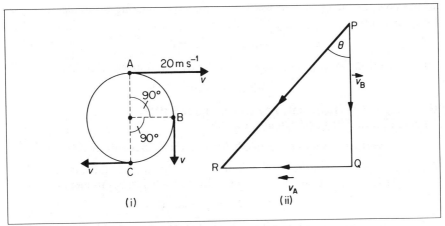

Figure 1.9 _Vector subtraction_

Components of Vectors

In mechanics and other branches of physics, we often need to find the _component_ of a vector in a certain direction.

The component is the 'effective part' of the vector in that direction. We can illustrate it by considering a picture held up by two strings OP and OQ each at an angle of 60° to the vertical, Figure 1.10 (i). If the force or tension in OP is 6 N, its _vertical_ component S acting upwards at P helps to support the weight W of the picture, which acts vertically downwards. The upward component T of the 6 N force in OQ acting in the direction QT, also helps to support W.

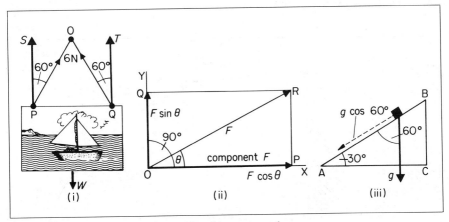

Figure 1.10 _Components of vectors_

Figure 1.10 (ii) shows how the component value of a vector F can be found in a direction OX. If OR represents F to scale, we draw a rectangle OPRQ which has OR as its diagonal. Now F is the sum of the vectors OP and OQ. _The vector OQ has no effect in a direction OX at 90° to itself._ So the effective part or component of F in the direction OX is the vector OP.

If θ is the angle between F and the direction OX, then

$$\text{OP/OR} = \cos\theta \quad \text{or} \quad \text{OP} = \text{OR}\cos\theta = F\cos\theta$$

So the component of any vector F in a direction making an angle θ to F is always given by

$$\textbf{component} = \textbf{\textit{F}}\cos\theta$$

In a direction OY *perpendicular* to OX, F has a component $F\cos(90° - \theta)$

which is $F\sin\theta$

This component is represented by OQ in Figure 1.10 (ii).
From Figure 1.10 (i), we can now see that

vertical component S of 6 N force in OP $= 6\cos 60° = 3$ N

This is also the value of the vertical component T of the 6 N force in OQ. Since the weight W of the picture is balanced by the two vertical components, then

weight $W = 3\,\text{N} + 3\,\text{N} = 6\,\text{N}$

The acceleration due to gravity, g, acts vertically downwards. In free fall, an object has an acceleration g. An object sliding freely down an inclined plane ABC, however, has an acceleration due to gravity equal to the *component* of g down the plane, Figure 1.10 (iii). If the plane is inclined at 60° to the vertical, the acceleration down the plane is then $g\cos 60°$ or $9\cdot8\cos 60°\,\text{m s}^{-2}$, which is $4\cdot9\,\text{m s}^{-2}$.

Since $\cos 60° = \sin 30°$, we can say that the acceleration down the plane is also given by $g\sin 30°$, where the angle made by the plane with the horizontal is 30°.

Projectiles

Consider an object O thrown forward from the top of a cliff OA with a horizontal velocity u of $15\,\text{m s}^{-1}$, Figure 1.11. Since u is horizontal, it has no component in a *vertical* direction. Similarly, since g acts vertically, it has no component in a *horizontal* direction.

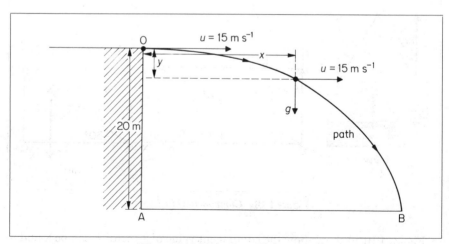

Figure 1.11 *Motion under gravity*

We may thus treat vertical and horizontal motion independently. Consider the vertical motion from O. If OA is 20 m, the ball has an initial vertical velocity of zero and a vertical acceleration of g, which is $9\cdot8\,\text{m s}^{-2}$ ($10\,\text{m s}^{-2}$ approximately). Thus, from $s = ut + \frac{1}{2}at^2$, *the time t to reach the bottom of the cliff is* given, using $g = 10\,\mathrm{m\,s^{-2}}$, by

$$20 = \tfrac{1}{2} \times 10 \times t^2 = 5t^2 \quad \text{or} \quad t = 2\,\mathrm{s}$$

So far as the horizontal motion is concerned, the ball continues to move forward with a constant horizontal velocity of $15\,\mathrm{m\,s^{-1}}$ since g has no component horizontally. In 2 seconds, therefore,

horizontal distance AB = distance from cliff = $15 \times 2 = 30\,\mathrm{m}$

Generally, in a time t the ball falls a *vertical* distance, y say, from O given by $y = \tfrac{1}{2}gt^2$. In the same time the ball travels a *horizontal* distance, x say, from O given by $x = ut$, where u is the velocity of $15\,\mathrm{m\,s^{-1}}$. If t is eliminated by using $t = x/u$ in $y = \tfrac{1}{2}gt^2$, we obtain $y = gx^2/2u^2$. This is the equation of a *parabola*. It is the path OB in Figure 1.11. In our discussion air resistance has been ignored.

Motion of Projectiles and their Range

In Figure 1.12, a ball at O on the ground is thrown with a velocity u at an angle α to the horizontal. We consider the vertical and horizontal motion *separately* in motion of this kind and use components.

Vertical motion. The vertical component of u is $u\cos(90° - \alpha)$ or $u\sin\alpha$; the acceleration $a = -g$. When the projectile reaches the ground at B, the *vertical* distance s travelled is *zero*. So, from $s = ut + \tfrac{1}{2}at^2$, we have

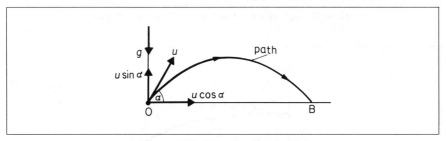

Figure 1.12 *Motion of projectiles*

$$0 = u\sin\alpha \,.\, t - \tfrac{1}{2}gt^2$$

Thus
$$t = \frac{2u\sin\alpha}{g} \qquad \qquad \text{.} \quad \text{.} \quad \text{.} \quad \text{.} \quad \text{.} \quad (1)$$

Horizontal motion. Since g acts vertically, it has no component in a horizontal direction. So the ball moves in a horizontal direction with an unchanged or constant velocity $u\cos\alpha$ because this is the component of u horizontally. So

Range R = OB = velocity × time

$$= u\cos\alpha \times \frac{2u\sin\alpha}{g} = \frac{2u^2\sin\alpha\cos\alpha}{g}$$

$$R = \frac{u^2\sin 2\alpha}{g}$$

The *maximum range* is obtained when $\sin 2\alpha = 1$, or $2\alpha = 90°$. So $\alpha = 45°$ for maximum range with a given velocity of throw u. In this case, the range is u^2/g.

At the maximum height A of the path, the *vertical* velocity of the ball is zero. So, applying $v = u + at$ in a vertical direction, the time t to reach A is given by

$$0 = u \sin \alpha - gt \quad \text{or} \quad t = u \sin \alpha / g$$

From (1), we see that this is half the time to reach B.

Example on Projectiles

A small ball A, suspended from a string OA, is set into oscillation, Figure 1.13. When the ball passes through the lowest point of the motion, the string is cut. If the ball is then moving with the velocity 0.8 m s^{-1} at a height 5 m above the ground, find the horizontal distance travelled by the ball. (Assume $g = 10 \text{ m s}^{-2}$.)

(*Analysis* (i) Horizontal distance = horizontal velocity (constant) × time (ii) Find time from vertical distance travelled, using g.)

When A is at the lowest point of the oscillation, it is moving horizontally with velocity 0.8 m s^{-1}. The ball lands at B on the ground.

To find the time of travel, consider the vertical motion. In this case the vertical distance s travelled is 5 m; the initial vertical velocity $u = 0$; and $a = g = 10 \text{ m s}^{-2}$. From $s = ut + \frac{1}{2}at^2$, we have

$$5 = \frac{1}{2} \times 10 \times t^2$$

So

$$t^2 = 1 \quad \text{or} \quad t = 1 \text{ s}$$

To find the horizontal distance travelled, consider the horizontal motion. The velocity is 0.8 m s^{-1} and this is constant. So

$$\text{horizontal distance to B} = 0.8 \text{ m s}^{-1} \times 1 \text{ s} = 0.8 \text{ m}$$

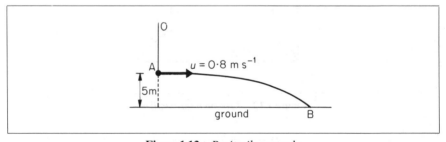

Figure 1.13   *Projectile example*

Exercises 1A

Linear Motion, Free-fall, Graphs

(*Assume $g = 10 \text{ m s}^{-2}$ or 10 N kg^{-1} unless otherwise given*)

1   A car moving with a velocity of 10 m s^{-1} accelerates uniformly at 1 m s^{-2} until it reaches a velocity of 15 m s^{-1}. Calculate   (i) the time taken,   (ii) the distance travelled during the acceleration,   (iii) the velocity reached 100 m from the place where the acceleration began.

2   A ball is thrown vertically upwards with an initial speed of 20 m s^{-1}. Calculate (i) the time taken to return to the thrower,   (ii) the maximum height reached.

3   A ball is thrown vertically upwards and caught by the thrower on its return. Sketch a graph of *velocity* (taking the upward direction as positive) against *time* for the whole of its motion, neglecting air resistance. How, from such a graph, would you obtain an estimate of the height reached by the ball? (*L.*)

4   A ball is dropped from a height of 20 m and rebounds with a velocity which is 3/4 of the velocity with which it hit the ground. What is the time interval between the first and second bounces?

5 A ball is thrown forward horizontally from the top of a cliff with a velocity of $10\,\mathrm{m\,s^{-1}}$. The height of the cliff above the ground is $45\,\mathrm{m}$. Calculate (i) the time to reach the ground, (ii) the distance from the cliff of the ball on hitting the ground, (iii) the direction of the ball to the horizontal just before it hits the ground.

6 A tennis ball is dropped from the hand, falls to the ground and bounces back at half the speed with which it hit the ground. Draw a velocity-time graph of its motion. Mark the point on the graph which corresponds to the ball hitting the ground.

Indicate how, from the graph,
(a) the distance the ball falls, and
(b) the distance the ball rises, can be found. (*L.*)

7 A projectile is fired with a velocity of $320\,\mathrm{m\,s^{-1}}$ at an angle of 30° to the horizontal. Find (i) the time to reach its greatest height, (ii) its horizontal range.

With the same velocity, what is the maximum possible range?

8 A small smooth object slides from rest down a smooth inclined plane inclined at 30° to the horizontal. What is (i) the acceleration down the plane, (ii) the time to reach the bottom if the plane is $5\,\mathrm{m}$ long?

The object is now thrown up the plane with an initial velocity of $15\,\mathrm{m\,s^{-1}}$.
(iii) How long does the object take to come to rest? (iv) How far up the plane has the object then travelled?

9 A stone attached to a string is whirled round in a horizontal circle with a constant speed of $10\,\mathrm{m\,s^{-1}}$. Calculate the difference in the *velocity* when the stone is (i) at opposite ends of a diameter, (ii) in two positions A and B, where angle AOB is 90° and O is the centre of the circle.

10 Two ships A and B are $4\,\mathrm{km}$ apart. A is due west of B. If A moves with a uniform velocity of $8\,\mathrm{km\,h^{-1}}$ due east and B moves with a uniform velocity of $6\,\mathrm{km\,h^{-1}}$ due south, calculate (i) the magnitude of the velocity of A relative to B, (ii) the closest distance apart of A and B.

11 Define *uniform acceleration*. State, for each case, *one* set of conditions sufficient for a body to describe
(a) a parabola,
(b) a circle.

A projectile is fired from ground level, with velocity $500\,\mathrm{m\,s^{-1}}$ at 30° to the horizontal. Find its horizontal range, the greatest vertical height to which it rises, and the time to reach the greatest height. What is the least speed with which it could be projected in order to achieve the same horizontal range? (The resistance of the air to the motion of the projectile may be neglected.) (*O.*)

12 A lunar landing module is descending to the Moon's surface at a steady velocity of $10\,\mathrm{m\,s^{-1}}$. At a height of $120\,\mathrm{m}$, a small object falls from its landing gear. Taking the Moon's gravitational acceleration as $1\cdot6\,\mathrm{m\,s^{-2}}$, at what speed, in $\mathrm{m\,s^{-1}}$, does the object strike the Moon?

A 202   B 22   C 19·6   D 16·8   E 10   (*AEB*, 1980.)

Laws of Motion, Force and Momentum

In the last section, we discussed velocity and acceleration. If you kick a moving ball, or hit a ball with a tennis racket, you can see that the force produces a change in velocity or acceleration. The first part of the next section will deal with the force on objects such as cars, for example, and the acceleration produced. After this section, we shall discuss momentum *of moving objects such as aeroplanes or trains, which is defined as* 'mass × velocity' *of a moving object.*

Newton's Laws of Motion

In 1687 Sir ISAAC NEWTON published a work called *Principia*, in which he set out clearly the Laws of Mechanics. He gave three 'laws of motion':

Law 1   Every body continues to be in a state of rest or to move with uniform velocity unless a resultant force acts on it.

Law 2   The change of momentum per second is proportional to the applied force and the momentum change takes place in the direction of the force.

Law 3   Action and reaction are always equal and opposite.

Inertia, Mass

Newton's first law expresses the idea of *inertia*. The inertia of a body is its reluctance to start moving, and its reluctance to stop once it has begun moving. Thus an object at rest begins to move only when it is pushed or pulled, i.e., when a *force* acts on it. An object O moving in a straight line with constant velocity will change its direction or move faster only if a new force acts on it, Figure 1.14 (i).

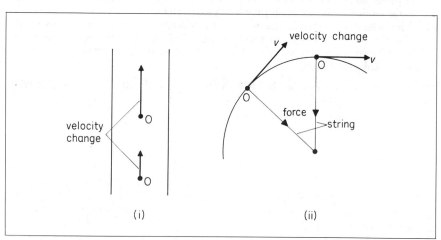

Figure 1.14   *Velocity changes:*   (i) *magnitude*,   (ii) *direction*

Passengers in a bus or car move forward when the vehicle stops suddenly. They continue in their state of motion until brought to rest by friction or collision. The use of safety belts reduces the shock.

Figure 1.14 (ii) illustrates a velocity change when an object O is whirled at constant speed by a string. This time the magnitude of the velocity v is constant but its direction changes. So a force due to the string acts on the object O.

'Mass' is a measure of the inertia of a body. If an object changes its direction or its velocity slightly when a large force acts on it, its inertial mass is high. The mass of an object is constant all over the world; it is the same on the earth as on the moon. Mass is measured in kilogram (kg) by means of a chemical balance, where it is compared with standard masses based on the International Prototype Kilogram.

Force, The Newton

When an object X is moving it is said to have an amount of *momentum* given, by definition, by

$$momentum = mass\ of\ X \times velocity \qquad . \qquad . \qquad . \qquad (1)$$

Thus a runner of mass 50 kg moving with a velocity of $10\,\mathrm{m\,s^{-1}}$ has a momentum of $500\,\mathrm{kg\,m\,s^{-1}}$. If another runner collides with X his velocity alters, and so the momentum of X alters.

From Newton's second law, a *force F* acts on X which is equal to its change in momentum per second. Using (1), it follows that if m is the mass of X,

$$F \propto m \times \text{change in velocity per second}$$

But the change in velocity per second is the *acceleration a* produced by the force.

$$\therefore F \propto ma$$

so
$$F = kma \qquad . \qquad . \qquad . \qquad . \qquad . \qquad (i)$$

where k is a constant.

With SI units, the **newton** (N) is the unit of force. It is defined as the force which gives a mass of 1 kg an acceleration of $1\,\mathrm{m\,s^{-2}}$. Substituting $F = 1\,\mathrm{N}$, $m = 1\,\mathrm{kg}$ and $a = 1\,\mathrm{m\,s^{-2}}$ in the expression for F in (i), we obtain $k = 1$. Hence, with units as stated, $k = 1$.

$$\therefore F = ma \qquad . \qquad . \qquad . \qquad . \qquad . \qquad (2)$$

which is a standard equation in dynamics. Thus if a mass of 0·2 kg is acted upon by a force F which produces an acceleration a of $4\,\mathrm{m\,s^{-2}}$, then, since $m = 0\cdot2\,\mathrm{kg}$,

$$F = ma = 0\cdot2\,(\mathrm{kg}) \times 4\,(\mathrm{m\,s^{-2}}) = 0\cdot8\,\mathrm{N}$$

Weight and Mass

The *weight* of an object is defined as the *force* acting on it due to gravitational pull, or gravity. So the weight of an object can be measured by attaching it to a spring-balance and noting the spring extension, as the extension is proportional to the force on it (p. 134).

Suppose the weight of an object of mass m is denoted by W. If the object is released so that it falls freely to the ground, its acceleration is g. Now $F = ma$. Consequently the force acting on it, or its weight, is given by

$$W = mg$$

If the mass is 1 kg, then, since $g = 9\cdot8\,\mathrm{m\,s^{-2}}$, the weight $W = 1 \times 9\cdot8 = 9\cdot8\,\mathrm{N}$.

The weight of a 5 kg mass is thus $5 \times 9 \cdot 8$ N or 49 N. Note that the weight of a 100 g (0·1 kg) mass is about 1 N; the weight of an average-sized apple is about 1 N.

Gravitational Field Strength

The space round the earth where the mass of an object experiences a gravitational pull or force due to gravity is called the *gravitational field* of the earth. Molecules of air, or this book or the reader, are all in the earth's gravitational field.

We can see that on the surface of the earth, the value of g may be expressed as about $9 \cdot 8$ N kg^{-1}. The *force per unit mass* in a gravitational field is called the *gravitational field strength*. On the moon's surface this is only about $1 \cdot 6$ N kg^{-1}, so a mass of 1 kg has a gravitational pull on it of $1 \cdot 6$ N.

The reader should note carefully that the *mass* of an object is constant all over the world, but its *weight* is a *force* whose magnitude depends on the value of g. The acceleration due to gravity, g, depends on the distance of the place considered from the centre of the earth; it is slightly greater at the poles than at the equator, since the earth is not a perfectly spherical shape. It therefore follows that the weight of an object differs in different parts of the world. On the moon, which is smaller than the earth and has a smaller density, an object would have the same mass as on the earth but it would weigh about one-sixth of its weight on the earth, as the acceleration of free fall on the moon is about $g/6$. For this reason astronauts tend to 'float' on the moon's surface.

Experimental Investigation of $F = ma$

An experimental investigation of $F = ma$ can be carried out by accelerating a trolley, mass m, down a friction-compensated inclined plane by a constant force F due to a stretched piece of elastic, and measuring the acceleration a with a ticker tape. Details of the experiment can be obtained from O-level texts, such as the author's *Principles of Physics* (Collins Educational, London). We assume the reader is familiar with the method. Figure 1.15 shows the results obtained. When the force is increased in the ratio $1 : 2 : 3$, experiment shows that the acceleration increases in the ratio $2 \cdot 4 : 4 \cdot 8 : 7 \cdot 3$, which is approximately $1 : 2 : 3$. Thus $a \propto F$ with constant mass.

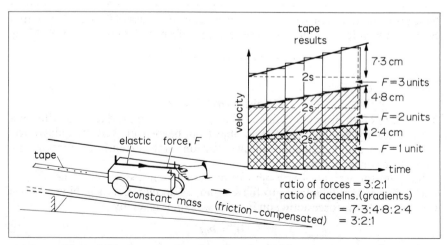

Figure 1.15   *Investigation of acceleration and force (mass constant)*

The mass m can also be varied by placing similar trolleys on top of each other and the force F can be kept constant. One experiment shows that with 1, 2 and 3 trolleys the accelerations decrease in the ratio $7.5 : 4.9 : 2.5$, which is $3 : 2 : 1$ approximately. Thus $a \propto 1/m$ when F is constant.

Applications of $F = ma$

The following examples illustrate the application of $F = ma$. It should be carefully noted that (i) if more than one force acts on a moving object, then F is the *resultant* force on the object, (ii) F must be in newtons (N) and m in kilograms (kg). Since F is a vector, the direction of the forces must be drawn in the diagram. Remember mass (m) is a scalar so this has no direction.

Examples

1 *Two forces* A force of 200 N pulls a sledge of mass 50 kg and overcomes a constant frictional force of 40 N. What is the acceleration of the sledge?

$$\text{Resultant force } F = 200 - 40 = 160 \text{ N}$$

From $F = ma$,

$$\therefore 160 = 50 \times a$$

$$\therefore a = 3.2 \text{ m s}^{-2}$$

2 *Lift problem* An object of mass 2·00 kg is attached to the hook of a spring-balance, and the balance is suspended vertically from the roof of a lift. What is the reading on the spring-balance when the lift is (i) ascending with an acceleration of 0.2 m s^{-2}, (ii) descending with an acceleration of 0.1 m s^{-2}, (iii) ascending with a uniform velocity of 0.15 m s^{-1} ($g = 10 \text{ m s}^{-2}$ or 10 N kg^{-1}).

Suppose T is the tension (force) in the spring-balance in N, Figure 1.16 (i).

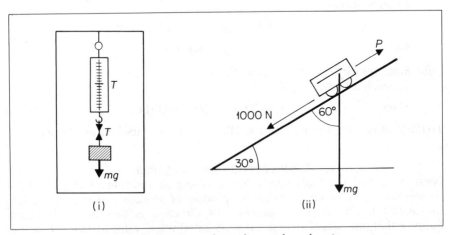

Figure 1.16 *Examples on force and acceleration*

(i) The object is acted on by two forces:
(a) the tension T in newtons in the spring balance which acts upwards,
(b) its weight mg or 20 N, which acts downwards.

Since the object accelerates *upwards*, T is greater than 20 N. So the net force, F, acting on the object $= T - 20$ N. Now

$$F = ma$$

where a is the acceleration in m s^{-2}, $0 \cdot 2$ m s^{-2}.

$$\therefore T - 20 = 2 \times a = 2 \times 0 \cdot 2$$

$$\therefore T = 20 \cdot 4 \text{ N} \qquad . \qquad . \qquad . \qquad . \qquad . \qquad (1)$$

(ii) When the lift *descends* with an acceleration of $0 \cdot 1$ m s^{-2}, its weight, 20 N, is now greater than T_1, the new tension in the spring-balance.

$$\therefore \text{ resultant force } = 20 - T_1$$

$$\therefore F = 20 - T_1 = ma = 2 \times 0 \cdot 1$$

$$\therefore T_1 = 20 - 0 \cdot 2 = 19 \cdot 8 \text{ N}$$

(iii) When the lift moves with constant velocity, the acceleration is zero. Since the resultant force is zero, the reading on the spring-balance is exactly equal to the weight, 20 N.

3 *Inclined plane* A car of mass 1000 kg is moving up a hill inclined at 30° to the horizontal. The total frictional force on the car is 1000 N, Figure 1.16 (ii).

Calculate the force P due to the engine when the car is
(a) accelerating at 2 m s^{-2},
(b) moving with a steady velocity of 15 m s^{-1}.

(*Analysis* (a) Use $F = ma$. (b) There are 3 forces on the car—its weight mg, P and the frictional force. (c) Since the car is moving on an incline, we need to find the component of the weight mg, down the incline.)

(a) Weight of car $= mg = 10\,000$ N
Component downhill $= mg \cos \theta = 10\,000 \cos 60° = 5000$ N

So   resultant force uphill, $F = P - 5000 - 1000$

From $F = ma$,

$$P - 5000 - 1000 = 1000 \times 2 = 2000$$

So $P = 8000$ N

(b) Since velocity is steady, acceleration $a = 0$
So resultant force $F = 0$

Then $P = 5000 + 1000 = 6000$ N

To the Student If required, Exercise 1B, page 31, has questions on $F = ma$.

Linear Momentum, Impulse

Newton defined the force acting on an object as the rate of change of its momentum, the momentum being the product of its mass and velocity (p. 19). *Momentum is thus a vector quantity*; its direction is that of the velocity. Suppose that the mass of an object is m, its initial velocity due to a force F acting on it for a time t is v. Then

$$\text{change of momentum} = mv - mu$$

and hence

$$F = \frac{mv - mu}{t}$$

$$\therefore F \times t = mv - mu = \textit{momentum change} \quad . \qquad . \qquad . \qquad (1)$$

The quantity $F \times t$ (force × time) is known as the *impulse* of the force on the object. From (1) it follows that the units of momentum are the same as those of Ft, that is, *newton second* (N s). From 'mass × velocity', alternative units are 'kg m s^{-1}'. The impulse ($F \times t$) is the 'time effect' of a force on an object.

Force and Momentum Change

A person of mass 50 kg who is jumping from a height of 5 metres will land on the ground with a velocity $= \sqrt{2gh} = \sqrt{2 \times 10 \times 5} = 10 \, \text{m s}^{-1}$, assuming $g = 10 \, \text{m s}^{-2}$ approx. If he does not flex his knees on landing, he will be brought to rest very quickly, say in $\frac{1}{10}$th second. The force F acting is then given by

$$F = \frac{\text{momentum change}}{\text{time}}$$

$$= \frac{50 \times 10}{\frac{1}{10}} = 5000 \, \text{N}$$

This is a force of about 10 times the person's weight and the large force has a severe effect on the body.

Suppose, however, that the person flexes his knees and is brought to rest much more slowly on landing, say in 1 second. Then, from above, the force F now acting is 10 times less than before, or 500 N. Consequently, much less damage is done to the person on landing.

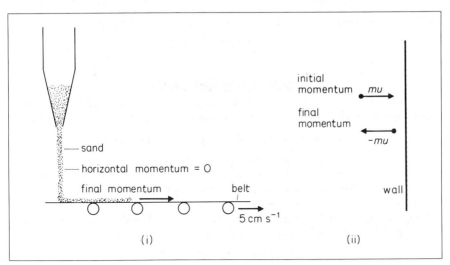

Figure 1.17   *Linear momentum changes*

Momentum is a *vector*. So we must always take account of its direction.

Suppose sand is allowed to fall vertically at a steady rate of 100 g s^{-1} on to a horizontal conveyor belt moving at a steady velocity of 5 cm s^{-1}, Figure 1.17 (i). The initial horizontal velocity of the sand is zero. The final horizontal velocity is 5 cm s^{-1}. Now in one second in a horizontal direction,

$$\text{mass} = 100 \, \text{g} = 0 \cdot 1 \, \text{kg, velocity gained} = 5 \, \text{cm s}^{-1} = 5 \times 10^{-2} \, \text{m s}^{-1}$$

$$\therefore \text{momentum change per second } \textit{horizontally} = 0 \cdot 1 \times 5 \times 10^{-2}$$

$$= 5 \times 10^{-3} \, \text{newton}$$

$$= \text{force on belt}$$

Observe that this is a case where the *mass* changes with time and the velocity gained is constant. In terms of the calculus, force is the rate of change of momentum mv, which is $v \times dm/dt$, and dm/dt is $100\,\text{g s}^{-1}$ in this numerical example.

Consider a molecule of mass m in a gas, which strikes the wall of a vessel repeatedly with a velocity u and rebounds with a velocity $-u$, Figure 1.17 (ii). Since momentum is a vector quantity, the momentum change = final momentum − initial momentum = $mu - (-mu) = 2mu$. If the containing vessel is a cube of side l, the molecule repeatedly takes a time $2l/u$ to make a collision with the same side as it moves to-and-fro across the vessel. So.

$$\text{number of collisions per second, } n = \frac{1}{2l/u} = \frac{u}{2l}$$

The average force on wall = $n \times$ one momentum change

$$= \frac{u}{2l} \times 2mu = \frac{mu^2}{l}$$

The total gas pressure is the average force per unit area on the walls of the container due to all the gas molecules and is discussed in *Heat*.

Suppose a ball of mass $0 \cdot 1\,\text{kg}$ hits a smooth wall normally with a velocity of $10\,\text{m s}^{-1}$ four times per second, rebounding each time with a velocity of $10\,\text{m s}^{-1}$. Then each time, momentum change = $0 \cdot 1\,[10 - (-10)] = 0 \cdot 1 \times 20$.

So　　　　average force on wall = $4 \times$ momentum change per second

$$= 4 \times 0 \cdot 1 \times 20 = 8\,\text{N}$$

Force due to Water Flow

When water from a horizontal hose-pipe strikes a wall at right angles, a force is exerted on the wall. Suppose the water comes to rest on hitting the wall. Then

$$\text{force} = \text{momentum change per second of water}$$

$$= (\text{mass} \times \text{velocity change}) \text{ per second}$$

---

force = mass per second × velocity change

Suppose the water flows out of the pipe at $2\,\text{kg s}^{-1}$ and its velocity changes from $5\,\text{m s}^{-1}$ to zero on hitting the wall. Then

$$\text{force } F = 2 \times (5 - 0) = 10\,\text{N}$$

Example on Force due to Water Flow

A hose ejects water at a speed of $20\,\text{cm s}^{-1}$ through a hole of area $100\,\text{cm}^2$. If the water strikes a wall normally, calculate the force on the wall in newtons, assuming the velocity of the water normal to the wall is zero after collision.

The volume of water per second striking the wall $= 100 \times 20 = 2000\,\text{cm}^3$

$\therefore$ mass per second striking wall $= 2000\,\text{g s}^{-1} = 2\,\text{kg s}^{-1}$

Velocity change of water on striking wall $= 20 - 0 = 20\,\text{cm s}^{-1} = 0 \cdot 2\,\text{m s}^{-1}$

$\therefore$ momentum change per second $= 2\,(\text{kg s}^{-1}) \times 0 \cdot 2\,(\text{m s}^{-1}) = 0 \cdot 4\,\text{N}$

$\therefore$ Force on wall $= 0 \cdot 4\,\text{N}$

Newton's Third Law, Action and Reaction

Newton's third law—action and reaction are equal and opposite—means that if a body A exerts a force (action) on a body B, then B will exert an equal and opposite force (reaction) on A.

These forces are produced between objects by direct contact when they touch, or by gravitational forces, for example, when they are apart. Thus if a ball is kicked upwards, the force on the ball by the kicker is equal and opposite to the force on the kicker by the ball. The initial upward acceleration of the ball is usually very much greater than the downward acceleration of the kicker because the mass of the ball is much less than that of the kicker.

As the ball falls downwards towards the ground, the force of attraction on the ball by the Earth is equal and opposite to the force of attraction on the Earth by the ball. The upward acceleration of the Earth is not noticeable since its mass is so large.

In the case of a rocket, the downward force on the burning gases from the exhaust is equal to the upward force on the rocket. This is one application of the reaction force. Another is a water-sprinkler which spins backwards as the water is thrown forwards.

Action-reaction forces therefore always occur in pairs. It should be noted that the two forces act on *different* bodies. So only *one* of the forces is used in discussing the motion of one of the two bodies. In the case of a man standing in a lift moving upwards, for example, the upward reaction of the floor *on the man* is the force we need to take into account in applying $F = ma$ to the motion of the *man*. The equal downward force on the floor is *not* required.

This principle is also illustrated in the next example.

Example on Truck and Trailer

Figure 1.18 shows a truck A of mass 1000 kg pulling a trailer of mass 3000 kg. The frictional force on A is 1000 N, on B it is 2000 N, and the truck engine exerts a force of 8000 N.

Calculate (i) the acceleration of the truck and trailer, (ii) the tension T in the tow-bar connecting A and B.

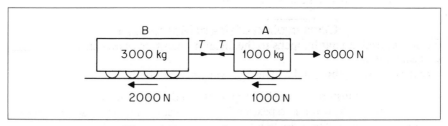

Figure 1.18 *Example on truck and trailer*

For B only   From $F = ma$, where F is the resultant force

$$T - 2000 = 3000\,a\ . \qquad . \qquad . \qquad . \qquad . \qquad . \qquad (1)$$

For A only $8000 - 1000 - T = 1000\,a\ . \qquad . \qquad . \qquad . \qquad . \qquad . \qquad (2)$

Adding (1) and (2) to eliminate T, then

$$8000 - 1000 - 2000 = 4000\,a$$

So $5000 = 4000\,a$ and $a = 1.25\,\mathrm{m\,s^{-2}}$

From (1), $T = (3000 \times 1.25) + 2000 = 5750\,\mathrm{N}$

Force due to Rotating Helicopter Blades

When helicopter blades are rotating, they strike air molecules in a downward direction. The momentum change per second of the air molecules produces a *downward* force and by the Law of Action and Reaction, an equal *upward* force is exerted by the molecules on the helicopter blades. This upward force helps to keep the helicopter hovering in the air because it can balance the downward weight of the machine. This is illustrated in the Example which follows.

Example on Rotating Helicopter Blades

A helicopter of mass 500 kg hovers when its rotating blades move through an area of 30 m² and gives an average speed v to the air.

Estimate v assuming the density of air is $1.3 \, kg \, m^{-3}$ and $g = 10 \, N \, kg^{-1}$.

(*Analysis*   (i) The reaction of the downward force on the air = weight of helicopter,   (ii) downward force = momentum change per second of air swept down,   (iii) mass of air per second moving downwards = volume per second × density = area swept by blades × velocity of air × density.)

Volume of air per second moving downwards = area × velocity v = $30v$

So   mass of air per second downwards = $30v \times 1.3 = 39v$

∴ momentum change per second of air = mass per second × velocity change

$$= 39v \times v = 39v^2$$

So   reaction force upwards = $39v^2$ = helicopter weight 5000 N

$$v^2 = \frac{5000}{39}$$

$$v = \sqrt{\frac{5000}{39}} = 11 \, m \, s^{-1} \, (\text{approx.})$$

At lift-off, the fuselage of the helicopter would turn round the opposite way to the rotation of the blades, from Newton's law of Action and Reaction. A vertical rotor on the tail provides a counter-thrust.

Conservation of Linear Momentum

We now consider what happens to the linear momentum of objects which *collide* with each other.

Experimentally, this can be investigated by several methods:

1   Trolleys in collision, with ticker-tapes attached to measure velocities.
2   Linear air-track, using perspex models in collision and stroboscopic photography for measuring velocities.

As an illustration of the experimental results, the following measurements were taken in trolley collisions (Figure 1.19):

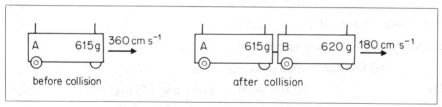

before collision   after collision

Figure 1.19   *Linear momentum experiment*

Before collision
Mass of trolley A = 615 g; initial velocity = 360 cm s^{-1}.

After Collision
A and B coalesced and both moved with velocity of 180 cm s^{-1}.

Thus the total linear momentum of A and B before collision = 0.615 (kg) $\times$ 3.6 (m s^{-1}) $+ 0 = 2.20$ kg m s^{-1} (approx.). The total momentum of A and B after collision = $1.235 \times 1.8 = 2.20$ kg m s^{-1} (approx.).
   Within the limits of experimental accuracy, it follows that *the total momentum of A and B before collision = the total momentum after collision.*
   Similar results are obtained if A and B are moving with different speeds after collision, or in opposite directions before collision.

Principle of Conservation of Linear Momentum

These experimental results can be shown to follow from Newton's second and third laws of motion (p. 18).

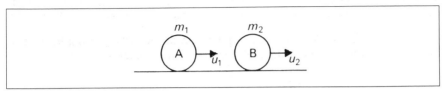

Figure 1.20   *Conservation of linear momentum*

Suppose that a moving object A, of mass m_1 and velocity u_1, collides with another object B, of mass m_2 and velocity u_2, moving in the same direction, Figure 1.20. By Newton's law of action and reaction, the force F exerted by A on B is equal and opposite to that exerted by B on A. Moreover, the time t during which the force acted on B is equal to the time during which the force of reaction acted on A. Thus the magnitude of the impulse, Ft, on B is equal and *opposite* to the magnitude of the impulse on A. From equation (1), p. 22, the impulse is equal to the change of momentum. It therefore follows that the *change* in the total momentum of the two objects is *zero*, i.e., the total momentum of the two objects is constant although a collision had occurred. Thus if A moves with a reduced velocity v_1 after collision, and B then moves with an increased velocity v_2,

$$m_1 u_1 + m_2 u_2 = m_1 v_1 + m_2 v_2$$

The principle of the conservation of linear momentum states that, if no external forces act on a system of colliding objects, the total momentum of the objects in a given direction before collision = total momentum in same direction after collision.

Examples on Conservation of Momentum

1   An object A of mass 2 kg is moving with a velocity of 3 m s^{-1} and collides head on with an object B of mass 1 kg moving in the opposite direction with a velocity of 4 m s^{-1}, Figure 1.21. After collision both objects stick, so that they move with a common velocity v. Calculate v.

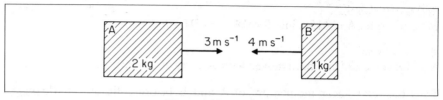

Figure 1.21 *Example*

Total momentum before collision of A and B in the direction of A

$$= 2 \times 3 - 1 \times 4 = 2 \, \text{kg m s}^{-1}$$

Note that momentum is a vector and the momentum of B is of opposite sign to A.

After collision, momentum of A and B in the direction of A $= 2v + 1v = 3v$

$$\therefore 3v = 2$$

$$\therefore v = \tfrac{2}{3} \, \text{m s}^{-1}$$

2   A bullet of mass 20 g travelling horizontally at $100 \, \text{m s}^{-1}$, embeds itself in the centre of a block of wood mass 1 kg which is suspended by light vertical strings 1 m in length. Calculate the maximum inclination of the strings to the vertical. (Assume $g = 9 \cdot 8 \, \text{m s}^{-2}$).

(*Analysis*   (i) The angle of swing θ depends on the velocity v of the bullet plus block,   (ii) v can be found using the conservation of momentum.)

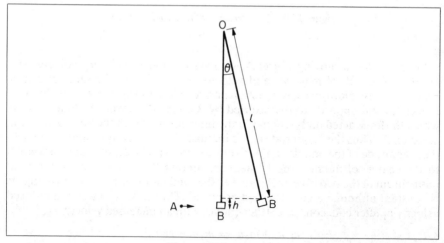

Figure 1.22 *Example*

Suppose A is the bullet, B is the block suspended from a point O, and θ is the maximum inclination to the vertical, Figure 1.22. If $v \, \text{m s}^{-1}$ is the common velocity of block and bullet when the bullet is brought to rest relative to the block, then, from the principle of the conservation of momentum, since $20 \, \text{g} = 0 \cdot 02 \, \text{kg}$,

$$(1 + 0 \cdot 02)v = 0 \cdot 02 \times 100$$

$$\therefore v = \frac{2}{1 \cdot 02} = \frac{100}{51} \, \text{m s}^{-1}$$

The vertical height risen by block and bullet is given by $v^2 = 2gh$, where $g = 9 \cdot 8 \, \mathrm{m\,s^{-2}}$ and $h = l - l \cos \theta = l(1 - \cos \theta)$

$$\therefore v^2 = 2gl(1 - \cos \theta)$$

$$\therefore \left(\frac{100}{51}\right)^2 = 2 \times 9 \cdot 8 \times 1(1 - \cos \theta)$$

$$\therefore 1 - \cos \theta = \left(\frac{100}{51}\right)^2 \times \frac{1}{2 \times 9 \cdot 8} = 0 \cdot 1962$$

$$\therefore \cos \theta = 0 \cdot 8038, \text{ or } \theta = 37° \text{ (approx.)}$$

3   A snooker ball X of mass $0 \cdot 3 \, \mathrm{kg}$, moving with velocity $5 \, \mathrm{m\,s^{-1}}$, hits a stationary ball Y of mass $0 \cdot 4 \, \mathrm{kg}$. Y moves off with a velocity of $2 \, \mathrm{m\,s^{-1}}$ at 30° to the initial direction of X, Figure 1.23.
   Find the velocity v of X and its direction after hitting Y.

(_Analysis_   We need two equations to find v and θ for X. So apply the momentum conservation   (i) along the initial direction of X,   (ii) perpendicular to the initial direction, direction Z.)

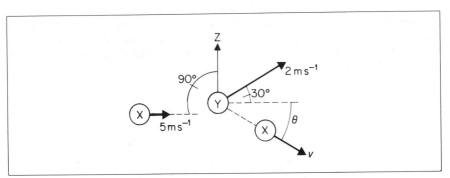

Figure 1.23   _Example on collision_

(i) In initial direction of X, from conservation of momentum,

$$0 \cdot 3v \cos \theta + 0 \cdot 4 \times 2 \cos 30° = 0 \cdot 3 \times 5$$

So $$0 \cdot 3 \, v \cos \theta = 1 \cdot 5 - 0 \cdot 8 \cos 30° = 0 \cdot 8 \qquad . \quad . \quad (1)$$

(ii) Along Z, 90° to initial X direction, initial momentum $= 0$.

So in this direction,

$$0 \cdot 4 \times 2 \sin 30° - 0 \cdot 3 \times v \sin \theta = 0$$

or $$0 \cdot 3 \, v \sin \theta = 0 \cdot 4 \qquad . \quad . \quad . \quad . \quad . \quad (2)$$

Dividing (2) by (1),

$$\frac{\sin \theta}{\cos \theta} = \tan \theta = \frac{0 \cdot 4}{0 \cdot 8} = 0 \cdot 5$$

So $$\theta = 27° \text{ (approx.)}$$

Also, from (2) $$v = \frac{0 \cdot 4}{0 \cdot 3 \sin 27°} = 3 \, \mathrm{m\,s^{-1}} \text{ (approx.)}$$

Inelastic and Elastic Collisions

In collisions, the total momentum of the colliding objects is always conserved. Usually, however, their total kinetic energy is not conserved. Some of it is changed to heat or sound energy, which is not recoverable. Such collisions are said to be *inelastic*. For example, when a lump of putty falls to the ground, the total momentum of the putty and earth is conserved, that is, the putty loses momentum and the earth gains an equal amount of momentum. But all the kinetic energy of the putty is changed to heat and sound on collision.

Inelastic collision = collision where total kinetic energy is *not* conserved (total momentum always conserved in *any* type of collision.)

If the total kinetic energy is conserved, the collision is said to be *elastic*. Gas molecules make elastic collisions. The collision between two smooth snooker balls is approximately elastic. Electrons may make elastic or inelastic collisions with atoms of a gas. As proved on p. 38, the kinetic energy of a mass m moving with a velocity v has kinetic energy equal to $\frac{1}{2}mv^2$.

Elastic collision = collision when total kinetic energy is conserved.

As an illustration of the mechanics associated with elastic collisions, consider a sphere A of mass m and velocity v incident on a stationary sphere B of equal mass m, Figure 1.24 (i). Suppose the collision is elastic, and after collision let A move with a velocity v_1 at an angle of 60° to its original direction and B move with the velocity v_2 at an angle θ to the direction of v.

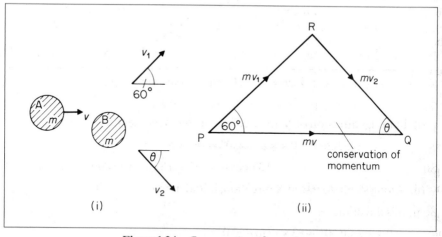

Figure 1.24   *Conservation of momentum*

Since momentum is a vector (p. 22), we may represent the momentum mv of A by the line PQ drawn in the direction of v, Figure 1.24 (ii). Likewise, PR represents the momentum mv_1 of A after collision. *Since momentum is conserved, the vector RQ must represent the momentum mv_2 of B after collision*, that is,

$$\overrightarrow{mv} = \overrightarrow{mv_1} + \overrightarrow{mv_2}$$

Hence

$$\overrightarrow{v} = \overrightarrow{v_1} + \overrightarrow{v_2}$$

or PQ represents v in magnitude, PR represents v_1 and RQ represents v_2. But if the collision is elastic,

$$\tfrac{1}{2}mv^2 = \tfrac{1}{2}mv_1{}^2 + \tfrac{1}{2}mv_2{}^2$$
$$\therefore v^2 = v_1{}^2 + v_2{}^2$$

Consequently, triangle PRQ is a right-angled triangle with angle R equal to 90°.

$$\therefore v_1 = v \cos 60° = \frac{v}{2}$$

Also, $\theta = 90° - 60° = 30°$, and $v_2 = v \cos 30° = \dfrac{\sqrt{3}v}{2}$

Momentum and Explosive Forces

There are numerous cases where momentum changes are produced by *explosive* forces. An example is a bullet of mass $m = 50\,\text{g}$ say, fired from a rifle of mass $M = 2\,\text{kg}$ with a velocity v of $100\,\text{m\,s}^{-1}$. Initially, the total momentum of the bullet and rifle is zero. From the principle of the conservation of linear momentum, when the bullet is fired the total momentum of bullet and rifle is still zero, since no external force has acted on them. Thus if V is the velocity of the rifle,

$$mv\,(\text{bullet}) + MV\,(\text{rifle}) = 0$$

$$\therefore MV = -mv \quad \text{or} \quad V = -\frac{m}{M}v$$

The momentum of the rifle is thus *equal and opposite* to that of the bullet. Further, $V/v = -m/M$. Since $m/M = 50/2000 = 1/40$, it follows that $V = -v/40 = 2.5\,\text{m\,s}^{-1}$. This means that the rifle moves back or *recoils* with a velocity only about $\frac{1}{40}$th that of the bullet.

If it is preferred, one may also say that the explosive force produces the same numerical momentum change in the bullet as in the rifle. Thus $mv = MV$, where V is the velocity of the rifle in the *opposite* direction to that of the bullet.

The joule (J) is the unit of energy (p. 35). As we see later,

the kinetic energy, E_1, of the bullet $= \tfrac{1}{2}mv^2 = \tfrac{1}{2}.0{\cdot}05.100^2 = 250\,\text{J}$

the kinetic energy, E_2, of the rifle $= \tfrac{1}{2}MV^2 = \tfrac{1}{2}.2.2{\cdot}5^2 = 6{\cdot}25\,\text{J}$

Thus the total kinetic energy produced by the explosion $= 256{\cdot}25\,\text{J}$. The kinetic energy E_1 of the bullet is then $250/256{\cdot}25$, or about 98%, of the total energy. This is explained by the fact that the kinetic energy depends on the *square* of the velocity. The high velocity of the bullet thus more than compensates for its small mass relative to that of the rifle. See also p. 38.

_____ Exercises 1B _____

Force and Momentum

(*Assume* $g = 10\,\text{m\,s}^{-2}$ *or* $10\,\text{N\,kg}^{-1}$ *unless otherwise stated*)

1 A car of mass $1000\,\text{kg}$ is accelerating at $2\,\text{m\,s}^{-2}$. What resultant force acts on the car? If the resistance to the motion is $1000\,\text{N}$, what is the force due to the engine?

2 A box of mass $50\,\text{kg}$ is pulled up from the hold of a ship with an acceleration of $1\,\text{m\,s}^{-2}$ by a vertical rope attached to it. Find the tension in the rope.

What is the tension in the rope when the box moves up with a uniform velocity of $1\,\mathrm{m\,s^{-1}}$?

3 A lift moves   (i) up and   (ii) down with an acceleration of $2\,\mathrm{m\,s^{-2}}$. In each case, calculate the reaction of the floor on a man of mass $50\,\mathrm{kg}$ standing in the lift.

4 A ball of mass $0{\cdot}2\,\mathrm{kg}$ falls from a height of $45\,\mathrm{m}$. On striking the ground it rebounds in $0{\cdot}1\,\mathrm{s}$ with two-thirds of the velocity with which it struck the ground. Calculate (i) the momentum change on hitting the ground,   (ii) the force on the ball due to the impact.

5 A ball of mass $0{\cdot}05\,\mathrm{kg}$ strikes a smooth wall normally four times in 2 seconds with a velocity of $10\,\mathrm{m\,s^{-1}}$. Each time the ball rebounds with the same speed of $10\,\mathrm{m\,s^{-1}}$. Calculate the average force on the wall.

Draw a sketch showing how the momentum varies with time over the 2 seconds.

6 The mass of gas emitted from the rear of a toy rocket is initially $0{\cdot}1\,\mathrm{kg\,s^{-1}}$. If the speed of the gas relative to the rocket is $50\,\mathrm{m\,s^{-1}}$, and the mass of the rocket is $2\,\mathrm{kg}$, what is the initial acceleration of the rocket?

7 A ball A of mass $0{\cdot}1\,\mathrm{kg}$, moving with a velocity of $6\,\mathrm{m\,s^{-1}}$, collides directly with a ball B of mass $0{\cdot}2\,\mathrm{kg}$ at rest. Calculate their common velocity if both balls move off together.

If A had rebounded with a velocity of $2\,\mathrm{m\,s^{-1}}$ in the opposite direction after collision, what would be the new velocity of B?

8 A bullet of mass $20\,\mathrm{g}$ is fired horizontally into a suspended stationary wooden block of mass $380\,\mathrm{g}$ with a velocity of $200\,\mathrm{m\,s^{-1}}$. What is the common velocity of the bullet and block if the bullet is embedded (stays inside) the block?

If the block and bullet experience a constant opposing force of $2\,\mathrm{N}$, find the time taken by them to come to rest.

9 A hose directs a horizontal jet of water, moving with a velocity of $20\,\mathrm{m\,s^{-1}}$, on to a vertical wall. The cross-sectional area of the jet is $5 \times 10^{-4}\,\mathrm{m^2}$. If the density of water is $1000\,\mathrm{kg\,m^{-3}}$, calculate the force on the wall assuming the water is brought to rest there.

10 A cable-operated lift of total mass $500\,\mathrm{kg}$ moves upwards from rest in a vertical shaft. The graph below shows how its velocity varies with time, Figure 1A.

(a) For the period of time indicated by DE, determine   (i) the distance travelled, (ii) the acceleration of the lift.

(b) Calculate the tension in the cable during the interval   (i) OA,   (ii) BC. Assume that the cable has negligible mass compared with that of the lift, and that friction between the lift and the shaft can be ignored. (*JMB*.)

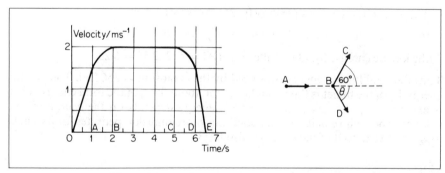

Figure 1A   **Figure 1B**

11 In a nuclear collision, an alpha-particle A of mass 4 units is incident with a velocity v on a stationary helium nucleus B of 4 mass units, Figure 1B. After collision, A moves in the direction BC with a velocity $v/2$, where BC makes an angle of $60°$ with the initial direction AB, and the helium nucleus moves along BD.

Calculate the velocity of rebound of the helium nucleus along BD and the angle θ made with the direction AB. (A solution by drawing is acceptable.)

12 A large cardboard box of mass $0{\cdot}75\,\mathrm{kg}$ is pushed across a horizontal floor by a

force of 4·5 N. The motion of the box is opposed by   (i) a frictional force of 1·5 N between the box and the floor, and   (ii) an air resistance force kv^2, where $k = 6·0 \times 10^{-2}\,\text{kg}\,\text{m}^{-1}$ and v is the speed of the box in m s^{-1}.

Sketch a diagram showing the directions of the forces which act on the moving box. Calculate maximum values for
  (a) the acceleration of the box,
  (b) its speed. (L.)

13 (a) A car of mass 1000 kg is initially at rest. It moves along a straight road for 20 s and then comes to rest again. The speed-time graph for the movement is (Figure 1C):   (i) What is the total distance travelled?   (ii) What resultant force acts on the car during the part of the motion represented by CD?   (iii) What is the momentum of the car when it has reached its maximum speed? Use this momentum value to find the constant resultant accelerating force.   (iv) During the part of the motion represented by OB on the graph, the constant resultant force found in (iii) is acting on the moving car although it is moving through air. Sketch a graph to show how the driving force would have to vary with time to produce this constant acceleration. Explain the shape of your graph.
  (b) If, when travelling at this maximum speed, the 1000-kg car had struck and remained attached to a stationary vehicle of mass 1500 kg, with what speed would the interlocked vehicles have travelled immediately after collision?
Calculate the kinetic energy of the car just prior to this collision and the kinetic energy of the interlocked vehicles just afterwards. Comment upon the values obtained.

Explain how certain design features in a modern car help to protect the driver of a car in such a collision. (L.)

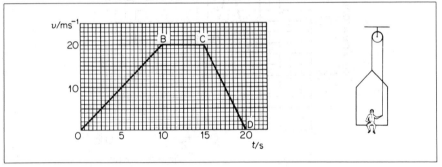

Figure 1C   **Figure 1D**

14 Answer the following questions making particular reference to the physical principles concerned:
  (a) explain why the load on the back wheels of a motor car increases when the vehicle is accelerating;
  (b) the diagram, Figure 1D, shows a painter in a crate which hangs alongside a building. When the painter who weighs 1000 N pulls on the rope the force he exerts on the floor of the crate is 450 N. If the crate weighs 250 N find the acceleration. (_JMB._)

15 (a) State Newton's laws of motion. Explain how the _newton_ is defined from these laws.
  (b) A rocket is propelled by the emission of hot gases. It may be stated that both the rocket and the emitted hot gases each gain kinetic energy and momentum during the firing of the rocket.
    Discuss the significance of this statement in relation to the laws of conservation of energy and momentum, explaining the essential difference between these two quantities.
  (c) A bird of mass 0·5 kg hovers by beating its wings of effective area 0·3 m^2. (i) What is the upward force of the air on the bird? ($g = 9·8\,\text{N}\,\text{kg}^{-1}$)   (ii) What

is the downward force of the bird on the air as it beats its wings? (iii) Estimate the velocity imparted to the air, which has a density of $1\cdot3\,\mathrm{kg\,m^{-3}}$, by the beating of the wings.

Which of Newton's laws is applied in each of (i), (ii) and (iii) above? (*L.*)

16 In an elastic head-on collision, a ball of mass $1\cdot0\,\mathrm{kg}$ moving at $4\cdot0\,\mathrm{m\,s^{-1}}$ collides with a stationary ball of mass $2\cdot0\,\mathrm{kg}$. Calculate the velocities of the balls after the collision indicating the directions in which they are then travelling. (*AEB.*)

17 Explain what is meant by a *force*. Define the SI unit in which it is measured.

Distinguish carefully the conditions under which
(a) linear momentum is conserved and
(b) kinetic energy is conserved.

A gun fires a shell with the horizontal component of its velocity equal to $200\,\mathrm{m\,s^{-1}}$. At the highest point in its flight, the shell explodes into three fragments. Two of these fragments, which have equal mass, fly off with equal speeds of $300\,\mathrm{m\,s^{-1}}$ relative to the ground, one along the flight direction of the shell at the instant of fragmentation and the other perpendicular to it and in a horizontal plane. Find the magnitude and direction of the velocity of the third fragment immediately after the explosion, assuming its mass is three times that of each of the other two fragments. Neglect air resistance.

Describe and explain qualitatively the subsequent motion of the three fragments. (*O. & C.*)

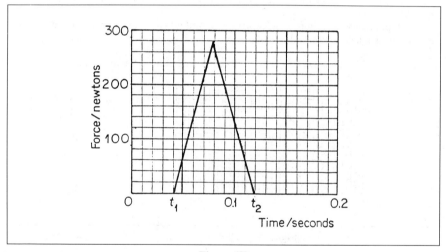

Figure 1E

18 A ball of mass $0\cdot16\,\mathrm{kg}$ moving with speed $35\,\mathrm{m\,s^{-1}}$ is struck by a bat and travels away from the bat in the opposite direction. What is the momentum of the ball before being struck?

The graph in Figure 1E shows the force which acts on the ball during its period of contact $t_1 t_2$ with the bat.

Calculate (i) the average force applied to the ball during its contact with the bat, (ii) the impulse applied to the ball, (iii) the speed of the ball after it has been struck by the bat. (*L.*)

Work, Energy, Power

In this section we deal with the important topics of work, energy and power, which are applied to the performance of all kinds of engines or machines, such as in cars, aeroplanes or the human body. The efficiency, for example, of any machine can be calculated from the ratio work (or power) out/work (or power) in.

Work

When an engine pulls a train with a constant force of 50 units through a distance of 20 units in its own direction, the engine is said by definition to do an amount of *work* equal to 50×20 or 1000 units, the product of the force and the distance. Thus if W is the amount of work,

---

$W = $ force $\times$ distance moved in direction of force

Work is a *scalar* quantity; it has no property of direction but only magnitude. When the force is one newton and the distance moved is one metre, then the work done is one *joule* (J). Thus a force of 50 N, moving through a distance of 10 m in its own direction, does 50×10 or 500 J of work.

The force to raise steadily a mass of 1 kg is equal to its weight, which is about 10 N (see p. 19). Thus if the mass of 1 kg is raised vertically through 1 m, then, approximately, work done $= 10\,(\text{N}) \times 1\,(\text{m}) = 10\,\text{J}$.

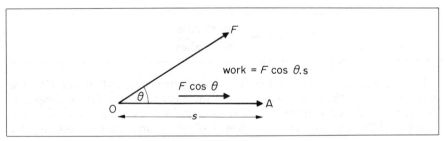

Figure 1.25   *Work and displacement*

Before leaving the topic of 'work', the reader should note carefully that we have assumed the force to move an object in its own direction. Suppose, however, that a force F pulls an object a distance s along a line OA acting at an angle θ to it, Fig. 1.25. The component of F along OA is $F \cos \theta$ (p. 14), and this is the effective part of F pulling along the direction OA. The component of F along a direction perpendicular to OA has no effect along OA. Consequently

work done $= F \cos \theta \times s$

In general, the work done by a force is equal to the product of the force and the displacement in the direction of the force.

Energy and Work

An engine does work when it pulls a train along a horizontal track. As a result of the work done, *energy* is transferred to the train. Assuming no energy losses, the amount of energy of the moving train is equal to the work done. The moving train has *kinetic energy*, or mechanical energy. Some of the chemical energy of the fuel used by the engine, or some of the electrical energy used by the engine if it is driven electrically, is thus transferred to mechanical energy.

When you wind a watch, you do some work. The work done is equal to the energy transferred to the moving parts of the watch. If a spring is wound up, the molecules in the metal spring are now closer together than before and they have gained molecular or 'elastic' energy. When you walk upstairs, you do work in moving your weight upwards. You have then gained mechanical *potential* energy, which is equal to the work done. The potential energy gained comes from the transfer of some chemical energy in your body.

Suppose an elastic band is pulled out steadily through a small distance of 1 cm or 0·01 m, and the force exerted increases steadily from zero to 10 N. Since the force is proportional to the extension of the band (p. 134), the work done = average force × distance moved. So

$$\text{work done} = \tfrac{1}{2}(0+10)\,\text{N} \times 0\cdot01\,\text{m}$$
$$= 5 \times 0\cdot01 = 0\cdot05\,\text{J}$$

This is the energy gained by the stretched elastic.

Power

When an engine does work quickly, it is said to be operating at a high *power*, if it does work slowly it is said to be operating at a lower power. 'Power' is defined as the *work done per second*, or energy spent per second. So

$$\textbf{power} = \frac{\textbf{work done}}{\textbf{time taken}}$$

The practical unit of power, the SI unit, is 'joule per second' or *watt* (W); the watt is defined as the rate of working at 1 joule per second. A 100 W lamp uses 100 J per second. A car engine of 1·2 kW uses 1200 J s^{-1}, since 1 kW (kilowatt) = 1000 W. 1 MW (megawatt) = 1 million (10^6) W, a unit used in industry for power.

Suppose a car is moving steadily with a velocity of v metres per second and the force due to the engine is F newtons. Then

$$\text{engine power} = \text{work done per second}$$
$$= F \times \text{distance per second} = F \times v$$

Power of engine $= \boldsymbol{F \times v}$

Examples on Power

1 (a) A car of mass 1000 kg moving on a horizontal road with a steady velocity of 10 m s^{-1} has a total frictional force on it of 400 N. Find the power due to the engine, Figure 1.26 (i).
   (b) The car now climbs a hill at an angle of 8° to the horizontal, Figure 1.26 (ii).
      Assuming the frictional force stays constant at 400 N, what engine power is now needed to keep the car moving at 10 m s^{-1}?

(*Analysis* (i) Power = $F \times v$, so we need F due to engine. (ii) In climbing the

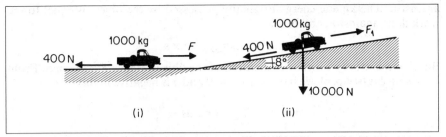

Figure 1.26   *Motion on inclined plane*

hill, the new force F_1 now has to overcome the component of the weight down the hill.)

(a) Since velocity is steady, acceleration = 0. So resultant force on car = 0. So if F is engine force,

$$F - 400 = 0 \quad \text{or} \quad F = 400 \, \text{N}$$

$$\therefore \text{power} = F \times v = 400 \times 10 = 4000 \, \text{W} = 4 \, \text{kW}$$

(b) Since velocity uphill is steady, acceleration uphill = 0 = resultant force uphill.

Now component of weight downhill = $10\,000 \sin 8°$ (or $\cos 82°$) = 1390 N. So if new engine force uphill is F_1,

$$F_1 - 400 - 1390 = 0 \quad \text{or} \quad F_1 = 1790 \, \text{N}$$

$$\therefore \text{new power} = F_1 \times v = 1790 \times 10 = 17\,900 \, \text{W} = 17·9 \, \text{kW}$$

2   Sand drops vertically at the rate of $2 \, \text{kg s}^{-1}$ on to a conveyor belt moving horizontally with a velocity of $0·1 \, \text{m s}^{-1}$. Calculate   (i) the extra power needed to keep the belt moving,   (ii) the rate of change of kinetic energy of the sand. Why is the power twice as great as the rate of change of kinetic energy?

(i) Force required to keep belt moving = rate of increase of horizontal momentum of sand = mass per second $(dm/dt) \times$ velocity change = $2 \times 0·1 = 0·2$ newton.

$$\therefore \text{power} = \text{work done per second} = \text{force} \times \text{rate of displacement}$$

$$= \text{force} \times \text{velocity} = 0·2 \times 0·1 = 0·02 \, \text{watt}$$

(ii) Kinetic energy of sand = $\frac{1}{2}mv^2$

$$\therefore \text{rate of change of energy} = \frac{1}{2}v^2 \times \frac{dm}{dt}, \text{ since } v \text{ is constant}$$

$$= \frac{1}{2} \times 0·1^2 \times 2 = 0·01 \, \text{watt}$$

Thus the power supplied is twice as great as the rate of change of kinetic energy. The extra power is due to the fact that the sand does not immediately assume the velocity of the belt, so that the belt at first moves relative to the sand. The extra power is needed to overcome the friction between the sand and belt.

Kinetic Energy

An object is said to possess *energy* if it can do work. When an object possesses energy because it is moving, the energy is said to be *kinetic*, e.g., a flying stone can break a window. Suppose that an object of mass m is moving with a velocity u, and is gradually brought to rest in a distance s by a constant force F acting against it. The kinetic energy originally possessed by the object is equal to the work done against F, and hence

$$\text{kinetic energy} = F \times s$$

But $F = ma$, where a is the deceleration of the object. Hence $F \times s = mas$. From $v^2 = u^2 + 2as$ (see p. 6), we have, since $v = 0$ and a is negative in this case,

$$0 = u^2 - 2as \quad \text{or} \quad as = \frac{u^2}{2}$$

$$\therefore \; \textbf{\textit{kinetic energy}} = \textbf{\textit{mas}} = \tfrac{1}{2}\textbf{\textit{mu}}^2$$

When m is in kg and u is in m s^{-1}, then $\tfrac{1}{2}mu^2$ is in *joule* J. Thus a car of mass 1000 kg, moving with a velocity of 36 km h^{-1} or 10 m s^{-1}, has an amount W of kinetic energy given by

$$W = \tfrac{1}{2}mu^2 = \tfrac{1}{2} \times 1000 \times 10^2 = 50\,000\,\text{J}$$

Kinetic Energies due to Explosive Forces

Suppose that, due to an explosion or nuclear reaction, a particle of mass m breaks away from the total mass concerned and moves with velocity v, and that the mass M left moves with velocity V in the opposite direction.

The kinetic energy E_1 of the mass m is given by

$$E_1 = \tfrac{1}{2}mv^2 = \frac{(mv)^2}{2m} = \frac{p^2}{2m} \qquad \cdot \qquad \cdot \qquad \cdot \qquad \cdot \qquad (1)$$

where p is the momentum mv of the mass. Similarly, the kinetic energy E_2 of the mass M is given by

$$E_2 = \tfrac{1}{2}MV^2 = \frac{p^2}{2M} \qquad \cdot \qquad \cdot \qquad \cdot \qquad \cdot \qquad (2)$$

because numerically the momentum $MV = mv = p$, from the conservation of momentum. Dividing (1) by (2), we see that

$$\frac{E_1}{E_2} = \frac{1/m}{1/M} = \frac{M}{m}$$

Hence the energy is *inversely*-proportional to the masses of the particles, that is, the smaller mass, m say, has the larger energy. Thus if E is the total energy of the two masses, the energy of the smaller mass $= ME/(M+m)$.

Suppose a bullet of mass $m = 50\,\text{g}$ is fired from a rifle of mass $M = 2\,\text{kg} = 2000\,\text{g}$ and that the total kinetic energy produced by the explosion is 2050 J. Since the energy is shared inversely as the masses,

$$\text{kinetic energy of bullet} = \frac{2000}{2000+50} \times 2050\,\text{J}$$

$$= \frac{2000}{2050} \times 2050\,\text{J} = 2000\,\text{J}$$

So $\qquad \text{kinetic energy of rifle} = 50\,\text{J}$

An α-particle has a mass of 4 units and a radium nucleus a mass of 228 units. If disintegration of a stationary thorium nucleus, mass 232, produces an α-particle and radium nucleus, and a release of energy of 4·05 MeV, where 1 MeV =

1.6×10^{-13} J, then

$$\text{energy of } \alpha\text{-particle} = \frac{228}{(4+228)} \times 4.05 = 3.98 \text{ MeV}$$

So the α-particle travels a relatively long distance before coming to rest compared with the comparatively heavy radium nucleus, which moves back or recoils a small distance.

Gravitational Potential Energy

A mass held stationary above the ground has energy, because, when released, it can raise another object attached to it by a rope passing over a pulley, for example. A coiled spring also has energy, which is released gradually as the spring uncoils. The energy of the weight or spring is called _potential energy_, because it arises from the position or arrangement of the body and not from its motion. In the case of the weight, the energy given to it is equal to the work done by the person or machine which raises it steadily to that position against the force of attraction of the earth. So this is _gravitational potential energy_. In the case of the spring, the energy is equal to the work done in displacing the molecules from their normal equilibrium positions against the forces of attraction of the surrounding molecules. So this is _molecular potential energy_.

If the mass of an object is m, and the object is held stationary at a height h above the ground, the energy released when the object falls to the ground is equal to the work done

$$= \text{force} \times \text{distance} = \text{weight of object} \times h$$

Suppose the mass m is 5 kg, so that the weight is 5×9.8 N or 49 N, and h is 4 metre. Then

$$\text{gravitational potential energy, p.e.} = 49 \text{ (N)} \times 4 \text{ (m)} = 196 \text{ J}$$

Generally, when a mass m is moved through a height h,

change in gravitational potential energy = _mgh_

where m is in kg, h is in metre, $g = 9.8 \text{ N kg}^{-1}$.

This formula assumes that 'g' is constant throughout the height h. Near the earth's surface 'g' is fairly constant. But if a mass such as a space vehicle is sent up from the earth to an orbit high above the earth, then the gravitational field strength varies appreciably throughout the height h and 'mgh' can not be used to find the gain in potential energy. We see later how the gain is calculated in this case.

Conservative Forces

If a ball of weight W is raised steadily from the ground to a point X at a height h above the ground, the work done is $W . h$. The potential energy, p.e., of the ball relative to the ground is thus $W . h$. Now whatever route is taken from ground level to X, the work done is the same—if a _longer_ path is chosen, for example, the component of the weight in the particular direction must then be overcome and so the force required to move the ball is correspondingly smaller. The p.e. of the ball at X is thus independent of the route to X. This implies that if the ball is taken in a closed path round to X again, _the total work done against the force of gravity is zero_. Work or energy has been expended on one part of the closed path, and regained on the remaining part.

When the work done in moving round a closed path in a field to the original point is zero, the forces in the field are called *conservative forces*. The earth's gravitational field is an example of a field containing conservative forces, as we now show.

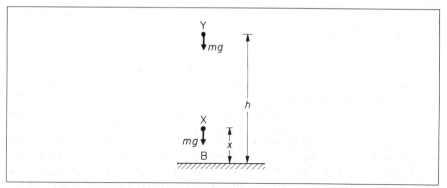

Figure 1.27   *Mechanical energy in a gravitational field*

Suppose the ball falls from a place Y at a height h to another X at a height of x above the ground, Figure 1.27. Then, if W is the weight of the ball and m its mass,

$$\text{the potential energy p.e. at X} = Wx = mgx$$

and   the kinetic energy k.e. at X $= \frac{1}{2}mv^2 = \frac{1}{2}m \cdot 2g(h-x) = mg(h-x)$

using $v^2 = 2as = 2g(h-x)$. Hence

$$\text{p.e.} + \text{k.e.} = mgx + mg(h-x) = mgh$$

Thus at any point such as X, the total mechanical energy of the falling ball is equal to the original energy at Y. The mechanical energy is hence constant or conserved. This is the case for a conservative field.

Principles of Conservation of Energy

When any object falls in the earth's gravitational field, a small part of the energy is used up in overcoming the resistance of the air. This energy is dissipated or lost as heat—it is not regained in moving the body back to its original position.

Although energy may be transferred from one form to another, such as from mechanical energy to heat energy as in this last example.

---

the total energy in a closed system is always constant.

If an electric motor is supplied with 1000 J of energy, for example, 850 J of mechanical energy, 140 J of heat energy and 10 J of sound energy may be produced. This is called the ***Principle of the Conservation of Energy*** and is one of the key principles in science.

Momentum and Energy in Gravitational Attraction

Consider a ball B held stationary at a height above the earth E. Relative to each other, the momentum and kinetic energy of B and E are both zero.

Suppose the ball is now released. The gravitational force of E on B accelerates the ball. So its momentum increases as it falls. From the law of conservation of momentum, the equal and opposite gravitational force of B on E produces an equal but opposite momentum on the earth. The earth is so heavy, however, that its velocity V towards B is extremely small. For example, suppose the mass m is 0.2 kg and the mass M of E is 10^{25} kg, and the velocity v of the ball B is $10\,\mathrm{m\,s}^{-1}$ at an instant. Then, from the conservation of momentum,

$$MV = mv$$

or $$V = \frac{m}{M} \times v = \frac{0.2}{10^{25}} \times 10 = 2 \times 10^{-25}\,\mathrm{m\,s}^{-1}$$

So the earth's velocity V is extremely small.

The total energy of B and E remains constant while the ball B is falling, since there are no external forces acting on the system. When B hits the ground the force (action) of B on E is equal and opposite to the force (reaction) of E on B, and the forces act for the same short time. So the total momentum of B and E is conserved. Thus when the ball rebounds, B moves upward with a momentum change equal and opposite to that of E. As B continues to rise its velocity and momentum decreases. So the momentum of E decreases. When B reaches its maximum height its momentum is zero. So the momentum of E is then zero.

Although momentum is conserved on collision with the earth, some mechanical energy is transformed to heat and sound. Hence the total mechanical energy of B and E is less after collision. As we showed on page 31, the ratio of the kinetic energy of the ball to the earth after collision is *inversely*-proportional to their masses. So with the above figures,

$$\frac{\text{kinetic energy of earth}}{\text{kinetic energy of ball}} = \frac{0.2}{10^{25}} = 2 \times 10^{-26}$$

Hence the kinetic energy of the earth after collision is extremely small. Practically all the kinetic energy is transferred to the ball.

Motion, Momentum and Energy Graphs in Gravitational Field

Figure 1.28 shows roughly the variation with *time t* of the speed, velocity, acceleration, distance, momentum and energy (kinetic and potential) of a ball thrown vertically upwards from the ground and then bouncing once on returning to the ground. See page 42.

Note that (i) speed is a scalar but velocity is a vector, (ii) the deceleration of the rising ball and the acceleration of the falling ball are both numerically g; but on hitting the ground and rising, the acceleration changes at contact with the ground and becomes opposite in direction as shown, (iii) momentum is a vector, (iv) energy is a scalar.

Dimensions

By the *dimensions* of a physical quantity we mean the way it is related to the fundamental quantities mass, length and time; these are usually denoted by M, L and T respectively. An area, length × breadth, has dimensions $L \times L$ or L^2; a volume has dimensions L^3; density, which is mass/volume, has dimensions M/L^3 or ML^{-3}; an angle has no dimensions, since it is the ratio of two lengths.

As an area has dimensions L^2, the *unit* may be written in terms of the metre as m^2. Similarly, the dimensions of a volume are L^3 and hence the unit is m^3. Density has dimensions ML^{-3}. The density of mercury is thus written as $13\,600\,\mathrm{kg\,m}^{-3}$. If some physical quantity has dimensions $ML^{-1}T^{-1}$, its unit may be written as $\mathrm{kg\,m}^{-1}\,\mathrm{s}^{-1}$.

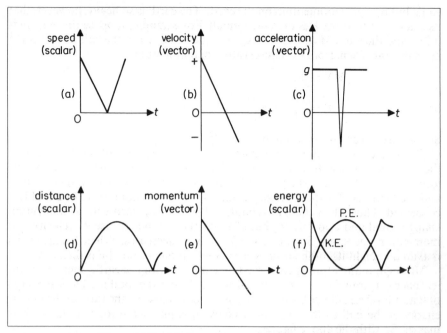

Figure 1.28   *Motion, momentum and energy graphs*

The following are the dimensions of some quantities in Mechanics:

Velocity. Since velocity $= \dfrac{\text{displacement}}{\text{time}}$, its dimensions are L/T or LT^{-1}.

Acceleration. The dimensions are those of velocity/time, i.e., L/T^2 or LT^{-2}.

Force. Since force $=$ mass $\times$ acceleration, its dimensions are MLT^{-2}.

Work or Energy. Since work $=$ force $\times$ distance, its dimensions are ML^2T^{-2}.

Example

In the gas equation $\left(p + \dfrac{a}{V^2}\right)(V - b) = RT$, what are the dimensions of the constants a and b?

p represents pressure, V represents volume. The quantity a/V^2 must represent a pressure since it is added to p. The dimensions of $p = [\text{force}]/[\text{area}] = \text{MLT}^{-2}/\text{L}^2 = \text{ML}^{-1}\text{T}^{-2}$; the dimensions of $V = \text{L}^3$. Hence

$$\frac{[a]}{\text{L}^6} = \text{ML}^{-1}\text{T}^{-2} \quad \text{or} \quad [a] = \text{ML}^5\text{T}^{-2}$$

The constant b must represent a volume since it is subtracted from V. Hence

$$[b] = \text{L}^3$$

Application of Dimensions, Simple Pendulum

We can often use dimensions to solve problems. As an example, suppose a small mass is suspended from a long thread so as to form a simple pendulum. We may reasonably suppose that the period, T, of the oscillations depends only on the mass m, the length l of the thread, and the acceleration, g, of free fall at the place concerned. Suppose then that

$$T = km^x l^y g^z. \qquad . \qquad . \qquad . \qquad . \qquad . \qquad \text{(i)}$$

where x, y, z, k are unknown numbers. The dimensions of g are LT^{-2} from p. 42. Now the dimensions of both sides of (i) must be the same.

$$\therefore T \equiv M^x L^y (LT^{-2})^z$$

Equating the indices of M, L, T on both sides, we have

$$x = 0$$
$$y + z = 0$$

and

$$-2z = 1$$
$$\therefore z = -\tfrac{1}{2}, \quad y = \tfrac{1}{2}, \quad x = 0$$

Thus, from (i), the period T is given by

$$T = kl^{\frac{1}{2}} g^{-\frac{1}{2}}$$

or

$$T = k\sqrt{\frac{l}{g}}$$

We cannot find the magnitude of k by the method of dimensions, since it is a number. A complete mathematical investigation shows that $k = 2\pi$ in this case, and hence $T = 2\pi\sqrt{l/g}$. (See also p. 87.) Note that T is independent of the mass m.

Velocity of Transverse Wave in a String

As another illustration of the use of dimensions, consider a wave set up in a stretched string by plucking it. The velocity, c, of the wave depends on the tension, F, in the string, its length l, and its mass m, and we can therefore suppose that

$$c = kF^x l^y m^z. \qquad . \qquad . \qquad . \qquad . \qquad . \qquad \text{(i)}$$

where x, y, z are numbers we hope to find by dimensions and k is a constant.

The dimensions of velocity, c, are LT^{-1}, the dimensions of tension, F, are MLT^{-2}, the dimensions of length, l, is L, and the dimension of mass, m, is M. From (i), it follows that

$$LT^{-1} \equiv (MLT^{-2})^x \times L^y \times M^z$$

Equating powers of M, L, and T on both sides,

$$\therefore 0 = x + z \qquad . \qquad . \qquad . \qquad . \qquad . \qquad \text{(i)}$$
$$1 = x + y \qquad . \qquad . \qquad . \qquad . \qquad . \qquad \text{(ii)}$$

and

$$-1 = -2x \qquad . \qquad . \qquad . \qquad . \qquad \text{(iii)}$$
$$\therefore x = \tfrac{1}{2}, \quad z = -\tfrac{1}{2}, \quad y = \tfrac{1}{2}$$
$$\therefore c = k \, . \, F^{\frac{1}{2}} l^{\frac{1}{2}} m^{-\frac{1}{2}}$$

or

$$c = k\sqrt{\frac{Fl}{m}} = k\sqrt{\frac{F}{m/l}} = k\sqrt{\frac{\text{Tension}}{\text{mass per unit length}}} = k\sqrt{\frac{T}{\mu}}$$

where μ is the mass per unit length. A complete mathematical investigation shows that $k = 1$.

The method of dimensions can thus be used to find the relation between quantities when the mathematics is too difficult. It has been extensively used in hydrodynamics, for example.

_____ Exercises 1C _____

Energy, Power, Dimensions

(Assume $g = 10\,m\,s^{-2}$ or $10\,N\,kg^{-1}$ unless otherwise stated)

1   An object A of mass 10 kg is moving with a velocity of 6 m s^{-1}. Calculate its kinetic energy and its momentum.
    If a constant opposing force of 20 N suddenly acts on A, find the time it takes to come to rest and the distance through which it moves.

2   An object A moving horizontally with kinetic energy of 800 J experiences a constant horizontal opposing force of 100 N while moving from a place X to a place Y, where XY is 2 m. What is the energy of A at Y?
    In what further distance will A come to rest if this opposing force continues to act on it?

3   A ball of mass 0·1 kg is thrown vertically upwards with an initial speed of 20 m s^{-1}. Calculate   (i) the time taken to return to the thrower,   (ii) the maximum height reached,   (iii) the kinetic and potential energies of the ball half-way up.

4   A 4 kg ball moving with a velocity of 10·0 m s^{-1} collides with a 16 kg ball moving with a velocity of 4·0 m s^{-1} (i) in the same direction,   (ii) in the opposite direction. Calculate the velocity of the balls in each case if they coalesce on impact, and the loss of energy resulting from the impact. State the principle used to calculate the velocity.

5   A ball of mass 0·1 kg is thrown vertically upwards with a velocity of 20 m s^{-1}. What is the potential energy at the maximum height? What is the potential energy of the ball when it reaches three-quarters of the maximum height while moving upwards?

6   A stone is projected vertically upwards and eventually returns to the point of projection. Ignoring any effects due to air resistance draw sketch graphs to show the variation with time of the following properties of the stone: (i) velocity,   (ii) kinetic energy,   (iii) potential energy,   (iv) momentum,   (v) distance from point of projection,   (vi) speed. (*AEB*, 1982.)

7   A stationary mass explodes into two parts of mass 4 units and 40 units respectively. If the larger mass has an initial kinetic energy of 100 J, what is the initial kinetic energy of the smaller mass? Explain your calculation.

8   What is an *elastic* and an *inelastic* collision? Give one example of each type. A bullet of mass 10 g is fired vertically with a velocity of 100 m s^{-1} into a block of wood of mass 190 g suspended by a long string above the gun. If the bullet comes to rest in the block, through what height does the block move?

9   A car of mass 1000 kg moves at a constant speed of 20 m s^{-1} along a horizontal road where the frictional force is 200 N. Calculate the power developed by the engine.
    If the car now moves up an incline at the same constant speed, calculate the new power developed by the engine. Assume that the frictional force is still 200 N and that sin $\theta = 1/20$, where θ is the angle of the incline to the horizontal.

10   Which of the following are   (i) scalars,   (ii) vectors? Obtain the dimensions of each.
    A momentum   B work   C speed   D force   E energy   F weight   G mass   H acceleration.

11   A horizontal force of 2000 N is applied to a vehicle of mass 400 kg which is initially at rest on a horizontal surface. If the total force opposing motion is constant at 800 N, calculate   (i) the acceleration of the vehicle,   (ii) the kinetic energy of the vehicle 5 s after the force is first applied,   (iii) the total power developed 5 s after the force is first applied. (*AEB*, 1985.)

12   The volume per second of a liquid flowing through a horizontal pipe of length *l* is given by $kpa^x/l\eta$, where k is a constant, p is the excess pressure (force per unit area),

a is the radius of the pipe and η is a frictional quantity of dimensions $ML^{-1}T^{-1}$. By dimensions, find the number x.

13   The period of vibration t of a liquid drop is given by $t = ka^x p^y \gamma^z$, where k is a constant, a is the radius of the drop, ρ is the density of the liquid and γ is the surface tension of dimensions MT^{-2}.

By dimensions, find the values of the indices x, y, z and the relation for t.

14   Explain what is meant by *kinetic energy*, and show that for a particle of mass m moving with velocity v, the kinetic energy is $\frac{1}{2}mv^2$.

A steel ball is

(a) projected horizontally with velocity v, at a height h above the ground,

(b) dropped from a height h and bounces on a fixed horizontal steel plate.

Neglecting air resistance and using suitable sketch graphs, explain how the kinetic energy of the ball varies in (a) with its height above the ground, and in (b) with its height above the plate. (*JMB.*)

15   Sand falls at a rate of $0.15\,kg\,s^{-1}$ on to a conveyor belt moving horizontally at a constant speed of $2\,m\,s^{-1}$. Calculate

(a) the extra force necessary to maintain this speed,

(b) the rate at which work is done by this force,

(c) the change in kinetic energy per second of the sand on the belt.

Account for the difference between your answers to (b) and (c). (*JMB.*)

16   A railway truck of mass 4×10^4 kg moving at a velocity of $3\,m\,s^{-1}$ collides with another truck of mass 2×10^4 kg which is at rest. The couplings join and the trucks move off together. What fraction of the first truck's initial kinetic energy remains as kinetic energy of the two trucks after the collision? Is energy conserved in a collision such as this? Explain your answer briefly. (*L.*)

17   A body moving through air at a high speed v experiences a retarding force F given by $F = kA\rho v^x$, where A is the surface area of the body, ρ is the density of the air and k is a numerical constant. Deduce the value of x.

A sphere of radius 50 mm and mass 1.0 kg falling vertically through air of density $1.2\,kg\,m^{-3}$ attains a steady velocity of $11.0\,m\,s^{-1}$. If the above equation then applies to its fall what is the value of k in this case? (*L.*)

18   A ball falls freely to Earth from a height H and rebounds to a height $h(<H)$. Discuss the linear momentum and energy changes that occur during   (i) the fall, (ii) the rebound, with reference to the principles of conservation of momentum and energy. The mass of the Earth, though very large compared with that of the ball, should not be taken as infinite. (Detailed mathematical treatment is not required.)

In a pile-driver a mass m falls freely from height H on to a vertical post of mass M and does not rebound. The ground exerts a constant force F opposing the motion of the post into the ground. The post is driven in a distance d. Find an expression for F (the motion of the earth due to the impact may be neglected). (*O. & C.*)

19   An α-particle having a speed of $1.00 \times 10^6\,m\,s^{-1}$ collides with a stationary proton which gains an initial speed of $1.60 \times 10^6\,m\,s^{-1}$ in the direction in which the α-particle was travelling.

What is the speed of the α-particle immediately after the collision? How much kinetic energy is gained by the proton in the collision?

It is known that this collision is perfectly elastic. Explain what this means. (Mass of α-particle $= 6.64 \times 10^{-27}$ kg.   Mass of proton $= 1.66 \times 10^{-27}$ kg.) (*L.*)

20   Explain what is meant by an *elastic collision* and a *completely inelastic collision*.

How much heat energy is produced

(a) when 0.5 kg of putty falls from a height of 2 m onto a rigid floor,

(b) when a ball of mass 0.5 kg falls from the same height onto the same floor and bounces to a height of 1 m?

5 g of fine sand is poured at a steady rate over 20 s onto the pan of a direct-reading chemical balance from a height of 0.25 m. Draw a graph showing how the reading of the balance varies with time, starting from the instant that pouring commences.

How, if at all, would the graph change if   (i) the height were doubled,   (ii) the time of pouring halved? (*W.*)

21   (a) A particle of mass m, initially at rest, is acted upon by a constant force until its velocity is v. Show that the kinetic energy of the particle is $\frac{1}{2}mv^2$.

(b) A train of mass 2.0×10^5 kg moves at a constant speed of 72 km h^{-1} up a straight incline against a frictional force of 1.28×10^4 N. The incline is such that the train rises vertically 1·0 m for every 100 m travelled along the incline. Calculate (i) the rate of increase per second of the potential energy of the train, (ii) the necessary power developed by the train. (*JMB.*)

22 Explain what is meant by *energy* and distinguish between *potential energy* and *kinetic energy*.

Define *linear momentum* and state the law of conservation of linear momentum. Describe an experiment to verify this law.

A rocket of mass M is moving at constant speed in free space and initially has kinetic energy E. An explosive charge of negligible mass divides it into three parts of equal mass $M/3$ in such a way that one part moves in the same direction as the parent rocket with kinetic energy $E/3$, and the other two portions move off at an angle 60° to this direction. Determine the total energy W imparted to the parts of the rocket in the explosion. (*O. & C.*)

23 Define *linear momentum*.

Describe an experiment which can be performed to investigate inelastic collisions between two bodies moving in one dimension. Explain how the velocities of the bodies can be measured. Summarise the results which would be obtained in terms of kinetic energies and momenta of the bodies before and after impact. How would these summarised results differ if the collision were perfectly elastic?

Two identical steel balls B and C lie in a smooth horizontal straight groove so that they are touching. A third identical ball A moves at a speed v along the groove and collides with B. Assuming that the collisions are all perfectly elastic explain why it is impossible for

(a) A to stop while B and C move off together at speed $v/2$,

(b) A to stop while B and C move off together at speed $v/\sqrt{2}$. (*L.*)

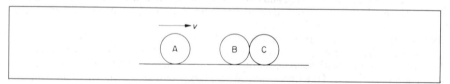

Figure 1E

24 As shown in the diagram, two trolleys P and Q of masses 0·50 kg and 0·30 kg respectively are held together on a horizontal track against a spring which is in a state of compression. When the spring is released the trolleys separate freely and P moves to the left with an initial velocity of 6 m s^{-1}. Calculate

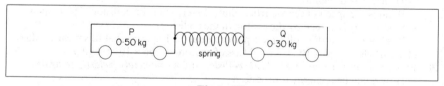

Figure 1F

(a) the initial velocity of Q,

(b) the initial total kinetic energy of the system.

Calculate also the initial velocity of Q if trolley P is held still when the spring under the same compression as before is released. (*JMB.*)

25 Define linear momentum and state the principle of conservation of linear momentum. Explain briefly how you would attempt to verify this principle by experiment.

Sand is deposited at a uniform rate of 20 kilogram per second and with negligible kinetic energy on to an empty conveyor belt moving horizontally at a constant speed of 10 metre per minute. Find

(a) the force required to maintain constant velocity,

(b) the power required to maintain constant velocity, and

(c) the rate of change of kinetic energy of the moving sand.

Why are the latter two quantities unequal? (*O. & C.*)

26 What do you understand by the *conservation of energy*? Illustrate your answer by reference to the energy changes occurring

(a) in a body whilst falling to and on reaching the ground,

(b) in an X-ray tube.

The constant force resisting the motion of a car, of mass 1500 kg, is equal to one-fifteenth of its weight. If, when travelling at 48 km per hour, the car is brought to rest in a distance of 50 m by applying the brakes, find the additional retarding force due to the brakes (assumed constant) and the heat developed in the brakes. (*JMB.*)

2
Circular motion, Gravitation, Simple Harmonic Motion

In the previous chapter we discussed the motion of an object moving in a straight line. There are many cases of objects moving in a curve or circular path about some point, such as bicycles or cars turning round corners or racing cars going round circular tracks. The earth and other planets move round the Sun in roughly circular paths. In place of speed in linear motion, we then have to use 'angular speed'. This helps to find the 'period' or time to go once round the circle. We shall also find that objects moving at constant speed round a circle have an acceleration towards the centre of the circle.

Circular Motion

Angular Speed

Consider an object moving in a circle with a *uniform speed* round a fixed point O as centre, Figure 2.1.

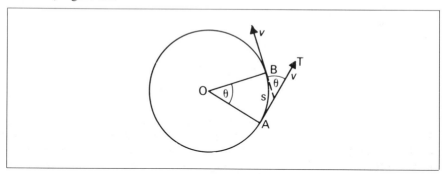

Figure 2.1 *Circular motion*

If the object moves from A to B so that the radius OA moves through an angle θ, its *angular speed*, ω, about O is defined as the *change of the angle per second*. So if t is the time taken by the object to move from A to B,

$$\omega = \frac{\theta}{t} \qquad . \qquad . \qquad . \qquad . \qquad . \qquad . \qquad (1)$$

Angular speed is usually expressed in 'radian per second' (rad s^{-1}). From (1),

$$\theta = \omega t \qquad . \qquad . \qquad . \qquad . \qquad . \qquad . \qquad (2)$$

which is analogous to the formula 'distance = uniform velocity × time' for motion in a straight line. It will be noted that the time T to describe the circle once, known as the *period* of the motion, is given by

$$T = \frac{2\pi}{\omega} \qquad . \qquad . \qquad . \qquad . \qquad . \qquad . \qquad (3)$$

since 2π radians is the angle in 1 revolution.

If s is the length of the arc AB, then $s/r = \theta$, by definition of an angle in radians.

$$\therefore s = r\theta$$

Dividing by t, the time taken to move from A to B,

$$\therefore \frac{s}{t} = r\frac{\theta}{t}$$

But $s/t =$ the *speed*, v, of the rotating object, and θ/t is the angular velocity.

$$\therefore v = r\omega \ . \qquad . \qquad . \qquad . \qquad . \qquad (4)$$

Example

A model car moves round a circular track of radius 0·3 m at 2 revolutions per second.
   What is
(a) the angular speed ω,
(b) the period T,
(c) the speed v of the car? Find also
(d) the angular speed of the car if it moves with a uniform speed of $2\,\mathrm{m\,s}^{-1}$ in a circle of radius 0·4 m.

(a) For 1 revolution, angle turned $\theta = 2\pi$ rad (360°). So

$$\omega = 2 \times 2\pi = 4\pi \text{ rad s}^{-1}$$

(b) Period $T =$ time for 1 rev $= \dfrac{2\pi}{\omega} = \dfrac{2\pi}{4\pi} = 0\text{·}5\,\mathrm{s}$. (Or, $T = 1\,\mathrm{s}/2\,\mathrm{rev} = 0\text{·}5\,\mathrm{s}$.)

(c) Speed $v = r\omega = 0\text{·}3 \times 4\pi = 1\text{·}2\pi = 3\text{·}8\,\mathrm{m\,s}^{-1}$
(d) From $v = r\omega$

$$\omega = \frac{v}{r} = \frac{2\,\mathrm{m\,s}^{-1}}{0\text{·}4\,\mathrm{m}} = 5\,\mathrm{rad\,s}^{-1}$$

Acceleration in a Circle

When a stone is attached to a string and whirled round at constant speed in a circle, one can feel the force (pull) in the string needed to keep the stone moving in its circular path. Although the stone is moving with a constant speed, the presence of the force implies that the stone has an *acceleration*.

The force on the stone acts *towards the centre* of the circle. We call it a *centripetal force*. The direction of the acceleration is in the same direction as the force, that is, towards the centre. We now show that if v is the uniform speed in the circle and r is the radius of the circle,

$$\textbf{acceleration towards centre} = \frac{v^2}{r} \qquad . \qquad . \qquad . \qquad (1)$$

or, since $v = r\omega$,

$$\textbf{acceleration towards centre} = \frac{r^2\omega^2}{r} = r\omega^2 \ . \qquad . \qquad . \qquad (2)$$

The dimensions of v are LT^{-1} and of r is L. So v^2/r has the dimensions LT^{-2}, which is an acceleration. Also, the dimension of ω is T^{-1}, so $r\omega^2$ has the dimensions LT^{-2}, which is an acceleration.

Proof of v^2/r or $r\omega^2$

To obtain an expression for the acceleration towards the centre, consider an object moving with a constant speed v round a circle of radius r, Figure 2.2 (i). At A, its *velocity* v_A is in the direction of the tangent AC; a short time Δt later at B, its velocity v_B is in the direction of the tangent BD. Since their directions are different, the velocity v_B is different from the velocity v_A, although their magnitudes are both equal to v. Thus a velocity change or acceleration has occurred from A to B.

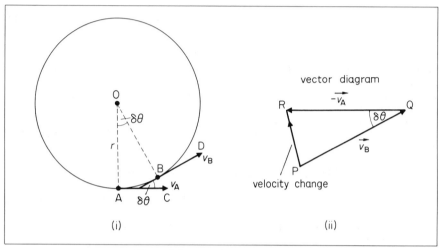

Figure 2.2   *Acceleration in a circle*

The velocity change from A to B $= \overrightarrow{v_B} - \overrightarrow{v_A} = \overrightarrow{v_B} + (-\overrightarrow{v_A})$. The arrows denote vector quantities. In Figure 2.2 (ii), PQ is drawn to represent $\overrightarrow{v_B}$ in magnitude (v) and direction (BD); QR is drawn to represent $(-\overrightarrow{v_A})$ in magnitude (v) and direction (CA). Then, as shown on p. 12,

$$\text{velocity change} = \overrightarrow{v_B} + (-\overrightarrow{v_A}) = \text{PR}$$

When Δt is small, the angle AOB or $\Delta\theta$ is small. Thus angle PQR, equal to $\Delta\theta$, is small. PR then points towards O, the centre of the circle. *The velocity change or acceleration is thus directed towards the centre.*

The magnitude of the acceleration, a, is given by

$$a = \frac{\text{velocity change}}{\text{time}} = \frac{\text{PR}}{\Delta t}$$

$$= \frac{v \cdot \Delta\theta}{\Delta t}$$

since $\text{PR} = v \cdot \Delta\theta$. In the limit, when Δt approaches zero, $\Delta\theta/\Delta t = d\theta/dt = \omega$, the angular speed. But $v = r\omega$ (p. 49). Hence, since $a = v\omega$,

$$a = \frac{v^2}{r} \quad \text{or} \quad r\omega^2 \quad . \quad . \quad . \quad . \quad . \quad (5)$$

Thus an object moving in a circle of radius r with a constant speed v has an acceleration towards the centre equal to v^2/r or $r\omega^2$.

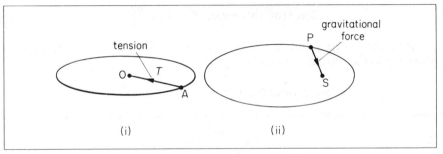

Figure 2.3 *Examples of centripetal forces*

Centripetal Forces

The centripetal force F required to keep an object of mass m moving in a circle of radius $r = ma = mv^2/r$. As already stated, it acts towards the centre of the circle. When a stone A is whirled in a horizontal circle of centre O by means of a string, the tension T provides the centripetal force, Figure 2.3(i). For a racing car moving round a circular track, the friction at the wheels provides the centripetal force. Planets such as P, moving in a circular orbit round the sun S, have a centripetal force due to gravitational attraction between S and P (p. 59), Figure 2.3(ii).

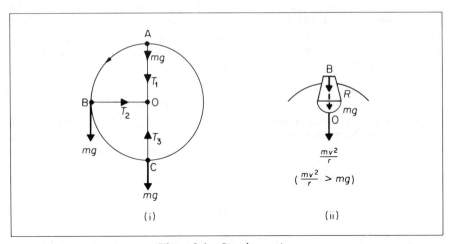

Figure 2.4 *Circular motion*

Figure 2.4(i) shows an object of mass m whirled with a high speed v in a *vertical* circle of centre O by a string of length r. At A, the top of the motion, suppose T_1 is the tension (force) in the string. Then, since the weight mg acts downwards towards the centre O,

$$\text{force towards centre, } F = T_1 + mg = \frac{mv^2}{r}$$

So
$$T_1 = \frac{mv^2}{r} - mg \qquad . \qquad . \qquad . \qquad . \qquad . \qquad (1)$$

At the point B, where OB is horizontal, suppose T_2 is the tension in the string. The weight mg acts vertically downwards and has no component in the horizontal direction BO. So

$$\text{force towards centre, } F = T_2 = \frac{mv^2}{r} \qquad . \qquad . \qquad . \qquad (2)$$

At C, the lowest point of the motion, the weight mg acts in the *opposite* direction to the tension T_3 in the string. So

$$\text{force towards centre, } F = T_3 - mg = \frac{mv^2}{r}$$

So

$$T_3 = \frac{mv^2}{r} + mg \qquad . \qquad . \qquad . \qquad . \qquad . \qquad (3)$$

From (1), (2) and (3), we see that
(a) the *maximum* tension is given by (3) and
(b) this occurs at the bottom C of the circle. Here the tension T_3 must be greater than mg by mv^2/r to make the object keep moving in a circular path.

The *minimum* tension is given in (1) and this occurs at A, the top of the motion. Here part of the required centripetal force is provided by the weight mg and the rest by T_1.

If some water is placed in a bucket B attached to the end of a string, the bucket can be whirled in a vertical plane without any water falling out. When the bucket is vertically above the point of support O, the weight mg of the water is less than the required force mv^2/r towards the centre and so the water stays in, Figure 2.4 (ii). The reaction R of the bucket base on the water provides the rest of the force mv^2/r. If the bucket is whirled slowly and $mg > mv^2/r$, part of the weight provides the force mv^2/r. The rest of the weight causes the water to accelerate downward and hence to leave the bucket.

Motion of Car (or Train) Round Banked Track

Suppose a car (or train) is moving round a banked track in a circular path of horizontal radius r, Figure 2.5 (i). If the only forces at the wheels A, B are the normal reaction forces R_1, R_2 respectively, that is, there is no side-slip or strain at the wheels, the force towards the centre of the track is $(R_1 + R_2) \sin \theta$, where θ is the angle of inclination of the plane to the horizontal.

$$\therefore (R_1 + R_2) \sin \theta = \frac{mv^2}{r} \qquad . \qquad . \qquad . \qquad . \qquad (i)$$

The car does not move in a vertical direction. So, for vertical equilibrium,

$$(R_1 + R_2) \cos \theta = mg \ . \qquad . \qquad . \qquad . \qquad . \qquad (ii)$$

Dividing (i) by (ii),

$$\therefore \tan \theta = \frac{v^2}{rg} \qquad . \qquad . \qquad . \qquad . \qquad (iii)$$

Thus for a given velocity v and radius r, the angle of inclination of the track for no side-slip must be $\tan^{-1}(v^2/rg)$. As the speed v increases, the angle θ increases, from (iii). A racing-track is made saucer-shaped because at high speeds the cars can move towards a part of the track which is steeper and sufficient to prevent side-slip, Figure 2.5 (ii). The outer rail of a curved railway track is raised above the inner rail so that the force towards the centre is largely provided by the component of the reaction at the wheels. It is desirable to bank a road at corners for the same reason as a racing track is banked.

An aeroplane with wings banked at an angle θ to the horizontal will make an

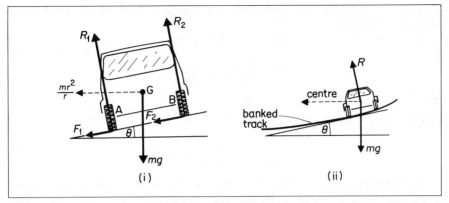

Figure 2.5   *Car on banked track*

aeroplane move with a speed v in a horizontal circular path of radius r, where $\tan \theta = v^2/rg$.

Conical Pendulum

Suppose a small object A of mass m is tied to a string OA of length l and then whirled round in a *horizontal* circle of radius r, with O fixed directly above the centre B of the circle, Figure 2.6. If the circular speed of A is constant, the string turns at a constant angle θ to the vertical. This is called a conical pendulum.

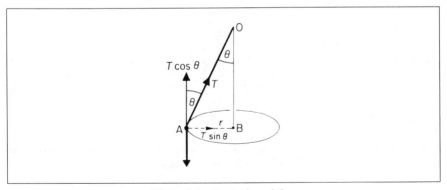

Figure 2.6   *Conical pendulum*

Since A moves with a constant speed v in a circle of radius r, there must be a centripetal force mv^2/r acting towards the centre B. The horizontal component, $T \sin \theta$, of the tension T in the string provides this force along AB. So

$$T \sin \theta = \frac{mv^2}{r} \qquad \cdot \qquad \cdot \qquad \cdot \qquad \cdot \qquad \cdot \qquad (1)$$

Also, since the mass does not move vertically, its weight mg must be counter-balanced by the vertical component $T \cos \theta$ of the tension. So

$$T \cos \theta = mg \qquad \cdot \qquad \cdot \qquad \cdot \qquad \cdot \qquad \cdot \qquad (2)$$

Dividing (1) by (2), then $\qquad\qquad \boldsymbol{\tan \theta} = \dfrac{v^2}{rg}$

A similar formula for θ was obtained for the angle of banking of a track which prevented side-slip.

If $v = 2\,\mathrm{m\,s^{-1}}, r = 0.5\,\mathrm{m}$ and $g = 10\,\mathrm{m\,s^{-2}}$, then

$$\tan\theta = \frac{v^2}{rg} = \frac{2^2}{0.5 \times 10} = 0.8$$

So

$$\theta = 39°$$

If $m = 2.0\,\mathrm{kg}$, it follows from (2) that

$$T = \frac{mg}{\cos\theta} = \frac{2 \times 10}{\cos 39°} = 25.7\,\mathrm{N}$$

A pendulum suspended from the ceiling of a train does not remain vertical while the train goes round a circular track. Its bob moves *outwards* away from the centre and the string becomes inclined at an angle θ to the vertical, as shown in Figure 2.6. In this case the centripetal force is provided by the horizontal component of the tension in the string, as we have already explained.

Motion of Bicycle Rider round Circular Track

When a person on a bicycle rides round a circular racing track, the necessary centripetal force mv^2/r is actually provided by the frictional force F at the ground, Figure 2.7. The force F has a moment about the centre of gravity G

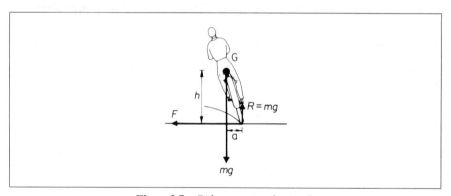

Figure 2.7 *Rider on a circular track*

equal to $F.h$ which tends to turn the rider outwards. When the rider leans inwards as shown, this is counterbalanced by the moment $R.a$ about G. $R = mg$ since there is no vertical motion, so the moment is $mg.a$. Thus, provided no slipping occurs, $F.h = mg.a$.

$$\therefore \frac{a}{h} = \tan\theta = \frac{F}{mg}$$

where θ is the angle of inclination to the vertical. Now $F = mv^2/r$

$$\therefore \tan\theta = \frac{v^2}{rg}$$

When F is greater than the limiting friction μR where μ is the coefficient of friction, skidding occurs. In this case $F > \mu mg$, or $mg\tan\theta > \mu mg$. Thus $\tan\theta > \mu$ is the condition for skidding.

Examples on Circular Motion

1 A stone of mass 0·6 kg, attached to a string of length 0·5 m, is whirled in a vertical circle at a constant speed.

If the maximum tension in the string is 30 N, calculate
(a) the speed of the stone,
(b) the maximum number of revolutions per second it can make. ($g = 10\,\mathrm{m\,s^{-2}}$)

(*Analysis* (i) Speed $= r\omega$, where $r = 0.5$ m, (ii) resultant force towards centre $= mr\omega^2$, (iii) maximum tension in string when mass at bottom of circle.)
(a) Suppose T newtons $=$ maximum tension in string. Then, at lowest point of circle,

$$T - mg = mr\omega^2$$

So $$30 - 0.6 \times 10 = 0.6 \times 0.5 \times \omega^2$$

Then $$\omega^2 = \frac{24}{0.3} = 80$$

$$\omega = \sqrt{80} = 9\,\mathrm{rad\,s^{-1}}\text{ (approx.)}$$

So $$v = r\omega = 0.5 \times 9 = 4.5\,\mathrm{m\,s^{-1}}$$

(b) $$\text{Period } T = \frac{2\pi}{\omega} = \frac{2\pi}{9}\,\mathrm{s}$$

So number of revs per second $= \dfrac{1}{2\pi/9} = \dfrac{9}{2\pi} = 1.5\,\mathrm{rev\,s^{-1}}$

2 A model aeroplane X has a mass of 0·5 kg and has a control wire OX of length 10 m attached to it when it flies in a horizontal circle with its wings horizontal, Figure 2.8. The wire OX is then inclined at 60° to the horizontal and fixed to a point O and X takes 2 s to fly once round its circular path.

Calculate
(a) the tension T in the control wire,
(b) the upward force on X due to the air.

(*Analysis* (i) Force F towards centre of circle $= mr\omega^2$, (ii) $F =$ horizontal component of T, (iii) upward force due to air $=$ weight of X $+$ downward component of T.)

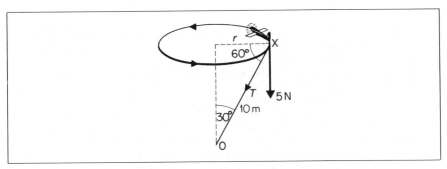

Figure 2.8 *Example on circular motion*

(a) $$\text{Angular velocity } \omega = 2\pi/2 = \pi\,\mathrm{rad\,s^{-1}}$$

For motion in horizontal circle,

$$F = mr\omega^2, \text{ where } F = T\cos 60°, \ r = 10\sin 30° = 5\,\text{m}$$

So
$$T\cos 60° = 0{\cdot}5 \times 5 \times \pi^2$$

$$T = \frac{2{\cdot}5\pi^2}{0{\cdot}5} = 50\,\text{N (approx.)}$$

(b) Upward force due to air = weight of $X + T\cos 30°$

$$= 5\,\text{N} + 50\cos 30°\,\text{N}$$

$$= 48\,\text{N (approx.)}$$

_____ **Exercises 2A** _____

Circular Motion

(Assume $g = 10\,\text{m s}^{-2}$ or $10\,\text{N kg}^{-1}$ unless otherwise given)

1 An object of mass 4 kg moves round a circle of radius 6 m with a constant speed of $12\,\text{m s}^{-1}$. Calculate   (i) the angular speed,   (ii) the force towards the centre.

2 An object of mass 10 kg is whirled round a horizontal circle of radius 4 m by a revolving string inclined to the vertical. If the uniform speed of the object is $5\,\text{m s}^{-1}$, calculate   (i) the tension in the string,   (ii) the angle of inclination of the string to the vertical.

3 A racing-car of 1000 kg moves round a banked track at a constant speed of $108\,\text{km h}^{-1}$. Assuming the total reaction at the wheels is normal to the track, and the horizontal radius of the track is 100 m, calculate the angle of inclination of the track to the horizontal and the reaction at the wheels.

4 An object of mass 8·0 kg is whirled round in a vertical circle of radius 2 m with a constant speed of $6\,\text{m s}^{-1}$. Calculate the maximum and minimum tensions in the string.

5 Calculate the force necessary to keep a mass of 0·2 kg moving in a horizontal circle of radius 0·5 m with a period of 0·5 s. What is the direction of the force?

6 Calculate the mean angular speed of the Earth assuming it takes 24·0 h to rotate about its axis.
   An object of mass 2·00 kg is   (i) at the Poles,   (ii) at the Equator. Assuming the Earth is a perfect sphere of radius $6{\cdot}4 \times 10^6$ m, calculate the change in weight of the mass when taken from the Poles to the Equator. Explain your calculation with the aid of a diagram.

7 A stone is rotated steadily in a horizontal circle with a period T by a string of length l. If the tension in the string is constant and l increases by 1%, find the percentage change in T.

8 A mass of 0·2 kg is whirled in a horizontal circle of radius 0·5 m by a string inclined at 30° to the vertical. Calculate   (i) the tension in the string,   (ii) the speed of the mass in the horizontal circle.

9 An object of mass 0·5 kg is rotated in a horizontal circle by a string 1 m long. The maximum tension in the string before it breaks is 50 N. What is the greatest number of revolutions per second of the object?

10 A mass of 0·4 kg is rotated by a string at a constant speed v in a vertical circle of radius 1 m. If the minimum tension of the string is 3 N, calculate   (i) v,   (ii) the maximum tension,   (iii) the tension when the string is just horizontal.

11 What force is necessary to keep a mass of 0·8 kg revolving in a horizontal circle of radius 0·7 m with a period of 0·5 s? What is the direction of this force? (Assume that $\pi^2 = 10$.) (*L.*)

12 A spaceman in training is rotated in a seat at the end of a horizontal rotating arm of length 5 m. If he can withstand accelerations up to $9g$, what is the maximum number of revolutions per second permissible? The acceleration of free fall (g) may be taken as $10\,\text{m s}^{-2}$. (*L.*)

13 Define the terms
(a) *acceleration*, and
(b) *force*.
  Show that the acceleration of a body moving in a circular path of radius *r* with uniform speed *v* is v^2/r, and draw a diagram to show the direction of the acceleration.
  A small body of mass *m* is attached to one end of a light inelastic string of length *l*. The other end of the string is fixed. The string is initially held taut and horizontal, and the body is then released. Find the values of the following quantities when the string reaches the vertical position:
(a) the kinetic energy of the body,
(b) the speed of the body,
(c) the acceleration of the body, and
(d) the tension in the string. (*O. & C.*)

14 Explain what is meant by *angular speed*. Derive an expression for the force required to make a particle of mass *m* move in a circle of radius *r* with uniform angular speed ω.
  A stone of mass 500 g is attached to a string of length 50 cm which will break if the tension in it exceeds 20 N. The stone is whirled in a vertical circle, the axis of rotation being at a height of 100 cm above the ground. The angular speed is very slowly increased until the string breaks. In what position is this break most likely to occur, and at what angular speed? Where will the stone hit the ground? (*C.*)

15 A special prototype model aeroplane of mass 400 g has a control wire 8 m long attached to its body. The other end of the control line is attached to a fixed point. When the aeroplane flies with its wings horizontal in a horizontal circle, making one revolution every 4 s, the control wire is elevated 30° above the horizontal. Draw a diagram showing the forces exerted on the plane and determine
(a) the tension in the control wire,
(b) the lift on the plane.
  (Assume acceleration of free fall, $g = 10\,\text{m s}^{-2}$ and $\pi^2 = 10$.) (*AEB*, 1982.)

16 (a) Explain why a particle moving with constant speed along a circular path has a radial acceleration. The value of such an acceleration is given by the expression v^2/r, where *v* is the speed and *r* is the radius of the path. Show that this expression is dimensionally correct.
(b) Explain, with the aid of clear diagrams, the following.
  (i) A mass attached to a string rotating at a constant speed in a horizontal circle will fly off at a tangent if the string breaks.   (ii) A cosmonaut in a satellite which is in a free circular orbit around the earth experiences the sensation of weightlessness even though he is influenced by the gravitational field of the earth.
(c) A pilot 'banks' the wings of his aircraft so as to travel at a speed of $360\,\text{km h}^{-1}$ in a horizontal circular path of radius 5·0 km. At what angle should he bank his aircraft in order to do this? (*L.*)

17 (a) Explain why a particle of mass *m* moving in a circular path of radius *r* at constant speed *v* must experience a force. Derive an expression from first principles for the magnitude *F* of this force.
(b) A racing car of mass 500 kg starts from rest and accelerates at $6·0\,\text{m s}^{-2}$ along a straight horizontal road for a distance of 150 m. It then enters at constant speed a horizontal circular curve of radius 200 m.
  (i) What is its speed through the curve?   (ii) What is the magnitude and direction of the resultant horizontal force acting on the racing car while it is rounding the curve?   (iii) If, while on the curve, the racing car accelerates forwards at $3·0\,\text{m s}^{-2}$, what is the resultant horizontal force acting on the car at the time it begins to accelerate forwards, and in which direction does it act? Illustrate your answer with a diagram.   (iv) State *two* parameters that limit the safe speed at which the racing car can travel around a horizontal curve of a given radius.   (v) Show that by suitable banking the road can be made perfectly safe for racing cars cornering at a particular speed. Calculate the banking angle needed for the speed at which the racing car entered the curve of radius 200 m. (*O.*)

18 (a) Write down an expression for the force required to maintain the motion of a body of mass m moving with constant speed v in a circle of radius R. In which direction does the force act?

(b) Figure 2A shows a toy runway. After release from a point such as X, a small model car runs down the slope, 'loops the loop', and travels on towards Z. The The radius of the loop is 0·25 m.

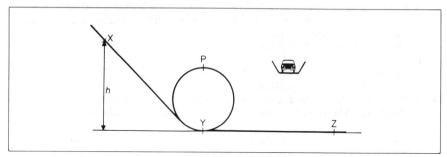

Figure 2A

(i) Ignoring the effect of friction outline the energy changes as the model moves from X to Z.   (ii) What is the minimum speed with which the car must pass point P at the top of the loop if it is to remain in contact with the runway? (iii) What is the minimum value of h which allows the speed calculated in (ii) to be achieved? The effect of friction can again be ignored. (Assume that the acceleration of free fall $g = 10\,\mathrm{m\,s^{-2}}$.) (*AEB*, 1984.)

Gravitation

In this section we shall show how Newton's universal law of gravitation is applied to the motion of planets round the sun, to satellites round the earth and to a moon satellite launched from the earth. Television pictures are now relayed from one part of the world to another by a satellite in a so-called parking or Clarke orbit.

Kepler's Laws

Kepler (1571–1630) had studied for many years the records of observations on the planets made by TYCHO BRAHE, and he discovered three laws now known by his name. *Kepler's laws* state:

Law 1   The planets describe ellipses about the sun as one focus.
Law 2   The line joining the sun and the planet sweeps out equal areas in equal times.
Law 3   The squares of the periods of revolution of the planets are proportional to the cubes of their mean distances from the sun.

Newton's Investigation on Planetary Motion

About 1666, at the early age of 24, Newton investigated the motion of a planet moving in a circle round the sun S as centre, Figure 2.9 (i). The force acting on the planet of mass m is $mr\omega^2$, where r is the radius of the circle and ω is the angular speed of the motion (p. 49). Since $\omega = 2\pi/T$, where T is the period of the motion,

$$\text{force on planet} = mr\left(\frac{2\pi}{T}\right)^2 = \frac{4\pi^2 mr}{T^2}$$

This is equal to the force of attraction of the sun on the planet. *Assuming an inverse-square law*, then, if k is a constant,

$$\text{force on planet} = \frac{km}{r^2}$$

$$\therefore \frac{km}{r^2} = \frac{4\pi^2 mr}{T^2}$$

$$\therefore T^2 = \frac{4\pi^2}{k} r^3$$

$$\therefore \boldsymbol{T^2 \propto r^3}$$

---

since k, π are constants.

Now Kepler had announced that the squares of the periods of revolution of the planets are proportional to the cubes of their mean distances from the sun (see above). Newton thus suspected that *the force between the sun and the planet was inversely proportional to the square of the distance between them.*

Motion of Moon round Earth

Newton now tested the inverse-square law by applying it to the case of the moon's motion round the earth, Figure 2.9 (ii). The moon has a period of

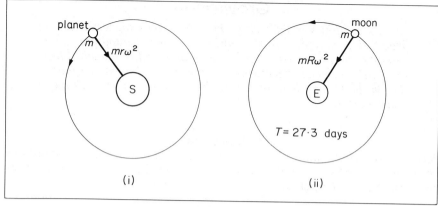

Figure 2.9 *Satellites*

revolution, T, about the earth of approximately 27·3 days, and the force on it $= mR\omega^2$, where R is the radius of the moon's orbit and m is its mass.

$$\therefore \text{force} = mR\left(\frac{2\pi}{T}\right)^2 = \frac{4\pi^2 mR}{T^2}$$

If the moon were at the earth's surface, the force of attraction on it due to the earth would be mg, where g is the acceleration due to gravity, Figure 2.9 (ii). Assuming that the force of attraction varies as the inverse square of the distance between the earth and the moon, then, by ratio,

$$= \frac{4\pi^2 mR}{T^2} : mg = \frac{1}{R^2} : \frac{1}{r_E^2}$$

where r_E is the radius of the earth.

$$\therefore \frac{4\pi^2 R}{T^2 g} = \frac{r_E^2}{R^2}$$

$$\therefore g = \frac{4\pi^2 R^3}{r_E^2 T^2} \quad . \qquad . \qquad . \qquad . \qquad . \qquad (1)$$

Newton substituted the then known values of R, r_E, and T, but was disappointed to find that the answer for g was not near to the observed value, $9\cdot8 \text{ m s}^{-2}$. Some years later, he heard of a new estimate of the radius of the earth, and we now know that r_E is about $6\cdot4 \times 10^6$ m. The radius R of the moon's orbit is about $60\cdot1 r_E$ and the period T of the moon is about 27·3 days or $27\cdot3 \times 24 \times 3600$ s. So

$$g = \frac{4\pi^2 R^3}{r_E^2 T^2} = \frac{4\pi^2 \times (60\cdot1 r_E)^3}{r_E^2 T^2} = \frac{4\pi^2 \times 60\cdot1^3 r_E}{T^2}$$

$$= \frac{4\pi^2 \times 60\cdot1^3 \times 6\cdot4 \times 10^6}{(27\cdot3 \times 24 \times 3600)^2} = 9\cdot9 \text{ m s}^{-2}$$

The result is very close to the measured value of g.

Newton's Law of Gravitation, G

Newton saw that a universal law could be stated for the attraction between any two particles of matter. He suggested that: *The force of the attraction between two given particles is inversely proportional to the square of their distance apart.*

From this law it follows that the force of attraction, F, between two particles of masses m and M respectively, at a distance r apart, is given by

$$F = G\frac{mM}{r^2}. \qquad \cdot \qquad \cdot \qquad \cdot \qquad \cdot \qquad (2)$$

where G is a universal constant known as the **gravitational constant**. This expression for F is **Newton's law of gravitation**. It is a universal law.

From (2), $G = Fr^2/mM$. So G can be expressed in $\text{N m}^2 \text{ kg}^{-2}$. Careful measurement shows that $G = 6.67 \times 10^{-11} \text{ N m}^2 \text{ kg}^{-2}$. The dimensions of G are

$$[G] = \frac{\text{MLT}^{-2} \times \text{L}^2}{\text{M}^2} = \text{L}^3\text{M}^{-1}\text{T}^{-2}$$

So the unit of G may also be expressed as $\text{m}^3 \text{ kg}^{-1} \text{ s}^{-2}$.

A celebrated experiment to measure G was carried out by C. V. Boys in 1895, using a method similar to one of the earliest determinations of G by Cavendish in 1798. Two identical balls, a, b, of gold, 5 mm in diameter, were suspended by a long and a short fine quartz fibre respectively from the ends, C, D, of a highly-polished bar CD, Figure 2.10. Two large identical lead spheres, A, B, 115 mm in diameter, were brought into position near a, b respectively. As a result of the attraction between the masses, two equal but opposite forces acted on CD. The bar was thus deflected, and the angle of deflection, θ, was measured by a lamp and scale method by light reflected from CD. The high sensitivity of the quartz fibres enabled the small deflection to be big enough to be measured accurately. The small size of the apparatus allowed it to be screened considerably from air convection currents.

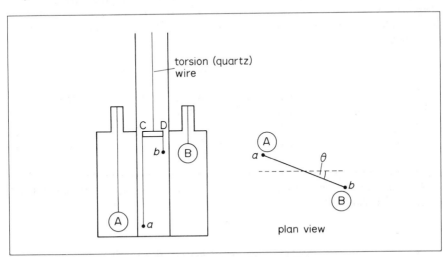

Figure 2.10   *Experiment on G*

Calculation for G

Suppose d is the distance between a, A, or b, B, when the deflection is θ. Then if m, M are the respective masses of a, A,

$$\text{torque of couple on CD} = G\frac{mM}{d^2} \times \text{CD}$$

But
$$\text{torque} = c\theta$$

where c is the torque in the torsion wire per unit radian of twist.

$$\therefore G\frac{mM}{d^2} \times CD = c\theta$$

$$\therefore G = \frac{c\theta d^2}{mM \times CD} \qquad . \qquad . \qquad . \qquad . \qquad . \qquad (1)$$

The constant c was determined by allowing CD to oscillate through a small angle and then observing its period of oscillation, T, which was of the order of 3 minutes. If I is the known moment of inertia of the system about the torsion wire, then

$$T = 2\pi\sqrt{\frac{I}{c}}$$

Gravitational Force on Masses, Relation between *g* and *G*

On the earth's surface, an object of mass m has a gravitational force of mg on it, where g is the acceleration of free fall. So a mass of 1 kg has a weight of $1g$ or $10\,N$, assuming g is $10\,m\,s^{-2}$ at the earth's surface.

To find the gravitational force on masses on the earth or outside it, it is legitimate to consider that the whole mass M_E of the earth is concentrated at its centre. Assuming the earth is a sphere of radius r_E, a mass m on the surface is at a distance r_E from the mass M_E. If the same mass is taken above the earth to a distance $2r_E$ from the centre, the force between M_E and m is reduced to $1/2^2$ or $1/4$, since the force between given masses is inversely-proportional to the square of their distance apart. So now

$$\text{gravitational force} = \frac{1}{4} \times 10\,N = 2\!\cdot\!5\,N$$

For a mass m on the earth's surface of radius r_E, gravitational force $= GMm/r_E^2 = mg$. Cancelling m on both sides, then

$$g = \frac{GM}{r_E^{\,2}}$$

As it is widely used, this relation between g and G should be memorised. From it, GM can be replaced in any formula by gr_E^2.

Variation of Acceleration of Free Fall

For points *outside* the earth, the gravitational force obeys an inverse-square law. So the acceleration of free fall, g', $\propto 1/r^2$, where r is the distance to the centre of the earth, Figure 2.11. The maximum value of g' is obtained at the earth's surface, where $r = r_E$.

Inside the earth, the value of g' is *not* inversely-proportional to the square of the distance from the centre. Assuming a uniform earth density, which is not true in practice, theory shows that g' varies linearly with the distance from the centre, as shown in Figure 2.11.

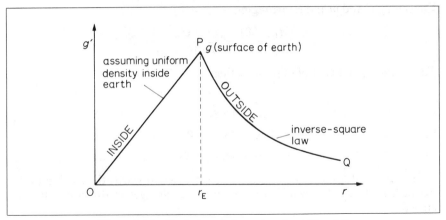

Figure 2.11   *Variation of g', acceleration of free fall*

Since the gravitational force F on a mass m is given generally by $F = mg'$, then $g' = F/m$. We see that g' can be expressed in 'newtons per kilogram' (N kg^{-1}). The *force per unit mass* in the gravitational field of the earth is called its *gravitational field strength*. We see that, on the earth, $g = 9.8 \text{ m s}^{-2} = 9.8 \text{ N kg}^{-1}$.

Examples on Gravitation

1   *Earth-Moon system*
The mass of the earth is 81 times that of the moon and the distance from the centre of the earth to that of the moon is about 4.0×10^5 km.

Calculate the distance from the centre of the earth where the resultant gravitational force becomes zero when a spacecraft is launched from the earth to the moon. Draw a sketch showing roughly how the gravitational force on the spacecraft varies in its journey.

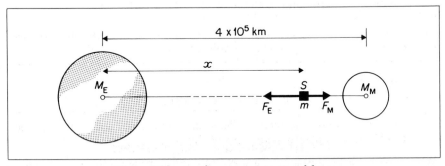

Figure 2.12   *Earth–moon gravitational force*

(*Analysis*   (i) The gravitational force on the spacecraft S due to the earth is opposite in direction to that of the moon,   (ii) $F = GMm/r^2$.)

Suppose the spacecraft S is a distance x in km from the centre of the earth and a distance $(4 \times 10^5 - x)$ from the moon when the resultant force is zero, Figure 2.12. If m is the spacecraft mass, then

$$\frac{GM_E m}{x^2} = \frac{GM_M m}{(4 \times 10^5 - x)^2}$$

Cancelling G and m and re-arranging,

$$\frac{M_E}{M_M} = \frac{81}{1} = \frac{x^2}{(4 \times 10^5 - x)^2}$$

Taking the square root of both sides, then

$$9 = \frac{x}{4 \times 10^5 - x}$$

So
$$10x = 9 \times 4 \times 10^5$$

$$x = 3 \cdot 6 \times 10^5 \, \text{km}$$

The resultant force F on m due to the earth acts towards the earth until S is reached. It then acts towards the moon. So F changes in direction after S is passed.

2 *Variation of* g

A man can jump $1 \cdot 5$ m on earth. Calculate the approximate height he might be able to jump on a planet whose density is one-quarter that of the earth and whose radius is one-third that of the earth.

Suppose the man of mass m leaps a height h on the earth and a height h_1 on the planet. Assuming he can give himself the same initial kinetic energy on the two planets, the potential energy gained is the same at the maximum height. So

$$mg_1 h_1 = mgh$$

where g_1 and g are the respective gravitational intensities on the planet and earth. So

$$h_1 = \frac{g}{g_1} \times h \qquad . \qquad . \qquad . \qquad . \qquad . \qquad (1)$$

But for the earth, $g = GM/r_E^2$ (p.62) $= G \cdot \frac{4}{3}\pi r_E^3 \rho_E / r_E^2 = G \cdot \frac{4}{3}\pi r_E \rho_E$, where ρ_E is the density of the earth. Similarly, $g_1 = G \cdot \frac{4}{3}\pi r_1 \rho_1$, where r_1, ρ_1 are the respective radius and density of the planet. So

$$\frac{g}{g_1} = \frac{r_E \rho_E}{r_1 \rho_1} = 4 \times 3 = 12$$

From (1), we have

$$h_1 = 12 \times 1 \cdot 5 \, \text{m} = 18 \, \text{m}$$

Force on Astronaut, Weightlessness

When a rocket is fired to launch a spacecraft and astronaut into orbit round the earth, the initial thrust must be very high owing to the large initial acceleration required. This acceleration, a, is of the order of $15g$, where g is the gravitational acceleration at the earth's surface.

Suppose S is the reaction of the couch to which the astronaut is initially strapped, Figure 2.13 (i). Then, from $F = ma$, $S - mg = ma = m \cdot 15g$, where m is the mass of the astronaut. Thus $S = 16mg$. This force is 16 times the weight of the astronaut and so, initially, he experiences a large force.

In orbit, however, the state of affairs is different. This time the acceleration of the spacecraft and astronaut are both g' in magnitude, where g' is the acceleration due to gravity at the particular height of the orbit, Figure 2.13 (ii). If

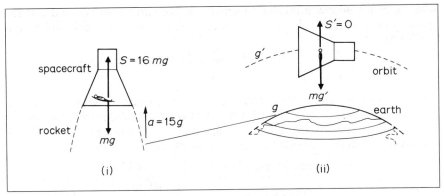

Figure 2.13 *Weight and weightlessness*

S' is the reaction of the surface of the spacecraft in contact with the astronaut, then, for circular motion,

$$F = mg' - S' = ma = mg'$$

Thus $S' = 0$. The astronaut now experiences no reaction at the floor when he walks about, for example, and so he experiences the sensation of being 'weightless' although he has a gravitational force mg' acting on him.

At the earth's surface we feel the reaction at the ground and are thus conscious of our weight. Inside a lift which is falling fast, the reaction at our feet diminishes. If the lift falls freely, the acceleration of objects inside is the same as that outside and hence the reaction on them is zero. This produces the sensation of 'weightlessness'. In orbit, as in Figure 2.13 (ii), objects inside a spacecraft are also in 'free fall' because they have the same acceleration g' as outside the spacecraft.

Earth Satellites

Satellites can be launched from the earth's surface to circle the earth. They are kept in their orbit by the gravitational attraction of the earth. Consider a satellite of mass m which just circles the earth of mass M close to its surface in an orbit 1, Figure 2.14. Then, if r_E is the radius of the earth,

$$\frac{mv^2}{r_E} = G\frac{Mm}{r_E^2} = mg$$

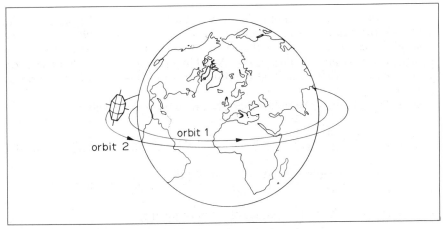

Figure 2.14 *Orbits round earth*

where g is the acceleration due to gravity at the earth's surface and v is the speed of m in its orbit. Thus $v^2 = r_E g$, and hence, using $r_E = 6\cdot4 \times 10^6$ m and $g = 9\cdot8$ m s^{-2},

$$v = \sqrt{r_E g} = \sqrt{6\cdot4 \times 10^6 \times 9\cdot8} = 8 \times 10^3 \text{ m s}^{-1} \text{ (approx.)}$$

$$= 8 \text{ km s}^{-1}$$

The speed v in the orbit is thus about 8 km s^{-1}. In practice, the satellite is carried by a rocket to the height of the orbit and then given an impulse, by firing jets, to deflect it in a direction parallel to the tangent of the orbit (see p. 67). Its velocity is boosted to 8 km s^{-1} so that it stays in the orbit. The period in orbit

$$= \frac{\text{circumference of earth}}{v} = \frac{2\pi \times 6\cdot4 \times 10^6 \text{ m}}{8 \times 10^3 \text{ m s}^{-1}}$$

$$= 5000 \text{ seconds (approx.)} = 83 \text{ min}$$

Parking Orbits

Consider now a satellite of mass m circling the earth in the plane of the equator in orbit 2 concentric with the earth, Figure 2.14. Suppose the direction of rotation is the same as the earth and the orbit is at a distance R from the centre of the earth. Then if v is the speed in orbit,

$$\frac{mv^2}{R} = \frac{GMm}{R^2}$$

But $GM = gr_E^2$, where r_E is the radius of the earth.

$$\therefore \frac{mv^2}{R} = \frac{mgr_E^2}{R^2}$$

$$\therefore v^2 = \frac{gr_E^2}{R}$$

If T is the period of the satellite in its orbit, then $v = 2\pi R/T$

$$\therefore \frac{4\pi^2 R^2}{T^2} = \frac{gr_E^2}{R}$$

$$\therefore T^2 = \frac{4\pi^2 R^3}{gr_E^2} \quad . \quad . \quad . \quad . \quad . \quad \text{(i)}$$

If the period of the satellite in its orbit is exactly equal to the period of the earth as it turns about its axis, which is 24 hours, *the satellite will stay over the same place on the earth* while the earth rotates. This is sometimes called a 'parking orbit'. Relay satellites can be placed in parking orbits, so that television programmes can be transmitted continuously from one part of the world to another.

Since $T = 24$ hours, the radius R can be found from (i). Its value is

$$R = \sqrt[3]{\frac{T^2 gr_E^2}{4\pi^2}} \quad \text{and} \quad g = 9\cdot8 \text{ m s}^{-2}, r_E = 6\cdot4 \times 10^6 \text{ m}$$

$$\therefore R = \sqrt[3]{\frac{(24 \times 3600)^2 \times 9\cdot8 \times (6\cdot4 \times 10^6)^2}{4\pi^2}} = 42\,400 \text{ km}$$

The height above the earth's surface of the parking orbit

$$= R - r_E = 42\,400 - 6400 = 36\,000 \text{ km}$$

In the orbit, assuming it is circular the speed of the satellite

$$= \frac{2\pi R}{T} = \frac{2\pi \times 42\,400\,\text{km}}{24 \times 3600\,\text{s}} = 3\cdot1\,\text{km s}^{-1}$$

The satellite, with the necessary electronic equipment inside, rises vertically from the equator when it is fired. At a particular height the satellite is given a horizontal momentum by firing rockets on its surface and the satellite then turns into the required orbit. This is illustrated in the next example.

Example on Satellite in Orbit

A satellite is to be put into orbit 500 km above the earth's surface. If its vertical velocity after launching is $2000\,\text{m s}^{-1}$ at this height, calculate the magnitude and direction of the impulse required to put the satellite directly into orbit, if its mass is 50 kg. Assume $g = 10\,\text{m s}^{-2}$, radius of earth, $r_E = 6400\,\text{km}$.

Suppose u is the velocity required for orbit, radius R. Then, with usual notation,

$$\text{Force on satellite} = \frac{mu^2}{R} = \frac{GmM}{R^2} = \frac{gr_E^2 m}{R^2}, \text{ as } \frac{GM}{r_E^2} = g$$

$$\therefore u^2 = \frac{gr_E^2}{R}$$

Now $r_E = 6400\,\text{km}$, $R = 6900\,\text{km}$, $g = 10\,\text{m s}^{-2}$

$$\therefore u^2 = \frac{10 \times (6400 \times 10^3)^2}{6900 \times 10^3}$$

$$\therefore u = 7700\,\text{m s}^{-1} \text{ (approx.)}$$

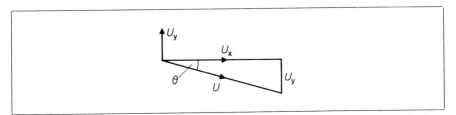

Figure 2.15   *Example on satellite*

At this height, vertical momentum

$$U_y = mv = 50 \times 2000 = 100\,000\,\text{kg m s}^{-1}$$

Horizontal momentum required $U_x = mu = 50 \times 7700 = 385\,000\,\text{kg m s}^{-1}$

$$\therefore \text{impulse needed, } U, = \sqrt{U_y^2 + U_x^2} = \sqrt{100\,000^2 + 385\,000^2} \text{ (Figure 2.15)}$$

$$= 4\cdot0 \times 10^5\,\text{kg m s}^{-1}$$

Direction. The angle θ made by the total impulse with the horizontal or orbit tangent is given by $\tan\theta = U_y/U_x = 100\,000/385\,000 = 0\cdot260$. Thus $\theta = 14\cdot6°$.

Mass and Density of Earth

At the earth's surface the force of attraction on a mass m is mg, where g is the acceleration due to gravity. Now it can be shown in this case that we can assume that the mass, M, of the earth is concentrated at its centre, if it is a sphere (p. 62). Assuming that the earth is spherical of radius r_E, it then follows that the force of attraction of the earth on the mass m is GmM/r_E^2. So

$$G\frac{mM}{r_E^2} = mg$$

$$\therefore g = \frac{GM}{r_E^2}$$

$$\therefore M = \frac{gr_E^2}{G}$$

Now, $g = 9\cdot8\,\mathrm{m\,s^{-1}}$, $r_E = 6\cdot4 \times 10^6\,\mathrm{m}$, $G = 6\cdot7 \times 10^{-11}\,\mathrm{N\,m^2\,kg^{-2}}$

$$\therefore M = \frac{9\cdot8 \times (6\cdot4 \times 10^6)^2}{6\cdot7 \times 10^{-11}} = 6\cdot0 \times 10^{24}\,\mathrm{kg}$$

The volume of a sphere is $4\pi r^3/3$, where r is its radius. So the mean density, ρ, of the earth is approximately given by

$$\rho = \frac{M}{V} = \frac{gr_E^2}{4\pi r_E^3 G/3} = \frac{3g}{4\pi r_E G}$$

By substituting known values of g, G and r_E, the mean density of the earth is found to be about $5500\,\mathrm{kg\,m^{-3}}$. The density of the earth is actually non-uniform and may approach a value of $10\,000\,\mathrm{kg\,m^{-3}}$ towards the interior.

Mass of Sun

The mass M_S of the sun can be found from the period of a satellite and its distance from the sun. Consider the case of the earth. Its period T is about 365 days or $365 \times 24 \times 3600$ seconds. Its distance r_S from the centre of the sun is about $1\cdot5 \times 10^{11}\,\mathrm{m}$. If the mass of the earth is m, then, for circular motion round the sun,

$$\frac{GM_S m}{r_S^2} = mr_S\omega^2 = \frac{mr_S 4\pi^2}{T^2}$$

$$\therefore M_S = \frac{4\pi^2 r_S^3}{GT^2} = \frac{4\pi^2 \times (1\cdot5 \times 10^{11})^3}{6\cdot7 \times 10^{-11} \times (365 \times 24 \times 3600)^2} = 2 \times 10^{30}\,\mathrm{kg}$$

In the equation $GM_S m/r_S^2 = mr_S\omega^2$ above, we see that the mass m of the satellite cancels on both sides and does not appear in the final equation for ω. So ω, the angular speed in the orbit, is *independent* of the mass of the satellite. The angular speed ω (and the period) depends only on the value of r_S, the orbit distance from the sun. This is true for all planets, that is

the angular speed of a planet depends only on the radius of the orbit and is independent of the mass of the planet.

Gravitational Potential

The *potential*, V, at a point due to the gravitational field of the earth is defined as numerically equal to the work done in taking a unit mass from infinity to that point. The potential at infinity is conventionally taken as *zero*. Points in electric fields have 'electric potential', as we see later.

For a point outside the earth, assumed spherical, we can imagine the whole mass M of the earth concentrated at its centre. The force of attraction on a unit mass outside the earth is thus GM/r^2, where r is the distance from the centre. The work done by the gravitational force in moving a distance Δr towards the earth = force × distance = $GM \cdot \Delta r/r^2$. Hence the potential at a point distant a from the centre greater than r is given by

$$V_a = \int_\infty^a \frac{GM}{r^2} dr = -\frac{GM}{a} \qquad . \qquad . \qquad . \qquad (1)$$

if the potential at infinity is taken as zero by convention. The negative sign indicates that the potential at infinity (zero) is *higher* than the potential close to the earth.

On the earth's surface, of radius r_E, we therefore obtain

$$V = -\frac{GM}{r_E} \qquad . \qquad . \qquad . \qquad . \qquad (2)$$

For large distances from the earth, for example, when a rocket travels from the earth to the moon, the change in potential energy of a mass can only be calculated by using *mass* $\times (GM/a - GM/b)$, where b and a are the distances from the centre of the earth. For small distances above the earth, however, the gravitational force on a mass is fairly constant. So the change in potential energy in this case can be calculated using *force* × *distance* or *mgh*.

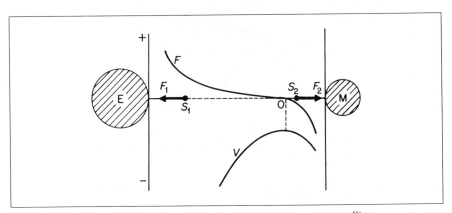

Figure 2.15A *Force (F) and potential (V) for moon satellite*

Figure 2.15 A shows roughly how the resultant force F on a satellite varies after it is launched from the earth E towards the moon M. The direction of F changes from F_1 at S_1, where the Earth's gravitational pull is greater than that of the moon, to F_2 at S_2 near the moon, where the pull of this planet is now stronger than that of the earth. At O, the gravitational pull of the earth is balanced by that of the moon.

The potential energy V of the satellite is the sum of its negative potential values due to the earth and the moon and is shown roughly in Figure 2.15 A. The maximum value of V occurs just below O. Here the resultant force F is zero. Since $F = -dV/dr$, the gradient dV/dr of the potential curve is then zero.

Velocity of Escape

Suppose a rocket of mass m is fired from the earth's surface Q so that it just escapes from the gravitational influence of the earth. Then work done $= m \times$ potential difference between infinity and Q

$$= \dot{m} \times \frac{GM}{r_E}$$

$\therefore$ kinetic energy of rocket $= \tfrac{1}{2}mv^2 = m \times \dfrac{GM}{r_E}$

$$v = \sqrt{\frac{2GM}{r_E}} = \text{velocity of escape}$$

Now
$$GM/r_E{}^2 = g$$

$$\therefore v = \sqrt{2gr_E}$$

$$\therefore v = \sqrt{2 \times 9{\cdot}8 \times 6{\cdot}4 \times 10^6} = 11 \times 10^3 \,\text{m s}^{-1} = 11 \,\text{km s}^{-1}\ (\text{approx.})$$

With an initial velocity, then, of about $11 \,\text{km s}^{-1}$, a rocket will completely escape from the gravitational attraction of the earth. It can be made to travel towards the moon, for example, so that eventually it comes under the gravitational attraction of this planet. At present, 'soft' landings on the moon have been made by firing retarding or retro rockets.

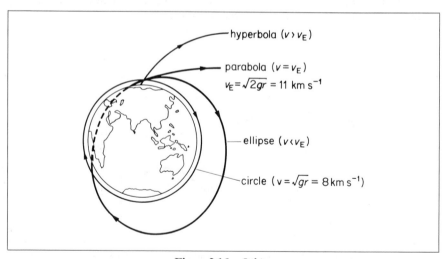

Figure 2.16 *Orbits*

Summarising, with a velocity of about $8 \,\text{km s}^{-1}$, a satellite can describe a circular orbit close to the earth's surface (p. 66). With a velocity greater than $8 \,\text{km s}^{-1}$ but less than $11 \,\text{km s}^{-1}$, a satellite describes an elliptical orbit round the earth. Its maximum and minimum height in the orbit depends on its particular velocity. Figure 2.16 illustrates the possible orbits of a satellite launched from the earth.

The molecules of air at normal temperatures and pressures have an average velocity of the order of $480 \,\text{m s}^{-1}$ or $0{\cdot}48 \,\text{km s}^{-1}$ which is much less than the velocity of escape. Many molecules move with higher velocity than $0{\cdot}48 \,\text{km s}^{-1}$

but gravitational attraction keeps the atmosphere round the earth. The gravitational attraction of the moon is much less than that of the earth and this accounts for the lack of atmosphere round the moon.

P.E. and K.E. of Satellite

A satellite of mass m in orbit round the earth has both kinetic energy, k.e., and potential energy, p.e. The k.e. $= \frac{1}{2}mv^2$, where v is the speed in the orbit. Now for circular motion in an orbit of radius r_0, if M is the mass of the earth,

$$\text{force towards centre} = \frac{mv^2}{r_0} = G\frac{Mm}{r_0{}^2}$$

$$\therefore \text{k.e.} = \tfrac{1}{2}mv^2 = G\frac{Mm}{2r_0} \qquad . \qquad . \qquad . \qquad . \qquad (1)$$

Assuming the zero of potential energy in the earth's field is at infinity (p. 69),

$$\text{p.e. of mass in orbit} = -G\frac{Mm}{r_0} . \qquad . \qquad . \qquad . \qquad (2)$$

So, from (1), the potential energy of the mass in orbit is numerically *twice* its kinetic energy and of opposite sign.

From (1) and (2),

$$\text{total energy in orbit} = -\frac{GMm}{r_0} + \frac{GMm}{2r_0}$$

$$= -\frac{GMm}{2r_0} \qquad . \qquad . \qquad . \qquad . \qquad (3)$$

Owing to friction in the earth's atmosphere, the satellite energy diminishes and the radius of the orbit decreases to r_1 say. The total energy in this orbit, from above, is $-GMm/2r_1$. Since this is *less* than the initial energy in (3), it follows that

$$\frac{GMm}{2r_1} > \frac{GMm}{2r_0}$$

From (1), these two quantities are the kinetic energy values in the respective orbits of radius r_1 and r_0. Hence the kinetic energy of the satellite *increases* when it falls to an orbit of smaller radius, that is, the satellite speeds up. This apparent anomaly is explained by the fact that the potential energy decreases by twice as much as the kinetic energy increases, from (2). Thus on the whole there *is* a loss of energy, as we expect.

Example on Energy of Satellite

A satellite of mass 1000 kg moves in a circular orbit of radius 7000 km round the earth, assumed to be a sphere of radius 6400 km. Calculate the total energy needed to place the satellite in orbit from the earth, assuming $g = 10\,\text{N kg}^{-1}$ at the earth's surface.

To launch the satellite, mass m, from the earth's surface of radius r_E into an orbit of radius r_0,

energy needed W = increase in potential energy and kinetic energy

$$= \frac{GMm}{r_E} - \frac{GMm}{r_0} + \tfrac{1}{2}mv^2$$

$$= \frac{GMm}{r_E} - \frac{GMm}{2r_0}$$

from equation (3) of the previous section. But $GM/r_E^2 = g$, or $GM/r_E = gr_E$.

So

$$W = mgr_E - \frac{mgr_E^2}{2r_0} = mg\left(r_E - \frac{r_E^2}{2r_0}\right)$$

$$= 1000 \times 10\left(6\cdot4 \times 10^6 - \frac{6\cdot4^2 \times 10^{12}}{2 \times 7 \times 10^6}\right)$$

$$= 3\cdot5 \times 10^{10}\,\text{J}$$

_____ **Exercises 2B** _____

Gravitation

(Assume $g = 10\,N\,kg^{-1}$ unless otherwise stated)

1   The gravitational force on a mass of 1 kg at the earth's surface is 10 N. Assuming the earth is a sphere of radius R, calculate the gravitational force on a satellite of mass 100 kg in a circular orbit of radius $2R$ from the centre of the earth.

2   Assuming the earth is a uniform sphere of mass M and radius R, show that the acceleration of free fall at the earth's surface is given by $g = GM/R^2$.
    What is the acceleration of a satellite moving in a circular orbit round the earth of radius $2R$?

3   A planet of mass m moves round the sun of mass M in a circular orbit of radius r with an angular speed ω. Show   (i) that ω is independent of the mass m of the planet,   (ii) that in a circular orbit of radius $4r$ round the sun, the angular speed decreases to $\omega/8$.

4   Obtain the dimensions of G.
    The period of vibration T of a star under its own gravitational attraction is given by $T = 2\pi/\sqrt{G\rho}$, where ρ is the mean density of the star. Show that this relation is dimensionally correct.

5   A satellite X moves round the earth in a circular orbit of radius R. Another satellite Y of the same mass moves round the earth in a circular orbit of radius $4R$. Show that   (i) the speed of X is twice that of Y,   (ii) the kinetic energy of X is greater than that of Y,   (iii) the potential energy of X is less than that of Y.
    Has X or Y the greater total energy (kinetic plus potential energy)?

6   Find the period of revolution of a satellite moving in a circular orbit round the earth at a height of $3\cdot6 \times 10^6$ m above the earth's surface. Assume the earth is a uniform sphere of radius $6\cdot4 \times 10^6$ m, the earth's mass is 6×10^{24} kg and G is $6\cdot7 \times 10^{-11}\,N\,m^2\,kg^{-1}$.

7   If the acceleration of free fall at the earth's surface is $9\cdot8\,m\,s^{-2}$, and the radius of the earth is 6400 km, calculate a value for the mass of the earth.
    ($G = 6\cdot7 \times 10^{-11}\,N\,m^2\,kg^{-2}$.) Give the theory.

8   Assuming the mean density of the earth is $5500\,kg\,m^{-3}$, that G is $6\cdot7 \times 10^{-11}\,N\,m^2\,kg^{-2}$, and that the earth's radius is 6400 km, find a value for the acceleration of free fall at the earth's surface. Derive the formula used.

9   Two binary stars, masses 10^{20} kg and 2×10^{20} kg respectively, rotate about their common centre of mass with an angular speed ω. Assuming that the only force on a star is the mutual gravitational force between them, calculate ω. Assume that the distance between the stars is 10^6 km and that G is $6\cdot7 \times 10^{-11}\,N\,m^2\,kg^{-2}$.

10   A preliminary stage of spacecraft *Apollo 11*'s journey to the moon was to place it in an earth parking orbit. This orbit was circular, maintaining an almost constant distance of 189 km from the earth's surface. Assuming the gravitational field strength in this orbit is $9\cdot4\,N\,kg^{-1}$, calculate
    (a) the speed of the spacecraft in this orbit and
    (b) the time to complete one orbit. (Radius of the earth = 6370 km.) (*L.*)

11   Explorer 38, a radio-astronomy research satellite of mass 200 kg, circles the earth
      in an orbit of average radius $3R/2$ where R is the radius of the earth. Assuming the
      gravitational pull on a mass of 1 kg at the earth's surface to be 10 N, calculate the
      pull on the satellite. (*L.*)

12   A satellite of mass 66 kg is in orbit round the earth at a distance of $5.7R$ above its
      surface, where R is the value of the mean radius of the earth. If the gravitational
      field strength at the earth's surface is 9.8 N kg^{-1}, calculate the centripetal force
      acting on the satellite.
         Assuming the earth's mean radius to be 6400 km, calculate the period of the
      satellite in orbit in hours. (*L.*)

13   (a) Explain what is meant by *gravitational field strength*. In what units is it
         measured?
            Starting with Newton's law of gravitation, derive an expression for g, the
         acceleration of free fall on the surface of the earth, stating clearly the meaning
         of each symbol used. (Assume that the earth may be considered as a point mass
         located at its centre.)
      (b) g may be found by measuring the acceleration of a freely falling body. Outline
         how you would measure g in this way, indicating the measurements needed and
         how you would calculate a value for g from them.
      (c) At one point on the line between the earth and the moon, the gravitational field
         caused by the two bodies is zero. Briefly explain why this is so.
            If this point is 4×10^4 km from the moon, calculate the ratio of the mass of
         moon to the mass of the earth. (Distance from earth to moon $= 4.0 \times 10^5$ km.)
         (*L.*)

14   Explain what is meant by the *constant of gravitation*. Describe a laboratory
      experiment to determine it, showing how the result is obtained from the
      observations.
         A proposed communication satellite would revolve round the earth in a circular
      orbit in the equatorial plane, at a height of 35 880 km above the earth's surface. Find
      the period of revolution of the satellite in hours, and comment on the result.
         (Radius of earth $= 6370$ km, mass of earth $= 5.98 \times 10^{24}$ kg, constant of
      gravitation $= 6.66 \times 10^{-11} \text{ N m}^2 \text{ kg}^{-2}$.) (*JMB.*)

15   (a) Explain what is meant by the terms: (i) *gravitational intensity g*;
         (ii) *gravitational potential V*.
      (b) A uniform spherical planet has a mass M and a radius R. Derive expressions in
         terms of these quantities and the gravitational constant G for values at the
         surface of the planet of: (i) the gravitational intensity g;   (ii) the gravitational
         potential V.
      (c) A small satellite is in a stable circular orbit of radius 7000 km around a planet of
         mass 5.7×10^{24} kg and radius 6500 km. [Take the gravitational constant G to be
         $6.7 \times 10^{-11} \text{ N m}^2 \text{ kg}^{-2}$.]
            Calculate: (i) the orbital speed of the satellite;   (ii) the orbital period of the
         satellite;   (iii) the velocity of escape from the surface of the planet.
      (d) By what factor would the velocity of escape be reduced if the linear dimensions of
         the planet were 10^3 smaller (i.e. radius $= 6.5$ km), its mean density remaining
         unchanged?
            In the light of your answer explain why many small planets do not have
         gaseous atmospheres. (*O.*)

16   (a) Describe how the unit of force is defined from Newton's Laws of Motion.
         Why is it necessary to introduce the dimensional constant G in Newton's Law of
         Gravitation? Find the dimensions of G in terms of mass M, length L and time T.
      (b) Derive an expression for the acceleration g due to gravity on the surface of the
         earth in terms of G, the radius of the earth R and its density ρ.
            The maximum vertical distance through which a fully-dressed astronaut can
      jump on the earth is 0.5 m. Estimate the maximum vertical distance through which
      he can jump on the moon, which has a mean density two-thirds that of the earth and
      a radius one-quarter that of the earth, stating any assumptions made. Determine
      the ratio of the time duration of his jump on the moon to that of his jump on the
      earth. (*O. & C.*)

17 (a) The gravitational field strength, g_0, on the surface of the earth is $9.81 \, \text{N kg}^{-1}$. Explain what this means.

Using Newton's law of gravitation show that $gr^2 = \text{constant}$ where g is the gravitational field strength at a distance r from the centre of the earth. ($r \geqslant r_0$, where r_0 is the radius of the earth.)

The gravitational field strength at the surface of the moon is $1.67 \, \text{N kg}^{-1}$. At what point on a line from the earth to the moon will the net gravitational field strength due to the earth and the moon be zero? Sketch a rough graph showing how this net gravitational field strength varies along the line between the surface of the earth and the surface of the moon.

(b) The gravitational potential on the surface of the earth is $-63 \, \text{MJ kg}^{-1}$. Explain what this means. If the gravitational potential on the surface of the moon is $-3 \, \text{MJ kg}^{-1}$, what is the gravitational potential difference between the surface of the earth and the surface of the moon?

The moon's surface is at a higher gravitational potential than the earth's surface, yet in returning to the earth from the moon, a spacecraft needs to use its rocket engines initially to propel it towards the earth. Why is this?

(Distance from the centre of the earth to the centre of the moon = 400 000 km. Radius of the earth = 6400 km. Radius of the moon = 1740 km.) (*L.*)

18 (a) State Newton's law of gravitation and explain how this law was established.

(b) Use Newton's law to deduce expressions for:
(i) the period T of a satellite in circular orbit of radius r about the earth in terms of the mass m_E of the earth and the gravitational constant G; (ii) the gravitational intensity g at this orbit in terms of the orbital radius r, the gravitational intensity g_0 at the earth's surface, and the radius r_E of the earth (assumed to be a uniform sphere).

(c) A satellite of mass 600 kg is in a circular orbit at a height of 2000 km above the earth's surface. [Take the radius of the earth to be 6400 km, and the value of g_0 to be $10 \, \text{N kg}^{-1}$.]
Calculate the satellite's: (i) orbital speed; (ii) kinetic energy; (iii) gravitational potential energy.

(d) Explain why any resistance to the forward motion of an artificial satellite in space results in an increase in its forward speed. (*O.*)

19 Describe, briefly, a method for the measurement of the gravitational constant G.

(a) Express the acceleration due to gravity, g, at the surface of the earth, in terms of G, the mass M of the earth and the radius R of the earth, assuming the earth is a uniform sphere. The effect of the earth's rotation may be neglected.

(b) Express the period T of a satellite in a circular orbit round the earth, in terms of the radius r of the orbit, g at the surface of the earth and R.

(c) For communication purposes it is desirable to have a satellite which stays vertically above one point on the earth's surface. Explain why the orbit of such a satellite (i) must be circular, and (ii) must lie in the plane of the equator. Find the radius of this orbit. (Radius of earth = 6400 km.) (*O. & C.*)

20 What do you understand by the *intensity of gravity (gravitational field strength)* and the *gravitational potential* at a point in the earth's gravitational field? How are they related?

Taking the earth to be uniform sphere of radius 6400 km, and the value of g at the surface to be $10 \, \text{m s}^{-2}$, calculate the total energy needed to raise a satellite of mass 2000 kg to a height of 800 km above the ground and to set it into circular orbit at that altitude.

Explain briefly how the satellite is set into orbit once the intended altitude has been reached, and also what would happen if this procedure failed to come into action. (*O.*)

Simple Harmonic Motion

In the last section we discussed circular motion. Now we consider simple harmonic motion, which has applications in many different branches of physics and is therefore important.

When the bob of a pendulum moves to-and-fro through a small angle, the bob is said to be moving with *simple harmonic motion*. The prongs of a sounding tuning fork, and the layers of air near it, are moving to-and-fro with simple harmonic motion. Light waves can be considered due to simple harmonic variations of electric and magnetic forces.

Simple harmonic motion is closely associated with circular motion. An example is shown in Figure 2.17. This illustrates an arrangement used to convert the circular motion of a disc D into the to-and-fro or simple harmonic motion of a piston P. The disc is driven about its axle O by a peg Q fixed near its rim. The vertical motion drives P up and down. Any horizontal component of the motion merely causes Q to move along the slot S. Thus the simple harmonic motion of P is the *projection* on the vertical line YY' of the circular motion of Q.

The projection of Q on YY' is the *foot* of the perpendicular from Q to the diameter passing through YY'. Figure 2.18 shows how the distance y from O of the projection varies as Q moves round the circular disc D with constant angular speed ω. In this rough sketch the horizontal axis represents angle of rotation or time, as the angle turned is proportional to the time. On one side of

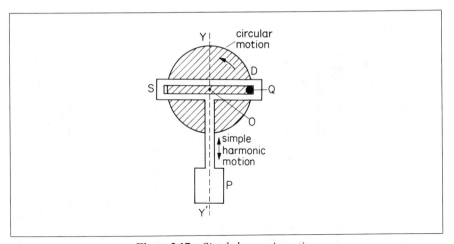

Figure 2.17 *Simple harmonic motion*

O, y has positive values; on the other side of O it has negative values. The graph of y against time t is a *simple harmonic* curve or *sine* (*sinusoidal*) *curve* as we see shortly. The maximum value of y is called the *amplitude*. One complete set of values of y is called one *cycle* because the graph repeats itself after one cycle.

Formulae in Simple Harmonic Motion

Consider an object moving round a circle of radius r and centre Z with a uniform angular speed ω, Figure 2.19. As we have just seen, if CZF is a fixed diameter, the *foot* of the perpendicular from the moving object to this diameter moves from Z to C, back to Z and across to F, and then returns to Z, while the object moves once round the circle from O in an anti-clockwise direction. The

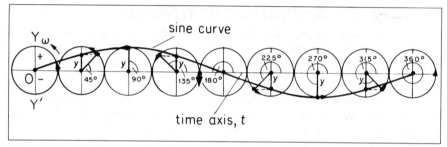

Figure 2.18 *Simple harmonic and circular motion. The diagram shows eight positions of a particle moving round a circle through 360° at constant angular speed. The distances y from O of the foot of the projection on YOY' all lie on a sine curve as shown*

to-and-fro motion along CZF of the foot of the perpendicular may be defined as *simple harmonic motion.*

Suppose the object moving round the circle is at A at some instant, where angle OZA = θ, and suppose the foot of the perpendicular from A to CZ is M. The acceleration of the object at A is $\omega^2 r$, and this acceleration is directed along the radius AZ (see p. 49). Hence the acceleration of M towards Z

$$= \omega^2 r \cos AZC = \omega^2 r \sin \theta$$

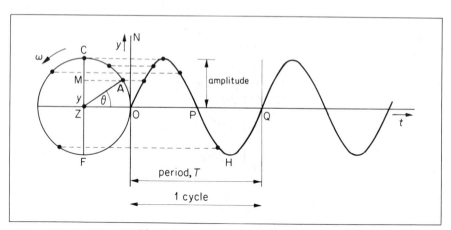

Figure 2.19 *Simple harmonic curve*

But $r \sin \theta = MZ = y$ say.

$$\therefore \text{ acceleration of M towards Z} = \omega^2 y$$

Now ω^2 is a constant.

$$\therefore \textit{acceleration of M towards Z} \propto \textit{distance of M from Z}$$

If we wish to express mathematically that the acceleration is always directed towards Z in simple harmonic motion, we must say

$$\text{acceleration towards Z} = -\omega^2 y \qquad . \qquad . \qquad . \qquad (1)$$

The minus indicates, of course, that the object begins to decelerate as it passes the centre, Z, of its motion. As we see later in discussing cases of simple harmonic motion, this is due to an opposing force. If the minus were omitted from equation (1) the latter would imply that the acceleration increases *in the direction* of y increasing, and the object would then never return to its original position.

We can now form a definition of **simple harmonic motion.**

It is the motion of a particle *whose acceleration is always*   (i) *directed towards a fixed point,*   (ii) *directly proportional to its distance from that point.*

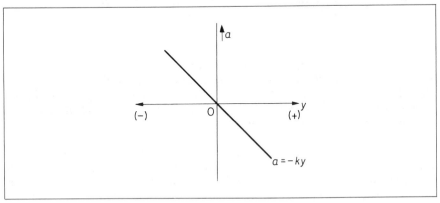

Figure 2.19A   *Graph of acceleration* a *against displacement* y *for s.h.m.*

Mathematically, if ω^2 is a constant

Acceleration $a = -\omega^2 y$

---

where y is the distance from the fixed point.

The straight-line graph in Figure 2.19A shows how the acceleration a varies with displacement y from a fixed point for simple harmonic motion. The line has a *negative* gradient, since $a = -ky$, where k is a positive constant.

Period, Amplitude, Sine Curve

The time taken for the foot of the perpendicular to move from C to F and back to C is known as the *period* (T) of the simple harmonic motion. In this time, the object moving round the circle goes exactly once round the circle from C; and since ω is the angular speed and 2π radians (360°) is the angle described, the period T is given by

$$T = \frac{2\pi}{\omega} \quad . \qquad . \qquad . \qquad . \qquad . \qquad (1)$$

The distance ZC, or ZF, is the maximum distance from Z of the foot of the perpendicular, and is known as the *amplitude* of the motion. It is equal to r, the radius of the circle. So maximum acceleration, $a_{max} = -\omega^2 r$.

We have now to consider the variation with time, t, of the distance, y, from Z of the foot of the perpendicular. The distance $y = ZM = r \sin \theta$. But $\theta = \omega t$, where ω is the angular speed.

$$\therefore y = r \sin \omega t \quad . \qquad . \qquad . \qquad . \qquad . \qquad (2)$$

The graph of y against t is shown in Figure 2.19; ON represents the y-axis and OQ the t-axis. Since the angular speed of the object moving round the circle is constant, θ is proportional to the time t. So at X, the angle θ or ωt is equal to 90° or $\pi/2$ in radians; at P, the angle θ is 180° or π radians; and at Q, the angle θ is 360° or 2π radians. The simple harmonic graph is therefore a sine (sinusoidal) curve.

A cosine curve such as $y = r\cos\omega t$, has the same waveform as a sine curve. So this also represents simple harmonic motion. But as $y = r$ when θ or ωt is zero, the cosine curve starts at a maximum value instead of zero as in a sine curve.

The complete set of values of y from O to Q is known as a cycle. The number of cycles per second is called the *frequency*. The unit '1 cycle per second' is called '1 *hertz (Hz)*'. The mains frequency in Great Britain is 50 Hz or 50 cycles per second.

Velocity in Simple Harmonic Motion

If y is the displacement at an instant, then the velocity v at this instant is dy/dt, the rate of change of displacement (p. 4). Now $y = r\sin\omega t$, Figure 2.20. To find v or dy/dt from this graph we take the *gradient* of the curve at the time t considered. Figure 2.20 shows how v varies with time, t.

The velocity–time (v–t) graph is a cosine curve. At $t = 0$, v has a maximum value. So $v = A\cos\omega t$ where A is the amplitude or maximum value of v. Now A is the gradient of the y–t graph at $t = 0$. We can see by drawing different graphs of y against t that A depends on both r, the maximum value of y, and ω, the angular velocity (or number of cycles per second). Since $dy/dt = \omega r\cos\omega t = v$, we see that $A = \omega r$. So the velocity v is given by

$$v = \omega r \cos \omega t \qquad . \qquad . \qquad . \qquad . \qquad . \qquad (1)$$

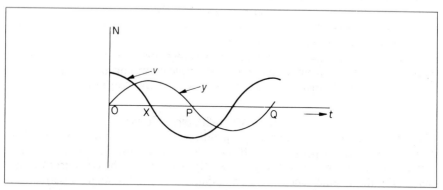

Figure 2.20   *Graph of velocity* v *and displacement* y *against time* t *in s.h.m.*

We can also express the velocity v in terms of y and r. From $y = r\sin\omega t$ and $v = r\omega\cos\omega t$, we have $\sin\omega t = y/r$ and $\cos\omega t = v/r\omega$. Now $\sin^2\omega t + \cos^2\omega t = 1$, from trigonometry. So

$$\frac{v^2}{r^2\omega^2} + \frac{y^2}{r^2} = 1$$

Simplifying
$$v = \pm\omega\sqrt{r^2 - y^2} \qquad . \qquad . \qquad . \qquad . \qquad . \qquad (2)$$

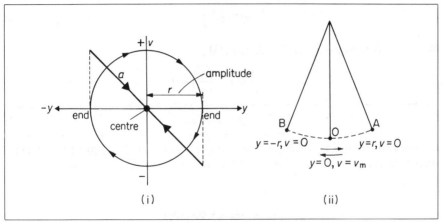

Figure 2.21 *Velocity* v *and displacement* y *in s.h.m.*

Figure 2.21 (i) shows the variation of v with *displacement* y. It is an ellipse. We can understand why this graph is obtained by considering the motion of a bob at the end of an oscillating simple pendulum, Figure 2.21 (ii). At the centre O ($y = 0$), the velocity v is a maximum. At the end A of the oscillation ($y = r$), $v = 0$. At the other end B ($y = -r$), $v = 0$. Note that v has an opposite direction on each half of the cycle. From (2), it follows that the maximum velocity v_m, when $y = 0$, is given numerically by

$$v_m = \omega r \qquad . \qquad . \qquad . \qquad . \qquad . \qquad (3)$$

When the velocity is a maximum ($y = 0$), the acceleration $a = 0$, since $a = -\omega^2 y$. When the velocity is zero ($y = r$), the acceleration a is a maximum.

S.H.M. Equations—Alternative Derivation

As we shall now show, all the equations used in s.h.m. can be derived by calculus without using the circle. With the usual notation,

$$\text{acceleration, } a = \frac{dv}{dt} = \frac{dy}{dt} \cdot \frac{dv}{dy} = v \frac{dv}{dy}$$

Now by definition of s.h.m., $a = -\omega^2 y$ (p. 76).

$$\therefore v \frac{dv}{dy} = -\omega^2 y$$

Integrating,
$$\therefore \frac{v^2}{2} = -\omega^2 \frac{y^2}{2} + c \qquad . \qquad . \qquad . \qquad . \qquad (1)$$

where c is a constant. Now $v = 0$ when $y = r$, the amplitude. So $c = \omega^2 r^2/2$, from (1). Substituting for c in (1) and simplifying,

$$\therefore v = \omega \sqrt{r^2 - y^2}$$

$$\therefore \frac{dy}{dt} = \omega \sqrt{r^2 - y^2} \qquad . \qquad . \qquad . \qquad . \qquad (2)$$

$$\therefore \frac{1}{\omega} \int \frac{dy}{\sqrt{r^2 - y^2}} = \int dt$$

$$\therefore \frac{1}{\omega}\sin^{-1}\left(\frac{y}{r}\right) = t + C \qquad \cdot \qquad \cdot \qquad \cdot \qquad \cdot \qquad (3)$$

When $t = 0$, then $y = 0$; so $C = 0$, from (3).

$$\therefore \frac{1}{\omega}\sin^{-1}\left(\frac{y}{r}\right) = t$$

$$\therefore y = r \sin \omega t \qquad \cdot \qquad \cdot \qquad \cdot \qquad \cdot \qquad \cdot \qquad (4)$$

When t increases to $t + 2\pi/\omega$, $y = r\sin(\omega t + 2\pi) = r\sin\omega t$, which is the same displacement value as at t. Hence the *period* T of the motion $= 2\pi/\omega$.

Learn these Results:

(1) If the acceleration a of an object $= -\omega^2 y$, where y is the distance or displacement of the object from a fixed point, the motion is simple harmonic motion. The graph of a against y is a straight line through the origin with a negative gradient. Maximum acceleration, $a_{max} = -\omega^2 r$, where r is amplitude.

(2) The *period, T*, of the motion $= 2\pi/\omega$, where T is the time to make a complete to-and-fro movement or cycle. The *frequency; f,* $= 1/T$ and its unit is 'Hz'. Note that $\omega = 2\pi/T = 2\pi f$.

(3) The amplitude, r, of the motion is the maximum distance on either side of the centre of oscillation.

(4) The velocity at any instant, v, $= \pm\omega\sqrt{r^2 - y^2}$; the maximum velocity $= \omega r$. The graph of the variation of v with displacement y is an ellipse.

Example on S.H.M.

A steel strip, clamped at one end, vibrates with a frequency of 20 Hz and an amplitude of 5 mm at the free end, where a small mass of 2 g is positioned. Find
(a) the velocity of the end when passing through the zero position,
(b) the acceleration at maximum displacement,
(c) the maximum kinetic energy of the mass.

(a) When the end of the strip passes through the zero position $y = 0$, the speed is a maximum v_m given by

$$v_m = \omega r$$

Now $\omega = 2\pi f = 2\pi \times 20$, and $r = 0.005$ m

$$\therefore v_m = 2\pi \times 20 \times 0.005 = 0.628 \text{ m s}^{-1}$$

(b) The acceleration $= -\omega^2 r$, where r is the amplitude

$$\therefore \text{acceleration} = (2\pi \times 20)^2 \times 0.005$$

$$= 79 \text{ m s}^{-2}$$

(c) $m = 2\text{ g} = 2 \times 10^{-3}\text{ kg}, v_m = 0.628 \text{ m s}^{-1}$

$$\therefore \text{maximum k.e.} = \tfrac{1}{2}mv_m{}^2 = \tfrac{1}{2} \times (2 \times 10^{-3}) \times 0.628^2 = 3.9 \times 10^{-4} \text{ J (approx.)}$$

S.H.M. and *g*

If a small coin is placed on a horizontal platform connected to a vibrator, and the amplitude is kept constant as the frequency is increased from zero, the coin will be heard 'chattering' at a particular frequency f_0. At this stage the reaction of the table with the coin becomes zero at some part of every cycle, so that it loses contact periodically with the surface, Figure 2.22.

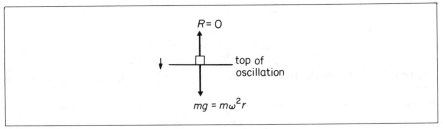

Figure 2.22   *s.h.m. of coin on surface*

The maximum acceleration in s.h.m. occurs at the end of the oscillation because the acceleration is directly proportional to the displacement. Thus maximum acceleration $= \omega^2 r$, where *r* is the amplitude and ω is $2\pi f_0$.

The coin will lose contact with the table when it is moving *down* with acceleration *g*, Figure 2.22. Suppose the amplitude *r* is 0·08 m. Then

$$(2\pi f_0)^2 r = g$$

$$\therefore 4\pi^2 f_0{}^2 \times 0\cdot08 = 9\cdot8$$

$$\therefore f_0 = \sqrt{\frac{9\cdot8}{4\pi^2 \times 0\cdot08}} = 1\cdot8\,\text{Hz}$$

Oscillating System—Spring and Mass

We now consider some oscillating systems in mechanics. A mass attached to a spring is a standard and useful case. For example, the body or chassis of a car is a mass attached to springs underneath and the oscillation needs study to provide a comfortable ride when travelling over ridges in the road surface. Also, when atoms or molecules in crystals vibrate, the molecular forces between the small masses can be represented by a 'spring'. In radio oscillators, one part of the basic arrangement can be considered to behave like a 'mass' and another as a 'spring'.

Suppose that one end of a spring S of negligible mass is attached to a smooth object A, and that S and A are laid on a horizontal smooth table, Figure 2.23. If the free end of S is attached to the table and A is pulled slightly to extend the

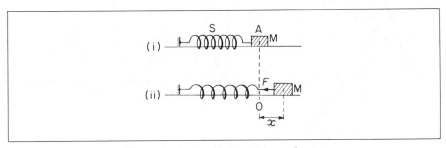

Figure 2.23   *Oscillating spring and mass*

spring and then released, the system vibrates with simple harmonic motion. The centre of oscillation O is the position of A at the end of the spring corresponding to its natural length, that is, when the spring is neither extended nor compressed.

Suppose the extension x of the spring is directly proportional to the force F in the spring (Hooke's law). F acts in the opposite direction to x, so $F = -kx$, where k is known as the *force constant* of the spring or 'force per unit extension'. If m is the mass of A, the acceleration a is given by $F = ma$. So

$$ma = -kx$$

Thus
$$a = -\frac{k}{m}x = -\omega^2 x$$

where $\omega^2 = k/m$. So the motion of A is simple harmonic and the period T is given by

$$T = \frac{2\pi}{\omega} = 2\pi\sqrt{\frac{m}{k}}$$

Potential and Kinetic Energy Exchanges in Oscillating Systems

The energy of the stretched spring is *potential energy*, p.e.—its molecules are continually displaced or compressed relative to their normal distance apart. The p.e. for an extension $x = \int F.\,dx = \int kx.\,dx = \frac{1}{2}kx^2$.

The energy of the mass is *kinetic energy*, k.e., or $\frac{1}{2}mv^2$, where v is the velocity. Now from $x = r\sin\omega t$, $v = dx/dt = \omega r\cos\omega t$

∴ total energy of spring plus mass $= \frac{1}{2}kx^2 + \frac{1}{2}mv^2$

$$= \frac{1}{2}kr^2\sin^2\omega t + \frac{1}{2}m\omega^2 r^2\cos^2\omega t$$

But $\omega^2 = k/m$, or $k = m\omega^2$

∴ total energy $= \frac{1}{2}m\omega^2 r^2(\sin^2\omega t + \cos^2\omega t) = \frac{1}{2}m\omega^2 r^2 = constant$

Thus the total energy of the vibrating mass and spring is constant. When the k.e. of the mass is a maximum (energy $= \frac{1}{2}m\omega^2 r^2$ and mass passing through the

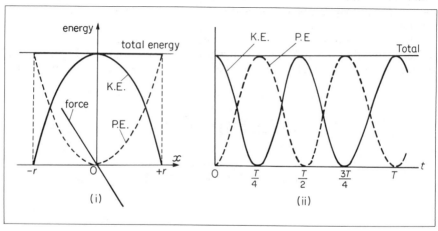

Figure 2.24 *Energy in s.h.m.*

centre of oscillation), the p.e. of the spring is then zero ($x = 0$). Conversely, when the p.e. of the spring is a maximum (energy $= \frac{1}{2}kr^2 = \frac{1}{2}m\omega^2 r^2$ and mass at the end of the oscillation), the k.e. of the mass is zero ($v = 0$). Figure 2.24 (i) shows the variation of p.e. and k.e. with displacement x; the force F extending the spring, also shown, is directly proportional to the displacement from the centre of oscillation. Figure 2.24 (ii) shows how the p.e. and k.e. vary with time t; the curves are simple harmonic or sine curves and T is the period.

The constant interchange of energy between potential and kinetic energies is essential for producing and maintaining oscillations, whatever their nature. In the case of the oscillating bob of a simple pendulum, for example, the bob loses kinetic energy after passing through the middle of the swing, and then stores the energy as potential energy as it rises to the top of the swing. The reverse occurs as it swings back. In the case of oscillating layers of air when a sound wave passes, kinetic energy of the moving air molecules is converted to potential energy when the air is compressed. In the case of electrical oscillations, a coil L and a capacitor C in the circuit constantly exchange energy; this is stored alternately in the magnetic field of L and the electric field of C.

Oscillation of Mass Suspended from Helical Spring

Consider a helical spring or an elastic thread PA suspended from a fixed point P, Figure 2.25. When a mass m is placed on it, the spring stretches to O by a length e given by

$$mg = ke \quad . \qquad . \qquad . \qquad . \qquad . \qquad . \qquad \text{(i)}$$

where k is the force constant (force per unit extension) of the spring, since the tension in the spring is then mg. If the mass is pulled down a little and then released, it vibrates up-and-down above and below O. Suppose at an instant that B is at a distance x below O. The tension T of the spring at B is then

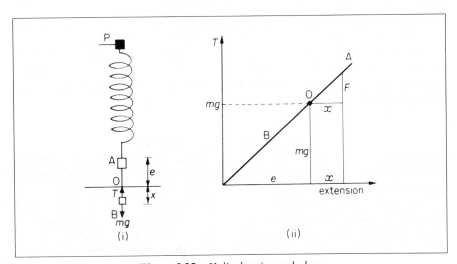

Figure 2.25 *Helical spring and s.h.m.*

equal to $k(e+x)$. Hence the resultant force F downwards $= mg - k(e+x) = mg - ke - kx = -kx$, since $ke = mg$ from (i). From $F = ma$,

$$\therefore -kx = ma$$

$$\therefore a = -\frac{k}{m}x = -\omega^2 x$$

where $\omega^2 = k/m$. Thus the motion is simple harmonic about O, and the period T is given by

$$T = \frac{2\pi}{\omega} = 2\pi\sqrt{\frac{m}{k}} \qquad . \qquad . \qquad . \qquad . \qquad . \qquad (1)$$

Also, since $mg = ke$, it follows that $m/k = e/g$.

$$\therefore T = 2\pi\sqrt{\frac{e}{g}} \qquad . \qquad . \qquad . \qquad . \qquad . \qquad (2)$$

Figure 2.25 (ii) shows the straight-line variation of the tension T in the spring with the extension, assuming Hooke's law (p. 134). The point O on the line corresponds to the extension e when the weight mg is on the spring and $T = mg$. When the mass is pulled down and released as in Figure 2.25 (i), the tension values vary along the straight line AOB. So at a displacement x from O, the *resultant* force $F(T - mg)$ on m is proportional to x. From $F = ma$, the acceleration a of m is proportional to x. So the motion is simple harmonic about O. Also, from Figure 2.25 (ii), $F/x = mg/e$. Hence $F/m = a = gx/e$. So $\omega^2 = g/e$ and the period $= 2\pi/\omega = 2\pi\sqrt{e/g}$, as deduced in (2).

From (1), it follows that $T^2 = 4\pi^2 m/k$. Consequently a graph of T^2 against m should be a straight line through the origin. In practice, when the load m is varied and the corresponding period T is measured, a straight line graph is obtained when T^2 is plotted against m, thus verifying indirectly that the motion of the load was simple harmonic. The graph does not pass through the origin, however, owing to the mass and the movement of the various parts of the spring. This has not been taken into account in the foregoing theory.

From (1), the period of oscillation T depends on the mass m and the force constant k of the spring. Since m and k are constants, it follows that if the same mass and spring are taken to the moon, the period of oscillation would be the same. The period of oscillation T of a simple pendulum of length l would change, however, if it were taken to the moon, as $T = 2\pi\sqrt{l/g}$ and the moon's gravitational intensity is about $g/6$.

Springs in Series and Parallel

Consider a helical spring of force constant k where $F = k \times$ extension. A mass m of weight mg then extends the spring by a length e given by $mg = ke$, and the period of oscillation of the mass is $T = 2\pi\sqrt{e/g}$ as previously obtained.

Suppose two identical helical springs are connected in *series*, each of force constant k, Figure 2.26 (i). The same weight mg will extend the springs twice as much as for a single spring since the total length is twice as much. So the extension is now $2e$. The period of oscillation of the mass m is therefore given by

$$T_1 = 2\pi\sqrt{\frac{2e}{g}}$$

So
$$T_1 = \sqrt{2}T$$

The mass therefore oscillates with a longer period at the end of the two springs.

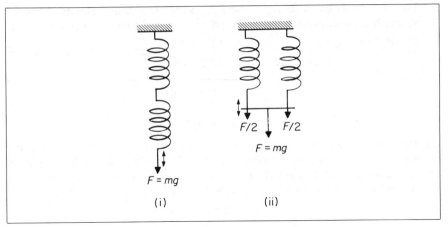

Figure 2.26 *Springs in series and parallel*

Now suppose the two springs are placed in *parallel* and the mass m is attached at the middle of a short horizontal connecting bar, Figure 2.26 (ii). This time the force on each spring is $mg/2$. So the extension is half as much as for a single spring, or $e/2$. So the period T_2 of the system is given by

$$T_2 = 2\pi \sqrt{\frac{e/2}{g}} = 2\pi \sqrt{\frac{e}{2g}} = \frac{1}{\sqrt{2}} T$$

The period of the parallel system is therefore less than for a single spring. Also, from above,

$$\frac{T_1}{T_2} = \frac{\sqrt{2}T}{T/\sqrt{2}} = \sqrt{4} = 2$$

Example on Spring-mass Oscillations

A small mass of 0·2 kg is attached to one end of a helical spring and produces an extension of 15 mm or 0·015 m. The mass is now pulled down 10 mm and set into vertical oscillation of amplitude 10 mm. What is

(a) the period of oscillation,
(b) the maximum kinetic energy of the mass,
(c) the potential energy of the spring when the mass is 5 mm below the centre of oscillation? ($g = 9\cdot8\,\mathrm{m\,s^{-2}}$.)

(*Analysis* (a) Since $T = 2\pi\sqrt{m/k}$, we need to find k, (b) max k.e. $= \frac{1}{2}mv_m{}^2 = \frac{1}{2}mr^2\omega^2$, (c) p.e. $= \frac{1}{2}kx^2$.)

(a) The force constant k of the spring in $\mathrm{N\,m^{-1}}$ is given by

$$k = \frac{mg}{e} = \frac{0\cdot2 \times 9\cdot8}{0\cdot015}$$

As we have previously shown,

$$T = 2\pi \sqrt{\frac{m}{k}} = 2\pi \sqrt{\frac{0\cdot2 \times 0\cdot015}{0\cdot2 \times 9\cdot8}}$$

$$= 2\pi \sqrt{\frac{0\cdot015}{9\cdot8}} = 0\cdot25\,\mathrm{s}$$

(b) The maximum k.e. $= \frac{1}{2}mv_m^2$, where v_m is the maximum velocity. Now for simple harmonic motion, $v_m = r\omega$ where $r = $ amplitude $= 10\,\text{mm} = 0.01\,\text{m}$. So, since $\omega = \sqrt{k/m} = \sqrt{9.8/0.015}$ from above

$$\text{maximum k.e.} = \frac{1}{2} \times 0.2 \times r^2\omega^2$$

$$= \frac{1}{2} \times 0.2 \times 0.01^2 \times \frac{9.8}{0.015}$$

$$= 6.5 \times 10^{-3}\,\text{J}$$

(c) The potential energy of the spring is given generally by $\frac{1}{2}kx^2$, where k is the force constant and x is the extension from its *original* length. The centre of oscillation is 15 mm below the unstretched length, so 5 mm below the centre of oscillation corresponds to an extension x of 20 mm or 0.02 m. Since

$$k = (0.2 \times 9.8)/0.015$$

$$\text{Potential energy of spring} = \frac{1}{2}kx^2 = \frac{\frac{1}{2} \times 0.2 \times 9.8}{0.015} \times 0.02^2\,\text{J}$$

$$= 2.6 \times 10^{-2}\,\text{J}$$

Simple Pendulum

We shall now study another case of simple harmonic motion. Consider a *simple pendulum*, which consists of a small mass m attached to the end of a length l of wire, Figure 2.27. If the other end of the wire is attached to a fixed point P and the mass is displaced slightly, it oscillates to-and-fro along the arc of a circle of centre P. We shall now show that the motion of the mass about its original position O is simple harmonic motion.

Suppose that the vibrating mass is at B at some instant, where $OB = y$ and angle $OPB = \theta$. At B, the force pulling the mass towards O is directed along the tangent at B, and is equal to $mg \sin \theta$. The tension, T, in the wire has no

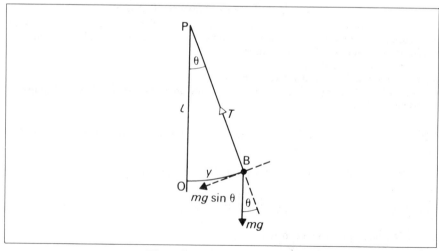

Figure 2.27 *Simple pendulum*

component in this direction, since PB is perpendicular to the tangent at B. Thus, since force $= $ mass $\times$ acceleration,

$$-mg \sin \theta = ma$$

where a is the acceleration along the arc OB; the minus indicates that the force is towards O, while the displacement, y, is measured along the arc from O in the opposite direction. *When θ is small, $\sin \theta = \theta$ in radians; also $\theta = y/l$.* Hence,

$$-mg\,\theta = -mg\frac{y}{l} = ma$$

$$\therefore a = -\frac{g}{l}\,y = -\omega^2 y$$

where $\omega^2 = g/l$. Since the acceleration is proportional to the distance y from a fixed point, the motion of the vibrating mass is simple harmonic motion (p. 77). Further, the period $T = 2\pi/\omega$.

$$\therefore T = \frac{2\pi}{\sqrt{g/l}} = 2\pi\sqrt{\frac{l}{g}} \qquad \cdot \qquad \cdot \qquad \cdot \qquad \cdot \qquad (1)$$

At a given place on the earth, where g is constant, the formula shows that the period T depends only on the length, l, of the pendulum. Moreover, the period remains constant even when the amplitude of the vibration diminishes owing to the resistance of the air. This result was first obtained by Galileo, who noticed a swinging lantern, and timed the oscillations by his pulse as clocks had not yet been invented. He found that the period remained constant although the swings gradually diminished in amplitude.

On the moon, g is about one-sixth that on the earth. From (1), we see that a pendulum of given length on the moon would have a period over twice as long as on the earth.

Example on Simple Harmonic Motion

A small bob of mass 20 g oscillates as a simple pendulum, with amplitude 5 cm and period 2 seconds. Find the speed of the bob and the tension in the supporting thread, when the speed of the bob is a maximum.

(*Analysis* (i) maximum speed $= \omega r$ and ω can be found from $T = 2\pi/\omega$. (ii) Use $F - mg = mv^2/r$ to find tension F.)

The speed, v, of the bob is a maximum when it passes through its original position. With the usual notation (see p. 79), the maximum v_m is given by

$$v_m = \omega r$$

where r is the amplitude of 0·05 m. Since $T = 2\pi/\omega$

$$\therefore \omega = \frac{2\pi}{T} = \frac{2\pi}{2} = \pi \qquad \cdot \qquad \cdot \qquad \cdot \qquad \cdot \qquad (1)$$

$$\therefore v_m = \omega r = \pi \times 0{\cdot}05 = 0{\cdot}16\,\mathrm{m\,s^{-1}}$$

Suppose F is the tension in the thread. The net force towards the centre of the circle along which the bob moves is then given by $(F - mg)$. The acceleration towards the centre of the circle, which is the point of suspension, is v_m^2/l, where l is the length of the pendulum.

$$\therefore F - mg = \frac{mv_m^2}{l}$$

$$\therefore F = mg + \frac{mv_m^2}{l} \qquad \cdot \qquad \cdot \qquad \cdot \qquad \cdot \qquad (2)$$

From $T = 2\pi\sqrt{l/g}$, $l = gT^2/4\pi^2 = 9\cdot8 \times 2^2/4\pi^2$. Also, $m = 20\,g = 0\cdot02\,kg$. So, from (2),

$$F = 0\cdot02 \times 9\cdot8 + \frac{0\cdot02 \times (0\cdot05\pi)^2 \times \pi^2}{9\cdot8}$$

$$= 19\cdot65 \times 10^{-2}\,N$$

Oscillations of a Liquid in a U-Tube

If the liquid on one side of a U-tube T is depressed by blowing gently down that side, the levels of the liquid will oscillate for a short time about their respective initial positions O, C, before finally coming to rest, Figure 2.28.

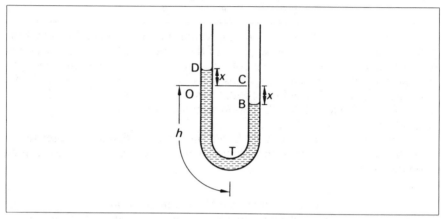

Figure 2.28   *s.h.m. of liquid*

At some instant, suppose that the level of the liquid on the left side of T is at D, at a height x above its original (undisturbed) position O. The level B of the liquid on the other side is then at a depth x below its original position C. So the excess pressure on the whole liquid, as shown on p. 105,

$$= \text{excess height} \times \text{liquid density} \times g = 2x\rho g$$

Since pressure = force per unit area,

force on liquid = pressure × area of cross-section of the tube = $2x\rho g \times A$

where A is the cross-sectional area of the tube. The mass of liquid in the U-tube = volume × density = $2hA\rho$, where $2h$ is the total length of the liquid in T. So, from $F = ma$ the acceleration, a, towards O or C is given by

$$-2x\rho gA = 2hA\rho a$$

The minus indicates that the force towards O is opposite to the displacement measured from O at that instant.

$$\therefore a = -\frac{g}{h}x = -\omega^2 x$$

where $\omega^2 = g/h$. So the motion of the liquid about O (or C) is simple harmonic, and the period T is given by

$$T = \frac{2\pi}{\omega} = 2\pi\sqrt{\frac{h}{g}}$$

In practice the oscillations are heavily damped owing to friction, which we have ignored.

Combining Two Perpendicular S.H.M.s, Lissajous Figures

As we see later in the cathode-ray oscilloscope (p. 778), electrons can move under two simple harmonic forces of the same frequency *at right-angles* to each other. In this case the electron has an *x*-motion say given by $x = a \sin \omega t$ and a perpendicular or *y*-motion given by $y = b \sin(\omega t + \theta)$, where a and b are the respective amplitudes and θ is the phase angle between the oscillations in the *x*- and *y*-directions respectively.

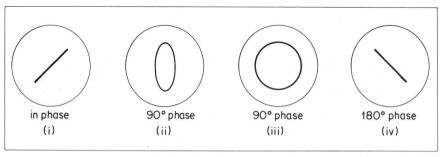

| in phase | 90° phase | 90° phase | 180° phase |
| (i) | (ii) | (iii) | (iv) |

Figure 2.29 *Lissajous figures*

If the oscillations are *in phase*, then $\theta = 0$. So $x = a \sin \omega t$ and $y = b \sin \omega t$. Then $\sin \omega t = x/a = y/b$. So $y = bx/a$. This is the equation of a *straight line* and this is the resultant motion of the electrons as shown on the oscilloscope screen, Figure 2.29 (i). This so-called 'Lissajous figure' is used to find out when two oscillations have the same phase (see p. 782).

If the two oscillations are 90° or $\pi/2$ out of phase, then $x = a \sin \omega t$ and $y = b \sin(\omega t + 90°) = b \cos \omega t$. So $\sin \omega t = x/a$ and $\cos \omega t = y/b$. Since, by trigonometry, $\sin^2 \omega t + \cos^2 \omega t = 1$, then $x^2/a^2 + y^2/b^2 = 1$. This is the equation of an *ellipse*, Figure 2.29 (ii). If the amplitudes a and b are equal, the equation is that of a *circle*, Figure 2.29 (iii). If the phase difference is 180° or π, then a straight line with a negative gradient is obtained, Figure 2.29 (iv). As explained in the cathode-ray oscilloscope section, two perpendicular oscillations which have a frequency ratio 2:1 produce a figure like the number eight on the screen.

_____ **Exercises 2C** _____

Simple Harmonic Motion

(Assume $g = 10 \, m \, s^{-2}$ or $10 \, N \, kg^{-1}$)

1 An object moving with simple harmonic motion has an amplitude of 0·02 m and a frequency of 20 Hz. Calculate   (i) the period of oscillation,   (ii) the acceleration at the middle and end of an oscillation,   (iii) the velocities at the corresponding instants.

2 A body of mass 0·2 kg is executing simple harmonic motion with an amplitude of 20 mm. The maximum force which acts upon it is 0·064 N. Calculate
(a) its maximum velocity,
(b) its period of oscillation. (*L.*)

3   A steel strip, clamped at one end, vibrates with a frequency of 50 Hz and an amplitude of 8 mm at the free end. Find
   (a) the velocity of the end when passing through the zero position,
   (b) the acceleration at the maximum displacement.
      Draw a sketch showing how the velocity and the acceleration vary with the displacement of the free end.

4   Some of the following graphs refer to simple harmonic motion, where v is the velocity, a is the acceleration, E_k is the kinetic energy, E is the total energy and x is the displacement from the mean (zero) position. Which graphs are correct?

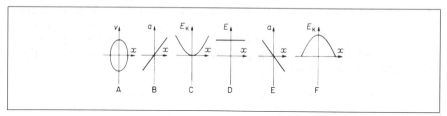

Figure 2B

5   A spring of force constant k of $5\,\mathrm{N\,m^{-1}}$ ($F = -kx$) is placed horizontally on a smooth table. One end of the spring is fixed and a mass X of $0.20\,\mathrm{kg}$ is attached to the free end. X is displaced a distance of 4 mm along the table and then released. Show that the motion of X is simple harmonic, and calculate   (i) the period,   (ii) the maximum acceleration,   (iii) the maximum kinetic energy,   (iv) the maximum potential energy of the spring.

6   A simple pendulum has a period of $4.2\,\mathrm{s}$. When the pendulum is shortened by 1 m, the period is $3.7\,\mathrm{s}$. From these measurements, calculate the acceleration of free fall g and the original length of the pendulum.
      If the pendulum is taken from the earth to the moon where the acceleration of free fall is $g/6$, what relative change, if any, occurs in the period T?

7   The bob of a simple pendulum moves simple harmonically with amplitude $8.0\,\mathrm{cm}$ and period $2.00\,\mathrm{s}$. Its mass is $0.50\,\mathrm{kg}$. The motion of the bob is undamped.
      Calculate maximum values for
   (a) the speed of the bob,
   (b) the kinetic energy of the bob. (*L.*)

8   Define *simple harmonic motion*. State one condition for simple harmonic motion. A spring is extended 10 mm when a small weight is attached to its free end. The weight is now pulled down slightly and released. Show that its motion is simple harmonic and calculate the period.

9   Define
   (a) *displacement*,
   (b) *amplitude*,
   (c) *angular frequency*,
   of a simple harmonic motion and give an expression relating them, explaining all symbols used.
      A student is under the impression that ω, the angular frequency of oscillation of a simple pendulum is dependent solely upon the length l of the pendulum and the mass m of its bob. Show, by dimensional analysis, that this cannot be correct. Derive from first principles the correct equation,

$$\omega^2 = g/l$$

where g is the acceleration of free fall.
      A small spherical mass is hung from the end of an elastic string of natural length $40.0\,\mathrm{cm}$ and when the pendulum so formed is set swinging with small amplitude, 20 oscillations are completed in $26.0\,\mathrm{s}$. The bob is then replaced by one of the same size but of a different mass and the new time for 20 oscillations is $26.4\,\mathrm{s}$. Account for this change and calculate the ratio of the masses. (*C.*)

10 (a) Define *simple harmonic motion*. Give *three* examples of systems which vibrate with approximately simple harmonic motion. How does the displacement of a simple harmonic motion vary with time?

What is meant by the *phase difference* between two simple harmonic motions of the same frequency? Illustrate your answer graphically, by considering the variation of the displacement with time of two motions vibrating with simple harmonic motion of the same frequency but which have phase differences of (i) 90° and (ii) 180°.

(b) At what points in a simple harmonic motion are (i) the acceleration, (ii) the kinetic energy and (iii) the potential energy of the system each at (1) a maximum, and (2) a minimum?

Sketch the graphs showing how (iv) the kinetic energy, (v) the potential energy, and (vi) the sum of the kinetic and potential energies for a simple harmonic oscillator each vary with displacement.

(c) Calculate the period of oscillation of a simple pendulum of length 1·8 m, with a bob of mass 2·2 kg. What assumption is made in this calculation? ($g = 9·8\,\mathrm{m\,s^{-2}}$)

If the bob of this pendulum is pulled aside a horizontal distance of 20 cm and released, what will be the values of (i) the kinetic energy and (ii) the velocity of the bob at the lowest point of the swing? (*L.*)

11 (a) State the conditions necessary for the motion of an oscillating body to be simple harmonic. Give **one** reason why the vertical oscillations of a body suspended from a spring may **not** satisfy these conditions.

(b) The vertical oscillations of a body on a spring are started by holding the body at a point where the spring is at its natural length and then releasing it.

State and explain briefly the effect of increasing the mass of the body on the value of each of the following quantities (i) the time period of the oscillation, (ii) the amplitude of the oscillation, (iii) the total energy of the oscillating system. (*AEB*, 1984.)

12 State the relationship between the force on a body and the distance of the body from a fixed position when the body is executing simple harmonic motion about that position.

Show that a body of mass m suspended by a light elastic string for which the ratio of tension to extension is λ will execute simple harmonic motion when given a small vertical displacement from its equilibrium position. Find the period of the motion for the case $m = 0·1\,\mathrm{kg}$ and $\lambda = 20\,\mathrm{N\,m^{-1}}$.

A second 0·1 kg mass is attached to the first by a light inextensible wire and hangs below it. The system is allowed to come to rest, and at time $t = 0$ the wire is cut. Calculate the position, velocity and acceleration of the first 0·1 kg mass at time $t = 1·05\,\mathrm{s}$, assuming no resistance to motion.

Give expressions for the kinetic and potential energy of the system at time t. Show that the total energy is independent of time. Outline qualitatively what would happen to the total energy of such a system set oscillating in the laboratory. (*O. & C.*)

13 (a) The displacement y of a mass vibrating with simple harmonic motion is given by $y = 20\sin 10\pi t$, where y is in millimetres and t is in seconds. What is (i) the amplitude, (ii) the period, (iii) the velocity at $t = 0$?

(b) A mass of 0·1 kg oscillates in simple harmonic motion with an amplitude of 0·2 m and a period of 1·0 s. Calculate its maximum kinetic energy. Draw a sketch showing how the kinetic energy varies with (i) the displacement, (ii) the time.

14 A mass X of 0·1 kg is attached to the free end of a vertical helical spring whose upper end is fixed and the spring extends by 0·04 m. X is now pulled down a small distance 0·02 m and then released. Find (i) its period, (ii) the maximum force acting on it during the oscillations, (iii) its kinetic energy when X passes through its mean position.

15 The displacement of a particle vibrating with simple harmonic motion of angular speed ω is given by $y = a\sin\omega t$ is the time. What does a represent? Sketch a graph of the *velocity* of the particle as a function of time starting from $t = 0\,\mathrm{s}$.

A particle of mass 0·25 kg vibrates with a period of 2·0 s. If its greatest displacement is 0·4 m what is its maximum kinetic energy? (*L.*)

16 Explain what is meant by *simple harmonic motion.*

Show that the vertical oscillations of a mass suspended by a light helical spring are simple harmonic and describe an experiment with the spring to determine the acceleration due to gravity.

A small mass rests on a horizontal platform which vibrates vertically in simple harmonic motion with a period of 0·50 s. Find the maximum amplitude of the motion which will allow the mass to remain in contact with the platform throughout the motion. (*L.*)

17 Define *simple harmonic motion.* Explain what is meant by the *amplitude*, the *period* and the *phase* of such a motion.

A simple pendulum of length 1·5 m has a bob of mass 2·0 kg.
(a) State the formula for the period of small oscillations and evaluate it in this case.
(b) If, with the string taut, the bob is pulled aside a horizontal distance of 0·15 m from the mean position and then released from rest, find the kinetic energy and the speed with which it passes through the mean position.
(c) After 50 complete swings, the maximum horizontal displacement of the bob has become only 0·10 m. What fraction of the initial energy has been lost?
(d) Estimate the maximum horizontal displacement of the bob after a further 50 complete swings. (Take g to be $10\,\mathrm{m\,s^{-2}}$.) (*O.*)

18 (a) What is meant by *simple harmonic motion*?

The equation $x = a\sin 2\pi f t$ can represent the motion of a body executing simple harmonic motion where x represents the displacement of the body from a fixed point at time t. Sketch two cycles of the motion beginning at $t = 0$, clearly labelling the axes of the graph. Use the graph to explain the physical meanings of a and f.

Explain how you could obtain from the graph the speed of the body at any instant.

(b) In order to check the timing of a camera shutter a student set up a simple pendulum of length 99·3 cm so that the bob swung in front of a horizontal metre scale. The bob was observed to swing between the 40·0 cm and 60·0 cm marks at its extreme positions. The camera was mounted directly in front of the scale, set for an exposure time (time for which the shutter is open) of 1/50 s and a photograph taken. The resulting photograph showed the bob to have moved from the 51·0 cm mark to the 51·6 cm mark while the shutter was open.

What is the percentage error in the exposure time indicated on the camera? (The period of oscillation of a simple pendulum of length *l* may be taken as

$$T = 2\pi\sqrt{\frac{l}{g}}$$

where g is the acceleration of free fall.) (*L.*)

19 Define simple harmonic motion and write down the appropriate equation to describe it. Give **two** examples of physical phenomena which exhibit such motion explaining carefully why the motion is simple harmonic in each case.

Derive an equation of motion of a mass suspended at the lower end of a helical spring and state in which way the period depends on   (i) the spring constant, (ii) the mass,   (iii) the amplitude.

A mass of 0·1 kg is set vibrating on the end of a spring of spring constant $4\,\mathrm{N\,m^{-1}}$. What is the period of oscillation?

In practice it is found that after 100 oscillations the amplitude is half the original amplitude and after another 100 oscillations it is a quarter the original. Compare this with radioactive decay and hence suggest a modification to the equation of motion derived above. What is the 'half-life' of this system? (*W.*)

20 Find, stating clearly any assumptions or conditions, an expression for the period of oscillation of a simple pendulum.

Such a pendulum is of length 1 m; the bob of mass 0·2 kg, is drawn aside through an angle of 5° and released from rest. The subsequent motion is described by

$$x = a\sin(\omega t + \varepsilon)$$

where x is the displacement of the bob (in metres) and t the time (measured in seconds from the instant of release). Find values for a, ω and ε.

What is the maximum velocity and maximum acceleration experienced by the bob?

What are the maximum and minimum values for the tension in the string, and where in the motion do these occur?

The angular amplitude reduces to 4° in 100 s. Find the mean loss of energy per cycle. (*W.*)

3

Forces in Equilibrium, Forces in Fluids

Forces in Equilibrium

We now consider forces which are in equilibrium, for example, forces which keep bridges stationary. As we shall see, forces are added together by a vector method. The effect of a force in a certain direction (its resolved component) depends on the size of the force and the angle between the force and the direction. This is often needed in problems of equilibrium.

Forces not only push or pull but also have a turning-effect or moment about an axis. In cases of equilibrium the moments have also to be considered. Objects are in equilibrium only if certain rules apply, as we discuss, and examples will show you how to use these rules.

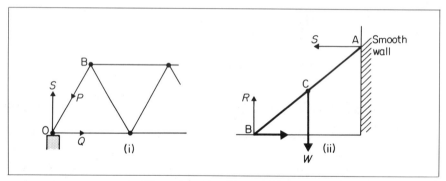

Figure 3.1 *Equilibrium of forces*

Figure 3.1 shows two examples of equilibrium. In Figure 3.1 (i) is part of a bridge structure. Here the joint O resting on brickwork is in equilibrium under three forces—P and Q are the forces or *tensions* in the metal beams, S is the *reaction force* on O due to the brickwork.

Figure 3.1 (ii) shows a uniform ladder AB with the top end A resting against a smooth wall and the bottom B resting on the rough ground. The forces on the ladder are its weight W acting at the midpoint C of the ladder, the frictional force F at the ground which stops the ladder sliding outwards, the *normal reaction R* of the ground at B acting at 90° to the ground and the normal reaction S at the wall acting at 90° to the wall. There is no frictional force here as the wall is smooth.

Adding Forces, Parallelogram of Forces

A force is a *vector* quantity (p. 4). So it can be represented in size and direction by a straight line drawn to scale. The sum or *resultant R* of two forces P and Q can be added by one of two vector methods.

(1) Figure 3.2 (i) shows two forces P and Q acting at 60° to each other. To add them, draw a line *ab* to represent P and from *b* draw a line *bc* to represent

Q—note that ab is parallel to P and bc is parallel to Q. Join ac. Then ac is the resultant R of P and Q in magnitude and direction. Note that the arrows on ab and bc follow each other.

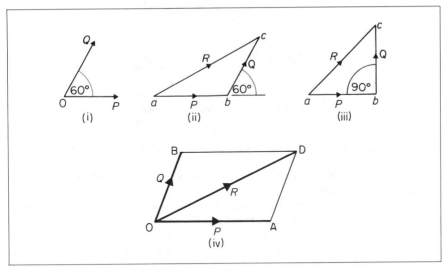

Figure 3.2 *Adding forces (vectors)*

We can find the size of R, and its direction θ to P either by accurate drawing or by calculation (using trigonometry in triangle abc). In branches of Physics, P and Q are often at 90° to each other. In this case the vector triangle abc is a right-angled triangle. Figure 3.2 (iii). Applying Pythagoras' theorem, then $R^2 = P^2 + Q^2$.

So
$$R = \sqrt{P^2 + Q^2}$$

Also, the angle θ R makes with P is given by $\tan\theta = bc/ab = Q/P$, so knowing P and Q, θ can be found.

(2) *Parallelogram of forces* The resultant R of P and Q can also be found by drawing a parallelogram of the forces. In Figure 3.2 (iv), draw OA to represent P and OB to represent Q at the angle, say 60°, between P and Q as in Figure 3.2 (i). Then complete the parallelogram OBDA. The resultant R is represented by *the diagonal OD through O*.

This gives the same result for R as in the previous method, since AD represents Q. If P and Q are 90° to each other, the parallelogram becomes a rectangle.

Resolved Components

We often need to find the effect of a force F in a particular direction at an angle θ to F. This is called the 'resolved component' or simply *component* of F in this direction.

In Figure 3.3, OD represents F and OX the direction at an angle θ in which the component of F is required. Using OD as the diagonal, we complete the *rectangle OADB*. Then the forces P, represented by OA, and Q, represented by OB, together represent F since F is their resultant. But Q is 90° to OX and so can have no effect in this direction. So P is the component of F in the direction

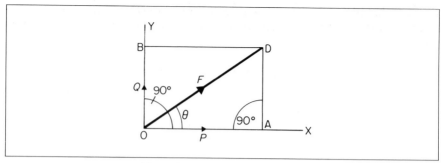

Figure 3.3 *Resolved components*

OX. Similarly, since *P* has no effect in the perpendicular direction OY, *Q* is the component of *F* in this direction.

We can find a general formula for a component. From the right-angle triangle ODA, $\cos \theta = P/F$. So

$$P = F \cos \theta$$

The cosine formula for the component should be memorised by the student. So the component *Q* in the direction OY is given by

$$Q = F \cos (90° - \theta) = F \sin \theta$$

So a force of 20 N has a component in a direction 60° to itself of $20 \cos 60° = 20 \times 0.5 = 10$ N In a direction 30° to itself, its component is $20 \cos 30° = 20 \times 0.87 = 17.4$ N. Remember that a force has no component at 90° to itself ($\cos 90° = 0$).

Example on Components

Figure 3.4 shows a stationary car of weight *W* on a road sloping at 30° to the horizontal. The frictional force on the car is 4000 N acting up the road BC.

What is (i) the weight *W*, (ii) the normal reaction force *R* of the road surface on the car?

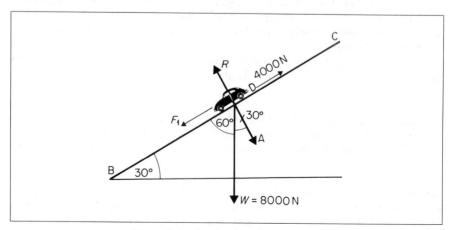

Figure 3.4 *Application of components*

(i) The weight W of the car acts vertically downwards. Since it makes an angle $60°$ to the road BC,

$$\text{component } F_1 \text{ down BC} = W \cos 60° = 0.5\,W$$

So
$$0.5\,W = \text{frictional force} = 4000\,\text{N}$$

$$\therefore W = 4000/0.5 = 8000\,\text{N}$$

(ii) The car does *not* move in a direction AD at $90°$ to the road. So R must balance exactly the component F_2 of the weight in the direction DA. So

R = component of 8000 N along DA, which makes an angle of $30°$ with 8000 N
  $= 8000 \cos 30° = 8000 \times 0.87 = 6960\,\text{N}$.

Forces in Equilibrium, Triangle and Polygon of Forces

We now consider the relations between forces *in equilibrium*. Figure 3.5 (i) shows forces P, Q and S acting on a joint O of a bridge structure which are in equilibrium.

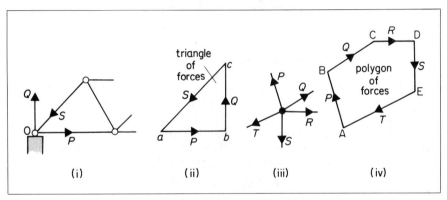

Figure 3.5 *Triangle and polygon of forces*

In Figure 3.5 (ii), we draw ab to represent P and bc to represent Q. Then ac represents the resultant (sum) of P and Q. For equilibrium, S must balance exactly the resultant. So S acts along ca opposite to ac, and has the same size as ac. In other words, the side ca of the triangle abc represents S.

This general result is stated in this way:

If three forces are in equilibrium, they can be represented in magnitude and direction by the three sides of a triangle taken in order.

This is called the *triangle of forces* theorem. Note that   (i) the arrows showing the directions of the forces follow each other (or are in order) round the triangle,   (ii) if the triangle does not close after drawing three forces acting on an object, then the forces are *not* in equilibrium.

We can extend this result to many forces in equilibrium. In Figure 3.5 (iii), five forces P, Q, R, S, T are in equilibrium at O. Starting with AB to represent P, all the forces taken in order form a *closed polygon* ABCDE, Figure 3.5 (iv). If the polygon did not close, the forces would not be in equilibrium.

Example on Triangle of Forces

A uniform ladder of weight 200 N and length 12 m is placed at an angle of 60° to the horizontal, with one end B leaning against a smooth wall and the other end A on the ground. Calculate the reaction force R of the wall at B and the force F of the ground at A.

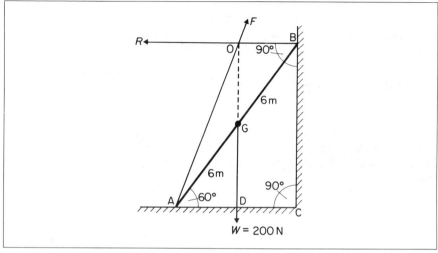

Figure 3.6   *Triangle of forces application*

The force R at B acts perpendicularly to the smooth wall. The weight W of the uniform ladder acts at its midpoint G. The forces W and R meet at O, as shown. So the force F at A must pass through O to balance their resultant.

The *triangle of forces* can be used to find the unknown forces R, F. Since DA is parallel to R, AO is parallel to F, and OD is parallel to W, the triangle of force is represented by AOD. By means of a scale drawing, R and F can be found, since

$$\frac{W(200)}{\text{OD}} = \frac{F}{\text{AO}} = \frac{R}{\text{DA}}$$

The result is $R = 58$ N and $F = 208$ N.

Moments

When the steering-wheel of a car is turned, the applied force is said to exert a *moment*, or turning-effect about the axle attached to the wheel. The magnitude of the moment of a force F about a point O is defined as *the product of the force F and the perpendicular distance OA from O to the line of action of F*, Figure 3.7.

Moment = force × *perpendicular* distance from axis

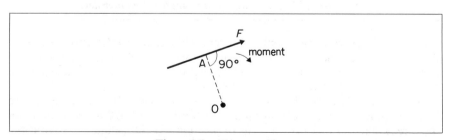

Figure 3.7   *Moment of force*

So $$\text{moment} = F \times AO$$

The magnitude of the moment is expressed in *newton metre* (N m) when F is in newton and AO is in metre. We can take an anticlockwise moment as positive in sign and a clockwise moment as negative in sign.

Poni   icaoponi a

Moments and Equilibrium

The resultant of a number of forces *in equilibrium* is zero. So the moment of the resultant about any point is zero. It therefore follows that

---

the algebraic sum of the moments of all the forces about any point is zero when those forces are in equilibrium.

This means that the total clockwise moment of the forces about any point = the total anticlockwise moment of the remaining forces about the same point.

As the following Examples show, this rule or *principle of moments* is widely used to find unknown forces when an object is in equilibrium.

Examples on Moments

1   A horizontal rod AB is suspended at its ends by two strings, Figure 3.8 (i). The rod is 0·6 m long and its weight of 3 N acts at G where AG is 0·4 m and BG is 0·2 m.
Find the tensions X and Y in the strings.

The forces X, Y and 3 N are parallel forces. So
$$X + Y = 3 . \qquad . \qquad . \qquad . \qquad . \qquad (1)$$

Clockwise moments about G = anticlockwise moments about G. Since 3 N has no moment about G,

$$X \times 0.4 = Y \times 0.2, \quad \text{so, cancelling,} \quad 2X = Y \quad . \quad . \quad (2)$$

From (1), $X + 2X = 3$,   so $X = 1$ N.   Then $Y = 2X = 2$ N

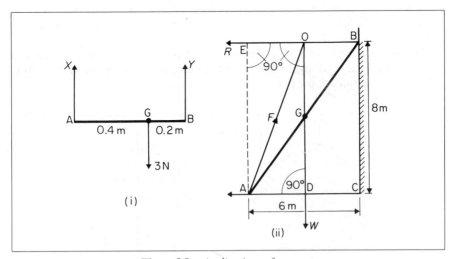

Figure 3.8   *Applications of moments*

2   In Figure 3.8 (ii), a ladder AB rests against a smooth wall at B and a rough ground at A. The weight W of 100 N acts at the midpoint G of the ladder, R is the normal reaction

force of the wall at B and F is the total force at the ground at A. The height BC = 8 m and
AC = 6 m.

Calculate R using the principle of moments and hence find F.

Take moments about A so as to eliminate F. Then, using perpendicular
distances,

$$R \times AE = W \times AD$$

or $\qquad R \times 8 = 100 \times 3 \quad (AE = BC = 8\,m, AD = \tfrac{1}{2}AC = 3\,m)$

So $\qquad R = \dfrac{100 \times 3}{8} = 37 \cdot 5\,N$

To find F, we see that F must be equal and opposite to the *resultant* of R and
W. Now R and W meet at 90° at O. So, from page 95,

$$\text{resultant} = F = \sqrt{R^2 + W^2}$$
$$= \sqrt{37 \cdot 5^2 + 100^2} = 107\,N$$

Couple and its Moment or Torque

There are many examples in practice where two forces, acting together, exert a
moment or turning-effect on some object. As a very simple case, suppose two
strings are tied to a wheel at X, Y, and *two equal and opposite forces*, F, are
exerted tangentially to the wheel, Figure 3.9 (i). If the wheel is pivoted at its
centre, O, it begins to rotate about O in an anticlockwise direction. The total
moment about O is then $(F \times OY) + (F \times OX) = F \times XY$.

Two equal and opposite forces whose lines of action do not coincide are said
to form a *couple*. The two forces always have a turning-effect, or moment, called
a *torque*, which is given by

$\qquad$ ***torque = one force × perpendicular distance between forces*** $\qquad$. $\qquad$ (1)

Since XY is perpendicular to each of the forces F in Figure 3.9 (i), the torque on
the wheel = $F \times XY = F \times$ diameter of wheel. So if $F = 10\,N$ and the diameter is
0·4 m, the torque = 4 N m.

The moving coil or armature in an electric motor is made to spin round by a
couple. When it is working, two parallel and equal forces, in opposite directions,
act on opposite sides of the coil.

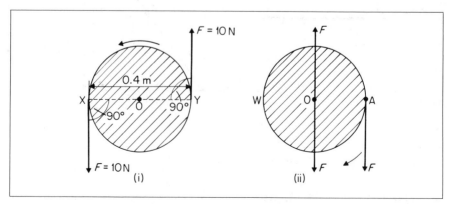

Figure 3.9 *Couples and moments*

In Figure 3.9 (ii), a wheel W is rotated about its centre O by a tangential force *F* at A on one side. To find the whole effect of *F* on the wheel, we can put two other forces *F at O* which are parallel to *F* at A but opposite in direction. This does not disturb the mechanics because the two forces at O would cancel and leave *F* at A. But the upward force *F* at O and the opposite parallel force *F* at A form a *couple* of moment *F* × OA. This is actually the moment of *F* at A.

In addition, however, we are left with the downward force *F* at O. So when *F* is applied at A to turn the wheel, it produces a couple *plus* an equal force *F* at O, the centre of the wheel.

Examples on Couples

1 In Figure 3.10 (i), a beam AB is acted on by a force of 3 N at A and a parallel force in the opposite direction of 4 N at B. What is the effect on the beam if it is on a smooth table?

The force of 4 N at B can be considered as a force of 3 N and a force of 1 N. The 3 N force at B and the 3 N force at A together form a *couple*. So the beam rotates. The force of 1 N left over at B would also make the beam move forward. So the total effect on the beam is a rotation *and* a forward movement.

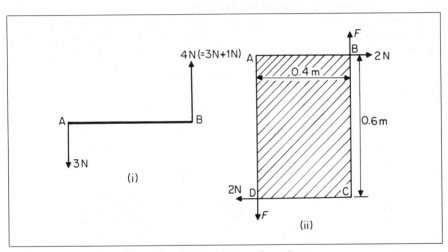

Figure 3.10   *Action of couples*

2 In Figure 3.10 (ii), two parallel forces *F* act in opposite directions along the sides AD and CB of a rectangular horizontal plate ABCD. Two equal and opposite forces of 2 N act along the sides CD and AB.

Calculate *F* if the plate does not rotate and the sides of the plate are 0·4 m and 0·6 m as shown.

Since the plate does not rotate, the moments of the two couples must be equal and opposite. So

$$F \times 0{\cdot}4 = 2 \times 0{\cdot}6$$

and
$$F = 2 \times 0{\cdot}6/0{\cdot}4 = 3\,\text{N}$$

Conditions for Equilibrium

We conclude by listing the conditions which apply when any object is in equilibrium.

(1) With three or more non-parallel forces acting on the object, a closed triangle or a closed polygon can be drawn to represent the forces in magnitude and direction.

(2) The algebraic sum of the moments of all the forces about any point is zero.

(3) The algebraic sum of the resolved components of all the forces in any direction is zero.

Centre of Mass

Consider a smooth uniform rod on a horizontal surface with negligible friction such as ice. If a force is applied to the rod near one end, the rod will rotate as it accelerates. If it is applied at the centre, *it will accelerate without rotation*. The *centre of mass* of an object may be defined as the point at which an applied force produces acceleration but no rotation.

With two separated masses m_1 and m_2 connected by a rigid rod of negligible mass, the centre of mass C is at a distance x_1 from m_1 and a distance x_2 from m_2 given numerically by $m_1 x_1 = m_2 x_2$. Figure 3.11 (i). So if the mass m_1 is 1 kg and the mass m_2 is 2 kg, and the length of the rod is 3 m, the centre of mass is 2 m from the 1 kg mass and 1 m from the 2 kg mass.

In a molecule of sodium chloride, the sodium atom has a relative atomic mass of about 23·0 and the chlorine atom one of about 35·5. If the separation of the atoms is a, the centre of mass C has a distance x from the sodium atom given by

$$23 \times x = 35 \cdot 5 \times (a - x)$$

Solving for x,
$$x = \frac{35 \cdot 5a}{58 \cdot 5} = 0 \cdot 6a$$

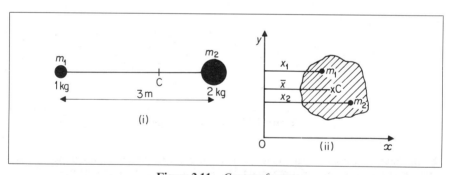

Figure 3.11 *Centre of mass*

Figure 3.11 (ii) shows the particles of masses $m_1, m_2, \ldots$ which together form the object of total mass M. If $x_1, x_2, \ldots$ are the respective x-co-ordinates of the particles relative to axes Ox, Oy, then generally the co-ordinate $\bar{x}$ of the centre of mass C is defined by

$$\bar{x} = \frac{m_1 x_1 + m_2 x_2 + \ldots}{m_1 + m_2 + \ldots} = \frac{\Sigma mx}{M}$$

Similarly, the distance $\bar{y}$ of the centre of mass C from Ox is given by

$$\bar{y} = \frac{\Sigma my}{M}$$

Centre of Gravity

Every particle is attracted towards the centre of the earth by the force of gravity.

The *centre of gravity* of a body is the point where the *resultant* force of attraction or *weight* of the body acts or appears to act.

In the simple case of a ruler, the centre of gravity is the point of support when the ruler is balanced.

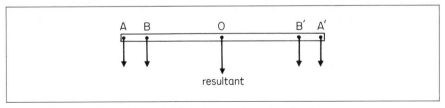

Figure 3.12 *Centre of gravity of rod*

An object can be considered to consist of many small particles. The forces on the particles due to the attraction of the earth are all parallel since they act vertically, and hence their resultant is the sum of all the forces. The resultant is the *weight* of the whole object, of course. In the case of a rod of uniform cross-sectional area, the weight of a particle A at one end, and that of a corresponding particle A' at the other end, have a resultant which acts at the mid-point O of the rod, Figure 3.12. Similarly, the resultant of the weight of a particle B, and that of a corresponding particle B', have a resultant acting at O. In this way, i.e., by symmetry, it follows that the resultant of the weights of all the particles of the rod acts at O. Hence the centre of gravity of a uniform rod is at its mid-point.

The centre of gravity, c.g., of the curved surface of a hollow cylinder acts at the mid-point of the cylinder axis. This is also the position of the c.g. of a uniform solid cylinder. The c.g. of a triangular plate of lamina is two-thirds of the distance along a median from corresponding point of the triangle. The c.g. of a uniform right solid cone is three-quarters along the axis from the apex.

Ideally, a *simple pendulum* has all its mass or weight concentrated at the centre of the swinging spherical bob. If the suspension has appreciable weight, the centre of gravity of the bob-suspension system would be higher than the centre of the spherical bob. The system would then no longer be a 'simple pendulum'.

Centre of Gravity and Centre of Mass

If the earth's field is uniform at all parts of an object, then the *weight* of a small mass m of it is typically mg. Thus, by moments, the distance of the centre of gravity from an axis Oy is given by

$$\frac{\Sigma mg \times x}{\Sigma mg} = \frac{\Sigma mx}{\Sigma m} = \frac{\Sigma mx}{M}$$

The gravitational field strength, g, cancels in numerator and denominator. It therefore follows that the centre of mass *coincides* with the centre of gravity. However, if the earth's field is *not* uniform at all parts of the object, the weight of a small mass m_1 of it is then $m_1 g_1$ say and the weight of a small mass m_2 at another part is $m_2 g_2$. Clearly, the centre of gravity does not now coincide with the centre of mass. A very long or very large object has different values of g at various parts of it.

(See Exercises 3, p. 110 for questions on Forces in Equilibrium)

Forces in Fluids

Pressure

Liquids and gases are called *fluids*. Unlike solid objects, fluids can flow.

If a piece of cork is pushed below the surface of a pool of water and then released, the cork rises to the surface again. The liquid thus exerts an upward force on the cork and this is due to the *pressure* exerted on the cork by the surrounding liquid. Gases also exert pressures. For example, when a thin closed metal can is evacuated, it usually collapses with a loud bang. The surrounding air exerts a pressure on the outside which is no longer counter-balanced by the pressure inside, and so there is a resultant force.

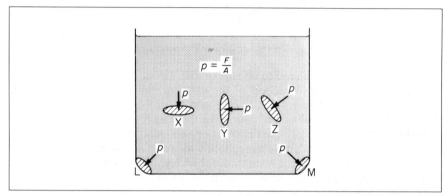

Figure 3.13   *Pressure in liquid*

Pressure is defined as the *average force per unit area* at the particular region of liquid or gas. In Figure 3.13, for example, X represents a small horizontal area, Y a small vertical area, and Z a small inclined area, all inside a vessel containing a liquid. The pressure p acts normally to the planes of X, Y or Z. In each case

$$\text{average pressure, } p = \frac{F}{A}$$

where F is the normal force due to the liquid on one side of an area A of X, Y or Z. Similarly, the pressure p on the sides L or M of the curved vessel acts normally to L and M and has magnitude F/A. In the limit, when the area is very small, $p = \mathrm{d}F/\mathrm{d}A$.

At a given point in a liquid, the pressure can act in any direction. *Pressure is a scalar*, not a vector. The direction of the force on a particular surface, however, is normal to the surface.

Formula for Pressure

Observation shows that the pressure increases with the depth, h, below the liquid surface and with its density ρ.

To obtain a formula for the pressure, p, suppose that a horizontal plate X of area A is placed at a depth h below the liquid surface, Figure 3.14. By drawing vertical lines from points on the perimeter of X, we can see that the force on X due to the liquid is equal to the weight of liquid of height h and uniform cross-section A. Since the volume of this liquid is Ah, the mass of the liquid $= Ah \times \rho$.

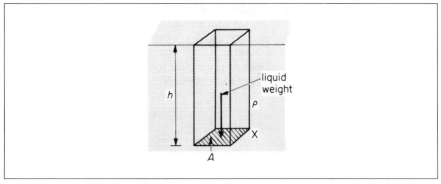

Figure 3.14 *Pressure and depth*

$$\therefore \text{weight} = Ah\rho g \text{ newton}$$

where g is 9·8, h is in m, A is in m², and ρ is in kg m⁻³.

$$\therefore \text{pressure, } p, \text{ on } X = \frac{\text{force}}{\text{area}} = \frac{Ah\rho g}{A}$$

$$\therefore p = h\rho g \qquad . \qquad . \qquad . \qquad . \qquad . \qquad (1)$$

When h, ρ, g have the units already mentioned, the pressure p is in *newton metre*⁻² (N m⁻² or in *pascal* (Pa), where 1 Pa = 1 N m⁻².

Pressure is also expressed in terms of the pressure due to a height of mercury (Hg). One unit is the *torr* (after Torricelli):

$$1 \text{ torr} = 1 \text{ mmHg} = 133\cdot3 \text{ Pa (approx.)}$$

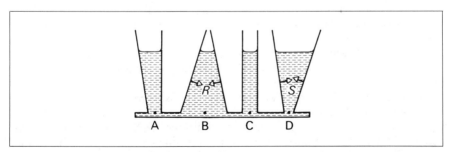

Figure 3.15 *Pressure and cross-section*

From $p = h\rho g$ it follows that *the pressure in a liquid is the same at all points on the same horizontal level in it.* Experiment also gives the same result. Thus a liquid filling the vessel shown in Figure 3.15 rises to the same height in each section if ABCD is horizontal. The cross-sectional area of B is greater than that of D. But the force on B is the sum of the weight of water above it together with the downward component of reaction R of the sides of the vessel, whereas the force on D is the weight of water above it *minus* the upward component of the reaction S of the sides of the vessel. This explains why the pressure in a vessel is independent of the cross-sectional area of the vessel such as those in Figure 3.15.

Atmospheric Pressure

A *barometer* is an instrument for measuring the pressure of the atmosphere, which is required in weather-forecasting, for example. An accurate form of barometer consists basically of a vertical barometer tube about a metre long containing mercury, with a vacuum at the closed top, Figure 3.16(i). The other end of the tube is below the surface of mercury contained in a vessel B.

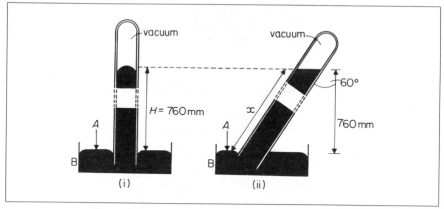

Figure 3.16   *Barometer*   *(i) vertical and*   *(ii) inclined*

The pressure on the surface of the mercury in B is atmospheric pressure, *A*; and since the pressure is transmitted through the liquid, the atmospheric pressure supports the column of mercury in the tube. Suppose the column is a vertical height *H* above the level of the mercury in B. Now the pressure, *p*, at the bottom of a column of liquid of vertical height *H* and density ρ is given by $p = H\rho g$ (p. 105). Thus if $H = 760$ mm $= 0.76$ m and $\rho = 13\,600$ kg m^{-3},

$$p = H\rho g = 0.76 \times 13\,600 \times 9.8 = 1.013 \times 10^5 \text{ Pa}$$

The pressure at the bottom of a column or mercury 760 mm high for a particular mercury density and value of *g* is known as *standard pressure* or *one atmosphere*.

By definition, 1 atmosphere $= 1.01325 \times 10^5$ Pa. *Standard temperature and pressure (s.t.p.)* **is 0°C and 760 mmHg pressure.**

The *bar* is 10^5 Pa and is nearly equal to one atmosphere.

It should be noted that the pressure *p* at a place X below the surface of a liquid is given by $p = H\rho g$, where *H* is the *vertical* distance of X below the surface. In Figure 3.16(ii), a very long barometer tube is inclined at an angle of 60° to the vertical. The length of the mercury along the slanted side of the tube is *x* mm say. If the atmospheric pressure here is the same as in Figure 3.16(i), this means that the *vertical* height to the mercury surface is still 760 mm. So

$$x \cos 60° = 760$$

and

$$x = \frac{760}{\cos 60°} = \frac{760}{0.5} = 1520 \text{ mm}$$

Variation of Atmospheric Pressure with Height

The density of a liquid varies very slightly with pressure. The density of a gas, however, varies appreciably with pressure. Thus at sea-level the density of the atmosphere is about 1·2 kg m^{-3}; as 1000 m above sea-level the density is about

$1\cdot1\,\text{kg}\,\text{m}^{-3}$; and at 5000 m above sea-level it is about $0\cdot7\,\text{kg}\,\text{m}^{-3}$. Standard atmospheric pressure is the pressure at the base of a column of mercury 760 mm high, a liquid which has a density about $13\,600\,\text{kg}\,\text{m}^{-3}$. Suppose air has a constant density of about $1\cdot2\,\text{kg}\,\text{m}^{-3}$. Then the height of an air column of this density which has a pressure equal to standard atmospheric pressure

$$= \frac{760}{1000} \times \frac{13\,600}{1\cdot2}\,\text{m} = 8\cdot6\,\text{km}$$

In fact, the air 'thins' the higher one goes, as explained above. The height of the air is thus much greater than $8\cdot6\,\text{km}$.

Density

As we have seen, the pressure in a fluid depends on the density of the fluid.

The *density* of a substance is defined as its *mass per unit volume*. So

$$\text{density, } \rho = \frac{\text{mass of substance}}{\text{volume of substance}} \qquad . \qquad . \qquad . \qquad (1)$$

The density of copper is about $9\cdot0\,\text{g}\,\text{cm}^{-3}$ or $9\cdot0 \times 10^3\,\text{kg}\,\text{m}^{-3}$; the density of aluminium is $2\cdot7\,\text{g}\,\text{cm}^{-3}$ or $2\cdot7 \times 10^3\,\text{kg}\,\text{m}^{-3}$; the density of water at 4°C is $1\,\text{g}\,\text{cm}^{-3}$ or $1000\,\text{kg}\,\text{m}^{-3}$.

Substances which float on water have a density less than $1000\,\text{kg}\,\text{m}^{-3}$ (p. 108). For example, ice has a density of about $900\,\text{kg}\,\text{m}^{-3}$; cork has a density of about $250\,\text{kg}\,\text{m}^{-3}$. Steel, of density $7800\,\text{kg}\,\text{m}^{-3}$, will float on mercury, whose density is about $13\,600\,\text{kg}\,\text{m}^{-3}$ at 0°C.

Archimedes' Principle

An object immersed in a fluid experiences a resultant upward force owing to the pressure of fluid on it. This upward force is called the *upthrust* of the fluid on the object.

ARCHIMEDES stated that *the upthrust is equal to the weight of fluid displaced by the object*, and this is known as *Archimedes' Principle*.

Thus if an iron cube of volume $400\,\text{cm}^3$ is totally immersed in water of density $1\,\text{g}\,\text{cm}^{-3}$, the upthrust on the cube = weight of $400 \times 1\,\text{g} = 4\,\text{N}$. If it is totally immersed in oil of density $0\cdot8\,\text{g}\,\text{cm}^{-3}$, the upthrust on it = weight of $400 \times 0\cdot8\,\text{g} = 3\cdot2\,\text{N}$.

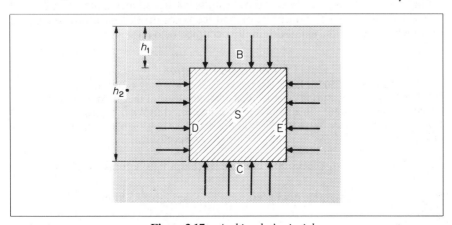

Figure 3.17 *Archimedes' principle*

Figure 3.17 shows why Archimedes' Principle is true. If S is a solid immersed in a liquid, the pressure on the lower surface C is greater than on the upper surface B, since the pressure at the greater depth h_2 is more than that at h_1. The pressure on the remaining surfaces D and E act as shown. The *force* on each of the four surfaces is calculated by summing the values of *pressure × area* over every part, remembering that vector addition is needed to sum forces. With a simple *rectangular-shaped solid* and the sides, D, E vertical, it can be seen that (i) the resultant horizontal force is zero, (ii) the upward force on C = pressure × area $A = h_2 \rho g A$, where ρ is the liquid density, and the downward force on B = pressure × area $A = h_1 \rho g A$. Thus

$$\text{resultant force on solid} = \text{upward force (upthrust)} = (h_2 - h_1)\rho g A$$

But
$$(h_2 - h_1)A = \text{volume of solid, } V$$

$$\therefore \text{ } upthrust = V \rho g = mg, \text{ where } m = V\rho$$

$$\therefore \text{ } \pmb{upthrust = weight\ of\ liquid\ displaced}$$

With a solid of irregular shape, taking into account horizontal and vertical components of forces, the same result is obtained. The upthrust is the weight of *liquid* displaced whatever the nature of the object immersed, or whether it is hollow or not. This is due primarily to the fact that the pressure on the object depends on the liquid in which it is placed.

Flotation

When an object *floats* in a fluid, the upthrust = the weight of the object, for equilibrium.

In *air*, for example, a balloon of constant volume 5000 m^3 and mass 4750 kg rises to an altitude where the upthrust is $4750\,g$ newtons, where g is the acceleration due to gravity at this height. From Archimedes' Principle, the upthrust = the weight of air displaced = $5000\,\rho g$ newtons, where ρ is the density of air at this height. Thus

$$5000\,\rho g = 4750\,g$$

$$\therefore \rho = 0.95 \text{ kg m}^{-3}$$

If a block of ice of volume 1 m^3 and mass 900 kg floats in *water* of density 1000 kg m^{-3}, the mass of water displaced is 900 kg, from Archimedes' Principle. Thus the volume of water displaced by the ice is 0.9 m^3. So the block floats with 0.1 m^3 above the water surface. If the ice, mass 900 kg, all melts, the water formed has a volume of 0.9 m^3. So all the melted ice takes up *exactly* the whole of the space which the solid ice had originally occupied below the water.

Example on Flotation

An ice cube of mass 50.0 g floats on the surface of a strong brine solution of volume 200.0 cm^3 inside a measuring cylinder. Calculate the level of the liquid in the measuring cylinder (i) before and (ii) after all the ice is melted. (iii) What happens to the level if the brine is replaced by 200.0 cm^3 water and 50.0 g of ice is again added? (Assume density of ice, brine = $900, 1100 \text{ kg m}^{-3}$ or $0.9, 1.1 \text{ g cm}^{-3}$ respectively.)

(i) Floating ice displaces 50 g of brine since upthrust equals weight of ice.

$$\therefore \text{ volume displaced} = \frac{\text{mass}}{\text{density}} = \frac{50}{1.1} = 45.5 \text{ cm}^3$$

$$\therefore \text{ level on measuring cylinder} = 245.5 \text{ cm}^3$$

(ii) 50 g of ice forms 50 g of water when all of it is melted.

$\therefore$ level on measuring cylinder *rises* to $250\cdot0\,\text{cm}^3$

(iii) *Water.* Initially, volume of water displaced $= 50\,\text{cm}^3$, since upthrust $= 50$ g.

$\therefore$ level on cylinder $= 250\cdot0\,\text{cm}^3$

If 1 g of ice melts, volume displaced is $1\,\text{cm}^3$ less. But volume of water formed is $1\,\text{cm}^3$. Thus the net change in water level is zero. Hence the water level remains unchanged as the ice melts.

Stokes' Law, Terminal Velocity

If we move through a pool of water we experience a resistance to our motion. This shows that there is a *frictional force* in liquids. We say this is due to the **viscosity** of the liquid.

When a small object, such as a small steel ball, is released in a viscous liquid like glycerine it accelerates at first, but its velocity soon reaches a steady value known as the *terminal velocity*. In this case the viscous (frictional) force, F, acting upwards, and the upthrust, U, due to the liquid on the object A, are together equal to its weight, mg acting downwards, or $F + U = mg$, Figure 3.18 (i). Since the resultant force on the object is zero it now moves with a constant (terminal) velocity. An object dropped from an aeroplane at first increases its speed v, but soon reaches its terminal speed. Figure 3.18 (ii) shows that variation of v with time as the terminal velocity v_0 is reached.

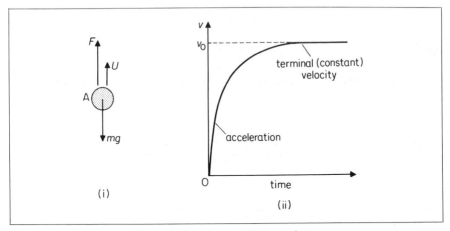

Figure 3.18   *Motion of falling sphere*

Suppose a sphere of radius a is dropped into a viscous liquid of coefficient of viscosity η, and its velocity at an instant is v. The frictional force, F, can be partly found by the method of dimensions. Thus suppose $F = ka^x\eta^yv^z$, where k is a constant. The dimensions of F are MLT^{-2}; the dimensions of a is L; the dimensions of η are $\text{ML}^{-1}\text{T}^{-1}$; and the dimensions of v are LT^{-1}.

$$\therefore \text{MLT}^{-2} \equiv \text{L}^x \times (\text{ML}^{-1}\text{T}^{-1})^y \times (\text{LT}^{-1})^z$$

Equating indices of M, L, T on both sides,

$$\therefore y = 1$$

$$x - y + z = 1$$

$$-y - z = -2$$

Hence $z = 1$, $x = 1$, $y = 1$. Consequently $F = k\eta av$. In 1850 STOKES showed mathematically that the constant k was 6π, and he arrived at the formula

$$F = 6\pi\eta av \qquad . \qquad . \qquad . \qquad . \qquad . \qquad (1)$$

Stokes' formula can be used to measure the coefficients of viscosity of very viscous liquids such as glycerine. It was also used by Millikan in his famous oil-drop experiment to measure the charge on an electron (p. 759). Here a small oil-drop falls through air with a terminal velocity and this enables the radius of the drop to be found.

The following Example shows how Stokes' formula is applied.

Example on Terminal Velocity

A small oil-drop falls with a terminal velocity of $4.0 \times 10^{-4}\,\mathrm{m\,s^{-1}}$ through air. Calculate the radius of the drop.

What is the new terminal velocity for an oil drop of half this radius? (Viscosity of air $= 1.8 \times 10^{-5}\,\mathrm{N\,s\,m^{-2}}$, density of oil $= 900\,\mathrm{kg\,m^{-3}}$, $g = 10\,\mathrm{m\,s^{-2}}$; neglect density of air.)

At terminal (steady) velocity,

frictional force on drop = weight − upthrust due to air

From Archimedes' principle, upthrust = weight of air displaced = mass of air $\times g$ = volume of sphere $(4\pi a^3/3) \times$ density of air $(\sigma) \times g$

So
$$6\pi\eta a v_0 = \frac{4}{3}\pi a^3 \rho g - \frac{4}{3}\pi a^3 \sigma g$$

where ρ is the oil density and σ that of the air. Neglecting σ, and simplifying,

$$a = \left(\frac{9v_0\eta}{2\rho g}\right)^{\frac{1}{2}}$$

$$= \left(\frac{9 \times 4 \times 10^{-4} \times 1.8 \times 10^{-5}}{2 \times 900 \times 10}\right)^{\frac{1}{2}}$$

$$= 1.9 \times 10^{-6}\,\mathrm{m}$$

The terminal velocity $v_0 \propto a^2$ from above. So when the radius a is decreased to one-half,

new terminal velocity $= \frac{1}{4} \times 4.0 \times 10^{-4} = 1.0 \times 10^{-4}\,\mathrm{m\,s^{-1}}$

_____ Exercises 3 _____

Forces in Equilibrium

1 Figure 3A shows three forces acting at a point. The lines drawn represent the forces roughly in magnitude and direction. Which diagram best represents equilibrium?

2 In Figure 3B (i) and (ii), calculate the torque acting on the rod AB, 0·4 m long.
    In Figure 3B (i), what work would be done if the torque remained constant and AB is rotated through 2 revolutions?

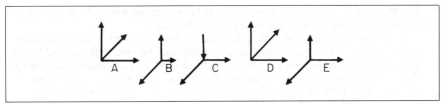

Figure 3A

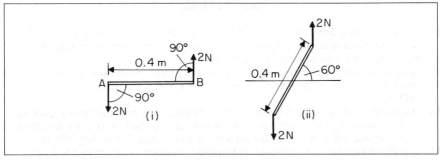

Figure 3B

3 The foot of a uniform ladder is on a rough horizontal ground, and the top rests against a smooth vertical wall. The weight of the ladder is 400 N, and a man weighing 800 N stands on the ladder one-quarter of its length from the bottom. If the inclination of the ladder to the horizontal is 30°, find the reaction at the wall and the total force at the ground.

4 A rectangular plate ABCD has two forces of 100 N acting along AB and DC in opposite directions. If AB = 3 m, BC = 5 m, what is the moment of the couple or torque acting on the plate? What forces acting along BC and AD respectively are required to keep the plate in equilibrium?

5 A flat plate is cut in the shape of a square of side 20·0 cm, with an equilateral triangle of side 20·0 cm adjacent to the square. Calculate the distance of the centre of mass from the apex of the triangle.

6 The dimensions of torque are the same as those of energy. Explain why it would nevertheless be inappropriate to measure torque in joules. State an appropriate unit. (*C.*)

7 A trap-door 120 cm by 120 cm is kept horizontal by a string attached to the mid-point of the side opposite to that containing the hinge. The other end of the string is tied to a point 90 cm vertically above the hinge. If the trap-door weight is 50 N, calculate the tension in the string and the reaction at the hinge.

8 Three forces in one plane act on a rigid body. What are the conditions for equilibrium?

The plane of a kite of mass 6 kg is inclined to the horizon at 60°. The thrust of the air acting normally on the kite acts at a point 25 cm above its centre of gravity, and the string is attached at a point 30 cm above the centre of gravity. Find the thrust of the air on the kite, and the tension in the string. (*C.*)

9 State and explain the conditions under which a rigid body remains in equilibrium under the action of a set of coplanar forces. Describe an experiment to determine the position of the centre of gravity of a flat piece of cardboard cut into the shape of a triangle. Use the conditions of equilibrium you have stated to justify your practical method.

A simple model of the ammonia molecule (NH_3) consists of a pyramid with the hydrogen atoms at the vertices of the equilateral base and the nitrogen atom at the apex. The N–H distance is 0·10 nm and the angle between two N–H bonds is 108°. Find the centre of gravity of the molecule. (Atomic weights: H = 1·0; N = 14; 1 nm = 1×10^{-9} m.) (*O.*)

10 Summarise the various conditions which are being satisfied when a body remains in equilibrium under the action of three non-parallel forces.

A wireless aerial attached to the top of a mast 20 m high exerts a horizontal force upon it of 600 N. The mast is supported by a stay-wire running to the ground from a point 6 m below the top of the mast, and inclined at 60° to the horizontal. Assuming that the action of the ground on the mast can be regarded as a single force, draw a diagram of the forces acting on the mast, and determine by measurement or by calculation the force in the stay-wire. (*C.*)

Forces in Fluids

11 An alloy of mass 588 g and volume 100 cm^3 is made of iron of density 8·0 g cm^{-3} and aluminium of density 2·7 g cm^{-3}. Calculate the proportion   (i) by volume, (ii) by mass of the constituents of the alloy.

12 A string supports a solid iron object of mass 180 g totally immersed in a liquid of density 800 kg m^{-3}. Calculate the tension in the string if the density of iron is 8000 kg m^{-3}.

13 A hydrometer floats in water with 6·0 cm of its graduated stem unimmersed, and in oil of density 0·8 g cm^{-3} with 4·0 cm of the stem unimmersed. What is the length of stem unimmersed when the hydrometer is placed in a liquid of density 0·9 g cm^{-3}?

14 A uniform capillary tube contains air trapped by a mercury thread 40 mm long. When the tube is placed horizontally as in Figure 3C (i), the length of the air column is 36 mm. When placed vertically, with the open end of the tube downwards, the length of air column is now x mm, Figure 3C (ii). Calculate x if the atmospheric pressure is 760 mmHg, assuming that the air obeys Boyle's law, pV = constant.

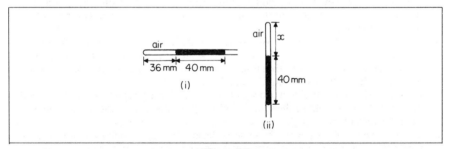

Figure 3C

15 A barometer tube, 960 mm long above the mercury in the reservoir, contains a little air above the mercury column inside it. When vertical, Figure 3D (i), the mercury column is 710 mm above the mercury in the reservoir. When inclined at 30° to the horizontal, Figure 3D (ii), the mercury column is now 910 mm along the barometer tube.

Assuming the air obeys Boyle's law, pV = constant, calculate the atmospheric pressure.

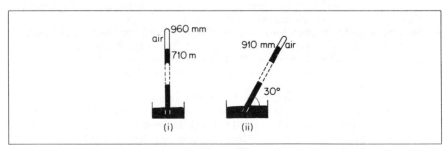

Figure 3D

16 State the principle of Archimedes and use it to derive an expression for the resultant force experienced by a body of weight W and density σ when it is totally immersed in a fluid of density ρ.

A solid weighs 237·5 g in air and 12·5 g when totally immersed in a liquid of density 0·9 g cm^{-3}. Calculate
(a) the density of the solid,
(b) the density of a liquid in which the solid would float with one-fifth of its volume exposed above the liquid surface. (*L.*)

17 State
(a) the laws of fluid pressure and
(b) the principle of Archimedes.

Show how (b) is a consequence of (a). Describe a simple experiment which verifies Archimedes' Principle.

The volume of a hot-air balloon is 600 m^3 and the density of the surrounding air is 1·25 kg m^{-3}. The balloon just hovers clear of the ground when the burner has heated the air inside to a temperature at which its density is 0·80 kg m^{-3}.
(a) What is the total mass of the balloon, including the hot air inside it?
(b) What is the total mass of the envelope of the balloon and its load?
(c) Find the acceleration with which the balloon will start to rise when the density of the air inside is reduced to 0·75 kg m^{-3}. (Take g to be 10 m s^{-2}.) (*O.*)

4
Further Topics in Mechanics and Fluids

Rotational Dynamics

So far we have considered the equations of linear motion and other dynamical formulae connected with a particle or small mass m. *We now consider the dynamics of large rotating objects such as spinning wheels in machines, for example.*

We shall find that dynamical formulae in rotational dynamics are similar to those in linear or translational dynamics. For example, the kinetic energy of a mass m *moving with a velocity* v *is $\frac{1}{2}mv^2$ and the rotational kinetic energy of a spinning object is $\frac{1}{2}I\omega^2$, where* I *is called the 'moment of inertia' of the object about its axis and ω is the angular velocity.*

Torque and Angular Acceleration

In linear motion, an object changing steadily from a velocity u to a velocity v in a time t has an acceleration a given by a = velocity change/time = $(v-u)/t$.

In rotational motion, a wheel spinning about its centre may increase its angular velocity from ω_0 to ω in a time t. The *angular acceleration* α is given by

$$\alpha = \frac{\omega - \omega_0}{t}$$

So $$\omega = \omega_0 + \alpha t \qquad . \qquad . \qquad . \qquad . \qquad . \qquad (1)$$

This is analogous to the linear relation $v = u + at$.

To make the wheel spin faster, a *couple* or *torque* T is applied to the wheel. We have already met forces which make a couple or a torque in a previous chapter. The turning-effect or torque T of a force F applied tangentially to a wheel of radius r spinning about its centre is given by $T = F \times r$ and the unit of T is N m (newton metre).

In linear dynamics, a force F produces an acceleration given by $F = ma$, where m is the mass of the object. In an analogous way, we soon see that a torque T, applied to a rotating wheel, gives it an angular acceleration given by

$$T = I\alpha \qquad . \qquad . \qquad . \qquad . \qquad . \qquad (2)$$

where I is the *moment of inertia* of the wheel about its axis of rotation, which we now explain.

Moment of Inertia I

Consider a large rigid object X rotating about an axis O when a torque T acts on the object, Figure 4.1. At the instant shown, a small mass m_1 of X, distant r_1 from O, has a linear acceleration a perpendicular to r_1 given by $a = r_1\alpha$, where α is the angular acceleration about O. So the force F on m_1 is given by

$$F = m_1 a = m_1 r_1 \alpha$$
$$\therefore \text{ torque } T \text{ about } O = F \times r_1 = m_1 r_1{}^2 \alpha$$

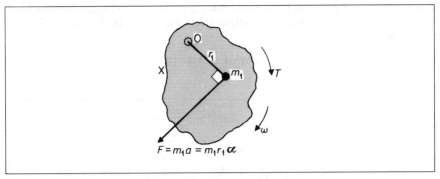

Figure 4.1 *Moment of inertia*

Adding together all the torques on the masses which make up the object X, then

$$\text{total torque } T = m_1 r_1{}^2\alpha + m_2 r_2{}^2\alpha + \ldots$$
$$= \Sigma(m_1 r_1{}^2)\alpha = I\alpha$$

where $I = \Sigma(m_1 r_1{}^2)$, the sum of all the products 'mr^2' for the masses m of the object and the square of their distances, r^2, from the axis of rotation. So, as in equation (2), $T = I\alpha$.

We call I the 'moment of inertia about the axis'. Calculation shows that a uniform rod of mass M and length l has a moment of inertia $I = Ml^2/12$ when it rotates about an axis at one end perpendicular to the rod. A sphere of mass M and radius r has a moment of inertia $I = 2Mr^2/5$ when it spins about an axis through its centre. The unit of I is 'kg m^2' and you should note that the value of I depends not only on the mass and dimensions of the object but also on the position of the axis of rotation.

Examples on Torque and Angular Acceleration

1   A heavy flywheel of moment of inertia 0.3 kg m^2 is mounted on a horizontal axle of radius 0.01 m and negligible mass compared with the flywheel. Neglecting friction, find   (i) the angular acceleration if a force of 40 N is applied tangentially to the axle,   (ii) the angular velocity of the flywheel after 10 seconds from rest.

(i) Torque $T = 40\,(\text{N}) \times 0.01\,(\text{m}) = 0.4\,\text{N m}$

From $$T = I\alpha$$

$$\text{angular acceleration } \alpha = \frac{T}{I} = \frac{0.4}{0.3} = 1.3\,\text{rad s}^{-2}$$

(ii) After 10 seconds, angular velocity $\omega = \alpha t$

$$= 1.3 \times 10 = 13\,\text{rad s}^{-1}$$

2   The moment of inertia of a solid flywheel about its axis is 0.1 kg m^2. It is set in rotation by applying a tangential force of 20 N with a rope wound round the circumference, the radius of the wheel being 0.1 m. Calculate the angular acceleration of the flywheel. What would be the angular acceleration if a mass of 2 kg were hung from the end of the rope? (*O. & C.*)

$$\text{Torque } T = I\alpha, \text{ and } T = 20 \times 0.1\,\text{N m}$$

So $$\text{angular acceleration } \alpha = \frac{T}{I} = \frac{20 \times 0.1}{0.1} = 20\,\text{rad s}^{-2}$$

If a mass m of 2 kg, or weight 20 N assuming $g = 10\,\mathrm{m\,s^{-2}}$, is hung from the end of the rope, it moves down with an acceleration a, Figure 4.2. In this case, if F is the tension in the rope,

$$mg - F = ma \qquad . \qquad . \qquad . \qquad . \qquad . \qquad (1)$$

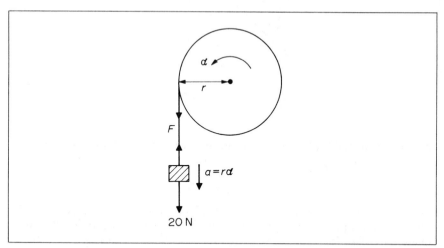

Figure 4.2 *Example on torque and angular acceleration*

For the flywheel, $\qquad\qquad F.r = \text{torque} = I\alpha$. $\qquad$. $\qquad$. $\qquad$. $\qquad$ (2)

where r is the radius of the wheel and α the angular acceleration about the centre. Now the mass descends a distance given by $r\theta$, where θ is the angle the flywheel has turned. Hence the acceleration $a = r\alpha$. Substituting in (1),

$$\therefore mg - F = mr\alpha$$

Multiplying by r,

$$\therefore mgr - F.r = mr^2\alpha . \qquad . \qquad . \qquad . \qquad (3)$$

Adding (2) and (3),

$$\therefore mgr = (I + mr^2)\alpha$$

$$\therefore \alpha = \frac{mgr}{I + mr^2} = \frac{2 \times 10 \times 0.1}{0.1 + 2 \times 0.1^2}$$

$$= 16.7\,\mathrm{rad\,s^{-2}}$$

Angular Momentum and Relation to Torque

In linear or straight-line motion, an important property of a moving object is its linear momentum (p. 22). When an object spins or rotates about an axis, its *angular momentum* plays an important part in its motion.

Consider a particle A of a rigid object rotating about an axis O, Figure 4.3. The momentum of A = mass × velocity = $m_1 v = m_1 r_1 \omega$. The 'angular momentum' of A about O is defined as the *moment of the momentum* about O. Its magnitude is thus $m_1 v \times p$, where p is the perpendicular distance from O to the direction of v. So angular momentum of A = $m_1 v p = m_1 r_1 \omega \times r_1 = m_1 r_1^2 \omega$.

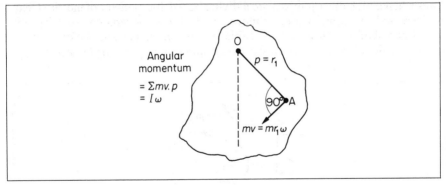

Figure 4.3 *Angular momentum*

∴ **total angular momentum of whole body** $= \Sigma m_1 r_1{}^2 \omega = \omega \Sigma m_1 r_1{}^2$

$$= I\omega$$

where I is the moment of inertia of the body about O.

Angular momentum is analogous to 'linear momentum', mv, in the dynamics of a moving particle. In place of m we have I, the moment of inertia; in place of v we have ω, the angular velocity.

Torque × Time

In linear dynamics, a force F acting on an object for a time t produces a momentum change given by

$F \times t$ **(impulse) = momentum change**

In an analogous way, a torque T acting for a time t on a rotating object produces an angular momentum change given by

$T \times t$ **= angular momentum change**

So if I is the moment of inertia about the axis concerned, and ω_1 and ω_2 are the initial and final angular velocities produced by a steady torque T, then

$$T \times t = I\omega_2 - I\omega_1$$

As an illustration, suppose a wheel of moment of inertia about its centre of $2\,\text{kg m}^2$ is spinning with an angular velocity of $15\,\text{rad s}^{-1}$. If it is brought to rest by a steady braking torque T in 5 s, the value of T is given by

$$T \times 5 = I\omega_2 - I\omega_1 = (2 \times 15) - 0 = 30$$

So $$T = 30/5 = 6\,\text{N m}$$

Conservation of Angular Momentum

The *conservation of angular momentum*, which corresponds to the conservation of linear momentum, states that *the angular momentum about an axis of a given rotating body or system of bodies is constant, if no external torque acts about that axis.*

Thus when a high diver jumps from a diving board, his moment of inertia, I, can be decreased by curling his body more, in which case his angular velocity ω is increased, Figure 4.4. He may then be able to turn more somersaults before striking the water. Similarly, a dancer on skates can spin faster by folding her arms.

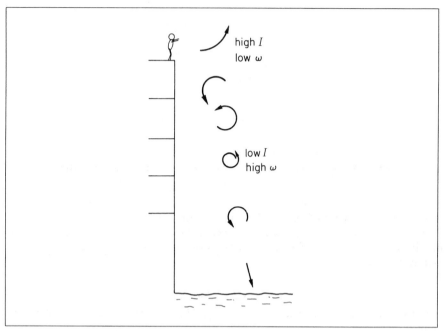

Figure 4.4 *Conservation of angular momentum*

The earth rotates about an axis passing through its geographic north and south poles with a period of 1 day. If it is struck by meteorites, then, since action and reaction are equal, no external couple acts on the earth and meteorites. Their total angular momentum is thus conserved. Neglecting the angular momentum of the meteorites about the earth's axis before collision compared with that of the earth, then

angular momentum of *earth plus meteorites* after collision = angular momentum of *earth* before collision

Since the effective mass of the earth has increased after collision, the moment of inertia has increased. So the earth will slow down slightly. Similarly, if a mass is dropped gently on to a turntable rotating freely at a steady speed, the conservation of angular momentum leads to a reduction in the speed of the table.

Angular momentum, and the principle of the conservation of angular momentum, have wide applications in physics. They are used in connection with enormous rotating masses such as the earth, as well as minute spinning particles such as electrons, neutrons and protons found inside atoms.

Examples on Conservation of Angular Momentum

1 A ballet dancer spins about a vertical axis at 1 revolution per second with arms outstretched. With her arms folded, her moment of inertia about the vertical axis decreases by 60%. Calculate the new rate of revolution.

Suppose I is the initial moment of inertia about the vertical axis and ω is the initial angular velocity corresponding to 1 rev s^{-1}. The new moment of inertia $I_1 = 40\%$ of $I = 0.4\,I$.

Suppose the new angular velocity is ω_1. Then, from the conservation of angular momentum,

$$I_1\omega_1 = I\omega$$

So
$$\omega_1 = \frac{I}{I_1}\omega = \frac{1}{0.4}\omega$$

Since angular velocity $\propto$ number of revs per second, the new number n of revs per second is given by

$$n = \frac{1}{0.4} \times 1 \text{ rev s}^{-1} = 2.5 \text{ rev s}^{-1}$$

2 A disc of moment of inertia 5×10^{-4} kg m^2 is rotating freely about axis O through its centre at 40 r.p.m., Figure 4.5. Calculate the new revolutions per minute (r.p.m.) if some wax W of mass 0.02 kg is dropped gently on to the disc 0.08 m from its axis.

Initial angular momentum of disc $= I\omega = 5 \times 10^{-4}\omega$

where ω is the angular velocity corresponding to 40 r.p.m.

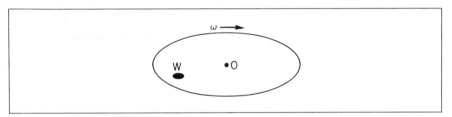

Figure 4.5 *Example on conservation of angular momentum*

When the wax of mass 0.02 kg is dropped gently on to the disc at a distance r of 0.08 m from the centre O, the disc slows down. Suppose the angular velocity is now ω_1. The total angular momentum about O of disc plus wax W

$$= I\omega_1 + mr^2\omega_1 = 5 \times 10^{-4}\omega_1 + 0.02 \times 0.08^2\omega_1$$
$$= 6.28 \times 10^{-4}\omega_1$$

From the conservation of angular momentum for the disc and wax about O

$$6.28 \times 10^{-4}\omega_1 = 5 \times 10^{-4}\omega$$
$$\therefore \frac{\omega_1}{\omega} = \frac{500}{628} = \frac{n}{40}$$

where n is the r.p.m. of the disc, because the angular velocity is proportional to the r.p.m.

$$\therefore n = \frac{500}{628} \times 40 = 32 \text{ (approx.)}$$

Central Forces and Conservation of Angular Momentum

Rotating objects sometimes have a force on them directed towards a particular point or axis. Figure 4.6 (i) shows a planet P moving in an *elliptical orbit* round the

Sun S under gravitational attraction. The force F on P is always directed towards S as it moves. So the moment of F about S is always *zero*.

Since the so-called central force or external force on P has no moment about S, the angular momentum about S is *constant*. At X, the nearest point to the Sun, the angular momentum about S $= mv_1r_1$, where m is the mass of the planet, v_1 is the velocity at X and SX $= r_1$. At Y, the furthest distance from the Sun, the angular momentum about S $= mv_2r_2$, where v_2 is the new velocity at Y and SY $= r_2$. So

$$mv_1r_1 = mv_2r_2$$

$$\therefore \frac{v_1}{v_2} = \frac{r_2}{r_1}$$

Since r_2 is greater than r_1, the velocity v_1 is greater than v_2. So the velocity of the planet P increases as it approaches the nearest distance to the Sun.

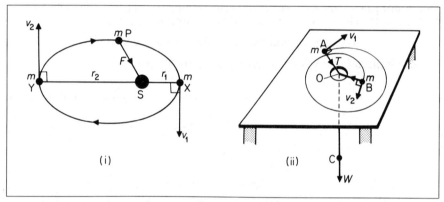

Figure 4.6 *Central forces and conservation of angular momentum*

Figure 4.6 (ii) shows a mass m at A on a *smooth table*. It is connected by a string AOC through a hole O in the table to a weight W, hanging down below the table, so that OC is vertical.

With the string OA taut, the mass m is pushed at right angles to OA with an initial velocity v_1 of $6 \, \text{m s}^{-1}$. The length OA is then 0.4 m. After the mass m rotates, the mass reaches a position at B on the table where OB is 0.3 m and the velocity of the mass has changed to a value v_2.

The tension T in the string may change as the mass rotates but T is always directed towards O on the table. So as the mass rotates, the external or central force T has no moment about O. So the angular momentum about O is *constant*. This helps us to find the velocity v_2 at B. We have, using angular momentum,

$$mv_1 \times 0.4 \, \text{(at A)} = mv_2 \times 0.3 \, \text{(at B)}$$

So
$$v_1 \times 0.4 = v_2 \times 0.3$$

and
$$v_2 = \frac{v_1 \times 0.4}{0.3} = \frac{6 \times 0.4}{0.3}$$

$$= 8 \, \text{m s}^{-1}$$

So the velocity increases to $8 \, \text{m s}^{-1}$ when the string shortens to 0.3 m, to keep the angular momentum constant.

Rotational Kinetic Energy, Work done by Torque

We now consider the *kinetic energy* of a rotating object. In Figure 4.7 (i), the rotational kinetic energy of the object X about O

$$= \text{sum of kinetic energy of all its individual masses}$$

$$= \tfrac{1}{2}m_1v_1^2 + \tfrac{1}{2}m_2v_2^2 + \ldots$$

$$= \tfrac{1}{2}m_1r_1^2\omega^2 + \tfrac{1}{2}m_2r_2^2\omega^2 + \ldots$$

$$= \tfrac{1}{2}(m_1r_1^2 + \tfrac{1}{2}m_2r_2^2 + \ldots)\omega^2 = \tfrac{1}{2}I\omega^2$$

So   **rotational kinetic energy** $= \tfrac{1}{2}I\omega^2$.   .   .   (3)

If $I = 2 \, \text{kg m}^2$ and $\omega = 3 \, \text{rad s}^{-1}$, then

$$\text{rotational kinetic energy} = \tfrac{1}{2} \times 2 \times 3^2 = 9 \, \text{J}$$

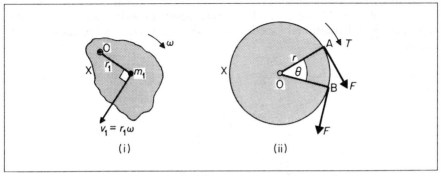

Figure 4.7   (i) *Rotational kinetic energy* (ii) *Work done by torque*

The *work done W* by a constant torque can be found from Figure 4.7 (ii). Here a force F is applied tangentially to a wheel X of radius r and X rotates through an angle θ as shown, while F stays tangential to the wheel. Then

$$\text{work done} = \text{force} \times \text{distance AB}$$

$$= F \times r\theta = F \cdot r \times \theta = \text{torque} \times \theta$$

So   **work done W = torque $\times$ angle of rotation** .   .   .   (4)

When the torque is in N m and θ is in radians, the work done is in J. Suppose a torque of constant value 6 N m rotates a wheel through 4 revolutions. Since the angle $\theta = 4 \times 2\pi = 8\pi \, \text{rad}$,

$$\text{work done } W = 6 \, (\text{N m}) \times 8\pi \, (\text{rad}) = 151 \, \text{J}$$

Consider a wheel rotating about its centre with an angular velocity of $15 \, \text{rad s}^{-1}$ and with a moment of inertia I of $2 \, \text{kg m}^2$ about its centre. If a steady braking torque T of 6 N m brings the wheel to rest in an angle of rotation θ, then

$$\text{work done by torque} = \text{change in kinetic energy}$$

So $T \times \theta = \tfrac{1}{2}I\omega^2 - 0$

$$\therefore 6 \times \theta = \tfrac{1}{2} \times 2 \times 15^2$$

$$\therefore \theta = 37 \cdot 5 \, \text{rad}$$

Since 1 revolution $= 2\pi$ rad,

$$\text{number of revs} = \frac{37\cdot5}{2\pi} = 6 \text{ (approx.)}$$

So the wheel comes to rest after about 6 revolutions when the braking torque is applied.

Kinetic Energy of a Rolling Object

When an object such as a cylinder or ball rolls on a plane, the object is rotating as well as moving down the plane. So it has rotational energy in addition to translational energy.

Consider a uniform cylinder C rolling along a plane without slipping, Figure 4.8. The forces on C are

(a) its weight Mg acting at its central axis O,

(b) the frictional force F at the plane which prevents slipping.

The force which produces linear acceleration and translational kinetic energy down the plane $= Mg \sin \alpha - F$. The *torque* about O which produces angular acceleration and rotational kinetic energy $= F.r$, where r is the radius of the cylinder.

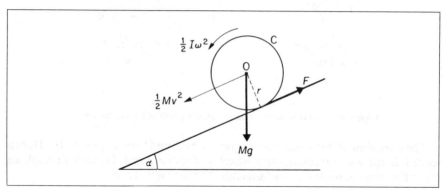

Figure 4.8 *Energy and acceleration of object rolling down plane*

Since energy is a scalar quantity (one with no direction), we can add the translational and rotational kinetic energies to obtain the total energy of the cylinder. So at a given instant,

---

$$\text{total kinetic energy} = \tfrac{1}{2}Mv^2 + \tfrac{1}{2}I\omega^2$$

---

where I is the moment of inertia about the axis O, ω is the angular velocity about O and v is the translational velocity down the plane. If the cylinder does not slip, then $v = r\omega$. So

$$\text{total kinetic energy} = \tfrac{1}{2}Mv^2 + \tfrac{1}{2}I\left(\frac{v}{r}\right)^2$$

$$= \tfrac{1}{2}v^2\left(M + \frac{I}{r^2}\right)$$

Suppose the cylinder rolls from rest through a distance s along the plane. The loss of potential energy $= Mgs \sin \alpha =$ gain in kinetic energy $= \tfrac{1}{2}v^2(M + I/r^2)$ from above. So

$$v^2 = \frac{2Mgs\sin\alpha}{M+(I/r^2)}$$

But $v^2 = 2as$ where a is the *linear acceleration* down the plane. So

$$2as = \frac{2Mgs\sin\alpha}{M+(I/r^2)}$$

Thus

$$a = \frac{Mg\sin\alpha}{M+(I/r^2)} \qquad . \qquad . \qquad . \qquad . \qquad . \qquad (1)$$

A uniform *solid* cylinder of mass M and radius r has a moment of inertia $I = Mr^2/2$ about its axis. So $I/r^2 = M/2$. Substituting in (1), we find that the acceleration down the plane $a = 2g\sin\alpha/3$. A uniform *hollow* cylinder open at both ends has a moment of inertia about its axis given by $I = Mr^2$, where M is the mass and r is the radius. From (1), we find that the acceleration down the plane, $a, = g\sin\alpha/2$. So the solid cylinder would have a greater acceleration down the plane than a hollow cylinder of the same mass. If no other tests were available, we could distinguish between a solid cylinder and a hollow cylinder closed at both ends, both of the same mass, by allowing them to roll from rest down an inclined plane. Starting from the same place the cylinder which reaches the bottom first would be the solid cylinder.

Measurement of Moment of Inertia of Flywheel

The moment of inertia of a flywheel W about a horizontal axle A can be determined by passing one end of some string through a hole in the axle, winding the string round the axle, and attaching a mass M to the other end of the string, Figure 4.9. The length of string is such that M reaches the floor, when released, at the same instant as the string is completely unwound from the axle.

M is released, and the number of revolutions, n, made by the wheel W up to the occasion when M strikes the ground is noted. The further number of revolutions n_1 made by W until it comes finally to rest, and the time t taken, are also observed by means of a chalk-mark W.

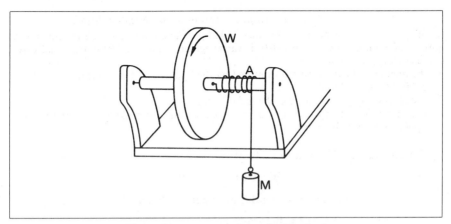

Figure 4.9 *Moment of inertia of flywheel*

Now the loss in potential energy of M = gain in kinetic energy of M + gain in kinetic energy of flywheel + work done against friction.

$$\therefore Mgh = \tfrac{1}{2}Mr^2\omega^2 + \tfrac{1}{2}I\omega^2 + nf \quad . \qquad . \qquad . \qquad (i)$$

where h is the distance M has fallen, r is the radius of the axle, ω is the angular velocity, I is the moment of inertia, and f is the energy per turn expended against friction. Since the energy of rotation of the flywheel when the mass M reaches the ground = work done against friction in n_1 revolutions, then

$$\tfrac{1}{2}I\omega^2 = n_1 f$$

$$\therefore f = \tfrac{1}{2}\frac{I\omega^2}{n_1}$$

Substituting for f in (i),

$$\therefore Mgh = \tfrac{1}{2}Mr^2\omega^2 + \tfrac{1}{2}I\omega^2\left(1 + \frac{n}{n_1}\right) \qquad . \qquad . \qquad . \qquad \text{(ii)}$$

Since the angular velocity of the wheel when M reaches the ground is ω, and the final angular velocity of the wheel is zero after a time t, the average angular velocity $= \omega/2 = 2\pi n_1/t$. Thus $\omega = 4\pi n_1/t$. Knowing ω and the magnitude of the other quantities in (ii), the moment of inertia I of the flywheel can be calculated.

Summary

We conclude with a summary showing a comparison between formulae in rotational and linear motion.

Linear Motion	Rotational Motion
1. Velocity, v	Angular velocity, $\omega = v/r$
2. Momentum $= mv$	Angular momentum $= I\omega$
3. Energy $= \tfrac{1}{2}mv^2$	Rotational energy $= \tfrac{1}{2}I\omega^2$
4. Force, F, $= ma$	Torque, T, $= I\alpha$
5. $F \times t =$ momentum change	$T \times t =$ angular momentum change
6. $F \times s =$ work done $=$ k.e. change	$T \times \theta =$ work done $=$ k.e. change
7. Conservation of linear momentum on collision, if no external forces	Conservation of angular momentum on collision, if no external torques.

Example on Torque, Angular Momentum, Kinetic Energy

A uniform circular disc of moment of inertia $0{\cdot}2\,\text{kg}\,\text{m}^2$ and radius $0{\cdot}15\,\text{m}$ is mounted on a horizontal cylindrical axle of radius $0{\cdot}015\,\text{m}$ and negligible mass. Neglecting frictional losses in the bearings, calculate

(a) the angular velocity acquired from rest by the application for 12 seconds of a force of 20 N tangential to the axle,

(b) the kinetic energy of the disc at the end of this period,

(c) the time required to bring the disc to rest if a constant braking force of 1 N were applied tangentially to its rim.

(a) Torque due to 20 N tangential to axle

$$= 20 \times 0{\cdot}015 = 0{\cdot}3\,\text{N}\,\text{m}$$

Torque $\times t =$ angular momentum change

$$\therefore 0{\cdot}3 \times 12 = 0{\cdot}2 \times \omega$$

$$\omega = 0{\cdot}3 \times 12/0{\cdot}2 = 18\,\text{rad}\,\text{s}^{-1}$$

(b) K.E. of disc after 12 seconds $= \tfrac{1}{2}I\omega^2$

$$= \tfrac{1}{2} \times 0{\cdot}2 \times 18^2 = 32{\cdot}4\,\text{J}$$

(c) Decelerating torque $= 1 \times 0{\cdot}15 = 0{\cdot}15\,\text{N m}$

$$\text{Torque} \times t = \text{angular momentum change}$$

$$\therefore 0{\cdot}15 \times t = 0{\cdot}2 \times 18$$

$$\therefore t = 0{\cdot}2 \times 18/0{\cdot}15 = 24\,\text{s}$$

(*See Exercises 4, p. 130, for questions on Rotational Dynamics*).

Fluids in Motion

Streamlines and Velocity

A stream or river flows slowly when it runs through open country and faster through narrow openings or constrictions. As shown shortly, this is due to the fact that water is practically an incompressible fluid, that is, changes of pressure cause practically no change in fluid density at various parts.

Figure 4.10 shows a tube of water flowing steadily between X and Y, where X has a bigger cross-sectional area A_1 than the part Y, of cross-sectional area A_2. The *streamlines* of the flow represent the directions of the velocities of the particles of fluid and the flow is uniform or laminar. Assuming the liquid is incompressible,

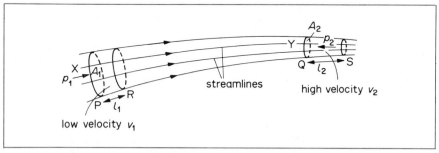

Figure 4.10 *Bernoulli's equation*

then, if it moves from PQ to RS, the volume of liquid between P and R is equal to the volume between Q and S. So $A_1 l_1 = A_2 l_2$, where l_1 is PR and l_2 is QS, or $l_2/l_1 = A_1/A_2$. Hence l_2 is greater than l_1. Consequently the *velocity* of the liquid at the narrow part of the tube, where the streamlines are closer together, is greater than at the wider part Y, where the streamlines are further apart. For the same reason, slow-running water from a tap can be made into a fast jet by placing a finger over the tap to narrow the exit.

Pressure and Velocity, Bernoulli's Principle

About 1740, Bernoulli obtained a relation between the pressure and velocity at different parts of a moving incompressible fluid. If the viscosity of the fluid is negligibly small, there are no frictional forces to overcome. In this case the work done by the pressure difference per unit volume of a fluid flowing along a pipe steadily is equal to the gain in kinetic energy per unit volume plus the gain in potential energy per unit volume.

Now the work done by a pressure in moving a fluid through a distance = force × distance moved = (pressure × area) × distance moved = pressure × volume moved, assuming the area is constant at a particular place for a short time of flow. At the beginning of the pipe where the pressure is p_1, the work done per unit volume on the fluid is thus p_1; at the other end, the work done per unit volume by the fluid is likewise p_2. Hence the net work done *on* the fluid per unit volume = $p_1 - p_2$.

The kinetic energy per unit volume = $\frac{1}{2}$ mass per unit volume × velocity² = $\frac{1}{2}\rho \times$ velocity², where ρ is the density of the fluid. Thus if v_2 and v_1 are the final and initial velocities respectively at the end and the beginning of the pipe, the kinetic energy gained per unit volume = $\frac{1}{2}\rho(v_2^2 - v_1^2)$. Further, if h_2 and h_1 are the respective heights measured from a fixed level at the end and beginning of the pipe, the potential energy gained per unit volume = mass per unit volume $\times$ $g \times (h_2 - h_1) = \rho g(h_2 - h_1)$.

So from the conservation of energy,

$$p_1 - p_2 = \tfrac{1}{2}\rho(v_2{}^2 - v_1{}^2) + \rho g(h_2 - h_1)$$
$$\therefore\ p_1 + \tfrac{1}{2}\rho v_1{}^2 + \rho g h_1 = p_2 + \tfrac{1}{2}\rho v_2{}^2 + \rho g h_2$$
$$\therefore\ p + \tfrac{1}{2}\rho v^2 + \rho g h = \textbf{constant}$$

where p is the pressure at any part and v is the velocity there. So for streamline motion of an incompressible non-viscous fluid,

---

the sum of the pressure at any part plus the kinetic energy per unit volume plus the potential energy per unit volume there is always constant.

This is known as *Bernoulli's Principle*.

Bernoulli's Principle shows that at points in a moving fluid where the potential energy change $\rho g h$ is very small, or zero as in flow through a horizontal pipe, the pressure is low where the velocity is high. Conversely, the pressure is high where the velocity is low. The principle has wide applications.

Example on Bernoulli Equation

As a numerical illustration, suppose the area of cross-section A_1 of X in Figure 4.10 is $4\,\text{cm}^2$, the area A_2 of Y is $1\,\text{cm}^2$, and water flows past each section in laminar flow at the rate of $400\,\text{cm}^3\,\text{s}^{-1}$. Then

$$\text{at X, speed } v_1 \text{ of water} = \frac{\text{vol. per second}}{\text{area}} = 100\,\text{cm s}^{-1} = 1\,\text{m s}^{-1}$$

$$\text{at Y, speed } v_2 \text{ of water} = 400\,\text{cm s}^{-1} = 4\,\text{m s}^{-1}$$

The density of water, $\rho = 1000\,\text{kg m}^{-3}$. So, if p is the pressure difference,

$$\therefore\ p = \tfrac{1}{2}\rho(v_2{}^2 - v_1{}^2) = \tfrac{1}{2} \times 1000 \times (4^2 - 1^2) = 7\cdot5 \times 10^3\,\text{N m}^{-2}$$

If h is in metres, $\rho = 1000\,\text{kg m}^{-3}$ for water, $g = 9\cdot8\,\text{m s}^{-2}$, then, from $p = h\rho g$

$$h = \frac{7\cdot5 \times 10^3}{1000 \times 9\cdot8} = 0\cdot77\,\text{m (approx.)}$$

The pressure head h is thus equivalent to $0\cdot77\,\text{m}$ of water.

Applications of Bernoulli's Principle

1. A suction effect is experienced by a person standing close to the platform at a station when a fast train passes. The fast-moving air between the person and train produces a decrease in pressure and the excess air pressure on the other side pushes the person towards the train.

2. *Filter pump.* A filter pump has a narrow section in the middle, so that a jet of water from the tap flows faster here, Figure 4.11 (i). This causes a drop in pressure near it and air therefore flows in from the side tube to which a vessel is connected. The air and water together are forced through the bottom of the filter pump.

A similar principle is used in the engine *carburettor* for vehicles. At one stage of its cycle, the engine draws in air. This rushes past the fine nozzle of a pipe connected to the petrol tank and lowers the air pressure at the nozzle, Figure 4.11 (ii). Some petrol is then forced out of the tank by atmospheric pressure through the nozzle

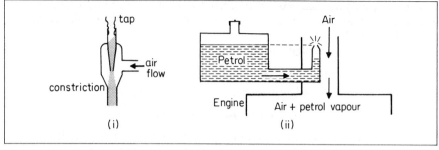

Figure 4.11 *Principle of* (i) *filter pump* (ii) *carburettor*

in a fine spray. The petrol vapour mixes with the air and so provides the air-petrol mixture required for the engine.

3. *Aerofoil lift.* The curved shape of an aerofoil creates a faster flow of air over its top surface than the lower one, Figure 4.12. This is shown by the closeness of the streamlines above the aerofoil compared with those below. From Bernoulli's Principle, the pressure of the air below is greater than that above, and this produces the lift on the aerofoil.

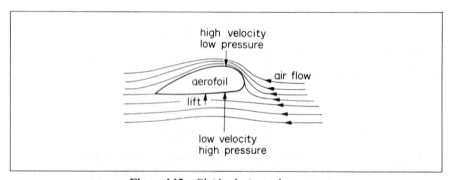

Figure 4.12 *Fluid velocity and pressure*

4. *Venturi meter.* This meter measures the volume of gas or liquid per second flowing through gas pipes or oil pipes.

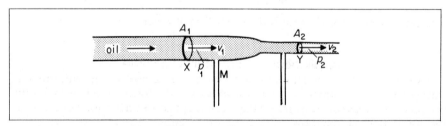

Figure 4.13 *Principle of Venturi meter*

Figure 4.13 shows the principle. A manometer M is connected between a wide section X, area A_1, and a narrower section Y, area A_2, of the horizontal pipe carrying a steady flow of oil, for example. Since the velocity v_2 at Y is greater than the velocity v_1 at X, the pressure p_2 at Y is less than the pressure p_1 at X. The manometer then has a difference in levels H of a liquid of density ρ' say.

Suppose Q is the volume per second of oil flowing at X or at Y. Then

$Q = A_1 v_1 = A_2 v_2$. Also, from the Bernoulli Principle, if ρ is the density of the oil,

$$p_1 + \tfrac{1}{2}\rho v_1{}^2 = p_2 + \tfrac{1}{2}\rho v_2{}^2$$

So $$p_1 - p_2 = H\rho' g = \tfrac{1}{2}\rho(v_2{}^2 - v_1{}^2) \ . \qquad . \qquad . \qquad . \qquad (1)$$

But $v_2 = Q/A_2$ and $v_1 = Q/A_1$. Substituting for v_2 and v_1 in (1),

$$H\rho' g = \tfrac{1}{2}\rho\left(\frac{Q^2}{A_2{}^2} - \frac{Q^2}{A_1{}^2}\right) = \tfrac{1}{2}\rho Q\left(\frac{A_1{}^2 - A_2{}^2}{A_1{}^2 A_2{}^2}\right)$$

So $$Q = \sqrt{\frac{2H\rho' g A_1{}^2 A_2{}^2}{\rho(A_1{}^2 - A_2{}^2)}} \qquad . \qquad . \qquad . \qquad . \qquad . \qquad (2)$$

This enables Q to be found. Since $Q \propto \sqrt{H}$, an experiment can be carried out to calibrate the difference in levels H of the manometer, using (2), in terms of volume per second rate of flow.

Measurement of Fluid Velocity, Pitot-static Tube

The velocity at a point in a fluid flowing through a horizontal tube can be measured by the application of the Bernoulli equation on p. 127. In this case h is zero and so

$$p + \tfrac{1}{2}\rho v^2 = \text{constant}$$

Here p is the static pressure at a point in the fluid, that is, the pressure unaffected by its velocity. The pressure $p + \tfrac{1}{2}\rho v^2$ is the total or dynamic pressure, that is, the pressure which the fluid would exert if it is brought to rest by striking a surface placed normally to the velocity at the point concerned.

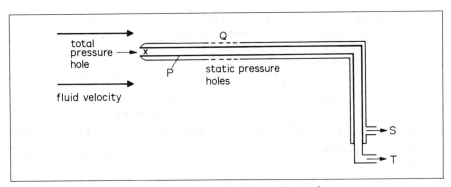

Figure 4.14 *Principle of Pitot-static tube*

Figure 4.14 illustrates the principle of a *Pitot-static tube*. The inner or Pitot tube P, named after its inventor, has an opening X at one end *normal* to the fluid velocity. A manometer connected to T would measure the total pressure $p + \tfrac{1}{2}\rho v^2$. The outer or static tube has holes Q in its side which are *parallel* to the fluid velocity. A manometer connected to S would measure the static pressure p. The difference in pressure, $h\rho' g$, in the two sides of a single manometer joined respectively to T and S would hence be equal to $\tfrac{1}{2}\rho v^2$. Thus v can be calculated from $v = \sqrt{2h\rho' g/\rho}$. In practice, corrections are applied to the manometer readings to take account of differences from the simple theory outlined.

Example on Fluid Motion in Pipe

(i) Water flows steadily along a horizontal pipe at a volume rate of $8 \times 10^{-3}\,\text{m}^3\,\text{s}^{-1}$. If the area of cross-section of the pipe is $40\,\text{cm}^2$ ($40 \times 10^{-4}\,\text{m}^2$), calculate the flow velocity of the water. (ii) Find the total pressure in the pipe if the static pressure in the horizontal pipe is 3.0×10^4 Pa, assuming the water is incompressible, non-viscous and its density is $1000 \, \text{kg m}^{-3}$. (iii) What is the new flow velocity if the total pressure is 3.6×10^4 Pa?

(i) Velocity of water $= \dfrac{\text{volume per second}}{\text{area}}$

$$= \frac{8 \times 10^{-3}}{40 \times 10^{-4}} = 2 \, \text{m s}^{-1}$$

(ii) Total pressure $=$ static pressure $+ \frac{1}{2}\rho v^2$

$$= 3.0 \times 10^4 + \frac{1000 \times 2^2}{2}$$

$$= 3.0 \times 10^4 + 0.2 \times 10^4 = 3.2 \times 10^4 \, \text{Pa}$$

(iii) $\frac{1}{2}\rho v^2 =$ total pressure $-$ static pressure

So $\frac{1}{2} \times 1000 \times v^2 = 3.6 \times 10^4 - 3.0 \times 10^4 = 0.6 \times 10^4$

$$v = \sqrt{\frac{0.6 \times 10^4}{500}} = 3.5 \, \text{m s}^{-1}$$

_____ **Exercises 4** _____

Rotational Dynamics

1 A disc of moment of inertia $10 \, \text{kg m}^2$ about its centre rotates steadily about the centre with an angular velocity of $20 \, \text{rad s}^{-1}$. Calculate (i) its rotational energy, (ii) its angular momentum about the centre, (iii) the number of revolutions per second of the disc.

2 A constant torque of $200 \, \text{N m}$ turns a wheel about its centre. The moment of inertia about this axis is $100 \, \text{kg m}^2$. Find (i) the angular velocity gained in 4 s, (ii) the kinetic energy gained after 20 revs.

3 A flywheel has a kinetic energy of 200 J. Calculate the number of revolutions it makes before coming to rest if a constant opposing couple of $5 \, \text{N m}$ is applied to the flywheel.
   If the moment of inertia of the flywheel about its centre is $4 \, \text{kg m}^2$, how long does it take to come to rest?

4 A constant torque of $500 \, \text{N m}$ turns a wheel which has a moment of inertia $20 \, \text{kg m}^2$ about its centre. Find the angular velocity gained in 2 s and the kinetic energy gained.

5 A ballet dancer spins with $2.4 \, \text{rev s}^{-1}$ with her arms outstretched, when the moment of inertia about the axis of rotation is I. With her arms folded, the moment of inertia about the same axis becomes $0.6 \, I$. Calculate the new rate of spin.
   State the principle used in your calculation.

6 A disc rolling along a horizontal plane has a moment of inertia $2.5 \, \text{kg m}^2$ about its centre and a mass of 5 kg. The velocity along the plane is $2 \, \text{m s}^{-1}$.
   If the radius of the disc is 1 m, find (i) the angular velocity, (ii) the total energy (rotational and translation) of the disc.

7 A wheel of moment of inertia $20 \, \text{kg m}^2$ about its axis is rotated from rest about its centre by a constant torque T and the energy gained in 10 s is 360 J. Calculate (i) the angular velocity at the end of 10 s, (ii) T, (iii) the number of revolutions made by the wheel before coming to rest if T is removed at 10 s and a constant opposing torque of $4 \, \text{N m}^{-1}$ is then applied to the wheel.

8 A uniform rod of length 3 m is suspended at one end so that it can move about an axis perpendicular to its length. The moment of inertia about the end is $6 \, \text{kg m}^2$ and

the mass of the rod is 2 kg. If the rod is initially horizontal and then released, find the angular velocity of the rod when   (i) it is inclined at 30° to the horizontal,   (ii) reaches the vertical.

9   A recording disc rotates steadily at 45 rev min⁻¹ on a table. When a small mass of 0·02 kg is dropped gently on the disc at a distance of 0·04 m from its axis and sticks to the disc, the rate of revolution falls to 36 rev min⁻¹. Calculate the moment of inertia of the disc about its centre.
Write down the principle used in your calculation.

10   A disc of moment of inertia 0·1 kg m² about its centre and radius 0·2 m is released from rest on a plane inclined at 30° to the horizontal. Calculate the angular velocity after it has rolled 2 m down the plane if its mass is 5 kg.

11   A flywheel with an axle 1·0 cm in diameter is mounted in frictionless bearings and set in motion by applying a steady tension of 2 N to a thin thread wound tightly round the axle. The moment of inertia of the system about its axis of rotation is $5·0 \times 10^{-4}$ kg m². Calculate
   (a) the angular acceleration of the flywheel when 1 m of thread has been pulled off the axle,
   (b) the constant retarding couple which must then be applied to bring the flywheel to rest in one complete turn, the tension in the thread having been completely removed. (*JMB.*)

12   Define the moment of inertia of a body about a given axis. Describe how the moment of inertia of a flywheel can be determined experimentally.
   A horizontal disc rotating freely about a vertical axis makes 100 r.p.m. A small piece of wax of mass 10 g falls vertically on to the disc and adheres to it at a distance of 9 cm from the axis. If the number of revolutions per minute is thereby reduced to 90, calculate the moment of inertia of the disc. (*JMB.*)

13   Write down an expression for the angular momentum of a point mass m moving in a circular path of radius r with constant angular velocity ω. Extend this result to a system of several point masses each rotating about a common axis with the same angular velocity and show how this leads to the concept of the moment of inertia I of a rigid body.
   Show that the quantity $\frac{1}{2}I\omega^2$ is the kinetic energy of rotation of a rigid body rotating about an axis with angular velocity ω.
   Describe how you would determine by experiment the moment of inertia of a flywheel.
   The atoms in the oxygen molecule O_2 may be considered to be point masses separated by a distance of $1·2 \times 10^{-10}$ m. The molecular speed of an oxygen molecule at s.t.p. is 460 m s⁻¹. Given that the rotational kinetic energy of the molecule is two-thirds of its translational kinetic energy, calculate its angular velocity at s.t.p. assuming that molecular rotation takes place about an axis through the centre of, and perpendicular to, the line joining the atoms. (*O. & C.*)

14   (a) For a rigid body rotating about a fixed axis, explain with the aid of a suitable diagram what is meant by *angular velocity, kinetic energy* and *moment of inertia*.
   (b) In the design of a passenger bus, it is proposed to derive the motive power from the energy stored in a flywheel. The flywheel, which has a moment of inertia of $4·0 \times 10^2$ kg m², is accelerated to its maximum rate of rotation $3·0 \times 10^3$ revolutions per minute by electric motors at stations along the bus route.
   (i) Calculate the maximum kinetic energy which can be stored in the flywheel.
   (ii) If, at an average speed of 36 kilometres per hour, the power required by the bus is 20 kW, what will be the maximum possible distance between stations on the level? (*JMB.*)

15   (a) Explain what is meant by   (i) a *couple*,   (ii) the *moment of a couple*. Show that a force acting along a given line can always be replaced by a force of the same magnitude acting along a parallel line, together with a couple.
   (b) A flywheel of moment of inertia 0·32 kg m² is rotated steadily at 120 rad s⁻¹ by a 50 W electric motor.   (i) Find the kinetic energy and angular momentum of the flywheel.   (ii) Calculate the value of the frictional couple opposing the rotation.   (iii) Find the time taken for the wheel to come to rest after the motor has been switched off. (*O.*)

16 A flywheel rotates about a horizontal axis fitted into friction free bearings. A light string, one end of which is looped over a pin on the axle, is wrapped ten times round the axle and has a mass of 1·5 kg attached to its free end. Discuss the energy changes as the mass falls. If the moment of inertia of the wheel and axle is 0·10 kg m² and the diameter of the axle 5·0 cm, calculate the angular velocity of the flywheel at the instant when the string detaches itself from the axle after ten revolutions. (*AEB*, 1983.)

Fluid Motion

17 An open tank holds water 1·25 m deep. If a small hole of cross-section area 3 cm² is made at the bottom of the tank, calculate the mass of water per second initially flowing out of the hole. ($g = 10 \, \text{m s}^{-2}$, density of water $= 1000 \, \text{kg m}^{-3}$.)

18 A lawn sprinkler has 20 holes each of cross-section area $2·0 \times 10^{-2} \, \text{cm}^2$ and is connected to a hose-pipe of cross-section area 2·4 cm². If the speed of the water in the hose-pipe is $1·5 \, \text{m s}^{-1}$, estimate the speed of the water as it emerges from the holes.

19 Show that the term $\frac{1}{2}\rho v^2$ which enters into the Bernoulli equation has the same dimensions as pressure p.
    A fluid flows through a horizontal pipe of varying cross-section. Assuming the flow is streamline and applying the Bernoulli equation $p + \frac{1}{2}\rho v^2 = \text{constant}$, show that the pressure in the pipe is greatest where the cross-section area is greatest.

20 Water flows along a horizontal pipe of cross-section area 48 cm² which has a constriction of cross-section area 12 cm² at one place. If the speed of the water at the constriction is $4 \, \text{m s}^{-1}$, calculate the speed in the wider section.
    The pressure in the wider section is $1·0 \times 10^5$ Pa. Calculate the pressure at the constriction. (Density of water $= 1000 \, \text{kg m}^{-3}$.)

21 Water flows steadily along a uniform flow tube of cross-section 30 cm². The static pressure is $1·20 \times 10^5$ Pa and the total pressure is $1·28 \times 10^5$ Pa.
    Calculate the flow velocity and the mass of water per second flowing past a section of the tube. (Density of water $= 1000 \, \text{kg m}^{-3}$.)

22 (a) Distinguish between *static pressure*, *dynamic pressure* and *total pressure* when applied to streamline (laminar) fluid flow and write down expressions for these three pressures at a point in the fluid in terms of the fluid velocity v, the fluid density ρ, pressure p, and the height h, of the point with respect to a datum.
    (b) Describe, with the aid of a labelled diagram, the Pitot-static tube and explain how it may be used to determine the flow velocity of an incompressible, non-viscous fluid.
    (c) The static pressure in a horizontal pipeline is $4·3 \times 10^4$ Pa, the total pressure is $4·7 \times 10^4$ Pa, and the area of cross-section is 20 cm². The fluid may be considered to be incompressible and non-viscous and has a density of $10^3 \, \text{kg m}^{-3}$.
    Calculate   (i) the flow velocity in the pipeline,   (ii) the volume flow rate in the pipeline. (*JMB*.)

5

Elasticity, Molecular Forces, Solid Materials

Metals and other solids can be classified as crystalline, glassy, amorphous or polymeric. All these solid materials are widely used in engineering and industry. In this chapter we start with metals and show how they react to forces which stretch them. We then consider the microscopic or molecular behaviour of metals and the ideas of dislocations and slip planes in a more detailed account of solid materials and their uses.

Elasticity

Elasticity of Metals

A bridge used by traffic is subjected to loads or forces of varying amounts. Before a steel bridge is constructed, therefore, samples of the steel are sent to a research laboratory. Here they undergo tests to find out whether the steel can withstand the loads likely to be put on them.

Figure 5.1 illustrates a simple laboratory method of investigating the property of steel we are discussing. Two long thin steel wires, P, Q, are suspended beside each other from a rigid support B, such as a girder at the top of the ceiling. The wire P is kept taut by a weight A attached to its end and carries a scale M graduated in millimetres. The wire Q carries a vernier scale V which is alongside the scale M.

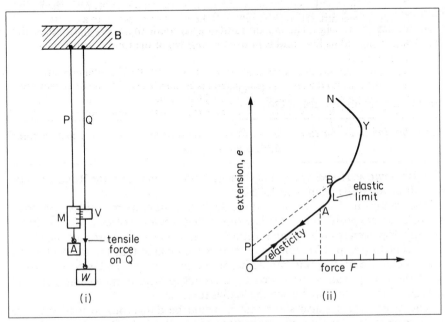

Figure 5.1 (i) *Elasticity experiment and* (ii) *result—extension against load*

When a load W such as 10 N is attached to the end of Q, the wire increases in length by a small amount which can be read from the change in the reading on the vernier V. If the load is taken off and the reading in V returns to its original value, the wire is said to be *elastic* for loads from zero to 10 N, a term adopted by analogy with an elastic thread. When the load W is increased to 20 N the extension (increase in length) is obtained from V again; and if the reading on V returns to its original value when the load is removed the wire is said to be elastic at least for loads from zero to 20 N. The load or force which stretches a wire is called a *tensile force*.

Proportional and Elastic Limits

The extension of a thin wire such as Q for increasing loads or forces F may be found by experiment to be as follows:

Force (N)	0	10	20	30	40	50	60	70	80
Extension (mm)	0	0·14	0·20	0·42	0·56	0·70	0·85	1·01	1·19

When the extension, e, is plotted against the force F in the wire, a graph is obtained which is a *straight line* OA, followed by a curve ABY rising slowly at first and then very sharply, Figure 5.1 (ii). Up to A, about 50 N, the results show that the extension increased by 0·014 mm per N added to the wire. A, then, is the *proportional limit*.

Along OA, and up to L just beyond A, the wire returned to its original length when the load was removed. The force at L is called the *elastic limit*. Along OL the metal is said to undergo changes called *elastic deformation*. Later we show that any energy stored in the metal during elastic deformation is recovered when the load is removed.

Beyond the elastic limit L, however, the wire has a permanent extension such as OP when the force is removed at B, for example, Figure 5.1 (ii). The extension increases rapidly along the curve ABY as the force on the wire is further increased and at Y the wire thins and breaks at N.

As we see later, when the elastic limit is exceeded, the energy stored in the metal is transferred to heat and is not recovered when the load is removed.

Hooke's Law

From the straight line graph OA, we deduce that

the extension is proportional to the force or tension in a wire if the proportional limit is not exceeded.

This is known as *Hooke's law*, after ROBERT HOOKE, founder of the Royal Society, who discovered the relation in 1676.

The extension of a wire is due to the displacement of its molecules from their mean (average) positions. So the law shows that when a molecule of the metal is slightly displaced from its mean position the restoring force is proportional to its displacement (see p. 150). One may therefore conclude that the molecules of a solid metal are undergoing simple harmonic motion (p. 77). Up to the elastic limit the energy gained or stored by a stretched wire is molecular potential energy, which is recovered when the load is removed.

The measurements also show that it would be dangerous to load the wire with weights greater than the magnitude of the elastic limit, because the wire then suffers a permanent strain. Similar experiments in the research laboratory enable scientists to find the maximum load which a steel bridge, for example, should carry for safety. Rubber samples are also subjected to similar experiments, to find the maximum safe tension in rubber belts used in machinery. See Figure 5A below.

Yield Point, Ductile and Brittle Substances, Breaking Stress

Careful experiments show that, for mild steel and iron for example, the planes of the atoms begin to 'slide' across each other soon after the load exceeds the elastic limit, that is, *plastic deformation* begins. This is indicated by the slight 'kink' at B beyond L in Figure 5.1 (ii), and it is called the *yield point* of the wire. The change from an elastic to a plastic stage is often shown by a sudden increase in the extension. In the plastic stage, the energy gained by the stretched wire is dissipated as heat and unlike the elastic stage, the energy is not recovered completely when the load is removed.

As the load is increased further the extension increases rapidly along the curve YN and the wire then becomes increasingly thinner and breaks at N. The *breaking stress* of the wire is the corresponding force per unit area of the narrowest cross-section of the wire.

Substances such as those just described, which lengthen considerably and undergo plastic deformation until they break, are known as *ductile* substances. Lead, copper and wrought iron are ductile. Other substances, however, break just after the elastic limit is reached; they are known as *brittle* substances. Glass and high carbon steels are brittle.

Brass, bronze, and many alloys appear to have no yield point. These materials increase in length beyond the elastic limit as the load is increased without the sudden appearance of a plastic stage.

The strength and ductility of a metal, its ability to flow, depend on defects in the metal crystal lattice. This is discussed later (p. 157).

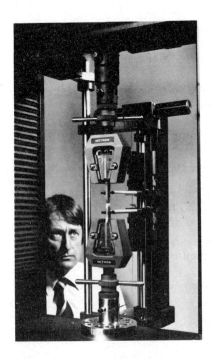

Figure 5A *The photograph shows a metal sample at the point of failure following a tensile test on an Instron Model 1185 Universal Materials Testing Machine. For precise measurement of extension, the system is fitted with an Automatic Extensometer. The load range is 0·1 N to 100 kN and the machine is used for a wide range of materials. (Courtesy of Instron Limited)*

Tensile Stress and Tensile Strain, Young Modulus

We have now to consider the technical terms used in the subject of elasticity of wires. When a force or tension F is applied to the end of a wire of original cross-sectional area A, Figure 5.2 (i),

$$\text{the } \textit{tensile stress} = \textit{force per unit area} = \frac{F}{A}. \qquad (1)$$

If the extension of the wire is e, and its original length is l,

$$\text{the } \textit{tensile strain} = \textit{extension per unit length} = \frac{e}{l} \qquad (2)$$

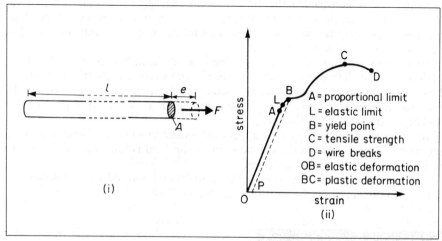

Figure 5.2 (i) *Tensile stress and tensile strain* (ii) *Stress against strain, ductile material*

Suppose a 2 kg mass is attached to the end of a vertical wire of length 2 m and diameter 0·64 mm, and the extension is 0·60 mm. Then

$$F = 2 \times 9{\cdot}8 \text{ N}, \ A = \pi \times 0.032^2 \text{ cm}^2 = \pi \times 0{\cdot}032^2 \times 10^{-4} \text{ m}^2$$

$$\therefore \text{ tensile stress} = \frac{2 \times 9{\cdot}8}{\pi \times 0{\cdot}032^2 \times 10^{-4}} = 6 \times 10^7 \text{ N m}^{-2} \qquad (3)$$

and $$\text{tensile strain} = \frac{0{\cdot}6 \times 10^{-3} \text{ m}}{2 \text{ m}} = 0{\cdot}3 \times 10^{-3}. \qquad (4)$$

It will be noted that 'stress' has units such as 'N m^{-2}'; 'strain' has no units because it is the ratio of two lengths. Figure 5.2 (ii) shows the general stress–strain graph for a ductile material.

Under elastic conditions, a *modulus of elasticity* of the wire, called the **Young modulus (E),** is defined as the ratio

$$E = \frac{\text{tensile stress}}{\text{tensile strain}} \qquad \cdot \qquad \cdot \qquad \cdot \qquad \cdot \qquad (5)$$

So
$$E = \frac{F/A}{e/l}$$

Using (3) and (4), when the elastic limit is not exceeded,

$$E = \frac{6 \times 10^7}{0 \cdot 3 \times 10^{-3}}$$

$$= 2 \cdot 0 \times 10^{11} \text{ N m}^{-2} \text{ (or Pa)}$$

Dimensions of Young Modulus

As stated before, the 'strain' of a wire has no dimensions of mass, length, or time, since, by definition, it is the ratio of two lengths. Now

$$\text{dimensions of stress} = \frac{\text{dimensions of force}}{\text{dimensions of area}}$$

$$= \frac{MLT^{-2}}{L^2} = ML^{-1}T^{-2}$$

∴ dimensions of the Young modulus, E,

$$\frac{\text{dimensions of stress}}{\text{dimensions of strain}} = ML^{-1}T^{-2}$$

Determination of Young Modulus

The magnitude of the Young modulus for a material in the form of a wire can be found with the apparatus illustrated in Figure 5.1 (i), p. 133, to which the reader should now refer. The following practical points should be specially noted, remembering that the elastic limit must not be exceeded:

(1) The use of *two wires*, P, Q, *of the same material and length*, eliminates the correction for (i) the yielding of the support when loads are added to Q, (ii) changes of temperature.

(2) The wire is made *thin* so that a moderate load of several kilograms produces a large tensile stress. The wire is also made *long* so that a measurable extension is produced.

(3) Both wires should be free of kinks, otherwise the increase in length cannot be accurately measured. The wires are straightened by attaching suitable weights to their ends, as shown in Figure 5.1 (i).

(4) A vernier scale is necessary to measure the extension of the wire since this is always small. The 'original length' of the wire is measured from the top B *to the vernier V* by a ruler, since an error of 1 millimetre is negligible compared with an original length of several metres. For very accurate work, the extension can be measured by using a spirit level between the two wires, and adjusting a vernier screw to restore the spirit level to its original reading after a load is added.

(5) The diameter of the wire must be found by a micrometer screw gauge at several places, and the average value then calculated. The area of cross-section, A, $= \pi r^2$, where r is the radius.

(6) The readings on the vernier are also taken when the load is gradually removed in steps of 1 kilogram; they should be very nearly the same as the readings on the vernier when the weights were added, showing that the elastic limit was not exceeded.

Calculation and Magnitude of Young Modulus

From the measurements, a graph can be plotted of the force F in newtons against the average extension e in metres. A straight line graph AB passing through the origin is drawn through all the points, Figure 5.3.

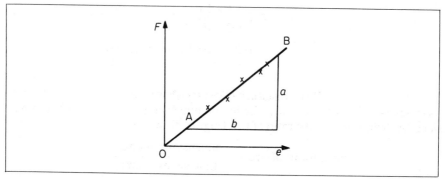

Figure 5.3 *Calculation of E*

Now

$$E = \frac{F/A}{e/l} = \frac{F}{e} \times \frac{l}{A}$$

with the usual notation. The value of F/e is the gradient, a/b, of the straight line AB and this can be found. So knowing F/e, the original length l of the wire and the cross-section area A ($\pi d^2/4$, where d is the diameter of the wire), E can be calculated.

Mild steel (0·2% carbon) has a Young modulus value of about $2·0 \times 10^{11} \, \mathrm{N\,m^{-2}}$, copper has a value about $1·2 \times 10^{11} \, \mathrm{N\,m^{-2}}$; and brass a value about $1·0 \times 10^{11} \, \mathrm{N\,m^{-2}}$.

The breaking stress (tenacity) of cast-iron is about $1·5 \times 10^8 \, \mathrm{N\,m^{-2}}$; the breaking stress of mild steel is about $4·5 \times 10^8 \, \mathrm{N\,m^{-2}}$.

At Royal Ordnance and other Ministry of Supply factories, tensile testing is carried out by placing a sample of the material in a machine known as an *extensometer*, which applies stresses of increasing value along the length of the sample and automatically measures the slight increase in length. When the elastic limit is reached, the pointer on the dial of the machine flickers, and soon after the yield point is reached the sample becomes thin at some point and then breaks. A graph showing the load against extension is recorded automatically by a moving pen while the sample is undergoing test.

Example on Loaded Wire

Find the maximum load which may be placed on a steel wire of diameter 1·0 mm if the permitted strain must not exceed $\frac{1}{1000}$ and the Young modulus for steel is $2·0 \times 10^{11} \, \mathrm{N\,m^{-2}}$.

We have
$$\frac{\text{max. stress}}{\text{max. strain}} = 2 \times 10^{11}$$

$$\therefore \text{max. stress} = \tfrac{1}{1000} \times 2 \times 10^{11} = 2 \times 10^8 \, \mathrm{N\,m^{-2}}$$

Now area of cross-section in $m^2 = \frac{\pi d^2}{4} = \frac{\pi \times 1·0^2 \times 10^{-6}}{4}$

and
$$\text{stress} = \frac{\text{load } F}{\text{area}}$$

$$\therefore F = \text{stress} \times \text{area} = 2 \times 10^8 \times \frac{\pi \times 1 \cdot 0^2 \times 10^{-6}}{4} \, \text{N}$$

$$= 157 \, \text{N}$$

Force in Bar due to Contraction or Expansion

When a bar is heated, and then prevented from contracting as it cools, a considerable force is exerted at the ends of the bar. We can derive a formula for the force if we consider a bar of Young modulus E, a cross-sectional area A, a linear expansivity of magnitude α, and a decrease in temperature of $\theta°\text{C}$. Then, if the original length of the bar is l, the decrease in length e if the bar were free to contract $= \alpha l \theta$ since, by definition, α is the change in length per unit length per degree temperature change.

Now
$$E = \frac{F/A}{e/l}$$

$$\therefore F = \frac{EAe}{l} = \frac{EA\alpha l\theta}{l}$$

$$\therefore F = EA\alpha\theta$$

As an illustration, suppose a steel rod of cross-sectional area $2 \cdot 0 \, \text{cm}^2$ is heated to $100°\text{C}$, and then prevented from contracting when it is cooled to $10°\text{C}$. The linear expansivity of steel $= 12 \times 10^{-6} \, \text{K}^{-1}$ and Young modulus $= 2 \cdot 0 \times 10^{11} \, \text{N m}^{-2}$. Then

$$A = 2 \, \text{cm}^2 = 2 \times 10^{-4} \, \text{m}^2, \quad \theta = 90°\text{C}$$
$$\therefore F = EA\alpha\theta = 2 \times 10^{11} \times 2 \times 10^{-4} \times 12 \times 10^{-6} \times 90 \, \text{N}$$
$$= 43\,200 \, \text{N}$$

Energy Stored in a Wire

Suppose that a wire has an original length l and is stretched by a length e when a force F is applied at one end. If the elastic limit is not exceeded, the extension is directly proportional to the applied load (p. 134). Consequently the force *in the wire* has increased uniformly in magnitude from zero to F, and so the average force in the wire while stretching was $F/2$. Now

$$\text{work done} = \text{force} \times \text{distance}$$

$$\therefore \textbf{work} = \textbf{average force} \times \textbf{extension}$$

$$= \tfrac{1}{2}Fe \qquad \cdot \quad \cdot \quad \cdot \quad \cdot \quad \cdot \qquad (1)$$

This is the amount of energy stored in the wire. It is the gain in molecular potential energy of the molecules due to their displacement from their mean positions. The formula $\frac{1}{2}Fe$ gives the energy in *joules* when F is in newtons and e is in metres.

Further, since $F = EAe/l$,

$$\text{energy } W = \tfrac{1}{2}EA\frac{e^2}{l}$$

Suppose that a vertical wire, suspended from one end, is stretched by attaching a weight of 20 N to the lower end. If the weight extends the wire by 1 mm or 1×10^{-3} m, then

$$\text{energy gained by wire} = \tfrac{1}{2}Fe = \tfrac{1}{2} \times 20 \times 1 \times 10^{-3}$$
$$= 10^{-2}\,\text{J} = 0{\cdot}01\,\text{J}$$

The gravitational potential energy (mgh) lost by the weight in dropping a distance of 1 mm $= 20 \times 1 \times 10^{-3}\,\text{J} = 0{\cdot}02\,\text{J}$. Half of this energy, 0·01 J, is the molecular energy gained by the wire; the remainder is the energy dissipated as heat when the weight comes to rest after vibrating at the end of the wire.

Graph of *F* Against *e* and Energy Measurement

The energy in the wire when it is stretched can also be found from the graph of F against e, Figure 5.4. Suppose the wire extension is e_1 when a force F_1 is applied.

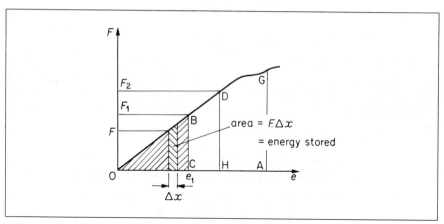

Figure 5.4 *Energy in stretched wire*

At some stage before the extension e_1 is reached suppose that the force in the wire is F and that the wire now extends by a very small amount Δx, as shown. Then over this small extension,

$$\text{energy in wire} = \text{work done} = F \,.\, \Delta x$$

Now $F \,.\, \Delta x$ is represented by the small *area* between the axis of e and the graph, shown shaded in Figure 5.4. So **the total work done between zero extension and e_1 is the area OBC between the graph and the axis of e.** The area of the triangle OBC $= \tfrac{1}{2}$ base × height $= \tfrac{1}{2}F_1e_1$, which is in agreement with our formula on p. — for the energy stored in the wire.

The area result is a general one. It can be used for both the linear (elastic) and the non-linear (non-elastic) parts of the force F against extension e graph. So in Figure 5.4, the work done when the force F_1 (extension e_1) is increased to F_2 (extension e_2) is the area of the trapezium BDHC. If the extension occurs from O to A, which is beyond the elastic limit, the work done is still equal to the area of OGA.

It should be noted that the energy in the wire is equal to the area between the graph and the e-axis because F is plotted vertically and e is plotted horizontally. If e is plotted vertically and F is plotted horizontally, the energy in the wire would then be the area between the graph and the *vertical* or e-axis.

Energy per Unit Volume of Wire

When the elastic limit is not exceeded, the energy per unit volume of a stretched wire is given by a simple formula, as we now see.

The energy stored $= \frac{1}{2}Fe$ and the volume of the wire $= Al$, where A is the cross-section area and l is the length of the wire. So

$$\text{energy per unit volume} = \frac{1}{2}\frac{F.e}{A.l} = \frac{1}{2}\times\left(\frac{F}{A}\right)\times\left(\frac{e}{l}\right)$$

So $\qquad\qquad$ **energy per unit volume $= \frac{1}{2}$ stress $\times$ strain**

So if the stress in a wire is $2 \times 10^7 \, N\,m^{-2}$ and the strain is 10^{-2}, then

$$\text{energy per unit volume} = \frac{1}{2}\times 2 \times 10^7 \times 10^{-2}$$
$$= 10^5 \, J\,m^{-3}$$

Examples on Young Modulus

1 A uniform steel wire of length $4\,m$ and area of cross-section $3 \times 10^{-6}\,m^2$ is extended $1\,mm$. Calculate the energy stored in the wire if the elastic limit is not exceeded. (Young modulus $= 2{\cdot}0 \times 10^{11}\,N\,m^{-2}$.)

(*Analysis* Energy stored $= \frac{1}{2}F \times e$)

$$\text{Stretching force } F = EA\frac{e}{l}$$

So $\qquad\qquad$ energy stored $= \frac{1}{2}Fe = \frac{1}{2}\dfrac{EAe^2}{l}$

$$= \frac{1}{2}\times\frac{2 \times 10^{11} \times 3 \times 10^{-6}\times(1 \times 10^{-3})^2}{4}\,J$$
$$= 0{\cdot}075\,J$$

2 Two vertical wires X and Y, suspended at the same horizontal level, are connected by a light rod XY at their lower ends, Figure 5.5. The wires have the same length l and cross-sectional area A. A weight of $30\,N$ is placed at O on the rod, where XO:OY $= 1:2$. Both wires are stretched and the rod XY then remains horizontal.

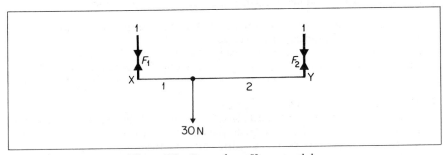

Figure 5.5 *Example on Young modulus*

If the wire X has a Young modulus E_1 of $1{\cdot}0 \times 10^{11}\,\mathrm{N\,m^{-2}}$, calculate the Young modulus E_2 of the wire Y assuming the elastic limit is not exceeded for both wires.

(*Analysis*: (i) Since the rod remains horizontal, extension e_1 of X = extension e_2 of Y. (ii) Forces F_1 and F_2 on wires can be found by moments. (iii) Use $e = Fl/EA$)

By moments about X, $F_2 \times 3 = 30 \times 1$, so $F_2 = 10\,\mathrm{N}$

So force at X, $F_1 = 30 - 10 = 20\,\mathrm{N}$

Since rod XY remains horizontal when wires are stretched, extension e_1 of X = extension e_2 of Y = e

Now, from
$$F = EAe/l, \quad e = Fl/EA$$

So
$$\frac{F_1 l}{E_1 A} = \frac{F_2 l}{E_2 A}$$

$$\therefore E_2 = \frac{F_2}{F_1} \times E_1 = \frac{10}{20} \times 1{\cdot}0 \times 10^{11}$$

$$= 5{\cdot}0 \times 10^{10}\,\mathrm{N\,m^{-2}}\ (\text{or Pa})$$

3 A rubber cord of a catapult has a cross-sectional area of $2\,\mathrm{mm^2}$ and an initial length of $0{\cdot}20\,\mathrm{m}$, and is stretched to $0{\cdot}24\,\mathrm{m}$ to fire a small object of mass 10 g ($0{\cdot}01$ kg). Calculate the initial velocity of the object when it is released.

Assume the Young modulus for rubber is $6 \times 10^8\,\mathrm{N\,m^{-2}}$ and that the elastic limit is not exceeded.

(*Analysis*: (i) Kinetic energy of object $= \frac{1}{2}mv^2 =$ energy stored in rubber. (ii) Energy stored $= \frac{1}{2}F.e$.)

Force stretching rubber, $F = EA\dfrac{e}{l} = \dfrac{6 \times 10^8 \times 2 \times 10^{-6} \times 0{\cdot}04}{0{\cdot}20} = 240\,\mathrm{N}$

since $A = 2\,\mathrm{mm^2} = 2 \times 10^{-6}\,\mathrm{m^2}$ and $e = 0{\cdot}24 - 0{\cdot}20 = 0{\cdot}04\,\mathrm{m}$

$$\therefore \text{energy stored in rubber} = \tfrac{1}{2}F.e = \tfrac{1}{2} \times 240 \times 0{\cdot}04 = 4{\cdot}8\,\mathrm{J}$$

$$\text{Kinetic energy of object} = \tfrac{1}{2}mv^2 = \tfrac{1}{2} \times 0{\cdot}01 \times v^2$$

$$\therefore \tfrac{1}{2} \times 0{\cdot}01 \times v^2 = 4{\cdot}8$$

$$\therefore v = \sqrt{\frac{4{\cdot}8 \times 2}{0{\cdot}01}} = 31\,\mathrm{m\,s^{-1}}$$

_____ **Exercises 5A** _____

Young Modulus, Strain, Energy

1 The speed c of longitudinal waves in a wire is given by the expression $c = \sqrt{\dfrac{E}{\rho}}$

where E is the Young modulus for the material of the wire and ρ is its density. Show that this equation is dimensionally correct.

The extension e, of a wire of cross-sectional area A and of initial length L, is measured for various extending forces F and a graph of F against e is plotted. How would you find a value of c from this graph? What other quantity would you need to measure? (*L.*)

2 Various masses, m, are added to a vertically suspended spring so that small extensions, x, are produced. Sketch the form of the graph obtained if values of m are plotted against values of x. How would you find from this graph
(a) the value of the force per unit extension for the spring, and
(b) the energy stored in the spring for a particular value of the extension? (*L.*)

3   A wire 2 m long and cross-sectional area $10^{-6}\,\text{m}^2$ is stretched 1 mm by a force of 50 N in the elastic region. Calculate   (i) the strain,   (ii) the Young modulus, (iii) the energy stored in the wire.

4   Figure 5A shows the variation of F, the load applied to two wires X and Y, and their extension e. The wires are both iron and have the same length.   (i) Which wire has

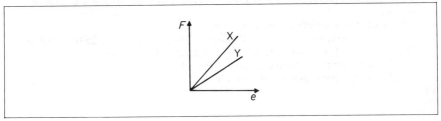

Figure 5A

the smaller cross-section?   (ii) Explain how you would use the graph for X to obtain a value for the Young modulus of iron, listing the additional measurements needed.

5   Define *tensile stress, tensile strain, Young modulus*. What are the units and dimensions of each?

A force of 20 N is applied to the ends of a wire 4 m long, and produces an extension of 0·24 mm. If the diameter of the wire is 2 mm, calculate the stress on the wire, its strain, and the value of the Young modulus.

6   What force must be applied to a steel wire 6 m long and diameter 1·6 mm to produce an extension of 1 mm? (Young modulus for steel $= 2\cdot0 \times 10^{11}\,\text{N}\,\text{m}^{-2}$.)

7   Find the extension produced in a copper wire of length 2 m and diameter 3 mm when a force of 30 N is applied. (Young modulus for copper $= 1\cdot1 \times 10^{11}\,\text{N}\,\text{m}^{-2}$.)

8   A spring is extended by 30 mm when a force of 1·5 N is applied to it. Calculate the energy stored in the spring when hanging vertically supporting a mass of 0·20 kg if the spring was unstretched before applying the mass. Calculate the loss in potential energy of the mass. Explain why these values differ. (*L.*)

9   Define *elastic limit* and *Young modulus* and describe how you would find the values for a copper wire.

What stress would cause a wire to increase in length by one-tenth of one per cent if the Young modulus for the wire is $12 \times 10^{10}\,\text{N}\,\text{m}^{-2}$? What force would produce this stress if the diameter of the wire is 0·56 mm? (*L.*)

10   In an experiment to measure the Young modulus for steel a wire is suspended vertically and loaded at the free end. In such an experiment,
(a) why is the wire long and thin,
(b) why is a second steel wire suspended adjacent to the first?

Sketch the graph you would expect to obtain in such an experiment showing the relation between the applied load and the extension of the wire. Show how it is possible to use the graph to determine
(a) Young modulus for the wire,
(b) the work done in stretching the wire.

If the Young modulus for steel is $2\cdot00 \times 10^{11}\,\text{N}\,\text{m}^{-2}$, calculate the work done in stretching a steel wire 100 cm in length and of cross-sectional area 0·030 cm² when a load of 100 N is slowly applied without the elastic limit being reached. (*JMB.*)

11   Define the terms *tensile stress* and *tensile strain* and explain why these quantities are more useful than *force* and *extension* for a description of the elastic properties of matter.

Describe the apparatus you would use and the measurements you would perform to investigate the relation between the tensile stress applied to a wire and the strain it produces.

A cylindrical copper wire and a cylindrical steel wire, each of length 1·5 m and diameter 2 mm, are joined at one end to form a composite wire 3 m long. The wire is loaded until its length becomes 3·003 m. Calculate the strains in the copper and steel wires and the force applied to the wire.

(Young modulus for copper $= 1·2 \times 10^{11}\,\mathrm{N\,m^{-2}}$; for steel $2·0 \times 10^{11}\,\mathrm{N\,m^{-2}}$.) (*O. & C.*)

12 State Hooke's law and describe, with the help of a rough graph, the behaviour of a copper wire which hangs vertically and is loaded with a gradually increasing load until it finally breaks. Describe the effect of gradually reducing the load to zero

(a) before,

(b) after the elastic limit has been reached.

A uniform steel wire of density $7800\,\mathrm{kg\,m^{-3}}$ weighs 16 g and is 250 cm long. It lengthens by 1·2 mm when stretched by a force of 80 N. Calculate

(a) the value of the Young modulus for the steel,

(b) the energy stored in the wire. (*JMB.*).

13

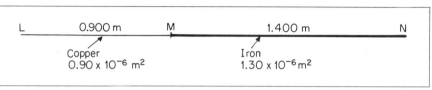

Figure 5B

A copper wire LM is fused at one end, M, to an iron wire MN, Figure 5B. The copper wire has length 0·900 m and cross-section $0·90 \times 10^{-6}\,\mathrm{m^2}$. The iron wire has length 1·400 m and cross-section $1·30 \times 10^{-6}\,\mathrm{m^2}$. The compound wire is stretched; its total length increases by 0·0100 m. Calculate

(a) the ratio of the extensions of the two wires,

(b) the extension of each wire,

(c) the tension applied to the compound wire.

(The Young modulus of copper $= 1·30 \times 10^{11}\,\mathrm{N\,m^{-2}}$. The Young modulus of iron $= 2·10 \times 10^{11}\,\mathrm{N\,m^{-2}}$.) (*L.*)

14 Explain the terms *stress, strain, modulus of elasticity* and *elastic limit*. Derive an expression in terms of the tensile force and extension for the energy stored in a stretched rubber cord which obeys Hooke's law.

The rubber cord of a catapult has a cross-sectional area 1·0 mm^2 and a total unstretched length 10·0 cm. It is stretched to 12·0 cm and then released to project a missile of mass 5·0 g. From energy considerations, or otherwise, calculate the velocity of projection, taking the Young modulus for the rubber as $5·0 \times 10^8\,\mathrm{N\,m^{-2}}$. State the assumptions made in your calculation. (*L.*)

15 What is meant by saying that a substance is 'elastic'?

A vertical brass rod of circular section is loaded by placing a 5 kg weight on top of it. If its length is 50 cm, its radius of cross-section 1 cm, and the Young modulus of the material $3·5 \times 10^{10}\,\mathrm{N\,m^{-2}}$, find

(a) the contraction of the rod,

(b) the energy stored in it. (*C.*)

16 (a) Define *stress, strain, Young modulus*.

(b) The formula for the velocity v of compressional waves travelling along a rod made of material of Young modulus E and density ρ is $v = (E/\rho)^{\frac{1}{2}}$. Show that this formula is dimensionally consistent.

(c) A uniform wire of unstretched length 2·49 m is attached to two points A and B which are 2·0 m apart and in the same horizontal line. When a 5 kg mass is attached to the midpoint C of the wire, the equilibrium position of C is 0·75 m below the line AB. Neglecting the weight of the wire and taking the Young modulus for its material to be $2 \times 10^{11}\,\mathrm{N\,m^{-2}}$, find (i) the strain in the wire, (ii) the stress in the wire, (iii) the energy stored in the wire. (*O.*)

17 Define the Young modulus of elasticity. Describe an accurate method of determining it. The rubber cord of a catapult is pulled back until its original length

has been doubled. Assuming that the cross-section of the cord is 2 mm square, and that Young modulus for rubber is 10^7 N m^{-2} calculate the tension in the cord. If the two arms of the catapult are 6 cm apart, and the unstretched length of the cord is 8 cm what is the stretching force? (O. & C.)

18   A copper wire and a length of rubber are each subjected to linear stress until they break. Sketch labelled graphs of stress against strain to show the behaviour of each material. Write brief notes on the important differences between the behaviour of these two materials.

The table below shows how the extension e of a 10 m length of a certain nylon climbing rope depends on the applied force F.

e/m	0	1·9	2·8	3·4	3·8	4·1	4·3
F/kN	0	2·0	4·0	6·0	8·0	10·0	12·0

(a) Draw a graph of applied force against extension.
   A climber of mass 70 kg, attached to a 10 m length of this rope, can withstand a force from the rope of no more than 6·5 kN without the risk of serious injury.
(b) Read off from your graph the extension which would be produced in the rope for a force of 6·5 k N.
(c) Use the graph to find the energy stored in the rope if it were stretched by this amount.
(d) If the upper end of the rope were securely anchored, through what vertical distance could the climber fall freely (before the rope started to stretch) without risk of injury from the force of the rope when his fall was arrested? (C.)

19   (a) Figure 5C is a graph showing how the extension of a steel wire of length 1·2 m and area of cross-section 0·012 mm^2 alters as a stretching force is applied. (i) Use the graph to calculate the Young modulus for steel. (ii) Draw a labelled diagram of an experimental arrangement suitable for obtaining such a set of results.

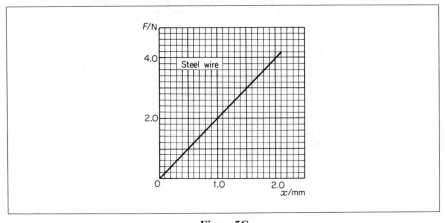

Figure 5C

(b) Figure 5D shows the results of a similar experiment done with a copper wire. In this case the wire has been stretched until it breaks.   (i) The graph drawn in this instance is a stress-strain curve. Explain *one* advantage of representing the results in this way.   (ii) Account in molecular terms for the behaviour of the wire as it is stretched from A to B.   (iii) The copper wire used was 2·0 m long and 0·25 mm^2 in cross-section. Calculate the tension

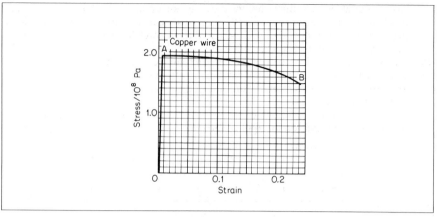

Figure 5D

in the wire at A and an approximate value for the work done in producing a strain of 0·1.

(c) A length of rubber cord is suspended from a rigid support and stretched by means of weights attached to its lower end.   (i) Sketch a stress-strain curve to represent the behaviour of such a cord as it is first loaded then unloaded. (ii) Suppose the cord were continuously stretched and relaxed at a rapid rate. What might you notice? How would this be explained by the stress-strain graph? (*L*.)

20 (a) What is meant by   (i) elastic behaviour,   (ii) plastic behaviour of a wire when it is stretched?

(b) (i) Describe how you would investigate the elastic and plastic properties of a soft copper wire under increasing tensile loads up to its breaking point. (ii) Draw a graph of the results you would expect to obtain, and label its principal features.   (iii) Interpret the graph in terms of the forces between atoms in the wire and of their arrangement in the wire.

(c) A lift in a skyscraper has a total mass of 8000 kg when loaded. It is hung from light cables made of steel of breaking stress $0.50 \times 10^9 \, \text{N m}^{-2}$. These cables will support a static load of 72 000 kg before they break. [Take the Young modulus for steel to be 2.0×10^{11} Pa.]

Calculate: (i) the total cross-sectional area of the lift cables,   (ii) the static extension of the cables when the lift is at rest at ground-floor level, if the height to the winding gear at the top of the building is 350 m,   (iii) the elastic strain energy stored in the cables when the lift is at rest at ground-floor level.

The lift now ascends at a steady speed of $8.0 \, \text{m s}^{-1}$.   (iv) Calculate the power needed to raise the lift,   (v) how could the lift system be designed to reduce significantly this large power requirement? (*O*.)

Molecular Forces

Particle Nature of Matter, Molecules

Matter is made up of many millions of molecules, which are particles whose dimensions are about 3×10^{-10} m. Evidence for the existence of molecules is given by experiments demonstrating *Brownian motion*, with which we assume the reader is familiar. One example is the random motion of smoke particles in air, which can be observed by means of a microscope. This is due to continuous bombardment of a tiny smoke particle by numerous air molecules all round it. The air molecules move with different velocities in different directions. The resultant force on the smoke particle is therefore unbalanced, and irregular in magnitude and direction. Larger particles do not show Brownian motion when struck on all sides by air molecules. The resultant force is then relatively negligible.

More evidence of the existence of molecules is supplied by the successful predictions made by the *kinetic theory of gases*. This theory assumes that a gas consists of millions of separate particles or molecules moving about in all directions (p. 683). *X-ray diffraction patterns* of crystals also provide evidence for the particle nature of matter (p. 869). The symmetrical patterns of spots obtained are those which one would expect from a three-dimensional grating or lattice formed from particles. A smooth continuous medium would not give a diffraction pattern of spots.

Size and Separation of Molecules

The size of atoms and molecules can be estimated in several different ways. By allowing an oil drop to spread on water, for example, an upper limit of about 5×10^{-9} m is obtained for the size of an oil molecule. X-ray diffraction experiments enable the interatomic spacing between atoms in a crystal to be accurately found. The results are of the order of a few angstrom units, such as 3 Å or 3×10^{-10} m or 0.3 nm (nanometre).

A simple calculation shows the order of magnitude of the enormous number of molecules present in a small volume. One gram of water occupies 1 cm^3. One mole has a mass of 18 g, and thus occupies a volume of 18 cm^3 or 18×10^{-6} m^3. Assuming the diameter of a molecule is 3×10^{-10} m, its volume is roughly $(3 \times 10^{-10})^3$ or 27×10^{-30} m^3. Hence the number of molecules in one mole $= 18 \times 10^{-6}/(27 \times 10^{-30}) = 7 \times 10^{23}$ approximately.

The *Avogadro constant*, N_A, is the number of molecules in one mole of a substance. Accurate values show that $N_A = 6.02 \times 10^{23}$ mol^{-1}, or 6.02×10^{26} kmol^{-1}, where 'kmol' represents a kilomole, 1000 moles.

The order of separation of molecules in liquids is about the same as in solids. We can calculate the separation of gas molecules at standard pressure from the fact that a mole of any gas occupies about 22·4 litres or 22.4×10^{-3} m^3 at s.t.p. Since one mole contains about 6×10^{23} molecules, then, roughly, taking the cube root of the volume per molecule,

$$\text{average separation} = \sqrt[3]{\frac{22.4 \times 10^{-3}}{6 \times 10^{23}}} \text{ m}$$

$$= 33 \times 10^{-10} \text{ m (approx.)}$$

This is about 10 times the separation of molecules in solids or liquids.

The lightest atom is hydrogen. Since about 6×10^{23} hydrogen molecules have a mass of 2 g or 2×10^{-3} kg, and each hydrogen molecule consists of two atoms, then

$$\text{mass of hydrogen atom} = \frac{2 \times 10^{-3}}{2 \times 6 \times 10^{23}} = 1{\cdot}7 \times 10^{-27}\,\text{kg (approx.)}$$

Heavier atoms have masses in proportion to their relative atomic masses.

Example on Molecular Separation

Estimate the order of separation of atoms in aluminium metal, given the density is $2700\,\text{kg m}^{-3}$, the relative atomic mass is 27 and the Avogadro constant is $6 \times 10^{23}\,\text{mol}^{-1}$.

1 mole of aluminium has a mass of 27 g. From the density value,

$$1\,\text{m}^3 \text{ of aluminium has} \frac{2700 \times 10^3}{27} \text{ or } 10^5 \text{ moles}$$

Now 1 mole contains 6×10^{23} molecules or atoms of aluminium

So $\qquad\qquad 6 \times 10^{23} \times 10^5$ atoms occupy a volume of $1\,\text{m}^3$

Hence $\qquad\quad$ volume per atom $= \dfrac{1}{6 \times 10^{28}} = 1{\cdot}7 \times 10^{-29}\,\text{m}^3$

The volume occupied per atom is of the order d^3, where d is the separation of the atoms. So, approximately,

$$d = \sqrt[3]{1{\cdot}7 \times 10^{-29}}\,\text{m}$$
$$= 2{\cdot}6 \times 10^{-10}\,\text{m}$$

Intermolecular Forces

The forces which exist between molecules can explain many of the bulk properties of solids, liquids and gases. These intermolecular forces arise from two main causes:

(1) The _potential energy_ of the molecules, which is due to interactions with surrounding molecules (this is principally electrical in origin).

(2) The _thermal energy_ of the molecules—this is the kinetic energy of the molecules and it depends on the temperature of the substance concerned.

We shall see later that the particular state or phase in which matter appears—that is, solid, liquid or gas—and the properties it then has, are determined by the relative magnitudes of these two energies.

Potential Energy and Force

In bulk, matter consists of numerous molecules. To simplify the situation, Figure 5.6 shows the variation of the mutual potential energy V between two molecules at a distance r apart.

Along the part BCD of the curve, the potential energy V is negative. Along the part AB, the potential energy V is positive. Generally, V can be written approximately as

$$V = \frac{a}{r^p} - \frac{b}{r^q} \qquad \cdot \qquad \cdot \qquad \cdot \qquad \cdot \qquad \cdot \quad (1)$$

where p and q are powers of r, and a and b are constants. The positive term with the constant a indicates a repulsive force and the negative term with the constant b an attractive force, as discussed shortly.

There are different kinds of bonds or forces between atoms and molecules in solids, depending on the nature of the solid. In an *ionic solid*, for example sodium chloride, V can be approximated by

$$V = \frac{a}{r^9} - \frac{b}{r} \qquad . \qquad . \qquad . \qquad . \qquad (2)$$

The force F between molecules is generally given by $F = -dV/dr$, the negative *potential gradient* in the field (see p. 203). From (2), it follows that, for the two ions,

$$F = \frac{9a}{r^{10}} - \frac{b}{r^2} \qquad . \qquad . \qquad . \qquad . \qquad (3)$$

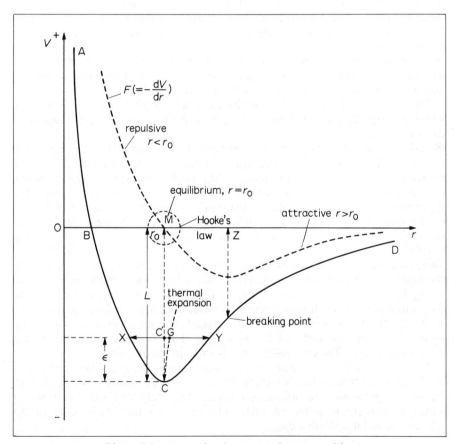

Figure 5.6   *Intermolecular potential energy and force*

The +ve term in (3) indicates a *repulsive* force since the force acts in the direction of increasing r. This is the force along ABC in Figure 5.6. The −ve term in (3) indicates an *attractive* force since the force acts oppositely to the direction of increasing r. This force acts along CD in Figure 5.6. As shown, it decreases with increasing separation, r, of the two molecules.

Properties of Solids from Molecular Theory

Several properties of a model solid can be deduced or calculated from the potential–separation $(V - r)$ graph or the force–separation $(F - r)$ graph.

Equilibrium spacing of molecules. The value of r when the potential energy V is a minimum corresponds to the stable or equilibrium spacing between the molecules. At the absolute zero, where the thermal energy is zero, this corresponds to C in Figure 5.6, or a separation $r = r_0$. At this separation or spacing, the repulsive and attractive forces balance, that is, $F = 0$. Hence, from (3), r_0 is given, for the ions concerned, by

$$r_0 = \left(\frac{9a}{b}\right)^{1/8}$$

The value of r_0 for solids is about 2 to 5×10^{-10} m.

If the separation r of the molecules is slightly increased from r_0, the attractive force between them will restore the molecules to their equilibrium position after the external force is removed. If the separation is decreased from r_0, the repulsive force will restore the molecules to their equilibrium position after the external force is removed. So the molecules of a solid *oscillate* about their equilibrium or mean position.

Elasticity and Hooke's law. Near the equilibrium position r_0, the graph of F against r approximates to a straight line, Figure 5.6. This means that the extension is proportional to the applied force (Hooke's law, p. 134).

The 'force constant', k, between the molecules is given by $F = -k(r - r_0)$, where r is slightly greater than r_0. So $k = -\mathrm{d}F/\mathrm{d}r = -$ gradient of tangent to curve at $r = r_0$.

Breaking strain. So long as the restoring force increases with increasing separation from $r = r_0$, the molecules will remain bound together. This is the case from $r = r_0$ to $r = OZ$ in Figure 5.6. Beyond a separation $r = OZ$, however, the restoring force *decreases* with increasing separation. OZ is therefore the separation between the molecules at the *breaking point* of the solid (see p. 136). It corresponds to the value of r for which $\mathrm{d}F/\mathrm{d}r = 0$, that is, to the point below Z on the $V - r$ curve. The *breaking strain* = extension/r_0 = $(x - r_0)/r_0$, where $x = OZ$.

Thermal expansion. Molecules remain stationary at absolute zero, since their thermal energy is then zero. This corresponds to the point C of the energy curve in Figure 5.6. At a higher temperature, the molecules have some energy, ε, above the minimum value, as shown. Hence they oscillate between points such as X and Y. Since the $V - r$ curve is not symmetrical, the mean position G of the oscillation is on the right of C', as shown. This corresponds to a greater separation than r_0. Thus the solid *expands* when its thermal energy is increased.

At a slightly higher temperature, the mean position moves further to the right of G and so the solid expands further. When the energy equals CM (latent heat, L), the energy enables the molecules to break completely the bonds of attraction which keep them in a bound state. The molecules then have little or no interaction and now form a *gas*.

Latent Heat of Vaporisation

Inside a liquid, molecules continually break and reform bonds with neighbours. The 'latent heat of vaporisation L' of a liquid is the energy to break all the bonds between its molecules.

Suppose ε is the energy to separate a particular molecule X from its nearest neighbour, that is, the energy per pair of molecules. If there are n nearest neighbours per molecule, and we neglect the effect of the other molecules, then the energy to break the bonds between X and its neighbours is $n\varepsilon$.

With a mole of liquid, there are N_A molecules inside it, where N_A is the Avogadro constant. The number of pairs of molecules is $\frac{1}{2}N_A$. So the energy required to break the bonds of all the molecules at the boiling point is roughly $\frac{1}{2}N_A n\varepsilon$. Thus the latent heat of vaporisation per mol, $L = \frac{1}{2}N_A n\varepsilon$. In molecular terms, it corresponds roughly to the energy difference between C and D in the $V - r$ curve in Figure 5.6, assuming C is about the equilibrium separation for two liquid molecules.

Bonds Between Atoms and Molecules

The atoms and molecules in solids, liquids and gases are held together by so-called *bonds* between them. There are different types of bonds. All are due to electrostatic forces which arise from the $+$ve charge on the nucleus of an atom and its surrounding electrons which carry $-$ve charges.

Briefly, the different types of bonds are:
(a) *Ionic bonds.* Sodium chloride in the solid state consists of positive sodium ions and negative chlorine ions held together by electrostatic attraction between the opposite charges.
(b) *Covalent bonds.* The electron in one atom of a hydrogen molecule, H_2, for example, wanders to the other atom, and the two atoms then attract each other as a result of their unlike charges. These covalent bonds, which are due to shared electrons between atoms, are very strong.
(c) *Metallic bond.* In solid metals such as sodium or copper, one or more electrons in the outermost part of the atom may leave and occupy the orbit of another atom. These so-called 'free' electrons wander through the metal crystal structure, which consists of fixed $+$ve ions. The metallic bond is similar to a covalent bond except that electrons are not attached to any particular atoms; it keeps the metal in its solid state. The metallic bond is not as strong as the ionic and covalent bonds.
(d) *Van der Waals' bond.* Over a long time-interval, the 'centre' of an electron cloud round the nucleus is at the nucleus itself. At any instant, however, more electrons may appear on one side of the nucleus than the other. In this case the 'centre' of the electron cloud, or $-$ve charge, is slightly displaced from the $+$ve charge on the nucleus. The two charges now form an 'electric dipole'. A dipole attracts the electrons in neighbouring atoms, forming other dipoles.

The electric dipoles have weak forces between them, called 'van der Waals' forces because similar attractive forces were predicted by van der Waals in connection with the molecules of gases. Solid neon, an inert element, is kept in this state by these bonds; the low melting point of solid neon shows that the bonds are weak.

_____ Exercises 5B _____

Molecular Forces and Energy

1 Figure 5E shows (i) the variation PADE of potential energy V between two molecules with their separation r, and, (ii) the variation QBLC of the force F between the molecules with their separation r.
  (a) Explain how the $F - r$ curve is obtained from the $V - r$ curve,
  (b) State which part of the $F - r$ curve corresponds to a repulsive force and which part to an attractive force. Describe briefly one experiment which shows the repulsive force and one experiment which shows the attractive force.

2 Explain how you would use the $V-r$ curve in Figure 5E
   (a) to obtain the equilibrium separation of the molecules,
   (b) to find the energy needed to completely separate two molecules initially at the equilibrium separation,
   (c) to show that a solid usually expands when its thermal energy is increased.

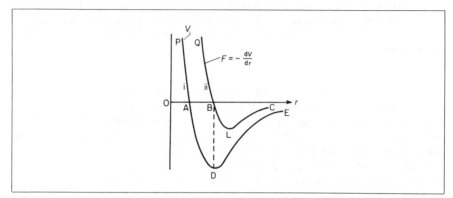

Figure 5E

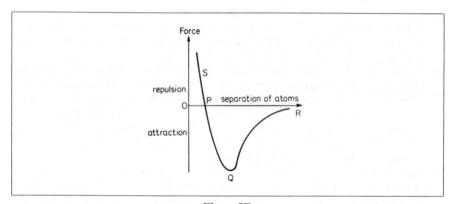

Figure 5F

3 Using the $F-r$ curve in Figure 5E, explain how you would
   (a) find the force constant k ($F = -kx$, where x is the extension of a wire in the elastic region)
   (b) account for Hooke's law of elasticity,
   (c) obtain a value for the breaking force of a solid.

4 The force between two molecules may be regarded as an attractive force which increases as their separation decreases and a repulsive force which is only important at small separations and which there varies very rapidly. Draw sketch graphs
   (a) for force-separation,
   (b) for potential energy-separation. On each graph mark the equilibrium distance and on (b) indicate the energy which would be needed to separate two molecules initially at the equilibrium distance.
   With the help of your graphs discuss briefly the resulting motion if the molecules are displaced from the equilibrium position. (*JMB*.)

5 The graph (Figure 5F) represents the relationship between the interatomic forces which exist in a material and the separation of the atoms. What point on the graph corresponds to the separation when the material is not subjected to any stress? Use the graph to explain   (i) why energy is stored in a material when it is compressed and when it is extended, and,   (ii) why, and over what region, the material can be expected to obey Hooke's law. (*AEB*, 1982.)

6

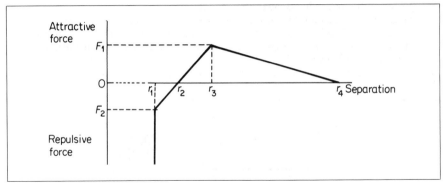

Figure 5G

The graph above shows a much simplified model of the force between two atoms plotted against their distance of separation, Figure 5G. Express (i) the maximum restoring force between the atoms when they are pulled apart, (ii) the equilibrium separation of the atoms, and, (iii) the energy required to separate the two atoms in terms of the forces and distances given on the graph.

With the aid of the graph explain why solids show resistance to both stretching and compressing forces. Explain over what region you would expect Hooke's law to apply. What would this model predict about the elastic limit and the yield point for a material whose atoms followed the model? (*L.*)

7 In the model of a crystalline solid the particles are assumed to exert both attractive and repulsive forces on each other. Sketch a graph of the potential energy between two particles as a function of the separation of the particles. Explain how the shape of the graph is related to the assumed properties of the particles.

The force F, in N, of attraction between two particles in a given solid varies with their separation d, in m, according to the relation

$$F = \frac{7 \cdot 8 \times 10^{-20}}{d^2} - \frac{3 \cdot 0 \times 10^{-96}}{d^{10}}$$

State, giving a reason, the resultant force between the two particles at their equilibrium separation. Calculate a value for this equilibrium separation.

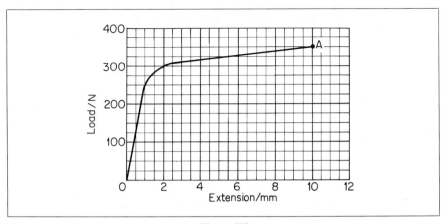

Figure 5H

The graph displays a load against extension plot for a metal wire of diameter $1 \cdot 5$ mm and original length $1 \cdot 0$ m, Figure 5H. When the load reached the value at A the wire broke. From the graph deduce values of

(a) the stress in the wire when it broke,
(b) the work done in breaking the wire,
(c) the Young modulus for the metal of the wire.

Define *elastic* deformation. A wire of the same metal as the above is required to support a load of 1·0 kN without exceeding its elastic limit. Calculate the minimum diameter of such a wire. (*O. & C.*)

8 (a) Sketch a graph which shows how the force between two atoms varies with the

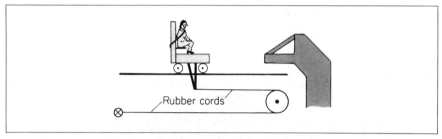
Rubber cords

Figure 5I

distance between their centres. With reference to this graph explain why (i) any reversible change in volume of a solid is always a small fraction of the unstressed volume, (ii) a metal wire obeys Hooke's law for small extensions.

(b) (i) Sketch a graph which shows how the length of a rubber cord varies with the tension in the cord as the tension increases from zero until the cord breaks. Account for the shape of the curve you have drawn in terms of the molecular structure of rubber. (ii) In what important way does the structure of a metal at the molecular level differ from that of a rubber? How do you account for the large extension of a rubber cord compared with the extension of a mild steel wire of the same dimensions and acted on by the same force?

(c) Figure 5I shows a trolley of total mass 560 kg which is used for testing seat belts. The trolley runs on rails and is attached to six identical, parallel rubber cords whose unstretched lengths are 40 m each. When the trolley is pulled back far enough to extend the cords by 21 m each and then released, it reaches a speed of 15 m s^{-1} just as the cords begin to slacken. If the cords are assumed to obey Hooke's law over the full range of their extension in this application, if the system is assumed to be free of friction and the Young modulus for the rubber is $2 \cdot 2 \times 10^7$ N m^{-2}, calculate (i) the maximum force applied to each cord, (ii) the area of cross-section of each cord when stretched. (*L.*)

Solid Materials

Industry uses many different kinds of solid materials. For example, metals such as iron, steel and aluminium; glassy solids such as perspex and window glass; and organic materials called polymers for making rubbers, plastics and resins.

Classification of Solids

Since the atoms of a solid occupy fixed positions, it is convenient to classify them according to the way their three-dimensional structure is built.

For many solids, and in particular metals, the *crystalline state* is the preferred one. Here the atoms are arranged in a regular, repetitive manner forming a three-dimensional lattice. With such an ordered packing system, the greatest number of atoms may be arranged in the smallest volume and the potential energy of the system tends to a minimum for stability.

Many solids, however, such as organic materials like rubber, are unable to adopt a structure such as a crystal state which has long-range order. *Polymers*, for example, are organic solids with very large and irregular molecules which are not capable of forming large-scale regular structures.

Other solids such as glass have no ordered structure on account of the way they are made. In making glass, molten material is cooled and its viscosity increases. The disordered liquid structure is then 'frozen in'. Solids which have their atoms arranged in a completely irregular structure are called *amorphous solids*. However, most non-crystalline solids do show some short-range order.

The way in which atoms are arranged in solids will obviously have a strong effect on the physical properties of the material. For example, diamond and graphite are both different structural forms of solid carbon. Diamond is transparent, hard and non-conducting but graphite is black, soft and conducting. Further, the nature of any solid structure will be largely determined by the nature of the interatomic forces. These are summarised below and were discussed on page 151.

Type	*Strength*	*Nature*	*Example*
Ionic	Strong	Electron transfer	Sodium salt
Covalent	Strong	Electron sharing	diamond
Metallic	Fairly strong	Electron sharing	copper
Van der Waals	Weak	Dipole interaction	solid neon

Solids can be classified into **(i) crystalline**—ordered structure,
(ii) amorphous—irregular structure, **(iii) glassy**—disordered structure,
(iv) polymer—organic irregular structure

Crystalline Solids

The simplest crystalline systems are those in pure metals. Here we are dealing with identical atoms linked through the fairly strong metallic bond (p. 151).

In any crystal, a particular grouping of atoms is repeated many times, like the pattern in some wallpapers or textiles. The unit of pattern is known as a *unit cell* and the whole structure is called a 'space lattice', with atoms or ions at the lattice corners. As shown in Figure 5.7 many atomic planes such as P, Q, R, S, T can be drawn through the crystal which are rich in atoms.

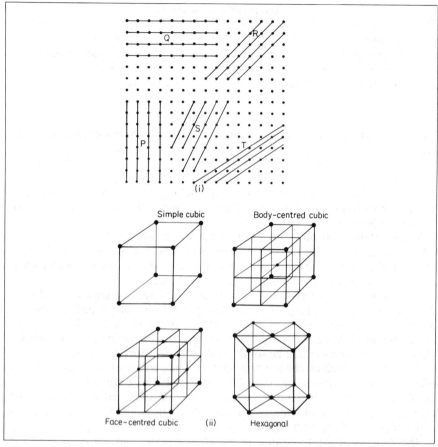

Figure 5.7 (i) *Atomic planes in crystal* (ii) *Types of crystal structure*

There are different types of unit cells. Figure 5.7 shows four types of *structure*. The *simple cubic* has an atom or ion at the eight corners of a cube. More common in nature is the *body-centred cube* (BCC), which has one atom at the centre of each cube in addition to the eight atoms at the corners, and the *face-centred cube* (FCC), which has one atom at the centre of the six faces of the cube in addition to eight atoms at the corners. Alternatively, crystals may have layers of atoms arranged with *hexagonal* rather than cubic symmetry. This appears to be a very efficient way of packing layers of atoms. The Table below compares the packing efficiencies of these crystal structures. Here the atoms are considered to be touching spheres and calculations of the volumes occupied by the atoms as a fraction of the available volume give a good guide to the efficiency of packing.

Structure	Packing fraction	Occurrence
Simple cubic	0·52	very rare
Body-centred cubic (BCC)	0·68	fairly common
Face-centred cubic (FCC)	0·74	very common
Hexagonal close packed (HCP)	0·74	very common

In crystalline solids such as metals, the atoms are grouped in a lattice structure with many planes rich in atoms.

Imperfections in Crystals

Real crystals are rarely perfect. Although less than one crystal site (place) in ten thousand may be imperfect, the existence of lattice defects can have considerable influence on the mechanical and electrical properties of a material. The industrial development of electronics was due to the control of the electrical properties of silicon and other semiconductors by adding impurity atoms in very low concentrations. Small impurities are also responsible for the characteristic colours of many gemstones. As we see later, the mechanical properties of solids are determined to a great extent by imperfections.

Broadly, crystal defects can be classified into two groups; either imperfections in the *occupation* of sites or in the *arrangement* of sites.

Imperfections in Occupation of Sites

Here the main imperfections are commonly *point defects*, especially
(a) the presence of foreign atoms and
(b) the existence of vacancies (unoccupied lattice sites).

Foreign atoms may exist among the 'host' atoms in several ways. They may be grouped in clusters in the host crystal or they may be dispersed through the crystal as single atoms. If sufficiently small, they may exist on *interstitial* sites, that is, they may occupy non-lattice sites between the host atoms. For example, pure iron can be made into steel, an engineering material widely used, by adding carbon, whose atoms reside between the iron atoms.

If foreign atoms are comparable in size to host atoms and valency requirements are met, they can exist as substitutes for host atoms on a lattice site. For example, one type of brass consists of 70% copper and 30% zinc whose atoms exist as substitutes for copper atoms in a cubic lattice.

Vacancies exist in all crystals and occur naturally when the crystal solidifies. They may also be present by diffusion into the crystal from the surface. Irradiation by α-particles may create a vacancy by knocking a host atom off its usual site and this is important to nuclear engineers.

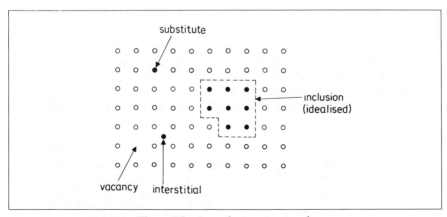

Figure 5.8 *Imperfections in crystals*

Figure 5.8 shows diagrammatically how these imperfections occur.

Imperfections in Arrangements of Sites, Dislocations

Imperfections in crystals can occur in the arrangements of sites, where atoms in the lattice structure should exist. These imperfections can be large and affect thousands of millions of atoms in the crystal.

From the viewpoint of the mechanical behaviour of a solid, the most important defects are the *dislocations* in the crystal. These are defects along a line of atoms. An *edge* dislocation can be considered as an extra part of a layer of atoms either removed from, or inserted into, a perfect crystal structure, Figure 5.9 (i).

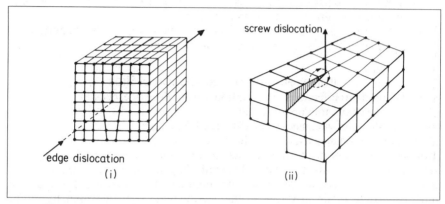

Figure 5.9   (i) *Dislocation is* edge *of plane of atoms*   (ii) *A screw dislocation*

Another type of dislocation is called a *screw* dislocation. Here the atoms are displaced so that they can be imagined to be on the spiral of a screw. The crystal then has the appearance of having been partially sliced and the two exposed faces displaced vertically, Figure 5.9 (ii). Details of the dislocations are given on page 161.

The atomic bonds along the line of a dislocation are strained. Where the dislocations meet a surface, these strained bonds can be made visible by etching the surface of the crystal and studying the surface in a microscope.

Crystals are imperfect. The most important defects are *dislocations*.

Polycrystalline Materials—Grains and Boundaries

Many crystal materials are *polycrystalline*, that is, they exist as a large collecting of tiny crystals all pointing in different (random) directions. Each tiny crystal is known as a *grain* and are connected together at the *grain boundaries*.

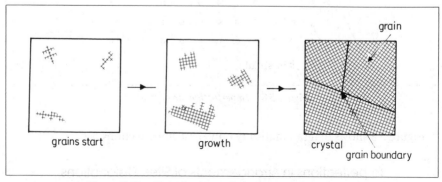

Figure 5.10   *Growth of grain boundaries (diagrammatic)*

The existence of grains appears to be due to the solidification process in manufacture which begins simultaneously at several places. The surface is the first to solidify or freeze. When the solid is completely solidified, the structure consists of grains pointing in different directions, as shown diagrammatically in Figure 5.10. A *single crystal* metal specimen has no grain boundaries and so is stronger than a polycrystalline metal.

Mechanical Behaviour of Solids

An engineer must select the right material to use in a project. He must then be confident that the final structure will perform safely and within the design limits which have been set. To achieve this, the engineer must know and understand how the materials available will respond or react to the stresses they may meet.

The section on Elasticity, page 136, discusses the strain in solid metals when stresses are applied. If required, the reader should refer to this chapter for topics such as Hooke's law, the Young modulus, yield point, breaking stress and the energy stored in a strained material, all of which are important for a full understanding of the mechanical behaviour of solids.

Here it may be useful to recall that if Hooke's law is obeyed, that is, the proportional limit is not reached, the material is said to deform *elastically* and the energy stored is recovered on removing the load. In this case, the strain is typically less than $\frac{1}{2}\%$ for metals.

In elastic deformation, when Hooke's law is obeyed and the atoms undergo small displacements, the energy stored is fully recovered when the load is removed.

Plastic Deformation, Breaking Stress

Beyond the elastic limit, a material undergoes *plastic deformation*. This occurs by movement of dislocations in the solid, discussed shortly. Figure 5.11 (i) shows a typical stress-strain curve for a specimen. If B is the elastic limit, the region BE corresponds to plastic deformation. In this case the energy stored in the solid is *not* recovered but is transferred to heat. Unloading from D, only the elastic energy is recovered and so DF is parallel to OA.

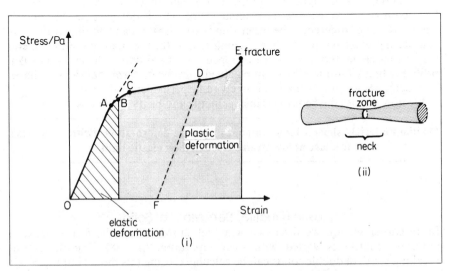

Figure 5.11 (i) *Stress-strain curve* (ii) *Breaking at fracture*

E corresponds to the point of *fracture*. The strain at fracture may be as high as 50% for a metal such as steel. Fracture usually occurs at E after a 'neck' has been produced because this has minimum cross-section area and so maximum stress, Figure 5.11 (ii). During necking, the material appears to 'flow' like a viscous liquid. The *breaking stress* at E (or *ultimate tensile stress*) is taken as a measure of the *tensile strength* of a material. The Young modulus value is a measure of the *stiffness* of the material and is the value of the gradient of the line OA (stress/strain). The Table shows the tensile strength of some materials.

Material	Tensile strength/10^8 Pa
Steel	4–10
Cast iron	1
Aluminium	0·8
Glass	1
Rubber	0·2

Plastic deformation is due to movement of dislocations. The energy in plastic deformation is converted to heat.

Stiffness = Young modulus. Tensile strength = breaking point stress.

Ductile and Brittle Materials

Materials which show a large amount of plastic deformation under stress are called *ductile*. Ductility is an important property and allows metals to be drawn into wires. Metals may be *malleable* or beaten into sheets and rolled and shaped. Metals are very useful engineering materials because many have high tensile strength, malleability and ductility.

Materials such as glass and ceramics fracture (break) close to the elastic limit without any appreciable plastic flow. They are called *brittle* materials. In these cases fracture occurs at low strains.

The absence of any significant plastic flow means that the fractured pieces may be fitted together to recreate the original shape. This is not the case with ductile materials. The difference between the two classes can be illustrated by considering what happens when a china teapot (brittle material) and a metal teapot (ductile material) are both dropped on a hard floor. The china pot will probably break but the pieces can be glued together to form the original shape. The metal pot will not break but is likely to be permanently dented.

We now discuss the behaviour inside ductile and brittle materials

Ductile materials show a large amount of plastic deformation. Brittle materials fracture at low strains close to their elastic limit.

Plastic (Ductile) Behaviour of Solids

In an earlier section we discussed a model showing how the forces between atoms or molecules varied with their separation (p. 148). The theoretical breaking stress of ductile solids can be calculated using this model but the result is many times greater than the value obtained in practice.

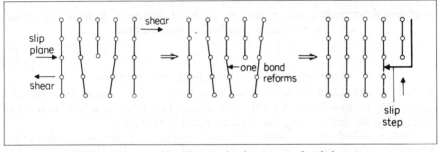

Figure 5.12 *Bond breaking and reforming at edge dislocation*

The reason for the relatively small value of breaking stress is due to a process called *slip*. Slip is due to a movement of dislocations throughout the crystal. Here *one bond at a time* is broken, so the process occurs at a much lower stress than that calculated theoretically, Figure 5.12. With a large number of dislocations in operation we can account for the plastic behaviour of ductile materials at the stresses observed. Large scale slip, requiring a whole plane of atoms to move bodily relative to an adjacent plane, would involve the simultaneous breaking of a very large number of bonds and require very much larger stresses than that obtained in practice.

A slip plane is generally one in, or on, which the atoms are most closely packed. Several such planes, pointing in different directions, may exist, depending on the crystal structure. Slipping preserves the crystal structure and results in a permanent extension of length, which is not recovered when the stress is removed, see Figure 5.13. Slip 'steps' at the surface, often visible to the naked eye, provide an indication of the mechanism involved. Slip steps may be several thousand atoms in height and occur close together to form slip bands.

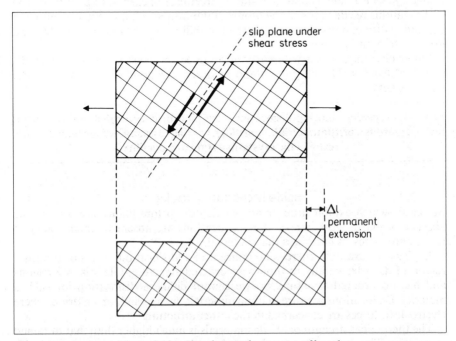

Figure 5.13 *Slip plane and extension of length*

In plastic deformation, *slip* occurs due to the movement of dislocations. Bonds between atoms are broken one at a time.

Work Hardening, Annealing, Iron and Steel

As we have seen, the ductility of metals is due to the existence of dislocations and their ability to move under a local shear stress. Materials with a large number of very mobile dislocations are therefore expected to be ductile.

Under repeated stress, however, dislocations move and intersect, and become entangled. They now *pin* each other and so become immobile (fixed). Dislocations between the pinning points can produce more dislocations, which also become immobilised due to intersections. This explains why copper wire, for example, can be broken by flexing it to-and-fro in the hand. The many dislocations produced by the repeated stress on the wire all become pinned and immobilised. Ductile behaviour is then not possible and the wire fractures in a brittle manner.

This is an example of *work hardening* a metal, a process which increases the strength of the metal but at the expense of ductility because plastic flow is then considerably reduced. *Cold rolling*, when the thickness of a metal is reduced by pressure, usually strengthens a metal by work hardening.

During cold working processes, the crystal structure is plastically deformed and there is lower ductility and toughness (see p. 161). The crystal grain boundaries (p. 158) are particularly affected. *Annealing* helps to restore the metal to its ductile state. In this process the metal is heated to a high temperature (below its melting point) and maintained at this temperature for a length of time. This increases the thermal vibrations of the crystal lattice and relaxes the internal strains. In this way the solid is *recrystallised* and returns to a ductile state.

Pure iron is usually too ductile for use in load-bearing applications such as bridges. Steel is made by adding a small percentage of carbon to pure iron. The carbon atoms reside between the atoms of the iron lattice and are very effective in pinning dislocations and reducing their mobility. So steel has less ductility and greater strength than pure iron.

Other elements, added with carbon, produce specialist steels. For example, stainless steel has 20% chromium and posseses very good corrosion resistance and hardness.

In *work hardening*, moving dislocations are pinned or entangled and the metal may fracture as a brittle material. *Annealing*, heating to a suitable high temperature, restores the crystalline state and ductility.

Brittle Materials, Cracks

Materials which exhibit little or no plastic flow before failure are *brittle*. The absence of plastic flow implies that dislocations are absent or their ability to move under stress is much less than in metals.

In glass, for example, there is no concept of a 'dislocation'. A dislocation is a region of disorder within an otherwise ordered structure. Glass is amorphous and has no ordered structure. So we cannot define a dislocation for such a material. Dislocations are rare in ionic materials such as sodium chloride, where electrostatic forces are concerned in the lattice structure.

The theoretical strength of brittle materials is much higher than that obtained in practice. The explanation for the low breaking stress was due to Griffith. He suggested that microscopic flaws or *cracks* in the surface (or just below) act as *stress concentrators*. Figure 5.14 illustrates the uniformly distributed stress in unflawed materials and the concentration of stress at the tip of the crack in flawed materials.

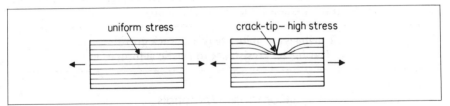

Figure 5.14   *Unflawed solid (uniform stress). Flawed solid with crack (high stress at tip).*

It can be shown that the stress at the crack tip is greater than the nominal uniform applied stress by a factor $k = 2\sqrt{l/r}$ approximately, where l is the crack length and r is the crack tip radius, which is of atomic dimensions. A scratch about 10^{-3} mm deep and a tip radius of 2×10^{-10} m would cause the stress at the tip to be about 140 times the nominal (applied) value. As the crack gets longer, the stress lines become more concentrated at the tip. Crack growth may then suddenly accelerate, leading to the characteristic sudden failure of brittle materials.

Griffith showed by experiment that freshly drawn glass fibres, with no surface flaws or cracks, came very close to their theoretical high strength. The fibres quickly lose their strength, however, as the cooling process introduces microscopic flaws which reduces the strength to normal values.

The sensitivity of brittle materials to cracks is shown by the cutting of glass. A fine line is scratched on the surface with a glass cutter. Slight pressure on either side of the crack causes the glass to fracture cleanly along the line. Most brittle materials are much stronger in compression because the surface cracks will then tend to be closed and unable to spread. Pillars made of cement, for example, are strong in compression. Cracks develop if the cement is in tension and it then breaks.

Cracks in the surface, or just below, produce high stress concentration at the tip. Rapid crack growth produces brittle behaviour (sudden fracture). Brittle materials are stronger in compression than in tension as this prevents cracks spreading.

Toughness and Hardness

The *toughness* of a material is a measure of its ability to resist crack growth. This is not to be confused with 'strength'. Plasticene, for example, is a tough material but not strong. Glass is much stronger than plasticene but not as tough.

Metals will contain surface cracks of microscopic dimensions. In general, however, metals are tough. This is due to the ability of dislocations to move and blunt the crack tip. The stress concentration is then relieved and the crack does not spread.

The *hardness* of a material is a measure of its resistance to plastic deformation. An 'indenter' such as a hardened metal sphere is pressed into the surface of the test material for a certain time and the 'hardness' is calculated by dividing the applied force by the contact area left in the surface by the indentation. A ductile material will produce a large area indentation and have low hardness.

The stress below the indenter is compressive, so it is possible to measure the hardness of brittle materials such as glass without causing brittle fracture.

Most cracks which occur in metals are due to *fatigue*, when the metal undergoes failure after many cycles of normal stress. Aircraft accidents can occur with metal fatigue starting at a rivet hole, for example. Metal *creep* is the gradual growth of plastic strain under a static, not cyclic, load.

Toughness = ability to resist crack growth
Hardness = resistance to plastic deformation

Composite Materials

In engineering design, it is often difficult to meet the mechanical properties needed by using a single material. *Composite materials*, where two materials are used, have wide application. Reinforced concrete, for example, is made by setting the concrete round steel wires or mesh. This improves the tensile (tension) properties of the concrete for use in structures or buildings. Concrete itself is a mixture of cement, sand and small stones and so is an example of a *particle composite*. Figure 5.15 shows pre-stressed concrete.

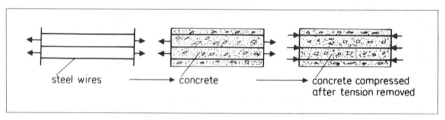

Figure 5.15 *Pre-stressed concrete*

In fibre-reinforced materials, which are used in the plastics industry, long straight fibres are embedded in a tough *matrix*. Polymers and metals have been used for matrix materials. Glass-fibre reinforced plastic (GRP), used as construction materials for many years, have glass fibres embedded in a matrix such as polyester resin. Carbon-fibre reinforced plastic (CFRP) is also used.

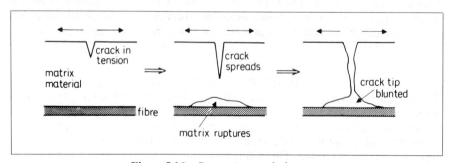

Figure 5.16 *Preventing cracks by matrix*

As we saw previously, *cracks* which spread in a material will weaken it. The principle of the toughening process in fibre-reinforced materials is illustrated in Figure 5.16. The matrix separates from the fibres and the tip of the crack is blunted by it and stopped from spreading. The matrix also helps to transfer the load to the fibres as these are bonded to the matrix surface. The GRP is used for making small boats and canoes, storage tanks and some car bodies. The CFRP is used in the aircraft industry as it has excellent strength/weight ratio property, like the GRP, and is much stiffer than steel.

Car windscreens are also made from composites. Thin sheets of laminated glass, bonded by resin, can prevent pieces of glass flying about dangerously after an accident. On impact, the cracks in the glass spread parallel to the surface but not through the windscreen, owing to the crack-blunting at the glass-resin boundary. Laminated glass is an example of a *layer composite*.

Composite materials improve mechanical properties. Glass-fibre and carbon-fibre reinforced materials prevent cracks spreading, are strong in relation to their weight and may be stiffer than steel.

Polymers, Structure and Mechanical Properties

We now discuss a class of organic materials generally described as *polymers*, which have wide application.

Polymers occur naturally in materials such as rubber, resin, cotton-wool and wood. They are used widely in the plastics industry. In the home, for example, there may be plastic dustbins, washing-up bowls, light fittings and wrapping paper, in addition to nylon socks. Plastics also make good thermal and electrical insulators, have low density, great toughness and resist corrosion. They are easy to mould and cheap to produce, which is a considerable advantage.

Their disadvantages are a low Young modulus and low tensile strength, making them unsuitable for many load-bearing applications. Their mechanical properties depend considerably on their temperature and they also tend to melt at relatively low temperatures accompanied by dangerous fumes.

Structure of Polymers

The basic structure of a polymer can be illustrated by considering polyethylene, better known as *polythene*. By a chemical process called *polymerisation*, the double bonds of a large number of ethylene molecules (C_2H_4) are broken to allow them to form a giant molecule, the polymer. Figure 5.17 illustrates the formation of a polyethylene $(CH_2)_n$ molecule. The basic unit ethylene is called a *monomer* or *mer*.

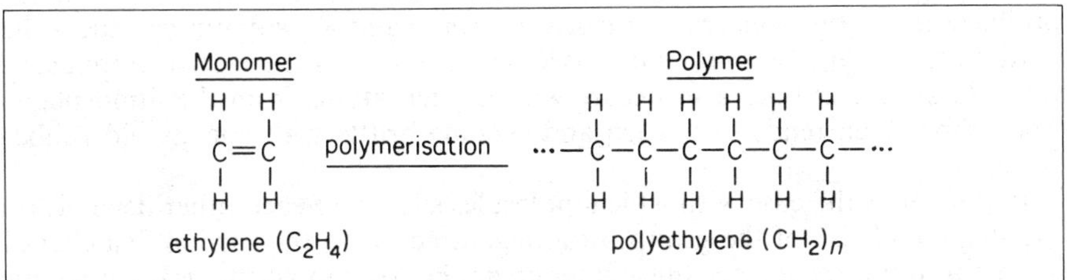

Figure 5.17 *Polymerisation*

The polymer molecule may contain thousands of carbon atoms so that n is very large. A wide range of materials can be made starting with different monomers. The Table illustrates how polyvinyl chloride (PVC), which uses chloride atoms, and polystyrene, which uses benzene rings, are made from their monomers, see Figure 5.18.

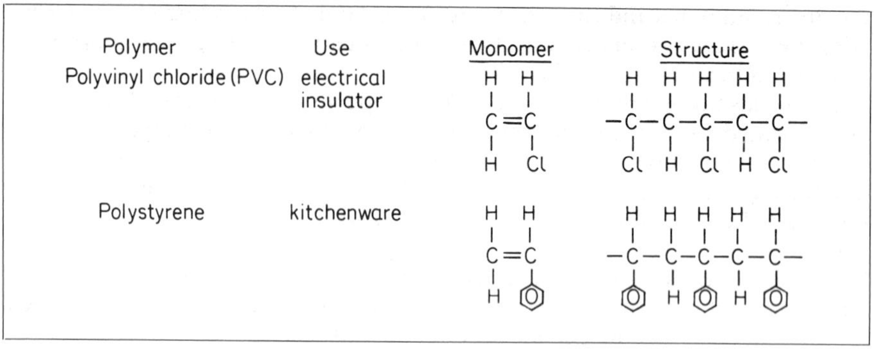

Polymer	Use	Monomer	Structure
Polyvinyl chloride (PVC)	electrical insulator	H H | | C=C | | H Cl	H H H H H | | | | | —C—C—C—C—C— | | | | | Cl H Cl H Cl
Polystyrene	kitchenware	H H | | C=C | | H ◎	H H H H H | | | | | —C—C—C—C—C— | | | | | ◎ H ◎ H ◎

Figure 5.18   *Polymer molecule from monomer*

***Polymers*, widely used in the plastics industry, are made chemically with monomers. They consist of very long chains of carbon atoms bonded to hydrogen and other atoms.**

Branching and Cross-linking of Polymer Molecules

During chemical manufacture, polymer molecules may form *branches* along them or become *cross-linked* like the rungs of a rope ladder, Figure 5.19. Polyethylene molecules, for example, usually contain more than a thousand carbon atoms and may have about seventy branches. As branching or cross-linking develops, freedom of movement of the molecules becomes less because they entangle.

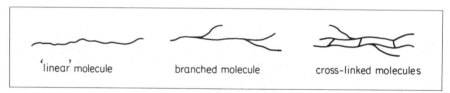

'linear' molecule   branched molecule   cross-linked molecules

Figure 5.19   *Linear, branched and cross-linked molecules*

Cross-linking is used in the manufacture of some materials. Natural rubber molecules, for example, can be cross-linked with sulphur atoms. 1% sulphur produces a soft but solid rubber called *vulcanite* and 4% sulphur produces the hard material called *ebonite*, both used as insulators in the electrical industry. Cross-linking in rubber also occurs with oxygen atoms from the atmosphere. The rubber then tends to harden and become brittle with age, as old rubber bands show.

In polymers, the ease with which molecules slide over each other depends on the shape and size of the groups of atoms attached to the carbon 'backbone'. With polystyrene, the mechanical interference between molecules is large, so this material is inflexible and glasslike. In PVC, the chlorine atoms produce interference, so PVC is much less flexible than polythene. PVC, however, can be made flexible for use as electrical insulators round copper wires by adding a 'plasticiser', which acts as an internal lubricant separating the PVC molecules.

Polymer molecules can be linear, branched or cross-linked. Cross-linking with other atoms is made in the manufacture of some materials.

Thermosetting and Thermoplastic Polymers

Polymers which form cross-links between their long chains of molecules in manufacture are called *thermosetting polymers*. At ordinary temperatures the cross-links keep these polymers solid and rigid. When heated, however, the agitation of the molecules destroys the cross-links and chemical decomposition occurs. So thermosetting polymers can not be re-moulded to a new shape by heating. Bakelite, melamine and epoxide resin are examples of thermosets.

Polymers with few cross-links are called *thermoplastic polymers*. When heated, these polymers become softer and can be re-moulded, unlike thermosetting polymers. Polythene, polystyrene, polyvinyl chloride (PVC) and nylon are examples of thermoplastics. The mechanical properties of these materials are

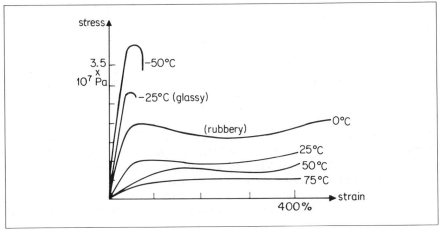

Figure 5.20   *Stress-strain graphs (diagrammatic) for Polythene at different temperatures*

more sensitive to temperature change. Figure 5.20 shows the stress-strain graph of polythene at different temperatures. At lower temperatures polythene has a 'glassy' behaviour (stiff and fracturing at low strain); at higher temperatures it has a 'rubbery' behaviour (less stiff and able to stretch much more).

Thermosetting **polymers have cross-links between chains of molecules. They are rigid and can not be re-moulded by heating.**
Thermoplastic **polymers have no cross-links. They become soft on reheating and can be re-moulded.**

Comparison of Mechanical Properties

The Table shows some of the mechanical properties of a wide range of materials, including plastics. Compared with metals and brittle materials such as glass, the plastics have much lower values of Young modulus (less stiff) and very large stretching or elongation. The elongation is a measure of the permanent extension of a length of the specimen after failure.

		Young modulus 10^9 Pa	Tensile strength 10^6 Pa	Elongation %
Metals	Steel	200	250	35
	Copper	120	150	45
	Aluminium	70	60–120	45
Woods	Oak (parallel to grain)	5–9	21	
Brittle	Glass	71	100 (about)	0 (about)
	Concrete	20–40	4	
Thermosets	Bakelite	6–8	50	0·6
	Melamine	9	70	
	Epoxide resin	1–5	30–80	
Thermoplastics	Perspex	3·4	55–70	2–10
	PVC (rigid)	2·5	60	2
	Polystyrene	3·5	40	2·5
	Nylon		70	60–300
Rubber	Natural	1(25% elongn)	32	850

In the majority of non-polymeric solids, the elastic strains are very small. Larger stresses result in either rapid brittle fracture (glass, for example) or plastic deformation (metals) which rarely produce elongations higher than 50%.

In plastics, the low Young modulus and enormous elongations occur because the long irregular molecules can uncoil and straighten out under tensile stresses. This process can be done without straining the individual bonds within the

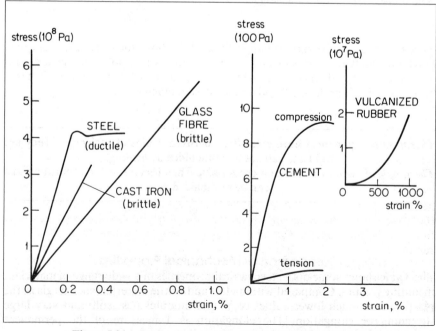

Figure 5.21 *Stress-strain graphs of some industrial materials*

molecules and so it can take place at much lower stresses than metals, for example. In general, then, due to uncoiling of molecules, polymeric materials do not obey Hooke's law. The molecules in a cross-linked thermoset are much more difficult to uncoil than those in a thermoplastic. For example, bakelite (a thermoset) is much stiffer and shows less elongation than polythene (a thermoplastic). Figure 5.21 shows roughly stress-strain curves for various materials used in engineering.

Metals and glass have high Young modulus and low elastic strain. Plastics have lower Young modulus (less stiff), large elongation and do not obey Hooke's law like metals do.

Rubber, Hysteresis

Unlike other materials, rubber shows an increase in Young modulus when its temperature rises. The increase in temperature produces more agitation of the molecules and increases their tendency to coil up. It is then more difficult to uncoil, or produce more order in the molecules, by tensile forces. So the Young modulus increases.

This also explains why rubber *contracts* when heated. The molecules become more coiled and the rubber shortens.

Figure 5.22 shows the stress-strain curve for a specimen of rubber when it is first loaded within the elastic limit, OAB, and then unloaded, BCA. BCA does

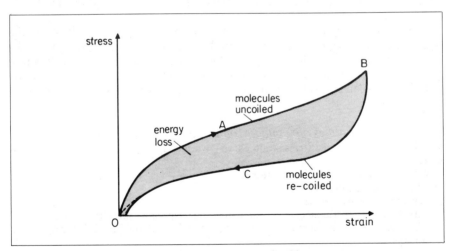

Figure 5.22 *Hysteresis of Rubber*

not coincide with OAB and this is called elastic *hysteresis*. It means that, under the same stress, rubber molecules do not re-coil in the same way as they were first coiled. There may be a small permanent elongation when the stress is finally removed.

Metals do not show significant hysteresis within the elastic limit—under stress, the strain is immediately the same whether loading or unloading takes place.

The area under OAB represents the work done per unit volume in stretching the rubber. The area under BCO represents the energy given up by the rubber on contracting. So the shaded area or *hysteresis loop* represents the energy 'lost' as heat during the loading-unloading cycle.

Effects of Hysteresis

The hysteresis of rubber enables it to convert mechanical energy to heat. It is therefore used as shock absorber material, for example.

A rolling car tyre is taken through many cycles of loading and unloading during a journey. A large area hysteresis loop for such a rubber is not desirable because the heat produced may lead to dangerously high tyre temperatures. It would also increase petrol consumption due to conversion of mechanical energy to heat. In practice, therefore, tyres are made with synthetic rubbers, which have small-area hysteresis loops. Materials with small hysteresis effects are called *resilient*.

Conventional plastics, notably thermoplastics, show hysteresis effects and are about ten times less resilient than steel, for example. Plastic gear wheels are less noisy than metal ones as they are able to dissipate vibrational energy more effectively as heat.

Rubber molecules are coiled. Under increased tension the molecules uncoil and high strain (800%) can be produced. On removing the load a hysteresis loop is obtained. The 'lost' energy is converted to heat and is proportional to the area of the hysteresis loop. *Resilient* materials have low hysteresis.

Wood

Wood contains a natural polymer based on the cellulose molecule. The grain of the wood is the line of the cellulose fibres. Wood is a composite material—the fibres are bonded together in a *lignin* matrix which consists of carbohydrates and is non-polymer.

Wood has strength and stiffness parallel to the grain but is weak across the grain. It is also a much weaker material in compression than in tension. In *plywood*, the wood is made stronger by glueing together alternately sheets whose grains go in perpendicular directions. The plywood is then equally strong in the two directions and so is less likely to warp than the single wood with grains in one direction.

Examples on Young Modulus and Solid Materials

1 Figure 5.23 shows the stress-strain curve for a metal alloy. Fracture occurred at an extension of 15%.

With reference to the diagram, estimate
(a) the Young modulus of the alloy,
(b) the ultimate tensile stress,
(c) the elongation (permanent extension remaining in specimen after fracture),
(d) the extension and breaking stress at fracture had the material been brittle rather than ductile.

(a) The Young modulus is the gradient of the linear (elastic) part of the stress-strain curve. So

$$\text{Young modulus, } E = \frac{\text{stress}}{\text{strain}} = \frac{300 \times 10^6 \, N\,m^{-2}}{0.005}$$

$$= 6 \times 10^{10} \, N\,m^{-2} \text{ (or Pa)}$$

(At stress $300 \times 10^6 \, N\,m^{-2}$, extension $= 0.5\%$, so strain $= 0.005$)
(b) Ultimate tensile stress = maximum stress material can withstand without fracture. So

$$\text{Ultimate tensile stress} = 380 \times 10^6 \, N\,m^{-2}$$

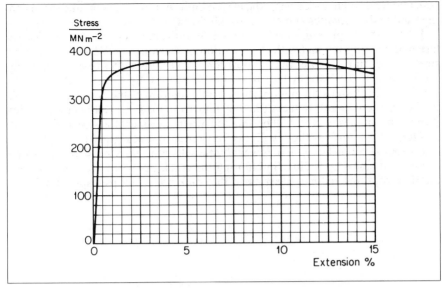

Figure 5.23 *Stress-strain curve*

(c) At fracture, extension is 15% but on fracturing, the *elastic* deformation, 0·5%, is recovered. So

$$\text{elongation} = 14{\cdot}5\%$$

(d) If the metal were brittle it would have fractured at, or just after, the end of the elastic deformation region, so no plastic deformation would have occurred. So

$$\text{extension at fracture} = 0{\cdot}5\% \text{ (brittle)}$$

and $$\text{breaking stress} = 300\text{--}320 \, \text{MN m}^{-2}$$

2   Figure 5.24 (i) shows a cross-section through a reinforced concrete beam. It is supported at its ends and vertically loaded in the middle.

From the properties of concrete and of steel, explain the purpose of the steel reinforcement rods and why the steel rods are placed as shown.

In this loading situation, the top and bottom faces of the beam are in compression and tension respectively, Figure 5.24 (ii). Brittle materials such as

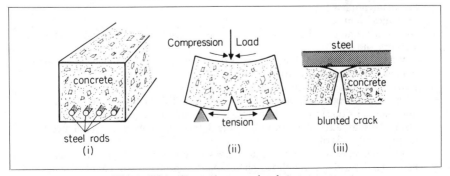

Figure 5.24 *Example on steel rods in concrete*

concrete tend to fail at, or near, the elastic limit. A surface crack, under tension, then suddenly elongates and runs through the material.

To stop this happening, steel rods are laid in the concrete close to the *bottom* face and running *parallel* with the direction of tension. The Young modulus of steel is many times greater than that of concrete, that is, stiffer than concrete. So the tension in the bottom face is supported largely by the steel than by the concrete. This lessens the possibility of brittle fracture of the concrete, increases the beam strength and reduces the 'sag' under the load. Also, any cracks which do elongate in the concrete will be 'blunted' when they meet the steel rods, Figure 5.24 (iii). This improves the toughness (resistance to crack growth) of the beam. The stress at a crack tip increases with crack length and also increases with a decrease in crack-tip radius. So blunting the crack tip, when the steel-concrete interface ruptures, reduces the stress at the crack tip and stops it spreading.

_____ Exercises 5C _____

1 The measured strengths of brittle and ductile solids are very much smaller than calculations based on the forces between individual atoms in the solid.
   Explain why this is the case.
2 The breaking strain of rubber may be as high as 800%, yet most metals break at strains which rarely are greater than 50%. With reference to the molecular or atomic structure of these materials, account for the difference in the breaking strain.
3 What is meant by the following terms:
   (a) stiffness,
   (b) strength,
   (c) toughness, of a solid material.
   Explain briefly how these properties might be affected if the material were fibre reinforced.
4 In terms of their structure and likely physical properties, distinguish between a *thermoplastic* and a *thermoset*. Name one example of each type of plastic.
5 Rubber shows large hysteresis. What is meant by 'hysteresis'?
   Give one application of the use of rubber where a large hysteresis would be
   (a) an advantage and
   (b) a disadvantage.
6 Aluminium and glass have almost the same values of the Young modulus, tensile strength and density.
   Why is glass, which is much cheaper to produce, not used in place of aluminium in load-bearing applications?
7 Explain the difference between *elastic deformation* and *plastic deformation*.
   Plastic deformation is the result of a process called *slipping*. What do you understand by this term and how does it take place in a typical metal.
8 Using the same axes, sketch approximate stress-strain curves for (1) a metal, (2) glass and, (3) rubber.
   With reference to the stress-strain curves, describe and explain what would happen if three identical hollow spheres made respectively from a metal, glass and rubber were dropped from a large height onto a hard surface.

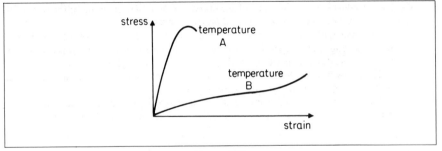

Figure 5.25

9   Figure 5.25 shows the stress-strain curves for a thermoplastic at two different
temperatures A and B. Which is the higher temperature? At which temperature
would the plastic behaviour be best described as
(a) glassy,
(b) rubbery?
    Explain why this material shows a transition from glassy to rubbery behaviour as
the temperature changes.

10 'Metal fatigue is thought to be caused by a slip mechanism in which the direction of
slip changes during the *stress cycle*. Point defects (e.g. an atom missing from the
lattice or an extra atom situated at an off-lattice point) interfere with the normal
movement of *edge-dislocations*. Normal slip is therefore inhibited over large parts of
a grain, and the slip becomes localised.'
(a) Explain the meaning of the term *metal fatigue*, making reference to the idea of a
    stress cycle.
(b) With the help of sketches, explain what is meant by an *edge-dislocation* and
    describe how it is able to move through a crystal. Suggest (i) how the presence of
    point defects might inhibit the mechanism of slip, and (ii) how the localised
    inhibition of slip might lead to fracture.
(c) Fatigue cracks often occur at points of weakness such as sharp corners. Describe
    briefly how an engineer might set about studying the stress in the region of
    points of weakness.
(d) Suggest how placing a metal under compression might increase the fatigue
    strength of the metal. (*L*.)

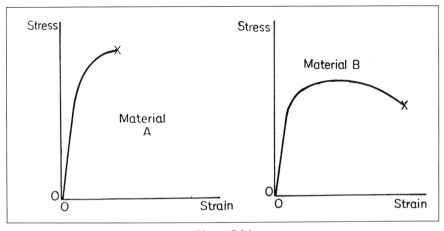

Figure 5.26

11 (a) Distinguish between (i) polycrystalline, and (ii) single crystal specimens of
       metals. Illustrate your answer with appropriate sketches.

(b) By considering the behaviour of a metal during elongation, explain the terms *strain*, *tensile stress*, *yield stress* and *breaking* stress.

Test pieces of two metal alloys of identical size and shape were subjected to tensile strength tests up to fracture (X), and the stress-strain diagrams (drawn to the same scale) which were obtained are shown in Figure 5.26.

Explain which of the two materials (i) has the greater tensile strength, (ii) is the more ductile, (iii) exhibits greater toughness.

(c) What is a composite material? Distinguish between layer and particle composite materials. Give an example of each type and state its engineering application. (*L.*)

Part 2
Electricity

6
Electrostatics

In this chapter we begin with an account of the more important phenomena about static (stationary) electric charges. We then show that charges in an electrostatic field are analogous to masses in a gravitational field—they have forces acting on them and have electric potential energy. The ideas here are widely used in many branches of electricity, for example, in solid state physics which deals with diodes and transistors, in the electron microscope, and in the theory of the atom.

General Phenomena

If a rod of ebonite is rubbed with fur, or a fountain-pen with a coat-sleeve, it gains the power to attract light bodies, such as pieces of paper or tin-foil or a piece of cork. The discovery that rubbed amber could attract silk was mentioned by THALES (640–548 B.C.). The Greek work for amber is *elektron*, and a body made attractive by rubbing is said to be 'electrified' or *charged*. This branch of Electricity, the earliest discovered, is called *Electrostatics*.

Conductors and Insulators, Positive and Negative Charges

A metal rod can be charged by rubbing with fur or silk, but only if it were held in a handle of glass or amber. The rod could not be charged if it were held directly in the hand. This is because electric charges can move along the metal and pass through the human body to the earth. The human body, metals and water are examples of *conductors*. Glass and amber are examples of *insulators*.

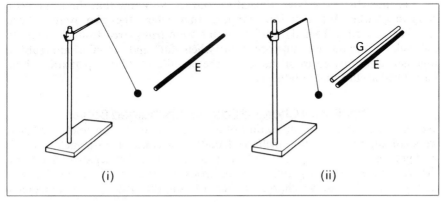

Figure 6.1 *Demonstrating that an electrified glass or acetate rod tends to oppose effect of electrified ebonite or polythene rod*

A suspended piece of cork is attracted to an electrified ebonite rod E, Figure 6.1 (i). But when we bring an electrified glass rod G towards the ebonite rod, the cork falls away, Figure 6.1 (ii). So the charge on the glass rod *opposes* the effect of the charge on the ebonite rod.

Benjamin Franklin, a pioneer of electrostatics, gave the name of 'positive electricity' to the charge on a glass rod rubbed with silk, and 'negative electricity' to that on an ebonite rod rubbed with fur. Rubbed by a duster, a cellulose acetate rod obtains a positive charge and a polythene rod obtains a negative charge.

Experiment shows that two positive, or two negative, charges repel each other but a positive and a negative charge attract each other. So a fundamental law of electrostatics is:

Like (similar) charges repel. Unlike charges attract.

Electrons and Electrostatics

Towards the end of the nineteenth century Sir J. J. Thomson discovered the existence of the *electron* (p. 757). This is a particle of very low mass—it is about 1/1840th of the mass of the hydrogen atom—and experiments show that it carries a tiny quantity of *negative* charge. Later experiments showed that electrons are present in all atoms.

The detailed structure of atoms is complicated, but generally, electrons exist round a very tiny core or nucleus carrying *positive* charge. Normally, atoms are electrically neutral, that is, there is no surplus of charge on them. Consequently the total negative charge on the electrons is equal to the positive charge on the nucleus. In insulators, all the electrons appear to be firmly 'bound' to the nucleus under the attraction of the unlike charges. In metals, however, some of the electrons appear to be relatively 'free'. These electrons play an important part in electrical phenomena concerning metals as we shall see.

Charge Transfer by Friction

The theory of electrons (negatively charged particles) gives a simple explanation of charging by friction. If the silk on which a glass rod has been rubbed is brought near to a charged and suspended ebonite rod it repels it; the silk must therefore have a negative charge. We know that the glass has a positive charge. We therefore suppose that when the two were rubbed together, electrons from the surface atoms were *transferred* from the glass to the silk. Likewise we suppose that when fur and ebonite are rubbed together, electrons go from the fur to the ebonite. Similar explanations hold for rubbed acetate and polythene.

Attraction of Charged Body for Uncharged Bodies

We can now explain the attraction of a charged body for an uncharged one; we shall suppose that the uncharged body is a conductor—a metal. If it is brought near to a charged polythene rod, say, then the negative charge on the rod repels the negative free electrons in the metal to its far end (Figure 6.2). A positive charge is then left on the near end of the metal.

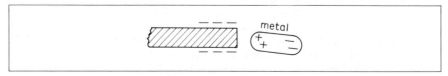

Figure 6.2 *Attraction by charged body*

Since this is nearer than the negative charge on the far end, it is attracted more strongly than the negative charge is repelled. On the whole, therefore, the metal is attracted. If the uncharged body is not a conductor, the mechanism by which it is attracted is more complicated; we shall leave this to a later chapter.

Electrostatics Today

The discovery of the electron led to a considerable increase in the practical importance of electrostatics. In cathode-ray tubes and in electron micro-scopes, for example, electrons are moved by electrostatic forces. The problems of preventing sparks and the breakdown of insulators are essentially electrostatic. These problems occur in high voltage electrical engineering. Later, we shall also describe an electrostatic generator used to provide a million volts or more for X-ray work and nuclear bombardment. Such generators work on principles of electrostatics discovered over a hundred years ago.

Gold-leaf Electroscope

One of the earliest instruments used for testing positive and negative charges consisted of a metal rod A to which gold leaves L were attached (Figure 6.3).

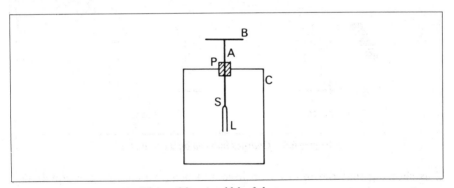

Figure 6.3   *A gold-leaf electroscope*

The rod was fitted with a circular disc or cap B, and was insulated with a plug P from a metal case C which screened L from outside influences other than those brought near to B.

When B is touched by a polythene rod rubbed with a duster, some

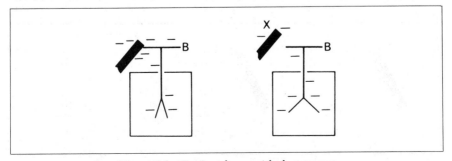

Figure 6.4   *Testing charge with electroscope*

negative charge on the rod passes to the cap and L; and since like charges repel, the leaves open or diverge, Figure 6.4(i). If an unknown charge X is now brought near to B, an increased divergence implies that X is negative, Figure 6.4(ii). A positive charge is tested in a similar way; the electroscope is first given a positive charge and an increased divergence indicates a positive charge.

Induction

We shall now show that it is possible to obtain charges, called *induced charges*, without any contact with another charge. An experiment on *electrostatic induction* as the phenomenon is called, is shown in Figure 6.5(i).

Two insulated metal spheres A, B are arranged so that they touch one another and a negatively charged polythene rod C is brought near to A. The spheres are now separated, and then the rod is taken away. Tests with a charged piece of cork now show that A has a positive charge and B a negative charge, Figure 6.5(ii). If the spheres are placed together so that they touch, it is found that they now have no effect on charged cork held near. Their charges must therefore have neutralised each other completely,

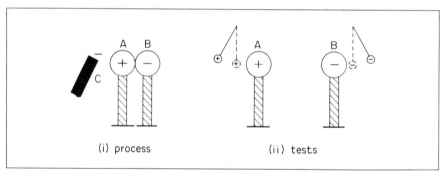

(i) process   (ii) tests

Figure 6.5 *Charges induced on a conductor*

thus showing that the induced positive and negative charges are equal. This is explained by the movement of electrons from A to B when the rod C is brought near, Figure 6.5(i). B has then a negative charge and A an equal positive charge.

Charging by Induction

Figure 6.6 shows how a conductor can be given a permanent charge by induction, without dividing it in two. We first bring a charged polythene rod

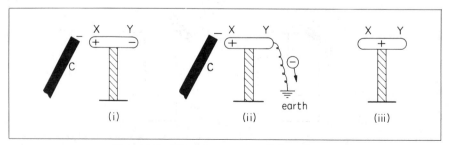

(i)   (ii)   (iii)

Figure 6.6 *Charging permanently by induction*

C, say, near to the conductor XY, (i); next we connect the conductor to earth by touching it momentarily, (ii). Finally we remove the polythene. We then find that the conductor is left with a positive charge, (iii). If we use a charged acetate rod, we find that the conductor is left with a negative charge. The charge left, called the induced charge, has always the *opposite* sign to the inducing charge.

This phenomenon of induction can again be explained by the movement of electrons. In Figure 6.6(i), the inducing charge C repels electrons to Y, leaving an equal positive charge at X as shown. When we touch the conductor XY, electrons are repelled from it to earth, as shown in Figure 6.6(ii), and a positive charge is left on the conductor. If the inducing charge is positive, then the electrons are attracted up *from the earth to the conductor*, which then becomes negatively charged.

The Action of Points, Van de Graaff Generator

Sometimes in experiments with an electroscope connected to other apparatus by a wire, the leaves of the electroscope gradually collapse, as though its charge were leaking away. This behaviour can often be traced to a sharp point on the wire—if the point is blunted, the leakage stops. Charge leaks away from a sharp point through the sir, being carried by molecules away from the point. This is explained later (p. 194).

Points are used to collect the charges produced in *electrostatic generators*. These are machines for continuously separating charges by induction, and thus building up very great charges and potential differences. Figure 6.7 is a simplified diagram of one such machine, due to Van de Graaff. A hollow metal sphere S is supported on an insulating tube T, many metres high. A silk

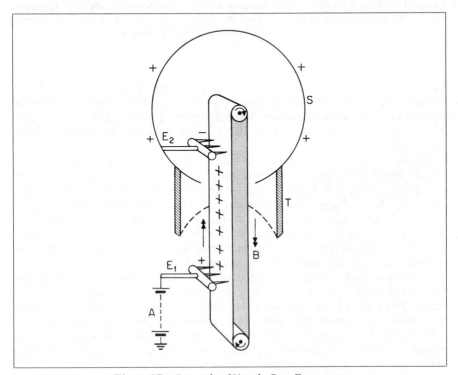

Figure 6.7   *Principle of Van de Graaff generator*

Figure 6.8 *Van de Graaff electrostatic generator at Aldermaston, England. The dome is the high-voltage terminal. The insulated rings are equipotentials, and provide a uniform potential gradient down the column. Beams of protons or deuterons, produced in the dome, are accelerated down the column to bombard different materials at the bottom, thereby producing nuclear reactions which can be studied*

belt B runs over the pulleys shown, of which the lower is driven by an electric motor. Near the bottom and top of its run, the belt passes close to the electrodes E, which are sharply pointed combs, pointing towards the belt. The electrode E_1 is made about 10 000 volts positive with respect to the earth by a battery A.

As shown later, the high electric field at the points ionises the air there; and so positive charges are repelled on to the belt, which carries it up into the sphere. There it induces a negative charge on the points of electrode E_2 and a positive charge on the sphere to which the blunt end of E_2 is connected. The high electric field at the points ionises the air there, and negative charges, repelled to the belt, discharges the belt before it passes over the pulley. In this way the sphere gradually charges up positively, until its potential is about a million volts relative to the earth.

Large machines of this type are used with high-voltage X-ray tubes, and for atom-splitting experiments. They have more elaborate electrode systems, stand about 15 m high, and have 4 m spheres. They can produce potential differences up to 5 000 000 volts and currents of about 50 microamperes. The *electrical energy* which they deliver comes from the work done by the motor in drawing the positively charged belt towards the positively charged sphere, which repels it.

In all types of high-voltage equipment sharp corners and edges must be avoided, except where points are deliberately used as electrodes. Otherwise, electric discharges called 'corona' discharges may occur at the sharp places.

Ice-pail Experiment

A famous experiment on electrostatic induction was made by FARADAY in 1843. In it he used the ice-pail from which it takes its name; but it was a modest pail, 27 cm high—not a bucket.

He stood the pail on an insulator, and connected it to a gold-leaf electroscope, as in Figure 6.9 (i). He next held a metal ball on the end of a long silk thread, and charged it positively. Then he lowered the ball into the pail, without letting it touch the sides or bottom, Figure 6.9 (ii). A positive charge was induced on the outside of the pail and the leaves, and made the leaves diverge. Once the ball was well inside the pail, Faraday found that the divergence of the leaves did not change when he moved the ball about—

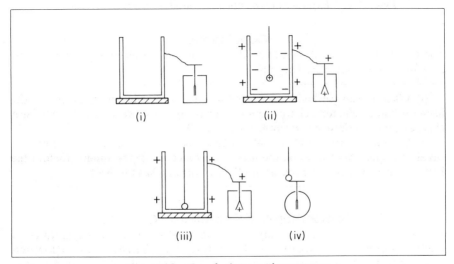

Figure 6.9 *Faraday's ice-pail experiment*

nearer to or farther from the walls or the bottom. This showed that the amount of the induced positive charge did not depend on the position of the ball, once it was well inside the pail.

Faraday then allowed the ball to touch the pail, and noticed that the leaves of the electroscope still did not move (Figure 6.9 (iii)). When the ball touched the pail, therefore, no charge was given to, or taken from, the outside of the pail. Faraday next lifted the ball out of the pail, and tested it for charge with another electroscope. He found that the ball had *no charge whatever*, Figure 6.9 (iv). The induced negative charge on the inside of the pail must therefore have been equal in magnitude to the original positive charge on the ball.

Faraday's experiment does not give these simple results unless the pail—or whatever is used in place of it—very nearly surrounds the charged ball, Figure 6.10 (i). If, for example, the ball is allowed to touch the pail before it is well inside, as in Figure 6.10 (iii), then it does not lose all its charge.

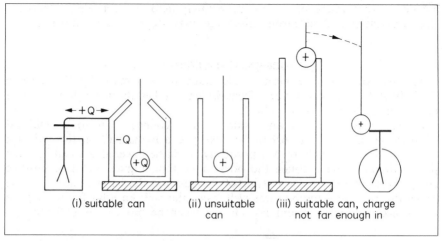

Figure 6.10 *Experimental conditions in Faraday's ice-pail experiment*

Conclusions

The conclusions to be drawn from the experiment therefore apply, strictly, to a *hollow closed conductor*. They are:

(i) **When a charged body is enclosed in a hollow conductor it induces on the inside of that conductor a charge equal but opposite to its own; and on the outside a charge equal and similar to its own, Figure 6.9 (i).**

(ii) **The *total* charge inside a hollow conductor is always zero: either there are equal and opposite charges on the inside walls and within the volume (before the ball touches), or there is no charge at all (after the ball has touched).**

Comparison and Collection of Charges

Faraday's ice-pail experiment gives us a method of comparing quantities of electric charges. The experiment shows that if a charged body is lowered well inside a tall narrow can, then it gives to the outside of the can a charge equal to its own. If the can is connected to the cap of an electroscope, the divergence of the leaves is a measure of the charge on the body. Thus we can compare the magnitudes of charges, without removing them from the bodies which carry them. We merely lower those bodies, in turn, into a tall insulated can, connected to an electroscope.

Sometimes we may wish to discharge a conductor completely, without letting its charge run to earth. We can do this by letting the conductor touch the bottom of a tall can on an insulating stand. The *whole* of the body's charge is then transferred to the outside of the can.

Charges Produced by Separation; Lines of Force

The ice-pail experiment suggests that a positive electric charge, for example, is always accompanied by an equal negative charge. Faraday repeated his experiment with a nest of hollow conductors, insulated from one another, and showed that equal and opposite charges were induced on the inner and outer walls of each (Figure 6.11).

Faraday also showed that equal and opposite charges are produced when a body is electrified by rubbing. He fitted an ebonite rod with a fur cap,

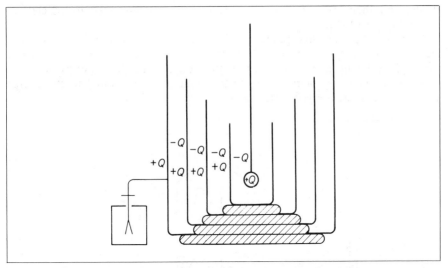

Figure 6.11 *Extension of ice-pail experiment on induced charges*

which he rotated by a silk thread or string wrapped round it and then compared the charges produced with an ice-pail and electroscope.

The idea that charges always occur in equal opposite pairs affects our drawing of lines of force diagrams. Lines of force radiate outwards from a positive charge, and inwards to a negative one. From any positive charge, therefore, we draw lines of force ending on an equal negative charge, as illustrated in Figure 6.12.

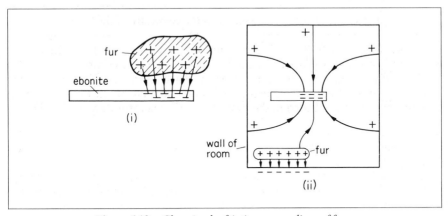

Figure 6.12 *Charging by friction—some lines of force*

Distribution of Charge; Surface Density

By using a can connected to an electroscope we can find how electricity is distributed over a charged conductor of any form—pear-shaped, for example.

We take a number of small leaves of tin-foil, all of the same area, but differently shaped to fit closely over the different parts of the conductor, and mounted on polythene handles, Figure 6.13 (i). These are called *proof-planes*. We first charge the body, press a proof-plane against the part which it fits, and then lower the proof-plane into a can connected to an electro-scope, Figure 6.13 (ii).

After noting the divergence of the leaves we discharge the can and electroscope by touching one of them, and repeat the observation with a proof-plane fitting a different part of the body. Since the proof-planes have equal areas, each of them carries away a charge proportional to the charge per unit area of the body, over the region which it touched.

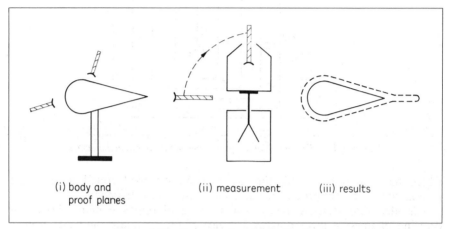

Figure 6.13 *Investigating charge distribution*

The charge per unit area over a region of the body is called the *surface-density* of the charge in that region. We find that

---

the surface-density increases with the curvature of the body,

---

as shown in Figure 6.13 (iii). The distance of the dotted line from the outline of the body is roughly proportional to the surface-density of charge.

Generally, a charged conductor with a sharp point (such as a lightning conductor) has a high surface density of charge at that point. For this reason pointed conductors are used in the Van de Graaff generator described earlier.

The Electrostatic Field

Law of Force Between Two Charges

The magnitude of the force between two electrically charged bodies was studied by COULOMB in 1875. He showed that, if the bodies were small compared with the distance between them, then the force F was inversely proportional to the square of the distance r,

$$F \propto \frac{1}{r^2} \qquad . \qquad . \qquad . \qquad . \qquad . \qquad (1)$$

This result is known as the *inverse square law*, or Coulomb's law.

Fundamental Law of Force

The SI unit of charge is the *coulomb* (C.). The *ampere* (A), the unit of current, is defined later (p. 338). The coulomb is defined as that quantity of charge which passes a section of a conductor in one second when the current flowing is one ampère.

By measuring the force F between two charges when their respective magnitudes Q and Q' are varied, we find that F is proportional to the product QQ'.

Together with the inverse-square law in (1), we can therefore write

$$F = k\frac{QQ'}{r^2} \qquad . \qquad . \qquad . \qquad . \qquad (2)$$

where k is a constant. With practical units for charge and r, k is written as $1/4\pi\varepsilon_0$, where ε_0 is a constant called the *permittivity* of free space if we suppose the charges are situated in a vacuum. So

$$F = \frac{1}{4\pi\varepsilon_0} \frac{QQ'}{r^2} \qquad . \qquad . \qquad . \qquad . \qquad (3)$$

In this expression, F is measured in newton (N), Q in coulomb (C) and r in metre (m). Now, from (3),

$$\varepsilon_0 = \frac{QQ'}{4\pi F r^2}$$

Hence the units of ε_0 are coulomb^2 newton^{-1} metre^{-2} ($C^2\,N^{-1}\,m^{-2}$). Another unit of ε_0, more widely used, is *farad metre*$^{-1}$ (F m^{-1}). See p. 219.

We shall see later that ε_0 has the numerical value of $8\cdot854 \times 10^{-12}$, and $1/4\pi\varepsilon_0$ then has the value 9×10^9 approximately.

So in free space we can write (3) as approximately

$$F = 9 \times 10^9 \frac{QQ'}{r^2} \qquad . \qquad . \qquad . \qquad . \qquad (4)$$

which is useful to simplify calculations as we shall see.

Permittivity, Relative Permittivity

So far we have considered charges in a vacuum. If charges are situated in other media such as water, then the force between the charges is *reduced*. Equation (3) is true only in a vacuum. In general, we write

$$F = \frac{1}{4\pi\varepsilon}\frac{QQ'}{r^2} \qquad . \qquad . \qquad . \qquad . \qquad (1)$$

where ε is the *permittivity* of the medium. The permittivity of air at normal pressure is only about 1·005 times that, ε_0, of a vacuum. For most purposes, therefore we may assume that value of ε_0 for the permittivity of air. The permittivity of water is about eighty times that of a vacuum. Thus the force between charges situated in water is eighty times less than if they were situated the same distance apart in a vacuum.

For this reason common salt (sodium chloride) dissolves in water. The electrostatic forces of attraction between the positive sodium ions and the negative chlorine ions, which keep the solid crystal structure in equilibrium, are reduced considerably by the water and the solid structure collapses.

The *relative permittivity*, ε_r, of a medium is the ratio of its permittivity ε to that of a vacuum, ε_0. So

$$\varepsilon_r = \varepsilon/\varepsilon_0$$

Although ε and ε_0 have dimensions, ε_r is a number and has no dimensions.

Examples on Force Between Charges

Figure 6.14 shows three small charges A, B and P in a line. The charge at A is positive, that at B is negative and that at P is positive. The values are those shown.
(a) Calculate the force on the charge at P due to A and B.
(b) At what point X on the line AB could there be *no* force on the charge P due to A and B if P were placed there?

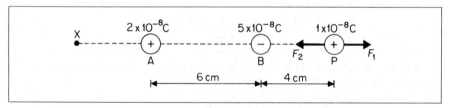

Figure 6.14 *Force on charges*

(a) The distance from A to P is 10 cm or 0·1 cm. So charge at A *repels* charge at P with a force F_1 given by

$$F_1 = 9 \times 10^9 \frac{Q_1 Q_2}{r^2}$$

$$= \frac{9 \times 10^9 \times 2 \times 10^{-8} \times 1 \times 10^{-8}}{0·1^2}$$

$$= 1·8 \times 10^{-4}\,\text{N}$$

This distance from B to P is 4 cm or 4×10^{-2} m. So charge at B *attracts* charge at P with a force F_2 given by

$$F_2 = 9 \times 10^9 \frac{Q_1 Q_2}{r^2}$$

$$= \frac{9 \times 10^9 \times 5 \times 10^{-8} \times 1 \times 10^{-8}}{(4 \times 10^{-2})^2}$$

$$= 2 \cdot 8 \times 10^{-3} \, \text{N}$$

So resultant force towards B

$$= F_2 - F_1 = 2 \cdot 8 \times 10^{-3} - 1 \cdot 8 \times 10^{-4}$$

$$= 2 \cdot 8 \times 10^{-3} - 0 \cdot 18 \times 10^{-3} = 2 \cdot 62 \times 10^{-3} \, \text{N}$$

(b) If the charge at P were taken to a point X to the *left* of A on the line AB, there would be no force on the charge.

In this case the smaller charge at P would repel the positive charge at X and the negative charge at B would attract the positive charge at X. Although the charge at A is smaller than the charge at B, it is *nearer* the charge at X. So at some point such as X the two forces would be equal and opposite and cancel each other.

2   In Figure 6.15, two small equal charges $2 \times 10^{-8} \text{C}$ are placed at A and B, one positive and the other negative. AB is 6 cm.

Find the force on a charge $+1 \times 10^{-8} \text{C}$ placed at P, where P is 4 cm from the line AB along the perpendicular bisector XP.

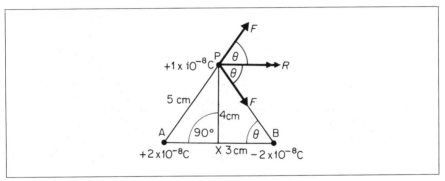

Figure 6.15   *Resultant force on charge*

From triangle APX, using Pythagoras, $\text{AP} = \sqrt{4^2 + 3^2} = \sqrt{25} = 5 \text{cm} = 5 \times 10^{-2} \text{m}$. The charge at A repels the charge at P with a force F given by

$$F = \frac{9 \times 10^9 \times 2 \times 10^{-8} \times 1 \times 10^{-8}}{(5 \times 10^{-2})^2}$$

$$= 7 \cdot 2 \times 10^{-4} \text{N, in a direction AP}$$

The charge at B attracts the charge at P with a force also equal to F, because $\text{BP} = 5 \text{cm} = \text{AP}$. But this force acts in the direction PB. So the *resultant* force R is along the bisector PE of the angle between the two forces, as shown.

To find R, we can use the components of the two forces F along PE (p. 96).

Then $\qquad\qquad\qquad R = F \cos \theta + F \cos \theta = 2F \cos \theta$

Now $\qquad\qquad\qquad \cos \theta = \text{BX/BP} = 3/5$

So $\qquad\qquad\qquad R = 2F \cos \theta = 2 \times 7 \cdot 2 \times 10^{-4} \times 3/5$

$$= 8 \cdot 64 \times 10^{-4} \, \text{N}$$

Electric Field-strength or Intensity, Field Patterns of Lines of Force

An 'electric field' can be defined as a region where an electric force is experienced. As in magnetism, electric fields can be mapped out by electrostatic lines of force, which may be defined as a line such that the tangent to it is in the direction of the force on a small *positive* charge at that point. Arrows on the lines of force show the direction of the force on a positive charge; the force on a negative charge is in the opposite direction. Figure 6.16 shows the lines of force, also called *electric flux*, in some electrostatic fields of charges.

The force exerted on a charged body in an electric field depends on the charge of the body and on the *strength* or *intensity* of the field. If we wish to explore the variation in strength of an electric field, then we must place a test charge Q' at the point concerned which is small enough not to upset the field by its introduction. The strength E of an electrostatic field at any point is defined as *the force per unit charge* which it exerts at that point. Its direction is that of the force exerted on a *positive* charge.

From this definition,

$$E = \frac{F}{Q'} \qquad F = EQ' . \qquad . \qquad . \qquad . \qquad . \qquad (1)$$

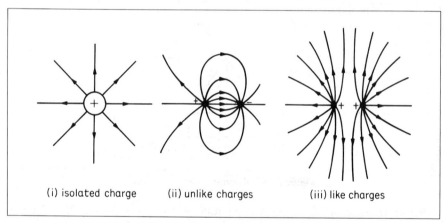

(i) isolated charge   (ii) unlike charges   (iii) like charges

Figure 6.16 *Field pattern of electric lines of force*

Since F is measured in newtons and Q' in coulombs, it follows that field-strength E *has units of newton per coulomb* $(N C^{-1})$. We shall see later that a more practical unit of E is *volt metre^{-1}* $(V m^{-1})$ (see p. 204).

Field-strength E due to Point Charge

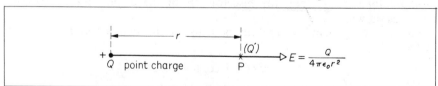

Figure 6.17 *Electric field-strength due to point charge*

We can easily find an expression for the strength E of the electric field due to a point charge Q situated in a vacuum (Figure 6.17). We start from the equation for the force between two such charges:

$$F = \frac{1}{4\pi\varepsilon_0}\frac{QQ'}{r^2}$$

If the test charge Q' is situated at the point P in Figure 6.17, the electric field-strength at that point is given by

$$E = \frac{F}{Q'} = \frac{Q}{4\pi\varepsilon_0 r^2} \qquad . \qquad . \qquad . \qquad . \qquad . \qquad (2)$$

The direction of the field is radially outward if the charge Q is positive (Figure 6.16(i)); it is radially inward if the charge Q is negative. If the charge were surrounded by a material of permittivity ε then,

$$E = \frac{Q}{4\pi\varepsilon r^2} \qquad . \qquad . \qquad . \qquad . \qquad . \qquad (3)$$

Flux from a Point Charge

We have already shown how electric fields can be described by lines of force. From Figure 6.16(i) it can be seen that the density of the lines increases near the charge where the field-strength is high. The field-strength E at a point can thus be represented by *the number of lines per unit area* or *flux density* through a surface perpendicular to the lines of force at the point considered. The *flux* through an area perpendicular to the lines of force is the name given to the product of $E \times area$, where E is the field-strength *normal* to the area at that place and is illustrated in Figure 6.18(i).

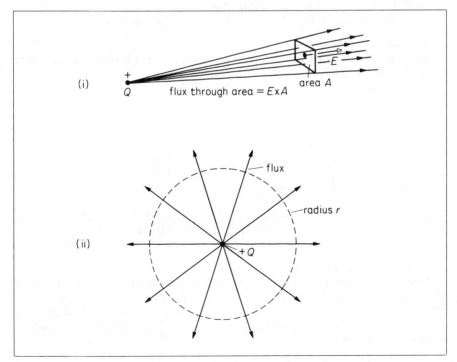

Figure 6.18 *Flux from a point charge*

Consider a sphere of radius r drawn in space concentric with a point charge, Figure 6.18 (ii). The value of E at this place is given by $E = Q/4\pi\varepsilon r^2$. The total normal flux through the sphere is,

$$E \times \text{area} = E \times 4\pi r^2$$

$$= \frac{Q}{4\pi\varepsilon r^2} \times 4\pi r^2 = \frac{Q}{\varepsilon}$$

So $\qquad\qquad E \times \text{area} = \dfrac{\textbf{charge inside sphere}}{\textbf{permittivity}}$ . . . . (1)

This demonstrates the important fact that the total flux crossing normally any sphere drawn outside and concentrically around a point charge is a constant. It does not depend on the distance from the charged sphere.

It should be noted that this result is only true if the inverse square law is true. To see this, suppose some other force law were valid, i.e. $E = Q/4\pi\varepsilon r^n$. Then the total flux through the area

$$= \frac{Q}{4\pi\varepsilon r^n} \times 4\pi r^2 = \frac{Q}{\varepsilon} r^{(2-n)}$$

This is only independent of r if $n = 2$.

Field due to Charged Sphere and Plane Conductor

Equation (1) can be shown to be generally true. Thus the total flux passing normally through any *closed* surface whatever its shape, is always equal to Q/ε, where Q is the total charge enclosed by the surface. This relation, called *Gauss's Theorem*, can be used to find the value of E in other common cases.

(1) *Outside a charged sphere*

The flux across a spherical surface of radius r, concentric with a small sphere carrying a charge Q (Figure 6.19 (i)), is given by,

$$\text{Flux} = \frac{Q}{\varepsilon}$$

$$\therefore E \times 4\pi r^2 = \frac{Q}{\varepsilon}$$

$$\therefore E = \frac{Q}{4\pi\varepsilon r^2}$$

This is the same answer as that for a point charge. This means that *outside* a charged sphere, the field behaves as if all the charge on the sphere were concentrated at the centre.

(2) *Inside a charged empty sphere*

Suppose a spherical surface A is drawn *inside* a charge sphere, as shown in Figure 6.19 (i). Inside this sphere there are no charges and so Q in equation (1) above is zero. This result is independent of the radius drawn, provided that it is less than that of the charged sphere. Hence from (1), **E must be zero everywhere inside a charged sphere**.

Figure 6.19 (ii) shows how E varies with the distance r from the *centre* of the sphere of radius r_0. E is zero from $r = 0$ to $r = r_0$. Beyond $r = r_0$, $E \propto 1/r^2$.

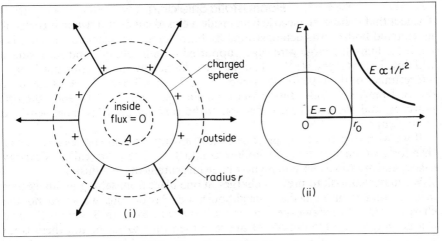

Figure 6.19 *Electric field of a charged sphere*

(3) *Outside a charged plane conductor*

Now consider a charged *plane* conductor S, with a surface charge density of σ coulomb metre^{-2}. Figure 6.20 shows a plane surface P, drawn outside S, which is parallel to S and has an area A metre^2. Applying equation (1),

$$\therefore E \times \text{area} = \frac{\text{Charge inside surface}}{\varepsilon}$$

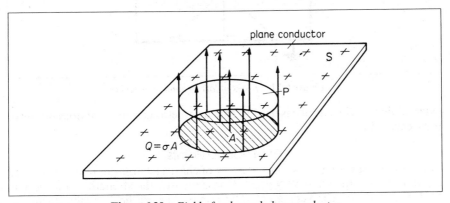

Figure 6.20 *Field of a charged plane conductor*

Now by symmetry, the intensity in the field must be perpendicular to the surface. Further, the charges which produce this field are those in the projection of the area P on the surface S, i.e. those within the shaded area A in Figure 6.20. The total charge here is thus σA coulomb.

$$\therefore E \cdot A = \frac{\sigma A}{\varepsilon}$$

$$\therefore E = \frac{\sigma}{\varepsilon}$$

Electrostatic Shielding

The fact that there is no electric field inside a closed conductor, when it contains no charged bodies, was demonstrated by Faraday in a spectacular manner. He made for himself a large wire cage, supported it on insulators, and sat inside it with his electroscopes. He then had the cage charged by an induction machine — a forerunner of the type we described on p. 181 — until painful sparks could be drawn from its outside. Inside the cage Faraday sat in safety and comfort, however, and there was no deflection to be seen on even his most sensitive electroscope.

If we wish to protect any persons or instruments from intense electric fields, therefore, we enclose them in hollow conductors. These are called 'Faraday cages', and are widely used in high-voltage measurements in industry.

We may also wish to prevent charges in one place from setting up an electric field beyond their immediate neighbourhood. To do this we surround the charges with a Faraday cage, and connect the cage to earth (Figure 6.21). The charge induced on the outside of the cage then runs to earth, and there is no

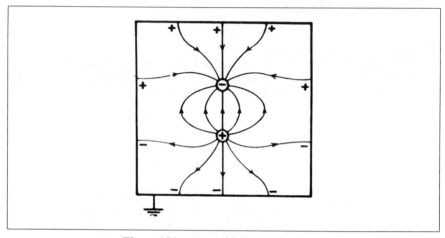

Figure 6.21 *Lines of force round charges*

external field. (When a cage is used to shield something *inside* it, it does not have to be earthed.)

Field Round Points

On p. 186 we saw that the surface-density of charge (charge per unit area) round a point of a conductor is very great. Consequently, the strength of the electric field near the point is very great. The intense electric field breaks down the insulation of the air, and sends a stream of charged molecules away from the point. The mechanism of the breakdown, which is called a 'corona discharge', is complicated, and we shall not discuss it here.

Corona breakdown starts when the electric field-strength E in air is about 3 million volt metre^{-1}. The corresponding surface-density of charge is about 2.7×10^{-5} coulomb metre^{-2} from $E = \sigma/\varepsilon_0$.

Example on Electron Motion in Strong Field

An electron of charge $e = 1.6 \times 10^{-19}$ C is situated in a uniform electric field of intensity λ or field-strength $120\,000$ V m^{-1}. Find the force on it, its acceleration, and the time it takes to travel 20 mm from rest (electron mass, $m = 9.1 \times 10^{-31}$ kg).

Force on electron $F = EQ = Ee$

Now $E = 120\,000\,\text{V m}^{-1}$.

$$\therefore F = 1{\cdot}6 \times 10^{-19} \times 1{\cdot}2 \times 10^{5}$$
$$= 1{\cdot}92 \times 10^{-14}\,\text{N}$$

Acceleration, $$a = \frac{F}{m} = \frac{1{\cdot}92 \times 10^{-14}}{9{\cdot}1 \times 10^{-31}}$$
$$= 2{\cdot}12 \times 10^{16}\,\text{m s}^{-2}$$

Time for 20 mm or 0·02 m travel is given by

$$s = \tfrac{1}{2}at^{2}$$

$$\therefore t = \sqrt{\frac{2s}{a}} = \sqrt{\frac{2 \times 0{\cdot}02}{2{\cdot}12 \times 10^{16}}}$$

$$= 1{\cdot}37 \times 10^{-9}\,\text{s}$$

The extreme shortness of this time is due to the fact that the ratio of charge-to-mass for an electron is very great:

$$\frac{e}{m} = \frac{1{\cdot}6 \times 10^{-19}}{9{\cdot}1 \times 10^{-31}} = 1{\cdot}8 \times 10^{11}\,\text{C kg}^{-1}$$

In an electric field, the charge e determines the force on an electron, while the mass m determines its inertia. Because of the large ratio e/m, the electron moves almost instantaneously, and requires very little energy to displace it. Also it can respond to changes in an electric field which take place even millions of times per second. Thus it is the large value of e/m for electrons which makes electronic tubes, for example, useful in electrical communication and remote control, explained later.

Electric Potential

Potential in Fields

When an object is held at a height above the earth it is said to have gravitational *potential energy*. A heavy body tends to move under the force of attraction of the earth from a point of great height to one of less, and we say that points in the earth's gravitational field have potential values depending on their height.

Electric potential is analogous to gravitational potential, but this time we think of points in an electric field. Thus in the field round a positive charge, for example, a positive charge moves from points near the charge to points further away. Points round the charge are said to have an 'electric potential'.

Potential Difference, Work, Energy of Charges

In mechanics we are always concerned with differences of height; if a point A on a hill is h metre higher than a point B, and our weight is w newton, then we do wh joule of work in climbing from B to A, Figure 6.22 (i). Similarly in electricity we are often concerned with differences of potential; and we define these also in terms of work.

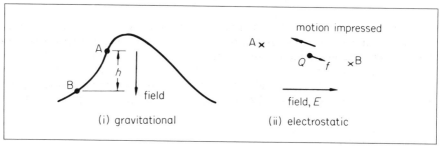

Figure 6.22   *Work done, in gravitational and electrostatic fields*

Let us consider two points A and B in an electrostatic field E, and let us suppose that the force on a positive charge Q has a component f in the direction AB, Figure 6.22 (ii). Then if we move a positively charged body from B to A, we do work against this component of the field E. *We define the potential difference between A and B as the work done in moving a unit positive charge from B to A.* We denote it by the symbol V_{AB}.

Potential difference V_{AB} = work per coulomb in moving charge from B to A

The work done will be measured in joules (J). The unit of potential difference is called the volt and may be defined as follows: *The potential difference between two points A and B is one volt if the work done in taking one coulomb of positive charge from B to A is one joule.*

---

1 volt = 1 joule per coulomb (1 V = 1 J/C)

From this definition, if a charge of Q coulomb is moved through a p.d. of V volt, then the work done W, in joule, is given by

$$W = QV . \qquad . \qquad . \qquad . \qquad . \qquad . \qquad (1)$$

Potential and Energy

Let us consider two points A and B in an electrostatic field, A being at a higher potential than B. The potential difference between A and B we denote as usual by V_{AB}. If we take a positive charge Q from B to A, we do work on it of amount QV_{AB}: the charge gains this amount of potential energy. If we now let the charge go back from A to B, it loses that potential energy: work is done on it by the electrostatic force, in the same way as work is done on a falling stone by gravity. This work may become kinetic energy, if the charge moves freely, or external work if the charge is attached to some machine, or a mixture of the two.

The work which we must do in first taking the charge from B to A does *not* depend on the path along which we carry it, just as the work done in climbing a hill does not depend on the route we take. If this were not true, we could devise a perpetual motion machine, in which we did less work in carrying a charge from B to A via X than it did for us in returning from A to B via Y, Figure 6.23.

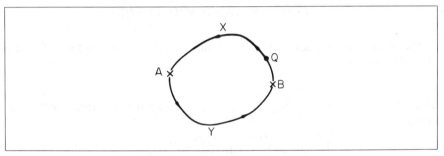

Figure 6.23 *A closed path in an electrostatic field*

The fact that the potential differences between two points is a constant, independent of the path chosen between the points, is a most important property of potential in general. This property can be conveniently expressed by saying that the work done in carrying a charge round any closed path in an electrostatic field, such as BXAYB in Figure 6.23 is zero.

As we also stress later, it shows that potential has magnitude but no direction. So

---

electric potential is a scalar.

The Electron-Volt

The kinetic energy gained by an electron which has been accelerated through a potential difference of 1 volt is called an *electron-volt* (eV). Since the energy gained in moving a charge Q through a p.d. $V = QV$,

$\therefore 1\,eV = $ electronic charge $\times 1 = (1\cdot6 \times 10^{-19} \times 1)\,$joule $= 1\cdot6 \times 10^{-19}\,$J

The electron-volt is a useful unit of energy in atomic physics. For example, the work necessary to extract a conduction electron from tungsten is $4\cdot52\,eV$. This quantity determines the magnitude of the thermionic emission from the metal at a given temperature (p. 762); it is analogous to the latent heat of evaporation of a liquid. An electron in an X-ray tube moving through a p.d. of $50\,000\,V$ will gain energy equal to $50\,000\,eV$.

Potential Difference due to Point Charge

We can now calculate the potential difference between two points in the field of a *single point positive charge*, Q in Figure 6.24. For simplicity we will assume that the points, A and B, lie on a line of force at distances a and b respectively from the charge. When a unit positive charge is at a distance r from the charge Q in free space the force f on it is

$$f = \frac{Q \times 1}{4\pi\varepsilon_0 r^2}$$

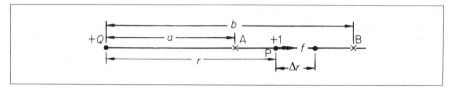

Figure 6.24 *Calculation of potential*

The work done in taking the charge from B to A, against the force f over a short distance Δr is

$$\Delta W = f\Delta r$$

Over the whole distance AB, therefore, the work done by the force on the unit charge is

$$\int_A^B \Delta W = \int_{r=a}^{r=b} f\,dr = \int_a^b \frac{Q}{4\pi\varepsilon_0 r^2}\,dr$$

$$= -\left[\frac{Q}{4\pi\varepsilon_0 r}\right]_a^b = \frac{Q}{4\pi\varepsilon_0 a} - \frac{Q}{4\pi\varepsilon_0 b}$$

This, then, is the value of the work which an external agent must do to carry a unit positive charge from B to A. The work per coulomb is the potential difference V_{AB} between A and B.

$$\therefore V_{AB} = \frac{Q}{4\pi\varepsilon_0}\left(\frac{1}{a} - \frac{1}{b}\right) \qquad . \qquad . \qquad . \qquad . \qquad (1)$$

V_{AB} will be in volt if Q is in coulomb, a and b are in metres and ε_0 is taken as $8.85 \times 10^{-12}\,\mathrm{F\,m^{-1}}$ or $1/4\pi\varepsilon_0$ as $9 \times 10^9\,\mathrm{m\,F^{-1}}$ approximately (see p. 187).

Example on Potential Difference and Work Done

Two positive point charges, of 12 and 8 microcoulomb respectively, are 10 cm apart. Find the work done in bringing them 4 cm closer. (Assume $1/4\pi\varepsilon_0 = 9 \times 10^9\,\mathrm{m\,F^{-1}}$.)

Suppose the 12 μC charge is fixed in position. Since 6 cm = 0·06 m and 10 cm = 0·1 m, then the potential difference between points 6 and 10 cm from it is given by (1).

$$\therefore V = \frac{12 \times 10^{-6}}{4\pi\varepsilon_0}\left(\frac{1}{0\cdot06} - \frac{1}{0\cdot1}\right)$$

$$= 12 \times 10^{-6} \times 9 \times 10^9 (16\tfrac{2}{3} - 10)$$

$$= 720\,000\,\mathrm{V}$$

(Note the very high potential difference due to quite small charges.)

The work done in moving the $8\,\mu C$ charge from 10 cm to 6 cm away from the $12\,\mu C$ charge is given by, using $W = QV$,

$$W = 8 \times 10^{-6} \times V$$
$$= 8 \times 10^{-6} \times 720\,000 = 5 \cdot 8\,J$$

Zero Potential, Potential at a Point

Instead of speaking continually of potential differences between pairs of points, we may speak of the potential at a single point—provided we always refer it to some other, agreed, reference point. This procedure is analogous to referring the heights of mountains to sea-level.

For practical purposes we generally choose as our zero reference point the electric potential of the surface of the *earth*. Although the earth is large it is all at the same potential, because it is a good conductor of electricity; if one point on it were at a higher potential than another, electrons would flow from the lower to the higher potential. As a result, the higher potential would fall, and the lower would rise; the flow of electricity would cease only when the potentials became equal.

In general it is difficult to calculate the potential of a point relative to the earth. This is because the electric field due to a charged body near a conducting surface is complicated, as shown by the lines of force diagram in Figure 6.25. In theoretical calculations, therefore, we often find it convenient to consider charges so far from the earth that the effect of the earth on their field is negligible; we call these 'isolated' charges.

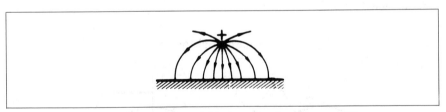

Figure 6.25 *Electric field of positive charge near earth*

So we define the potential at a point A as

the work done per coulomb in bringing a positive charge from infinity to A.

Potential due to Point Charge and to Charged Sphere
Point charge

Equation (1), p. 198, gives the potential difference between two points A and B in the field of an isolated point charge Q:

$$V_{AB} = \frac{Q}{4\pi\varepsilon_0}\left(\frac{1}{a} - \frac{1}{b}\right)$$

If B is at infinity, then, since b is very much greater than a, $1/b$ is negligible compared with $1/a$. So the potential at A is:

$$V_A = \frac{Q}{4\pi\varepsilon_0 a}$$

So at a distance r from a point charge Q, the potential V is:

$$V = \frac{Q}{4 \pi \varepsilon_0 r} \qquad \cdot \qquad \cdot \qquad \cdot \qquad \cdot \qquad \cdot \qquad (1)$$

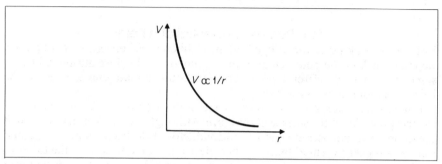

Figure 6.26 *Variation of* V *with* r

Figure 6.26 shows how *V* varies with the distance *r* from the point charge *Q*. The curve does not fall as rapidly as the curve of *E*, the field-strength, with *r*, since $E \propto 1/r^2$.

When *Q* is a positive charge, *V* is positive. This means that work is done by an *external* force in moving a positive charge to the point concerned. When *Q* is a negative charge, *V* is *negative*. This means that the *field itself* does work or loses energy when a positive charge is moved to the point concerned, since it is now attracted by *Q*.

Charged sphere

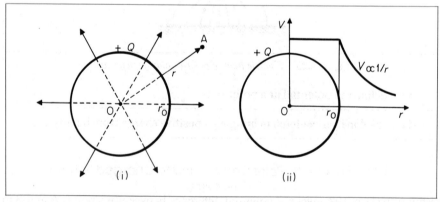

Figure 6.27 *Potential due to charged sphere*

Suppose the charge on a spherical conductor is $+Q$ and the radius of the sphere is r_0, Figure 6.27 (i). The lines of force spread out radially from the surface and we can imagine them starting from a charge *Q* concentrated at the centre point O. We have just seen that the potential at a distance *r* from a point charge Q is $V = Q/4 \pi \varepsilon_0 r$. *Outside the sphere*, then, the potential at a point such as A distance *r* from the centre is

$$V = \frac{Q}{4 \pi \varepsilon_0 r} \qquad \cdot \qquad \cdot \qquad \cdot \qquad \cdot \qquad \cdot \qquad (1)$$

At the *surface*, where $r = r_0$, the radius, the potentials is

$$V = \frac{Q}{4\pi\varepsilon_0 r_0} \qquad \cdot \qquad \cdot \qquad \cdot \qquad \cdot \qquad \cdot \qquad (2)$$

Inside the sphere, the electric field-strength $E = 0$ (p. 192). So no work is done when a charge is taken from *any* point inside to a point on the surface S. Therefore there is no potential difference between any point inside and S. But the potential of S $= Q/4\pi\varepsilon_0 r_0$. So the potential at any point inside the sphere is

$$V = \frac{Q}{4\pi\varepsilon_0 r_0} \qquad \cdot \qquad \cdot \qquad \cdot \qquad \cdot \qquad \cdot \qquad (3)$$

Note that *all* points inside the sphere have this same potential value, because $E = 0$ for all points inside, as we saw earlier.

Figure 6.27 (ii) shows the variation of the potential V due to a charged sphere with the distance r measured from the *centre* of the sphere. Note that V is constant ($= Q/4\pi\varepsilon_0 r_0$) from $r = 0$ to $r = r_0$.

Example on Potential Variation due to Charges

In Figure 6.28, a positively-charged sphere C is near a long insulated conductor AB. Draw sketches to show how the potential V all round C varies with the distance from C measured along AB and beyond (i) before and (ii) after AB is placed in position.

Draw a sketch showing the new variation of V with distance if AB is earthed.

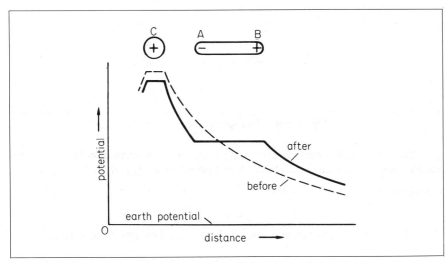

Figure 6.28 *Potential near positive charge before and after bringing up uncharged conductor*

(i) *Before* AB is placed in position, the variation of V with distance r is similar to the isolated spherical charged conductor in Figure 6.27 (ii).

(ii) *After* AB is placed in position, there are now induced equal and opposite charges at A and B as shown. Since the charge on A is nearer C than B and opposite to that on C, the potential between C and A is now *less* than before.

Further, the potential of a conductor such as AB is constant. So this part of the graph is a horizontal straight line. Beyond the +ve charge at B, the potential decreases as shown.

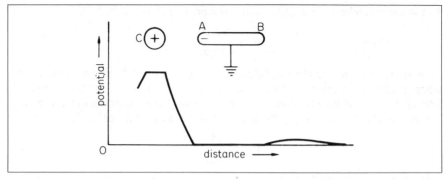

Figure 6.29 *Potential near positive charge in the presence of an earthed conductor*

Earthed conductor AB. There is now only a $-$ve charge at the end A and the potential falls rapidly to zero as shown in Figure 6.29. Beyond B, C has a small potential.

Example on Potential Energy

In Figure 6.30, an alpha-particle A of charge $+3\cdot2 \times 10^{-19}$ C and mass $6\cdot8 \times 10^{-27}$ kg is travelling with a velocity v of $1\cdot0 \times 10^{7}$ m s^{-1} directly towards a nitrogen nucleus N which has a charge of $+11\cdot2 \times 10^{-19}$ C.

Calculate the closest distance of approach of A to N, assuming that A is initially a very long way from N compared with the closest distance of approach.

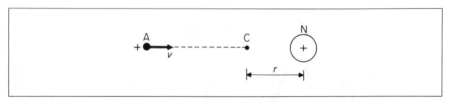

Figure 6.30 *Example on potential energy*

(*Analysis* The alpha-particle loses kinetic energy as it approaches the nitrogen nucleus and this is transferred to electrical potential energy in the field of N.)

Initial kinetic energy of A $= \frac{1}{2}mv^2 = \frac{1}{2} \times 6\cdot8 \times 10^{-27} \times (1\cdot0 \times 10^7)^2$

$$= 3\cdot4 \times 10^{-13} \text{ J}$$

If r is the closest distance of approach at C, potential energy at r due to N

$$= \frac{Q_1 Q_2}{4\pi\varepsilon_0 r} = \frac{9 \times 10^9 \times 3\cdot2 \times 10^{-19} \times 11\cdot2 \times 10^{-19}}{r}$$

$$= \frac{3\cdot2 \times 10^{-27}}{r}$$

assuming $1/4\pi\varepsilon_0 = 9 \times 10^9$ and the initial potential energy of A is zero.

So $\qquad\qquad 3\cdot4 \times 10^{-13} = \dfrac{3\cdot2 \times 10^{-27}}{r}$

$$\therefore r = \frac{3\cdot2 \times 10^{-27}}{3\cdot4 \times 10^{-13}} = 9\cdot4 \times 10^{-15} \text{ m}$$

Potential Gradient and Field-strength (Intensity)

We shall now see how potential difference is related to field-strength or intensity. Suppose A, B are two neighbouring points on a line of force, so close together that the electric field-strength between them is constant and equal to E (Figure 6.31). If V is the potential at A, $V + \Delta V$ is that at B, and the respective distances of A, B from the origin are x and $x + \Delta x$, then

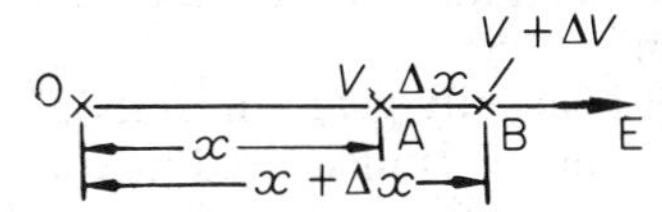

$$V_{AB} = \text{potential difference between A, B}$$

$$= V_A - V_B = V - (V + \Delta V) = -\Delta V$$

Figure 6.31 _Field-strength and potential gradient_

The work done in taking a unit charge from B to A

$$= \text{force} \times \text{distance} = E \times \Delta x = V_{AB} = -\Delta V$$

Hence

$$E = -\frac{\Delta V}{\Delta x}$$

or, in the limit,

$$E = -\frac{dV}{dx} . \qquad . \qquad . \qquad . \qquad . \qquad . \qquad (1)$$

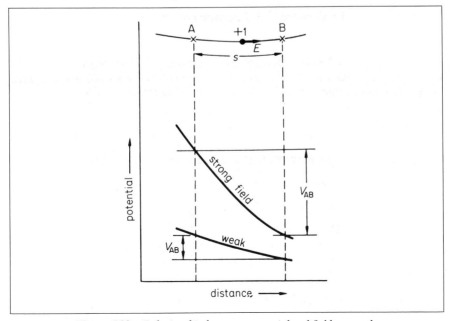

Figure 6.32 _Relationship between potential and field-strength_

The quantity dV/dx is the rate at which the potential rises with distance, and is called the **potential gradient**. Equation (1) shows that the strength of the electric field is equal to the negative of the potential gradient.

Potential Variation in Fields, Unit of *E*

Strong and weak fields in relation to potential are illustrated in Figure 6.32.

In Figure 6.33 the electric field-strength $= V/h$, the potential gradient, and this is uniform in magnitude in the middle of the plates. At the edge of the plates the field becomes non-uniform.

We can now see why E is usually given in units of 'volt per metre' ($V\,m^{-1}$). From (1), $E = -(dV/dx)$. Since V is measured in volts and x in metres, then E will be in volt per metre ($V\,m^{-1}$). From the original definition of $E(= F/Q)$, the

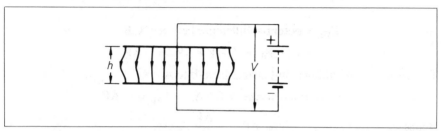

Figure 6.33   *Electric field between parallel plates; in middle,* $E = V/h$

units of E were newton coulomb^{-1} ($N\,C^{-1}$). To show these two units are equivalent, we have 1 joule $=$ 1 newton $\times$ 1 metre from mechanics and so

$$1 \text{ volt} = 1 \text{ joule coulomb}^{-1}$$

$$= 1 \text{ newton metre coulomb}^{-1}$$

$$\therefore 1 \text{ volt metre}^{-1} = 1 \text{ newton coulomb}^{-1}$$

Examples on Potential Gradient and Field-strength

1   An oil drop of mass 2×10^{-14} kg carries a charge Q. The drop is stationary between two parallel plates 20 mm apart which a p.d. of 500 V between them. Calculate Q.

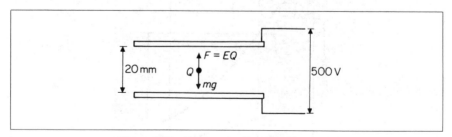

Figure 6.34   *Oil drop in electric field*

Since the drop is stationary,

upward force on charge, $F = $ weight of drop, mg

Now $\qquad F = EQ = \dfrac{V}{d}Q$, since $E =$ potential gradient $= \dfrac{V}{d}$

So $\qquad \dfrac{V}{d}Q = mg$

$$\therefore Q = \frac{mgd}{V} = 2 \times 10^{-14} \times 10 \times 20 \times 10^{-3}$$

$$= 8 \times 10^{-18}\,\mathrm{C}$$

2 An electron is liberated from the lower of two large parallel metal plates separated by a distance $h = 20\,\mathrm{mm}$. The upper plate has a potential of $+2400\,\mathrm{V}$ relative to the lower. How long does the electron take to reach it? (Assume charge-mass ratio, e/m, for electron $= 1.8 \times 10^{11}\,\mathrm{C\,kg^{-1}}$.)

Between large parallel plates, close together, the electric field is uniform except near the edges of the plates, as shown in Figure 6.33. Except near the edges, therefore, the potential gradient between the plates is uniform; its magnitude is V/h, where $h = 0.02\,\mathrm{m}$, so

$$\text{electric field-strength } E = \text{potential gradient}$$

$$= 2400/0.02\,\mathrm{V\,m^{-1}}$$

$$= 1.2 \times 10^5\,\mathrm{V\,m^{-1}}$$

Force on electron of charge e is given by $F = Ee$.

$$\text{Acceleration, } a = \frac{F}{m} = \frac{Ee}{m}$$

$$= 1.2 \times 10^5 \times 1.8 \times 10^{11}$$

$$= 2.16 \times 10^{16}\,\mathrm{m\,s^{-2}}$$

Then, from $s = \frac{1}{2}at^2$,

$$t = \sqrt{\frac{2s}{a}} = \sqrt{\frac{2 \times 20 \times 10^{-3}}{2.16 \times 10^{16}}}$$

$$= 1.4 \times 10^{-9}\,\mathrm{s}$$

Equipotentials

We have already said that the earth must have the same potential all over, because it is a conductor. In any conductor there can be no differences of potential. Otherwise these would set up a potential gradient or electric field and electrons would then redistribute themselves throughout the conductor, under the influence of the field, until they had destroyed the field. This is true whether the conductor has a net charge, positive or negative, or whether it is uncharged.

Any surface or volume over which the potential is constant is called an *equipotential*. The space inside a hollow charged conductor has the same potential as the surface at all points and so is an equipotential volume. The surface of a conductor of *any* shape is an equipotential surface.

Equipotential surfaces can be drawn throughout any space in which there is an electric field. Figure 6.35 (i) shows the field of an isolated point charge Q.

At a distance r from the charge, the potential is $Q/4\pi\varepsilon_0 r$; a sphere of radius r and centre at Q is therefore an equipotential surface, of potential $Q/4\pi\varepsilon_0 r$. In fact, all spheres centred on the charge are equipotential surfaces, whose potentials are inversely proportional to their radii, Figure 6.35 (i). Values proportional to the potentials are shown.

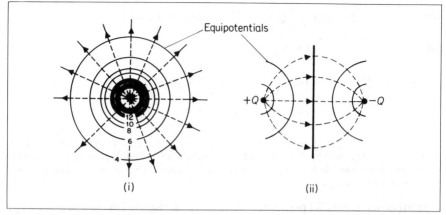

Figure 6.35 *Equipotentials and lines of force round* (i) *a point charge* (ii) *two opposite charges*

Figure 6.35 (ii) shows two equal and opposite point charges $+Q$ and $-Q$, and some typical lines of force and equipotentials in their field. The equipotential lines meet the lines of force at 90°.

An equipotential surface has the property that, along any direction lying in the surface, there is no electric field; for there is no potential gradient. *Equipotential surfaces are therefore always at right angles to lines of force*, as shown in Figure 6.35. Since conductors are always equipotentials, if any conductors appear in an electric-field diagram the lines of force must always be drawn to meet them at right angles.

Potential due to a System of Charges

When we consider the electric field due to more charges than one, we see the advantages of the idea of potential over the idea of field-strength. If we wish to find the field-strength E at the point P in Figure 6.36, due to the two positive charges Q_1 and Q_2, we have first to find the force exerted by each on a unit charge at P, and then to add these forces by a *vector* method such as the parallelogram method, shown in Figure 6.36.

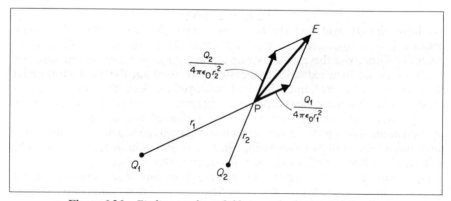

Figure 6.36 *Finding resultant field-strength of two point charges*

On the other hand, if we wish to find the potential at P, we merely calculate the potential due to each charge, and *add the potentials algebraically*, since potentials are scalar. So potential at P, $V = Q_1/4\pi\varepsilon_0 r_1 + Q_2/4\pi\varepsilon_0 r_2$.

When we have plotted the equipotentials, they turn out to be more useful than lines of force. A line of force diagram appeals to the imagination, and helps us to see what would happen to a charge in the field. But it tells us little about the strength of the field—at the best, if it is more carefully drawn than most, we can only say that the field is strongest where the lines are closest. But equipotentials can be labelled with the values of potential they represent; and from their spacing we can find the actual value of the potential gradient, and hence the field-strength. The direction of the field-strength is always at right angles to the equipotential curves.

Comparison of Static and Current Phenomena

Broadly speaking, we may say that in electrostatic phenomena we meet small quantities of charge, but great differences of potential. On the other hand in the phenomena of current electricity discussed later, the potential differences are small but the amounts of charge transported by the current are great. Sparks and shocks are common in electrostatics, because they require great potential differences; but they are rarely dangerous, because the total amount of energy available is usually small. On the other hand, shocks and sparks in current electricity are rare, but, when the potential difference is great enough to cause them, they are likely to be dangerous.

These quantitative differences make problems of insulation much more difficult in electrostatic apparatus than in apparatus for use with currents. The high potentials met in electrostatics make leakage currents relatively great, and the small charges therefore tend to disappear rapidly. Any wood, for example, ranks as an insulator for current electricity, but a conductor in electrostatics. In electrostatic experiments we sometimes wish to connect a charges body to earth; all we have then to do is to touch it.

Comparison between Electrostatic and Gravitational Fields

We conclude with a comparison between electrostatic (electric) fields and gravitational fields. Scientists consider that gravitational forces are the weakest in the universe and electric forces are much stronger.

Like charges repel and unlike charges attract in electric fields, so electric forces may be repulsive or attractive. In the gravitational field, masses attract each other and no repulsive force has yet been detected. So electric potential may be positive or negative but gravitational potential is only negative (zero potential is at infinity in both cases). Further, the Earth is such a large sphere, that the gravitational field strength, g, near the Earth's surface is fairly uniform for a height above the surface which is small compared to its radius.

The following Table summarises some other points:

		Electric	**Gravitational**
1	Field-strength Unit	F/Q $N\,C^{-1}$	F/m $N\,kg^{-1}$
2	Force formula Force direction	$F = Q_1 Q_2 / 4\pi\varepsilon_0 r^2$ attractive/repulsive	$F = Gm_1 m_2 / r^2$ attractive only
3	Strength outside isolated sphere	$E = \pm Q / 4\pi\varepsilon_0 r^2$	$E = GM/r^2$
4	Potential outside isolated sphere	$V = \pm Q / 4\pi\varepsilon_0 r$	$V = -GM/r$

_____ Exercises 6 _____

(*Where necessary, assume* $\varepsilon_0 = 8.85 \times 10^{-12}\, F\, m^{-1}$)

1 What is the *potential gradient* between two parallel plane conductors when their separation is 20 mm and a p.d. of 400 V is applied to them? Calculate the force on an oil drop between the plates if the drop carries a charge of 8×10^{-19} C.

2 Using the same graphical axes in each case, draw sketches showing the variation of potential (i) inside and outside an isolated hollow spherical conductor A which has a positive charge, (ii) between A and an insulated sphere B brought near to A, (iii) between A and B if B is now earthed.

3 A charged oil drop remains stationary when situated between two parallel horizontal metal plates 25 mm apart and a p.d. of 1000 V is applied to the plates. Find the charge on the drop if it has a mass of 5×10^{-15} kg. (Assume $g = 10\,\mathrm{N\,kg^{-1}}$.)
   Draw a sketch of the electric field between the plates and state if the field is everywhere uniform.

4 How do
   (a) the magnitude of the gravitational field, and
   (b) the magnitude of the electrostatic field, vary with distance from a point mass and a point charge respectively?
      Sketch a graph illustrating the variation of electrostatic field-strength E with distance r from the *centre* of a uniformly solid metal sphere of radius r_0 which is positively charged. Explain the shape of your graph (i) for $r > r_0$, and, (ii) for $r < r_0$. (L.)

5 Define
   (a) electric intensity,
   (b) difference of potential.
      How are these quantities related?
      A charged oil-drop of radius 1.3×10^{-6} m is prevented from falling under gravity by the vertical field between two horizontal plates charged to a difference of potential of 8340 V. The distance between the plates is 16 mm, and the density of oil is $920\,\mathrm{kg\,m^{-3}}$. Calculate the magnitude of the charge on the drop ($g = 9.81\,\mathrm{m\,s^{-2}}$). (O. & C.)

6 Show how (i) the surface density, (ii) the intensity of electric field, (iii) the potential, varies over the surface of an elongated conductor charged with electricity. Describe experiments you would perform to support your answer in cases (i) and (ii).
      Describe and explain the action of points on a charged conductor; and give two practical applications of the effect. (L.)

7 Describe, with the aid of a labelled diagram, a Van de Graaff generator, explaining the physical principles of its action.
      The high voltage terminal of such a generator consists of a spherical conducting shell of radius 0.50 m. Estimate the maximum potential to which it can be raised in air for which electrical breakdown occurs when the electric intensity exceeds $3 \times 10^6\,\mathrm{V\,m^{-1}}$.
      State two ways in which this maximum potential could be increased. (*JMB.*)

8 Define *potential at a point* in an electric field.
      Sketch a graph illustrating the variation of potential along a radius from the centre of a charged isolated conducting sphere to infinity.
      Assuming the expression for the potential of a charged isolated conducting sphere in air, determine the change in the potential of such a sphere caused by surrounding it with an earthed concentric thin conducting sphere having three times its radius. (*JMB.*)

9 Two plane parallel conducting plates 15.0 mm apart are held horizontal, one above the other, in air. The upper plate is maintained at a positive potential of 1500 V while the lower plate is earthed. Calculate the number of electrons which must be attached to a small oil drop of mass 4.90×10^{-15} kg, if it remains stationary in the air between the plates. (Assume that the density of air is negligible in comparison with that of oil.)

If the potential of the upper plate is suddenly changed to -1500 V what is the initial acceleration of the charged drop? Indicate, giving reasons, how the acceleration will change. (*L.*)

10 Describe carefully Faraday's ice-pail experiments and discuss the deductions to be drawn from them. How would you investigate experimentally the charge distribution over the surface of a conductor? (*C.*)

11 What is an *electric field*? With reference to such a field define *electric potential*.

Two plane parallel conducting plates are held horizontal, one above the other, in a vacuum. Electrons having a speed of 6.0×10^6 m s^{-1} and moving normally to the plates enter the region between them through a hole in the lower plate which is earthed. What potential must be applied to the other plate so that the electrons just fail to reach it? What is the subsequent motion of these electrons? Assume that the electrons do not interact with one another.

(Ratio of charge to mass of electron is 1.8×10^{11} C kg^{-1}.) (*JMB.*)

12 An isolated conducting spherical shell of radius 0.10 m, in vacuo, carries a positive charge of 1.0×10^{-7} C. Calculate
(a) the electric field-strength,
(b) the potential, at a point on the surface of the conductor.
   Sketch a graph to show how one of these quantities varies with distance along a radius from the centre to a point well outside the spherical shell. Point out the main features of the graph. (*JMB.*)

13 (a) A charged oil drop falls at constant speed in the Millikan oil drop experiment when there is no p.d. between the plates. Explain this.
(b) Such an oil drop, of mass 4.0×10^{-15} kg, is held stationary when an electric field is applied between the two horizontal plates. If the drop carries 6 electric charges each of value 1.6×10^{-19} C, calculate the value of the electric field-strength. (*L.*)

14 (a) A conductor carrying a negative charge has an insulating handle. Describe how you would use it to charge (i) negatively, (ii) positively a thin spherical conducting shell which is isolated and initially uncharged.
   In each case explain why the procedure you describe produces the desired result.
(b) For one of these cases, sketch graphs showing how the electric field-strength and the electric potential vary along a line outwards from the centre of the shell, when the charging device has been removed.

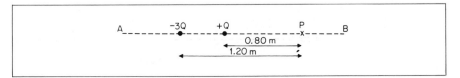

Figure 6A

(c) Figure 6A shows an arrangement of two point charges in air, Q being $0.30 \, \mu$C.
(i) Find the electric field-strength and the electric potential at P. (ii) Find the point on AB between the two charges at which the electric potential is zero.
(iii) Explain why the potential on AB on the left of the $-3Q$ charge is always negative.

Take $\varepsilon_0 = 8.8 \times 10^{-12}$ F m^{-1} or $= \dfrac{1}{36\pi} \times 10^{-9}$ F m^{-1} (*JMB.*)

15 (a) Figure 6B shows a hollow metal sphere supported on an insulating stand. In (i) a large positive charge is near to the sphere; in (ii) the sphere is earthed; in (iii) the earth connection has been removed and finally in (iv) the positive charge has been removed.
   Sketch the distribution of charge on the sphere which you would expect at each of the four stages.

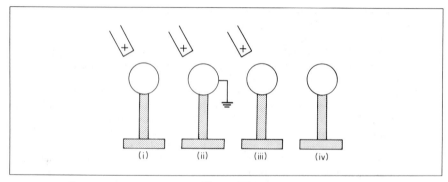

Figure 6B

(b) A large, hollow, metal sphere is charged positively and insulated from its surroundings. Sketch graphs of (i) the electric field-strength, and (ii) the electric potential, from the centre of the sphere to a distance of several diameters. (*AEB*, 1985.)

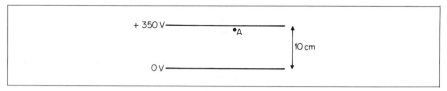

Figure 6C

16 Two point charges Q_1 and Q_2 are situated as shown in Figure 6C. Q_1 is a positive charge and Q_2 is a negative charge; the magnitude of Q_1 is greater than Q_2. A third point charge, which is positive, is now placed in such a position, X, that it experiences no resultant electrostatic force due to Q_1 and Q_2. Explain carefully why X must lie somewhere on the line AB which passes through Q_1 and Q_2. Copy the diagram and indicate clearly in which section of the line AB the point X must lie. Give reasons for your answer. Explain why the position X would be unchanged if the magnitude or the sign of the third charge were altered. (*L*.)

17 (a) Define the terms *potential* and *field-strength* at a point in an electric field. Figure 6D shows two horizontal parallel conducting plates in a vacuum.

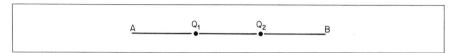

Figure 6D

A small particle of mass 4×10^{-12} kg, carrying a positive charge of $3 \cdot 0 \times 10^{-14}$ C is released at A close to the upper plate. What *total* force acts on this particle?

Calculate the kinetic energy of the particle when it reaches the lower plate.

(b) Figure 6E shows a positively charged metal sphere and a nearby uncharged metal rod.

Figure 6E

Explain why a redistribution of charge occurs on the rod when the charged metal sphere is brought close to the rod.

Copy this diagram and show on it the charge distribution on the rod. Sketch a few electric field lines in the region between the sphere and the rod.

Sketch graphs which show how (i) the potential relative to earth, and (ii) the field-strength vary along the axis of the rod from the centre of the charged sphere to a point beyond the end of the rod furthest from the sphere. How is graph (i) related to graph (ii)?

How will the potential distribution along this axis be changed if the rod is now earthed? (*L.*)

18 Define the *electric potential V* and the *electric field-strength E* at a point in an electrostatic field. How are they related? Write down an expression for the electric field-strength at a point close to a charged conducting surface, in terms of the surface density of charge.

Corona discharge into the air from a charged conductor takes place when the potential gradient at its surface exceeds $3 \times 10^6 \text{ V m}^{-1}$; a potential gradient of this magnitude also breaks down the insulation afforded by a solid dielectric. Calculate the greatest charge that can be placed on a conducting sphere of radius 20 cm supported in the atmosphere on a long insulating pillar; also calculate the corresponding potential of the sphere. Discuss whether this potential could be achieved if the pillar of insulating dielectric was only 50 cm long. (Take ε_0 to be $8.85 \times 10^{-12} \text{ F m}^{-1}$.) (*O.*)

7

Capacitors

*Capacitors are important components in the electronics and tele-
communications industries. They are essential, for example, in
radio and television receivers and in transmitter circuits. We
shall describe how charges and energy are stored in capacitors,
the series and parallel circuit arrangements of capacitors and the
charge and discharge of a capacitor through a resistor which
occurs in many practical circuits.*

A capacitor is a device for storing charge. The earliest capacitor was invented—almost accidentally—by van Musschenbroek of Leyden, in about 1746,
and became known as a Leyden jar. One form of it is shown in Figure 7.1 (i); J is

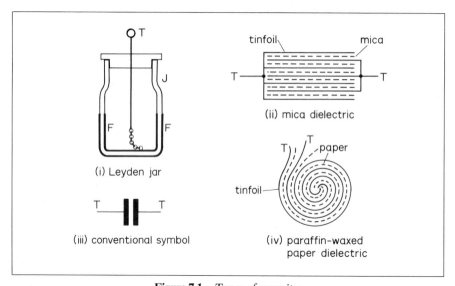

Figure 7.1 *Types of capacitor*

a glass jar, FF are tin-foil coatings over the lower parts of its walls, and T is a
knob connected to the inner coating. Modern forms of capacitor are shown at
(ii) and (iv) in the figure. Essentially

---

all capacitors consist of two metal plates separated by an insulator.

The insulator is called the *dielectric*; in some capacitors it is polystyrene, oil or
air. Figure 7.1 (iii) shows the circuit symbol for such a capacitor; T, T are
terminals joined to the plates.

Charging and Discharging Capacitor
Figure 7.2 (i) shows a circuit which may be used to study the action of a
capacitor. C is a large capacitor such as 500 microfarad (see later), R is

a large resistor such as 100 kilohms ($10^5\,\Omega$), A is a current meter reading 100 − 0 − 100 microamperes (100 μA), K is a two-way key, and D is a 6 V d.c. supply.

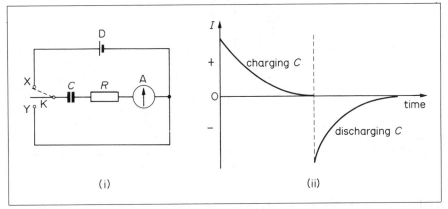

Figure 7.2 *Charging and discharging capacitor*

When the battery is connected to C by contact at X, the current I in the meter A is seen to be initially about 60 μA. Then as shown in Figure 7.2 (ii), it slowly decreases to zero. Thus current flows to C for a short time when the battery is connected to it, even though the capacitor plates are separated by an insulator.

We can disconnect the battery from C by opening X. If contact with Y is now made, so that in effect the plates of C are joined together through R and A, the current in the meter is observed to be about 60 μA initially in the opposite direction to before and then slowly decreases to zero, Figure 7.2 (ii). This flow of current shows that C *stored charge when it was connected to the battery* originally.

Generally, a capacitor is *charged* when a battery or p.d. is connected to it. When the plates of the capacitor are joined together, the capacitor becomes *discharged*. Large values of C and R in the circuit of Figure 7.2 (i) help to slow the current flow, so that we can see the charging and discharging which occurs, as explained more fully later.

We can also show that a charged capacitor has stored *energy* by connecting the terminals by a piece of wire. A spark, a form of light and heat, passes just as the wire makes contact.

Charging and Discharging Processes

When we connect a capacitor to a battery, electrons flow from the negative terminal of the battery on to the plate A of the capacitor connected to it (Figure 7.3). At the same rate, electrons flow from the other plate B of the capacitor towards the positive terminal of the battery. Equal positive and negative charges thus appear on the plates, and oppose the flow of electrons which causes them. As the charges accumulate, the potential difference between the plates increases, and the charging current falls to zero when the potential difference becomes equal to the battery voltage V_0. The charges on the plates B and A are now $+Q$ and $-Q$, and the capacitor is said to gave stored a charge Q in amount.

When the battery is disconnected and the plates are joined together by a wire, electrons flow back from plate A to plate B until the positive charge on B is completely neutralised. A current thus flows for a time in the wire, and at the end

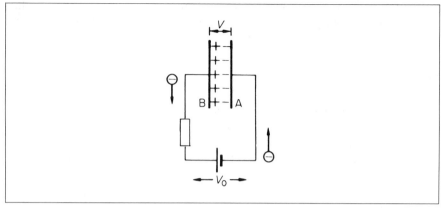

Figure 7.3 *A capacitor charging (resistance is shown because some is always present, even if only that of the connecting wires)*

of the time the charges on the plates become zero. So the capacitor is *discharged*. Note that a charge Q flows from one plate to the other during the discharge.

Capacitors and A.C. Circuits

Capacitors are widely used in alternating current and radio circuits, because they can transmit alternating currents. To see how they do so, let us consider the circuit of Figure 7.4, in which the capacitor may be connected across either of

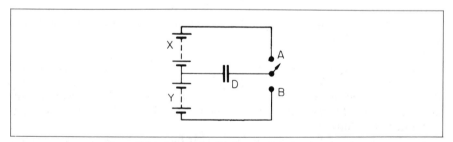

Figure 7.4 *Reversals of voltage applied to capacitor*

the batteries X, Y. When the key is closed at A, current flows from the battery X, and charges the plate D of the capacitor positively. If the key is now closed at B instead, current flows from the battery Y; the plate D loses its positive charge and becomes negatively charged. Thus if the key is rocked rapidly between A and B, current surges backwards and forwards along the wires connected to the capacitor. An alternating voltage, as we shall see later, is one which reverses many times a second. When such a voltage is applied to a capacitor, therefore, an alternating current flows in the connecting wires.

Variation of Charge with P.D., Vibrating Reed Switch

Figure 7.5 shows a circuit which may be used to investigate how the charge Q stored on a capacitor C varies with the p.d. V applied. The d.c. supply D can be altered in steps from a value such as 10 V to 25 V, C is a capacitor consisting of two large square metal plates separated by small pieces of polythene at the

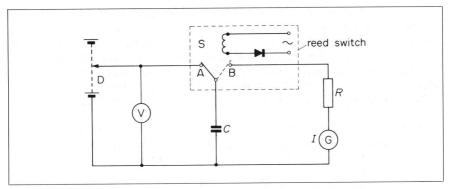

Figure 7.5   *Variation of* Q *with* V—*vibrating reed switch*

corners, G is a sensitive current meter such as a light beam galvanometer, and *R* is a protective high resistor in series with G.

The capacitor can be charged and discharged rapidly by means of a *vibrating reed switch* S. The vibrator charges *C* by contact with A and discharges *C* through G by contact with B. When the vibrator frequency *f* is made suitably high, such as several hundred hertz, a steady current *I* flows in G. Its magnitude is given by

---

$$I = \textbf{charge per second} = fQ$$

---

where *Q* is the charge on the capacitor each time it is charged, since *f* is the number of times per second it is charged. Thus, for a given value of *f*, *the charge Q is proportional to the current I* in G.

When *V* is varied and values of *I* are observed, results show that $I \propto V$. So experiment shows that, for a given capacitor, $Q \propto V$.

Capacitance Definition and Units

Since $Q \propto V$, then Q/V is a constant for the capacitor. The ratio of the charge on either plate to the potential difference between the plates is called the <u>*capacitance*</u>, *C*, of the capacitor:

$$C = \frac{Q}{V} \qquad . \qquad . \qquad . \qquad . \qquad . \qquad . \qquad (1)$$

So
$$Q = CV \qquad . \qquad . \qquad . \qquad . \qquad . \qquad . \qquad (2)$$

and
$$V = \frac{Q}{C} \qquad . \qquad . \qquad . \qquad . \qquad . \qquad . \qquad (3)$$

When *Q* is in coulomb (C) and *V* in volt (V), then capacitance *C* is in farad (F). One farad (1 F) is the capacitance of an extremely large capacitor. In practical circuits, such as in radio receivers, the capacitance of capacitors used are therefore expressed in *microfarad* (µF). One microfarad is one millionth part of a farad, that is $1\,\mu F = 10^{-6}\,F$. It is also quite usual to express small capacitors, such as those used on record players, in picofarad (pF). A picofarad is one millionth part of a microfarad, that is $1\,pF = 10^{-6}\,\mu F = 10^{-12}\,F$.

$$1\,\mu F = 10^{-6}\,F \qquad 1\,pF = 10^{-12}\,F$$

Comparison of Capacitances, Measurement of C

The vibrating reed circuit shown in Figure 7.5 can be used to compare large capacitances (of the order of microfarads) or to compare small capacitances. With large capacitances, a meter with a suitable range of current of the order of milliamperes may be required. With smaller capacitances, a sensitive galvanometer may be more suitable, as the current flowing is then much smaller. In both cases suitable values for the applied p.d. V and the frequency f must be chosen.

Suppose two large, or two small, capacitances, C_1 and C_2, are to be compared. Using C_1 first in the vibrating reed circuit, the current flowing is I_1 say. When C_1 is replaced by C_2, suppose the new current is I_2.

Now $Q_1 = C_1 V$ and $Q_2 = C_2 V$; hence $Q_1/Q_2 = C_1/C_2 = C_1/C_2$. But from p. 215, $Q \propto I$. Thus $Q_1/Q_2 = I_1/I_2$.

$$\therefore \frac{C_1}{C_2} = \frac{I_1}{I_2}$$

Thus the ratio C_1/C_2 can be found from the current readings I_1 and I_2.

An *unknown capacitor* C can also be found using the vibrating reed circuit. Suppose I is the current measured in G when the applied p.d. is V. Using a low voltage from the a.c. mains for the switch, C is charged and discharged 50 times per second, the mains frequency. Since the current I is the charge flowing per second, then

$$I = 50\,CV$$

So

$$C = \frac{I}{50\,V}$$

With I in amperes and V in volts, then C is in *farads*.

Ballistic Galvanometer Method

Large capacitances, of the order of microfarads, can also be compared with the aid of a *ballistic galvanometer*. In this instrument, as explained later, the first 'throw' or deflection is proportional to the quantity of charge (Q) passing through it.

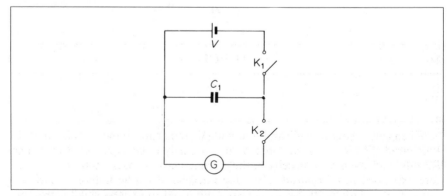

Figure 7.6 *Comparison of capacitances—ballistic galvanometer*

The circuit required is shown in Figure 7.6. The capacitor of capacitance C_1 is charged by a battery of e.m.f. V, and then discharged through the ballistic galvanometer G. The corresponding first deflection θ_1 is observed. The capacitor is now replaced by another of capacitance C_2, charged again by the battery, and the new deflection θ_2 is observed when the capacitor is discharged.

Now

$$\frac{Q_1}{Q_2} = \frac{\theta_1}{\theta_2}$$

$$\therefore \frac{C_1 V}{C_2 V} = \frac{C_1}{C_2} = \frac{\theta_1}{\theta_2}$$

If C_2 is a standard capacitor, whose value is known, then the capacitance of C_1 can be found.

Factors Determining Capacitance

As we have seen, a capacitor consists of two metal plates separated by an insulator called a 'dielectric'. We can now find out by experiment what factors influence capacitance.

Distance between plates. Figure 7.7 (i) shows two parallel metal plates X and Y

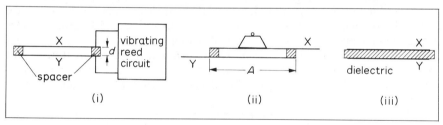

Figure 7.7   *Factors affecting capacitance*

separated by a distance d equal to the thickness of the polythene spacers shown. The capacitance C can be varied by separating the plates a distance $2d$ and then $3d$ and $4d$, using more spacers.

The capacitance can be found each time using the vibrating reed circuit described before. As we have shown, the current I in G is proportional to C for a given applied p.d. V. Experiment shows that, allowing for error, $C \propto 1/d$, where d is the *separation* between the plates. So halving the separation will double the capacitance.

Area between plates. By placing a weight on the top plate X and moving sideways, Figure 7.7 (ii), the area A of overlap, or common area between the plates, can be varied while d is kept constant. Alternatively, pairs of plates of different area can be used which have the same separation d. By using the vibrating reed circuit, experiment shows that $C \propto A$.

Dielectric. Let us now replace the air between the plates by completely filling the space with a 'dielectric' such as polystyrene or polythene sheets or glass, Figure 7.7 (iii). In this case the area A and distance d remain constant. The vibrating reed experiment then shows that the capacitance has *increased* appreciably when the dielectric is used in place of air.

Some Practical Capacitors

As we have just seen, the simplest capacitor consists of two flat parallel plates with an insulating medium between them. Practical capacitors have a variety of forms but basically they are all forms of parallel-plate capacitors.

A capacitor in which the effective area of the plates can be adjusted is called a *variable capacitor*. In the type shown in Figure 7.8, the plates are semicircular

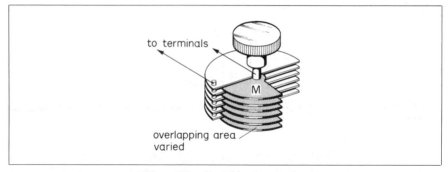

Figure 7.8   *Variable air capacitor*

although other shapes may be used, and one set can be swung into or out of the other. The capacitance is proportional to the area of overlap of the plates. The plates are made of brass or aluminium, and the dielectric may be air or oil or mica. The *variable air capacitor* is used in radio receivers for tuning to the different wavelengths of commercial broadcasting stations.

Figure 7.1 (ii), p. 212, shows a *multiple capacitor* with a mica dielectric. The capacitance is *n* times the capacitance between two successive plates where *n* is the number of dielectrics between all the plates. The whole arrangement is sealed into a plastic case.

Figure 7.1 (iv), p. 212, shows a *paper* capacitor—it has a dielectric of paper impregnated with paraffin wax or oil. Unlike the mica capacitor, the papers can be rolled and sealed into a cylinder of relatively small volume. To increase the stability and reduce the power losses, the paper is now replaced by a thin layer of *polystyrene*.

Electrolytic capacitors are widely used. Basically, they are made by passing a direct current between two sheets of aluminium foil, with a suitable electrolyte or liquid conductor between them, Figure 7.9. A very thin film of aluminium oxide is then formed on the anode plate, which is on the *positive* side of the d.c. supply as shown. This film is an insulator. It forms the dielectric between the two plates, the electrolyte being a good conductor, Figure 7.9 (i). Since the

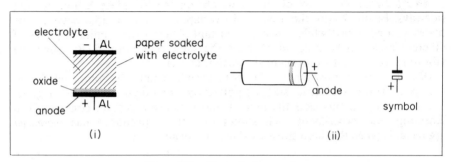

Figure 7.9   *Electrolytic capacitor*

dielectric thickness *d* is so very small, and $C \propto 1/d$, the capacitance value can be very high. Several thousand microfarads may easily be obtained in a capacitor of small volume. To maintain the oxide film, the anode terminal is marked in red or by a + sign, Figure 7.9 (ii). This terminal must be connected to the positive side of the circuit in which the capacitor is used, otherwise the oxide film will break down. It is represented by the unblacked rectangle in the symbol for the electrolytic capacitor shown in Figure 7.9 (ii).

Parallel Plate Capacitor

We now obtain a formula for the capacitance of a parallel-plate capacitor which is widely used.

Suppose two parallel plates of a capacitor each have a charge numerically equal to Q, Figure 7.10. The surface density σ is then Q/A where A is the area of either plate, and the field-strength between the plates, E, is given, from p. 193, by

$$E = \frac{\sigma}{\varepsilon} = \frac{Q}{\varepsilon A}$$

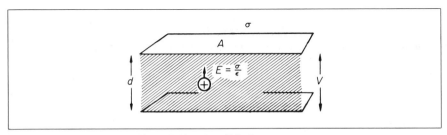

Figure 7.10 *Parallel-plate capacitor*

Now E is numerically equal to the potential gradient V/d, p. 203.

$$\therefore \frac{V}{d} = \frac{Q}{\varepsilon A}$$

$$\therefore \frac{Q}{V} = \frac{\varepsilon A}{d}$$

$$\therefore C = \frac{\varepsilon A}{d} \, . \qquad . \qquad . \qquad . \qquad . \qquad . \qquad (1)$$

It should be noted that this formula for C is approximate, as the field becomes non-uniform at the edges. See Figure 6.33, p. 204.

Thus a capacitor with parallel plates, having a vacuum (or air, if we assume the permittivity of air is the same as a vacuum) between them, has a capacitance given by

$$C = \frac{\varepsilon_0 A}{d}$$

where C = capacitance in farad (F), A = area of overlap of plates in metre^2, d = distance between plates in metre and $\varepsilon_0 = 8 \cdot 854 \times 10^{-12}$ farad metre^{-1}.

Capacitance of Isolated Sphere

Suppose a sphere of radius r metre situated in air is given a charge of Q coulomb. We assume, as on p. 200, that the charge on a sphere gives rise to potentials *on and outside* the sphere as if all the charge were concentrated at the *centre*. From p. 200, the surface of the sphere thus has a potential relative to that 'at infinity' (or, in practice, to that of the earth) given by:

$$\therefore V = \frac{Q}{4\pi\varepsilon_0 r}$$

$$\therefore \frac{Q}{V} = 4\pi\varepsilon_0 r$$

$$\therefore \textbf{Capacitance, } C = 4\pi\varepsilon_0 r \qquad . \qquad . \qquad . \qquad . \qquad (2)$$

The other 'plate' of the capacitor is the earth.

Suppose $r = 10\,\text{cm} = 0\cdot1\,\text{m}$. Then

$$C = 4\pi\varepsilon_0 r = 4\pi \times 8\cdot85 \times 10^{-12} \times 0\cdot1\,\text{F}$$

$$= 11 \times 10^{-12}\,\text{F (approx.)} = 11\,\text{pF}$$

Concentric Spheres

Faraday used two concentric spheres to investigate the relative permittivity (p. 221) of liquids. Suppose a, b are the respective radii of the inner and outer spheres, Figure 7.11. Let $+Q$ be the charge given to the inner sphere and let the outer sphere be earthed, with air between them.

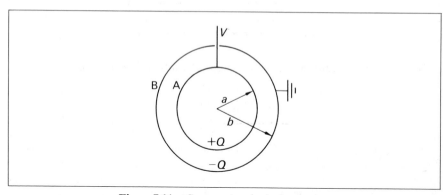

Figure 7.11   *Concentric spherical capacitor*

The induced charge on the outer sphere is $-Q$ (see p. 164). The potential V_a of the inner sphere = potential due to $+Q$ plus potential due to $-Q$ = $+\dfrac{Q}{4\pi\varepsilon_0 a} - \dfrac{Q}{4\pi\varepsilon_0 b}$, since the potential due to the charge $-Q$ is $-Q/4\pi\varepsilon_0 b$ everywhere inside the larger sphere (see p. 201).

But $V_b = 0$, as the outer sphere is earthed.

$$\therefore \text{ potential difference, } V = V_a - V_b = \frac{1}{4\pi\varepsilon_0}\left(\frac{Q}{a} - \frac{Q}{b}\right)$$

$$\therefore V = \frac{Q}{4\pi\varepsilon_0}\left(\frac{b-a}{ab}\right)$$

$$\therefore \frac{Q}{V} = \frac{4\pi\varepsilon_0 ab}{b-a}$$

$$C = \frac{4\pi\varepsilon_0 ab}{b-a} \qquad . \qquad . \qquad . \qquad . \qquad . \qquad (3)$$

As an example, suppose $b = 10\,\text{cm} = 0{\cdot}1\,\text{m}$ and $a = 9\,\text{cm} = 0{\cdot}09\,\text{m}$.

$$\therefore C = \frac{4\pi\varepsilon_0 ab}{b-a}$$

$$= \frac{4\pi \times 8{\cdot}85 \times 10^{-12} \times 0{\cdot}1 \times 0{\cdot}09}{(0{\cdot}1 - 0{\cdot}09)}\,\text{F}$$

$$= 100\,\text{pF (approx.)}$$

Note that the inclusion of a nearby second plate to the capacitor increases the capacitance. For an *isolated* sphere of radius 10 cm, the capacitance was 11 pF (p. 220).

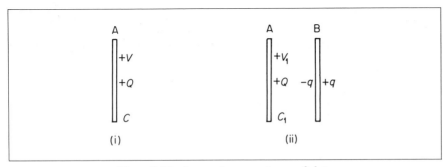

Figure 7.12 *Increasing capacitance of plate*

The same effect is obtained for a metal plate A which has a charge $+Q$, Figure 7.12 (i). If the plate is isolated, A will then have some potential V relative to earth and its capacitance $C = Q/V$.

Now suppose that another metal plate B is brought near to A, as shown, Figure 7.12 (ii). Induced charges $-q$ and $+q$ are then obtained on B. Now the charge $-q$ is nearer A than the charge $+q$. This *lowers* the potential V to a value V_1. So the value of C changes from $C = Q/V$ to $C_1 = Q/V_1$ and since V_1 is less than V, the new capacitance is *greater* than C.

If B is *earthed*, only the negative charge $-q$ is left on B. This lowers the potential of A more than before. So the capacitance C is again *increased*.

Relative Permittivity (Dielectric Constant) and Dielectric Strength

The ratio of the capacitance with and without the dielectric between the plates is called the *relative permittivity* (or *dielectric constant*) of the material used. The expression 'without a dielectric' strictly means 'with the plates in a vacuum'; but the effect of air on the capacitance of a capacitor is so small that for most purposes it may be neglected. The relative permittivity of a substance is denoted by the letter ε_r. So

$$\varepsilon_r = \frac{C_d}{C_v}$$

where C_d is the capacitance with a dielectric completely filling the space between the plates and C_v is the capacitance with a vacuum between the plates. An experiment to measure relative permittivity is given on page 224.

The following table gives the value of relative permittivity, and also of *dielectric strength*, for various substances. The strength of a dielectric is the potential gradient at which its insulation breaks down, and a spark passes through it. A solid dielectric is ruined by such a breakdown, but a liquid or gaseous one heals up as soon as the applied potential difference is reduced.

Water is not suitable as a dielectric in practice, because it is a good insulator only when it is very pure, and to remove all matter dissolved in it is almost impossible.

PROPERTIES OF DIELECTRICS

Substance	Relative permittivity	Dielectric strength, kilovolts per mm
Glass	5–10	30–150
Mica	6	80–200
Ebonite	2·8	30–110
Ice*	94	—
Paraffin wax	2	15–50
Paraffined paper	2	40–60
Methyl alcohol*.	32	—
Water*	81	—
Air (*normal pressure*)	1·0005	—

* Polar molecules (see p. 223).

Action of Dielectric

We regard a molecule as a collection of atomic nuclei, positively charged, and surrounded by a cloud of negative electrons. When a dielectric is in a charged capacitor, its molecules are in an electric field; the nuclei are urged in the direction of the field, and the electrons in the opposite direction, Figure 7.13 (i). Thus each molecule is distorted, or *polarized*: one end has an excess of positive

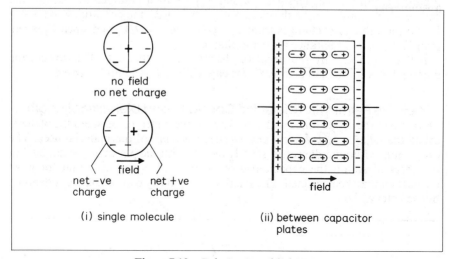

(i) single molecule

(ii) between capacitor plates

Figure 7.13 *Polarization of dielectric*

charge, the other an excess of negative. At the surfaces of the dielectric, therefore, charges appear, as shown in Figure 7.13 (ii). These charges are of opposite sign to the charges on the plates. So they *reduce* the electric field strength E between the plates. Since E = potential difference/plate separation (V/d), the potential difference between the plates is reduced. From $C = Q/V$, where Q is the charge on the plates and V is the p.d. between the plates, it follows that C is *increased*.

If the capacitor is connected to a battery, then its potential difference is constant; but the surface charges on the dielectric still increase its capacitance. They do so because they offset the charges on the plates, and so enable greater charges to accumulate there before the potential difference rises to the battery voltage.

Some molecules, we believe, are permanently polarized: they are called *polar molecules*. Water has polar molecules. The effect of this, in a capacitor, is to increase the capacitance in the way already described. The increase is, in fact, much greater than that obtained with a dielectric which is polarized merely by the action of the field.

ε_0 and its Measurement

We can now see how the unit of ε_0 may be stated in a more convenient manner and how its magnitude may be measured.

Unit. From $C = \dfrac{\varepsilon_0 A}{d}$, we have $\varepsilon_0 = \dfrac{Cd}{A}$.

Thus the unit of $\varepsilon_0 = \dfrac{\text{farad} \times \text{metre}}{\text{metre}^2}$

$$= \text{farad metre}^{-1}, \text{F m}^{-1} \text{ (see also p. 187)}$$

Measurement. In order to find the magnitude of ε_0, the circuit in Figure. 7.14 is used.

C is a parallel plate capacitor, which may be made of sheets of glass or perspex coated with aluminium foil. The two conducting surfaces are placed facing inwards, so that only air is present between these plates. The area A of the plates in metre^2, and the separation d in metres, are measured. P is a high tension supply capable of delivering about 200 V, and G is a calibrated sensitive galvanometer. S is a *vibrating reed switch* unit, energised by a low a.c. voltage

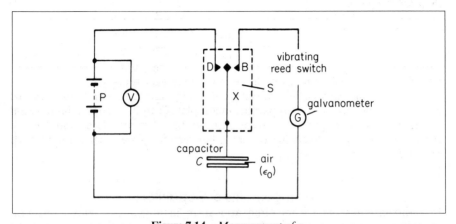

Figure 7.14 *Measurement of ε_0*

from the mains. When operating, the vibrating bar X touches D and then B, and the motion is repeated at the mains frequency, fifty times a second.

As explained previously, when the circuit is on, the vibrating reed switch charges and discharges the capacitor 50 times per second. The average steady current I in G is then read.

Charged once, the charge Q on C is

$$Q = CV = \frac{\varepsilon_0 VA}{d}$$

The capacitor is discharged fifty times per second. Since the current is the charge flowing per second,

$$\therefore I = \frac{\varepsilon_0 VA . 50}{d} \text{ ampere}$$

$$\therefore \varepsilon_0 = \frac{Id}{50 \, VA} \text{ farad metre}^{-1}$$

The following results were obtained in one experiment:

$$A = 0.0317 \, \text{m}^2, d = 1.0 \, \text{cm} = 0.010 \, \text{m}, V = 150 \, \text{V}, I = 0.21 \times 10^{-6} \, \text{A}$$

$$\therefore \varepsilon_0 = \frac{Id}{50VA}$$

$$= \frac{0.21 \times 10^{-6} \times 0.01}{50 \times 150 \times 0.0317}$$

$$= 8.8 \times 10^{-12} \, \text{F m}^{-1}$$

As very small currents are concerned, care must be taken to make the apparatus of high quality insulating material, otherwise leakage currents will lead to serious error.

Relative Permittivity of Glass and Oil

The same method can be used to find the relative permittivity of various solid materials such as *glass*. If the glass completely fills the space between the two plates, and the current in G is I with the glass and I_0 with air between the plates, then, for the glass,

$$\varepsilon_r = \frac{C_{glass}}{C_{air}} = \frac{I}{I_0}$$

So ε_r for glass can be found from the ratio of the two currents.

If ε_r of an insulating liquid such as an *oil* is required, a similar method can be used. This time, however, two parallel metal plates can be used in a large vessel as the capacitor. If I_0 is the current with air between the plates and I is the current when oil completely fills the space between the plates, then ε_r for oil is the ratio I/I_0.

Arrangements of Capacitors

In radio circuits, capacitors often appear in arrangements whose resultant capacitances must be known. To derive expressions for these, we need the equation defining capacitance in its three possible forms:

$$C = \frac{Q}{V}, \quad V = \frac{Q}{C}, \quad Q = CV$$

In Parallel. Figure 7.15 shows three capacitors, having all their left-hand

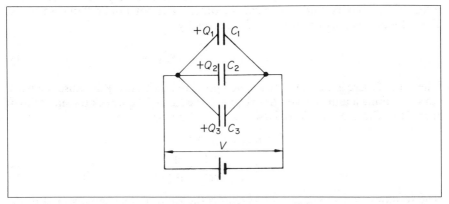

Figure 7.15   *Capacitors in parallel*

plates connected together, and also all their right-hand plates. They are said to be connected in parallel across the same potential difference V. The charges on the individual capacitors are respectively

$$\left. \begin{array}{l} Q_1 = C_1 V \\ Q_2 = C_2 V \\ Q_3 = C_3 V \end{array} \right\} \qquad . \qquad . \qquad . \qquad . \qquad . \qquad (1)$$

The total charge on the system of capacitors is

$$Q = Q_1 + Q_2 + Q_3 = (C_1 + C_2 + C_3)V$$

So the system is equivalent to a single capacitor, of capacitance

$$C = \frac{Q}{V} = C_1 + C_2 + C_3$$

Thus when capacitors are connected in parallel, their resultant capacitance is the *sum* of their individual capacitances. It is greater than the greatest individual one.

In Series. Figure 7.16 shows three capacitors having the right-hand plate of

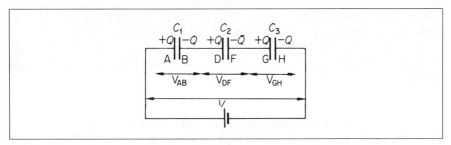

Figure 7.16   *Capacitors in series*

one connected to the left-hand plate of the next, and so on—connected in series. When a cell is connected across the ends of the system, a charge Q is transferred from the plate H to the plate A, a charge $-Q$ being left on H. This charge induces a charge $+Q$ on plate G; similarly, charges appear on all the other capacitor plates, as shown in the figure. (The induced and inducing charges are equal because the capacitor plates are very large and very close together; in effect, either may be said to enclose the other.) The potential differences across the individual capacitors are, therefore, given by

$$V_{AB} = \frac{Q}{C_1}, \quad V_{DF} = \frac{Q}{C_2}, \quad V_{GH} = \frac{Q}{C_3} \qquad \cdot \qquad \cdot \qquad \cdot \qquad (2)$$

The sum of these is equal to the applied potential difference V because the work done in taking a unit charge from H to A is the sum of the work done in taking it from H to G, from F to D, and from B to A. Therefore

$$V = V_{AB} + V_{DF} + V_{GH}$$

$$= Q\left(\frac{1}{C_1} + \frac{1}{C_2} + \frac{1}{C_3}\right) \qquad \cdot \qquad \cdot \qquad \cdot \qquad (3)$$

The resultant capacitance of the system is the ratio of the charge stored to the applied potential difference, V. The charge stored is equal to Q, because, if the battery is removed, and the plates HA joined by a wire, *a charge Q will pass through that wire*, and the whole system will be discharged. The resultant capacitance is therefore given by

$$C = \frac{Q}{V}, \quad \text{or} \quad \frac{1}{C} = \frac{V}{Q}$$

so, by equation (3),

$$\frac{1}{C} = \frac{1}{C_1} + \frac{1}{C_2} + \frac{1}{C_3} \qquad \cdot \qquad \cdot \qquad \cdot \qquad (4)$$

Thus, to find the resultant capacitance of capacitors in series, we must add the reciprocals of their individual capacitances. The resultant is less than the smallest individual.

Comparison of Series and Parallel Arrangements. Let us compare Figures 7.15 and 7.16. In Figure 7.16, where the capacitors are in *series*, all the capacitors carry the same charge, which is equal to the charge carried by the system as a *whole*, Q. So to find the charge Q on each capacitor, use

$$Q = CV$$

where C is the *resultant* or *total* capacitance given by the $1/C$ formula in (4). The potential difference applied to the system, however, is divided amongst the capacitors, in inverse proportion to their capacitances (equations (2)).

In Figure 7.15, where the capacitors are in *parallel*, they all have the same potential difference. The charge stored is divided amongst them, in direct proportion to the capacitances (equations (1)).

Examples on Capacitors in Series and Parallel

1 In Figure 7.17(i), C_1 (3 μF) and C_2 (6 μF) are in series across a 90 V d.c. supply. Calculate the charges on C_1 and C_2 and the p.d. across each.

Total capacitance C is given by $1/C = 1/C_1 + 1/C_2$

$$\frac{1}{C} = \frac{1}{3} + \frac{1}{6} = \frac{3}{6}$$
$$\therefore C = 6/3 = 2\,\mu F$$

The charges on C_1 and C_2 are the same and equal to Q on C.

So $\qquad Q = CV = 2 \times 10^{-6} \times 90 = 180 \times 10^{-6}\,C$

Then $\qquad V_1 = Q/C_1 = 180 \times 10^{-6}/3 \times 10^{-6} = 60\,V$

and $\qquad V_2 = Q/C_2 = 180 \times 10^{-6}/6 \times 10^{-6} = 30\,V$

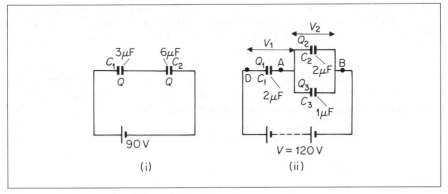

Figure 7.17 *Examples on capacitors*

2 Find the charges on the capacitors in Figure 7.17(ii) and the potential differences across them.

Capacitance between A and B,

$$C' = C_2 + C_3 = 3\,\mu F$$

Overall capacitance B to D, since C_1 and C' are in series, is, from $1/C = 1/C_1 + 1/C'$,

$$C = \frac{C_1 C'}{C_1 + C'} = \frac{2 \times 3}{2 + 3} = 1 \cdot 2\,\mu F$$

Charge stored in this capacitance C

$$= Q_1 = Q_2 + Q_3 = CV = 1 \cdot 2 \times 10^{-6} \times 120$$
$$= 144 \times 10^{-6}\,C$$
$$\therefore V_1 = \frac{Q_1}{C_1} = \frac{144 \times 10^{-6}}{2 \times 10^{-6}} = 72\,V$$

So $\qquad V_2 = V - V_1 = 120 - 72 = 48\,V$

$$Q_2 = C_2 V_2 = 2 \times 10^{-6} \times 48 = 96 \times 10^{-6}\,C$$
$$Q_3 = C_3 V_2 = 10^{-6} \times 48 = 48 \times 10^{-6}\,C$$

Measuring Charge and Capacitance

To measure a charge, a capacitor C_i such as $0 \cdot 01\,\mu F$ or $0 \cdot 1\,\mu F$ is first connected to a digital voltmeter V with an electronic amplifier, which has a very high input impedance or resistance, Figure 7.18.

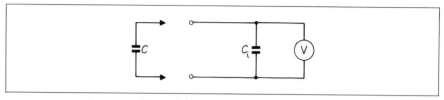

Figure 7.18   *Measuring Q and C*

The charge Q on a capacitor C (or on an insulated metal sphere) is then transferred to C_i as shown and the voltmeter reading V is taken. Suppose this is $0\cdot3$ V and C_i is $0\cdot1$ μF. Then if all the charge on C is transferred to C_i,

$$Q = C_i V = 0\cdot1 \times 10^{-6} \times 0\cdot3 = 3 \times 10^{-8}\,C$$

We can now see how much of the charge Q on C is transferred to the uncharged capacitor C_i. If Q_i is the charge on C_i, the charge left on $C = Q - Q_i$. Now on contact, the capacitors have the same p.d. V. So

$$V = \frac{Q_i}{C_i} = \frac{Q - Q_i}{C}$$

Simplifying,
$$Q_i = \frac{C_i}{C + C_i} Q$$

So if $C_i = 20 \times C$, then $Q_i = (20/21) \times Q = 95\%$ of Q. Therefore C_i must be very large compared with C in order to transfer practically all the charge to C_i.

A capacitor C can be measured by charging it to a suitable known value V, and then transferring the charge Q as we have just described. Then $C = Q/V$.

Energy of a Charged Capacitor

A charged capacitor is a store of electrical energy, as we may see from the vigorous spark it can give on discharge. This can also be shown by charging a large electrolytic capacitor C, such as $10\,000$ μF, to a p.d. of 6 V, and then discharging it through a small or toy electric motor A, Figure 7.19 (i). A small

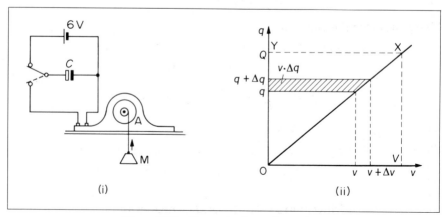

Figure 7.19   *Energy in charged capacitor*

mass M such as $10\,g$, suspended from a thread tied round the motor wheel, now rises as the motor functions. Some of the stored energy in the capacitor is thus transferred to gravitational potential energy of the mass; the remainder is transferred to kinetic energy and heat in the motor.

To find the energy stored in the capacitor, we note that since q (charge) is proportional to v (p.d. across the capacitor) at any instant, the graph OX showing how q varies with v is a straight line, Figure 7.19 (ii). We may therefore consider that the final charge Q on the capacitor moved from one plate to the other through an *average* p.d. equal to $\frac{1}{2}(0 + V)$, since there is zero p.d. across the plates at the start and a p.d. V at the end. So

$$\text{work done, } W = \text{energy stored} = \text{charge} \times \text{p.d.} = Q \times \tfrac{1}{2}V$$

So
$$W = \tfrac{1}{2}QV$$

From $Q = CV$, other expressions for the energy stored are

$$W = \tfrac{1}{2}CV^2 = \frac{Q^2}{2C}$$

$$\textbf{Energy } W = \tfrac{1}{2}CV^2 = \tfrac{1}{2}\frac{Q^2}{C} = \tfrac{1}{2}QV$$

If C is measured in farad, Q in coulomb and V in volt, then the formulae will give the energy W in joules.

Alternative Proof of Energy Formulae

We can also calculate the energy stored in a charged capacitor by a calculus method.

At any instant of the charging process, suppose the charge on the plates is q and the p.d. across the plates is then v. If an additional tiny charge Δq now flows from the negative to the positive plate, we may say that the charge Δq has moved through a p.d. equal to v. So

$$\text{work done in displacing the charge } \Delta q = v \,.\, \Delta q$$

and
$$\text{total work done} = \text{energy stored} = \int_0^Q v \,.\, dq$$

where the limits are $q = Q$, final charge, and $q = 0$, as shown. To integrate, we substitute $v = q/C$. Then

$$\text{energy stored } W = \int_0^Q \frac{q \,.\, dq}{C} = \frac{1}{C}\left[\frac{q^2}{2}\right]_0^Q = \frac{Q^2}{2C}$$

Using $Q = CV$, other expressions for W are
$$W = \tfrac{1}{2}CV^2 \quad \text{or} \quad W = \tfrac{1}{2}QV$$

Energy and Q-V Graph, Heat Produced in Charging

Figure 7.19 (ii) shows the variation of the charge q on the capacitor and its corresponding p.d. v while the capacitor is charged to a final value q. The small shaded area shown $= v \,.\, \Delta q$. So the area represents the small amount of work done or energy stored during a change from q to $q + \Delta q$. It therefore follows that the total energy stored by the capacitor is represented by the area of the triangle OXY. This area $= \tfrac{1}{2}QV$, as previously obtained.

If a high resistor R is included in the charging circuit, the rate of charging is slowed. When the charging current ceases to flow, however, the final charge Q on the capacitor is the same as if negligible resistance was present in the circuit, since the whole of the applied p.d. V is the p.d. across the capacitor when the current in the resistor is zero. Thus the energy stored in the capacitor is $\frac{1}{2}QV$ *whether the resistor is large or small.*

It is important to note that the energy in the capacitor comes from the battery. This supplies an amount of energy equal to QV during the charging process. Half of the energy, $\frac{1}{2}QV$, goes to the capacitor. The other half is transferred to *heat* in the circuit resistance. If this is a high resistance, the charging current is low and the capacitor gains its final charge after a long time. If it is a low resistance, the charging current is higher and the capacitor gains its final charge after a long time. If it is a low resistance, the charging current is higher and the capacitor gains its final charge in a quicker time. In *both* cases, however, the total amount of heat produced is the same, $\frac{1}{2}QV$.

Connected Capacitors, Loss of Energy

Consider a capacitor C_1 of $2\,\mu F$ charged to a p.d. of $50\,V$, and a capacitor C_2 of $3\,\mu F$ charged to a p.d. of $100\,V$, Figure 7.20 (i). Then

$$\text{charge } Q_1 \text{ on } C_1 = C_1 V_1 = 2 \times 10^{-6} \times 50 = 10^{-4}\,C$$

and $\qquad$ $$\text{charge } Q_2 \text{ on } C_2 = C_2 V_2 = 3 \times 10^{-6} \times 100 = 3 \times 10^{-4}\,C$$

$$\therefore \text{ total charge} = 4 \times 10^{-4}\,C \quad . \qquad . \qquad . \qquad . \qquad (1)$$

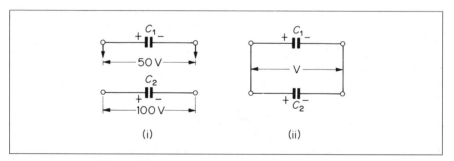

Figure 7.20 *Loss of energy in connected capacitors*

Suppose the capacitors are now joined with plates of like charges connected together, Figure 7.20 (ii). Then some charge will flow from C_1 to C_2 until the p.d. across each capacitor becomes *equal* to some value V. Further, since charge is conserved, the total charge on C_1 and C_2 after connection = the total charge before connection. Now after connection,

$$\text{total charge} = C_1 V + C_2 V = (C_1 + C_2)V = 5 \times 10^{-6}\,V \quad . \qquad (2)$$

Hence, from (1),

$$5 \times 10^{-6}\,V = 4 \times 10^{-4}$$

$$\therefore V = 80\,V$$

$\therefore$ total energy of C_1 and C_2 after connection

$$= \tfrac{1}{2}(C_1 + C_2)V^2$$

$$= \tfrac{1}{2} \times 5 \times 10^{-6} \times 80^2 = 0 \cdot 016\,J \quad . \qquad . \qquad . \qquad . \qquad . \qquad (3)$$

The total energy of C_1 and C_2 *before* connection

$$= \tfrac{1}{2}C_1 V_1{}^2 + \tfrac{1}{2}C_2 V_2{}^2$$
$$= \tfrac{1}{2} \times 2 \times 10^{-6} \times 50^2 + \tfrac{1}{2} \times 3 \times 10^{-6} \times 100^2$$
$$= 0{\cdot}0025 + 0{\cdot}015 = 0{\cdot}0175\,\text{J} \qquad . \qquad . \qquad . \qquad . \qquad . \qquad (4)$$

Comparing (4) with (3), we can see that a *loss of energy* occurs when the capacitors are connected. This loss of energy is converted to *heat* in the connecting wires.

The heat is produced by *flow of current* in the wires connecting the two capacitors when they are joined.

When two capacitors are connected together, in calculations always use:
1   **After connection, the p.d. V across both capacitors is the *same*.**
2   **The total charge before connection = the total charge after connection.**

Discharge in C-R Circuit

We now consider in more detail the *discharge* of a capacitor C through a resistor R, which is widely used in electronic circuits. Suppose the capacitor is initially charged to a p.d. V_0 so that its charge is then $Q = CV_0$. At a time t after the discharge through R has begun, the current I flowing $= V/R$ where V is then the p.d. across C, Figure 7.21 (i). Now

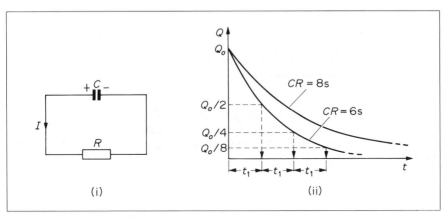

Figure 7.21   *Discharge in C–R circuit*

$$V = \frac{Q}{C} \text{ and } I = -\frac{dQ}{dt} \text{ (the minus shows } Q \text{ decreases with increasing } t\text{).}$$

Hence, from $I = V/R$, we have

$$-\frac{dQ}{dt} = \frac{1}{CR}Q$$

Integrating, $\therefore \displaystyle\int_{Q_0}^{Q} \frac{dQ}{Q} = -\frac{1}{CR}\int_0^t dt$

$$\therefore \ln\left(\frac{Q}{Q_0}\right) = -\frac{t}{CR}$$

$$\therefore Q = Q_0 e^{-t/CR}. \qquad . \qquad . \qquad . \qquad . \qquad (1)$$

Hence Q decreases exponentially with time t, Figure 7.21 (ii). Since the p.d. V across C is proportional to Q, it follows that $V = V_0 e^{-t/CR}$. Further, since the current I in the circuit is proportional to V, then $I = I_0 e^{-t/CR}$, where I_0 is the initial current value, V_0/R.

From (1), Q decreases from Q_0 to half its value, $Q_0/2$, in a time t given by

$$e^{-t/CR} = \tfrac{1}{2} = 2^{-1}$$

$$\therefore t = CR \ln 2$$

Similarly, Q decreases from $Q_0/2$ to half this value, $Q_0/4$, in a time $t = CR \ln 2$. This is the same time from Q_0 to $Q_0/2$. Thus the time for a charge to diminish to half its initial value, no matter what the initial value may be, is always the same. See Fig. 7.21 (ii). This is true for fractions other than one-half. It is typical of an exponential variation or 'decay' which also occurs in radioactivity (p. 887).

Time Constant

The *time constant* T of the discharge circuit is defined as CR seconds, where C is in farad and R is in ohm. Thus if $C = 4\,\mu F$ and $R = 2\,M\Omega$, then $T = (4 \times 10^{-6}) \times (2 \times 10^6) = 8$ seconds. Now, from (1), if $t = CR$, then

$$Q = Q_0 e^{-1} = \frac{1}{e} Q_0$$

So the time constant may be defined as the time for the charge to decay to $1/e$ times its initial value ($e = 2 \cdot 72$ approximately, so that $1/e = 0 \cdot 37$ approx.). If the time constant CR is high, then the charge will diminish slowly; if the time constant is small, the charge will diminish rapidly. See Figure 7.21 (ii).

Charging C through R

Consider now the *charging* of a capacitor C through a resistance R in series, and suppose the applied battery has an e.m.f. E and a negligible internal resistance, Figure 7.22 (i).

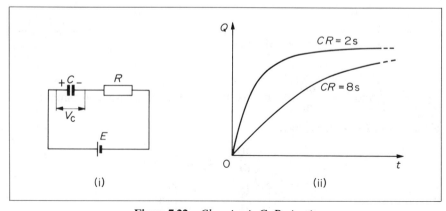

Figure 7.22 *Charging in C–R circuit*

At the instant of making the circuit, there is no charge on C and hence no p.d. across it. So the p.d. across $R = E$, the applied circuit p.d. Thus the initial current flowing, $I_0 = E/R$. Suppose I is the current flowing after a time t. Then, if V_C is the p.d. now across C,

$$I = \frac{E - V_C}{R}$$

Now $I = dQ/dt$ and $V_C = Q/C$. Substituting in the above equation and simplifying,

$$\therefore CR \frac{dQ}{dt} = CE - Q = Q_0 - Q$$

where $Q_0 = CE =$ final charge on C, when no further current flows.
Integrating,

$$\therefore \frac{1}{CR} \int_0^t dt = \int_0^Q \frac{dQ}{Q_0 - Q}$$

$$\therefore \frac{t}{CR} = -\ln\left(\frac{Q_0 - Q}{Q_0}\right)$$

$$\therefore Q = Q_0(1 - e^{-t/CR}) \qquad \cdot \quad \cdot \quad \cdot \quad \cdot \qquad (2)$$

As in the case of the discharge circuit, the *time constant* T is defined as CR seconds with C in farad and R in ohm. If T is high, it takes a long time for C to reach its final charge, that is, C charges slowly. If T is small, C charges rapidly. See Figure 7.22 (ii). The voltage V_C follows the same variation as Q, since $V_C \propto Q$.

Rectangular Pulse Voltage and C–R Circuit
We can apply our results to find how the voltages across a capacitor C and resistor R vary when a *rectangular pulse voltage*, shown in Figure 7.23 (i), is applied to a C–R series circuit. This type of circuit is used in analogue computers.

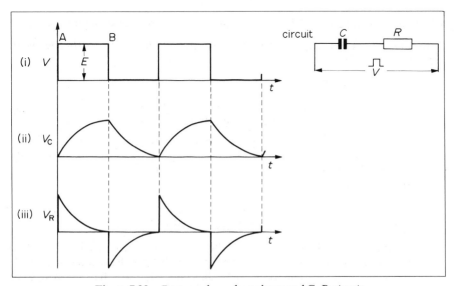

Figure 7.23 *Rectangular pulse voltage and C–R circuit*

On one half of a cycle, the p.d. is constant along AB at a value E say. We can therefore consider that this is similar to the case of *charging* a C–R circuit by a battery of e.m.f. E. The p.d. V_C across the capacitor hence *rises* along exponential curve, Figure 7.23 (ii). During the same time, the p.d. across R, V_R, falls as shown in Figure 7.23 (iii), since $V_R = E - V_C$; that is, the curves for V_R and V_C together *add up to* the straight line graph AB in Figure 7.23 (i).

Similarly, during the time when $V = 0$, the curves for V_C and V_R add up to zero.

Examples on Capacitors

1 Energy

A capacitor of capacitance C is fully charged by a 200 V battery. It is then discharged through a small coil of resistance wire embedded in a thermally insulated block of specific heat capacity $2.5 \times 10^2 \,\mathrm{J\,kg^{-1}\,K^{-1}}$ and of mass 0.1 kg. If the temperature of the block rises by 0.4 K, what is the value of C? (*L.*)

(*Analysis*   Energy (heat) through coil = energy in capacitor.)

$$\text{Energy in capacitor} = \tfrac{1}{2}CV^2 = \tfrac{1}{2} \times C \times 200^2 = 20\,000\,C$$

$$\text{Energy through coil} = mc\theta = 0.1 \times 2.5 \times 10^2 \times 0.4 = 10\,\mathrm{J}$$

So

$$20\,000\,C = 10$$

$$C = \frac{10}{20\,000} = \frac{1}{2000}\,\mathrm{F}$$

$$= 500\,\mu\mathrm{F}$$

2 Vibrating reed switch, Parallel-plate capacitor

In a vibrating reed experiment, two parallel plates have an area $0.12\,\mathrm{m}^2$ and are separated 2 mm by a dielectric. The battery of 150 V charges and discharges the capacitor at a frequency of 50 Hz, and a current of $20\,\mu\mathrm{A}$ is produced. Calculate the relative permittivity of the dielectric if the permittivity of free space is $8.9 \times 10^{-12}\,\mathrm{F\,m^{-1}}$.

What is the new capacitance if the dielectric is half withdrawn from the plates?

(*Analysis*   (i) Use $I = 50\,CV$,   (ii) $C \propto A$, common area between plates.)

Suppose C is the capacitance between the plates. Then, with the usual notation,

$$\text{current } I = 50CV$$

So

$$C = \frac{I}{50V} = \frac{20 \times 10^{-6}}{50 \times 150} = \frac{4 \times 10^{-8}}{15} \qquad . \quad . \quad . \quad \text{(i)}$$

But

$$C = \frac{\varepsilon_r \varepsilon_0 A}{d} = \frac{\varepsilon_r \times 8.9 \times 10^{-12} \times 0.12}{2 \times 10^{-3}} \qquad . \quad . \quad \text{(ii)}$$

So, from (i) and (ii),

$$\varepsilon_r = \frac{4 \times 10^{-8} \times 2 \times 10^{-3}}{15 \times 8.9 \times 10^{-12} \times 0.12} = 5$$

If the dielectric is half withdrawn, the common area of each of the two capacitors formed is now $0.5A$. One capacitor, with air dielectric, has a capacitance given by $0.5\,\varepsilon_0 A/d$. The other, with dielectric of $\varepsilon_r = 5$, has a capacitance given by $2.5\varepsilon_0 A/d$. These capacitances are in parallel, so adding, total capacitance, $C = \dfrac{3\varepsilon_0 A}{d}$

$$= \frac{3 \times 8 \cdot 9 \times 10^{-12} \times 0 \cdot 12}{2 \times 10^{-3}} = 1 \cdot 6 \times 10^{-9}\,\text{F}$$

3 Connected capacitors

The plates of a parallel plate air capacitor consisting of two circular plates, each of 10 cm radius, placed 2 mm apart, are connected to the terminals of an electrostatic voltmeter. The system is charged to give a reading of 100 on the voltmeter scale. The space between the plates is then filled with oil of dielectric constant 4·7 and the voltmeter reading falls to 25. Calculate the capacitance of the voltmeter. You may assume that the voltage recorded by the voltmeter is proportional to the scale reading.

(*Analysis* (i) Total charge is constant,   (ii) p.d. is same for both capacitors after connection.)

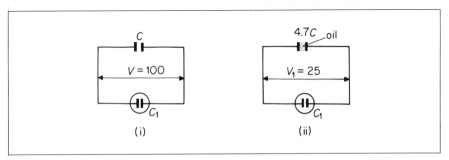

Figure 7.24 *Example on capacitors*

Suppose V is the initial p.d. across the air capacitor and voltmeter, and let C_1 be the voltmeter capacitance and C the plates capacitance, Figure 7.24 (i).

Then   total charge $= CV + C_1 V = (C + C_1)V$.   .   .   (i)

When the plates are filled with oil the capacitance increases to $4 \cdot 7C$, and the p.d. fall to V_1, Figure 7.24 (ii). But the total charge remains constant.

$$\therefore 4 \cdot 7 C V_1 + C_1 V_1 = (C + C_1)V, \quad \text{from (i)}$$

$$\therefore (4 \cdot 7C + C_1)V_1 = (C + C_1)V$$

$$\therefore \frac{4 \cdot 7C + C_1}{C + C_1} = \frac{V}{V_1} = \frac{100}{25} = 4$$

$$\therefore 0 \cdot 7C = 3C_1$$

$$\therefore C_1 = \frac{0 \cdot 7C}{3} = \frac{7}{30}C$$

Now $C = \varepsilon_0 A/d$, where A is in metre^2 and d is in metre.

$$\therefore C = \frac{8 \cdot 85 \times 10^{-12} \times \pi \times (10 \times 10^{-2})^2}{2 \times 10^{-3}}\,\text{F}$$

$$= 1 \cdot 4 \times 10^{-10}\,\text{F (approx.)}$$

$$\therefore C_1 = \frac{7}{30} \times 1 \cdot 4 \times 10^{-10}\,\text{F} = 3 \cdot 3 \times 10^{-11}\,\text{F}$$

_____ Exercises 7 _____

1 A capacitor charged from a 50 V d.c. supply is discharged across a charge-measuring instrument and found to have carried a charge of $10 \, \mu C$. What was the capacitance of the capacitor and how much energy was stored in it? (*L.*)

2 A 300 V battery is connected across capacitors of $3 \, \mu F$ and $6 \, \mu F$
   (a) in parallel,
   (b) in series.
   Calculate the charge and energy stored in each capacitor in (a) and (b).

3 A parallel-plate capacitor with air as the dielectric has a capacitance of $6 \times 10^{-4} \, \mu F$ and is charged by a 100 V battery. Calculate
   (a) the charge,
   (b) the energy stored in the capacitor,
   (c) the energy supplied by the battery.
   What accounts for the difference in the answers for (b) and (c)?
      The battery connections are now removed, leaving the capacitor charged, and a dielectric of relative permittivity 3 is then carefully placed between the plates. What is the new energy stored in the capacitor?

4

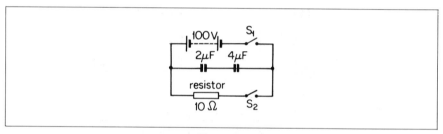

Figure 7A

   If S_2 is left open and S_1 is closed, calculate the quantity of charge on each capacitor, Figure 7A.
   If S_1 is now opened and S_2 is closed, how much charge will flow through the $10 \, \Omega$ resistor?
   If the entire process were repeated with the $10 \, \Omega$ resistor replaced by one of much larger resistance what effect would this have on the flow of charge? (*L.*)

5 (a) Define *capacitance*. Describe briefly the structure of (i) a variable air capacitor, (ii) an electrolytic capacitor, and (iii) a simple paper capacitor.
   (b) The circuit in Figure 7B (i) shows a capacitor C and a resistor R in series. The applied voltage V varies with time as shown in Figure 7B (ii). The product CR is

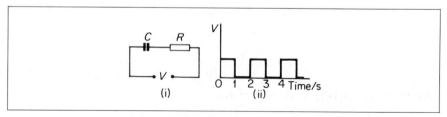

Figure 7B

   of the order 1 s. Sketch graphs showing the way the voltages across C and R vary with time.
      If the product CR were made considerably smaller than 1 s what would be the effect on the graphs?

(c) A capacitor of capacitance $4\,\mu F$ is charged to a potential of 100 V and another of capacitance $6\,\mu F$ is charged to a potential of 200 V. These capacitors are now joined, with plates of like charge connected together. Calculate   (i) the potential across each after joining,   (ii) the total electrical energy stored before joining, and,   (iii) the total electrical energy stored after joining. Explain why the energies calculated in (ii) and (iii) are different. (L.)

6   Three $1\cdot0$-μF capacitors are
(a) connected in series to a $2\cdot0$-V battery,
(b) connected in parallel with each other and a $2\cdot0$-V battery.
Calculate the charge on each of the capacitors in each of cases (a) and (b).
   Account, *without* calculation, for the difference in energy stored in each capacitor in cases (a) and (b). (L.)

7

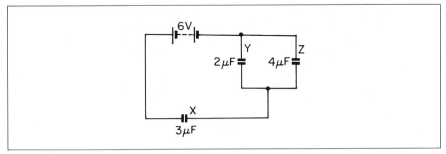

Figure 7C

   Examine the circuit above (Figure 7C) and calculate
(a) the potential difference across capacitor X,
(b) the charge on the plates of capacitor Y,
(c) the energy associated with the charge stored in capacitor Z. (L.)

8   Explain what is meant by *dielectric constant (relative permittivity)*. State two physical properties desirable in a material to be used as the dielectric in a capacitor.
   A sheet of paper 40 mm wide and $1\cdot5 \times 10^{-2}$ mm thick between metal foil of the same width is used to make a $2\cdot0\,\mu F$ capacitor. If the dielectric constant (relative permittivity) of the paper is $2\cdot5$, what length of paper is required? ($\varepsilon_0 = 8\cdot85 \times 10^{-12}\,F\,m^{-1}$.) (*JMB.*)

9

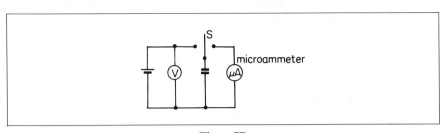

Figure 7D

   In the circuit shown in Figure 7D, S is a vibrating reed switch and the capacitor consists of two flat metal plates parallel to each other and separated by a small air-gap. When the number of vibrations per second of S is n and the potential difference between the battery terminals is V, a steady current I is registered on the microammeter.
(a) Explain this and show that $I = nCV$, where C is the capacitance of the parallel plate arrangement.
(b) Describe how you would use the apparatus to determine how the capacitance C

depends on   (i) the area of overlap of the plates,   (ii) their separation, and show how you would use your results to demonstrate the relationships graphically.

(c) Explain how you could use the measurements made in (b) to obtain a value for the permittivity of air.

(d) In the above arrangement, the microammeter records a current I when S is vibrating. A slab of dielectric having the same thickness as the air-gap is slid between the plates so that one-third of the volume is filled with dielectric. The current is now observed to be $2I$. Ignoring the edge effects, calculate the relative permittivity of the dielectric. (*JMB.*)

10

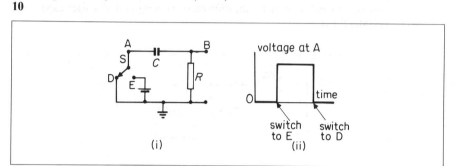

Figure 7E

In the Figure 7E circuit, C is a capacitor and R is a high resistor. By operating the switch S the voltage at A is made to vary with time as shown in the diagram. Sketch the voltage-time graph you would expect to obtain at B and explain its form. (*L.*)

11 Derive an expression for the energy stored in a capacitor C when there is a potential difference V between the plates. If C is in microfarad and V is in volt, express the result in joule.

Show that when a battery is used to charge a capacitor through a resistor, the heat dissipated in a circuit is equal to the energy stored in the capacitor.

Describe the structure of a 1 microfarad capacitor and describe an experiment to compare the capacitance of two capacitors of this type. (*JMB.*)

12 In an experiment to investigate the discharge of a capacitor through a resistor, the circuit shown in Figure 7F was set up. The battery had an e.m.f. of 10 V and negligible internal resistance. The switch was first closed and the capacitor allowed to charge fully. The switch was then opened (at time $t = 0$), and Figure 7G shows how the milliammeter reading subsequently changed with time.

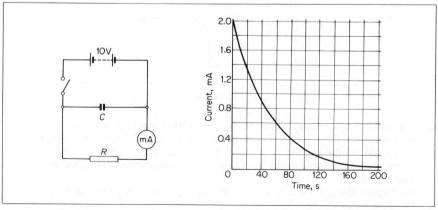

Figure 7F   **Figure 7G**

(a) Use the graph to estimate the initial charge on the capacitor. Explain how you arrived at your answer.
(b) Use your answer to (a) to estimate the capacitance of C.
(c) Calculate the resistance of R. (*AEB*, 1984.)

13 (a) Describe a method for measuring the relative permittivity of a material. Your account should include a labelled circuit diagram, brief details of the procedure and the method used to calculate the result.

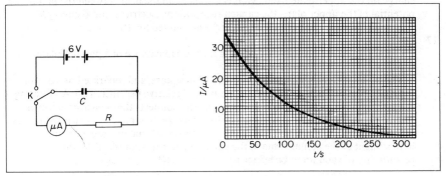

Figure 7H	**Figure 7I**

(b) In the capacitor above, Figure 7H, the capacitor C is first fully charged by using the two-way switch K. The capacitor C is then discharged through the resistor R. The graph (Figure 7I) shows how the current in the resistor R changes with time. Use the graph to help you answer the following questions.

Calculate the resistance R. (The resistance of the microammeter can be neglected.)

Find an approximate value for the charge on the capacitor plates at the beginning of the discharging process and hence calculate (i) the energy stored by the capacitor at the beginning of the discharging process, and (ii) the capacitance C. (*L.*)

14 Define electric field-strength and potential at a point in an electric field.

Explain what is meant by the *relative permittivity* of a material. How may its value be determined experimentally?

A capacitor of capacitance $9 \cdot 0 \, \mu F$ is charged from a source of e.m.f. 200 V. The capacitor is now disconnected from the source and connected in parallel with a second capacitor of capacitance $3 \cdot 0 \, \mu F$. The second capacitor is now removed and discharged. What charge remains on the $9 \cdot 0 \, \mu F$ capacitor? How many times would the process have to be performed in order to reduce the charge on the $9 \cdot 0 \, \mu F$ capacitor to below 50% of its initial value? What would the p.d. between the plates of the capacitor now be? (*L.*)

15 Define the capacitance of a parallel plate capacitor. Write down an expression for this capacitance and explain why your expression is only approximately correct.

A potential difference of 600 V is established between the top cap and the case of a calibrated electroscope by means of a battery which is then removed, leaving the electroscope isolated. When a parallel plate capacitor with air dielectric is connected across the electroscope, one plate to the top cap and the other plate to the case, the p.d. across the electroscope is found to drop to 400 V. If the capacitance of the parallel plate capacitor is $1 \cdot 0 \times 10^{-11} \, F$, calculate
(a) the capacitance of the electroscope;
(b) the change in electrical energy which results from the sharing of the charge. Explain why the total energy is different after sharing.

If the space between the parallel plates of the capacitor were then filled with material of relative permittivity 2, what would then be the potential of the electroscope? (*JMB.*)

16 Two horizontal parallel plates, each of area $500 \, cm^2$, are mounted 2 mm apart in a vacuum. The lower plate is earthed and the upper one is given a positive charge of

0·05 µC. Neglecting edge effects, find the electric field-strength between the plates and state in what direction the field acts.

Deduce values for
(a) the potential of the upper plate,
(b) the capacitance between the two plates,
(c) the electrical energy stored in the system.

If the separation of the plates is doubled, keeping the lower plate earthed and the charge on the upper plate fixed, what is the effect on the field between the plates, the potential of the upper plate, the capacitance and the electrical stored energy?

Discuss how the change in energy can be accounted for. (*O. & C.*)

17 Define *potential, capacitance.*

Obtain from first principles a formula for the capacitance of a parallel-plate capacitor.

The plates of such a capacitor are each 0·4 m square, and separated by 10^{-3} m, the space between being filled with a medium of relative permittivity 5. A vibrating contact, with frequency 50 second^{-1}, repeatedly connects the capacitor across a 120-volt battery and then discharges it through a galvanometer whose resistance is of the order of 50 ohm. Calculate the current recorded, and explain why this is independent of the actual value of the galvanometer resistance. (Take the permittivity of vacuum to be $8·85 \times 10^{-12}$ F m^{-1}.) (*O.*)

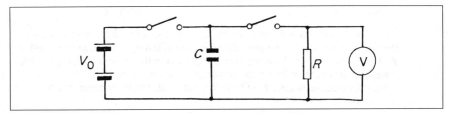

Figure 7J

18 (i) Figure 7J shows the apparatus used by a student to measure the capacitance C of a capacitor. The resistor R is a wire-wound resistor of value 100·0 kΩ. The time constant for the circuit is about 2 minutes.

What do you understand by the term *time constant*?

The potential difference V across the capacitor varies with the discharge time t according to the equation

$$\ln V = \ln V_0 - t/CR.$$

State what measurements you would make in order to obtain a value for C. How would you use a graph to find the result?

How might the voltmeter introduce into the experiment (1) a random error, and (2) a systematic error? Explain why the voltmeter should be read at approximately 15-s intervals for a period of two or three minutes.

(ii) A student has a 360 kΩ resistor and two capacitors. One capacitor is known to have a capacitance of 300 µF. The two capacitors in parallel discharge through the resistor with a time constant of 180 s. Calculate a value for the capacitance of the second capacitor.

What will be the time constant if the two capacitors are connected in series with the resistor? (*L.*)

19 A charged capacitor of capacitance 100 µF is connected across the terminals of a voltmeter of resistance 100 kΩ. When time $t = 0$, the reading on the voltmeter is 10·0 V. Calculate
(a) the charge on the capacitor at $t = 0$,
(b) the reading on the voltmeter at $t = 20·0$ s,
(c) the time which must elapse, from $t = 0$, before 75% of the energy stored in the capacitor at $t = 0$ has been dissipated. (*JMB.*)

8
Current Electricity

We begin current electricity with a study of conduction in metals and the formula for current in terms of the drift velocity of charges. Series and parallel circuits, and ammeters and voltmeters, are then fully discussed, followed by ohmic and non-ohmic conductors and the formulae for electrical energy and power. Finally, we discuss the complete circuit with e.m.f. and internal resistance of batteries and their terminal p.d., and the general Kirchhoff laws.

Ohm's and Joule's Laws: Resistance and Power

Discovery of Electric Current

By the middle of the eighteenth century, electrostatics was a well-established branch of physics. Machines had been invented which could produce by friction great amounts of charge, giving sparks and electric shocks. The momentary current (as we would now call it) carried by the spark or the body was called a 'discharge'.

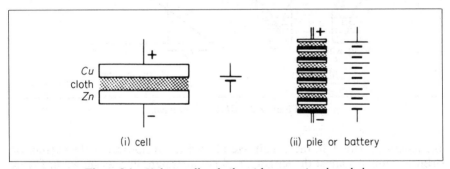

Figure 8.1 *Voltaic cell and pile, with conventional symbols*

In 1799 Volta discovered how to obtain from two metals a continuous supply of electricity: he placed a piece of cloth soaked in brine between copper and zinc plates, Figure 8.1 (i). The arrangement is called a *voltaic cell,* and the metal plates its 'poles'; the copper is known as the positive pole, the zinc as the negative. Volta increased the power by building a pile of cells, with the zinc of one cell resting on the copper of the other, Figure 8.1 (ii). From this pile he obtained sparks and shocks similar to those given by electrostatic machines.

Shortly after, it was found that water was decomposed into hydrogen and oxygen when connected to a voltaic pile. This was the earliest discovery of the chemical effect of an electric current. The heating effect was also soon found, but the magnetic effect, the most important effect, was discovered some twenty years later.

Ohm's Experiment on Resistance

The properties of an electric circuit, as distinct from the effects of a current, were first studied by Ohm in 1826. He set out to find how the length of wire in a circuit affected the current through it—in modern language, he investigated *electrical resistance*. In his first experiment he used voltaic piles as sources of current, but he found that the current which they gave varied considerably, and he later replaced them by thermocouples (p. 273). The voltaic pile or battery and the thermocouple are 'electrical generators'. As we see later, a battery transfers chemical energy to electrical energy and a thermocouple transfers heat energy to electrical energy. These, and other, electrical generators produce a *potential difference* (p.d.), *V*, at their terminals. When a length of wire is joined to the terminal an electric current, *I*, flows along the wire whose magnitude depends on the magnitude of *V*.

Using a constant p.d. from a thermocouple made of copper (Cu) and bismuth (Bi) wires, Ohm passed currents through various lengths of brass wire, 0·37 mm in diameter, and observed the current in a galvanometer G, Figure 8.2 (i). He found that the current *I* in his experiments was almost inversely proportional to

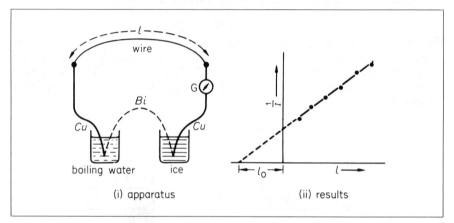

(i) apparatus   (ii) results

Figure 8.2   *Ohm's experiment*

the length of wire, *l*, in the circuit. He plotted the reciprocal of the current (in arbitrary units) against the length *l*, and got a straight line, as shown in Figure 8.2 (ii). So

$$I \propto \frac{1}{l_0 + l}$$

where l_0 is the intercept of the line on the axis of length. Ohm explained this result by supposing, naturally, that the thermocouples and galvanometer, as well as the wire, offered resistance to the current. He interpreted the constant l_0 as the length of wire equal in resistance to the galvanometer and thermocouples.

Conduction in Metals, Heating Effect of Current

The conduction of electricity in metals is due to *free electrons*. Free electrons have thermal energy which depends on the metal temperature, and they wander randomly through the metal from atom to atom.

When a battery is connected across the ends of the metal, an electric field is set up. The electrons are now accelerated by the field, so they gain velocity and energy. When they 'collide' with an atom vibrating about its fixed mean position (called a 'lattice site'), they give up some of their energy to it.

The amplitude of the vibrations is then increased and the temperature of the metal rises.

The electrons are then again accelerated by the field and again give up some energy. Although their movement is erratic, on the average the electrons drift in the direction of the field with a mean speed we calculate shortly. This drift constitutes an 'electric current'. It will be noted that heat is generated by the collision of electrons whichever way they flow. Thus the heating effect of a current—called *Joule heating* (p. 257)—is irreversible, that is, it still occurs when the current in a wire is reversed.

Drift Velocity of Electrons

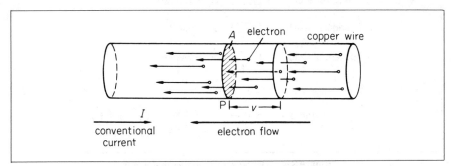

Figure 8.3 *Theory of metallic conduction*

A simple calculation enables the average drift speed to be estimated. Figure 8.3 shows a part of a copper wire of cross-sectional area A through which a current I is flowing. We suppose that there are n electrons per unit volume, and that each electron carries a charge e. Now in one second all those electrons within a distance v to the right of the plane at P, that is, in a volume Av, will flow through this plane, as shown. This volume contains nAv electrons and hence a charge $nAve$. Thus a charge of $nAve$ per second passes P, and so the current I is given by

---

$$I = nAve \qquad . \qquad . \qquad . \qquad . \qquad . \qquad (1)$$

To find the order of magnitude of v, suppose $I = 10\,\text{A}$, $A = 1\,\text{mm}^2 = 10^{-6}\,\text{m}^2$, $e = 1{\cdot}6 \times 10^{-19}\,\text{C}$, and $n = 10^{28}$ electrons m^{-3}. Then, from (1),

$$v = \frac{I}{nAe} = \frac{10}{10^{28} \times 10^{-6} \times 1{\cdot}6 \times 10^{-19}}$$

$$= \frac{1}{160}\,\text{m s}^{-1}\,(\text{approx.})$$

This is a surprisingly slow drift compared with the average thermal speeds, which are of the order of several hundred metres per second (p. 687).

Resistance

The *resistance R* of a conductor is *defined* as the ratio V/I, where V is the p.d. across the conductor and I is the current flowing in it. Thus if the same p.d. V is applied to two conductors A and B, and a smaller current I flows in A, then the resistance of A is greater than that of B. We write, then,

$$\frac{V}{I} = R \qquad . \qquad . \qquad . \qquad . \qquad . \qquad (2)$$

The unit of potential difference, V is the *volt*, symbol V; the unit of current, I, is the *ampere*, symbol A; the unit of resistance, R, is the *ohm*, symbol Ω. The ohm is thus the resistance of a conductor through which a current of one ampere flows when a potential difference (p.d.) of one volt is maintained across it. Figure 8.4 shows some symbols which may be used for different types of resistors, and for ammeters, voltmeters and galvanometers (sensitive current-measuring meters).

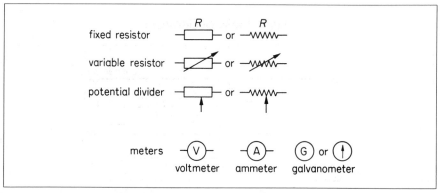

Figure 8.4 *Symbols for resistors and meters*

From the above equation, it also follows that

$$V = IR \quad \text{and} \quad I = \frac{V}{R} \qquad . \qquad . \qquad . \qquad . \qquad (3)$$

Smaller units of current are the milliampere (one-thousandth of an ampere or 10^{-3} A), symbol mA, and the microampere (one-millionth of an ampere or 10^{-6} A), symbol μA. Smaller units of p.d. are the millivolt (10^{-3} V) and the microvolt (10^{-6} V). A small unit of resistance is the microhm ($1/10^6$ or 10^{-6} Ω); larger units are the kilohm ($1000\,\Omega$), symbol kΩ, and the megohm (10^6 ohms), symbol MΩ.

Conductance is defined as the ratio I/V, and is therefore the inverse of resistance, or $1/R$ in numerical value. The unit of conductance is the *siemens*, symbol S.

Series Resistors

The resistors of an electric circuit may be arranged in series, so that the charges carrying the current flow through each in turn (Figure 8.5); or they may be arranged in parallel, so that the flow of charge divides between them as in Figure 8.6, p. 245.

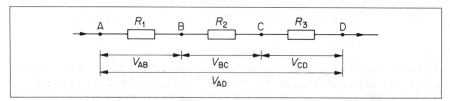

Figure 8.5 *Resistances in series*

Figure 8.5 shows three passive resistors in series, carrying a current I. If V_{AD} is the potential difference across the whole system, the electrical energy supplied to the system per second is IV_{AD} (p. 259). This is equal to the electrical energy per second in all the resistors.

So $$IV_{AD} = IV_{AB} + IV_{BC} + IV_{CD}$$

from which $$V_{AD} = V_{AB} + V_{BC} + V_{CD} \qquad \cdot \quad \cdot \quad \cdot \quad (1)$$

The individual potential differences are given, from previous, by

$$V_{AB} = IR_1, \ V_{BC} = IR_2, \ V_{CD} = IR_3 \qquad \cdot \quad \cdot \quad \cdot \quad (2)$$

So, by equation (1),

$$V_{AD} = IR_1 + IR_2 + IR_3$$
$$= I(R_1 + R_2 + R_3) \qquad \cdot \quad \cdot \quad \cdot \quad (3)$$

And the effective resistance of the system is

$$R = \frac{V_{AD}}{I} = R_1 + R_2 + R_3 \qquad \cdot \quad \cdot \quad \cdot \quad (4)$$

Summarising:

(i) *Current same through all resistors.*
(ii) *Total potential difference = sum of individual potential differences (equation (1)).*
(iii) *Individual potential differences directly proportional to individual resistances (equation (2)).*
(iv) *Total resistance = sum of individual resistances.*

Resistors in Parallel

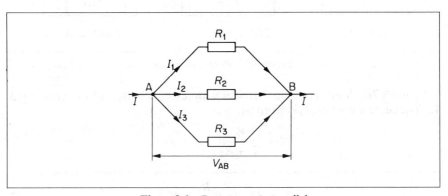

Figure 8.6 *Resistances in parallel*

Figure 8.6 shows three passive resistors connected in parallel, between the points A, B. A passive device is one which produces no energy. A current I enters the system at A and leaves at B, setting up a potential difference V_{AB} between those points. The current branches into I_1, I_2, I_3, through the three elements, and

$$I = I_1 + I_2 + I_3 \qquad \cdot \quad \cdot \quad \cdot \quad \cdot \quad (5)$$

Now
$$I_1 = \frac{V_{AB}}{R_1}, \quad I_2 = \frac{V_{AB}}{R_2}, \quad I_3 = \frac{V_{AB}}{R_3}$$

$$\therefore I = V_{AB}\left(\frac{1}{R_1} + \frac{1}{R_2} + \frac{1}{R_3}\right)$$

$$\therefore \frac{I}{V_{AB}} = \frac{1}{R} = \frac{1}{R_1} + \frac{1}{R_2} + \frac{1}{R_3} \quad . \quad . \quad . \quad . \quad (6)$$

where R is the effective resistance (V_{AB}/I) of the system.

Summarising:

(i) *Potential difference same across each resistor.*
(ii) *Total current = sum of individual currents* (equation (5)).
(iii) *Individual currents inversely proportional to individual resistances.*
(iv) *Effective resistance less than least individual resistance* (equation (6)).

The Potential Divider

Two resistances in series are often used to provide a known fraction of a given p.d. The arrangement is known as a 'potential divider'.

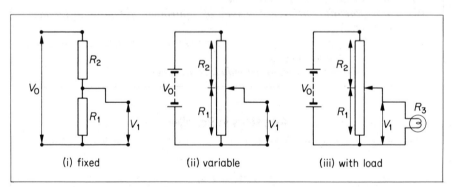

Figure 8.7 *Potential divider*

Figure 8.7 (i) shows a potential divider with resistances R_1 and R_2 across a p.d. V_0. The current flowing, I, is given by

$$I = \frac{V_0}{R_1 + R_2}$$

$$\therefore V_1 = IR_1 = \frac{R_1}{R_1 + R_2} V_0 \quad . \quad . \quad . \quad . \quad (7)$$

So the fraction of V_0 obtained across R_1 is $R_1/(R_1 + R_2)$. If R_1 is $10\,\Omega$ and R_2 is $1000\,\Omega$, then

$$V_1 = \frac{10}{10 + 1000} V_0 = \frac{10}{1010} V_0 = \frac{1}{101} V_0$$

A resistor with a sliding contact can similarly be used, as shown in Figure 8.7 (ii), to provide a continuously variable potential difference, from zero to the full supply value V_0. This is a convenient way of controlling the voltage applied to a load such as a lamp, Figure 8.7 (iii). The resistance of the load, R_3, however, acts in parallel with the resistance R_1. So equation (7) is no longer true, and the voltage V_1 must be measured with a voltmeter. It can be calculated, as in the following example, if R_3 is known. But if the load is a lamp its resistance varies greatly with the current through it, because its temperature varies.

Example on Potential Divider

A load of $2000\,\Omega$ is connected, via a potential divider of resistance $4000\,\Omega$, to a $10\,\text{V}$ supply, Figure 8.8. What is the potential difference across the load when the slider is
(a) one-quarter,
(b) half-way up the divider?

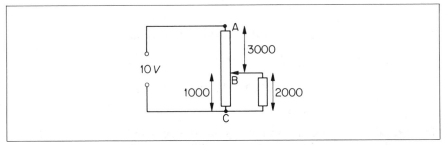

Figure 8.8   *A loaded potential divider*

(a) Since
$$\frac{1}{R_{BC}} = \frac{1}{2000} + \frac{1}{1000}$$

$$R_{BC} = \frac{2000 \times 1000}{2000 + 1000} = \frac{2000}{3}\,\Omega$$

$$\therefore R_{AC} = R_{AB} + R_{BC} = 3000 + \frac{2000}{3} = \frac{11\,000}{3}\,\Omega$$

$$\therefore V_{BC} = \frac{R_{BC}}{R_{AC}} V_{AC}$$

$$= \frac{2000/3}{11\,000/3} \times 10 = \frac{2}{11} \times 10$$

$$= 1{\cdot}8\,\text{V}$$

If the load were removed, V_{BC} would be $(1000/4000)$ of $10\,\text{V}$ or $2{\cdot}5\,\text{V}$.
(b) It is left for the reader to show similarly that $V_{BC} = 3{\cdot}3\,\text{V}$ if the slider is half-way up the divider. Without the load it would be $5\,\text{V}$.

Conversion of a Milliammeter into a Voltmeter

We will now see how to use a milliammeter as a voltmeter. Let us suppose that we have a moving-coil instrument which requires 5 milliamperes ($5\,\text{mA}$ or $5 \times 10^{-3}\,\text{A}$ or $0{\cdot}005\,\text{A}$) for full-scale deflection (f.s.d.). And let us suppose that the resistance of its coil, r, is $20\,\Omega$, Figure 8.9. Then, when it is fully deflected, the potential difference across it is

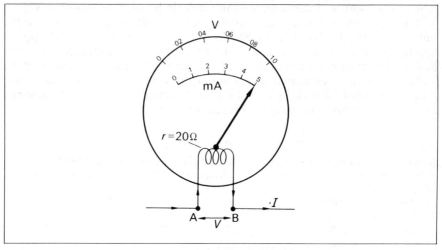

Figure 8.9   *P.d. across moving-coil ammeter*

$$V = rI$$

$$= 20 \times 5 \times 10^{-3} = 100 \times 10^{-3}\,\text{V}$$

$$= 0{\cdot}1\,\text{V}$$

So if the coil resistance is constant, the instrument can be used as a *voltmeter*, giving full-scale deflection for a potential difference of 0·1, or 100 mV. Its scale could be engraved as shown at the top of Figure 8.9.

The potential differences to be measured in the laboratory are usually greater than 100 mV, however. To measure such a potential difference we insert a resistor *R in series* with the coil, as shown in Figure 8.10. If we wish to measure

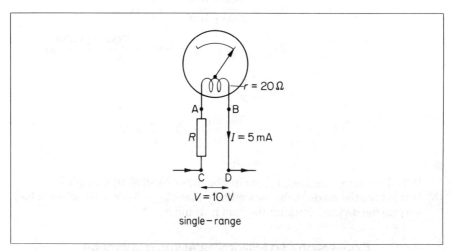

Figure 8.10   *Milliammeter converted to 0–10 V voltmeter*

up to 10 V we must choose the resistance *R* so that, when 10 V is applied between the terminals CD, then a *full-scale* current of 5 mA (5×10^{-3} A) flows through the moving coil.

Now
$$V = (R+r)I$$
$$\therefore 10 = (R+20) \times 5 \times 10^{-3}$$

or
$$R+20 = \frac{10}{5 \times 10^{-3}} = 2 \times 10^3 = 2000\,\Omega$$

$$\therefore R = 2000 - 20$$
$$= 1980\,\Omega \qquad . \quad . \quad . \quad . \quad . \quad . \quad (8)$$

The resistance R is called a *multiplier*. Many voltmeters contain a series of multipliers of different resistances. The range of potential difference measured by the meter can then be varied, for example, from $0-10\,\text{mV}$ to $0-150\,\text{V}$.

Conversion of a Milliammeter into an Ammeter

Moving-coil instruments give full-scale deflection for currents smaller than those generally met in the laboratory. If we wish to measure a current of the order of an ampere or more we connect a low resistance S, called a *shunt*, across the terminals of a moving-coil meter. Figure 8.11. The shunt diverts most of the current to be measured, I, away from the coil—hence its name. Let us suppose

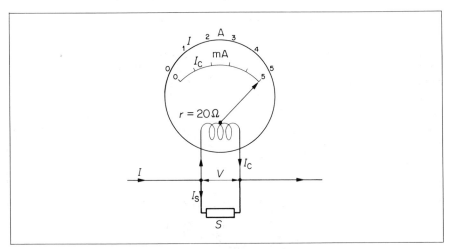

Figure 8.11 *Conversion of milliammeter to ammeter*

that, as before, the coil of the meter has a resistance r of $20\,\Omega$ and is fully deflected by a current, I_C, of $5\,\text{mA}$ ($0\cdot005\,\text{A}$).

Suppose we wish to shunt it so that a full-scale deflection is obtained for a current I of $5\,\text{A}$, and the meter is converted to a range $0-5\,\text{A}$. Then the shunt resistance S must shunt a current I_S of $(5-0\cdot005)\,\text{A}$ or $4\cdot995\,\text{A}$ through itself.

The potential difference across the shunt is the *same* as that across the coil, which is

$$V = rI_C = 20 \times 0\cdot005 = 0\cdot1\,\text{V}$$

The resistance of the shunt must therefore be

$$S = \frac{V}{I_S} = \frac{0\cdot1}{4\cdot995} = 0\cdot020\,02\,\Omega \qquad . \quad . \quad . \quad (9)$$

The ratio of the current measured to the current through the coil is

$$\frac{I}{I_C} = \frac{5}{5 \times 10^{-3}} = 1000$$

This ratio is the same whatever the current I, because it depends only on the resistances S and r; the reader may easily show that its value is $(S+r)/S$. The deflection of the coil is therefore proportional to the measured current, as indicated in Figure 8.11, and the shunt is said to have a 'power' of 1000 when used with this instrument.

The resistance of shunts and multipliers are always given with four-figure accuracy. The moving-coil instrument itself has an error of the order of 1%. A similar error in the shunt or multiplier would therefore double the error in the instrument as a whole. On the other hand, there is nothing to be gained by making the error in the shunt less than about 0·1%, because at that value it is swamped by the error of the moving system.

Multimeters

A *multimeter* is an instrument which is adapted for measuring both current and voltage. It has a shunt R as shown, and a series of voltage multipliers R', Figure 8.12. The shunt is connected permanently across the coil, and the resistances in

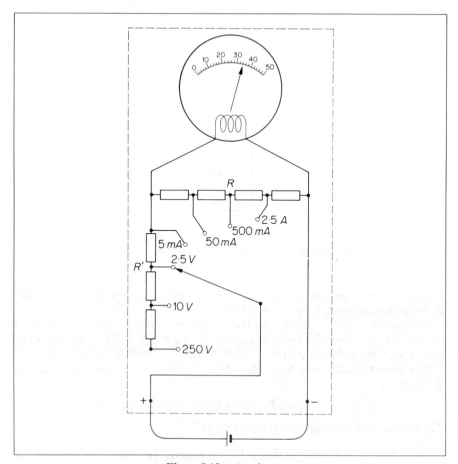

Figure 8.12   *A multimeter*

R' are adjusted to give the desired full-scale voltages with the shunt in position. A switch or plug enables the various full-scale values of current or voltage to be chosen, but the user does the mental arithmetic. The instrument shown in the figure is reading 1·7 volts; if it were on the 10-volt range, it would be reading 6·4.

The terminals of a meter, multimeter or otherwise, are usually marked + and −; the pointer is deflected to the right when current passes through the meter from + to −.

Use of Voltmeter and Ammeter

A moving-coil voltmeter is a current-operated instrument. It can be used to measure potential differences if we assume that the current which it draws is always proportional to the potential difference applied to it as the current varies. Since its action depends on Ohm's law, a moving-coil voltmeter cannot be used in any experiment to demonstrate that law.

We use moving-coil voltmeters as they are more sensitive and more accurate than other forms of voltmeters. The current which they take does, however, sometimes complicate their use. To see how it may do so, let us suppose that we wish to measure a resistance R of about $100\,\Omega$. As shown in Figure 8.13 (i), we connect it in series with a cell, a milliammeter, and a variable resistance; across it we place the voltmeter. We adjust the current until the voltmeter reads, say, $V_1 = 1\,\text{V}$; let us suppose that the milliammeter then reads $I = 12\,\text{mA}$. The value of the resistance then appears to be

$$R = \frac{V_1}{I} = \frac{1}{12 \times 10^{-3}} = \frac{10^3}{12}$$

$$= 83\,\Omega\ (\text{approx.})$$

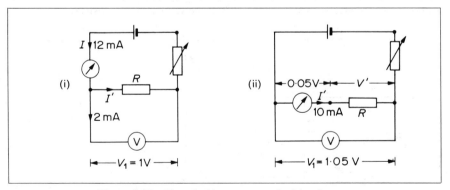

Figure 8.13 *Use of* (i) *ammeter and* (ii) *voltmeter*

But the milliammeter reading *includes the current drawn by the voltmeter*. If that is 2 mA, then the current through the resistor, I', is only 10 mA and its resistance is actually

$$R = \frac{V_1}{I'} = \frac{1}{10 \times 10^{-3}} = \frac{1}{10^{-2}}$$

$$= 100\,\Omega$$

The current drawn by the voltmeter has made the resistance appear 17% lower than its true value.

To try and avoid this error, we might connect the voltmeter as shown in Figure 8.13(ii): across both the resistor and the milliammeter. But its reading would then include the potential difference across the milliammeter. Let us suppose that this is 0·05 V when the current through the milliammeter is 10 mA. Then the potential difference V' across the resistor would be 1 V, and the voltmeter would read 1·05 V. The resistance would appear to be

$$R = \frac{1·05}{10 \times 10^{-3}} = \frac{1·05}{10^{-2}}$$

$$= 105 \, \Omega$$

Thus the voltage drop across the milliammeter would make the resistance appear 5% higher than its true value.

To reduce errors, in *low-resistance* circuits the voltmeter should be connected as in Figure 8.13(i), so that its reading does not include the voltage drop across the ammeter. But in *high-resistance* circuits the voltmeter should be connected as in Figure 8.13(ii) so that the ammeter does not carry its current.

As we see later, using a *potentiometer* to measure p.d. in Figure 8.13(i) is equivalent to using a voltmeter of infinitely-high resistance. In this case no current is drawn by the potentiometer, so this increases the accuracy of measuring R.

Ohm's Law

Ohm investigated how the current I in a given metal varied with the p.d. V across it and came to a conclusion about their relationship, stated later, called *Ohm's law*. Let us consider an experiment which can easily be done with modern apparatus. As shown in Figure 8.14(i), we connect in series the following apparatus: (i) one or more accumulators, S, (ii) a milliammeter A reading to 15 milliamperes, (iii) a wire-wound resistor Q of the order of 50 ohms, (iv) a suitable variable resistance or *rheostat P* of the same order of resistance.

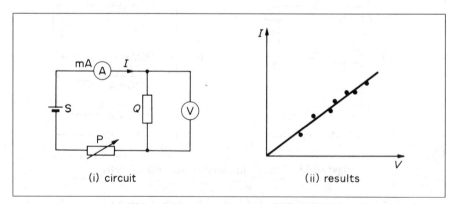

(i) circuit   (ii) results

Figure 8.14 *Demonstration of Ohm's law*

Across the resistor Q we connect a device to measure the potential difference V across it, such as a potentiometer (p. 280) whose calibration does not depend on Ohm's law, otherwise the experiment would not be valid. The milliammeter calibration likewise must not depend on Ohm's law. By adjusting the resistor we vary the current I through the circuit, and at each value of I we measure V. On plotting V against I we get a straight line through the origin, as in Figure 8.14(ii); this shows that the potential difference across the resistor Q is proportional to the current through it:

$$V \propto I$$

This relation was found by Ohm to hold for many conductors. So their resistance R, which is the ratio V/I, is a constant independent of V or I. This is known as *Ohm's law*. Taking into account that resistance depends on temperature and other physical conditions such as mechanical strain, Ohm's law for these type of conductors can be stated as follows:

Under constant physical conditions, the resistance V/I is a constant independent of V or I and their directions.

Ohmic and Non-ohmic Conductors

Ohm's law is obeyed by the most important class of conductors, metals. These are called *ohmic conductors*. In this type of conductor the current I is reversed in direction when the p.d. V is reversed but the magnitude of I is unchanged. The characteristic or I–V graph is thus a straight line *passing through the origin*, as shown in Figure 8.15 (i). An electrolyte such as copper sulphate solution with copper electrodes obeys Ohm's law, Figure 8.15 (ii).

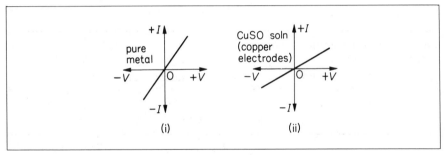

Figure 8.15 *Characteristics of ohmic conductors*

Non-ohmic conductors are those which do not obey Ohm's law ($V \propto I$). Many useful components in the electrical industry must be non-ohmic; for example, a non-ohmic component is essential in a radio receiver circuit. A non-ohmic characteristic or I–V graph may have a curve instead of a straight line; or it may not pass through the origin as in the ohmic characteristic; or it may conduct poorly or not at all when the p.d. is reversed ($-V$). Figure 8.16 illustrates the non-ohmic characteristics of a junction (semiconductor) diode, neon gas, a diode valve, and the electrolyte dilute sulphuric acid with tungsten electrodes where,

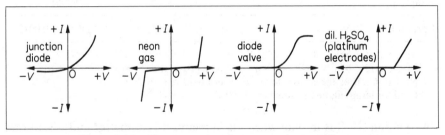

Figure 8.16 *Characteristics of some non-ohmic conductors*

unlike Figure 8.16 (ii), an e.m.f. is produced at the electrodes by the chemicals liberated there.

Resistivity

Ohm showed, by using wires of different length and diameter, that the resistance of a wire, R, is proportional to its length, l, and inversely proportional to its cross-sectional area A. The truth of this can easily be demonstrated today by experiments with a Wheatstone bridge, discussed later, and suitable lengths of wire. We have, then, for a given wire,

$$R \propto \frac{l}{A}$$

we may therefore write

$$R = \rho \frac{l}{A} \qquad \qquad \qquad (1)$$

where ρ is a constant for the material of the wire. It is called the *resistivity* of that material. So

$$\rho = R \frac{A}{l} \qquad \qquad \qquad (2)$$

and resistivity has units

$$\text{ohm} \times \frac{\text{metre}^2}{\text{metre}} = \text{ohm} \times \text{metre or } \Omega \, \text{m}$$

RESISTIVITIES

Substance	Resistivity ρ, Ω m (at 20°C)	Temperature coefficient α, K^{-1}
Aluminium	$2{\cdot}82 \times 10^{-8}$	$0{\cdot}0039$
Brass	$c. \, 8 \times 10^{-8}$	$c. \, 0{\cdot}0015$
Constantan[1]	$c. \, 49 \times 10^{-8}$	$0{\cdot}00001$
Copper	$1{\cdot}72 \times 10^{-8}$	$0{\cdot}0043$
Iron	$c. \, 9{\cdot}8 \times 10^{-8}$	$0{\cdot}0056$
Manganin[2]	$c. \, 44 \times 10^{-8}$	$c. \, 0{\cdot}00001$
Mercury	$95{\cdot}77 \times 10^{-8}$	$0{\cdot}00091$
Nichrome[3]	$c. \, 100 \times 10^{-8}$	$0{\cdot}0004$
Silver	$1{\cdot}62 \times 10^{-8}$	$c \, 0{\cdot}0039$
Tungsten[4]	$5{\cdot}5 \times 10^{-8}$	$0{\cdot}0058$
Carbon (graphite)	$33 \text{ to } 185 \times 10^{-8}$	$-0{\cdot}0006 \text{ to } -0{\cdot}0012$

[1] Also called Eureka; 60% Cu, 40% Ni.
[2] 84% Cu, 12% Mn, 4% Ni; used for resistance boxes and shunts.
[3] Ni–Cu–Cr; used for electric fires—does not oxidize at 1000°C.
[4] Used for lamp filaments—melts at 3380°C.

In equation (1), R is in ohm when l is in metre, A is in metre^2 and ρ is in ohm metre.

The resistivity of a metal is increased by even small amounts of impurity; and alloys, such as Constantan, may have resistivities far greater than any of their constituents as the Table of resistivities shows.

The *temperature coefficient* α is the fractional increase in resistivity per kelvin temperature rise from the resistivity value at 0°C.

(see p. 293).

Examples on Circuits

1   A cell C, having an e.m.f. 2·2 V and negligible internal resistance, is connected to the combination of resistors shown in Figure 8.17. What is the effective value of the resistance connected across the terminals of the cell? What are the values of the currents i_1, i_2 and i_3?

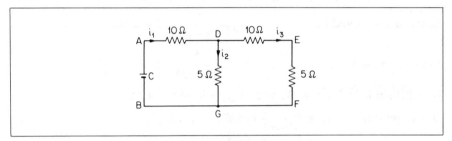

Figure 8.17   *Example on circuits*

Resistance along DEF $= 10 + 5 = 15\,\Omega$

Since DEF is in parallel with the $5\,\Omega$ resistor between D, G (G and F are connected together), the combined resistance R is given by

$$\frac{1}{R} = \frac{1}{15} + \frac{1}{5} = \frac{4}{15}, \text{ so } R = 15/4 = 3\cdot75\,\Omega$$

Thus total resistance between terminals of C $= 10 + 3\cdot75 = 13\cdot75\,\Omega$ (1)

So
$$i_1 = \frac{E}{R} = \frac{2\cdot2}{13\cdot75} = 0\cdot16 \text{ A} \quad . \quad . \quad . \quad . \quad (2)$$

Also, since DEF ($15\,\Omega$) is in parallel with DG ($5\,\Omega$),

$$i_2 = \frac{15}{5+15} \times 0\cdot16 \text{ A} = 0\cdot12 \text{ A}$$

and
$$i_3 = 0\cdot16 - 0\cdot12 = 0\cdot04 \text{ A}$$

2 In the circuit shown in Figure 8.18, calculate the p.d. between B and D, assuming the battery of 12 V has negligible internal resistance.

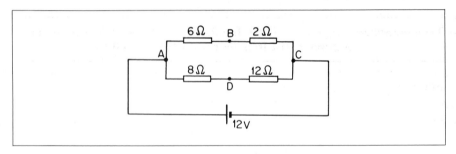

Figure 8.18   *Example on potential difference*

Has B or D the higher potential?

$$\text{P.d. across ABC} = 12\,\text{V} = \text{p.d. across ADC}$$

So p.d. across A and B, $V_{AB} = \dfrac{6}{(6+2)} \times 12\,\text{V} = 9\,\text{V}$

and p.d. across A and D, $V_{AD} = \dfrac{8}{(8+12)} \times 12\,\text{V} = 4\!\cdot\!8\,\text{V}$

By subtraction, p.d. across B and D, $V_{BD} = 9 - 4\!\cdot\!8 = 4\!\cdot\!2\,\text{V}$

Higher potential $V_{AB} = V_A - V_B = 9\,\text{V}$.   .   .   .   .   .   (1)

$$V_{AD} = V_A - V_D = 4\!\cdot\!8\,\text{V}$$.   .   .   .   .   .   (2)

Subtracting (2) from (1),

$$V_D - V_A = 4\!\cdot\!2\,\text{V}$$

So the potential of D is higher than that of B by 4·2 V.

Heat and Power

Electrical Heating, Joule's Laws

In 1841 Joule studied the heating effect of an electric current by passing it through a coil of wire in a jar of water, Figure 8.19. He used various currents, measured by an early form of galvanometer G, and various lengths of wire, but always the same mass of water. The rise in temperature of the water, in a given time, was then proportional to the heat developed by the current in that time. Joule found that the heat produced in a given time, with a given wire, was proportional to I^2, where I is the current flowing. If H is the heat produced per second, then

$$H \propto I^2 \qquad . \qquad . \qquad . \qquad . \qquad . \qquad (1)$$

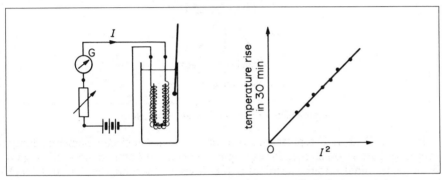

Figure 8.19 *Joule's experiment on heating effect of current*

Joule also made experiments on the heat produced by a given current in different wires. He found that the rate of heat production was proportional to the *resistance* of the wire:

$$H \propto R \qquad . \qquad . \qquad . \qquad . \qquad . \qquad (2)$$

Relationships (1) and (2) together give

$$H \propto I^2 R \qquad . \qquad . \qquad . \qquad . \qquad . \qquad (3)$$

How Current Produces Heat in Metals

Heat is a form of energy. The heat produced per second by a current in a wire is therefore a measure of the energy which it liberates in one second, as it flows through the wire.

The heat is produced, we suppose, by the free electrons as they move through the metal. On their way they collide frequently with atoms. *At each collision they lose some of their kinetic energy*, and give it to the atoms which they strike. Thus, as the current flows through the wire, it increases the kinetic energy of vibration of the metal atoms: it generates heat in the wire. The electrical resistance of the metal is due, we say, to its atoms obstructing the drift of the electrons past them: it is analogous to mechanical friction. As the current flows through the wire, the energy lost per second by the electrons is the electrical power supplied by the battery which maintains the current. That power comes, as we shall see later, from the chemical energy liberated by these actions within the battery.

Potential Difference and Energy

On p. 196 we defined the potential difference V_{AB} between two points, A and B, as the work done by an external agent in taking a unit positive charge from B to A, Figure 8.20(i). This definition applies equally well to points in an electrostatic field and to points on a conductor carrying a current.

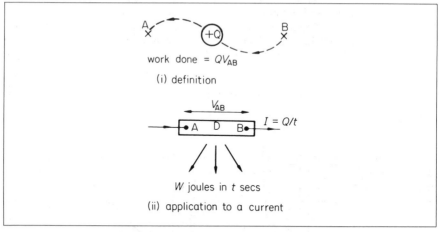

Figure 8.20 *Potential difference and energy*

In Figure 8.20(ii), D represents any electrical device or circuit element: a lamp, motor, or battery on charge, for example. A current of I ampere flows through it from the terminal A to the terminal B. If it flows for t second, the charge Q which it carries from A to B is, since a current is the quantity of electricity per second flowing,

$$Q = It \text{ coulomb} \qquad . \qquad . \qquad . \qquad . \qquad . \qquad (1)$$

Let us suppose that the device D liberates a total amount of energy W joules in the time t; this total may be made up of heat, light, sound, mechanical work, chemical transformation, and any other forms of energy. Then W is the amount of *electrical energy* given up by the charge Q in passing through the device D from A to B. From the definition of p.d.,

$$\therefore W = QV_{AB} \qquad . \qquad . \qquad . \qquad . \qquad . \qquad (2)$$

where V_{AB} is the potential difference between A and B in volts.

The work, in all its forms, which the current I does in t seconds as it flows through the device, is therefore

$$W = IV_{AB}t \qquad . \qquad . \qquad . \qquad . \qquad . \qquad (3)$$

by equations (1) and (2). W, the energy produced, is in joules if I is in amperes, V_{AB} in volts and t in seconds.

Electrical Power

The energy liberated per second in the device is defined as its electrical *power*. The electrical power, P, supplied is given, from above, by

$$P = \frac{W}{t} = \frac{IV_{AB}t}{t}$$

(Figure 8.21 (i)).

or $$P = IV_{AB} \qquad . \qquad . \qquad . \qquad . \qquad . \qquad (1)$$

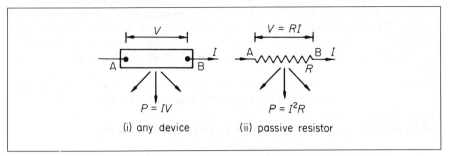

Figure 8.21 *Power equations*

When an electric current flows through a wire or 'passive' resistor, all the power which it conveys to the wire appears as *heat*. If I is the current, R is the resistance, then $V_{AB} = IR$, Figure 8.21 (ii)

$$\therefore P = I^2R \qquad . \qquad . \qquad . \qquad . \qquad . \qquad (2)$$

Also, $$P = \frac{V_{AB}^{2}}{R} \qquad . \qquad . \qquad . \qquad . \qquad (3)$$

The power, P, is in *watts* (W) when I is in ampere, R is in ohm, and V_{AB} is in volt. 1 kilowatt (kW) = 1000 watts.

The formulae for power, $P = I^2R$ or V^2/R, is true only when all the electrical power supplied is dissipated as heat. As we shall see, the formulae do not hold when part of the electrical energy supplied is converted into mechanical work, as in a motor, or into chemical energy, as in an accumulator being charged. A device which converts all the electrical energy supplied to it into heat is called a 'passive' resistor; it may be a wire, or a strip of carbon, or a liquid which conducts electricity but is not decomposed by it. Since the joule (J) is the unit of heat, it follows that, for a resistor, the heat H in it in joules is given by

$$H = IVt$$
or by $$H = I^2Rt \qquad . \qquad . \qquad . \qquad . \qquad . \qquad (4)$$
or by $$H = \frac{V^2t}{R}$$

The units of I, V, R are ampere (A), volt (V), ohm (Ω) respectively.

High-tension (High-voltage) Transmission

When electricity has to be transmitted from a source, such as a power station, to a distant load, such as a factory, the two must be connected by cables. These cables have resistance, which is in effect added to the internal resistance of the generator. Power is wasted in them as heat. If r is the total resistance of the cables, and I the supply current, the power wasted is I^2r.

The power delivered to the factory is IV, where V is the potential difference at the factory. Economy requires the waste power, I^2r, to be small; but it also requires the cables to be thin, and therefore cheap to buy and erect. The thinner the cables, however, the higher their resistance r. Thus the most economical way to transmit the power is to make the current, I, as small as possible; this means making the potential difference V as *high* as possible. When large amounts of power are to be transmitted, therefore, very high voltages are used: 400 000 volts on the main lines of the British grid, 23 000 volts on subsidiary lines.

These voltages are much too high to be brought into a house, or even a factory. They are stepped down by transformers, in a way which we shall describe later; stepping-down in that way is possible only with alternating current, which is one of the main reasons why alternating current is so widely used.

Summary of Formulae Related to Power

In any device whatever

Electrical power consumed = power developed in other forms,

$$P = IV$$

watts = amperes × volts

In a passive resistor

(i)
$$V = IR \quad I = \frac{V}{R} \quad R = \frac{V}{I}$$

(ii) **Power consumed = heat developed per second, in watts.**

$$P = I^2 R = IV = \frac{V^2}{R}$$

(iii) **Heat developed in time t:**

Electrical energy consumed = heat developed in joules

$$I^2 Rt = IVt = \frac{V^2}{R}t$$

Commercial unit = kilowatt hour (kWh) = kilowatt × hour

$$= 3\cdot 6 \times 10^6 \text{ joule}$$

Example on Heating Effect of Current

An electric heating element to dissipate 480 watts on 240 V mains is to be made from Nichrome ribbon 1 mm wide and thickness 0·05 mm. Calculate the length of ribbon required if the resistivity of Nichrome is $1\cdot 1 \times 10^{-6}$ ohm metre.

Power,
$$P = \frac{V^2}{R}$$

$$\therefore R = \frac{V^2}{P} = \frac{240^2}{480} = 120\,\Omega$$

The area A of cross-section of the ribbon $= 1 \times 0\cdot 05 \text{ mm}^2 = 0\cdot 05 \times 10^{-6} \text{ m}^2$.

From
$$R = \frac{\rho l}{A}$$

$$\therefore l = \frac{R \cdot A}{\rho} = \frac{120 \times 0\cdot 05 \times 10^{-6}}{1\cdot 1 \times 10^{-6}} = 5\cdot 45 \text{ m}$$

Electromotive Force

E.M.F. and Internal Resistance

An electrical generator provides energy and power. This is considered later. Here we consider the current and potential difference, p.d., in circuits connected to a generator such as a battery.

If a high resistance voltmeter is connected across the terminals of a dry battery B, the meter may read about 1·5 V, Figure 8.22 (i). Since practically no current flows from the battery in this case we say it is on 'open circuit'. The p.d. across the terminals of a battery (or any other generator) on open circuit is called its *electromotive force* or *e.m.f.*, symbol *E*. We define e.m.f. in terms of energy later (p. 266).

When a resistor is connected to the battery, the current flows through the resistor and through the *internal resistance*, *r*, of the battery to complete the circuit flow.

The e.m.f. of a battery depends on the nature of the chemicals used and not on its size. A tiny battery has the same e.m.f. as a large battery made of the same chemicals. The internal resistance of the tiny battery, however, is much less than the large battery. Provided only a small current is taken from a battery, its e.m.f. and internal resistance are fairly constant.

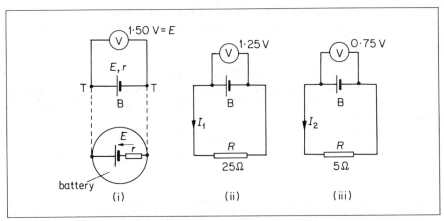

Figure 8.22 *E.m.f. and internal resistance*

Any electrical generator, then, has two important properties, an e.m.f. *E* and an internal resistance *r*. As shown in Figure 8.22 (i), *E* and *r* may be represented separately in a diagram, though in practice they are inseparable between the terminals T, T.

Circuit Principles, Terminal p.d.

In Figure 8.22 (ii), a resistor of 25 Ω is connected to the battery B so that a current I_1 flows in the circuit. The voltmeter reading across the battery terminals, or terminal p.d., may then be 1·25 V, although the e.m.f. is 1·5 V. When the resistor is replaced by one of 5 Ω, Figure 8.22 (iii), a larger current I_2 flows and the voltmeter reading or terminal p.d. is now 0·75 V.

To understand why the terminal p.d. varies when a current flows from a battery, it is important to realise that the voltmeter is connected across the *external* or outside resistance in Figure 8.22 (ii). So 1·25 V is the p.d. across the

$25\,\Omega$ resistor. Now the e.m.f., $1\cdot5\,V$, maintains the current in the *whole* circuit, that is, through the external *and* internal resistance r. So we deduce that the p.d. across the internal resistance $r = 1\cdot5 - 1\cdot25 = 0\cdot25\,V$.

Similarly, in Figure 8.22 (iii) $0\cdot75\,V$ is the p.d. across the external resistance $5\,\Omega$. So in this case the p.d. across the internal resistance $r = 1\cdot5 - 0\cdot75 = 0\cdot75V$. A common error is to think that the voltmeter across the terminals reads the e.m.f. This is not the case here as there is a p.d. across the internal resistance when a current flows, and the voltmeter can only read the p.d. across the external resistance R, which is the terminal p.d.

In Figure 8.22 (ii), the p.d. across the $25\,\Omega$ external resistor R is $1\cdot25\,V$ and the p.d. across the internal resistance r is $0\cdot25\,V$. Since the same current flows in R and r it follows that $R = 5r$, or $r = R/5 = 5\,\Omega$.

Similarly, the p.d. across the external resistor R of $5\,\Omega$ in Figure 8.22 (iii) is $0\cdot75\,V$ and that across the internal resistance r is $0\cdot75\,V$. So $r = R = 5\,\Omega$, as previously calculated.

You should now see that as the external resistance R increases, the terminal p.d. increases. When R is an infinitely-high value, so that $I = 0$, the terminal p.d. is equal to the e.m.f. E.

Summarising:
1. **E = p.d. across the *whole* circuit, R plus r.**
2. **Terminal p.d. V, when current flows = p.d. across R *only*.**

Circuit Formulae

In Figure 8.22 (ii), a battery of e.m.f. E and internal resistance r is joined to an external resistor R, and a current I flows in the circuit. The p.d. across $R = IR$ and the p.d. across $r = Ir$. So

$$E = IR + Ir \qquad . \qquad . \qquad . \qquad . \qquad . \qquad (1)$$

or
$$I = \frac{E}{R+r} \qquad . \qquad . \qquad . \qquad . \qquad (2)$$

Note carefully that when the e.m.f. E is used to find the current I, the resistance $(R + r)$ of the *whole* circuit is required.

On the other hand, the terminal p.d., V = p.d. across external resistor R. So

$$\text{terminal p.d. } V = IR = \frac{ER}{R+r}$$

Further, from (1),
$$V = E - Ir \qquad . \qquad . \qquad . \qquad . \qquad . \qquad (3)$$

This is a useful formula for the terminal p.d. when the e.m.f. E and internal resistance r are known. For example, suppose a current of $0\cdot5\,A$ flows from a battery of e.m.f. E of $3\,V$ and internal resistance $4\,\Omega$. Then

$$\text{terminal p.d. } V = E - Ir = 3 - (0\cdot5 \times 4) = 1\,V$$

The internal resistance r can be found from the e.m.f. E, the terminal p.d. V and the current I. From (1),

$$Ir = E - IR = E - V$$

So
$$r = \frac{E-V}{I} \qquad \cdot \quad \cdot \quad \cdot \quad \cdot \quad \cdot \quad (4)$$

If I is needed, we may use $I = V/R$.

Example on Circuit Calculation

A battery of e.m.f. 1·50 V has a terminal p.d. of 1·25 V when a resistor of 25 Ω is joined to it. Calculate the current flowing, the internal resistance r and the terminal p.d. when a resistor of 10 Ω replaces the 25 Ω resistor.

We have
$$I = \frac{V}{R} = \frac{1·25}{25} = 0·05 \text{ A}$$

Also,
$$r = \frac{\text{p.d.}}{\text{current}} = \frac{E-V}{I}$$

$$= \frac{1·50 - 1·25}{0·05} = \frac{0·25}{0·05} = 5 \Omega$$

When the external resistor is 10 Ω, the current I flowing is
$$I = \frac{E}{R+r} = \frac{1·50}{10+5} = 0·1 \text{ A}$$

So   terminal p.d. $V = IR = 0·1 \times 10 = 1 \text{ V}$

Terminal p.d. with Current in Opposition to E.M.F.

So far we have considered the terminal p.d. when the battery e.m.f. maintains the current. Suppose, however, that a current is passed through a battery in *opposition* to its e.m.f., a case which occurs in re-charging an accumulator, for example.

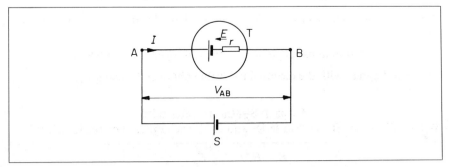

Figure 8.23 *Terminal p.d.*

Figure 8.23 shows a supply S sending a current I through a battery T in opposition to its e.m.f. E. The terminal p.d. V_{AB} must be greater than E in this case. Since the net p.d., $V_{AB} - E$, across the terminals must maintain the current I in r, then the net p.d. = Ir. So
$$V_{AB} - E = Ir$$

Hence
$$V_{AB} = E + Ir$$

In contrast, when the battery e.m.f. itself maintains a current, so that the current is in the same direction as the e.m.f., then the terminal p.d. $V = E - Ir$, as we have already seen.

Example on Terminal p.d.

Figure 8.24 shows a circuit with two batteries in opposition to each other. One has an

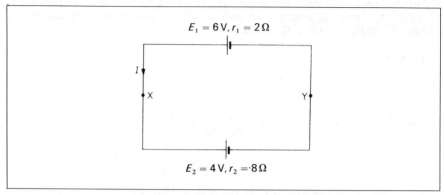

Figure 8.24 *Example*

e.m.f. E_1 of 6 V and internal resistance r_1 of $2\,\Omega$ and the other an e.m.f. E_2 of 4 V and internal resistance r_2 of $8\,\Omega$. Calculate the p.d. V_{XY} across XY.

Net e.m.f. in circuit $= E_1 - E_2 = 6 - 4 = 2$ V

So   current, $I = \dfrac{E_1 - E_2}{r_1 + r_2} = \dfrac{6 - 4}{2 + 8} = 0{\cdot}2$ A

The e.m.f. E_1 is in the same direction as the current. So

terminal p.d., $V_{XY} = E_1 - Ir_1 = 6 - (0{\cdot}2 \times 2) = 5{\cdot}6$ V

If we consider the battery of e.m.f. E_2, we see that I flows in *opposition* to E_2. In this case,

terminal p.d., $V_{XY} = E_2 + Ir_2 = 4 + (0{\cdot}2 \times 8) = 5{\cdot}6$ V

This result agrees with the terminal p.d. value obtained by using E_1.

Cells in Series and Parallel

When cells or batteries are in series and assist each other, then the total e.m.f.

$$E = E_1 + E_2 + E_3 + \ldots \qquad\qquad . \qquad . \qquad . \quad (1)$$

and the total internal resistance

$$r = r_1 + r_2 + r_3 + \ldots \qquad\qquad . \qquad . \qquad . \quad (2)$$

where E_1, E_2 are the individual e.m.f.s and r_1, r_2 are the corresponding internal resistances. If one cell, e.m.f. E_2 say, is turned round 'in opposition' to the others, then $E = E_1 - E_2 + E_3 + \ldots$; but the total internal resistance remains unaltered.

When *similar cells are in parallel*, the total e.m.f. $= E$, the e.m.f. of any one of them. The internal resistance r is here given by

$$\frac{1}{r} = \frac{1}{r_1} + \frac{1}{r_1} + \dots \qquad \cdot \quad \cdot \quad \cdot \quad \cdot \quad (3)$$

where r_1 is the internal resistance of each cell. If different cells are in parallel, there is no simple formula for the total e.m.f. and the total internal resistance, and any calculations involving circuits with such cells are dealt with by applying Kirchhoff's laws, discussed later.

Examples on Circuits

1 Two similar cells A and B are connected in series with a coil of resistance $9.8\,\Omega$. A voltmeter of very high resistance connected to the terminals of A reads 0.96 V and when connected to the terminals of B it reads 1.00 V, Figure 8.25. Find the internal resistance of each cell. (Take the e.m.f. of a cell as 1.08 V.) (L.)

The p.d. across both cells $= 0.96 + 1.00 = 1.96$ V

$$= \text{p.d. across } 9.8\,\Omega$$

$$\therefore \text{ current flowing, } I, = \frac{V}{R} = \frac{1.96}{9.8} = 0.2 \text{ A}$$

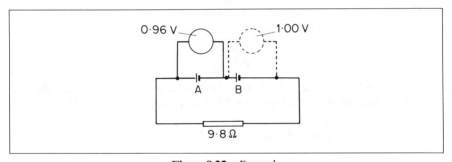

Figure 8.25 *Example*

Now terminal p.d. across each cell $= E - Ir$

$$\therefore \text{ for cell A, } 0.96 = 1.08 - 0.2r, \text{ or } r = 0.6\,\Omega$$

$$\text{for cell B, } 1.00 = 1.08 - 0.2r, \text{ or } r = 0.4\,\Omega$$

2 What is meant by the *electromotive force* of a cell?
   A voltmeter is connected in parallel with a variable resistance, R, which is in series with an ammeter and a cell. For one value of R the meters read 0.3 A and 0.9 V. For another value of R the readings are 0.25 A and 1.0 V. Find the values of R, the e.m.f. of the cell, and the internal resistance of the cell. What assumptions are made about the resistance of the meters in the calculation?
   If in this experiment the ammeter had a resistance of $10\,\Omega$ and the voltmeter a resistance of $100\,\Omega$ and R was $2\,\Omega$, what would the meters read? (L.)

(i) The voltmeter reads the terminal p.d. across the cell if the resistances of the meters are neglected. Thus, with the usual notation (Fig. 8.26 (i)),

$$E - Ir = 0.9, \text{ or } E - 0.3r = 0.9 \qquad \cdot \quad \cdot \quad \cdot \quad (i)$$

and
$$E - 0.25r = 1.0 \qquad \cdot \quad \cdot \quad \cdot \quad \cdot \quad (ii)$$

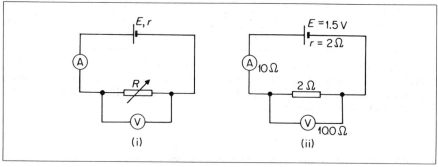

Figure 8.26 *Example*

Subtracting (i) from (ii),

$$0{\cdot}05r = 0{\cdot}1, \text{ i.e. } r = 2\,\Omega$$

Also, from (i),

$$E = 0{\cdot}3r + 0{\cdot}9 = 0{\cdot}6 + 0{\cdot}9 = 1{\cdot}5\text{ V}$$

Further,

$$R_1 = \frac{V}{I} = \frac{0{\cdot}9}{0{\cdot}3} = 3\,\Omega$$

and

$$R_2 = \frac{1{\cdot}0}{0{\cdot}25} = 4\,\Omega$$

(ii) If the voltmeter has $100\,\Omega$ resistance and is in parallel with the $2\,\Omega$ resistance, the combined resistance R is given by (Figure 8.26 (ii))

$$\frac{1}{R} = \frac{1}{2} + \frac{1}{100} = \frac{51}{100}, \text{ or } R = \frac{100}{51}\,\Omega$$

$$\therefore \text{ current, } I = \frac{E}{\text{Total resistance}}$$

$$= \frac{1{\cdot}5}{\dfrac{100}{51} + 10 + 2} = 0{\cdot}11\text{ A}$$

Also, voltmeter reading $V = IR = 0{\cdot}11 \times \dfrac{100}{51} = 0{\cdot}21$ V

Electromotive Force and Energy

We can now get a definition of electromotive force E from energy principles.

We can define the e.m.f. E of a battery or any other generator as the *total energy per coulomb* it delivers round a circuit joined to it.

So if a device has an electromotive force E, then, in passing a charge Q round a circuit joined to it, it liberates an amount of electrical energy equal to QE. If a charge Q is passed through the source against its e.m.f., then the work done against the e.m.f. is QE. The above definition of e.m.f. does not depend on any assumptions about the nature of the generator.

If a device of e.m.f. E passes a steady current I for a time t, then the charge that it circulates is

$$Q = It$$

So, from the definition of E,

total electrical energy liberated, W, $= QE = IEt$ . . (1)

and **total electrical power generated, $P = \dfrac{W}{t} = EI$** . . (2)

We can also define e.m.f. in terms of power and current, and therefore in a way suitable for dealing with circuit problems. From equation (2)

$$P = EI$$

or $$E = \frac{P}{I}$$

So *the e.m.f. of a device is the ratio of the electrical power which it generates to the current which it delivers.* If current is forced through a device in opposition to its e.m.f., then equation (2) gives the power used in overcoming the e.m.f.

Electromotive force resembles potential difference in that both can be defined as the ratio of power to current. The unit of e.m.f. is therefore 1 watt per ampere, or 1 volt; and *the e.m.f. of a source, in volt, is numerically equal to the power which it generates when it delivers a current of* 1 *ampere.*

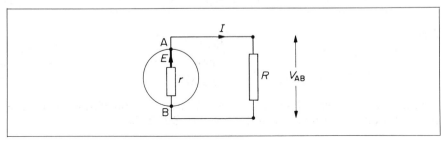

Figure 8.27 *A complete circuit*

Current Formula

We can apply the definition of E in terms of power to the circuit in Figure 8.27. The total power supplied by the source is EI. The power delivered to the external resistor is called the *output power*. Its value $= IV_{AB} = I \times IR = I^2R$. The power delivered to the internal resistance $r = I^2r$. So

$$EI = I^2R + I^2r \qquad . \qquad . \qquad . \qquad . \quad (1)$$

Dividing by I, $$E = IR + Ir \qquad . \qquad . \qquad . \qquad . \quad (2)$$

and $$I = \frac{E}{R+r} \qquad . \qquad . \qquad . \qquad . \quad (3)$$

The same results in (2) and (3) were obtained earlier in the chapter.

Output Power and Efficiency

As we have just seen, the power delivered to the external resistor R, often called the *load* of a battery or other generator, is given by $P_{out} = IV_{AB}$ in Figure 8.27. The power supplied by the source or generator $P_{gen} = IE$. The difference between the power generated and the output is the power wasted as heat in the source: $I^2 r$. The ratio of the power output to the power generated is the efficiency, η, of the circuit as a whole:

$$\eta = \frac{P_{out}}{P_{gen}} \qquad \qquad (1)$$

So
$$\eta = \frac{P_{out}}{P_{gen}} = \frac{IV_{AB}}{IE} = \frac{V_{AB}}{E}$$

Now $V_{AB} = IR = ER/(R+r)$. So

$$\eta = \frac{R}{R+r} \qquad \qquad (2)$$

This shows that the efficiency tends to unity (or 100 per cent) as the load resistance R tends to *infinity*. For high efficiency the load resistance must be several times the internal resistance of the source. When the load resistance is equal to the internal resistance, the efficiency is 50 per cent. (See Figure 8.28 (i)).

Power Variation, Maximum Power

Now let us consider how the *power output* varies with the load resistance. Since the power output $= IV_{AB} = I \times IR$, then

$$P_{out} = I^2 R$$

Also
$$I = \frac{E}{R+r}$$

so
$$P_{out} = \frac{E^2 R}{(R+r)^2}$$

If we take fixed values of E and r, and plot P_{out} as a function of R, we find that it passes through a maximum when $R = r$, Figure 8.28 (i). So

---

power output to R is a maximum when $R = r$, internal resistance.

We shall explain this result shortly. Physically, this result means that the power output is very small when R is either very large or very small, compared with r. When R is very large, the terminal potential difference, V_{AB}, approaches a constant value equal to the e.m.f. E (Figure 8.28 (ii)); as R is increased the current falls, and the power IV_{AB} falls with it. When R is very small, the current approaches the constant value E/r, but the potential difference (which is equal to IR) falls steadily with R; the power output therefore falls likewise. Consequently the power output is greater for a moderate value of R; the mathematics shows that this value is actually $R = r$.

To prove $R = r$ for maximum power, we have, from before,

$$P_{out} = \frac{E^2 R}{(R+r)^2}$$

Now
$$(R+r)^2 = R^2 + 2Rr + r^2 = (R-r)^2 + 4Rr$$

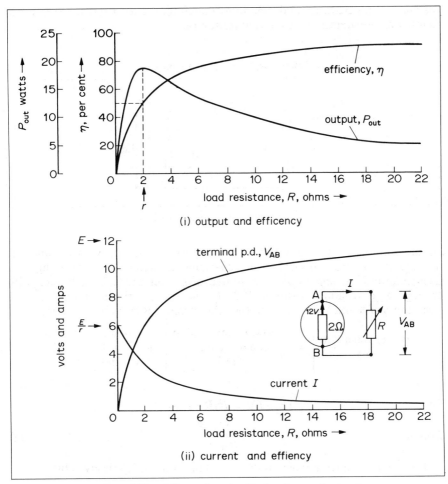

Figure 8.28 *Effects of varying load resistance in circuit*

So
$$P_{out} = \frac{E^2 R}{(R-r)^2 + 4Rr} = \frac{E^2}{(R-r)^2/R + 4r}$$

When $R = r$, the denominator of the fraction is *least* and so P_{out} is then a maximum. We see that the maximum output power $= E^2/4r$, which agrees with the power value when $R = r$.

Examples of Loads in Electrical Circuits

Loading for greatest *power output* is common in communication engineering. For example, the last transistor in a receiver delivers electrical power to the loudspeaker, which the speaker converts into mechanical power as sound waves.

To get the loudest sound, the speaker resistance (or impedance) is 'matched' to the internal resistance (or impedance) of the transistor, so that maximum power is delivered to the speaker.

The loading on a dynamo or battery is generally adjusted, however, for high *efficiency*, because that means greatest economy. Also, if a large dynamo were used with a load not much greater than its internal resistance, the current would be so large that the heat generated in the internal resistance would ruin the machine. With batteries and dynamos, therefore, the load resistance is made many times greater than the internal resistance.

Load Not a Passive Resistor

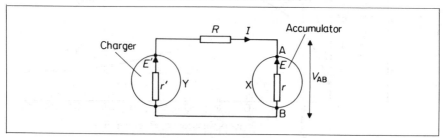

Figure 8.29 *Accumulator charging*

As an example of a load which is not a passive resistor, we shall take an accumulator being charged. The charging is done by connecting the accumulator X in *opposition* to a source of greater e.m.f., Y in Figure 8.29, through a controlling resistor R. If E, E' and r, r' are the e.m.f. and internal resistances of X and Y respectively, then the current I is given by the equation:

$$\left.\begin{array}{c}\text{power generated}\\ \text{in Y}\end{array}\right\} = \left\{\begin{array}{c}\text{power converted to}\\ \text{chemical energy}\\ \text{in X}\end{array}\right\} + \left\{\begin{array}{c}\text{power dissipated}\\ \text{as heat in all}\\ \text{resistances}\end{array}\right.$$

$$E'I \quad = \quad EI \quad + \quad I^2R + I^2r' + I^2r \quad (1)$$

So
$$(E' - E)I = I^2(R + r' + r),$$

from which
$$I = \frac{E' - E}{R + r' + r} \qquad . \qquad . \qquad . \qquad . \qquad (2)$$

The potential difference across the accumulator itself, V_{AB}, is given by

$$\left.\begin{array}{c}\text{power delivered}\\ \text{to X}\end{array}\right\} = \left\{\begin{array}{c}\text{power converted to}\\ \text{chemical energy}\end{array}\right\} + \left\{\begin{array}{c}\text{power dissipated}\\ \text{as heat}\end{array}\right.$$

So
$$IV_{AB} \quad = \quad IE \quad + \quad I^2r$$

and
$$V_{AB} = E + Ir \qquad . \qquad . \qquad . \qquad . \qquad (3)$$

Equation (3) shows that, when current is driven through a generator in opposition to its e.m.f., then the potential difference across the generator is equal to the *sum* of its e.m.f. and the voltage drop across its internal resistance. This result follows at once from energy considerations, as we have just seen.

Kirchhoff's Laws

A 'network' is usually a complicated system of electrical conductors. KIRCHHOFF (1824–87) extended Ohm's law to networks, and gave two laws, which together enabled the current in any part of the network to be calculated.

The *first law* refers to any point in the network, such as A in Figure 8.30 (i); it states that the total current flowing into the point is equal to the total current flowing out of it:

$$I_1 = I_2 + I_3$$

The law follows from the fact that electric charges do not accumulate at the points of a network. It is often put in the form that

the algebraic sum of the currents at a junction of a circuit is zero,

where a current, I, is reckoned positive if it flows towards the point, and negative if it flows away from it. Thus at A in Figure 8.30 (i),

$$I_1 - I_2 - I_3 = 0$$

Kirchhoff's first law gives a set of equations which help towards solving of the network. In practice we can shorten the work by putting the first law straight into the diagram, as shown in Figure 8.30 (ii) for example, since

$$\text{current along AC} = I_1 - I_g$$

Kirchhoff's second law connects the e.m.f. and p.d. in a complete circuit. It refers to any *closed loop*, such as AYCA in Figure 8.30 (ii). It states that

round such a loop, *the algebraic sum of the e.m.f.s is equal to the algebraic sum of all the p.d.s in that circuit.*

So, going clockwise round the loop AYCA in Figure 8.30 (ii),

$$E_2 = R_{AC}(I_1 - I_g) - R_g I_g$$

Note carefully that a p.d. is *positive* if it is in the *same* direction as the net e.m.f. Since I_g is opposite to E_2, the p.d. $R_g I_g$ is *negative*.

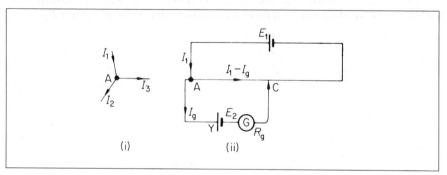

Figure 8.30 *Kirchhoff's laws*

Example on Kirchhoff's laws

Figure 8.31 shows a network in which the currents I_1, I_2, can be found from Kirchhoff's laws.

From the first law, the current in the $8\,\Omega$ wire is $(I_1 + I_2)$, assuming I_1, I_2 are the currents through the cells.

Taking closed circuits formed by each cell with the $8\,\Omega$ wire, we have, from the second law,

$$E_1 = 6 = 3I_1 + 8(I_1 + I_2) = 11I_1 + 8I_2$$

and

$$E_2 = 4 = 2I_2 + 8(I_1 + I_2) = 8I_1 + 10I_2$$

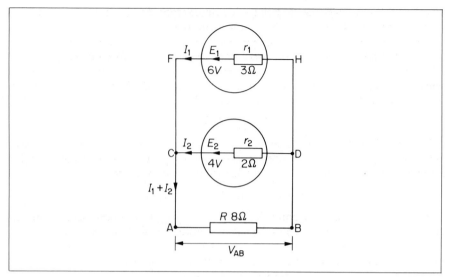

Figure 8.31 *Load R across cells in parallel*

Solving the two equations, we find

$$I_1 = \frac{14}{23} = 0.61 \text{ A}, \ I_2 = -\frac{2}{23} = -0.09 \text{ A}$$

The minus sign indicates that the current I_2 flows in the opposite direction to that shown in the diagram; so it flows against the e.m.f. of the generator E_2.

The Thermoelectric Effect

Seebeck Effect

The heating effect of the current transfers electrical energy into heat, but we have not so far described any mechanism which converts heat into electrical energy. This was discovered by SEEBECK in 1822.

In his experiments he connected a plate of bismuth between copper wires leading to a galvanometer, as shown in Figure 8.32 (i). He found that if one of the bismuth-copper junctions was heated, while the other was kept cool, then a

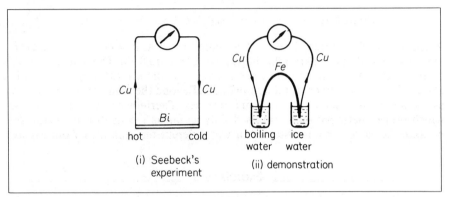

Figure 8.32 *The thermoelectric effect*

current flowed through the galvanometer. The direction of the current was from the copper to the bismuth at the cold junction. We can easily repeat Seebeck's experiment, using copper and iron wires and a galvanometer capable of indicating a few microamperes, Figure 8.32 (ii).

Thermocouples

Seebeck went on to show that a current flowed, without a battery, in any circuit containing two different metals, with their two junctions at different temperatures. Currents obtained in this way are called *thermoelectric currents*, and a pair of metals, with their junctions at different temperatures, are said to form a *thermocouple*. The following is a list of metals, such that any two of them form a thermocouple, then the current will flow from the higher to the lower in the list, across the cold junction:

Antimony, Iron, Zinc, Lead, Copper, Platinum, Bismuth.

Thermoelectric currents often appear when they are not wanted. They may occur from small differences in purity of two samples of the same metal, and from small differences of temperature—due, perhaps, to the warmth of the hand. They can cause a great deal of trouble in circuits used for precise measurements, or for detecting other small currents, not of thermal origin. As sources of electrical energy, thermoelectric currents are neither convenient nor economical, but they have been used in solar batteries with semiconductor materials. Thermocouples are used in the measurement of temperature, and of other quantities, such as radiant energy, which can be measured by a temperature rise.

Variation of Thermoelectric E.M.F. with Temperature

Later we shall see how thermoelectric e.m.f.s are measured. When the cold junction of a given thermocouple is kept constant at 0°C, and the hot junction

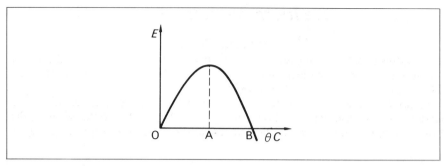

Figure 8.33 *Thermoelectric e.m.f. variation with temperature*

temperature $\theta°C$ is varied, the e.m.f. E is found to vary as $E = a\theta + b\theta^2$, where a, b are constants. This is a parabola-shaped curve (Figure 8.33). The temperature A corresponding to the maximum e.m.f. is known as the *neutral temperature*; it is about 250°C for a copper-iron thermocouple. Beyond the temperature B, known as the *inversion temperature*, the e.m.f. reverses. Thermoelectric thermometers, which utilise thermocouples, are used only as far as the neutral temperature, because the same e.m.f. is obtained at two different temperatures, from Figure 8.33.

_____ **Exercises 8** _____

Circuits

1 A battery of e.m.f. 4 V and internal resistance $2\,\Omega$ is joined to a resistor of $8\,\Omega$. Calculate the terminal p.d.
    What additional resistance in series with the $8\,\Omega$ resistor would produce a terminal p.d. of 3·6 V?

2 A battery of e.m.f. 24 V and internal resistance r is connected to a circuit having two parallel resistors of $3\,\Omega$ and $6\,\Omega$ in series with an $8\,\Omega$ resistor, Figure 8A (i). The current flowing in the $3\,\Omega$ resistor is then 0·8 A. Calculate   (i) the current in the $6\,\Omega$ resistor,   (ii) r,   (iii) the terminal p.d. of the battery.

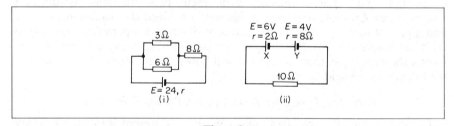

Figure 8A

3 A battery X of e.m.f. 6 V and internal resistance $2\,\Omega$ is in series with a battery Y of e.m.f. 4 V and internal resistance $8\,\Omega$ so that the two e.m.f.s act in the same direction, Figure 8A (ii). A $10\,\Omega$ resistor is connected to the batteries. Calculate the terminal p.d. of each battery.
    If Y is reversed so that the e.m.f.s now oppose each other, what is the new terminal p.d. of X and Y?

4 Two resistors of $1200\,\Omega$ and $800\,\Omega$ are connected in series with a battery of e.m.f. 24 V and negligible internal resistance, Figure 8B (i). What is the p.d. across each resistor?
    A voltmeter V of resistance $600\,\Omega$ is now connected firstly across the $1200\,\Omega$ resistor as shown, and then across the $800\,\Omega$ resistor. Find the p.d. recorded by the voltmeter in each case.

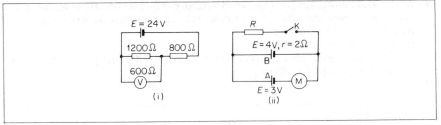

Figure 8B

5 In Figure 8B (ii), A has an e.m.f. of 3 V and negligible internal resistance and B has an e.m.f. of 4 V and internal resistance 2·0 Ω. With the switch K open, what current flows in the meter M?
   When K is now closed, no current flows in M. Calculate the value of R.
   (*Hint*. When no current flows in M, p.d. across B = p.d. across A.)

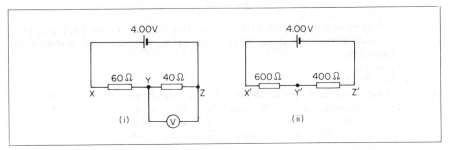

Figure 8C

6 The 4·00-V cell in the circuits shown above has zero internal resistance, Figure 8C. An accurately calibrated voltmeter connected across YZ records 1·50 V. Calculate
   (a) the resistance of the voltmeter,
   (b) the voltmeter reading when it is connected across Y'Z'.
   What do your results suggest concerning the use of voltmeters? (*L.*)

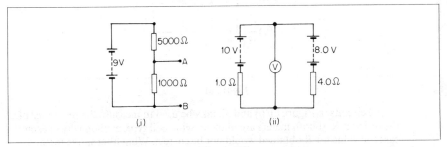

Figure 8D

7 (i) What is the final potential difference between *A* and *B* in the circuit in Figure 8D (i):
   (a) in the circuit as shown,
   (b) if an additional 500 Ω resistor were connected from *A* to *B*,
   (c) if the 500 Ω were replaced by a 2 μF capacitor? For what purpose would the circuit in (a) be useful?
   (ii) In the circuit of Figure 8D (ii), the batteries have negligible internal resistance and the voltmeter V has a very high resistance. What would be the reading of the voltmeter? (*L.*)

8 A laboratory power supply is known to have an e.m.f. of 1000 V. However, when a voltmeter of resistance 10 kΩ is connected to the output terminals of the supply, a reading of **only 2 V** is obtained.
(a) Explain this observation,
(b) Calculate (i) the current flowing in the meter, (ii) the internal resistance of the power supply. (*AEB*, 1984.)

9

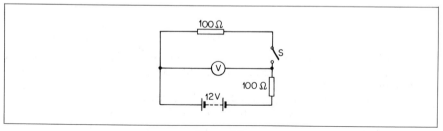

Figure 8E

A 12-V battery of negligible internal resistance is connected as shown in Figure 8E. The resistance of the voltmeter is 100 Ω. What reading will the voltmeter show when the switch, S is
(a) open, and
(b) closed? (*L.*)

10 (a) Define *the volt* and *the ohm*. Water in a barrel may be released at a variable rate by an adjustable tap at the bottom. If this system is compared to an electric circuit what in the system would be analogous to (i) the potential difference, (ii) the charge flowing, (iii) the current, and (iv) the resistance, in the circuit?

Give *two* examples, illustrated by appropriate graphs, of conductors or components which do not obey Ohm's law.

(b)

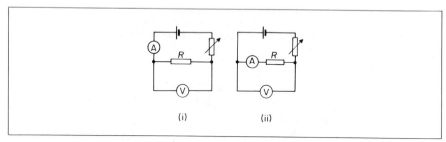

(i)   (ii)

Figure 8F

The circuits in Figure 8F (i) and (ii) may be used to measure the resistance of the resistor, *R*. If both meters are of the moving coil type, explain why in each case the value for *R* obtained would not be correct. What alternative method would you use to obtain a better value for *R*? (No circuit details are required.)

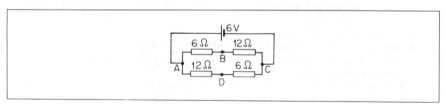

Figure 8G

(c) In Figure 8G, what is the potential difference between the points B and D? What resistor could you add to the 12 Ω resistor in branch ADC in order to make the potential difference between B and D zero? (*L.*)

11 Copper wire of cross-section area 2·0 mm² carries a current of 1·5 A. Find the drift velocity of the electrons, assuming $9·0 \times 10^{28}$ conduction electrons per metre³ and the electron charge $1·6 \times 10^{-19}$ C.

Why is the drift velocity slow?

12 Twelve cells each of e.m.f. 2 V and of internal resistance $\frac{1}{2}$ Ω are arranged in a battery of *n* rows and an external resistance of $\frac{3}{8}$ Ω is connected to the poles of the battery. Determine the current flowing through the resistance in terms of *n*.

Obtain numerical values of the current for the possible values which *n* may take and draw a graph of current against *n* by drawing a smooth curve through the points. Give the value of the current corresponding to the maximum of the curve and find the internal resistance of the battery when the maximum current is produced. (*L.*)

13 Describe with full experimental details an experiment to test the validity of Ohm's law for a metallic conductor.

An accumulator of e.m.f. 2 V and of negligible internal resistance is joined in series with a resistance of 500 Ω and an unknown resistance *X* Ω. The readings of a voltmeter successively across the 500 Ω resistance and *X* are 2/7 and 8/7 V respectively. Comment on this and calculate the value of *X* and the resistance of the voltmeter. (*JMB.*)

14 Explain clearly the difference between e.m.f. and potential difference.

Write down the Kirchhoff network laws and point out that each is essentially a statement of a conservation law.

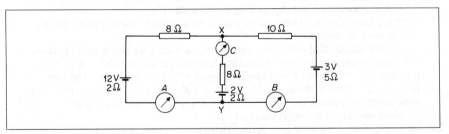

Figure 8H

Find, for the above circuit, Figure 8H, (i) the readings on the ammeters *A*, *B* and *C* (assumed to have effectively zero resistances), (ii) the potential difference between X and Y, (iii) the power dissipated as heat in the circuit, (iv) the power delivered by the 12 V cell.

Account carefully for the difference between (iii) and (iv). (*W.*)

15 State Ohm's law, and describe the experiments you would make in order to verify it. The positive poles A and C of two cells are connected by a uniform wire of resistance 4 ohms and their negative poles B and D by a uniform wire of resistance 6 Ω. The middle point of BD is connected to earth. The e.m.f.s of the cells AB and CD are 2 V and 1 V respectively, their resistances 1 Ω and 2 Ω respectively. Find the potential at the middle point of AC. (*O. & C.*)

Power. Heating Effect

16 The maximum power dissipated in a 10 000 Ω resistor is 1 W. What is the maximum current?

17 Two heating coils A and B, connected in parallel in a circuit, produce powers of 12 W and 24 W respectively. What is the ratio of their resistances, R_A/R_B, when used?

18 A heating coil of power rating 10 W is required when the p.d. across it is 20 V.

Calculate the length of nichrome wire needed to make the coil if the cross-sectional area of the wire used is $1 \times 10^{-7}\,\text{m}^2$ and the resistivity of nichrome is $1 \times 10^{-6}\,\Omega\,\text{m}$.

What length of wire would be needed if its diameter was half that previously used?

19 The running temperature of the filament of a 12-V, 48-W tungsten filament lamp is 2700°C and the average temperature coefficient of resistance for tungsten from 0°C to 2700°C is $6.4 \times 10^{-3}\,\text{K}^{-1}$. Calculate the resistance of the filament at 0°C. (*L.*)

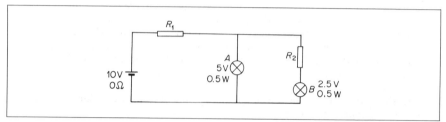

Figure 8I

20 In the above circuit, Figure 8I, what must be the values of R_1 and R_2 for the two lamps A and B to be operated at the ratings indicated? If lamp A burns out, what would be the effect on lamp B? Give reasons. (*W.*)

21 A thin film resistor in a solid-state circuit has a thickness of 1 μm and is made of nichrome of resistivity $10^{-6}\,\Omega\,\text{m}$. Calculate the resistance available between opposite edges of a $1\,\text{mm}^2$ area of film

(a) if it is square shaped,

(b) if it is rectangular, 20 times as long as it is wide. (*C.*)

22 State the laws of the development of heat when an electric current flows through a wire of uniform material.

An electrical heating coil is connected in series with a resistance of $X\,\Omega$ across the 240 V mains, the coil being immersed in a kilogram of water at 20°C. The temperature of the water rises to boiling-point in 10 minutes. When a second heating experiment is made with the resistance X short-circuited, the time required to develop the same quantity of heat is reduced to 6 minutes. Calculate the value of X. (Heat losses may be neglected.) (*L.*)

23 (a) Explain what is meant by (i) the electrical resistance of a conductor, and (ii) the resistivity of the material of a conductor.

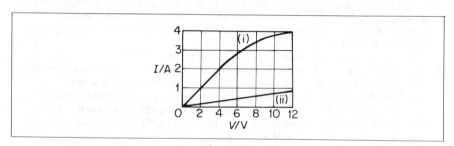

Figure 8J

(b) The graphs in Figure 8J show how the current varies with applied potential difference across (i) a 12 V, 36 W filament lamp, and (ii) a metre length of nichrome wire of cross-section $0.08\,\text{mm}^2$. Using the graphs, find the ratio of the values of the electrical resistance of the filament lamp to the nichrome wire (1) when the potential difference across them is 12 V, and (2) when the potential difference across them is 0.5 V.

How does the resistance of the filament lamp change as the current increases? Suggest a physical explanation for this change.

(c) The resistivity of copper is about $1.8 \times 10^{-8}\,\Omega\,m$ at 20°C. Show, using the information in (b) above, that the resistivity of nichrome is approximately 60 times this value. Explain why, in a domestic circuit containing a fire element and connecting cable, only the element becomes appreciably hot. (*L.*)

24 Indicate, by means of graphs, the relation between the current and voltage
(a) for a uniform manganin wire;
(b) for a water voltameter;
(c) for a diode valve.

How do you account for the differences between the three curves?

An electric hot plate has two coils of manganin wire, each 20 metres in length and $0.23\,mm^2$ cross-sectional area. Show that it will be possible to arrange for three different rates of heating, and calculate the wattage in each case when the heater is supplied from 200 V mains. The resistivity of manganin is $4.6 \times 10^{-7}\,\Omega\,m$. (*O. & C.*)

25 Describe an experiment for determining the variation of the resistance of a coil of wire with temperature.

An electric fire dissipates 1 kW when connected to a 250 V supply. Calculate to the nearest whole number the percentage change that must be made in the resistance of the heating element in order that it may dissipate 1 kW on a 200 V supply. What percentage change in the length of the heating element will produce this change of resistance if the consequent increase in the temperature of the wire causes its resistivity to increase by a factor 1.05? The cross-sectional area may be assumed constant. (*JMB.*)

26 Describe an instrument which measures the strength of an electric current by making use of its heating effect. State the advantages of this method.

A surge suppressor is made of a material whose conducting properties are such that the current passing through is directly proportional to the fourth power of the applied voltage. If the suppressor dissipates energy at a rate of 6.0 W when the potential difference across it is 240 V, estimate the power dissipated when the potential difference rises to 1200 V. (*C.*)

More Circuits

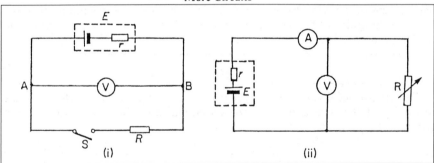

Figure 8K

27 (a) State Kirchhoff's second law.
(b) (i) In Figure 8K (i), E is a cell of source (internal) resistance r and the resistance of R is $4.0\,\Omega$. With the switch S open, the high resistance voltmeter V reads 10.0 V and with S closed the voltmeter reads 8.0 V. Show that $r = 1.0\,\Omega$.
(ii) If R were replaced by a cell of e.m.f. 4.0 V and source resistance $1.0\,\Omega$ with its negative terminal connected to B, what would be the reading of the voltmeter with S closed? (*JMB.*)

28 Figure 8K (ii) shows a cell of e.m.f. E and internal resistance r connected in series with a variable resistor R. The current I in the resistor is measured with an ammeter and the potential difference V between its ends is measured with a voltmeter. E, I, V and r are related as follows: $V = E - Ir$.

You are asked to use this circuit to find the values of E and r from a graph.
(a) State what measurements you would make.
(b) What graph would you plot? Sketch the graph you would expect to obtain and label the axes clearly.
(c) How would you use the graph to find E and r? (*L.*)

Measurements by Potentiometer and Wheatstone Bridge

In this chapter we discuss the accurate measurement of potential difference and resistance and their applications. As we see later, this enables the National Physical Laboratory to check the readings on current meters, for example, made by commercial instrument makers. Since the measurements are based on a comparison, standards of potential difference and of resistance will be needed, as discussed later.

The Potentiometer

Pointer instruments are useless for very accurate measurements: the best of them have an intrinsic error of about 1% of full scale. Where greater accuracy than this is required, elaborate measuring circuits are used.

One of the most useful of these circuits is the *potentiometer*. It consists of a uniform wire, AB in Figure 9.1 (i), about a metre long. An accumulator X, sometimes called a *driver cell*, keeps a steady current I in AB. Since the wire is uniform, its resistance per centimetre, R, is constant; the potential difference across 1 cm of the wire, RI, is therefore also constant. The potential difference between the end A of the wire, and any point C upon it, is thus proportional to the length of wire l between A and C:

$$V_{AC} \propto l . \qquad . \qquad . \qquad . \qquad . \qquad (1)$$

We can also see this relation is true from $V_{AC} = IR_{AC} = I\rho l/A$, where ρ is the resistivity of the wire and A its cross-sectional area. So if I, ρ and A keep constant, $V_{AC} \propto l$.

Comparison of E.M.F.s

To illustrate the use of the potentiometer, suppose we take a cell, Y in Figure 9.1 (ii), and join its positive terminal to the point A (to which the positive terminal of X is also joined). We connect the negative terminal of Y through a sensitive galvanometer to a slider S, which we can press on to any point in the wire.

Let us suppose that the cell Y has an e.m.f. E, which is less than the potential difference V_{AB} across the whole of the wire. Then if we press the slider on B, a current I' will flow through Y in opposition to its e.m.f., Figure 9.1 (iii). This current will deflect the galvanometer G—let us say to the *right*. If we now press the slider on A, the cell Y will be connected straight across the galvanometer, and will deliver a current in the direction of its e.m.f., Figure 9.1 (iii). The galvanometer will therefore show a deflection to the *left*. If the deflections at A and B are not opposite, then either the e.m.f. of Y is greater than the potential difference across the whole wire, or we have connected the circuit wrongly. The

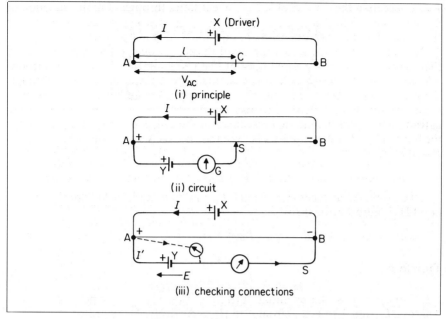

Figure 9.1 *The potentiometer*

commonest mistake in connecting up is not joining both positive poles of X and Y to A.

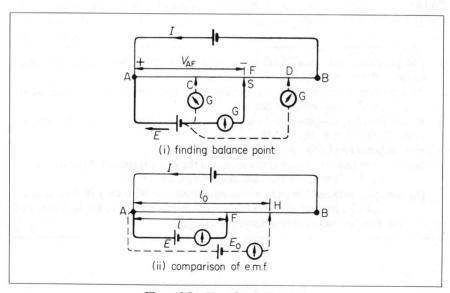

Figure 9.2 *Use of potentiometer*

Now let us suppose that we place the slider on to the wire at a point a few centimetres from A, then at a point a few centimetres farther on, and so forth. (We do not run the slider continuously along the wire, because the scraping would destroy its uniformity.)

When the slider is at a point C near A (Figure 9.2(i)) the potential difference

V_{AC} is less than the e.m.f. E of Y. So current flows through G in the direction of E, and G may deflect to the left. When the slider is at D near B, V_{AD} is greater than E, current flows through G in opposition to E, and G deflects to the right.

By trial and error (but no scraping of the slider) we can find a point F such that, when the slider is pressed upon it, the galvanometer shows *no* deflection. *The potential difference* V_{AF} *is then equal to the e.m.f.* E; no current flows through the galvanometer because E and V_{AF} act in opposite directions in the galvanometer circuit, Figure 9.2 (i). Because no current flows, the resistance of the galvanometer, and the internal resistance of the cell, cause no voltage drop. So the full e.m.f. E therefore appears between the points A and S, and is balanced by V_{AF}, that is,

$$E = V_{AF}$$

If we now take another cell of e.m.f. E_0, and balance it in the same way, at a point H (Figure 9.2 (ii)), then

$$E_0 = V_{AH}$$

Therefore
$$\frac{E}{E_0} = \frac{V_{AF}}{V_{AH}}$$

But, from previous, the potential differences V_{AF}, V_{AH} are proportional to the lengths l, l_0 from A to F, and from A to H, respectively. Therefore

$$\frac{E}{E_0} = \frac{l}{l_0} \qquad . \qquad . \qquad . \qquad . \qquad . \qquad . \qquad (2)$$

So the ratio of the e.m.f.s is proportional to the ratio of the balancing lengths and can therefore be calculated.

---

1 **The potentiometer uses a null (no-deflection) method. So it does not depend on the accuracy of an instrument reading.**
2 **At a potentiometer balance, no current flows from the cell. So the p.d. at the cell terminals = the e.m.f. of the cell.**
3 **If the balance lengths are l_1 and l_2 for two cells of e.m.f.s E_1 and E_2 respectively, then $E_1/E_2 = l_1/l_2$.**
4 **If no balance can be found on the wire, then**
   (a) **the +ve pole of the cell is not joined to the same terminal of the wire as the +ve pole of the driver (potentiometer) cell, or**
   (b) **The p.d. between the ends of the potentiometer wire may be *less* than the e.m.f. to be measured. So only a *small* resistance is needed in series with the wire for more balance-length readings.**

Accuracy of Potentiometer

The following points should be noted:

(1) When the potentiometer is used to compare the e.m.f.s of cells, no errors are introduced by the internal resistances, because no current flows through the cells at the balance-points.

(2) The potentiometer is more accurate than the moving-coil voltmeter for measuring e.m.f. The moving-coil voltmeter has a resistance and this lowers the p.d. between the terminals of the cell when it is connected. In contrast, since no

current flows from the cell when a balance is found, the potentiometer may be considered to be a voltmeter with an *infinitely-high resistance*, which is the ideal voltmeter.

(3) The accuracy of a potentiometer is limited by the non-uniformity of the slide-wire, the uncertainty of the balance-point, and the error in measuring the length *l* of wire from the balance-point to the end A. With even crude apparatus, the balance-point can be located to within about 0·5 mm; if the length *l* is 50 cm, or 500 mm, then the error in locating the balance-point is 1 : 1000. If the wire has been carefully treated, its non-uniformity may introduce an error of about the same magnitude. The overall error is then about ten times less than that of a pointer instrument.

(4) The *precision* with which the balance-point of a potentiometer can be found depends on the *sensitivity* of the galvanometer. With a very sensitive galvanometer a very small current can be detected.

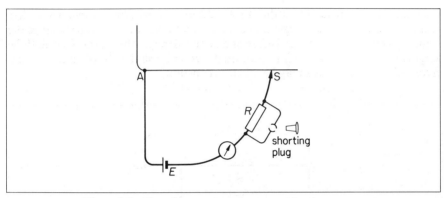

Figure 9.3   *Use of protective resistance with galvanometer*

A moving-coil galvanometer must be protected by a series resistance *R* of several thousand ohms, which is shorted out when the balance is nearly reached, Figure 9.3. A series resistance is preferable to a shunt, because it reduces the current drawn from the cell under test when the potentiometer is unbalanced. Looking for the balance-point then causes less change in the chemical condition of the cell, and therefore in its e.m.f. The actual magnitude of *R* does not matter as no current flows through *R* at a balance.

It is important to realise that the accuracy of a potentiometer does not depend on the accuracy of the galvanometer, but only on its sensitivity. The galvano-meter is used not to measure a current but merely to show one when the potentiometer is off balance. It is said to be used as a null-indicator, and the potentiometer method of measurement is called a null method.

(5) The current through the potentiometer wire must be steady—it must not change appreciably between the finding of one balance-point and the next. The accumulator which provides it should therefore be neither freshly charged nor nearly run-down; when an accumulator is in either of those conditions its e.m.f. falls with time.

Errors in potentiometer measurements may be caused by non-uniformity of the wire, and by the resistance of its connection to the terminal at A. This resistance is added to the resistance of the length *l* of the wire between A and the balance-point, and if it is appreciable it makes equation (2) not true.

1 **No current is taken from the cell at a balance. So the potentiometer acts like a perfect voltmeter of infinitely-high resistance.**

2 **The magnitude of the series resistance protecting the galvanometer does not matter because no current flows in this part of the circuit at a balance.**

Uses of Potentiometer, E.M.F. and Internal Resistance

All the uses of the potentiometer depend on the fact that it can measure potential difference accurately, and without drawing current from the circuit under test.

(a) *E.m.f. measurement* If one of the cells in Figure 9.2 (ii) is a *standard cell* of known e.m.f., say E_0, such as a Weston cadmium cell, then the unknown e.m.f. of the other, E is given by equation:

$$\frac{E}{E_0} = \frac{l}{l_0} \qquad . \qquad . \qquad . \qquad . \qquad (1)$$

Equation (1) is true only if the current I through the potentiometer wire has remained constant. The easiest way to check that it has done so is to balance the standard cell against the wire before and after balancing the unknown cell. If the lengths to the balance-point are equal—within the limits of experimental error—then the current I may be taken as constant.

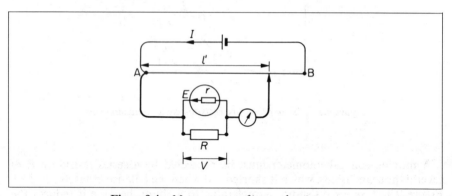

Figure 9.4 *Measurement of internal resistance*

(b) The *internal resistance of a cell*, r, can be found with a potentiometer by balancing first its e.m.f., E, when the cell is on open circuit. Suppose the balance length is l. A known resistance R is then connected to the cell, as shown in Figure 9.4. The terminal p.d. V is now balanced by a smaller length l' than l since a current flows from the cell. Now

$$V = IR = \frac{E}{R+r}R$$

So

$$\frac{V}{E} = \frac{R}{R+r} \qquad . \qquad . \qquad . \qquad . \qquad (2)$$

But

$$\frac{V}{E} = \frac{l'}{l} \qquad . \qquad . \qquad . \qquad . \qquad (3)$$

where l and l' are the lengths of potentiometer wire required to balance E and V. From equations (2) and (3), $l/l' = (R+r)/R$. So r can be found from

$$r = \left(\frac{l}{l'} - 1\right) R$$

Also, since $r\left(\dfrac{1}{R}\right) = \dfrac{l}{l'} - 1$, we can vary R and measure l' for each value of R. A graph of l/l' against $1/R$ is a straight line whose *gradient* is equal to r.

Measurement of Current

A current can be measured on a potentiometer by measuring the potential difference V which it sets up across a standard known resistance R in Figure 9.5 (i), and then using $I = V/R$. A low resistance R is chosen so that it does not disturb the circuit in which it is placed.

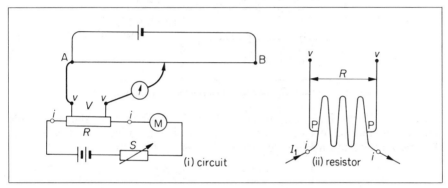

Figure 9.5 *Calibration of ammeter with potentiometer*

Figure 9.5 (i) shows how an ammeter M can be calibrated by a potentiometer. The rheostat S is adjusted until the required ammeter reading is obtained and the p.d. V between the terminals v, v of R is balanced on the potentiometer wire. Suppose this gives a balance length l. The e.m.f. E_0 of a standard cell is now balanced on the wire (see Figure 9.6 (ii)). If this balance length is l_0, then

$$\frac{V}{E_0} = \frac{l}{l_0}$$

So V can be found since E_0, l_0 and l are known and the true current I is then calculated from $I = V/R$.

The resistance of the wires connecting the potential terminals to the points PP, and to the potentiometer circuit, do not affect the result, because at the balance-point the current through them is zero.

Figure 9.5 (ii) shows in detail the standard resistance used. It consists of a broad strip of alloy, such as manganin, whose resistance varies very little with temperature (p. 254). The current is led in and out at the terminals i, i. The terminals v, v are connected to fine wires soldered to points PP on the strip; they are called the potential terminals. The marked value R of the resistance is the value between the points PP.

Calibration of Voltmeter

Figure 9.6 (i) shows how a potentiometer can be used to calibrate a voltmeter. A standard cell is first used to find the p.d. per cm or volt per cm of the wire (Figure 9.6 (ii)). If its e.m.f. E_0 is balanced by a length l_0, then

$$\text{volt per cm} = \frac{E_0}{l_0} \quad . \qquad . \qquad . \qquad . \qquad . \qquad (1)$$

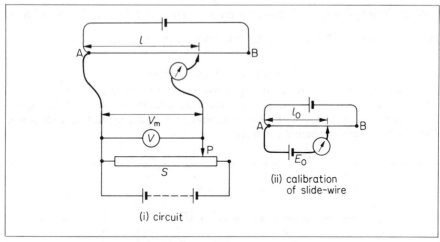

Figure 9.6 *Calibration of voltmeter with potentiometer*

Different voltages V_m are now applied to the voltmeter by the adjustable potential divider or rheostat S, Figure 9.6 (i), which has a high resistance. If l is the length of potentiometer wire which balances a p.d. V_m then

$$V_m = l \times (\text{volt/cm of wire})$$

$$= l\frac{E_0}{l_0} \quad . \qquad . \qquad . \qquad . \qquad . \qquad . \qquad (2)$$

The value of V_m is the true value of the p.d. across the voltmeter terminals. If the voltmeter reading is V_{obs}, then the correction to be added to it is $V_m - V_{obs}$. This is plotted against V_{obs}, as in Figure 9.7, and this provides a correction curve for the voltmeter readings when the meter is used.

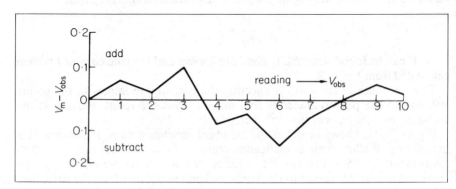

Figure 9.7 *Correction curve of voltmeter*

Comparison of Resistances

A potentiometer can be used to compare resistances, by comparing the potential differences across them when they are carrying the same current I_1, Figure 9.8.

This method is particularly useful for very *low resistances*, because, as we have just seen, the resistances of the connecting wires do not affect the result of the experiment. It can, however, be used for higher resistances. With low resistances the ammeter A′ and rheostat P are necessary to adjust the current to a value which will neither exhaust the accumulator Y, nor overheat the resistors, and a series resistor (not shown) is needed with X in the potentiometer circuit.

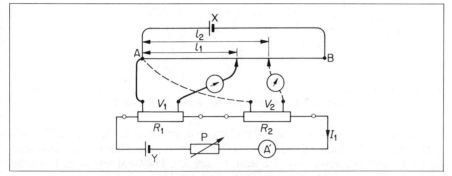

Figure 9.8 *Comparison of resistances with potentiometer*

No standard cell is required. The potential difference across the first resistor, $V_1 = R_1 I_1$, is balanced against a length l_1 of the potentiometer wire, as shown by the full lines in the figure. *Both* potential terminals of R_1 are then disconnected from the potentiometer, and those of R_2 are connected in their place. If l_2 is the length to the new balance-point, then

$$\frac{l_1}{l_2} = \frac{V_1}{V_2} = \frac{R_1 I_1}{R_2 I_1} = \frac{R_1}{R_2}$$

This result is true only if the current I_1 is constant, as well as the potentiometer current. The accumulator Y, as well as X, must therefore be in good condition. To check the constancy of the current I_1, the ammeter $A′$ is not accurate enough. The reliability of the experiment as a whole can be checked by balancing the potential V_1 a second time, after V_2. If the new value of l_1 differs from the original then at least one of the accumulators is running down and must be replaced.

Measurement of Thermoelectric E.M.F.

The e.m.f.s of the thermojunctions (p. 273) are small—of the order of a millivolt. If we tried to measure such an e.m.f. on a simple potentiometer we should find the balance-point very near one end of the wire, so that the end-error would be serious.

Figure 9.19 shows a potentiometer circuit for measuring thermoelectric e.m.f. A suitable high resistance R, produced by two resistance boxes R_1, R_2, is needed in series with the wire. Suppose the wire has a resistance of $3 \cdot 0\,\Omega$ and we assume that the accumulator D has an e.m.f. 2 V and negligible internal resistance. If the p.d. across the whole wire needs to be, say, 4 mV or 0·004 V, then the p.d. across R is $2 - 0 \cdot 004 = 1 \cdot 996$ V. Since the p.d. across resistors in a series circuit is proportional to the resistance, then R is given by

$$\frac{R}{3} = \frac{1 \cdot 996}{0 \cdot 004}$$

So $R = 1497\,\Omega$ on calculation.

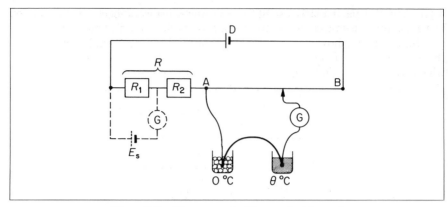

Figure 9.9 *Measurement of thermoelectric e.m.f.*

Calibration experiment. First, we find the 'p.d. per ohm' in the circuit. For this purpose a standard cell, of e.m.f. $E_s = 1\cdot018$ V say, is placed across one resistance R_1 with a galvanometer G in one lead, Figure 9.9. The total resistance from the boxes R_1 and R_2 must be kept constant at $1497\,\Omega$ so that the potentiometer current is constant. Initially, then, $R_1 = 750\,\Omega$ and $R_2 = 747\,\Omega$, for example, and by taking out a resistor from one box and replacing an equal resistor in the other box, a balance in G can soon be obtained. Suppose $R_1 = 779\,\Omega$ in this case. Then the p.d. across the 100 cm length of potentiometer wire, $3\cdot0\,\Omega$, is

$$\frac{3}{779} \times 1\cdot018 = 0\cdot003\,92 \text{ V} = 3\cdot92 \times 10^{-3} \text{ V}$$

Thermoelectric e.m.f. After removing the standard cell, we now proceed to measure the thermoelectric e.m.f. E of a thermocouple at various temperatures t°C of the hot junction, the other junction being kept constant at 0°C, Figure 9.9. Suppose the balance length on the wire at a particular temperature is 62·4 cm. Then, from above,

$$E = \frac{62\cdot4}{100} \times 3\cdot92 \times 10^{-3} = 2\cdot45 \times 10^{-3} \text{ V}$$

Use of the standard cell overcomes the error in assuming that the accumulator e.m.f. is 2 V. If the thermoelectric e.m.f. is not required accurately, then we can assume that the p.d. across 100 cm length of potentiometer wire is 4 m V when the series resistance R is $1497\,\Omega$, as calculated above, and not use the calibration part of the experiment.

Thermoelectric E.M.F. and Temperature

Figure 9.10 shows the results of measuring the e.m.f. E when the cold junction is at 0°C and the hot junction is at various temperatures θ in °C. The curves approximate to parabolas:

$$E = a\theta + b\theta^2 \qquad . \qquad . \qquad . \qquad . \qquad (1)$$

Since the same value of E is obtained at *two* different temperatures θ, the thermocouple is never used for measuring temperature greater than the value corresponding to its maximum e.m.f.

To find the values of a and b, we see from (1) that

$$E/\theta = a + b\theta \qquad . \qquad . \qquad . \qquad . \qquad (2)$$

A graph of E/θ against θ is then a straight line whose *gradient* is the value of b. The value of a is the *intercept* on the E/θ-axis.

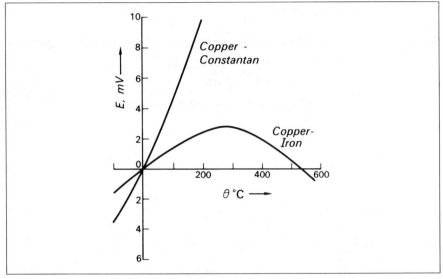

Figure 9.10 *E.m.f.s of thermocouples (reckoned positive when into copper at the cold junction)*

THERMOELECTRIC E.M.F.s
(E in microvolt when θ is in °C and cold junction at 0°C)

Junction	a	b	Range for a and b, °C	Limits of use, °C
Cu/Fe . . .	14	-0.02	0–100	*See 1*
Cu/Constantan[2] .	41	0.04	-50 to $+300$	-200 to $+300$
Pt/Pt – Rh[3] . .	6.4	0.006	0–200	0–1700
Chromel[4]/Alumel[5] .	41	0.001	0–900	0–1300

[1] Simple demonstrations. [2] See p. 254.
[3] 10% Rh; used only for accurate work or very high temperatures.
[4] 90% Ni, 10% Cr.
[5] 94% Ni, 3% Mn. 2% Al, 1% Si.

Example on Potentiometer

In the circuit shown, the e.m.f. E_s of a standard cell is 1·02 V and this is balanced by the p.d. across a resistance of 2040 Ω in series with a potentiometer wire AB. If AB is 1·00 m long and has a resistance of 4 Ω, calculate the length AC on it which balances the e.m.f. 1·2 mV of the thermocouple XY, Figure 9.11.

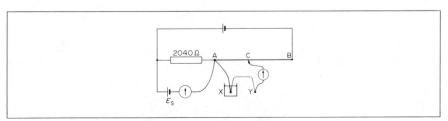

Figure 9.11 *Example*

Since 1.02 V is the p.d. across $2040\,\Omega$, and the $4\,\Omega$ wire AB is in series with $2040\,\Omega$, then

$$\text{p.d. across AB} = \frac{4}{2040} \times 1.02 \text{ V} = \frac{4}{2000} \text{V} = 2\,\text{mV}$$

So thermocouple e.m.f., $1.2\,\text{mV}$, is balanced by a length AC on AB (1 m) given by

$$\frac{AC}{AB} = \frac{1.2\,\text{mV}}{2\,\text{mV}} = \frac{3}{5}$$

$$\therefore AC = \tfrac{3}{5} \times AB = \tfrac{3}{5} \times 1.00\,\text{m} = 0.60\,\text{m}$$

Wheatstone Bridge: Measurement of Resistance

Wheatstone Bridge Circuit

About 1843 Wheatstone designed a circuit called a 'bridge circuit' which gave an accurate method for measuring resistance. We shall deal later with the practical aspects. In Figure 9.12, X is the unknown resistance, and P, Q, R are resistance

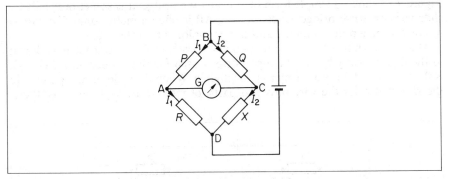

Figure 9.12   *Wheatstone bridge*

boxes. One of these—usually R—is adjusted until the galvanometer G between A and C shows no deflection, a so-called 'balance' condition. In this case the current I_g in G is zero. Then, as we shall show,

$$\frac{P}{Q} = \frac{R}{X}$$

so

$$X = \frac{Q}{P} R$$

Wheatstone Bridge Proof

At balance, since no current flows through the galvanometer, the points A and C must be at the same potential, Figure 9.12. Therefore

$$V_{AB} = V_{CB} \text{ and } V_{AD} = V_{CD}$$

So

$$\frac{V_{AB}}{V_{AD}} = \frac{V_{CB}}{V_{CD}} \quad . \quad . \quad . \quad . \quad . \quad \text{(i)}$$

Also, since $I_g = 0$, P and R carry the same current, I_1, and X and Q carry the same current, I_2. There

$$\frac{V_{AB}}{V_{AD}} = \frac{I_1 P}{I_1 R} = \frac{P}{R}$$

and

$$\frac{V_{CB}}{V_{CD}} = \frac{I_2 Q}{I_2 X} = \frac{Q}{X} \quad . \quad . \quad . \quad . \quad \text{(ii)}$$

From equations (i) and (ii),

$$\frac{P}{R} = \frac{Q}{X}$$

So

$$X = \frac{Q}{P} R$$

Exactly the same relationship between the four resistances is obtained if the galvanometer and cell positions are interchanged. Further analysis of the circuit shows that the bridge is most sensitive when the galvanometer is connected between the junction of the highest resistances and the junction of the lowest resistances.

The Slide-wire (Metre) Bridge

Figure 9.13 shows a simple form of Wheatstone bridge; it is sometimes called a slide-wire or metre bridge, since the wire AB is often a metre long. The wire is uniform, as in a potentiometer, and can be explored by a slider S.

The unknown resistance X and a known resistance R are connected as shown in the figure; heavy brass or copper strip is used for the connections AD, FH, KB, whose resistances are generally negligible. When the slider is at a point C in the wire it divides the wire into two parts, of resistances R_{AC} and R_{CB}; these,

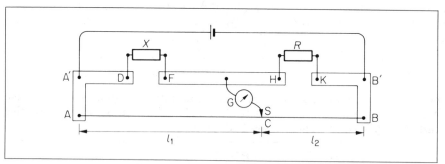

Figure 9.13 *Slide-wire (metre) bridge*

with X and R, form a Wheatstone bridge. (The galvanometer and battery are interchanged relative to the circuits we have given earlier; that enables the slider S to be used as the galvanometer key. We have already seen that the interchange does not affect the condition for balance in G.) The connections are checked by placing S first on A, then on B. The balance-point is found by trial and error—not by scraping S along AB. At balance,

$$\frac{X}{R} = \frac{R_{AC}}{R_{CB}}$$

Since the wire is uniform, the resistances R_{AC} and R_{CB} are proportional to the lengths of wire, l_1 and l_2. Therefore

$$\frac{X}{R} = \frac{l_1}{l_2} \qquad . \qquad . \qquad . \qquad . \qquad . \qquad (1)$$

The resistance R should be chosen so that the balance-point C comes fairly near to the centre of the wire—within, say, its middle third. If either l_1 or l_2 is small, the resistance of its end connection AA' or BB' in Figure 9.13 is not negligible in comparison with its own resistance; equation (1) then does not hold. Some idea of the accuracy of a particular measurement can be got by interchanging R and X, and balancing again. If the new ratio agrees with the old within about 1%, then their average may be taken as the value of X.

Since the galvanometer G is a sensitive current-reading meter, a high protective resistor (not shown) is required in series with it until a *near* balance is found on the wire. At this stage the high resistor is shunted or removed and the final balance-point found.

The lowest resistance which a bridge of this type can measure with reasonable accuracy is about 1 ohm. Resistances lower than about 1 ohm cannot be measured accurately on a Wheatstone bridge, because of the resistances of the wires connecting them to the X terminals, and of the contacts between those wires and the terminals to which they are, at each end, attached. This is the reason why the potentiometer method is more satisfactory for comparing and measuring low resistances.

Lorenz devised a method of measuring resistance without using a standard resistance. This *absolute method* is described on page 353.

Temperature Coefficient of Resistance

We have already seen that the resistance of a wire varies with its temperature. If we put a coil of fine copper wire into a water bath, and use a Wheatstone bridge to measure its resistance at various moderate temperatures θ, we find that the resistance, R_θ, increases with the temperature, Figure 9.14. We may therefore define a *temperature coefficient of resistance*, α, such that

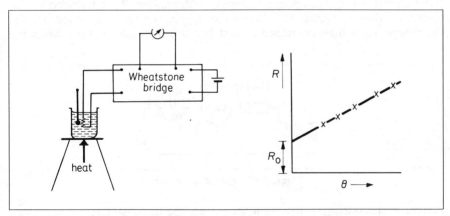

Figure 9.14 *Measurement of temperature coefficient*

$$R_\theta = R_0(1 + \alpha\theta) \quad . \quad . \quad . \quad . \quad . \quad (1)$$

where R_0 is the resistance at 0°C. In words, starting with the resistance at 0°C,

$$\alpha = \frac{\text{increase of resistance per K rise of temperature}}{\text{resistance at 0°C}}$$

If R_1 and R_2 are the resistances at θ_1°C and θ_2°C, then, from (1),

$$\frac{R_1}{R_2} = \frac{1 + \alpha\theta_1}{1 + \alpha\theta_2} \quad . \quad . \quad . \quad . \quad (2)$$

Values of α for pure metals are of the order of 0.004 K^{-1}. They are much less for alloys than for pure metals, a fact which makes alloys useful materials for resistance boxes and shunts.

Equation (1) represents the change of resistance with temperature fairly well, but not as accurately as it can be measured. More accurate equations are given later in the Heat section of this book, where resistance thermometers are discussed.

Thermistors

A *thermistor* is a heat-sensitive resistor usually made from semiconductors. One type of thermistor has a high positive temperature coefficient of resistance. So when it is placed in series with a battery and a current meter and warmed, the current is observed to decrease owing to the rise in resistance. Another type of thermistor has a *negative* temperature coefficient of resistance, that is, its resistance rises when its temperature is decreased, and falls when its temperature is increased. Thus when it is placed in series with a battery and a current meter and warmed, the current is observed to increase owing to the decrease in resistance.

Thermistors with a high negative temperature coefficient are used for resistance thermometers in very low temperature measurement of the order of 10 K, for example. The higher resistance at low temperature enables more accurate measurement to be made.

Thermistors with negative temperature coefficient may be used to safeguard against current surges in circuits where this could be harmful, for example, in a radio circuit where heaters are in series. A thermistor, T, is included in the circuit, as shown, Figure 9.15. When the supply voltage is switched on, the thermistor has a high resistance at first because it is cold. It thus limits the

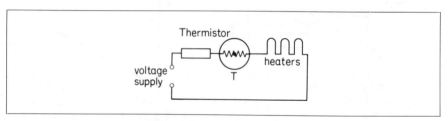

Figure 9.15 *Use of thermistor*

current to a moderate value. As it warms up, the thermistor resistance drops appreciably and an increased current then flows through the heaters. Thermistors are also used in transistor receiver circuits to compensate for excessive rise in collector current.

Example on Temperature Coefficient

How would you compare the resistances of two wires A and B, using
(a) a Wheatstone bridge method, and
(b) a potentiometer?

For each case draw a circuit diagram and indicate the method of calculating the result.

In an experiment carried out at 0°C, A was 1·20 m of Nichrome wire of resistivity $100 \times 10^{-8} \, \Omega$ m and diameter 1·20 mm, and B a German silver wire 0·80 mm diameter and resistivity $28 \times 10^{-8} \, \Omega$ m. The ratio of the resistances A/B was 1·20. What was the length of the wire B?

If the temperature coefficient of Nichrome is $0·000\,40 \, \mathrm{K}^{-1}$ and of German silver is $0·000\,30 \, \mathrm{K}^{-1}$, what would the ratio of resistances become if the temperature were raised by 100 K? (*L*.)

With usual notation,

for A, $$R_1 = \frac{\rho_1 l_1}{A_1}$$

and for B,
$$R_2 = \frac{\rho_2 l_2}{A_2}$$

$$\therefore \frac{R_1}{R_2} = \frac{\rho_1}{\rho_2} \cdot \frac{l_1}{l_2} \cdot \frac{A_2}{A_1} = \frac{\rho_1}{\rho_2} \cdot \frac{l_1}{l_2} \cdot \frac{d_2^{\,2}}{d_1^{\,2}}$$

where d_2, d_1 are the respective diameters of B and A.

$$\therefore 1{\cdot}20 = \frac{100}{28} \times \frac{1{\cdot}20}{l_2} \times \frac{0{\cdot}8^2}{1{\cdot}20^2}$$

$$\therefore l_2 = \frac{100 \times 1{\cdot}20 \times 0{\cdot}64}{1{\cdot}20 \times 28 \times 1{\cdot}44} = 1{\cdot}59\,\text{m} \qquad . \qquad . \qquad . \qquad (i)$$

When the temperature is raised by 100 K, the resistance increases according to the relation $R_\theta = R_0(1 + \alpha\theta)$. Thus

new Nichome resistance, $R_A = R_1(1 + \alpha . 100) = R_1 \times 1{\cdot}04$

and new German silver resistance, $R_B = R_2(1 + \alpha' . 100) = R_2 \times 1{\cdot}03$

$$\therefore \frac{R_A}{R_B} = \frac{R_1}{R_2} \times \frac{1{\cdot}04}{1{\cdot}03} = 1{\cdot}20 \times \frac{1{\cdot}04}{1{\cdot}03} = 1{\cdot}21 . \qquad . \qquad . \qquad (ii)$$

_____ **Exercises 9** _____

Potentiometer

1 The e.m.f. of a battery A is balanced by a length of 75·0 cm on a potentiometer wire. The e.m.f. of a standard cell, 1·02 V, is balanced by a length of 50·0 cm. What is the e.m.f. of A?

   Calculate the new balance length if A has an internal resistance of 2 Ω and a resistor of 8 Ω is joined to its terminals.

2 A 1·0 Ω resistor is in series with an ammeter M in a circuit. The p.d. aross the resistor is balanced by a length of 60·0 cm on a potentiometer wire. A standard cell of e.m.f. 1·02 V is balanced by a length of 50·0 cm. If M reads 1·10 A, what is the error in the reading?

3 The driver cell of a potentiometer has an e.m.f. of 2 V and negligible internal resistance. The potentiometer wire has a resistance of 3 Ω. Calculate the resistance needed in series with the wire if a p.d. of 5 mV is required across the whole wire.

   The wire is 100 cm long and a balance length of 60 cm is obtained for a thermocouple e.m.f. E. What is the value of E?

4 In a potentiometer experiment, a balance length cannot be found. Write down *two* possible reasons, with explanations.

5 Explain the reasons for the following procedures in potentiometer experiments:
   (a) The positive pole of a battery whose e.m.f. is required is connected to the same terminal of the potentiometer wire as the positive pole of the driver cell.
   (b) The protective resistor is removed before a final balance point is determined.
   (c) A rheostat is sometimes included in the potentiometer circuit with the driver cell but its resistance must not be too high.
   (d) A standard cell is needed in an experiment to calibrate an ammeter but not in an experiment to measure the internal resistance of a cell.
   (e) In comparing the resistances of two resistors A and B, the resistors are placed in series in a circuit.

6 (a) Figure 9A (i), in which AB is a uniform resistance wire, is a simple potentiometer circuit. Explain why a point X may be found on the wire which gives zero galvanometer deflection.

   When the circuit was first set up it was impossible to find a balance point. State and explain two possible causes of this.

How would you use the circuit to compare the e.m.f.s of two cells with minimum error? Why is this circuit not suitable for the comparison of an e.m.f. of a few millivolts with an e.m.f. of about a volt?

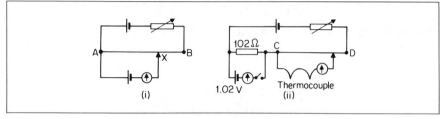

Figure 9A

(b) The second circuit, Figure 9A (ii), may be used to measure the e.m.f. of a thermocouple provided that the resistance of CD is known. Describe how you would use it. If the resistance of CD were $2.00\,\Omega$, its length were 1.00 m and the balance length were 79 cm, what would be the e.m.f. of the thermocouple? (*L.*)

7 A simple potentiometer circuit is set up as in Figure 9B, using a uniform wire AB, 1.0 m long, which has a resistance of $2.0\,\Omega$. The resistance of the 4-V battery is negligible. If the variable resistor R were given a value of $2.4\,\Omega$, what would be the length AC for zero galvanometer deflection?

If R were made $1.0\,\Omega$ and the 1.5 V cell and galvanometer were replaced by a voltmeter of resistance $20\,\Omega$, what would be the reading of the voltmeter if the contact C were placed at the mid-point of AB? (*L.*)

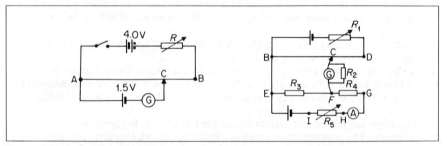

Figure 9B **Figure 9C**

8 (a) Figure 9C shows a potentiometer circuit arranged to compare the values of two low resistance resistors. (i) Which resistors are to be compared? (ii) What are the functions of the remaining three resistors? (iii) During the experiment what connection changes would need to be made? (iv) When the circuit was initially set up the galvanometer was found to be deflected in the same direction wherever along the wire BD the sliding contact C was placed. Suggest *two* possible reasons for this. (v) For the purpose of this experiment explain whether or not it is necessary to calibrate the potentiometer with a standard cell.

(b) A potentiometer may be regarded as equivalent to a voltmeter. Illustrate this by drawing the *basic* potentiometer circuit, marking the points which correspond to the positive and negative terminals of the equivalent voltmeter.

Describe how this basic circuit may be developed in order to measure the internal resistance of a cell. How may the observations be displayed in the form of a straight line graph and how could the internal resistance be found from this graph? (*L.*)

9 (a) Describe, with a circuit diagram, a potentiometer circuit arranged (i) to compare the e.m.f. of a cell with that of a standard cell, and (ii) to measure

accurately a steady direct current of approximately 1 A. What factors determine the accuracy of the current measurement?

(b) A 12 V, 24 W tungsten filament bulb is supplied with current from n cells connected in series. Each cell has an e.m.f. of 1·5 V and internal resistance 0·25 Ω. What is the value of n in order that the bulb runs at its rated power?

An additional resistance R is introduced into the circuit so that the potential difference across the bulb is 6 V. Why is the power dissipated in the bulb not 6 W? Is it greater or less than 6 W? (*O. & C.*)

10 The slide connection J in the circuit shown below (Figure 9D) is moved along the 100-cm potentiometer wire AB to find the point C at which the centre zero galvanometer registers zero current.

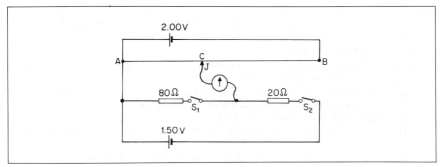

Figure 9D

(a) Both cells have negligible internal resistance. Calculate the length AC (i) with switches S_1 and S_2 both closed, (ii) with switch S_1 open and switch S_2 closed.

(b) The 1·50-V cell develops an internal resistance of a few ohms. Identify and explain, without calculations, any effect on the two balance lengths determined in (a). (*L.*)

11 The circuit in Figure 9E is being used to measure the e.m.f. of a thermocouple T. AB is a uniform wire of length 1·00 m and resistance 2·00 Ω. With K_1 closed and K_2 open, the balance length is 90·0 cm. With K_2 closed and K_1 open, the balance length is 45 cm. What is the e.m.f. of the thermocouple?

What is the value of R if the resistance of the driver cell is negligible? (*L.*)

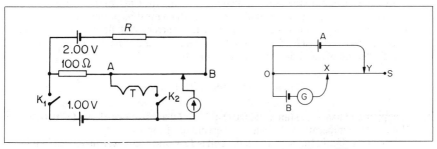

Figure 9E   **Figure 9F**

12 State *Ohm's law*. Discuss two examples of non-ohmic conductors.

Cells A and B and a galvanometer G are connected to a slide wire OS by two sliding contacts X and Y as shown in Figure 9F. The slide wire is 1·0 m long and has a resistance of 12 Ω. With OY 75 cm, the galvanometer shows no deflection when OX is 50 cm. If Y is moved to touch the end of the wire at S, the value of OX which gives no deflection is 62·5 cm. The e.m.f. of cell B is 1·0 V.

Calculate

(a) the p.d. across OY when Y is 75 cm from O (with the galvanometer balanced),

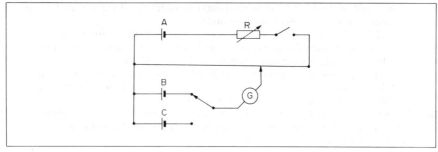

Figure 9G

(b) the p.d. across OS when Y touches S (with the galvanometer balanced),
(c) the internal resistance of cell A,
(d) the e.m.f. of cell A. (*C.*)

13 The circuit diagram in Figure 9G represents the slide wire potentiometer used for the comparison of the e.m.f.s of the cells B and C.

(a) What is the main advantage of using a potentiometer for this purpose?
(b) What is the main quality required of the driver cell A?
(c) What is the purpose of the rheostat R?
(d) If, in practice, a balance point could not be found for cell B suggest two possible reasons.
(e) Outline the experimental procedure, which you would adopt for comparing the e.m.f.s.

Draw a circuit diagram to show how a potentiometer may be adapted to measure an e.m.f. of a few millivolts. Explain how you would standardise this potentiometer using a standard cell. Indicate the approximate values of the components used, if the potentiometer wire has a resistance of $5\,\Omega$ (*L.*)

14 Describe and explain how a potentiometer is used to test the accuracy of the 1 V reading of a voltmeter.

A potentiometer consists of a fixed resistance of $2030\,\Omega$ in series with a slide wire of resistance $4\,\Omega\,\text{metre}^{-1}$. When a constant current flows in the potentiometer circuit a balance is obtained when

(a) a Weston cell of e.m.f. 1.018 V is connected across the fixed resistance and 150 cm of the slide wire and also when
(b) a thermocouple is connected across 125 cm of the slide wire only.

Find the current in the potentiometer circuit and the e.m.f. of the thermocouple.

Find the value of the additional resistance which must be present in the above potentiometer circuit in order that the constant current shall flow through it, given that the driver cell is a lead accumulator of e.m.f. 2 V and of negligible resistance and the length of the slide wire is 2 metres. (*L.*)

Wheatstone Bridge, Resistance

15 A copper coil has a resistance of $20.0\,\Omega$ at $0°C$ and a resistance of $28.0\,\Omega$ at $100°C$. What is the temperature coefficient of resistance of copper?

Used in a circuit, the p.d. across the coil is 12 V and the power produced in it is 6 W. What is the temperature of the coil?

16 A tungsten coil has a resistance of $12.0\,\Omega$ at $15°C$. If the temperature coefficient of resistance of tungsten is $0.004\,\text{K}^{-1}$, calculate the coil resistance at $80°C$.

17 A heating coil is to be made, from nichrome wire, which will operate on a 12 V supply and will have a power of 36 W when immersed in water at 373 K. The wire available has an area of cross-section of $0.10\,\text{mm}^2$. What length of wire will be required? (Resistivity of nichrome at 273 K $= 1.08 \times 10^{-6}\,\Omega\,\text{m}$. Temperature coefficient of resistivity of nichrome $= 8.0 \times 10^{-5}\,\text{K}^{-1}$.) (*L.*)

18 Describe how you would measure the temperature coefficient of resistance of a metal.

Give a short account of the platinum resistance thermometer.

A steady potential difference of 12 V is maintained across a wire which has a resistance of $3 \cdot 0\,\Omega$ at 0°C; the temperature coefficient of resistance of the material is $4 \times 10^{-3}\,K^{-1}$. Compare the rates of production of heat in the wire at 0°C and at 100°C.

The wire is embedded in a body of constant heat capacity $600\,J\,K^{-1}$. Neglecting heat losses, and taking the thermal conductivity of the body to be large, find the time taken to increase the temperature of the body from 0°C to 100°C. (*O.*)

19 Define temperature coefficient of resistance. Describe how you would measure the average temperature coefficient of resistance for an iron wire across the temperature range 0°C to 100°C by a potentiometer method, using an iron wire of resistance about $4\,\Omega$ at 0°C and $6 \cdot 5\,\Omega$ at 100°C wound on an insulating former and provided with copper leads. State approximate values for the circuit components which you would use. (*L.*)

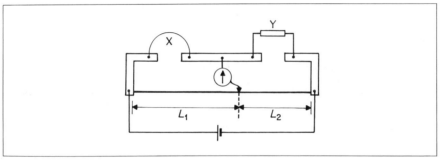

Figure 9H

20 (a) In the Wheatstone bridge arrangement shown above (Figure 9H) X is the resistance of a length of constantan wire and Y is the resistance of a standard resistor. Derive from first principles the equation relating X, Y, L_1 and L_2 when no current flows through the galvanometer, stating any assumptions you make.

(b) A student plans to use the apparatus described in (a) to determine the resistance of several different lengths of constantan wire of diameter $0 \cdot 50\,mm$. If a $2\,\Omega$ standard resistor is available, suggest the range of lengths he might use given that the resistivity of constantan wire is approximately $5 \times 10^{-7}\,\Omega\,m$. Give reasons for your answers. (*JMB.*)

21 What do you understand by *temperature coefficient of resistance*?

Describe fully how you would use two equal resistors, one calibrated variable resistor, and other apparatus, to measure the temperature coefficient of resistance of copper, by means of a Wheatstone bridge circuit. Derive from first principles the equation which is satisfied when your bridge is 'balanced'.

To a good approximation, the resistivity of copper near room temperature is proportional to its absolute temperature. Calculate the temperature coefficient of resistance of copper, and explain your calculation. (Take 0°C as 273 K.) (*C.*)

22 Derive the balance condition for a Wheatstone bridge. Describe a practical form of Wheatstone bridge and explain how you would use it to determine the resistance of a resistor of nominal value $20\,\Omega$.

An electric fire element consists of $4 \cdot 64\,m$ of nichrome wire of diameter $0 \cdot 500\,mm$, the resistivity of nichrome at 15°C being $112 \times 10^{-8}\,\Omega\,m$. When connected to a 240 V supply the fire dissipates $2 \cdot 00\,kW$ and the temperature of the element is 1015°C. Determine a value for the mean temperature coefficient of resistance of nichrome between 15°C and 1015°C. (*L.*)

23 Describe the Wheatstone bridge circuit and deduce the condition for 'balance'. State clearly the fundamental electrical principles on which you base your argument. Upon what factors do
(a) the sensitivity of the bridge,

(b) the accuracy of the measurement made with it, depend?

Using such a circuit, a coil of wire was found to have a resistance of $5\,\Omega$ in melting ice. When the coil was heated to $100°C$, a $100\,\Omega$ resistor had to be connected in parallel with the coil in order to keep the bridge balanced at the same point. Calculate the temperature coefficient of resistance of the coil. (*C.*)

24 (a) Define the volt and use your definition to derive an expression for the power dissipated (in watt) in a resistor of resistance R (in ohm) when a current I (in amp) flows through it.

(b) Explain how you would use a metre bridge to determine the resistivity of a metal in the form of a wire.

(c) Discuss the process of conduction in a metal. Derive an expression for the current flowing in a wire in terms of the number of free electrons per unit volume n, the area of cross section of the wire A, the electronic charge e, and the average drift velocity of the electrons v. (*AEB*, 1982.)

10

Magnetic Field and Force on Conductor

In this chapter we introduce magnetic fields. We shall see the difference in pattern of the lines of force or magnetic flux round magnets, and round current-carrying straight and coiled conductors.

The force on a current-carrying conductor in a magnetic field is due to the 'interaction' between two magnetic fields. We discuss in detail how the force on a conductor is used in the moving-coil meter and the electric motor. Finally, we consider the force on moving charges in a magnetic field and show how it is applied in the Hall effect.

Magnetism

Natural magnets were known some thousands of years ago, and in the eleventh century A.D. the Chinese invented the magnetic compass. This consisted of a magnet, floating on a buoyant support in a dish of water. The respective ends of the magnet, where iron filings are attracted most, are called the north and south poles.

In the thirteenth century the properties of magnets were studied by Peter Peregrinus. He showed that

like poles repel and *unlike poles attract.*

His work was forgotten, however, and his results were rediscovered in the sixteenth century by Dr. Gilbert, who is famous for his researches in magnetism and electrostatics.

Ferromagnetism

About 1823 STURGEON placed an iron core into a coil carrying a current, and found that the magnetic effect of the current was increased enormously. On switching off the current the iron lost nearly all its magnetism. Iron, which can be magnetised strongly, is called a *ferromagnetic* material. Steel, made by adding a small percentage of carbon to iron, is also ferromagnetic. It retains its magnetism, however, after removal from a current-carrying coil, and is more difficult to magnetise than iron.

Nickel and cobalt are the only other ferromagnetic elements in addition to iron, and are widely used for modern magnetic apparatus. A modern alloy for permanent magnets, called *alnico*, has the composition 54 per cent iron, 18 per cent nickel, 12 per cent cobalt, 6 per cent copper, 10 per cent aluminium. It retains its magnetism extremely well, and, by analogy with steel, is therefore said to be magnetically very hard. Alloys which are very easily magnetised, but do not retain their magnetism, are said to be magnetically soft. An example is *mumetal*, which contains 76 per cent nickel, 17 per cent iron, 5 per cent copper, 2 per cent chromium.

Magnetic Fields

The region round a magnet, where a magnetic force is experienced, is called a *magnetic field.* The appearance of a magnetic field is quickly obtained by iron filings, and accurately plotted with a small compass, as the reader knows. The *direction* of a magnetic field is taken as the direction of the force on a *north* pole if placed in the field.

Figure 10.1 shows a few typical fields. The field round a bar-magnet is 'non-uniform', that is, its strength and direction vary from place to place, Figure 10.1 (i). The earth's field locally, however, is uniform, Figure 10.1 (ii). A bar of soft iron placed north–south becomes magnetised by induction by the earth's field, and the lines of force become concentrated in the soft iron, Figure 10.1 (iii). The *tangent* to a line of force at a point gives the direction of the magnetic field at that point.

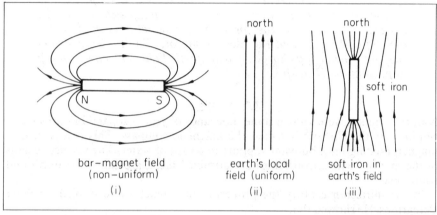

bar–magnet field (non–uniform) (i)

earth's local field (uniform) (ii)

soft iron in earth's field (iii)

Figure 10.1 *Magnetic fields*

Oersted's Discovery

The magnetic effect of the electric current was discovered by OERSTED in 1820. Like many others, Oersted suspected a relationship between electricity and magnetism, and was deliberately looking for it. In the course of his experiments, he happened to lead a wire carrying a current over, but parallel to, a compass-

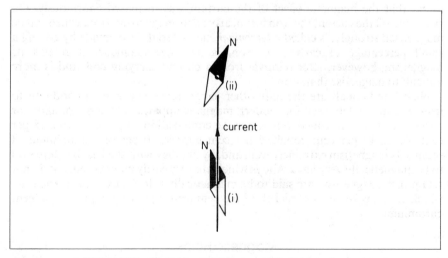

Figure 10.2 *Deflection of compass needle by electric current*

needle, as shown in Figure 10.2 (i); the needle was deflected. Oersted then found that if the wire was led under the needle, it was deflected in the opposite sense, Figure 10.2 (ii).

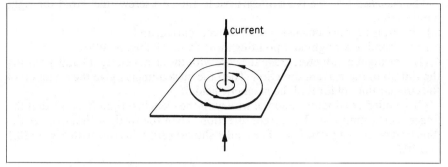

Figure 10.3   *Magnetic field of long straight conductor*

From these observations he concluded that the magnetic field was *circular* round the wire. We can see this by plotting the lines of force of a long vertical wire, as shown in Figure 10.3. To get a clear result a strong current is needed, and we must work close to the wire, so that the effect of the earth's field is negligible. It is then seen that the lines of force are *circles*, concentric with the wire.

Directions of Current and Field; Rules

The relationship between the direction of the lines of force and of the current is expressed in Maxwell's *corkscrew rule*: if we imagine ourselves driving a corkscrew in the direction of the current, then the direction of rotation of the corkscrew is the direction of the lines of force. Figure 10.4 illustrates this rule, the small, heavy circle representing the wire, and the large light one a line of force. At (i) the current is flowing into the paper; its direction is indicated by a cross, which stands for the tail of an arrow moving away from the reader. At (ii) the current is flowing out of the paper; the dot in the centre of the wire stands for the point of an approaching arrow.

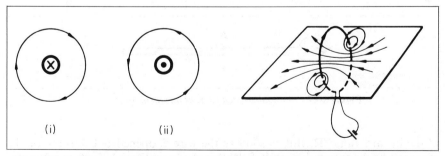

Figure 10.4   *Illustrating corkscrew rule*   **Figure 10.5**   *Magnetic field of narrow coil*

If we plot the magnetic field of a circular coil carrying a current, we get the result shown in Figure 10.5. Near the circumference of the coil, the lines of force are closed loops, which are not circular, but whose directions are still given by the corkscrew rule, as in Figure 10.5. Near the centre of the coil, the lines are almost straight and parallel. Their direction here is again given by the corkscrew rule, but the current and the lines of force are interchanged, that is, if we turn the screw in the direction of the current, then its point travels in the direction of the lines.

The *clenched fist rule* is an alternative to the corkscrew rule: Hold the right hand so that

(a) the fist is tightly clenched with the fingers curled, and

(b) the thumb is straight and pointing away from the fingers. With

(1) a straight conductor, grasp the wire with the clenched right hand, pointing the thumb in the current direction. Then the curled fingers give the direction of the circular lines of force of the magnetic field. If

(2) a coiled conductor, hold the wire with the clenched right hand so that the fingers curl round it in the current direction. Then the straight thumb gives the direction of the magnetic field. The reader should verify this rule with Figure 10.4 and 10.6.

The Solenoid

The magnetic field of a long cylindrical coil is shown in Figure 10.6. Such a coil is called a *solenoid*; it has a field similar to that of a bar-magnet, whose poles are indicated in the figure. If an iron or steel core were put into the coil, it would become magnetised with the polarity shown.

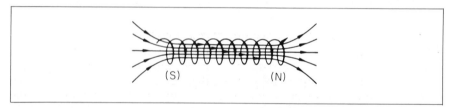

Figure 10.6 *Magnetic field of solenoid*

If the terminals of a battery are joined by a wire which is simply doubled back on itself, as in Figure 10.7, there is no magnetic field at all. Each element of the outward run, such as AB, in effect cancels the field of the corresponding element

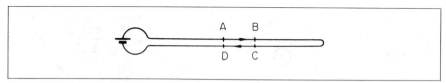

Figure 10.7 *A doubled-back current has no magnetic field*

of the inward run, CD. But as soon as the wire is opened out into a loop, its magnetic field appears, Figure 10.8. Within the loop, the field is strong, because all the elements of the loop give magnetic fields in the same sense, as we can see by applying the corkscrew or other rule to each side of the square ABCD. Outside the loop, for example at the point P, corresponding elements of the loop give opposing fields (for example, DA opposes BC); but these elements are at different distances from P (DA is farther away than BC). So there is a resultant field at P, but it is weak compared with the field inside the loop. A magnetic field

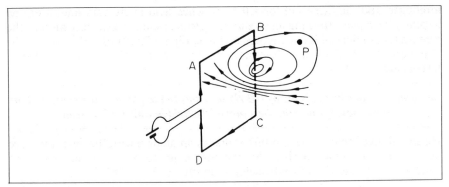

Figure 10.8 *An open loop of current has magnetic field*

can thus be set up either by wires carrying a current, or by the use of permanent magnets.

Force on Conductor, Fleming's Left hand Rule

When a conductor carrying a current is placed in a magnetic field due to some source other than itself, it experiences a mechanical force. To demonstrate this, a

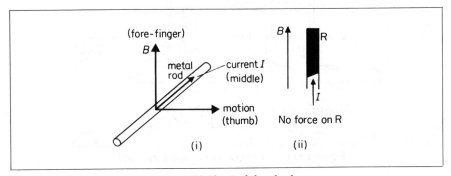

Figure 10.9 *Force on current in magnetic field*

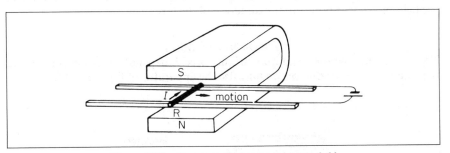

Figure 10.10 *Left-hand rule*

short brass rod R is connected across a pair of brass rails, as shown in Figure 10.9. A horseshoe magnet is placed so that the rod lies in the vertically upward field between its N, S poles. When we pass a current I through the rod, from an accumulator, the rod rolls along the rails.

The relative directions of the current, the applied field, and the motion are shown in Figure 10.10 (i). They are the same as those of the middle finger, the forefinger, and the thumb of the *left* hand when held all at right angles to one another. If we place the magnet so that its field B lies in the *same* direction as the current I, then the rod R experiences no force, Figure 10.10 (ii).

Experiments like this were first made by Ampère in 1820. As a result of them, he concluded that

---

the force on a conductor is *always* at right angles to the plane which contains both the conductor and the direction of the field in which it is placed.

He also showed that, if the conductor makes an angle α with the field, the force on it is proportional to $\sin\alpha$. So the maximum force is exerted when the conductor is *perpendicular* to the field, when $\sin\alpha = 1$.

Dependence of Force on Physical Factors

Since the magnitude of the force on a current-carrying conductor is given by

$$F \propto \sin\alpha \qquad . \qquad . \qquad . \qquad . \qquad . \qquad (1)$$

where α is the angle between the conductor and the field, it follows that F is zero when the conductor is parallel to the field direction. This defines the direction of the magnetic field. To find which way it points, we can apply Fleming's rule to the case when the conductor is placed at right angles to the field. The direction of the field then corresponds to the direction of the forefinger.

Variation of F with I

To investigate how the magnitude of the force F depends on the current I and the length l of the conductor, we may use the apparatus of Figure 10.11.

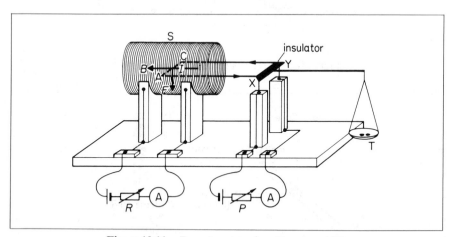

Figure 10.11 *Experiment to show* F *varies with* I

Here the conductor AC is situated in the field B of a solenoid S. The current flows into, and out of, the wire via the pivot points Y and X. The scale pan T is placed at the same distance from the pivot as the straight wire AC, which is perpendicular to the axis of the coil. The frame is first balanced with no current flowing in AC. A current is then passed, and the extra weight needed to restore the frame to a horizontal position is equal to the force on the wire AC. By varying the current in AC with the rheostat P, for example, by doubling or halving the circuit resistance, it may be shown that:

$$F \propto I \quad . \qquad . \qquad . \qquad . \qquad . \qquad . \qquad (2)$$

If different frames are used so that the length, l, of AC is changed, it can be shown that, with constant current and field,

$$F \propto l \quad . \qquad . \qquad . \qquad . \qquad . \qquad . \qquad (3)$$

Effect of B

The magnetic field due to the solenoid will depend on the current flowing in it. If this current is varied by adjusting the rheostat R, it can be shown that the larger the current *in the solenoid*, S, the larger is the force F. It is reasonable to suppose that a larger current in S produces a stronger magnetic field. Thus the force F increases if the magnetic field strength is increased. The magnetic field is represented by a *vector* quantity which is given the symbol B and is defined shortly. This is called the *flux density* in the field. We assume that:

$$F \propto B \qquad . \qquad . \qquad . \qquad . \qquad . \qquad (4)$$

Magnitude of F

From the results expressed in equations (1) to (4), we obtain

$$F \propto BIl \sin \alpha$$

or

$$F = kBIl \sin \alpha \qquad . \qquad . \qquad . \qquad . \qquad . \qquad (5)$$

where k is a constant.

In the SI system of units, the unit of B is the tesla (T). One tesla may be defined as the flux density of a uniform field when the force on a conductor 1 metre long, placed perpendicular to the field and carrying a current of 1 ampere, is 1 newton. Substituting $F = 1$, $B = 1$, $l = 1$ and $\sin \alpha = \sin 90° = 1$ in (5), then $k = 1$. So in Figure 10.12 (i), with the above units,

$$F = BIl \sin \alpha \qquad . \qquad . \qquad . \qquad . \qquad . \qquad (6)$$

When the whole length of the conductor is *perpendicular* to the field B, Figure 10.12 (ii), then, since $\alpha = 90°$ in this case,

$$F = BIl \qquad . \qquad . \qquad . \qquad . \qquad . \qquad . \qquad (7)$$

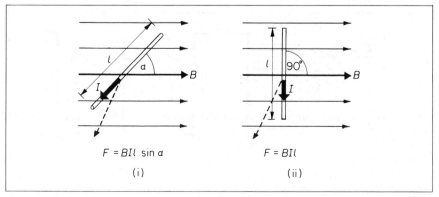

$$F = BIl \sin \alpha$$

(i)

$$F = BIl$$

(ii)

Figure 10.12 *Magnitude of* F *which acts towards reader*

It may be noted that the apparatus of Figure 10.11 can be used to determine the flux density B of the field in the solenoid. In this case, $\alpha = 90°$ and $\sin \alpha = 1$. So measurement of F, I and l enables B to be found from (7).

It may help the reader if we now summarize the main points about B:

---

1 When a current-carrying conductor XY is turned in a uniform magnetic field of flux density B until no force acts on it, then XY points in the direction of B.

2 When a straight conductor of length l carrying a current I is placed perpendicular to a uniform field and a force F acts on the conductor, then the magnitude B of the flux density is *defined* by

$$B = \frac{F}{Il}$$

Since F, I and l can all be measured, B can be calculated.

3 B is a vector. So its component in a direction at an angle θ to B is $B \cos \theta$.

Example on Force on Conductor

A wire carrying a current of 10 A and 2 metres in length is placed in a field of flux density 0·15 T. What is the force on the wire if it is placed
(a) at right angles to the field,
(b) at 45° to the field,
(c) along the field.

From (6) $\qquad\qquad F = BIl \sin \alpha$

(a) $\qquad\qquad\qquad F = 0{\cdot}15 \times 10 \times 2 \times \sin 90°$

$\qquad\qquad\qquad\qquad = 3\,\text{N}$

(b) $\qquad\qquad\qquad F = 0{\cdot}15 \times 10 \times 2 \times \sin 45°$

$\qquad\qquad\qquad\qquad = 2{\cdot}12\,\text{N}$

(c) $\qquad\qquad\qquad F = 0$, since $\sin 0° = 0$

Interaction of Magnetic Fields

The force on a conductor in a magnetic field can be accounted for by the interaction between magnetic fields.

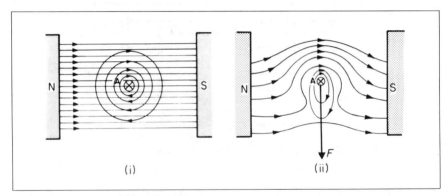

Figure 10.13 *Interaction of magnetic fields*

Figure 10.13 (i) shows a section A of a vertical conductor carrying a downward current. The field pattern consists of circles round A as centre (p. 303). When the conductor is in the uniform horizontal field B due to the poles N, S, the magnetic flux (lines) due to B, which consists of straight parallel lines, passes on either side of A. The two fields interact. As shown, the resultant field has a *greater* flux density above A in Figure 10.13 (ii) and a *smaller* flux density below A. The conductor moves from the region of greater flux density to smaller flux density. So A moves downwards as shown. As the reader should verify, the direction of the force F on the conductor is given by Fleming's left hand rule.

If a current-carrying conductor is placed in the *same* direction as a uniform magnetic field, the flux-density on both sides of the conductor is the same, as the reader should verify. The conductor is now not affected by the field, that is, no force acts on it in this case.

Torque on Rectangular Coil in Uniform Field

A rectangular coil of insulated copper wire is used in the moving-coil meter, which we discuss shortly. Industrial measurements of current and p.d. are made mainly with moving-coil meters.

Consider a rectangular coil situated with its plane *parallel* to a uniform magnetic field of flux density B. Suppose a current I is passed into the coil, Figure 10.14 (i). Viewed from above, the coil appears as shown in Figure 10.14 (ii).

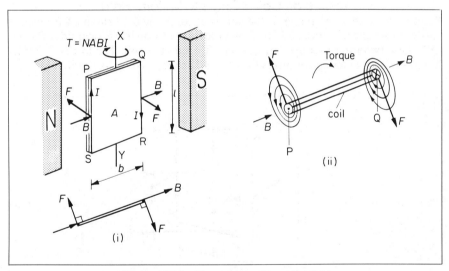

Figure 10.14 *Torque on coil in radial field*

The side PS of length l is perpendicular to B. So the force on it is given by $F = BIl$. If the coil has N turns, the length of the conductor is increased N times and so the force on the side PS, F, $= BIlN$.

The force on the opposite side QR is also given by $F = BIlN$, but its direction is *opposite* to that on PS. There are no forces on the sides PQ and SR although they carry currents because PQ and SR are parallel to the field B.

The two forces F on the sides PS and QR tend to turn the coil about an axis XY passing through the middle of the coil. The two forces together are called a *couple* and their moment (turning-effect) or *torque* T is given, by definition, by

$$T = F \times p$$

where p is the *perpendicular* distance between the two forces. See p. 100. Now from Figure 10.14 (i), $p = b$, the width PQ or SR of the coil. So.

$$T = F \times p = BIlN \times b$$

But $l \times b =$ area A of the coil. So

---

torque $T = BANI$ . . . . . (1)

The unit of torque (force × distance) is newton metre, symbol N m. In using $T = BANI$, B must be in units of T (tesla), A in m^2 and I in A.

If there were no opposition to the torque, the coil PQRS would turn round and settle with its plane normal to B, that is, facing the poles N, S in Figure 10.14 (i). As we see later, springs can control the amount of rotation of the coil.

Figure 10.14 (ii) is a plan view PQ of the rectangular coil with its plane in the same direction as the uniform magnetic field of the magnet N, S. As we explained previously, the magnetic field of the current in the straight sides PS, QR of the coil interacts with the field of the magnet. Figure 10.14 (ii) shows roughly the appearance of the resultant field round the vertical sides of the conductors whose tops are P and Q respectively. The current is downward in Q and upward towards the reader in P. The forces F act from the dense to the less dense flux and together they produce a torque on the coil.

Torque on Coil at Angle to Uniform Field

Suppose now that the plane of the coil is at an angle θ to the field B when it carries a current I. Figure 10.15 (i) shows the forces F_1 on its vertical sides PS and QR; these two forces set up a torque which rotate the coil. The forces F_2 on its horizontal sides merely compress the coil and are resisted by its rigidity.

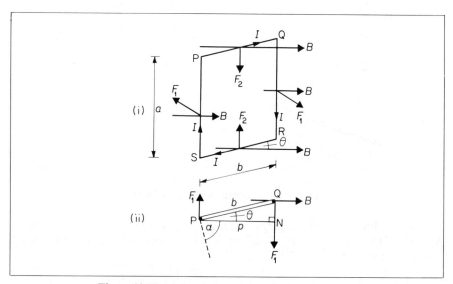

Figure 10.15   *Torque on coil at angle to uniform field*

The forces F_1 on the sides PS and QR are still given by $F_1 = BIlN$ because PS and QR are perpendicular to B. But now the forces F_1 are not separated by a perpendicular distance b, the coil breadth. The perpendicular distance p is less than b and is given by (Figure 10.15 (ii))

$$p = b \cos \theta$$

So this time

$$\text{torque } T = F_1 \times p = BIlN \times b \cos \theta$$

So
$$\boldsymbol{T = BANI \cos \theta} \quad . \quad . \quad . \quad . \quad . \quad (2)$$

When the plane of the coil is *parallel* to B, then $\theta = 0°$ and $\cos \theta = 1$. So the torque $T = BANI$ as we have already shown. If the plane of the coil is *perpendicular* to B, then $\theta = 90°$ and $\cos \theta = 0$. So the torque $T = 0$ in this case.

If α is the angle between B and the *normal* to the plane of the coil, then $\theta = 90° - \alpha$. From (2), the torque T is then given by

$$T = BANI \sin \alpha \quad . \quad . \quad . \quad . \quad . \quad (3)$$

Magnetism is due to circulating and spinning electrons inside atoms. The moving charges are equivalent to electric currents. Consequently, like a current-carrying coil, permanent magnets also have a torque acting on them when they are placed with their axis at an angle to a magnetic field. Like the coil, they turn and settle in equilibrium with their axis along the field direction. Thus the magnetic compass needle will point magnetic north–south in the direction of the Earth's magnetic field. By analogy with the torque on a magnet in a magnetic field, the current-carrying coil is said to have a magnetic moment equal to NIA, from (3).

Example on Torque

A vertical rectangular coil of sides 5 cm by 2 cm has 10 turns and carries a current of 2 A. Calculate the torque on the coil when it is placed in a uniform horizontal magnetic field of 0·1 T with its plane
(a) parallel to the field,
(b) perpendicular to the field,
(c) 60° to the field.

The area A of the coil $= 5 \times 10^{-2} \text{ m} \times 2 \times 10^{-2} \text{ m} = 10^{-3} \text{ m}^2$

So (a) $\quad\quad$ torque $T = BANI = 10 \times 10^{-3} \times 0·1 \times 2$

$$= 2 \times 10^{-3} \text{ N m}$$

(b) Here $\quad\quad T = 0$

(c) $\quad\quad\quad T = BANI \cos 60° \text{ or } BANI \sin 30°$

$$= 2 \times 10^{-3} \times 0·5 = 10^{-3} \text{ N m}$$

The Moving-coil Meter

All current measurements except the most accurate are made today with a moving-coil meter. In this instrument a rectangular coil of fine insulated copper wire is suspended in a strong magnetic field, Figure 10.16 (i). The field is set up between soft iron pole-pieces, NS, attached to a powerful permanent magnet.

The pole-pieces are curved to form parts of a cylinder coaxial with the suspension of the coil. And between them lies a cylindrical core of soft iron, C. It is supported on a brass pin, T in Figure 10.16 (ii), which is placed so that it does not foul the coil. As the diagram shows, the magnetic field B is *radial* to the core and pole-pieces, over the region in which the coil can swing. In this case the

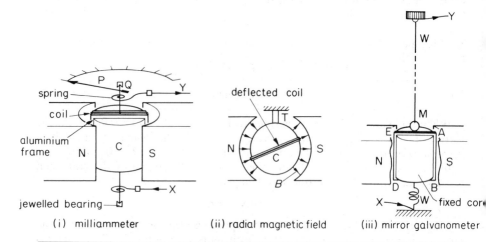

(i) milliammeter   (ii) radial magnetic field   (iii) mirror galvanometer

Figure 10.16 *Moving-coil meters*

deflected coil *always* comes to rest with its plane *parallel* to the field in which it is then situated, as shown in Figure 10.16 (ii).

The moving-coil milliammeter or ammeter have hair-springs and jewelled bearings. The coil is wound on a rigid but light aluminium frame, which also carries the pivots. The pivots are insulated from the former if it is aluminium, and the current is led in and out through the springs. The framework, which carries the springs and jewels, is made from brass or aluminium—if it were steel it would affect the magnetic field. An aluminium pointer, P, shows the deflection of the coil; it is balanced by a counterweight, Q, Figure 10.16 (i).

In the more sensitive instruments, the coil is suspended on a phosphor-bronze wire, WM, which is kept taut, Figure 10.16 (iii). The current is led into and out of the coil EABD through the suspension, at X and Y, and the deflection of the coil is shown by a beam of light, reflected by a mirror M to a scale in front of the instrument.

Theory of Moving-coil Instrument

The rectangular coil is situated in the radial field B. When a current is passed into it, the coil rotates through an angle θ which depends on the strength of the springs. *No matter where the coil comes to rest*, the field B in which it is situated always lies along the *plane* of the coil because the field is radial. As we have previously seen, the torque T on the coil is then always given by $BANI$. So the torque $T \propto I$, since B, A, N are constant.

In equilibrium, the deflecting torque T on the coil is equal to the opposing torque due to the elastic forces in the spring. The opposing torque $= c\theta$, where c is a constant of the springs which depends on its elasticity under twisting forces and on its dimensions. So

$$BANI = c\theta$$

and
$$I = \frac{c}{BAN}\theta \qquad . \qquad . \qquad . \qquad . \qquad (1)$$

Equation (1) shows that the deflection θ is proportional to the current I. So the scale showing current values is a *uniform* one, that is, equal divisions along the calibrated scale represent equal steps in current. This is an important advantage of the moving coil meter. It can be accurately calibrated and its subdivisions read accurately.

If the radial field were not present, for example, if the soft iron cylinder were removed, the torque would then by $BANI \cos \theta$ (p. 311) and I would be proportional to $\theta/\cos \theta$. The scale would then be *non-uniform* and difficult to calibrate or to read accurately.

The pointer type of instrument (Fig. 10.16(i)) usually has a scale calibrated directly in milliamperes or microamperes. Full-scale reading on such an instrument corresponds to deflection θ of 90° to 120°; it may represent a current of 50 microamperes to 15 milliamperes, according to the strength of the hair springs, the geometry of the coil, and the strength of the magnetic field. The less sensitive models are more accurate, because their pivots and springs are more robust, and therefore are less affected by dust, vibration, and hard use.

Summary. A moving-coil meter has:
  (1) a rectangular coil,   (2) springs   (3) a radial magnetic field which produces a linear (uniform) scale,   (4) a current given by $BANI$ (deflection torque) $= c\theta$ (opposing spring torque)

Sensitivity of Current Meter

The *sensitivity* of a current meter is the *deflection per unit current*, or θ/I. Small currents must be measured by a meter which gives an appreciable deflection. From $BANI = c\theta$, we have $\theta/I = BAN/c$. So greater sensitivity is obtained with a stronger field B, a low value of c, that is, *weak* springs, and a greater value of N and A. The size and number of turns of a coil would increase the resistance of the meter, which is not desirable. The elastic constant c of the springs can be varied, however.

When a galvanometer is of the suspended-coil type (Figure 10.16(iii)), its sensitivity is generally expressed in terms of the displacement of the spot of light reflected from the mirror on to the scale. A Scalamp or Edspot, a form of light beam galvanometer, may give a deflection of 25 mm per microampere.

All forms of moving-coil galvanometer have one disadvantage: they are easily damaged by overload. A current much greater than that which the instrument is intended to measure will burn out its hair-springs or suspension.

Sensitivity of Voltmeter

The sensitivity of a voltmeter is the deflection per unit p.d., or θ/V, where θ is the deflection produced by a p.d. V.

If the resistance of a moving coil meter is R, the p.d. V across its terminals when a current I flows through it is given by $V = IR$. From our expression for I given previously,

$$V = \frac{cR}{BAN}\theta$$

So   **voltage sensitivity** $= \dfrac{\theta}{V} = \dfrac{BAN}{cR}$

So unlike the current sensitivity, the voltage sensitivity depends on the resistance R of the meter coil.

Example on Sensitivity of Meter

A moving coil meter X has a coil of 20 turns and a resistance $10\,\Omega$. Another moving coil meter Y has a coil of 10 turns and a resistance of $4\,\Omega$. If the area of each coil, the strength of the springs and the field B are the same in each meter, which has
(a) the greater current sensitivity,
(b) the greater voltage sensitivity?

(a) The current sensitivity is given by

$$\frac{\theta}{I} = \frac{BAN}{c}$$

Since the sensitivity $\propto N$, with A, c and B constant, then X (20 turns) has a greater sensitivity than Y (10 turns).
(b) The voltage sensitivity $= BAN/cR$. So with A, B, c constant,

$$\text{sensitivity} \propto \frac{N}{R}$$

Now $N/R = 20/10 = 2$ numerically for X, and $N/R = 10/4 = 2.5$ for Y. So Y has the greater voltage sensitivity.

As we showed on p. 247, a moving-coil milliammeter can be converted to a voltmeter by adding a suitable high resistance in *series* with the meter, and to an ammeter by adding a suitable low resistance in *parallel* with the meter to act as a shunt. See pp. 249–251.

Multimeters, widely used in the radio and electrical industries, are moving-coil meters which can read potential differences or currents on the same scale, by switching to series or shunt resistors at the back of the meter.

The Wattmeter

The wattmeter is an instrument for measuring electrical power. In construction and appearance it resembles a moving-coil voltmeter or ammeter, but it has no permanent magnet. Instead it has two fixed coils, FF in Figure 10.17, which set up the magnetic field in which the suspended coil, M, moves.

When the instrument is in use, the coils FF are connected in series with the device X whose power consumption is to be measured. The magnetic field B, set up by FF, is then proportional to the current I drawn by X:

$$B \propto I$$

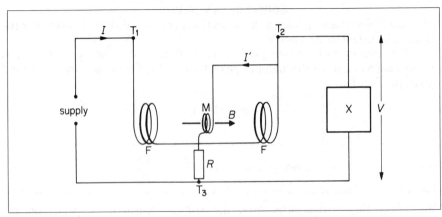

Figure 10.17 *Principle of wattmeter*

The moving-coil M is connected across the device X. In series with M is a high resistance R, similar to the multiplier of a voltmeter; M is, indeed, often called the volt-coil. The current I' through the volt-coil is small compared with the main current I, and is proportional to the potential difference V across the device X:

$$I' \propto V$$

The torque acting on the moving-coil is proportional to the current through it, and to the magnetic field in which it is placed:

$$T \propto BI'$$

Therefore $$T \propto IV$$

So the torque on the coil is proportional to the product of the current through the device X, and the voltage across it. The torque is therefore proportional to the power consumed by X, and the power can be measured by the deflection of the coil.

The diagram shows that, because the volt-coil draws current, the current through the fixed coils is a little greater than the current through X. As a rule, the error arising from this is negligible; if not, it can be allowed for as when a voltmeter and ammeter are used separately.

Force on Charges Moving in Magnetic Fields

We now consider the forces acting on charges moving through a magnetic field. The forces are used to focus the moving electrons on to the screen of a television receiver using a magnet. The forces due to the Earth's magnetic field make electrical particles bunch together near the North pole of the Earth and produce a glow in the sky called Northern Lights.

As we explained earlier, an electric current in a wire can be regarded as a drift of electrons in the wire, superimposed on their random thermal motions. If the electrons in the wire drift with average velocity v, and the wire lies at right angles to the field, then the force on *each* electron, as we soon show, is given by

$$F = Bev \qquad . \qquad . \qquad . \qquad . \qquad . \qquad (1)$$

Generally, the force F on a charge Q moving at right angles to a field of flux density B is given by

$$F = BQv \qquad . \qquad . \qquad . \qquad . \qquad . \qquad (2)$$

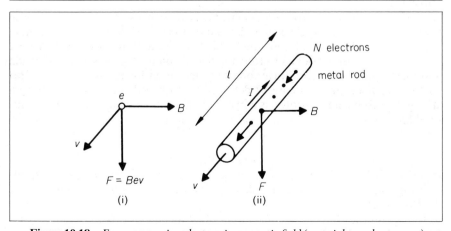

Figure 10.18 *Force on moving electron in magnetic field (v at right angles to page)*

If B is in tesla (T), e or Q is in coulomb (C) and v in metre second^{-1} (m s^{-1}), then F will be in newton (N) (Figure 10.18 (i)).

The proof of equation (1) can be obtained as follows. Suppose a current I flows in a straight conductor of length l when it is perpendicular to a uniform field of flux density B. From p. 243, $I = nvAe$, where n is the number of electrons per unit volume, v is the drift velocity of the electrons, A is the area of cross-section of the conductor and e is the electron charge. Then the force F' on the conductor is given by

$$F' = BIl = BnevAl = Bev \times nAl$$

Now Al is the volume of the wire. So nAl is the number N of electrons in the conductor. Figure 10.18 (ii).

So $\qquad$ force on one electron, $F = \dfrac{F'}{N} = Bev$

Generally, a charge Q moving *perpendicular* to a magnetic field B with a velocity v has a force on it given by

$$F = BQv$$

If the velocity v and the field B are inclined to each other at an angle θ,

$$F = BQv \sin \theta$$

Force Direction, Energy in Magnetic Field

It should be carefully noted that the force F acts *perpendicular* to v and to B. This means that F is a *deflecting force*, that is, it changes the direction of motion of the moving charge when the charge enters the field B but does not alter the magnitude of v.

Further, since F is perpendicular to the direction of motion or displacement of the charge, *no work* is done by F as the charge moves in the field. So *no energy* is gained by a charge when it enters a magnetic field and forces act on it.

The *direction* of F is given by Fleming's left hand rule. The middle finger points in the direction of the conventional current or direction of motion of a *positive* charge. If a *negative* charge moves from X to Y, the middle finger points in the opposite direction, Y to X, since this is the equivalent positive charge movement.

An electron moving across a magnetic field experiences a force whether it is in a wire or not—for example, it may be one of a beam of electrons in a vacuum tube. Because of this force, a magnetic field can be used to *focus* or deflect an electron beam, instead of an electrostatic field as on p. 766. Magnetic deflection and focusing are common in cathode ray tubes used for television. In nuclear energy machines, protons may be deflected and whirled round in a circle by a strong magnetic field. A proton is a hydrogen nucleus carrying a positive charge (p. 898).

Hall Effect

In 1879, Hall found that an e.m.f. is set up *transversely* or *across* a current-carrying conductor when a perpendicular magnetic field is applied. This is called the *Hall effect*.

To explain the Hall effect, consider a slab of metal carrying a current, Figure 10.19. The flow of electrons is in the opposite direction to the conventional current. If the metal is placed in a magnetic field B at right angles to the face AGDC of the slab and directed out of the plane of the paper, a force Bev then acts on each electron in the direction from CD to AG. *Thus electrons collect along the side AG of the metal*, which will make AG negatively charged and lower its potential with respect to CD. So a potential difference or e.m.f. opposes the electron flow. The flow ceases when the e.m.f. reaches a particular value V_H called the *Hall voltage* as shown in Figure 10.19, which may be measured by using a high impedance voltmeter.

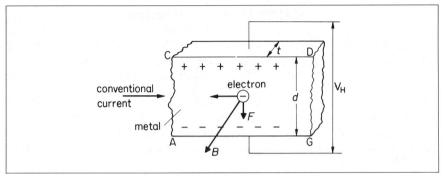

Figure 10.19   *Hall voltage*

Magnitude of Hall Voltage

Suppose V_H is the magnitude of the Hall voltage and d is the width of the slab. Then the electric field intensity E set up across the slab is numerically equal to the potential gradient and hence $E = V_H/d$. So the force on each electron $= Ee = V_He/d$.

The force, which is directed upwards from AG to CD, is equal to the force produced by the magnetic field when the electrons are in equilibrium.

$$\therefore Ee = Bev$$

$$\therefore \frac{V_He}{d} = Bev$$

$$\therefore V_H = Bvd \qquad . \qquad . \qquad . \qquad . \qquad . \qquad (1)$$

From p. 243, the drift velocity of the electrons is given by

$$I = nevA \qquad . \qquad . \qquad . \qquad . \qquad . \qquad (2)$$

where n is the number of electrons per unit volume and A is the area of cross-section of the conductor. In this case $A = td$ where t is the thickness. Hence, from (2),

$$v = \frac{I}{netd}$$

Substituting in (1),

$$\therefore V_H = \frac{BI}{net} \qquad . \qquad . \qquad . \qquad . \qquad . \qquad (3)$$

We now take some typical values for copper to see the order of magnitude of V_H. Suppose $B = 1$ T, a field obtained by using a large laboratory electromagnet.

For copper, $n \simeq 10^{29}$ *electrons* per metre³, and the charge on the electron is 1.6×10^{-19} coulomb. Suppose the specimen carries a current of 10 A and that its thickness is about 1 mm or 10^{-3} m. Then

$$V_{\mathrm{H}} = \frac{1 \times 10}{10^{29} \times 1.6 \times 10^{-19} \times 10^{-3}} = 0.6\,\mu\mathrm{V} \text{ (approx.)}$$

This e.m.f. is very small and would be difficult to measure. The importance of the Hall effect becomes apparent when semiconductors are used, as we now see.

Hall Effect in Semiconductors

In semiconductors, the charge carriers which produce a current when they move may be positively or negatively charged (see p. 788). The Hall effect helps us to find the sign of the charge carried. In Figure 10.19, p. 317, suppose that electrons were not responsible for carrying the current, and that the current was due to the movement of positive charges in the *same* direction as the conventional current. The magnetic force on these charges would also be *downwards*, in the same direction as if the current were carried by electrons. This is because the sign *and* the direction of movement of the charge carriers have both been reversed. Thus AG would now become *positively* charged, and the polarity of the Hall voltage would be reversed.

Experimental investigation of the polarity of the Hall voltage hence tells us whether the current is predominantly due to the drift of positive charges or to the drift of negative charges. In this way it was shown that the current in a metal such as copper is due to movement of negative charges, but that in impure semiconductors such as germanium or silicon, the current may be predominantly due to movement of either negative or positive charges (p. 788).

The magnitude of the Hall voltage V_{H} in metals was shown as above to be very small. In semiconductors it is much larger because the number n of charge carriers per metre³ is much *less* than in a metal and $V_{\mathrm{H}} = BI/net$. Suppose that n is about 10^{25} per metre³ in a semiconductor, and $B = 1$ T, $t = 10^{-3}$ m, $e = 1.6 \times 10^{-19}$ C, as above. Then

$$V_{\mathrm{H}} = \frac{1 \times 10}{10^{25} \times 1.6 \times 10^{-19} \times 10^{-3}} = 6 \times 10^{-3}\,\mathrm{V} \text{ (approx.)} = 6\,\mathrm{mV}$$

The Hall voltage is thus much more measurable in semiconductors than in metals.

Use of Hall Effect

Apart from its use in semiconductor investigations, a *Hall probe* may be used to measure the flux density B of a magnetic field. A simple Hall probe is shown in Figure 10.20. Here a wafer of semiconductor has two contacts on opposite sides which are connected to a high impedance voltmeter, V. A current, generally less than one ampere, is passed through the semiconductor and is measured on the ammeter, A. The 'araldite' glue prevents the wires from being detached from the wafer. Now, from (3) on p. 317.

$$V_{\mathrm{H}} = \frac{BI}{net}$$

$$\therefore B = \frac{V_{\mathrm{H}} net}{I}$$

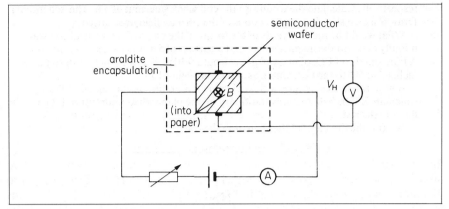

Figure 10.20 *Measurement of* B *by Hall voltage*

Now *net* is a constant for the given semiconductor, which can be determined previously. Thus from the measurement of V_H and I, B can be found.

In practice, the voltmeter scale is calibrated in teslas(T) by the manufacturer and so the flux density B of the magnetic field is read directly from the scale. Note that the direction of B must be *perpendicular* to the semiconductor probe when measuring B. Later we shall use the Hall probe to measure the flux density B round a straight current-carrying conductor and inside a current-carrying solenoid (p. 324).

Summary

1   **With B perpendicular to a conductor S, a Hall voltage is obtained on the sides of S normal to the current flowing through S.**
2   **Hall voltage $V_H = BI/net$.**
3   **The Hall voltage is used**
     (a) in semiconductors to find whether the current flow is due mainly to positive or negative charges,
     (b) to measure n, the charge density,
     (c) as a basis of a Hall probe, for measuring the flux density B of a magnetic field.

Exercises 10

1   A vertical straight conductor X of length 0·5 m is situated in a uniform horizontal magnetic field of 0·1 T.   (i) Calculate the force on X when a current of 4 A is passed into it. Draw a sketch showing the directions of the current, field and force. (ii) Through what angle must X be turned in a vertical plane so that the force on X is halved?
2   A straight horizontal rod X, of mass 50 g and length 0·5 m, is placed in a uniform horizontal magnetic field of 0·2 T perpendicular to X. Calculate the current in X if the force acting on it just balances its weight. Draw a sketch showing the directions of the current, field and force. ($g = 10\,\mathrm{N\,kg^{-1}}$.)
3   A narrow vertical rectangular coil is suspended from the middle of its upper side with its plane parallel to a uniform horizontal magnetic field of 0·02 T. The coil has 10 turns, and the lengths of its vertical and horizontal sides are 0·1 m and 0·05 m

respectively. Calculate the torque on the coil when a current of 5 A is passed into it. Draw a sketch showing the directions of the current, field and torque.

What would be the new value of the torque if the plane of the vertical coil was initially at 60° to the magnetic field and a current of 5 A was passed into the coil?

4   A horizontal rod PQ, of mass 10 g and length 0·10 m, is placed on a smooth plane inclined at 60° to the horizontal, as shown in Figure 10A.

A uniform vertical magnetic field of value B is applied in the region of PQ. Calculate B if the rod remains stationary on the plane when a current of 1·73 A flows in the rod.

What is the direction of the current in the rod?

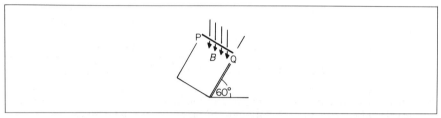

Figure 10A

5   An electron beam, moving with a velocity of $10^6\,\mathrm{m\,s^{-1}}$, moves through a uniform magnetic field of 0·1 T which is perpendicular to the direction of the beam. Calculate the force on an electron if the electron charge is $-1\cdot6 \times 10^{-19}$ C. Draw a sketch showing the directions of the beam, field and force.

6   A current of 0·5 A is passed through a rectangular section of a semiconductor 4 mm thick which has majority carriers of negative charges or free electrons. When a magnetic field of 0·2 T is applied perpendicular to the section, a Hall voltage of 6·0 mV is produced between the opposite edges.

Draw a diagram showing the directions of the field, charge carriers and Hall voltage, and calculate the number of charge carriers per unit volume.

7   Figure 10B represents a cylindrical aluminium bar A resting on two horizontal aluminium rails which can be connected to a battery to drive a current through A. A magnetic field, of flux density 0·10 T, acts perpendicularly to the paper and into it. In which direction will A move if the current flows?

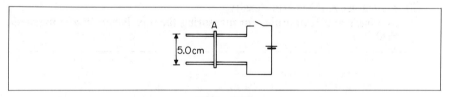

Figure 10B

Calculate the angle to the horizontal to which the rails must be tilted to keep A stationary if its mass is 5·0 g, the current in it is 4·0 A and the direction of the field remains unchanged. (Acceleration of free fall, $g = 10\,\mathrm{m\,s^{-2}}$.) (*L.*)

8   Describe an experiment to show that a force is exerted on a conductor carrying a current when it is placed in a magnetic field. Give a diagram showing the directions of the current, the field, and the force.

A rectangular coil of 50 turns hangs vertically in a uniform magnetic field of magnitude 10^{-2} T, so that the plane of the coil is parallel to the field. The mean height of the coil is 5 cm and its mean width 2 cm. Calculate the strength of the current that must pass through the coil in order to deflect it 30° if the torsional constant of the suspension is 10^{-9} newton metre per degree. Give a labelled diagram of a moving-coil galvanometer. (*L.*)

9 Describe with the aid of diagrams the structure and mode of action of a moving coil galvanometer having a linear scale and suitable for measuring small currents. If the coil is rectangular, derive an expression for the deflecting couple acting upon it when a current flows in it, and hence obtain an expression for the current sensitivity (defined as the deflection per unit current).

If the coil of a moving galvanometer having 10 turns and of resistance $4\,\Omega$ is removed and replaced by a second coil having 100 turns and of resistance $160\,\Omega$ calculate
(a) the factor by which the current sensitivity changes and
(b) the factor by which the voltage sensitivity changes.
Assume that all other features remain unaltered. (*JMB.*)

10 Define the coulomb. Deduce an expression for the current I in a wire in terms of the number of free electrons per unit volume, n, the area of cross-section of the wire, A, the charge on the electron, e, and its drift velocity, v.

A copper wire has 1.0×10^{29} free electrons per cubic metre, a cross-sectional area of $2.0\,\text{mm}^2$ and carries a current of $5.0\,\text{A}$. Calculate the force acting on each electron if the wire is now placed in a magnetic field of flux density $0.15\,\text{T}$ which is perpendicular to the wire. Draw a diagram showing the directions of the electron velocity, the magnetic field and this force on an electron.

Explain, without experimental detail, how this effect could be used to determine whether a slab of semiconducting material was n-type or p-type. (Charge on electron $= -1.6 \times 10^{-19}\,\text{C}$.) (*L.*)

11 (a) A moving coil meter posses a square coil mounted between the poles of a strong permanent magnet. The torque on the coil is $4.2 \times 10^{-9}\,\text{N m}$ when the current is $100\,\mu\text{A}$.   (i) The meter is designed so that whatever the deflection of the coil, the magnetic flux density is always parallel to the plane of the coil. Explain, with the aid of a labelled diagram how this is achieved.   (ii) The restoring springs bring the coil to rest after it has turned through a certain angle. If the restoring couple per unit angular displacement applied by the springs is $3.0 \times 10^{-9}\,\text{N m}$ per radian, through what angle, in radian, will the coil turn when a current of $100\,\mu\text{A}$ flows?   (iii) Explain what is meant by the *current sensitivity* of such a meter. If the pointer on the instrument is $7.0\,\text{cm}$ long, what length of arc on the scale would correspond to a change in current of $2\,\mu\text{A}$?   (iv) The instrument indicates full scale deflection for a current of $100\,\mu\text{A}$. What current produces full scale deflection if the number of turns in the coil is doubled?

Increasing the number of turns also increases the resistance of the coil. Explain whether or not this change affects the sensitivity of the meter.
(b) A moving coil meter has a resistance of $1000\,\Omega$ and gives a full scale deflection for a current of $100\,\mu\text{A}$.   (i) What value resistor would be required to convert it to an ammeter reading up to $1.00\,\text{A}$? Draw a circuit diagram showing where the resistor would be connected. What form might this resistor have?   (ii) Draw a diagram showing the additional circuitry needed for the moving coil meter to be adapted to measure alternating currents. Mark clearly on the diagram the connecting points for the meter and for the a.c. supply.

What is the relationship between the steady current registered by the meter and the current from the a.c. supply? (*L.*)

12 Draw a labelled sketch showing the construction of a moving-coil galvanometer. Deduce an expression for the angle of deflection in terms of the current and any other relevant quantities.

Discuss the factors that determine the sensitivity of the galvanometer.

You are provided with two identical meters of f.s.d. $50\,\text{mA}$ and resistance $100\,\Omega$. Describe how to convert one of them to an ammeter reading up to $1\,\text{A}$ and the other to a voltmeter reading up to $200\,\text{V}$.

They are to be used to check the power consumption of a lamp rated $100\,\text{W}$ and $200\,\text{V}$. Two circuits can be arranged, with the voltmeter connected   (i) across the lamp only or   (ii) across the lamp and the ammeter.
(a) Show that when the power is determined from the readings on the meters both methods give the wrong answer.
(b) Which, if either, is the more accurate? (*W.*)

13 Write down a formula for the magnitude of the force on a straight current-carrying wire in a magnetic field, explaining clearly the meaning of each symbol in your formula.

Derive an expression for the couple on a rectangular coil of n turns and dimensions $a \times b$ carrying a current I when placed in a uniform magnetic field of flux density B at right angles to the sides of the coil of length a and at an angle θ to the sides of length b. Describe briefly how you would demonstrate experimentally that the couple on a plane coil in a uniform field depends only on its area and not on its shape.

A circular coil of 50 turns and area $1\cdot25 \times 10^{-3}\,\mathrm{m}^2$ is pivoted about a vertical diameter in a uniform horizontal magnetic field and carries a current of 2 A. When the coil is held with its plane in a north–south direction, it experiences a couple of $0\cdot04\,\mathrm{N\,m}$. When its plane is east–west, the corresponding couple is $0\cdot03\,\mathrm{N\,m}$. Calculate the magnetic flux density. (Ignore the earth's magnetic field.) (*O. & C.*)

14 A strip of metal $1\cdot2$ cm wide and $1\cdot5 \times 10^{-3}$ cm thick carries a current of $0\cdot50$ A along its length. If it is assumed that the metal contains 5×10^{22} free electrons per cm^3, calculate the mean drift velocity of these electrons ($e = 1\cdot6 \times 10^{-19}$ C).

The metal foil is placed normal to a magnetic field of flux density B. Explain why, in these circumstances, you might expect a p.d. to be developed across the foil. By equating the magnetic and electric forces acting on an electron when the p.d. has been established, derive an expression for the p.d. in terms of B, the current I, the electron charge e, the number of electrons per unit volume N and the thickness of the foil t. Illustrate your answer with a clear diagram. (*JMB.*)

15 Describe a moving-coil type of galvanometer and deduce a relation between its deflection and the steady current passing through it.

A galvanometer, with a scale divided into 150 equal divisions, has a current sensitivity of 10 divisions per milliampere and a voltage sensitivity of 2 divisions per millivolt. How can the instrument be adapted to serve

(a) as an ammeter reading to 6 A,

(b) as a voltmeter in which each division represents 1 V? (*L.*)

16 Explain the origin of the Hall effect. Include a diagram showing clearly the directions of the Hall voltage and other relevant vector quantities for a specimen in which electron conduction predominates.

A slice of indium antimonide is $2\cdot5$ mm thick and carries a current of 150 mA. A magnetic field of flux density $0\cdot5$ T, correctly applied, produces a maximum Hall voltage of $8\cdot75$ mV between the edges of the slice. Calculate the number of free charge carriers per unit volume, assuming they each have a charge of $-1\cdot6 \times 10^{-19}$ C. Explain your calculation clearly.

What can you conclude from the observation that the Hall voltage in different conductors can be positive, negative or zero? (*C.*)

11
Magnetic Fields of Current-Carrying Conductors

In this chapter we shall deal more fully with the magnetic fields due to currents in the main types of conductor, the solenoid, the straight conductor (wire) and a narrow circular coil.

Solenoids are widely used, particularly with soft iron inside, in the electrical and radio industries. The straight conductor can be used in a basic current-measuring meter and is used to define the ampere. Two narrow circular coils are used as so-called Helmholtz coils to provide a uniform magnetic field in experiments.

We shall first state the values of the flux density B of each of the three conductors and show how they are applied.

Experiments to verify these formulae for B will also be given and a formal proof of the formulae will be found at the end of the chapter.

Solenoid

Solenoids, or relatively long coils of wire, are widely used in industry. For example, solenoids are used in telephone earpieces to carry the speech current and in magnetic relays used in telecommunications.

The magnetic field inside an infinitely-long solenoid is constant in magnitude. A form of coil which gives a very nearly uniform field is shown in Figure 11.1 (i). It is a solenoid of N turns and length L metre wound on a circular support

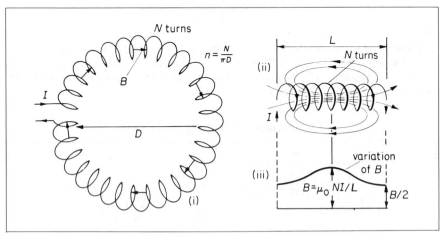

Figure 11.1 *A toroid and solenoid*

instead of a straight one, and is called a *toroid*. If its average diameter D is several times its core diameter, then the turns of wire are almost equally spaced around its inside and outside circumferences; their number per metre is therefore

$$n = \frac{N}{L} = \frac{N}{\pi D} \qquad \qquad (1)$$

The magnetic field within a toroid is very nearly uniform, because the coil has no ends. The coil is equivalent to an infinitely long solenoid. If I is the current, the flux density B at all points within it is given by

$$B = \mu_0 nI . \qquad \qquad (2)$$

μ_0 is a constant known as the *permeability of free space* which has the value $4\pi \times 10^{-7}\,\mathrm{H\,m^{-1}}$ (H is a unit called a 'henry' and is discussed later). The constant μ_0 is necessary to make the units correct, that is, B is then in teslas (T) when I is in amperes (A) and l is in metres (m).

Solenoids of Finite Length

In practice, solenoids cannot be made infinitely long. But if the length L of a solenoid is about ten times its diameter, the field near its middle is fairly uniform, and has the value given by equation (2). Figure 11.1 (ii) shows a solenoid of length L and N turns, so that $n = N/L$. The flux density in the *middle* of the coil is given approximately by

$$B = \mu_0 nI = \mu_0 \frac{NI}{L} \qquad \qquad (3)$$

If a long solenoid is imagined cut at any point R near the middle, the two solenoids on each side have the same field B at their respective centres since each has the same number of turns per unit length as the long solenoid. So each solenoid contributes equally to the field at R. Hence each solenoid provides a field $B/2$ at their end R. We therefore see that the field at the *end* of any long solenoid is *half* that at the centre and is given by

$$B = \frac{1}{2}\mu_0 \frac{NI}{L} \qquad \qquad (4)$$

Figure 11.1 (iii) shows roughly the variation of B along the solenoid.

As we explained on p. 303, the direction of B inside the solenoid can be found from the 'corkscrew rule' or the 'clenched fist rule'. The reader should verify the directions of B shown in Figure 11.1 (i) and (ii).

Experiment for *B* using Hall Probe

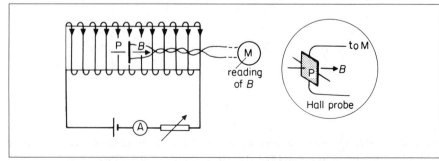

Figure 11.2   *B measured by Hall probe inside solenoid*

Figure 11.2 shows how the flux density B in the middle of a long solenoid can be investigated. S is a 'Slinky' (loose) coil, with its N turns uniformly spaced in a length L. P is a Hall probe in the middle of S and placed so that the flux density B is *normal* to P. As shown on p. 318, the Hall voltage produced at P is proportional to the value of B and this can be read directly in tesla (T) on the meter M.

In the experiment, the uniform spacing of S is varied by pulling out the coil more and the total length L of the coil and the value of B in the middle are measured each time. The number of turns per metre length is given by $n = N/L$, so $n \propto 1/L$ as N is constant. A graph of B against $1/L$ produces a straight line passing through the origin, so showing that $B \propto n$. The same circuit can be used to verify $B \propto I$, the current in the solenoid, for a given value of n.

Effect on *B* of Relative Permeability

As we have stated, the constant μ_0 in the formula for flux density B is called the permeability of free space (or vacuum) and has the value $4\pi \times 10^{-7}\,\mathrm{H\,m^{-1}}$. The permeability of air at normal pressure is only very slightly different from that of a vacuum. So we can consider the permeability of air to be practically $4\pi \times 10^{-7}\,\mathrm{H\,m^{-1}}$.

If the solenoid is wound round soft iron, so that this material is now the core of the solenoid, the permeability is increased considerably. The name 'relative permeability', symbol μ_r, is given to the number of times the permeability has increased relative to that of free space or air. So if $\mu_r = 1000$, the value of B in the solenoid is 1000 times as great as with an air core. Generally, the permeability μ of an iron core would be given by

$$\mu = \mu_r \mu_0$$

Note that μ_r is a number and has no units, unlike μ_0 and μ.

Long Straight Conductor

We now consider the magnetic field of a long straight current-carrying conductor. A submarine cable carrying messages is an example of such a conductor.

All round a straight current-carrying wire, the field pattern consists of circles concentric with the wire. Figure 11.3 (i) shows the field round one section of the

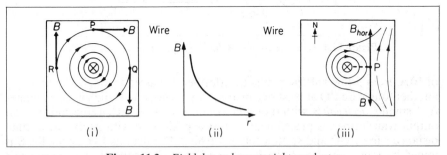

Figure 11.3 *Field due to long straight conductor*

conductor. Maxwell's corkscrew rule gives the field direction: If a right-handed corkscrew is turned so that the point moves along the current direction, the field direction is the same as the direction of turning.

The direction of B is along the tangent to a circle at the point concerned. So at P due north of the wire, B points east for a downward current. At a point due east, B points south and at a point due west, B points north.

At a point distance r from an infinitely-long wire, the value of B is given by

$$B = \frac{\mu_0 I}{2\pi r}$$

So for a given current, $B \propto 1/r$, Figure 11.3 (ii).

The earth's horizontal magnetic field B_{hor} is about $4 \times 10^{-5}\,\mathrm{T}$ and acts due north. When this cancels exactly the magnetic field of the current, a *neutral point* is obtained in the combined field of the earth and the current. Since the field due to the current must be due south, the neutral point P in Figure 11.3 (iii) is due *east* of the wire. Suppose the current is 5 A. The distance r of the neutral point from the wire is then given by

$$\frac{\mu_0 I}{2\pi r} = B_{hor} = 4 \times 10^{-5}$$

So
$$r = \frac{\mu_0 I}{2\pi \times 4 \times 10^{-5}} = \frac{4\pi \times 10^{-7} \times 5}{2\pi \times 4 \times 10^{-5}}$$

$$= 0{\cdot}025\,\mathrm{m} = 25\,\mathrm{mm}$$

Variation of *B* using a Search Coil

An apparatus suitable for finding the variation of B with distance r from a long straight wire CD is shown in Figure 11.4. Alternating current (a.c.) of the order

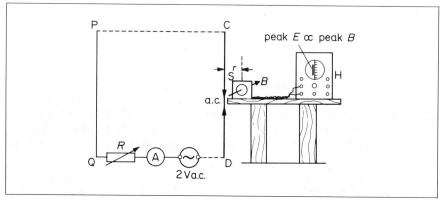

Figure 11.4   *Investigation of* B *due to long straight conductor CD*

of 10 A, from a low voltage mains transformer, is passed through CD by using another long wire PQ at least one metre away, a rheostat R and an a.c. ammeter A. A small *search coil* S, with thousands of turns of wire, such as the coil from an output transformer, is placed near CD. It is positioned with its axis at a small distance r from CD and so that the flux from CD enters its face normally. S is joined by long twin flex to the Y-plates of an oscilloscope H and the greatest sensitivity, such as $5\,\mathrm{mV/cm}$, is used.

When the a.c. supply is switched on, the varying flux through S produces an induced alternating e.m.f. E. The peak (maximum) value of E can be found by switching off the time-base and measuring the length of the line trace, Figure 11.4. See p. 781. Now the peak value of the magnetic flux density B is proportional to the peak value of E, as shown later. So the length of the trace gives a measure of the peak value of B.

The distance r of the coil CD is then increased and the corresponding length of the trace is measured. The length of the trace plotted against $1/r$ gives a straight line graph passing through the origin. Hence $B \propto 1/r$. A similar method can be used for investigating the field B inside of a solenoid.

Forces between Currents

In 1821, Ampere discovered by experiment that current-carrying conductors exert a force on each other. For example, when the currents in two long neighbouring straight conductors X and Y are in the same direction, there is a force of *attraction* between them, Figure 11.5 (i). If the currents flow in opposite directions, there is a *repulsive* force between them, Figure 11.5 (ii). Each conductor has a force on it due to the magnetic field of the other, from the law of Action and Reaction.

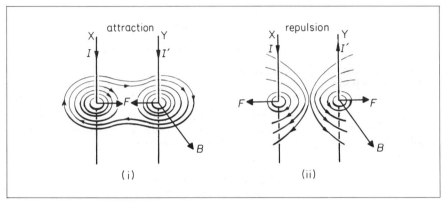

Figure 11.5   *Forces between currents*

Figure 11.5 (i) shows the resultant magnetic flux round two long straight vertical conductors X, Y in a horizontal plane when the currents are both downwards. The lines tend to pull the conductors towards each other. In Figure 11.5 (ii), the currents are in opposite directions. Here the lines tend to push the conductors apart.

Fleming's left hand rule confirms the direction of the forces. At Y, the flux-density B due to the conductor X is perpendicular to Y (the flux due to X alone consists of circles with X as centre and at Y the *tangent* to the circular line is perpendicular to Y). So, from Fleming's rule, the force F on Y in Figure 11.5 (i) is towards X. From the law of action and reaction, the force F on X is towards Y and equal to that on Y. So the conductors *attract* each other.

In Figure 11.5 (ii), the current I' in Y is *opposite* to that in Figure 11.5 (i). From Fleming's left hand rule, the force F on Y is now away from X and so the force is *repulsive*.

Magnitude of Force, The Ampere

If two long straight conductors X and Y lie parallel and close together at a distance r apart, and carry currents I, I' respectively as in Figure 11.5 (i), then the current I is in a magnetic field of flux density B equal to $\mu_0 I'/2\pi r$ due to the current I' (p. 326). The force *per metre* length, F, on X is hence given by

$$F = BIl = BI \times 1 = \frac{\mu_0 I'}{2\pi r} \times I \times 1$$

$$\therefore F = \frac{\mu_0 II'}{2\pi r} \qquad \cdot \qquad \cdot \qquad \cdot \qquad \cdot \qquad \cdot \qquad (1)$$

From the law of Action and Reaction, this would also be the force per metre on the other conductor Y.

Nowadays the ampere is *defined* in terms of the force between conductors.

It is *that current, which flowing in each of two infinitely-long parallel straight wires of negligible cross-sectional area separated by a distance of* 1 *metre* in vacuo, *produces a force* between the wires of 2×10^{-7} *newton metre*$^{-1}$.

Taking $I = I' = 1$ A, $r = 1$ metre, $F = 2 \times 10^{-7}$ newton metre^{-1}, then, from (1),

$$2 \times 10^{-7} = \frac{\mu_0 \times 1 \times 1}{2\pi \times 1}$$

$$\therefore \mu_0 = 4\pi \times 10^{-7} \text{ henry metre}^{-1}$$

which is the value used in formulae with μ_0.

It may be noted that the electrostatic force of repulsion between the negative charges of the moving electrons in the two wires is completely neutralised by the attractive force on them by the positive charges on the stationary metal ions in the wires. Thus the force between the two wires is only the *electromagnetic* force, which is due to the magnetic fields of the moving electrons.

Example on Force between Conductors

A long straight conductor X carrying a current of 2 A is placed parallel to a short conductor Y of length 0·05 m carrying a current of 3 A, Figure 11.6. The two conductors are 0·10 m apart. Calculate (i) the flux density due to X at Y, (ii) the approximate force on Y.

(i) Due to X, $$B = \frac{\mu_0 I}{2\pi r} = \frac{4\pi \times 10^{-7} \times 2}{2\pi \times 0\cdot 10}$$

$$= 4 \times 10^{-6} \text{ T}$$

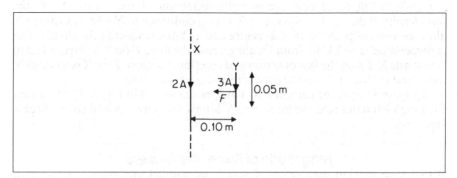

Figure 11.6 *Force between conductors*

(ii) On Y, length $l = 0\cdot 05$ m

force $F = BIl = 4 \times 10^{-6} \times 3 \times 0\cdot 05$

$$= 6 \times 10^{-7} \text{ N}$$

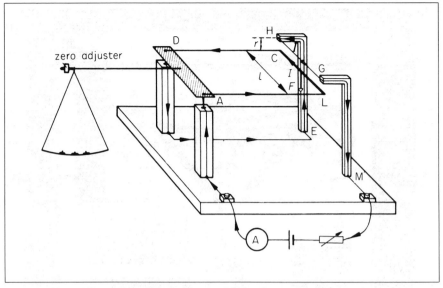

Figure 11.7 *Laboratory form of ampere balance*

Absolute Determination of Current, Ampere Balance

A simple laboratory form of an *ampere balance*, which measures current by measuring the force between current-carrying conductors, is shown in Figure 11.7.

With no current flowing, the zero screw is adjusted until the plane of ALCD is horizontal. The current I to be measured is then switched on so that it flows through ALCD and EHGM in series and HG repels CL. The mass m necessary to restore balance is then measured, and mg is the force between the conductors since the respective distances of CL and the scale pan from the pivot are equal. The equal lengths l of the straight wires CL and HG, and their separation r, are all measured.

From equation (1) above,

$$\text{force per metre} = \frac{4\pi \times 10^{-7} I^2}{2\pi r}$$

$$\therefore mg = \frac{4\pi \times 10^{-7} I^2 l}{2\pi r}$$

$$\therefore I = \sqrt{\frac{mgr}{2 \times 10^{-7} l}}$$

In this expression I will be in ampere if m is in kilogram, $g = 9.8\ \text{m s}^{-2}$ and l and r are measured in metre.

Figure 11.7A Ampere balance. *Current balance at the National Physical Laboratory. One large coil (bottom left) has been lowered so that the small suspended coil above it, at the end of the beam on the left, can be seen. To measure current, the large coils (left and right) and the two suspended coils above them are all connected in series, in such a way that one suspended coil is repelled upwards and the other is attracted downwards when the current flows. Equilibrium is restored by adding masses on one of the scale pans. These are placed on or lifted off the scale pan by rods controlled by the knobs outside the case.* (Crown copyright, Courtesy of National Physical Laboratory)

Narrow Circular Coil

The third of our typical conductors is the narrow circular coil.

Figure 11.8(i) shows the magnetic field pattern round a narrow vertical circular coil C carrying a current I, in the horizontal (perpendicular) plane passing through the middle of the coil. In the middle M of the coil, the field is uniform for a short distance either side. Here the field value B is given by

$$B = \frac{\mu_0 I}{2r}$$

where r is the radius in metres.

Figure 11.8(ii) shows how B varies as we move from the centre of the coil along a line perpendicular to the plane of the coil. The field value decreases continuously. Helmholtz, an eminent scientist of the 19th century, showed that two narrow circular coils of the same radius and carrying the same current could provide a *uniform* magnetic field between them. For this purpose they are placed facing each other at a distance apart equal to their radius R. As shown in

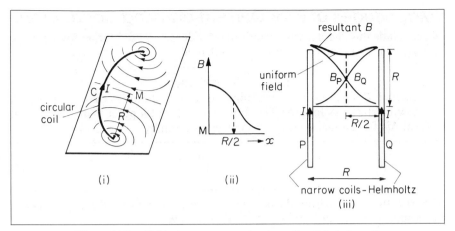

Figure 11.8   *Fields due to narrow circular coils. Helmholtz coils*

Figure 11.8 (iii), the resultant magnetic field B round a point half-way between the coils P and Q is fairly uniform for some distance on either side of the point. The flux density B of the uniform field is given approximately by

$$B = 0{\cdot}72\frac{\mu_0 NI}{R}$$

where N is the number of turns in each coil, I is the current in amperes and R is the radius in metres.

Helmholtz coils were used by Sir J. J. Thomson to obtain a uniform magnetic field of known value in a famous experiment to find the charge–mass ratio of an electron (see p. 769).

Magnitudes of B for Current-carrying Conductors

We conclude this chapter with proofs of the values of B used earlier for a narrow circular coil, a straight conductor and a solenoid.

Law of Biot and Savart

To calculate B for any shape of conductor, Biot and Savart gave a law which can now be stated as follows: The flux density ΔB at a point P due to a small element Δl of a conductor carrying a current is given by

$$\Delta B \propto \frac{I\Delta l \sin \alpha}{r^2} \qquad . \qquad . \qquad . \qquad . \qquad . \qquad (1)$$

where r is the distance from the point P to the element and α is the angle between the element and the line joining it to P, Figure 11.9.

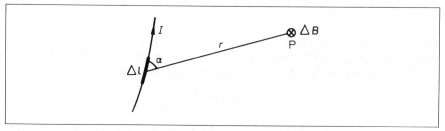

Figure 11.9 *Biot and Savart law*

The formula in (1) cannot be proved directly, as we cannot experiment with an infinitesimally small conductor. We believe in its truth because the deductions for large practical conductors turn out to be true.

The constant of proportionality in equation (1) depends on the medium in which the conductor is situated. In air (or, more exactly, in a vacuum), we write

$$\Delta B = \frac{\mu_0}{4\pi} \frac{I\Delta l \sin \alpha}{r^2} \qquad . \qquad . \qquad . \qquad . \qquad (2)$$

The value of μ_0, from p. 328, is

$$\mu_0 = 4\pi \times 10^{-7}$$

and its unit is *'henry per metre'* ($H\,m^{-1}$) as will be shown later.

B for Narrow Coil

The formula for the value of B at the centre of a narrow circular coil can be immediately deduced from (2). Here the radius r is constant for all the elements

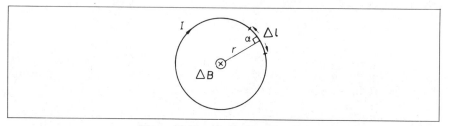

Figure 11.10 *Field of circular coil*

Δl, and the angle α is constant and equal to 90°, Figure 11.10. If the coil has N turns, the length of wire in it is $2\pi rN$, and the field at its centre is therefore given, if the current is I, by

$$B = \int dB = \frac{\mu_0}{4\pi} \int_0^{2\pi rN} \frac{Idl \sin 90°}{r^2}$$

$$= \frac{\mu_0 I}{4\pi r^2} \int_0^{2\pi rN} dl = \frac{\mu_0 I}{4\pi r^2} 2\pi rN$$

$$= \frac{\mu_0 NI}{2r} \qquad . \quad . \quad . \quad . \quad . \quad . \quad . \quad (1)$$

From (1), $B \propto I$ where r and N are constant, $B \propto 1/r$ when I and N are constant, and $B \propto N$ when I and r are constant.

B along Axis of a Narrow Circular Coil

We will now find the magnetic field at a point anywhere on the axis of a narrow circular coil (P in Figure 11.11). We consider an element Δl of the coil, at right angles to the plane of the paper. This sets up a field ΔB at P, in the plane of the paper, and at right angles to the radius vector r. If β is the angle between r and

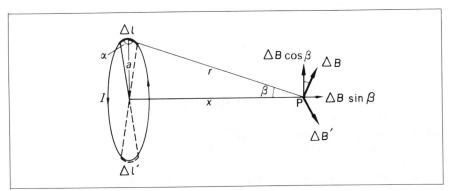

Figure 11.11 *Field on axis of flat coil*

the axis of the coil, then the field ΔB has components $\Delta B \sin \beta$ along the axis, and $\Delta B \cos \beta$ at right angles to the axis. If we now consider the element $\Delta l'$ diametrically opposite to Δl, we see that it sets up a field $\Delta B'$ equal in magnitude to ΔB. This also has a component, $\Delta B' \cos \beta$, at right angles to the axis; but this component acts in the opposite direction to $\Delta B \cos \beta$ and therefore cancels it. By considering elements such as Δl and $\Delta l'$ all round the circumference of the coil, we see that the field at P can have no component at right angles to the axis. Its value along the axis is

$$B = \int dB \sin \beta$$

From Figure 11.11, we see that the length of the radius vector r is the same for all points on the circumference of the coil, and that the angle α is also constant, being 90°. Thus, if the coil has a single turn, and carries a current I,

$$\Delta B = \frac{\mu_0 I \, \Delta l \sin \alpha}{4\pi r^2} = \frac{\mu_0 I}{4\pi r^2} \Delta l$$

And, if the coil has a radius a, then

$$B = \int dB \sin \beta = \int_0^{2\pi a} \frac{\mu_0 I}{4\pi r^2} dl \sin \beta$$

$$B = \frac{\mu_0 I a \sin \beta}{2r^2} \qquad . \qquad . \qquad . \qquad . \qquad . \qquad (i)$$

When the coil has more than one turn, the distance r varies slightly from one turn to the next. But if the width of the coil is small compared with all its other dimensions, we may neglect it, and write,

$$B = \frac{\mu_0 N I a \sin \beta}{2r^2} \qquad . \qquad . \qquad . \qquad . \qquad . \qquad (ii)$$

where N is the number of turns.

Equation (ii) can be put into a variety of forms, by using the facts that

$$\sin \beta = \frac{a}{r}$$

and $$r^2 = x^2 + a^2$$

where x is the distance from P to the centre of the coil. Thus

$$B = \frac{\mu_0 N I a^2}{2r^3} = \frac{\mu_0 N I a^2}{2(x^2 + a^2)^{3/2}} \qquad . \qquad . \qquad . \qquad . \qquad (1)$$

Helmholtz Coils

The field along the axis of a single coil varies with the distance x from the coil. In order to obtain a *uniform* field, Helmholtz used two coaxial parallel coils of equal radius R, separated by a distance R. In this case, when the same current flows around each coil in the same direction, the resultant field B is uniform for some distance on either side of the point on their axis midway between the coils. See p. 331.

The magnitude of the resultant field B at the midpoint can be found from our previous formula for a single coil. We now have $a = R$ and $x = R/2$. Thus, for the two coils,

$$B = 2 \times \frac{\mu_0 N I R^2}{2(R^2/4 + R^2)^{3/2}} = \left(\frac{4}{5}\right)^{3/2} \times \frac{\mu_0 N I}{R}$$

$$B = 0.72 \frac{\mu_0 N I}{R} \text{ (approx.)}$$

B on Axis of a Long Solenoid

We may regard a solenoid as a long succession of narrow coils; if it has n turns per metre, then in an element Δx of it there are $n\Delta x$ coils, Figure 11.12. At a point P on the axis of the solenoid, the field due to these is, by equation (ii),

$$\Delta B = \frac{\mu_0 I a \sin \beta}{2r^2} n\Delta x$$

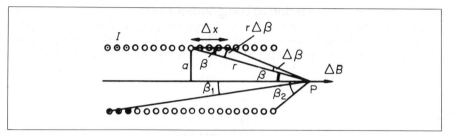

Figure 11.12   *Field on axis of solenoid*

in the notation which we have used for the flat coil. If the element Δx subtends an angle $\Delta \beta$ at P, then, from the figure,

$$r\,\Delta\beta = \Delta x \sin \beta$$

so

$$\Delta x = \frac{r\,\Delta\beta}{\sin \beta}$$

Also,

$$a = r \sin \beta$$

Thus

$$\Delta B = \frac{\mu_0 I r \sin^2 \beta}{2r^2}\, n\frac{r\,\Delta\beta}{\sin \beta}$$

$$= \frac{\mu_0 n I}{2} \sin \beta \Delta\beta$$

If the radii of the coil, at its ends, subtend the angles β_1 and β_2 at P, then the field at P is

$$B = \int_{\beta}^{\beta_2} \frac{\mu_0 n I}{2} \sin \beta \mathrm{d}\beta$$

$$= \frac{\mu_0 n I}{2}\left[-\cos \beta\right]_{\beta_1}^{\beta_2}$$

$$= \frac{\mu_0 n I}{2}(\cos \beta_1 - \cos \beta_2) \qquad . \qquad . \qquad . \qquad . \qquad (1)$$

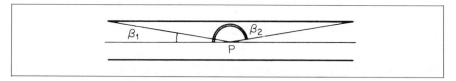

Figure 11.13   *A very long solenoid*

If the point P inside a very long solenoid—so long that we may regard it as infinite—then $\beta_1 = 0$ and $\beta_2 = \pi$, as shown in Figure 11.13. Then, by equation (1):

$$B = \frac{\mu_0 n I}{2}\left[-\cos \beta\right]_{0}^{\pi}$$

so

$$B = \mu_0 n I . \qquad . \qquad . \qquad . \qquad . \qquad . \qquad (2)$$

The quantity nI is often called the 'ampere-turns per metre'.

B due to Long Straight Wire

In Figure 11.14, AC represents part of a long straight wire. P is taken as a point so near it that, from P, the wire looks infinitely long—it subtends very nearly 180°. An element XY of this wire, of length Δl, makes an angle α with the radius

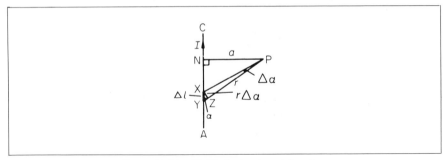

Figure 11.14   *Field of a long straight wire*

vector, r, from P. It therefore contributes to the magnetic field at P an amount

$$\Delta B = \frac{\mu_0 I \Delta l \sin \alpha}{4\pi r^2} \qquad . \qquad . \qquad . \qquad . \qquad \text{(i)}$$

when the wire carries a current I. If a is the perpendicular distance, PN, from P to the wire, then

$$PN = PX \sin \alpha \quad \text{or} \quad a = r \sin \alpha$$

so

$$r = \frac{a}{\sin \alpha} \quad . \qquad . \qquad . \qquad . \qquad . \qquad \text{(ii)}$$

Also, if we draw XZ perpendicular to PY, we have

$$XZ = XY \sin \alpha = \Delta l \sin \alpha$$

If Δl subtends an angle $\Delta \alpha$ at P, then

$$XZ = r \Delta \alpha = \Delta l \sin \alpha$$

From (i)

$$\therefore \Delta B = \frac{\mu_0 I \, \Delta l \sin \alpha}{4\pi r^2} = \frac{\mu_0 I r \, \Delta \alpha}{4\pi r^2} = \frac{\mu_0 I \, \Delta \alpha}{4\pi r}$$

From (ii),

$$\therefore \Delta B = \frac{\mu_0 I \sin \alpha \Delta \alpha}{4\pi a}$$

When the point Y is at the bottom end A of the wire, $\alpha = 0$; and when Y is at the top C of the wire, $\alpha = \pi$. Therefore the total magnetic field at P is

$$B = \frac{\mu_0}{4\pi} \int_0^\pi \frac{I \sin \alpha \Delta \alpha}{a} = \frac{\mu_0 I}{4\pi a} \left[-\cos \alpha \right]_0^\pi$$

$$\therefore B = \frac{\mu_0 I}{2\pi a} \qquad . \qquad . \qquad . \qquad . \qquad \text{(1)}$$

Equation (1) shows that the magnetic field of a long straight wire, at a point near it, is inversely proportional to the distance of the point from the wire. The result was discovered experimentally by Biot and Savart, and led to their general formula in (i) which we used to derive equation (1).

Ampère's Theorem

In the calculation of magnetic flux density B, we have used so far only the Biot and Savart law. Another law useful for calculating B is *Ampère's theorem*.

Ampère showed that if a *continuous closed line or loop* is drawn round one or more current-carrying conductors, and B is the flux density in the direction of an element dl of the loop, then for free space

$$\oint \frac{B}{\mu_0} \cdot dl = I$$

where the symbol $\oint$ represents the integral taken completely round the closed loop and I is the total current enclosed by the loop. So we can write

$$\oint B \cdot dl = \mu_0 I \qquad . \qquad . \qquad . \qquad . \qquad . \qquad (1)$$

The proof of (1) is outside the scope of this book.

We now apply the theorem to two special cases of current-carrying conductors.

1 Straight wire

Figure 11.15 shows a circular loop L of radius r, drawn concentrically round a straight wire carrying a current I. The flux lines are circles and so, at every part

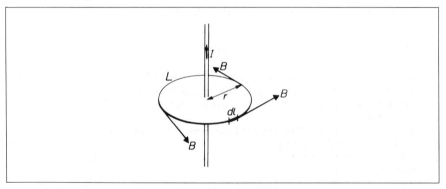

Figure 11.15 B *due to straight wire*

of a closed line, B is directed along the tangent to the circle at that part. Further, by symmetry, B has the same value everywhere along the line.

So
$$\oint B \cdot dl = B \oint dl = B \cdot 2\pi r$$

since B is constant. Hence, from (1),

$$B \cdot 2\pi r = \mu_0 I$$

and so
$$B = \frac{\mu_0 I}{2\pi r}$$

This agrees with the result derived earlier.

2 Toroid (Solenoid)

Consider the closed loop M indicated by the broken line in Figure 11.16. Again B is everywhere the same at M and is directed along the loop at every point.

So
$$\oint B \cdot dl = B \oint dl = BL$$

where L is the total length of the loop M. Hence, from (1),

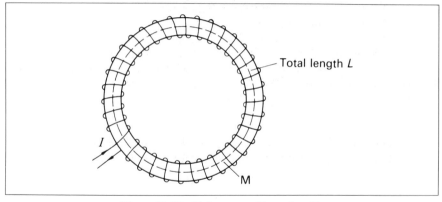

Figure 11.16 B *due to toroid or solenoid*

$$BL = \mu_0 NI$$

So
$$B = \frac{\mu_0 NI}{L} = \mu_0 nI$$

where N is the total number of turns, and n is the number of turns per metre. This agrees with the result previously obtained on p. 335.

_____ Exercises 11 _____

1 A vertical conductor X carries a downward current of 5 A.
   (a) Draw the pattern of the magnetic flux in a horizontal plane round X.
   (b) What is the flux density due to the current alone at a point P 10 cm due east of X.
   (c) If the earth's horizontal magnetic flux density has a value 4×10^{-5} T, calculate the resultant flux density at P.
   Is the resultant flux density at a point 10 cm due north of X greater *or* less than at P? Explain your answer.

2 A horizontal wire, of length 5 cm and carrying a current of 2 A, is placed in the middle of a long solenoid at right angles to its axis. The solenoid has 1000 turns per metre and carries a steady current I. Calculate I if the force on the wire is vertically downwards and equal to 10^{-4} N.

3 Two vertical parallel conductors X and Y are 0·12 m apart and carry currents of 2 A and 4 A respectively in a downward direction. Figure 11A (i).
   (a) Draw the resultant flux pattern between X and Y.
   (b) Ignoring the earth's magnetic field, find the distance from X of a point where the magnetic fields due to X and Y neutralise each other.

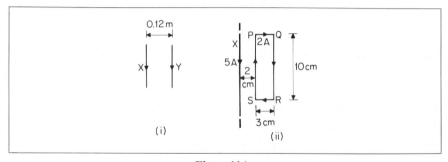

Figure 11A

(c) Calculate the force per metre on X and on Y, and show their directions in a sketch.

4 In Figure 11A (ii), X is a very long straight conductor carrying a current of 5 A. A metal rectangle PQRS is suspended with PS 2 cm from X as shown. The dimensions of PQRS are 10 cm by 3 cm, and a current of 2 A flows in the coil. Calculate the resultant force on PQRS in magnitude and direction.

5 Two very long thin straight parallel wires each carrying a current in the same direction are separated by a distance d. With the aid of a diagram which indicates the current directions, account for the force on each wire and show on the diagram the direction of one of the forces.

Write down an expression for the magnitude of the force per unit length of wire and hence define the ampere. Why is the electrostatic force between charges ignored in the definition? ($\mu_0 = 4\pi \times 10^{-7}\,\mathrm{H\,m^{-1}}$.) (*JMB.*)

6 Define the *ampere*. Write down expressions for   (i) the magnitude of the flux density B at a distance of d from a very long straight conductor carrying a current I, and   (ii) the mechanical force acting on a straight conductor of length l carrying a current I at right angles to a uniform magnetic field of flux density B.

Show how these two expressions may be used to deduce a formula for the force per unit length between two long straight parallel conductors *in vacuo* carrying currents I_1 and I_2 separated by a distance d.

A horizontal straight wire 5 cm long weighing $1 \cdot 2\,\mathrm{g\,m^{-1}}$ is placed perpendicular to a uniform horizontal magnetic field of flux density $0 \cdot 6$ T. If the resistance of the wire is $3 \cdot 8\,\Omega\,\mathrm{m^{-1}}$, calculate the p.d. that has to be applied between the ends of the wire to make it just self-supporting. Draw a diagram showing the direction of the field and the direction in which the current would have to flow in the wire ($g = 9 \cdot 8\,\mathrm{m\,s^{-2}}$). (*C.*)

7 State the law of force acting on a conductor carrying an electric current in a magnetic field. Indicate the direction of the force and show how its magnitude depends on the angle between the conductor and the direction of the field.

Sketch the magnetic field due solely to two long parallel conductors carrying respectively currents of 12 and 8 A in the same direction. If the wires are 10 cm apart, find where a third parallel wire also carrying a current must be placed so that the force experienced by it shall be zero. (*L.*)

8 Define the *ampere*.

Two long vertical wires, set in a plane at right angles to the magnetic meridian, carry equal currents flowing in opposite directions. Draw a diagram showing the pattern, in a horizontal plane, of the magnetic flux due to the currents alone—that is, for the moment ignoring the earth's magnetic field.

Next, taking into account the earth's magnetic field, discuss the various situations that can give rise to neutral points in the plane of the diagram.

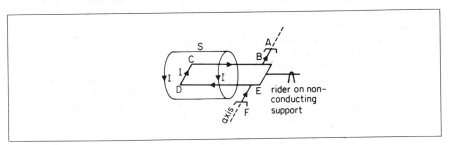

Figure 11B

Figure 11B shows a simple form of current balance. The 'long' solenoid S, which has 2000 turns per metre, is in series with the horizontal rectangular copper loop ABCDEF, where BC = 10 cm and CD = 3 cm. The loop, which is freely pivoted on the axis AF, goes well inside the solenoid, and CD is perpendicular to the axis of the solenoid. When the current is switched on, a rider of mass 0·2 g placed 5 cm from the axis is needed to restore equilibrium. Calculate the value of the current, I. (O.)

9 (a) A long straight wire of radius a carries a steady current. Sketch a diagram showing the lines of magnetic flux density (B) near the wire and the relative directions of the current and B. Describe, with the aid of a sketch graph, how B varies along a line from the surface of the wire at right-angles to the wire.

(b) Two such identical wires R and S lie parallel in a horizontal plane, their axes being 0·10 m apart. A current of 10 A flows in R in the opposite direction to a current of 30 A in S. Neglecting the effect of the earth's magnetic flux density calculate the magnitude and state the direction of the magnetic flux density at a point P in the plane of the wires if P is (i) midway between R and S, (ii) 0·05 m from R and 0·15 m from S. The permeability of free space, $\mu_0 = 4\pi \times 10^{-7}\,\mathrm{H\,m^{-1}}$. (*JMB*.)

10 Define the *ampere*.

Draw a labelled diagram of an instrument suitable for measuring a current absolutely in terms of the ampere, and describe the principle of it.

A very long straight wire PQ of negligible diameter carries a steady current I_1. A square coil ABCD of side l with n turns of wire also of negligible diameter is set up with sides AB and DC parallel to and coplanar with PQ; the side AB is nearest to PQ and is at a distance d from it. Derive an expression for the resultant force on the coil when a steady current I_2 flows in it, and indicate on a diagram the direction of this force when the current flows in the same direction in PQ and AB.

Calculate the magnitude of the force when $I_1 = 5\,\mathrm{A}$, $I_2 = 3\,\mathrm{A}$, $d = 3\,\mathrm{cm}$, $n = 48$ and $l = 5\,\mathrm{cm}$. (O. & C.)

11 Draw a sketch showing clearly the direction of the magnetic flux density at a point due to a long straight wire carrying a steady current. Mark the direction of the current in the wire.

The formula for the magnitude of the flux density B is given below. Describe an experiment to test both the formula and the direction.

A long straight wire in a uniform magnetic field carries a steady current. Show on a sketch the directions of the field, the current and the force experienced by a small element of the wire. Assume that the field is in some arbitrary direction with respect to the wire.

Two infinitely long parallel wires 0·5 m apart each carry a current of 2 A in the same direction. Find the force per metre on each wire and deduce the directions of these forces.

Two flat coils each of 20 turns have a mean radius of 30 cm. They are mounted coaxially and are 1 cm apart. Find an *approximate* value for the force between them when a current of 5 A flows in the same sense through each coil. Comment briefly on whether the approximate value is greater or less than the true value, and whether the approximation gets better or worse as the distance between the coils is increased.

($B = \mu_0 I/2\pi a$ where I is the current, a the perpendicular distance from the wire and $\mu_0 = 4\pi \times 10^{-7}\,\mathrm{H\,m^{-1}}$.) (*W*.)

12 (a) Figure 11C shows a conducting circular coil of radius 10 cm mounted in a north–south vertical plane. A current in the coil generates a magnetic flux density B_0 in an easterly direction at its centre and a magnetic flux density B at a point along the line perpendicular to the plane of the coil and passing through its centre. If θ is the angle shown in the diagram it can be shown that

$$B = B_0 \cos^3 \theta$$

Explain how you would verify this relation experimentally. Give details of the apparatus you would use and the measurements you would make.

(b) Write down an expression for the force F on a straight wire of length l and which carries current I in a uniform magnetic field of flux density B.

Use this expression to derive a relation for the couple on a rectangular coil with sides of length a and b, with N turns and carrying current I, mounted with its plane parallel to a uniform magnetic field of flux density B.

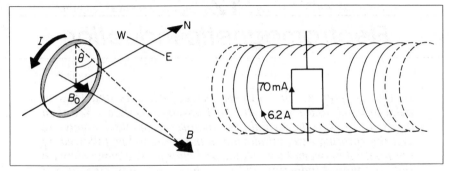

Figure 11C   **Figure 11D**

A square coil of side 1·2 cm and with 20 turns of fine wire is mounted centrally inside, and with its plane parallel to the axis of, a long solenoid which has 50 turns per cm, Figure 11D. The current in the coil is 70 mA and the current in the solenoid is 6·2 A.

Calculate   (i) the magnetic flux density in the solenoid,   (ii) the couple on the square coil. (Permeability of vacuum $= 4\pi \times 10^{-7}\,\mathrm{H\,m^{-1}}$.) (L.)

12

Electromagnetic Induction

In this chapter we first discuss the experiments and laws of induced e.m.f. and current due to Faraday and Lenz. We then apply the laws to the straight conductor, the simple dynamo coil and the rotating disc, all moving in magnetic fields, followed by the relation between flux linkage and charge. We conclude with self and mutual induction and with the variation of current and p.d. in an inductor-resistor circuit.

Faraday's Discovery

After Ampere and others had investigated the magnetic effect of a current Faraday tried to find its opposite. He tried to produce a current by means of a magnetic field. He began work on the problem in 1825 but did not succeed until 1831.

The apparatus with which he worked is represented in Figure 12.1; it consists of two coils of insulated wire, A, B, wound on a wooden core. One coil was connected to a galvanometer, and the other to a battery. No current flowed through the galvanometer, as in all Faraday's previous attempts. But when he disconnected the battery Faraday happened to notice that the galvanometer

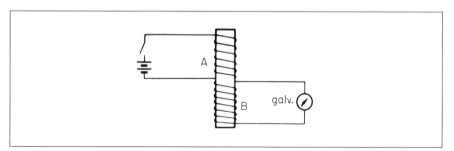

Figure 12.1 *Faraday's experiment on induction*

needle gave a kick. And when he connected the battery back again, he noticed a kick in the opposite direction. However often he disconnected and reconnected the battery, he got the same results. The 'kicks' could hardly be all accidental—they must indicate momentary currents. Faraday had been looking for a steady current—that was why it took him six years to find it.

Conditions for Generation of Induced Current

The results of Faraday's experiments showed that a current flowed in coil B of Figure 12.1 only while the magnetic field due to coil A was changing—the field building up as the current in A was switched on, decaying as the current in A was switched off. And the current which flowed in B while the field was decaying was in the opposite direction to the current which flowed while the field was building up. Faraday called the current in B an *induced current*. He found that it could be made much greater by winding the two coils on an iron core, instead of a wooden one. This historic apparatus can be seen at the Royal Institution, London.

Once he had realised that an induced current was produced only by a *change* in the magnetic field inducing it. Faraday was able to find induced currents wherever he had previously looked for them. In place of the coil A he used a magnet, and showed that as long as the coil and the magnet were at rest, there was no induced current, Figure 12.2 (i). But when he moved either the coil or the magnet an induced current flowed *as long as the motion continued*, Figure 12.2 (ii), (iii). If the current flowed one way when the north pole of the magnet was approaching the end X of the coil, it flowed the other way when the north pole was moving away from X.

Since a flow of current implies the presence of an e.m.f., Faraday's experiments showed that an e.m.f. could be induced in a coil by moving it relatively to a

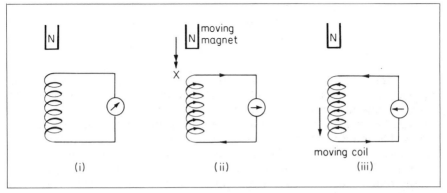

Figure 12.2 *Induced current by moving magnet or moving coil*

magnetic field, Figure 12.2 (iii). In discussing induction it is more fundamental to deal with the e.m.f. than the current, because the current depends on both the e.m.f. and the resistance.

Summarising, *relative motion* is needed between a magnet and a coil to produce induced currents. The induced current increases when the relative velocity increases and when a soft iron core is used inside the coil.

Direction of Induced Current or E.M.F.: Lenz's Law of Energy

Before considering the magnitude of an induced e.m.f., let us investigate its direction. To do so we must first see which way the galvanometer deflects when a current passes through it in a known direction: we can find this out with a battery and a megohm resistor, Figure 12.3 (i). We then take a coil whose direction of winding we know, and connect this to the galvanometer. In turn we plunge each pole of a magnet into and out of the coil; and we get the results shown in Figure 12.3 (ii), (iii), (iv) for the currents flowing in the coil.

These results were generalised into a simple rule by Lenz in 1835. He said that

_____

the induced current flows always in such a direction as to oppose the change causing it.

_____

For example, in Figure 12.3 (ii), the clockwise current flowing in the coil makes this end an S pole. So it repels the approaching S pole. In Figure 12.3 (iii), the

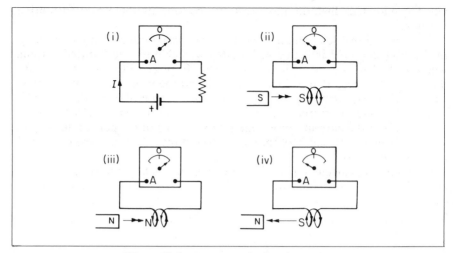

Figure 12.3 *Direction of induced currents*

induced anticlockwise current makes the end of the coil an N pole. So the approaching N pole is repelled. In Figure 12.3 (iv), the induced clockwise current in the coil now attracts the N pole moving away from it.

Lenz's law is a beautiful example of the Principle of the Conservation of Energy. The induced current sets up a force on the magnet, which the mover of the magnet must overcome. The work done in overcoming this force provides the electrical energy of the current. (This energy is dissipated as heat in the coil.)

If the induced current flowed in the opposite direction to that which it actually takes, then it would speed up the motion of the magnet. So the current would continuously increase the kinetic energy of the magnet. So both mechanical *and* electrical energy would be produced, without any agent having to do work. The system would be a perpetual motion machine and this is impossible. So the induced current always flows in a direction to *oppose* the motion and the electrical energy comes from the mechanical energy required to overcome the force opposing the motion.

The direction of the induced e.m.f., E, is the same as that of the current, as in Figure 12.4 (i). If we wished to reword Lenz's law, substituting e.m.f. for current, we would have to speak of the e.m.f.s *tending* to oppose the change...etc., because there can be no opposing force unless the circuit is closed and a current

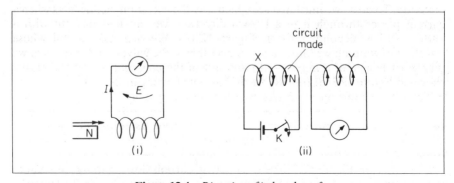

Figure 12.4 *Direction of induced e.m.f.*

can flow. If the terminals of a coil are *not* closed, and a flux change is made in the coil, an e.m.f. is produced between the terminals but no current flows.

In Figure 12.4 (ii), a coil X connected to a battery is placed near a coil Y. When the circuit in X is made be pressing the switch K, the current in the face of the coil near Y flows anticlockwise when viewed from Y. This is similar to bringing a N-pole suddenly near Y. So the induced current in Y is anticlockwise, as shown. If the current in X is switched off, this is similar to removing a N-pole suddenly from Y. So the current in Y is now clockwise, that is, in the opposite direction to before. The induced e.m.f. in Y, which follows the direction of the current, therefore reverses when the current in X is switched on and off.

Magnitude of E.M.F., Faraday's Law

Accurate experiments on induction are difficult to do with simple apparatus; but rough-and-ready experiments will show on what factors the magnitude depends. We require coils of the same diameter but different numbers of turns, coils of the same number of turns but different diameters, and two similar magnets, which

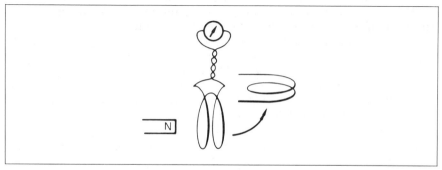

Figure 12.5   *E.m.f. induced by turning coil*

we can use singly or together. If we use a high-resistance galvanometer, the current will not vary much with the resistance of the coil in which the e.m.f. is induced, and we can take the deflection as a measure of the e.m.f. There is no need to plunge the magnet into and out of the coil: we can get just as great a deflection by simply turning the coil through a right angle, so that its plane changes from parallel to perpendicular to the magnet, or vice versa, Figure 12.5. We find that the induced e.m.f. increases with: (i) the speed with which we turn the coil,   (ii) the area of the coil,   (iii) the strength of the magnetic field (two magnets give a greater e.m.f. than one),   (iv) the number of turns in the coil.

To generalise these results and to build up useful formulae, we use the idea of *magnetic flux*, or field lines, passing through a coil. Figure 12.6 shows a coil, of area A, whose normal makes an angle θ with a uniform field of flux density B. The component of the field at right angles to the plane of the coil is $B \cos \theta$, and we say that the magnetic flux Φ through the coil is

$$\Phi = AB \cos \theta \qquad . \qquad . \qquad . \qquad . \qquad . \qquad (1)$$

If either the strength B of the field is changed, or the coil is turned so as to change the angle θ, then the flux through the coil changes.

Results (i) to (iii) above, therefore, show that the e.m.f. induced in a coil increases with the *rate of change of the magnetic flux* through it. More accurate

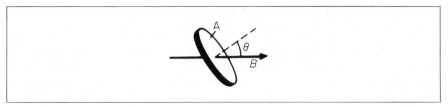

Figure 12.6 *Magnetic flux*

experiments show that the induced e.m.f. is actually *proportional* to the rate of change of flux through the coil. This result is sometimes called *Faraday's*, or *Neumann's, law*: The induced e.m.f. is proportional to the *rate of change of magnetic flux linking the coil or circuit*.

The unit of magnetic flux Φ is the *weber* (Wb). So the unit of B, the flux density or flux per unit area, is the *weber per metre*^{2} (Wb m^{-2}) or *tesla* (T).

Flux Linkage

If a coil has more than one turn, then the flux through the whole coil is the sum of the fluxes through the individual turns. We call this the *flux linkage* through the whole coil. If the magnetic field is uniform, the flux through one turn is given, from (1), by $AB \cos \theta$. If the coil has N turns, the total flux linkage Φ is given by

$$\Phi = BAN \cos \theta \qquad \qquad \qquad (2)$$

From Faraday's or Neumann's law, the e.m.f. induced in a coil is proportional to the rate of change of the flux linkage, Φ. Hence

$$E \propto \frac{d\Phi}{dt}$$

or
$$E = -k \frac{d\Phi}{dt} \qquad \qquad \qquad (3)$$

where k is a positive constant. The minus sign expresses Lenz's law. It means that the induced e.m.f. is in such a direction that, if the circuit is closed, the induced current *opposes* the change of flux. Note that an induced e.m.f. exists across the terminals of a coil when the flux linkage changes, even though the coil is on 'open circuit'. A current, of course, does not flow in this case.

On p. 350, it is shown that $E = -k\,d\Phi/dt$ is consistent with the expression $F = BIl$ for the force on a conductor only if $k = 1$. We may therefore say that

$$E = -\frac{d\Phi}{dt} \qquad \qquad \qquad (4)$$

where Φ is the flux linkage in webers, t is in seconds, and E is in volts.

From (4), it follows that one weber is the flux linking a circuit if the induced e.m.f. is one volt when the flux is reduced uniformly to zero in one second.

1 Lenz's law states that the induced current *opposes* the motion or change producing it
2 Faraday's (Neumann's) Law states that the induced e.m.f. is directly proportional to the rate of change of magnetic flux linking the circuit or coil
3 $\qquad \qquad \qquad E = -d\Phi/dt$

Example on e.m.f. due to Flux Change

(a) A narrow coil of 10 turns and area $4 \times 10^{-2}\,\text{m}^2$ is placed in a uniform magnetic field of flux density B of $10^{-2}\,\text{T}$ so that the flux links the turns normally. Calculate the average induced e.m.f. in the coil if it is removed completely from the field in $0.5\,\text{s}$.

(b) If the same coil is *rotated* about an axis through its middle so that it turns through 60° in $0.2\,\text{s}$ in the field B, calculate the average induced e.m.f.

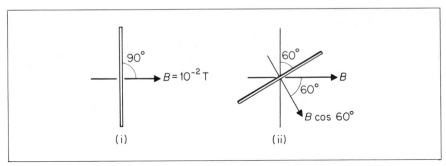

Figure 12.7 *Example*

(a) Flux linking coil initially $= NAB = 10 \times 4 \times 10^{-2} \times 10^{-2}$

$$= 4 \times 10^{-3}\,\text{Wb (Figure 12.7 (i))}$$

So average induced e.m.f. $= \dfrac{\text{flux change}}{\text{time}} = \dfrac{4 \times 10^{-3}}{0.5}$

$$= 8 \times 10^{-3}\,\text{V}$$

(b) When the coil is initially perpendicular to B, flux linking coil $= NAB$, Figure 12.7 (ii). When the coil is turned through 60°, the flux density normal to the coil is now $B\cos 60°$. So

flux change through coil $= NAB - NAB\cos 60°$

$$= 4 \times 10^{-3} - 4 \times 10^{-3} \times 0.5$$

So average induced e.m.f. $= \dfrac{\text{flux change}}{\text{time}} = \dfrac{2 \times 10^{-3}}{0.2}$

$$= 10^{-2}\,\text{V}$$

E.M.F. Induced in Moving Straight Conductor

Generators at power stations produce high induced voltages by rotating long *straight conductors*. Figure 12.8 (i) shows a simple apparatus for demonstrating that an e.m.f. may be induced in a straight rod or wire, when it is moved across a magnetic field. The apparatus consists of a rod AC resting on rails XY, and lying between the poles NS of a permanent magnet. The rails are connected to a galvanometer G.

If we move the rod to the left, so that it *cuts across* the field B of the magnet, a current I flows as shown. If we move the rod to the right, the current reverses. We notice that the current flows only while the rod is moving, and so we conclude that the motion of the rod AC induces an e.m.f. E in it.

By turning the magnet into a vertical position (Figure 12.8 (ii)) we can show that no e.m.f. is induced in the rod when it moves *parallel* to the field B. We conclude that an e.m.f. is induced in the rod only when it *cuts across* the field.

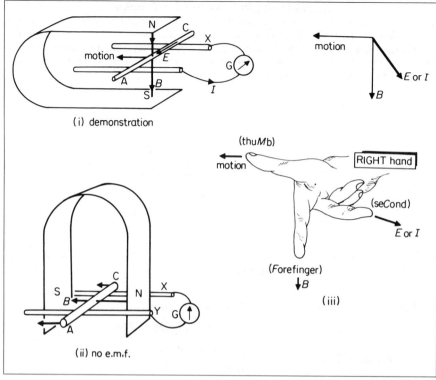

Figure 12.8 *E.m.f. induced in moving rod. Fleming Right-hand rule.*

And, whatever the direction of the field, no e.m.f. is induced when we slide the rod parallel to its own length. The induced e.m.f. is greatest when we move the rod at right angles, both to its own length and to the magnetic field. These results may be summarised in *Fleming's right-hand rule*:

If we extend the thumb and first two fingers of the right *hand, so that they are all at right angles to one another, then the directions of field, motion, and induced e.m.f. are related as in Figure 12.8 (iii).*

Students should remember that the *right* hand rule is used for induced current or e.m.f. but the *left* hand rule refers to the *force* on a conductor.

To show E.M.F. ∝ Rate of Change

The variation of the magnitude of the e.m.f. in a rod with the speed of 'cutting' magnetic flux can be demonstrated with the apparatus in Figure 12.9 (i).

Here AC is a copper rod, which can be rotated by a wheel W round one pole N of a long magnet. Brush contacts at X and Y connect the rod to a galvanometer G and a series resistance R. When we turn the wheel, the rod AC cuts across the field B of the magnet, and an e.m.f. is induced in it. If we turn the wheel steadily, the galvanometer gives a steady deflection, showing that a steady current is flowing round the circuit.

To find how the current and e.m.f. depend on the speed of the rod, we keep the circuit resistance constant, and vary the rate at which we turn the wheel. We time the revolutions with a stop-watch, and find that the deflection θ is

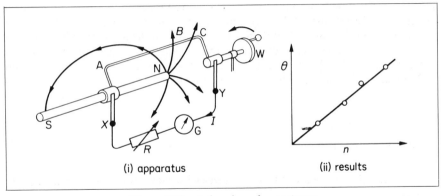

(i) apparatus   (ii) results

Figure 12.9   *Induced e.m.f. experiment*

proportional to the number of revolutions per second, n, Figure 12.9(ii). It follows that the induced e.m.f. is directly proportional to the speed of the rod.

Calculation of E.M.F. in Straight Conductor

Consider the circuit shown in Figure 12.10. PQ is a straight wire touching the two connected parallel wires QR, PS and free to move over them. All the conductors are situated in a uniform vertical magnetic field of flux density B, perpendicular to the horizontal plane of PQRS.

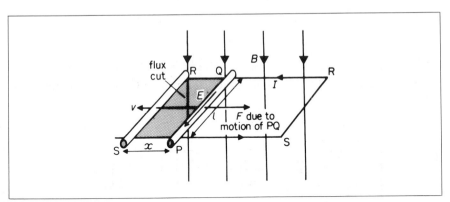

Figure 12.10   *Calculation of induced e.m.f.*

Suppose PQ, length l, moves with uniform velocity to a position SR. If the distance moved is x in a time t, then

$$\text{flux cut, } \Phi = B \times \text{area PQRS} = Blx$$

So, numerically, induced e.m.f. $E = \Phi/t = Blx/t = Blv$, since velocity $v = x/t$. So

$$E = Blv$$

The induced e.m.f. E produces a current I which flows round the circuit. A *force* will now act on the wire PQ due to the current flowing and to the presence of the magnetic field. From Fleming's left hand rule, we find that the force F acts in the *opposite* direction to the motion of the rod PQ, as shown. So, as Lenz's law states, the induced current flows in a direction so as to oppose the motion of PQ.

If the current flowing is I, and the length of PQ is l, the force on PQ is BIl. This is equal to the force moving PQ because PQ is not accelerating. From the principle of conservation of energy, work done per second by force moving PQ = electrical energy produced per second. So

$$BIl \times v = EI$$
$$\therefore Blv = E$$

as already deduced from flux changes.

When a straight conductor of length *l* moves with constant velocity *v* in a magnetic field *B*,
1   the induced e.m.f. $E = Blv$ when *l* and *v* are both 90° to *B*
2 $E = 0$ when *l* or *v* is parallel to *B*

Examples on Induced e.m.f. in Straight Conductors

1   A train travels at $30\,\mathrm{m\,s^{-1}}$ due east.

Calculate the induced e.m.f. between the ends of a horizontal axle CD of the train which is $1\cdot5\,\mathrm{m}$ long, assuming the Earth's magnetic field strength is $6 \times 10^{-5}\,\mathrm{T}$ and acts downwards at 65° to the horizontal. Which end of CD is at a higher potential?

The induced e.m.f. along CD will be due to the *vertical* component B of the Earth's magnetic field.

$$B = 6 \times 10^{-5} \sin 65° = 5\cdot4 \times 10^{-5}\,\mathrm{T}$$

With the train and CD moving due east. Figure 12.11 (i),

$$\text{induced e.m.f. } E = Blv = 5\cdot4 \times 10^{-5} \times 1\cdot5 \times 30$$
$$= 2\cdot4 \times 10^{-3}\,\mathrm{V}$$

Using Fleming's Right hand rule, E acts from D to C. So C is at the higher potential.

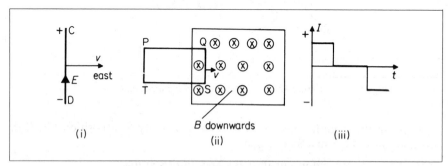

Figure 12.11   *Induced e.m.f. in straight conductors*

2   A horizontal metal frame PQST moves with uniform velocity v of $0\cdot2\,\mathrm{m\,s^{-1}}$ into a uniform field B of $10^{-2}\,\mathrm{T}$ acting vertically downwards. Figure 12.11 (ii). PT = $0\cdot1\,\mathrm{m}$ and PQ = $0\cdot2\,\mathrm{m}$ and the resistance R of the frame is $5\,\Omega$. The sides QS and PT enter the field in a direction normal to the field boundary as shown.

What current flows in the metal frame when
(a) QS just enters the field,

(b) when the whole frame is moving through the field,
(c) when QS just moves out of the field on the other side? Draw a sketch graph showing the variation of current.

(a) The sides PQ and TS move parallel to v, so no induced e.m.f. is obtained in these sides. For the moving side QS in the field,

$$E = Blv = 10^{-2} \times 0.1 \times 0.2 = 2 \times 10^{-4} \, V$$

So $$I = E/R = 2 \times 10^{-4}/5 = 4 \times 10^{-5} \, A$$

(b) With the whole frame PQST moving through the field, the flux through PQST in constant. So no induced e.m.f. is obtained. Alternatively, the induced e.m.f. in QS and PT act in opposite directions round the frame, so their resultant e.m.f. is zero.

(c) When QS just leaves the field on the other side, the induced e.m.f. in PT moving through the field is the same as in (a) and so the current I has the same value. But the direction of I round the frame is now *opposite* to the current in (a). So the graph of I with time t is that shown roughly in Figure 12.11 (iii).

Induced E.M.F. and Force on Moving Electrons

The e.m.f. induced in a wire moving through a magnetic field is due to the motion of electrons inside the metal, as we now explain.

When we move the wire downwards across the field B as in Figure 12.12, each

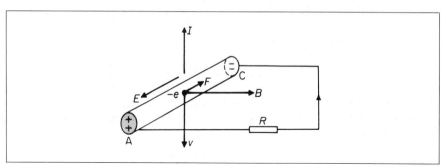

Figure 12.12 *Forces on a moving electron*

electron moves downwards across the field. A downward movement of electron charge $-e$ is equivalent to an *upward* movement of positive charge or conventional current I, as shown. Applying Fleming's left hand rule (in which the force F is at right angles to the velocity v of the wire and to B), we see that F *drives the electrons along the wire* from A to C. So if the wire is not connected in a closed circuit, electrons will pile up at C. Thus the end C will gain a negative charge and A will be left with an equal positive charge. After a time the charge at C will oppose further electron movement along the wire and so the drift stops.

The charges between A and C produce an *electromotive force E*, Figure 12.13. As in a battery, A is the 'positive pole' of the wire generator and C is the 'negative pole'. So when an external resistor R is joined to A and C, the conventional current flows in it as shown.

Homopolar or Disc Generator

Another type of generator, which gives a very steady e.m.f., is illustrated in

Figure 12.13 (i). It consists of a copper disc which rotates between the poles of a magnet. Connections are made from its axle X and circumference Y to a galvanometer G. We assume for simplification that the magnetic field B is uniform over the radius XY.

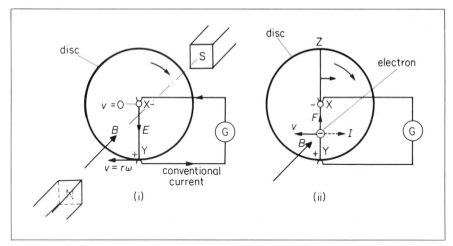

Figure 12.13   *Disc generator*

The radius XY continuously cuts the magnetic flux between the poles of the magnet. For this straight conductor, the velocity at the end X is zero and that at the other end Y is $r\omega$, where ω is the angular velocity of the disc. Since the velocity varies uniformly from X to Y,

$$\text{average velocity of XY, } v = \frac{1}{2}(0+r\omega) = \frac{r\omega}{2}$$

Now the induced e.m.f. E in a straight conductor of length l and moving with velocity v normal to a field B is given by $E = Blv$. Since in this case $l = r$ and $v = \frac{1}{2}r\omega$, then

$$E = B \times r \times \tfrac{1}{2}r\omega = \tfrac{1}{2}Br^2\omega \quad . \qquad . \qquad . \qquad . \qquad (1)$$

As $\omega = 2\pi f$, where f is the number of revolutions per second of the disc, we can say that

$$E = B.\pi r^2.f \quad . \qquad . \qquad . \qquad . \qquad . \qquad (2)$$

The direction of E is given by Fleming's *right* hand rule. Applying the rule, we find that E acts from X to Y so that Y is at the *higher* potential, as shown.

We can understand the origin of the e.m.f. by considering an electron between X and Y, Figure 12.13 (ii). When the disc rotates, the electron moves to the left as shown. The equivalent conventional current I is then to the right. Applying Fleming's left hand rule, we find that the force F on the electron drives it to X. So X obtains a negative charge and Y a positive charge. The radius is thus a generator with Y as its positive pole.

Diameter of Disc, E.M.F. with Axle

As the disc rotates clockwise, Figure 12.13 (ii), the radius XY moves to the left at the same time as the radius XZ moves to the right. If the magnetic field covered the whole disc, the induced e.m.f. in the two radii would be in *opposite* directions. So the resultant e.m.f. between the ends of the diameter YZ would be *zero*. $B . \pi r^2 . f$, the e.m.f. between the centre and rim of the disc, is the maximum e.m.f. which can be obtained from the dynamo.

If the disc had a radius r_1 and an axle at the centre of radius r_2, the area swept out by a rotating radius of the metal disc $= \pi r_1{}^2 - \pi r_2{}^2 = \pi(r_1{}^2 - r_2{}^2)$. In this case the induced e.m.f. would be $E = B . \pi(r_1{}^2 - r_2{}^2) . f$.

Generators of this kind are called *homopolar* because the e.m.f. induced in the moving conductor is always in the same direction. They are sometimes used for electroplating, where only a small voltage is required, but they are not useful for most purposes, because they give too small an e.m.f. The e.m.f. of a commutator dynamo can be made large by having many turns in the coil but the e.m.f. of a homopolar dynamo is limited to that induced in one radius of the disc.

Lorenz Absolute Method for Resistance

Lorenz devised a method of measuring resistance in which no electrical quantities are needed, that is, this is an *absolute method*. It is therefore adopted for measuring resistance in national physical laboratories. In contrast, when measuring resistance by V/I, one relies upon the accuracy of the voltmeter and ammeter used. In measuring resistance by a Wheatstone bridge method, one relies upon the accuracy of the standard resistance provided.

Figure 12.14 shows the principle of the method. A long coil A with n turns per metre is placed in series with the resistance R so that each carries a current I. A circular metal disc D is placed with its plane perpendicular to the magnetic field B inside the coil. By means of brushes, connections are made to R from the axle or centre of the disc and the circumference or edge. A galvanometer G is included in one lead.

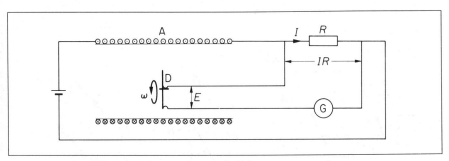

Figure 12.14   *Absolute method for measuring resistance*

As explained before, the disc acts as a generator (dynamo) when it is rotated steadily in the field B. By varying the angular velocity ω, the induced e.m.f. E between the centre and circumference is used to balance the constant p.d. IR across R. The galvanometer then shows no deflection.

The e.m.f. $E = \omega r^2 B/2$, where r is the radius of the disc (p. 352). The field value $B = \mu_0 nI = 4\pi nI \times 10^{-7}$ (p. 324). Since there is a balance,

$$IR = \frac{Br^2\omega}{2} = \frac{4\pi nIr^2\omega \times 10^{-7}}{2}$$

$$\therefore R = 2\pi nr^2\omega \times 10^{-7} \, \Omega$$

So by measuring ω in rad s^{-1}, r in metre and n, the resistance R can be calculated in ohms.

Example on Rotating Disc

A circular metal disc is placed with its plane perpendicular to a uniform magnetic field of flux density B. The disc has a radius of 0·20 m and is rotated at $5\,\text{rev s}^{-1}$ about an axis through its centre perpendicular to its plane. The e.m.f. between the centre and the rim of the disc is balanced by the p.d. across a $10\,\Omega$ resistor when carrying a current of 1·0 mA. Calculate B.

The induced e.m.f. $E = B \cdot \pi r^2 \cdot f = B \times \pi \times 0 \cdot 2^2 \times 5 = 0 \cdot 2\pi B$

and p.d. across $10\,\Omega$, $V = IR = 1 \times 10^{-3} \times 10 = 10^{-2}\,\text{V}$

$$\therefore 0 \cdot 2\pi B = 10^{-2}$$

$$\therefore B = 1 \cdot 6 \times 10^{-2}\,\text{T}$$

The Dynamo Generator

Faraday's discovery of electromagnetic induction was the beginning of electrical engineering. Nearly all the commercial electric current used today is generated by induction, in machines which contain coils moving continuously in a magnetic field.

Figure 12.15 illustrates the principle of such a machine, which is called a *dynamo*, or *generator*. A coil DEFG, shown for simplicity as having only one turn, rotates on a shaft, (not shown), between the poles NS of a horseshoe magnet. The ends of the coil are connected to flat brass rings R, which are supported on the shaft by discs of insulating material, also not shown. Contact with the rings is made by small blocks of carbon H, supported on springs, and shown connected to a lamp L.

As the coil rotates, the flux linking it changes, and a current I is induced in it which flows, through the carbon blocks H, to the lamp L. The magnitude (which

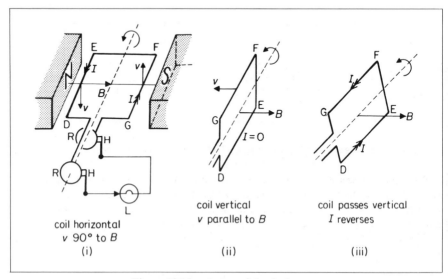

Figure 12.15 *Action of simple dynamo*

we study shortly) and the direction of the current are not constant. Thus when the coil is in the position shown, the side ED is moving downwards through the lines of force, and GF is moving *upwards*. Half a revolution later, ED and GF will have interchanged their positions, and ED will be moving upwards. Consequently, applying Fleming's right-hand rule (p. 348), the current round the coil must *reverse* as ED changes from downward to upward motion, Figure 12.15 (iii). The actual direction of the current at the instant shown on the diagram is indicated by the double arrows, using Fleming's rule. By applying this rule, it can be seen that *the current reverses* every time the plane of the coil passes the vertical position.

Note that when the coil is vertical, Figure 12.15 (ii), the velocity v of ED and GF are both *parallel* to the field B. So at this instant the induced current is zero.

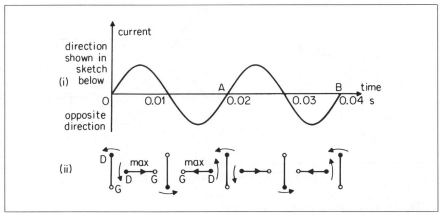

Figure 12.16 *Current generation by dynamo of Figure 12.15 plotted against time and coil position*

We shall see shortly that the magnitude of the e.m.f. and current varies with time as shown in Figure 12.16 (i). This diagram also shows the corresponding position of DG in Figure 12.16 (ii), which should be verified by the reader.

This type of current is called an *alternating current* (a.c.). A complete alternation, such as from A to B in the figure, is called a 'cycle'; and the number of cycles which the current goes through in one second is called its 'frequency'. The frequency of the current represented in the figure is that of most domestic supplies in Britain—50 Hz (cycles per second). Thus from A to B, which is one cycle, the time taken $(0.04-0.02) = 0.02\,s = 1/50\,s$. So the frequency is 50 Hz.

When the dynamo coil is horizontal, the e.m.f. is a maximum. When the coil is vertical, the e.m.f. is zero.

E.M.F. in Dynamo

We can now calculate the e.m.f. in the rotating coil. If the coil of N turns has an area A, and its normal makes an angle θ with the magnetic field B, as in Figure 12.17 (i), then the flux linkage with the coil $= NA \times$ component of B normal to coil.

So
$$\Phi = NAB\cos\theta$$

Figure 12.17 (ii) shows how the flux linkage Φ varies with the angle θ starting from $\theta = 0$, when the coil is vertical (V). Since $\theta = \omega t$, then $\theta \propto t$. So the horizontal axis can also represent the time t, as indicated. When $\theta = 90°$ the coil is horizontal (H) and no flux links the coil. As the coil rotates further the flux linking the face same reverse and so Φ becomes negative as shown.

We can now find the induced e.m.f. E. If the coil turns with a steady angular velocity ω or $d\theta/dt$, then the e.m.f. induced in the coil is given by $E = -d\Phi/dt = -gradient$ of the $\Phi - t$ graph in Figure 12.17 (ii). Figure 12.17 (iii) shows the negative gradient variation found from Figure 12.17 (ii). This is the variation of E with time t.

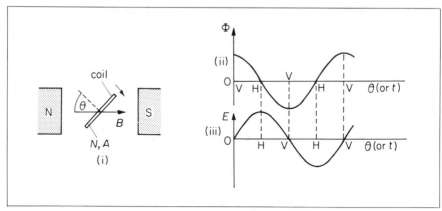

Figure 12.17   *Coil inclined to magnetic field*

We can calculate E exactly as follows.

$$E = -\frac{d\Phi}{dt}$$

$$= -NAB\frac{d}{dt}(\cos\theta)$$

$$= NAB\sin\theta\frac{d\theta}{dt} \qquad . \qquad . \qquad . \qquad . \qquad . \qquad (1)$$

$d\theta/dt$ is ω, the angular velocity and $\omega = 2\pi f$ where f is the number of revolutions per second. Also, in a time t, $\theta = \omega t = 2\pi ft$. So, from (1), we can write

$$E = \omega NAB\sin\omega t \qquad . \qquad . \qquad . \qquad . \qquad . \qquad (2)$$

or $\qquad\qquad\qquad E = 2\pi fNAB\sin 2\pi ft \qquad . \qquad . \qquad . \qquad . \qquad (3)$

Thus the e.m.f. varies *sinusoidally* with time, that is, like a sine wave, the frequency being f cycles per second.

The maximum (peak) value or amplitude of E occurs when $\sin 2\pi ft$ reaches the value 1. If the maximum value is denoted by E_0, it follows that

$$E_0 = 2\pi f NAB$$

and $\qquad\qquad\qquad E = E_0\sin 2\pi ft \qquad . \qquad . \qquad . \qquad . \qquad . \qquad (4)$

The e.m.f. E sends an alternating current of a similar sine equation through a resistor connected across the coil.

Dynamo E.M.F. from Energy Principles

The e.m.f. in a simple dynamo can also be found from energy principles.

At a time t, suppose the normal to the plane of the coil makes an angle θ with the field B, Figure 12.17 (i). The torque (moment of couple) acting on the coil is then given by $NABI \sin \theta$ (p. 311), where I is the current flowing if the ends of the coil are connected to an external resistor.

The work done by a torque in rotation through an angle = torque × angle of rotation (p. 121). So if the coil is rotated through a small angle $\Delta\theta$ in a time Δt,

$$\text{mechanical work done per second by torque} = \text{torque} \times \Delta\theta/\Delta t$$

$$= NABI \sin \theta \times \omega$$

since $\omega = \Delta\theta/\Delta t$. But if E is the induced e.m.f. in the coil, the electrical energy per second generated in the coil $= EI$. So

$$EI = NABI \sin \theta \times \omega$$

or $$E = \omega NAB \sin \theta = \omega NAB \sin \omega t$$

This result for E agrees with the calculation using $E = -\,d\Phi/dt$

A simple dynamo rotating at a constant angular velocity ω (or f rev/s) in a uniform field B has an alternating e.m.f. (a.c. voltage) given by $E = E_0 \sin \omega t$, where E_0 = maximum e.m.f. $= \omega NAB = 2\pi f NAB$.

Alternators

Generators of alternating current are often called *alternators*. In all but the smallest, the magnetic field of an alternator is provided by an electromagnet called a field-magnet or *field*, as shown in Figure 12.18. It has a core of cast steel, and is fed with direct current from a separate d.c. generator. The rotating coil, called the *armature*, is wound on an iron core, which is shaped so that it can turn within the pole-pieces of the field-magnet. With the field magnet, the armature core forms a system which is almost wholly iron, and can be strongly magnetised by a small current through the field winding. The field in which the armature turns is much stronger than if the coil had no iron core, and the e.m.f. is proportionately greater. In the small alternators used for bicycle lighting the

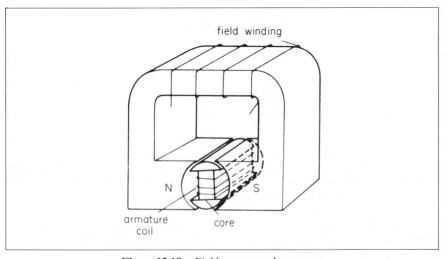

Figure 12.18   *Field magnet and armature*

armature is stationary, and the field is provided by permanent magnets, which rotate around it. In this way rubbing contacts, for leading the current into and out of the armature, are avoided.

When no current is being drawn from a generator, the power required to turn its armature is merely that needed to overcome friction, since no electrical energy is produced. But when a current is drawn, the power required increases, to provide the electrical power. The current, flowing through the armature winding, causes the magnetic field to set up a couple which opposes the rotation of the armature, and so demands the extra power.

The Transformer

A *transformer* is a device for stepping up—or down—an alternating voltage. It has primary and secondary windings but no make-and-break, Figure 12.19. It has an iron core, which is made from E-shaped laminations, interleaved so that the magnetic flux does not pass through air at all. In this way the greatest flux is obtained with a given current.

When an alternating e.m.f. E_p is connected to the primary winding, it sends an alternating current through it. This sets up an alternating flux in the core of magnitude BA, where B is the flux density and A is the cross-sectional area. This

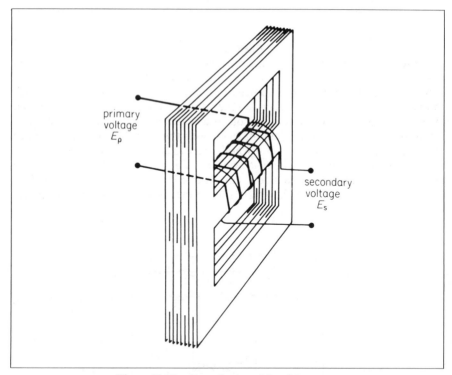

Figure 12.19 *Transformer with soft iron core*
(*Power losses in use:* (*i*) *heat losses in coils* (*ii*) *eddy-current losses in core* (*reduced by laminations, p. 361*), (*iii*) *magnetic hysteresis* (*reduced by special iron alloys with low 'magnetic friction'*) (*iv*) *magnetic flux leakage from coils.*)

induces an alternating e.m.f. in the secondary E_s. If N_p, N_s are the number of turns in the primary and secondary coils, their linkages with the flux Φ are:

$$\Phi_p = N_pAB \quad \Phi_s = N_sAB$$

The magnitude of the e.m.f. induced in the secondary is,

$$E_s = \frac{d\Phi_s}{dt} = N_s A \frac{dB}{dt}$$

The changing flux also induces a back-e.m.f. in the primary, whose magnitude is

$$E_p = \frac{d\Phi_p}{dt} = N_p A \frac{dB}{dt}$$

The voltage applied to the primary, from the source of current, is used simply in overcoming the back-e.m.f. E_p, if we neglect the resistance of the wire. Therefore it is equal in magnitude to E_p. (This is analogous to saying, in mechanics, that action and reaction are equal and opposite.) Consequently we have

$$\frac{e.m.f.\ induced\ in\ secondary}{voltage\ applied\ to\ primary} = \frac{E_s}{E_p} = \frac{N_s}{N_p} \qquad . \qquad . \qquad . \qquad (1)$$

So the transformer steps voltage up or down according to its 'turns-ratio'.

$$\frac{\textbf{secondary voltage}}{\textbf{primary voltage}} = \frac{\textbf{number of secondary turns}}{\textbf{number of primary turns}} \ .$$

The relation in (1) is only true when the secondary is on open circuit. When a load is connected to the secondary winding, a current flows in it. Thus the power drawn from the secondary is drawn, in turn, from the supply to which the primary is connected. So now a *greater* primary current is flowing than before the secondary was loaded.

Transformers are used to step up the voltage generated at a power station from 23 000 to 400 000 volts for high-tension transmission (p. 259). After transmission they are used to step it down again to a value safer for distribution (240 volts in houses). Inside a house a transformer may be used to step the voltage down from 240 V to 4 V, for ringing bells. Transformers with several secondaries are used in television receivers, where several different voltages are required.

D.C. Generators

Figure 12.20 (i) is a diagram of a *direct-current* (d.c.) generator or dynamo. Its essential difference from an alternator is that the armature winding is connected to a *commutator* instead of slip-rings.

The commutator consists of two half-rings of copper C, D, insulated from one another, and turning with the coil. Brushes BB, with carbon tips, press against the commutator and are connected to the external circuit. The commutator is arranged so that it *reverses* the connections from the coil to the circuit at the instant when the e.m.f. reverses in the coil.

Figure 12.20 (ii) shows several positions of the coil and commutator, and the e.m.f. observed at the terminals XY. This e.m.f. varies in magnitude, but it acts always in the same way round the circuit connected to XY and so it is a varying direct e.m.f. The average value in this case can be shown to be $2/\pi$ of the maximum e.m.f. E_0, given in equation (4), p. 356.

In practice, as in an alternator, the armature coil is wound with insulated wire on a soft iron core, and the field-magnet is energised by a current. This current is provided by the dynamo itself. The steel of the field-magnet has always a small residual magnetism, so that as soon as the armature is turned an e.m.f. is induced in it. This then sends a current through the field winding, which increases the field and the e.m.f. The e.m.f. rapidly builds up to its working value.

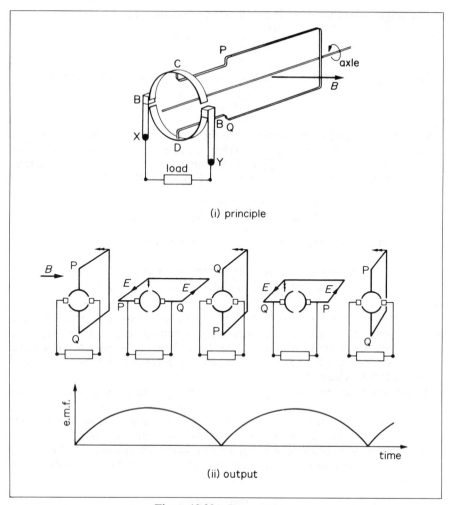

Figure 12.20 *D.c. generator*

Most consumers of direct current wish it to be steady, not varying as in Figure 12.20. A reasonably steady e.m.f. is given by an armature with many coils, inclined to one another, and a commutator with a correspondingly large number of segments. The coils are connected to the commutator in such a way that their e.m.f.s add round the external circuit.

Applications of Alternating and Direct Currents

Direct currents are less easy to generate than alternating currents, and alternating e.m.f.s are more convenient to step up and to step down, and to distribute over a wide area. The national grid system, which supplies electricity to the whole country, is therefore fed with alternating current. Alternating current is just as suitable for heating as direct current, because the heating effect of a current is independent of its direction. It is also equally suitable for lighting, because filament lamps depend on the heating effect, and gas-discharge lamps—neon, sodium, mercury—run as well on alternating current as on direct.

Small motors, of the size used in vacuum-cleaners and common machine-tools, run satisfactorily on alternating current, but large ones, as a general rule, do not. Direct current is therefore used on most electric railway systems. These systems either have their own generating stations, or convert alternating current from the grid into direct current. One way of converting alternating current into direct current is to use a *rectifier*, whose principle we shall describe later.

Eddy-currents and Power Losses

The core of the armature of a dynamo is built up from thin sheets of soft iron insulated from one another by an even thinner film of oxide, as shown in Figure 12.21 (i). These are called *laminations*, and the armature is said to be laminated.

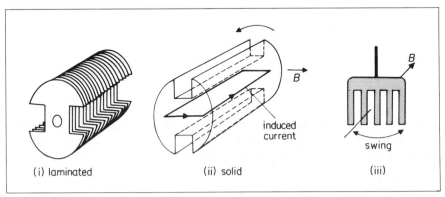

Figure 12.21 *Armature cores. Eddy currents*

If the armature were solid, then, since iron is a conductor, currents would be induced in it by its motion across the magnetic field, Figure 12.21 (ii). These currents would absorb power by opposing the rotation of the armature, and they would dissipate that power as heat, which would damage the insulation of the winding. But when the armature is laminated, these currents cannot flow, because the induced e.m.f. acts at right angles to the laminations, and therefore to the insulation between them. The magnetisation of the core, however, is not affected, because it acts along the laminations. Thus the induced currents, called *eddy-currents*, are practically eliminated, while the desired e.m.f.—in the armature coil—is not.

Eddy-currents, by Lenz's law, always tend to oppose the motion of a solid conductor in a magnetic field. The opposition can be shown in many ways. One of the most impressive is to make a chopper with a thick copper blade, and to try to slash it between the poles of a stronger electromagnet; then to hold it delicately and allow it to drop between them. The resistance to the motion in the case of the fast-moving chopper can be felt.

If a rectangular metal plate of aluminium is set swinging between the poles of a strong magnet, it soon comes to rest. The eddy-currents circulating inside the metal oppose the motion. But if many deep slots are cut into the metal, as in a comb, the pendulum now keeps oscillating for a much longer time before coming to rest. Figure 12.21 (iii). The eddy-currents are considerably reduced in this case as they cannot flow across the many air gaps formed by the slots.

Damping of Moving-Coil Meters

Sometimes eddy-currents can be made use of—for example, in damping a meter. When a current is passed through the coil of an ammeter, a couple acts on the coil which sets it swinging. If the swings are opposed only by the viscosity of the air, they decay very slowly and are said to be naturally damped, Figure 12.22. The pointer then takes a long time to come to its final steady deflection θ.

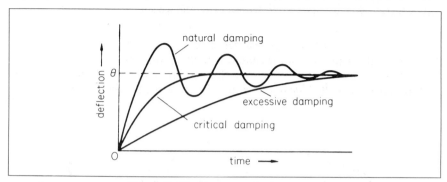

Figure 12.22   *Damping of galvanometer*

To bring the pointer more rapidly to rest, the damping must be increased. One way of increasing the damping is to wind the coil on a *metal* frame or former made of aluminium. Then, as the coil swings, the field of the permanent magnet induces eddy-currents in the former, and these, by Lenz's law, oppose the motion. They therefore slow down the turning of the coil towards its eventual position, and *also stop its swings about that position*. So in the end the deflected coil comes to rest sooner than if it were not damped.

Galvanometer coils which are wound on insulating formers can be damped by short-circuiting a few of their turns, or by joining the galvanometer terminals with connecting wire so that the whole coil is short-circuited. The meter can then be carried safely from one place to another without excessive swinging of the coil, which might otherwise damage the instrument.

If the coil is overdamped, as shown in Figure 12.22, it may take almost as long to come to rest as when it is undamped. The damping which is just sufficient to prevent 'overshoot' is called 'critical' damping.

Electric Motors

If a simple direct-current dynamo, with a split-ring commutator, is connected to a battery it will run as a *motor*, Figure 12.23. Current flows round the armature coil, and the magnetic field exerts a couple on this, as in a moving-coil meter. The

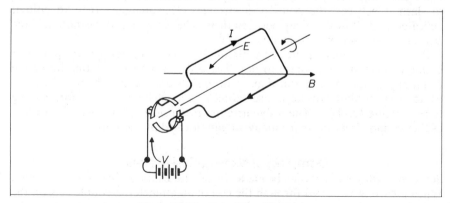

Figure 12.23   *Principle of d.c. motor*

commutator reverses the current just as the sides of the coil are changing from upward to downward movement and vice versa. Thus the couple on the armature is always in the same direction, and so the shaft turns continuously.

The armature of a motor is laminated, in the same way and for the same reason, as the armature of a dynamo.

Back-e.m.f. in Motor

When the armature of a motor rotates, an e.m.f. is induced in its windings. By Lenz's law this e.m.f. opposes the current which is making the coil turn. It is therefore called a *back-e.m.f.* If its magnitude is E, and V is the potential difference applied to the armature by the supply, then the armature current is

$$I_a = \frac{V - E}{R_a} \qquad . \qquad . \qquad . \qquad . \qquad . \qquad (1)$$

Here R_a is the resistance of the armature, which is generally small—of the order of 1 ohm.

The back-e.m.f. E is proportional to the strength of the magnetic field, and to the speed of rotation of the armature. When the motor is first switched on, the back-e.m.f. is zero: it rises as the motor speeds up. In a large motor the starting current would be much too great and destroy the armature coil. To limit it, a variable resistance is therefore inserted in series with the armature, and this is gradually reduced to zero as the motor gains speed.

When a motor is running, the back-e.m.f. in its armature E is not much less than the supply voltage V. For example, a motor running off the mains ($V = 240$ V say) might develop a back-e.m.f. $E = 230$ V. If the armature had a resistance of 1 ohm, the armature current would then be 10 A (equation (1)). When the motor was switched on, the armature current would be 240 A if no starting resistor were used.

The following example shows how the back e.m.f. is taken into account.

Example on Motor and Speed

A motor has an armature resistance of $4 \cdot 0 \, \Omega$. On a 240 V supply and a light load, the motor speed is 200 rev min^{-1} and the armature current is 5 A. Calculate the motor speed at a full load when the armature current is 20 A.

Generally,
$$I = \frac{240 - e_1}{4}$$

where e_1 is the back-e.m.f. So at $I = 5$ A,

$$240 - e_1 = 5 \times 4 = 20$$

and
$$e_1 = 240 - 20 = 220 \text{ V}$$

Suppose e_2 is the new back-e.m.f. when $I = 20$ A. Then, from the equation for I,

$$240 - e_2 = 20 \times 4 = 80$$

So
$$e_2 = 240 - 80 = 160 \text{ V}$$

Now the speed of the motor is proportional to the back-e.m.f.

So
$$\text{full load speed} = \frac{160}{220} \times 200 \text{ rev min}^{-1}$$

$$= 145 \text{ rev min}^{-1} \text{ (approx.)}$$

Back-e.m.f. and Power

If V is the supply voltage to a motor and I_a is the current, the power supplied to the motor is $I_a V$. Part of this power is dissipated in the resistance R_a of the motor coil and this is equal to $I_a^2 R$. The rest of the power is the *mechanical power* developed in the motor.

If E_b is the back-e.m.f. in the motor, then

$$I_a = \frac{V - E_b}{R_a}$$

So

$$V = I_a R_a + E_b$$

Multiplying by I_a, then

$$I_a V = I_a^2 R + I_a E_b$$

$$\therefore I_a E_b = I_a V - I_a^2 R_a$$

So

$$I_a E_b = \textit{mechanical power developed in motor.}$$

In the Example on the motor just done, the supply voltage was 240 V and the armature current was 5 A. So the power supplied $= I_a V = 5 \times 240 = 1200$ W. The heat per second in the resistance $R_a = I_a^2 R_a = 5^2 \times 4 = 100$ W. So the mechanical power developed in the motor $= 1200 - 100 = 1100$ W.

Series- and Shunt-Wound Motors

The field winding of a motor may be connected in series or in parallel with the armature. If it is connected in series, it carries the armature current, which is large, Figure 12.24. The field winding therefore has few turns of thick wire, to keep down its resistance and so waste little power in it as heat. The few turns are enough to magnetise the iron, because the current is large.

Series motors are used where great torque is required in starting—for example, in cranes. They develop a great starting torque because the armature current flows through the field coil. At the start the armature back-e.m.f. is small, and the current is great—as great as the starting resistance will allow. The field-magnet is therefore very strongly magnetised. The torque on the armature is proportional to the field and to the armature current; since both are great at the start, the torque is very great.

If the field coil is connected in parallel with the armature, as in Figure 12.25, the motor is said to be 'shunt-wound'. The field winding has many turns of fine wire to keep down the current which it consumes.

Shunt-wound motors are used for driving machine-tools, and in other jobs where a steady speed is required. A shunt motor keeps a nearly steady speed for the following reason. If the load is increased, the speed falls a little; the back-

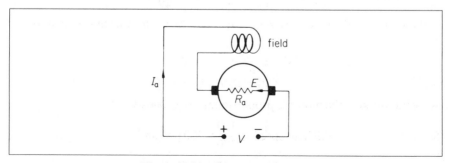

Figure 12.24   *Series-wound motor*

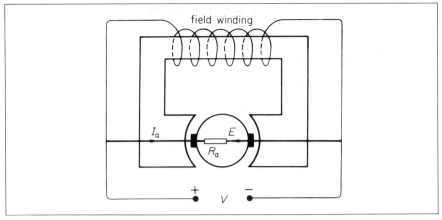

Figure 12.25 *Current and voltages in shunt-wound motor*

e.m.f. then falls in proportion to the speed, and the current rises, enabling the motor to develop more power to overcome the increased load. A series motor does not keep such a steady speed as a shunt motor.

_____ Exercises 12A _____

Electromagnetic Induction

1 A bar magnet M, with its S pole at the bottom, is dropped vertically through a horizontal flat coil C, Figure 12A (i). Draw a sketch showing the direction of the induced current
   (a) just before M passes through C,
   (b) just after M has passed completely through C.

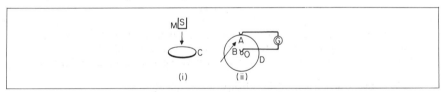

Figure 12A

2 Figure 12A (ii) shows a vertical copper disc D rotating clockwise in a uniform horizontal magnetic field *B* directed normally towards D. A galvanometer G is connected to contacts at O and A. The radius OA can be considered as a straight conductor moving in the field. Copy the diagram and show in your sketch the direction of the induced current flowing through G. Has A or O the higher potential?

3 A horizontal rod PQ of length 1·5 m is perpendicular to a uniform horizontal field *B* of 0·1 T, Figure 12B (i). Calculate the induced e.m.f., if any, in PQ when the rod is

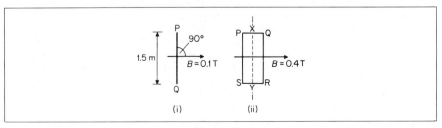

Figure 12B

moved through the field with a uniform velocity of $4 \, \text{m s}^{-1}$
(a) in the direction of B,
(b) perpendicular to B and upwards. Which end of PQ has the higher potential?

4 Figure 12B(ii) shows a vertical rectangular coil PQRS with its plane parallel to a uniform horizontal magnetic field B of $0.4 \, \text{T}$. The coil has 5 turns, PS is 10 cm long and SR is 5 cm long. Calculate the average induced e.m.f. in the coil, if any,
(a) when it is moved sideways in the direction of B with a velocity of $2 \, \text{m s}^{-1}$,
(b) when it is rotated through 90° about the vertical axis XY in $0.1 \, \text{s}$.
   If the resistance of the coil PQRS is $10 \, \Omega$ and its terminals are connected, what charge circulates in the coil in case (b)?

5 In Figure 12B(ii) the coil PQRS is rotated about the axis XY at $50 \, \text{rev s}^{-1}$. Calculate the maximum e.m.f. induced in the coil.
   What is the instantaneous e.m.f. in the coil when its plane is (i) parallel to the direction of B, (ii) 60° to B, (iii) 90° to B?

6 Explain, using the case of a N pole of a magnet approaching a coil, why Lenz's law is a consequence of the law of the conservation of energy.
   How is Lenz's law applied to explain why an induced e.m.f. is obtained in a straight conductor cutting flux in a magnetic field?

7 Figure 12C(i) shows a horizontal metal rod XY moving with a uniform velocity v perpendicularly to a uniform magnetic field B acting into the paper. By considering the force on electron, charge $-e$, explain why an induced e.m.f. is obtained along XY.

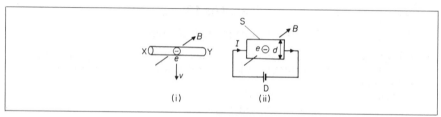

Figure 12C

8 Figure 12C(ii) shows an n-type semiconductor S, with a battery D connected so that a current I flows through it. When a strong magnetic field B is applied perpendicularly to the plane of S, explain why an e.m.f. (Hall voltage) E is obtained.
   Show that
(a) $E = Bvd$, where v is the drift velocity of the n-charges,
(b) $E = BI/net$, where n is the number of n-charges per metre^3 and t is the thickness of S.

9 A closed square coil consisting of a single turn of area A rotates at a constant angular speed, ω, about a horizontal axis through the mid-points of two opposite sides. The coil rotates in a uniform horizontal magnetic flux density, B, which is directed perpendicularly to the axis of rotation.
(a) Give an expression for the flux linking the coil when the normal to the plane of the coil is at an angle α to the direction of B.
(b) If at time $t = 0$ the normal to the plane of the coil is in the same direction as that of B, show that the e.m.f. E, induced in the coil is given by $E = BA\omega \sin \omega t$.
(c) With the aid of a diagram, describe the positions of the coil relative to B when E is (i) a maximum (ii) zero. Explain your answer. (*JMB.*)

10 (a) Show, by considering the force on an electron, that a potential difference will be established between the ends of a metal rod which is moving in a direction at right angles to a magnetic field. Draw a diagram in which the direction of motion of the rod is shown, the direction of the magnetic field is stated and the polarity of the ends of the conductor is shown.
(b) How would you show that an induced current in a conductor moving in a magnetic field is such a direction as to oppose the motion of the conductor? Explain why this follows from conservation of energy.

(c) The primary of a transformer is connected to a constant voltage a.c. supply and the secondary is on open circuit. Discuss the factors which determine the current flowing in the primary winding. (*L.*)

11 (a) A wire of length *l* is horizontal and oriented North–South. It moves East with velocity *v* through the earth's magnetic field which has a downward vertical component of flux density *B*. Write down an expression for the potential difference between the two ends of the wire. Which end of the wire is at the more positive potential?

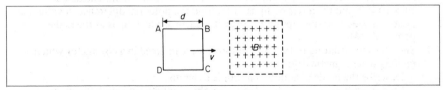

Figure 12D

(b) A horizontal square frame ABCD, of side *d*, moves with velocity *v* parallel to sides AB, DC from a field-free region into a region of uniform magnetic field of flux density *B*, Figure 12D. The boundaries of the field are parallel to the sides BC, AD of the frame and the field is directed vertically downward. Write down expressions for the electromotive force induced in the frame (i) when side BC has entered the field but side AD has not, (ii) when the frame is entirely within the field region, (iii) when side BC has left the field but side AD has not.

For each position derive an expression for the magnitude and direction of the current in the frame and the resultant force acting on the frame due to the current. The total resistance of the wire frame is *R*, and its self-inductance may be neglected. (*O. & C.*)

12 State the *laws of electromagnetic induction*. Show how Lenz's law is consistent with the principle of conservation of energy.

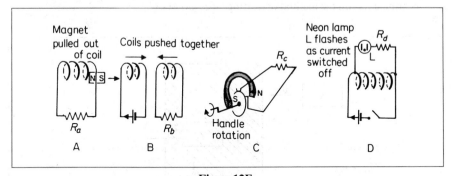

Figure 12E

Draw four arrows, labelled *A*, *B*, *C* and *D*, showing the directions of the currents induced in the resistors in the experiments illustrated. Explain how the e.m.f. arises in cases *C* and *D*, Figure 12E.

A copper disc of area *A* rotates at frequency *f* at the centre of a long solenoid of turns per unit length *n* and carrying a current *I*. The plane of the disc is normal to the flux. The rotation rate is adjusted so that the e.m.f. generated between the centre of the copper disc and its rim is 1% of the potential difference across the ends of the solenoid. Deduce an expression for the e.m.f. generated between the centre of the copper disc and its rim. Hence find the resistance of the solenoid in terms of μ_0, *A*, *f* and *n*. (*C.*)

13 Define *electromotive force* and state the *laws of electromagnetic induction*. Using the definition and the laws, derive an expression for the e.m.f. induced in a conductor moving in a magnetic field.

When a wheel with metal spokes 1·2 m long is rotated in a magnetic field of flux density 5×10^{-5} T normal to the plane of the wheel, an e.m.f. of 10^{-2} V is induced between the rim and the axle. Find the rate of rotation of the wheel. (*L.*)

14 State Lenz's law and describe how you would demonstrate it using a solenoid with two separate superimposed windings with clearly visible turns, a cell with marked polarity, and a centre-zero galvanometer. Illustrate your answer with diagrams.

A metal aircraft with a wing span of 40 m flies with a ground speed of 1000 km h^{-1} in a direction due east at constant altitude in a region of the northern hemisphere where the horizontal component of the earth's magnetic field is $1·6 \times 10^{-5}$ T and the angle of dip is 71·6°. Find the potential difference in volts that exists between the wing tips and state, with reasons, which tip is at the higher potential (*JMB.*)

15 State the laws relating to the electromotive force induced in a conductor which is moving in a magnetic field.

Describe the mode of action of a simple dynamo.

Find in volts the e.m.f. induced in a straight conductor of length 20 cm, on the armature of a dynamo and 10 cm from the axis when the conductor is moving in a uniform radial field of 0·5 T and the armature is rotating at 1000 r.p.m. (*L.*)

16 State Lenz's law of electromagnetic induction and describe, with explanation, an experiment which illustrates its truth.

Describe the structure of a transformer suitable for supplying 12 V from 240-V mains and explain its action. Indicate the energy losses which occur in the transformer and explain how they are reduced to a minimum.

When the primary of a transformer is connected to the a.c. mains the current in it
(a) is very small if the secondary circuit is open, but
(b) increases when the secondary circuit is closed. Explain these facts. (*L.*)

17 State the laws of electromagnetic induction and describe experiments you would perform to illustrate the factors which determine the magnitude of the induced current set up in a closed circuit.

A simple electric motor has an armature of 0·1 Ω resistance. When the motor is running on a 50 V supply the current is found to be 5 A. Explain this and show what bearing it has on the method of starting large motors. (*L.*)

18

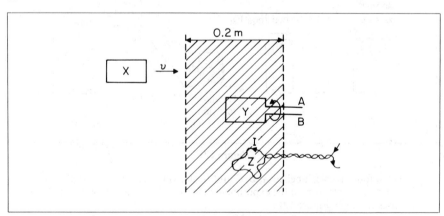

Figure 12F

In Figure 12F a uniform magnetic field of 2·0 T exists, normal to the paper, in the shaded region. Outside this region the magnetic field is zero. X, Y and Z are three loops of wire.
(a) Loop X is a rigid rectangle in the plane of the paper with dimensions 50×100 mm and resistance 0·5 Ω. It is pulled through the field at a constant velocity of 20 mm s^{-1} as shown. Its leading edge enters the field at time $t = 0$. Sketch graphs, giving explanations, of how the following vary with time from $t = 0$ to $t = 20$ s: (i) the magnetic flux linking loop X, (ii) the induced e.m.f.

around the loop, (iii) the current in the loop. Neglect any effect of self-inductance.

(b) Loop Y of the same dimensions as X is mounted on a shaft within the field region. It is rotated at a constant frequency f to provide a source of alternating e.m.f. At $t = 0$ the loop is in the plane of the paper. (i) Give an expression for the e.m.f. between the terminals A and B at any instant. (ii) Calculate the frequency of rotation required to give an output of 3 V r.m.s.

(c) Loop Z, made of a fixed length 0·3 m of flexible wire, rests on a smooth surface in the plane of the paper within the field region. A direct current I in the wire causes it to take up a circular shape. (i) Explain this observation. (ii) What would you expect to happen if the current I was reversed in direction? (*O. & C.*)

19 (a) Magnetic fields can be described in terms of field lines (lines of force). Use this concept to distinguish between *magnetic flux density* (magnetic induction) and *magnetic flux*.

(b) A coil of cross-sectional area 0·001 6 m² and length 50 cm, having 400 turns, is to be used to produce a uniform magnetic field of value 1·51 mT. It is calculated that this can be done if there is a current of 1·5 A in the coil. (i) State where the magnetic field will be uniform. (ii) Show how the value of the current is calculated. (iii) Calculate the total flux through the coil for this current value. (Permeability of free space, $\mu_0 = 4\pi \times 10^{-7}\,\mathrm{H\,m^{-1}}$.)

(c) A coil of 8 turns is now wound around the centre portion of the above coil and its ends are connected to the Y plates of a cathode ray oscilloscope. A 50-Hz alternating current is passed through the 400-turn coil. (i) Explain briefly why an e.m.f. will be induced in the 8-turn coil. (ii) Explain how, with the time base switched off, you would use the oscilloscope to measure the e.m.f. induced in the coil. (iii) If the waveform of the 50-Hz alternating current is as shown in Figure 12G.

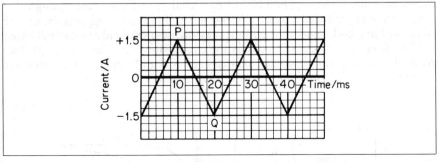

Figure 12G

(1) Calculate the change in the flux linkage through the 8-turn coil during the time interval between the points P and Q on the diagram; also calculate the induced e.m.f. during this time interval.

(2) Sketch a graph showing how the induced e.m.f. varies with time over two cycles of current change, beginning at $t = 0$ and using the same time scale as above. (*L.*)

Self-induction

The phenomenon called *self-induction* was discovered by the American, Joseph Henry, in 1832. It is used in the *inductor*, a component widely use in communication and radio circuits.

When a current flows through a coil, it sets up a magnetic field. And that field threads the coil which produces it, Figure 12.26 (i). If the current I through the coil is changed—by means of a variable resistance, for example—the flux linked with the turns of the coil changes. An e.m.f. is therefore induced in the coil. By

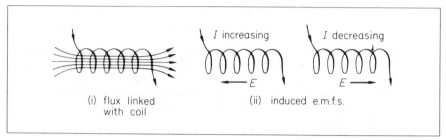

Figure 12.26 *Self-induction*

Lenz's law the direction of the induced e.m.f. will be such as to oppose the change of current. So the e.m.f. will be against the current if it is increasing, but in the same direction if it is decreasing, Figure 12.26 (ii).

Back-e.m.f.

When an e.m.f. is induced in a circuit by a change in the current through that circuit, the e.m.f. induced is called a *back-e.m.f.* Self-induction opposes the growth of current in a coil, and so the current may increase gradually to its final value.

This effect can be demonstrated by the circuit shown in Figure 12.27 (i). Two parallel arrangements are connected to a battery B and a key K. One consists of an iron-cored coil L with many turns in series with a small lamp A_1. The other

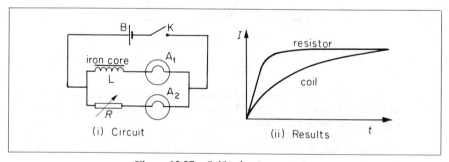

Figure 12.27 *Self-induction experiment*

has a variable resistor R in series with a similar lamp A_2. Initially R is adjusted so that the two lamps have the same brightness in their respective circuits with steady current flowing.

With the circuit open as shown in Figure 12.27 (i), K is closed, so that B is now connected. The lamp A_2 with R is seen to become bright almost immediately but the lamp A_1 with L increases slowly to full brightness. The induced or back-e.m.f. in the coil L *opposes* the growth of current so the glow in the lamp filament in A_1 increases slowly. The resistor R, however, has a negligible back-e.m.f. So its lamp A_2 glows fully bright as soon as K is closed. See Figure 12.27 (ii).

Induced e.m.f. Across Contacts

Just as self-induction opposes the rise of an electric current when it is switched on, so also it opposes the decay of the current when it is switched off. When the circuit is broken, the current starts to fall very rapidly, and a correspondingly great e.m.f. is induced, which tends to maintain the current.

This e.m.f. is often great enough to break down the insulation of the air between the switch contacts, and produce a spark. To do so, the e.m.f. must be about 350 volts or more, because air will not break down—not over any gaps, narrow or wide—when the voltage is less than that value. The e.m.f. at break may be much greater than the e.m.f. of the supply which maintained the current. A spark can easily be obtained, for example, by breaking a circuit consisting of an iron-cored coil and accumulator.

Non-inductive Coils

In some circuits containing coils, self-induction is a nuisance. To minimise their

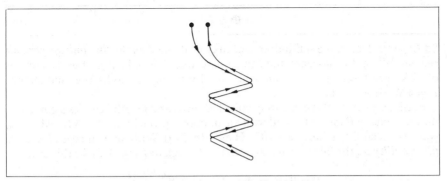

Figure 12.28   *Non-inductive winding*

self-inductance, the coils of resistance boxes are wound so as to set up extremely small magnetic fields. As shown in Figure 12.28, the wire is doubled-back on itself before being coiled up. Every part of the coil is then travelled by the same current in opposite directions, and so its resultant magnetic field is negligible. Such a coil is said to be *non-inductive*.

Self-inductance, L

To discuss the effects of self-induction more fully, we define the property of a coil called its *self-inductance*. By definition,

$$self\text{-}inductance = \frac{back\text{-}e.m.f.\ induced\ in\ coil\ by\ a\ changing\ current}{rate\ of\ change\ of\ current\ through\ coil}$$

Self-inductance is denoted by the symbol L. Numerically, we may therefore write its definition as

$$L = \frac{E_{back}}{dI/dt}$$

or

$$E_{back} = L\frac{dI}{dt} \qquad . \qquad . \qquad . \qquad . \qquad . \qquad (1)$$

The unit of self-induction is the henry (H). *A coil has a self-inductance of 1 henry if the back-e.m.f. in it is 1 volt, when the current through it is changing at the rate of 1 ampere per second.* Equation (1) then becomes:

$$E_{back}(\text{volts}) = L(\text{henrys}) \times \frac{dI}{dt}(\text{ampere/second})$$

The iron-cored coils used for smoothing the rectified supply current to a television receiver are usually very large and have an inductance L of about 50 H.

An air-cored coil may have a small inductance of 0·001 H or 1 millihenry (1 mH).

Resistance, R, of coil $= V/I =$ opposition to *steady* current
Inductance, L, of coil $= E/(dI/dt) =$ opposition to *varying* current

---

L for Coil

Since the induced e.m.f. $E = d\Phi/dt = L\,dI/dt$, numerically, it follows by integration from zero that

---

$$\Phi = LI$$

So $L = \Phi/I$. Hence the self-inductance may be defined as the *flux linkage per unit current*. When Φ is in weber and I in ampere, then L is in henry. Thus if a current of 2A produces a flux linkage of 4 Wb in a coil, the inductance $L = 4\,\text{Wb}/2\,\text{A} = 2\,\text{H}$.

Earlier we saw that when a long coil of N turns and length l carries a current I, the flux density B inside the coil with an air core is given by $B = \mu_0 NI/l$, where μ_0 is the permeability of air, $4\pi \times 10^{-7}\,\text{H m}^{-1}$ (p. 324). With an iron core of *relative permeability* μ_r, the flux density is given by $B = \mu_r\mu_0 NI/l$ (p. 325). In this case

$$\text{flux linkage } \Phi = NAB = \frac{\mu_r\mu_0 N^2 AI}{l}$$

$$\therefore L = \frac{\Phi}{I} = \frac{\mu_r\mu_0 N^2 A}{l} \qquad \qquad (1)$$

This formula may be used to find the approximate value of the inductance of a coil. L is in henry when A is in metre^2, l in metre and μ_0 is $4\pi \times 10^{-7}$ henry metre^{-1}. Note that L depends on N^2, the *square* of the number of turns.

From (1), $\mu_0 = Ll/\mu_r N^2 A$. Now the unit of L is henry, H, the unit of l/A is metre^{-1}, m^{-1}, and μ_r and N^2 are numbers. So

$$\text{unit of } \mu_0 = \text{H m}^{-1}, \text{ or } \mu_0 = 4\pi \times 10^{-7}\,\text{H m}^{-1}$$

Energy Stored

When the current in a coil is interrupted by breaking the circuit, a spark passes across the gap and energy is liberated in the form of heat and light. The energy has been stored in the *magnetic field of the coil*, just as the energy of a charged capacitor is stored in the electrostatic field between its plates (p. 466). When the current in the coil is first switched on, the back-e.m.f. opposes the rise of current. The current flows against the back-e.m.f. and therefore does *work* against it (p. 247). When the current becomes steady, there is no back-e.m.f. and no more work done against it. The total work done in bringing the current to its final value is stored in the magnetic field of the coil. It becomes liberated when the current collapses because then the induced e.m.f. tends to maintain the current, and to do external work of some kind.

To calculate the energy stored in a coil, suppose that the current through it is rising at a rate dI/dt ampere per second. Then, if L is its self-inductance in henrys, the back-e.m.f. across it is given numerically by

$$E = L\frac{dI}{dt}$$

The total work W done against the back-e.m.f. in bringing the current from zero to a steady value I_0 is therefore

$$W = \int EI\, dt = \int_0^{I_0} LI\frac{dI}{dt}\, dt = \int_0^{I_0} LI \cdot dI$$

So
$$W = \tfrac{1}{2}LI_0{}^2$$

This is the energy stored in the magnetic field of the coil.

E.M.F. across Contacts at Break

To calculate the e.m.f. induced at break is, in general, a complicated business. But we can easily do it for one important practical circuit. To prevent sparking

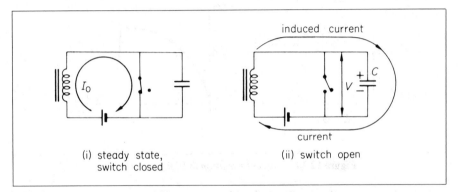

Figure 12.29 *Prevention of sparking by capacitor*

at the contacts of a switch in an inductive circuit, such as a relay used in telecommunications, a capacitor is often connected across the switch, Figure 12.29 (i). When the circuit is broken, the collapsing flux through the coil tends to maintain the current because the current can continue to flow for a brief time by charging the capacitor, Figure 12.29 (ii). Consequently the current does not decay as rapidly as it would without the capacitor, and the back-e.m.f. never rises as high. If the capacitance of the capacitor is great enough, the potential difference across it (and therefore across the switch) never rises high enough to cause a spark.

To find the value to which the potential difference does rise, we assume that all the energy originally stored in the magnetic field of the coil is now stored in the electrostatic field of the capacitor.

If C is the capacitance of the capacitor in farad, and V_0 the final value of potential difference across it on volt, then the energy stored in it is $\tfrac{1}{2}CV_0{}^2$ joule (p. 229). Equating this to the original value of the energy stored in the coil, we have

$$\tfrac{1}{2}CV_0{}^2 = \tfrac{1}{2}LI_0{}^2$$

Let us suppose that a current of 1 ampere is to be broken, without sparking, in a circuit of self-inductance 1 henry and to prevent sparking, the potential difference across the capacitor must not rise above 350 volt. The least capacitance that must be connected across the switch is therefore given by

$$\tfrac{1}{2}C \times 350^2 = \tfrac{1}{2} \times 1 \times 1^2$$

So
$$C = \frac{1}{350^2} = 8 \times 10^{-6}\,\text{F} = 8\,\mu\text{F}$$

A capacitor of capacitance $8\,\mu\text{F}$, and able to withstand 350 volts, would therefore be required.

Current in *L* and *R* Series Circuit

Consider a coil of inductance $L = 2\,\text{H}$ and resistance $R = 5\,\Omega$ connected to a 10 V battery of negligible internal resistance, with a switch S in the circuit, Figure 12.30 (i).

When the switch is closed so that current flows, part of the 10 V is needed to maintain the current in R and the rest of the p.d. is needed to maintain the growth of the current against the back-e.m.f. E_b due to the inductance L.

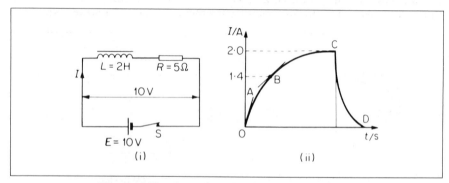

Figure 12.30 *Current variation in L, R series circuit*

1. At the instant the switch is closed (time $t = 0$), there is no current in the circuit. So there is no p.d. across R, from $V = IR$. Hence the whole of the $10\,\text{V} = E_b$, the back-e.m.f. So

$$E_b = 10 = L\frac{dI}{dt} = 2\frac{dI}{dt}$$

Hence
$$\frac{dI}{dt} = \frac{10}{2} = 5\,\text{A s}^{-1}$$

In Figure 12.30 (ii), OA represents the rate of change of current with time at $t = 0$ and the line has a gradient of $5\,\text{A s}^{-1}$.

2. Suppose the current I rises to a value 1·4 A, which is represented by B in Figure 12.30 (ii). The p.d. across R is then given by

$$V = IR = 1·4 \times 5 = 7\,\text{V}$$

So
$$\text{back-e.m.f. } E_b = 10 - 7 = 3\,\text{V}$$

Hence
$$L\frac{dI}{dt} = 3 \quad \text{and} \quad \frac{dI}{dt} = \frac{3}{2} = 1·5\,\text{A s}^{-1}$$

So the current rise, shown by the gradient at B, decreases as time goes on.

3. When the current is finally established and is constant, there is no flux

change in the coil and hence no back-e.m.f. In this case the whole of the 10 V maintains the current in R. So if I_0 is the final or steady current, $V = 10 = IR$. Hence

$$I_0 = \frac{V}{R} = \frac{10}{5} = 2\,\text{A}$$

This is the current value corresponding to C in Figure 12.30 (ii). The graph shows how I increases to its final value after a time t. It is an exponential graph (see below).

Generally, we see that $E = V + L\,dI/dt = IR + L\,dI/dt$. When we solve this differential equation (see *Scholarship Physics* by the author), the result is

$$I = I_0(1 - e^{-Rt/L})$$

So the curve OC in Figure 12.30 (ii) is exponential.

When the circuit is broken, the flux in the coil decreases rapidly and the current falls quickly along CD in Figure 12.30 (ii). A spark may then be obtained across the switch due to the high voltage, as we previously explained.

Mutual Induction

We have already seen that an e.m.f. may be induced in one circuit by a changing current in another (Figure 12.1, p. 342). The phenomenon is often called *mutual induction*, and the pair of circuits which show it are said to have mutual inductance. The *mutual inductance*, M, between two circuits is defined by the equation:

$$\left.\begin{array}{l}\text{e.m.f. induced in B, by}\\ \text{changing current in A}\end{array}\right\} = M \times \left\{\begin{array}{l}\text{rate of change of}\\ \text{current in A}\end{array}\right.$$

across the switch due to the high voltage, as we previously explained:

The same value of M would be obtained if we changed the current in B and observed the e.m.f. induced in A. So, from above,

$$E_B = M\,dI_A/dt \qquad . \qquad . \qquad . \qquad . \qquad . \qquad (1)$$

Further, since $E_B = d\Phi_B/dt$ numerically from Faraday's law, then if Φ_B is the flux change in B due to a current change I_A in A, then we can write

$$\Phi_B = MI_A \qquad . \qquad . \qquad . \qquad . \qquad . \qquad (2)$$

Either equation (1) or equation (2) can be used for defining or calculating M. So if 0·02 V is the induced e.m.f. in B when the current in A changes at $2\,\text{A s}^{-1}$, then

$$M = E_B/(dI_A/dt) = 0{\cdot}02/2 = 0{\cdot}01\,\text{H}$$

Also, if the flux linking a coil B is 0·04 Wb when the current in a neighbouring coil A is 2 A, then

$$M = \Phi_B/I_A = 0{\cdot}04/2 = 0{\cdot}02\,\text{H}$$

The greatest value of mutual inductance M occurs when the two coils A and B are wound over each other on a soft iron core, as in the case of the primary and secondary coils of a commercial (soft iron) transformer. In this case it can be shown that $M = \sqrt{L_A L_B}$, where L_A and L_B are the respective self-inductances of the separate coils A and B.

_____ **Exercises 12B** _____

Self and Mutual Induction

1 Define *self inductance*.

A 12-V battery of negligible internal resistance is connected in series with a coil of resistance $1.0\,\Omega$ and inductance L. When switched on the current in the circuit grows from zero. When the current is 10 A the rate of growth of the current is $500\,A\,s^{-1}$. What is the value of L? (*L*.)

2 A circuit contains an iron-cored inductor, a switch and a d.c. source arranged in series. The switch is closed and, after an interval, reopened. Explain why a spark jumps across the switch contacts.

In order to prevent sparking a capacitor is placed in parallel with the switch. The energy stored in the inductor at the instant when the circuit is broken is $2.00\,J$ and to prevent sparking the voltage across the contacts must not exceed 400 V. Assuming there are no energy losses due to resistance in the circuit, calculate the minimum capacitance required. (*AEB*, 1982.)

3 In the circuit shown in Figure 12H below, L has a very large inductance and negligible resistance and R has a resistance of $600\,\Omega$. The four cells are each of e.m.f. $1.5\,V$.

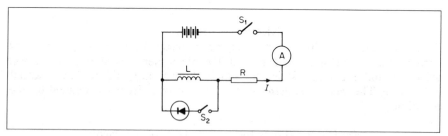

Figure 12H

(a) (i) Sketch a graph showing how the current, I, varies with time, t, when switch S_2 is left open and switch S_1 is closed. Draw a matching graph to show how the p.d., V, across L varies with time over the *same* interval. (ii) Account for the shapes of the graphs. (iii) If S_2 is closed before S_1 is closed, I and V vary with time exactly as before. Explain why the silicon diode makes no difference to I and V.

(b) (i) With S_2 open, S_1 is opened to switch off a steady current I. Explain why sparking occurs at the switch contacts. (ii) With S_2 closed, S_1 is opened to switch off a steady current I. Explain why no sparking occurs. (*L*.)

4 (a) What is meant by the statement that a solenoid has an inductance of 2 H?

A $2.0\,H$ solenoid is connected in series with a resistor, so that the total resistance is $0.50\,\Omega$, to a $2.0\,V$ d.c. supply. Sketch the graph of current against time when the current is switched on. What is (i) the final current, (ii) the initial rate of change of current with time, (iii) the rate of change of current with time when the current is $2.0\,A$?

Explain why an e.m.f. greatly in excess of $2.0\,V$ will be produced when the current is switched off.

(b) A long air-cored solenoid has 1000 turns of wire per metre and a cross-sectional area of $8.0\,cm^2$. A secondary coil, of 2000 turns, is wound around its centre, and connected to a ballistic galvanometer, the total resistance of coil and galvanometer being $60\,\Omega$. The sensitivity of the galvanometer is 2.0 divisions per microcoulomb. If a current of $4.0\,A$ in the primary solenoid were switched off, what would be the deflection of the galvanometer? (Permeability of free space $= 4\pi \times 10^{-7}\,H\,m^{-1}$.) (*L*.)

5 Explain what is meant by the mutual inductance of two coils. If you were provided with a calibrated cathode ray oscilloscope (or a high resistance millivoltmeter) and a means of producing a steadily increasing current, how would you measure the

mutual inductance of two coils? (Assume that normal laboratory equipment is also available.)

Five turns of wire are wound closely about the centre of a long solenoid of radius 20 mm. If there are 500 turns per metre in the solenoid, calculate the mutual inductance of the two coils. Show your reasoning. (The permeability of free space is $4\pi \times 10^{-7}\,H\,m^{-1}$.) (L.)

6   State what is meant by
(a) self induction, and
(b) mutual induction.

Describe one experiment in each case to illustrate these effects.

In the circuit shown (Figure 12I) A and B have equal ohmic resistance but A is of negligible self inductance whilst B has a high self inductance.

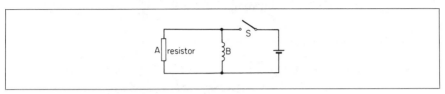

Figure 12I

Describe and explain how the currents through A and B change with time
(i) when the switch S is closed, and   (ii) when it is opened. Illustrate your answers graphically.

Describe briefly *two* applications of self inductors. (L.)

7   Describe the phenomena of self induction and mutual induction.

Describe the construction and explain the action of a simple form of a.c. transformer.

In an a.c. transformer in which the primary and secondary windings are perfectly coupled and in which a negligible primary current flows when there is no load in the secondary, a current of 5 A (r.m.s.) was observed to flow in the primary under an applied voltage of 100 V (r.m.s.) when the secondary was connected to resistors only. If the primary contains 100 turns and the secondary 25 000 turns, calculate
(a) the voltage
(b) the current in the secondary, stating any simplifiying assumptions you make.
(O. & C.)

8   A choke of large self inductance and small resistance, a battery and a switch are connected in series. Sketch and explain a graph illustrating how the current varies with time after the switch is closed. If the self inductance and resistance of the coil are 10 H and 5 Ω respectively and the battery has an e.m.f. of 20 V and negligible resistance, what are the greatest values after the switch is closed of
(a) the current,
(b) the rate of change of current? (*JMB*.)

Flux Linkage and Charge Relation

We have already seen that an electromotive force is induced in a circuit when the magnetic flux linked with it changes. If the circuit is closed, a current flows, and electric charge, Q, is carried round the circuit. As we shall now show, there is a simple relationship between the charge and the change of flux.

Consider a closed circuit of total resistance R ohm, which has a total flux linkage Φ with a magnetic field, Figure 12.31. If the flux linkages start to change,

$$\text{induced e.m.f., } E = -\frac{d\Phi}{dt}$$

$$\therefore \text{ current, } I = \frac{E}{R} = -\frac{1}{R}\frac{d\Phi}{dt} \qquad \cdot \qquad \cdot \qquad \cdot \qquad (1)$$

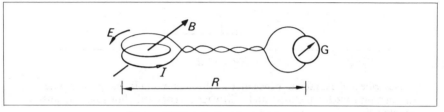

Figure 12.31 *Coil with changing flux*

In general, the flux linkage will not change at a steady rate, and the current will not be constant. But, throughout its change, charge is being carried round the circuit. In a time t from zero, the charge carried round the circuit is

$$Q = \int_0^t I\,dt$$

From (1),

$$\therefore Q = -\frac{1}{R}\int_0^t \frac{d\Phi}{dt}\,dt$$

$$= -\frac{1}{R}\int_{\Phi_0}^{\Phi_t} d\Phi$$

where Φ_0 is the number of linkages at $t = 0$, and Φ_t is the number of linkages at time t. Thus

$$Q = -\frac{\Phi_t - \Phi_0}{R} = \frac{\Phi_0 - \Phi_t}{R}$$

The quantity $\Phi_0 - \Phi_t$ is positive if the linkages Φ have decreased, and negative if they have increased. But as a rule we are interested only in the magnitude of the change, and so we may write

$$Q = \frac{\textit{change of flux linkage}}{R} \qquad \cdot \qquad \cdot \qquad \cdot \qquad (2)$$

Equation (2) shows that the charge circulated is proportional to the change of flux-linkages, and is *independent of the time*.

Ballistic Galvanometer

From equation (2), we see that if the charge Q flowing is measured by a *ballistic galvanometer* G, as shown in Figure 12.31, then we have a measure of the change in the flux linkage, Φ.

Ballistics is the study of the motion of an object, such as a bullet, which is set off by a blow, and then allowed to move freely without friction. A ballistic galvanometer is one used to measure an electrical blow, or impulse, for example, the charge Q which circulates when a capacitor is discharged through it.

A galvanometer which is intended to be used ballistically has   (i) a heavier coil than one which is not, and   (ii) has as little damping as possible—an insulating former, no short-circuited turns, no shunt. The mass of its coil makes it swing slowly. In the example above, for instance, the capacitor has discharged, and the charge has finished circulating, while the galvanometer coil is just beginning to turn. The galvanometer coil continues to turn, however: and as it does so it twists the controlling spring. The coil stops turning when its kinetic energy, which it gained from the forces set up by the current, has been converted into potential energy of the spring. The coil then swings back, as the spring untwists, and it continues to swing back and forth for some time. Eventually it comes to rest, but only because of the damping due to the viscosity of the air, and to the spring. Theory shows that, if the damping is negligible, *the first deflection of the galvanometer is proportional to the quantity of charge, Q, that passed through its coil, as it began to move.* This first deflection, θ, is often called the 'throw' of the galvanometer; We then have

$$Q = k\theta, \quad . \quad \quad . \quad \quad . \quad \quad . \quad \quad . \quad \quad . \quad (1)$$

where k is a constant of the galvanometer.

Equation (1) is true only if all the energy given to the coil is spent in twisting the suspension. If an appreciable amount of energy is used to overcome damping—that is, dissipated as heat by eddy currents—then the galvanometer is not ballistic, and θ is not proportional to Q.

To calibrate the ballistic galvanometer, a capacitor of known capacitance, for example, $2\,\mu\text{F}$, is charged by a battery of known e.m.f. such as 50 volt, and then discharged through the instrument. Suppose the deflection is 200 divisions. The charge $Q = CV = 100$ microcoulomb, and so the galvanometer sensitivity is 2 divisions per microcoulomb.

Measurement of Flux Density

Figure 12.32 illustrates the principle of measuring the flux density B in the field

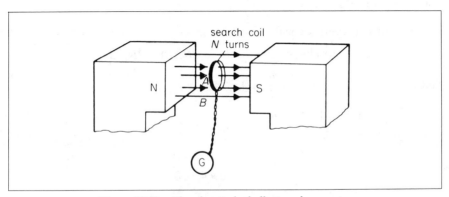

Figure 12.32 *Flux density by ballistic galvanometer*

between the poles of a powerful magnet such as a loudspeaker magnet. A small coil, called a *search coil*, with a known area and number of turns, is connected to a ballistic galvanometer G. It is positioned at right angles to the field to be measured, as shown, so that the flux enters the coil face normally.

The coil is then pulled completely out of the field by moving it smartly downwards, for example, and the throw θ produced in the galvanometer is observed. The charge Q which passes round the circuit is proportional to θ, from above.

Suppose B is the field-strength in tesla (T), A is the area of the coil in m^2 and N is the number of turns. Then

$$\text{change of flux-linkages} = NAB$$

$$\therefore \text{quantity, } Q, \text{ through galvanometer} = \frac{NAB}{R}$$

where R is the *total* resistance of the galvanometer and search coil. But

$$Q = c\theta$$

where c is the quantity per unit deflection of the ballistic galvanometer.

$$\therefore \frac{NAB}{R} = c\theta$$

$$\therefore B = \frac{Rc\theta}{NA} \qquad \cdot \quad \cdot \quad \cdot \quad \cdot \quad \cdot \quad (1)$$

The constant c is found by discharging a capacitor through the galvanometer (see p. 216). By knowing c, θ, R, N and A, the flux density B can be calculated from (1).

In this way, by using a suitable so-called 'search' coil connected to a ballistic galvanometer, the flux-density at various points between the poles of a large horseshoe magnet can be compared.

Example on Measuring B

A long solenoid carries a current which produces a flux-density B as its centre. A narrow coil Y of 10 turns and mean area $4.0 \times 10^{-5} \, m^2$ is placed in the middle of the solenoid so that the flux links its turns normally and the ends of Y are connected. If a charge of $1.6 \times 10^{-6} \, C$ circulates through Y when the current in the solenoid is reversed, and the resistance of Y is $0.2 \, \Omega$, calculate B.

When current reverses, the flux reverses. So

$$\text{flux change } \Phi = NAB - (-NAB) = 2NAB$$

Since

$$\frac{\Phi}{R} = Q$$

$$\therefore \frac{2 \times 10 \times 4 \times 10^{-5} B}{0.2} = 1.6 \times 10^{-6}$$

$$\therefore B = \frac{1.6 \times 10^{-6} \times 0.2}{2 \times 10 \times 4 \times 10^{-5}}$$

$$= 4 \times 10^{-4} \, T$$

_____ **Exercises 12C** _____

Charge and Flux Linkage

1 A flat search coil containing 50 turns each of area $2.0 \times 10^{-4} \, m^2$ is connected to a galvanometer; the total resistance of the circuit is $100 \, \Omega$. The coil is placed so that its plane is normal to a magnetic field of flux density $0.25 \, T$.
   (a) What is the change in magnetic flux linking the circuit when the coil is moved to a region of negligible magnetic field?
   (b) What charge passes through the galvanometer? (*C*.)

2 (i) A flat coil of N turns and area A is placed so that its plane is perpendicular to a magnetic field of flux density B. The coil is suddenly removed from the field. If the coil is in a closed circuit of resistance R prove, from first principles, that the charge q which is circulated is given by

$$q = \frac{NBA}{R}$$

(ii) A ballistic galvanometer has charge sensitivity $100 \, mm \, \mu C^{-1}$ and a resistance of $100 \, \Omega$. A square search coil of negligible resistance, of 25 turns, having sides of length $10 \, mm$ is in series with the galvanometer. When the coil is removed from a magnetic field the deflection on the galvanometer is $250 \, mm$. Calculate the magnetic flux density.   (iii) If the search coil is removed from the field in $0.5 \, s$, how much heat is generated by the circulating charge?   (iv) How would a current balance rather than the search coil be used to measure the flux density? (*W.*)

3 Describe and account for two constructional differences between a moving-coil galvanometer used to measure current and the ballistic form of the instrument.
   An electromagnet has plane-parallel pole faces. Give details of an experiment, using a search coil and ballistic galvanometer of known sensitivity, to determine the variation in the magnitude of the magnetic flux density along a line parallel to the pole faces and mid-way between them. Indicate in qualitative terms the variation you would expect to get.
   A coil of 100 turns each of area $2.0 \times 10^{-3} \, m^2$ has a resistance of $12 \, \Omega$. It lies in a horizontal plane in a vertical magnetic flux density of $3.0 \times 10^{-3} \, Wb \, m^{-2}$. What charge circulates through the coil if its ends are short-circuited and the coil is rotated through $180°$ about a diametral axis? (*JMB.*)

13

A.C. Circuits

A.C. circuits are needed for understanding the action and design of radio and television circuits. We start with the root-mean-square value of a.c. The single components L, C and R in a.c. circuits are then discussed, and the analogy with d.c. circuits is stressed. This is followed by series L, R and C, R circuits and the important series resonance L, C, R circuit and its application in radio reception. As we show, power in a.c. circuits depends on the phase difference between current and voltage.

Measurement of A.C.

If an alternating current (a.c.) is passed through a moving-coil meter, the pointer does not move. The coil is urged clockwise and anticlockwise at the frequency of the current—50 times per second if it is drawn from the British grid—and does not move at all. In a sensitive instrument the pointer may be seen to vibrate with a small amplitude.

The relation between current I and pointer deflection θ in a moving-coil meter is $I \propto \theta$. This is unsuitable for measuring alternating current as the deflection reverses on the negative half of the cycle. Instruments for measuring alternating currents must be so made that the pointer deflects the same way when the current flows through the instrument in either direction. As we shall see, a suitable law of deflection is $\theta \propto I^2$, a square-law deflection.

Hot-wire Instrument, Mean-square Value of Current

One type of 'square law' instrument is the hot-wire ammeter, Figure 13.1. In it the current flows through a fine resistance-wire XY, which it heats. The wire

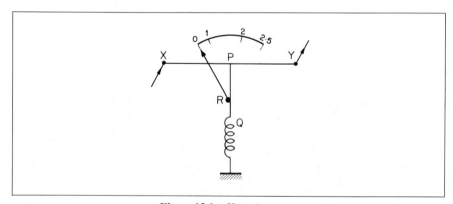

Figure 13.1 *Hot wire meter*

warms up to such a temperature that it loses heat—mainly by convection—at a rate equal to the average rate at which heat is developed in the wire. The rise in temperature of the wire makes it expand and sag. The sag is taken up by a second fine wire PQ, which is held taut by a spring. The wire PQ passes round a pulley R attached to the pointer of the instrument, which rotates the pointer. The deflection of the pointer is roughly proportional to the average rate at which heat is developed in the wire XY. It is therefore roughly proportional to the average value of the *square* of the alternating current, and the scale is a square-law (non-uniform) one as shown.

Root-mean-square Value of A.C. Sinusoidal (Sine Wave) A.C.

Earlier we saw that an alternating current I varied sinusoidally; that is, it could be represented by the equation $I = I_m \sin \omega t$, where I_m was the peak (maximum) value of the current. In commercial practice, alternating currents are always measured and expressed in terms of their *root-mean-square* (*r.m.s.*) value.

Consider two resistors of equal resistance R, one carrying an alternating current and the other a direct current. Suppose both are dissipating the same power P, as heat. The root-mean-square (r.m.s.) value of the alternating current, I_r, is then defined as equal to the direct current, I_d. Thus:

> *the root-mean-square value of an alternating current is defined as that value of steady current which would dissipate heat at the same rate in a given resistance.*

Since the power dissipated by the direct current is

$$P = I_d{}^2 R$$

our definition means that, in the a.c. circuit,

$$P = I_r{}^2 R \qquad . \qquad . \qquad . \qquad . \qquad (1)$$

Whatever the wave-form of the alternating current, if I is its value at any instant, the power which it delivers to the resistance R at that instant is $I^2 R$. Consequently, the average power P is given by

$$P = \text{average value of } (I^2 R)$$
$$= \text{average value of } (I^2) \times R$$

since R is a constant. Therefore, by equation (1), taking the average value over a cycle,

$$I_r{}^2 = \text{average value of } (I^2) \qquad . \qquad . \qquad . \qquad . \qquad (2)$$

The average value of (I^2) is called the *mean-square* current. Figure 13.2 (i)

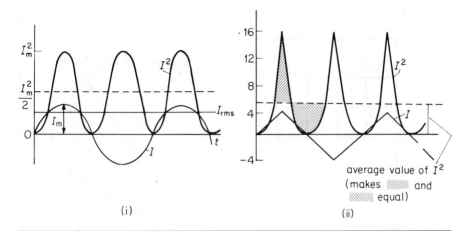

Figure 13.2   *Mean-square values*

shows a sinusoidal (sine variation) current I from the a.c. mains and the way its I^2 values vary. The values are *positive* on the negative half cycle. Since this graph of I^2 is a symmetrical one, the mean or average value of I^2 is $I_m^2/2$, where I_m is the *maximum* or *peak* value of the current. So in this case, the *root-mean-square* (r.m.s.) value, I_r, of the current is given by taking the square root of $I_r^2/2$ and therefore

$$I_r = \frac{I_m}{\sqrt{2}} = 0.71\,I_m \quad . \qquad . \qquad . \qquad . \qquad (3)$$

If the r.m.s. value I_r is known, the peak value of the current I_m is calculated from

$$I_m = \sqrt{2}I_r$$

In Britain, the a.c. mains supply is 240 V (r.m.s.). So the peak or maximum value of the voltage is

$$V_m = \sqrt{2}V_r = 1.41 \times 240 = 338 \text{ V}$$

This means that an electrical appliance is unsuitable for use on the a.c. mains if it cannot withstand a voltage of about 338 V.

Other A.C. Waveforms, Square Wave A.C.

The root-mean-square (r.m.s.) value of a varying current or voltage depends on its waveform.

Figure 13.2 (ii) shows an alternating current I which is not sinusoidal and the way its I^2 values vary with time. Unlike the sine-wave current in Figure 13.2 (i), the mean-square value is *not* half-way between the zero and the peak or maximum value. The mean-square value corresponds to the value which, for a cycle, makes the areas equal on both sides, as shown in Figure 13.2 (ii). In this case it can be seen that the mean-square value is *less* than $I_m^2/2$, where I_m is the maximum value of the current. So the r.m.s. is less than $I_m/\sqrt{2}$.

Figure 13.3 (i) shows one form of a *square wave* alternating current. Unlike

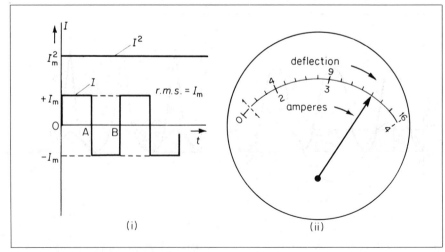

(i)   (ii)

Figure 13.3   (i) *Square wave*   (ii) *Scale of a.c. ammeter*

sinusoidal a.c., this has a constant positive current I_m for half a cycle OA and a constant negative current I_m for the other half of the cycle AB. The square of the current is positive on both halves of the cycle and equal to $I_m{}^2$ throughout. So the root-mean-square value is I_m. The power delivered to a pure resistance R would therefore be $I_m{}^2 R$.

A.C. Meter and Scale

We can see that for measuring alternating current, we require a meter whose deflection measures not the current through it but the average value of the square of the current. As we have already seen, hot-wire meters have just this property.

For convenience, such meters are scaled to read amperes, not (amperes)^2, as in Figure 13.3 (ii). The scale reading is then proportional to the square-root of the deflection, and indicates directly the root-mean-square value of the current, I_r. An a.c. meter of the hot-wire type can be calibrated by using direct current. This follows at once from the definition of the r.m.s. value of current as the value of direct current which produces the same heat per second in a resistor.

Moving-coil meters with semiconductor diode rectifiers are widely used in multimeters for measuring alternating current and voltage, as described later (p. 792). They work in a different way to a hot-wire meter and give much more accurate readings of a.c. current or voltage.

A.C. with a Capacitor C

In many radio circuits, resistors, capacitors, and inductors or coils are present. An alternating current can flow through a resistor, but it is not obvious at first that it can flow through a capacitor. This can be demonstrated, however, by connecting a capacitor of 1 μF or more in series with a mains filament lamp of low rating such as 25 W. The lamp lights up, showing that a current is flowing through it. Direct current cannot flow through a capacitor because an insulating medium is between the plates. So with a mixture of a.c. and d.c., only the a.c. flows through a circuit with a capacitor.

The current flows because the capacitor plates are being continually charged,

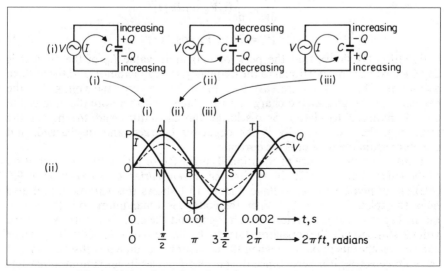

Figure 13.4   *Flow of a.c. through capacitor, frequency 50 Hz*

discharged, and charged the other way round by the alternating voltage of the mains, Figure 13.4(i). The current thus flows round the circuit, and can be measured by an a.c. milliammeter inserted in any of the connecting wires.

Figure 13.4(ii) shows how the alternating voltage V varies with time, t. Since the charge Q on the capacitor plates is given at any instant by $Q = CV$, and C is constant, the graph of Q is in phase (in step) with that of V as shown.

The current I is the rate of change of Q with time, that is, $I = dQ/dt$. So the value of I at any instant is the corresponding *gradient* of the $Q-t$ graph. At O, the gradient value OP is a maximum, so I is then a maximum. From O to A, the gradient of the $Q-t$ graph decreases to zero. So I decreases to zero at N. From A to B, the gradient of the $Q-t$ curve is *negative* and so I is negative from N to R. In this way we see that the $I-t$ graph is PNRST.

So I and V are 90° out of phase, *with I leading V by 90°*.

If V is made bigger, we see that the gradient at O is bigger. So I_m increases when V_m increases. Also, if the frequency f of V is doubled, for example, so that there are now two cycles between O and B, the gradient at O becomes greater. So I_m increases when both f and V_m increase.

Calculation for *I*

To find the exact variation of I with time t, suppose the amplitude or peak of the voltage V applied to the capacitor C is V_m and its frequency is f. Then, assuming a sinusoidal voltage variation, the instantaneous voltage at any time t is

$$V = V_m \sin 2\pi ft$$

If C is the capacitance of the capacitor, then the charge Q on its plates is

$$Q = CV$$

so
$$Q = CV_m \sin 2\pi ft$$

The current, I, flowing at any instant, is equal to the rate at which charge is accumulating on the capacitor plates. Thus

$$I = \frac{dQ}{dt} = \frac{d}{dt}(CV_m \sin 2\pi ft)$$

$$= 2\pi f CV_m \cos 2\pi ft \quad . \quad \quad . \quad \quad . \quad \quad . \quad (1)$$

Equation (1) shows that the peak or maximum value I_m of the current is $2\pi f CV_m$; so I_m is proportional to the frequency, the capacitance, and the voltage amplitude. These results are easy to explain. The greater the voltage, or the capacitance, the greater the charge on the plates, and therefore the greater the current required to charge or discharge the capacitor. And the higher the frequency, the more rapidly is the capacitor charged and discharged, and therefore again the greater is the current.

A more puzzling feature of equation (1) is the factor giving the time variation of the current, $\cos 2\pi ft$. It shows that *the current varies a quarter-cycle or 90° ($\pi/2$) out of phase with the voltage*. Figure 13.4 shows this variation, and also helps to explain it physically. When the voltage is a maximum, so is the charge on the capacitor. It is therefore not charging and the current is zero. When the voltage starts to fall, the capacitor starts to discharge. The rate of discharging, or current, reaches its maximum when the capacitor is completely discharged and the voltage across it is zero. Since the current I passes its maximum a quarter-cycle ahead of the voltage V, we see that I leads V by 90° ($\pi/2$).

For a capacitor C
I leads V by 90°

Reactance of C

The *reactance* of a capacitor is its opposition in ohms to the passage of alternating current. We do not use the term 'resistance' in this case because this is the opposition to direct current.

The reactance, symbol X_C, is defined by

$$X_C = \frac{V_m}{I_m}$$

where V_m and I_m are the peak or maximum of the a.c. voltage and current. Since the ratio $V_m/I_m = V_r/I_r$, where V_r and I_r are the r.m.s. voltage and current respectively, we can also define reactance X_C by the ratio V_r/I_r. We shall omit the suffix r when using r.m.s. values and so

$$X_C = \frac{V}{I}$$

Here X_C is in ohms when V is in volts (r.m.s.) and I is in amperes (r.m.s.). As we have just seen, the amplitude or peak value of the current through a capacitor is given by

$$I_m = 2\pi f C V_m$$

The reactance of the capacitor is therefore

$$X_C = \frac{V_m}{I_m} = \frac{1}{2\pi f C}$$

X_C is in ohms when f is in Hz (cycles per second), and C in farads.

For convenience we often write $\omega = 2\pi f$. The quantity ω is called the angular frequency of the current and voltage. It is expressed in radians per second. Then an alternating voltage, for example, may be written as

$$V = V_m \sin \omega t$$

The reactance of a capacitor can therefore also be written as

$$X_C = \frac{1}{\omega C}$$

Calculations of Reactance X_C

As an illustration, suppose a capacitor C of $0.1\,\mu F$ is used on the mains frequency of 50 Hz. Then the reactance is

$$X_C = \frac{1}{2\pi f C} = \frac{1}{2 \times 3.14 \times 50 \times 0.1 \times 10^{-6}}$$

$$= \frac{10^6}{2 \times 3.14 \times 50 \times 0.1} = 32\,000\,\Omega \text{ (approx.)}$$

From the formula for reactance we note that $X_C \propto 1/C$ for a given frequency. So if a $1\,\mu F$ capacitor is used on the 50 Hz mains, its reactance is 10 times *less*

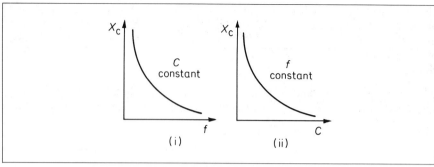

Figure 13.5 *Reactance of capacitor*

than that of $0.1 \, \mu\text{F}$, which is $32\,000\,\Omega/10$ or $3200\,\Omega$. Figure 13.5 (ii) shows how X_C varies with C.

Also, since $X_C \propto 1/f$ for a given capacitor, at $f = 1000\,\text{Hz}$ a capacitor of $1\,\mu\text{F}$ has 20 times *less* reactance than at $f = 50\,\text{Hz}$. So $X_C = 3200\,\Omega/20 = 160\,\Omega$. See Figure 13.5 (i).

Since $X_C = V/I$, where V and I are both r.m.s. (or peak) values, we can see that

$$I = \frac{V}{X_C} \quad \text{and} \quad V = IX_C$$

These are similar formulae to d.c. circuit formulae. The difference with d.c. circuits is that we must always consider the phase difference between V and I, and V and I are 90° out of phase as we have previously shown.

Example on Capacitor Reactance

A capacitor C of $1\,\mu\text{F}$ is used in a radio circuit where the frequency is $1000\,\text{Hz}$ and the current flowing is $2\,\text{mA}$ (r.m.s.). Calculate the voltage across C.

What current flows when an a.c. voltage of $20\,\text{V}$ r.m.s., $f = 50\,\text{Hz}$ is connected to this capacitor?

(i) Reactance, $X_C = \dfrac{1}{2\pi f C} = \dfrac{1}{2\pi \times 1000 \times 10^{-6}} = 159\,\Omega$ (approx.)

$$\therefore V = IX_C = \frac{2}{1000} \times 159 = 0.32\,\text{V (approx.)}$$

(ii) When $20\,\text{V}$ r.m.s., $f = 50\,\text{Hz}$, is connected to C, the reactance of C changes. Since $X_C \propto 1/f$,

$$X_C \text{ at } f = 50\,\text{Hz is 20 times } X_C \text{ at } f = 1000\,\text{Hz}$$

So $\qquad\qquad X_C = 20 \times 159\,\Omega = 3180\,\Omega$

$$\therefore I = \frac{V}{X_C} = \frac{20}{3180} = 6.3 \times 10^{-3}\,\text{A r.m.s}$$

For a capacitor C   *V lags on I by 90°*

$$X_C = \frac{1}{\omega C} = \frac{1}{2\pi f C}$$

A.C. through an Inductor

Since a coil is made from conducting wire, we have no difficulty in seeing that an alternating current can flow through it. However, if the coil has appreciable self-inductance, the current is less than would flow through a non-inductive coil of the same resistance. We have already seen how self-inductance opposes changes of current; it must therefore oppose an alternating current, which is continuously changing.

Let us suppose that the resistance of the coil is negligible, a condition which can be satisfied in practice. We can simplify the theory by considering first the current, and then finding the potential difference across the coil. Suppose the current is

$$I = I_m \sin 2\pi f t \quad . \qquad . \qquad . \qquad . \qquad . \qquad (1)$$

where I_m is its peak value, Figure 13.6. If L is the inductance of the coil, the changing current sets up a back-e.m.f. in the coil, of magnitude

$$E = L\frac{dI}{dt}$$

To maintain the current, the applied supply voltage must be equal to the back-e.m.f. The voltage applied to the coil must therefore be given by

$$V = L\frac{dI}{dt}$$

Since L is constant, it follows that V is proportional to dI/dt.

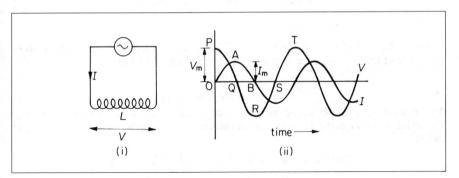

Figure 13.6 *Flow of a.c. through a coil*

Figure 13.6 (ii) shows how the current I varies with time t. The values of dI/dt are the *gradients* of the $I-t$ graph at the time concerned. At O the gradient is a maximum; so the maximum voltage of V, or V_m, occurs at O and is represented by OP as shown. From O to A, the gradient of the $I-t$ graph decreases to zero. So the voltage V decreases from P to Q. From A to B the gradient of the $I-t$ graph is *negative* (downward slope). So the voltage decreases along QR. We now see that V *leads* I by 90° ($\pi/2$).

We can find the value of V_m from the equation $V = LdI/dt$. From (1),

$$I = I_m \sin 2\pi f t$$

So, by differentiation with respect to t,

$$V = L\frac{dI}{dt}$$

$$= L\frac{d}{dt}(I_m \sin 2\pi f t) = 2\pi f L I_m \cos 2\pi f t$$

So $\qquad V_{\mathrm{m}} = \text{maximum (peak) voltage} = 2\pi f L I_{\mathrm{m}}$

Hence the reactance of the inductor is

$$X_{\mathrm{L}} = \frac{V_{\mathrm{m}}}{I_{\mathrm{m}}} = 2\pi f L$$

X_{L} is in ohms when f is in Hz, and L is in henrys (H). An *iron-cored* coil has a high inductance L such as 20 H. Used on the mains frequency f of 50 Hz, its reactance $X_{\mathrm{L}} = 2\pi f L = 2 \times 3\cdot14 \times 50 \times 20 = 6280\,\Omega$. Since this type of inductor provides a high reactance, it is sometimes called a 'choke'.

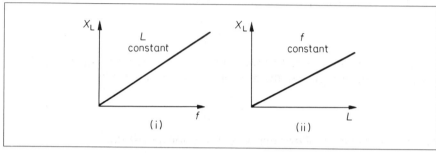

Figure 13.7   *Variation of reactance* X_L

Since $X_{\mathrm{L}} = 2\pi f L$, it follows that $X_{\mathrm{L}} \propto f$ for a given inductance, Figure 13.7 (i), and that $X_{\mathrm{L}} \propto L$ for a given frequency, Figure 13.7 (ii).

Example on Reactance X_{L}

An inductor of 2 H and negligible resistance is connected to a 12 V mains supply, $f = 50$ Hz. Find the current flowing. What current flows when the inductance is changed to 6 H?

$$\text{Reactance, } X_{\mathrm{L}} = 2\pi f L = 2\pi \times 50 \times 2 = 628\,\Omega$$

$$\therefore I = \frac{V}{X_{\mathrm{L}}} = \frac{12}{628}\,\mathrm{A} = 19\,\mathrm{mA}\ (\text{approx.})$$

When the inductance is increased to 6 H, its reactance X_{L} is increased 3 times since $X_{\mathrm{L}} \propto L$ for a given frequency. So the current is reduced to 1/3rd of its value. So now $I = 6$ mA (approx.).

For an inductor L

V leads on I by 90°
$X_{\mathrm{L}} = \omega L = 2\pi f L$

Phasor Diagrams

In the Mechanics section of this book, it is shown that a quantity which varies sinusoidally with time may be represented as the projection of a rotating vector (p. 76). These quantities are called *phasors*, as the phase angle must also be represented. Alternating currents and voltages may therefore be represented as phasors. Figure 13.8 shows, on the left, the phasors representing the current

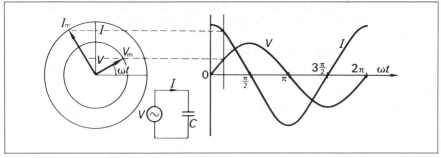

Figure 13.8 *Phasor diagram for capacitor*

through a capacitor, and the voltage across it. Since the current leads the voltage by $\pi/2$, the current vector I is displaced by 90° ahead of the voltage vector V.

Figure 13.9 shows the phasor diagram for a pure inductor. In drawing it, the voltage has been taken as $V = V_m \sin \omega t$, and the current drawn lagging $\pi/2$

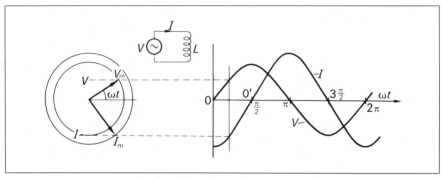

Figure 13.9 *Phasor diagram for pure inductance*

behind it. This enables the diagram to be readily compared with that for a capacitor. To show that it is essentially the same as Figure 13.6 (ii), we have only to shift the origin by $\pi/2$ to the right, from 0 to 0′.

When an alternating voltage is connected to a pure resistance R, the current I at any instant $= V/R$, where V is the voltage at that instant. So I is zero when V is zero and I is a maximum when V is a maximum. Hence I and V are in phase,

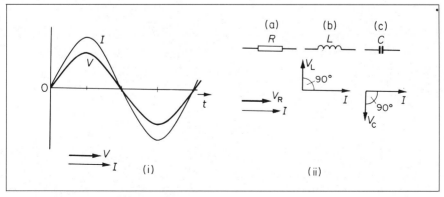

Figure 13.10 *Phasor diagrams for R, L, C*

Figure 13.10 (i). Since the phase angle is zero, we draw the vector or phasor I in the same direction as V, where the phasors represent either peak values or r.m.s. values.

Figure 13.10 (ii) summarises the phasor diagrams for

(a) a pure resistance R,
(b) a pure inductance L and
(c) a pure capacitance C.

They should be memorised by the reader for use in all kinds of a.c. circuits.

Series Circuits

L and R in Series

Consider an inductor L in series with resistance R, with an alternating voltage V (r.m.s.) of frequency f connected across both components, Figure 13.11 (i).

The sum of the respective voltages V_L and V_R across L and R is equal to V. But the voltage V_L leads by 90° on the current I, and the voltage V_R is in phase with I (see p. 391). Thus the two voltages can be drawn to scale as shown in Figure

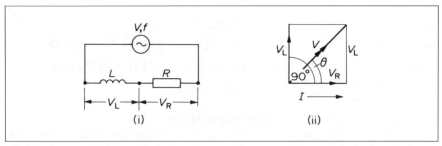

Figure 13.11   *Inductance and resistance in series*

13.11 (ii), and hence, by Pythagoras' theorem, it follows that the vector sum V is given by

$$V^2 = V_L{}^2 + V_R{}^2$$

But $V_L = IX_L$, $V_R = IR$.

$$\therefore V^2 = I^2 X_L{}^2 + I^2 R^2 = I^2(X_L{}^2 + R^2)$$

$$\therefore I = \frac{V}{\sqrt{X_L{}^2 + R^2}} \qquad . \qquad . \qquad . \qquad . \qquad . \qquad \text{(i)}$$

Also, from Figure 13.11 (ii), the current I lags on the applied voltage V by an angle θ given by

$$\tan \theta = \frac{V_L}{V_R} = \frac{IX_L}{IR} = \frac{X_L}{R} \qquad . \qquad . \qquad . \qquad . \qquad \text{(ii)}$$

From (i), it follows that the 'opposition' Z to the flow of alternating current is given in ohms by

$$Z = \frac{V}{I} = \sqrt{X_L{}^2 + R^2} \qquad . \qquad . \qquad . \qquad . \qquad \text{(iii)}$$

This 'opposition', Z, is known as the *impedance* of the circuit.

Example on L and R in Series

An iron-cored coil of 2 H and 50 Ω resistance is placed in series with a resistor of 450 Ω, and a 100 V, 50 Hz, a.c. supply is connected across the arrangement. Find
(a) the current flowing in the coil,
(b) its phase angle relative to the voltage supply,
(c) the voltage across the coil.

(a) The reactance $X_L = 2\pi f L = 2\pi \times 50 \times 2 = 628\,\Omega$.
Total resistance $R = 50 + 450 = 500\,\Omega$.

$$\therefore \text{ circuit impedance } Z = \sqrt{X_L{}^2 + R^2} = \sqrt{628^2 + 500^2} = 803\,\Omega$$

$$\therefore I = \frac{V}{Z} = \frac{100}{803}\,\text{A} = 12\cdot 5\,\text{mA (approx.)}$$

(b)
$$\tan\theta = \frac{X_L}{R} = \frac{628}{500} = 1\cdot 256$$

So
$$\theta = 515\cdot 5°$$

(c) For the coil,
$$X_L = 628\,\Omega \text{ and } R = 50\,\Omega$$

So coil impedance
$$Z = \sqrt{X_L{}^2 + R^2} = \sqrt{628^2 + 50^2} = 630\,\Omega$$

Thus
$$\text{voltage across coil } V = IZ = 12\cdot 5 \times 10^{-3} \times 630$$
$$= 7\cdot 9\,\text{V (approx.)}$$

C and R in Series

A similar analysis enables the impedance to be found of a capacitance C and resistance R in series, Figure 13.12 (i). In this case the voltage V_C across the

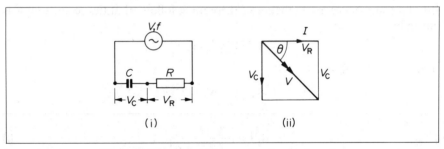

(i)   (ii)

Figure 13.12 *Capacitance and resistance in series*

capacitor lags by 90° on the current I (see p. 386), and the voltage V_R across the resistance is in phase with the current I. As the vector sum is V, the applied voltage, it follows by Pythagoras' theorem that

$$V^2 = V_C{}^2 + V_R{}^2 = I^2 X_C{}^2 + I^2 R^2 = I^2 (X_C{}^2 + R^2)$$

$$\therefore I = \frac{V}{\sqrt{X_C{}^2 + R^2}} \qquad \qquad \text{(i)}$$

Also, from Figure 13.12 (ii), the current I leads on V by an angle θ given by

$$\tan\theta = \frac{V_C}{V_R} = \frac{IX_C}{IR} = \frac{X_C}{R} \qquad \qquad \text{(ii)}$$

It follows from (i) that the impedance Z of the $C-R$ series circuit is

$$Z = \frac{V}{I} = \sqrt{X_C{}^2 + R^2}$$

It should be noted that although the impedance formula for a C–R series circuit is of the same mathematical form as that for a L–R series circuit, the current in the C–R series case *leads* on the applied voltage but the current in the L–R series case *lags* on the applied voltage.

For L–R series:

$$\text{impedance } Z = \sqrt{X_L^2 + R^2}, \quad \tan\theta = X_L/R$$

For C–R series:

$$\text{impedance } Z = \sqrt{X_C^2 + R^2}, \quad \tan\theta = X_C/R$$

L, C, R in Series

The most general series circuit is the case of L, C, R in series, Figure 13.13 (i). As we see later, it is widely used in radio. The phasor diagram has V_L leading by 90° on V_R, V_C lagging by 90° on V_R, with the current I in phase with V_R, Figure 13.13 (ii). If V_L is greater than V_C, their resultant is $(V_L - V_C)$ in the direction of V_L, as shown. Thus, from Pythagoras' theorem for triangle ODB, the applied voltage V is given by

$$V^2 = (V_L - V_C)^2 + V_R^2$$

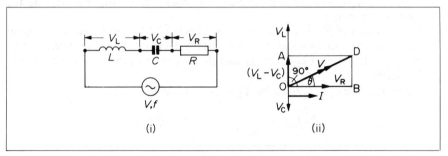

Figure 13.13   L, C, R *in series*

But $V_L = IX_L$, $V_C = IX_C$, $V_R = IR$.

$$\therefore V^2 = (IX_L - IX_C)^2 + I^2 R^2 = I^2[(X_L - X_C)^2 + R^2]$$

$$\therefore I = \frac{V}{\sqrt{(X_L - X_C)^2 + R^2}} \qquad . \quad . \quad . \quad . \quad \text{(i)}$$

Also, I lags on V by an angle θ given by

$$\tan\theta = \frac{DB}{OB} = \frac{V_L - V_C}{V_R} = \frac{IX_L - IX_C}{IR} = \frac{X_L - X_C}{R} \qquad . \quad . \quad \text{(ii)}$$

Resonance in the L, C, R Series Circuit

From (i), it follows that the impedance Z of the circuit is given by

$$Z = \sqrt{(X_L - X_C)^2 + R^2}$$

The impedance varies as the frequency, f, of the applied voltage varies, because X_L and X_C both vary with frequency. Since $X_L = 2\pi f L$, then $X_L \propto f$, and thus the variation of X_L with frequency is a straight line passing through the origin, Figure 13.14 (i). Also, since $X_C = 1/2\pi f C$, then $X_C \propto 1/f$, and thus the variation of X_C with frequency is a curve approaching the two axes, Figure 13.14 (i). The resistance R is independent of frequency, and is thus represented by a line

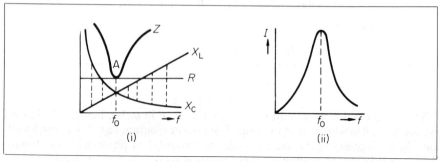

Figure 13.14   *Resonance curves*

parallel to the frequency axis. The difference $(X_L - X_C)$ is represented by the dotted lines shown in Figure 13.14 (i), and it can be seen that $(X_L - X_C)$ decreases to zero for a particular frequency f_0, and thereafter increases again. Thus, from $Z = \sqrt{(X_L - X_C)^2 + R^2}$, the impedance diminishes and then increases as the frequency f is varied.

The variation of Z with f is shown in Figure 13.14 (i), and since the current $I = V/Z$, the current varies as shown in Figure 13.14 (ii). Thus the current has a maximum value at the frequency f_0, and this is known as the *resonant frequency* of the circuit.

The magnitude of f_0 is given by $X_L - X_C = 0$, or $X_L = X_C$.

$$\therefore 2\pi f_0 L = \frac{1}{2\pi f_0 C} \quad \text{or} \quad 4\pi^2 LC f_0^2 = 1$$

$$\therefore f_0 = \frac{1}{2\pi\sqrt{LC}}$$

At frequencies above and below the resonant frequency, the current is less than the maximum current, see Figure 13.14 (ii), and the phenomenon is thus basically the same as the forced and resonant vibrations obtained in Sound or Mechanics.

At resonance: **(1)** $f_0 = 1/2\pi\sqrt{LC}$ **(2)** $X_L = X_C$ **(3) impedance** $Z = R$
(4) maximum current $I = V/R$ **(5)** I **and** V **are in phase**

Tuning in Radio Receivers

The series resonance circuit is used for tuning a radio receiver. In this case the incoming waves of frequency f say from a distant transmitting station induces a varying voltage in the aerial, which in turn induces a voltage V of the same frequency in a coil and capacitor circuit in the receiver, Figure 13.15 (i). When the capacitance C is varied the resonant frequency is changed; and at one setting of C the resonant frequency becomes f, the frequency of the incoming waves. The maximum current is then obtained, and the station is now heard very loudly.

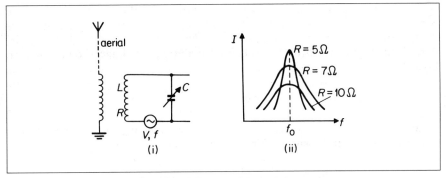

Figure 13.15 *Tuning a receiver*

The sharpness of the resonance is an important matter in radio reception of transmitting stations. If it is not sharp, other transmitting stations may produce a current I or 'response' in the circuit of about the same value as the station required. Considerable 'interference' then occurs. If the resistance R in the L, C, R series is small, then the resonance is sharp, as illustrated roughly in Figure 13.15 (ii). An inductor (L, R) and a capacitor (C) in series form an L, C, R series circuit, and the resistance R is low if the inductor coil is made with the *minimum* wire needed. In this case the resonance is sharp.

Parallel Circuits

We now consider briefly the principles of a.c. parallel circuits. In d.c. parallel circuits, the currents in the individual branches are added arithmetically to find their total. In a.c. circuits, however, we add the currents by vector methods, taking into account the phase angle between them.

L, R in Parallel

In the parallel circuit in Figure 13.16 (i), the supply current I is the vector sum of

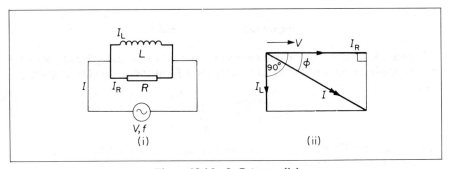

Figure 13.16   L, R *in parallel*

I_L and I_R, Figure 13.16 (ii) shows the vector addition. I_L is 90° out of phase with I_R, since I_R is in phase with V, and I_L lags 90° behind V. So

$$I^2 = I_R{}^2 + I_L{}^2 = \left(\frac{V}{R}\right)^2 + \left(\frac{V}{X_L}\right)^2$$

Then
$$I = V\sqrt{\frac{1}{R^2} + \frac{1}{X_L{}^2}}$$

Also, from Figure 13.16 (ii), I lags behind the applied voltage V by an angle φ given by

$$\tan \varphi = \frac{I_L}{I_R} = \frac{R}{X_L}$$

Similar analysis shows that when an a.c. voltage V r.m.s. is applied to a parallel C, R circuit, the supply I is given by

$$I = V\sqrt{\frac{1}{R^2} + \frac{1}{X_C{}^2}} \quad \text{and} \quad \tan \varphi = \frac{R}{X_C}$$

where I now leads V by the angle φ.

L, C in Parallel, Coil-capacitor Resonance

A parallel arrangement of coil (L, R) and capacitor (C) is widely used in transistor oscillators and in radio-frequency amplifier circuits. To simplify matters, let us assume that the resistance of the coil is negligible compared with its reactance. We then have effectively an inductor L in parallel with a capacitor C, Figure 13.17 (i).

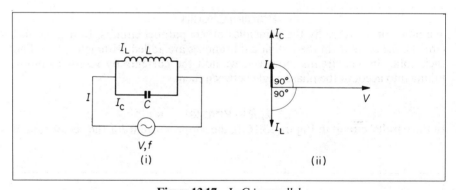

Figure 13.17 L, C *in parallel*

Figure 13.17 (ii) shows the two currents in the components. I_L lags by 90° on V but I_C leads by 90° on V. If I_C is greater than I_L at the particular frequency f, then

$$I = I_C - I_L = \frac{V}{X_C} - \frac{V}{X_L}$$

Since I leads by 90° on V in this case, we say that the circuit is 'net capacitive'. If, however, I_L is greater than I_C, then

$$I = I_L - I_C = \frac{V}{X_L} - \frac{V}{X_C}$$

Since I lags by 90° on V in this case, the circuit is 'net inductive'.

Suppose V, L and C are kept constant and the frequency f of the supply is varied from a low to a high value. The magnitude and phase of I then varies according to the relative magnitudes of $X_L (2\pi f L)$ and $X_C (1/2\pi f C)$, as shown in

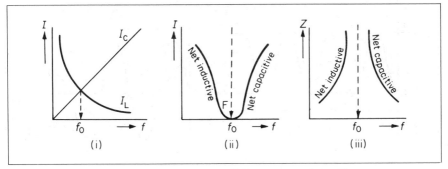

Figure 13.18   L, C *in parallel—variation of* Z *and* I

Figure 13.18 (i). A special case occurs when $X_L = X_C$. Then $I_L = I_C$ and so $I = 0$, Figure 13.18 (ii). At this frequency f_0, we have $X_L = X_C$, so

$$2\pi f_0 L = \frac{1}{2\pi f_0 C}, \quad \text{or} \quad f_0 = \frac{1}{2\pi\sqrt{LC}}$$

Figure 13.18 (ii) shows how the current I varies with the frequency f. Figure 13.18 (iii) shows how the *impedance* Z of the parallel L, C circuit varies with frequency f. Since $Z = V/I$, and I is zero at the frequency f_0, it follows that Z is infinitely high at f_0. The parallel inductor-capacitor circuit has therefore a *resonant frequency* (f_0) so far as its impedance Z is concerned.

In practice, when the resistance R of the coil is taken into account, a similar variation of Z with frequency f is obtained. The maximum value of Z is now finite and theory shows that $Z = L/CR$. The resonant frequency f_0 is practically still given by $f_0 = 1/2\pi\sqrt{LC}$.

For a particular frequency, a high impedance is often needed as a 'load' in certain radio circuits. A parallel coil-capacitor circuit is then used, tuned to the frequency wanted. In contrast, the *series* L, C, R circuit gives a maximum *current* I at resonance in a radio tuning circuit.

Power in A.C. Circuits

Resistance R. The power absorbed is usually $P = IV$. In the case of a resistance, $V = IR$, and $P = I^2R$. The variation of power is shown in Figure 13.19 (i), where I_m = the peak (maximum) value of the current. On p. 363 we explained the reason for choosing the root-mean-square value of alternating current. So the average power absorbed in R is given by

$$P = I^2R$$

where I is the r.m.s. value.

The power P in a resistor can also be written as

$$P = IV = \frac{I_m V_m}{2}$$

as shown in Figure 13.19 (i).

Inductance L. In the case of a pure inductor, the voltage V across it leads by $90°$ on the current I. Thus if $I = I_m \sin \omega t$, then $V = V_m \sin(90° + \omega t) = V_m \cos \omega t$. Hence, at any instant,

$$\text{power absorbed} = IV = I_m V_m \sin \omega t . \cos \omega t = \tfrac{1}{2} I_m V_m \sin 2\omega t$$

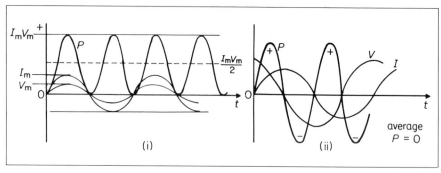

Figure 13.19 *Power in R, L and C*

The variation of power, P, with time t is shown in Figure 13.19 (ii); it is a sine curve with an average of zero. *Hence no power is absorbed in a pure inductance.* This is explained by the fact that on the first quarter of the current cycle, power is absorbed ($+$) in the magnetic field of the coil (see p. 372). On the next quarter-cycle the power is returned ($-$) to the generator, and so on.

Capacitance. With a pure capacitance, the voltage V across it lags by 90° on the current I (p. 386). Thus if $I = I_m \sin \omega t$,

$$V = V_m \sin(\omega t - 90°) = -V_m \cos \omega t$$

Hence, numerically,

$$\text{power at an instant, } P = IV = I_m V_m \sin \omega t \cos \omega t = \frac{I_m V_m}{2} \sin 2\omega t$$

Thus, as in the case of the inductance, *the power absorbed in a cycle is zero,* Figure 13.19 (ii). This is explained by the fact that on the first quarter of the cycle, energy is stored in the electrostatic field of the capacitor. On the next quarter the capacitor discharges, and the energy is returned to the generator.

Formulae for A.C. Power, Power Factor

It can now be seen that, if I is the r.m.s. value of the current in amps in a circuit containing a resistance R ohms, the power absorbed is I^2R watts. Care should be taken to exclude the inductances and capacitances in the circuit, as no power is absorbed in them. So if a current of 2 A r.m.s. flows in a circuit containing a coil of 2 H and resistance 10 Ω in series with a capacitor of 1 μF, the power absorbed in the circuit $= I^2R = 2^2 \times 10 = 40$ W.

If the voltage V across a circuit leads by an angle θ on the current I, the voltage

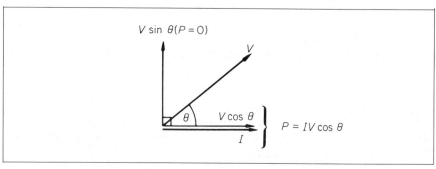

Figure 13.20 *Power absorbed*

can be resolved into a component $V \cos \theta$ in phase with the current, and a voltage $V \sin \theta$ perpendicular to the current, Figure 13.20. The former component, $V \cos \theta$, represents that part of the voltage across the total *resistance* in the circuit, and hence the power absorbed is

$$P = IV \cos \theta$$

The component $V \sin \theta$ is that part of the applied voltage across the total inductance and capacitance. Since the power absorbed here is zero, it is sometimes called the 'wattless component' of the voltage.

The *power factor* of an a.c. circuit is defined as the ratio

$$power\ absorbed/IV$$

since IV is the maximum power which would be absorbed if the whole circuit was a resistance. So

$$\textbf{power factor} = \frac{IV \cos \theta}{IV} = \cos \theta$$

If the circuit has considerable reactance (X_L or X_C) compared to the amount of resistance (R), then the phase angle θ is nearly 90°. So $\cos \theta$, and hence the power factor, is very small. This is because a pure inductor L and a pure capacitor C absorb no power as we have seen.

Examples on A.C. Circuits

1 A circuit consists of a capacitor of $2\,\mu\text{F}$ and a resistor of $1000\,\Omega$. An alternating e.m.f. of 12 V (r.m.s.) and frequency 50 Hz is applied. Find   (1) the current flowing,   (2) the voltage across the capacitor,   (3) the phase angle between the applied e.m.f. and current, (4) the average power supplied.

The reactance X_C of the capacitor is given by

$$X_C = \frac{1}{2\pi f C} = \frac{1}{2\pi \times 50 \times 2 \times 10^{-6}} = 1590\,\Omega \text{ (approx.)}$$

$\therefore$ total impedance $Z = \sqrt{R^2 + X_C{}^2} = \sqrt{1000^2 + 1600^2} = 1880\,\Omega$ (approx.)

(1) $\therefore$ current, $I = \dfrac{V}{Z} = \dfrac{12}{1880} = 6\cdot4 \times 10^{-3}\,\text{A}$

(2) voltage across C, $V_C = IX_C = \dfrac{12}{1880} \times 1590 = 10\cdot2\,\text{V (approx.)}$

(3) The phase angle θ is given by

$$\tan \theta = \frac{X_C}{R} = \frac{1590}{1000} = 1\cdot59$$

$$\therefore \theta = 58° \text{ (approx.)}$$

(4) Power supplied $= I^2 R = \left(\dfrac{12}{1880}\right)^2 \times 1000 = 0\cdot04\,\text{W (approx.)}$

2 A capacitor of capacitance C, a coil of inductance L and resistance R, and a lamp are placed in series with an alternating voltage V. Its frequency f is varied from a low to a high value while the magnitude of V is kept constant. Describe and explain how the brightness of the lamp varies.

If $V = 0.01$ V (r.m.s.) and $C = 0.4\,\mu\text{F}$, $L = 0.4\,\text{H}$, $R = 10\,\Omega$, calculate (i) the resonant frequency, (ii) the maximum current, (iii) the voltage across C at resonance, neglecting the lamp resistance. What is the effect of reducing the resistance R to $5\,\Omega$?

When f is varied, the impedance Z of the circuit decreases to a minimum value (resonance) and then increases. Z is a minimum when $X_L = X_C$, so that $Z = R$ at resonance. Since the *current* flowing in the circuit increases to a maximum and then decreases, the brightness of the lamp increases to a maximum at resonance and then decreases.

(i)   Resonant frequency $f_0 = \dfrac{1}{2\pi\sqrt{LC}} = \dfrac{1}{2\pi\sqrt{0.4 \times 0.4 \times 10^{-6}}}$

$$= \dfrac{10^3}{2\pi \times 0.4} = 400 \text{ Hz (approx.)}$$

(ii)   Maximum current $I = \dfrac{V}{R} = \dfrac{0.01}{10} = 0.001$ A (r.m.s.)

(iii)   Voltage across $C = IX_C = 0.001 \times \dfrac{1}{2\pi \times 400 \times 0.4 \times 10^{-6}}$

$$= \dfrac{0.001 \times 10^6}{2\pi \times 400 \times 0.4} = 1 \text{ V}$$

When R is reduced to $5\,\Omega$, the maximum current I is doubled, since $I = V/R$. Also, the sharpness of resonance is considerably increased.

Exercises 13

1   Figure 13A reresents alternating currents of different wave shapes, each of peak (amplitude) value 3.0 A. (i) is a sinusoidal a.c., (ii) is a square wave and (iii) is a rectangular wave. Calculate the r.m.s. value of the current in each case.

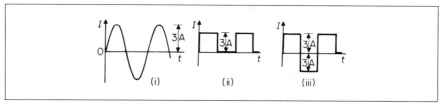

Figure 13A

2   An alternating voltage of 10 V r.m.s. and frequency 50 Hz is applied to (i) a resistor of $5\,\Omega$, (ii) an inductor of 2 H, and (iii) a capacitor of 1 μF. Determine the r.m.. current flowing in each case and draw a phasor diagram of the current and voltage for each.

3   An alternating current of 0.2 A r.m.s. and frequency $100/2\pi$ Hz flows in a circuit consisting of a series arrangement of a resistor R of $20\,\Omega$, an inductor L of 0.15 H and a capacitor C of $500\,\mu$F. Calculate the a.c. voltage (i) across each component, (ii) across R and L together, (iii) across L and C together, (iv) the total voltage across L, C, R.
    What power is dissipated in each componnt?

4   A coil of inductance L and negligible resistance is in series with a resistance R. A supply voltage of 40 V (r.m.s.) is connected to them. If the voltage across L is equal to that across R, calculate
   (a) the voltage across each component,

(b) the frequency f of the supply,

(c) the power absorbed in the circuit, if $L = 0.1$ H and $R = 40\,\Omega$.

5 An inductor L of negligible resistance is connected in *parallel* with a capacitor C and an a.c. voltage V of constant r.m.s. value is connected across the arrangement. With the aid of a phasor diagram, explain why the current I drawn from the a.c. supply is zero at particular frequency.

Draw a sketch showing the variation of I with frequency f when f is varied from a very low value to a very high value.

6 If a sinusoidal current, of peak value 5 A, is passed through an a.c. ammeter the reading will be $5/\sqrt{2}$ A. Explain this.

What reading would you expect if a square-wave current, switching rapidly between $+0.5$ and -0.5 A, were passed through the instrument? (*L.*)

7 (a) The *impedance* of a circuit containing a capacitor C and a resistor R connected to an alternating voltage supply is given by $\sqrt{R^2 + X^2}$, where X is the *reactance* of the capacitor. Define the two terms in italics.

The current in the circuit leads the voltage by a phase angle θ, where $\tan\theta = X/R$. Explain, using a vector diagram, why this is so.

An inductor is put in series with the capacitor and resistor and a source of alternating voltage of constant value but variable frequency. Sketch a graph to show how the current will vary as the frequency changes from zero to a high value.

(b) An alternating voltage of 10 V r.m.s. and 5.0 kHz is applied to a resistor, of resistance $4.0\,\Omega$, in series with a capacitor of capacitance $10\,\mu F$. Calculate the r.m.s. potential differences across the resistor and the capacitor. Explain why the sum of these potential differences is not equal to 10 V. (Assume $\pi^2 = 10$.) (*L.*)

8 A coil having inductance and resistance is connected to an oscillator giving a fixed sinusoidal output voltage of 5.00 V r.m.s. With the oscillator set at a frequency of 50 Hz, the r.m.s. current in the coil is 1.00 A and at a frequency of 100 Hz, the r.m.s. current is 0.625 A.

(a) Explain why the current through the coil changes when the frequency of the supply is changed.

(b) Determine the inductance of the coil.

(c) Calculate the ratio of the powers dissipated in the coil in the two cases. (*JMB.*)

9 An inductor and a capacitor are connected one at a time to a variable-frequency power source. State how, and explain in non-mathematical terms why, the current through the inductor and the capacitor varies as the frequency is varied.

A circuit is set up containing an inductor, a capacitor, a lamp and a variable-frequency source with the components arranged in series. Explain why, as the frequency of the supply is varied, the lamp is found to increase in brightness, reach a maximum and then become less bright. Explain why the inductor is heated by the passage of the current while the capacitor remains cool. (*L.*)

10 (a) A flat coil of wire is rotated at constant angular velocity in a uniform field between the poles of a magnet. Prove the e.m.f. generated varies sinusoidally with time.

Explain what is meant by the peak value of the e.m.f.

Why was the concept of a root-mean-square introduced into a.c. theory? State the numerical relation between the root-mean-square value and the peak value.

(b) Calculate the reactance of a pure inductance of 1 mH carrying a.c. of frequency 0.5 MHz and that of an inductance of 10 H carrying a.c. of frequency 50 Hz.

(i) Comment on the average power dissipated in each. (ii) What is the advantage of a variable inductance for controlling a.c. rather than a variable resistance?

(c) Why would the larger inductance have an iron core? The cores of large transformers are found to become warm in use. Explain why this is so, and point out how this would be minimised in practice. (*W.*)

11 A lamp, which may be regarded as a non-inductive resistor, is rated at 2 A, 220 W. In order to operate the lamp from the 240 V, 50 Hz mains, an inductor is placed in series with it. If the resistance of the inductor is $5.0\,\Omega$ what should the value of its inductance be? (*L.*)

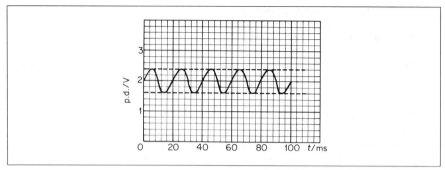

Figure 13B

12 A potential difference varying with time t as shown in the Figure 13B is applied to a capacitor of reactance $1.0\,k\Omega$.
  (a) Use the graph to calculate   (i) the peak value of the current in the circuit, and (ii) the capacitance of the capacitor.
  (b) Sketch a graph, using the same origin and time scale as in the diagram above, to show how the current varies with time over the first 40 ms.
  (c) Calculate the new peak value of the current if the frequency of the applied p.d. were increased by a factor of 10^2. (*L.*)

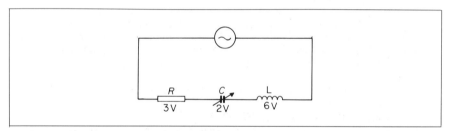

Figure 13C

13 A constant voltage source in the circuit illustrated supplies a sinusoidal alternating e.m.f., Figure 13C. The voltages marked against the other components are the peak values developed across each for a particular value of the capacitance of C.
  (a) Determine   (i) the peak value of the applied e.m.f.,   (ii) the phase angle between the applied e.m.f. and the current.
  (b) If the variable capacitor C is now adjusted until the voltage across R is a maximum, the circuit is said to be resonant. Explain why the current in the circuit has its maximum value when this is so. (*L.*)
14 A series circuit consists of an inductor L, a resistor R and a capacitor C driven by an a.c. source of constant peak voltage E_0 and variable frequency f. Connections are made from C and L to the Y plates of a double-beam oscilloscope so that the time variation of the potential differences across each of these components, V_C and V_L, can be simultaneously displayed.
  Sketch the displays you would see   (i) when f is very small,   (ii) at resonance, (iii) when f is very large. Give brief explanations in **each** case.
  At a certain frequency the peak p.d.'s across the three components are found to be: $(V_R)_0 = 100\,V$, $(V_L)_0 = 150\,V$, $(V_C)_0 = 100\,V$.
  (a) Is the frequency greater than, equal to, or less than the resonant frequency?
  (b) Find the peak source e.m.f.
  (c) Find the phase difference between the e.m.f. and the current. Which leads?
  (d) Find the peak resonance current, and the power dissipated in the circuit at resonance if $R = 10\,k\Omega$. (*W.*)
15 (a) Explain why a moving coil ammeter cannot be used to measure an alternating current even if the frequency is low. Draw a diagram of a bridge rectifier circuit

which could be used with such an ammeter and explain its action. (You are not required to explain the mode of operation of an individual diode in the bridge.)

(b) An alternating voltage is connected to a resistor and an inductor in series. By using a vector diagram, and explaining the significance of each vector, derive an expression for the impedance of the circuit.

A 50 V, 50 Hz a.c. supply is connected to a resistor, of resistance 40 Ω, in series with a solenoid whose inductance is 0·20 H. The p.d. between the ends of the resistor is found to be 20 V. What is the resistance of the wire of the solenoid? (Assume $\pi^2 = 10$) (L.)

16 Explain what is meant by the *peak value* and *root-mean-square value* of an alternating current. Establish the relation between these quantities for a sinusoidal waveform.

What is the r.m.s. value of the alternating current which must pass through a resistor immersed in oil in a calorimeter so that the initial rate of rise of temperature of the coil is three times that produced when a direct current of 2 A passes through the resistor under the same conditions? (*JMB.*)

17 A pure capacitor C is connected across an a.c. source. Draw sketch-graphs on the same time axis to show how the current, the p.d. across C and the source e.m.f. vary with time.

Write down, in terms of appropriate quantities which must be defined, an expression for the power drawn from the source at any instant. Show, with the aid of a sketch-graph, that the average power dissipated is zero.

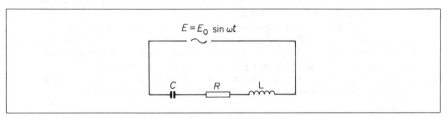

Figure 13D

For the circuit given (Figure 13D), draw a typical vector diagram representing V_C, V_R and V_L, the peak potential differences across C, R and L respectively, together with E_0. Express V_C, V_R and V_L in terms of the quantities shown in the above diagram together with I_0, the peak current.

Deduce from your vector diagram expressions for   (i) the phase difference ϕ between the applied e.m.f. and the current,   (ii) the impedance of the circuit, (iii) the value of I_0 at resonance.

Describe briefly how you would use an oscilloscope to measure ϕ and V_C. (*W.*)

18 (a) Define the *impedance* of a coil carrying an alternating current. Distinguish between the *impedance* and *resistance* of a coil and explain how they are related. Describe and explain how you would use a length of insulated wire to make a resistor having an appreciable resistance but negligible inductance.

(b) Outline how you would determine the impedance of a coil at a frequency of 50 Hz using a resistor of known resistance, a 50 Hz a.c. supply and a suitable measuring instrument. Show how to calculate the impedance from your measurements.

(c) A coil of inductance L and resistance R is connected in series with a capacitor of capacitance C and a variable frequency sinusoidal oscillator of negligible impedance. Sketch qualitatively how the current in the circuit varies with the applied frequency and account for the shape of the curve. Sketch on the same axes the curve you would expect for a considerably larger value of R, the values of L and C remaining unchanged, taking care to indicate which curve refers to the larger value of R. (*JMB.*)

19 (a) A sinusoidal alternating potential difference of which the peak value is 20 V is connected across a resistor of resistance 10 Ω. What is the mean power dissipated in the resistor?

(b) A sinusoidal alternating potential difference is to be rectified using the circuit in Figure 13E (i), which consists of a diode D, a capacitor C and a resistor R. Sketch the variation with time of the potential difference between A and B which you would expect and explain why it has that form.

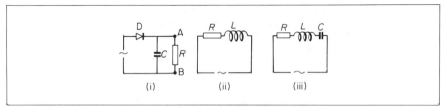

(i)   (ii)   (iii)

Figure 13E

(c) A sinusoidal alternating difference of a constant amplitude is applied across the resistor R and inductor L, Figure 13E (ii). Explain why the amplitude of the current through the circuit decreases as the frequency of the alternating potential difference is increased.

(d) A sinusoidal alternating potential difference of constant frequency and amplitude is applied to the circuit in Figure 13E (iii). Describe and explain how the amplitude of the current through the circuit changes as the capacitance C is increased slowly from a very small value to a very large value. (*O. & C.*)

20 Define the *impedance* of an a.c. circuit.

A $2.5\,\mu F$ capacitor is connected in series with a non-inductive resistor of $300\,\Omega$ across a source of p.d. of r.m.s. value 50 V alternating at $1000/2\pi$ Hz. Calculate
(a) the r.m.s. values of the current in the circuit and the p.d. across the capacitor,
(b) the mean rate at which energy is supplied by the source. (*JMB.*)

21 A source of a.c. voltage is connected by wires of negligible resistance across a capacitor. Explain, without the use of mathematical expressions, why
(a) a current flows;
(b) the current is not in phase with the voltage;
(c) the size of the current depends upon the frequency of the supply voltage;
(d) the power output of the source is zero.

If a resistor, of resistance R, is connected in series with a capacitor, of capacitance C, to an a.c. voltage of frequency f, derive an expression for the phase difference between the voltage and the current.

If the supply voltage were 10 V, the frequency $1.0\,kHz$ and the capacitance $2.0\,\mu F$, what value of R in the circuit would allow a current of 0.10 A to flow? (*L.*)

22 Explain what is meant in an alternating current circuit by
(a) reactance,
(b) impedance,
(c) resonance.

Describe an experiment by which (c) may be demonstrated.

A coil of self-inductance of 0.200 H and resistance $50.0\,\Omega$ is to be supplied with a current of 1.00 A from a 240 V, 50 Hz, supply and it is desired to make the current in phase with the potential difference of the source. Find the values of the components that must be put in series with the coil. Illustrate the conditions in the circuit with a phasor (vector) diagram. (*L.*)

23 An electrical appliance is operated from a 240 V, 50 Hz supply. Sketch a graph, with suitable values marked on the axes, to illustrate how the potential difference across the appliance varies with time.

A current $I_0 \cos \omega t$ flows in a circuit containing a pure capacitor of capacitance C. Starting from the definition of capacitance, show that the potential difference across the capacitor has a maximum value $I_0/\omega C$. Sketch graphs to show the variation with time of the current in the circuit and of the p.d. across the capacitor, using the same time axes for both.

A pure variable resistor is connected in series with a pure fixed capacitor. With the aid of a phasor diagram, or otherwise, explain what happens to

(a) the impedance of the circuit and

(b) the phase angle between the current and the p.d. across the combination as the resistance is increased from zero.

If the capacitor has a capacitance of $1 \cdot 6 \, \mu F$, determine the value of the resistance so that the phase angle is 45° at a frequency of 50 Hz. (C.)

24 Explain what is meant by the *reactance* of an inductor or capacitor.

An alternating potential difference is applied (i) to an inductor, (ii) to a capacitor. Describe and explain the phase lag or lead between the current and the applied potential difference in each case.

Calculate the reactance of an inductor L of inductance 100 mH and of a capacitor C of capacitance $2 \, \mu F$, both at a frequency of 50 Hz. At what frequency f_0 are their reactances equal in magnitude?

The inductor L and capacitor C are connected in parallel and an alternating potential difference of constant amplitude and variable frequency f is applied, Figure 13F.

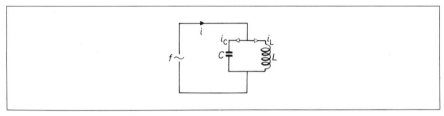

Figure 13F

(a) What is the phase relationship between i_L and i_C?

(b) What are the relative magnitudes of i_L and i_C when $f = f_0$ and what is then the value of i?

(c) What are the relative magnitudes of i, i_L and i_C when f is very much greater than f_0? Explain briefly your conclusions in each case. (O. & C.)

Part 3

*Geometrical Optics. Waves.
Wave Optics. Sound Waves*

14
Geometrical Optics

Reflection, Refraction, Principles of Optical Fibres

LIGHT is an electromagnetic wave and later we shall discuss the wave theory of light in detail. In this first chapter in Optics, however, we are mainly concerned with geometrical (ray) optics. So we start with the effect on light rays when they meet mirrors (reflection) and when they travel from one medium such as air to another such as glass (refraction). In particular, total internal reflection occurs when rays meet the boundary between two different media at an angle of incidence greater than the critical angle.

The principles of optical fibres are discussed at the end of the chapter. These very thin strands of glass are now used in telecommunications to transmit signals along them. We shall show how the laws of refraction and total internal reflection are applied in optical fibres.

Light Energy and Light Beams

Light is a form of energy. We know this is the case because plants and vegetables grow when they absorb sunlight. Also, electrons are emitted by certain metals when light is incident on them, showing that there was some energy in the light. This phenomenon is the basis of the *photoelectric cell* (p. 848). Substances like wood or brick which allow no light to pass through them are called 'opaque' substances. Unless an opaque object is perfectly black, some of the light falling on it is reflected. A 'transparent' substance, like glass, is one which allows some of the light energy incident on it to pass through. The rest of the energy is absorbed and (or) reflected.

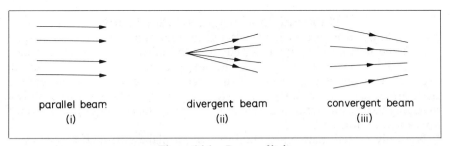

parallel beam (i) divergent beam (ii) convergent beam (iii)

Figure 14.1 *Beams of light*

A *ray* of light is the direction along which the light energy or light waves travel. Although rays are represented in diagrams by straight lines, in practice a ray has a finite width. A *beam* of light is a collection of rays. A searchlight emits a *parallel beam* of light. The rays from a point on a very distant object like the sun are substantially parallel, Figure 14.1 (i). A lamp emits a *divergent beam* of light;

while a source of light behind a lens, as in a projection lantern, can provide a *convergent beam*, Figure 14.1 (ii), (iii).

Reflection by Plane Mirrors, Reversibility of Light

When we see an object, rays of light enter the eye and produce the sensation of vision. In Figure 14.2 (i), rays from a small (point) object O are reflected by a plane mirror so that the angle of incidence i = the angle of reflection r (law of reflection). The rays enter the eye of an observer at D. We always see images in the direction *in which the rays enter the eye*. So the image of O appears to be at I, behind the mirror.

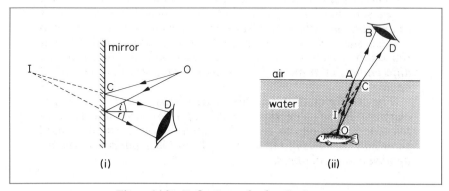

Figure 14.2 *Reflection and refraction images*

In Figure 14.2 (ii), the rays OA and OC from a point object O on a fish change their direction at the boundary with air. They travel along AB and CD. So an observer sees the image of O at I, higher in the water than O.

Light rays never change their light path. So if the ray of light CD is reversed in direction to travel along DC, it will travel along CO in the water. This is known as the *principle of reversibility of light*. We shall need to use this principle later. It follows from a general law in Optics due to Fermat. This states that light travels between two points such as O and D in the minimum (or maximum) time. So the light path between O and D is the same in either direction.

Virtual and Real Images in Plane Mirrors

As was shown in Figure 14.2, an object O in front of a mirror has an image I behind the mirror. The rays reflected from the mirror do not actually pass through I, but only *appear* to do so. The image cannot be received on a screen because the image is behind the mirror, Figure 14.3 (i). This type of image is therefore called a *virtual* image. You can see that the light beam from O is a diverging beam. After reflection from the mirror it is still a diverging beam which appears to come from I.

Not only virtual images are obtained with a plane mirror. If a *convergent* beam is incident on a plane mirror M, the reflected rays pass through a point I *in front of* M, Figure 14.3 (ii). If the incident beam converges to the point O, then O is called a 'virtual' object. I is called a *real* image because it can be received on a screen. Figure 14.3 (i) and (ii) should now be compared. In the former, a real object (divergent beam) gives rise to a virtual image; in the latter, a virtual object (convergent beam) gives rise to a real image. In each case the image and object are at equal distances from the mirror. A plane mirror produces an image which is the same size as the object.

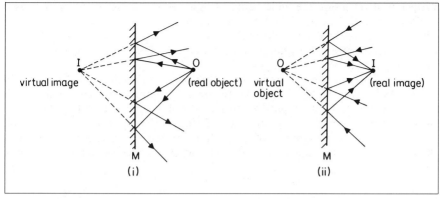

Figure 14.3 *Virtual and real image in plane mirror*

Curved Mirrors, Spherical and Paraboloid

Curved mirrors are widely used as driving mirrors in cars. Make-up and dentists mirrors are curved mirrors. The largest telescope in the world uses an enormous curved mirror to collect light from distant stars. British Telecom use large curved reflectors in suitable parts of the country to transmit and receive radio signals, which are electromagnetic waves like light waves (p. 414).

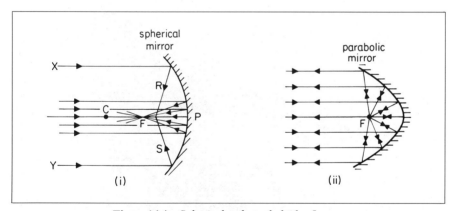

Figure 14.4 *Spherical and paraboloid reflectors*

Figure 14.4 (i) shows a *concave mirror* P. Its surface is part of a sphere of centre C. When a *narrow* parallel beam of rays from a distant object such as the sun is incident on the middle of P, all the rays are reflected to one point or *focus* F.

When a *wide* beam of light XY, parallel to the principal axis, is incident on a concave spherical mirror, reflected rays such as R and S do not pass through a single point, as was the case with a narrow beam. In the same way, if a small lamp is placed at the focus F of a concave spherical mirror, those rays from the lamp which strike the mirror at points well away from the pole P will be reflected in different directions and not as a parallel beam. In this case the reflected beam diminishes in intensity as its distance from the mirror increases.

So a concave spherical mirror is useless as a searchlight mirror. For this reason a mirror whose section is the shape of a parabola (the path of a ball thrown forward into the air) is used in searchlights. A paraboloid mirror has the property of reflecting the wide beam of light from a lamp at its focus F as a perfectly parallel beam. The intensity of the reflected beam is practically undiminished as the distance from the mirror increases, Figure 14.4 (ii). For the same reason, motor headlamp reflectors and those used in torches are paraboloid in shape.

British Telecom use aerials in the shape of a paraboloid dish to send and receive radio signals. See Plate 14A (below). A communications satellite high above the Earth sends a parallel beam of radio signals to all parts of the dish. This is reflected to a receiver at the focus, like light waves. One aerial reflector dish has a diameter of 32 m. It is steered by mechanisms to point directly at the communications satellite and so to receive maximum power from it.

Plate 14A   *British Telecom aerial dish at Madley, Hereford. International radio signals, received from Earth satellites in geostationary orbits above the equator, are reflected by the paraboloid dish to a sensitive receiver at the focus.*

Refraction at Plane Surfaces

Laws of Refraction

When a ray of light AO is incident at O on the plane surface of a glass medium, some of the light is reflected from the surface along OC in accordance with the laws of reflection. The rest of the light travels along a new direction, OB, in the glass, Figure 14.5. The light is said to be 'refracted' on entering the glass. The *angle of refraction, r,* is the angle made by the refracted ray OB with the normal at O.

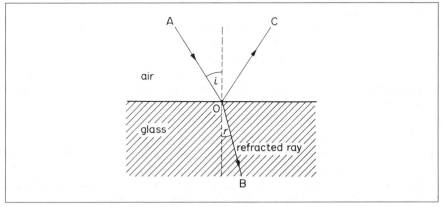

Figure 14.5   *Refraction at plane surface*

SNELL, a Dutch professor, discovered in 1620 that the sines of the angles of incidence and refraction have a constant ratio to each other. The *laws of refraction* are:

---

1   *The incident and refracted rays, and the normal at the point of incidence, all lie in the same plane.*

2   *For two given media,* $\dfrac{\sin i}{\sin r}$ *is a constant, where i is the angle of incidence and r is the angle of refraction* (**Snell's law**).

Refractive Index

The constant ratio $\sin i / \sin r$ is known as the *refractive index*, symbol n, for the two given media. As the value of n depends on the colour of the light used, it is usually given as the value for a particular yellow wavelength emitted from sodium vapour. If the medium containing the incident ray is denoted by 1, and that containing the refracted ray by 2, the refractive index can be denoted by $_1n_2$.

Scientists have drawn up tables of refractive indices when the incident ray is travelling in a vacuum and is then refracted into the medium for example, glass or water. The values obtained are known as the *absolute* refractive indices of the media; and as a vacuum is always the first medium, the subscripts for the absolute refractive index, n, can be dropped. An average value for the magnitude of n for glass is about 1·5, n for water is about 1·33, and n for air at normal pressure is about 1·00028. In fact the refractive index of a medium is only very slightly altered when the incident light is in air instead of a vacuum. So experiments to determine the absolute refractive index n are usually performed with the light incident from air into the medium. We can take $_{air}n_{glass}$ as equal to $_{vacuum}n_{glass}$ for most practical purposes.

Light is refracted because it has different speeds in different media. The Wave Theory of Light, discussed later, shows that the refractive index $_1n_2$ for two given media 1 and 2 is given by

$$_1n_2 = \frac{speed\ of\ light\ in\ medium\ 1\ (c_1)}{speed\ of\ light\ in\ medium\ 2\ (c_2)} \qquad . \qquad . \qquad . \qquad (1)$$

This is a *definition* of refractive index which can be used instead of the ratio $\sin i/\sin r$. An alternative definition of the absolute refractive index, n, of a medium 1 is then

$$n = \frac{\textbf{speed of light in a vacuum, } c}{\textbf{speed of light in medium 1, } c_1} \qquad . \qquad . \qquad . \qquad (2)$$

In practice the velocity of light in air can replace the velocity in a vacuum in this definition.

Relations between Refractive Indices

1   Consider a ray of light, AO, refracted from *glass to air* along the direction OB. The refracted ray OB is bent away from the normal, Figure 14.6. The refractive index from glass to air, $_gn_a$, is given by $\sin x/\sin y$ where x is the angle of incidence in the glass and y is the angle of refraction in the air.

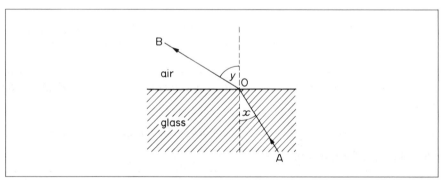

Figure 14.6   *Refraction from glass to air*

From the principle of the reversibility of light (p. 412), it follows that a ray travelling along BO in air is refracted along OA in the glass. The refractive index from air to glass, $_an_g$, is given by $\sin y/\sin x$. But $_gn_a = \sin x/\sin y$.

$$\therefore {}_gn_a = \frac{1}{_an_g} \qquad . \qquad . \qquad . \qquad . \qquad . \qquad (3)$$

If $_an_g$ is 1·5, then $_gn_a = 1/1·5 = 0·67$. Similarly, if the refractive index from air to water is 4/3, the refractive index from water to air is 3/4.

2   Consider a ray AO incident in air on a plane glass boundary, then refracted from the glass into a water medium, and finally emerging along a direction CD into air. *If the boundaries of the media are parallel, the emergent ray CD is parallel*

to the incident ray AO, although there is a relative displacement, Figure 14.7. Thus the angles made with the normals by AO, CD are equal, and we shall denote them by i_a.

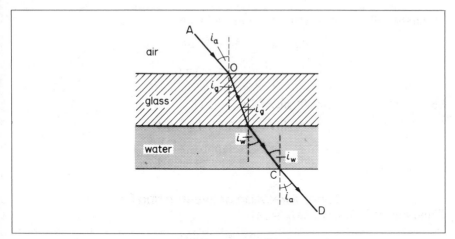

Figure 14.7   *Refraction at parallel plane surfaces*

Suppose i_g, i_w are the angles made with the normals by the respective rays in the glass and water media. Then, by definition, $_gn_w = \sin i_g / \sin i_w$.

But
$$\frac{\sin i_g}{\sin i_w} = \frac{\sin i_g}{\sin i_a} \times \frac{\sin i_a}{\sin i_w}$$

and
$$\frac{\sin i_g}{\sin i_a} = {}_gn_a, \text{ and } \frac{\sin i_a}{\sin i_w} = {}_an_w$$

$$\therefore {}_gn_w = {}_gn_a \times {}_an_w \qquad . \qquad . \qquad . \qquad . \qquad . \qquad \text{(i)}$$

We can derive this relation more simply from the definition of refractive index n in terms of the speed of light (p. 416). Assuming the speed of light in air is practically the same as the speed c in a vacuum, then

$$_gn_a \times {}_an_w = \frac{c_g}{c} \times \frac{c}{c_w} = \frac{c_g}{c_w} = {}_gn_w$$

Also, as $_gn_a = \dfrac{1}{_an_g}$, we can write:

$$_gn_w = \frac{_an_w}{_an_g}$$

Using $_an_w = 1.33$ and $_an_g = 1.5$, it follows that $_gn_w = \dfrac{1.33}{1.5} = 0.89$.

We see from equation (i) above that, for different media 1, 2 and 3,

$$_1n_3 = {}_1n_2 \times {}_2n_3 \qquad . \qquad . \qquad . \qquad . \qquad . \qquad \text{(4)}$$

The order of the suffixes enables this formula to be easily memorised.

Example on Refractive Index

A film of oil, refractive index 1·20, lies on water of refractive index 1·33. A light ray is incident at 30° in the oil on the oil–water boundary.

Calculate the angle of refraction in the water.

Using the suffix o for oil, w for water and a for air,

$$_on_w = {_on_a} \times {_an_w} = (1/{_an_o}) \times {_an_w}$$

$$= \frac{1}{1·20} \times 1·33 = 1·11$$

So

$$\sin 30°/\sin r = {_on_w} = 1·11$$

$$\therefore \sin r = \sin 30°/1·11 = 0·45$$

$$r = 27° \text{ (approx.)}$$

General Relation between *n* and Sin *i*

From Figure 14.7, $\sin i_a/\sin i_g = {_an_g}$

$$\therefore \sin i_a = {_an_g} \sin i_g \quad . \qquad . \qquad . \qquad . \qquad . \qquad \text{(i)}$$

Also, $\sin i_w/\sin i_a = {_wn_a} = 1/{_an_w}$

$$\therefore \sin i_a = {_an_w} \sin i_w . \quad . \qquad . \qquad . \qquad . \qquad \text{(ii)}$$

From (i) and (ii),

$$\sin i_a = {_an_g} \sin i_g = {_an_w} \sin i_w$$

If the equations are re-written in terms of the absolute refractive indices of air (n_a), glass (n_g), and water (n_w), we have

$$n_a \sin i_a = n_g \sin i_g = n_w \sin i_w$$

since $n_a = 1$. This relation shows that when a ray is refracted from one medium to another, *the boundaries being parallel,*

$$\boldsymbol{n \sin i = constant} \quad . \qquad . \qquad . \qquad . \qquad . \qquad \text{(5)}$$

---

where *n* is the absolute refractive index of a medium and *i* is the angle made by the ray with the normal in that medium.

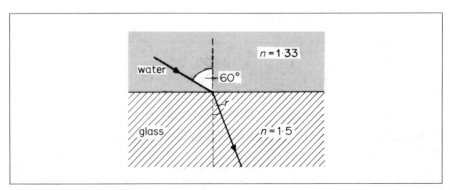

Figure 14.8   *Refraction from water to glass*

This relation, used later in fibre optics, also applies to the case of light passing directly from one medium to another. Suppose a ray is incident on a water–glass boundary at an angle of 60°, Figure 14.8. Then, applying '$n \sin i$ is a constant,' we have

$$1{\cdot}33 \sin 60° \text{ (water)} = 1{\cdot}5 \sin r \text{ (glass)} \qquad . \qquad . \qquad . \qquad \text{(iii)}$$

where r is the angle of refraction in the glass, and $1{\cdot}33$, $1{\cdot}5$ are the respective values of n_w and n_g. So $\sin r = 1{\cdot}33 \sin 60°/1{\cdot}5 = 0{\cdot}7679$, from which $r = 50{\cdot}1°$.

Total Internal Reflection

If a ray AO in glass is incident at a small angle α on a glass–air plane boundary, part of the incident light is reflected along OE in the glass, while the rest of the light is refracted away from the normal at an angle β into the air. The reflected ray OE is weak, but the refracted ray OL is bright, Figure 14.9 (i). This means that most of the incident light energy is transmitted, and only a little is reflected.

When the angle of incidence, α, in the glass is increased, the angle of emergence, β, is increased at the same time. At some angle of incidence C in the glass the refracted ray OL travels along the *glass–air boundary*, making the angle of refraction 90°, Figure 14.9 (ii). The reflected ray OE is still weak in intensity, but as the angle of incidence in the glass is increased slightly the reflected ray suddenly becomes bright, and no refracted ray is seen. Figure 14.9 (iii) shows what happens. Since *all* the incident light energy is now reflected, *total internal reflection* is said to take place in the glass at O.

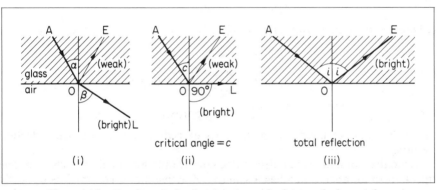

Figure 14.9 *Total internal reflection at a perfectly smooth glass surface*

Critical Angle Values

When the angle of refraction in air is 90°, a critical stage is reached at the point of incidence O. The angle of incidence *in the glass* is known as the *critical angle* for glass and air, Figure 14.9 (ii). Since '$n \sin i$ is a constant' (p. 418), we have

$$n \sin C \text{ (glass)} = 1 \times \sin 90° \text{ (air)}$$

where n is the refractive index of the glass. As $\sin 90° = 1$, then

$$n \sin C = 1$$

or,
$$\sin C = \frac{1}{n} . \qquad . \qquad . \qquad . \qquad . \qquad . \qquad \text{(8)}$$

Crown glass has a refractive index of about 1·51 for yellow light, and thus the critical angle for glass to air is given by $\sin C = 1/1·51 = 0·667$. Consequently $C = 41·5°$. Thus if the incident angle in the glass is greater than C, for example 45°, total internal reflection occurs, Figure 14.9 (iii).

The refractive index of glass for blue light is greater than that for red light (p. 418). Since $\sin C = 1/n$, we see that the critical angle for blue light is *less* than for red light.

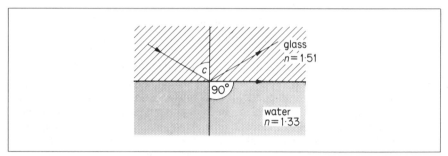

Figure 14.10 *Critical angle for water and glass*

Total reflection may also occur when light in glass ($n_g = 1·51$, say) is incident on a boundary with water ($n_w = 1·33$). Applying '$n \sin i$ is a constant' to the critical case, Figure 14.10, we have

$$n_g \sin C = n_w \sin 90°$$

where C is the critical angle. As $\sin 90° = 1$

$$n_g \sin C = n_w$$

$$\therefore \sin C = \frac{n_w}{n_g} = \frac{1·33}{1·51} = 0·889$$

So $\qquad\qquad\qquad\qquad \therefore C = 63°$ (approx.)

So if the angle of incidence in the glass exceeds 63°, total internal reflection occurs.

Note that total internal reflection can occur only when light travels from one medium to another which has a *smaller* refractive index, i.e. which is optically less dense. It cannot occur when light travels from one medium to another optically denser, for example from air to glass, or from water to glass. In this case a refracted ray is always obtained.

_____ Exercises 14A _____

Refraction at Plane Surface, Critical Angle

1 A ray of light is incident at 60° in air–glass plane surface. Find the angle of refraction in the glass (n for glass = 1·5).

2 A ray of light is incident in water at an angle of 30° on a water–air plane surface. Find the angle of refraction in the air (n for water = 4/3).

3 A ray of light is incident in water at an angle of (i) 30°, (ii) 70° on a water–glass plane surface. Calculate the angle of refraction in the glass in each case ($_a n_g = 1·5$, $_a n_w = 1·33$).

4 Calculate the critical angle for (i) an air–glass surface, (ii) an air–water

surface, (iii) a water–glass surface; draw diagrams in each case illustrating the total reflection of a ray incident on the surface ($_an_g = 1.5$, $_an_w = 1.33$).

5 State the conditions under which total reflection will occur for light entering normally one face of an isosceles right-angle prism of glass of $n = 1.5$ but not in the case when light enters similarly a similar thin hollow prism full of water of $n = 1.33$.

6 Explain the meaning of critical angle and total internal reflection. Describe fully
   (a) one natural phenomenon due to total internal reflection,
   (b) one practical application of it.
   Light from a luminous point on the lower face of a rectangular glass slab, 2·0 cm thick, strikes the upper face and the totally reflected rays outline a circle of 3·2 cm radius on the lower face. What is the refractive index of the glass? (*JMB.*)

7 Figure 14A shows a narrow parallel horizontal beam of monochromatic light from a laser directed towards the point A on a vertical wall. A semicircular glass block G is placed symmetrically across the path of the light and with its straight edge vertical. The path of the light is unchanged.
   The glass block is rotated about the centre, O, of its straight edge and the bright spot where the beam strikes the wall moves down from A to B and then disappears.

$$OA = 1.50 \, m \qquad AB = 1.68 \, m$$

   (a) Account for the disappearance of the spot of light when it reaches B.
   (b) Find the refractive index of the material of the glass block G for light from the laser.
   (c) Explain whether AB would be longer or shorter if a block of glass of higher refractive index was used. (*L.*)

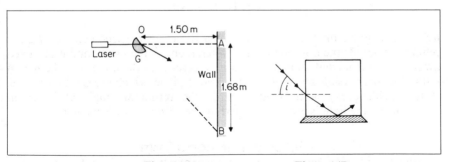

Figure 14A   **Figure 14B**

8 (a) For light travelling in a medium of refractive index n_1 and incident on the boundary with a medium of refractive index n_2, explain what is meant by total internal reflection and state the circumstances in which it occurs.
   (b) A cube of glass of refractive index 1·500 is placed on a horizontal surface separated from the lower face of the cube by a film of liquid, as shown in Figure 14B. A ray of light from outside and in a vertical plane parallel to one face of the cube strikes another vertical face of the cube at an angle of incidence $i = 48° 27'$ and, after refraction, is totally reflected at the critical angle at the glass–liquid interface. Calculate   (i) the critical angle at the glass–liquid interface and   (ii) the angle of emergence of the ray from the cube. (*JMB.*)

Refraction Through Prisms

Glass prisms are used in many optical instruments, for example, prism binoculars. They are also used for separating the colours of the light emitted by glowing objects, which would then give an accurate knowledge of their chemical composition. A prism of glass is used to measure the refractive index of glass very accurately.

The angle between the inclined plane surfaces XDFZ, XDEY is known as the *angle of the prism*, or the *refracting angle*, the line of intersection XD of the planes is known as the *refracting edge*, and any plane in the prism perpendicular to XD, such as PQR, is known as a *principal section* of the prism, Figure 14.11. A ray of

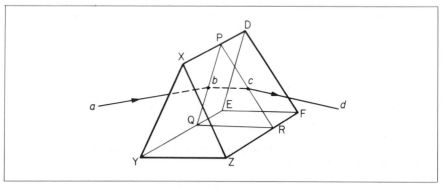

Figure 14.11   *Prism*

light *ab*, incident on the prism at *b* in a direction perpendicular to XD, is refracted towards the normal along *bc* when it enters the prism, and is refracted away from the normal along *cd* when it emerges into the air. From the law of refraction, the rays *ab*, *bc*, *cd* all lie in the same plane, which is PQR in this case. If the incident ray is directed towards the refracting angle at X, as in Figure 14.11, the light is always deviated by the prism towards its base.

Refraction through a Prism

Consider a ray HM incident in air on a prism of refracting angle A, and suppose the ray lies in the principal section PQR, Figure 14.12. Then, if i_1, r_1 and i_2, r_2

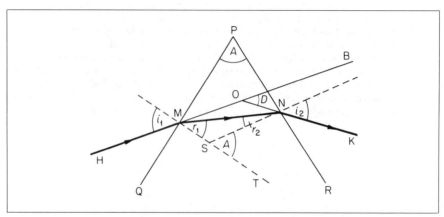

Figure 14.12   *Refraction through prism*

are the angles of incidence and refraction at M, N as shown, and n is the prism refractive index,

$$\sin i_1 = n \sin r_1 \quad . \quad . \quad . \quad . \quad . \quad \text{(i)}$$

$$\sin i_2 = n \sin r_2 \quad . \quad . \quad . \quad . \quad . \quad \text{(ii)}$$

Further, as MS and NS are normals to PM and PN respectively, angle MPN + angle MSN = 180°, considering the quadrilateral PMSN. But angle NST + angle MSN = 180°.

$$\therefore \text{angle NST} = \text{angle MPN} = A$$

$$\therefore A = r_1 + r_2 \quad . \quad . \quad . \quad . \quad . \quad \text{(iii)}$$

as angle NST is the exterior angle of triangle MSN. Memorise equation (iii) as this is often needed in prism refraction.

In Figure 14.12, D is the *angle of deviation* of the ray HM produced by the prism. The angle of deviation at M = angle OMN = $i_1 - r_1$; the angle of deviation at N = angle MNO = $i_2 - r_2$. Since the deviations at M, N are in the same direction, the total deviation, D (angle BOK), is

$$D = (i_1 - r_1) + (i_2 - r_2) \quad . \quad . \quad . \quad . \quad \text{(iv)}$$

Equations (i)–(iv) are the general relations which hold for refraction through a prism.

We can now illustrate refraction through a prism by examples.

Examples on Refraction through a Prism

1 A glass prism with a refracting angle A has a refractive index 1·6. A ray PO is incident normally on one side and strikes the other side at Q, Figure 14.13 (i).

Calculate the least value of A for the ray not to be refracted at Q into the air.

For ray at Q not to be refracted into air, angle of incidence in glass = C, critical angle.

Now
$$\sin C = \frac{1}{n} = \frac{1}{1·6} = 0·625$$

So
$$C = 39° \text{ (approx.)}$$

Now, by geometry, $A = C$ in this special case

So
$$\text{minimum value of } A = 39°$$

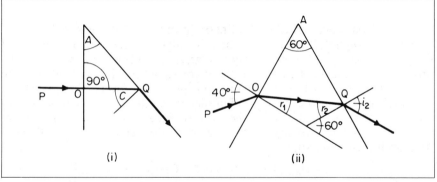

Figure 14.13   *Refraction through prism*

2 A ray PO is incident at 40° at O on a glass prism of angle $A = 60°$ and refractive index $n = 1.5$, Figure 14.13 (ii).

Calculate the angle i_2 at which the ray comes out at Q into the air and the deviation of the ray PO produced by the prism.

The angle of refraction r_1 at O in the glass is given by

$$\frac{\sin 40°}{\sin r_1} = 1.5$$

So
$$\sin r_1 = \frac{\sin 40°}{1.5} = 0.429$$

$$\therefore r_1 = 25° \text{ (approx.)} \qquad . \qquad . \qquad . \qquad . \qquad . \qquad (1)$$

At Q, the angle of incidence r_2 in the glass is found from $r_1 + r_2 = A$.

So
$$r_2 = A - r_1 = 60° - 25° = 35°$$

At Q,
$$\frac{\sin i_2}{\sin 35°} = 1.5$$

So
$$\sin i_2 = 1.5 \times \sin 35° = 0.860$$

$$\therefore i_2 = 59°$$

Also, deviation of PO = deviation at O$(40° - r_1)$ + deviation at Q$(i_2 - r_2)$

$$= (40° - 25°) + (59° - 35°)$$

$$= 39°$$

3 A glass prism has a refractive index $n = 1.5$. What is the largest angle A of the prism if light incident at 90° on one face *just* emerges (comes out) at the other face into the air after refraction through the prism.

The light just comes out at the other face if the angle of refraction in the air $= 90°$.

So the angles in the glass on *both* sides $= C$, the critical angle value. Since $\sin C = 1/n = 1/1.5 = 0.6667$, $C = 42°$ (approx.).

So
$$A = r_1 + r_2 = C + C = 2C = 84°$$

If the prism angle is greater than 84°, no light can be refracted through the prism.

Maximum Deviation by Prism, Angle of Incidence 90°

Maximum deviation, D_{max}, occurs when the angle of incidence on the face of the prism is 90°. In this case the ray has 'grazing incidence' on the face of the prism, as shown in Figure 14.14 (i), and it comes out making an angle i to the normal at the second face. From the principle of the reversibility of light, it follows that a ray making an angle of incidence i on the face of the prism comes out making an angle of 90° to the normal, Figure 14.14 (ii). So maximum deviation occurs for two angles of incidence on the face of a prism, 90° and i. Also, from Figure 14.14 (i),

$$D_{max} = d_1 + d_2 = (90° - C) + (i - r)$$

since the angle in the glass is the critical angle C when the angle in the air is 90°.

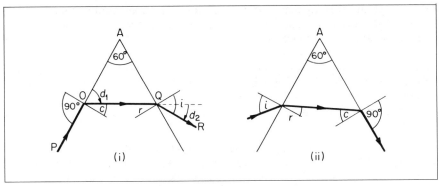

Figure 14.14 *Maximum deviation*

Example on Refraction through Prism for Incident Angle 90°

Calculate the angle of emergence i, and the deviation, when light is incident at 90° on the face of a 60° prism of refractive index 1·5.

In Figure 14.14 (i), the incident ray PO is refracted at the critical angle C along OQ in the glass. From $\sin C = 1/n = 1/1·5 = 0·6667$, then $C = 41·8°$.

Since $A = 60° = C + r$, where r is the angle of incidence at Q, then $r = 60° - 41·8° = 18·2°$. From $\sin i/\sin r = n$, the angle of emergence i is given by

$$\frac{\sin i}{\sin 18·2°} = 1·5$$

or $$\sin i = 1·5 \times \sin 18·2° = 0·4685$$

So $$i = 27·9°$$

Deviation, $$D_{max} = d_1 + d_2 = 90° - C + i - r$$
$$= 90° - 41·8° + 27·9° - 18·2°$$
$$= 57·9°$$

Minimum Deviation

Experiment shows that as the angle of incidence i on a glass prism is increased from zero, the deviation D begins to decrease continuously to some *minimum*

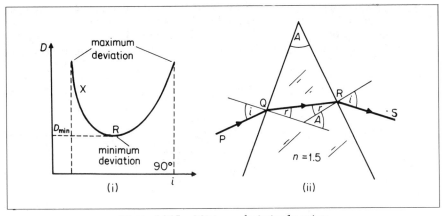

Figure 14.15 *Minimum deviation by prism*

value D_{min}, and then increases to a maximum as i is increased further to 90°. A graph of D plotted against i has the appearance of the curve X, which has a minimum value at R, Figure 14.15 (i).

Experiment and theory show that *the minimum deviation, D_{min}, of the light occurs when the ray passes symmetrically through the prism*. Suppose the ray is PQRS in Figure 14.15 (ii). Then the incident angle, i, at Q is equal to the angle of emergence, i, into the air at R for this special case.

An example will illustrate how to find the minimum deviation.

Example on Minimum Deviation

In Figure 14.15 (ii), the glass prism has an angle $A = 60°$ and a refractive index $n = 1.5$. Calculate the angle of incidence i for minimum deviation, and the value of the minimum deviation, assuming the ray passes symmetrically through the prism in this case.

Since the angle i is the same at Q and R, the angles r in the glass at Q and R are the same.

So
$$r + r = A = 60°$$

$$\therefore r = \frac{60°}{2} = 30°$$

At Q,
$$\frac{\sin i}{\sin 30°} = n = 1.5$$

So
$$\sin i = 1.5 \times \sin 30° = 0.75$$

$$\therefore i = 49° \text{ (approx.)}$$

Then
$$\begin{aligned} \text{minimum deviation} &= \text{deviation at Q} + \text{deviation at R} \\ &= (49° - 30°) \quad + (49° - 30°) \\ &= 38° \end{aligned}$$

Dispersion

White light has a band of wavelengths of different colours. This is called the *spectrum* of white light. The longest wavelength is red light, which has a wavelength in air of about 700 nm (700×10^{-9} m or 0.7μm). The shortest wavelength is violet, which has a wavelength in air of about 450 nm (450×10^{-9} m or 0.45μm).

In a vacuum (and practically in air), all the colours travel at the same speed. In a medium such as glass, however, the colours travel at different speeds—red has the fastest speed and violet the slowest. On the wave theory, refraction is due to the change in speed of light when it enters a different medium. So when a ray AO of white light is incident at O on a glass prism, the colours are refracted in

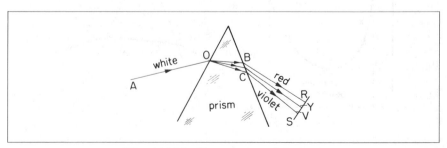

Figure 14.16 *Dispersion in a glass prism*

different directions such as OBR and OCS, Figure 14.16. The glass prism has therefore separated or *dispersed* the white light into its various colours or wavelengths, as Newton first discovered in 1666. After leaving the glass a band or spread of impure colours are formed on a white screen S. The spectrum of white light consists of (bands of) red, orange, yellow, green, blue, indigo and violet. The separation of the colours by the prism is known as *dispersion*. As we see shortly, the wavelengths in a light signal produced in telecommunications is dispersed when it travels along a glass optical fibre.

The sun and the hot tungsten filament of a lamp have a continuous spectrum of visible wavelengths. Hot gases such as hydrogen and krypton have visible wavelengths which form a *line spectrum*. The light from a laser has practically one wavelength, so it is a *monochromatic* light source. The topic of Spectra is discussed more fully in a later chapter.

The Spectrometer

The spectrometer is an optical instrument which is mainly used to study the light from different sources. Using dispersion by a glass prism, it can be used to investigate the different wavelengths from a light source. It can measure accurately the deviation of light by a prism and the refractive index of the glass prism. It can also measure accurately the wavelength of light using a diffraction grating (p. 552).

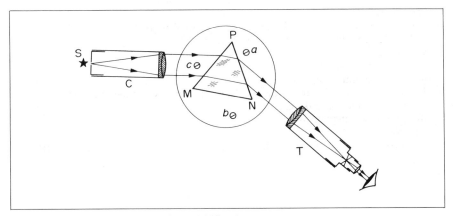

Figure 14.17 *Spectrometer*

The instrument consists essentially of a *collimator*, C, a *telescope*, T, and a *table*, on which a prism PMN, for example, can be placed. The lenses in C, T are achromatic lenses (p. 461). The collimator is fixed, but the table and the telescope can be rotated round a circular scale graduated in half-degrees (not shown) which has a common vertical axis with the table, Figure 14.17. A vernier is also provided for this scale. The *source* of *light*, S, used in the experiment is placed in front of a narrow slit at one end of the collimator, so that the prism is illuminated by light from S.

Before the spectrometer can be used, however, three adjustments must be made: (1) The collimator C must be adjusted so that parallel light emerges from it; (2) the telescope T must be adjusted so that parallel rays entering it are brought to a focus at cross-wires near its eyepiece; (3) the refracting edge of the prism must be parallel to the axis of rotation of the telescope, that is, the table must be 'levelled' using the screws *a*, *b*, *c*.

Details of experiments with a spectrometer will be found in *Advanced Level Practical Physics* by Nelkon and Ogborn (Heinemann).

_____ Exercises 14B _____

Refraction through Prisms

1 A ray of light is refracted through a prism of angle 70°. If the angle of refraction in the glass at the first face is 28°, what is the angle of incidence in the glass at the second face?

2 A prism of glass of refractive index 1·63 has an angle A between two of its faces. If a ray of light is incident normally on one face of the prism, for what range of values of A will the ray emerge from the second face? (*C.*)

3 A narrow beam of light is incident normally on one face of an equilateral prism (refractive index 1·45) and finally emerges from the prism. The prism is now surrounded by water (refractive index 1·33). What is the angle between the directions of the emergent beam in the two cases? (*L.*)

4 A is the vertex of a triangular glass prism, the angle at A being 30°. A ray of light OP is incident at P on one of the faces enclosing the angle A, in a direction such that the angle $OPA = 40°$. Show that, if the refractive index of the glass is 1·50, the ray cannot emerge from the second face.

5 The refracting angle of a prism is 62·0° and the refractive index of the glass for yellow light is 1·65°. Find the smallest possible angle of incidence of a ray of this yellow light which is transmitted without total internal reflection. Explain what happens if white light is used instead, and the angle of incidence is varied about this minimum.

6 A ray passes symmetrically through a glass prism of angle 60° and refractive index 1·6, when the deviation is a minimum.
Calculate
(a) the angle of incidence,
(b) the minimum deviation.

7 Draw a sketch showing the dispersion by a glass prism when the source is
(a) a hot gas such as hydrogen,
(b) a hot tungsten filament,
(c) the sun.
   The sun's spectrum is crossed by fine dark lines. What is the cause of the lines?

Optical Fibres in Communications

Monomode and Multimode Fibres

As we shall see shortly, light signals can travel along very fine long glass fibres roughly the same diameter as a human hair. *Optical fibres*, as they are called, are replacing the copper cables previously used in telecommunications. The fibre is a very fine glass rod of diameter about 125 µm (125×10^{-6} m). After manufacture it has a central glass *core* surrounded by a glass coating or *cladding* of smaller refractive index than the core, Plate 14B.

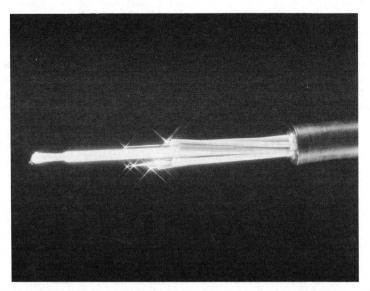

Figure 14B Optical fibres *are being used in telephone and other transmitting cables by British Telecom in a new network. The fibres are hair thin strands of specially coated glass. They can transmit a laser or other light beam from one end to the other as a result of repeated total internal reflections at the glass boundary, even if the fibre is bent or twisted. Each fibre can carry as many as 2000 telephone conversations, with less signal loss than in conventional telephone cables.* (By courtesy of The Post Office).

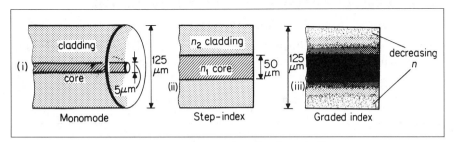

Figure 14.18 *Monomode and multimode fibres*

The fibres are classified into two main types.

(a) The *monomode* fibre has a very narrow core of diameter about 5 µm (5×10^{-6} m) or less, so the cladding is relatively big, Figure 14.18 (i).

(b) The *multimode* fibre has a core of relatively large diameter such as 50 µm. In one form of multimode fibre the core has a constant refractive index n_1

such as 1·52 from its centre to the boundary with the cladding, Figure 14.18 (ii). The refractive index then changes to a lower value n_2 such as 1·48 which remains constant throughout the cladding. This is called a *step-index* multimode fibre, in the sense that the refractive index 'steps' from 1·52 to 1·48 at the boundary with the cladding.

As we discuss later, to transmit light signals more efficiently a multimode fibre is made whose refractive index decreases smoothly from the middle to the outer surface of the fibre, Figure 14.18 (iii). There is now no noticeable boundary between the core and the cladding. This is called a *graded index* multimode fibre.

Optical Paths in Fibres

We shall now see what happens when a light signal enters one end of an optical fibre.

Figure 14.19 shows a step-index fibre. With a large angle of incidence, a ray OA entering one end at O is refracted into the core along OP and then refracted along PQ in the cladding. At Q, the fibre surface, the ray passes into the air. In this case only a very small amount of light, due to reflection, passes along the fibre.

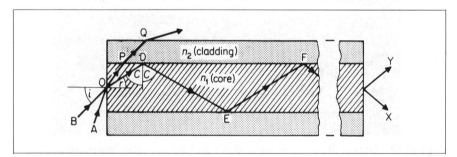

Figure 14.19 *Light path by total internal reflection—multiple reflections*

With a smaller angle of incidence, however, a ray such as BO is refracted in the core along OD and meets the boundary between the core and cladding *at their critical angle, C*. In this case, since $n \sin i$ is constant (p. 418),

$$n_1 \sin C = n_2 \sin 90° = n_2$$

where n_1 is the core refractive index and n_2 the slightly smaller cladding refractive index. If $n_1 = 1·52$ and $n_2 = 1·48$, it follows that

$$\sin C = \frac{n_2}{n_1} = \frac{1·48}{1·52} = 0·974$$

and so $$C = 77° \text{ (approx.)}$$

The ray OD is now *totally* reflected at D along DE, where it again meets the core-cladding boundary at the critical angle. At E, therefore, it is totally reflected along EF.

In this way, by total reflection, a ray of light entering one end of a fibre can travel along the fibre by *multiple reflections* with a fairly high light intensity. At the other end of the fibre the ray emerges in a direction X (odd number of multiple reflections) or a direction Y (even number of multiple reflections).

Maximum Angle of Incidence

The maximum angle of incidence in air for which *all* the light is totally reflected at the core-cladding fibre is the angle i in Figure 14.19. To calculate i, we have

$$1 \times \sin i = n_1 \sin r \text{ (refraction from air to core)} \qquad . \qquad . \qquad (1)$$

and $\qquad n_1 \sin C = n_2 \sin 90° = n_2$ (refraction from core to cladding) $\quad . \qquad (2)$

Also, $\qquad\qquad\qquad\qquad r = 90° - C$, so $\sin r = \cos C$

From (1), $\cos C = \sin i/n_1$; from (2), $\sin C = n_2/n_1$
   Using the trigonometrical relation $\sin^2 C + \cos^2 C = 1$, then

$$\frac{n_2^2}{n_1^2} + \frac{\sin^2 i}{n_1^2} = 1$$

Simplifying,

$$\sin i = \pm\sqrt{n_1^2 - n_2^2}$$

With $n_1 = 1.52$ and $n_2 = 1.48$, calculation shows that $i = 20°$ (approx.). So an incident beam from air, making an angle of incidence not more than $20°$ will be transmitted along the fibre with appreciable intensity.

Losses of Power, Dispersion

When a light signal travels along fibres by multiple reflections, some light is absorbed due to impurities in the glass. Some is scattered at groups of atoms which collect together at places such as joints when fibres are joined together. Careful manufacture can reduce the power loss by absorption and scattering.

   The information received at the other end of a fibre can be in error due to *dispersion* or spreading of the light signal. No light signal is perfectly mono-chromatic. A narrow band of wavelengths is present in the spectrum of the light signal. As we saw when we considered dispersion in a glass prism (p. 426), the various wavelengths are refracted in different directions when the light signal enters the glass fibre and the light spreads.

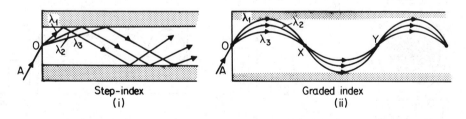

Figure 14.20 *Light paths in step-index and graded index fibres*

Figure 14.20 (i) shows the light paths followed by three wavelengths λ_1, λ_2 and λ_3. λ_1 meets the core-cladding at the critical angle and λ_2 and λ_3 at slightly greater angles. All the rays travel along the fibre by multiple reflections as previously explained. But the light paths have different lengths. So the wave-lengths reach the other end of the fibre *at different times*. The signal received is therefore faulty or distorted.

   Figure 14.20 (i) shows a step-index fibre. Its disadvantage can be considerably reduced by using a *graded index* fibre (p. 430). Figure 14.20 (ii) shows roughly what happens in this case. Each wavelength still takes a different path and at some layer in the glass, different for each, the rays are totally reflected. But unlike the step index fibre, all the rays come to a focus at X as shown, and then again at Y, and so on. We can see this is possible because the speed is inversely-proportional to the refractive index (speed $= c/n$). So the wavelength λ_1 travels a longer path than λ_2 or λ_3 but at a greater speed. Fermat's principle (p. 412) states that light takes the minimum (or maximum) time to travel between points such as O and X, so the time of travel is the same whichever path is taken.

In spite of the different dispersion, then, all the wavelengths arrive at the other end of the fibre at the same time. With a step index fibre, the overall time difference may be about 33 ns (33×10^{-9} s) per km length of fibre. Using a graded index fibre, the time difference is reduced to about 1 ns per km.

Figure 14.21 shows diagrammatically the monomode and multimode fibres, the light paths through them and their effect on an input light pulse, where I is the light intensity and t is the time.

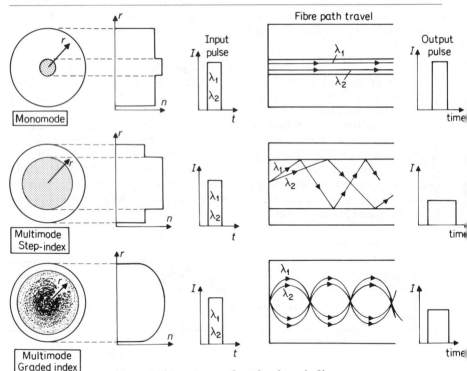

Figure 14.21 *Monomode and multimode fibres*

Light Signal Transmission, Conversion to Sound

A gallium phosphide (GaP) light-emitting diode (LED) can be used as a light source with a graded index fibre. Its light intensity is weak, however, and the absorption and scattering in a long fibre makes this an unsatisfactory source in practice.

A gallium-arsenide (GaAs) semiconductor laser is a much better light source than the LED, though it is more expensive. It has a relatively high light intensity and a much narrower band of wavelengths about a mean value such as 1·3 μm, which reduces dispersion problems. With a laser light source, British Telecom prefer to use a *monomode* fibre for long distance transmission. The monomode fibre with a very thin core of diameter 5 μm or less can now be manufactured with precision. Using a narrow band of wavelengths of mean value 1·3 μm, the light travels straight along the core with only one mode or path. See Figure 14.21.

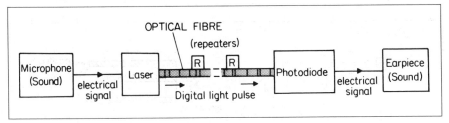

Figure 14.22 *Transmission and reception of sound information* (diagrammatic)

Figure 14.22 shows in block form how sound energy is transmitted along a fibre and reconverted at the other end to sound energy. Sound information such as speech is converted to an electrical audio signal by a microphone and this is made to modulate light from a laser. The information is then carried along the fibre as a train of light pulses in digital form. Signal losses occur by absorption and scattering, so amplifiers or boosters called *repeaters* (R) are placed at places along the fibre cable. Between Nottingham and Sheffield, repeaters are placed every 50 km of cable. Copper cables would need many more amplifiers per 50 km length than optical fibres due to greater power losses.

At the other (receiving) end of the cable, a photodiode converts the digital light pulses into a corresponding electrical signal. Technical problems of noise carried by the incoming signal are overcome by using special types of transistors and the audio currents are reconverted to sound in the earpiece of the listener in the telephone system. By a system called *time division multiplexing,* many thousands of telephone calls can be transmitted along one optical fibre by using light pulses in digital form.

Further telecommunication details are outside the scope of this book and should be obtained from specialist books on telecommunications.

_____ **Exercises 14C** _____

Optical Fibres

1 (a) Explain in terms of a wave model how a beam of light is refracted as it crosses an interface between two transparent media. Hence derive Snell's law of refraction in terms of the speeds of light in the media. (*see Chap. 18.*)

   (b) Describe the phenomenon of total internal reflection and explain what is meant by the critical angle. How is the critical angle related to the speeds of light in the media involved?

   (c) A portion of a straight glass rod of diameter d and refractive index n is bent into an arc of a circle of mean radius R, and a parallel beam of light is shone down it, as shown in Figure 14C. (i) Derive an expression in terms of R and d for the angle of incidence i of the central ray C on reaching the glass-to-air surface at the circular arc. (ii) Show that the smallest value of R which will allow *all* the light to pass around the arc is given by

$$R = \frac{d(n+1)}{2(n-1)}$$

(iii) Use this result to explain why glass fibres, rather than rods, are used to carry optical signals around sharp corners.

(d) A glass fibre of refractive index 1·5 and diameter 0·50 mm is bent into a semi-circular arc of mean radius 4·0 mm, and a beam of light is shone along it. (i) Show that no light escapes from the sides of the fibre. (ii) Show by a suitable calculation that if the fibre is immersed in oil of refractive index 1·4 some light will escape. (iii) Suggest an application for such a device. (*O.*)

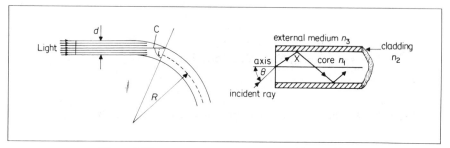

Figure 14C   **Figure 14D**

2 What do you understand by *angle of refraction* and *refractive index*?

A ray of light crosses the interface between two transparent media of refractive indices n_A and n_B. Give a formula relating the directions of the ray on the two sides of the interface. Show the angles you use in your formula on a diagram. Hence, deduce the conditions necessary for total internal reflection to take place at an interface.

In the simple 'light pipe' shown in Figure 14D, a ray of light may be transmitted (with little loss) along the core by repeated internal reflection. The diagram shows a cross-section through the diameter of the 'pipe' with a ray incident in that plane. The core, cladding and external medium have refractive indices n_1, n_2 and n_3 respectively. Show that total internal reflection takes place at X provided that the angle θ is smaller than a value θ_m given by the expression $\sin \theta_m = \sqrt{(n_1{}^2 - n_2{}^2)}/n_3$. Explain why the pipe does not work for rays for which $\theta > \theta_m$. (Reminder: $\sin^2 \theta + \cos^2 \theta = 1$.) (*C.*)

3 Draw sketches showing the different light paths through a monomode and a multimode fibre. Why is the monomode fibre preferred in telecommunications?

4 The refractive index of the core and cladding of an optical fibre are 1·6 and 1·4 respectively. Calculate
  (a) the critical angle at the interface,
  (b) the (maximum) angle of incidence in the air of a ray which enters the fibre and is then incident at the critical angle on the interface.

5 A short pulse of white light is sent out at one end of an optical fibre 4 km long. (i) Calculate the time interval between the red and blue light emerging at the other end, given the speed of light in air is 3×10^8 m s^{-1} and the refractive index of blue and red light are respectively 1·53 and 1·50. (ii) If the pulse of white light is a train of short duration square waves of intensity against time, draw a labelled sketch of the pulse arriving at the other end of the fibre. With a telecommunication optical fibre, how is this disadvantage overcome? (iii) What is the frequency of the white light pulses at one end if the red and blue pulses at the other end are just separated?

15
Lenses and Optical Instruments

In this chapter we deal first with refraction through converging and diverging lenses and the different images obtained. We then apply the lens equation to calculate image positions and magnification. Next, we consider the astronomical telescope in normal adjustment and show that its magnifying power is the ratio of the focal lengths of its two lenses. The eye-ring and resolving power then follow, and the radio telescope is discussed. The chapter concludes with the simple microscope and its magnifying power.

Converging and Diverging Lenses

A *lens* is an object, usually made of glass, bounded by one or two spherical surfaces. Figure 15.1 (i) illustrates three types of *converging* lenses, which are thicker in the middle than at the edges. Figure 15.1 (ii) shows three types of *diverging* lenses, which are thinner in the middle than at the edges.

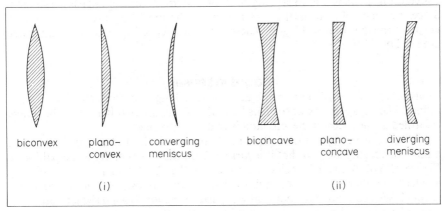

| biconvex | plano-convex | converging meniscus | biconcave | plano-concave | diverging meniscus |

(i)   (ii)

Figure 15.1   (i) *Converging lens*   (ii) *Diverging lens*

The *principal axis* of a lens is the line joining the centres of curvature of the two surfaces, and passes through the middle of the lens. Experiments with a ray-box show that a thin converging lens brings an incident parallel beam of rays to a *principal focus*, F, on the other side of the lens when the beam is narrow and incident close to the principal axis, Figure 15.2 (i). On account of the convergent beam contained with it, the lens is better described as a 'converging' lens. If a similar parallel beam is incident on the other (right) side of the lens, it converges to a focus F', which is at the same distance from the lens as F when the lens is thin. To distinguish F from F' the latter is called the 'first principal focus'; F is known as the 'second principal focus'.

When a narrow parallel beam, close to the principal axis, is incident on a thin diverging lens, experiment shows that a beam is obtained which appears to diverge from a point F on the same side as the incident beam, Figure 15.2 (ii). F is known as the principal 'focus' of the diverging lens.

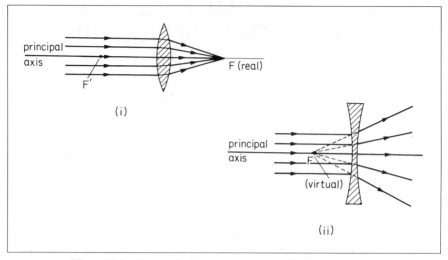

Figure 15.2 *Focus of (i) converging, and (ii) diverging lenses*

Signs of Focal Length, *f*

From Figure 15.2 (i), it can be seen that a converging lens has a real focus. By convention (p. 438), the focal length, *f*, of a *converging* lens is *positive* in sign. Since the focus of a diverging lens is virtual, the focal length of such a lens is negative in sign, Figure 15.2 (ii). These signs are needed when optical formulae are used (p. 438).

Images in Lenses

Converging lens. (i) When an object is a very long way from this lens, i.e., at infinity, the rays arriving at the lens from the object are parallel. Thus the image is formed at the focus of the lens, and is real and inverted.

(ii) Suppose an object OP is placed at O perpendicular to the principal axis of a thin converging lens, so that it is *farther* from the lens than its principal focus, Figure 15.3 (i). A ray PC incident on the middle, C, of the lens is very slightly displaced by its refraction through the lens, as the opposite surfaces near C are parallel. We therefore consider that PC passes *straight through* the lens, and this is true for any ray incident on the middle of a thin lens.

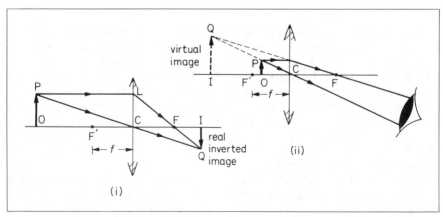

Figure 15.3 *Images in converging lenses*

A ray PL parallel to the principal axis is refracted so that it passes through the focus F. Thus the image, Q, of the top point P of the object is formed below the principal axis, and hence the whole image IQ is real and inverted. In making accurate drawings the lens should be represented by a straight line, as illustrated in Figure 15.3, as we are only concerned with thin lenses and a narrow beam incident close to the principal axis.

(iii) *Image same size as object.* When an object is placed at a distance $2f$ from the lens, the real inverted image has the same size as the object and is also a distance $2f$ from the lens on the other side. So if a converging lens has a focal length 10 cm, an object 20 cm $(2f)$ from the lens forms an image of the same size at 20 cm from the lens on the other side.

If the object is *further* than 20 cm from the lens, the real inverted image moves nearer the lens and becomes *smaller* than the object. If the object is nearer than 20 cm but greater than 10 cm (f) from the lens, the image moves back and becomes bigger than the object.

(iv) The least distance between an object and a real image formed by a lens is $4f$ $(2f + 2f)$. To form a real image on a screen, the distance between the object and the screen must be at least $4f$. So if a lens has a focal length 10 cm, and a screen is placed 30 cm (less than 4×10 cm) from the object, the lens can not form an image on the screen.

(v) The image formed by a converging lens is always real and inverted until the object is placed *nearer* the lens than its focal length, Figure 15.3 (ii). In this case the rays from the top point P *diverge* after refraction through the lens, and hence the image Q is *virtual*. The whole image, IQ, is erect (the same way up as the object) and magnified, besides being virtual, and hence the converging lens can be used as a simple 'magnifying glass' (see p, 450).

Images in Converging lens:
   When the object is
   1   **at distance $2f$ from lens, image is real, inverted and same size as object.**
   2   **Between $2f$ and f, image is real, inverted and bigger than object.**
   3   **Further than $2f$, the image is real, inverted and smaller than object.**
   4   **Nearer than f, image is upright, magnified and virtual (magnifying glass).**

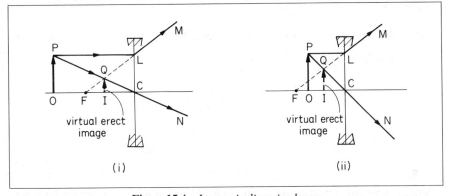

Figure 15.4   *Images in diverging lenses*

Diverging lens. In the case of a converging lens, the image is sometimes real and sometimes virtual. In a diverging lens, the image is always virtual; in addition, the image is always erect and diminished. Figure 15.4 (i), (ii) illustrate the formation of two images. A ray PL appears to diverge from the focus F after refraction through the lens, a ray PC passes straight through the middle of the lens and emerges along CN, and hence the emergent beam from P appears to diverge from Q on the same side of the lens as the object. The image IQ is thus *virtual*.

Lens Equation and Magnification Formula

Provided a sign rule is used for the distances, the equation

$$\frac{1}{v} + \frac{1}{u} = \frac{1}{f}. \tag{1}$$

is the relation between the object distance u from the lens, the image distance v and the focal length f. The 'Real is Positive' sign rule, which we shall use, is: (1) give a *plus*($+$) sign for *real* object and image distances, (2) give a *minus*($-$) sign for *virtual* object and image distances.

The sign rule also applies to focal lengths. A converging lens has a real focus. So $f = +10$ cm for a converging lens of focal length 10 cm. A diverging lens has a virtual focus. So $f = -20$ cm for a diverging lens of focal length 20 cm.

The linear (transverse) magnification m produced by a lens is defined as the ratio *height of image/height of object*. Numerically,

$$m = \frac{v}{u}. \tag{2}$$

Applications of Lens Equation and Magnification Formula

The following examples illustrate how to apply the lens equation $1/v + 1/u = 1/f$ and the magnification formula $m = v/u$. The case of a virtual object should be carefully noted.

Examples on Lenses

1 *Converging lens. Real object*
An object is placed 12 cm from a converging lens of focal length 18 cm. Find the position of the image.

Since the lens is converging, $f = +18$ cm. The object is real, and therefore $u = +12$ cm. Substituting in $\dfrac{1}{v} + \dfrac{1}{u} = \dfrac{1}{f}$,

$$\therefore \frac{1}{v} + \frac{1}{(+12)} = \frac{1}{(+18)}$$

$$\therefore \frac{1}{v} = \frac{1}{18} - \frac{1}{12} = -\frac{1}{36}$$

$$\therefore v = -36 \text{ cm}$$

Since v is negative in sign the image is *virtual*, and it is 36 cm from the lens. See Figure 15.3 (ii).

The magnification,

$$m = \frac{v}{u} = \frac{-36}{12} = -3$$

So the object is magnified 3 times and the minus shows it is upright (magnifying glass).

2 *Converging lens. Virtual object*

A beam of light, converging to a point 10 cm behind a converging lens, is incident on the lens. Find the position of the point image if the lens has a focal length of 40 cm.

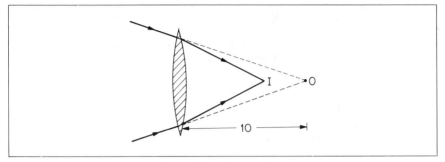

Figure 15.5 *Virtual object*

If the incident beam converges to the point O, then O is a *virtual object*, Figure 15.5. Thus $u = -10$ cm. Also, $f = +40$ cm, since the lens is converging. Substituting in $\dfrac{1}{v} + \dfrac{1}{u} = \dfrac{1}{f}$,

$$\frac{1}{v} + \frac{1}{(-10)} = \frac{1}{(+40)}$$

$$\therefore \frac{1}{v} = \frac{1}{40} + \frac{1}{10} = \frac{5}{40}$$

$$\therefore v = \frac{40}{5} = 8$$

Since v is positive in sign the image is *real*, and it is 8 cm from the lens. The image is I in Figure 15.5.

If the beam of light formed an object of finite size at O, and a real image of this object at I, then

$$\text{magnification} = \frac{v}{u} = \frac{8}{-10} = -0{\cdot}8$$

So the image is smaller than the object.

Diverging lens. Suppose a beam converges to a point 10 cm behind a diverging lens of focal length 40 cm, so $f = -40$ cm. Then $u = -10$ cm (virtual object). So

$$1/v + 1/(-10) = 1/(-40)$$

Solving, $v = 40/3 = 13{\cdot}3$ cm. So a real image is formed 13·3 cm behind the lens.

3 A luminous object and a screen are placed on an optical bench and a converging lens is placed between them to throw a sharp image of the object on the screen; the linear magnification of the image is found to be 2·5. The lens is now moved 30 cm nearer the screen and a sharp image again formed. Calculate the focal length of the lens.

If O, I are the object and screen positions respectively, Figure 15.6, and L_1, L_2 are the two positions of the lens, then $OL_1 = IL_2$, because the u and v values are interchanged. Suppose $OL_1 = x = L_2I$.

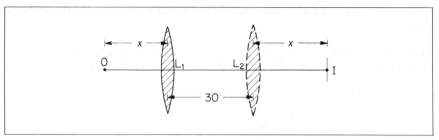

Figure 15.6 *Example*

For the lens in the position L_1, $u = OL_1 = x$, and $v = L_1I = 30 + x$.

But magnification, $m = \dfrac{v}{u} = 2\cdot5$

So
$$\frac{30 + x}{x} = 2\cdot5$$

$$\therefore x = 20\,\text{cm}$$

$$\therefore u = OL_1 = 20\,\text{cm}$$

$$v = L_1I = 30 + x = 50\,\text{cm}$$

Substituting in $\dfrac{1}{v} + \dfrac{1}{u} = \dfrac{1}{f}$,

$$\therefore \frac{1}{20} + \frac{1}{50} = \frac{1}{f}$$

from which
$$f = 14\cdot3\,\text{cm}$$

4 A slide projector has a converging lens of focal length 20·0 cm and is used to magnify the area of a slide, 5 cm², to an area of 0·8 m² on a screen.

Calculate the distance of the slide from the projector lens.

The ratio *area* of image/*area* of object $= 0\cdot8\,\text{m}^2/5\,\text{cm}^2$

$$= \frac{8000\,\text{cm}^2}{5\,\text{cm}^2} = 1600$$

So linear magnification $m =$ square root of area ratio $= 40$

$$\therefore \frac{v}{u} = 40, \text{ and } v = 40\,u$$

From the lens equation $\dfrac{1}{v} + \dfrac{1}{u} = \dfrac{1}{f}$, since u and v are both real and $+$ve,

$$\frac{1}{40\,u} + \frac{1}{u} = \frac{1}{+20}$$

Solving,
$$u = \frac{41}{2} = 20\cdot5\,\text{cm}$$

_____ Exercises 15A _____

Lenses

1 An object is placed   (i) 12 cm,   (ii) 4 cm from a converging lens of focal length 6 cm. Calculate the image position and the magnification in each case, and draw sketches illustrating the formation of the image.

2 What do you know about the image obtained with a diverging lens? The image of a real object in a diverging lens of focal length 10 cm is formed 4 cm from the lens. Find the object distance and the magnification. Draw a sketch to illustrate the formation of the image.

3 The image obtained with a converging lens is upright and three times the length of the object. The focal length of the lens is 20 cm. Calculate the object and image distances.

4 Used as a magnifying glass, the image of an object 4 cm from a converging lens is five times the object length. What is the focal length of the lens?

5 A slide of dimensions 2 cm by 2 cm produces a clear image of area 6400 cm^2 on a projector screen. Calculate the focal length of the projector lens if the screen is 82 cm from the slide.

6 An object placed 20 cm from a converging lens forms a magnified clear image on a screen. When the lens is moved 20 cm towards the screen, a smaller clear image is formed on the screen. Calculate the focal length of the lens.

7 A beam of light converges to a point 9 cm behind   (i) a converging lens of focal length 12 cm,   (ii) a diverging lens of focal length 15 cm. Find the image position in each case, and draw sketches illustrating them.

8 Draw a ray diagram to show how a converging lens produces an image of finite size of the moon clearly focused on a screen. If the moon subtends an angle of 9.1×10^{-3} radian at the centre of the lens, which has a focal length of 20 cm, calculate the diameter of this image. With the screen removed, a second converging lens of focal length 5·0 cm is placed coaxial with the first and 24 cm from it on the side remote from the moon. Find the position, nature and size of the final image. (*JMB.*)

9 A converging lens of 6 cm focal length is mounted at a distance of 10 cm from a screen placed at right angles to the axis of the lens. A diverging lens of 12 cm focal length is then placed coaxially between the converging lens and the screen so that an image of an object 24 cm from the converging lens is focused on the screen. What is the distance between the two lenses? Before commencing the calculation state the sign convention you will employ. (*JMB.*)

10 A lamp and a screen are 80 cm apart and a converging lens placed midway between them produces a focused image on the screen.

A thin diverging lens is placed 10 cm from the lamp, between the lamp and the converging lens. When the lamp is moved back so that it is 30 cm from the diverging lens, the focused image reappears on the screen. What is the focal length of the diverging lens? (*L.*)

11 Light from an object passes through a thin converging lens, focal length 20 cm, placed 24 cm from the object and then through a thin diverging lens, focal length 50 cm, forming a real image 62·5 cm from the diverging lens. Find
(a) the position of the image due to the first lens,
(b) the distance between the lenses,
(c) the magnification of the final image.

Optical Instruments

When a telescope or microscope is used to look at an object, the image we see depends on the eye. We therefore need to know some basic points about vision.

Firstly, the image formed by the eye lens L must appear on the retina R at the back of the eye if the object is to be clearly seen, Figure 15.7. Secondly, the normal eye can focus an object at infinity (the 'far point' of the normal eye). In this case the eye is relaxed or said to be 'unaccommodated'. Thirdly, the eye can see an object in greatest detail when it is placed at a certain distance D from the eye, known as the *least distance of distinct vision*, which is about 25 cm, for a normal eye. The point at a distance D from the eye is known as its 'near point'.

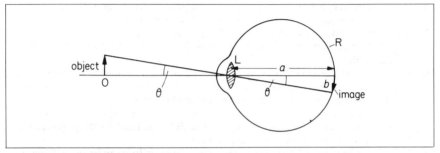

Figure 15.7   *Length of image on retina, and visual angle*

Visual Angle

Consider an object O placed some distance from the eye, and suppose θ is the angle in radians subtended by it at the eye, Figure 15.7. Since the opposite angles at L are equal, the length b of the image on the retina is given by $b = a\theta$, where a is the distance from R to L. But a is a constant; so $b \propto \theta$. We thus arrive at the important conclusion that *the length of the image formed by the eye is proportional to the **angle** subtended at the eye by the object*. This angle is known as the *visual angle*; the greater the visual angle, the greater is the apparent size of the object.

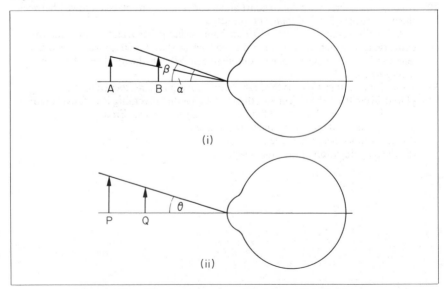

Figure 15.8   *Relation between visual angle and length of image*

Figure 15.8 (i) illustrates the case of an object moved from A to B, and viewed by the eye in both positions. At B the angle β subtended at the eye is greater than the visual angle α subtended at A. So the object appears larger at B than at A, although its physical size is the same. Figure 15.8 (ii) illustrates the case of two objects, at P, Q respectively, which subtend the same visual angle θ at the eye. The objects then appear to be of equal size, although the object at P is physically bigger than that at Q. Of course, an object is not clearly seen if it is brought closer to the eye than the near point.

Angular Magnification of Telescopes

Telescopes and microscopes are instruments designed to increase the visual angle, so that the object viewed can be made to appear much larger with their help. Before they are used the object may subtend a small angle α at the eye; when they are used the final images should subtend an increased angle β at the eye. The *angular magnification, M*, of the instrument is defined as the ratio

$$M = \frac{\beta}{\alpha} \qquad . \qquad . \qquad . \qquad . \qquad . \qquad (1)$$

This is also popularly known as the *magnifying power* of the instrument. It should be carefully noted that we are concerned with visual angles in the theory of optical instruments, and not with the physical sizes of the object and the image obtained.

Telescopes are instruments used for looking at distant objects. High power telescopes are used at astronomical observatories. In 1609 Galileo made a telescope through which he saw the satellites of Jupiter and the rings of Saturn. The telescope led the way for great astronomical discoveries, particularly by Kepler. Newton also designed telescopes. He was the first to suggest the use of curved mirrors for telescopes, as we see later.

If α is the angle subtended at the unaided eye by a *distant* object, and β is the angle subtended at the eye by its final image when a telescope is used, the angular magnification M (also called the 'magnifying power') of the telescope is given by

$$M = \frac{\beta}{\alpha}$$

Astronomical Telescope in Normal Adjustment

An astronomical telescope made from lenses consists of an *objective* of long focal length and an *eyepiece* of short focal length, for a reason given on p. 444. Both lenses are converging. *The telescope is in normal adjustment when the final image is formed at infinity.* The eye is then relaxed or unaccommodated when viewing the image. The unaided eye is also relaxed when a distant object viewed can be considered to be at infinity.

The objective lens O collects parallel rays from the distant object. So it forms an image I at its focus F_0. Figure 15.9 shows three of the many non-axial rays a from the *top* point of the object, which pass through the top point T of the image. The three rays b from the foot of the object would pass through the foot of I (not shown). As the final image is at infinity, I must be at the focus F_e of the eyepiece. So F_e and F_0 are at the *same* place.

To draw the final image, take one lens at a time.

(a) For O, draw a central ray a straight through C_1 to T, the top of the objective

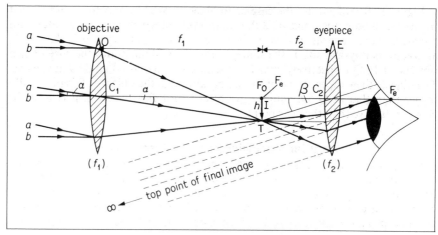

Figure 15.9 *Telescope in* normal *adjustment*

image below F_0. Then draw the other two rays a to pass through T, as shown.

(b) For E, draw a line from T to pass straight through C_2 and another line from T parallel to the principal axis to pass through F_e. The lines emerging from E are parallel; so the final image is at infinity.

(c) Now continue the rays passing through T from O so that they meet the lens E; then draw each refracted ray *parallel to* TC_2 because they must pass through the top of the image of T when produced back. Note carefully that the two lines first drawn from T to E to find the image position are construction lines and *not* actual light rays, and so should not have arrows on them.

To find the *angular magnification M* of the telescope, assume that the eye is closed to the eyepiece. Since the telescope length is very small compared with the distance of the object from either lens, we can take the angle α subtended at the unaided eye by the object as that subtended at the *objective* lens, as shown. Since I is distance f_1 from C_1, where f_1 is the focal length of O, we see that $\alpha = h/f_1$, where h is the length of I. Also, the angle β subtended at the eye when the telescope is used is given by h/f_2, where f_2 is the focal length of the eye-piece E. So

$$M = \frac{\beta}{\alpha} = \frac{h/f_2}{h/f_1}$$

$$\therefore M = \frac{f_1}{f_2} \qquad . \qquad . \qquad . \qquad . \qquad . \qquad (2)$$

Thus the angular magnification is equal to the ratio of the focal length of the objective (f_1) to that of the eyepiece (f_2). For high angular magnification, it follows from (2) that the objective should have a *long* focal length and the eyepiece a *short* focal length. Note that the separation of the lenses is $f_1 + f_2$.

> **Telescope in normal adjustment = Final image at infinity**
> **Then** $M = f_1$ **(objective)**$/f_2$ **(eyepiece)**
> **and separation of (distance between) lenses** $= f_1 + f_2$

Examples on Telescopes

1 An astronomical telescope has an objective of focal length 120 cm and an eyepiece of focal length 5 cm. What is
(a) the angular magnification (magnifying power),
(b) the separation of the two lenses?

(a)
$$M = \frac{f_1}{f_2} = \frac{120}{5} = 24$$

(b)
$$\text{separation} = f_1 + f_2 = 120 + 5 = 125 \, \text{cm}$$

2 An astronomical telescope has an objective focal length of 100 cm and an eyepiece focal length of 5 cm, Figure 15.10. With the eye close to the eyepiece, an observer sees clearly the final image of a star at a distance 25 cm from the lens.
Calculate
(a) the separation between the lenses,
(b) the angular magnification (magnifying power M.)

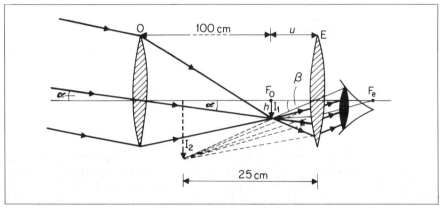

Figure 15.10 *Telescope with image at 25 cm from eye* (not to scale)

(a) The objective lens O forms an image I_1 of the star at its focus F_0 since parallel rays are incident on the lens. F_0 is 100 cm from O.
   The eyepiece E, $f = 5$ cm, forms a magnified and virtual image of I_1 at I_2, which is 25 cm from E. Suppose u is the distance of I_1 from E. Then, for lens E, $v = -25$ cm (virtual image) and $f = +5$ cm. From $1/v + 1/u = 1/f$,

$$\frac{1}{-25} + \frac{1}{u} = \frac{1}{+5}$$

Solving
$$u = \frac{25}{6} = 4 \cdot 2 \, \text{cm}$$

So   separation OE of lenses $= 100 + 4 \cdot 2 = 104 \cdot 2$ cm

(b) In Figure 15.10, final image I_2 subtends an angle β at the eye close to E. If h is the height of I_1, then

$$M = \frac{\beta}{\alpha} = \frac{h/u}{h/100} = \frac{100}{u}$$

$$= \frac{100}{4 \cdot 2} = 24$$

Eye-Ring of Telescope

When an object is viewed by an optical instrument, only those rays from the object which are bounded by the perimeter or edge of the objective lens enter the instrument. The lens thus acts as a *stop* to the light from the object. With a given objective, the best position of the eye is one where it collects as much light as possible from that passing through the objective.

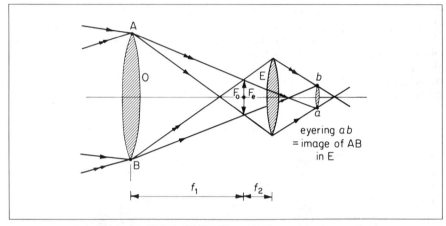

Figure 15.11 *Eye-ring position*

Figure 15.11 shows the rays from the field of view which are refracted at the *boundary* of the objective O to form an image at F_0 or F_e with the telescope in normal adjustment. These rays are again refracted at the boundary of the eyepiece E to form a small image ab. From the ray diagram, we see that a is the image of A on the objective and b is an image of B on the objective. *So ab is the image of the objective AB in the eyepiece.*

The small circular image ab is called the *eye–ring*. It is the best position for the eye. Here the eye can collect the maximum amount of light entering the objective from outside so that it has a *wide field of view*. If the eye were placed closer to the eyepiece than the eye-ring the observer would have a smaller field of view.

If the telescope is in normal adjustment, the distance u of the objective from the eyepiece E, focal length f_2, is $(f_1 + f_2)$. From the lens equation, the eye-ring distance v from E is given by

$$\frac{1}{v} + \frac{1}{+(f_1+f_2)} = \frac{1}{+(f_2)}$$

from which

$$v = \frac{f_2}{f_1}(f_1 + f_2)$$

Now the objective diameter: eye-ring diameter $= AB:ab = u:v$

$$= (f_1+f_2): \frac{f_2}{f_1}(f_1+f_2)$$

$$= f_1/f_2$$

But the angular magnification of the telescope $= f_1/f_2$ (p. 444). So the angular magnification, M, is also given by

$$M = \frac{\text{diameter of objective}}{\text{diameter of eye-ring}} \qquad . \qquad . \qquad . \qquad (3)$$

the telescope being in normal adjustment.

The relation in (3) provides a simple way of measuring M for a telescope.

Resolving Power

If two distant objects are close together, it may not be possible to see their images apart through a telescope even though the lenses are perfect and produce high magnifying power. This is due to the phenomenon of diffraction and is explained later (p. 544). Here we can state that the *smallest* angle θ subtended at a telescope by two distant objects which can just be seen separated is given approximately by

$$\theta = \frac{1 \cdot 22\lambda}{D}$$

where λ is the mean wavelength of the light from the distant objects and D is the diameter of the *objective* lens.

θ is called the *resolving power* of the telescope. The smaller the value of θ, *the greater* is the resolving power because two distant objects which are closer together can then be seen separated through the telescope. Note that the formula for θ only depends on the diameter of the objective and *not* on its focal length, and that it does not concern the eyepiece. As we have seen, the focal lengths of the objective and eyepiece affect the angular magnification of a telescope but high angular magnification does not produce high resolving power. Higher resolving power is obtained by using an objective lens of greater diameter.

So if the objective of a telescope has a diameter of 200 mm, and the mean wavelength of the light from distant stars is 6×10^{-7} m, the resolving power

$$\theta = \frac{1 \cdot 22 \times 6 \times 10^{-7}}{0 \cdot 200} = 4 \times 10^{-6} \text{ rad (approx.)}$$

This means that the two stars which subtend this angle at the telescope objective can just be seen separated or resolved.

Reflector Telescope

The astronomical telescope so far discussed has a lens objective and is therefore a *refractor* telescope. A *reflector* telescope, with a large curved mirror as its objective, was first suggested by Newton.

The construction of the Hale telescope at Mount Palomar, the largest telescope in the world, is one of the most fascinating stories of scientific skill and invention. The major feature of the telescope is a *parabolic mirror*, 5 metres across, which is made of pyrex, a low expansion glass. The glass itself took more than six years to grind and polish, and the front of the mirror is coated with aluminium, instead of being covered with silver, as it lasts much longer. The huge size of the mirror enables enough light from very distant stars and planets to be collected and brought to a focus for them to be photographed. Special cameras are incorporated in the instrument. This method has the advantage that plates can be exposed for hours, if necessary, to the object to be studied, enabling records to be made. It is used to obtain useful information about the building-up and breaking-down of the elements in space, to investigate astronomical theories

Plate 15A Radcliffe Refractor Telescope. *The larger telescope has an objective of 60 cm diameter and focal length about 7 m. The resolving power is of the order 10^{-6} rad. This telescope acts as a camera whereas the smaller telescope shown in use, which has a 50 cm objective, acts as a guide telescope.*

Plate 15B Allen Reflector Telescope. *This has an objective mirror of 60 cm diameter and focal length about 2 m. A Cassegrain mirror is at the top of the tube, which has a length much shorter than the refractor telescope.* (Photographs of the Allen and Radcliffe telescopes of the University of London Observatory are reproduced by courtesy of the University of London Observatory, Department of Physics and Astronomy, University College, London.)

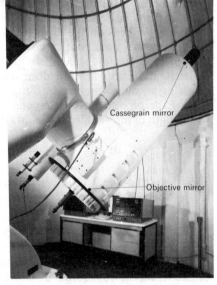

of the universe, and to photograph planets such as Mars. The Hale telescope was built on the top of Mount Palomar, California, where the air is particularly free of mist and other hindrances to night vision.

Besides the main parabolic mirror O, which is the telescope objective, seven other mirrors are used in the 5 metre telescope. Some are plane, Figure 15.12 (i), while others are convex, Figure 15.12 (ii), and they are used to bring the light to a more convenient focus, where the image can be photographed, or magnified several hundred times by an eyepiece E for observation. The various methods of focusing the image were suggested respectively by *Newton, Cassegrain,* and *Coudé,* the last thing being a combination of the other two methods, Figure 15.12 (iii).

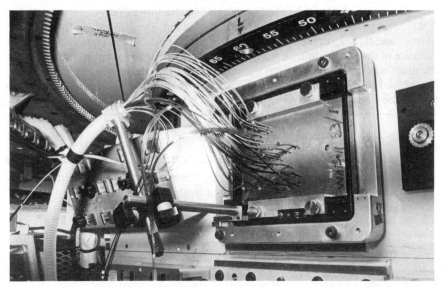

Plate 15C *Fifty optical fibres plugged into an aperture plate at the Cassegrain focus of the 4-metre Anglo-Australian Telescope. The output of the fibre bundle feeds the spectrograph slit, allowing simultaneous spectroscopy of fifty separate objects.*
Courtesy of Peter Gray, Epping Laboratory, Anglo-Australian Observatory

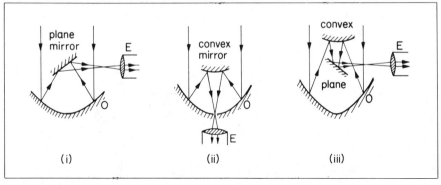

Figure 15.12 (*i*) *Newton reflector* (*ii*) *Cassegrain reflector* (*iii*) *Coudé reflector*

The reflecting telescope is free from the coloured images produced by refraction at the glass lenses of the refractor telescope. This so-called 'chromatic aberration' of the lens makes the image seen indistinct. The image is also brighter than in a refractor telescope, where some loss of light occurs by reflection at the lens' surfaces and by absorption. The large diameter of the mirror, which is the telescope objective, also produces high resolving power.

Radio waves from galaxies in outer space are detected by *radio telescopes*. These consist of a concave aerial 'dish' of metal rods which reflect the radio waves to a sensitive detector at the focus of the dish. See p. 414. The signal received is then amplified and recorded automatically. The resolving power of the telescope is increased by moving several widely-spaced dishes along rails, while pointing them skywards in the same direction. This effectively increases the diameter of the telescope objective.

Simple Microscope or Magnifying Glass

A microscope is an instrument used for viewing *near* objects. When it is in normal use, therefore, the image formed by the microscope is usually at the least distance of distinct vision, *D*, from the eye, i.e., at the near point of the eye. With the unaided eye (that is, without the instrument), the object is seen clearest when it is placed at the near point. So the angular magnification of a microscope in *normal* use is given by

$$M = \frac{\beta}{\alpha}$$

where β is the angle subtended at the eye by the image at the near point, and α is the angle subtended at the unaided eye by the object at the near point.

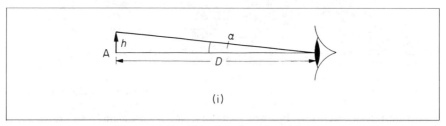

(i)

Figure 15.13   *Visual angle with unaided eye*

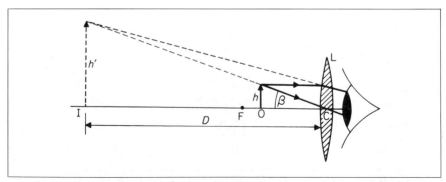

Figure 15.14   *Simple microscope, or magnifying glass*

Suppose an object of length *h* is viewed at the near point A by the unaided eye, Figure 15.13. The visual angle, α, is then h/D in radian measure. Now suppose that a converging lens L is used as a magnifying glass to view the same object. An erect, magnified image is obtained when the object O is nearer to L than its focal length (p. 437), and the observer moves the lens until the image at I is situated at his or her near point. If the observer's eye is close to the lens at C, the distance IC is then equal to *D*, the least distance of distinct vision, Figure 15.14. Thus the new visual angle β is given by h'/D, where h' is the length of the virtual image. We can see that β is greater than α by comparing Figure 15.13 with Figure 15.14.

The angular magnification, *M*, of this simple microscope can be found in terms of *D* and the focal length *f* of the lens. From definition, $M = \beta/\alpha$.

But
$$\beta = \frac{h'}{D}, \quad \alpha = \frac{h}{D}$$

$$\therefore M = \frac{h'}{D} \bigg/ \frac{h}{D} = h'/h \qquad . \qquad . \qquad . \qquad (1)$$

Now h'/h is the 'linear magnification' produced by the lens, and is given by $h'/h = v/u$, where v is the image distance CI and u is the object distance CO (see p. 438). Since $1/v + 1/u = 1/f$, with the usual notation, we have

$$1 + \frac{v}{u} = \frac{v}{f}, \quad \text{or} \quad \frac{v}{u} = \frac{v}{f} - 1$$

by multiplying throughout by v. Since the image is virtual, $v = CI = -D$, where D is the _numerical_ value of the least distance of distinct vision,

$$\therefore \frac{v}{u} = \frac{v}{f} - 1 = -\frac{D}{f} - 1$$

$$\therefore \frac{h'}{h} = -\frac{D}{f} - 1$$

$$\therefore M = -\left(\frac{D}{f} + 1\right) \qquad . \qquad . \qquad . \qquad (2)$$

from (1) above. So numerically, $\quad M = \left(\frac{D}{f} + 1\right)$

If the magnifying glass has a focal length of 5 cm, $f = +5$ as it is converging; also, if the least distance of distinct vision is 25 cm, $D = 25$ numerically. Substituting in (2),

$$M = -\left(\frac{25}{5} + 1\right) = -6$$

Thus the angular magnification is 6. The position of the object O is given by substituting $v = -25$ and $f = +5$ in the lens equation $1/v + 1/u = 1/f$, from which the object distance u is found to be $+4·2$ cm.

From the formula for M in (2), it follows that a lens of _short_ focal length is required for high angular magnification.

When an object OA is viewed through a converging lens acting as a _magnifying glass_, various coloured virtual images, corresponding to I_R, I_V for red and violet rays for example, are formed, Figure 15.15. The top point of each image lies on the line CA. So each image subtends the same angle at the eye close

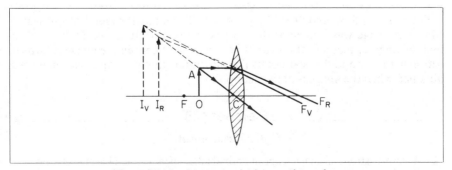

Figure 15.15 _Dispersion with magnifying glass_

to the lens, so that the colours received by the eye will practically overlap. Thus the virtual image seen in a magnifying glass is almost free of chromatic aberration. A little colour is observed at the edges as a result of spherical aberration. A *real* image formed by a lens, however, has chromatic aberration, as explained on p. 461.

Magnifying Glass with Image at Infinity

We have just considered the normal use of the simple microscope, where the image formed is at the near point of the eye and the eye is accommodated (p. 442). When the image is formed at infinity, however, which is not a normal use of the microscope, the eye is undergoing the least strain and is then unaccommodated (p. 442). In this case the object must be placed at the focus, F, of the lens, Figure 15.16.

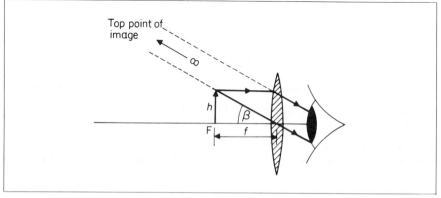

Figure 15.16 *Final image at infinity*

Suppose that the focal length of the lens is f. The visual angle β now subtended at the eye is then h/f if the eye is close to the lens, and hence the angular magnification, M, is given by

$$M = \frac{\beta}{\alpha} = \frac{h/f}{h/D}$$

as $\alpha = h/D$, see Figure 15.13.

$$\therefore M = \frac{D}{f} \qquad . \qquad . \qquad . \qquad . \qquad . \qquad (3)$$

When $f = +5$ cm and $D = 25$ cm, $M = 5$. The angular magnification was 6 when the image was formed at the near point (p. 451). It can easily be verified that the angular magnification varies between 5 and 6 when the image is formed between infinity and the near point. The maximum angular magnification is thus obtained when the image is at the near point.

Exercises 15B

Optical Instruments

1 A simple astronomical telescope in normal adjustment has an objective of focal length 100 cm and an eyepiece of focal length 5 cm.   (i) Where is the final image

formed? (ii) Calculate the angular magnification. (ii) How would you increase the *resolving power* of the telescope?

2 Draw a ray diagram showing how the image of a distant star is formed at the least distance of distinct vision of an observer using a simple astronomical telescope. In your sketch show the principal focus of the two lenses.

The same telescope is now required to produce the image of the star on a photographic plate beyond the eyepiece. What adjustment is required? Draw a diagram to explain your answer.

3 What is the *eye-ring* of a telescope? Draw a ray diagram showing how the eye-ring is formed in a simple astronomical telescope and explain why this telescope has a wide field of view.

Calculate the distance of the eye-ring from the eyepiece of a simple astronomical telescope in normal adjustment whose objective and eyepiece have focal lengths of 80 cm and 10 cm respectively.

4 Calculate the position of the eye-ring for an astronomical telescope consisting of two thin converging lenses, an objective of focal length 1·0 m and an eyepiece of focal length 20 mm, placed 1·02 m apart.

Explain the advantage of placing the eye at the eye-ring position when using the telescope. (*L*.)

5 Draw a sketch of a *reflector telescope* and show with a ray diagram how an observer sees the final image of a distant star.

State (i) the advantages of a reflector telescope over a refractor telescope, (ii) how the resolving power of the reflector telescope can be increased, (iii) the purpose of a radio reflector telescope.

6 Explain the term *angular magnification* as related to an optical instrument. Describe, with the aid of a ray diagram, the structure and action of an astronomical telescope. Derive an expression for its angular magnification when used so that the final image is at infinity. With such an instrument what is the best position for the observer's eye? Why is this the best position?

Even if the lenses in such an instrument are perfect it may not be possible to produce clear separate images of two points which are close together. Explain why this is so. Keeping the focal lengths of the lenses the same, what could be changed in order to make the separation of the images more possible? (*L*.)

7 A refracting telescope has an objective of focal length 1·0 m and an eyepiece of focal length 2·0 cm. A real image of the sun, 10 cm in diameter, is formed on a screen 24 cm from the eyepiece. What angle does the sun subtend at the objective? (*L*.)

8 Draw a diagram showing the passage of rays through a simple astronomical refracting telescope when it is used to view a distant extended object such as the moon, and is adjusted so that the final image is at infinity. Using the diagram, show how the magnifying power of the telescope is related to the focal lengths of the objective and eyepiece lenses.

The objective of a telescope has a diameter of 100 mm. Estimate the approximate angular separation of two stars which can just be resolved by the telescope.

What are the advantages of using a reflecting (rather than a refracting) objective in an astronomical telescope? (*O. & C.*)

9 Explain the essential features of the astronomical telescope. Define and deduce an expression for the magnifying power of this instrument.

A telescope is made of an object glass of focal length 20 cm and an eyepiece of 5 cm, both converging lenses. Find the magnifying power in accordance with your definition in the following cases:

(a) when the eye is focused to receive parallel rays, and
(b) when the eye sees the image situated at the nearest distance of distinct vision which may be taken as 25 cm. (*L*.)

10 An astronomical telescope may be constructed using as objective either
(a) a converging lens, or
(b) a concave mirror.

Draw diagrams to illustrate the optical system of both types of telescope. Include in each diagram at least three rays reaching the instrument from an off-axial direction.

Define the magnifying power of a telescope. A telescope consists of two thin converging lenses of focal lengths 0·3 m and 0·03 m separated by 0·33 m. It is focused on the moon, which subtends an angle of 0·5° at the objective. Starting from first principles, find the angle subtended at the observer's eye by the image of the moon formed by the instrument.

Explain why one would expect this image to be coloured. Suggest how this defect might be rectified. (*O. & C.*)

11 What is the *eye-ring* of a telescope?

For an astronomical telescope in normal adjustment deduce expressions for the size and position of the eye-ring in terms of the diameter of the object glass and the focal lengths of the object glass and eye-lens.

Discuss the importance of   (i) the magnitude of the diameter of the object glass, (ii) the structure of the object glass,   (iii) the position of the eye. (*L.*)

12 Show, by means of a ray diagram, how an image of a distant extended object is formed by an astronomical refracting telescope in normal adjustment (i.e. with the final image at infinity).

A telescope objective has focal length 96 cm and diameter 12 cm. Calculate the focal length and minimum diameter of a simple eyepiece lens for use with the telescope, if the magnifying power required is × 24, and all the light transmitted by the objective from a distant point on the telescope axis is to fall on the eyepiece. Derive any formulae you use. (*O. & C.*)

13 An astronomical telescope consisting of an objective focal length 60 cm and an eyepiece of focal length 3 cm is focused on the moon so that the final image is formed at the minimum distance of distinct vision (25 cm) from the eyepiece. Assuming that the diameter of the moon subtends an angle of $\frac{1}{2}°$ at the objective, calculate

(a) the angular magnification,

(b) the actual size of the image seen.

How, with the same lenses, could an image of the moon, 10 cm in diameter, be formed on a photographic plate? (*C.*)

14 Explain, with the aid of a ray diagram, how a simple astronomical telescope employing two converging lenses may form an apparently enlarged image of a distant extended object. State with reasons where the eye should be placed to observe the image.

A telescope constructed from two converging lenses, one of focal length 250 cm, the other of focal length 2 cm, is used to observe a planet which subtends an angle of 5×10^{-5} radian. Explain how these lenses would be placed for normal adjustment and calculate the angle subtended at the eye of the observer by the final image.

How would you expect the performance of this telescope for observing a star to compare with one using a concave mirror as objective instead of a lens, assuming that the mirror had the same diameter and focal length as the lens. (*O & C.*)

15 A converging lens is used to cast an image of the full moon on to a screen.

(a) Assuming the moon to be in the centre of the field of view of the lens, draw a ray diagram showing clearly how the image is produced. Explain your method of construction.

(b) The moon is $3·5 \times 10^3$ km in diameter and is $3·8 \times 10^5$ km away from the Earth. (i) Calculate the angle in radian measure that it subtends at the eye of an observer on Earth.   (ii) If the lens has a focal length of 0·30 m, what is the diameter of the image of the moon produced on the screen?   (iii) If the observer views the image on the screen from a distance of 0·25 m, what angle will it subtend at his eye?   (iv) Calculate the angular magnification that has been achieved by using this lens.

(c) Explain carefully the effect on the angular magnification of using a lens of much longer focal length.

A much greater angular magnification can be achieved by using two converging lenses, one as an objective and the other as an eyepiece. If a converging lens of focal length 0·050 m were placed 0·050 m beyond the screen (now removed) and the observer viewed the moon through both the eyepiece and the objective lenses, what would be the new angular magnification?

Apart from greater magnification, name *one* other difference between the image produced in this case and when only one lens was used. (*L*.)

16 A converging lens of focal length 5 cm is used as a magnifying glass. If the near point of the observer is 25 cm from the eye and the lens is held close to the eye, calculate (i) the distance of the object from the lens,   (ii) the angular magnification.

What is the angular magnification when the final image is formed at infinity?

17 Explain what is meant by the magnifying power of a magnifying glass.

Derive expressions for the magnifying power of a magnifying glass when the image is

(a) 25 cm from the eye and

(b) at infinity.

In each case draw the appropriate ray diagram. (*JMB*.)

16
Further Topics in Optics

In this chapter we shall consider three optical topics: (1) the compound microscope and its magnifying power, (2) the lens camera f-number and depth of field, (3) defects of lenses (chromatic and spherical aberrations) and the way to reduce these defects.

Compound Microscope

From the formula $M = -\left(\dfrac{D}{f} + 1\right)$ for the magnifying power of a single lens (p. 451), we see that M is greater numerically the smaller the focal length of the lens. As it is impracticable to decrease f beyond a certain limit, owing to the mechanical difficulties of grinding a lens of short focal length (great curvature), *two* separated lenses are used to obtain a high angular magnification. This forms a *compound* microscope. The lens nearer to the object is called the *objective*; the lens through which the final image is viewed is called the *eyepiece*. The objective and the eyepiece are both converging, and both have small focal lengths for a reason explained later.

When the microscope is used, the object O is placed at a slightly *greater* distance from the objective than its focal length. In Figure 16.1, F_o is the focus of this lens. An inverted real image is then formed at I_1 in the microscope tube, and the eyepiece is adjusted so that a large virtual image is formed by it at I_2. Thus I_1 is *nearer* to the eyepiece than the focus F_e of this lens. It can now be seen that the eyepiece acts as a simple magnifying glass, used for viewing the image formed at I_1 by the objective.

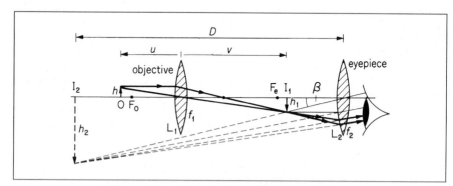

Figure 16.1 *Compound microscope in* normal *use*

To draw the final image I_2, we first draw construction lines from the top of I_1 to the eyepiece as shown in Figure 16.1. The actual rays, shown by heavy lines in Figure 16.1, can then be drawn as we explained for the telescope on p. 443, to which the reader should refer.

Figure 16.1 illustrates only the basic principle of a compound microscope. The single lens objective shown would produce a real image of the object which is coloured (see *chromatic aberration*, p. 461). The single lens eyepiece would produce a virtual image fairly free of colour (p. 452). In practice, both the objective and eyepiece of microscopes are made of several lenses which together reduce chromatic aberration as well as spherical aberration.

The best position for the eye is at *the image of the objective in the eyepiece* or *eye-ring*. All the rays from the object pass through this image. See p. 446. Suppose the objective is 16 cm from L_2, which has a focal length of 2 cm. The image distance, v, in L_2 is given by $\dfrac{1}{v} + \dfrac{1}{(+16)} = \dfrac{1}{(+2)}$, from which $v = 2\cdot3$ cm. Thus the eye-ring is a short distance from the eyepiece, and in practice the eye should be farther from the eyepiece than in Figure 16.1. This is arranged in commercial microscopes by having a circular opening fixed at the eye-ring distance from the eyepiece, so that the observer's eye has automatically the best position when it is placed close to the opening.

Angular Magnification with Microscope in Normal Use

When the microscope is in normal use the image at I_2 is formed at the least distance of distinct vision, D, from the eye (p. 450). Suppose that the eye is close to the eyepiece, as shown in Figure 16.1. The visual angle β subtended by the image at I_2 is then given by $\beta = h_2/D$, where h_2 is the height of the image. With the unaided eye, the object subtends a visual angle given by $\alpha = h/D$, where h is the height of the object, see Figure 15.13.

$$\therefore \text{ angular magnification, } M = \frac{\beta}{\alpha}$$

$$= \frac{h_2/D}{h/D} = \frac{h_2}{h}$$

Now $\dfrac{h_2}{h}$ can be written as $\dfrac{h_2}{h_1} \times \dfrac{h_1}{h}$, where h_1 is the length of the intermediate image formed at I_1.

$$\therefore M = \frac{h_2}{h_1} \cdot \frac{h_1}{h} \qquad . \qquad . \qquad . \qquad . \qquad \text{(i)}$$

The ratio h_2/h_1 is the linear magnification of the 'object' at I_1 produced by the *eyepiece*, and we have shown on p. 451 that the linear magnification is also given by $v/f_2 - 1$, where v is the image distance from the lens and f_2 is the focal length. Since $v = -D$ where D is the numerical value of the least distance of distinct vision, it follows that

$$\frac{h_2}{h_1} = \frac{D}{f_2} - 1 = -\left(\frac{D}{f_2} + 1\right) \qquad . \qquad . \qquad . \qquad \text{(ii)}$$

Also, the ratio h_1/h is the linear magnification of the object at O produced by the *objective* lens. Thus if the distance of the image I_1 from this lens is denoted by v, we have

$$\frac{h_1}{h} = \frac{v}{f_1} - 1 \qquad . \qquad . \qquad . \qquad . \qquad \text{(iii)}$$

$$\therefore M = \frac{h_2}{h_1} \cdot \frac{h_1}{h} = -\left(\frac{D}{f_2} + 1\right)\left(\frac{v}{f_1} - 1\right) \qquad . \qquad . \qquad \text{(4)}$$

It can be seen that if f_1 and f_2 are small, M is large. Thus the angular magnification is high if the focal lengths of the objective and the eyepiece are small.

Example on Compound Microscope

A model of a compound microscope is made up of two converging lenses of 3 cm and 9 cm focal length at a fixed separation of 24 cm. Where must the object be placed so that the final image may be at infinity? What will be the magnifying power if the microscope as thus arranged is used by a person whose nearest distance of distinct vision is 25 cm? State what is the best position for the observer's eye and explain why.

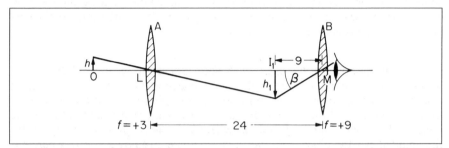

Figure 16.2 *Example on compound microscope*

(i) Suppose the objective A is 3 cm focal length, and the eyepiece B is 9 cm focal length, Figure 16.2. If the final image is at infinity, the image I_1 in the objective must be 9 cm from B, the focal length of the eyepiece, see p. 452. So the image distance LI_1, from the objective A $= 24 - 9 = 15$ cm. The object distance OL is thus given by

$$\frac{1}{(+15)} + \frac{1}{u} = \frac{1}{(+3)}$$

from which
$$u = OL = 3\tfrac{3}{4}\,cm$$

(ii) The angle β subtended at the observer's eye is given by $\beta = h_1/9$, where h_1 is the height of the image at I_1, Figure 16.2. Without the lenses, the object subtends an angle α at the eye given by $\alpha = h/25$, where h is the height of the object, since the least distance of distinct vision is 25 cm.

$$\therefore \text{ magnifying power } M = \frac{\beta}{\alpha} = \frac{h_1/9}{h/25} = \frac{25}{9} \times \frac{h_1}{h}$$

But
$$\frac{h_1}{h} = \frac{LI_1}{LO} = \frac{15}{3\tfrac{3}{4}} = 4$$

$$\therefore M = \frac{25}{9} \times 4 = 11 \cdot 1$$

The best position of the eye is at the eye-ring, which is the image of the objective A in the eyepiece B.

Lens Camera; f-number

The photographic camera consists essentially of a *lens system* L, a *light-sensitive film* F at the back, a *focusing* device for adjusting the distance of the lens from F, and an *exposure* arrangement which provides the correct exposure for a given lens aperture, Figure 16.3. The lens system may contain an achromatic doublet and separated lenses which together reduce considerably chromatic and spherical aberration (p. 461). An *aperture* or *stop* of diameter d is provided so that the light is incident centrally on the lens, thus diminishing distortion (p. 462).

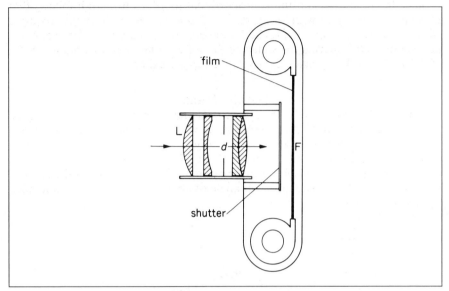

Figure 16.3 *Photographic camera*

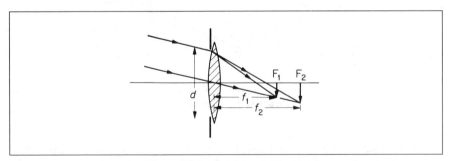

Figure 16.4 *Brightness of image*

The amount of luminous flux falling on the image in a camera is proportional to the area of the lens aperture, or to d^2, where d is the diameter of the aperture. The area of the image formed is proportional to f^2, where f is the focal length of the lens, since the length of the image formed is proportional to the focal length, as illustrated by Figure 16.4. It therefore follows that the luminous flux per unit area of the image, or *brightness B*, of the image, is proportional to d^2/f^2. The time of exposure, t, for activating the chemicals on the given negative is inversely proportional to B. Hence

$$t \propto \frac{f^2}{d^2} \qquad . \qquad . \qquad . \qquad . \qquad . \qquad . \qquad \text{(i)}$$

The *relative aperture* of a lens is defined as the ratio d/f, where d is the diameter of the aperture and f is the focal length of the lens. The aperture is usually expressed by its f-*number*. If the aperture is f-4, this means that the diameter d of the aperture is $f/4$, where f is the focal length of the lens. An aperture of f-8 means a diameter d equal to $f/8$, which is a smaller aperture than $f/4$.

Since the time t of exposure is proportional to f^2/d^2, from (i) it follows that the exposure required with an aperture f-8 ($d = f/8$) is 16 times that required with an aperture f-2 ($d = f/2$). The f-numbers on a camera are 2, 2·8, 3·5, 4, 4·8, for example. On squaring the values of f/d for each number, we obtain 4, 8, 12, 16, 20, or 1, 2, 3, 4, 5, which are the relative exposure times.

Depth of Field

An object will not be seen by the eye until its image on the retina covers at least the area of a single cone, which transmits along the optic nerve light energy just sufficient to produce the sensation of vision. As a basis of calculation in photography, a circle of finite diameter about 0·25 mm viewed 250 mm away will just be seen by the eye as a fairly sharp point, and this is known as the *circle of least confusion*. It corresponds to an angle of about 1/1000th radian subtended by an object at the eye.

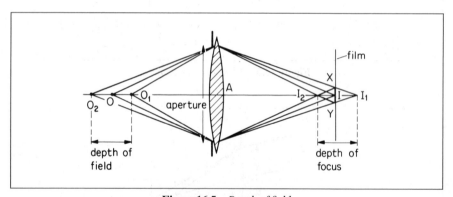

Figure 16.5   *Depth of field*

On account of the lack of resolution of the eye, a camera can take clear pictures of objects at different distances. Consider a point object O in front of a camera lens A which produces a point image I on a film, Figure 16.5. If XY represents the diameter of the circle of least confusion round I, the eye will see all points in the circle as reasonably sharp points. Now rays from the lens aperture to the edge of XY meet at I_1 beyond I, and also at I_2 in front of I. The point images I_1, I_2 correspond to point objects O_1, O_2 on either side of O, as shown. Consequently the images of all objects between O_1, O_2 are seen clearly on the film.

The distance $O_1\ O_2$ is therefore known as the *depth of field*. The distance $I_1\ I_2$ is known as the *depth of focus*. The depth of field depends on the lens aperture. If the aperture is made smaller, and the diameter XY of the circle of least confusion is unaltered, it can be seen from Figure 16.5 that the depth of field increases. If the aperture is made larger, the depth of field decreases.

Defects of Lenses

Chromatic Aberration, Achromatic Lenses

When white light from an object is refracted by a lens, a coloured image is formed. This is because the glass refracts different colours such as red, *r*, and blue, *b*, to a different focus (Figure 16.6). The coloured images are formed at slightly different places and this is called the *chromatic aberration* (colour defect) of a single lens.

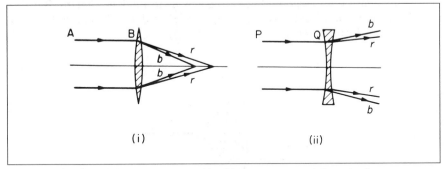

Figure 16.6 *Dispersion produced by converging and diverging lens*

A converging lens deviates an incident ray such as AB towards its principal axis, Figure 16.6(i). A diverging lens, however, deviates a ray PQ away from its principal axis, Figure 16.6(ii). The dispersion between two colours produced by a converging lens can thus be neutralised by placing a suitable diverging lens beside it. Two such lenses which together eliminate the chromatic aberration of a single lens are called an *achromatic* combination of lenses. Figure 16.7 illustrates an achromatic lens combination, known as an *achromatic doublet*. The biconvex lens is made of crown glass, while the diverging lens is made of

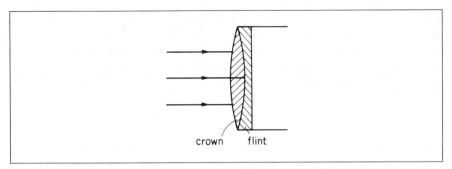

Figure 16.7 *Achromatic doublet in telescopic objective*

flint glass and is a plano-concave lens. So that the lenses can be cemented together with Canada balsam, the radius of curvature of the curved surface of the plano-concave lens is made numerically the same as that of one surface of the converging lens. The achromatic combination acts as a converging lens when used as the objective lens in a high-quality telescope or microscope.

It should be noted that chromatic aberration would occur if the diverging and converging lenses were made of the *same* material, as the two lenses together would then constitute a single thick lens of one material.

Spherical Aberration

If a *wide* parallel beam of light is incident on a lens experiment it shows that the rays are not all brought to the same focus, Figure 16.8. It therefore follows that the image of an object is distorted if a wide beam of light falls on the lens, and this is known as *spherical aberration*. The aberration may be reduced by surrounding the lens with an opaque disc having a hole in the middle, so that light is incident only on the middle of the lens as in the lens camera. But this method reduces the brightness of the image since it reduces the amount of light energy passing through the lens.

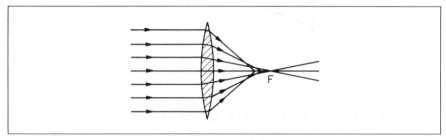

Figure 16.8   *Spherical aberration*

As rays converge to a single focus for small angles of incidence, spherical aberration can be diminished if the angles of incidence on the lens' surfaces are reduced. In general, then, the *deviation* of the light by a lens should be shared as equally as possible by its surfaces, as each angle of incidence would then be as small as possible. A practical method of reducing spherical aberration is to use *two* lenses, when four surfaces are obtained, and to share the deviation equally between the lenses. The lenses are usually plano-convex.

In the compound microscope, the slide or other object is placed close to the objective (p. 456). A large angle is then subtended by the object at the lens and so the angle of incidence of rays is large. Correction for spherical aberration is hence more important for the objective of this instrument than chromatic aberration. The reverse is the case for the objective of a refractor telescope. A compound microscope of good quality has several lenses which help to correct the aberrations.

_____ Exercises 16 _____

Compound Microscope

1   Draw a labelled ray diagram to illustrate the action of a compound microscope.
    State and explain how you would arrange simple converging lenses, one of focal length 2 cm and one of focal length 5 cm, to act as a compound microscope with magnifying power (angular magnification) × 42, the final image being 25 cm from the eye lens.
    Assume that the focal lengths quoted, and your calculations, relate to the image formed when the object is illuminated by monochromatic red light. Without further calculation, state and explain the changes in the position and size of the image formed by the objective and in the apparent size of the final image (i.e.

the angle it subtends at the centre of the eye lens) which would occur on changing the illumination of the object through the spectral range from red to violet, the setting of the microscope remaining unchanged. (*O. & C.*)

2 A point object is placed on the axis of, and 3·6 cm from, a thin converging lens of focal length 3·0 cm. A second thin converging lens of focal length 16·0 cm is placed coaxial with the first and 26·0 cm from it on the side remote from the object. Find the position of the final image produced by the two lenses.

Why is this not a suitable arrangement for a compound microscope used by an observer with normal eyesight?

For such an observer wishing to use the two lenses as a compound microscope with the eye close to the second lens decide, by means of a suitable calculation, where the second lens must be placed relative to the first. (*JMB.*)

3 Draw the path of two rays, from a point on an object, passing through the optical system of a compound microscope to the final image as seen by the eye.

If the final image formed coincides with the object, and is at the least distance of distinct vision (25 cm) when the object is 4 cm from the objective, calculate the focal lengths of the objective and eye lenses, assuming that the magnifying power of the microscope is 14. (*L.*)

4 Give a detailed description of the optical system of the compound microscope, explaining the problems which arise in the design of an objective lens for a microscope.

A compound microscope has lenses of focal length 1 cm and 3 cm. An object is placed 1·2 cm from the object lens; if a virtual image is formed 25 cm from the eye, calculate the separation of the lenses and the magnification of the instrument. (*O. & C.*)

Lens Camera

5 A camera has a lens, of focal length 120 mm, which can be moved along its principal axis towards and away from the film. If the camera is to be able to form perfect images of objects from infinite distance down to 1·00 m from the camera, through what distance must it be possible to move the lens? (*L.*)

6 A convex camera lens is used to form an image of an object 1·00 m away from it on a film 0·050 m from the lens. What is the focal length of the lens?

If the camera is used to photograph a distant object, how far from the film would the clear image be formed? What type of lens should be placed close to the first lens in order to enable the distant object to be focused on the film if the separation of the first lens and film cannot be changed in this camera? What is the focal length of this added lens? (*L.*)

7 Under certain conditions a suitable setting for a camera is: exposure time 1/125 second, aperture $f/5\cdot6$. If the aperture is changed to $f/16$ what would be the new exposure time in order to achieve the same film image density? What other effect would this change in f-number produce? (*L.*)

8 (a) Define (i) linear magnification, and (ii) magnifying power. Why is the latter appropriate in considering optical instruments such as telescopes?

(b) Draw a ray diagram to illustrate the action of an astronomical telescope consisting of two convex (converging) lenses, the instrument being in normal adjustment. On your diagram indicate the positions of the principal foci of the lenses.

(c) When a single convex lens is used as a magnifying glass, and the eye is placed close to the lens, the angles subtended by object and image are approximately the same. This being the case, explain why the magnifying glass produces a magnified image.

When white light is refracted on passing through a lens it undergoes dispersion and each colour produces a separate image. Why, then, is a series of coloured images not observed when the eye is placed close to a magnifying glass?

(d) A camera is set at $f5\cdot6$, 1/120 s. If the aperture is changed to $f16$, to what value should the exposure time be set to achieve the same exposure? What other effect would the change of aperture have? (*AEB*, 1982.)

Aberrations of Lenses

9 A white object in front of a converging lens produces an inverted coloured image. Using the lens focus for red and blue rays, draw a sketch showing how the red and blue images are formed. What is this lens defect called?

10 What is *spherical aberration* of a lens? Draw a sketch to illustrate your answer and explain why this produces an unclear image. How can the aberration be reduced?

11 An *achromatic doublet* is used as a telescope objective. Draw a sketch of the doublet and explain how it reduces the colour defect due to a single lens.

12 When a compound microscope is used, the tiny object viewed is close to the objective lens. Using a ray diagram, explain why it is particularly important to reduce spherical aberration.

13 A parallel beam of white light is incident on a converging glass lens which has a focal length of 20·0 cm for yellow light. A white screen is placed 20·0 cm from the lens on the other side of the beam. With the aid of a diagram, explain the change in the appearance of the coloured image seen on the screen when the screen is moved
(a) towards the lens, and
(b) away from the lens.

Miscellaneous Questions

14

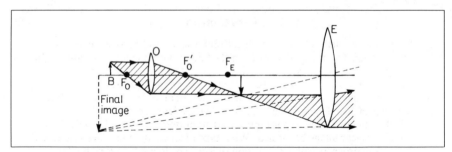

Figure 16A

The diagram shows the paths of two rays of light from the tip of an object B through the objective O, and the eyelens E of a compound microscope. The final image is at the near point of an observer's eye when the eye is close to E. F_O and F_O' are the principal foci of O and F_E is one of the principal foci of E. The diagram is *not drawn to scale*.
(a) Explain why
(i) the object is placed to the left of F,
(ii) the eyepiece is adjusted so that the intermediate image is to the right of F_E.
(b) In this arrangement, the focal lengths of O and E are 10 mm and 60 mm respectively. If B is 12 mm from O and the final image is 300 mm from E, calculate the distance apart of O and E. (*JMB*.)

17

Oscillations and Waves

In this chapter we first study the general properties of oscillations, together with resonance and phase difference. We then consider different types of waves—longitudinal and transverse waves, and progressive and stationary or standing waves. Matter waves such as sound waves, and electromagnetic waves such as light or radio waves, are then compared and discussed, together with their speeds.

S.H.M.

Simple harmonic motion (s.h.m.) occurs when the force acting on an object or system is directly proportional to its displacement x from a fixed point and is always directed towards this point. If the object moves with s.h.m., the variation of the displacement x with time t is a sine relation given by

$$x = a \sin \omega t \qquad . \qquad . \qquad . \qquad . \qquad . \qquad (1)$$

Here a is the greatest displacement from the mean or equilibrium position and is the *amplitude* of the motion, Figure 17.1. The constant $\omega = 2\pi f$ where f is the *frequency* of oscillation or number of cycles per second. The period T of the motion, or time to undergo one complete cycle, is equal to $1/f$, so that $\omega = 2\pi/T$.

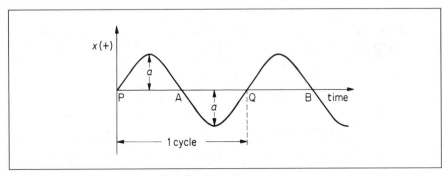

Figure 17.1 *Sine curve*

The small oscillation of a pendulum bob or a vibrating layer of air is a *mechanical oscillation*, so that x is a displacement from a mean fixed position. Later, *electrical oscillations* are considered; x may then represent the instantaneous charge on the plates of a capacitor when the charge alternates about a mean value of zero. In an *electromagnetic wave*, x may represent the component of the electric or magnetic field vectors at a particular place.

Energy in S.H.M.

On p. 82, it was shown that the sum of the potential and kinetic energies of a body moving with s.h.m. is *constant* and equal to the total energy in the vibration. Further, it was shown that the time averages of the potential energy (p.e.) and kinetic energy (k.e.) are equal; each is half the total energy. In any mechanical oscillation, *there is a continuous interchange or exchange of energy from p.e. to k.e. and back again.*

For vibrations to occur, therefore, an agency is needed which can have and store p.e. and another which can have and store k.e. This was the case for a mass oscillating on the end of a spring, as we saw on p. 83. The mass stores k.e. and the spring stores p.e.; and interchange occurs continuously from one to the other as the spring is compressed and released alternately. In the oscillations of a simple pendulum, the mass stores k.e. as it swings downwards from the end of an oscillation, and this is changed to p.e. as the height of the bob increases above its mean position.

Electrical Oscillations

So far we have dealt with mechanical oscillations and energy. The energy in electrical oscillations takes a different form. There are still two types of energy. One is the energy stored in the electric field, and the other that stored in the magnetic field. To obtain electrical oscillations, an inductor (a coil) is used to produce the magnetic field and a capacitor to produce the electric field. This is discussed more fully on page 396.

Suppose the capacitor is charged and there is no current at this moment, Figure 17.2 (i). A p.d. then exists across the capacitor and an electric field is present between the plates. At this instant all the energy is stored in the electric field, and since the current is zero there is no magnetic energy. Because of the p.d. a current will begin to flow and magnetic energy will begin to be stored in the inductor. Thus there will be a change from electric to magnetic energy. The p.d. is the agency which causes the transfer of energy.

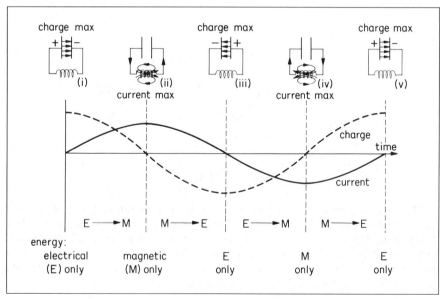

Figure 17.2 *Electrical oscillations—energy exchanges*

One quarter of a cycle later the capacitor will be fully discharged and the current will be at its greatest, so that the energy is now entirely stored in the magnetic field, Figure 17.2 (ii). The current continues to flow for a further quarter-cycle until the capacitor is fully charged in the opposite direction, when the energy is again completely stored in the electric field, Figure 17.2 (iii). The current then reverses and the processes occur in reverse order, Figure 17.2 (iv), after which the original state is restored and a complete oscillation has taken place, Figure 17.2 (v). The whole process then repeats, giving continuous oscillations.

Electrical oscillations are produced by exchange of energy between a capacitor, which stores electrical energy, and a coil or inductor, which stores magnetic energy.

Phase of Vibrations

Consider an oscillation given by $x_1 = a \sin \omega t$. Suppose a second oscillation has the same amplitude, a, and angular frequency, ω, but is out of step and reaches the end of its oscillation a fraction, β, of the period T later than the first one. The second oscillation thus *lags behind* the first by a time βT, and so its displacement x_2 is given by

$$x_2 = a \sin \omega(t - \beta T)$$

$$= a \sin (\omega t - \varphi) \quad . \quad . \quad . \quad . \quad . \quad (2)$$

where $\varphi = \omega \beta T = 2\pi \beta T / T = 2\pi \beta$. If the second oscillation *leads* the first by a time βT, the displacement is given by

$$x_2 = a \sin (\omega t + \varphi) \quad . \quad . \quad . \quad . \quad . \quad (3)$$

φ is known as the *phase angle* of the oscillation. It represents the *phase difference* between the oscillations $x_1 = a \sin \omega t$ and $x_2 = a \sin (\omega t - \varphi)$.

Graphs of displacement against time are in Figure 17.3. Curve 1 represents $x_1 = a \sin \omega t$. Curve 2 represents $x_2 = a \sin (\omega t + \pi/2)$, so that its phase lead is

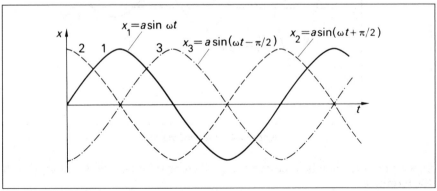

Figure 17.3 *Phase difference* (i) $1,2 = \pi/2$ or $90°$, (ii) $2,3 = \pi$ or $180°$

$\pi/2$ or $90°$; this is a *lead* of one quarter of a period. Curve 3 represents $x_3 = a \sin (\omega t - \pi/2)$ so that its phase lag is $\pi/2$ or $90°$, this is a lag of one quarter of a period on curve 1. If the phase difference is 2π, the oscillations are effectively in phase.

Note that if the phase difference is π or $180°$, the displacement of one oscillation reaches a positive maximum value at the same instant as the other oscillation reaches a *negative* maximum value. The two oscillations are thus sometimes said to be 'antiphase'. This is the case for curves 2 and 3 in Figure 17.3.

Damped Vibrations

In practice, the amplitude of vibration in simple harmonic motion does not remain constant but becomes progressively smaller. Such a vibration is said to be *damped*. The decrease in amplitude is due to loss of energy; for example, the amplitude of the bob of a simple pendulum diminishes slowly owing to the viscosity (friction) of the air. This is shown by curve 1 in Figure 17.4.

The general behaviour of mechanical systems subject to various amounts of damping may be conveniently investigated using a coil of a ballistic galvano-meter (p. 379). If a resistor is connected to the terminals of a ballistic galvano-meter when the coil is swinging, the induced e.m.f. due to the motion of the coil in the magnetic field of the galvanometer magnet causes a current to flow through the resistor. This current, by Lenz's Law (p. 343), opposes the motion of the coil and so causes damping. The smaller the value of the resistor, the greater is the degree of damping. The galvanometer coil is set swinging by discharging a capacitor through it. The time period, and the time taken for the amplitude to be reduced to a certain fraction of its original value, are then measured. The

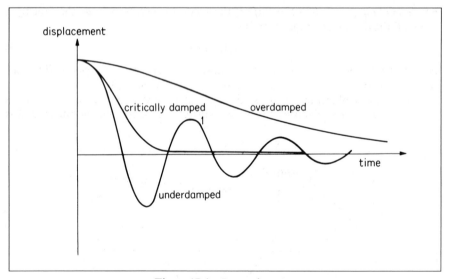

Figure 17.4 *Damped motion*

experiment can then be repeated using different values of resistor connected to the terminals.

It is found that as the damping is increased the time period increases and the oscillations die away more quickly. As the damping is increased further, there is a value of resistance which is just sufficient to prevent the coil from vibrating past its rest position. This degree of damping, called the *critical damping, reduces the motion to rest in the shortest possible time*. If the resistance is lowered further, to increase the damping, no vibrations occur but the coil takes a longer time to settle down to its rest position. Graphs showing the displacement against time for 'underdamped', 'critically damped', and 'overdamped' motion are shown in Figure 17.4.

When it is required to use a galvanometer as a current-measuring instrument, rather than ballistically to measure charge, it is generally critically damped. The return to zero is then as rapid as possible.

These results, obtained for the vibrations of a damped galvanometer coil, are quite general. All vibrating systems have a certain critical damping, which brings the motion to rest in the shortest possible time.

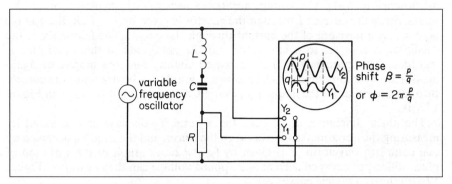

Figure 17.5 *Demonstration of oscillations*

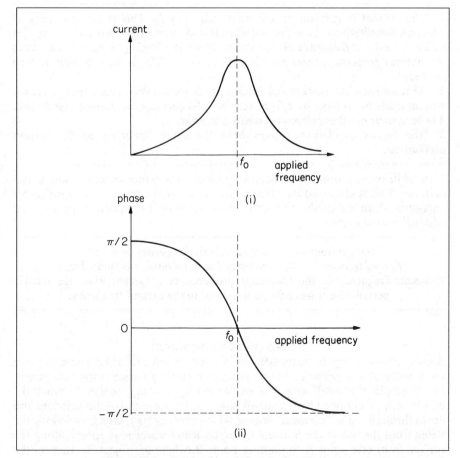

Figure 17.6 *Amplitude and phase in forced vibrations*

Forced Oscillations, Resonance

In order to keep a system, which has a degree of damping, in continuous oscillatory motion, some outside periodic force must be used. The frequency of this force is called the *forcing frequency*. In order to see how systems respond to a forcing oscillation, we may use an electrical circuit comprising a coil L, capacitor C and resistor R, shown in Figure 17.5.

The applied oscillating voltage is displayed on the Y_2 plates of a double-beam oscilloscope (p. 781). The voltage across the resistor R is displayed on the Y_1 plates. Since the current I through the resistor is given by $I = V/R$, the voltage across R is a measure of the current through the circuit. The frequency of the oscillator is now set to a low value and the amplitude of the Y_1 display is recorded. The frequency is then increased slightly and the amplitude again measured. By taking many such readings, a graph can be drawn of the current through the circuit as the frequency is varied. A typical result is shown in Figure 17.6 (i).

The phase difference, φ, between the Y_1 and Y_2 displays can be found by measuring the horizontal shift p between the traces, and the length q occupied by one complete waveform. φ is given by $(p/q) \times 2\pi$. A graph of the variation of phase difference between current and applied voltage can then be drawn. Figure 17.6 (ii) shows a typical curve.

The following observations may be made:

---

1 The current is greatest at a certain frequency f_0. This is the frequency of undamped oscillations of the system, when it is allowed to oscillate *on its own*. f_0 is called the *natural frequency* of the system. When the forcing frequency is equal to the natural frequency, *resonance* is said to occur. The largest current is then produced.
2 At resonance, the current and voltage are in phase. Well below resonance, the current leads the voltage by $\pi/2$; at very high frequencies the current lags by $\pi/2$. The behaviour of other resonant systems is similar.
3 The forced oscillations always have the same frequency as the forcing oscillations.

In addition to resonance in electrical circuits, resonance occurs in sound and in optics. This is discussed later (see p. 601). It should be noted that considerable energy is absorbed at the resonant frequency from the system supplying the external periodic force.

Natural **frequency of a system is the frequency on its own.**
Forced **frequency is the frequency due to an outside periodic force.**
Resonant **frequency is the frequency produced in a system when the outside periodic force has a frequency equal to the natural frequency.**

Waves and Wave-motion

A *wave* allows energy to be transferred from one point to another some distance away without any particles of the medium travelling between the two points. For example, if a small weight is suspended by a string, energy to move the weight may be obtained by repeatedly shaking the other end of the string up and down through a small distance. Waves, which carry energy, then travel along the string from the top to the bottom. Likewise, *water waves* may spread along the surface from one point A to another point B, where an object floating on the water will be disturbed by the wave. No particles of water at A actually travel to

B in the process. The energy in the electromagnetic spectrum, comprising X-rays and light waves, for example, may be considered to be carried by *electromagnetic waves* from the radiating body to the absorber. Again, *sound waves* carry energy from the source to the ear by disturbance of the air (p. 473).

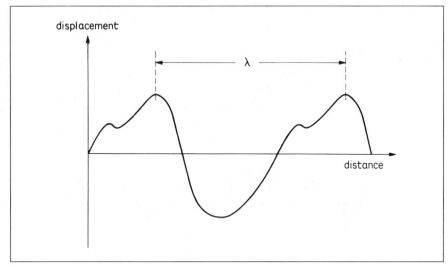

Figure 17.7 *Wave and wavelength*

If the source or origin of the wave oscillates with a frequency f, then each point in the medium concerned oscillates with the same frequency. A snapshot of the wave profile or waveform may appear as in Figure 17.7 at a particular instant. The source repeats its motion f times per second, so a repeating *waveform* is observed spreading out from it.

The distance between corresponding points in successive waveforms, such as two successive crests or two successive troughs, is called the *wavelength*, λ. Each time the source vibrates once, the waveform moves forward a distance λ. So in one second, when f vibrations occur, the wave moves forward a distance $f\lambda$. Hence the velocity c of the waves, which is the distance the profile moves in one second, is given by:

$$c = f\lambda$$

This equation is true for all wave motion, whatever its origin, that is, it applies to sound waves, electromagnetic waves and mechanical waves.

Transverse Waves

A wave which is propagated by vibrations *perpendicular* to the direction of travel of the wave is called a *transverse* wave. Examples of transverse waves are waves on plucked strings and on water. Electromagnetic waves, which include light waves, are also transverse waves.

The propagation of a transverse wave is illustrated in Figure 17.8. Each particle vibrates perpendicular to the direction of propagation with the same amplitude and frequency, and the wave is shown successively at $t = 0$, $T/4$, $T/2$, $3T/4$, in Figure 17.8, where T is the period.

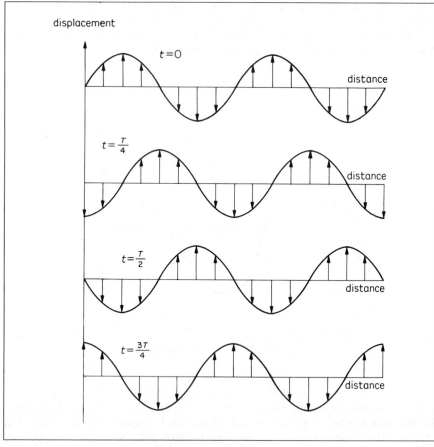

Figure 17.8 *Progressive transverse wave*

Longitudinal Waves

In contrast to a transverse wave, a *longitudinal* wave is one in which the vibrations occur in the *same* direction as the direction of travel of the wave. Figure 17.9 illustrates the propagation of a longitudinal wave. The row of dots shows the actual positions of the particles whereas the *graph* shows the *displacement* of the particles from their equilibrium positions. The positions at time $t = 0$, $t = T/4$, $t = T/2$ and $t = 3T/4$ are shown. The diagram for $t = T$ is, of course, the same as $t = 0$. With displacements to R (right) and to L (left), see graph axis, note that   (i) the displacements of the particles cause regions of high density (*compressions* C) and of low density (*rarefactions* R) to be formed along the wave.   (ii) These regions move along with the speed of the wave, as shown by the broken diagonal line.   (iii) Each particle vibrates about its mean position with the same amplitude and frequency.   (iv) The regions of greatest compression are one-quarter wavelength (90° phase) ahead of the greatest displacement in the direction of the wave. Compare the *pressure graph* at $t = 3T/4$ with the displacement graph. This result is useful in sound waves.

The most common example of a longitudinal wave is a *sound* wave. This is propagated by alternate compressions and rarefactions of the air.

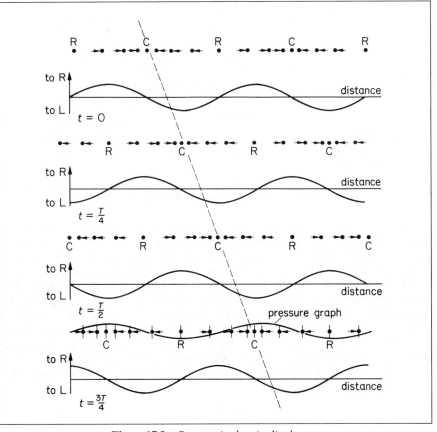

Figure 17.9 *Progressive longitudinal wave*

A *transverse* wave is one in which the direction of the oscillations is perpendicular to the direction of the wave. Light and all other electromagnetic waves, and water waves, are transverse.

A *longitudinal* wave is one in which the direction of the oscillations is in the same direction as the wave. Sound is a longitudinal wave.

Progressive Waves

Both the transverse and longitudinal waves described above are *progressive*. This means that the wave profile moves along with the speed of the wave. If a snapshot is taken of a progressive wave, it repeats at equal distances. The repeat distance is the *wavelength* λ. If one point is taken, and the profile is observed as it passes this point, then the profile is seen to repeat at equal intervals of time. The repeat time is the *period, T*.

The vibrations of the particles in a progressive wave are of the same amplitude and frequency. But *the phase of the vibrations changes for different points along the wave*. This can be seen by considering Figures 17.8 and 17.9. The phase difference may be demonstrated by the following experiment, in which sound waves of the order of 1000 to 2000 Hz may be used.

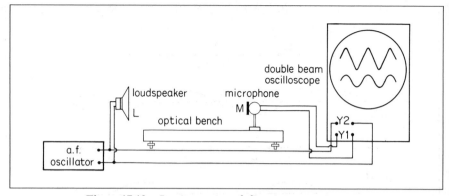

Figure 17.10 *Demonstration of phase in progressive wave*

An audio-frequency (af) oscillator is connected to the loudspeaker L and to the Y_2 plates of a double-beam oscilloscope, Figure 17.10. A microphone M, mounted on an optical bench, is connected to the Y_1 plates. When M is moved away from or towards L, the two traces on the screen are as shown in Figure 17.11 (i) at one position. This occurs when the distance LM is equal to a whole number of wavelengths, so that the signal received by M is in phase with that sent out by L. When M is now moved further away from L through distance $\lambda/4$, where λ is the wavelength, the appearance on the screen changes to that shown in Figure 17.11 (ii). The resultant phase change is $\pi/2$, so that the signal now arrives a quarter of a period later. When M is moved a distance $\lambda/2$ from its 'in-phase' position, the signal arrives half a period later, a phase change of π, Figure 17.11 (iii).

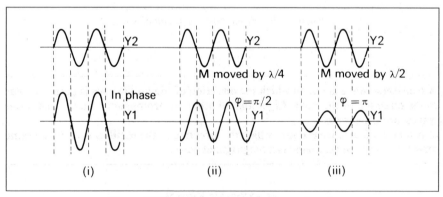

Figure 17.11 *Phase difference and wavelength*

Speed of Sound in Free Air

The speed of sound in free air can be found from this experiment. Firstly, a position of the microphone M is obtained when the two signals on the screen are in phase, as in Figure 17.10. The reading of the position of M on the optical bench is then taken. M is now moved slowly until the phase of the two signals on the screen is seen to change through $\pi/2$ to π and then to be in phase again. The shift of M is then measured. It is equal to λ, the wavelength. From several measurements the average value of λ is found, and the velocity of sound is calculated from $v = f\lambda$, where f is the frequency obtained from the oscillator dial. Another method for finding the speed of sound in free air is given on p. 488.

Progressive Wave Equation

An equation can be formed to represent generally the displacement y of a vibrating particle in a medium in which a *wave* passes. Suppose the wave moves from left to right and that a particle at the origin O then vibrates according to the equation $y = a \sin \omega t$, where t is the time and $\omega = 2\pi f$ (p. 465).

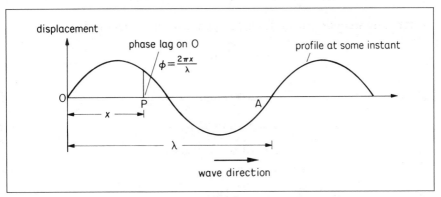

Figure 17.12 *Progressive wave equation*

At a particle P at a distance x from O to the right, the phase of the vibration will be different from that at O, Figure 17.12. A distance λ from O corresponds to a phase difference of 2π (p. 474). Thus the phase difference φ at P is given by $(x/\lambda) \times 2\pi$ or $2\pi x/\lambda$. Hence the displacement of any particle at a distance x from the origin is given by

$$y = a \sin (\omega t - \varphi)$$

or
$$y = a \sin\left(\omega t - \frac{2\pi x}{\lambda} \right). \qquad . \qquad . \qquad . \qquad . \qquad (4)$$

Since $\omega = 2\pi f = 2\pi v/\lambda$, where v is the velocity of the wave, this equation may be written:

$$y = a \sin\left(\frac{2\pi v t}{\lambda} - \frac{2\pi x}{\lambda} \right)$$

or
$$y = a \sin \frac{2\pi}{\lambda}(vt - x) \qquad . \qquad . \qquad . \qquad . \qquad . \qquad (5)$$

Also, since $\omega = 2\pi/T$, equation (4) may be written:

$$y = a \sin 2\pi \left(\frac{t}{T} - \frac{x}{\lambda} \right). \qquad . \qquad . \qquad . \qquad . \qquad (6)$$

Equations (5) or (6) represent a *plane-progressive wave*. The negative sign in the bracket indicates that, since the wave moves from left to right, the vibrations at points such as P to the right of O will *lag* on that at O. A wave travelling in the *opposite direction*, from right to left, arrives at P before O. Thus the vibration at

P *leads* that at O. Consequently a wave travelling in the opposite direction is given by

$$y = a \sin 2\pi \left(\frac{t}{T} + \frac{x}{\lambda} \right) \quad . \quad . \quad . \quad . \quad (7)$$

that is, the sign in the bracket is now a plus sign.

As an illustration of calculating the constants of a wave, suppose a wave is represented by

$$y = a \sin \left(2000\pi t - \frac{\pi x}{0 \cdot 17} \right)$$

where t is in seconds, y in m. Then, comparing it with equation (5),

$$y = a \sin \frac{2\pi}{\lambda} (vt - x)$$

we have

$$\frac{2\pi v}{\lambda} = 2000\pi \text{ and } \frac{2\pi}{\lambda} = \frac{\pi}{0 \cdot 17}$$

$$\therefore \lambda = 2 \times 0 \cdot 17 = 0 \cdot 34 \, \text{m}$$

and

$$v = 1000\lambda = 1000 \times 0 \cdot 34$$

$$= 340 \, \text{m s}^{-1}$$

$$\therefore \text{ frequency, } f, = \frac{v}{\lambda} = \frac{340}{0 \cdot 34} = 1000 \, \text{Hz}$$

$$\therefore \text{ period, } T, = \frac{1}{f} = \frac{1}{1000} \, \text{s}$$

If two layers of the wave are 1·8 m apart, they are separated by 1·8/0·34 wavelengths, or by $5\frac{10}{34}\lambda$. Their *phase difference* for a separation λ is 2π; and hence, for a separation $10\lambda/34$, omitting 5λ from consideration, we have:

$$\text{phase difference} = \frac{10}{34} \times 2\pi = \frac{10\pi}{17} \text{ radians}$$

Since y represents the displacement of a particle as the wave travels, the *velocity v* of the particle at any instant is given by dy/dt. From equation (6),

$$v = \frac{dy}{dt} = \frac{2\pi a}{T} \cos 2\pi \left(\frac{t}{T} - \frac{x}{\lambda} \right)$$

So the graph of v against x is 90° out of phase with the graph of y against x. Some microphones used in broadcasting are 'velocity' types, in the sense that the audio current produced is proportional to the velocity of the particles of air. Other microphones, such as those used in the telephone handset, may be 'pressure' types—the audio current is here proportional to the pressure changes in the air.

In progressive waves, the amplitude may be constant and neighbouring points are out of phase with each other.

Principle of Superposition

When two waves travel through a medium, their combined effect at any point can be found by the *Principle of Superposition*. This states that *the resultant displacement at any point is the sum of the separate displacements due to the two waves.*

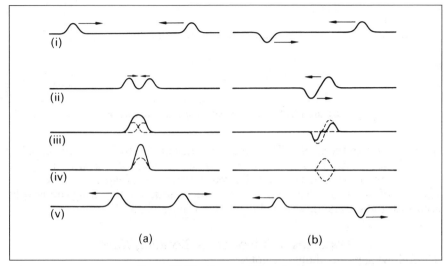

Figure 17.13   *Superposition of waves*

The principle can be illustrated by means of a long stretched spring ('Slinky'). If wave pulses are produced at each end simultaneously, the two waves pass through the wire. Figure 17.13(a) shows the stages which occur as the two pulses pass each other. In Figure 17.13(a)(i), they are some distance apart and are approaching each other, and in Figure 17.13(a)(ii) they are about to meet. In Figure 17.13(a)(iii), the two pulses, each shown by broken lines, are partly overlapping. The resultant is the sum of the two curves. In Figure 17.13(a)(iv), the two pulses exactly overlap and the greatest resultant is obtained. The last diagram shows the pulses receding from one another.

The diagrams in Figure 17.13(b) show the same sequence of events (i)–(v) but the pulses are now equal and opposite. The Principle of Superposition is widely used in discussion of wave phenomena such as interference, as we shall see (p. 515).

Stationary or Standing Waves

We have already discussed progressive waves and their properties. Figure 17.14 shows an apparatus which produces a different kind of wave (see also p. 592). If the weights on the scale-plan are suitably adjusted, a number of *stationary vibrating loops* are seen on the string when one end is set vibrating. This time the wave-like profile on the string does *not* move along the medium, which is the string, and the wave is therefore called a *stationary* (or *standing*) *wave*. In Figure 17.14 a wave travelling along the string to one end is reflected here. So the stationary wave is due to the superposition of *two waves* of equal frequency and amplitude travelling in *opposite* directions along the string. This is discussed more fully on p. 592.

The motion of the string when a stationary wave is produced can be studied by using a stroboscope (strobe). This instrument gives a flashing light whose frequency can be varied. The apparatus is set up in a darkened room and

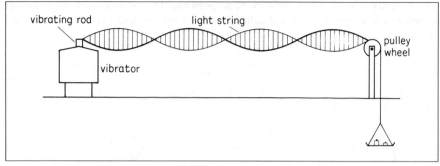

Figure 17.14 *Demonstration of stationary wave*

illuminated with the strobe. When the frequency of the strobe is nearly equal to that of the string, the string can be seen moving up and down slowly. Its observed frequency is equal to the difference between the frequency of the strobe and that of the string. Progressive stages in the motion of the string can now be seen and studied, and these are illustrated in Figure 17.15.

Properties of Stationary or Standing Waves

The following points should be noted:

1   There are points such as B where the displacement is permanently zero. These points are called *nodes* of the stationary wave.
2   At points between successive nodes the vibrations *are in phase*.

This last property of the stationary wave is in sharp contrast to the progressive

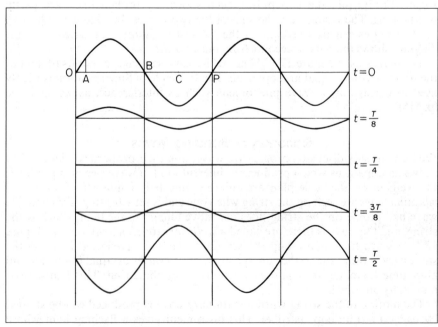

Figure 17.15 *Changes in motion of stationary wave*

wave, where the phase of points near each other are all different. Thus when one point of a stationary wave is at its maximum displacement, *all* points are then at their maximum displacement. When a point (other than a node) has zero displacement, *all* points then have zero displacement.

3 **Each point along the wave has a *different amplitude* of vibration from neighbouring points. Points such as C in Figure 17.15 which have the greatest amplitude are called *antinodes*.**

Again this is different from the case of a progressive wave, where every point vibrates with the same amplitude.

4 **The wavelength is equal to the distance OP, Figure 17.15. So the wavelength λ is *twice* the *distance between successive nodes or successive antinodes*. The distance between successive nodes or antinodes is $\lambda/2$; the distance between a node and a neighbouring antinode is $\lambda/4$.**

Stationary Longitudinal Waves

In Sound, stationary longitudinal waves can be set up in a pipe closed at one end (closed pipe). We shall study this in more detail in a later chapter. Here we may note that all the possible frequencies obtained from the pipe are subject to the condition that the closed end must be a displacement *node* of the stationary wave formed, since the air cannot move here, and the open end must be a displacement *antinode* as the air is most free to move here. Figure 17.16 (i) shows the stationary wave formed for the lowest possible frequency f_0 and other possible or allowed frequencies $3f_0$ and $5f_0$.

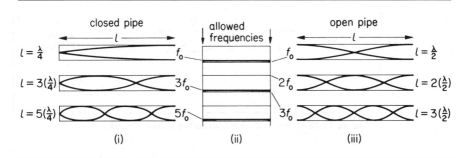

Figure 17.16 *Stationary waves in pipes*

Figure 17.16 (ii) shows the possible frequencies, $f_0, 2f_0, 3f_0, \ldots$, for the case of the pipe open at both ends. Here the two ends of the pipes must be antinodes and so the possible stationary waves are those shown.

Stationary Transverse Waves

Figure 17.17 shows the possible frequencies for stationary transverse waves produced by plucking in the middle a string fixed at both ends. Here the ends must always be displacement nodes and the middle an antinode.

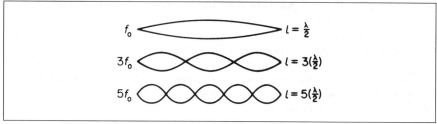

Figure 17.17 *Stationary waves in strings*

Pressure in Stationary Wave

Consider the instant corresponding to curve 1 of the displacement graph of a stationary wave, Figure 17.18(i). At the node *a*, the particles on either side produce a compression (increase of pressure), from the direction of their displacement. At the same instant the pressure at the antinode *b* is normal and that at the node *c* is a rarefaction (decrease in pressure). Figure 17.18(ii) shows the pressure variation along the stationary wave—the displacement nodes are the pressure antinodes. So the closed end of the pipe in Figure 17.16(i) is a *pressure antinode.*

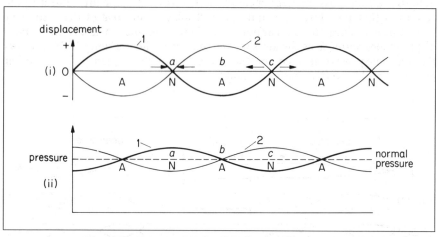

Figure 17.18 *Pressure variation due to stationary wave*

A stationary or standing wave is one in which some points are permanently at rest (*nodes*), others between these points are vibrating with varying amplitude and the maximum amplitude is midway between the nodes (*antinodes*). Points between successive nodes are in phase with each other. (In a progressive wave, neighbouring points are out of phase with each other.)

In air or other gases, a pressure antinode occurs at a displacement node and a pressure node at a displacement antinode.

Stationary Light Waves

Light waves have extremely short wavelengths of the order of 5×10^{-7} m or 0·0005 mm. In 1890 Wiener succeeded in detecting stationary light waves.

He deposited a very thin photographic film, about one-twentieth of the wavelength of light, on glass and placed it in a position XY inclined at the extremely small angle of about 4′ to a plane mirror CD. Figure 17.19 is an exaggerated sketch for clarity. When the mirror was illuminated normally by monochromatic light and the film was developed, bright and dark bands were seen. These were respectively antinodes A and nodes N of the stationary light waves formed by reflection at the mirror (see Figure 17.20).

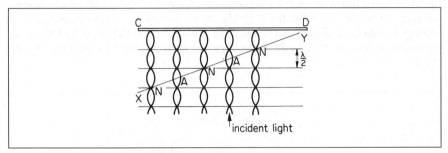

Figure 17.19 *Stationary light waves*

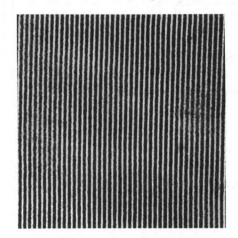

Figure 17.20 *Stationary light waves due to the mercury line of wavelength 546 nm*

Stationary Waves in Aerials

Stationary waves, due to oscillating electrons, are produced in *aerials* tuned to incoming radio waves, or aerials transmitting radio waves. Figure 17.21 illustrates the stationary wave obtained on a vertical metal rod acting as a 'quarter-

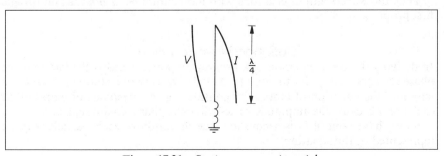

Figure 17.21 *Stationary waves in aerials*

wave' aerial. Here the electrons cannot move at the top of the rod, so this is a current (I) node. The current antinode is near the other end of the rod. The current node corresponds to a voltage (V) antinode, as shown (compare 'displacement' and 'pressure' for the case of the closed pipe on p. 479).

Stationary Waves in Electron Orbits

Moving electrons have wave properties (see p. 875). If we consider a circular orbit of the simplest atom, the hydrogen atom, there must be a complete number of such waves in the orbit for a stable atom; otherwise some of the waves or energy would be radiated as the electron rushed round the orbit and the atom would then lose its energy. So stationary waves are formed in the orbit. Figure 17.22.

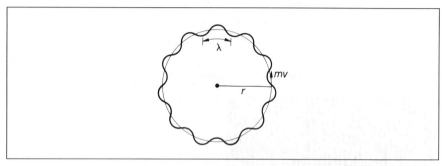

Figure 17.22 *Stationary waves in electron orbit*

So if the radius is r and there are n waves of wavelength λ, we must have

$$2\pi r = n\lambda$$

Now Bohr suggested that the angular momentum about the centre $= nh/2\pi$ (p. 862), where n is an integer and h is the Planck constant.

$$\therefore\ mv \times r = \frac{nh}{2\pi}$$

From above,
$$r = \frac{n\lambda}{2\pi} = \frac{nh}{2\pi mv}$$

$$\therefore \lambda = \frac{h}{mv}$$

Thus the wavelength of electrons with momentum mv is h/mv, as de Broglie first proposed (p. 876).

Stationary Wave Equation

In deriving the wave equation of a progressive wave, we used the fact that the phase changes from point to point (p. 475). In the case of a stationary wave, we may find the equation of motion by considering the *amplitude* of vibration at each point because the amplitude varies while the phase remains constant.

As we have seen, if ω is a constant, the vibration of each particle may be represented by the equation

$$y = Y \sin \omega t \qquad . \qquad . \qquad . \qquad . \qquad (8)$$

where Y is the amplitude of the vibration at the point considered. Y varies along the wave with the distance x from some origin. If we suppose the origin to be at an antinode, then the origin will have the greatest amplitude, A, say. Now the wave repeats at every distance λ, and it can be seen that the amplitudes at differing points vary sinusoidally with their particular distance x. An equation representing the changing amplitude Y along the wave is thus:

$$Y = A \cos \frac{2\pi x}{\lambda} = A \cos kx \qquad . \qquad . \qquad . \qquad (9)$$

where $k = 2\pi/\lambda$. When $x = 0$, $Y = A$; when $x = \lambda$, $Y = A$. When $x = \lambda/2$, $Y = -A$. This equation hence correctly describes the variation in amplitude along the wave, as shown in Figure 17.15. Hence the equation of motion of a stationary wave is, with equation (8),

$$y = A \cos kx \sin \omega t . \qquad . \qquad . \qquad . \qquad (10)$$

From equation (10), $y = 0$ at all times when $\cos kx = 0$. Thus $kx = \pi/2, 3\pi/2, 5\pi/2, \ldots$, in this case. This gives values of x corresponding to $\lambda/4, 3\lambda/4, 5\lambda/4, \ldots$ These points are *nodes* since the displacement at a node is always zero. Thus equation (10) gives the correct distance, $\lambda/2$, between nodes.

A stationary wave can be considered as produced by the superposition of *two progressive waves, of the same amplitude and frequency, travelling in opposite directions*, as we now show.

Mathematical Proof of Stationary Wave Properties

The properties of the stationary wave just deduced can be obtained by a mathematical treatment. Suppose $y_1 = a \sin 2\pi \left(\dfrac{t}{T} - \dfrac{x}{\lambda} \right)$ is a plane-progressive wave travelling in one direction along the x-axis (p. 475). Then $y_2 = a \sin 2\pi \left(\dfrac{t}{T} + \dfrac{x}{\lambda} \right)$ represents a wave of the same amplitude and frequency travelling in the opposite direction. The resultant displacement, y, is hence given by

$$y = y_1 + y_2 = a \left[\sin 2\pi \left(\frac{t}{T} - \frac{x}{\lambda} \right) + \sin 2\pi \left(\frac{t}{T} + \frac{x}{\lambda} \right) \right]$$

from which
$$y = 2a \sin \frac{2\pi t}{T} . \cos \frac{2\pi x}{\lambda} \qquad . \qquad . \qquad . \qquad (i)$$

using the transformation of the sum of two sine functions to a product.

$$\therefore y = Y \sin \frac{2\pi t}{T} \qquad . \qquad . \qquad . \qquad . \qquad (ii)$$

where
$$Y = 2a \cos \frac{2\pi x}{\lambda} . \qquad . \qquad . \qquad . \qquad (iii)$$

From (ii), Y is the magnitude of the *amplitude* of vibration of the various layers; and from (iii) it also follows that the amplitude is a maximum and equal to $2a$ at $x = 0$, $x = \lambda/2$, $x = \lambda$, and so on. These points are thus antinodes, and consecutive antinodes are hence separated by a distance $\lambda/2$. The amplitude Y is zero when $x = \lambda/4$, $x = 3\lambda/4$, $x = 5\lambda/4$, and so on. These points are thus nodes, and they are hence midway between consecutive antinodes.

Wave Properties, Reflection

Any wave motion can be *reflected*. The reflection of light waves is discussed later.

Like light waves, sound waves are reflected from a plane surface so that the angle of incidence is equal to the angle of reflection. This can be demonstrated by placing a tube T_1 in front of a plane surface AB and blowing a whistle gently at S, Figure 17.23. Another tube T_2, directed towards N, is place on the other side of the normal NQ, and moved until a sensitive microphone, connected to a cathode-ray oscilloscope, is considerably affected at R, showing that the reflected wave is in the direction NR. It will then be found that angle RNQ = angle SNQ.

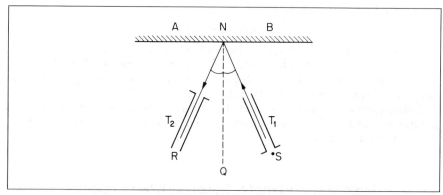

Figure 17.23   *Reflection of sound*

It can also be shown that sound waves come to a focus when they are incident on a curved concave mirror. A surface shaped like a parabola reflects sound waves to long distances if the source of sound is placed at its focus (see also p. 413). The famous whispering gallery of St. Paul's is a circular-shaped chamber whose walls repeatedly reflect sound waves round the gallery, so that a person talking quietly at one end can be heard distinctly at the other end.

Electromagnetic waves of about 21 cm wavelength from outer space are now detected by radio-telescopes. The waves are reflected by a large parabolic 'dish' to a sensitive receiver (see p. 414). A demonstration of the reflection of 3 cm electromagnetic waves is shown on p. 490.

Refraction

Waves can also be *refracted*, that is, their direction changes when they enter a new medium. This is due to the change in velocity of the waves on entering a different medium. Refraction of light is discussed on p. 415.

Sound waves can be refracted as well as reflected. TYNDALL placed a watch in front of a balloon filled with carbon dioxide, which is heavier than air, and found that the sound was heard at a definite place on the other side of the balloon. The sound waves thus converged to a focus on the other side of the balloon, which therefore has the same effect on sound waves as a converging lens has on light waves (see p. 507). If the balloon is filled with hydrogen, which is lighter than air, the sound waves diverge on passing through the balloon. In this case the gas balloon acts similarly to a diverging lens when light waves are incident on it (see p. 507).

The refraction of sound explains why sounds are easier to hear at night than during day-time. In the day-time, the upper layers of air are colder than the layers near the earth. Now sound travels faster the higher the temperature (see p. 496), and sound waves are hence refracted in a direction away from the earth. The intensity of the sound waves thus diminishes. At night-time, however, the layers of air near the earth are colder than those higher up, and hence sound waves are now refracted towards the earth, with a consequent increase in intensity.

For a similar reason, a distant observer O hears a sound from a source S more easily when the wind is blowing towards him than away from him, Figure 17.24. When the wind is blowing towards O, the bottom of the sound wavefront is

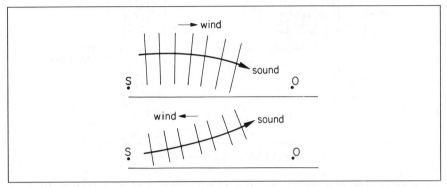

Figure 17.24 *Refraction of sound*

moving slower than the upper part, and hence the wavefronts turn towards the observer, who therefore hears the sound easily. When the wind is blowing in the opposite direction the reverse is the case, and the wavefronts turn upwards away from the ground and O. The sound intensity thus diminishes. This phenomenon shows how wavefronts may change direction due to variation in wind velocity.

Radio (electromagnetic) waves are refracted in the ionosphere (a layer of electrons and ions) high above the earth when they are transmitted from one side of the globe to the other. A demonstration of the refraction of microwaves, 3 cm electromagnetic waves, is shown on p. 490.

Diffraction

Waves can also be 'diffracted'. *Diffraction* is the name given to the spreading of waves when they pass through apertures or around obstacles.

The general phenomenon of diffraction may be illustrated by using water waves in a ripple tank, with which we assume the reader is familiar. Figure 17.25 (i) shows the effect of widening the aperture and Figure 17.25 (ii) the effect of shortening the wavelength and keeping the same width of opening. In certain circumstances in diffraction, reinforcement of the waves, or complete cancellation occurs in particular directions from the aperture, as shown in Figure 17.25 (i) and (ii). These patterns are called 'diffraction bands' (p. 540).

Generally, the smaller the width of the aperture in relation to the wavelength, the greater is the spreading or diffraction of the waves. This explains why we cannot see round corners. The wavelength of *light waves* is about 6×10^{-7} m (p. 521). This is so short that no appreciable diffraction is obtained around obstacles of normal size. With very small obstacles or narrow apertures, how-

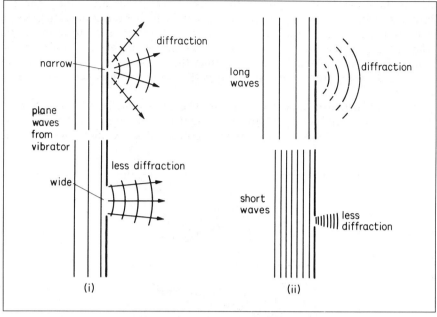

Figure 17.25   *Diffraction of waves*

ever, diffraction of light may be appreciable (see p. 540). *Electromagnetic waves* can be diffracted, as shown on p. 490.

Sound waves are diffracted round wide openings such as doorways because their wavelength is comparable with the width of the opening. For example, the wavelength for a frequency of, say, 680 Hz is about 0·5 m and the width of a door may be about 0·8 m. Generally, the diffraction increases with longer wavelength. For this reason, the low notes of a band marching away out of sight round a corner are heard for a longer time than the high notes. Similarly, the low notes of an orchestra playing in a hall can be heard through a doorway better than the high notes by a listener outside the hall.

Interference

When two or more waves of the same frequency overlap, the phenomenon of *interference* occurs. Interference is easily demonstrated in a ripple tank. Two sources, A and B, of the same frequency are used. These produce circular waves which spread out and overlap, and the pattern seen on the water surface is shown in Figure 17.26.

The interference pattern can be explained from the Principle of Superposition (p. 476). If the oscillations of A and B are in phase, crests from A will arrive at the same time as crests from B at any point on the line RS. Hence by the Principle of Superposition there will be reinforcement or a large wave along RS. Along XY, however, crests from A will arrive before corresponding crests from B. In fact, every point on XY is half a wavelength, λ, nearer to A than to B, so that crests from A arrive at the same time as troughs from B. Thus, by the Principle of Superposition, the resultant is *zero*. Generally, reinforcement (constructive interference) occurs at a point C when the path difference $AC - BC = 0$ or λ or 2λ, and cancellation (destructive interference) when $AC - BC = \lambda/2$ or $3\lambda/2$ or $5\lambda/2$.

Interference of *light waves* is discussed in detail on p. 515. An experiment to

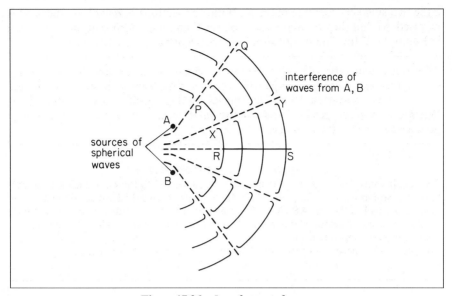

Figure 17.26 *Interference of waves*

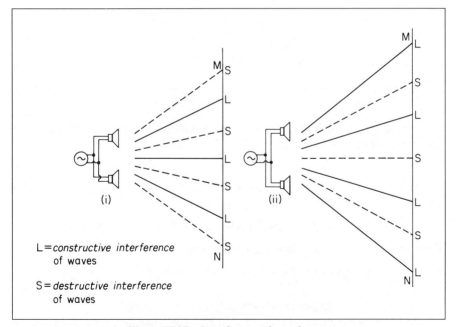

Figure 17.27 *Interference of sound waves*

demonstrate the interference of *electromagnetic waves* (microwaves) is given on p. 490. The interference of *sound waves* can be demonstrated by connecting two loudspeakers in parallel to an audio-frequency oscillator, Figure 17.27 (i). As the ear or microphone is moved along the line MN, alternate loud (L) and soft (S) sounds are heard according to whether the receiver of sound is on a line of reinforcement (constructive interference) or cancellation (destructive inter-ference) of waves. Figure 17.27 (i) indicates the positions of loud and soft sounds if the two speakers oscillate in phase. If the connections to *one* of the speakers is reversed, so that they oscillate out of phase, then the pattern is altered as shown in Figure 17.27 (ii). The reader should try to account for this difference.

All waves can be reflected, refracted (due to change of speed) and diffracted (they spread through openings comparable to their wavelength). Interference occurs where two waves overlap. From the Principle of Superposition, two crests arriving together at a place produce constructive interference but a crest and a trough produce destructive interference.

Example on Sound Interference

Two small loudspeakers A, B, 1·00 m apart, are connected to the same oscillator so that both emit sound waves of frequency 1700 Hz in phase, Figure 17.28. A sensitive detector, moving parallel to the line AB along PQ 2·40 m away, detects a maximum wave at P on the perpendicular bisector MP of AB and another maximum wave when it first reaches a point Q directly opposite to B.

Calculate the speed c of the sound waves in air from these measurements.

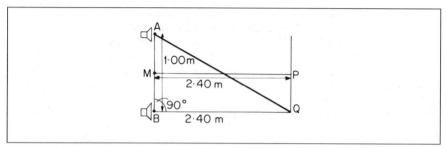

Figure 17.28 *Example*

There is constructive interference of the sound waves at P and Q. Since Q is the first maximum after P, where AP = BP, it follows that

$$AQ - BQ = \lambda \qquad . \qquad . \qquad . \qquad . \qquad . \qquad (1)$$

where λ is the wavelength of the sound waves. Now BQ = 2·40 m, AB = 1·00 m and angle ABQ = 90°. So

$$AQ = \sqrt{BQ^2 + AB^2} = \sqrt{2 \cdot 40^2 + 1 \cdot 00^2} = 2 \cdot 60 \text{ m}$$

From (1), $\lambda = 2 \cdot 60 - 2 \cdot 40 = 0 \cdot 20 \text{ m}$

So   wave speed $c = f\lambda = 1700 \times 0 \cdot 20 = 340 \text{ m s}^{-1}$

Measurement of Speed of Sound

Figure 17.29 shows one method of measuring the speed of sound in free air by an interference method.

Sound waves of constant frequency, such as 1500 Hz, travel from a loud-speaker L towards a vertical board M. Here the waves are reflected and interfere with the incident waves. As explained before, the two waves travelling in opposite directions produce a *stationary wave* between the board M and L.

A small microphone, positioned in front of the board, is connected to the Y-plates of an oscilloscope. As the microphone is moved back from M towards L, the amplitude of the waveform seen on the screen increases to a *maximum* at one position A, as shown. This is an antinode of the stationary wave. When the

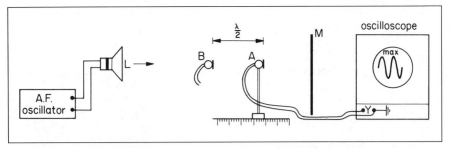

Figure 17.29 *Velocity of sound in free air—interference method*

microphone is moved on, the amplitude diminishes to a minimum (a node) and then increases to a maximum again at a position B, the next antinode. The distance between successive antinodes is $\lambda/2$ (p. 479). Thus by measuring the average distance d between successive maxima, the wavelength λ can be found. Knowing the frequency f of the note from the loudspeaker, the velocity c of the sound wave can be calculated from $c = f\lambda$.

Velocity of Sound by Lissajous' Figures

The velocity of sound can be measured by a different method using an oscilloscope. In this case, the loudspeaker L is connected to the X-plates of the oscilloscope and the microphone A to the Y-plates. The board M in Figure 17.29 is completely removed.

When A faces L, the oscilloscope beam is affected
(a) *horizontally* by a voltage V_L due to the sound waves from L,
(b) *vertically* by a voltage V_A due to sound waves arriving at A.
These two sets of waves together produce a resultant waveform on the screen, whose geometrical form is called a *Lissajous figure*.

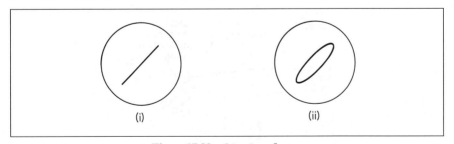

Figure 17.30 *Lissajous figures*

If V_L and V_A are exactly in phase, a straight inclined line is seen on the screen. Figure 17.30 (i). If V_L and V_A are out of phase, an ellipse is seen, Figure 17.30 (ii). By moving the microphone, the average distance between successive positions when a sloping straight line in the same direction as the original appears on the screen can be measured. This distance is λ, the wavelength. The velocity of sound v is then calculated from $v = f\lambda$, where f is the known frequency of the sound. This method is more accurate than the method described earlier, as the Lissajous' straight line position of the microphone can be found with far greater accuracy than the maxima (or minima) positions required in Figure 17.29.

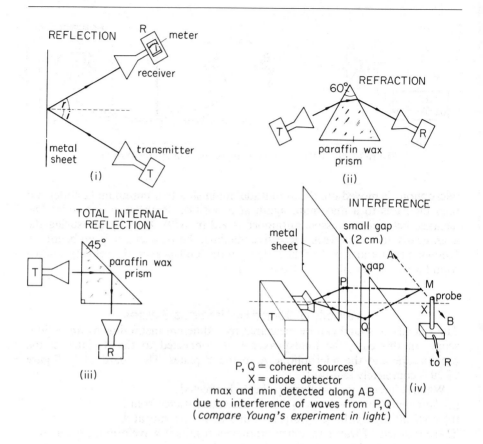

REFLECTION

meter

receiver

metal sheet

transmitter

T

(i)

REFRACTION

60°

T

R

paraffin wax prism

(ii)

TOTAL INTERNAL REFLECTION

45°

paraffin wax prism

T

R

(iii)

INTERFERENCE

metal sheet

small gap (2 cm)

gap

A

P

M

probe

T

X

B

Q

to R

P, Q = coherent sources
X = diode detector
max and min detected along A B
due to interference of waves from P, Q
(*compare Young's experiment in light*)

(iv)

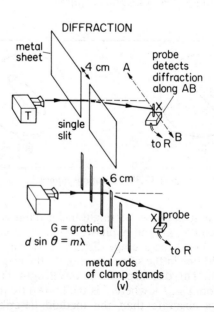

DIFFRACTION

metal sheet

4 cm   A

probe detects diffraction along AB

T

single slit

X

B

to R

6 cm

G = grating
$d \sin \theta = m\lambda$

X   probe

to R

metal rods of clamp stands

(v)

Figure 17.31 *Experiments with microwaves:* (i) *reflection,* (ii) *refraction,* (iii) *total internal reflection,* (iv) *interference,* (v) *diffraction*

Wave Properties of Electromagnetic Waves

Electromagnetic waves, like all waves, can undergo reflection, refraction, interference and diffraction. In laboratory demonstrations, *microwaves* of about 3 cm wavelength may be used. These are radiated from a horn waveguide T and are received by a similar waveguide R or by a smaller *probe* X. The detected wave then produces a deflection in a connected meter. Some experiments which can be performed in a school laboratory are illustrated in Figure 17.31 (i)–(v).

Polarisation of Waves

A transverse wave due to vibrations in *one plane* is said to be *plane-polarised*. Figure 17.32 shows a plane-polarised wave due to vibrations in the vertical plane *yOx* and another plane-polarised wave due to vibrations in the perpendicular plane *zOx*. Both waves travel in the direction O*x*.

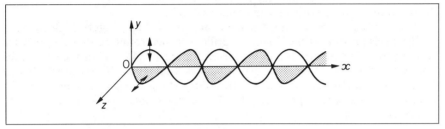

Figure 17.32 *Plane-polarised waves*

Consider a horizontal rope AD attached to a fixed point D at one end, Figure 17.33 (i). Transverse waves due to vibrations in many different planes can be set up along A by holding the end A in the hand and moving it up and down in all directions perpendicular to AD, as illustrated by the arrows in the plane X. Suppose we repeat the experiment but this time we have two parallel slits B and C between A and D as shown. A wave then emerges along BC, but unlike the waves along the part AB of the rope (not shown), which are due to vibrations in many different planes, the wave along BC is due only to vibrations parallel to the slit B. This plane-polarised wave passes through the parallel slit C. But when C is turned so that it is *perpendicular* to B, as shown in Figure 17.33 (ii), no wave is now obtained beyond C.

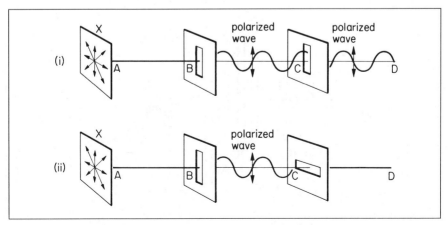

Figure 17.33 *Transverse waves and polarisation*

It is important to note that *no polarisation can be obtained with longitudinal waves*. Figure 17.33 illustrates how transverse waves can be distinguished by experiment from longitudinal waves. If the rope AD is replaced by a thick elastic cord, and longitudinal waves are produced along AD, then turning the slit C round from the position shown in Figure 17.33 (i) to that shown in Figure 17.32 (ii) makes no difference to the wave—it travels through B and C undisturbed. Since sound waves are longitudinal waves, no polarisation of sound waves can be produced. As we shall see in a later chapter, because light waves are transverse waves they can be polarised. The phenomena of interference and diffraction occur both with sound and light waves, but only the phenomenon of polarisation can distinguish between waves which may be longitudinal or transverse.

Figure 17.34 illustrates an experiment on polarisation carried out with *electromagnetic waves*. Here a grille of parallel metal rods is rotated between
(a) a source T of 3 cm electromagnetic waves or microwaves,
(b) detector, a probe, with a meter connected to it.
When the rods are horizontal the meter reading is high, Figure 17.34 (i). So a wave travels past the grille. When the grille is turned round so that the rods are vertical, there is no deflection in the meter, Figure 17.34 (ii). Thus the wave does not travel past the grille. This experiment is analogous to that illustrated in Figure 17.33 for mechanical waves travelling along a rope. It shows that the electromagnetic waves produced by T are *plane-polarised* and so they are transverse waves.

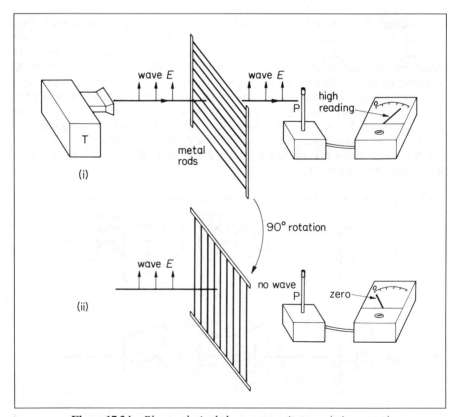

Figure 17.34   *Plane-polarised electromagnetic waves (microwaves)*

We can also see that the electromagnetic waves from T are plane-polarised by placing T in front of the probe P *without* using the grille, so that the meter joined to P indicates a high reading. If T is now rotated about its axis through 90°, the meter reading falls to zero. The probe P detects vibrations in one plane, so that when T is rotated through 90° the waves are not detected as they are plane-polarised.

Transverse waves can be polarised. Vibrations in one plane only (plane-polarised waves) can be produced from transverse waves. So all electromagnetic waves such as light, radio waves and X-rays can be polarised.
   Longitudinal waves such as sound can not be polarised. Transverse and longitudinal waves can only be distinguished by *polarisation* and not by reflection, refraction, interference or diffraction.

Velocity of Waves

We now list, for convenience, the velocity c of waves of various types, some of which are considered more fully in other sections of the book:

1. *Transverse wave on string*

$$c = \sqrt{\frac{T}{\mu}} \qquad . \quad . \quad . \quad . \quad . \quad (1)$$

where T is the tension and μ is the mass per unit length.

2. *Sound waves in gas*

$$c = \sqrt{\frac{\gamma p}{\rho}} \qquad . \quad . \quad . \quad . \quad . \quad (2)$$

where p is the pressure, ρ is the density and γ is the ratio of the molar heat capacities of the gas (p. 672).

3. *Longitudinal waves in solid*

$$c = \sqrt{\frac{E}{\rho}} \qquad . \quad . \quad . \quad . \quad (3)$$

where E is Young modulus and ρ is the density.

4. *Electromagnetic waves*

$$c = \sqrt{\frac{1}{\mu\varepsilon}} \qquad . \quad . \quad . \quad . \quad . \quad (4)$$

where μ is the permeability and ε is the permittivity of the medium. In free space $\mu_0 = 4\pi \times 10^{-7}\ \text{H m}^{-1}$, $\varepsilon_0 = 8\cdot85 \times 10^{-12}\ \text{F m}^{-1}$, so $c = 3\cdot0 \times 10^8\ \text{m s}^{-1}$.

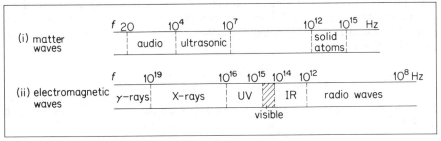

Figure 17.35   *Frequency spectrum*

Figure 17.35(i) shows roughly the range of frequencies in the spectrum of *matter waves*—waves due to vibrations in solids, liquids and gases including sound waves. Sound frequencies in air range from about 20 to 20000 Hz for those detected by the human ear but very much higher particle frequencies can be obtained in solids.

Figure 17.35(ii) shows roughly the range of frequencies in the spectrum of *electromagnetic waves*. The various waves are discussed later. The frequency range detected by the eye is about 4×10^{14} to 7×10^{14} Hz, a range factor of about 2. The human ear, however, has a range factor of about 1000.

Velocity of Sound in a Medium

When a sound wave travels in a medium, such as a gas, a liquid, or a solid, the particles in the medium are subjected to varying stresses, with resulting strains (p. 472). The velocity of a sound wave is thus partly governed by the *modulus of elasticity*, E, of the medium, which is defined by the relation

$$E = \frac{\text{stress}}{\text{strain}} = \frac{\text{force per unit area}}{\text{change in length (or volume)/original length (or volume)}} \quad \text{(i)}$$

The velocity, c, also depends on the density, ρ, of the medium, and it can be shown that

$$c = \sqrt{\frac{E}{\rho}} . \quad . \quad . \quad . \quad . \quad (1)$$

When E is in newton per metre^2 (N m^{-2}) and ρ in kg m^{-3}, then c is in metre per second (m s^{-1}). The relation (1) was first obtained by Newton.

For a solid, E is Young modulus of elasticity. The magnitude of E for steel is about $2 \times 10^{11} \text{ N m}^{-2}$, and the density ρ of steel is 7800 kg m^{-3}. So the velocity of sound in steel is given by

$$c = \sqrt{\frac{E}{\rho}} = \sqrt{\frac{2 \times 10^{11}}{7800}} = 5060 \text{ m s}^{-1}$$

For a liquid, E is the bulk modulus of elasticity. Water has a bulk modulus of $2 \cdot 04 \times 10^9 \text{ N m}^{-2}$, and a density of 1000 kg m^{-3}. The calculated velocity of sound in water is thus given by

$$c = \sqrt{\frac{2 \cdot 04 \times 10^9}{1000}} = 1430 \text{ m s}^{-1}$$

The proof of the velocity formula requires advanced mathematics, and is beyond the scope of this book. It can partly be verified by the method of dimensions, however. Thus since density, ρ, = mass/volume, the dimensions of ρ are given by ML^{-3}. The dimensions of force (mass × acceleration) are MLT^{-2}, the dimensions of area are L^2; and the denominator in (i) has zero dimensions since it is the ratio of two similar quantities. So the dimensions of modulus of elasticity, E, are given by

$$\frac{\text{ML}}{\text{T}^2\text{L}^2} \text{ or } \text{ML}^{-1}\text{T}^{-2}$$

Suppose the velocity, c, = $kE^x\rho^y$, where k is a constant. The dimensions of c are LT^{-1}

$$\therefore \text{LT}^{-1} \equiv (\text{ML}^{-1}\text{T}^{-2})^x \times (\text{ML}^{-3})^y$$

using the dimensions of E and ρ obtained above. Equating the respective indices of M, L, T on both sides, then

$$x + y = 0 \quad . \quad . \quad . \quad . \quad . \quad \text{(ii)}$$

$$-x - 3y = 1 \quad . \quad . \quad . \quad . \quad . \quad \text{(iii)}$$

$$-2x = -1 \quad . \quad . \quad . \quad . \quad . \quad \text{(iv)}$$

From (iv), $x = 1/2$; from (ii), $y = -1/2$. Thus, as $c = kE^x\rho^y$,

$$c = kE^{\frac{1}{2}}\rho^{-\frac{1}{2}}$$

$$\therefore c = k\sqrt{\frac{E}{\rho}}$$

It is not possible to find the magnitude of k by the method of dimensions, but a rigid proof of the formula by calculus shows that $c = \sqrt{E/\rho}$, so $k = 1$.

Velocity of Sound in a Gas, Laplace's Correction

Changes in pressure and volume occur at different places in a gas when a sound wave travels along the gas. If the pressure–volume changes take place *isothermally* (constant temperature), the velocity of the sound wave is given by $c = \sqrt{p/\rho}$, where p is the pressure and ρ is the density of the gas.

For air at normal conditions, $p = 1.01 \times 10^5$ Pa and $\rho = 1.29 \, \text{kg m}^{-3}$. So

$$c = \sqrt{\frac{1.01 \times 10^5}{1.29}} = 280 \, \text{m s}^{-1}$$

This theoretical value was first obtained by Newton and it is well below the experimental value of about $340 \, \text{m s}^{-1}$.

About a century later, Laplace suggested that the pressure–volume changes at different places in the air took place *adiabatically*, that is, no heat enters or leaves the gas while the wave travels along. In this case the speed of sound formula changes, if γ is the ratio of the molar heat capacities of the gas p. 672), to

$$c = \sqrt{\frac{\gamma p}{\rho}} \quad . \quad . \quad . \quad . \quad . \quad \text{(2)}$$

The magnitude of γ for air is 1.40, and *Laplace's correction*, as it is known, then changes the value of the velocity in air at 0°C to

$$c = \sqrt{\frac{1.40 \times 1.01 \times 10^5}{1.29}} = 331 \, \text{m s}^{-1}$$

This is in good agreement with the experimental value.

Effect of Pressure and Temperature on Velocity of Sound in a Gas

Suppose that a mole of gas has a mass M and a volume V. The density is then M/V and so the velocity of sound, c, is

$$c = \sqrt{\frac{\gamma p}{\rho}} = \sqrt{\frac{\gamma p V}{M}}$$

But the molar gas equation is $pV = RT$, where R is the molar gas constant and

T is the absolute temperature. Hence

$$c = \sqrt{\frac{\gamma RT}{M}} . \qquad . \quad . \quad . \quad . \quad \text{(i)}$$

Since γ, M and R are constants for a given gas, it follows that

---

the velocity of sound in a gas is independent of the pressure if the temperature remains constant.

This has been verified by experiments which showed that the velocity of sound at the top of a mountain is about the same as at the bottom. It also follows from (i) that

---

the velocity of sound is proportional to the square root of its absolute temperature.

Thus if the velocity in air at 16°C is 338 m s^{-1} by experiment, the velocity, c, at 0°C is calculated from

$$\frac{c}{338} = \sqrt{\frac{273}{289}}$$

from which

$$c = 338\sqrt{\frac{273}{289}} = 328 \cdot 5 \text{ m s}^{-1}$$

Adiabatic and Isothermal Sound Waves in a Gas

Consider a progressive sound wave travelling in a gas. At an instant, suppose ABC represents a wavelength λ, with a compression at A and a rarefaction at B, Figure 17.36. Under compression the gas temperature at A is raised; at the same instant the gas temperature at B is lowered. Thus heat tends to flow from A to B. If the temperatures can be equalised in the time taken by the wave to reverse the pressure conditions or temperatures at A and B, shown by the broken line in Figure 17.36, then the pressure–volume changes in the gas will take place isothermally. If not, the changes will be adiabatic.

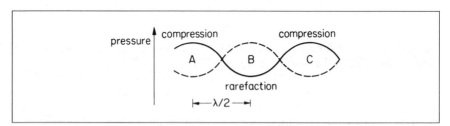

Figure 17.36 *Adiabatic and isothermal sound waves*

Consider, therefore, a time equal to half a period, $T/2$. Since $T = \lambda/c$, where c is the constant velocity of the wave, then $T/2 \propto \lambda$, the wavelength. From the formula for conduction of heat (p. 698), assuming a linear temperature gradient for simplification,

$$\text{heat flow from A to B} \propto \text{temperature gradient} \times \text{time}$$

$$\propto \frac{\text{temperature difference}}{\lambda/2} \times \lambda$$

$\propto$ temperature difference

Hence the *temperature change* of the mass of gas in the region AB

$$= \frac{\text{heat flowing}}{\text{mass} \times \text{sp. ht. capacity}}$$

$$\propto \frac{\text{temperature difference}}{\lambda} \propto \frac{1}{\lambda}$$

since the heat flowing is proportional to the temperature difference and the mass of gas between A and B is proportional to the wavelength.

At audio-frequencies, the relatively long wavelength produces a small temperature change. No equalisation of temperature then occurs between B and C by the time the conditions reverse. Thus the wave travels under *adiabatic* conditions. This is why Laplace's formula, $c = \sqrt{\gamma p/\rho}$, holds for the velocity of sound waves. As the wavelength decreases, however, the temperature change increases. Thus at ultrasonic frequencies, for example, the wave travels under conditions which are more isothermal than adiabatic and the velocity is then $c = \sqrt{p/\rho}$.

A more detailed analysis shows that isothermal conditions would be obtained if the frequency is extremely high, which is not the case for normal sound waves (see *Gases, Liquids and Solids* by D. Tabor (Penguin)).

In a gas, the speed of sound c is given by $c = \sqrt{\gamma p/\rho}$. For air, $\gamma = 1\cdot 4$, so

$$c = \sqrt{1\cdot 4p/\rho}$$

The speed $\propto \sqrt{T}$ but does not depend on the pressure.

_____ Exercises 17 _____

1 A small piece of cork in a ripple tank oscillates up and down as ripples pass it. If the ripples travel at $0\cdot 20\,\text{m s}^{-1}$, have a wavelength of 15 mm and an amplitude of $5\cdot 0$ mm, what is the maximum velocity of the cork? (*L.*)

2 A beam of electromagnetic waves of wavelength $3\cdot 0$ cm is directed normally at a grid of metal rods, parallel to each other and arranged vertically about $2\cdot 0$ cm apart. Behind the grid is a receiver to detect the waves. It is found that when the grid is in this position, the receiver detects a strong signal but that when the grid is rotated in a vertical plane through 90°, the detected signal strength falls to zero. What property of the wave gives rise to this effect? Account briefly in general terms for the effect described above. (*L.*)

3 If the velocity of sound in air is 340 metres per second, calculate   (i) the wavelength when the frequency is 256 Hz,   (ii) the frequency when the wavelength is $0\cdot 85$ m. Prove the velocity formula used in your calculations.

4 State and explain the differences between progressive and stationary waves.
   A progressive and a stationary simple harmonic wave each have the same frequency of 250 Hz and the same velocity of $30\,\text{m s}^{-1}$. Calculate   (i) the phase difference between two vibrating points on the progressive wave which are 10 cm apart,   (ii) the equation of motion of the progressive wave if its amplitude is $0\cdot 03$ metre,   (iii) the distance between nodes in the stationary wave.

5 Explain carefully what is meant by the term *resonance* and give *two* specific examples of its occurrence.
   There are several ways in which a metre rule, clamped at one end, may vibrate. With each of these ways is associated a resonant frequency. Draw sketches to illustrate *two* resonant modes for the rule and explain how the rule could be induced to vibrate in these modes. (*L.*)

6 Two waves of equal frequency and amplitude travel in opposite directions in a medium.
   (a) Why is the resultant wave called 'stationary'?
   (b) State *two* differences between a stationary and a progressive wave.
   (c) Explain briefly, using the principle of superposition, why nodes and antinodes of displacement are obtained in a stationary wave.
      If the amplitudes of the two waves are 3 units and 1 unit respectively, show by the principle of superposition that the ratio of the amplitudes of the stationary wave at an antinode and node respectively is $2:1$.

7 Two small loudspeakers A and B are 0·50 m apart. Figure 17A (i). Both emit sound waves of the same frequency f. A detector moves along a line CD 1·20 m from AB and parallel to it. A maximum wave is detected at C where OC is the perpendicular bisector of AB and again at D directly opposite A. Calculate f. (Velocity of sound $= 340\,\mathrm{m\,s^{-1}}$.)

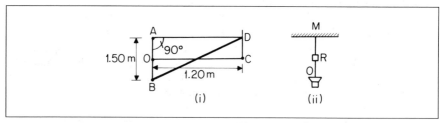

Figure 17A

8 A small source O of electromagnetic waves is placed some distance from a plane metal reflector M. Figure 17A (ii). A receiver R, moving between O and M along the line normal to the reflector, detects successive maximum and minimum readings on the meter joined to it.
   (a) Explain why these readings are obtained.
   (b) Calculate the frequency of the source O if the average distance between successive minima is 1·5 cm and the speed of electromagnetic waves in air $= 3\cdot0\times10^{8}\,\mathrm{m\,s^{-1}}$.

9 Explain what is meant by the statement that 'sound is propagated in air as longitudinal progressive waves', and outline the experimental evidence in favour of this statement. Compare the mode of propagation of sound in air with that of
   (a) waves travelling along a long metal rod, produced by tapping one end,
   (b) water waves.
      Two loudspeakers face each other at a separation of about 100 m and are connected to the same oscillator, which gives a signal of frequency 110 Hz. Describe and explain the variation of sound intensity along the line joining the speakers. A man walks along the line with a uniform speed of 2·0 m s⁻¹. What does he hear? (Speed of sound $= 330\,\mathrm{m\,s^{-1}}$.) (*O*.)

10 Explain the terms *damped oscillation, forced oscillation* and *resonance*. Give one example of each.
      Describe an experiment to illustrate the behaviour of a simple pendulum (or pendulums) undergoing forced oscillation. Indicate qualitatively the results you would expect to observe.
      What factors determine
   (a) the period of free oscillations of a mechanical system,
   (b) the amplitude of a system undergoing forced oscillation? (*O. & C.*)

11 What is the principle of superposition as applied to wave motion?
      Discuss as fully as you can the result of superposing two waves of equal amplitude
   (a) of the same frequency travelling in opposite directions,
   (b) of slightly different frequencies travelling in the same direction.
      Describe how you would demonstrate the validity of your conclusions in *one* of these cases.

Two plane sound waves of the same frequency travelling in opposite directions have different amplitudes. When an observer moves along the direction of travel of one of the waves, the amplitude of the sound he hears fluctuates by a factor of two. Explain this and find the ratio of the amplitudes of the two travelling waves. (O. & C.)

12 Write an equation that represents a progressive sinusoidal wave motion. With the aid of suitable diagrams, explain the meanings of the quantities appearing in your equation.

On axes immediately above each other, sketch two similar sinusoidal waves each of amplitude A and with phase differences such that, when superposed, the waves would produce

(a) maximum constructive interference,
(b) maximum destructive interference.

Give the value (magnitude and unit) of the phase difference in case (b).

If two such waves have exactly one-third of the phase difference relevant to case (b) and are superposed, find (by means of a phasor diagram or otherwise) (i) the amplitude of the resultant wave in terms of A, (ii) the ratio of the power carried by the resultant wave to the total power carried by the two component waves considered separately. Comment on your result in relation to the principle of conservation of energy. (C.)

13 (a) A small source S emits electromagnetic waves of wavelength about 3 cm which can be detected by an aerial A (a straight wire) connected to a meter measuring the intensity of the radiation. Initially the distance $SA = d$. When the distance between the source and the aerial is increased to $2d$, the meter reading falls to one-quarter of its original value. What conclusion can be drawn from this?

What would you expect the meter reading to be if the aerial were moved until $SA = 3d$?

If the source is rotated through 90° about the line SA, the meter reading falls to zero. Explain briefly the reason for this.

(b) A metal reflecting screen is now placed some distance beyond A with its plane perpendicular to the line SA. It is found that as the screen is moved slowly away from A, alternate maximum and minimum readings are shown on the meter. Explain briefly the reason for this. If the screen is displaced a distance of 8·7 cm between a first and a seventh minimum, calculate the wavelength and the frequency of the wave. (Speed of electromagnetic waves in air $= 3·0 \times 10^8\,\mathrm{m\,s^{-1}}$).

(c) The source of the electromagnetic waves is assumed to be *monochromatic*. Explain what monochromatic means. What would you have observed in (b) if the source had emitted simultaneously waves of wavelength 3 cm and 6 cm, of the same intensity? (L.)

14 A plane-progressive wave is represented by the equation

$$y = 0·1 \sin(200\pi t - 20\pi x/17)$$

where y is the displacement in millimetres, t is in seconds and x is the distance from a fixed origin O in metres (m).

Find (i) the frequency of the wave, (ii) its wavelength, (iii) its speed, (iv) the phase difference in radians between a point 0·25 m from O and a point 1·10 m from O, (v) the equation of a wave with double the amplitude and double the frequency but travelling exactly in the opposite direction.

15 The equation $y = a \sin(\omega t - kx)$ represents a plane wave travelling in a medium along the x-direction, y being the displacement at the point x at time t.

Deduce whether the wave is travelling in the positive x-direction or in the negative x-direction

If $a = 1·0 \times 10^{-7}\,\mathrm{m}$, $\omega = 6·6 \times 10^3\,\mathrm{s^{-1}}$ and $k = 20\,\mathrm{m^{-1}}$, calculate

(a) the speed of the wave,
(b) the maximum speed of a particle of the medium due to the wave. (JMB.)

Sound Waves

16 A source of sound of frequency 550 Hz emits waves of wavelength 600 mm in air

at 20°C. What is the velocity of sound in air at this temperature? What would be the wavelength of the sound from the source in air at 0°C? (*L.*)

17 If a detonator is exploded on a railway line an observer standing on the rail 2·0 km away hears two reports. Why is this so? What is the time interval between these reports?

(The Young modulus for steel $= 2·0 \times 10^{11}\,\mathrm{N\,m^{-2}}$. Density of steel $= 8·0 \times 10^3\,\mathrm{kg\,m^{-3}}$. Density of air $= 1·4\,\mathrm{kg\,m^{-3}}$. Ratio of the molar heat capacities of air $= 1·40$. Atmospheric pressure $= 10^5\,\mathrm{N\,m^{-2}}$.) (*L.*)

18 Describe how a *sound wave* passes through air, using graphs which illustrate and compare the variation of   (i) the *displacement* of the air particles,   (ii) the *pressure changes*, while the wave travels.

Using the same axes as the displacement graph, draw a sketch of the graph showing the variation of *velocity* of the air particles.

19 Describe the factors on which the velocity of sound in a gas depends. A man standing at one end of a closed corridor 57 m long blew a short blast on a whistle. He found that the time from the blast to the sixth echo was two seconds. If the temperature was 17°C, what was the velocity of sound at 0°C? (*C.*)

20 Write down an expression for the speed of sound in an ideal gas. Give a consistent set of units for the quantities involved.

Discuss the effect of changes of pressure and temperature on the speed of sound in air.

Describe an experimental method for finding a *reliable* value for the speed of sound in free air. (*JMB.*)

21 Describe an experiment to measure the velocity of sound in the open air. What factors may affect the value obtained and in what way may they do so?

It is noticed that a sharp tap made in front of a flight of stone steps gives rise to a ringing sound. Explain this and, assuming that each step is 0·25 m deep, estimate the frequency of the sound. (The velocity of sound may be taken to be $340\,\mathrm{m\,s^{-1}}$.) (*L.*)

18

Wave Theory of Light. Speed of Light

Historical

It has already been mentioned that light is a form of energy which stimulates our sense of vision. About 1660 Newton proposed that particles, or corpuscles, were emitted from a luminous object. The corpuscular theory of light was adopted by many scientists of the day owing to the authority of Newton, but HUYGENS, an eminent Dutch scientist, proposed about 1680 that light energy travelled from one place to another by means of a wave-motion.

If the *wave theory of light* was correct, light should bend round a corner, just as sound travels round a corner. The experimental evidence for the wave theory in Huygens' time was very small, and the theory was dropped for more than a century. In 1801, however, THOMAS YOUNG obtained evidence that light could produce wave effects such as interference, and he was among the first to see clearly the close analogy between sound and light waves. As the principles of the subject became understood other experiments were carried out which showed that light could spread round corners, and Huygens' wave theory of light was revived. Newton's corpuscular theory was rejected since it did not agree with experimental observations (see p. 510). The wave theory of light has played, and is still playing, an important part in the development of the subject.

In 1905 EINSTEIN suggested that the energy in light could be carried from place to place by 'particles' whose energy depended on the wavelength of the light. This was a return to a corpuscular theory, though it was completely different from that of Newton, as we see later. Experiments showed that Einstein's theory was true, and the particles of light energy are known as 'photons' (p. 846). It is now considered that *either* the wave theory *or* the particle theory of light can be used in a problem on light, depending on the circumstances of the problem. In this section we shall consider Huygens' wave theory, which led to many important applications in light.

Wavefronts

Consider a point source of light, S, in air, and suppose that a disturbance, or wave, starts at S and travels outwards. After a time t the wave has travelled a distance ct, where c is the speed of light in air, and the light energy has thus reached the surface of a sphere of centre S and radius ct, Figure 18.1. The surface of the sphere is called the *wavefront* of the light at this instant, and every point on it is vibrating 'in step' or *in phase* with every other point. As time goes on the wave travels farther and new wavefronts are obtained. These are all surfaces of spheres of centre S.

At points a long way from S, such as C or D, the wavefronts are parts of a sphere of very large radius, and so the wavefronts are then substantially *plane*. Light from the sun reaches the earth in plane wavefronts because the sun is so far away. Plane wavefronts are also produced by a converging lens when a point source of light is placed at its focus.

The significance of the wavefront, then, is that it shows how the light energy travels from one place in a medium to another. A *ray* is the name given to the direction along which the energy travels, and consequently a ray of light passing through a point is *perpendicular* to the wavefront at that point. The rays diverge near S, but they are approximately parallel a long way from S, as the curved wavefronts are then fairly plane, Figure 18.1.

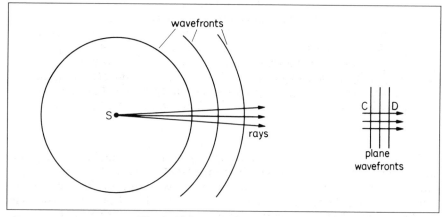

Figure 18.1 *Wavefronts and rays*

Huygens' Construction for the New Wavefront

Suppose that the wavefront from a centre of disturbance S had reached the surface AB in a medium at some instant, Figure 18.2. To obtain the position of the new wavefront after a further time t, Huygens said that *every point, A, ..., C, ..., E, ..., B, on AB becomes a new or 'secondary' centre of disturbance.* The wavelet from A then reaches the surface M of a sphere of radius ct and centre A, where c is the speed of light in the medium; the wavelet from C reaches the surface D of a sphere of radius ct and centre C; and so on for every point on AB. According to Huygens,

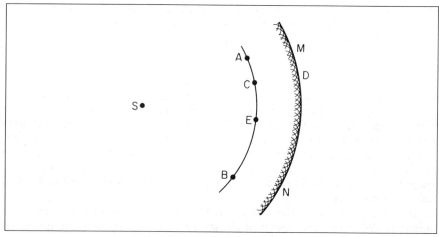

Figure 18.2 *Huygens' construction*

___

***the new wavefront is the surface MN which touches all the wavelets from the
secondary sources***

___

and in the case considered, it is the surface of a sphere of centre S.

In this simple example of drawing the new wavefront, the light travels in the
same medium. Huygens' Principle, however, is especially valuable for drawing
the new wavefront when the light travels from one medium to another, as we
soon show.

Reflection at Plane Surface

Suppose that a beam of parallel rays between HA and LC is incident on a plane
mirror, and imagine a plane wavefront AB which is normal to the rays, reaching
the mirror surface, Figure 18.3. At this instant the point A acts as a centre of
disturbance. Suppose we require the new wavefront at a time corresponding to
the instant when the disturbance at B reaches C. The wavelet from A reaches the
surface of a sphere of radius AD at this instant. When other points between AC

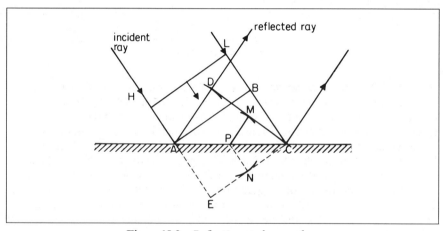

Figure 18.3 *Reflection at plane surface*

on the mirror, such as P, are reached by the disturbances starting at AB, wavelets
of smaller radius than AD such as PM are obtained at the instant we are
considering. The new wavefront is the surface CMD which touches all the
wavelets.

In the absence of the mirror, the plane wavefront AB would reach the position
EC in the time considered. Thus AD = AE = BC, and PN = PM, where PN is
perpendicular to EC. The triangles PMC, PNC are hence congruent, as PC is
common, angles PMC, PNC are each 90°, and PN = PM. Thus angle
PCM = angle PCN. But triangles ACD, AEC are congruent. Consequently
angle ACD = angle ACE = angle PCN = angle PCM, since EC is a plane. So
CMD is a *plane* surface.

Law of reflection. We can now deduce the law of reflection for the angles of
incidence and reflection. The triangles ABC, AEC are congruent, and triangles
ADC, AEC are congruent. The triangles ABC, ADC are hence congruent, and
therefore angle BAC = angle DCA. Now these are the angles made by the
wavefronts AB and CD respectively with the mirror surface AC. Since the
incident and reflected rays, for example HA, AD, are normal to the wavefronts,

these rays also make equal angles with AC. So the angles of incidence and reflection are equal.

Point Object

Consider now a *point object* O in front of a plane mirror M, Figure 18.4. A spherical wave spreads out from O, and at some time the wavefront reaches ABC. In the absence of the mirror the wavefront would reach a position DEF in a time t. But every point between D and F on the mirror acts as a secondary

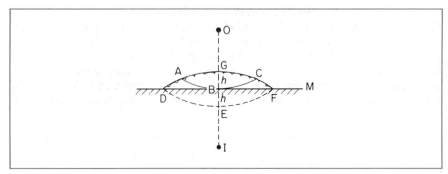

Figure 18.4 *Point object*

centre of disturbance and so wavelets are reflected back into the air. At the end of the time t, a surface DGF is drawn to touch all the wavelets, as shown. DGF is part of a spherical wave which advances into the air, and since it appears to have come from a point I as centre *below* the mirror, then I is a virtual image of O.

The sphere of which DGF is part has a chord DF. Suppose the distance from B, the midpoint of the chord, to G is h. The sphere of which DEF is part has the same chord DF, and the distance from B to E is also h. It follows, from the theorem of product of intersection of chords of a circle, that $DB \cdot BF = h(2r - h) = h(2R - h)$, where r is the radius OE and R is the radius IG. So $R = r$, or $IG = OE$, and hence $IB = OB$. The image and object are thus equidistant from the mirror.

Refraction at Plane Surface

Consider a beam of parallel rays between LO and PD incident on the plane surface of a water medium from air in the direction shown, and suppose that a plane wavefront has reached the position OA at a certain instant, Figure 18.5. Each point between O, A becomes a new centre of disturbance as the wavefront advances to the surface of the water, and the wavefront changes in direction when the disturbance enters the liquid because the speed is now less than in air.

Suppose that t is the time taken by the light to travel from A to D. The disturbance from O travels a distance OB, or $c_w t$, in water in a time t, where c_w is the speed of light in water. At the end of the time t, the wavefronts in the water from the other secondary centres between O, D reach the surfaces of spheres to each of which DB is a tangent. Thus DB is the new wavefront in the water, and the ray OB which is normal to the wavefront is consequently the refracted ray.

Since c is the speed of light in air, $AD = ct$. Now

$$\frac{\sin i}{\sin r} = \frac{\sin \text{LON}}{\sin \text{BOM}} = \frac{\sin \text{AOD}}{\sin \text{ODB}}$$

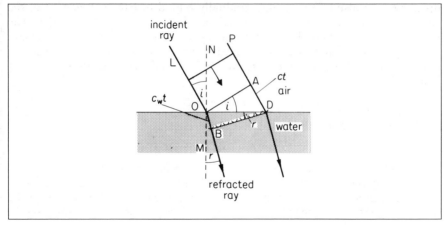

Figure 18.5 *Refraction at plane surface*

$$\therefore \frac{\sin i}{\sin r} = \frac{AD/OD}{OB/OD} = \frac{AD}{OB} = \frac{ct}{c_w t} = \frac{c}{c_w} \qquad . \qquad . \qquad . \qquad (i)$$

But c, c_w are constants for the given media.

$$\therefore \frac{\sin i}{\sin r} \text{ is a constant}$$

which is Snell's law of refraction (p. 415).

It can now be seen from (i) that the refractive index, *n*, of a medium is given by

$$n = \frac{c}{c_m},$$ **where *c* is the speed of light in air and c_m is the speed of light**

in the medium.

Newton's Corpuscular Theory of Light

Before the wave theory of light, Newton had suggested a corpuscular or *particle* theory of light. According to Newton, particles are emitted by a source of light, and they travel in a straight line until the boundary of a new medium is encountered.

In the case of *reflection at a plane surface*, Newton stated that at some very small distance from the surface M, represented by AB, the particles were acted

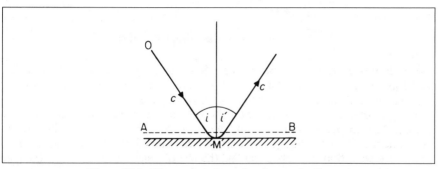

Figure 18.6 *Newton's corpuscular theory of reflection*

upon by a repulsive force, which gradually diminished the component of the velocity c in the direction of the normal and then reversed it, Figure 18.6. The *horizontal* component of the velocity remained unaltered, and hence the velocity of the particles of light as they moved away from M is again c. Since the horizontal components of the incident and reflected velocities are the same, it follows that

$$c \sin i = c \sin i' \qquad . \qquad . \qquad . \qquad . \qquad . \qquad (i)$$

where i' is the angle of reflection

$$\therefore \sin i = \sin i', \text{ or } i = i'$$

Thus the corpuscular theory explains the law of reflection at a plane surface.

To explain *refraction at a plane surface* when light travels from air to a denser medium such as water, Newton stated that a force of attraction acted on the particles as they approached beyond a line DE very close to the boundary N,

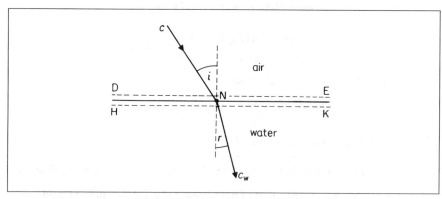

Figure 18.7 *Newton's corpuscular theory of refraction*

Figure 18.7. The vertical component of the velocity of the particles would be *increased* on entering the water, the horizontal component of the velocity remaining unaltered, and beyond a line HK close to the boundary the vertical component would remain constant at its increased value. The resultant velocity, c_w, of the particles in the water ought to be *greater* than the velocity, c, in air.

Suppose i, r are the angles of incidence and refraction respectively. Then, as the horizontal components of the velocity are unaltered,

$$c \sin i = c_w \sin r$$

$$\therefore \frac{\sin i}{\sin r} = \frac{c_w}{c}$$

$$\therefore n = \frac{c_w}{c} = \text{the refractive index}$$

Since n is greater than 1, the speed of light in water, c_w, should be greater than the speed in air, c, as was stated above. This is according to Newton's corpuscular theory. On the wave theory, however, $n = c/c_w$ (see p. 505); and hence the velocity of light in water is *less* than the velocity in air according to the wave theory. The corpuscular theory and wave theory are thus in conflict. In an experiment carried out about 150 years later, Foucault obtained a value for c_w which showed that the corpuscular theory of Newton could not be true (see p. 510).

Dispersion

The dispersion of colours produced by a medium such as glass is due to the difference in speeds of the various colours in the medium. Thus suppose a plane wavefront AC of white light is incident in air on a plane glass surface, Figure 18.8. In the time the light takes to travel in air from C to D, the red light from the

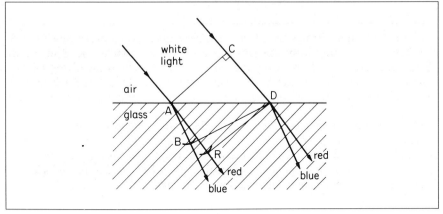

Figure 18.8 *Dispersion*

centre of disturbance A reaches a position shown by the wavelet at R. The blue light from A reaches another position shown by the wavelet at B, since the speed of blue light in glass is less than that of red light, so AB is less than AR. On drawing the new wavefronts DB, DR, it can be seen that the blue wavefront BD is refracted *more* in the glass than the red wavefront DR. The refracted blue ray is AB and the refracted red ray is AR, and so dispersion occurs.

Power of a Lens

We can now consider briefly the effect of lenses on the *curvature* of wavefronts. The curvature of a spherical wavefront is defined as $1/r$, where r is the radius of the wavefront surface.

When a plane wavefront is incident on a converging lens L, a spherical wavefront, S, of radius f emerges from L, where f is the focal length of the lens, Figure 18.9 (i). This is because the light travels faster in the air at the top L than in the glass at the middle of the lens. Parallel rays, which are normal to the

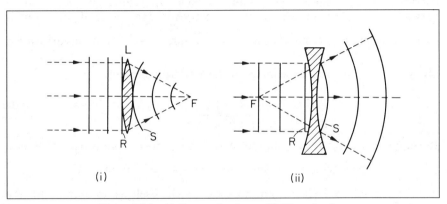

Figure 18.9 *(i) Converging lens (ii) Diverging lens*

wavefront, are thus refracted towards F, the focus of the lens. Now the curvature of a plane wavefront is zero, and the curvature of the spherical wavefront S is $1/f$. So the converging lens adds a curvature of $1/f$ to a wavefront incident on it. $1/f$ is defined as the *converging power* of the lens:

$$\text{Power } P = \frac{1}{f}$$

Figure 18.9 (ii) illustrates the effect of a *diverging* lens on a plane wavefront R. The front S emerging from the lens has a curvature opposite to S in Figure 18.9 (i), and it appears to be diverging from a point F behind the diverging lens, which is its focus. The curvature of the emerging wavefront is thus $1/f$, where f is the focal length of the lens, and the powers of the converging and diverging lens are opposite in sign.

The power of a converging lens is positive, since its focal length is positive, while the power of a diverging lens is negative. The unit of power is the *dioptre*, D, which is the power of a lens of 1 metre focal length. A lens of $+8$ dioptres, or $+8D$, is therefore a converging lens of focal length $1/8$ m or $12\cdot5$ cm, and a lens of $-4D$ is a diverging lens of $1/4$ m or 25 cm focal length.

The Lens Equation

Suppose that an object O is placed a distance u from a converging lens, Figure 18.10. The spherical wavefront A from O which reaches the lens has a radius of

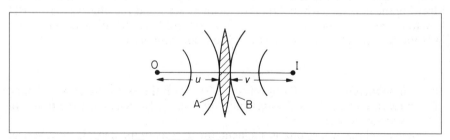

Figure 18.10   *Effect of lens on wavefront*

curvature u, and hence a curvature $1/u$. Since the converging lens adds a curvature of $1/f$ to the wavefront as we proved, the spherical wavefront B emerging from the lens into the air has a curvature $\left(\dfrac{1}{u} + \dfrac{1}{f}\right)$. But the curvature is also given by $\dfrac{1}{v}$, where v is the image distance IB from the lens.

$$\therefore \frac{1}{v} = \frac{1}{u} + \frac{1}{f}$$

It can be seen that the curvature of A is of an opposite sign to that of B; and taking this into account, the lens equation $\dfrac{1}{v} + \dfrac{1}{u} = \dfrac{1}{f}$ is obtained. A similar method can be used for a diverging lens, which is left as an exercise for the student.

Speed of Light

For many centuries the speed of light was thought to be infinitely large; from about the end of the seventeenth century, however, evidence began to be obtained which showed that the speed of light, though enormous, was a finite quantity. Galileo, in 1600, attempted to measure the speed of light by covering and uncovering a lantern at night, and timing how long the light took to reach an observer a few miles away. Owing to the enormous speed of light, however, the time was too small to measure, and the experiment was a failure. The first successful attempt to measure the speed of light was made by RÖMER, a Danish astronomer, in 1676.

Römer recorded the date and time of the eclipse of one of Jupiter's satellites. From his measurements, Römer found that the expected eclipse of the satellite was $16\frac{1}{2}$ minutes later than expected. He deduced that this was the time taken by the light to travel across the diameter of the earth's orbit. The diameter is about 3×10^{11} m. So with these figures,

$$\text{speed of light, } c = \frac{3 \times 10^{11}\,\text{m}}{16 \cdot 5 \times 60\,\text{s}} = 3 \times 10^8\,\text{m s}^{-1} \text{ (approx.)}$$

Measurement of Speed of Light, Rotating Mirror Method

In 1862 Foucault used a fairly accurate way of measuring the speed of light in a laboratory, Figure 18.11 shows only the basic principle.

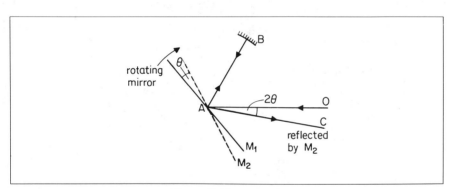

Figure 18.11 *Example*

A beam of light OA is incident on a plane mirror M_1 rotated at high speed and then reflected along AB to a concave mirror at B. This reflects the light back to M_1 but in this short time the mirror has rotated to a new position M_2. So the returning light is reflected along a different direction AC.

In the experiment, O was a bright light source and its displacement to C was measured. This gives a measure of θ, the angle of rotation of the mirror from M_1 to M_2. Now the time for the light to travel from A to B and back $= 2AB/c$, where c is the speed of light in air. In this time, the mirror rotated through θ radians. So if the mirror makes n revolutions per second,

$$\text{time} = \frac{\theta}{2\pi n} = \frac{2AB}{c}$$

$$\therefore c = 4\pi n AB/\theta$$

This relation enables c to be calculated.

Since AB was only about 20 metres, the experiment can be carried out in a laboratory. After measuring the speed of light in air, Foucault placed a long pipe filled with water along AB and measured the speed in water. He found that the speed in water was *less* than in air. Newton's 'corpuscular theory' of light predicted that light should travel faster in water than in air (p. 506), whereas the 'wave theory' of light predicted that light should travel slower in water than in air. The direct observation of the speed of light in water by Foucault's method showed that the corpuscular theory of Newton could not be true.

Michelson's Rotating Prism Method for Speed of Light

The speed of light, *c*, is a quantity which appears in many fundamental formulae in advanced Physics, especially in connection with the theories concerning particles such as electrons in atoms and calculations on nuclear energy. EINSTEIN has shown, for example, that the energy W released from an atom is given by $W = mc^2$ joules, where *m* is the decrease in mass of the atom in kilograms and *c* is the numerical value of the speed of light in metres per second. A knowledge of the magnitude of *c* is thus important. A. A. MICHELSON, an American physicist, spent many years of his life in measuring the speed of light, and the method he devised was considered as one of the most accurate.

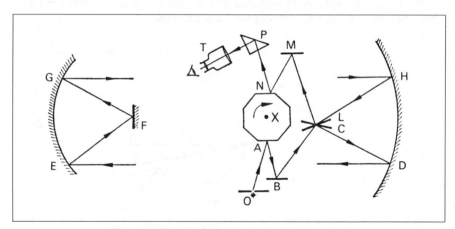

Figure 18.12 *Michelson's rotating prism method*

The essential features of Michelson's apparatus are shown in Figure 18.12. X is an equiangular octagonal steel prism which can be rotated at constant speed about a vertical axis through its centre. The faces of the prism are highly polished, and the light passing through a slit from a very bright source O is reflected at the surface A towards a plane mirror B. From B the light is reflected to a plane mirror L, which is placed so that the image of O formed by this plane mirror is at the focus of a large concave mirror HD. The light then travels as a parallel beam to another concave mirror GE a long distance away, and it is reflected to a plane mirror F at the focus of GE. The light is then reflected by the mirror, travels back to H, and is there reflected to a plane mirror C placed just below L and inclined to it as shown. From C the light is reflected to a plane mirror M, and is then incident on the face N of the octagonal prism opposite to A. The final image thus obtained is viewed through T with the aid of a totally reflecting prism P.

The image is seen by light reflected from the top surface of the octagonal prism

X. When the prism is rotated the image disappears at first, as the light reflected from A when the prism is just in the position shown in Figure 18.12 arrives at the opposite face to find this surface in some position inclined to that shown.When the speed of rotation is increased and suitably adjusted, however, the image reappears and is seen in the same position as when the prism X is at rest. *The light reflected from A now arrives at the opposite surface in the time taken for the prism to rotate through 45°, or $\frac{1}{8}$th of a revolution*, since in this case the surface on the left of N, for example, will occupy exactly the position of N when the light arrives at the upper surface of X.

Suppose d is the total distance in metres travelled by the light in its journey from A to the opposite face; the time taken is then d/c, where c is the speed of light. But this is the time taken by X to make $\frac{1}{8}$th of a revolution, which is $1/8m$ seconds if the number of revolutions per second is m.

$$\therefore \frac{1}{8m} = \frac{d}{c}$$

$$\therefore c = 8md \text{ metre per second}$$

Thus c can be calculated from a knowledge of m and d.

Michelson performed the experiment in 1926, and again in 1931, when the light path was enclosed in an evacuated tube 1.6 km long. Multiple reflections were obtained to increase the effective path of the light. A prism with 32 faces was also used, and Michelson's result for the speed of light *in vacuo* was $2.99\,774 \times 10^8 \text{ m s}^{-1}$.

Example on Speed of Light

A beam of light is reflected by a rotating mirror on to a fixed mirror, which sends it back to the rotating mirror from which it is again reflected, and then makes an angle of 18° with its original direction. The distance between the two mirrors is 10^4 m, and the rotating mirror is making 375 revolutions per second. Calculate the speed of light. (L.)

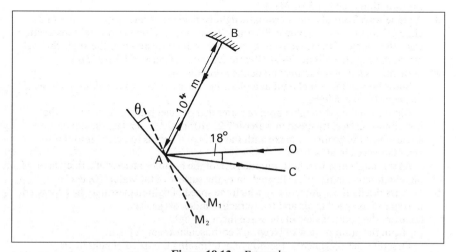

Figure 18.13 *Example*

Suppose OA is the original direction of the light, incident at A on the mirror in the position M_1, B is the fixed mirror, and AC is the direction of the light reflected from the rotating mirror when it reaches the position M_2, Figure 18.13.

The angle θ between M_1, M_2 is $\frac{1}{2} \times 18°$, since the angle of rotation of a mirror is half the angle of deviation of the reflected ray when the incident ray (BA in this case) is kept constant. Thus $\theta = 9°$.

$$\text{Time taken by mirror to rotate } 360° = \frac{1}{375}\text{s}$$

$$\therefore \text{ time taken to rotate } 9° = \frac{9}{360} \times \frac{1}{375}\text{s}$$

But this is also the time taken by the light to travel from A to B and back, which is given by $2 \times 10^4/c$, where c is the speed of light in m s^{-1}.

$$\therefore \frac{2 \times 10^4}{c} = \frac{9}{360} \times \frac{1}{375}$$

$$\therefore c = \frac{2 \times 10^4 \times 360 \times 375}{9} = 3 \times 10^8 \text{ m s}^{-1}$$

Exercises 18

Wave Theory

1 Using Huygens' principle of secondary wavelets explain, making use of a diagram, how a refracted wavefront is formed when a beam of light, travelling in glass, crosses the glass-air boundary. Show how the sines of the angles of incidence and refraction are related to the speeds of light in air and glass. (*L.*)

2 A parallel beam of monochromatic radiation travelling through glass is incident on the plane boundary between the glass and air. Using Huygens' principle draw diagrams (one in each case) showing successive positions of the wavefronts when the angle of incidence is
(a) 0°,
(b) 30°,
(c) 60°.
Indicate clearly and explain the constructions used. (The refractive index of glass for the radiation used is 1·5.) (*JMB.*)

3 A plane wavefront of monochromatic light is incident normally on one face of a glass prism, of refracting angle 30°, and is transmitted. Using Huygens' construction trace the course of the wavefront. Explain your diagram and find the angle through which the wavefront is deviated. (Refractive index of glass = 1·5.) (*JMB.*)

4 State *Snell's law of refraction* and define *refractive index.*
Show how refraction of light at a plane interface can be explained on the basis of the wave theory of light.
Light travelling through a pool of water in a parallel beam is incident on the horizontal surface. Its speed in water is $2·2 \times 10^8$ m s^{-1}. Calculate the maximum angle which the beam can make with the vertical if light is to escape into the air where its speed is $3·0 \times 10^8$ m s^{-1}.
At this angle in water, how will the path of the beam be affected if a thick layer of oil, of refractive index 1·5, is floated on to the surface of the water? (*O. & C.*)

5 Discuss briefly the arguments by which the speed of light in glass may be expressed in terms of its speed in air and the refractive index of the glass
(a) from the point of view of the wave theory of light,
(b) from the point of view of Newton's corpuscular theory of light.
Describe an experimental method of determining the speed of light in air.
Figure 18A represents a plane wavefront AB striking a plane surface in air. The refractive index of the glass is 1·5, and the speed of light in air is 3×10^8 m s^{-1}. The distance BC is 3 cm. Taking the time from the instant shown in the diagram, and considering only refraction,

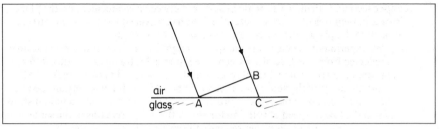

Figure 18A

(a) construct accurately the wavefront at time 10^{-10} s;
(b) draw the position of the wavefront at time 2×10^{-10} s;
(c) draw the position of the wavefront at time 5×10^{-11} s. (*O*.)

6 How did Huygens explain the reflection of light on the wave theory? Using Huygens' conceptions, show that a series of light waves diverging from a point source will, after reflection at a plane mirror, appear to be diverging from a second point, and calculate its position. (*C*.)

7 (a) Explain briefly Huygens' method for constructing wavefronts.
   A parallel beam of light is projected on to the surface of a plane mirror at an angle of incidence of about 70°. Draw a diagram showing clearly how Huygens' method can be used to determine the direction of the reflected beam.
   (b) What is meant by *critical angle*? Under what conditions will a wave be totally reflected on meeting a boundary between two media, both of which will allow passage of the wave?
   A beam of light travelling through a transparent medium A is incident on a plane interface into air at an angle of 20°. If the speed of light in the medium is 60% of that in air, calculate the angle of refraction in air.
   When the beam is shone through another transparent medium B and the incident angle is again 20°, it is found that the beam is just totally reflected at a plane interface with air. Calculate the speed of light in B as a percentage of the speed of light, *c*, in air. (*L*.)

8 What is Huygens' principle?
   Draw and explain diagrams which show the positions of a light wavefront at successive equal time intervals when
   (a) parallel light is reflected from a plane mirror, the angle of incidence being about 60°,
   (b) monochromatic light originating from a small source in water is transmitted through the surface of the water into the air.
   Describe an experiment, and add the necessary theoretical explanation, to show that in air the wavelength of blue light is less than that of red light. (*JMB*.)

9 Using Huygens' concept of secondary wavelets show that a plane wave of monochromatic light incident obliquely on a plane surface separating air from glass may be refracted and proceed as a plane wave. Establish the physical significance of the refractive index of the glass.
   In what circumstances does dispersion of light occur? How is it accounted for by the wave theory?
   If the wavelength of yellow light in air is 6.0×10^{-7} m, what is its wavelength in glass of refractive index 1.5? (*JMB*.)

Speed of Light

10 Write down two advantages and two disadvantages of
    (a) Foucault's rotating mirror method and
    (b) Michelson's rotating prism method of determining the speed of light.

11 Draw a diagram of Foucault's method of measuring the speed of light. How has the speed of light in water been shown to be less than in air? The radius of curvature of the curved mirror is 20 metres and the plane mirror is rotated at 20

revs per second. Calculate the angle in degrees between a ray incident on the plane mirror and then reflected from it after the light has travelled to the curved mirror and back to the plane mirror (speed of light $= 3 \times 10^8\,\mathrm{m\,s^{-1}}$).

12 (a) In an experiment to determine the speed of light in air, light from a point source is reflected from one face of a sixteen-sided mirror M, travels a distance d to a stationary mirror from which it returns and, after a second reflection from M, forms an image of the source on a screen. When M is rotated at certain speeds, the image is still seen in the same position. Explain how this can occur and show that, if the lowest speed of rotation for which the image remains in the same position is n (in revolutions per second), the speed of light, c, is given by $c = 32nd$.

(b) Using the above arrangement, an image is seen on the screen when the speed of rotation is 900 revolutions per second. The speed of rotation is gradually increased until at 1200 revolutions per second the image is again seen. If $c = 3 \cdot 00 \times 10^8\,\mathrm{m\,s^{-1}}$, calculate a value for d consistent with these figures. What is the lowest speed of rotation for which an image will be seen on the screen? (*JMB.*)

13 Describe an experiment to determine the speed of light in a vacuum or in air. Show how the result is calculated from the measurements made, estimate the errors to be expected in the measurements, and deduce the maximum possible error in the result.
What results would be obtained in a determination of the speed of light in water? How have the results of such experiments influenced views as to the nature of light? (*O. & C.*)

14 A beam of light after reflection at a plane mirror, rotating 2000 times per minute, passes to a distant reflector. It returns to the rotating mirror from which it is reflected to make an angle of 1° with its original direction. Assuming that the velocity of light is 300 000 km s^{-1}, calculate the distance between the mirrors. (*L.*)

15 Describe a method of measuring the speed of light. Explain precisely what observations are made and how the speed is calculated from the experimental data.
A horizontal beam of light is reflected by a vertical plane mirror A, travels a distance of 250 metres, is then reflected back along the same path and is finally reflected again by the mirror A. When A is rotated with constant angular velocity about a vertical axis in its plane, the emergent beam is deviated through an angle of 18 minutes. Calculate the number of revolutions per second made by the mirror.
If an atom may be considered to radiate light of wavelength 500 nm for a time of 10^{-10} second, how many cycles does the emitted wave train contain? (*O. & C.*)

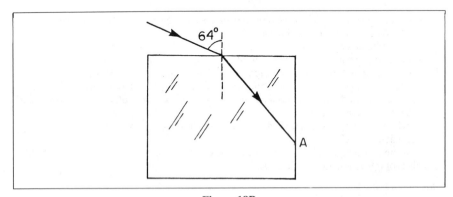

Figure 18B

16 A beam of white light enters a rectangular block of transparent material as shown in Figure 18B. If its speed in air is $3 \times 10^8\,\mathrm{m\,s^{-1}}$, at what speed must the light travel through the transparent material so that the refracted beam is just totally reflected at A.
In practice the beam would be dispersed as it enters the block. Why is this? (*L.*)

Interference of Light Waves

*We shall first discuss coherent sources, which are necessary for
the phenomenon of interference, and the use of path difference for
constructive and destructive interference. We then consider the
interference produced in the Young two-slit experiment, the
air-wedge experiment and Newton's rings together with their
applications, and conclude with the blooming of lenses for clearer
images.*

Coherent Sources

As we see later, light waves from a sodium lamp, for example, are due to energy
changes in the sodium atoms. The emitted waves occur in bursts lasting about
10^{-8} second. The light waves produced by the different atoms are out of phase
with each other, as they are emitted randomly and rapidly. We call such sources
of light waves as these atoms *incoherent sources* on account of the continual
change of phase.

Two sodium lamps X and Y both emit light waves of the same colour or
wavelength. But owing to the random emission of light waves from their atoms,
their resultant light waves are constantly out of phase. So X and Y are
incoherent sources. *Coherent* sources are those which emit light waves of the
same wavelength or frequency which are *always* in phase with each other or
have a *constant phase difference*. As we now show, two coherent sources can
together produce the phenomenon of interference.

Interference of Light Waves, Constructive Interference

Suppose two sources of light, A, B, have exactly the same wavelength and
amplitude of vibration, and that their vibrations are always in phase with each
other, Figure 19.1. The two sources A and B are therefore *coherent* sources.

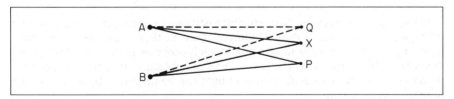

Figure 19.1 *Interference of waves*

*Their combined effect at a point is obtained by adding algebraically the dis-
placements at the point due to the sources individually.* This is known as the
Principle of Superposition. Thus their resultant effect at X, for example, is the
algebraic sum of the vibrations at X due to the source A alone and the vibrations
at X due to the source B alone. If X is equidistant from A and B, the vibrations at
X due to the two sources are *always* in phase as   (i) the distance AX travelled by
the wave starting from A is equal to the distance BX travelled by the wave
starting from B,   (ii) the sources A and B are assumed to have the same
wavelength and to be always in phase with each other.

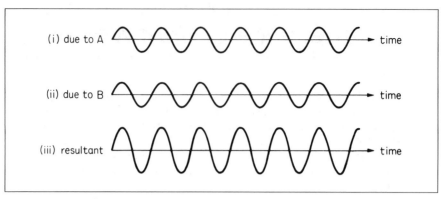

Figure 19.2 *Vibrations at X—constructive interference*

Figure 19.2 (i), (ii) illustrate the vibrations at X due to A and B, which have the same amplitude. The resultant vibration at X is obtained by adding the two curves, and has an amplitude *double* that of either curve, Figure 19.2 (iii). Now the energy of a vibrating source is proportional to the square of its amplitude (p. 83). Consequently the light energy at X is four times that due to A or B alone. A bright band of light is thus obtained at X. As A and B are coherent sources, the bright band is *permanent*. With wave crests and troughs arriving at X at the same time, we say that the bright band is due to *constructive interference* of the light waves from A and B at X.

If Q is a point such that BQ is greater than AQ by a whole number of wavelengths (Figure 19.1), the vibration at Q due to A is in phase with the vibration there due to B (see p. 467). A permanent bright band is then obtained at O.

Generally, a permanent bright band is obtained at any point Y if the *path difference*, **BY – AY**, is given by

$$BY - AY = m\lambda$$

where λ is the wavelength of the sources A, B, and $m = 0, 1, 2$ and so on.

We now see that permanent interference between two sources of light can only take place if they are *coherent* sources, that is they must have the same wavelength and be always in phase with each other or have a constant phase difference. This implies that the two sources of light must have the same colour. As we see later, two coherent sources of light can be produced by using a single primary source of light.

Destructive Interference

Consider now a point P in Figure 19.1 whose distance from B is half a wavelength longer than its distance from A, $AP - BP = \lambda/2$. The vibration at P due to B will then be 180° out of phase with the vibration there due to A (see p. 467), Figure 19.3 (i), (ii). The resultant effect at P is thus zero, as the displacements at any instant are equal and opposite to each other, Figure 19.3 (iii). No light is therefore seen at P. With a wave crest from A arriving at P at the same time as a wave trough from B, the permanent dark band here is said to be due to *destructive interference* of the waves from A and B.

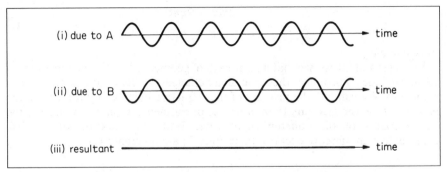

Figure 19.3 *Vibrations at P—destructive interference*

If the path difference, AP—BP, were $3\lambda/2$ or $5\lambda/2$, instead of $\lambda/2$, a permanent dark band would again be seen at P as the vibrations there due to A and B would be 180° out of phase.

Summarising:
If the path-difference is zero or a whole number of wavelengths, a bright band is obtained; if it is an odd number of half-wavelengths, a dark band is obtained.

From the principle of the conservation of energy, the total light energy from the sources A and B above must be equal to the light energy in all the bright bands of the interference pattern. The light energy missing from the dark bands is therefore found in the bright bands. It follows that the bright bands on a screen appear *brighter* than the screen when this is uniformly illuminated by A and B without forming an interference pattern.

Optical Path, Reflection of Waves

The phase of a wave arriving at a point is affected by the medium through which it travels. For example, part of its path may be in air and part in glass. Since the velocity of light is less in glass than in air, there are more waves in a given length in glass than in an equal length in air.

Suppose light travels a distance t in a medium of refractive index n. Then if λ is the wavelength in the medium, the phase difference Δ due to this path (p. 467) is

$$\Delta = \frac{2\pi t}{\lambda} \quad . \qquad . \qquad . \qquad . \qquad . \qquad (1)$$

If the wave travels from a vacuum (or air) to this medium, its frequency does not alter but its wavelength and velocity become smaller. Suppose λ_0 is the wavelength and c is the speed in a vacuum. Then if c_m is the speed in the medium,

$$\text{frequency} = \frac{c}{\lambda_0} = \frac{c_m}{\lambda}$$

$$\therefore \lambda = \frac{c_m}{c}\lambda_0 \qquad . \qquad . \qquad . \qquad . \qquad (2)$$

Substituting for λ from (2) in (1),

$$\therefore \Delta = \frac{2\pi ct}{c_m \lambda_0} = \frac{2\pi nt}{\lambda_0} . \qquad . \qquad . \qquad . \qquad . \qquad (3)$$

since $n = c/c_m$.

From (1) and (3), we see that a light path of geometric length t in a medium of refractive index n produces the same phase change as a light path of length nt in a vacuum. We call 'nt', the product of the refractive index and path length, the *optical path* in the medium. In interference phenomena, we always calculate the optical paths of the coherent light rays. With the notation on p. 516, constructive interference occurs if their optical path difference is $m\lambda$.

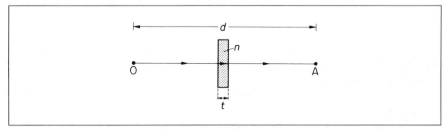

Figure 19.4 *Optical path*

As an illustration of optical path, suppose light travels from O to A, a distance d, in air, Figure 19.4. The optical path $= n_0 d = d$, since the refractive index n_0 of air is practically 1. Now suppose a thin slab of glass of thickness t and refractive index n is placed between O and A so that the light passes through a length t in the glass. The optical path between O and A is now

$$(d - t) + nt = d + (n - 1)t$$

since the light travels a distance $(d - t)$ in air and a distance t in glass.

Reflection of waves. Light waves may also undergo phase change by reflection at some point in their path. If the waves are reflected at a *denser* medium, for example, at an air–glass interface (boundary) after travelling in air, the reflected waves have a phase change of π or 180° compared to the incident waves, Figure 19.5 (i). This phase change also occurs with matter waves such as sound waves,

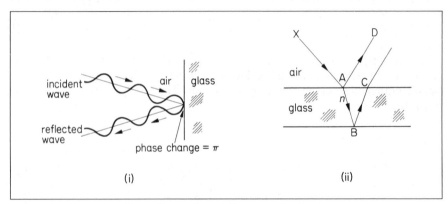

Figure 19.5 *Phase difference and reflection*

as shown in Figure 23.2, p. 592. A phase change of 2π or $360°$ is equivalent to a path length λ

$$\therefore t = \frac{\lambda}{2}$$

To take into account reflection at a *denser* medium, then, we must add (or subtract) $\lambda/2$ to the optical path.

Figure 19.5 (ii) shows an incident ray of light XA partly refracted at A from air to glass and then reflected at B, the glass–air interface. The optical path from A to C is $n(AB + BC)$; there is no phase change by reflection at B since this occurs at an interface with the *less* dense medium air. By contrast, a phase change equivalent to a path of $\lambda/2$ occurs when XA is reflected at A along AD, since this is reflection at a denser medium, glass.

Young's Two-Slit Experiment

From our previous discussion, two conditions are essential to obtain an interference phenomenon in light: (i) two coherent sources of light must be produced, (ii) the coherent sources must be very close to each other as the wavelength of light is very small, otherwise the bright and dark pattern produced some distance away would be too close to each other and no interference pattern would then be seen.

One of the first demonstrations of the interference of light waves was given by YOUNG in 1801. He placed a source, S, of monochromatic light in front of a narrow slit C, and arranged two very narrow slits A, B, close to each other, in front of C. Young then saw bright and dark bands on either side of O on a screen T, where O is on the perpendicular bisector of AB, Figure 19.6.

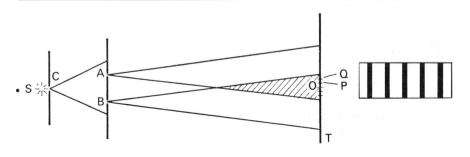

Figure 19.6 *Young's experiment—photographed fringes are shown on right*

Young's observations can be explained by considering the light from S illuminating the two slits A, B. Since the light diverging from A has exactly the same frequency as, and is always in phase with, the light diverging from B, A and B act as *two close coherent sources*. Interference thus takes place in the shaded region, where the light beams overlap, Figure 19.6. As AO = OB, a bright band is obtained at O. At a point close to O, such that BP − AP = $\lambda/2$, where λ is the wavelength of the light from S, a dark band is obtained. At a point Q such that BQ − AQ = λ, a bright band is obtained; and so on for either side of O. Young demonstrated that the bands or *fringes* were due to interference by covering A or B, when the fringes disappeared. Young's experiment is an example of inter-ference by *division of wavefront* from C at A and B. Diffraction at A and B controls the overall angular width of the bands and their intensities (see Figure 20.10, page 546).

Separation of Fringes

Suppose P is the position of the mth bright fringe, so that $BP - AP = m\lambda$, Figure 19.7. Let $OP = x_m$ = distance from P to O, the centre of the fringe system, where MO is the perpendicular bisector of AB. If a length PN equal to PA is described on PB, then $BN = BP - AP = m\lambda$. Now in practice AB is very small, and PM is

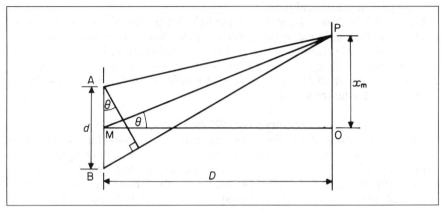

Figure 19.7   *Theory of Young's fringes* (exaggerated)

very much larger than AB. So AN meets PM practically at right angles. It then follows that

$$\text{angle } PMO = \text{angle } BAN = \theta \text{ say}$$

From triangle BAN,

$$\sin \theta = \frac{BN}{AB} = \frac{m\lambda}{d}$$

where $d = AB$ = the distance between the slits or their separation. From triangle PMO,

$$\tan \theta = \frac{PO}{MO} = \frac{x_m}{D}$$

where $D = MO$ = the distance from the screen to the slits. Since θ is very small, about 0·01 radian or 0·5° for 20 fringes when D is 1 metre, $\tan \theta = \sin \theta$.

$$\therefore \frac{x_m}{D} = \frac{m\lambda}{d}$$

$$\therefore x_m = \frac{mD\lambda}{d}$$

If Q is the neighbouring or $(m-1)$th bright fringe, it follows that

$$OQ = x_{m-1} = \frac{(m-1)D\lambda}{d}$$

$\therefore$ separation y between successive fringes $= x_m - x_{m-1} = \dfrac{\lambda D}{d}$. (i)

$$\therefore \lambda = \frac{dy}{D} . \qquad . \qquad . \qquad . \qquad . \qquad \text{(ii)}$$

Measurement of Wavelength by Young's Interference Fringes

A laboratory experiment to measure wavelength by Young's interference fringes is shown in Figure 19.8. Light from a small filament lamp is focused by a lens on to a narrow slit S, such as that in the collimator of a spectrometer. Two narrow slits A, B, about 0·5 millimetre apart, are placed a short distance in front of S, and the light coming from A, B is viewed in a low-powered microscope or eyepiece M about one metre away. Some coloured interference fringes are then

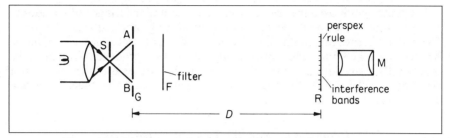

Figure 19.8 *Laboratory experiment on Young's interference fringes*

observed by M. A red and then a blue filter, F, placed in front of the slits, produces red and then blue fringes. Observation shows that the separation of the red fringes is more than that of the blue fringes. Now $\lambda = dy/D$, from (ii), where y is the separation of the fringes. It follows that the wavelength of red light is *longer* than that of blue light.

An approximate value of the wavelength of red or blue light can be found by placing a Perspex rule R in front of the eyepiece and moving it until the graduations are clearly seen, Figure 19.8. The average distance, y, between the fringes is then measured on R. The distance d between the slits can be found by magnifying the distance by a converging lens, or by using a travelling microscope. The distance D from the slits to the Perspex rule, where the fringes are formed, is measured with a metre rule. The wavelength λ can then be calculated from $\lambda = dy/D$; it is of the order 6×10^{-7} m. Further details of the experiment can be obtained from *Advanced Level Practical Physics* by Nelkon and Ogborn (Heinemann).

Measurements can also be made using a spectrometer, with the collimator and telescope adjusted for parallel light (p. 427). The narrow collimator slit is illuminated by sodium light, for example, and the double slits placed on the table. Young's fringes can be seen through the telescope after alignment. From the theory on p. 520, the angular separation of the fringes is λ/d. Thus by measuring the average angular separation of a number of fringes with the telescope, λ can be calculated from $\lambda = d\theta$, if d is known or measured.

The wavelengths of the extreme colours of the visible spectrum vary with the observer. This may be 4×10^{-7} m for violet and 7×10^{-7} m for red; an 'average' value for visible light is $5·5 \times 10^{-7}$ m, which is a wavelength in the green.

Appearance of Young's Interference Fringes

The experiment just outlined can also be used to demonstrate the following points:

1. If the source slit S is moved *nearer* the double slits the separation of the fringes is unaffected but their brightness increases. This can be seen from the formula y (separation) $= \lambda D/d$, since D and d are constant.

2. If the distance apart d of the slits is diminished, keeping S fixed, the separation of the fringes increases. This follows from $y = \lambda D/d$.

3. If the source slit S is *widened* the fringes gradually disappear. The slit S is then equivalent to a large number of narrow slits, each producing its own fringe system at different places. The bright and dark fringes of different systems therefore overlap, giving rise to uniform illumination. It can be shown that, to produce interference fringes which are recognisable, the slit width of S must be less than $\lambda D'/d$, where D' is the distance of S from the two slits A, B.

4. If one of the slits, A or B, is covered up, the fringes disappear.

5. If white light is used the central fringe is white, and the fringes either side are *coloured*. Blue is the colour nearer to the central fringe and red is farther away. The path difference to a point O on the perpendicular bisector of the two slits A, B is zero for all colours, and consequently each colour produces a bright fringe here. As they overlap, a white fringe is formed. Farther away from O, in a direction parallel to the slits, the shortest visible wavelengths, blue, produce a bright fringe first.

Examples on Young's Two-slit Experiment

1 In a Young's slits experiment, the separation between the first and fifth bright fringe is 2·5 mm when the wavelength used is $6·2 \times 10^{-7}$ m. The distance from the slits to the screen is 0·80 m. Calculate the separation of the two slits.

From previous, $\qquad \lambda = \dfrac{dy}{D}$, where d is the slit separation

$$\therefore d = \frac{\lambda D}{y} = \frac{6·2 \times 10^{-7} \times 0·8}{2·5 \times 10^{-3}/4}$$

$$= 8 \times 10^{-4}\,\text{m} = 0·8\,\text{mm}$$

2 In Figure 19.9, S_1 and S_2 are two coherent light sources in a Young's two-slit experiment separated by a distance 0·50 mm and O is a point equidistant from S_1 and S_2. O is on a screen A which is 0·80 m from the slits.

When a thin parallel-sided piece of glass G of thickness $3·6 \times 10^{-6}$ m is placed near S_1 as shown, the centre of the fringe system moves from O to a point P. Calculate OP if the wavelength of the monochromatic light from the two slits is $6·0 \times 10^{-7}$ m and the refractive index of the glass is 1·5.

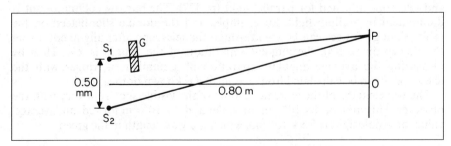

Figure 19.9 *Example on Young's two-slit experiment*

(*Analysis* (a) If P is the centre of the fringe system, the number of waves in S_1P = the number of waves in S_2P. (b) Since light travels slower in the glass G than in air, the number of wavelengths in G is *more* than in air of the same thickness as G)

In air, $\lambda_a = 6 \cdot 0 \times 10^{-7}$ m. In glass, $\lambda_g = 6 \cdot 0 \times 10^{-7}/1 \cdot 5 = 4 \times 10^{-7}$ m.

The *increase* in the number of waves in G when it replaces air of the same thickness

$$= \frac{3 \cdot 6 \times 10^{-6}}{4 \times 10^{-7}} - \frac{3 \cdot 6 \times 10^{-6}}{6 \times 10^{-7}} = 9 - 6 = 3$$

So   number of bright bands from O to P = 3

Now   separation of bright bands y is given by

$$y = \frac{D\lambda}{d} = \frac{0 \cdot 8 \times 6 \times 10^{-7}}{0 \cdot 5 \times 10^{-3}}$$

$$= 9 \cdot 6 \times 10^{-4}\,\text{m}$$

So $\text{OP} = 3 \times 9 \cdot 6 \times 10^{-4} = 2 \cdot 88 \times 10^{-3}\,\text{m} = 2 \cdot 88\,\text{mm}$

(*Alternatively*   Extra path difference due to G $= nt - t = (n-1)t$

So   extra number of wavelengths $= \dfrac{(n-1)t}{\lambda_a}$

$$= \frac{(1 \cdot 5 - 1) \times 3 \cdot 6 \times 10^{-6}}{6 \times 10^{-7}} = 3$$

As shown above, $\text{OP} = 3D\lambda/d = 2 \cdot 88 \times 10^{-3}\,\text{m}$)

Interference in Thin Wedge Films

A very thin wedge of an air film can be formed by placing a thin piece of foil or paper between two microscope slides at one end Y, with the slides in contact at the other end X, Figure 19.10. The wedge has then a very small angle θ, as shown. When the air-film is illuminated by monochromatic light from an extended source S, *straight* bright and dark fringes are seen which are parallel to the line of intersection X of the two slides.

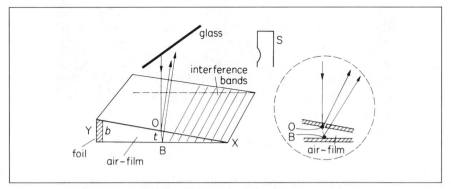

Figure 19.10   *Air-wedge fringes*

The light reflected down towards the wedge is partially reflected upwards from the *lower* surface O of the top slide, as shown inset in Figure 19.10. The rest of the light passes through the slide and some is reflected upward from the *top* surface B of the lower slide. The two trains of waves are coherent, since both have originated from the same centre of disturbance at O. So they produce an interference pattern if brought together by the eye or in an eyepiece.

The path difference is $2t$, where t is the small thickness of the air-film at O. At X, where the path difference is apparently zero, we would expect a bright fringe. But a *dark* fringe is observed at X. This is due to a phase change of 180°, equivalent to an extra path difference of $\lambda/2$, which occurs when a wave is reflected at a denser medium. See p. 518. The optical path difference between the two coherent beams is thus actually $2t + \lambda/2$. So, if the beams are brought together to interfere, a bright fringe is obtained when $2t + \lambda/2 = m\lambda$, or $2t = (m - \frac{1}{2})\lambda$. A dark fringe is obtained at a thickness t given by $2t = m\lambda$.

The bands or fringes are located at the air-wedge film, and the eye or microscope must be focused here to see them. The appearance of a fringe is the contour of all points in the wedge air film where the optical path difference is the same. If the wedge surfaces make perfect optical contact at one edge, the fringes are straight lines parallel to the line of intersection of the surfaces. If the glass surfaces are uneven, and the contact at one edge is not regular, the fringes are not perfectly straight. A particular fringe still shows the locus of all points in the air-wedge which have the same optical path difference in the air-film.

In *transmitted light*, the appearance of the fringes is complementary to those seen by reflected light, from the law of conservation of energy. The bright fringes thus correspond in position to the dark fringes seen by reflected light, and the fringe where the surfaces touch is now bright instead of dark.

The wedge air film is an example of interference by *division of amplitude*. Here part of the wave is transmitted at O and the remainder is reflected at O, so that the amplitude of the wave, which is a measure of its energy (p. 584), is 'divided' into two parts. This method of producing interference is basically different from producing interference by division of wavefront (p. 519).

Figure 19.11 *Interference bands in air-wedges. The angle of the air-wedge on the right is about $2\frac{1}{2}$ times less than that on the left, so that the separation of the bands is correspondingly greater*

Thickness of Thin Foil by Air Wedge, Crystal Expansion

If there is a bright fringe at Y at the edge of the foil, Figure 19.10, the thickness b of the foil is given by $2b = (m + \frac{1}{2})\lambda$, where m is the number of bright fringes between X and Y. If there is a dark band at Y, then $2b = m\lambda$. So by counting m, the thickness b can be found. The small angle θ of the wedge is given by b/a, where a is the distance XY, and by measuring a with a travelling microscope focused on the air-film, θ can be found.

The angle θ of the wedge can also be found from the separation s of the bright bands. In Figure 19.12, B_1 and B_2 are *consecutive* bright bands. So the extra path

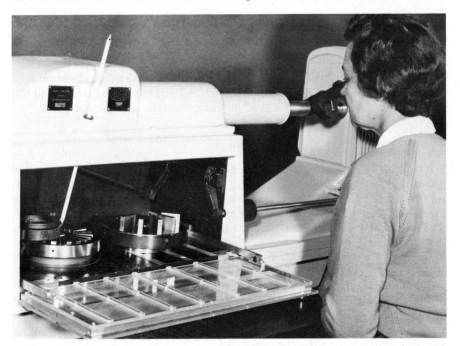

Plate 19A Interferometer measurement of length. *The photograph shows an interfero-meter used to measure the accuracy of block (slip) gauges, widely used for precision measurements in the engineering industry. As shown, the gauges are placed on turntables. Optical flats are used with them to form interference bands; a circular flat is shown ready for use above the gauges on the left. Any error in length is deduced from an accurately-known wavelength in the visible spectrum of cadmium, which is used to form the bands. Correction is needed for temperature changes, observed by the long mercury thermometer shown.*
(Crown copyright. Courtesy of National Physical Laboratory)

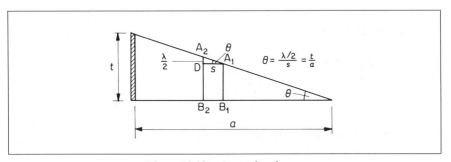

Figure 19.12 *Air-wedge theory*

difference in the air film from one band to the other is λ. The extra path difference is $2t_2 - 2t_1$, or $2A_2D$, where A_1D is the perpendicular from A_1 to A_2B_2. So $A_2D = \lambda/2$. Hence if s is the separation, B_1B_2, of the bands, which is equal to A_1D, then

$$\tan \theta = \frac{A_2D}{A_1D} = \frac{\lambda/2}{s} = \frac{\lambda}{2s}$$

Since θ is very small, $\tan \theta = \theta$ in radians. So

$$\theta = \frac{\lambda}{2s}$$

As an illustration, suppose an air wedge is illuminated normally by mono-chromatic light of wavelength $5 \cdot 8 \times 10^{-7}$ m (580 nm) and the separation of the bright bands is 0·29 mm or $2 \cdot 9 \times 10^{-4}$ m. Then, from $\theta = \lambda/2s$, we have

$$\theta = \frac{5 \cdot 8 \times 10^{-7}}{2 \times 2 \cdot 9 \times 10^{-4}} = 10^{-3} \, \text{rad}$$

If a *liquid wedge* is formed between the plates, the optical path difference becomes $2nt$, where the air thickness if t, n being the refractive index of the liquid. An optical path difference of λ now occurs for a change in t which is n times *less* than in the case of the air-wedge. The spacing of the bright and dark fringes is thus n times *closer* than for air. So measurement of the relative spacing enables n to be found.

The expansivity of a crystal can be found by forming an air-wedge of small angle between a fixed horizontal glass plate and the upper surface of the crystal, and illuminating the wedge by monochromatic light. When the crystal is heated a number of bright fringes, m say, cross the field of view in a microscope focused on the air-wedge. The increase in length of the crystal in an upward direction is $m\lambda/2$, since a change of λ represents a change in the thickness of the film is $\lambda/2$, and the expansivity can then be calculated.

Examples on Air Wedge

1 A wedge air film is formed by placing aluminium foil between two glass slides at a distance of 75 mm from the line of contact of the slides. When the air wedge is illuminated normally by light of wavelength $5 \cdot 60 \times 10^{-7}$ m, interference fringes are produced parallel to the line of contact which have a separation of 1·20 mm. Calculate the angle of the wedge and the thickness of the foil.

If s is the separation of the bands, then

$$\text{angle of wedge, } \theta = \frac{\lambda/2}{s} = \frac{\lambda}{2s}$$

$$= \frac{5 \cdot 6 \times 10^{-7}}{2 \times 1 \cdot 2 \times 10^{-3}} = 2 \cdot 3 \times 10^{-4} \, \text{rad}$$

If t is the thickness of the foil, then

$$\frac{t}{75 \times 10^{-3}} = \theta = 2 \cdot 3 \times 10^{-4}$$

So $\qquad\qquad t = 75 \times 10^{-3} \times 2 \cdot 3 \times 10^{-4} = 1 \cdot 7 \times 10^{-5} \, \text{m}$

2 Two optically flat glass plates, in contact along one edge, make a very small angle with each other. They are illuminated by red light of wavelength 750 nm and blue light of wavelength 450 nm. Looking down on the wedge the first place where it appears purple is 5·0 mm from the line of contact.

If red and blue light together produce purple light, find the angle between the plates.

The first place where the wedge appears purple corresponds to a thickness t of the air wedge where both red and blue light first form a coincident bright interference band. The colours then mix and produce purple light.

At the thickness t, the path difference for bright interference band of wavelength λ is given by $2t = (m-\frac{1}{2})\lambda$, from p. 524. So for $\lambda = 750\,\text{nm}$ and $\lambda = 450\,\text{nm}$,

$$2t = (m-\tfrac{1}{2})\,750\,\text{nm} = (m+\tfrac{1}{2})\,450\,\text{nm}$$

since if m is the whole number for the 750 nm wavelength, then $(m+1)$ is the whole number for the overlapping 450 nm (shorter) wavelength.

$$\therefore (m-\tfrac{1}{2})\,750 = (m+\tfrac{1}{2})\,450$$

Solving, $$m = 2$$

So $2t = (2-\tfrac{1}{2}) \times 750\,\text{nm} = 1125\,\text{nm}$, and $t = 562\cdot5\,\text{nm} = 5\cdot6 \times 10^{-7}\,\text{m}$ (approx.)

Hence angle of wedge $\theta = \dfrac{t}{5 \times 10^{-3}} = \dfrac{5\cdot6 \times 10^{-7}}{5 \times 10^{-3}} = 1\cdot1 \times 10^{-4}\,\text{rad}$ (approx.)

Newton's Rings

Newton discovered an example of interference which is known as 'Newton's rings'. In this case a lens L is placed on a sheet of plane glass H, L having a lower surface of very large radius of curvature so that a curved air wedge is formed

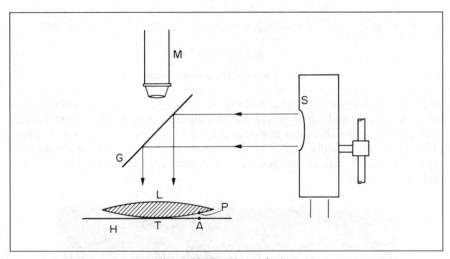

Figure 19.13   *Newton's rings*

with H, Figure 19.13. By means of a sheet of glass G, monochromatic light from a sodium lamp S, for example, is reflected downwards towards L; and when the light reflected upwards is observed through a microscope M focused on H, a series of bright and dark rings is seen. The circles have increasing radius, and are concentric with the point of contact T of L with H, see Figure 19.14.

Consider the air-film PA between A on the plate and P on the lower lens surface. Some of the incident light is reflected from P to the microscope, while the remainder of the light passes straight through to A, where it is also reflected to the microscope and brought to the same focus. The two rays of light have thus a net path difference of $2t$, where $t = $ PA. The same path difference is obtained at *all points round* T which have the same distance TA from T. So if $2t = m\lambda$, where

Figure 19.14 *Newton's rings, formed by interference of yellow light between converging lens and flat glass plate*

m is an integer and λ is the wavelength, we might expect a bright *ring* with centre T. Similarly, if $2t = (m + \frac{1}{2})\lambda$, we might expect a dark ring.

When a ray is reflected from an optically *denser* medium, however, a phase change of 180° occurs in the wave, which is equivalent to an extra path difference of $\lambda/2$ (see also p. 518). The truth of this statement can be seen by the presence of the dark spot at the centre, T, of the rings. At this point there is no geometrical path difference between the rays reflected from the lower surface of the lens and H, so that they should be in phase when they are brought to a focus and should form a bright spot. The dark spot means, therefore, that one of the rays suffers a phase change of 180°. Taking the phase change into account, it follows that

$$2t = m\lambda \text{ for a } \textit{dark} \text{ ring} \qquad . \qquad . \qquad . \qquad (1)$$

and
$$2t = (m + \tfrac{1}{2})\lambda \text{ for a } \textit{bright} \text{ ring} \qquad . \qquad . \qquad . \qquad (2)$$

where *m* is an integer. Young verified the phase change by placing oil of sassafras between a crown and a flint glass lens. This liquid had a refractive index greater than that of crown glass and less than that of flint glass, so that light was now reflected at an optically denser medium at each lens. A *bright* spot was then observed in the middle of the Newton's rings, showing that no net phase change had now occurred.

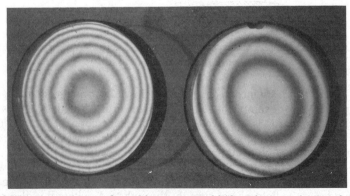

Figure 19.15 *Newton's rings formed by transmitted light. The central spot is bright since there is no phase change on transmission (compare the dark central spot in Newton's rings formed by reflected light). If the lens surface is imperfect the rings are distorted, as shown on the right*

The grinding of a lens surface can be tested by observing the appearance of the Newton's rings formed between it and a flat glass plate when monochromatic light is used. If the rings are not perfectly circular as in Figure 19.14, the grinding is imperfect (see Figure 19.15). As in the case of the wedge air film, Newton's rings is an example of interference by division of amplitude (p. 524).

Measurement of Wavelength by Newton's Rings

The radius r of a ring can be expressed in terms of the thickness, t, of the corresponding layer of air by simple geometry. Suppose TO is produced to D to meet the completed circular section of the lower surface PO of the lens of radius a, PO being perpendicular to the diameter TD through T, Figure 19.16. Then, from the well-known theorem concerning the segments of chords in a circle, TO. OD = QO. OP. But AT = r = PO, QO = OP = r, AP = t = TO, and OD = $2a-$OT = $2a-t$.

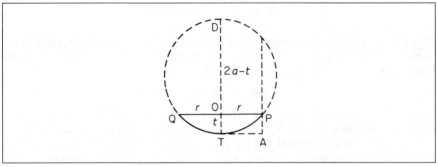

Figure 19.16 *Theory of radius of Newton's rings*

$$\therefore t\,(2a-t) = r \times r = r^2$$

$$\therefore 2at - t^2 = r^2$$

But t^2 is very small compared with $2at$, as a is large.

$$\therefore 2at = r^2$$

$$\therefore 2t = \frac{r^2}{a} \qquad . \qquad . \qquad . \qquad . \qquad . \qquad (i)$$

But
$$2t = (m+\tfrac{1}{2})\lambda \text{ for a bright ring}$$

$$\therefore \frac{r^2}{a} = (m+\tfrac{1}{2})\lambda \qquad . \qquad . \qquad . \qquad . \qquad (3)$$

The first bright ring obviously corresponds to the case of $m = 0$ in equation (3); the second bright ring corresponds to the case of $m = 1$. Thus the radius of the 15th bright ring is given from (3) by $r^2/a = 14\tfrac{1}{2}\lambda$, from which $\lambda = 2r^2/29a$. Knowing r and a, therefore, the wavelength λ can be calculated. Experiment shows that the rings become narrower when blue or violet light is used in place of red light, which proves, from equation (3), that the wavelength of violet light is shorter than the wavelength of red light. Similarly it can be proved that the wavelength of yellow light is shorter than that of red light and longer than the wavelength of violet light.

Visibility of Newton's Rings

When white light is used in Newton's rings experiment the rings are coloured, generally with violet at the inner and red at the outer edge. This can be seen from the formula $r^2 = (m+\frac{1}{2})\lambda a$, since $r^2 \propto \lambda$.

When Newton's rings are formed by sodium light, close examination shows that the clarity, or visibility, of the rings gradually diminishes as one moves outwards from the central spot, after which the visibility improves again. The variation in clarity is due to the fact that sodium light is not monochromatic but consists of *two wavelengths*, λ_2, λ_1, close to one another. These are (i) $\lambda_2 = 5.890 \times 10^{-7}$ m (589·0 nm) (D_2), (ii) $\lambda_1 = 5.896 \times 10^{-7}$ m (589·6 nm) (D_1). Each wavelength produces its own pattern of rings, and the ring patterns gradually separate as m, the number of the ring, increases. When $m\lambda_1 = (m+\frac{1}{2})\lambda_2$, the bright rings of one wavelength fall in the dark spaces of the other and the visibility is a minimum. In this case

$$5896m = 5890\,(m+\tfrac{1}{2})$$

$$\therefore m = \frac{5890}{12} = 490 \text{ (approx.)}$$

At a further number of ring m_1, when $m_1\lambda_1 = (m_1+1)\lambda_2$, the bright (and dark) rings of the two ring patterns coincide again, and the clarity, or visibility, of the interference pattern is restored. In this case

$$5896m_1 = 5890(m_1+1)$$

from which $m_1 = 980$ (approx.). Thus at about the 500th ring there is a minimum visibility, and at about the 1000th ring the visibility is a maximum.

It may be noted here that the fringes in films of varying thickness, such as Newton's rings and the air-wedge fringes, p. 523, appear to be formed in the film itself, and the eye must be focused on the film to see them. We say that the fringes are 'localised' at the film. With a thin film of uniform thickness, however, fringes are formed by parallel rays which enter the eye, and these fringes are therefore localised at infinity.

'Blooming' of Lenses

Whenever lenses are used, a small percentage of the incident light is reflected from each surface. In compound lens systems, as in telescopes and microscopes, this produces a background of unfocused light, so the clarity of the final image is reduced. There is also a reduction in the intensity of the image, since less light is transmitted through the lenses.

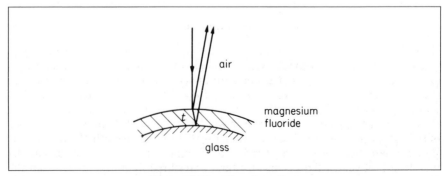

Figure 19.17   *Blooming of lens*

The amount of reflected light can be considerably reduced by evaporating a thin coating of a fluoride salt such as magnesium fluoride on to the surfaces, Figure 19.17. Some of the light, of average wavelength λ, is then reflected from the air-fluoride surface and the remainder penetrates the coating and is partially reflected from the fluoride-glass surface. Destructive interference occurs between the two reflected beams when there is a phase difference of 180°, or a path difference of $\lambda/2$, as the refractive index of the fluoride is less than that of glass. Thus if t is the required thickness of the coating and n' its refractive index, $2n't = \lambda/2$. Hence $t = \lambda/4n' = 6 \times 10^{-7}/(4 \times 1\cdot38)$, assuming λ is 6×10^{-7} m and n' is $1\cdot38$, from which $t = 1\cdot1 \times 10^{-7}$ m.

For best results n' should have a value equal to about $\sqrt{n}$, where n is the refractive index of the glass lens. The intensities of the two reflected beams are then equal, and hence complete interference occurs between them. No light is now reflected back from the lens. In practice, since complete interference is not possible simultaneously for every wavelength of white light, an average wavelength for λ, such as green-yellow, is chosen. The lens thus appears purple, a mixture of red and blue, since these colours in white light are reflected. 'Bloomed' lenses produce a marked improvement in the clarity of the final image in optical instruments.

Lloyd's Mirror

In 1834 LLOYD obtained interference fringes on a screen by using a plane mirror M, and illuminating it with light nearly at grazing incidence, coming from a slit O parallel to the mirror, Figure 19.18. A point such as A on the screen is illuminated   (i) by waves from O travelling along OA and   (ii) by waves from O travelling along OM and then reflected along MA, which appear to come from the virtual image I of O in the mirror. Since O and I are close coherent sources interference fringes are obtained on the screen.

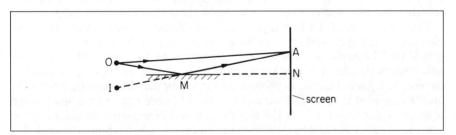

Figure 19.18   *Lloyd's mirror experiment*

Experiment showed that the fringe at N, which corresponds to the point of intersection of the mirror and the screen, was *dark*. Since ON = IN, this fringe might have been expected, before the experiment was carried out, to be bright. Lloyd concluded that a phase change of 180°, equivalent to half a wavelength, occurred by reflection at the mirror surface, which is a denser surface than air (see p. 518). Lloyd's mirror experiment is an example of interference by division of wavefront (p. 519).

Colours in Thin Films

The colours in thin films of oil or glass are due to interference from an extended source such as the sky or a cloud. Figure 19.19 illustrates interference between

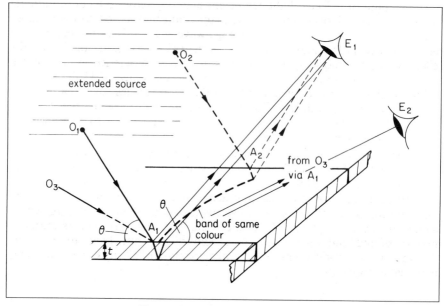

Figure 19.19   *Colours in thin films*

rays from points O_1, O_2 respectively on the extended source. Each ray is reflected and refracted at points such as A_1 or A_2, on the film, and enter the eye at E_1. Although O_1, O_2 are non-coherent, the eye will see the same colour of a particular wavelength λ if $2nt \cos r = (m - \frac{1}{2})\lambda$, the condition for constructive interference, or the complementary colour if $2nt \cos r = m\lambda$, the condition for destructive interference. If n is the refractive index of the oil film of thickness t and r is the angle of refraction in the film, the path difference between the rays reflected and refracted at A_1 (or at A_2) can be shown to be $2nt \cos r + \lambda/2$.

The separation of the two rays from A_1 or from A_2 must be less than the diameter of the eye-pupil for interference to occur, and this is the case only for thin films. The angle of refraction r is determined by the angle of incidence, or reflection, at the film. The particular colour seen thus depends on the position of the eye. At E_2, for example, a different colour will be seen from another point O_3 on the extended source. The variation of θ and hence r is small when the eye observes a particular area of the film. So a fringe of a particular colour is the contour of paths of *equal inclination* to the film such as $A_1 A_2$. Since the angle θ is constant round the perpendicular line from the eye to the film, the fringe or band of a particular colour is *circular*. So if $2nt \cos r = m\lambda$ for a blue colour, the eye sees a circular band of the complementary colour such as green-yellow.

Vertical Soap Film Colours

An interesting experiment on thin films, due to C. V. Boys, can be performed by illuminating a vertical soap film with monochromatic light. At first the film appears uniformly coloured. As the soap drains to the bottom, however, a wedge-shaped film of liquid forms in the ring, the top of the film being thinner than the bottom. The thickness of the wedge is constant in a horizontal direction, and thus horizontal bright and dark fringes are observed across the film. When the upper part of the film becomes extremely thin a *black* fringe is observed at the top (compare the dark central spot in Newton's rings experiment), and the film breaks shortly afterwards.

With white light, a succession of broad coloured fringes is first observed in the soap film. Each fringe contains colours of the spectrum, red to violet. The fringes widen as the film drains, and just before it breaks a black fringe is obtained at the top. The black fringe is due to the 180° (or π) phase change by reflection when the film is $\lambda/4$ thick and so destructive interference occurs.

For normal incidence of white light, a particular wavelength λ is seen where the optical path difference due to the film $= (m-\frac{1}{2})\lambda$ and m is an integer. Thus a red colour of wavelength $7\cdot0 \times 10^{-7}$ m is seen where the optical path difference is $3\cdot5 \times 10^{-7}$ m, corresponding to $m = 1$. No other colour is seen at this part of the thin film. Suppose, however, that another part of the film is much thicker and the optical path difference here is $21 \times 3\cdot5 \times 10^{-7}$ m. Then a red colour of wavelength $7\cdot0 \times 10^{-7}$ m, $m = 11$, an orange colour of wavelength about $6\cdot4 \times 10^{-7}$ m, $m = 12$, a yellow wavelength about $5\cdot9 \times 10^{-7}$ m, $m = 13$, and other colours of shorter wavelengths corresponding to higher integral values of m, are seen at the same part of the film. These colours all overlap and produce a white colour. If the film is thicker still, it can be seen that numerous wavelengths throughout the visible spectrum are obtained and the film then appears uniformly white.

_____ **Exercises 19** _____

1 In Young's two-slit experiment using red light, state the effect of the following procedure on the appearance of the fringes:
   (a) The separation of the slits is decreased.
   (b) The screen is moved closer to the slits.
   (c) The source slit is moved closer to the two slits.
   (d) Blue light is used in place of red light.
   (e) One of the two slits is covered up.
   (f) The source slit is made wider.

2 In a Young's two-slit experiment using light of wavelength $6\cdot0 \times 10^{-7}$ m, the slits were $0\cdot40$ mm apart and the distance of the slits to the screen was $1\cdot20$ m. Find the separation of the fringes.
   What is the angle in radians subtended by a central pair of bright fringes at the slits?

3 In an air-wedge experiment using white light, the following are observed:
   (a) A dark band is obtained where the two slides touch.
   (b) The bands nearest to the dark band are coloured blue.
   (c) The bands are straight and parallel.
   Explain each of these effects.

4 (a) A wedge-shaped film of air is formed between two, thin, parallel-sided, glass plates by means of a straight piece of wire. The two plates are in contact along one edge of the film and the wire is parallel to this edge.   (i) Draw and label a diagram of the experimental arrangement you would use to observe and make measurements on interference fringes produced with light incident normally on the film.   (ii) Explain the function of each part of the apparatus.
   (b) In such an experiment using light of wavelength 589 nm, the distance between the seventh and one hundred and sixty-seventh dark fringes was 26·3 mm and the distance between the junction of the glass plates and the wire was 35·6 mm. Calculate the angle of the wedge and the diameter of the wire. (*JMB.*)

5 (a) Describe, with the aid of a diagram, how you would produce and view interference fringes using a monochromatic light source, a double slit and any other essential apparatus. Describe the appearance of the fringes.
   (b) State the measurements you would make in order to determine the wavelength of the light and indicate the instrument you would use for each measurement, justifying your choice in each case.
   (c) Derive an expression for the wavelength in terms of the relevant measurements.
   (d) Describe the effect on the appearance of the fringes of reducing the slit

separation and discuss how this would affect the accuracy of your measurements in (b) above. (*N.*)

6 Two plane glass plates which are in contact at one edge are separated by a piece
of metal foil 12·50 cm from that edge. Interference fringes parallel to the line
of contact are observed in reflected light of wavelength $5·46 \times 10^{-7}$ m and are
found to be 1·50 mm apart. Find the thickness of the foil. (*L.*)

7 Explain why, for visible interference effects, it is normally necessary for the
light to come from a single source and to follow different optical paths. Include
a statement of the conditions required for complete destructive interference.

Explain *either* the colours seen in thin oil films *or* the colours seen in soap
bubbles.

A wedge-shaped film of air between two glass plates gives equally spaced dark
fringes, using reflected sodium light, which are 0·22 mm apart. When monochromatic light of another wavelength is used the fringes are 0·24 mm apart.
Explain why the two fringe spacings are different. (The incident light falls normally
on the air film in both cases.) Discuss what you would expect to see if the air
film were illuminated by both sources simultaneously. Calculate the wavelength
of the second source of light.
(Wavelength of sodium light = 589 nm (589×10^{-9} m).) (*L.*)

8 Describe how to set up apparatus to observe and make measurements on the
interference fringes produced by Young's slits. Explain how   (i) the wavelengths
of two monochromatic light sources could be compared,   (ii) the separation of
the slits could be deduced using a source of known wavelength. Establish any
formula required.

State, giving reasons, what you would expect to observe
(a) if a white light source were substituted for a monochromatic source,
(b) if the source slit were then displaced slightly at right angles to its length in
    the plane parallel to the plane of the Young's slits. (*L.*)

9 Describe, with the aid of a labelled diagram, how the wavelength of monochromatic light may be found using Young's slits. Give the theory of the
experiment.

State, and give physical reasons for the features which are common to this
method and to the method based on Lloyd's mirror.

In an experiment using Young's slits the distance between the centre of the
interference pattern and the tenth bright fringe on either side is 3·44 cm and the
distance between the slits and the screen is 2·00 m. If the wavelength of the light
used is $5·89 \times 10^{-7}$ m determine the slit separation. (*JMB.*)

10 A low-flying aircraft can adversely affect the quality of television reception in
the houses over which it is flying. The effects produced are caused by alternate
rises and falls in the strength of the signal received at the aerial. Suggest why
these occur. (*L.*)

11 With the aid of a clearly labelled diagram, explain how thin film interference
effects can be used to test the flatness of a surface. (*L.*)

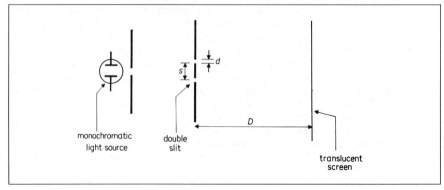

Figure 19A

12 Figure 19A (not drawn to scale) shows the apparatus used in an attempt to measure the wavelength of light using double slit interference.

(a) Explain how the apparatus produces double slit interference fringes on the translucent screen.

(b) If measurable fringes are to be seen on the screen, then (i) the slit separation s must be small, (ii) the distance D between the slits and the screen must be large. Explain each of these conditions.

(c) The light from the monochromatic source has a wavelength of $5 \cdot 0 \times 10^{-7}$ m and it is required to produce a fringe separation of $5 \cdot 0$ mm. Suggest suitable values for s and D justifying your answer. Describe the important features of the pattern you would hope to obtain.

(d) Describe and explain any changes in the pattern which would be brought about by each of the following changes made separately: (i) one slit is covered with an opaque material; (ii) the slits are made narrower although their separation remains the same. (*AEB*, 1984.)

13 Figure 19B shows a wavefront of light AB arriving at the surface of a flat mica sheet. CD marks the position of the same wavefront just inside the mica.

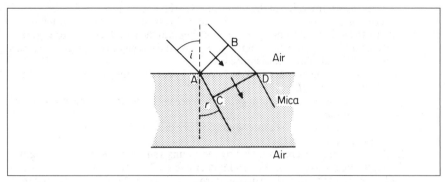

Figure 19B

(a) State Huygens' principle.

Explain how the refraction of light shown in Figure 19B is accounted for by Huygens' principle and one additional assumption. State this assumption. Derive the result

$$n = \frac{\sin i}{\sin r} = \frac{c}{c_m}$$

where n is the refractive index of mica and where c and c_m are the speeds of light in air and mica respectively.

(b) The thickness of the mica sheet is $4 \cdot 80$ µm. The refractive index of mica for light of wavelength 512 nm is $1 \cdot 60$. Calculate the ratio of the thickness of the sheet of mica to (i) the wavelength of the 512 nm light in air, and (ii) the wavelength of the same light in mica.

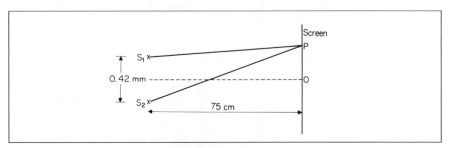

Figure 19C

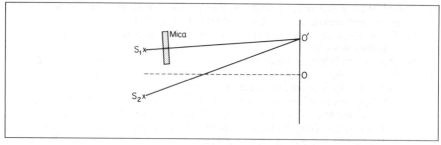

Figure 19D

(c) Figure 19C shows the positions of the slits and the screen in an optical two slits experiment. Light of wavelength 512 nm is used, the separation of the slits is 0·42 mm and the perpendicular distance from the line S_1S_2 to the screen at O is 75 cm. P marks the centre of the third bright fringe from O. Calculate the distance OP.

Write down the value of the distance $PS_2 - PS_1$.

The same mica sheet as was referred to in (b) above is mounted normally across the light path from S_1 without cutting into the light from S_2, see Figure 19D. Explain why the centre of the pattern on the screen moves from O to a new position O'. Calculate the distance OO'.

How would the distance OO' be affected by rotating the mica sheet about an axis parallel to S_1? (L.)

14 Explain what is meant by the term *path-difference* with reference to the interference of two-wave-motions.

Why is it not possible to see interference where the light beams from the headlamps of a car overlap?

Interference fringes were produced by the Young's slits method, the wavelength of the light being 6×10^{-7} m. When a film of material $3·6 \times 10^{-3}$ cm thick was placed over *one* of the slits, the fringe pattern was displaced by a distance equal to 30 times that between two adjacent fringes. Calculate the refractive index of the material. To which side are the fringes displaced?

(When a layer of transparent material whose refractive index is *n* and whose thickness is *d* is placed in the path of a beam of light, it introduces a path difference equal to $(n-1)d$.) (O. & C.)

15 What do you understand by (i) *interference*, (ii) *coherence* between two separate wave trains, (iii) *coherence* along one wave train?

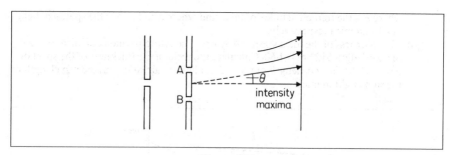

Figure 19E

In a 'Young's slits' experiment, the centres of the double slits are 0·25 mm apart and the wavelength of the light used is $6·0 \times 10^{-4}$ mm. Calculate the angle θ subtended at the slits by adjacent maxima of the fringe pattern (see Figure 19E).

Describe and explain what happens to these fringes if

(a) slit A is covered with a thin sheet of transparent material of high refractive index,

(b) the light emerging from slit A is reduced in intensity to half that emerging from the other slit B,

(c) A and B are each covered with a thin film of Polaroid and one of these films is slowly rotated,

(d) the distance between slit A and slit B is slowly increased. (*C.*)

16   What are the necessary conditions for interference of light to be observable? Describe with the aid of a labelled diagram how optical interference may be demonstrated using Young's slits. Indicate suitable values for all the distances shown.

How are the colours observed in thin films explained in terms of the wave nature of light? Why does a small oil patch on the road often show approximately circular coloured rings? (*L.*)

Newton's Rings

17   Draw a labelled diagram of the apparatus you would use to view Newton's rings. Explain why darkness is produced at certain points and state the conditions required for this.

Also explain why

(a) the interference fringes are circular;

(b) the centre of the system is normally black;

(c) the radii of the rings are proportional to the square roots of the natural numbers.

State with reasons, what you would expect to see if the spherical lens were replaced by a cylindrical one.

Newton's rings were produced using a plano-convex lens, made of glass of refractive index 1·48 resting on a flat glass plate. The diameter of the 10th dark ring from the centre was measured and found to be 3·36 mm. The diameter of the 30th dark ring was found to be 5·82 mm. The wavelength of the light used was 589 nm (589 × 10^{-9} m). What was the focal length of the lens? (*L.*)

18   A glass converging lens rests in contact with a horizontal plane sheet of glass. Describe how you would produce and view Newton's rings, using reflected sodium light.

Explain how the rings are formed and derive a formula for their diameters. Describe the measurements you would make in order to determine the radii of curvature of the faces of the lens, assuming that the wavelength of sodium light is known. Show how the result is derived from the observations.

How would the ring pattern change if

(a) the lens were raised vertically one quarter of a wavelength,

(b) the space between the lens and the plate were filled with water? (*JMB.*)

19   In the interference of light what is meant by the requirement of coherency? How is this usually achieved in practice?

A thin spherical lens of long focal length is placed on a flat piece of glass. How, using this arrangement, would you demonstrate interference by reflection?

Explain how these fringes are formed and describe their appearance. What would be the effect if the spherical lens were replaced by a cylindrical one?

Such a system using a spherical lens is illuminated with light of wavelength 600 nm. When the lens is carefully raised from the plate 50 extra fringes appear at and move away from the centre of the fringe system. By what distance was the lens raised? (*L.*)

20   Explain how Newton's rings are formed, and describe how you would demonstrate them experimentally. How is it possible to predict the appearance of the centre of the ring pattern when

(a) the surfaces are touching, and

(b) the surfaces are not touching?

In a Newton's rings experiment one surface was fixed and the other movable along the axis of the system. As the latter surface was moved the rings appeared to contract and the centre of the pattern, initially at its darkest, became alternately bright and dark, passing through 26 bright phases and finishing at its darkest again. If the wavelength of the light was $5 \cdot 461 \times 10^{-7}$ m, how far was the surface moved and did it approach, or recede from the fixed surface? Suggest one possible application of this experiment. (*O. & C.*)

21 (a) What conditions must be fulfilled if interference between two light beams is to be observed?

   (b) State *three* ways in which a beam of light from a laser differs from a parallel beam of light from a sodium lamp.

   (c) A thin plano-convex lens is placed with its curved face downwards on a plane glass plate and is illuminated normally by sodium light of wavelength 589 nm. A series of circular interference fringes is observed by reflected light. (i) Draw a diagram of a suitable experimental arrangement by which the fringes could be observed and measured.   (ii) Explain the formation of the fringes. Why is the centre of the pattern dark?   (iii) If the radius of the 20th dark ring from the centre is 4·99 mm, calculate the radius of curvature of the lens face, proving any formula you use in your calculation.   (iv) The air-space between the lens and the glass plate is now filled with water, of refractive index 1·33. Describe the changes in the fringe system, and calculate the new radius of the 20th dark ring from the centre.

   (d) Explain why colours are observed when a thin layer of transparent liquid such as petrol spreads over a water surface and is illuminated by daylight. (*O.*)

20
Diffraction of Light Waves

In this chapter we study the diffraction or spreading of waves through openings. We first deal with diffraction at a single slit and the variation of intensity of the image formed. This leads to the way in which the resolving power of optical and radio telescopes can be increased. The diffraction grating is then considered in detail and its use in measuring wavelengths in spectra is discussed. We conclude with a brief account of the use of diffraction in holography.

Diffraction Image

In 1665 GRIMALDI observed that the shadow of a very thin wire in a beam of light was much broader than he expected. This is due to diffraction.

Consider two points on the *same wavefront*, for example the two points A, B, on a plane wavefront, arriving at a narrow slit in a screen, Figure 20.1. A and B

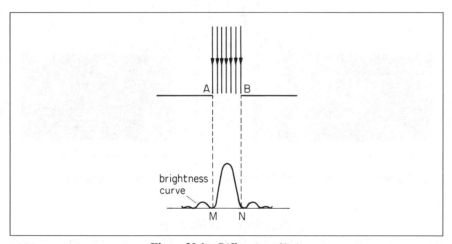

Figure 20.1   *Diffraction of light*

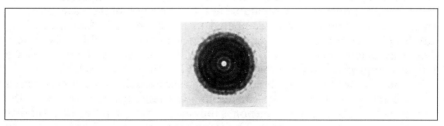

Figure 20.2   *Diffraction rings in the shadow of a small circular disc. The bright spot is at the centre of the geometrical shadow*

can be considered as secondary sources of light from Huygens' Principle in the wave theory of light (p. 502); and as they are on the same wavefront, A and B have identical amplitudes and frequencies and are in phase with each other. Consequently A and B are *coherent sources*. So we can expect to find an interference pattern on a screen in front of the slit, provided its width is small compared with the wavelength of light. In fact, for a short distance beyond the edges M, N, of the projection of AB, that is, in the geometrical shadow, observation shows that there are some alternate bright and dark fringes, see Figure 20.4.

Figure 20.3 *(below) The variation in intensity of the bright diffraction bands due to a single small rectangular aperture is shown roughly in the diagram. The central bright band, in which most of the light is concentrated, has maximum intensity in the direction of P, the middle of the band. The intensity of this band diminishes to a minimum at Q or Q_1, which are at an angle of diffraction θ to the direction of P*

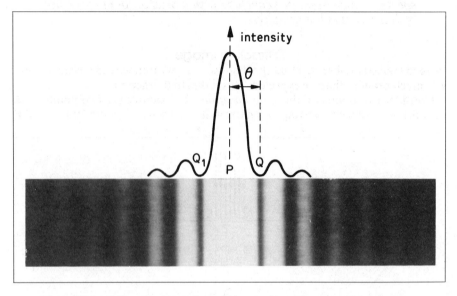

Figure 20.4 *Diffraction bands formed by a single small rectangular aperture*

Diffraction and Slit Width

Diffraction is the name given to the spreading of waves after they pass through small openings (or round small obstacles). The diffraction is appreciable when the width of the opening is comparable to the wavelength of the waves and very small when the width is large compared to the wavelength. Sound has long wavelengths and can diffract after passing through doorways. Light has a very short wavelength such as 6×10^{-7} m or 6×10^{-4} mm and so light waves are diffracted appreciably only through very small openings as we shall see.

If a source of white light is observed through the eyelashes a series of coloured images can be seen. These images are due to interference between sources on the same wavefront, and the phenomenon is thus an example of diffraction. Another example of diffraction was deduced by POISSON at a time when the wave theory was new. Poisson considered mathematically the combined effect of the wavefronts round a circular disc illuminated by a distant small source of light, and he came to the conclusion that the light should be visible beyond the disc in the middle of the geometrical shadow. Poisson thought this was impossible; but experiment confirmed his deduction as shown in Figure 20.2, and he became a supporter of the wave theory of light.

Diffraction at Single Slit

We now consider in more detail the image produced by diffraction at an opening or slit. The image of a distant star produced by a telescope objective is due to diffraction at a circular opening. On this account a study of the image has application in Astronomy, as we see later.

Suppose parallel plane wavefronts from a distant object are diffracted at a rectangular slit AB of width a, Figure 20.5 (i). This is called *Fraunhofer diffraction*. If the light passing through the slit is received on a screen S a long way from AB, we may consider that parallel wavefronts have travelled to S to form the image of the slit.

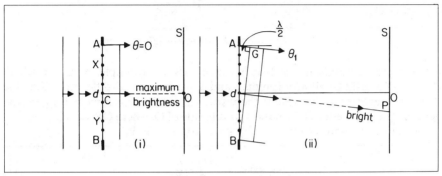

Figure 20.5 *Rectangular slit diffraction image*

Consider a plane wavefront which reaches the opening AB. All points on it between A and B are in phase, that is, they are coherent. These points act as secondary centres, sending out waves beyond the slit. Their combined effect at any distant point can be found by summing the numerous waves arriving there, from the Principle of Superposition. The mathematical treatment is beyond the scope of this book. The general effect, however, can be seen by a simplified treatment.

Central Bright Image

Consider first a point O on the screen which lies on the normal to the slit passing through its midpoint C, Figure 20.5 (i). O is the *centre* of the diffraction image. It corresponds to a direction $\theta = 0$, where θ is measured from the normal to AB. Now in this direction the waves from the secondary sources such as A, X, Y, B have no path difference. So all the waves arrive *in phase* at O. The centre of the diffraction image is therefore brightest.

In a direction very slightly inclined to $\theta = 0$, the waves from all the sources between A and B arrive slightly out of phase at the corresponding point P of the image near the centre, Figure 20.5 (ii). So the brightness decreases. In a particular direction θ_1, we reach the dark edge or boundary Q of the central image. As we soon show, this direction corresponds to a path difference of λ between the two sources at the edges A and B of the slit. Figure 20.6 shows the path difference AF of the waves from A and B in this case, which totals $\left(\dfrac{\lambda}{2} + \dfrac{\lambda}{2}\right)$ or λ.

To find the resultant amplitude of the waves arriving at Q, divide the wavefront AB into two halves. The top point A of the upper half CA, and the top point C of the lower half BC, send out waves to Q which have a path difference $\lambda/2$, from above. Thus the resultant amplitude at Q is zero. All other pairs of

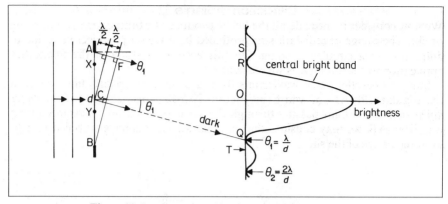

Figure 20.6 *Variation of brightness of diffraction image*

corresponding points in the two halves of AB, for example, X and Y where
CX = BY and the two bottom points C and B, also have a path difference $\lambda/2$. So the
brightness at Q is zero. This point, then, is one *edge* of the central bright fringe on the
screen. The other edge is R on the opposite side of O, where OR = OQ.

We can easily find the direction θ_1. From triangle BAF,

$$\sin \theta_1 = \frac{AF}{AB} = \frac{\lambda}{d} \qquad \qquad (1)$$

or if θ_1 is small, its value in radians is

$$\theta_1 = \frac{\lambda}{d} \qquad \qquad (2)$$

As we discuss later, the direction $\theta_1 = \lambda/d$ becomes important in finding the
resolving power of a telescope (p. 544).

Secondary Fringes

Secondary bright and dark fringes are also obtained on the screen beyond Q, as
shown in the photograph in Figure 20.4. Consider, for example, a point T which
lies in a direction θ_3 where the path difference of waves starting from A and B is
$3\lambda/2$, Figure 20.6. We can imagine the wavefront AB divided into *three* equal
parts. Now the waves from the extreme ends of the upper two parts have a path
difference λ. Thus, as we explained for the dark fringes at Q, these two parts of
the wavefront produce darkness at T. The third part produces a fringe of light at
T much less bright than the central fringe, which was due to the whole wavefront
between A and B. Calculation shows that the intensity at T is less than 5%
of the intensity at the middle of the central bright fringe. Thus most of the
light incident on AB is diffracted into the central bright fringe.

We can find the directions of the edge or minimum brightness of the secondary
bright fringes in the same way as the central bright fringe, by dividing the
diffracted wavefront at the slit into four or six, and so on, equal parts. The result
shows that the directions for the successive minima are given by

$$\sin \theta_2 = 2\lambda/d \quad \sin \theta_3 = 3\lambda/d \qquad \qquad (3)$$

**With plane wavefronts (parallel rays) incident normally on a rectangular opening
of width d:**

1 the central image has a maximum brightness in a direction $\theta = 0$ and a minimum brightness (edge) in a direction where $\sin \theta = \lambda/d$
2 the secondary images have minima brightness where $\sin \theta = 2\lambda/d$, $3\lambda/d$, and so on.

Rectilinear and Non-Rectilinear Propagation

As we have seen, the *angular width* of the central bright fringe is 2θ, where θ is the angle between the direction of maximum intensity and the direction of minimum intensity or the edge of the bright fringe in Figure 20.6. The angle θ is given by

$$\sin \theta = \frac{\lambda}{d}$$

where d = width of slit AB. The angle θ is the angular half-width of the central fringe.

When the slit is widened and d becomes large compared with λ, then $\sin \theta$ is very small and hence θ is very small. In this case the directions of the minimum and maximum intensities of the central fringe are very close to each other. Practically the whole of the light is thus confined to a direction immediately in front of the incident direction, that is, no spreading occurs. This explains the *rectilinear propagation of light* for wide openings. When the slit width d is very small and equal to 2λ, for example, then $\sin \theta = \lambda/d = 1/2$, or $\theta = 30°$. The light waves now spread round through $30°$ on either side of the slit, that is, the diffraction is appreciable.

These results are true for any wave phenomenon. In the case of an electromagnetic wave of 3 cm wavelength, a slit of these dimensions produces sideways spreading. Sound waves of a particular frequency 256 Hz have a wavelength of about 1·3 m. Consequently, sound waves spread round openings such as a doorway or an open window, which have comparable dimensions to their wavelengths. Light waves would not spread round these openings as they have very small wavelengths of the order 6×10^{-7} m.

Diffraction in Telescope Objective

When a parallel beam of light from a distant object such as a star S_1 enters a telescope objective L, the lens collects light through a circular opening. So a *diffraction* image of the star is formed round its principal focus, F. This is illustrated in the exaggerated diagram of Figure 20.7.

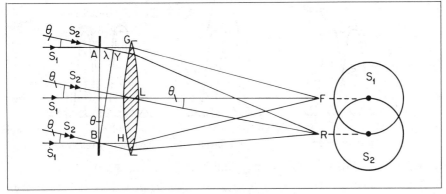

Figure 20.7 *Diffraction in telescope objective*

Consider an incident plane wavefront AB from the star S_1, and suppose for a moment that the aperture is rectangular. The diffracted rays such as AG, BH normal to the wavefront are incident on the lens in a direction parallel to the principal axis LF. The optical paths AGF, BHF are equal. This is true for all other diffracted rays from points between A, B which are parallel to LF, since the optical paths to an image produced by a lens are equal. The central part F of the star pattern is therefore bright.

Now consider those diffracted rays from all points between AB which enter the lens at an angle θ to the principal axis. This corresponds to a diffracted plane wavefront BY at an angle θ to AB. As explained previously, if $AY = \lambda$, then R is the dark edge of the central maximum of the diffraction pattern of the star S_1.

The angle θ corresponding to the edge R is given by

$$\sin \theta = \frac{\lambda}{D}$$

where D is the diameter of the lens aperture. This is the case where the aperture can be divided into a number of rectangular slits. For a *circular* opening such as a lens, or the concave mirror of the Palomar telescope, the formula becomes $\sin \theta = 1\cdot22\lambda/D$. As λ is small compared to D, then θ is small and so we may write $\theta = 1\cdot22\lambda/D$, where θ is in radians.

Resolving Power of Telescope

Suppose now that another distant star S_2 is at an angular distance θ from S_1, Figure 20.8. The maximum intensity of the central pattern of S_2 then falls on the minimum or edge of the central pattern of the star S_1, corresponding to R in Figure 20.8 (i). Experience shows that the two stars can then just be distinguished or *resolved*. Lord Rayleigh stated a criterion for the resolution of two objects, which is generally accepted: *Two objects are just resolved when the maximum intensity of the central pattern of one object falls on the first minimum or dark edge of the other.* Figure 20.8 (i) shows the two stars just resolved. The resultant intensity in the middle dips to about 0·8 of the maximum, and the eye is apparently sensitive to the change here. Figure 20.8 (ii) shows two stars S_1, S_2 unresolved, and Figure 20.8 (iii) the same stars completely resolved. See also Figure 20.9.

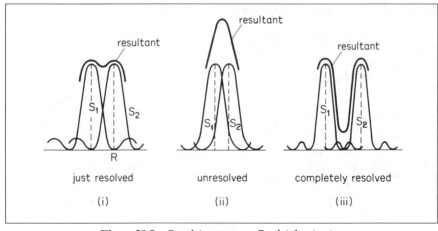

Figure 20.8 *Resolving power—Rayleigh criterion*

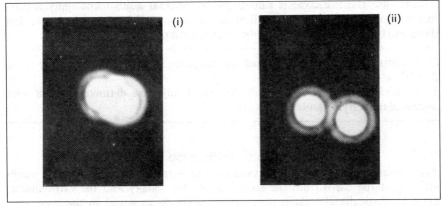

Figure 20.9 *Resolving: (i) Two sources just resolved according to Rayleigh's criterion (ii) Two sources completely resolved*

The angular distance θ between two distant stars just resolved is thus given by $\sin\theta = \theta = 1.22\,\lambda/D$, where D is the diameter of the objective. This is an expression for the *limit of resolution*, or *resolving power*, of a telescope. The limit of resolution or resolving power increases when θ is *smaller*, as two stars closer together can then be resolved. Consequently telescope objectives of *large diameter D* gives high resolving power.

The Yerkes Observatory has a large telescope objective diameter of about 1 metre. The angular distance θ between two stars which can just be resolved is thus given by

$$\theta = \frac{1.22\lambda}{D} = \frac{1.22 \times 6 \times 10^{-7}}{1} = 7.3 \times 10^{-7} \text{ radians}$$

assuming 6×10^{-7} m for the wavelength of light. The Mount Palomar telescope has a parabolic mirror objective of aperture 5 metres. The resolving power is thus five times as great as the Yerkes Observatory telescope. In addition to the advantage of high resolving power, a large aperture collects more light from the distant source and so provides an image of higher intensity or brightness.

Magnifying Power of Telescope and Resolving Power

If the width of the emergent beam from a telescope is greater than the diameter of the eye-pupil, rays from the outer edge of the objective do not enter the eye. Hence the full diameter D of the objective is not used. If the width of the emergent beam is less than the diameter of the eye-pupil, the eye itself, which has a constant aperture, may not be able to resolve the distant objects. Theoretically, the angular resolving power of the eye is $1.22\,\lambda/d$, where d is the diameter of the eye-pupil. In practice an angle of 1 minute is resolved by the eye, which is more than the theoretical value.

Now the angular magnification, or magnifying power, of a telescope is the ratio β/α, where β is the angle subtended at the eye by the final image and α is the angle subtended at the objective (p. 443). To make the fullest use of the diameter D of the objective, therefore, the magnifying power should be increased to the angular ratio given by, if D is in metre,

$$\frac{\text{resolving power of eye}}{\text{resolving power of objective}} = \frac{\pi/(180 \times 60)}{1.22 \times 6 \times 10^{-7}/D} = 400\,D \text{ (approx.)}$$

In this case the telescope is said to be in 'normal adjustment'. Any further increase in magnifying power will make the distant objects appear larger, but there will be no increase in definition or resolving power.

1 Smallest angle θ just resolved by telescope objective of diameter D is $\theta = 1\cdot22\lambda/D$.
2 Resolving power increases with increased objective diameter D, *not* with increased magnifying power M.

Radio Telescope

Radio telescopes are used to investigate and map the sources of radio waves arriving at the earth from the solar system, the galaxy and the extragalactic nebulae. Basically the radio telescope consists of an aerial in the form of a paraboloid-shaped metal surface or 'dish', which can be rotated to face any part of the sky. Distant radio waves, like distant light waves, are reflected towards a focus from all parts of the paraboloid surface (p. 414). The converging waves are then passed to a sensitive receiver and after detection the signals may be recorded on a paper chart or they may be recorded on tape for feeding into a computer for analysis.

The Jodrell Bank radio telescope has a dish about 75 m in diameter. The hydrogen line emitted from interstellar space has a wavelength of about 21 cm or 0·21 m. Thus the angular resolution

$$= \frac{1\cdot22\lambda}{D} = \frac{1\cdot22 \times 0\cdot21}{75} = 0\cdot0034 \, \text{rad} = 0\cdot2°$$

Larger telescopes can provide greater resolving power but the technical problems and cost make this unpractical. A technique using interferometry principles, in which the dishes need not be large, provides much greater resolving power. This has been developed particularly at the Mullard Radio Astronomy Observatory, Cambridge, England.

Radio Interferometers

The principle of the interferometer type of radio telescope is illustrated in Figure 20.10.

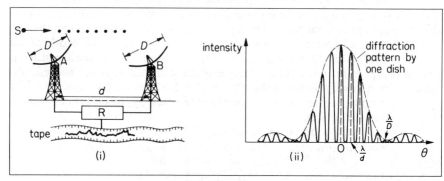

Figure 20.10 *Radio telescope—interferometer type*

Two dishes A and B, separated by a distance of say 5 km, are connected to the same receiver R, Figure 20.10 (i). Radiation from a moving source will reach both aerials or dishes. If the path difference is a whole number of wavelengths or zero, the two signals are in phase and a maximum resultant signal is obtained (constructive interference); if the signals are in antiphase, the resultant signal is a minimum or zero (destructive interference). Thus as a source S moves across the sky, the resultant signal varies in intensity. The principles concerned are similar to a Young's two-slit experiment, except that in this case we are dealing with two 'point receivers' of radiation instead of 'point emitters' of light. The mathematics of the interference is the same in both cases.

The variation in intensity with angle is shown roughly in Figure 20.10 (ii). It is similar to the variation in intensity of the interference fringes in a Young's two-slit experiment. Successive maxima or peaks are thus obtained at angular separations equal to λ/d, where d is the separation of the two dishes. So if $\lambda = 3\,\text{cm} = 0\cdot03\,\text{m}$ and $d = 5\,\text{km} = 5000\,\text{m}$,

$$\text{angular separation} = \frac{\lambda}{d} = \frac{0\cdot03}{5000} = 6 \times 10^{-6}\,\text{rad} = 0\cdot0003°\ (\text{approx.})$$

Further, the angular resolution θ, the angle from the central maximum to the first minimum (p. 544), is given by

$$\theta = \frac{\lambda}{2d} = 3 \times 10^{-6}\,\text{rad}$$

If only one dish were used, with a diameter $D = 13\,\text{m}$ for example, the single slit intensity pattern (p. 544) shows that the angular resolution is now

$$\theta = \frac{1\cdot22\lambda}{D} = \frac{1\cdot22 \times 0\cdot03}{13} = 3 \times 10^{-3}\,\text{rad (approx.)}$$

which is 1000 times less resolution than that obtained with the two dishes. Thus the angular diameter of a source, or of two sources, may be found more accurately by the interferometer method. Further, the method enables the direction of a moving source to be tracked more accurately—the fringe of the central maximum is much narrower with two dishes than with one and this helps to locate the source better.

The interferometer principle has been successfully applied to surveying regions of the sky and to mapping radio galaxies. The Mullard Radio Astronomy Observatory has a telescope consisting of eight parabolic reflectors or dishes, each about 13 m in diameter and mounted on rails about $1\frac{1}{4}$ km long, acting together as a 'grating interferometer'. The effective baseline of the telescope is 5 km. The dishes can be moved along the rails to different sets of positions to map a region of the sky. The signals received from the different aerials are recorded on tape and then synthesised by means of a computer. In this way a telescope of very large aperture (5 km) can be simulated. This is called *aperture synthesis*. With this telescope extremely accurate maps have been made of radio sources in outer space. The Nobel Prize was awarded in 1974 to Sir Martin Ryle, then director of the Mullard Radio Astronomy Observatory, and to Professor A. Hewish of Cambridge for their contributions to radio astronomy, especially aperture synthesis.

Increasing Number of Slits

On p. 540 we saw that the image of a single narrow rectangular slit is a bright central or principal maximum diffraction fringe, together with subsidiary maxima diffraction fringes which are much less bright. Suppose that parallel light is incident on two more parallel close slits, and the light passing through the slits is received by a telescope focused at infinity. Since each slit produces a similar diffraction effect in the same direction, the observed diffraction pattern will have an intensity variation identical to that of a single slit. This time, however, the pattern is crossed by a number of interference fringes, which are due to interference between slits (see *Young's experiment*, p. 519). The envelope of the intensity variation of the interference fringes, shown by the broken lines in Figure 20.11, follows the *diffraction pattern variation due to a single slit*. In general, if I_s is the intensity at a point due to interference between slits and I_d that due to diffraction of a *single* slit, then the resultant intensity I is given by $I = I_d \times I_s$. Hence if $I_d = 0$ at any point, then $I = 0$ at this point irrespective of the value of I_s.

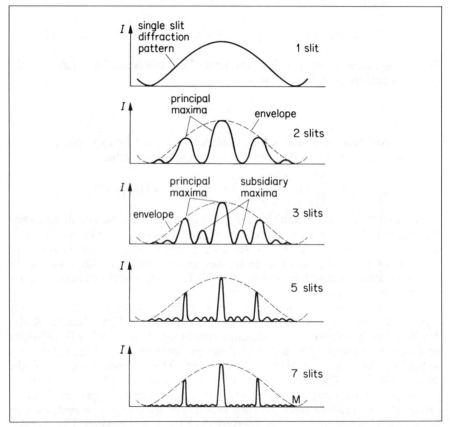

Figure 20.11 *Principal maxima with increasing slits*

As more parallel equidistant slits are introduced, the intensity and sharpness of the principal maxima increase and those of the subsidiary maxima decrease. The effect is illustrated roughly in Figure 20.11. With several hundred lines per millimetre, only a few sharp principal maxima are seen in directions discussed shortly. Their angular separation depends only on the distance between successive slits. The slit width affects the intensity of the higher order principal maxima; the narrower the slit, the greater is the diffraction of light into the higher orders. As shown, the single slit diffraction pattern modulates the maxima. In the direction corresponding to M (7 slits), the single slit intensity

$I_d = 0$. Hence, as stated above, the resultant intensity $I = 0$. This means that no light is diffracted in this direction from any of the slits. Figure 20.12 shows how the principal maxima diffraction images become *sharper* as the number of slits increases, which is one reason why the 'diffraction grating' was invented, as we now discuss.

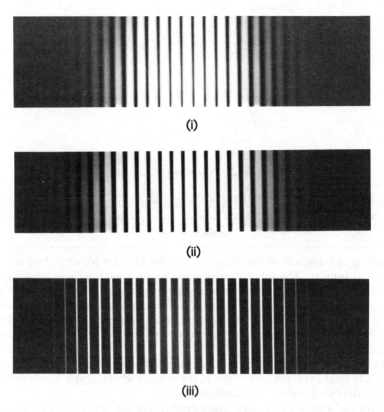

Figure 20.12 *Diffraction by gratings (i) 3 slits; (ii) 5 slits; (iii) 20 slits. Diffraction patterns by gratings with different numbers of slits, all of the same width and separation. As the number of slits increases, the bright lines become sharper*

Principal Maxima of Diffraction Grating

A *diffraction grating* is a large number of close parallel equidistant slits, ruled on glass or metal; it provides a very valuable way of studying spectra. If the width of a slit or clear space is a and the thickness of a ruled opaque line is b, the spacing d of the slits is $(a + b)$. Thus with a grating of 600 lines per millimetre, the spacing $d = 1/600$ millimetre $= 17 \times 10^{-7}$ m, or a few wavelengths of visible light.

The angular positions of the principal maxima produced by a diffraction grating can easily be found. Suppose X, Y are corresponding points in consecutive slits, where $XY = d$, and the grating is illuminated normally by monochromatic light of wavelength λ, Figure 20.13. In a direction θ, the diffracted rays XL, YM have a path difference XA of $d \sin \theta$. The diffracted rays from all other corresponding points in the two slits have a path difference of $d \sin \theta$ in the same direction. Other pairs of slits throughout the grating can be

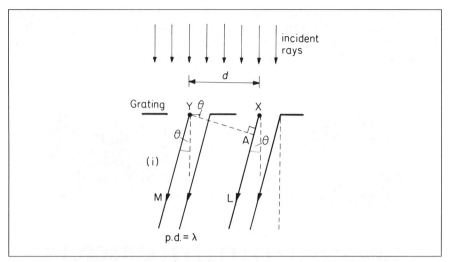

Figure 20.13   *Diffraction grating images*

treated in the same way. Hence bright or principal maxima are obtained when

$$d \sin \theta = m\lambda \qquad . \qquad . \qquad . \qquad . \qquad . \qquad \text{(i)}$$

where m is an integer, if all the diffracted parallel rays are collected by a telescope focused at infinity. The images corresponding to $m = 0, 1, 2,\ldots$ are said to be respectively of the zero, first, second... orders respectively. The zero order image is the image where the path difference of diffracted rays is zero, and corresponds to that seen directly opposite the incident beam on the grating. It should again be noted that all points in the slits are secondary centres on the same wavefront and therefore coherent sources.

Unlike the images obtained with glass prisms, the grating gives *two* sets of diffraction images on opposite sides of the normal where $d \sin \theta = m\lambda$. The angle between the two images of the same order is 2θ.

Figure 20.14 shows how the grating produces diffracted waves which interfere

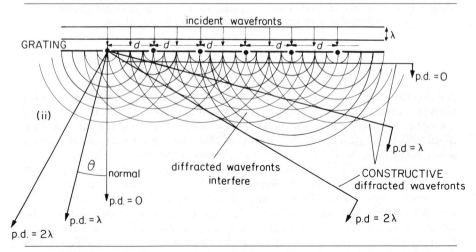

Figure 20.14   *Grating action—diffraction and interference of wavefronts*

constructively in directions given by $d \sin \theta = m\lambda$. Only six slits are shown and only six corresponding secondary sources in the slits, separated by a distance d. The diffracted wavefronts correspond to a path difference of λ and 2λ respectively. They travel at an angle θ to the normal given by $d \sin \theta = \lambda$ and 2λ respectively. In practice the build-up of the diffracted wavefronts is due to the effect of many slits, for example, 300 or more per millimetre, and to the numerous secondary sources within each slit.

Diffraction Images

The *first order* diffraction image is obtained when $m = 1$. So

$$d \sin \theta = \lambda$$

or
$$\sin \theta = \frac{\lambda}{d}$$

If the grating has 600 lines per millimetre ($600 \, \text{mm}^{-1}$), the spacing of the slits, d, is $1/600 \, \text{mm} = 1/(600 \times 10^3) \, \text{m}$. Suppose yellow light, of wavelength $\lambda = 5\cdot89 \times 10^{-7} \, \text{m}$, is used to illuminate the grating. Then

$$\sin \theta = \frac{\lambda}{d} = 5\cdot89 \times 10^{-7} \times 600 \times 10^3 = 0\cdot3534$$

$$\therefore \theta = 20\cdot7°$$

The *second order* diffraction image is obtained when $m = 2$. In this case $d \sin \theta = 2\lambda$.

$$\sin \theta = \frac{2\lambda}{d} = 2 \times 5\cdot89 \times 10^{-7} \times 600 \times 10^3 = 0\cdot7068$$

$$\therefore \theta = 45\cdot0°$$

If $m = 3$, $\sin \theta = 3\lambda/d = 1\cdot060$. Since the sine of an angle cannot be greater than 1, it is impossible to obtain a third order image with the diffraction grating for this wavelength.

With a grating of 1200 lines per mm the diffraction images of sodium light would be given by $\sin \theta = m\lambda/d = m \times 5\cdot89 \times 10^{-7} \times 12 \times 10^5 = 0\cdot7068 \, m$. Thus only $m = 1$ is possible here. As all the diffracted light is now concentrated in one image, instead of being distributed over several images, the first order image is very bright, which is an advantage.

Missing Orders in Principal Maxima

In a diffraction grating, the order m of the principal maxima obtained is given by $\sin \theta = m\lambda/d$. As we saw previously, however, the intensity of the principal maxima will be zero if the intensity due to a *single slit* of the grating is zero, as this modulates the over-all intensity variation (see Figure 20.11). Now the intensity of the single-slit pattern is zero where $\sin \theta = \lambda/a$, if a is the width of the slit. So the *missing order* m in the principal maxima obtained by a diffraction grating is given by

$$\sin \theta = \frac{m\lambda}{d} = \frac{\lambda}{a}$$

Thus
$$m = \frac{d}{a}$$

If the width a of the slit is one-quarter of the separation d of the grating lines, then $m = 4$ from above. The 4th order will be missing from the principal maxima of the grating. With a particular grating, missing orders occur only if d/a is a whole number.

Measurement of Wavelength by Diffraction Grating

The wavelength of monochromatic light can be accurately measured by a diffraction grating with a spectrometer.

The collimator C and telescope T of the instrument are first adjusted for parallel light, and the grating P is then placed on the table so that its plane is

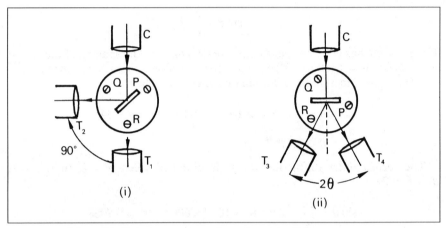

Figure 20.15 *Measurement of wavelength by diffraction grating*

perpendicular to two screws, Q, R, Figure 20.15 (i). To level the table so that the plane of P is parallel to the axis of rotation of the telescope, the telescope is first placed in the position T_1 directly opposite the illuminated slit of the collimator, and then rotated exactly through 90° to a position T_2. The table is now turned until the slit is seen in T_2 by *reflection* at P, and one of the screws Q, R turned until the slit image is in the middle of the field of view. The plane of P is now parallel to the axis of rotation of the telescope. The table is then turned through 45° so that the plane of the grating is exactly perpendicular to the light from C, and the telescope is turned to a position T_3 to receive the first diffraction image, Figure 20.15 (ii). If the lines of the grating are not parallel to the axis of rotation of the telescope, the image will not be in the middle of the field of view. The third screw is then adjusted until the image is central.

The readings of the first diffraction image are observed on both sides of the normal. The angular difference is 2θ, and the wavelength is calculated from $\lambda = d \sin \theta$, where d is the spacing of the slits, obtained from the number of lines per centimetre of the grating. If a second order image is obtained for a diffraction angle θ_1, then $\lambda = \frac{1}{2} d \sin \theta_1$.

Diffraction gratings with suitable values of d have been used to measure a wide range of wavelengths. For very short wavelengths such as X-rays, a grating of similar linear dimensions is provided by rows of atomic planes inside a crystal (see p. 869).

Resolving Power of Grating

The resolving power of a grating is a measure of how effectively it can separate or *resolve* two wavelengths in a given order of their spectrum. Using the Rayleigh criterion for resolving power (p. 544), it can be shown that the resolving power depends on the *total lines or width Nd* of the grating, where N is the number of lines and d is their spacing.

So a grating 5 cm wide with 3000 lines per cm has twice the resolving power of the same grating if it was masked to be only 2·5 cm wide. In both gratings the diffraction images of two close wavelengths would be formed at the same angles for a given order, since $d \sin \theta = m\lambda$ and d is the same. But the images with the 2·5 cm grating are not as sharp as with the 5 cm grating and the two close wavelengths may not be seen separated or resolved.

Note that $Nd = Nm\lambda/\sin \theta$, so the resolving power increases with the order m of the image.

Example on Gratings

A parallel beam of sodium light is incident normally on a diffraction grating. The angle between *the two first order spectra on either side* of the normal is 27° 42′. Assuming that the wavelength of the light is $5\cdot893 \times 10^{-7}$ m, find
(a) the number of rulings per mm on the grating, and
(b) the greatest number of bright images obtained.

(a) The first order spectrum occurs at an angle $\theta = \frac{1}{2} \times 27° \, 42′ = 13° \, 51′$

$$\therefore d = \frac{\lambda}{\sin \theta} = \frac{5\cdot893 \times 10^{-7}}{\sin 13° \, 51′} \, \text{m}$$

$$\therefore \text{number of rulings per metre} = \frac{1}{d} = \frac{\sin 13° \, 51′}{5\cdot893 \times 10^{-7}} = 406\,000$$

$$= 406 \text{ per millimetre}$$

(b) From $d \sin \theta = m\lambda$, the greatest number of images is obtained when $\theta = 90°$ and so $\sin \theta = 1$. In this case,

$$d \sin \theta = d = m\lambda$$

So

$$m = \frac{d}{\lambda} = \frac{(1/4\cdot06) \times 10^{-5}}{5\cdot893 \times 10^{-7}}$$

$$= \frac{100}{4\cdot06 \times 5\cdot893} = 4\cdot18$$

Since m is a whole number, the number of images $= 4$

Spectra with Grating

If white light is incident normally on a diffraction grating several coloured spectra are observed on either side of the normal, Figure 20.16 (i). The first order diffraction images are given by $d \sin \theta = \lambda$, and as violet has a shorter wavelength than red, θ is less for violet than for red. Consequently the spectrum colours on either side of the incident white light are violet to red. In the case of a spectrum produced by dispersion in a glass prism, the colours range from red, the least deviated, to violet, Figure 20.16 (ii). Second and higher order spectra are obtained with a diffraction grating on opposite sides of the normal, whereas only one spectrum is obtained with a glass prism. The angular spacing of the colours is also different in the grating and the prism.

If $d \sin \theta = m_1 \lambda_1 = m_2 \lambda_2$, where m_1, m_2 are integers, then a wavelength λ_1 in the m_1 order spectrum overlaps the wavelength λ_2 in the m_2 order. The extreme

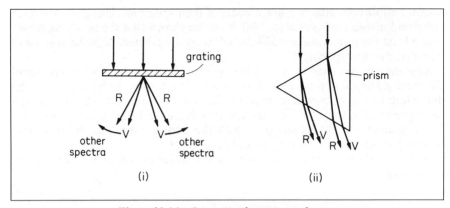

Figure 20.16 *Spectra with grating and prism*

violet in the visible spectrum has a wavelength about 3.8×10^{-7} m. The violet direction in the second order spectrum would thus correspond to $d \sin \theta = 2\lambda = 7.6 \times 10^{-7}$ m, and this would not overlap the extreme colour, red, in the first order spectrum, which has a wavelength about 7.0×10^{-7} m. In the second order spectrum, a wavelength λ_2 would be overlapped by a wavelength λ_3 in the third order if $2\lambda_2 = 3\lambda_3$. If $\lambda_2 = 6.9 \times 10^{-7}$ m (red), then $\lambda_3 = 2\lambda_2/3 = 4.6 \times 10^{-7}$ m (blue). Thus overlapping of colours occurs in spectra of higher orders than the first.

Holography

When an object is photographed by a camera, only the *intensity* of the light from its different points is recorded on the photographic film to form the image. Now the intensity is a measure of the mean square value of the amplitude of the original light wave from the object. Consequently the *phase* of the wave arriving at the film from the different points on the object is lost.

In *holography*, however, both the phase and the amplitude of the light waves are recorded on the film. The resulting photograph is called a *hologram*. As we see later, it is a speckled pattern of fine dots. Dr. D. Gabor, who laid the foundations of holography in 1948, gave this name from the Greek 'holos' meaning 'the whole', because it contains the whole information about the light wave, that is, its phase as well as its amplitude.

Gabor's method of producing a hologram and of reconstructing the original wave were difficult to put in practice in 1948 because sources which remain coherent only over very short path differences were then available. In 1962, however, the laser was invented. This gave a powerful source of light which remained coherent over long paths (see Figure 20.17) and the subject of holography then developed rapidly.

Making a Hologram

Figure 20.18 shows the basic principle of making a hologram. By means of a half-silvered mirror M, part of the coherent light from a laser is reflected towards the object O and the remainder passes straight through as shown. The photographic plate or film P is thus illuminated by
(a) light waves scattered or diffracted from O, called the 'object beam', and
(b) by a direct beam called the 'reference beam'.

The numerous points which make up the image on P are then formed by interference between the overlapping coherent waves of the object beam and the

Figure 20.17 *Newton's rings. Obtained simply by reflecting a helium–neon laser beam from the front and back surfaces of a low power lens and so demonstrating the coherency of a laser beam over long paths*

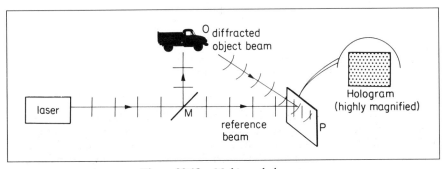

Figure 20.18 *Making a hologram*

reference beam. As shown inset diagrammatically, the hologram consists of a very large number of closely-spaced points, invisible to the naked eye but seen under a powerful microscope. The 'structure' of the hologram is like a diffraction grating; it has opaque and transparent regions very closely spaced.

In commercial broadcasting, audio-frequency currents are arranged to modulate the carrier radio wave from the transmitter. By analogy, in making a hologram the reference beam may be considered analogous to a 'carrier wave' which is 'modulated' at the photographic film by interference with the object beam. Thus, to simplify matters, suppose the object and reference beams arriving at the plate are plane waves represented by $y = O \sin(\omega t + \Delta)$ and $y = R \sin \omega t$ respectively, where Δ is their phase difference due to the path difference. The resultant amplitude at the plate is given, from vector addition, by $(O^2 + R^2 + 2OR \cos \Delta)^{1/2}$. The third term $2OR \cos \Delta$ contains (i) the phase angle Δ and (ii) the amplitude O of the object wave modulated by the amplitude R of the reference beam. After the plate is exposed and developed, the object wave, accompanied by the others due to interference, is 'recorded' on the emulsion. It may be noted that if the object moves very slightly (of the order of a fraction of a wavelength) while the photograph is taken, the diffracted waves from the object no longer produces a hologram. Rigid mounting of the object is therefore essential.

Reconstructing the Hologram Image

Figure 20.19 shows how the object wavefront or image is 'reconstructed'. The optical arrangement is simply reversed and the hologram H is illuminated by coherent light from the laser, which was the original reference beam. Now as a general principle, if an image produced by a grating is recorded on photographic film, then the image formed by illuminating the resulting grating on the negative will be identical to the original grating. In a similar way, we may consider the object as a '3-D grating' and its 3-D image is formed when the hologram is illuminated as in Figure 20.19.

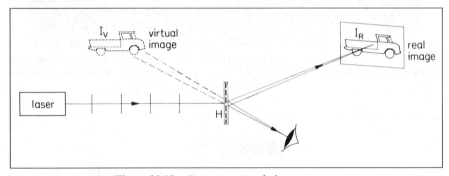

Figure 20.19 *Reconstructing hologram image*

The points on the hologram H thus act as a diffraction grating. The waves diffracted through H carry the phase and amplitude of the waves originally diffracted from the object O when the hologram was made. The object wavefronts have thus been 'reconstructed'. One of the diffracted beams forms a real image I_R, as shown. Another diffracted beam forms a virtual image I_V. This image can be seen on looking through the hologram H. The hologram thus acts like a 'window' through which the image can be seen.

One of the most remarkable features of the hologram is that by moving the head while looking through it, one can see more of the object originally hidden from view. Thus a three-dimensional (3D) view is recorded on a two-dimensional photographic film, see Figure 20.20. This is due to the fact that all parts of the object originally photographed have sent diffracted waves to the photographic film. Further, the 'grain' of a particular hologram is repeated regularly throughout the whole of the film. Thus if the hologram is cut into small pieces, although the quality is poor the whole of the image can be seen through one piece.

Figure 20.20 *Hologram of chess pieces* (*i*) *Focused on the castle* (*ii*) *Focused on the knight and taken from a different angle and showing the 3D nature of the image*

_____ Exercises 20 _____

1 A diffraction grating has 400 lines per mm and is illuminated normally by monochromatic light of wavelength 600 nm (6×10^{-7} m).
   Calculate
   (a) the grating spacing,
   (b) the angle to the normal at which the first order maximum is seen,
   (c) the number of diffraction maxima obtained.

2 A diffraction grating is illuminated normally by monochromatic light of wavelength λ. Show in a diagram using waves (i) where diffraction occurs, (ii) where interference occurs, (iii) the directions in which the first and second order diffraction maxima are seen, together with the corresponding path difference.

3 A plane diffraction grating is illuminated by a source which emits two spectral lines of wavelengths 420 nm (420×10^{-9} m) and 600 nm (600×10^{-9} m). Show that the 3rd order line of one of these wavelengths is diffracted through a greater angle than the 4th order of the other wavelength. (*L*.)

4 A spectral line of known wavelength 5.792×10^{-7} m emitted from a mercury vapour lamp is used to determine the spacing between the lines ruled on a plane diffraction grating. When the light is incident normally on the grating the third order spectrum, measured using a spectrometer, occurs at an angle of 60° 19′ to the normal. Calculate the grating spacing.
   Why is the value obtained using the third order spectrum likely to be more accurate than if the first order were used? (*L*.)

5 When a plane wavefront of monochromatic light meets a diffraction grating, the diffraction pattern produced is caused by the interference of beams diffracted through the various grating slits. Explain, with the aid of a diagram, exactly where and under what conditions diffraction occurs.

Explain why only beams diffracted in certain directions interfere constructively. (*L.*)

6 Light of wavelength λ falls normally on a transmission diffraction grating of spacing d. Show, with the aid of a suitable diagram, how light from the various slits reinforces at certain values of the diffraction angle, θ, and that then $n\lambda = d \sin \theta$, where n is an integer.

A spectrometer is used with the grating to determine the wavelength of monochromatic light. You may assume that all the initial adjustments for the spectrometer and for positioning the grating have been made, and that the grating spacing, d, is known. List the readings you would take in order to use the apparatus to find the best value of the wavelength.

A stationary ultrasonic (high frequency sound) plane wave of frequency 5.0×10^6 Hz is set up in a transparent tank of liquid. This produces a stationary pattern of density variation. Monochromatic light of wavelength 390 nm (in the liquid) passes through the tank at right angles to the ultrasonic wave. A diffraction pattern similar to that of a grating is produced in the light, the deviation of the sixth order ray being $1.0°$ (0.017 rad). Explain why a diffraction pattern is produced and calculate the speed of ultrasound in the liquid. (*L.*)

7 In the Young's double slit experiment *interference* occurs when *diffraction* of light takes place at each of the two slits.
   (a) Explain with the aid of a diagram the meaning of *interference* and *diffraction* in this case.
   (b) Explain how and why *sharper* and *more widely* spaced interference fringes may be obtained using a diffraction grating. (*AEB*, 1985.)

8 (a) What conditions are necessary in order that interference patterns between light from two sources may be observed?
   (b) Draw a labelled diagram showing the apparatus required to determine the wavelength of red light using a pair of slits. Indicate approximate values of the dimensions of the apparatus and state a measuring instrument suitable for each measurement required.

   How would you use the measured values of the dimensions of the apparatus to estimate the separation of the fringes produced by light of wavelength 500 nm?

   What part is played by diffraction in this experiment?

   How are the fringes produced in this experiment very different from those produced using a diffraction grating and the same source of light (i) when the grating spacing is the same as the slit separation, and (ii) when the grating spacing is much smaller than the slit separation?

   If the red light is replaced by blue light, how do the fringes produced in the experiment differ markedly from those produced using red light? (*L.*)

9 The wavelengths of the two yellow sodium lines are 589·0 nm and 589·6 nm. The eye, with the aid of the spectrometer telescope, can resolve an angle of 0·2 minutes. How many lines per metre must there be in a diffraction grating for the two sodium lines to be seen as just separate in the first-order in the spectrometer? Assume the diffraction angle is small so that $\sin \theta \simeq \theta$ in radians. (*W.*)

10 (a) A soap film is formed on a vertical wire frame and viewed in reflected white light. Initially a number of horizontal coloured fringes are seen. As the soap film

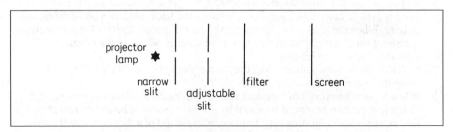

Figure 20A

gradually drains the fringe pattern changes until eventually, just before the film breaks, the top part appears black. (i) Explain how the coloured fringes are formed. (ii) Describe how the appearance of the fringes changes as the film drains. (iii) Explain why part of the film appears black just before it breaks.

(b) Figure 20A shows an experimental arrangement for demonstrating the diffraction of light through a narrow horizontal slit. The apparatus is used in a darkened room. (i) Sketch a graph to show the intensity pattern you would see on the screen when the apparatus is correctly adjusted. (ii) The width of the adjustable slit is 0·50 mm, the filter transmits yellow light of wavelength $6·0 \times 10^{-7}$ m and the screen is 1·0 m away from the adjustable slit. Calculate the positions of the first and second minima in the diffraction pattern. (iii) What would be the effect on the pattern of removing the filter? (*JMB.*)

11 Describe and give the theory of an experiment to compare the wavelengths of yellow light from a sodium and red light from a cadmium discharge lamp, using a diffraction grating. Derive the required formula from first principles.

White light is reflected normally from a soap film of refractive index 1·33 and then directed upon the slit of a spectrometer employing a diffraction grating at normal incidence. In the first-order spectrum a dark band is observed with minimum intensity at an angle of 18° 0' to the normal. If the grating has 500 lines per mm, determine the thickness of the soap film assuming this to be the minimum value consistent with the observations. (*L.*)

12 Describe how you would determine the wavelength of monochromatic light using a diffraction grating and a spectrometer. Give the theory of the method.

A filter which transmits only light between $6·3 \times 10^{-7}$ and $6·0 \times 10^{-7}$ m is placed between a source of white light and the slit of a spectrometer; the grating has 5000 lines to the centimetre; and the telescope has an objective of focal length 15 cm with an eyepiece of focal length 3 cm. Find the width in millimetres of the first-order spectrum formed in the focal plane of the objective. Find also the angular width of this spectrum seen through the eyepiece. (*O.*)

13 (a) Parallel monochromatic light is incident normally on a thin slit of width d and focused on to a screen. Derive the relationship between the wavelength λ and the angle of diffraction θ for the first minimum of intensity on the screen.

(b) If the light has a wavelength of 540 nm and is focused by a converging lens of focal length 0·50 m placed immediately in front of the slit which has a width of 0·10 mm, calculate the distance from the centre of the intensity distribution to the first minimum.

(c) Explain how diffraction effects, similar to those referred to above, limit the sharpness of the image produced by a telescope. (*JMB.*)

14 Give a labelled sketch and a brief description of the essential features of a spectrometer incorporating a plane diffraction grating. What part is played by (a) diffraction, and (b) interference, in the operation of the diffraction grating?

A source emitting light of two wavelengths is viewed through a grating spectrometer set at normal incidence. When the telescope is set at an angle of 20° to the incident direction, the second order maximum for one wavelength is seen superposed on the third order maximum for the other wavelength. The shorter wavelength is 400 nm. Calculate the longer wavelength and the number of lines per metre in the grating. At what other angles, if any, can superposition of two orders be seen using this source? (*O. & C.*)

15 A pure spectrum is one in which there is no overlapping of light of different wavelengths. Describe how you would set up a diffraction grating to display on a screen as close an approximation as possible to a pure spectrum. Explain the purpose of each optical component which you would use.

A grating spectrometer is used at normal incidence to observe the light from a sodium flame. A strong yellow line is seen in the first order when the telescope axis is at an angle of 16° 26' to the normal to the grating. What is the highest order in which the line can be seen?

The grating has 4800 lines per cm; calculate the wavelength of the yellow radiation.

What would you expect to observe in the spectrometer set to observe the first-order spectrum if a small but very bright source of white light is placed close to the sodium flame so that the flame is between it and the spectrometer? (*O. & C.*)

16 Explain what is meant by diffraction, and discuss the parts played by diffraction and by interference in the action of a diffraction grating. Describe briefly the diffraction patterns observed when a parallel beam of monochromatic light illuminates
(a) a fairly narrow slit,
(b) a very narrow slit,
(c) a straight edge.

A diffraction grating is set up on a spectrometer table so that parallel light is incident on it normally. For a light of wavelength $5 \cdot 5 \times 10^{-7}$ m, first order reinforcement is observed in directions making angles θ of 18° with the normal on each side of the normal.   (i) Find the value of the grating spacing d.   (ii) Calculate the wavelength of monochromatic light which would give a first order spectral line at $\theta = 25°$.   (iii) What would be the values of θ for second order spectral lines for each of these two wavelengths?   (iv) Is a third order spectrum possible for each? Explain. (*O.*)

17 What is *Huygens' theory*? Indicate with the aid of diagrams how the theory accounts for
(a) the reflection of waves from a plane surface, and
(b) the diffraction of waves by a grating.

Describe how you would demonstrate the diffraction of waves at a single slit. How does the width of the slit affect the results?

A grating has 500 lines per mm and is illuminated normally with monochromatic light of wavelength 589 nm. How many diffraction maxima may be observed? Calculate their angular positions. (*L.*)

21

Polarisation of Light Waves

In this chapter we show that light waves are transverse waves because they can be polarised. We first deal with polarisation by Polaroid, then with polarisation by reflection and by double refraction. We conclude with a brief account of applications of polarisation such as photoelasticity and saccharimetry.

Polarised Light

Polaroid is an artificial crystalline material which can be made in thin sheets. As we shall see shortly, it has the property of allowing only light waves due to vibrations in a particular plane to pass through.

Suppose two Polaroids, P and Q, are placed one behind the other in front of a window, and the light passing through P and Q is observed, Figure 21.1. When

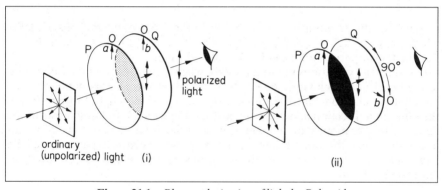

Figure 21.1   *Plane-polarisation of light by Polaroid*

the Polaroids have their axes *a* and *b* parallel, the light passing through Q appears slightly darker, Figure 21.1 (i). If Q is now rotated slowly about the line of vision with its plane parallel to P, the light passing through Q becomes darker and darker and disappears at one stage. In this case the axes *a* and *b* are perpendicular, Figure 21.1 (ii). When Q is rotated further the light reappears, and becomes brightest when the axes *a, b* are again parallel.

This simple experiment leads to the conclusion that light waves are *transverse* waves; otherwise the light emerging from Q could never be extinguished by simply rotating the Polaroid. The experiment, in fact, is analogous to that illustrated in Figure 17.33, where transverse mechanical waves were set up along a rope and plane-polarised waves were obtained by means of a slit. Polaroid, because of its internal molecular structure, transmits only those vibrations of light in a particular plane. Consequently

(a) plane-polarised light is obtained beyond the crystal P, and

(b) no light emerges beyond Q when its axis is *perpendicular* to that of P.

Vibrations in Unpolarised and Polarised Light

Figure 21.2 (i) is an attempt to represent diagrammatically the vibrations of ordinary or unpolarised light at a point A when a ray travels in a direction AB. X is a plane perpendicular to AB, and ordinary (unpolarised) light may be

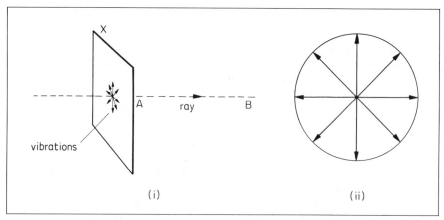

Figure 21.2 (i) *Vibrations occur in every plane perpendicular to ray* (ii) *Vibrations in ordinary light*

imagined as due to vibrations which occur in every one of the millions of planes which pass through AB and are perpendicular to X. As represented in Figure 21.2 (ii), the amplitudes of the vibrations are all equal.

Consider the vibrations in ordinary light when it is incident on the Polaroid P in Figure 21.1 (i). Each vibration can be resolved into two components, one in a direction parallel to *a*, the direction of easy transmission of light through P, and the other in a direction *m* perpendicular to *a*, Figure 21.3. Polaroid absorbs the light due to vibrations parallel to *m*, known as the *ordinary rays*, but allow light due to the other vibrations, known as the *extraordinary rays*, to pass through. So *plane-polarised light*, light due to vibrations in one plane, is produced, as illustrated in Figure 21.1.

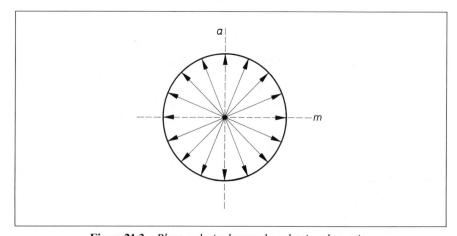

Figure 21.3 *Plane-polarised waves by selective absorption*

Polaroid thus absorbs light due to vibrations in a particular direction and allows light due to vibrations in a perpendicular direction to pass through. This 'selective absorption' is also shown by certain natural crystals such as tourmaline.

Theory and experiment show that the vibrations of light are *electromagnetic* in origin. This is discussed later at the end of the chapter.

1   Light consists of transverse waves
2   In ordinary (unpolarised) light, the vibrations are in every plane at right angles to the direction of the light
3   In plane-polarised light, the vibrations are only in one plane perpendicular to the direction of the light
4   A Polaroid crystal produces plane-polarised light in a particular allowed direction through the crystal. When another Polaroid is rotated in front of the first Polaroid, the light decreases in intensity. Darkness occurs when the two Polaroids have their allowed directions at 90° to each other.

Polarised Light by Reflection

In 1808 MALUS discovered that polarised light is obtained when ordinary light is reflected by a plane sheet of glass (p. 564). The most suitable angle of incidence is about 57°, Figure 21.4. If the reflected light is viewed through a Polaroid which is slowly rotated about the line of vision, the light is practically extinguished at one

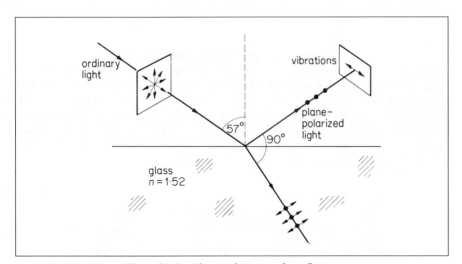

Figure 21.4   *Plane-polarisation by reflection*

position of the Polaroid. This proves that the light reflected by the glass is practically plane-polarised. Light reflected from the surface of a table becomes darker when viewed through a rotated Polaroid, showing it is partially plane-polarised.

The production of the polarised light by the glass is explained as follows. Each of the vibrations of the incident (ordinary) light can be resolved into a component parallel to the glass surface and a component perpendicular to the surface. The light due to the components parallel to the glass is largely reflected, but the remainder of the light, due mainly to the components perpendicular to the glass, is *refracted* into the glass. Thus the light reflected by the glass is partially plane-polarised.

Brewster's Law, Polarisation by Pile of Plates

The particular angle of incidence i on a transparent medium when the reflected light is almost completely plane-polarised is called the *polarising angle*. BREWSTER found that, in this case, $\tan i = n$, where n is the refractive index of the medium (*Brewster's law*). With crown glass of $n = 1·52$, $i = 57°$ (approx.) from $\tan i = n$. Since n varies with the colour of the light, white light can not be completely plane-polarised by reflection.

As $\sin i/\sin r = n$, where r is the angle of refraction, it follows from Brewster's law that $\cos i = \sin r$, or $i + r = 90°$. Thus the reflected and refracted beams are at 90° to each other at the polarising angle, as illustrated in Figure 21.4.

The refracted beam contains light mainly due to vibrations perpendicular to that reflected and is therefore partially plane-polarised. Since refraction and reflection occur at both sides of a glass plate, the transmitted beam contains a fair percentage of plane-polarised light. A *pile of plates* increases the percentage, and thus provides a simple way of producing plane-polarised light. They are mounted inclined in a tube so that the ordinary (unpolarised) light is incident at the polarising angle, and the transmitted light is then practically plane-polarised.

---

1   **Light beams reflected and refracted by glass are partially plane-polarised.**

2   **Maximum polarisation of the reflected beam occurs for an angle of incidence i given by $\tan i = n$, refractive index of glass (Brewster's law). The reflected and refracted rays are then at 90° to each other.**

Polarisation by Double Refraction

We have already considered two methods of producing polarised light. The first observation of polarised light, however, was made by BARTHOLINUS in 1669, who placed a crystal of Iceland spar on some words on a sheet of paper. To his surprise, two images were seen through the crystal. Bartholinus therefore gave the name of *double refraction* to the phenomenon, and experiments more than a century later showed that the crystal produced plane-polarised light when ordinary light was incident on it, see Figure 21.5.

Figure 21.5   *Double refraction. A ring with a spot in the centre, photographed through a crystal of Iceland spar. The two plane-polarised beams of light emerging from the crystal form two rings and two spots*

Iceland spar is a crystalline form of calcite (calcium carbonate) which cleaves in the form of a 'rhomboid' when it is lightly tapped; this is a solid whose opposite faces are parallelograms. When a beam of unpolarised light is incident on one face of the crystal, its internal molecular structure produces two beams of polarised light, E, O, whose vibrations are *perpendicular* to each other, Figure 21.6. If the incident direction AB is parallel to a plane known as the 'principal

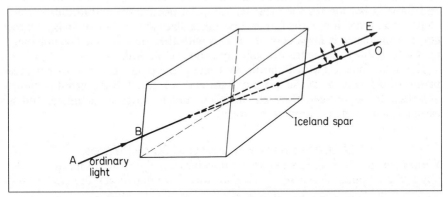

Figure 21.6 *Plane-polarised light by double refraction*

section' of the crystal, one beam O emerges parallel to AB, while the other beam E emerges in a different direction. As the crystal is rotated about the line of vision the beam E revolves round O. On account of this abnormal behaviour the rays in E are called 'extraordinary' rays; the rays in O are known as 'ordinary' rays (p. 562). Thus two images of a word on a paper, for example, are seen when an Iceland spar crystal is placed on top of it; one image is due to the ordinary rays, while the other is due to the extraordinary rays.

With the aid of an Iceland spar crystal Malus discovered the polarisation of light by reflection (p. 563). While on a visit to Paris he looked through the crystal at the light of the sun reflected from the windows of the Palace of Luxemburg, and observed that only one image was obtained for a particular position of the crystal when it was rotated slowly. The light reflected from the windows could not therefore be ordinary (unpolarised) light, and Malus found that it was plane-polarised.

Nicol Prism

We have seen that a Polaroid produces polarised light, and that the Polaroid can be used to detect light (p. 561). NICOL designed a form of Iceland spar crystal

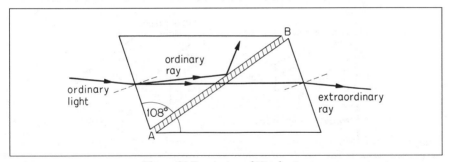

Figure 21.7 *Action of Nicol prism*

which was widely used for producing and detecting polarised light, and it is known as a *Nicol prism*. A crystal whose faces contain angles of 72° and 108° is broken into two halves along the diagonal AB, and the halves are cemented together by a layer of Canada balsam, Figure 21.7. The refractive index of the crystal for the ordinary rays is 1·66, and is 1·49 for the extraordinary rays; the refractive index of the Canada balsam is about 1·55 for both rays, since Canada balsam does not polarise light. A critical angle thus exists between the crystal and Canada balsam for the ordinary rays, but not for the extraordinary rays. Hence total reflection of the former rays takes place at the Canada balsam if the angle of incidence is large enough, as it is with the Nicol prism. The emergent light is then due to the extraordinary rays, and is polarised.

The prism is used like a Polaroid to detect plane-polarised light, namely, the prism is held in front of the beam of light and is rotated. If the beam is plane-polarised the light seen through the Nicol prism varies in intensity, and is extinguished at one position of the prism.

Differences Between Light and Sound Waves

We are now in a position to distinguish fully between light and sound waves. The physical difference, of course, is that light waves are due to varying electric and magnetic fields and is an *electromagnetic wave*, while sound waves are due to vibrating layers or particles of the medium concerned and is a *matter wave*. Light can travel through a vacuum, but sound cannot travel through a vacuum. Another very important difference is that the vibrations of the particles in sound waves are in the same direction as that along which the sound travels, whereas the vibrations in light waves are perpendicular to the direction along which the light travels. Sound waves are therefore *longitudinal* waves, whereas light waves are *transverse* waves. As we have seen, sound waves can be reflected and refracted, and can give rise to interference phenomena; but no polarisation phenomena can be obtained with sound waves since they are longitudinal waves, unlike the case of light waves.

Polarisation and Electric Vector

As we have just stated, light is a transverse electromagnetic wave and thus contains electric and magnetic fields. Experiment shows that only the electric field is concerned in the blackening of a photographic film when exposed to light. Hence we usually define *the direction of light vibrations to be that of the electric vector E*. The 'plane of polarisation' is then defined as the plane containing the light ray and *E*.

Figure 21.8 (i) shows *plane-polarised* light due to vertical vibrations or a

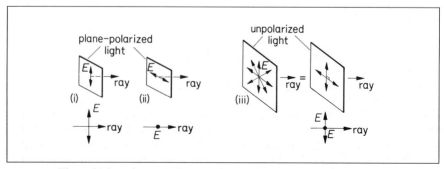

Figure 21.8 _Electric vectors in plane-polarised and unpolarised light_

vertical vector E. Figure 21.8 (ii) shows plane-polarised light due to horizontal vibrations; in this case a dot is used to indicate the vibrations or vector E perpendicular to the paper. Figure 21.8 (iii) shows *ordinary* or *unpolarised* light.

The associated field vectors E in ordinary light act in all directions in a plane perpendicular to the ray and vary in phase. Since each vector can be resolved into components in perpendicular directions, the total effect is equivalent to two perpendicular vectors equal in magnitude as explained on p. 562. So ordinary or unpolarised light can be represented by an arrow and a dot (representing an arrow going down into the page), as shown in Figure 21.8 (iii).

Polaroid Transmission and Light Intensity

Suppose a Polaroid A produces polarised light whose electric vector vibrates in a particular direction, say AY in Figure 21.9 (i). If another Polaroid B is placed in front of A so that its 'easy' direction of transmission BX is perpendicular to AY, the overlapping areas appear black because the electric vector has no component in a perpendicular direction and so no light is transmitted.

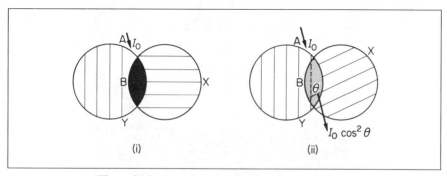

Figure 21.9   *Intensity variation by rotating Polaroid*

Suppose B is turned so that BX makes an angle θ with AY, Figure 21.9 (ii). If E_0 is the initial amplitude of the electric vector in A, the amplitude of the vector transmitted by B is the component E given by $E = E_0 \cos \theta$. The intensity is proportional to the square of the amplitude (p. 584). Since $E^2 = E_0^2 \cos^2 \theta$, the intensity I of the light transmitted by B is given by $I = I_0 \cos^2 \theta$, where I_0 is the light intensity incident from A on B, Figure 21.9 (ii). So if B is rotated so that $\theta = 60°$, then $I = I_0 \cos^2 60° = I_0/4$, since $\cos 60° = 1/2$. So the light intensity is reduced to one-quarter of that incident from A.

It may be noted that the intensity of the unpolarised light *incident on A is $2I_0$*, since, from Figure 21.9, one of the perpendicular vectors of the unpolarised light is absorbed by A.

Applications of Polarised Light

Polaroids are used in many practical applications of polarised light. For example, they are used in sunglasses to reduce the intensity of incident sunlight and to eliminate reflected light or glare.

Photoelasticity, or photoelastic stress analysis, utilises polarised light. Under mechanical stress, certain isotropic substances such as glass and celluloid become doubly refracting. Ordinary white light does not pass through crossed Polaroids (p. 561). However, if a celluloid model with a cut-out design is placed between the crossed Polaroids and then subjected to compression by a vice,

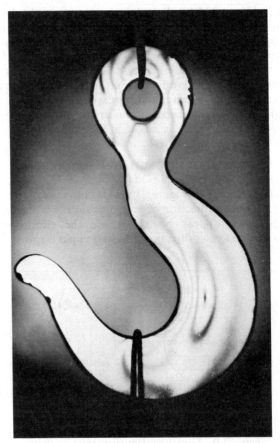

Figure 21A *Photoelasticity. The pattern in a perspex hook between polaroids when the hook is under stress. Similar patterns in transparent models help the engineer to investigate the stresses in mechanical structures subjected to loads or forces.* (Courtesy of Kodak Limited)

coloured bright and dark fringes can now be seen or projected on a screen. The fringes spread from places where the stress is most concentrated. Now the pattern of the fringe varies with the stress. So using a model, a study of the fringe pattern provides the engineer with valuable information on design. This is particularly useful with complicated shapes, where mathematical computation is difficult.

In *films*, it is possible to give the illusion of three-dimensions or 3-D by projecting two overlapping pictures, with slightly different views, on to the same screen. Each picture has been taken by light polarised respectively in perpendicular directions. The viewer is provided with special spectacles, with perpendicular Polaroids in the respective frames. The two pictures received simultaneously by the two eyes provide a 3-D view of the scene.

Saccharimetry is the measurement of the concentration of sugars such as cane sugar in solution. Due to the molecular structure of the sugar, these solutions rotate the plane of polarisation of plane-polarised light as the light passes through. Solids such as quartz produce the same effect, which is called *optical activity*. The rotation of the plane of polarisation when the incident light is viewed may be right-handed (clockwise) or left-handed (anticlockwise).

There are various types of saccharimeter. Here we are only concerned with the basic principle of their action. A saccharimeter usually consists of a nicol prism P, which produces plane-polarised light from an incident source S of

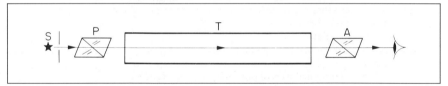

Figure 21.10 *Saccharimeter principle*

monochromatic light and is called the *polariser;* a tube T containing the solution; and a nicol A through which the emerging light is observed, called the *analyser,* Figure 21.10. Before the solution is poured into T, the analyser A is rotated until the plane-polarised light emerging from P is completely extinguished. T is then filled with the sugar solution. On looking through A, light can now be seen. A is then rotated until the light is again just extinguished and the angle of rotation θ is measured.

For a solution of a given substance in a given solvent, the amount of rotation θ depends on the length of light path travelled through the liquid, the temperature of the solution and the wavelength of the light. Tables provide the rotation in degrees when the sodium D-line is used, the temperature is 20°C, and the light travels a column of length 10 cm of the liquid which contains 1 gram of the active substance per millilitre of solution. This is called the *specific rotation* or *rotary power.* Knowing its value, the concentration of a sugar solution can be found by a polarimeter experiment and using the Tables.

_____ **Exercises 21** _____

1 With the aid of suitable diagrams, explain the difference between transverse waves which are polarised and those which are unpolarised. State one useful application of plane polarised waves, specifying the type of wave involved (e.g. electromagnetic). (*L.*)

2 What is meant by *plane of polarisation*? Explain why the phenomenon of polarisation is met with in dealing with light waves, but not with sound waves.
    Describe and explain the action of
    (a) a Nicol prism,
    (b) a sheet of Polaroid.
    How can a pair of Polaroid sheets and a source of natural light be used to produce a beam of light the intensity of which may be varied in a calculable manner? (*L.*)

3 Explain what is meant by the statement that a beam of light is *plane-polarised.* Describe *one* experiment in each instance to demonstrate
    (a) polarisation by reflection,
    (b) polarisation by double refraction,
    (c) polarisation by scattering.
    The refractive index of diamond for sodium light is 2·417. Find the angle of incidence for which the light reflected from diamond is completely plane polarised. (*L.*)

4 Explain the terms *linearly polarised light* and *unpolarised light.*
    A parallel beam of light is incident on the surface of a transparent medium of refractive index n at an angle of incidence i such that the reflected and refracted

beams are at right angles to each other. Assuming the laws of reflection and refraction, find the relation between n and i.

At this value of i, the Brewster angle, the reflected beam is found to be linearly polarised. Describe the apparatus and procedure you would use to find the refractive index of a material, using the Brewster angle. Has the method any advantage over other methods for measurement of n? (*O. & C.*)

5 Give an account of the action of
(a) a single glass plate.
(b) a Nicol prism, in producing plane-polarised light. State *one* disadvantage of *each* method.
Mention *two* practical uses of polarising devices. (*JMB.*)

6 Explain what is meant by *polarisation* and *interference*.
Describe
(a) an experiment to demonstrate polarisation and
(b) an application in which polarised light is involved. Explain why electromagnetic waves can be polarised but sound waves cannot.
A series of interference fringes is formed on a screen placed 1·6 m away from a double slit illuminated by a narrow line source. The slits are 1·0 mm apart and are illuminated by light of wavelength 640 nm (640×10^{-9} m). Calculate the spacing of the fringes. Explain why the fringes disappear if the line source is made too wide. (*L.*)

7 Give one piece of experimental evidence in each case, with a brief account of how that evidence is obtained, that light behaves
(a) as a wave,
(b) as an electromagnetic wave,
(c) as a stream of particles.
When an unpolarised beam of light travelling in air is incident on a glass plate at an angle of incidence 30° the reflected light is found to be *partially linearly polarised*. Explain the meaning of the term in italics and describe how you would verify this statement experimentally. (*O. & C.*)

8 A beam of plane-polarised light falls normally on a sheet of Polaroid, which is at first set so that the intensity of the transmitted light, as estimated by a photographer's light-meter, is a maximum. (The meter is suitably shielded from all other illumination.) Describe and explain the way in which you would expect he light-meter readings to vary as the Polaroid is rotated in stages through 180° about an axis at right angles to its plane.
How would you show experimentally
(a) that calcite is doubly refracting,
(b) that the two refracted beams are plane-polarised, in planes at right angles to one another, and
(c) that in general the two beams travel through the crystal with different velocities? (*O.*)

9 Explain what is meant by *plane-polarised* electromagnetic radiation. How may plane-polarised
(a) light and
(b) radio waves be produced and detected?
Two polarising sheets initially have their polarisation directions parallel. Through what angle must one sheet be turned so that the intensity of transmitted light is reduced to a third of the original transmitted intensity?
A vertical dipole is connected to a radio frequency generator. An identical dipole, similarly connected to the same generator, is arranged horizontally a quarter of a wavelength in front of the first so that both dipoles are normal to the horizontal line joining their midpoints. Describe as fully as you can the nature of the radiation at a distant point on this line. (*C.*)

10 Explain as fully as you can the nature of linearly polarised light.
When unpolarised light falls on the surface of a block of glass the reflected light is partially polarised. If the angle of incidence is $\tan^{-1} n$, where n is the refractive index of the glass, the reflected light is completely linearly polarised. Describe the apparatus you would use and the experiments you would perform in order to verify these statements for a sample of glass of known refractive index.

Why would it be necessary, if very accurate results were required, to use monochromatic light to verify the second statement?

Show that, when the condition for completely polarised light is satisfied, the reflected and refracted beams are at right angles to one another. (*O. & C.*)

11 A beam of electromagnetic waves of wavelength 3·0 cm is directed normally at a grid of metal rods, parallel to each other and arranged vertically about 2·0 cm apart. Behind the grid is a receiver to detect the waves. It is found that when the grid is in this position, the receiver detects a strong signal but that when the grid is rotated in a vertical plane through 90°, the detected signal strength falls to zero. What property of the wave gives rise to this effect? Account briefly in general terms for the effect described above. (*L.*)

12 Draw a labelled diagram of an optical arrangement that could be used to detect the presence of strain in a transparent material.

Describe briefly how regions of high strain in the material under test in this way are identified. (*L.*)

13 Draw a labelled sketch of a simple practical arrangement for demonstrating interference by Young's slits in the laboratory. How would you ensure clearly visible fringes?

Give a careful account of a simple theory which explains the formation of the fringes, and deduce an expression for the distance x between adjacent bright fringes in terms of the distance d between slits, the distance D from the slits to where the fringes are formed, and the wavelength λ of the light.

Describe and account for the effect on the fringes of

(a) increasing the widths of the individual slits, and

(b) using a white light source.

Pieces of Polaroid are arranged over the slits (using monochromatic light) and the planes of polarisation are originally parallel. Describe and account qualitatively for the effect on the fringes of slowly rotating one of the Polaroids through 180° in its plane while keeping the other fixed. (*W.*)

In the Waves chapter, we discussed some properties of sound waves. Here we deal with the factors which influence the pitch, loudness and quality (timbre) of sound waves, the phenomenon of beats and its uses, and the Doppler effect and its application in sound and light.

Characteristics of Notes
Notes may be similar to or different from each other in three respects: (i) *pitch*, (ii) *loudness*, (iii) *quality*. These three quantities define or 'characterise' a note.

Pitch and Frequency
Pitch is analogous to colour in light, which depends only on the wavelength or frequency of the light wave (p. 521). Similarly, the pitch of a note depends only on the frequency of the sound vibrations. A high frequency gives rise to a high-pitched note; a low frequency produces a low-pitched note. Thus the high-pitched whistle of a boy may have a frequency of several thousand Hz, whereas a low-pitched hum due to a.c. mains frequency when first switched on in a television receiver may be 100 Hz. The range of sound frequencies is about 15 to 20 000 Hz depending on the observer.

Musical Intervals
If a note of frequency 300 Hz, and then a note of 600 Hz, are sounded by a siren, the pitch of the higher note is recognised to be an upper octave of the lower note. A note of frequency 1000 Hz is recognised to be an upper octave of a note of frequency 500 Hz. So the *musical interval* between two notes is an upper octave if the ratio of their frequencies is 2:1. It can be shown that the musical interval between two notes depends on the *ratio* of their frequencies, and not on the actual frequencies.

Ultrasonics, Production and Use
There are sound waves of higher frequency than 20 000 Hz, which are inaudible to a human being. These are known as *ultrasonics*; and since velocity = wavelength × frequency, ultrasonics have short wavelengths compared with sound waves in the audio-frequency range.

In 1881 CURIE discovered that a thin plate of quartz increased or decreased in length if an electrical battery was connected to its opposite faces. By correctly cutting the plate, the expansion or contraction could be made to occur along the axis of the faces to which the battery was applied. When an alternating voltage of ultrasonic frequency was connected to the faces of such a quartz crystal the faces vibrated at the same frequency, and ultrasonic sound waves were produced.

Another method of producing ultrasonics is to place an iron or nickel rod inside a solenoid carrying an alternating current of ultrasonic frequency. Since the length of a magnetic specimen increases slightly when it is magnetised, ultrasonic sound waves are produced by the vibrations of the rod.

In recent years ultrasonics have been utilised for a variety of industrial purposes. They are used on board coasting vessels for depth sounding, the time taken by the wave to reach the bottom of the sea from the surface and back being

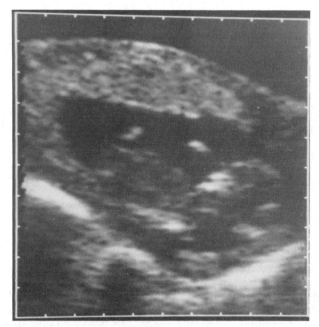

Plate 22A *Magnified photograph of 12 week foetus during pregnancy, taken by ultrasonics. Head and limbs (top) are easily identified.* Professor R. W. Shaw, Academic Dept, Obstetrics & Gynaecology and the Ultrasound Unit of the Royal Free Hospital, London

determined. Ultrasonics are also used to kill bacteria in liquids, and they are used extensively to locate faults and cracks in metal castings following a method similar to that of radar. Ultrasonic waves are sent into the metal under investigation, and the beam reflected from the fault is picked up on a cathode-ray tube screen together with the reflection from the other end of the metal. The position of the fault can then easily be located.

Ultrasonic Medical Use

Ultrasonic techniques are used in clinical practice in hospitals. In diagnostic ultrasonics, the frequencies are very high and the waves reflected from inside the patient are amplified and displayed on the screen of a receiver. The image shows the various soft tissues and their connecting links.

Ultrasonics are now used to study the development of the unborn baby. The waves are reflected back from the foetus. In this way the birth can be followed throughout the pre-natal period to check on any abnormalities, Plate 22A.

The chief advantage of diagnostic ultrasonics is the absence of danger to the patient. Radiography, which is the use of X-rays, could be very dangerous.

Beats

If two notes of nearly equal frequency are sounded together, a periodic rise and fall in intensity can be heard. This is known as the phenomenon of *beats*. The frequency of the beats is the number of intense or loud sounds heard per second.

Consider a layer of air some distance away from two pure notes of nearly equal frequency, say 48 and 56 Hz respectively, which are sounding. The variation of the displacement, y_1, of the layer due to one fork alone is shown in Figure 22.1 (i); the variation of the displacement y_2, of the layer due to the second fork alone is shown in Figure 22.1 (ii).

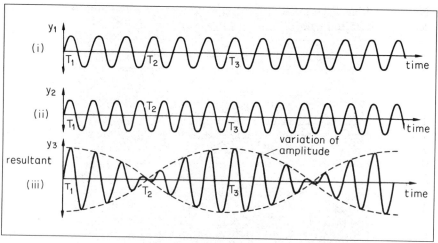

Figure 22.1 *Beats* (not to scale)

According to the Principle of Superposition (p. 515), the variation of the resultant displacement, y, of the layer is the algebraic sum of the two curves, which varies in amplitude in the way shown in Figure 22.1 (iii). To understand the variation of y, suppose that the displacements y_1, y_2 are in phase at some instant T_1, Figure 22.1. Since the frequency of the curve in Figure 22.1 (i) is 48 cycles per sec the variation y_1 undergoes 3 complete cycles in $\frac{1}{16}$th second. In the same time, the variation y_2 undergoes $3\frac{1}{2}$ cycles, since its frequency is 56 cycles per second. Thus y_1 and y_2 are 180° out of phase with each other at this instant, and their resultant y is then a minimum at some instant T_2. So $T_1 T_2$ represents $\frac{1}{16}$th of a second in Figure 22.1 (iii).

In $\frac{1}{8}$th of a second from T_1, y_1 has undergone 6 complete cycles and y_2 has undergone 7 complete cycles. The two waves are hence in phase again at T_3 where $T_1 T_3$ represents $\frac{1}{8}$th of a second, and their resultant at their instant is again a maximum, Figure 22.1 (iii). In this way it can be seen that loud sound is heard after every $\frac{1}{8}$ second, and thus the beat frequency is 8 cycles per second. This is the difference between the frequencies, 48, 56, of the two notes. We show soon that *the beat frequency is always equal to the difference of the two nearly equal frequencies.* From Figure 22.1 we see that beats are a phenomenon of interference of two sound waves travelling in the same direction to an observer.

Beat Frequency Formula

Suppose two sounding tuning-forks have frequencies f_1, f_2 cycles per second which are close to each other. At some instant the displacement of a particular layer of air near the ear due to each fork will be a maximum to the right. The resultant displacement is then a maximum, and a loud sound or beat is heard.

After this, the vibrations of air due to each fork go out of phase, and t seconds later the displacement due to each fork is again a maximum to the right, so that a loud sound or beat is heard again. One fork has then made exactly *one cycle* more than the other. But the number of cycles made by each fork in t seconds is $f_1 t$ and $f_2 t$ respectively. Assuming f_1 is greater than f_2,

$$\therefore f_1 t - f_2 t = 1$$

$$\therefore f_1 - f_2 = \frac{1}{t}$$

Now 1 beat has been made in t seconds, so that $1/t$ is the number of beats per second or beat frequency.

$$\therefore f_1 - f_2 = beat\ frequency$$

Uses of Beats

The phenomenon of beats can be used to measure the unknown frequency, f_1, of a note. For this purpose a note of known frequency f_2 is used to provide beats with the unknown note, and the frequency f of the beats is obtained by counting the number made in a given time. Since f is the difference between f_2 and f_1, it follows that $f_1 = f_2 - f$, or $f_1 = f_2 + f$. Thus suppose $f_2 = 1000\ Hz$, and the number of beats per second made with a tuning-fork of unknown frequency f_1 is 4. Then $f_1 = 1004$ or $996\ Hz$.

To decide which value of f_1 is correct, the end of the tuning-fork prong is loaded with a small piece of plasticine which diminishes the frequency a little, and the two notes are sounded again. If the beat frequency is *increased*, a little thought indicates that the frequency of the note must have been originally $996\ Hz$. If the beats are decreased, the frequency of the note must have been originally $1004\ Hz$. The tuning-fork must not be overloaded, as the frequency may decrease, if it was $1004\ Hz$, to a frequency such as $995\ Hz$, in which case the significance of the beats can be wrongly interpreted.

Beats are also used to 'tune' an instrument to a given note. As the instrument note approaches the given note, beats are heard. The instrument may be regarded as 'tuned' when the beats occur at a very slow rate.

Doppler Effect

The whistle of a train or a jet aeroplane appears to increase in pitch as it approaches a stationary observer; as the moving object passes the observer, the pitch changes and becomes lower. The apparent alteration in frequency was first predicted by DOPPLER in 1845. He stated that a change of frequency of the wave-motion should be observed when a source of sound or light was moving, and this is known as the *Doppler effect*.

The Doppler effect occurs when a source of sound or light moves *relative* to an observer. In light, the effect was observed when measurements were taken of the wavelength of the light from a moving star; they showed a marked variation. In sound, the Doppler effect can be demonstrated by placing a whistle in the end of a long piece of rubber tubing, and whirling the tube in a horizontal circle above the head while blowing the whistle. The open end of the tube acts as a moving source of sound, and an observer hears a rise and fall in pitch as the end approaches and recedes from him or her.

A complete calculation of the apparent frequency in particular cases is given shortly, but Figure 22.2 shows why a change of wavelength, and hence frequency, occurs when a source of sound is moving towards a stationary observer. At a certain instant the position of the moving source is at 4. At four successive seconds *before* this instant the source had been at the positions 3, 2, 1, 0 respectively. If c is the velocity of sound, the wavefront from the source when in the position 3 reaches the surface A of a sphere of radius c and centre 3 when the source is just at 4. In the same way, the wavefront from the source when it was in the position 2 reaches the surface B of a sphere of radius $2c$ and centre 2. The wavefront C corresponds to the source when it was in the position 1, and the wavefront D to the source when it was in the position O. So if the observer is on

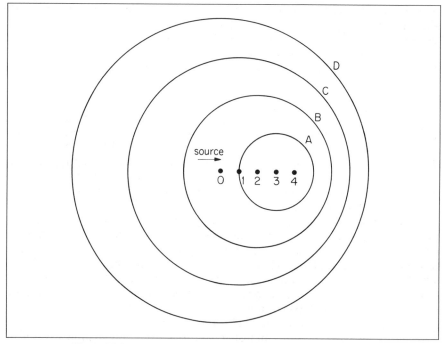

Figure 22.2 *Doppler effect: moving source, stationary observer*

the right of the source S, he receives wavefronts which are relatively more
crowded together than if S were stationary; the frequency of S thus appears to
increase.

When the observer is on the left of S, in which case the source is moving away
from him, the wavefronts are farther apart than if S were stationary. So the
observer receives correspondingly fewer waves per second. The apparent
frequency is therefore lowered.

Calculation of Apparent Frequency

Suppose c is the velocity of sound in air, u_s is the velocity of the source of sound
S, u_0 is the velocity of an observer O, and f is the true frequency of the source.

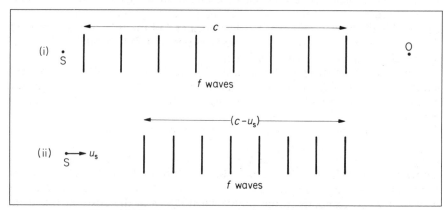

Figure 22.3 *Source moving towards stationary observer*

(i) *Source moving towards stationary observer*. If the source S were stationary, the f waves sent out in one second towards the observer O would occupy a distance c, and the wavelength would be c/f, Figure 22.3 (i). If S moves with a velocity u_s towards O, however, the f waves sent out occupy a distance $(c-u_s)$, because S has moved a distance u_s towards O in 1 s, Figure 22.3 (ii). So the wavelength λ' of the waves reaching O is now $(c-u_s)/f$.

But velocity of sound waves $= c$

$$\therefore \text{apparent frequency, } f' = \frac{\text{velocity of sound relative to O}}{\text{wavelength of waves reaching O}}$$

$$= \frac{c}{\lambda'} = \frac{c}{(c-u_s)/f}$$

$$\therefore f' = \frac{c}{c-u_s} f \qquad . \qquad . \qquad . \qquad . \qquad . \qquad (1)$$

Since $(c-u_s)$ is less than c, f' is greater than f; the apparent frequency thus appears to increase when a source is moving towards an observer.

(ii) *Source moving away from stationary observer*. In this case the f waves sent out towards O in 1 s occupy a distance $(c+u_s)$, Figure 22.4. The wavelength λ' of the waves reaching O is thus $(c+u_s)/f$, and hence the apparent frequency f' is given by

$$f' = \frac{c}{\lambda'} = \frac{c}{(c+u_s)/f}$$

$$\therefore f' = \frac{c}{c+u_s} \cdot f \qquad . \qquad . \qquad . \qquad . \qquad (2)$$

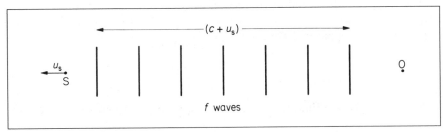

Figure 22.4 *Source moving away from stationary observer*

Since $(c+u_s)$ is greater than c, f' is less than f, and hence the apparent frequency decreases when a source moves away from an observer.

(iii) *Source stationary, and observer moving towards it*. Since the source is stationary, the f waves sent out by S towards the moving observer O occupies a

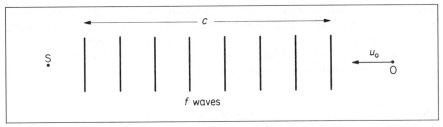

Figure 22.5 *Observer moving towards stationary source*

distance c, Figure 22.5. The wavelength of the waves reaching O is hence c/f, and thus unlike the cases already considered, the wavelength is unaltered.

The velocity of the sound waves relative to O is not c, however, as O is moving relative to the source. The velocity of the sound waves relative to O is given by $(c+u_0)$ in this case, and hence the apparent frequency f' is given by

$$f' = \frac{\text{velocity of sound relative to O}}{\text{wavelength of waves reaching O}}$$

$$= \frac{c+u_0}{c/f}$$

$$\therefore f' = \frac{c+u_0}{c}\cdot f \qquad . \qquad . \qquad . \qquad . \qquad . \qquad . \qquad (3)$$

Since $(c+u_0)$ is greater than c, f' is greater than f; thus the apparent frequency is increased.

(iv) *Source stationary, and observer moving away from it*, Figure 22.6. As in the case just considered, the wavelength of the waves reaching O is unaltered, and is given by c/f.

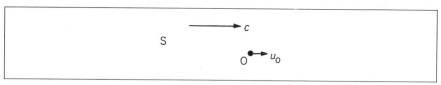

Figure 22.6 *Observer moving away from stationary source*

The velocity of the sound waves relative to $O = c-u_0$, and hence

$$\text{apparent frequency, } f', = \frac{c-u_0}{\text{wavelength}} = \frac{c-u_0}{c/f}$$

$$\therefore f' = \frac{c-u_0}{c}\cdot f \qquad . \qquad . \qquad . \qquad . \qquad (4)$$

Since $(c-u_0)$ is less than c, the apparent frequency f' appears to be decreased.

Source and Observer Both Moving

If the source and the observer are both moving, the apparent frequency f' can be found from the formula

$$f' = \frac{c'}{\lambda'}$$

where c' is the velocity of the sound waves relative to the observer, and λ' is the wavelength of the waves reaching the observer. This formula can also be used to find the apparent frequency in any of the cases considered before.

Suppose that the observer has a velocity u_0, the source a velocity u_s, and that both are moving in the *same* direction. Then

$$c' = c-u_0$$

and

$$\lambda' = (c-u_s)/f$$

as was deduced in case (i), p. 577.

$$\therefore f' = \frac{c'}{\lambda'} = \frac{c - u_0}{(c - u_s)/f} = \frac{c - u_0}{c - u_s} \cdot f \qquad . \qquad . \qquad . \qquad (i)$$

If the observer is moving towards the source, $c' = c + u_0$, and the apparent frequency f' is given by

$$f' = \frac{c + u_0}{c - u_s} \cdot f \qquad . \qquad . \qquad . \qquad . \qquad (ii)$$

From (i), it follows that $f' = f$ when $u_0 = u_s$, in which case there is no relative velocity between the source and the observer. It should also be noted that the motion of the observer affects only c', the velocity of the waves reaching the observer, while the motion of the source affects only λ', the wavelength of the waves reaching the observer.

The effect of the wind can also be taken into account in the Doppler effect. Suppose the velocity of the wind is u_w, in the direction of the line SO joining the source S to the observer O. Since the air has then a velocity u_w relative to the ground, and the velocity of the sound waves relative to the air is c, the velocity of the waves relative to ground is $(c + u_w)$ if the wind is blowing in the same direction as SO. All our previous expressions for f' can now be adjusted by replacing the velocity c in it by $(c + u_w)$. If the wind is blowing in the opposite direction to SO, the velocity c must be replaced by $(c - u_w)$.

Example on Doppler Principle

A car, sounding a horn producing a note of 500 Hz, approaches and then passes a stationary observer O at a steady speed of $20\,\mathrm{m\,s^{-1}}$. Calculate the change in pitch of the note heard by O (velocity of sound $= 340\,\mathrm{m\,s^{-1}}$).

Towards O. Velocity of sound relative to O, $c' = 340\,\mathrm{m\,s^{-1}}$

Wavelength of waves reaching O, $\lambda' = (340 - 20)/500$ m

$$\therefore \text{apparent frequency to O, } f' = \frac{c'}{\lambda'}$$

$$= \frac{340}{320} \cdot 500 = 531\,\mathrm{Hz} \qquad . \qquad (1)$$

Away from O. With the above notation, $c' = 340\,\mathrm{m\,s^{-1}}$

$$\text{and} \qquad \lambda' = (340 + 20)/500\,\mathrm{m}$$

$$\therefore \text{apparent frequency to O, } f'' = \frac{340}{(340 + 20)} \cdot 500$$

$$= 472\,\mathrm{Hz} \qquad . \qquad . \qquad . \qquad (2)$$

From (1) and (2), change in pitch $= \dfrac{f''}{f'} = 0 \cdot 9$ (approx.)

Reflection of Waves

Consider a source of sound A approaching a fixed reflector R such as a wall or bridge, for example. The reflected waves then appear to travel from R to A as if they came from the 'mirror image' A' of A in R, Figure 22.7.

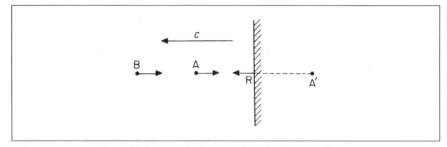

Figure 22.7 *Doppler's principle and reflection of waves*

As an illustration, suppose a car A approaches R with a velocity of $20\,\mathrm{m\,s^{-1}}$ when sounding a note of 1000 Hz from its horn, and that another car B behind A is travelling towards A with a velocity of $30\,\mathrm{m\,s^{-1}}$, Figure 22.7.

In B, the driver hears a note from R which has an apparent frequency $f' = c'/\lambda'$, where c' is the velocity of sound relative to B and λ' is the wavelength of the waves reaching B. If the velocity of sound c is $340\,\mathrm{m\,s^{-1}}$, then

$$c' = 340 + 30 = 370\,\mathrm{m\,s^{-1}}$$

and

$$\lambda' = (340 - 20)/1000 = 320/1000\,\mathrm{m}$$

$$\therefore f' = \frac{c'}{\lambda'} = \frac{370}{320} . 1000 = 1156\,\mathrm{Hz}$$

In B, the driver also hears a note directly from A. In this case,

$$c' = 340 + 30 = 370\,\mathrm{m\,s^{-1}}$$

and

$$\lambda' = (340 + 20)/1000 = 360/1000\,\mathrm{m}$$

$$\therefore f'' = \frac{c'}{\lambda'} = \frac{370}{360} . 1000 = 1028\,\mathrm{Hz}$$

With a *reflector moving* with velocity v directly towards a stationary sound source S of frequency f, the frequency f' of the waves reflected back to S is given by

$$f' = \frac{c+v}{c-v} f$$

where c is the velocity of sound in air. The frequency f_0 of the waves received by the moving reflector is $f_0 = (c+v)f/c$ and the frequency f' of the waves reflected back to S $= c f_0/(c-v) = (c+v)f/(c-v)$.

Doppler Effect in Circular Motion

Consider a sound signal of constant frequency carried by a car movig round a circular track.

At any instant, the velocity is directed along the tangent to the circle. So a distant observer O will hear a note which rises in frequency as the car approaches O in its circular path and decreases in frequency as the car goes away from O. The maximum frequency occurs when the velocity is directly towards O and the minimum frequency when the velocity is directly away from O.

If the observer O is *inside* the circle, O will hear the frequency rise when the component of the velocity is directed towards him or her. The frequency will decrease when the velocity component is directed away from O. If O is at the centre of the circle, the velocity has no component towards O. So a continuous note of the same frequency as the signal is heard in this case.

Doppler Principle in Light

The speed of distant stars and planets has been estimated from measurements of the wavelengths of the spectrum lines which they emit. Suppose a star or planet is moving with a velocity v away from the earth and emits light of wavelength λ. If the frequency of the vibrations is f cycles per second, then f waves are emitted in one second, where $c = f\lambda$ and c is the velocity of light *in vacuo*. Owing to the velocity v, the f waves occupy a distance $(c + v)$. Thus the *apparent wavelength* λ' to an observer on the earth in line with the star's motion is

$$\lambda' = \frac{c+v}{f} = \frac{c+v}{c} \cdot \lambda = \left(1 + \frac{v}{c}\right)\lambda$$

$$\therefore \ \lambda' - \lambda = \text{`shift' in wavelength} = \frac{v}{c}\lambda \qquad . \qquad . \qquad . \qquad \text{(i)}$$

and hence $\qquad \dfrac{\lambda' - \lambda}{\lambda} = \text{fractional change in wavelength} = \dfrac{v}{c} \qquad . \qquad . \qquad \text{(ii)}$

From (i), it follows that λ' is greater than λ when the star or planet is moving away from the earth, that is, there is a 'shift' or displacement *towards the red*. The position of a particular wavelength in the spectrum of the star is compared with that obtained in the laboratory, and the difference in the wavelengths, $\lambda' - \lambda$, the 'red shift', is measured. From (i), knowing λ and c, the velocity v can be calculated.

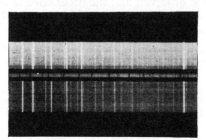

Figure 22.8 *Doppler shift. The central band of dark lines is the absorption spectrum of the star Eta Cephei. The bright lines in the wider bands above and below are the same lines in the emission spectrum of iron as obtained in the laboratory. Because the star is moving away from the earth, on account of the Doppler effect each dark line has a wavelength slightly greater than if the star were stationary*

If the star is moving *towards* the earth with a velocity u, the apparent wavelength λ'' is given by

$$\lambda'' = \frac{c-u}{f} = \frac{c-u}{c} \cdot \lambda = \left(1 - \frac{u}{c}\right)\lambda$$

$$\therefore \ \lambda - \lambda'' = \frac{u}{c}\lambda \qquad . \qquad . \qquad . \qquad . \qquad . \qquad \text{(iii)}$$

Since λ'' is less than λ, there is a displacement towards the blue in this case*.

*Equations (i)–(iii) apply for velocities much less than c, otherwise relativistic corrections are required. See *Introduction to Relativity*, Rosser (Butterworth).

In measuring the speed of a star, a photograph of its spectrum is taken. The spectral lines are then compared with the same lines obtained by photographing in the laboratory an arc or spark spectrum of an element present in the star If the lines are displaced towards the red, the star is receding from the earth; if displaced towards the violet, the star is approaching the earth. By this method the velocities of the stars have been found to be between about 10 km s^{-1} and 300 km s^{-1}.

The Doppler effect has also been used to measure the speed of rotation of the sun. Photographs are taken of the east and west edges of the sun; each contains absorption lines due to elements such as iron vaporised in the sun, and also some absorption lines due to oxygen in the earth's atmosphere. When the two photographs are put together so that the oxygen lines coincide, the iron lines in the two photographs are displaced relative to each other. In one case the edge of the sun approaches the earth, and in the other the opposite edge recedes from the earth. Measurements show a rotational speed of about 2 km s^{-1}. See Worked Example on page 583.

Measurement of Plasma Temperature

In very hot gases or plasma, used in thermonuclear fusion experiments, the temperature is of the order of millions of degrees Celsius. At these high temperatures molecules of the glowing gas are moving away and towards the observer with very high speeds and, owing to the Doppler effect, the wavelength λ of a particular spectral line is apparently changed. One edge of the line now corresponds to an apparently increased wavelength λ_1 due to molecules moving directly away from the observer, and the other edge to an apparent decreased wavelength λ_2 due to molecules moving directly towards the observer. The line is thus observed to be *broadened*.

From our previous discussion, if v is the velocity of the molecules,

$$\lambda_1 = \frac{c+v}{c} . \lambda$$

and

$$\lambda_2 = \frac{c-v}{c} . \lambda$$

$$\therefore \text{ breadth of line, } \lambda_1 - \lambda_2 = \frac{2v}{c} . \lambda \qquad . \qquad . \qquad . \qquad \text{(iii)}$$

The breadth of the line can be measured by a diffraction grating, and as λ and c are known, the velocity v can be calculated. By the kinetic theory of gases, the velocity v of the molecules is roughly the root-mean-square velocity, or $\sqrt{3RT/M}$, where T is the absolute temperature, R is the molar gas constant and M is the mass of one mole.

Examples on Doppler Effect in Sound and Light

1 A whistle giving out 500 Hz moves away from a stationary observer in a direction towards and perpendicular to a flat wall with a velocity of $1 \cdot 5 \text{ m s}^{-1}$. How many beats per second will be heard by the observer? (Take the velocity of sound as 336 m s^{-1} and assume there is no wind.)

The observer hears a note of apparent frequency f' from the whistle directly, and a note of apparent frequency f'' from the sound waves reflected from the wall.

Now $$f' = \frac{c'}{\lambda'}$$

where c' is the velocity of sound in air relative to the observer and λ' is the wavelength of the waves reaching the observer. Since

$$c' = 336 \, \text{m s}^{-1} \text{ and } \lambda' = \frac{336 + 1 \cdot 5}{500} \, \text{m}$$

$$\therefore f' = \frac{336 \times 500}{337 \cdot 5} = 497 \cdot 8 \, \text{Hz}$$

The note of apparent frequency f'' is due to sound waves moving towards the observer with a velocity of $1 \cdot 5 \, \text{m s}^{-1}$

$$\therefore f'' = \frac{c'}{\lambda'} = \frac{336}{(336 - 1 \cdot 5)/500}$$

$$= \frac{336 \times 500}{334 \cdot 5} = 502 \cdot 2 \, \text{Hz}$$

$$\therefore \text{beats per second} = f'' - f' = 502 \cdot 2 - 497 \cdot 8 = 4 \cdot 4$$

2 Two observers A and B are provided with sources of sound of frequency 500. A remains stationary and B moves away from him at a velocity of $1 \cdot 8 \, \text{m s}^{-1}$. How many beats per second are observed by A and by B, the velocity of sound being $330 \, \text{m s}^{-1}$?

Beats observed by A. A hears a note of frequency 500 due to its own source of sound. He also hears a note of apparent frequency f' due to the moving source B. With the usual notation,

$$f' = \frac{c'}{\lambda'} = \frac{330}{(330 + 1 \cdot 8)/500}$$

since the velocity of sound, c', relative to A is $330 \, \text{m s}^{-1}$ and the wavelength λ' of the waves reaching him is $(330 + 1 \cdot 8)/500 \, \text{m}$.

$$\therefore f' = \frac{330 \times 500}{331 \cdot 8} = 497 \cdot 3$$

$$\therefore \text{beats observed by A} = 500 - 497 \cdot 29 = 2 \cdot 71 \, \text{Hz}$$

Beats observed by B. The apparent frequency f' of the sound from A is given by

$$f' = \frac{c'}{\lambda'}$$

In this case $c' = $ velocity of sound relative to B $= 330 - 1 \cdot 8 = 328 \cdot 2 \, \text{m s}^{-1}$ and the wavelength λ' of the waves reaching B is unaltered. Since $\lambda' = 330/500 \, \text{m}$, it follows that

$$f' = \frac{328 \cdot 2}{330/500} = \frac{328 \cdot 2 \times 500}{330} = 497 \cdot 27$$

$$\therefore \text{beats heard by B} = 500 - 497 \cdot 27 = 2 \cdot 73 \, \text{Hz}$$

3 In the Sun's spectrum, a line of wavelength $589 \cdot 00 \, \text{nm}$ differs by $7 \cdot 8 \times 10^{-3} \, \text{nm}$ when opposite edges of the Sun are observed across the equatorial diameter.

Estimate the speed of rotation of the Sun, assuming speed of light, $c = 3 \cdot 00 \times 10^8 \, \text{m s}^{-1}$.

One edge moves with velocity v away from the observer. From the Doppler principle, the increased wavelength is given by, if λ is the actual wavelength,

$$\lambda' = \frac{c+v}{c}\lambda$$

The opposite edge moves towards the observer with velocity v. So the decreased wavelength is given by

$$\lambda'' = \frac{c-v}{c}\lambda$$

So   change in wavelength $= \lambda' - \lambda'' = \left(\dfrac{c+v}{c} - \dfrac{c-v}{c}\right)\lambda$

$$= 2\frac{v}{c}\lambda = 7{\cdot}8 \times 10^{-3}\,\text{nm}$$

$$\therefore v = \frac{7{\cdot}8 \times 10^{-3} \times 3 \times 10^8}{2 \times 589}$$

$$= 2 \times 10^3\,\text{m s}^{-1} = 2\,\text{km s}^{-1}$$

Intensity and Amplitude

The *intensity* of a sound at a place is defined as the energy per second flowing through one square metre held normally at that place to the direction along which the sound travels. At a distance r from a *small* source of sound, the energy passes through the surface area, $4\pi r^2$, of a surrounding sphere. So in this case, intensity $\propto 1/r^2$.

Suppose the displacement y of a vibrating layer of air is given by $y = a\sin\omega t$, where $\omega = 2\pi/T$ and a is the amplitude of vibration, see equation (1), p. 465. The velocity, v, of the layer is given by

$$v = \frac{dy}{dt} = \omega a \cos \omega t$$

and hence the kinetic energy, W, is given by

$$W = \tfrac{1}{2}mv^2 = \tfrac{1}{2}m\omega^2 a^2 \cos^2 \omega t \qquad . \qquad . \qquad . \qquad \text{(i)}$$

where m is the mass of the layer. The layer also has potential energy as it vibrates. Its total energy, W_0, which is constant, is therefore equal to the maximum value of the kinetic energy. From (i), it follows that

$$W_0 = \tfrac{1}{2}m\omega^2 a^2 \qquad . \qquad . \qquad . \qquad . \qquad \text{(ii)}$$

In 1 second, the air is disturbed by the wave over a distance c m, where c is the velocity of sound in m s^{-1}; and if the area of cross-section of the air is $1\,\text{m}^2$, the volume of air disturbed is $c\,\text{m}^3$. The mass of air disturbed per second is thus $c\rho$ kg, where ρ is the density of air in kg m^{-3}, and hence, from (ii),

$$W = \tfrac{1}{2}c\rho\omega^2 a^2 \qquad . \qquad . \qquad . \qquad . \qquad \text{(iii)}$$

It therefore follows that *the intensity of a sound due to a wave of given frequency is proportional to the square of its amplitude of vibration.*

At the beginning of this section we showed that the intensity due to a small source of sound at a distance r from it was proportional to $1/r^2$. But we have just proved that the intensity is also proportional to a^2, where a is the amplitude of vibration at this distance from the source. So $a^2 \propto 1/r^2$, or

$$a \propto \frac{1}{r}$$

Quality or Timbre

If the same note is sounded on the violin and then on the piano, an untrained listener can tell which instrument is being used, without seeing it. We say that the *quality* or *timbre* of the note is different in each case.

The waveform of a note is never simple harmonic in practice; the nearest approach is that obtained by sounding a tuning-fork, which thus produces what may be called a 'pure' note, Figure 22.9 (i). If the same note is played on a violin and piano respectively, the waveforms produced might be represented by Figure 22.9 (ii), (iii), which have the same frequency and amplitude as the waveform in Figure 22.9 (i). Now curves of the shape of Figure 22.9 (ii), (iii) can be analysed mathematically into the sum of a number of *simple harmonic* curves, whose

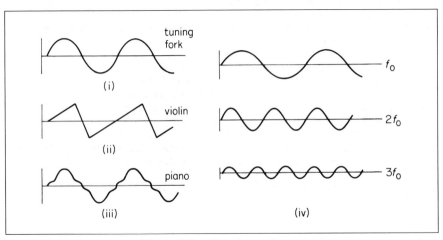

Figure 22.9   *Waveforms of notes*

frequencies are multiples of f_0, the frequency of the original waveform; the amplitudes of these curves diminish as the frequency increases, Figure 22.9 (iv), for example, might be an analysis of a curve similar to Figure 22.9 (iii), corresponding to a note on a piano. The ear is able to detect simple harmonic waves and therefore it registers the presence of notes of frequencies $2f_0$ and $3f_0$, in addition to f_0, when the note is sounded on the piano. The amplitude of the curve corresponding to f_0 is greatest, Figure 22.9 (iv), and the note of frequency f_0 is heard predominantly because the intensity is proportional to the square of the amplitude (p. 584). In the background, however, are the notes of frequencies $2f_0$, $3f_0$, which are called the *overtones*. The frequency f_0 is called the *fundamental*.

As the waveform of the same note is different when it is obtained from different instruments, it follows that the analysis of each will differ. For example, the waveform of a note of frequency f_0 from a violin may contain overtones of frequencies $2f_0$, $4f_0$, $6f_0$. The musical 'background' to the fundamental note is therefore different when it is sounded on different instruments, and hence *the overtones present in a note determine its quality or timbre*.

A *harmonic* is the name given to a note whose frequency is a simple multiple of the fundamental frequency f_0. So f_0 is called the 'first harmonic'; a note of frequency $2f_0$ is called the 'second harmonic', and so on. Certain harmonics of a note may be absent from its overtones; for example, the only possible notes obtained from an organ-pipe closed at one end are $f_0, 3f_0, 5f_0, 7f_0$, and so on (p. 595).

_____ Exercises 22 _____

Sound Waves

1 Describe the nature of the disturbance set up in air by a vibrating tuning-fork and show how the disturbance can be represented by a sine curve. Indicate on the curve the points of
(a) maximum particle velocity,
(b) maximum pressure.
   What characteristics of the vibration determine the pitch, intensity, and quality respectively of the note? (*JMB*.)

2 Define *frequency* and explain the term *harmonics*. How do harmonics determine the *quality* of a musical note?
   It is much easier to hear the sound of a vibrating tuning fork if it is
(a) placed in contact with a bench, or
(b) held over a *certain* length of air in a tube. Explain why this is so in both these cases and give two further examples of the phenomenon occurring in (b).
   Describe how you would measure the wavelength in the air in a tube of the note emitted by the fork. How would the value obtained be affected by changes in
(i) the temperature of the air and   (ii) the pressure of the air? (*L*.)

3 Distinguish between *longitudinal* and *transverse* wave motions, giving examples of each type. Find a relationship between the frequency, wavelength and velocity of propagation of a wave motion.
   Describe experiments to investigate quantitatively for sound waves the phenomena of
(a) reflection,
(b) refraction,
(c) interference. (*C*.)

4 Explain why sounds are heard very clearly at great distances from the source
(a) on still mornings after a clear night, and
(b) when the wind is blowing from the source to the observer.

5 Continuous sound waves of a single frequency are emitted from two small loudspeakers A and B, fed by the same signal generator and located as shown in

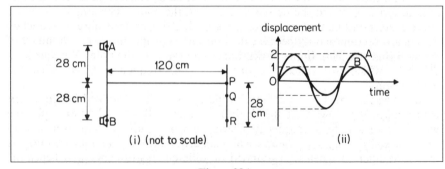

Figure 22A

Figure 22A (i), which is not to scale. A small sensitive microphone placed at P is connected (via a pre-amplifier) to a cathode ray oscilloscope. The graphs in Figure 22A (ii) show the traces that appear on the screen of the c.r.o. due to A alone, then B alone, and represent the variation of the displacement at P produced by each wave separately with time.

(a) Calculate the relative intensities of the waves emitted by A and B.
The width of the trace on the c.r.o. is 12·0 cm and the time base 'speed' is 50 μs cm^{-1}. Calculate the frequency of the waves.

(b) The two waves interfere at P. Using the superposition principle, construct on the graph (Figure 22A (ii)) the resultant displacement-time curve, using the same axes as for the original waves. What is the intensity of the resultant wave at P compared with that caused by B alone?

(c) Using axes on a separate sheet, construct the resultant displacement for a point Q where the waves from A and B arrive 180° out of phase with each other. (Assume that the amplitudes of the waves arriving at Q are the same as those arriving at P.) What is the intensity of the resultant wave compared with that caused by B alone in this case?

(d) Explain why, although the displacement of the resultant wave produced at P or Q varies with time, the sound intensity does not.

(e) A maximum of sound intensity occurs when the microphone is at R. From the dimensions given in the diagram determine the largest possible value for the wavelength of the sound waves.

(f) The experimental arrangement shown in Figure 22A (i) could be used to measure the speed of sound, c, in air. Explain briefly how you would use the apparatus to measure c. Explain how by altering the frequency of the sound a more reliable value of c might be obtained than by using only one frequency. (L.)

6 A thin, vertical rod is partially immersed in a large deep pool of water. It moves vertically with simple harmonic motion of small amplitude. Describe the waves produced on the water surrounding the rod. State and explain how (i) the wavelength and (ii) the amplitude of the waves depend on the distance from the rod of the point at which they are measured.

Describe and briefly explain what happens when a second rod, similar to the first and vibrating with the same frequency and amplitude, and in phase with it, is placed in the water at a distance d from the first rod.

Discuss the difficulties encountered in attempting to demonstrate similar behaviour for two sources of visible light and describe an experiment you would perform to achieve this. (O. & C.)

7 Describe a method for the accurate measurement of the velocity of sound in *free* air.
Indicate the factors which influence the velocity and how they are allowed for or eliminated in the experiment you describe.

At a point 20 m from a small source of sound the intensity is 0·5 microwatt cm^{-2}. Find a value for the rate of emission of sound energy from the source, and state the assumptions you make in your calculation. (*JMB.*)

8 Explain the origin of the beats heard when two tuning-forks of slightly different frequency are sounded together. Deduce the relation between the frequency of the beats and the difference in frequency of the forks. How would you determine which fork had the higher frequency?

A simple pendulum set up to swing in front of the 'seconds' pendulum ($T = 2$ s) of a clock is seen to gain so that the two swing in phase at intervals of 21 s. What is the time of swing of the simple pendulum? (*L.*)

9 What is meant by
(a) the *amplitude*,
(b) the *frequency* of a wave in air? What are the corresponding characteristics of the musical sound associated with the wave? How would you account for the difference in quality between two notes of the same pitch produced by two different instruments, e.g., by a violin and by an organ pipe?

What are 'beats'? Given a set of standard forks of frequencies 256, 264, 272, 280, and 288, and a tuning-fork whose frequency is known to be between 256 and 288, how would you determine its frequency accurately?

Doppler's Principle

10 An observer travels with constant velocity of $30 \, \text{m s}^{-1}$ towards a distant source of sound, which has a frequency of 1000 Hz. Calculate the apparent frequency of the sound heard by the observer. What frequency is heard after passing the source of sound? (Assume velocity of sound $= 330 \, \text{m s}^{-1}$.)

11 (a) A sound of frequency of 500 Hz is emitted from an alarm system in a vehicle. The frequency of the note as heard by a stationary observer changes as the vehicle accelerates away. What will be the value of this frequency when the vehicle reaches a speed of $10 \, \text{m s}^{-1}$? (Speed of sound in air $= 340 \, \text{m s}^{-1}$.)

(b) Assume the vehicle is capable of accelerating up to the speed of sound. Sketch a graph showing how the frequency of the sound heard by the stationary observer changes as the speed of the vehicle increases from 0 to $340 \, \text{m s}^{-1}$. (*L*.)

12 The wavelength of a particular line in the emission spectrum of a distant star is measured as 600·80 nm. The true wavelength is 600·00 nm.

(a) Is the star moving away from or towards the observer?

(b) Calculate the speed of the star. (Velocity of light $= 3.0 \times 10^{8} \, \text{m s}^{-1}$.)

13 An observer, travelling with a constant velocity of $20 \, \text{m s}^{-1}$, passes close to a stationary source of sound and notices that there is a change of frequency of 50 Hz as he passes the source. What is the frequency of the source?
(Speed of sound in air $= 340 \, \text{m s}^{-1}$.) (*L*.)

14 Deduce expressions for the frequency heard by an observer

(a) when he is stationary and a source of sound is moving towards him and

(b) when he is moving towards a stationary source of sound. Explain your reasoning carefully in each case.

Give an example of change of frequency due to motion of source or observer from some other branch of physics, explaining either, a use which is made of it, or a deduction from it.

A car travelling at $10 \, \text{m s}^{-1}$ sounds its horn, which has a frequency of 500 Hz, and this is heard in another car which is travelling behind the first car, in the same direction, with a velocity of $20 \, \text{m s}^{-1}$. The sound can also be heard in the second car by reflection from a bridge head. What frequencies will the driver of the second car hear? (Speed of sound in air $= 340 \, \text{m s}^{-1}$.) (*L*.)

15 An object, vibrating vertically with a frequency of 10 Hz, is moving in a horizontal straight line with a velocity of $2.0 \, \text{cm s}^{-1}$. It is producing waves, which travel with a speed of $12 \, \text{cm s}^{-1}$, on a water surface. Draw a diagram showing the instantaneous positions of the waves emitted during the previous half second.

Calculate the frequency of the waves in the direction of motion of the object.

A boy sitting on a swing which is moving to an angle of 30° from the vertical is blowing a whistle which has a frequency of $1.0 \, \text{kHz}$. The whistle is $2.0 \, \text{m}$ from the point of support of the swing. A girl stands in front of the swing. Calculate the maximum and minimum frequencies she will hear. (Speed of sound $= 330 \, \text{m s}^{-1}$, $g = 9.8 \, \text{m s}^{-2}$.) (*L*.)

16 (a) State the conditions necessary for 'beats' to be heard and derive an expression for their frequency.

(b) A fixed source generates sound waves which travel with a speed of $330 \, \text{m s}^{-1}$. They are found by a distant stationary observer to have a frequency of 500 Hz. What is the wavelength of the waves? From first principles find (i) the wavelength of the waves in the direction of the observer, and (ii) the frequency of the sound heard if (1) the source is moving towards the stationary observer with a speed of $30 \, \text{m s}^{-1}$, (2) the observer is moving towards the stationary source with a speed of $30 \, \text{m s}^{-1}$, (3) both source and observer move with a speed of $30 \, \text{m s}^{-1}$ and approach one another. (*JMB*.)

17 The Sun rotates about its centre. Observation of the light from the two edges at the ends of a diameter shows a Doppler shift of about 0·008 nm for a wavelength 600·000 nm.

Estimate the angular velocity of the Sun about its centre, given that its radius is $7.0 \times 10^{8} \, \text{m}$ and $c = 3.0 \times 10^{8} \, \text{m s}^{-1}$.

18 Describe
  (a) the *Doppler effect* and
  (b) *beats*, as observed with sound waves. Derive expressions for the apparent frequency of a sound signal heard by an observer in still air   (i) from a source of frequency f moving with velocity u towards a stationary observer and   (ii) from a stationary source of frequency f when the observer approaches the source with velocity v.

  An ultrasonic burglar alarm in still air transmits a signal at a frequency of 4.5×10^4 Hz, part of which is reflected by the burglar to a receiver alongside the transmitter.

  The burglar moves towards the transmitter at $1\,\mathrm{m\,s^{-1}}$; calculate   (iii) the frequency of the signal received by the burglar, and   (iv) the frequency detected by the receiver alongside the transmitter.

  Hence find   (v) the beat frequency between the signal reflected from the moving burglar and the original signal from the transmitter.

  The alarm is triggered by any beat frequency greater than 5 Hz; estimate   (vi) the minimum velocity of approach of a burglar to activate the alarm. (Velocity of sound in air $= 340\,\mathrm{m\,s^{-1}}$). (*O. & C.*)

23

Waves in Pipes and Strings

As we shall see, stationary or standing waves are formed in sounding pipes and strings. We first deal in detail with the waves produced in closed and open pipes and the range of frequencies produced and the way in which the speed of sound in pipes is measured. This is followed by waves in strings and the frequencies produced, and the application to measuring the a.c. mains frequency.

Introduction

The music from an organ, a violin, or a xylophone is due to vibrations in the air set up by oscillations in these instruments. In the organ, air is blown into a pipe, which sounds its characteristic note as the air inside it vibrates; in the violin, the strings are bowed so that they oscillate; and in a xylophone a row of metallic rods are struck in the middle with a hammer, which sets them into vibration.

Before considering each of the above cases in more detail, it would be best to consider the feature common to all of them. A violin string is fixed at both ends, A, B, and waves travel along m, n to each end of the string when it is bowed and are there reflected, Figure 23.1 (i).

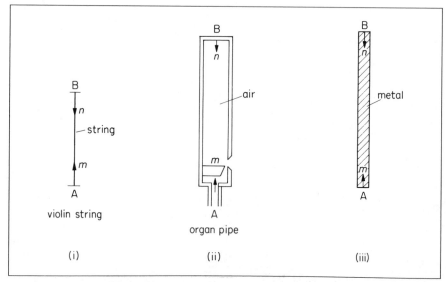

Figure 23.1 *Reflection of waves in instruments*

The vibrations of the particles of the string are hence due to *two waves of the same frequency and amplitude travelling in opposite directions*. A similar effect is obtained with an organ pipe closed at one end B, Figure 23.1 (ii). If air is blown into the pipe at A, a wave travels along the direction m and is reflected at B in the opposite direction n. The vibrations of the air in the pipe are thus due to two waves travelling in opposite directions. If a metal rod is fixed at its middle in a vice

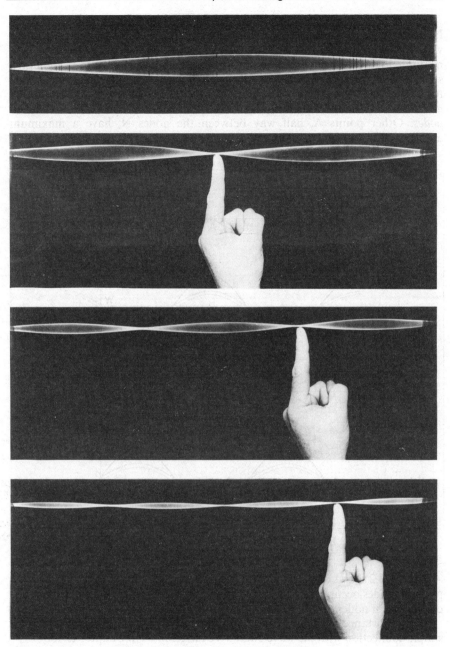

Figure 23A (a)–(d) *The photographs illustrate how 'stopping' with a light touch of a finger at different points of a vibrating cord produces successive harmonics. At the top, the mode of vibration corresponds to the fundamental frequency f_0 and the others to $2f_0$, $3f_0$, $4f_0$. As shown, a displacement node is produced at the stop in each case.*
(Courtesy of Prof. C. A. Taylor, Cardiff University)

and stroked at one end A, a wave travels along the rod in the direction m and is reflected at the other end B in the direction n, Figure 23.1 (iii). The vibrations of the rod, which produce a high-pitched note, are thus due to two waves travelling in opposite directions.

Stationary Waves and Wavelength

In Chapter 17 on Waves, we showed that a *stationary wave* is formed when two waves of equal amplitude and frequency travel in opposite directions in a medium. Figure 23.2 shows a plane-progressive sound wave *a* in air incident on a smooth wall W, together with the reflected wave *b*. When the displacements due to the two waves are added together at the times shown, where *T* is the period of *a* or *b*, the resultant wave S has points N of permanent zero displacement called *nodes*. Other points A, half way between the nodes N, have a maximum amplitude of vibration; they are called *antinodes*.

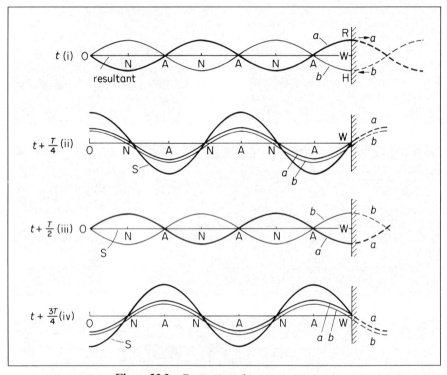

Figure 23.2 *Formation of stationary waves*

Figure 23.2 shows displacement nodes and antinodes. As we saw on p. 472, the pressure nodes occur at displacement antinodes; the pressure antinodes occur at displacement nodes.

The importance of the nodes and antinodes in a stationary wave lies in their simple connection with the wavelength. We have shown in Chapter 17 that

$$\text{\textit{the distance between consecutive nodes, NN}} = \frac{\lambda}{2} \qquad . \qquad . \qquad \text{(i)}$$

where λ is the wavelength of the progressive wave;

$$\text{\textit{the distance between consecutive antinodes, AA}} = \frac{\lambda}{2} \qquad . \qquad . \qquad \text{(ii)}$$

and

_____

the distance from a node to the next antinode, $NA = \dfrac{\lambda}{4}$.   (iii)

_____

Examples on Stationary Waves

Plane sound waves of frequency 100 Hz fall normally on a smooth wall. At what distances from the wall will the air particles have
(a) maximum,
(b) minimum amplitude of vibration? Give reasons for your answer. (The velocity of sound in air may be taken as $340\,\mathrm{m\,s^{-1}}$.) (*L.*)

A stationary wave is set up between the source and wall, due to the production of a reflected wave. The wall is a displacement node, since the air in contact with it cannot move; and other nodes are at equal distances, d, from the wall. Now if the wavelength is λ,

$$d = \frac{\lambda}{2}$$

Since

$$\lambda = \frac{v}{f} = \frac{340}{100} = 3\cdot4\,\mathrm{m}$$

$$\therefore d = \frac{3\cdot4}{2} = 1\cdot7\,\mathrm{m}$$

Thus minimum amplitude of vibration is obtained $1\cdot7$, $3\cdot4$, $5\cdot1\,\mathrm{m}\ldots$ from the wall.

The antinodes are midway between the nodes. So maximum amplitude of vibration is obtained $0\cdot85$, $2\cdot55$, $4\cdot25\,\mathrm{m},\ldots$ from the wall.

Waves in Pipes

Closed Pipe

A *closed* or *stopped organ pipe* consists essentially of a metal pipe closed at one end Q, and a blast of air is blown into it at the other end P, Figure 23.3 (i). A wave thus travels up the pipe to Q, and is reflected at this end down the pipe, so that a

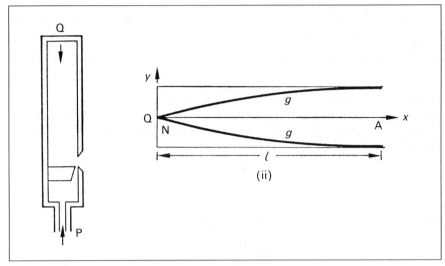

Figure 23.3 *(i) Closed (stopped) pipes* *(ii) Fundamental of closed (stopped) pipe*

stationary wave is obtained. The end Q of the closed pipe must be a node N, since the layer in contact with Q must be permanently at rest, and the open end A, where the air is free to vibrate, must be an antinode A. The simplest stationary wave in the air in the pipe is therefore represented by *g* in Figure 23.3 (ii). Here the pipe is positioned horizontally to show the relative displacement, *y*, of the layers at different distances, *x*, from the closed end Q; the axis of the stationary wave is Q*x*.

It can now be seen that the length *l* of the pipe is equal to the distance between a node N and a consecutive antinode A of the stationary wave. But NA = $\lambda/4$, where λ is the wavelength (p. 593).

$$\therefore \frac{\lambda}{4} = l \text{ or } \lambda = 4l$$

But the frequency, *f*, of the note is given by $f = c/\lambda$, where *c* is the velocity of sound in air.

$$\therefore f = \frac{c}{4l}$$

This is the frequency of the lowest note obtainable from the pipe, and it is known as its *fundamental*. We shall denote the fundamental frequency by f_0, so that

$$f_0 = \frac{c}{4l} \qquad . \qquad . \qquad . \qquad . \qquad . \qquad . \qquad (1)$$

Figure 23.3 (ii) shows the stationary wave of *displacement* of air molecules along the closed pipe. The *pressure* variation is also a stationary wave. But in contrast to Figure 23.3 (ii), the pressure node is at the open end since the air pressure here is constant and equal to the external atmospheric pressure; and the pressure antinode is at the closed end since here the layers of air are compressed.

Overtones of Closed Pipe

If a stronger blast of air is blown into the pipes, notes of higher frequency can be obtained which are simple multiples of the fundamental frequency f_0. Two possible cases of stationary waves are shown in Figure 23.4. In each, the closed

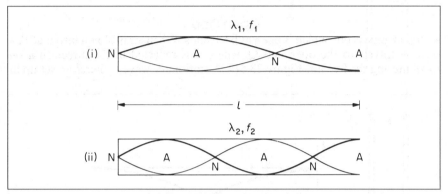

Figure 23.4 *Overtones in closed pipe*

end of the pipe is a node, and the open end is an antinode. In Figure 23.4 (i), however, the length l of the pipe is related to the wavelength λ_1 of the wave by

$$l = \frac{3}{4}\lambda_1$$

$$\therefore \lambda_1 = \frac{4l}{3}$$

The frequency f_1 of the note is thus given by

$$f_1 = \frac{c}{\lambda_1} = \frac{3c}{4l} \quad . \qquad . \qquad . \qquad . \qquad . \qquad (i)$$

But

$$f_0 = \frac{c}{4l}$$

$$\therefore f_1 = 3f_0 \quad . \qquad . \qquad . \qquad . \qquad . \qquad (ii)$$

As we previously explained, the stationary *pressure* wave in the air has pressure nodes at the displacement antinodes A in Figure 23.4 (i) and pressure antinodes at the displacement nodes N.

In Figure 23.4 (ii), when a note of frequency f_2 is obtained, the length l of the pipe is related to the wavelength λ_2 by

$$l = \frac{5\lambda_2}{4} \text{ or } \lambda_2 = \frac{4l}{5}$$

$$\therefore f_2 = \frac{c}{\lambda_2} = \frac{5c}{4l} \qquad . \qquad . \qquad . \qquad . \qquad . \qquad \text{(iii)}$$

$$\therefore f_2 = 5f_0 \qquad . \qquad . \qquad . \qquad . \qquad . \qquad . \qquad \text{(iv)}$$

By drawing other sketches of stationary waves, with the closed end as a node and the open end as an antinode, it can be shown that higher frequencies can be obtained which have frequencies of $7f_0$, $9f_0$, and so on. They are produced by blowing harder at the open end of the pipe. The frequencies obtainable at a closed pipe are hence f_0, $3f_0$, $5f_0$, and so on, i.e., the closed pipe gives only odd harmonics, and hence the frequencies $3f_0$, $5f_0$, etc. are possible *overtones*.

Open Pipe

An 'open' pipe is one which is open at both ends. When air is blown into it at P, a wave m travels to the open end Q, where it is reflected in the direction n on encountering the free air, Figure 23.5 (i). A stationary wave is therefore set up in

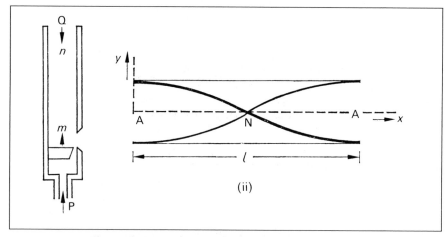

Figure 23.5   (i) *Open pipe*   (ii) *Fundamental of open pipe*

the air in the pipe, and as the two ends of the pipe are open, they must both be *antinodes*. The simplest type of wave is hence that shown in Figure 23.5 (ii), the x-axis of the wave being drawn along the middle of the pipe, which is horizontal. A node N is midway between the two antinodes.

The length l of the pipe is the distance between consecutive antinodes. But the distance between consecutive antinodes $= \lambda/2$, where λ is the wavelength (p. 593).

$$\therefore \frac{\lambda}{2} = l \text{ or } \lambda = 2l$$

Thus the frequency f_0 of the note obtained from the pipe is given by

$$f_0 = \frac{c}{\lambda} = \frac{c}{2l} \qquad . \qquad . \qquad . \qquad . \qquad . \qquad \text{(2)}$$

This is the frequency of the fundamental note of the pipe.

Overtones of Open Pipe

Notes of higher frequencies than f_0 can be obtained from the pipe by blowing harder. The stationary wave in the pipe has always an antinode A at each end, and Figure 23.6 (i) represents the case of a note of a frequency f_1.

The length l of the pipe is equal to the wavelength λ_1 of the wave in this case.

Thus
$$f_1 = \frac{c}{\lambda_1} = \frac{c}{l}$$

But
$$f_0 = \frac{c}{2l}, \text{ from (2) above}$$

$$\therefore f_1 = 2f_0. \quad . \quad . \quad . \quad . \quad . \quad \text{(i)}$$

In Figure 23.6 (ii), the length $l = 3\lambda_2/2$, where λ_2 is the wavelength in the pipe, so $\lambda_2 = 2l/3$. The frequency f_2 is thus given by

$$f_2 = \frac{c}{\lambda_2} = \frac{3c}{2l}$$

$$\therefore f_2 = 3f_0 \quad . \quad . \quad . \quad . \quad . \quad \text{(ii)}$$

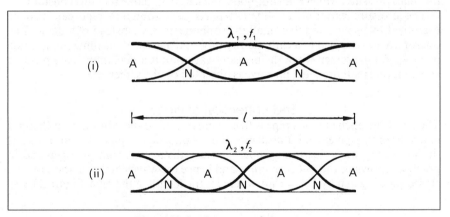

Figure 23.6 *Overtones of open pipes*

The frequencies of the overtones in the open pipe are thus $2f_0$, $3f_0$, $4f_0$, and so on, that is, all harmonics are obtainable. The frequencies of the overtones in the closed pipe are $3f_0$, $5f_0$, $7f_0$, and so on, and hence the *quality* of the same note obtained from a closed and an open pipe is different (see p. 585).

Detection of Nodes and Antinodes, and Pressure Variation, in Pipes

The *nodes and antinodes* in a sounding pipe have been detected by suspending inside it a very thin piece of paper with lycopodium or fine sand particles on it, Figure 23.7 (i). The particles can be heard vibrating on the paper at the antinodes, but they are still at the nodes.

The *pressure variation* in a sounding pipe has been examined by means of a sensitive flame, designed by Lord Rayleigh. The length of the flame can be made

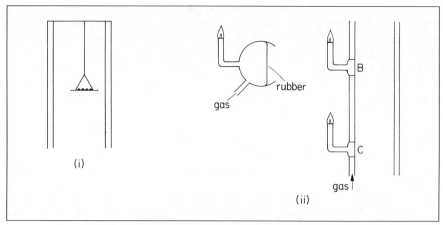

Figure 23.7 (*i*) *Detection of nodes and antinodes* (*ii*) *Detection of pressure*

sensitive to the pressure of the gas supplied, so that if the pressure changes the length of flame is considerably affected. Several of the flames can be arranged at different parts of the pipe, with a thin rubber or mica diaphragm in the pipe, such as at B, C, Figure 23.7 (ii). At a place of maximum pressure variation, which is a node (p. 480), the length of flame alters accordingly. At a place of constant (normal) pressure, which is an antinode, the length of flame remains constant.

The pressure variation at different parts of a sounding pipe can also be examined by using a suitable small microphone at B, C, instead of a flame. The microphone is coupled to a cathode-ray tube and a wave of maximum amplitude is shown on the screen when the pressure variation is a maximum. At a place of constant (normal) pressure, no wave is observed on the screen.

End-correction of Pipes

The air at the open end of a pipe is free to move, and so the vibrations at this end of a sounding pipe extend a little into the air outside the pipe. The antinode of the stationary wave due to any note is thus a distance e from the open end in practice, known as the *end-correction*, and hence the wavelength λ in the case of a closed pipe is given by $\lambda/4 = l + e$, where l is the length of the pipe, Figure 23.8 (i).

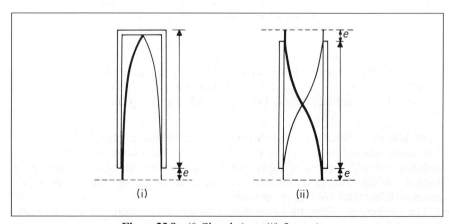

Figure 23.8 (*i*) *Closed pipe* (*ii*) *Open pipe*

In the case of an open pipe sounding its fundamental note, the wavelength λ is given by $\lambda/2 = l + e + e$, since *two* end-corrections are required, assuming the end-corrections are equal, Figure 23.8 (ii). Thus $\lambda = 2(l + 2e)$. See also p. 596.

The mathematical theory of the end-correction was developed independently by Helmholtz and Rayleigh. It is now generally accepted that $c = 0.58r$, or $0.6r$, where r is the radius of the pipe, so that the wider the pipe, the greater is the end-correction. It was also shown that the end-correction depends on the wavelength λ of the note, and tends to vanish for very short wavelengths.

Effect of Temperature, and End-correction, on Pitch of Pipes

The frequency, f_0, of the fundamental note of a closed pipe of length l and end-correction e is given by

$$f_0 = \frac{c}{\lambda} = \frac{c}{4(l+e)} \qquad \qquad \text{(i)}$$

with the usual notation, since $\lambda = 4(l+e)$. See above. Now the velocity of sound, c, in air at $\theta°C$ is related to its velocity c_0 at $0°C$ by

$$\frac{c}{c_0} = \sqrt{\frac{273+\theta}{273}} = \sqrt{1 + \frac{\theta}{273}} \qquad \qquad \text{(ii)}$$

since the velocity is proportional to the square root of T the kelvin temperature. Substituting for c in (i),

$$\therefore f_0 = \frac{c_0}{4(l+e)}\sqrt{1 + \frac{\theta}{273}} \qquad \qquad \text{(iii)}$$

From (iii), it follows that, with a given pipe, *the frequency of the fundamental increases as the temperature increases.* Also, for a given temperature and length of pipe, the frequency decreases as e increases. Now $e = 0.6r$ where r is the radius of the pipe. Thus *the frequency of the note from a pipe of given length is lower the wider the pipe*, the temperature being constant. The same results hold for an open pipe.

Resonance

If a diving springboard is bent and then allowed to vibrate freely, it oscillates with a frequency which is called its *natural frequency*. When a diver on the edge of the board begins to jump up and down repeatedly, the board is forced to vibrate at the frequency of the jumps; and at first, when the amplitude is small, the board is said to be undergoing *forced vibrations*. As the diver jumps up and down to gain increasing height for his dive, the frequency of the periodic downward force reaches a stage where it is practically the same as the natural frequency of the board. The amplitude of the board then becomes very large, and the periodic force is said to have set the board in *resonance* (see also p. 470).

A mechanical system which is free to move, like a wooden bridge or the air in pipes, has a natural frequency of vibration, f_0, which depends on its dimensions. When a periodic force of a frequency different from f_0 is applied to the system, the latter vibrates with a small amplitude and undergoes forced vibrations. When the periodic force has a frequency equal to the natural frequency f_0 of the system, the amplitude of vibration becomes a maximum, and the system is then set into resonance. Figure 23.9 is a typical curve showing the variation of amplitude with frequency. Some time ago it was reported in the newspapers that

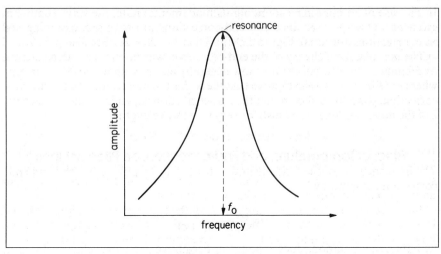

Figure 23.9 *Resonance curve*

a soprano who was broadcasting had broken a glass tumbler on the table of a listener when she had reached a high note. This is an example of resonance. The glass had a natural frequency equal to that of the note sung, and was thus set into a large amplitude of vibration sufficient to break it.

The phenomenon of resonance occurs in branches of Physics other than Sound and Mechanics. When an electrical circuit containing a coil and capacitor is 'tuned' to receive the radio waves from a distant transmitter, the frequency of the radio waves is equal to the natural frequency of the circuit and resonance is therefore obtained. A large current then flows in the electrical circuit (p. 396).

Sharpness of Resonance

As the resonance condition is approached, the effect of the frictional or *damping* forces on the amplitude increases. Damping prevents the amplitude from becoming excessively large at resonance. The lighter the damping, the sharper is the resonance, that is, the amplitude diminishes considerably at a frequency slightly different from the resonant frequency, Figure 23.10. A heavily-damped system has a fairly flat resonance curve. Tuning is therefore more difficult in a system which has light damping.

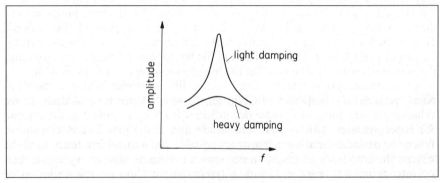

Figure 23.10 *Sharpness of resonance*

The effect of damping can be illustrated by attaching a simple pendulum carrying a very light bob, and one of the same length carrying a lead bob of equal size, to a horizontal string. The pendula are set into vibration by a third pendulum of equal length attached to the same string. It is then seen that the amplitude of the lead bob is much greater than that of the light bob. The damping of the light bob due to air resistance is much greater than for the lead bob.

See also page 468.

Resonance in a Tube or Pipe

If a person blows gently down a pipe closed at one end, the air inside vibrates freely, and a note is obtained from the pipe which is its fundamental (p. 594). A stationary wave then exists in the pipe, with a node N at the closed end and an antinode A at the open end, as explained previously.

If the prongs of a tuning-fork are held over the top of the pipe, the air inside it is set into vibration by the periodic force exerted on it by the prongs. In general, however, the vibrations are feeble, as they are *forced* vibrations, and the intensity of the sound heard is correspondingly small. But when a tuning-fork of the same frequency as the fundamental frequency of the pipe is held over the pipe, the air inside is set in *resonance* by periodic force, and the amplitude of the vibrations is large. A loud note, which has the same frequency as the fork, is then heard coming

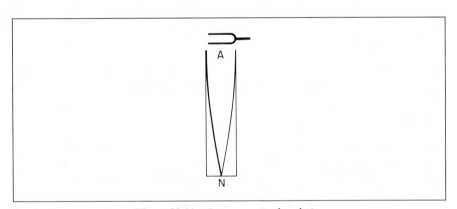

Figure 23.11 *Resonance in closed pipe*

from the pipe, and a stationary wave is set up with the top of the pipe acting as an antinode and the fixed end as a node, Figure 23.11. If a sounding tuning-fork is held over a pipe open at both ends, resonance occurs when the stationary wave in the pipe has antinodes at the two open ends, as shown by Figure 23.5; the frequency of the fork is then equal to the frequency of the fundamental of the open pipe. A similar case to the closed pipe, but using electrical oscillations, was discussed on p. 481.

Resonance Tube Experiment, Measurement of Velocity of Sound and 'End-Correction' of Tube

If a sounding tuning-fork is held over the open end of a tube T filled with water, resonance is obtained at some position as the level of water is gradually lowered, Figure 23.12 (i). The stationary wave set up is then as shown. If e is the end-correction of the tube (p. 598), and l is the length from the water level to the top of the tube, then

$$l + e = \frac{\lambda}{4} \quad . \qquad . \qquad . \qquad . \qquad . \qquad . \qquad \text{(i)}$$

But
$$\lambda = \frac{c}{f}$$

where f is the frequency of the fork and c is the velocity of sound in air.

$$\therefore l + e = \frac{c}{4f} \qquad . \qquad . \qquad . \qquad . \qquad . \qquad \text{(ii)}$$

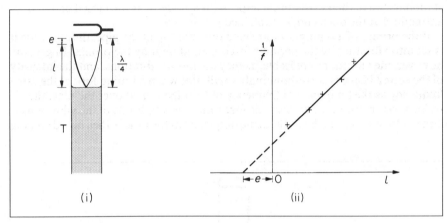

(i)   (ii)

Figure 23.12   *Resonance tube experiment*

If different tuning-forks of known frequency f are taken, and the corresponding values of l obtained when resonance occurs, it follows from equation (ii) that a graph of $1/f$ against l is a straight line, Figure 23.12 (ii). Now from equation (ii), the gradient of the line is $4/c$; thus c can be determined. Also, the negative intercept of the line on the axis of l is e, from equation (ii); hence the end-correction can be found.

If only one fork is available, and the tube is sufficiently long, another method for c and e can be adopted. In this case the level of the water is lowered further

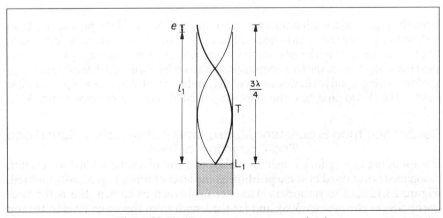

Figure 23.13   *Resonance at new water level*

from the position in Figure 23.12 (i), until resonance is again obtained at a level L_1, Figure 23.13. Since the stationary wave set up is that shown and the new length to the top from L_1 is l_1, it follows that

$$l_1 + e = \frac{3\lambda}{4} \qquad . \qquad . \qquad . \qquad . \qquad . \qquad \text{(iii)}$$

But
$$l + e = \frac{\lambda}{4}, \text{ from (ii)}$$

Subtracting,
$$l_1 - l = \frac{\lambda}{2}$$

$$\therefore \lambda = 2(l_1 - l)$$

$$\therefore c = f\lambda = 2f(l_1 - l) \qquad . \qquad . \qquad . \qquad . \qquad \text{(3)}$$

In this method for c, therefore, the end-correction e is eliminated. The magnitude of e can be found from equations (ii) and (iii). Thus, from (ii),

$$3l + 3e = \frac{3\lambda}{4}$$

But, from (iii),
$$l_1 + e = \frac{3\lambda}{4}$$

$$\therefore 3l + 3e = l_1 + e$$

$$\therefore 2e = l_1 - 3l$$

$$\therefore e = \frac{l_1 - 3l}{2} \qquad . \qquad . \qquad . \qquad . \qquad \text{(4)}$$

Hence e can be found from measurements of l_1 and l.

Velocity of Sound in Air by Dust Tube Method

Figure 23.14 illustrates another method for measuring the velocity of sound in air by means of stationary waves.

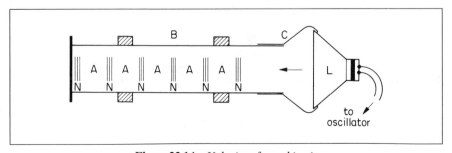

Figure 23.14   *Velocity of sound in air*

A measuring cylinder B is placed on its side and is arranged to lie horizontally on supports such as plasticene. The inside of the cylinder is coated lightly with lycopodium powder or cork dust along its length. A paper cone C, attached to a loudspeaker L, is fitted over the open end of B. By connecting a suitable oscillator to L, sound waves are produced which travel to the closed end of B and are reflected, so that stationary waves are formed in the air.

The frequency of the oscillator is varied. At a frequency of the order of a kilohertz or more depending on the length of the measuring cylinder, the dust suddenly settles into regularly spaced heaps at positions along the cylinder. These are nodes, N, of the stationary wave (zero displacement positions). Midway between the nodes are antinodes, A, where the air has maximum amplitude of vibration and so little dust settles there.

Measurement of the average distance NN between successive nodes $= \lambda/2$, where λ is the wavelength. So λ can be found. The velocity in air $c = f\lambda$, where f is known, and hence c can be calculated. This is an approximate method for c as

(a) the sound waves are damped by the sides of the tube and so this is not the velocity in free air (see p. 488),

(b) the distance NN cannot be measured to a high degree of accuracy. In the method outlined here, the dust and the tube must both be dry.

Examples on Waves in Pipes

1   A cylindrical pipe of length 28 cm closed at one end is found to be at resonance when a tuning fork of frequency 864 Hz is sounded near the open end. Determine the mode of vibration of the air in the pipe, and deduce the value of the end-correction. [Take the velocity of sound in air as $340 \, \mathrm{m \, s^{-1}}$.]

Let $\lambda =$ the wavelength of the sound in the pipe.

Then
$$\lambda = \frac{c}{f} = \frac{34\,000}{864} = 39 \cdot 35 \, \mathrm{cm}$$

If the pipe is resonating to its fundamental frequency f_0, the stationary wave in the pipe is that shown in Figure 23.12 (i) and the wavelength λ_0, is given by $\lambda_0/4 = 28 \, \mathrm{cm}$. Thus $\lambda_0 = 112 \, \mathrm{cm}$. Since $\lambda = 39 \cdot 35 \, \mathrm{cm}$, the pipe cannot be sounding its resonant frequency. The first overtone of the pipe is $3 f_0$, which corresponds to a wavelength λ_1 given by $3\lambda/4 = 28$ (see Figure 23.4).

$$\therefore \lambda_1 = \frac{112}{3} = 37\tfrac{1}{3} \, \mathrm{cm}$$

Consequently, allowing for the effect of an end correction, the pipe is sounding its *first overtone*.

Let $e =$ the end-correction in cm

Then
$$28 + e = \frac{3\lambda_1}{4}$$

But, accurately,
$$\lambda_1 = \frac{c}{f} = \frac{34\,000}{864} = 39 \cdot 35$$

$$\therefore 28 + e = \tfrac{3}{4} \times 39 \cdot 35$$

$$\therefore e = 1 \cdot 5 \, \mathrm{cm}$$

2   Explain, with diagrams, the possible states of vibration of a column of air in
(a) an open pipe,
(b) a closed pipe.

An open pipe 30 cm long and a closed pipe 23 cm long, both of the same diameter, are each sounding its first overtone, and these are in unison. What is the end-correction of these pipes? (L.)

Suppose c is the velocity of sound in air, and f is the frequency of the note. The wavelength, λ, is thus c/f.

When the open pipe is sounding its first overtone, the length of the pipe plus end-corrections $= \lambda$.

$$\therefore \frac{c}{f} = 30 + 2e \qquad . \qquad . \qquad . \qquad . \qquad . \qquad \text{(i)}$$

since there are two end-corrections.

When the closed pipe is sounding its first overtone,

$$\frac{3\lambda}{4} = 23 + e$$

$$\therefore \frac{3c}{4f} = 23 + e \qquad . \qquad . \qquad . \qquad . \qquad . \qquad \text{(ii)}$$

From (i) and (ii), it follows that

$$23 + e = \tfrac{3}{4}(30 + 2e)$$

$$\therefore 92 + 4e = 90 + 6e$$

$$\therefore e = 1 \text{ cm}$$

_____ Exercises 23A _____

Waves in Pipes

1 Write down in terms of wavelength, λ, the distance between   (i) consecutive nodes,   (ii) a node and an adjacent antinode,   (iii) consecutive antinodes. Find the frequency of the fundamental of a closed pipe 15 cm long if the velocity of sound in air is 340 m s^{-1}.

2 Discuss what is meant by the statement that *sound is a wave motion*. Use the example of the passage of a sound wave through air to explain the terms wavelength (λ), frequency (f), and velocity (v) of a wave. Show that $v = f\lambda$.

Explain the increase in loudness (or 'resonance') which occurs when a sounding tuning-fork is held near the open end of an organ pipe when the length of the pipe has certain values, the other end of the pipe being closed. Find the shortest length of such a pipe which resonates with a 440 Hz tuning-fork, neglecting end corrections. (Velocity of sound in air $= 350 \text{ m s}^{-1}$). (*O. & C.*)

3 Explain the conditions necessary for the creation of stationary waves in air.
    Describe how
    (a) the displacement,
    (b) the pressure vary at different points along a stationary wave in air and describe how these effects might be demonstrated experimentally.
    A tube is closed at one end and closed at the other by a vibrating diaphragm which may be assumed to be a displacement node. It is found that when the frequency of the diaphragm is 2000 Hz a stationary wave pattern is set up in the tube and the distance between adjacent nodes is then 8·0 cm. When the frequency is gradually reduced the stationary wave pattern disappears but another stationary wave pattern reappears at a frequency of 1600 Hz. Calculate   (i) the speed of sound in air,   (ii) the distance between adjacent nodes at a frequency of 1600 Hz,   (iii) the length of the tube between the diaphragm and the closed end,   (iv) the next lower frequency at which a stationary wave pattern will be obtained. (*JMB.*)

4 What are the chief characteristics of a progressive wave motion? Give your reasons for believing that sound is propagated through the atmosphere as a longitudinal

wave motion, and find an expression relating the velocity, the frequency, and the wavelength.

Neglecting end effects, find the lengths of

(a) a closed organ pipe, and

(b) an open organ pipe, each of which emits a fundamental note of frequency 256 Hz. (Take the speed of sound in air to be 330 m s^{-1}.) (*O.*)

5 (a) Explain in terms of the properties of a gas, but without attempting mathematical treatment, how the vibration of a sound source, such as a loudspeaker diaphragm, can be transmitted through the air around it.

Explain, also, the reflection which occurs when the vibration reaches a fixed barrier, such as a wall.

(b) Plane, simple harmonic, progressive sound waves of wavelength 1·2 m and speed 348 m s^{-1}, are incident normally on a plane surface which is a perfect reflector of sound. What statements can be made about the amplitude of vibration and about air pressure changes at points distant (i) 30 cm, (ii) 60 cm, (iii) 90 cm, (iv) 10 cm from the reflector? Justify your answers. (*O. & C.*)

6 Describe the motion of the air in a tube closed at one end and vibrating in its fundamental mode. An observer

(a) holds a vibrating tuning-fork over the open end of a tube which resounds to it,

(b) blows lightly across the mouth of the tube. Describe and explain the difference in the quality of the notes that he hears.

A uniform tube, 60·0 cm long, stands vertically with its lower end dipping into water. When the length above water is 14·8 cm, and again when it is 48·0 cm, the tube resounds to a vibrating tuning-fork of frequency 512 Hz. Find the lowest frequency to which the tube will resound when it is open at both ends. (*L.*)

7 Discuss the factors which determine the pitch of the note given by a 'closed' pipe. Explain why the fundamental frequency and the quality of the note from a 'closed' pipe differ from those of the note given under similar conditions by a pipe of the same length which is open at both ends. (*JMB.*)

8 What do you understand by

(a) forced vibrations,

(b) free vibrations, and

(c) resonance?

Illustrate your answer by giving three distinct examples, one for each of (a), (b) and (c).

Explain how a stationary sound wave may be set up in a gas column and how you would demonstrate the presence of nodes and antinodes. State what measurements would be required in order to deduce the speed of sound in air from your demonstration, and show how you would calculate your result.

The speed of sound, c, in an ideal gas is given by the formula $c = \sqrt{\gamma p/\rho}$, where p is the pressure, ρ is the density of the gas and γ is a constant. By considering this formula explain the effect of a change in (i) temperature, and (ii) pressure, on speed of sound. (*L.*)

9 Distinguish between the formation of an echo and the formation of a stationary sound wave by reflection, explaining the general circumstances in which each is produced.

Describe an experiment in which the velocity of sound in air may be determined by observations on stationary waves.

An organ pipe is sounded with a tuning-fork of frequency 256 Hz. When the air in the pipe is at a temperature of 15°C, 23 beats are heard in 10 seconds; when the tuning-fork is loaded with a small piece of wax, the beat frequency is found to decrease. What change of temperature of the air in the pipe is necessary to bring the pipe and the unloaded fork into unison? (*C.*)

10 What is meant by

(a) *a stationary wave motion* and

(b) *a node*?

Describe how the phenomenon of resonance may be demonstrated using a loudspeaker, a source of alternating voltage of variable frequency and a suitable tube open at one end and closed at the other. Explain how resonance occurs in the arrangement you describe, draw a diagram showing the position of the nodes in the tube in a typical case of resonance and state clearly the meaning of the diagram. How would you demonstrate the position of the nodes experimentally? (*O. & C.*)

11 Describe and give the theory of one experiment in each instance by which the velocity of sound may be determined,

(a) in free air,

(b) in the air in a resonance tube.

What effect, if any, do the following factors have on the velocity of sound in free air; frequency of the vibrations; temperature of the air; atmospheric pressure; humidity?

State the relationship between this velocity and temperature. (*L.*)

Waves in Strings

If a horizontal rope is fixed at one end, and the other end is moved up and down, a wave travels along the rope. The particles of the rope are then vibrating vertically, and since the wave travels horizontally, this is an example of a *transverse* wave (see p. 471). The waves propagated along the surface of the water when a stone is dropped into it are also transverse waves, as the particles of the water are moving up and down while the wave travels horizontally. A transverse wave is also obtained when a stretched string, such as a violin string, is plucked. Before we can study waves in strings, we require to know the velocity of transverse waves travelling along them.

Velocity of Transverse Waves Along a Stretched String

Suppose that a transverse wave is travelling along a thin string of length l and mass m under a constant tension T. If we assume that the string has no 'stiffness', that is, the string is perfectly flexible, the velocity c of the transverse wave along it depends only on the values of T, m, l. The velocity is given by

$$c = \sqrt{\frac{T}{m/l}}$$

$$\text{or } c = \sqrt{\frac{T}{\mu}} \qquad . \qquad . \qquad . \qquad . \qquad . \qquad (1)$$

where μ is the 'mass per unit length' of the string.

When T is in *newton* and m in *kilogram per metre*, then c is in *metre per second*.

The formula for c may be partly deduced by the method of dimensions, in which all the quantities concerned are reduced to the fundamental units of mass, M, length, L, and time, T. Suppose that

$$c = kT^x m^y l^z \qquad . \qquad . \qquad . \qquad . \qquad (i)$$

where k, x, y, z, are numbers. The dimensions of velocity c are LT^{-1}, the dimensions of tension T, a force, are MLT^{-2}, the dimension of m is M, and the dimension of l is L. As the dimensions on both sides of (i) must be equal, it follows that

$$LT^{-1} = (MLT^{-2})^x (M^y)(L^z)$$

Equating the indices of M, L, T on both sides, we have

for M, $\qquad\qquad\qquad x + y = 0$

for L, $\qquad\qquad\qquad x + z = 1$

for T, $\qquad\qquad\qquad 2x = 1$

$$\therefore x = \tfrac{1}{2}, z = \tfrac{1}{2}, y = -\tfrac{1}{2}$$

Thus, from (i)

$$c = kT^{\frac{1}{2}} m^{-\frac{1}{2}} l^{\frac{1}{2}}$$

$$\therefore c = k\sqrt{\frac{Tl}{m}} = k\sqrt{\frac{T}{m/l}}$$

A rigid mathematical treatment shows that the constant $k = 1$, so $c = \sqrt{\dfrac{T}{m/l}}$.

Since m/l is the 'mass per unit length' of the string, then

$$c = \sqrt{\frac{T}{\mu}}$$

where μ is the mass per unit length.

Modes of Vibration of Stretched String

If a wire is stretched between two points N, N and is plucked in the middle, a transverse wave travels along the wire and is reflected at the fixed end. A *stationary wave* is thus set up in the wire, and the simplest mode of vibration is one in which the fixed ends of the wire are nodes, N, and the middle is an antinode, A, Figure 23.15. Since the distance between consecutive nodes is

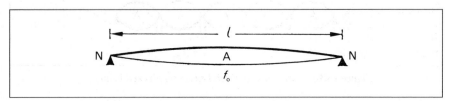

Figure 23.15 *Fundamental of stretched string*

$\lambda/2$, where λ is the wavelength of the transverse wave in the wire, it follows that

$$l = \frac{\lambda}{2}$$

where l is the length of the wire. Thus $\lambda = 2l$. The frequency f of the vibration is hence given by

$$f = \frac{c}{\lambda} = \frac{c}{2l}$$

where c is the velocity of the transverse wave. But $c = \sqrt{T/\mu}$, from previous.

$$\therefore f = \frac{1}{2l}\sqrt{\frac{T}{\mu}}$$

This is the frequency of the *fundamental* note obtained from the string; and if we denote the frequency by the usual symbol f_0, we have

$$f_0 = \frac{1}{2l}\sqrt{\frac{T}{\mu}} \qquad . \qquad . \qquad . \qquad . \qquad (2)$$

Overtones of Stretched String

The first overtone f_1 of a string plucked in the middle corresponds to a stationary wave shown in Figure 23.16, which has nodes at the fixed ends and an antinode in the middle. If λ_1 is the wavelength, it can be seen that

$$l = \frac{3}{2}\lambda_1$$

$$\text{or } \lambda_1 = \frac{2l}{3}$$

The frequency f_1 is thus given by

$$f_1 = \frac{c}{\lambda_1} = \frac{3c}{2l} = \frac{3}{2l}\sqrt{\frac{T}{\mu}} \qquad \cdot \qquad \cdot \qquad \cdot \qquad \cdot \qquad \cdot \qquad \text{(i)}$$

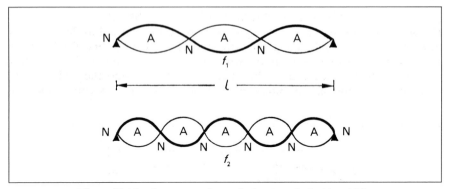

Figure 23.16 *Overtones of stretched string plucked in middle*

But the fundamental frequency, $f_0, = \dfrac{1}{2l}\sqrt{\dfrac{T}{\mu}}$, from equation (2).

$$\therefore f_1 = 3f_0$$

The second overtone f_2 of the string when plucked in the middle corresponds to a stationary wave shown in Figure 23.16. In this case $l = \dfrac{5}{2}\lambda_2$, where λ_2 is the wavelength.

$$\therefore \lambda_2 = \frac{2l}{5}$$

$$\therefore f_2 = \frac{c}{\lambda_2} = \frac{5c}{2l}$$

where f_2 is the frequency. But $c = \sqrt{T/\mu}$

$$\therefore f_2 = \frac{5}{2l}\sqrt{\frac{T}{\mu}} = 5f_0$$

The overtones are thus $3f_0$, $5f_0$, and so on.

Other notes than those considered above can be obtained by touching or 'stopping' the string lightly at its midpoint, for example, so that the latter becomes a node in addition to those at the fixed ends. If the string is plucked one-quarter of the way along it from a fixed end, the simplest stationary wave set up is that illustrated in Figure 23.17 (i) (see also page 591). Thus the wavelength $\lambda = l$, and hence the frequency f is given by

$$f = \frac{c}{\lambda} = \frac{c}{l} = \frac{1}{l}\sqrt{\frac{T}{\mu}}$$

$$\therefore f = 2f_0, \text{ since } f_0 = \frac{1}{2l}\sqrt{\frac{T}{\mu}}$$

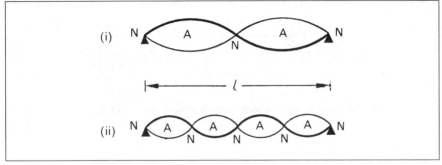

Figure 23.17 *Even harmonics in stretched string*

If the string is plucked one-eighth of the way from a fixed end, a stationary wave similar to that in Figure 23.17 (ii) may be set up. The wavelength, $\lambda' = l/2$, and hence the frequency

$$f' = \frac{c}{\lambda'} = \frac{2c}{l}$$

$$\therefore f' = \frac{2}{l}\sqrt{\frac{T}{\mu}} = 4f_0$$

Verification of the Laws of Vibration of a Fixed String, The Sonometer

As we have already shown (p. 609), the frequency of the fundamental of a stretched string is given by

$$f = \frac{1}{2l}\sqrt{\frac{T}{\mu}}$$

writing f for f_0. It thus, follows that:

---

1   $f \propto \dfrac{1}{l}$ *for a given tension (T) and string (μ constant).*

2   $f \propto \sqrt{T}$ *for a given length (l) and string (μ constant).*

3   $f \propto \dfrac{1}{\sqrt{\mu}}$ *for a given length (l) and tension (T).*

These are known as the 'laws of vibration of a fixed string'. The *sonometer* was designed to verify them.

The sonometer consists of a hollow wooden box Q, with a thin horizontal wire attached to A at one end, Figure 23.18. The wire passes over a grooved wheel H, and is kept taut by a mass M hanging down at the other end. Wooden bridges, B, C, can be placed beneath the wire so that a definite length of wire is obtained, and the length of wire can be varied by moving one of the bridges. The length of wire between B, C can be read from a fixed horizontal scale D, graduated in millimetres, on the box below the wire.

(1) *To verify $f \propto 1/l$ for a given tension (T) and mass per unit length (μ)*, the mass M is kept constant so that the tension, *T*, in the wire AH is constant. The length,

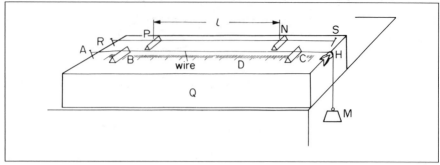

Figure 23.18   *Sonometer verification of f $\propto$ 1/l and $\sqrt{T}$*

l, of the wire between B, C is varied by moving C until the note obtained by plucking BC in the middle is the same as that produced by a sounding tuning-fork of known frequency *f*. If the observer lacks a musical ear, the 'tuning' can be recognised by listening for beats when the wire and the tuning-fork are both sounding, as in this case the frequencies of the two notes are nearly equal (p. 574). Alternatively, a small piece of paper in the form of an inverted *V* can be placed on the middle of the wire, and the end of the sounding tuning-fork then placed on the sonometer box. The vibrations of the fork are transmitted through the box to the wire, which vibrates in resonance with the fork if its length is 'tuned' to the note. The paper will then vibrate considerably and may be thrown off the wire.

Different tuning-forks of known frequency *f* are used, and the lengths, *l* of the wire are observed when they are tuned to the corresponding note. A graph of *f* against 1/*l* is then plotted, and is found to be a straight line within the limits of experimental error. Thus *f* $\propto$ 1/*l* for a given tension and mass per unit length of wire.

(2) *To verify f $\propto$ $\sqrt{T}$ for a given length and mass per unit length*, the length BC between the bridges is kept fixed, so that the length of wire is constant, and the mass M is varied to alter the tension. To obtain a measure of the frequency *f* of the note produced when the wire between B, C is plucked in the middle, a second wire, fixed to R, S on the sonometer, is utilised. This usually has a weight (not shown) attached to one end to keep the tension constant, Figure 23.18. The wire RS has bridges P, N beneath it, and N is moved until the note from the wire between P, N is the same as the note from the wire between B, C. Now the tension in PN is constant as the wire is fixed to R and S. Thus, since frequency, *f* $\propto$ 1/*l* for a given tension and wire, the frequency of the note from BC is proportional to 1/*l*, where *l* is the length of PN.

By varying the mass *M*, the tension *T* in BC is varied. A graph of 1/*l* against $\sqrt{T}$ is found to be a straight line passing through the origin. So *f* $\propto$ $\sqrt{T}$ for a given length of wire.

(3) *To verify f $\propto$ 1/$\sqrt{\mu}$ for a given length and tension*, wires of different material are connected to B, C, and the same mass *M* and the same length BC are taken. The frequency, *f*, of the note obtained from BC is again found by using the second wire RS in the way already described. The mass per unit length, μ, is the mass per metre length of wire, and is given by $\pi r^2 \rho$ kg m^{-1}, where *r* is the radius of the wire in m and ρ is its density in kg m^{-3}, as $(\pi r^2 \times 1)$ m^3 is the volume of 1 m of the wire.

When 1/*l* is plotted against 1/$\sqrt{\mu}$, the graph is found to be a straight line passing through the origin. So *f* $\propto$ 1/$\sqrt{\mu}$ for a given length and tension.

Measurement of Frequency of A.C. Mains

The frequency of the alternating current (a.c.) mains can be measured using a sonometer wire.

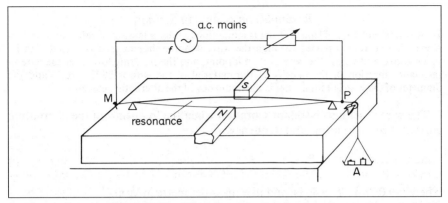

Figure 23.19 *Mains frequency by vibrating wire*

The alternating current is passed into the wire MP, and the poles N, S of a powerful magnet are placed on either side of the wire so that the magnetic field due to it is perpendicular to the wire, Figure 23.19. As a result of the magnetic effect of the current, a *force* acts on the wire which is perpendicular to the directions of both the magnetic field and the current, and so the wire is subjected to a transverse force. If the current is an alternating one of 50 Hz, the magnitude of the force varies at the rate of 50 Hz. By adjusting the tension in the sonometer wire by varying weights in a scale-pan A, a position can be reached when the wire is seen to be vibrating through a large amplitude; in this case the wire is *resonating* to the applied force, Figure 23.19.

The length l of wire between the bridges is now measured, and the tension T and the mass per unit length, μ, are also found. The frequency f of the alternating current is then calculated from

$$f = \frac{1}{2l}\sqrt{\frac{T}{\mu}}$$

Velocity of Longitudinal Waves in Wires

If a sonometer wire is stroked along its length by a rosined cloth, a high-pitched note is obtained. This note is due to *longitudinal* vibrations in the wire, and must be clearly distinguished from the note produced when the wire is plucked, which sets up *transverse* vibrations of the wire and a corresponding transverse wave. As we saw on p. 493, the velocity c of a longitudinal wave in a medium is

$$c = \sqrt{\frac{E}{\rho}}$$

where E is Young modulus for the wire and ρ is its density. The wavelength, λ, of the longitudinal wave is $2l$, where l is the length of the wire, since a stationary longitudinal wave is set up. Thus the frequency f of the note is given by

$$f = \frac{c}{\lambda} = \frac{1}{2l}\sqrt{\frac{E}{\rho}}$$

The frequency of the note may be obtained approximately with the aid of an audio oscillator, and so the velocity of sound in the wire, or its Young modulus, can be roughly calculated.

Examples on Waves in Strings

1 A sonometer wire of length 76 cm is maintained under a tension of value 40 N and an alternating current is passed through the wire. A horse-shoe magnet is placed with its poles above and below the wire at its midpoint, and the resulting forces set the wire in resonant vibration. If the density of the material of the wire is 8800 kg m^{-3} and the diameter of the wire is 1 mm, what is the frequency of the alternating current?

The wire is set into resonant vibration when the frequency of the alternating current is equal to its natural frequency, f.

Now
$$f = \frac{1}{2l}\sqrt{\frac{T}{\mu}} \qquad \qquad \qquad \text{(i)}$$

where $l = 0.76$ m, $T = 40$ N, and μ = mass per metre in kg m^{-1}

Also, mass of 1 metre = volume × density

$$= \pi r^2 \times 1 \times 8800 \text{ kg}$$

where radius r of wire $= \frac{1}{2}$ mm $= 0.5 \times 10^{-3}$ m

From (i), $\therefore f = \dfrac{1}{2 \times 0.76}\sqrt{\dfrac{40}{\pi \times 0.5^2 \times 10^{-6} \times 1 \times 8800}}$

$$= 50 \text{ Hz}$$

2 A piano string has a length of 2·0 m and a density of 8000 kg m^{-3}. When the tension in the string produces a strain of 1%, the fundamental note obtained from the string in transverse vibration is 170 Hz. Calculate the Young modulus value for the material of the string.

If E is the Young modulus, A is the cross-section area of the string, l is the length of the string and e is the extension due to a force (tension) T, then, from page 137,

$$T = EA\frac{e}{l} = EA \times \frac{1}{100}$$

since the strain $e/l = 1\% = 1/100$. So

$$\text{frequency, } f = \frac{1}{2l}\sqrt{\frac{T}{\mu}} = \frac{1}{2l}\sqrt{\frac{EA}{100\,A\rho}}$$

since μ = mass per unit length $= A \times 1 \times \rho = A\rho$. So cancelling A,

$$f = \frac{1}{2l}\sqrt{\frac{E}{100\rho}}$$

Squaring, $\qquad E = 4f^2l^2 \times 100\rho$

$$= 4 \times 170^2 \times 2^2 \times 100 \times 8000$$

$$= 3.7 \times 10^{11} \text{ N m}^{-2} \text{ or Pa}$$

_____ **Exercises 23B** _____

Waves in Strings

1   Describe the differences between stationary waves and progressive waves. Outline an experimental arrangement to illustrate the formation of a stationary wave in a string.
     Waves of wavelength λ, from a source S, reach a common point P by two different routes. At P the waves are found to have a phase difference $3\pi/4$ rad. Show graphically what this means. What is the minimum path difference between the two routes?
     A string fixed at both ends is vibrating in the lowest mode of vibration for which a point a quarter of its length from one end is a point of maximum vibration. The note emitted has a frequency of 100 Hz. What will be the frequency emitted when it vibrates in the next mode such that this point is again a point of maximum vibration? (*L*.)

2   Explain what is meant by the *wavelength*, the *frequency*, and the *speed* of a sinusoidal travelling wave and derive a relation between them.
     What is meant by a stationary wave? A stationary sinusoidal wave of period T is set up on a stretched string so that there are nodes only at the two ends of the string and at its midpoint. The displacement of each point of the string has its maximum value at $t = 0$. Show on a single sketch the shape taken by the string at times $t = 0$, $T/8$, $T/4$, $3T/8$ and $T/2$.
     A piano string 1·5 m long is made of steel of density $7·7 \times 10^3$ kg m^{-3} and Young's modulus 2×10^{11} N m^{-2}. It is maintained at a tension which produces an elastic strain of 1% in the string. What is the fundamental frequency of transverse vibration of the string? (*O. & C.*)

3   Describe the motion of the particles of a string under constant tension and fixed at both ends when the string executes transverse vibrations of
     (a) its fundamental frequency,
     (b) the first overtone (second harmonic). Illustrate your answer with suitable diagrams.
     A horizontal sonometer wire of fixed length 0·50 m and mass $4·5 \times 10^{-3}$ kg is under a fixed tension of $1·2 \times 10^2$ N. The poles of a horse-shoe magnet are arranged to produce a horizontal transverse magnetic field at the midpoint of the wire, and an alternating sinusoidal current passes through the wire. State and explain what happens when the frequency of the current is progressively increased from 100 to 200 Hz. Support your explanation by performing a suitable calculation. Indicate how you would use such an apparatus to measure the fixed frequency of an alternating current. (*JMB.*)

4   Explain the meaning of the following terms in relation to wave motion: displacement, amplitude, wavelength, frequency, phase.
     What is the nature of a wave motion in which
     (a) the amplitude is the same at all points, but the phase varies with position,
     (b) the phase is the same at all points, but the amplitude varies with position?
     The velocity of transverse waves along a string depends only on the tension F, the radius r and the density ρ of the material. Use the method of dimensions to determine the form of the dependence, and describe briefly how you would attempt to verify your result experimentally. (*O. & C.*)

5   What is meant by
     (a) a forced vibration,
     (b) resonance? Give an example of each from   (i) mechanics,   (ii) sound.
     Using the same axes sketch graphs showing how the amplitude of a forced vibration depends upon the frequency of the applied force when the damping of the system is
     (a) light,
     (b) heavy.
     Point out any special features of the graphs.
     A sonometer wire is stretched by hanging a metal cylinder of density 8000 kg m^{-3}

at the end of the wire. A fundamental note of frequency 256 Hz is sounded when the wire is plucked.

Calculate the frequency of vibration of the same length of wire when a vessel of water is placed so that the cylinder is totally immersed. (*JMB.*)

6   Distinguish between a *progressive* wave and a *stationary* wave. Explain in detail how you would use a sonometer to establish the relation between the fundamental frequency of a stretched wire and
(a) its length,
(b) its tension. You may assume a set of standard tuning-forks and set of weights in steps of half a kilogram to be available.

A pianoforte wire having a diameter of 0·90 mm is replaced by another wire of the same material but with diameter 0·93 mm. If the tension of the wire is the same as before, what is the percentage change in the frequency of the fundamental note? What percentage change in the tension would be necessary to restore the original frequency? (*L.*)

7   Describe experiments to illustrate the differences between
(a) *transverse* waves,
(b) *longitudinal* waves,
(c) *progressive* waves and
(d) *stationary* waves? To which classes belong   (i) the vibrations of a violin string, (ii) the sound waves emitted by the violin into the surrounding air?

A wire whose mass per unit length is 10^{-3} kg m^{-1} is stretched by a load of 4 kg over the two bridges of a sonometer 1 m apart. If it is struck at its middle point, what will be
(a) the wavelength of its subsequent fundamental vibrations,
(b) the fundamental frequency of the note emitted? If the wire were struck at a point near one bridge what further frequencies might be heard? (Do not derive standard formulae.) (Assume $g = 10$ m s^{-2}.) (*O. & C.*)

8   A uniform wire vibrates transversely in its fundamental mode. On what factors, other than the length does the frequency of vibration depend, and what is the form of the dependence for each factor?

Describe the experiment you would perform to verify the form of dependence for *one* factor.

A wire of diameter 0·040 cm and made of steel of density 8000 kg m^{-3} is under constant tension of 80 N. A fixed length of 50 cm is set in transverse vibration. How would you cause the vibration of frequency about 840 Hz to predominate in intensity? (*JMB.*)

9   Give an expression for the velocity of a transverse wave along a thin flexible string and show that it is dimensionally correct. Explain how reflection may give rise to transverse *standing waves* on a stretched string and use the expression for the velocity to drive the frequency of the fundamental mode of vibration.

A steel wire of length 40·0 cm and diameter 0·0250 cm vibrates transversely in unison with a tube, open at each end and of effective length 60·0 cm, when each is sounding its fundamental note. The air temperature is 27°C. Find the tension in the wire. (Assume that the velocity of sound in air at 0°C is 331 m s^{-1} and the density of steel is 7800 kg m^{-3}.) (*L.*)

Part 4
Heat

24

Introduction: Temperature, Heat, Energy

Temperature

We are interested in heat because it is the commonest form of energy, and because changes of temperature have great effects on our personal comfort, and on the properties of substances, such as water, which we use every day. Temperature is a scientific quantity which corresponds to primary sensations—hotness and coldness. These sensations are not reliable enough for scientific work, because they depend on contrast—the air in a thick-walled barn or church feels cool on a summer's day, but warm on a winter's day, although a thermometer may show that it has a lower temperature in the winter. A thermometer, such as the familiar mercury-in-glass instrument (Figure 24.1), is a device whose readings depend on hotness or coldness, and which we choose to consider more reliable than our senses. We are justified in considering it more reliable because different thermometers of the same type agree with one another better than different people do.

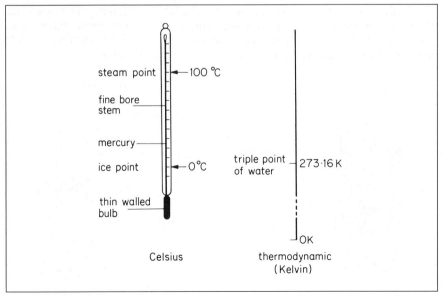

Figure 24.1 *Mercury-in-glass thermometer (left); °C and K scales*

Types of Thermometers

The temperature of an object is not a fixed number but depends on the type of thermometer used and on the temperature scale adopted, discussed shortly.

In general, thermometers use some measurable property of a substance which is sensitive to temperature change. The *constant-volume gas thermometer*, for example, uses the pressure change with temperature of a gas at constant volume. The *resistance thermometer* uses the change of electrical resistance of a pure

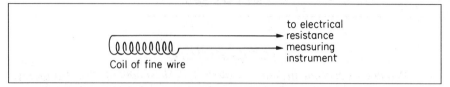

Figure 24.2 *A resistance thermometer; the wire is usually of pure platinum*

metal with temperature, Figure 24.2. The *mercury-in-glass* thermometer depends on the change in volume of mercury with temperature relative to that of glass. A *thermoelectric thermometer* depends on the electromotive force change with temperature of two metals joined together.

Thermodynamic Temperature Scale

The *thermodynamic* temperature scale is the standard temperature scale adopted for scientific measurement. Thermodynamic temperature is denoted by the symbol *T* and is measured in *kelvin*, symbol K. The kelvin is the SI unit of temperature or of temperature change.

The thermodynamic temperature scale uses one fixed point, the *triple point of water*. This is the temperature at which saturated water-vapour, pure water and melting ice are all in equilibrium. The triple point temperature is *defined* as 273·16 K. Figure 24.3 (i) shows an apparatus used for obtaining the triple point of water; distilled water (from which dissolved air has been driven out), water-vapour and ice are here in equilibrium. The kelvin is the fraction 1/273·16 of the thermodynamic temperature of the triple point of water.

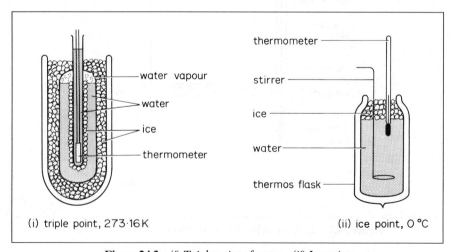

Figure 24.3   (i) *Triple point of water*   (ii) *Ice point*

On the thermodynamic scale, the ice point has a temperature 273·15 K. The slight difference from the triple point is due to the difference in pressure (4·6 mmHg at the triple point and 760 mmHg at the ice point, Figure 24.3 (ii)) and to the removal of dissolved air from the distilled water used for the triple point.

Using the constant-volume gas thermometer, for example, the gas pressure p_{tr} is measured at the triple point of water, $273 \cdot 16$ K. If the pressure is p at an unknown temperature T on the thermodynamic scale, then, by definition,

$$T = \frac{p}{p_{tr}} \times 273 \cdot 16 \text{ K}$$

With a platinum resistance thermometer, the resistance R of the metal is measured at the unknown temperature. The temperature T_{pt} on the thermodynamic scale is then given, if R_{tr} is the resistance at the triple point, by

$$T_{pt} = \frac{R}{R_{tr}} \times 273 \cdot 16 \text{ K}$$

Celsius Temperature Scale

The *Celsius temperature*, symbol θ, is now defined by $\theta = T - 273 \cdot 15$, where T is the thermodynamic temperature (see *The International Practical Temperature Scale of 1968*, National Physical Laboratory, HMSO). The ice point is a Celsius temperature of 0°C and the steam point, the temperature of steam at 760 mmHg pressure, is 100°C.

It should be noted that if different types of thermometers are used to measure temperature, they only agree at the fixed points—$273 \cdot 16$ K on the thermodynamic scale and 0°C and 100°C, for example, on the Celsius scale.

The temperature change or interval of one degree Celsius, 1°C, is exactly the same as the temperature interval 1 K on the thermodynamic scale. So '°C' may be replaced by 'K' in SI units, and '°C^{-1}' by 'K^{-1}'. Approximately, the temperature 0°C = 273 K and 100°C = 373 K. The absolute zero of temperature 0 K is approximately -273°C.

Heat and Energy

Heat and Temperature

If we run hot water into a lukewarm bath, we make it hotter; if we run in cold water, we make it cooler. The hot water, we say, gives *heat* to the cooler bath-water; but the cold water takes heat from the warmer bath-water. The quantity of heat which we can get from hot water depends on both the mass of water and on its temperature: a bucket-full at 80°C will warm the bath more than a cup-full at 100°C.

Roughly speaking, temperature is analogous to electrical potential, and heat is analogous to quantity of electricity. We can detect temperature changes, and whenever the temperature of a body rises, that body has gained heat. The converse is not always true; when a body is melting or boiling, it is absorbing heat from the flame beneath it, but its temperature is not rising.

Thermal Equilibrium, The Zeroth Law

When two bodies are in thermal contact and there is no net flow of heat between them, they are said to be in *thermal equilibrium*. Experimentally, it is found that when two bodies A and B are each in thermal equilibrium with a third body C, then A and B are also in thermal equilibrium with each other. This is called the 'Zeroth Law of Thermodynamics'—the number 'zero' was used because this law logically precedes the 'first' and 'second' laws of thermodynamics and, in fact, is assumed in the two laws.

The Zeroth Law leads to the conclusion that temperature is a well-defined physical quantity. For example, suppose we wish to determine whether two bodies A and B are in thermal equilibrium. In practice, we do this by bringing each in turn into contact with a third body, a thermometer T say. Experimentally, then, we bring A and T into thermal equilibrium, and B and T into thermal equilibrium. If the temperature reading is the same in the two cases, we deduce, but only with the help of the Zeroth Law, that A and B are in thermal equilibrium. The Law thus enables temperature to be defined as that property of a body which decides whether or not it is in thermal equilibrium with another body.

Heat and Energy

The idea of heat as a form of energy was developed particularly by BENJAMIN THOMPSON (1753–1814). He was an American who, after adventures in Europe, became a Count of the Holy Roman Empire, and war minister of Bavaria. He is now generally known as Count Rumford.

While supervising his arsenal, he noticed the great amount of heat which was liberated in the boring of cannon. The idea common at the time was that this heat was a fluid, pressed out of the chips of metal as they were bored out of the barrel. To measure the heat produced, Rumford used a blunt borer, and surrounded it and the end of the cannon with a wooden box, which was filled with water. From the mass of water, and the rate at which its temperature rose, he showed that the amount of heat liberated was in no way connected with the mass of metal bored away, and concluded that it depended only on the *work done* against friction. It followed that *heat was a form of energy*.

Rumford published the results of his experiments in 1798. No similar experiments were made until 1840, when Joule began his study of heat and other forms of energy. Joule measured the work done, and the heat produced, when water was churned, in an apparatus he designed for the experiment. He also measured the work done and heat produced when oil was churned, when air was compressed, when water was forced through fine tubes, and when cast iron bevel wheels were rotated one against the other. Always, within the limits of experimental error, he found that the heat liberated was *proportional* to the mechanical work done, and that the ratio of the two was the same in all types of experiment.

In other experiments, Joule measured the heat liberated by an electric current in flowing through a resistance; at the same time he measured the work done in driving the dynamo which generated the current. He obtained about the same ratio for work done to heat liberated as in his direct experiments. This work linked the ideas of heat, mechanical, and electrical energy. He also showed that the heat produced by a current is related to the chemical energy used up.

We can thus say

(a) that an object gains *energy* when its temperature rises,

(b) that *energy* (heat) passes from a warm to a cold object if they are placed in contact.

The metric unit of work or energy is the *joule*, J. Since experiment shows that heat is a form of energy, *the joule is the scientific unit of heat.* 'Heat per second' is expressed in 'joules per second' or *watts*, W.

The Conservation of Energy

As a result of all his experiments, Joule developed the idea that energy in any one form could be converted into any other. There might be a loss of useful energy in the process—for example, some of the heat from the furnace of a steam-engine is lost up the chimney, and some more down the exhaust—but no energy is destroyed. The work done by the engine, added to the heat lost as described and the heat developed as friction, is equal to the heat provided by the fuel burnt. The idea underlying this statement is called the *Principle of the Conservation of Energy*. It means that, if we start with a given amount of energy in any one form, we convert it in turn into all other forms. We may not always be able to convert it completely, but if we keep an accurate balance-sheet we shall find that the *total* amount of energy, expressed in any one form—say heat or work—is always the same, and is equal to the original amount.

The principle of the conservation of energy is often expressed concisely in mathematical form by the equation

$$\Delta Q = \Delta U + \Delta W$$

Here ΔQ is the *quantity of heat of energy* given to a system, ΔU is the consequent rise in *internal energy* of the system and ΔW is the *external work* done by the system. As we see later, the rise in internal energy, ΔU, of a system is shown by a temperature rise. If the system also expands, it does external work, ΔW, against the external forces. The principle of conservation of energy will be applied later to many cases of energy changes.

The conservation of energy applies to living organisms—plants and animals—as well as to inanimate systems. For example, we may put a man or a mouse into a box or a room, give him a treadmill to work, and feed him. His food is his fuel; if we burn a sample of it, we can measure its chemical energy, in heat units. And if we now add up the heat value of the work which the man does, and the heat which his body gives off, we find that their total is equal to the chemical energy of the food which the man eats. Because food is the source of man's energy, food value may be expressed in *kilojoules*. A man needs about 120 000 kilojoules per day.

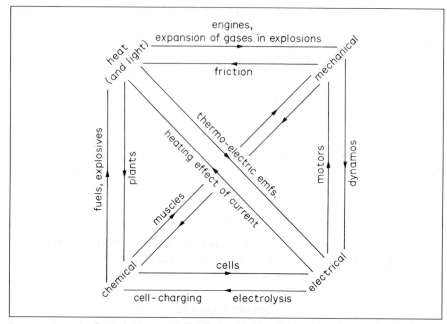

Figure 24.4 *Forms of energy, and their interconversions*

All the energy by which we live comes from the sun. The sun's ultra-violet rays are absorbed in the green matter of plants, and make them grow; the animals eat the plants, and we eat them—we are all vegetarians at one remove. The plants and trees of an earlier age decayed, were buried, and turned into coal. Even water-power comes from the sun—we would have no lakes if the sun did not evaporate the water and provide the rainfall which fills the lakes. The relationship between the principal forms of energy are summarised in Figure 24.4. It shows how different kinds of energy can be transferred from one form to another.

25

Thermometry

Realisation of Temperature Scale

In the previous chapter we discussed the general idea of a temperature scale. To establish such a scale we need: (i) some physical property of a substance—such as the volume of a gas or the electrical resistance of pure platinum—which increases continuously with increasing temperature, but is constant at constant temperature; (ii) fixed temperatures—fixed points— which can be accurately reproduced in the laboratory.

The Fixed Points

On the thermodynamic scale, the triple point of water is chosen as the fixed point and is defined at $273 \cdot 16$ K (p. 620). The absolute zero is 0 K, the ice point is $273 \cdot 15$ K and the steam point is $373 \cdot 15$ K.

Suppose P is the chosen temperature-measuring quantity, such as a gas pressure at constant volume or the electrical resistance of a pure metal. If P_{tr} is the value of P at the triple point and P_T is the value at an unknown temperature T on the absolute thermodynamic scale, then, by definition,

$$T = \frac{P_T}{P_{tr}} \times 273 \cdot 16 \text{ K}$$

The temperatures of melting ice and pure boiling water at 760 mmHg pressure were chosen as the fixed points on the Celsius scale. Nowadays, the temperature θ on the Celsius scale is defined by $\theta = T - 273 \cdot 15$.

If P is the chosen temperature-measuring quantity, its values P_0 at the ice point and P_{100} at the steam point determine the fundamental interval, 100°C, of the Celsius scale, $P_{100} - P_0$. And the temperature θ_P on the Celsius scale which corresponds to a value P_θ is given by

$$\theta_P = \frac{P_\theta - P_0}{P_{100} - P_0} \times 100$$

Gas Thermometry

In most accurate work, temperatures are measured by *gas thermometers*; for example, by the changes in pressure of a gas at constant volume.

At pressures of the order of one atmosphere, different gases give slightly different temperature scales, because none of them obeys the gas laws perfectly. But as the pressure is reduced, the gases approach closely to the ideal, and their temperature scales come together (see p. 743). By observing the departure of a gas from Boyle's law at moderate pressures it is possible to allow for its departure from the ideal. Temperatures measured with the gas in a constant volume thermometer can then be converted to the values which would be given by the same thermometer if the gas were ideal.

Gas thermometer temperatures are used as *standard* temperatures. Temperatures measured with other types of thermometers are converted to a gas thermometer temperature.

The Constant Volume Gas Thermometer

Figure 25.1 shows a constant volume hydrogen thermometer. B is a bulb of platinum–iridium, holding the gas. The volume is defined by the level of the index I in the glass tube A. The pressure is adjusted by raising or lowering the mercury reservoir R. A barometer CD is fitted directly into the pressure-measuring system; if H_1 is its height, and h the difference in level between the

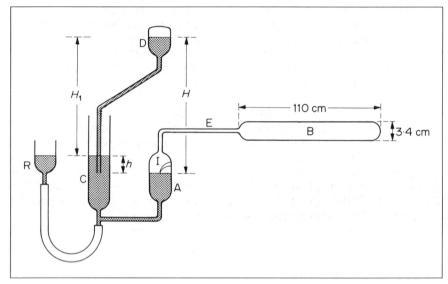

Figure 25.1 *Constant volume hydrogen thermometer* (not to scale)

mercury surfaces in A and C, then the pressure H of the hydrogen, in mm mercury is

$$H = H_1 + h$$

H is measured with a cathetometer (a travelling telescope with vernier).

The glass tubes A, C, D, all have the same diameter to prevent errors due to surface tension; and A and D are optically worked to prevent errors due to refraction (as in looking through common window-glass).

Observations made with a constant volume gas thermometer must be corrected for the following errors: (i) the expansion of the bulb B; (ii) the temperature of the gas in the tube E and A, which lies between the temperature of B and the temperature of the room; (iii) the temperature of the mercury in the barometer and manometer.

The expansion of the bulb can be estimated from its cubical expansivity, by using the temperature shown by the gas thermometer. Since the expansion appears only as a small correction to the observed temperature, the uncorrected value of the temperature may be used in estimating it.

The tube E is called the 'dead-space' of the thermometer. Its diameter is made small, about 0·7 mm, so that it contains only a small fraction of the total mass of gas. Its volume is known, and the temperatures at various points in it are measured with mercury thermometers. The effect of the gas in it is then allowed for in a calculation similar to that used to calculate the pressure of a gas in two bulbs at different temperatures (p. 660).

Plate 25A Gas Thermometry. *The photograph shows the whole assembly of the National Physical Laboratory constant volume gas thermometer, used to measure a low temperature in the range about 2 K to 27 K. The working parts such as the gas bulb are immersed in liquid helium in the stainless steel Dewar vessel on the left. The pressure of the gas is measured by a sensitive pressure balance top right. The value in Pa is recorded on the circular dial above the Dewar vessel.* (Crown copyright. Courtesy of the National Physical Laboratory)

A gas thermometer is a large awkward instrument, demanding much skill and time, and useless for measuring changing temperatures. In practice, gas thermometers are used only for calibrating electrical thermometers—resistance thermometers and thermocouples. The readings of these, when they are used to measure unknown temperatures, can then be converted into temperatures on the ideal gas scale. Helium gas is widely used in gas thermometers.

The Celsius temperature θ on the constant volume gas thermometer scale would be calculated from

$$\theta = \frac{p_\theta - p_0}{p_{100} - p_0} \times 100°C$$

where p_θ, p_0 and p_{100} are the respective gas pressures at the unknown temperature θ, the ice point and steam point.

Examples on Temperature Measurement

1 The pressure recorded by a constant volume gas thermometer at a kelvin temperature T is $4\cdot80 \times 10^4\,\mathrm{N\,m^{-2}}$. Calculate T if the pressure at the triple point, $273\cdot16\,\mathrm{K}$, is $4\cdot20 \times 10^4\,\mathrm{N\,m^{-2}}$.

$$T = \frac{p_T}{p_{tr}} \times 273\cdot16\,\mathrm{K}$$

$$= \frac{4\cdot80 \times 10^4}{4\cdot20 \times 10^4} \times 273\cdot16$$

$$= 312\,\mathrm{K}$$

2 A constant mass of gas maintained at constant pressure has a volume of 200.0 cm^3 at the temperature of melting ice, 273.2 cm^3 at the temperature of water boiling under standard pressure, and 525.1 cm^3 at the normal boiling-point of sulphur. A platinum wire has resistance of 2.000, 2.778 and $5.280 \, \Omega$ at the temperatures. Calculate the values of the boiling-point of sulphur given by the two sets of observations, and comment on the results.

On the gas thermometer scale, the boiling-point of sulphur is given by

$$\theta = \frac{V_\theta - V_0}{V_{100} - V_0} \times 100$$

$$= \frac{525.1 - 200.0}{273.2 - 200.0} \times 100$$

$$= 444.1°C$$

On the platinum resistance thermometer scale, the boiling-point is given by

$$\theta_p = \frac{R_\theta - R_0}{R_{100} - R_0} \times 100$$

$$= \frac{5.280 - 2.000}{2.778 - 2.000} \times 100$$

$$= 421.6°C$$

The temperatures recorded on the thermometers are therefore different. This is due to the fact that the variation of gas pressure with temperature at constant volume is different from the variation of the electrical resistance of platinum with temperature.

Electric Thermometers

Electrical thermometers have great advantages over other types. They are more accurate than any except gas thermometers, but are quicker in action and less cumbersome than those thermometers.

The measuring element of a *thermoelectric thermometer* is the welded junction of two fine wires. It is very small in size, and can therefore measure the temperature almost at a point. It causes very little disturbance wherever it is placed, because the wires leading from it are so thin that heat loss along them is usually negligible. It has a very small heat capacity, and can therefore follow a rapidly changing temperature. To measure such a temperature, however, the e.m.f. of the junction must be measured with a galvanometer, instead of a potentiometer, and some accuracy is then lost. The Celsius temperature θ_{th} on the thermoelectric thermometer scale would be calculated from

$$\theta_{th} = \frac{E_\theta - E_0}{E_{100} - E_0} \times 100°C$$

where E_θ, E_0 and E_{100} are the respective thermoelectric e.m.f.s at the unknown temperature θ, the ice point and the steam point.

The measuring element of a *resistance thermometer* is a spiral of fine wire. It has a greater size and heat capacity than a thermojunction, and cannot therefore measure a local or rapidly changing temperature. But, over the range from about room temperature to a few hundred degrees Celsius, it is more accurate.

The platinum-resistance scale differs appreciably from the mercury-in-glass scale, for example, as the following table shows:

Mercury-in-glass	0	50	100	200	300	°C
Platinum-resistance	0	50·25	100	197	291	°C

Resistance Thermometers

Resistance thermometers are usually made of platinum. The wire is wound on two strips of mica, arranged crosswise as shown in Figure 25.2 (i). The ends of the coil are attached to a pair of leads A, for connecting them to a Wheatstone bridge. A similar pair of leads B is near to the leads from the coil, and connected in the adjacent arm of the bridge (Figure 25.2 (ii)). At the end near the coil, the pair of leads B is short-circuited. If the two pairs of leads are identical, their resistances are equal, whatever their temperature. Thus if $P = Q$ the dummy pair, B, just compensates for the pair A going to the coil; and the bridge measures the resistance of the coil alone. The Celsius temperature θ_p on the platinum resistance thermometer scale would be calculated from

$$\theta_p = \frac{R_\theta - R_0}{R_{100} - R_0} \times 100°C$$

where R_θ, R_0 and R_{100} are the respective resistances at the unknown temperature θ, the ice point and the steam point.

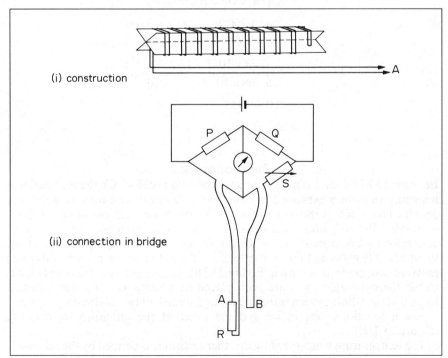

(i) construction

(ii) connection in bridge

Figure 25.2 *Platinum resistance thermometer* ($P = Q$ *and B compensates A so that* $S = R$)

The platinum resistance thermometer is used to measure temperature on the International Practical Temperature Scale between $-259.34°C$ and $630.74°C$. The platinum used in the coil must be strain-free and annealed pure platinum. Its purity and reliability are judged by the increase in its resistance from the ice point to the steam point. Thus if R_0 and R_{100} are the resistances of the coil at these points, then the coil is suitable if

$$\frac{R_{100}}{R_0} > 1.39250$$

Various formulae and tables are provided to obtain the temperature over the wide temperature range.

Example on Resistance Thermometer

The resistance R_t of a platinum wire at temperature $t°C$, measured on the gas scale, is given by $R_t = R_0(1 + at + bt^2)$, where $a = 3.800 \times 10^{-3}$ and $b = -5.6 \times 10^{-7}$. What temperature will the platinum thermometer indicate when the temperature on the gas scale is 200°C?

$$R_t = R_0(1 + at + bt^2)$$

$$\therefore R_{200} = R_0(1 + 200a + 200^2 b)$$

and
$$R_{100} = R_0(1 + 100a + 100^2 b)$$

$$\therefore \theta_p = \frac{R_{200} - R_0}{R_{100} - R_0} \times 100$$

$$= \frac{R_0(1 + 200a + 200^2 b) - R_0}{R_0(1 + 100a + 100^2 b) - R_0} \times 100$$

$$= \frac{200a + 200^2 b}{a + 100b} = \frac{200(a + 200b)}{a + 100b}$$

$$= 200 \frac{(3.8 \times 10^{-3} - 11.2 \times 10^{-5})}{3.8 \times 10^{-3} - 5.6 \times 10^{-5}}$$

$$\therefore \theta_p = 200 \times 0.985$$

$$= 197°C$$

Thermocouples

Between $630.74°C$ and the freezing point of gold ($1064.43°C$), the international temperature scale is expressed in terms of the electromotive force of a thermocouple. The wires of the thermocouple are platinum, and platinum–rhodium alloy (90% Pt.: 10% Rh.). Since the e.m.f. is to be measured on a potentiometer, care must be taken that thermal e.m.f.s are not set up at the junctions of the thermocouple wires and the copper leads to the potentiometer. To do this three junctions are made, as shown in Figure 25.3 (i). The junctions of the copper leads to the thermocouple wires are both placed in melting ice. The electromotive force E of the whole system is then equal to the e.m.f. of two platinum/platinum–rhodium junctions, one in ice and the other at the unknown temperature (Figure 25.3 (ii)).

The temperature θ measured by this thermocouple is defined by the relation

$$E = a + b\theta + c\theta^2$$

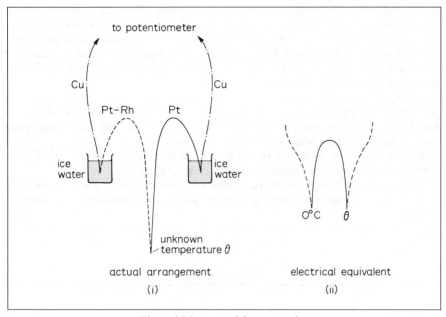

Figure 25.3 *Use of thermocouples*

where *a*, *b* and *c* are constants. The values of the constants are determined by measurements of *E* at the gold point (1064·43°C), the silver point (960·8°C), and the temperature of freezing antimony (about 630·74°C).

Other Thermocouples

Because of their convenience, thermocouples are used to measure temperatures outside their range on the international scale, when the highest accuracy is not required. The arrangement of three junctions and potentiometer may be used, but for less accurate work the potentiometer may be replaced by a galvanometer G, in the simpler arrangement of Figure 25.4 (i). The galvanometer scale may be calibrated to read directly in temperatures, the known melting-points of metals like tin and lead being used as subsidiary fixed points. For rough work,

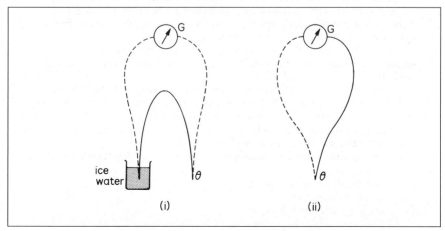

Figure 25.4 *Simple thermojunction thermometer*

particularly at high temperatures, the cold junction may be omitted (Figure 25.4 (ii)). An uncertainty of a few degrees in a thousand is often of no importance.

Owing to the small size and small heat capacity of their thermojunctions, thermocouples can be used to measure varying temperatures at a particular place of a metal surface, for example. In this case a hole would be bored for the thermojunction to be pushed in.

Summary: *Gas thermometer* temperatures are used as standard temperatures. At very low pressures all gases give the same temperature value. Very wide temperature range. Disadvantage—bulky and so difficult to use.

Resistance thermometers have a wide temperature range, about −200°C to 1200°C. Used to measure liquid temperatures accurately.

Thermoelectric thermometers have wide temperature range, about −250°C to 1500°C. Used to measure varying temperatures as the junction has a low heat capacity.

_____ Exercises 25 _____

1

	Steam point 100°C	Ice point 0°C	Room temperature
Resistance of resistance thermometer	75·000 Ω	63·000 Ω	64·992 Ω
Pressure recorded by constant volume gas thermometer	$1·10 \times 10^5$ $N m^{-2}$	$8·00 \times 10^4$ $N m^{-2}$	$8·51 \times 10^4$ $N m^{-2}$

Using the above data, which refers to the observations of a particular room temperature using two types of thermometer, calculate the room temperature on the scale of the resistance thermometer and on the scale of the constant volume gas thermometer.

Why do these values differ slightly? (*L.*)

2 (a) Explain how a temperature scale is defined.
   (b) Discuss the relative merits of (i) a mercury-in-glass thermometer, (ii) a platinum resistance thermometer, (iii) a thermocouple, for measuring the temperature of an oven which is maintained at about 300°C. (*JMB.*)

3 Tabulate various physical properties used for measuring temperature. Indicate the temperature range for which each is suitable.

Discuss the fact that the numerical value of a temperature expressed on the scale of the platinum resistance thermometer is not the same as its value on the gas scale except at the fixed points.

If the resistance of a platinum thermometer is 1·500 ohms at 0°C, 2·060 ohms at 100°C and 1·788 ohms at 50°C on the gas scale, what is the diference between the numerical values of the latter temperature on the two scales? (*JMB.*)

4 Describe the structure of a simple constant volume gas thermometer. Discuss how it would be used to establish a scale of temperature.

Explain why the same temperature measured on two different scales need not have the same value.

Discuss the circumstances in which
   (a) a gas thermometer and
   (b) a thermocouple might be used.

Why is it generally not sensible to use a thermoelectric e.m.f. as the physical property used to *define* a scale of temperature? (*L.*)

5 The table below gives data for two thermometers at three different temperatures (the ice-point, the steam-point and room temperature).

Type of thermometer	Property	Value of property		
		ice-point	steam-point	room temp
Gas	Pressure in mm Hg	760	1040	795
Thermistor	Current in mA	12·0	54·0	15·0

(a) Calculate the temperature of the room according to each thermometer.
(b) State why the thermometers disagree in their value for room temperature.
(c) Explain why a gas thermometer is seldom used for temperature measurement in the laboratory. (*AEB*, 1985.)

6 The resistance of the element of a platinum resistance thermometer is $2·00\,\Omega$ at the ice point and $2·73\,\Omega$ at the steam point. What temperature on the platinum resistance scale would correspond to a resistance value of $8·43\,\Omega$?

Measured on the gas scale, the same temperature corresponded to a value of 1020°C. Explain the discrepancy. (*L*.)

7 Explain what is meant by a change in temperature of 1°C on the scale of a platinum resistance thermometer.

Draw and label a diagram of a platinum resistance thermometer together with a circuit in which it is used.

Give *two* advantages of this thermometer and explain why, in its normal form, it is unsuited for measurement of varying temperatures.

The resistance R_t of platinum varies with the temperature t°C as measured by a constant volume gas thermometer according to the equation

$$R_t = R_0(1 + 8000\,\alpha t - \alpha t^2)$$

where α is a constant. Calculate the temperature on the platinum scale corresponding to 400°C on this gas scale. (*JMB*.)

8 (a) What is meant by a thermometric property? What qualities make a particular property suitable for use in a practical thermometer?

A Celsius temperature scale may be defined in terms of a thermometric property X by the following equation:

$$\theta = \frac{X - X_0}{X_{100} - X_0} \times 100°C \tag{1}$$

where X_0 is the value of the property at the ice point, X_{100} at the steam point, and X at some intermediate temperature. If X is plotted against θ a straight line always results no matter what thermometric property is chosen. Explain this.

(b) On the graph Figure 25A, line A shows how X varies with θ (following equation (1) above), line B shows how a second thermometric property Q varies with θ, the temperature measured on the X scale. (i) Describe, in principle, how you would conduct an experiment to obtain line B. (ii) If $\theta = 40°C$ recorded by an X-scale thermometer, what temperature would be recorded by a Q-scale thermometer? (iii) At what two temperatures will the X and Q scales coincide?

(c) The ideal gas scale of temperature is one based on the properties of an ideal gas. What is the particular virtue of this scale? Describe very briefly how readings on such a scale can be obtained using a thermometer containing a real gas. (*L*.)

9 (a) When bodies are in thermal equilibrium, their temperatures are the same.

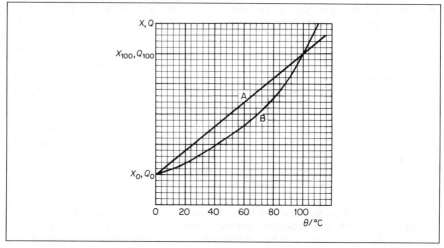

Figure 25A

Explain in energy terms the condition for two bodies to be in thermal equilibrium with one another.

(b) The temperature of a beaker of water is to be measured using a mercury-in-glass thermometer. (i) Why is it necessary to wait before taking the reading? (ii) Explain briefly how you might estimate the heat capacity (energy required per unit temperature rise) of a mercury-in-glass thermometer. (iii) If the beaker contains 120 g of water at 60°C, what temperature would be recorded by the mercury-in-glass thermometer if it was initially at 18°C and had a heat capacity of $30 \, \text{J K}^{-1}$?
(Assume the specific heat capacity of water to be $4200 \, \text{J kg}^{-1} \, \text{K}^{-1}$ and ignore the heat losses to the beaker and surroundings while the temperature is being taken.) (iv) Why, if a more accurate value of the temperature were required in this case, might you use a thermocouple? (v) Describe briefly how you would calibrate a thermocouple and use it to measure the temperature of the water. Show how you would calculate the temperature of the water from your readings. (*L.*)

10 (a) Define the following: (i) thermodynamic coordinates, (ii) thermodynamic state, (iii) thermal equilibrium. (*see chapters 27, 30*)

(b) State the zeroth law of thermodynamics and explain how the concept of temperature arises from the zeroth law.

(c) (i) Define the ideal gas temperature. (ii) A constant-volume gas thermometer is immersed in water at the triple point and the pressure P_3 recorded; then it is immersed in a thermostatic bath at θ and the pressure P_θ recorded. The experiment is repeated for different values of P_3. The table gives the results. Find the ideal gas temperature θ.

| P_3 (mm Hg) | 1000·0 | 750·00 | 500·00 | 250·00 |
| P_θ (mm Hg) | 1535·3 | 1151·76 | 767·82 | 383·95 |

Triple point of water, $T_{Tr} = 273 \cdot 16 \, \text{K}$. (*W.*)

26

Heat Capacity, Latent Heat

Heat Capacity, Specific Heat Capacity

The heat capacity *of a body, such as a lump of metal, is the quantity of heat required to raise its temperature by 1 degree. It is expressed in* joule per kelvin $(J\,K^{-1})$ *in SI units.*

The *specific heat capacity* of a substance is the heat required to raise the temperature of 1 kg of it through 1 degree; it is the heat capacity per kg of the substance. Specific heat capacities are expressed in *joule per kilogram per kelvin* and the symbol is c. The specific heat capacity of water, c_w, is about $4200\,\text{J}\,\text{kg}^{-1}\,\text{K}^{-1}$, or $4\cdot2\,\text{kJ}\,\text{kg}^{-1}\,\text{K}^{-1}$, where $1\,\text{kJ} = 1$ kilojoule $= 1000\,\text{J}$. The megajoule, MJ, is a larger unit than the kilojoule used in the gas or electrical industry. $1\,\text{MJ} = 10^6\,\text{J} = 1000\,\text{kJ}$.

From the definition of specific heat capacity, it follows that, for a particular object,

---

heat capacity, C = mass × specific heat capacity

The specific heat capacity of copper, for example, is about $400\,\text{J}\,\text{kg}^{-1}\,\text{K}^{-1}$. Hence the heat capacity of 5 kg of copper $= 5 \times 400 = 2000\,\text{J}\,\text{K}^{-1} = 2\,\text{kJ}\,\text{K}^{-1}$. If the copper temperature rises by 10°C, then heat gained $= 5 \times 400 \times 10 = 20\,000\,\text{J}$.

Generally, then, the heat Q gained (or lost) by an object is given by

---

$$Q = mc\theta$$

---

where m is the mass of the object, c its specific heat capacity and θ its temperature change.

Temperature Changes

From $Q = mc\theta$, the temperature change θ of an object of mass m which loses or gains a given quantity of heat Q is given by

$$\theta = \frac{Q}{mc}$$

where c is the specific heat capacity of the object. So for a given loss of heat Q to the surroundings, the temperature fall θ of a small mass of warm water in a room is greater than a large mass of water at the same temperature. We shall see later that the rate at which a hot solid or liquid cools depends on the nature and area of its surface, in addition to its temperature, mass and specific heat capacity.

When a thermoelectric thermometer is used, the thermometer junction is placed in contact with the object whose temperature is required. The junction has a small thermal capacity, mc, since its mass is small. The temperature of the junction thus quickly reaches the temperature of the object, which is an advantage of the thermometer.

Measurement of Specific Heat Capacity

Specific Heat Capacity of Solid by Electrical Method

The specific heat capacity of a metal can be found by an *electrical method*. Figure 26.1 shows a simple form of laboratory apparatus. M is a thick solid block of metal such as aluminium, with an electric heater element H completely inside a deep hole bored into the metal and a thermometer T inside another deep hole. Both H and T must make good thermal contact with the block. An insulating jacket J is placed round the metal.

In an experiment, suppose the voltmeter reads 12 V, the ammeter A reads 4·0 A and the block, mass 1·0 kg, rises by 16°C in 5 min or 300 s. Then (p. 259)

$$\text{heat supplied} = IVt = 12 \times 4 \times 300 = 14\,400\,\text{J}$$

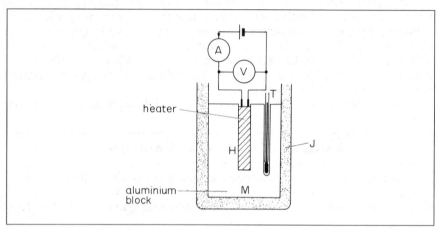

Figure 26.1 *Electrical method for specific heat capacity*

If c is the specific heat capacity of the metal, then, assuming negligible heat losses,

$$Q = mc\theta = 1 \times c \times 16 = 14\,400$$
$$\therefore c = 900\,\text{J}\,\text{kg}^{-1}\,\text{K}^{-1}$$

If θ is the temperature rise corrected for heat losses (p. 641), then, generally,

$$IVt = mc\theta$$

The electrical energy supplied, $\Delta Q = IVt$. From the principle of conservation of energy (p. 623), this is equal to the rise in internal energy ΔU of the metal $(mc\theta)$ plus the heat losses h by cooling *plus* the external work ΔW done against external atmospheric pressure by the metal when it expands on warming. Since metals expand very slightly in volume on warming, ΔW can be neglected in this experiment. So $IVt = mc\theta + h$.

If θ is not corrected for heat losses, we see from $IVt = mc\theta$ that the result for c is too high. The cooling correction is discussed later.

Specific Heat Capacity by Mechanical Method

Figure 26.2 illustrates one simple form of apparatus for measuring the specific heat capacity of a metal by the transfer of *mechanical energy to heat energy*.

It consists of a small solid metal cylinder A, insulated from the rest of the apparatus by a nylon bush. The metal temperature can be read on a short-range thermometer B placed inside a hole bored axially in A. A flexible nylon cord C is wound several times round the cylinder. One end of C is attached to a rubber band D. The other end is attached to a heavy weight W which hangs from it.

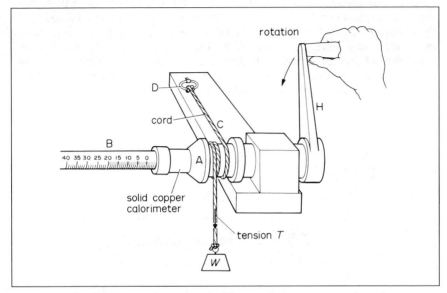

Figure 26.2 *Mechanical method for specific heat capacity of metal*

By means of a handle H, the cylinder is rotated so that the rubber band goes slack. The tension in the cord C is then equal to the weight W and mechanical energy is expended against this force when the cylinder is rotated. After a suitable number of revolutions, the final temperature of the metal cylinder is noted, a cooling correction having been applied (p. 642).

Suppose $W = 50$ N, the mass of the metal cylinder is 0·200 kg, its diameter is 25 mm or 0·025 m, the temperature rise is 10·0 K and the number of revolutions is 200. Then, since the circumference of a cylinder $= \pi \times$ diameter,

$$\text{mechanical energy expended} = \text{force} \times \text{distance}$$

$$= 50 \times (\pi \times 0{\cdot}025) \times 200 \text{ J}$$

From the principle of conservation of energy, we see that the mechanical energy ΔQ spent in turning the handle = the rise in internal energy ΔU of the metal *plus* the external work ΔW done against the external atmospheric pressure when the metal expands on warming. The metal expansion is very small and so ΔW can be neglected in the equation for c. So $\Delta Q = \Delta U$.

Assuming negligible heat losses, ΔU = heat gained by metal = $mc\theta$, where c is its specific heat capacity.

$$\therefore 0{\cdot}2 \times c \times 10 = 50 \times \pi \times 0{\cdot}025 \times 200$$

$$\therefore c = 390 \text{ J kg}^{-1} \text{ K}^{-1} \text{ (approx.)}$$

Specific Heat Capacity of a Liquid by Continuous Flow Method

Callendar and Barnes devised an electrical method for the specific heat capacity of a liquid in which only steady temperatures are measured. They used platinum resistance thermometers, which are more accurate than mercury ones but take more time to read. In the measurement of steady temperatures, however, this is no drawback. As we shall see shortly *the heat capacity of the apparatus is not required*, which is a great advantage of the method.

Figure 26.3 shows Callendar and Barnes' apparatus used to measure the specific heat capacity of water. Water from the constant-head tank K flows through the glass tube U, and can be collected as it flows out. It is heated by the spiral resistance wire R, which carried a steady electric current I. Its temperature, as it enters and leaves, is measured by the Thermometers T_1 and T_2. (In a simplified laboratory experiment, these may be mercury thermometers.) Surrounding the apparatus is a glass jacket G, which is evacuated, so that heat cannot escape from the water by conduction or convection.

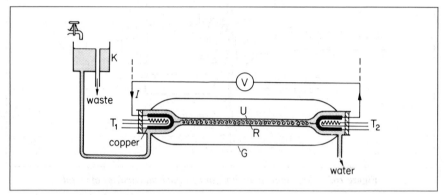

Figure 26.3   *Callendar and Barnes' apparatus (contracted several times in length relative to diameters)*

When the apparatus is running, it settles down eventually to a steady state, in which the heat supplied by the current is all carried away by the water. *None is then taken in warming the apparatus, because every part of it is at a constant temperature.* The mass of water m, which flows out of the tube in t seconds, is then measured. If the water enters at a temperature θ_1 and leaves at θ_2, then if c_w is its mean specific heat capacity,

$$\text{heat gained by water} = Q = mc_w(\theta_2 - \theta_1) \text{ joules}$$

The energy which liberates this heat is electrical. To find it, the current I, and the potential difference across the wire V, are measured with a potentiometer. If I and V are in amperes and volts respectively, then, in t seconds, energy supplied to the wire $= IVt$ joules.

If we ignore any heat losses and any external work, then, from the principle of conservation of energy,

$$mc_w(\theta_2 - \theta_1) = IVt$$

$$\therefore c_w = \frac{IVt}{m(\theta_2 - \theta_1)}$$

Elimination of Heat Losses

To get the highest accuracy from this experiment, the small heat losses due to radiation, and conduction along the glass, must be allowed for. These are determined by the temperatures θ_1 and θ_2. For a given pair of values of θ_1 and θ_2, and constant-temperature surroundings (not shown), let the heat lost per second be h. Then if m now represents the mass of liquid flowing *per second*,

$$\text{power supplied by heating coil} = mc_w(\theta_2 - \theta_1) + h$$

$$\therefore IV = mc_w(\theta_2 - \theta_1) + h \quad . \quad . \quad (1)$$

To allow for the loss h, the rate of flow of water is changed, to about half or twice its previous value. The current and voltage are then adjusted to bring θ_2 back to its original value, for example, if the rate of flow of the water is reduced, then I and V are reduced. The inflow temperature θ_1 is fixed by the temperature of the water in the tank. If I', V' are the new values of I, V and m' is the new mass of water flowing per second, then:

$$I'V' = m'c_w(\theta_2 - \theta_1) + h \quad . \quad . \quad . \quad (2)$$

Note that h in equation (2) is the same as in equation (1) because the *mean* temperature of the liquid is arranged to be at the same for each experiment.

On subtracting equation (2) from equation (1), we have

$$(IV - I'V') = (m - m')c_w(\theta_2 - \theta_1)$$

$$\therefore c_w = \frac{(IV - I'V')}{(m - m')(\theta_2 - \theta_1)} \quad . \quad . \quad . \quad (3)$$

When the temperature rise, $\theta_2 - \theta_1$, is made small, for example, $\theta_1 = 20\cdot0°\text{C}$, $\theta_2 = 22\cdot0°\text{C}$, then c_w may be considered as the specific heat capacity at $21\cdot0°\text{C}$, the mean temperature. If the inlet water temperature is now raised to say $\theta_1 = 40\cdot0°\text{C}$ and θ_2 is then $42\cdot0°\text{C}$, c_w is now the specific heat capacity at $41\cdot0°\text{C}$. In this way it was found that *the specific heat capacity of water varies with temperature*. The continuous flow method can be used to find the variation in specific heat capacity of any liquid in the same way.

The table below shows the relative variation of the specific heat capacity of water, taking the value at 15°C as 1·0000 in magnitude.

Specific heat capacity of water

Temperature (°C)	5	15	25	40	70	100
c_w	1·0047	1·0000	0·9980	0·9973	1·0000	1·0057

Examples on Constant Flow Method

1 To measure the specific heat capacity of a liquid, a p.d. of 6·0 V was applied to the heating coil in a constant flow calorimeter. When the rate of flow of liquid was halved, a new p.d. V was required to produce the same inlet and outlet temperatures. Calculate V, assuming heat losses are negligible.

$$\text{Heat supplied per second} = IV = \frac{V^2}{R}$$

where R is the resistance of the heating coil. Since R is constant when the temperature is constant,

$$\text{heat supplied per second} \propto V^2$$

When the rate of flow is halved, *half* the heat per second is required to produce the same outlet temperature, assuming negligible heat losses. So, by proportion,

$$\frac{V^2}{6^2} = \frac{1}{2}$$

So
$$V = \sqrt{\frac{6^2}{2}} = \sqrt{18} = 4 \cdot 2 \text{ V}$$

2 Water flows at the rate of $0 \cdot 1500 \text{ kg min}^{-1}$ through a tube and is heated by a heater dissipating $25 \cdot 2 \text{ W}$. The inflow and outflow water temperatures are $15 \cdot 2°C$ and $17 \cdot 4°C$ respectively. When the rate of flow is increased to $0 \cdot 2318 \text{ kg min}^{-1}$ and the rate of heating to $37 \cdot 8 \text{ W}$, the inflow and outflow temperatures are unaltered. Find (i) the specific heat capacity of water, (ii) the rate of loss of heat from the tube.

Suppose c_w is the specific heat of water in $\text{J kg}^{-1} \text{K}^{-1}$ and h is the heat lost in J s^{-1}. Then, since $1 \text{ W} = 1 \text{ J}$ per second,

$$25 \cdot 2 = \frac{0 \cdot 1500}{60} c_w (17 \cdot 4 - 15 \cdot 2) + h. \qquad . \qquad . \qquad . \qquad (1)$$

and
$$37 \cdot 8 = \frac{0 \cdot 2318}{60} c_w (17 \cdot 4 - 15 \cdot 2) + h. \qquad . \qquad . \qquad . \qquad (2)$$

Subtracting (1) from (2),

$$\therefore 37 \cdot 8 - 25 \cdot 2 = \frac{0 \cdot 2318 - 0 \cdot 1500}{60} c_w (17 \cdot 4 - 15 \cdot 2)$$

$$\therefore c_w = \frac{12 \cdot 6 \times 60}{0 \cdot 0818 \times 2 \cdot 2} = 4200 \text{ J kg}^{-1} \text{K}^{-1}$$

Substituting for c_w in (1),

$$\therefore h = 25 \cdot 2 - \frac{0 \cdot 15}{60} \times 4200 \times 2 \cdot 2 = 2 \cdot 1 \text{ J s}^{-1} = 2 \cdot 1 \text{ W}$$

Method of Mixtures

A common way of measuring specific heat capacities of solids (or liquids) is the method of mixtures.

As an illustration of a specific heat capacity determination, suppose a metal of mass $0 \cdot 2 \text{ kg}$ at $100°C$ is dropped into $0 \cdot 08 \text{ kg}$ of water at $15°C$ contained in a calorimeter of mass $0 \cdot 12 \text{ kg}$ and specific heat capacity $400 \text{ J kg}^{-1} \text{K}^{-1}$. The final temperature reached is $35°C$. Then, assuming negligible heat losses,

$$\text{heat capacity of calorimeter} = 0 \cdot 12 \times 400 = 48 \text{ J K}^{-1}$$

$$\text{heat capacity of water} = 0 \cdot 08 \times 4200 = 336 \text{ J K}^{-1}$$

$$\therefore \text{ heat gained by water} + \text{cal.} = (336 + 48) \times (35 - 15) \text{ J}$$

$$\text{and heat lost by hot metal} = 0 \cdot 2 \times c \times (100 - 35) \text{ J}$$

$$\therefore 0 \cdot 2 \times c \times 65 = 384 \times 20$$

$$\therefore c = \frac{384 \times 20}{0 \cdot 2 \times 65} = 590 \text{ J kg}^{-1} \text{K}^{-1} \text{ (approx.)}$$

Heat Losses

In a calorimetric experiment, some heat is always lost by leakage. Leakage of heat cannot be prevented, as leakage of electricity can, by insulation, because even the best insulator of heat still has appreciable conductivity (p. 701).

When convection is prevented, gases are the best thermal insulators. Hence calorimeters are often surrounded with a shield S and the heat loss due to conduction is made small by packing S with insulating material or by supporting the calorimeter on an insulating ring, or on threads. The loss by radiation is small at small excess temperatures over the surroundings. In some simple calorimetric experiments the final temperature of the mixture is reached quickly, so that the time for leakage is small. The total loss of heat is therefore negligible in laboratory experiments on the specific heats of metals, but not on the specific heat capacities of bad conductors, such as rubber, which give up their heat slowly. Here a 'cooling correction' must be added to the observed final temperature. When great accuracy is required, the loss of heat by leakage is always taken into account.

Newton's Law of Cooling

Newton was the first person to investigate the heat lost by a body in air. He found that *the rate of loss of heat is proportional to the excess temperature over the surroundings*. This result, called *Newton's law of cooling*, is approximately true in still air only for a temperature excess of about 20 K or 30 K; but it is true for all excess temperatures in conditions of forced convection of the air, i.e. in a draught. With natural convection Dulong and Petit found that the rate of loss of heat was proportional to $\theta^{5/4}$, where θ is the excess temperature, and this appears to be true for higher excess temperatures, such as from 50 K to 300 K.

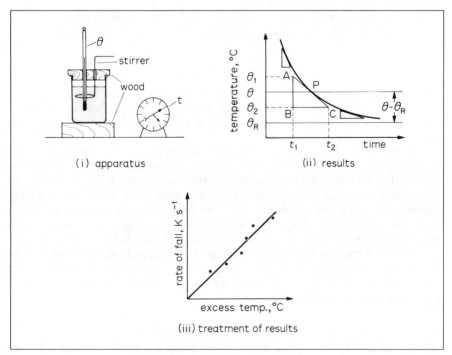

(i) apparatus

(ii) results

(iii) treatment of results

Figure 26.4 *Newton's law of cooling*

To demonstrate Newton's law of cooling, we plot a temperature (θ)-time (t) cooling curve for hot water in a calorimeter placed in a draught (Figure 26.4 (i)). If θ_R is the room temperature, then the excess temperature of the water is $(\theta - \theta_R)$. At various temperatures, such as θ in Figure 26.4 (ii), we draw tangents such as APC to the curve. The slope of the tangent, in degrees per second, gives us the rate of fall of temperature, when the water is at the temperature θ:

$$\text{rate of fall} = \frac{AB}{BC} = \frac{\theta_1 - \theta_2}{t_2 - t_1}$$

We then plot these rates against the excess temperature, $\theta - \theta_R$, as in Figure 26.4 (iii), and find a straight line passing through the origin. Since the heat lost per second by the water and calorimeter is proportional to the rate of fall of the temperature, Newton's law is thus verified.

Heat Loss and Temperature Fall

Besides the excess temperature, the rate of heat loss depends on the exposed area of the calorimeter, and on the nature of its surface: a dull surface loses heat a little faster than a shiny one, because it is a better radiator (p. 718). This can be shown by doing a cooling experiment twice, with equal masses of water, but once with the calorimeter polished, and once after it has been blackened in a candle-flame. In general, for any body with a uniform surface at a uniform temperature θ, we may write, if Newton's law is true,

$$\text{heat loss/second} = \frac{dQ}{dt} = kS(\theta - \theta_R) \qquad . \qquad . \qquad . \qquad (2)$$

where S is the area of the body's surface, θ_R is the temperature of its surroundings, k is a constant depending on the nature of the surface, and Q denotes the heat lost from the body.

When a body loses heat Q, its temperature θ falls; if m is its mass, and c its specific heat capacity, then its rate of fall of temperature, $d\theta/dt$, is given by

$$\frac{dQ}{dt} = -mc\frac{d\theta}{dt}$$

Now the mass of a body is proportional to its volume. The rate of heat loss, however, is proportional to the surface area of the body. The rate of fall of temperature is therefore proportional to the ratio of surface to volume of the body.

For bodies of similar shape, the ratio of surface to volume is inversely proportional to any linear dimension. If the bodies have surfaces of similar nature, therefore, the rate of fall of temperature is inversely proportional to the linear dimension; a small body cools faster than a large one. This is a fact of daily experience: a small coal which falls out of the fire can be picked up sooner than a large one; a tiny baby should be more thoroughly wrapped up than a grown man. In calorimetry by the method of mixtures, the fact that a small body cools faster than a large one means that the larger the specimen, the less serious is the heat loss in transferring it from its heating place to the calorimeter. It also means that the larger the scale of the whole apparatus, the less serious are the errors due to loss of heat from the calorimeter.

Cooling Correction

As we mentioned previously, a 'cooling correction' is needed in electrical heating experiments and the method of mixtures.

Newton's law can be used to make an approximation for the cooling correction where the heat gained by a solid, for example, is fairly rapid. Figure 26.5 shows the temperature rise from the initial temperature A to the observed

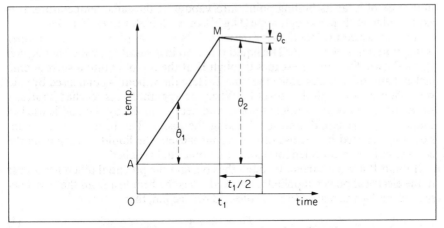

Figure 26.5 *Cooling correction (good conductor)*

maximum M when the solid gained heat, and the temperature fall from M when the solid cooled a few degrees.

We can see from the diagram that the average temperature excess θ_2 above the surroundings during cooling is about *twice* the average temperature excess θ_1 during heating. Hence, from Newton's law, the rate of cooling during the time t_1 of heating is about *half* that during cooling. So the cooling correction θ_c is *the temperature drop from the maximum M in half the time t_1.*

As an illustration, suppose that the temperature of the metal block in the electrical measurement of its specific heat capacity rose to an observed maximum of 21·5°C in 4·0 min. The cooling correction is then roughly the temperature drop in 2·0 min. If this drop is 0·2 K, the final temperature corrected for cooling is 21·7°C.

The approximate cooling correction described is suitable only for good conductors. It can thus be used in determining the specific heat capacity of a metal by the electrical method. In similar experiments with poor conductors such as rubber or glass, however, it may take a long time for the solid to reach its final temperature. In such cases a more accurate way of making the cooling correction is needed.

Specific Latent Heat

Vaporisation, Electrical Method for Specific Latent Heat

The *specific latent heat of vaporisation* of a liquid is the heat required to convert unit mass of it, at its boiling point, into vapour at the same temperature. It is expressed in joule per kilogramme ($J \, kg^{-1}$), or, with high values, in $kJ \, kg^{-1}$.

A modern electrical method for the specific latent heat of evaporation of water is illustrated in Figure 26.6. The liquid is heated in a vessel U by the heating coil R. As shown, the vapour escapes through H at the top of U into a surrounding jacket J and then passes down the tube T. Here the vapour is condensed by cold water flowing through the jacket K. When the apparatus has reached its *steady state*, the liquid is at its boiling-point, and heat supplied by the coil is used in evaporating the liquid, and in offsetting the losses. The heat losses are considerably reduced by the use of the vapour jacket. The liquid coming from the condenser is then collected for a measured time, and weighed.

If I and V are the current through the coil, and the potential difference across it, the electrical power supplied is IV. And if h is the heat lost from the vessel per second, and m the mass of liquid collected per second, then

$$IV = ml + h \qquad . \qquad . \qquad . \qquad . \qquad . \qquad (1)$$

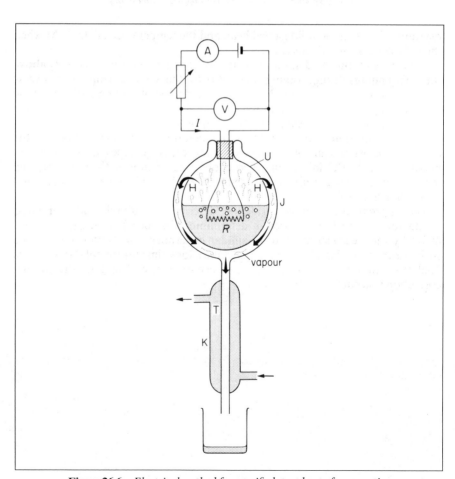

Figure 26.6 *Electrical method for specific latent heat of evaporation*

The heat losses h are determined by the temperature of the vessel, which is fixed at the boiling-point of the liquid. So they can be eliminated by a second experiment with a different rate of evaporation, as in the case of the continuous flow method, p. 638. If I', V' are the new current and potential difference, and if m' is the mass per second evaporated, then

$$I'V' = m'l + h$$

Hence by subtraction from equation (1)

$$l = \frac{(IV - I'V')}{(m - m')}$$

It may be noted that a much higher power supply is needed for determining the specific latent heat of vaporisation of water than for alcohol, as water has a much higher value of l. The specific latent heat of evaporation of water is about $l = 2260\,\text{kJ}\,\text{kg}^{-1}$ or $2\cdot26 \times 10^6\,\text{J}\,\text{kg}^{-1}$.

Method of Mixtures

To find the specific latent heat of vaporisation of water by mixtures, we pass steam into a calorimeter with water (Figure 26.7). On its way the steam passes

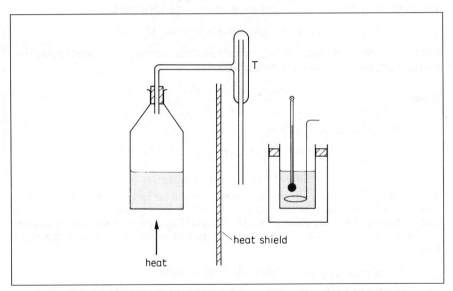

Figure 26.7 *Specific latent heat of evaporation of water by mixtures*

through a vessel, T in the figure, which traps any water carried over by the steam and is called a steam-trap. The mass m of condensed steam is found by weighing. If θ_1 and θ_2 are the initial and final temperatures of the water, the specific latent heat l is given by:

$$\left.\begin{array}{l}\text{heat given by steam}\\\text{condensing}\end{array}\right\} + \left\{\begin{array}{l}\text{heat given by condensed}\\\text{water cooling from}\\100°C \text{ to } \theta_2\end{array}\right\} = \left\{\begin{array}{l}\text{heat taken by}\\\text{calorimeter and}\\\text{water}\end{array}\right.$$

$$ml \qquad + \qquad mc_w(100 - \theta_2) \qquad = \qquad (m_1 c_w + C)(\theta_2 - \theta_1)$$

where m_1 is the mass of water in the calorimeter, c_w is the specific heat capacity of water and C is the heat capacity (mass × specific heat capacity) of the metal calorimeter.

Hence
$$l = \frac{(m_1 c_w + C)(\theta_2 - \theta_1)}{m} - c_w(100 - \theta_2)$$

The result for l is only approximate as the steam-trap is not efficient.

Fusion

The specific latent heat of fusion of a solid is the heat required to convert unit mass of it, at its melting-point, into liquid at the same temperature. It is expressed in joules per kilogramme ($J\,kg^{-1}$). High values can be more conveniently expressed in $kJ\,kg^{-1}$.

Ice is one of the substances whose specific latent heat of fusion we are likely to have to measure. To do so, place warm water, at a temperature of θ_1 a few degrees above room temperature, inside a calorimeter. Then add small lumps of ice, dried by blotting paper, until the temperature reaches a value θ_2 as much below room temperature as θ_1 was above. In this case a 'cooling correction' is not necessary. Weigh the mixture, to find the mass m of ice which has been added. Then the specific latent heat l is given by:

$$\left.\begin{array}{l}\text{heat given by calorimeter}\\ \text{and water in cooling}\end{array}\right\} = \left\{\begin{array}{l}\text{heat used in}\\ \text{melting ice}\end{array}\right\} + \left\{\begin{array}{l}\text{heat used in warming melted}\\ \text{ice from 0°C to }\theta_2\end{array}\right.$$

$$\therefore (m_1 c_w + C)(\theta_1 - \theta_2) = ml + mc_w(\theta_2 - 0)$$

where m_1 = mass of water and c_w = specific heat capacity, C = heat capacity of calorimeter, and θ_1 = initial temperature.

Hence
$$l = \frac{(m_1 c_w + C)(\theta_1 - \theta_2)}{m} - c_w \theta_2$$

A modern electrical method gives

$$l = 334\,kJ\,kg^{-1} \quad \text{or} \quad 3\cdot34 \times 10^5\,J\,kg^{-1}$$

Example on Specific Latent Heat

An electric kettle with a 2·0 kW heating element has a heat capacity of $400\,J\,K^{-1}$. 1·0 kg of water at 20°C is placed in the kettle. The kettle is switched on and it is found that 13 minutes later the mass of water in it is 0·5 kg. Ignoring heat losses calculate a value for the specific latent heat of vaporisation of water. (Specific heat capacity of water $= 4\cdot2 \times 10^3\,J\,kg^{-1}\,K^{-1}$.)

Total heat supplied $= 2000 \times 13 \times 60 = 1\cdot56 \times 10^6\,J$

Heat used for kettle $= C\theta = 400 \times (100 - 20) = 32\,000 = 0\cdot032 \times 10^6\,J$

Heat used to raise temperature of 1 kg of water from 20°C to 100°C

$$= mc\theta = 1 \times 4200 \times (100 - 20) = 0\cdot336 \times 10^6\,J$$

So total heat to change water at 100°C to steam at 100°C

$$= 1\cdot56 \times 10^6 - (0\cdot032 \times 10^6 + 0\cdot336 \times 10^6)\,J$$
$$= 1\cdot192 \times 10^6\,J$$

Since mass of water changed to steam $= 1\cdot0 - 0\cdot5 = 0\cdot5$ kg, then

$$l = \frac{1\cdot192 \times 10^6}{0\cdot5} = 2\cdot38 \times 10^6\,J\,kg^{-1}$$

_____ Exercises 26 _____

1   A metal cylinder of mass 0·5 kg is heated electrically by a 12 W heater in a room at 15°C. The cylinder temperature rises uniformly to 25°C in 5 min and later becomes constant at 45°C.   (i) What is the rate of loss of heat of the cylinder to the surroundings at 45°C? Explain your answer   (ii) Assuming the rate of heat loss is proportional to the excess temperature over the surroundings calculate the rate of loss of heat of the cylinder at 20°C.   (iii) Calculate the specific heat capacity of the metal, taking into account the loss of heat to the surroundings.

2   An electrical heater of 2 kW is used to heat 0·5 kg of water in a kettle of heat capacity 400 J K^{-1}. The initial water temperature is 20°C.
    Neglecting heat losses,   (i) how long will it take to heat the water to its boiling point, 100°C?   (ii) starting from 20°C, what mass of water is boiled away in 5 min? (Assume for water, specific heat capacity = 4200 J kg^{-1} K^{-1} and specific latent heat of vaporisation = 2 × 10^6 J kg^{-1}.)

3   In an electrical constant flow experiment to determine the specific heat capacity of a liquid, heat is supplied to the liquid at a rate of 12 W. When the rate of flow is 0·060 kg min^{-1} the temperature rise along the flow is 2·0 K. Use these figures to calculate a value for the specific heat capacity of the liquid.
    If the true value of the specific heat capacity is 5400 J kg^{-1} K^{-1}, estimate the percentage of heat lost in the apparatus. Explain briefly how, in practice, you would reduce or make allowance for this heat loss. (L.)

4   What is meant by the specific latent heat of vaporisation of a liquid? Explain how latent heat of vaporisation can be regarded as molecular potential energy.
    Calculate the potential energy per molecule released when 18 g of steam condenses to water at 100°C. (Specific latent heat of vaporisation of water = 2·26 × 10^6 J kg^{-1}. Mass of 1 mole of water = 18 g. Number of molecules in a mole of molecules = 6·02 × 10^{23}.) (L.)

5   In a constant flow calorimeter, being used for measuring the specific heat capacity of a liquid, a p.d. of 4·0 V was applied to the heating coil. The rate of flow was now doubled and, by adjusting the applied p.d., the same inlet and outlet temperatures were obtained. Assuming heat losses to be negligible calculate the new value of the applied p.d. (L.)

6   Describe how you might measure, by an electrical method, the specific heat capacity of copper provided in the form of a cylinder 4 cm long and 1 cm in diameter.
    Describe the procedure you would use to make an allowance for heat loss and explain how you would derive the specific heat capacity from your measurements.
    When a metal cylinder of mass 2·0 × 10^{-2} kg and specific heat capacity 500 J kg^{-1} K^{-1} is heated by an electrical heater working at constant power, the initial rate of rise of temperature is 3·0 K min^{-1}. After a time the heater is switched off and the initial rate of fall of temperature is 0·3 K min^{-1}. What is the rate at which the cylinder gains heat energy immediately before the heater is switched off? (N.)

7   In an experiment to measure the specific heat capacity of water, a stream of water flows at a steady rate of 5·0 g s^{-1} over an electrical heater dissipating 135 W, and a temperature rise of 5·0 K is observed. On increasing the rate of flow to 10·0 g s^{-1}, the same temperature rise is produced with a dissipation of 240 W. Explain why the power in the second case is not twice that needed in the first case, and deduce a value for the specific heat capacity of water.
    Discuss the advantages of using such a continuous flow method for measuring the specific heat capacity of a liquid. Describe how a temperature rise of the order of 5 K in the region between 0°C and 100°C may be measured to an accuracy of better than $\frac{1}{2}$%. (O. & C.)

8   A student using a continuous flow calorimeter obtains the following results:
    (a) Using water, which enters at 18·0°C and leaves at 22·0°C, the rate of flow is 20 g min^{-1}, the current in the heating element is 2·3 A and the potential difference across it is 3·3 V.

(b) Using oil, which flows in and out at the same temperatures as the water, the rate of flow is 70 g min⁻¹, the current is 2·7 A and the potential difference is 3·9 V.

Taking the specific heat capacity of water to be 4200 J kg⁻¹ K⁻¹, calculate, explaining your method clearly, (i) the rate of heat loss from the apparatus, (ii) the specific heat capacity of the oil.

Explain carefully how, using this same method, the specific heat capacity of the oil could be obtained without a knowledge of the specific heat capacity of water.

Explain why readings should only be taken when a steady state exists. How would you ensure that such a condition had been attained?

Explain in principle what steps you would take to increase the accuracy in the measurements of the values of (i) the rate of flow of the liquid, (ii) the temperature difference between the inflow and outflow, (iii) the potential difference across and the current in the heating element. (*L*.)

9 (a) The conservation of energy principle may be written in equation form as $\Delta Q = \Delta U + \Delta W$. Express this relationship in words, making clear the meaning of each term.

(b) Outline a continuous flow method for measuring the specific heat capacity of water. Show how the conservation of energy principle can be used to calculate the specific heat capacity. Explain one important advantage of your method.

(c) In an experiment to determine the specific heat capacity of aluminium, a cylindrical 1 kg block of aluminium was heated electrically by a 17·3 W immersion heater inserted into a hole in the centre of the block. The block was suspended in a draught-free room at 20°C. The temperature of the block at first rose steadily (10 K in 10 min), then, more slowly, finally stabilising at 85°C. (i) Explain, using the conservation of energy principle, why the temperature of the block stabilised, although the heater was still switched on. (ii) Assuming the rate of heat loss from the block was proportional to the excess temperature of the block above that of the room, calculate (1) the rate of heat loss from the block at 25°C, and (2) the specific heat capacity of aluminium (corrected for heat loss). (iii) When a similar experiment was performed under the same conditions with a cylindrical 1 kg block of iron, the final steady temperature was considerably higher than 85°C. Suggest the most likely reason for this. (*L*.)

10 Define *specific heat capacity*.

In an experiment to determine the specific heat capacity of a liquid, the liquid flows past an electric heating coil and in the steady state the inlet and outlet temperatures are 10·4°C and 13·5°C respectively. When the mass rate of flow of the liquid is $3·2 \times 10^{-3}$ kg s⁻¹ the power supplied to the coils is 27·4 W. The flow rate is then changed to $2·2 \times 10^{-3}$ kg s⁻¹ and, in order to maintain the same inlet and outlet temperatures the power supplied is adjusted to 19·3 W. Explain why two sets of data are obtained, and calculate the specific heat of the liquid.

Why are the temperatures made the same in each part of the experiment? What are the advantages of this method over the method of mixtures?

What is the rate of loss of heat in the above experiment? If in a further experiment the surrounding temperature is 11·95°C, the rate of loss of heat will be zero. Why is this so? (*L*.)

11 Give an account of an electrical method of finding the specific latent heat of vaporisation of a liquid boiling at about 60°C. Point out any cases of inaccuracy and explain how to reduce their effect.

Ice at 0°C is added to 200 g of water initially at 70°C in a vacuum flask. When 50 g of ice has been added and has all melted the temperature of the flask and contents is 40°C. When a further 80 g of ice has been added and has all melted the temperature of the whole becomes 10°C. Calculate the specific latent heat of fusion of ice, neglecting any heat lost to the surroundings.

In the above experiment the flask is well shaken before taking each temperature reading. Why is this necessary? (*C.*)

12 In the absence of bearing friction a winding engine would raise a cage weighing 1000 kg at $10 \, \text{m s}^{-1}$, but this is reduced by friction to $9 \, \text{m s}^{-1}$. How much oil, initially at 20°C, is required per second to keep the temperature of the bearings down to 70°C? (Specific heat capacity of oil $= 2100 \, \text{J kg}^{-1} \, \text{K}^{-1}$; $g = 9 \cdot 81 \, \text{m s}^{-1}$.) (*O. & C.*)

13 Figure 26A represents a laboratory apparatus for determining the specific latent heat of vaporisation of water. Draw a diagram of a suitable electric circuit in which the heater coil could be incorporated, and describe briefly how the experiment is carried out and how the result is calculated from the observations made.

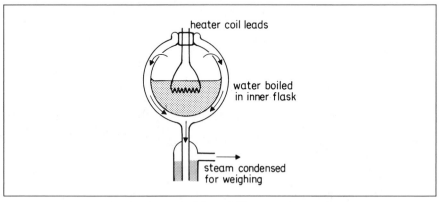

Figure 26A

A pupil performing the experiment finds that, when the heat supply is 16 W, it takes 30 minutes for the temperature of the water to rise from 20°C to 100°C, and that the rate of vaporisation is very slow even at the latter temperature. Estimate an upper limit to the value of the heat capacity of the inner flask and its contents. Calculate the mass of water collected after 30 minutes of steady boiling when the power supply is 60 W. (Take specific latent heat of vaporisation of water to be $2 \cdot 26 \times 10^{6} \, \text{J kg}^{-1}$.) (*O.*)

14 Give an account of a method of determining the specific latent heat of evaporation of water, pointing out the ways in which the method you describe achieves, or fails to achieve, high accuracy.

A 600 watt electric heater is used to raise the temperature of a certain mass of water from room temperature to 80°C. Alternatively, by passing steam from a boiler into the same initial mass of water at the same initial temperature the same temperature rise is obtained in the same time. If 16 g of water were being evaporated every minute in the boiler, find the specific latent heat of steam, assuming that there were no heat losses. (*O. & C.*)

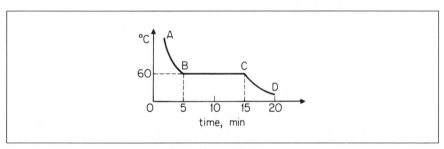

Figure 26B

15 Describe, giving the necessary theory, how the specific latent heat of a liquid may be determined by a method which involves a constant rate of evaporation. What are the advantages of this method over a method of mixtures?

In calorimetry experiments a cooling correction has to be applied for heat losses. Explain how this is done in the experiment you describe.

Figure 26B shows a cooling curve for a substance which, starting as a liquid, eventually solidifies. Explain the shape of the curve and use the following data to obtain a value for the specific latent heat of fusion of the substance: Room temperature $= 20°C$, slope of tangent to curve when temperature is $70°C = 10\,K\,min^{-1}$, specific heat capacity of liquid $= 2\cdot0 \times 10^3\,J\,kg^{-1}\,K^{-1}$, mass of liquid $= 1\cdot50 \times 10^{-2}\,kg$. (You may assume that Newton's law of cooling holds.) (*L.*)

27

Gas Laws, Thermodynamics, Heat Capacities

In this chapter we shall discuss the relationship between the temperature, pressure and volume of a gas. Unlike the case of a solid or liquid this can be expressed in very simple laws, called the Gas Laws, and reduced to a simple equation called the ideal Gas Equation. We shall then deal with the relation between heat energy, internal energy of a gas and work done by a gas in basic thermodynamics and discuss isothermal and adiabatic changes. We conclude with the principal heat capacities of a gas and their relationship.

The Gas Laws

Pressure and Volume: Boyle's Law

In 1660 ROBERT BOYLE—whose epitaph reads 'Father of Chemistry, and Nephew of the Earl of Cork'—published the results of his experiments on the 'natural spring of air'. He meant what we now call the relationship between the pressure of air and its volume.

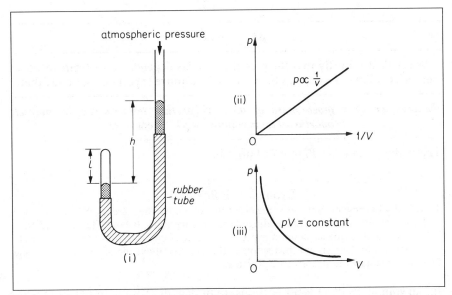

Figure 27.1 *Boyle's law apparatus*

We can repeat Boyle's experiment with the apparatus shown in Figure 27.1 (i), which contains dry air above mercury in a closed uniform tube. We set the open limb of the tube at various heights above and below the closed limb and measure the difference in level, h, of the mercury. When the mercury in the open limb is *below* that in the closed, we reckon h as negative. At each value of h we measure the corresponding length l of the air column in the closed limb. To find the pressure of the air we add the difference in level h to the height of the barometer, H; their sum gives the pressure p of the air in the closed limb:

$$p = g\rho(H+h)$$

where g is the acceleration of gravity and ρ is the density of mercury.

If A is the area of cross-section of the closed limb, the volume of the trapped air is

$$V = lA$$

To interpret our measurements we may either plot $H+h$, which is a measure of p, against $1/l$, or make a table of the product $(H+h)l$. We find that the plot is a straight line, and therefore

$$(H+h) \propto \frac{1}{l} \qquad . \qquad . \qquad . \qquad . \qquad . \qquad (1)$$

Alternatively, we find

$$(H+h)l = \text{constant} \quad . \qquad . \qquad . \qquad . \qquad . \qquad (2)$$

which means the same as (1).

Since g, ρ, and A are constants, the relationships (1) and (2) give

$$p \propto \frac{1}{V}$$

or $$pV = \text{constant}$$

We shall see shortly that the pressure of a gas depends on its temperature as well as its volume. To express the results of the above experiments, we say that

the pressure of a given mass of gas, at constant temperature, is inversely proportional to its volume, or $pV = $ constant.

This is *Boyle's law*. See Figure 27.1 (ii), (iii).

Examples on Boyle's law

1 A faulty barometer tube has some air at the top above the mercury. When the length of the air column is 250 mm, the reading of the mercury above the outside level is 750 mm. When the length of the air column is decreased to 200 mm, by depressing the barometer tube further into the mercury the reading of the mercury above the outside level becomes 746 mm. Calculate the atmospheric pressure.

Initial volume V_1 of air is proportional to the length 250 mm. Initial air pressure $p_1 = (A - 750)$ mmHg, where A is the atmospheric pressure. Also, new volume V_2 of air is proportional to length 200 mm, and new pressure of air, $p_2 = (A - 746)$ mmHg.

From Boyle's law, $\qquad p_1 V_1 = p_2 V_2$

So $\qquad\qquad 250 \times (A - 750) = 200 \times (A - 746)$

Thus $\qquad\qquad 5(A - 750) = 4(A - 746)$

So $\qquad\qquad 5A - 4A = A = 3750 - 2984 = 766\,\text{mmHg}$

2 A vacuum pump has a cylinder of volume v and is connected to a vessel of volume V to pump out air from the vessel. The initial pressure in the vessel is p. Calculate the reduced pressure after n strokes of the pump.

What number of strokes is needed to reduce the pressure from $2{\cdot}0 \times 10^5\,\text{Pa}$ to $1{\cdot}0 \times 10^3\,\text{Pa}$ if the closed vessel has a volume of $200\,\text{cm}^3$ and the pump cylinder is $20\,\text{cm}^3$?

On the first part of the stroke, the air in the vessel expands from V to $(V + v)$. So, from Boyle's law, the reduced pressure p_1 is given by

$$p_1(V + v) = pV$$

So $\qquad\qquad p_1 = p\left(\dfrac{V}{V+v}\right).$ $\qquad$. $\qquad$. $\qquad$. $\qquad$. $\qquad$. $\qquad$ (1)

On the other part of the stroke, the air drawn in is pushed out into the atmosphere by the pump, leaving a pressure p_1 in the vessel. After the second stroke, the reduced pressure p_2 is given, from (1), by

$$p_2 = p_1\left(\frac{V}{V+v}\right) = p\left(\frac{V}{V+v}\right)^2$$

So after n strokes, the reduced pressure p_n is given by

$$p_n = p\left(\frac{V}{V+v}\right)^n.$$ $\qquad$. $\qquad$. $\qquad$. $\qquad$. $\qquad$ (2)

Calculation Since $V = 200\,\text{cm}^3$, $v = 20\,\text{cm}^3$, $p = 2{\cdot}0 \times 10^5\,\text{Pa}$, $p_n = 1{\cdot}0 \times 10^3\,\text{Pa}$,

From (2), $\qquad 1{\cdot}0 \times 10^3 = 2{\cdot}0 \times 10^5\left(\dfrac{200}{200 + 20}\right)^n$

So $\qquad\qquad \dfrac{1}{200} = \left(\dfrac{200}{220}\right)^n = \left(\dfrac{10}{11}\right)^n$

Taking logs to base 10, $\log(1/200) = n(\log 10 - \log 11)$

$$\therefore\ -2{\cdot}301 = n(1 - 1{\cdot}041)$$

$$\therefore\ n = \frac{2{\cdot}301}{0{\cdot}041} = 56\ \text{strokes}$$

Volume and Temperature: Charles's Law

Measurements of the change in volume of a gas with temperature, at constant pressure, can be made with the apparatus shown in Figure 27.2.

Dry air is trapped by mercury in the closed limb C of the tube AC; a scale engraved upon C enables us to measure the length of the air column, l. The tube is surrounded by a water-bath W, which we can heat by passing in steam. After making the temperature uniform by stirring, we level the mercury in the limbs A and C, by pouring mercury in at A, or running it off at B. The air in C is then always at atmospheric (constant) pressure. We measure the length l and plot it against the temperature, θ (Figure 27.3).

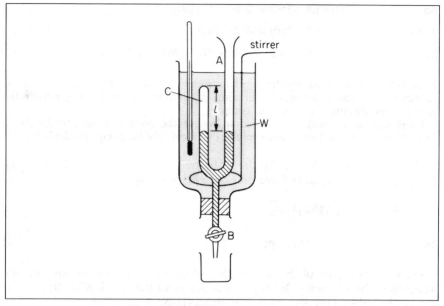

Figure 27.2 *Charles's law experiment*

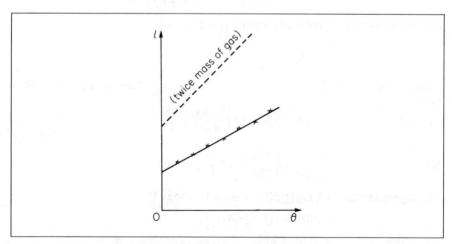

Figure 27.3 *Results of experiment*

If A is the constant cross-section of the tube, the volume of the trapped air is

$$V = lA$$

The cross-section A, and the distance between the divisions on which we read l, both increase with the temperature θ. But their increases are very small compared with the expansion of the gas. So we may say that the volume of the gas is proportional to the scale-reading of l. The graph then shows that the volume of the gas, at constant pressure, increases *uniformly* with its temperature as measured on the mercury-in-glass scale. A similar result is obtained with twice the mass of gas, as shown in Figure 27.3.

Expansivity of Gas (Constant Pressure)

The rate at which the volume of a gas increases with temperature can be defined by a quantity called its *expansivity at constant pressure*, α_p,

$$\alpha_p = \frac{\text{volume at } \theta°C - \text{volume at } 0°C}{\text{volume at } 0°C} \times \frac{1}{\theta}$$

Thus, if V is the volume at $\theta°C$, and V_0 the volume at $0°C$, then

$$\alpha_p = \frac{V - V_0}{V_0 \theta}$$

or $$V = V_0(1 + \alpha_p \theta)$$

Charles found that α_p had the same numerical value, $\frac{1}{273}$, for all gases. This so-called Charles's law states: *The volume of a given mass of any gas, at constant pressure, increases by $\frac{1}{273}$ of its value at $0°C$, for every degree Celsius rise in temperature.*

Absolute (Kelvin) Temperature

Charles's law shows that, if we plot the volume V of a given mass of any gas at constant pressure against its temperature θ, we shall get a straight line graph A as shown in Figure 27.4. If we produce this line backwards, it will meet the temperature axis at $-273°C$. This temperature is called the *absolute zero*. If a gas

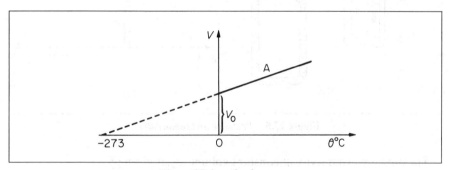

Figure 27.4 *Absolute zero*

is cooled, it liquefies before it reaches $-273°C$, and Charles's Law no longer holds; but that fact does not affect the form of the relationship between the volume and temperature at higher temperatures. We may express this relationship by writing

$$V \propto (273 + \theta)$$

The quantity $(273 + \theta)$ is called the *absolute temperature* of the gas, and is denoted by T. The idea of absolute temperature was developed by Lord Kelvin, and absolute temperatures, T, are therefore expressed in kelvin:

$$T(K) = (273 + \theta)(°C)$$

From Charles's law, we see that the volume of a given mass of gas at constant pressure is proportional to its absolute or kelvin temperature, since

$$V \propto (273 + \theta) \propto T$$

So, if a given mass of gas has a volume V_1 at $\theta_1°$C, and is heated at constant pressure to $\theta_2°$C, its new volume is given by

$$\frac{V_1}{V_2} = \frac{273+\theta_1}{273+\theta_2} = \frac{T_1}{T_2}$$

Pressure and Temperature

The effect of temperature on the *pressure* of a gas, at constant volume, can be investigated with the apparatus shown in Figure 27.5 (i). The bulb B contains air, which can be brought to any temperature θ by heating the water in the surrounding bath W. When the temperature is steady, the mercury in the closed limb of the tube is brought to a fixed level D, so that the volume of the air is fixed. The difference in level, h of the mercury in the open and closed limbs is then added to the height of the barometer, H, to give the pressure p of the gas in cm of mercury. If p, $(h+H)$, is plotted against the temperature, the plot is a straight line, Figure 27.5 (ii).

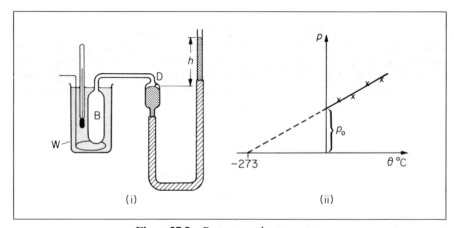

Figure 27.5 *Pressure and temperature*

The pressure expansivity at constant volume, α_V, is given by

$$\alpha_V = \frac{p-p_0}{p_0\theta}$$

where p_0 is the pressure at 0°C. The expansivity α_V, which expresses the change of pressure with temperature, at constant volume, has practically the same value for all gases: $\frac{1}{273}$ K^{-1}. It is thus numerically equal to the expansivity α_p. We may therefore say that, at constant volume, the pressure of a given mass of gas is proportional to its absolute or kelvin temperature T, since

$$p \propto (273+\theta) \propto T$$

$$\therefore \frac{p_1}{p_2} = \frac{273+\theta_1}{273+\theta_2} = \frac{T_1}{T_2}$$

Gas Laws: At constant pressure, $V \propto T$ (kelvin) or $V_1/V_2 = T_1/T_2$
At constant volume, $p \propto T$ (kelvin) or $p_1/p_2 = T_1/T_2$
At constant temperature, $p \propto 1/V$ or pV = constant (Boyle)

The Ideal Gas Equation

Figure 27.6 illustrates how we may find the general relationship between pressure, volume and temperature of a given mass of gas. This relationship is called the *ideal gas equation.*

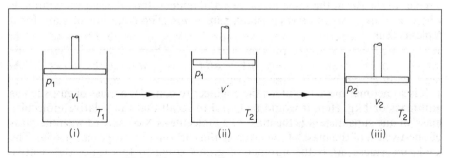

Figure 27.6   *Changing temperature and pressure of a gas*

At (i) we have the gas occupying a volume V_1 at a pressure p_1, and an absolute temperature T_1. We wish to calculate its volume V_2 at an absolute temperature T_2 and pressure p_2, as at (iii). We first raise the temperature to T_2 while keeping the pressure constant at p_1, as at (ii). If V is the volume of the gas at (ii) then, by Charles's law:

$$\frac{V'}{V_1} = \frac{T_2}{T_1} \qquad . \quad . \quad . \quad . \quad . \quad (1)$$

As shown in (iii), we now increase the pressure to p_2, while keeping the temperature constant at T_2. By Boyle's law,

$$\frac{V_2}{V'} = \frac{p_1}{p_2} . \qquad . \quad . \quad . \quad . \quad (2)$$

Eliminating V' between equations (1) and (2), we find

$$\frac{V_2}{V_1} = \frac{T_2}{T_1} . \frac{p_1}{p_2}$$

or

$$\frac{p_2 V_2}{T_2} = \frac{p_1 V_1}{T_1}$$

In general, therefore,

$$\frac{pV}{T} = R \qquad . \quad . \quad . \quad . \quad . \quad (3)$$

where **R** is a constant, so that

$$pV = RT . \qquad . \quad . \quad . \quad . \quad (4)$$

Ideal Gas Equation, The Gas Constant

Equation (4) is the *ideal gas equation* or *equation of state*. The magnitude of the constant in the equation depends on the nature of the gas—air, hydrogen, for example,—and on the amount—number of moles or mass—of the gas.

For *one mole* of a particular gas, an amount discussed shortly, the corresponding *molar gas constant* is given the symbol R. So if 1 mole occupies a volume V equal to V_m at a pressure p and absolute temperature T, we write

$$pV_m = RT. \qquad . \quad . \quad . \quad . \quad . \qquad (5)$$

n moles of the gas at the same pressure p and temperature T occupy a volume V where $V = nV_m$. So the gas constant here is nR. The equation of state for n moles is thus

$$pV = nRT \qquad . \quad . \quad . \quad . \quad . \qquad (5A)$$

It is sometimes more useful for the engineer to consider a mass of gas, such as unit mass or 1 kg. Here it would be useful to recall that the 'relative molecular mass' of any substance X is the mass of a molecule of X compared with the mass of one-twelfth of the mass of the atom of the carbon-12 isotope (see p. 659). The 'molecular mass' M is the number of grams equal to this ratio, the relative molecular mass. For the gases hydrogen, helium, oxygen and carbon dioxide, the molecular mass M is respectively about 2, 4, 32 and 44 g, or, in the SI unit of kilogram, 2×10^{-3}, 4×10^{-3}, 32×10^{-3}, and 44×10^{-3} kg respectively.

We shall denote the gas constant per unit mass by the symbol r. Thus if M is the molecular mass in *kg* (SI unit), $r = R/M$. So the equation of state for unit mass of gas, 1 kg, may be written

$$pV = \frac{R}{M}T = rT . \qquad . \quad . \quad . \quad . \qquad (6)$$

For a mass m kg of gas, the equation of state is

$$pV = m\frac{R}{M}T = mrT \qquad . \quad . \quad . \quad . \qquad (6A)$$

Since $m/M = n$, the number of moles, equation (6A) is identical to (5A).

For a mole of gas, the *density* $\rho = M/V_m$, or $V_m = M/\rho$. From (5), it follows that the equation of state can be written

$$\frac{p}{\rho} = \frac{R}{M}T \qquad . \quad . \quad . \quad . \quad . \qquad (7)$$

Units of Gas Constant

The unit of

$$\frac{pV}{T} \text{ or } \frac{\text{pressure} \times \text{volume}}{\text{temperature}} = \frac{N\,m^{-2} \times m^3}{K} = N\,m\,K^{-1}$$

$$= J\,K^{-1}$$

since 1 newton $\times$ 1 metre = 1 joule. So

unit of molar gas constant = $J\,mol^{-1}\,K^{-1}$

and   unit of gas constant per unit mass = $J\,kg^{-1}\,K^{-1}$

Avogadro's Hypothesis: Molar Gas Constant

Avogadro, with one simple-looking idea, illuminated chemistry as Newton illuminated mechanics. In 1811 he suggested that chemically active gases, such as oxygen, existed not as single atoms, but as pairs: he proposed to distinguish between an atom, O, and a molecule, O_2. In 1814 Avogadro also put forward another idea, now called *Avogadro's hypothesis*: that equal volumes of all gases, at the same temperature and pressure, contain equal numbers of molecules. The number of molecules in $1 \, cm^3$ of gas at s.t.p. is called Loschmidt's number; it is $2 \cdot 69 \times 10^{19}$.

The amount of a substance which contains as many elementary units as there are atoms in $0 \cdot 012 \, kg$ (12 g) of carbon-12 is called a *mole*, symbol 'mol'. The number of molecules per mole is the same for all substances. It is called the *Avogadro constant*, symbol N_A, and is equal to $6 \cdot 02 \times 10^{23} \, mol^{-1}$.

From Avogadro's hypothesis, it follows that the mole of *all* gases, at the same temperature and pressure, occupy equal volumes. Experiment confirms this; at s.t.p. 1 mole of any gas occupies $22 \cdot 4$ litres. Consequently, if V_m is the volume of 1 mole, then the ratio pV_m/T is the same for all gases. So the molar gas constant R, which is equal to this ratio, is the *same for all gases*.

At s.t.p.
$$V_m = 22 \cdot 4 \, \text{litres} = 22 \cdot 4 \times 10^{-3} \, m^3$$

$$p = 760 \, \text{mmHg} = 1 \cdot 013 \times 10^5 \, \text{Pa}$$

$$T = 273 \, K$$

$$\therefore R = \frac{pV_m}{T} = \frac{1 \cdot 013 \times 10^5 \times 22 \cdot 4 \times 10^{-3}}{273}$$

$$= 8 \cdot 31 \, J \, mol^{-1} \, K^{-1}$$

Gas Constant per Unit Mass

If the amount of gas is 1 kilogram, the gas constant is expressed as $J \, kg^{-1} \, K^{-1}$.

As an example, consider the gas oxygen, O_2, which has a molecular mass of 32 g. For 1 mole, 32 g, the gas constant $= R = 8 \cdot 31$ numerically. So for 1 kg (1000 g),

$$\text{gas constant} = \frac{1000}{32} \times 8 \cdot 31 = 260 \, J \, kg^{-1} \, K^{-1}$$

Applying the Gas Equation

We can apply the equation $pV = nRT$, where n is the number of moles, to calculate the mass of a gas. Suppose oxygen gas, contained in a cylinder of volume V of $1 \times 10^{-2} \, m^3$ has a temperature T of 300 K and a pressure p_1 of $2 \cdot 5 \times 10^5 \, \text{Pa}$. After some of the oxygen is used at constant temperature, the pressure falls to $1 \cdot 3 \times 10^5 \, \text{Pa}$, p_2.

From $pV = nRT$, the number of moles in the cylinder initially is given by

$$n_1 = \frac{p_1 V}{RT}$$

and finally by
$$n_2 = \frac{p_2 V}{RT}$$

Hence the number of moles used $= n_1 - n_2$

$$= \frac{(p_1 - p_2)V}{RT}$$

$$= \frac{(2 \cdot 5 - 1 \cdot 3) \times 10^5 \times 1 \times 10^{-2}}{8 \cdot 3 \times 300} = 0 \cdot 48$$

using the value $R = 8 \cdot 3 \, \text{J mol}^{-1} \, \text{K}^{-1}$.

But   molecular mass of oxygen $= 32 \, \text{g} = 32 \times 10^{-3} \, \text{kg}$

$\therefore$ mass of oxygen used $= 0 \cdot 48 \times 32 \times 10^{-3} \text{kg}$

$$= 0 \cdot 015 \, \text{kg (approx.)}$$

Example on Gas Equation

A cylinder containing 19 kg of compressed air at a pressure 9·5 times that of the atmosphere is kept in a store at 7°C. When it is moved to a workshop where the temperature is 27°C a safety valve on the cylinder operates, releasing some of the air. If the valve allows air to escape when its pressure exceeds 10 times that of the atmosphere, calculate the mass of air that escapes. (*L*.)

From $pV = nRT$, if A is the atmospheric pressure and V is the cylinder volume,

$$\text{initial number of moles of air, } n_1 = \frac{pV}{RT} = \frac{9 \cdot 5A \times V}{R \times 280} \quad . \quad . \quad (1)$$

and $$\text{final number of moles, } n_2 = \frac{pV}{RT} = \frac{10A \times V}{R \times 300} . \quad . \quad . \quad (2)$$

Suppose m is the mass of air left in the cylinder. Then, since the mass is proportional to the number of moles, it follows that $m/19 \, \text{kg} = n_2/n_1$. Dividing (2) by (1) and simplifying, then

$$\frac{m}{19 \, kg} = \frac{10 \times 280}{9 \cdot 5 \times 300}$$

So $$m = \frac{19 \times 10 \times 280}{9 \cdot 5 \times 300} = 18 \cdot 67 \, \text{kg}$$

So mass of air escaped $= 19 - 18 \cdot 67 = 0 \cdot 33 \, \text{kg}$

Connected Gas Containers

As we have seen, the number of moles, n, in a gas can be found from the relation

$$n = \frac{pV}{RT}$$

In a closed system of connected gas containers, some gas may flow out of one container when its temperature rises, to other containers. The *total* number of moles of gas in all the containers remains constant, however, no matter what changes take place in individual containers. Hence, in a closed system, the sum of the values of pV/RT *for all the containers is constant*. This is used in the following example.

Example on Connected Gas Containers

Two gas containers with volumes of 100 cm³ and 1000 cm³ respectively are connected by a tube of negligible volume, and contain air at a pressure of 1000 mm mercury. If the temperature of both vessels is originally 0°C, how much air will pass through the connecting tube when the temperature of the smaller is raised to 100°C? Give your answer in cm³ measured at 0°C and 760 mm mercury.

The *total* number of moles of air in the two containers remains constant, although some air is transferred from the hot to the cold container on heating.

From $n = pV/RT$,

$$\text{initially, total moles of air} = \frac{pV}{RT} = \frac{1000 \times 1100}{R \times 273}$$

$$\text{and finally, total moles of air} = \frac{p \times 100}{R \times 373} + \frac{p \times 1000}{R \times 273}$$

where p is the new pressure in both containers after heating.

Equating the two numbers of moles, which are unchanged by any transfer,

$$\therefore \frac{p \times 100}{R \times 373} + \frac{p \times 1000}{R \times 273} = \frac{1000 \times 1100}{R \times 273}$$

Cancelling R and simplifying, then

$$p = \frac{11\,000 \times 373}{4003} = 1025\,\text{mmHg}$$

As the question requires, we now have to convert the initial volume of $100\,\text{cm}^3$ of air in the smaller container at 0°C and 1000 mmHg to 0°C and 760 mmHg, and the final volume of $100\,\text{cm}^3$ at 100°C and 1025 mmHg to 0°C and 760 mmHg.

$$\text{Initial volume at 0°C and 760 mmHg} = 100 \times \frac{1000}{760} = 131 \cdot 6\,\text{cm}^3$$

$$\text{Final volume at 0°C and 760 mmHg} = 100 \times \frac{1025}{760} \times \frac{273}{373} = 98 \cdot 7\,\text{cm}^3$$

$\therefore$ volume of air at 0°C and 760 mmHg flowing out $= 33\,\text{cm}^3$ (approx.)

Mixture of Gases: Dalton's Law

Figure 27.7 shows an apparatus with which we can study the pressure of a mixture of gases. A is a bulb, of volume V_1, containing air at atmospheric pressure, p_1. C is another bulb, of volume V_2, containing carbon dioxide at a pressure p_2. The pressure p_2 is measured on the manometer M; in millimetres of mercury it is

$$p_2 = h + H$$

where H is the height of the barometer. (In the same units, the air pressure, $p_1 = H$.)

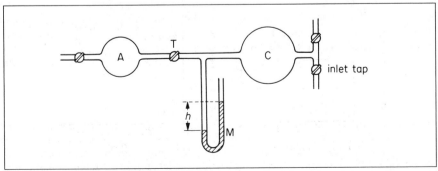

Figure 27.7 *Apparatus for demonstrating Dalton's law of partial pressure*

When the bulbs are connected by opening the tap T, the gases mix, and reach the same pressure, p; this pressure is given by the new height of the manometer. Its value is found to be given by

$$p = p_1 \frac{V_1}{V_1 + V_2} + p_2 \frac{V_2}{V_1 + V_2}$$

Now the quantity $p_1 V_1/(V_1 + V_2)$ is the pressure which the air originally in A would have, if it expanded to occupy A and C; for, if we denote this pressure by p', then $p'(V_1 + V_2) = p_1 V_1$. Similarly $p_2 V_2/(V_1 + V_2)$ is the pressure which the carbon dioxide originally in C would have, if it expanded to occupy A and C. Thus the total pressure of the mixture is the sum of the pressures which the individual gases exert, when they have expanded to fill the vessel containing the mixture.

The pressure of an individual gas in a mixture is called its *partial pressure*: it is the pressure which would be observed if that gas alone occupied the volume of the mixture, and had the same temperature as the mixture. The experiment described shows that *the pressure of a mixture of gases is the sum of the partial pressures of its constituents*. This statement was first made by Dalton, in 1801, and is called *Dalton's Law of Partial Pressures*.

So if air and water-vapour have a total pressure of $1{\cdot}01 \times 10^5$ Pa in a container, and the water-vapour alone has a pressure of $0{\cdot}01 \times 10^5$ Pa, then the air pressure in the mixture has a pressure $= 1{\cdot}01 \times 10^5 - 0{\cdot}01 \times 10^5 = 1{\cdot}00 \times 10^5$ Pa. The volume occupied by either the air or the water-vapour is the whole volume of the container.

Unsaturated and Saturated Vapours

An *unsaturated* vapour is a vapour which is not in contact with its own liquid in a closed space. So a mixture of air and water vapour in a flask, without any water present, is a mixture of air and unsaturated vapour.

A *saturated* vapour is a vapour in contact with, or in equilibrium with, its own liquid. So if a flask contains air and some water at 20°C, the water vapour above the water is saturated vapour at 20°C. The mixture, then, is air and saturated water vapour. Values of saturated vapour pressure (s.v.p.) vary from a very low value at 0°C to 760 mmHg ($1{\cdot}013 \times 10^5$ Pa) at 100°C. If the external atmospheric pressure is 760 mmHg, water will boil at 100°C. The bubbles inside the water then have sufficient vapour pressure to just overcome the outside atmospheric pressure and burst open at the surface.

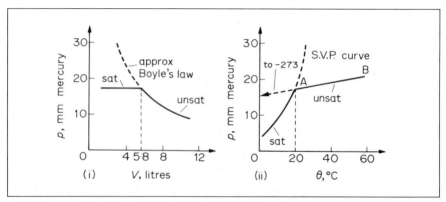

Figure 27.8 *Effect of volume and temperature on pressure of water vapour*

Gas Laws for Vapours

Saturated vapours do not obey Boyle's law: *their pressure is independent of their volume.* Unsaturated vapours obey Boyle's law roughly, as they also obey roughly Charles's law, Figure 27.8 (i). Vapours, saturated and unsaturated, are gases because they spread throughout their vessels; but we find it convenient to distinguish them by name from gases such as air, which obey Charles's and Boyle's laws closely.

Figure 27.8 (ii) shows the effect of heating a saturated vapour. More and more of the liquid evaporates, and the pressure rises very rapidly. As soon as all the liquid has evaporated, however, the vapour becomes unsaturated, and its pressure rises more steadily along a straight line AB. Well away from the saturated state, the unsaturated vapour obeys the relation $p \propto T$ at constant volume, which is a gas law.

Gas Laws Applied to Mixture of Vapour and Gas

The following examples illustrate how the gas laws are applied to the case of a vapour in a mixture with other gases. Note the following: (i) *Vapour unsaturated.* Here the gas laws give a good approximation to changes in pressure, volume and temperature; (ii) *Vapour saturated.* If the vapour is in a mixture with a gas such as air, we apply the gas laws to the *air* after using Dalton's law of partial pressures to allow for the saturated vapour. We do *not* apply the gas laws to the saturated vapour because its mass changes due to condensation or evaporation as conditions change, and the gas laws apply to constant mass of gas.

Examples on Saturated Vapours

1 A narrow tube of uniform bore, closed at one end, has some air entrapped by a small quantity of water. If the pressure of the atmosphere is 760 mmHg, the equilibrium vapour pressure of water at 12°C and at 35°C is 10·5 mmHg and 42·0 mmHg respectively, and the length of the air column at 12°C is 10 cm, calculate its length at 35°C.

For the given mass of air,

$$\frac{p_1 V_1}{T_1} = \frac{p_2 V_2}{T_2}$$

$$p_1 = 76 - 1\cdot05 = 74\cdot95 \text{ cm}, V_1 = 10, T = 273 + 12 = 285 \text{ K}$$

$$p_2 = 76 - 4\cdot2 = 71\cdot8 \text{ cm}, T_2 = 273 + 35 = 308 \text{ K}$$

$$\therefore \frac{74\cdot95 \times 10}{285} = \frac{71\cdot8 V_2}{308}$$

$$\therefore V_2 = \frac{74\cdot95 \times 10 \times 308}{285 \times 71\cdot8} = 11\cdot3$$

So length at 35°C $= 11\cdot3$ cm

2 A closed vessel contains air, saturated water-vapour, and an excess of water. The total pressure in the vessel is 760 mmHg when the temperature is 25°C; what will it be when the temperature has been raised to 100°C? (Saturation vapour pressure of water at 25°C is 24 mmHg.)

From Dalton's law, the pressure of the air at 25°C $= 760 - 24 = 736$ mmHg. Suppose the pressure is p mm at 100°C. Then, since pressure is proportional to absolute temperature for a fixed mass of air, we have

$$\frac{p}{736} = \frac{373}{298}$$

from which $\qquad\qquad p = 921\,\text{mmHg}$

Now the saturation vapour pressure of water at $100°C = 760\,\text{mmHg}$

$$\therefore \text{total pressure in vessel} = 921 + 760 = 1681\,\text{mmHg}$$

_____ **Exercises 27A** _____

Gas Laws

1 A container of gas has a volume of $0.10\,\text{m}^3$ at a pressure of $2.0 \times 10^5\,\text{N m}^{-2}$ and a temperature of 27°C.   (i) Find the new pressure if the gas is heated at constant volume to 87°C.   (ii) The gas pressure is now reduced to $1.0 \times 10^5\,\text{N m}^{-2}$ at constant temperature. What is the new volume of the gas?   (iii) The gas is cooled to $-73°C$ at constant pressure. Find the new volume of the gas.

2 Using a vertical (y) axis to represent pressure, p, and a horizontal axis to represent volume, V, illustrate by rough sketches the changes which take place in (i), (ii) and (iii) in question **1**.

3 A cylinder of gas has a mass of $10.0\,\text{kg}$ and a pressure of 8.0 atmospheres at 27°C. When some gas is used in a cold room at $-3°C$, the gas remaining in the cylinder at this temperature has a pressure of 6.4 atmospheres. Calculate the mass of gas used.

4 State Boyle's law and Charles's law and show how they may be combined to give the equation of state of an ideal gas.

   Two glass bulbs of equal volume are joined by a narrow tube and are filled with a gas at s.t.p. When one bulb is kept in melting ice and the other is placed in a hot bath, the new pressure is $877.6\,\text{mm}$ mercury. Calculate the temperature of the bath. (*L.*)

5 The formula $pv = mrT$ is often used to describe the relationship between the pressure p, volume v, and temperature T of a mass m of a gas, r being a constant. Referring in particular to the experimental evidence how do you justify
   (a) the use of this formula,
   (b) the usual method of calculating T from the temperature t of the gas on the centigrade (Celsius) scale?

   Two vessels of capacity 1.00 litre are connected by a tube of negligible volume. Together they contain $3.42 \times 10^{-4}\,\text{kg}$ of helium at a pressure of $1.07 \times 10^5\,\text{Pa}$ and temperature 27°C. Calculate   (i) a value for the constant r for helium,   (ii) the pressure developed in the apparatus if one vessel is cooled to 0°C and the other heated to 100°C, assuming that the capacity of each vessel is unchanged. (*JMB.*)

6 State *Boyle's law* and *Charles' law*, and show how they lead to the gas equation $PV = RT$. Describe an experiment you would perform to measure the thermal expansion coefficient of dry air.

   What volume of liquid oxygen (density $1140\,\text{kg m}^{-3}$) may be made by liquefying completely the contents of a cylinder of gaseous oxygen containing 100 litres of oxygen at 120 atmospheres pressure and 20°C? Assume that oxygen behaves as an ideal gas in this latter region of pressure and temperature.
   (1 atmosphere $= 1.01 \times 10^5\,\text{N m}^{-2}$; molar gas constant $= 8.31\,\text{J mol}^{-1}\,\text{K}^{-1}$; relative molecular mass of oxygen $= 32.0$.) (*O. & C.*)

7 State the First Law of Thermodynamics and explain what is meant by the *internal energy* of a system. What constitutes the internal energy of an ideal gas? Starting from the expression $P = \frac{1}{3}\rho c^2$, show that the internal energy of an ideal *monatomic* gas is $\frac{3}{2}PV$, and discuss the interpretation of temperature in the kinetic theory, (*see chapter 28.*)

Explain the following observations:
(a) when pumping up a bicycle tyre the pump barrel gets warm, and
(b) when a gas at high pressure in a container is suddenly released, the container cools.

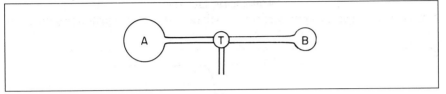

Figure 27A

Two bulbs, A of volume $100\,cm^3$ and B of volume $50\,cm^3$, are connected to a three way tap T which enables them to be filled with gas or evacuated. The volume of the tubes may be neglected, Figure 27A.

(c) Initially bulb A is filled with an ideal gas at 10°C to a pressure of $3\cdot0 \times 10^5$ Pa. Bulb B is filled with an ideal gas at 100°C to a pressure of $1\cdot0 \times 10^5$ Pa. The two bulbs are connected with A maintained at 10°C and B at 100°C. Calculate the pressure at equilibrium.

(d) Bulb A, filled at 10°C to a pressure of $3\cdot0 \times 10^5$ Pa, is connected to a vacuum pump with a cylinder of volume $20\,cm^3$. Calculate the pressure in A after one inlet stroke of the pump. The air in the pump is now expelled into the atmosphere. Calculate the pressure in A after the second inlet stroke. Calculate the number of strokes of the pump to reduce the pressure in A to $1\cdot0 \times 10^3$ Pa. The whole system is maintained at 10°C throughout the process. (*O. & C.*)

Thermodynamics

We now deal with the relation between work and heat energy, which is a branch of thermodynamics. In this book we are concerned only with basic principles.

Work Done by Gas

Consider some gas, at a pressure p, in a cylinder fitted with a piston (Figure 27.9).

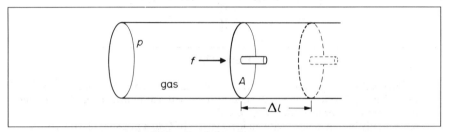

Figure 27.9 *Work done by gas in expansion*

If the piston has an area A, the force on it is

$$f = pA$$

If we allow the piston to move outwards a distance Δl, the gas will expand, and its pressure will fall. But by making the distance very short, we can make the fall in pressure so small that we may consider the pressure constant. The force f is then constant, and the work done is

$$\Delta W = f \cdot \Delta l = pA \cdot \Delta l$$

The product $A \cdot \Delta l$ is the increase in volume, ΔV, of the gas, so that

$$\Delta W = p \cdot \Delta V \qquad . \qquad . \qquad . \qquad . \qquad . \qquad (1)$$

The product of pressure and volume, in general, therefore represents work. If the pressure p is in $N\,m^{-2}$, and the area A is in m^2, the force f is in newton. And if the movement Δl is in m, the work $f \cdot \Delta l$ is in newton × metre or *joule* (J). The increase of volume, ΔV, is in m^3. Thus the product of pressure in $N\,m^{-2}$, and volume in m^3, represents work in joule.

From (1), **work done = pressure × volume change**

So if the volume of a gas at constant pressure of $10^5\,N\,m^{-2}$ expands by $0.01\,m^3$, then

$$\text{work done} = 10^5 \times 0.01 = 1000\,J$$

Latent Heat and Internal Energy, Molecular Potential Energy

The volume of 1 g of steam at 100°C and 760 mmHg pressure is $1672\,cm^3$. Therefore when 1 g of water turns into steam, it expands by $1671\,cm^3$; in doing so, it does work against the atmospheric pressure. The heat equivalent of this work is that part of the latent heat which must be supplied to the water to make it overcome atmospheric pressure as it evaporates; it is called the 'external latent heat'. The rest of the specific latent heat—the internal part—is the equivalent of the work done in separating the molecules, against their mutual attractions. This amount of energy, then, is a measure of the *increase in potential energy* of the molecules of water in the gaseous state over that in the liquid state, at the same temperature.

The work done, W, in the expansion of 1 g from water to steam is the product of the atmospheric pressure p and the increase in volume ΔV:

$$W = p \cdot \Delta V$$

Normal atmospheric pressure corresponds to a pressure $p = 1 \cdot 013 \times 10^5$ Pa,

So $\qquad W = p \cdot \Delta V = 1 \cdot 013 \times 10^5 \times 1671 \times 10^{-6} = 170 \, \text{J}$

The external specific latent heat is therefore

$$l_{ex} = 1 \cdot 013 \times 10^5 \times 1671 \times 10^{-6}$$
$$= 170 \, \text{J g}^{-1} = 170 \, \text{kJ kg}^{-1}$$

This result shows that the external part of the specific latent heat is much less than the internal part. Since the total specific latent heat of vaporisation l is $2270 \, \text{J g}^{-1}$, the internal part is

$$l_{in} = l - l_{ex} = 2270 - 170$$
$$= 2100 \, \text{J g}^{-1} = 2100 \, \text{kJ kg}^{-1}$$

For 1 mole of water, 18 g, the number of molecules is about 6×10^{23}, the Avogadro constant. So the gain in potential energy of a molecule of water when changing from liquid at 100°C to vapour at 100°C is

$$\frac{18 \times 2100}{6 \times 10^{23}} = 6 \cdot 3 \times 10^{-20} \, \text{J molecule}^{-1}$$

Internal Energy of Gas

The *internal energy* of an ideal gas is the kinetic energy of thermal motion of its molecules. As we see later, the magnitude depends on the temperature of the gas and on the number of atoms in its molecule.

The thermal motion of the molecules is a random motion—it is often called the thermal 'agitation' of the molecules. We must appreciate that the energy of the gas which we call internal energy is quite independent of any motion of the gas in bulk. When a cylinder of oxygen is being carried by an express train, its kinetic energy as a whole is greater than when it is standing on the platform; but the random motion of the molecules relative to the cylinder is unchanged—and so is the temperature of the gas.

The same is true of a liquid. In a water-churning experiment to convert mechanical energy into heat, baffles must be used to prevent the water from acquiring any mass-motion—all the work done must be converted into random motion, if it is to appear as heat. Likewise, the internal energy of a solid is the kinetic energy of the vibration of its atoms about their mean positions. Throwing a lump of metal through the air does not raise its temperature, but hitting it with a hammer does.

We shall use the symbol U for the internal energy. Although its absolute magnitude is not known, we are mainly concerned with *changes* in internal energy, denoted by ΔU. We can often calculate ΔU, as we soon show.

First Law of Thermodynamics

The First Law of Thermodynamics states that the *total energy in a closed system is constant*, that is, the energy is conserved in any transfer of energy from one form to another. This law is also known as the Principle of Conservation of Energy (p. 623).

When we warm a gas so that it expands, the heat ΔQ we give to it appears partly as an increase ΔU to its *internal energy*—and hence its temperature—and partly as the energy required for the *external work* done. ΔW. Thus from the First Law of Thermodynamics, we may write

$$\Delta Q = \Delta U + \Delta W \quad . \qquad . \qquad . \qquad . \qquad . \qquad (1)$$

If the expansion of the gas occurs reversibly, then no friction forces are present (p. 671). In the case, $\Delta W = p \cdot \Delta V$. Thus, from (1),

$$\Delta Q = \Delta U + p \cdot \Delta V \quad . \qquad . \qquad . \qquad . \qquad . \qquad (2)$$

Equation (2), then, is a mathematical statement of the First Law of Thermodynamics applied to the case of energy changes associated with a gas.

In the relation $\Delta Q = \Delta U + \Delta W$, ΔW is the external work done *by* the gas. We can express the relation in a different way if heat ΔQ is given to the gas and ΔW is the external work done *on* the gas. In this case, the increase in the internal energy of the gas, ΔU, is given by

$$\Delta U = \Delta Q + \Delta W \quad . \qquad . \qquad . \qquad . \qquad . \qquad (3)$$

The work done on the gas is $p \cdot \Delta V$ and so in this case

$$\Delta U = \Delta Q + p \cdot \Delta V \quad . \qquad . \qquad . \qquad . \qquad . \qquad (4)$$

In electrical heating of a wire, ΔW = work done on the electrons as they flow = energy supplied. From (3), when a lamp filament is *first* switched on, $\Delta W = \Delta U$. So the temperature rises. The filament reaches a constant temperature a short time later. So $\Delta U = 0$. $\Delta W = -\Delta Q$ = heat lost by radiation.

Internal Energy Changes, Ideal Gas

If the volume of a gas is kept *constant* when it is warmed, then no external work is done. So *all* the heat supplied goes in raising the internal energy of the gas. In this special case, then, $\Delta U = \Delta Q$, the heat supplied. If we have 1 mole of gas whose molar heat capacity at constant volume is C_V (discussed shortly), and ΔT is its temperature rise, then $\Delta Q = C_V \cdot \Delta T$. At constant volume, then, the change in internal energy is given by

$$\Delta U = C_V \cdot \Delta T \quad . \qquad . \qquad . \qquad . \qquad . \qquad (3)$$

It should be noted that this is always the change in internal energy of an ideal gas for a temperature change ΔT, no matter how the change has occurred. *The internal energy of an ideal gas is independent of its volume*—it depends only on its temperature.

A perfect, or ideal, gas is one which obeys Boyle's law exactly and whose internal energy is independent of its volume. No such gas exists, but at room temperature, and under moderate pressures, many gases approach the ideal closely enough for most purposes.

Suppose a gas in a vessel is thermally insulated and the gas expands. In this case *no heat* enters or leaves the system. The work done by the gas is then taken from its internal energy, with the result that the gas *cools*. If the gas is compressed instead of expanding, the work done on the gas increases the internal energy. So this time the gas temperature *rises*. In either case the change in internal energy $\Delta U = C_V \cdot \Delta T$.

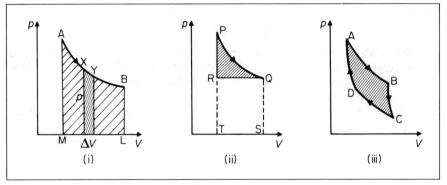

Figure 27.10 *Work done from* p — V *graphs*

Work Done from p— V Graphs

The work done by or on a gas can be found from a graph of pressure p against volume V when the work is done.

In Figure 27.10 (i), a gas has a pressure, volume and temperature represented by the point A and then does work and expands to a new volume and a new pressure and temperature represented by the point B. For a *very* small volume change V from X to Y, small amount of work done W = pressure × volume change, from our result on page 666.

So $\qquad W = p \cdot \Delta V = $ *area* of shaded strip below XY

From A to B, then, we can say that total work done = area below AB, ABLM

So **work done = Area between $p - V$ graph and volume-axis**

In Figure 27.10 (ii), a gas expands from P to Q, then it is compressed from Q to R while its pressure is kept constant (QR is a horizontal line), and finally the gas is compressed back to P while its volume is kept constant (RP is a vertical line. Now from P to Q, the work done *by* the gas is the area PQST. From Q to R, the done *on* the gas is the area QRTS. No work is done along PR (volume constant). So, by subtracting the two areas, as shown shaded,

---

net work done by gas = area enclosed, PQR

Figure 27.10 (iii) shows a gas taken round a so-called *cycle* of changes ABCD, from A and back to A. As before,

net work done by gas = area ABCD

Isothermal Changes

In a car, motor-cycle or aeroplane engine, gases expand and are compressed, and are heated and cooled, in ways more complicated than those already described. We now consider some chief ways in which such changes take place.

When a gas expands or is compressed at *constant temperature*, the gas is said to undergo an *isothermal expansion or compression*.

For a mole of gas, we have already seen that the pressure p, the volume V and the absolute temperature T are related by $pV = RT$. So if the temperature is constant,

$$pV = \text{constant}$$

the value of the constant depending on T and R. So the curve of pressure and volume represents a Boyle's law relation. Such a curve is called an *isothermal* for the given mass of gas. Figure 27.11 shows a family of isothermals for 1 g of air, each isothermal being labelled with the particular constant temperature. At the higher temperatures, the pressure and volume values are higher from $pV = RT$.

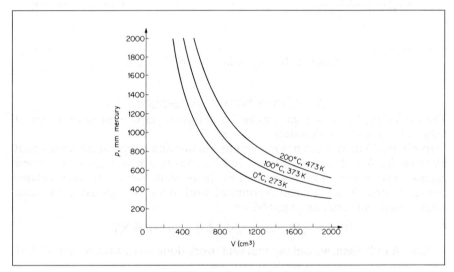

Figure 27.11   *Isothermals for 1 g of air*

Kinetic Theory in Isothermal Change

When a gas expands, it does work—for example, in driving a piston (Figure 27.9, p. 666). The molecules of the gas bombarded the piston, and if the piston moves they give up some of their kinetic energy to it. When a molecule bounces off the *moving* piston, it does so with a velocity less than that with which it struck. The change in velocity is small, because the piston moves much more slowly than the molecule; but there are many molecules striking the piston at any instant, and their total loss of kinetic energy is equal to the work done in driving the piston forward. The work done by a gas in expanding, therefore, is done at the expense of its internal energy. The temperature of the gas will consequently *fall* during expansion, unless heat is supplied to it.

Conversely, if a gas is compressed, its temperature rises. The molecules now rebound from the forward-moving piston with a velocity greater than their incident velocity. The total increase in kinetic energy of all the molecules is equal to the work done in moving the piston. In an isothermal compression or expansion, the gas must be held in a thin-walled, highly conducting vessel, surrounded by a constant temperature bath. And the expansion must take place slowly, so that heat can pass into the gas to maintain its temperature at every instant during the expansion.

External Work Done in Expansion

The heat taken in when a gas expands isothermally is the heat equivalent of the mechanical work done, because there is no change in internal energy of an ideal gas when its temperature is constant. As we have seen earlier, if the volume of the gas increases by a small amount ΔV, at the pressure p, then the work done is

$$\Delta W = p \, \Delta V$$

In an expansion from V_1 to V_2, the work done is equal to the area between the isothermal curve and the volume-axis from V_1 to V_2. Calculation shows that the work W is given by

$$W = RT \ln (V_2/V_1) = Q$$

where Q is the heat taken in during the expansion, (see p. 750).

Now let us consider an isothermal compression. When a gas is compressed, work is done on it by the compressing agent. To keep its temperature constant, therefore, heat must be withdrawn from the gas, to prevent the work done from increasing its internal energy. The gas must again be held in a thin well-conducting vessel, surrounded by a constant-temperature bath; and it must be compressed slowly.

Reversible Isothermal Change

Suppose a gas expands isothermally from p_1, V_1, T to p_2, V_2, T. If the change can be reversed so that the state of the gas is returned from p_2, V_2, T to p_1, V_1, T through exactly the same values of pressure and volume at every stage, then the isothermal change is said to be *reversible*. A reversible isothermal change is an ideal one. It requires conditions such as a light frictionless piston, so that the pressure inside and outside the gas can always be equalised and no work is done against friction; very slow expansion, so that no eddies are produced in the gas to dissipate the energy; and a constant temperature reservoir with very thin good-conducting walls, as we have seen. In a reversible isothermal change of 1 mole of gas, $pV = \text{constant} = RT$.

Reversible Adiabatic Change

Let us now consider a change of volume in which the conditions are opposite to isothermal; no heat is allowed to enter or leave the gas.

An expansion or contraction in which *no heat* enters or leaves the gas is called an *adiabatic expansion or contraction*.

In an adiabatic expansion, the external work is done wholly at the expense of the internal energy of the gas, and the gas therefore cools. In an adiabatic compression, all the work done on the gas by the compressing agent appears as an increase in its internal energy and therefore as a rise in its temperature. We have already discussed a reversible isothermal change. A *reversible adiabatic change* is an adiabatic change which can be exactly reversed in the sense explained on p. 673. As we noted there, a reversible change is an ideal case.

Adiabatic Curve and Equation

The curve relating pressure and volume for a given mass of a given gas for adiabatic changes is called an 'adiabatic'. In Figure 27.12, the curve shown is an adiabatic for 1 kg of air. It is steeper, at any point, than the isothermal through that point. The curve AB is the isothermal for the temperature $T_0 = 373 \text{ K}$, which cuts the adiabatic at the point p_0, V_0.

If the gas is adiabatically compressed from V_0 to V_1, its temperature rises to some value T_1. Its representative point p_1, V_1 now lies on the isothermal for T_1,

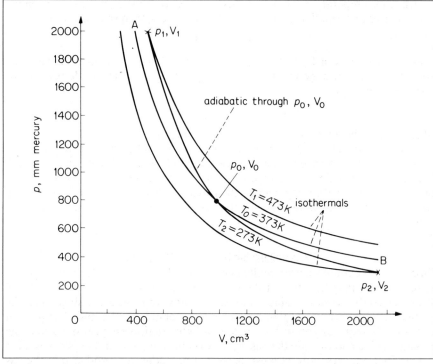

Figure 27.12 *Relationship between adiabatic and isothermals*

since $p_1 V_1 = RT_1$. Similarly, if the gas is expanded adiabatically to V_2, it *cools* to T_2 and its representative point p_2, V_2 lies on the isothermal for T_2. Thus the adiabatic through any point—such as p_0, V_0—is steeper than the isothermal.

Calculation shows that the equation for a reversible adiabatic $p-V$ change of a given mass of gas is

$$pV^\gamma = \text{constant} \qquad . \qquad . \qquad . \qquad . \qquad . \qquad (1)$$

where $\gamma = C_p/C_V$ = the ratio of the molar heat capacities of the gas, discussed later. For air, $\gamma = 1\cdot4$. So for an adiabatic change for this gas, $pV^{1\cdot4}$ = constant.

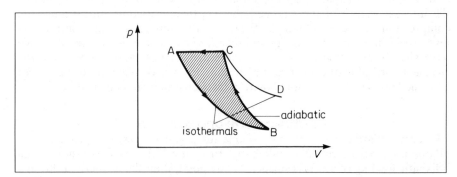

Figure 27.13 *Work done by gas*

In Figure 27.13, a gas expands *isothermally* from a pressure, volume and temperature, represented by A, to that represented by B. At B, the gas is now compressed *adiabatically* until the point C is reached. Since the temperature has increased, C lies on an isothermal CD of a higher temperature than AB. At C, the gas is compressed *at constant pressure* until the point A is again reached. The net amount of work done *on* the gas is equal to the enclosed area ABC, as we explained before.

Ideal and Real $p-V$ Curves

The condition for an adiabatic change is that no heat must enter or leave the gas. The gas must therefore be held in a thick-walled, badly conducting vessel; and the change of volume must take place rapidly, to give as little time as possible for heat to escape. However, in a rapid compression, for example, eddies may be formed, so that some of the work done appears as kinetic energy of the gas in bulk, instead of as random kinetic energy of its molecules. All the work done then does not go to increase the internal energy of the gas, and the temperature rise is less than in a truly adiabatic compression. If the compression is made slowly, then more heat leaks out, since no vessel has perfectly insulating walls.

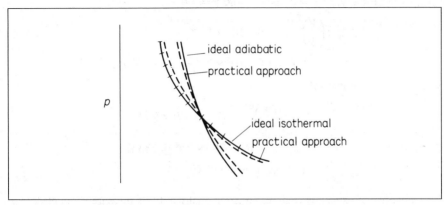

Figure 27.14 *Ideal and real* p $-$ V *curves for a gas*

Perfectly adiabatic changes are therefore impossible; and so, we have seen, are perfectly isothermal ones. Any practical expansion or compression of a gas must lie between isothermal and adiabatic. It may lie anywhere between them, but if it approximates to isothermal, the curve representing it will always be a little steeper than the ideal (Figure 27.14). If it approximates to adiabatic, the curve representing it will never be quite as steep as the ideal.

Equation for Temperature Change in an Adiabatic

If we wish to introduce the temperature, T, into an adiabatic change, we use the gas equation for one mole of gas,

$$pV = RT$$

Now

$$pV^\gamma = k, \text{ a constant}$$

Substituting $p = RT/V$, then

$$\frac{RT}{V} \cdot V^\gamma = k$$

So $RTV^{\gamma-1} = k$. But R is a constant.

So $$T \cdot V^{\gamma-1} = \text{constant}$$

The applications of $pV^{\gamma} = \text{constant}$ and $TV^{\gamma-1} = \text{constant}$ are illustrated in the examples which follow.

Examples on Adiabatic Change

1 An ideal gas at 17°C has a pressure of $1\cdot0 \times 10^5$ Pa, and is compressed (i) isothermally, (ii) adiabatically until its volume is halved, in each case reversibly. Calculate in each case the final pressure and temperature of the gas, assuming $\gamma = 1\cdot4$.

(i) Isothermally, $pV = \text{constant}$.

$$\therefore p \times \frac{V}{2} = 1\cdot0 \times 10^5 \times V$$

$$\therefore p = 2\cdot0 \times 10^5 \text{ Pa}$$

The temperature is constant at 17°C.
(ii) Adiabatically, $pV^{\gamma} = \text{constant}$, and $\gamma = 1\cdot4$.

$$\therefore p \times \left(\frac{V}{2}\right)^{1\cdot4} = 1\cdot0 \times 10^5 \times V^{1\cdot4}$$

$$\therefore p = 1\cdot0 \times 10^5 \times 2^{1\cdot4} = 2\cdot6 \times 10^5 \text{ Pa}$$

Since $TV^{\gamma-1} = \text{constant}$,

$$\therefore T \times \left(\frac{V}{2}\right)^{0\cdot4} = (273 + 17) \times V^{0\cdot4}$$

$$\therefore T = 290 \times 2^{0\cdot4} = 383 \text{ K}$$

$$\therefore \text{temperature} = 110°C$$

2 A quantity of oxygen is compressed isothermally until its pressure is doubled. It is then allowed to expand adiabatically until its original volume is restored. Find the final pressure in terms of the initial pressure. (The ratio of the molar heat capacities of oxygen is to be taken as $1\cdot40$.)

Let p_0, V_0 = the original pressure and volume of the oxygen.
Since $pV = \text{constant}$ for an isothermal change,

$$\therefore \text{new volume} = \frac{V_0}{2} \text{ when new pressure is } 2p_0$$

Suppose the gas expands adiabatically to its volume V_0, when the pressure is p.
Then
$$p \times V_0^{1\cdot4} = 2p_0 \times \left(\frac{V_0}{2}\right)^{1\cdot4}$$

$$\therefore p = 2p_0 \times \left(\frac{1}{2}\right)^{1\cdot4} = 0\cdot8 \, p_0$$

Heat and Mechanical Work in Engines

If one metal surface is rubbed against another, practically the whole of the mechanical work done against friction is transformed into heat (see p. 637). Thus the conversion from mechanical energy to heat can be almost 100% efficient. As we now show, however, a practical machine or *engine* which converts heat into mechanical energy, the opposite process, can never be 100% efficient.

If the engine operates in a *cycle*, so that it returns to its original state after a series of operations, then it can be made to do work continuously.

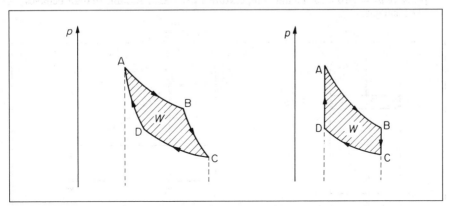

Figure 27.15 *Heat and work. Carnot cycle*

Consider a gas taken through a cycle of reversible pressure (*p*)—volume (*V*) changes. Two examples of a cycle are shown in Figure 27.15 (i) and (ii). Starting from the state represented by A, the gas expands reversibly from A to B. In Figure 27·15 (i), it then expands reversibly from B to C. Here the mechanical work done *by* the gas, which is the integral of $p.dV$ for the two stages of the cycle, is represented by the *area* ABCXYA between the curves and the *V*-axis (see p. 669). To restore the gas to its initial state at A the gas must now be compressed reversibly along CD and DA. In this case the work done *on* the gas is represented by the area CDAYXC. The gas now gives up a quantity of heat Q_2, whereas it takes in a quantity of heat Q_1 along AB and BC while doing work.

Since the gas returns to its initial state at A, there is no change in the internal energy of the gas at the end of a cycle. From the First Law of Thermodynamics, then, the net work done in a cycle = heat gained by gas = $Q_1 - Q_2$. The *net* work done *W* is represented by the area ABCD in both Figure 27.15 (i) and (ii).

The efficiency *E* of an engine is defined by

$$E = \frac{\text{work obtained}}{\text{energy supplied}} \times 100\%$$

$$= \frac{Q_1 - Q_2}{Q_1} \times 100\% = \left(1 - \frac{Q_2}{Q_1}\right) \times 100\%$$

It follows that an engine can never be 100% efficient, that is, all the heat supplied can *never* be transferred or converted into mechanical energy during a complete cycle. This is one statement of the *Second Law of Thermodynamics*.

In 1824, Carnot showed that the most efficient engine, that is, the engine which transfers the maximum amount of the heat supplied between two given temperatures to mechanical energy, was one working under *reversible* conditions (p. 671). Figure 27.15 (i) represents a *Carnot cycle* if AB is a reversible isothermal expansion and CD a reversible isothermal contraction, and BC, DA are respectively a reversible adiabatic expansion and contraction. Figure 27.15 (ii) represents two reversible adiabatic changes of a gas, and two reversible constant volume changes, known as the *Otto cycle*. This has maximum efficiency like the Carnot cycle because the changes are reversible. See also p. 752.

Exercises 27B

Thermodynamics, Isothermal and Adiabatic Changes

1 The cylinder in Figure 27B holds a volume $V_1 = 1000\,cm^3$ of air at an initial pressure $p_1 = 1\cdot10 \times 10^5$ Pa and temperature $T_1 = 300$ K. Assume that air behaves like ideal gas.

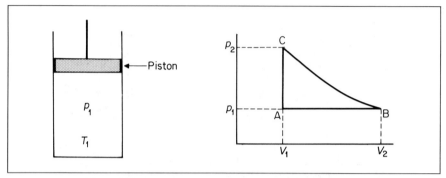

Figure 27B	**Figure 27C**

Figure 27C shows a sequence of changes imposed on the air in the cylinder.
(a) AB—the air is heated to 375 K at constant pressure. Calculate the new volume, V_2.
(b) BC—the air is compressed isothermally to volume V_1. Calculate the new pressure, p_2.
(c) CA—the air cools at constant volume to pressure p_1. State how a value for the work done on the air during the full sequence of changes may be found from the graph in Figure 27C. (*L.*)

2 Explain why the temperature of a gas is liable to change when it expands.
Explain what energy changes take place in
(a) *reversible isothermal*,
(b) *reversible adiabatic* alterations in the volume of a gas.
What conditions would be needed to produce such changes?
Gas in a cylinder, initially at a temperature of 17°C and a pressure of $1\cdot01 \times 10^5\,N\,m^{-2}$, is to be compressed to one-eighth of its volume. What would be the difference between the final pressures if the compression were done
(a) isothermally,
(b) adiabatically?
What would be the final temperature in the latter case? (Ratio of the molar heat capacities = 1·40.) (*L.*)

3 A litre of air, initially at 20°C and at 760 mm of mercury pressure, is heated at constant pressure until its volume is doubled. Find
(a) the final temperature,
(b) the external work done by the air in expanding,
(c) the quantity of heat supplied.
(Assume that the density of air at s.t.p. is $1\cdot293\,kg\,m^{-3}$ and that the specific heat capacity of air at constant volume is $714\,J\,kg^{-1}\,K^{-1}$.) (*L.*)

4 Distinguish between an *isothermal change* and an *adiabatic change*. In each instance state, for a reversible change of an ideal gas, the relation between pressure and volume.
A mass of air occupying initially a volume $2 \times 10^{-3}\,m^3$ at a pressure of 760 mm of mercury and a temperature of 20·0°C is expanded adiabatically and reversibly to twice its volume, and then compressed isothermally and reversibly to a volume of $3 \times 10^{-3}\,m^3$. Find the final temperature and pressure, assuming the ratio of the specific heat capacities of air to be 1·40. (*L.*)

5 Define the terms
   (a) *isothermal change*, and
   (b) *adiabatic change*, as applied to the expansion of a gas.
      Explain how these changes may be approximated to in practice. What would be the relationship between the *pressure* and *temperature* for each of them for an ideal gas?
      Sketch, using the same axes, the *pressure-volume* curves for each of (a) and (b) for the expansion of a gas from a volume V_1 and pressure p_1 to a volume V_2. How, from these graphs, could you calculate the work done in each of the expansions?
      Explain why the temperature falls during an adiabatic expansion and discuss whether or not the temperature fall would be the same for an ideal gas and a real gas. (*L.*)

6 (a) The first law of thermodynamics may be written $\delta Q = \delta U + \delta W$. Explain the meaning of this equation as applied to the heating of a gas. Use the equation to justify the fact that the molar heat capacity of a gas at constant pressure is greater than the molar heat capacity at constant volume.
   (b) Explain the meaning of the terms isothermal change and adiabatic change. What is meant by a reversible change?
      A mass of gas is expanded isothermally and then compressed adiabatically to its original volume. What further operation must be performed on the gas to restore it to its original state? Sketch a labelled $p - V$ graph to represent the series of operations. What quantity is represented by the area enclosed?
   (c) An ideal gas at an initial temperature of 15°C and pressure of $1 \cdot 10 \times 10^5$ Pa is compressed isothermally to one quarter of its original volume. What will be its final pressure and temperature? What would have been the pressure and temperature if the compression had been adiabatic? (Ratio of principal specific heat capacities of the gas = $1 \cdot 40$.) (*AEB*, 1982.)

Heat Capacities

Heat Capacities at Constant Volume and Constant Pressure

When we warm a gas, we may let it expand or not, as we please. If we do not let it expand—if we warm it in a closed vessel—then it does no external work, and all the heat we give it goes to increase its internal energy. *The heat required to warm one mole of gas through one degree, when its volume is kept constant, is called the molar heat capacity of the gas at constant volume.* It is denoted by C_V and is generally expressed in $J\,mol^{-1}\,K^{-1}$.

A similar definition applies to the *specific heat capacity* at constant volume, where the mass of gas warmed is 1 kg. The symbol used in this case is c_V and the unit is $J\,kg^{-1}\,K^{-1}$.

We can also warm a gas while keeping its pressure constant, and define the corresponding heat capacity. *The molar heat capacity of a gas at constant pressure is the heat required to warm one mole of it by one degree, when its pressure is kept constant.* It is denoted by C_p, and is expressed in the same units as C_V.

In the case of the specific heat capacity at constant pressure, which applies to a mass of 1 kg, the symbol used is c_p.

A change made at constant pressure is called an *isobaric* change.

It can be seen that $c_V = C_V/M$ and $c_p = C_p/M$, where M is the numerical value of the mass of one mole expressed in kg.

Any number of heat capacities can be defined for a gas, according to the mass and the conditions imposed upon its pressure and volume. For unit mass, 1 kg, of a gas, the heat capacities at constant pressure c_p, and at constant volume c_V, are called the *principal specific heat capacities.*

Molar Heat Capacities: their Difference

Figure 27.16 shows how we can find a relationship between the molar heat capacities of a gas. We first consider 1 mole of the gas warmed through 1 K at

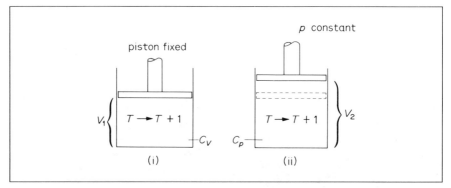

Figure 27.16 *Molar heat capacity at constant volume and pressure*

constant volume, (i). The heat required is C_V joules, and goes wholly to increase the internal energy U.

We next consider 1 mol warmed through 1 K at constant pressure, (ii). It expands from V_1 to V_2, and does an amount of external work given by

$$W = p \times \text{volume change} = p(V_2 - V_1)$$

Further, since the temperature rise of the gas is 1 K, and the internal energy of the gas is independent of volume, the rise in internal energy is C_V, the molar heat capacity at constant volume. Hence, from $\Delta Q = \Delta U + p.\Delta V$, the total amount of heat required to warm the gas at constant pressure is therefore

$$C_p = C_V + p(V_2 - V_1). \qquad . \qquad . \qquad . \qquad . \qquad (1)$$

We can simplify the last term of this expression by using the equation of state for one mole:

$$pV = RT$$

where T is the absolute temperature of the gas, and R is the molar gas-constant. If T_1 is the absolute temperature before warming, then

$$pV_1 = RT_1 \qquad . \qquad . \qquad . \qquad . \qquad . \qquad (2)$$

The absolute temperature after warming is $T_1 + 1$; therefore

$$pV_2 = R(T_1 + 1) \qquad . \qquad . \qquad . \qquad . \qquad . \qquad (3)$$

and on subtracting (2) from (3) we find

$$p(V_2 - V_1) = R$$

Equation (1) now gives $\qquad\qquad C_p = C_V + R$

or $\qquad\qquad\qquad\qquad \boldsymbol{C_p - C_V = R} \qquad . \qquad . \qquad . \qquad . \qquad (4)$

A similar expression to (4) can be derived for the difference in the principal specific heat capacities of a gas, $c_p - c_V$. Thus if r is the gas constant per unit mass,

$$c_p - c_V = r \qquad . \qquad . \qquad . \qquad . \qquad . \qquad (4A)$$

On p. 636, the specific heat capacity of a metal is measured at constant atmospheric pressure. So c_p was measured. The volume expansion of a metal at constant pressure is very small compared to that of a gas. So the external work done is very small. Hence it follows that there is not much difference between c_p and c_V for a metal.

Enthalpy

In engines and other machines such as refrigerator units, a gas may pass from a constant high pressure to a constant low pressure through a fine opening such as a needle or throttle valve, without any exchange of heat with the surroundings. So the change is adiabatic. Such experiments were first carried out many years ago using a porous plug with fine holes and this is known as a *Joule-Kelvin experiment.*

Suppose p_1 is the constant high pressure on one side of the fine opening and a volume V_1 of gas is pushed through. Then work done *on* the gas $= p_1 V_1$, since the volume of gas changes from V_1 to zero on the high-pressure side and work done $= p.\Delta V$.

If p_2 is the constant low pressure on the other side of the opening and the gas expands to a volume V_2, then work done *by* gas in expanding from zero to $V_2 = p_2 V_2$.

So net work done by gas $= p_2 V_2 - p_1 V_1$.

If the process takes place adiabatically, then the work done by the gas $=$ decrease in internal energy of gas $= U_1 - U_2$, where U_1 is the initial and U_2 is the final internal energy. So

$$U_1 - U_2 = p_2 V_2 - p_1 V_1$$

and $\qquad\qquad\qquad U_1 + p_1 V_1 = U_2 + p_2 V_2$

The quantity $(U + pV)$ for a gas, which remained constant as the gas passed through the fine opening is called its *enthalpy* or *total heat*, symbol H.

So
$$\Delta H = \Delta U + p\Delta V$$

The enthalpy is constant in a throttle process as we have seen and is therefore an important quantity in mechanical engineering and industrial chemistry.

With most gases the internal energy decreases and the gas cools when passing through a fine opening from a high to a low pressure side. Exceptions are hydrogen and helium, which increase in temperature on account of the way their values of pV change with pressure (see p. 743). With an ideal gas, $p_2 V_2 = p_1 V_1$ since pV is independent of pressure, and $U_2 = U_1$. So no temperature change occurs in this case. In a refrigerator unit, when the coolant liquid is pumped from the high pressure to the low pressure side through a valve, the liquid evaporates and produces a drop in temperature.

The *latent heat of vaporisation* of water $= \Delta U + p . \Delta V$ where ΔU is the internal energy gain and $p . \Delta V$ is the external work done (page 666). So the specific latent heat of vaporisation is the *enthalpy change per kg* from water to steam. The specific latent heat of fusion of ice is the enthalpy change per kg from ice to water.

Example on Work done by Gas and Internal Energy Change

At 27°C and a pressure of 1.0×10^5 Pa, an ideal gas has a volume of 0.04 m^3. It is heated at constant pressure until its volume increases to 0.05 m^3.

Find (i) the external work done, (ii) the new temperature of the gas, (iii) the change in internal energy of the gas if its mass is 45 g, its molar mass is 28 g and its molar heat capacity at constant volume is $0.6 \text{ J mol}^{-1} \text{ K}^{-1}$, (iv) the total heat given to the gas.

(i) External work done $= p \times$ volume change

$$= 1.0 \times 10^5 \times (0.05 - 0.04)$$

$$= 1000 \text{ J}$$

(ii) From the gas law $V \propto T$ (constant pressure), the new kelvin temperature T_2 is given by

$$\frac{V_2}{V_1} = \frac{T_2}{T_1}$$

So
$$\frac{0.05}{0.04} = \frac{T_2}{(273 + 27)} = \frac{T_2}{300}$$

Then
$$T_2 = 300 \times 0.05/0.04 = 375 \text{ K} (102°\text{C})$$

(iii) Number of moles of gas, $n, = 45/28$. So

$$\text{internal energy rise} = nC_V(T_2 - T_1)$$

$$= \frac{45}{28} \times 0.6 \times (375 - 300)$$

$$= 72 \text{ J}$$

(iv) Total heat given to gas, $\Delta Q = \Delta U + \Delta W$

$$= 72 + 1000$$

$$= 1072 \text{ J}$$

_____ Exercises 27C _____

Heat Capacities of Gases, Internal Energy, Work

1 For hydrogen, the molar heat capacities at constant volume and constant pressure are respectively $20.5\,J\,mol^{-1}\,K^{-1}$ and $28.8\,J\,mol^{-1}\,K^{-1}$. What does this mean? (i) Which heat capacity is related to internal energy assuming hydrogen is an ideal gas? Explain your answer.   (ii) From the values given, calculate the molar gas constant.   (iii) Is C_p always greater than C_V? Explain your answer.

2 From the values of C_V and C_p given in question **1**, calculate   (i) the heat needed to raise the temperature of 8 g of hydrogen from 10°C to 15°C at constant pressure,   (ii) the increase in internal energy of the gas,   (iii) the external work done. (Molar mass of hydrogen = 2 g.)

3 (a) The specific heat capacities of air are $1040\,J\,kg^{-1}\,K^{-1}$ measured at constant pressure and $740\,J\,kg^{-1}\,K^{-1}$ measured at constant volume. Explain briefly why the values are different.

   (b) A room of volume $180\,m^3$ contains air at a temperature of 16°C having a density of $1.13\,kg\,m^{-3}$. During the course of the day the temperature rises to 21°C. Calculate an approximate value for the amount of energy transferred to the air during the day. Assume that air can escape from the room but no fresh air enters. Explain your reasoning. (L.)

4 A gas has a volume of $0.02\,m^3$ at a pressure of $2 \times 10^5\,Pa(N\,m^{-2})$ and temperature of 27°C. It is heated at constant pressure until its volume increases to $0.03\,m^3$.

   Calculate   (i) the external work done,   (ii) the new temperature of the gas, (iii) the increase in internal energy of the gas if its mass is 16 g, its molar heat capacity at constant volume is $0.8\,J\,mol^{-1}\,K^{-1}$ and its molar mas is 32 g.

5 Why is the energy needed to raise the temperature of a given mass of gas by a certain amount greater if the pressure is kept constant than if the volume is kept constant? (L.)

6 What happens to the energy added to an ideal gas when it is heated

   (a) at constant volume, and

   (b) at constant pressure?

   Show from this that a gas can have a number of values of specific heat capacity. Deduce an expression for the difference between the specific heat capacities of a gas at constant pressure and at constant volume.

   If the ratio of the principal specific heat capacities of a certain gas is 1.40 and its density at s.t.p. is $0.090\,kg\,m^{-3}$ calculate the values of the specific heat capacity at constant pressure and at constant volume. (Standard atmospheric pressure = $1.01 \times 10^5\,N\,m^{-2}$.) (L.)

7 Explain why the values of the specific heat capacities of a gas when measured at constant pressure and at constant volume respectively are different. Derive an expression for the difference, for an ideal gas, in terms of its relative molecular mass M and the molar gas constant R.

   Given that the volume of a gas at s.t.p. is $2.24 \times 10^{-2}\,m^3\,mol^{-1}$ and that standard pressure is $1.01 \times 10^5\,N\,m^{-2}$, calculate a value for the molar gas constant R and use it to find the difference between the quantities of heat required to raise the temperature of $0.01\,kg$ of oxygen from 0°C to 10°C when

   (a) the pressure,

   (b) the volume of the gas is kept constant.

   (Relative molecular mass of oxygen = 32.) (O. & C.)

8 Explain why the specific heat capacity of a gas is greater if it is allowed to expand while being heated than if the volume is kept constant. Discuss whether it is possible for the specific heat capacity of a gas to be zero.

   When 1 g of water at 100°C is converted into steam at the same temperature 2264 J must be supplied. How much of this energy is used in forcing back the atmosphere? Explain what happens to the remainder of the energy. [1 g of water 100°C occupies $1\,cm^3$. 1 g of steam at 100°C and 760 mmHg occupies $1601\,cm^3$. Density of mercury = $13\,600\,kg\,m^{-3}$.] (C.)

9 Starting from the first law of thermodynamics, deduce, with careful explanation, the relationship for the difference between the principal molar heat capacities ($C_{p,M}$ and $C_{V,M}$) for an ideal gas

$$C_{p,M} - C_{V,M} = R.$$

What is
(a) an isothermal, and
(b) an adiabatic change of state?

A mole of ideal gas ($\gamma = 1{\cdot}67$) is at temperature 300 K and presure 10^5 Pa. The gas is first expanded adiabatically until its volume is doubled (step 1); then compressed isothermally to its original volume (step 2).   (i) Sketch these changes on a $p - V$ diagram.   (ii) Find the pressure after step 1.   (iii) Find the temperature after step 2.

It is found that, after step 2, an amount of energy 1·38 kJ has to be transferred at constant volume to regain the original temperature. Find $C_{v,M}$ using only the data given in the question. (*W.*)

10 Explain why the molar heat capacity of a gas at constant pressure is different from that at constant volume.

The density of an ideal gas is $1{\cdot}60$ kg m^{-3} at 27°C and $1{\cdot}00 \times 10^5$ newton metre^{-2} pressure and its specific heat capacity at constant volume is 312 J kg^{-1} K^{-1}. Find the ratio of the specific heat capacity at constant pressure to that at constant volume. Point out any significance to be attached to the result. (*JMB.*)

28
Kinetic Theory of Gases

In this chapter we deal with the kinetic theory of gases and show how it can explain all the gas laws of an ideal gas. We shall also see that the molecules have a range of speeds, as Maxwell first showed.

Gas Pressure, Assumptions

In the kinetic theory of gases, we explain the behaviour of gases by considering the motion of their molecules. We suppose that the pressure of a gas is due to the molecules bombarding the walls of its container. Whenever a molecule bounces off a wall, its momentum at right-angles to the wall is reversed; the force which it exerts on the wall is equal to the rate of change of its momentum. The average force exerted by the gas on the whole of its container is the average rate at which the momentum of its molecules is changed by collision with the walls. Since pressure is force per unit area, to find the pressure of the gas we must find this force, and then divide it by the area of the walls.

The following assumptions are made to simplify the calculation:

(a) **The attraction between the molecules is negligible.**
(b) **The volume of the molecules is negligible compared with the volume occupied by the gas.**
(c) **The molecules are like perfectly elastic spheres.**
(d) **The duration of a collision is negligible compared with the time between collisions.**

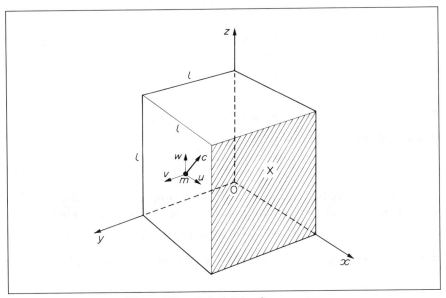

Figure 28.1 *Calculation of gas pressure*

Calculation of Pressure

Consider for convenience a cube of side l containing N molecules of gas each of mass m, Figure 28.1. A typical molecule will have a velocity c at any instant and this will have components of u, v, w respectively in the direction of the three perpendicular axes Ox, Oy, Oz as shown. So $c^2 = u^2 + v^2 + w^2$.

Consider the force exerted on the face X of the cube due to the component u. Just before impact, the momentum of the molecule due to u is mu. After impact, the momentum is $-mu$, since the momentum reverses. Thus

$$\text{momentum change on impact} = mu - (-mu) = 2mu$$

The time taken for the molecule to move across the cube to the opposite face and back to X is $2l/u$.

This is the time to make one impact. So the number of impacts per second n' on a given face $= 1 \div 2l/u = u/2l$. So at face X

$$\text{momentum change per second} = n' \times \text{one momentum change}$$

$$= \frac{u}{2l} \times 2mu = \frac{mu^2}{l}$$

$$\therefore \text{force on X} = \frac{mu^2}{l}$$

$$\therefore \text{pressure on X} = \frac{\text{force}}{\text{area}} = \frac{mu^2}{l \times l^2} = \frac{mu^2}{l^3} \qquad . \qquad . \qquad \text{(i)}$$

We now take account of the N molecules in the cube. Each has a different velocity and hence a component of different magnitude in the direction Ox. If these are represented by $u_1, u_2, u_3, \ldots u_N$, it follows from (i) that the total pressure on X, p, is given by

$$p = \frac{mu_1{}^2}{l^3} + \frac{mu_2{}^2}{l^3} + \frac{mu_3{}^2}{l^3} + \ldots + \frac{mu_N{}^2}{l^3}$$

$$= \frac{m}{l^3}(u_1{}^2 + u_2{}^2 + u_3{}^2 + \ldots + u_N{}^2) \qquad . \qquad . \qquad \text{(ii)}$$

Let the symbol $\overline{u^2}$ represent the average or mean value of all the squares of the components in the Ox direction, that is,

$$\overline{u^2} = \frac{u_1{}^2 + u_2{}^2 + u_3{}^2 + \ldots + u_N{}^2}{N}$$

Then
$$N\overline{u^2} = u_1{}^2 + u_2{}^2 + u_3{}^2 + \ldots + u_N{}^2$$

Hence, from (ii),

$$p = \frac{Nm\overline{u^2}}{l^3} \qquad . \qquad . \qquad . \qquad . \qquad \text{(iii)}$$

Now with a large number of molecules of varying speed in random motion, the mean square of the component speed in any one of the three axes in the *same*.

$$\therefore \overline{u^2} = \overline{v^2} = \overline{w^2}$$

But, for each molecule, $c^2 = u^2 + v^2 + w^2$, so that the mean square $\overline{c^2}$ is given by $\overline{c^2} = \overline{u^2} + \overline{v^2} + \overline{w^2}$.

$$\therefore \overline{u^2} = \tfrac{1}{3}\overline{c^2}$$

Hence, from (iii),

$$p = \tfrac{1}{3} \frac{Nm\overline{c^2}}{l^3}$$

The *number of molecules per unit volume*, $n = N/l^3$. Thus we may write

$$p = \tfrac{1}{3} nm\overline{c^2}* \qquad . \qquad . \qquad . \qquad . \qquad . \qquad (1)$$

If n is in molecules per metre^3, m in kilogram and c in metre per second, then the pressure p is in *newton per metre^2* ($N\,m^{-2}$) or *pascals* (Pa).

In our calculation, we assumed that molecules of a gas do not collide with other molecules as they move to-and-fro across the cube. If, however, we assume that their collisions are perfectly elastic, both the kinetic energy and the momentum are conserved in them. The average momentum with which all the molecules strike the walls is then not changed by their collisions with one another; what one loses, another gains. The important effect of collisions between molecules is to distribute their individual speeds; on the average, the fast ones lose speed to the slow. We suppose, then, that different molecules have different speeds, and that the speeds of individual molecules vary with time, as they make collisions with one another; but we also suppose that the average speed of all the molecules is constant. These assumptions are justified by the fact that the kinetic theory leads to conclusions which agree with experiment.

Root-Mean-Square (R.M.S.) Speed

In equation (1) the factor nm is the product of the number of molecules per unit volume and the mass of one molecule. It is therefore the total mass of the gas per unit volume, its density ρ. Thus the equation gives

$$p = \tfrac{1}{3} \rho \overline{c^2}* \qquad . \qquad . \qquad . \qquad . \qquad . \qquad (2)$$

or

$$\overline{c^2} = \frac{3p}{\rho} \qquad . \qquad . \qquad . \qquad . \qquad . \qquad (3)$$

If we substitute known values of p and ρ in equation (3), we can find $\overline{c^2}$. For hydrogen at s.t.p., $\rho = 0.09\,kg\,m^{-3}$ and $p = 1.013 \times 10^5$ Pa.

$$\therefore \overline{c^2} = \frac{3p}{\rho} = \frac{3 \times 1.013 \times 10^5}{9 \times 10^{-2}}$$

$$= 3.37 \times 10^6\,m^2\,s^{-2}$$

The square root of $\overline{c^2}$ is called the *root-mean-square speed*; it is of the same magnitude as the average speed, but not quite equal to it. See p. 692. Its value is

$$\sqrt{\overline{c^2}} = \sqrt{3.37} \times 10^3 = 1840\,m\,s^{-1}\ \text{(approx.)}$$

$$= 1.84\,km\,s^{-1}$$

From (3), note that $\qquad \sqrt{\overline{c^2}} = \sqrt{\dfrac{3p}{\rho}} = $ **r.m.s. speed**

Molecular speeds were first calculated in this way by Joule in 1848; they turn out to have a magnitude which is high, but reasonable. The value is reasonable

_____

* $\overline{c^2}$ may also be printed as $\langle c^2 \rangle$.

because it has the same order of magnitude as the speed of sound ($1.30 \, \text{km s}^{-1}$ in hydrogen at 0°C). The speed of sound is the speed with which the molecules of a gas pass on a disturbance from one to another, and this we expect to be of the same magnitude as the speeds of their natural motion.

Root-mean-square Speed and Mean Speed

It should be carefully noted that the pressure p of the gas depends on the 'mean square' of the speed. This is because

(a) the momentum change at a wall is proportional to u, as previously explained, and

(b) the number of impacts per second on a given face is proportional to u.

So the rate of change of momentum is proportional to $u \times u$ or to u^2. Further, the mean square speed is *not* equal to the square of the average speed. As an example, let us suppose that the speeds of six molecules are, 1, 2, 3, 4, 5, 6 units. Their mean speed $\bar{c}$ is given by

$$\bar{c} = \frac{1+2+3+4+5+6}{6} = \frac{21}{6} = 3.5$$

and its square is

$$(\bar{c})^2 = 3.5^2 = 12.25$$

Their mean square speed, however, is

$$\overline{c^2} = \frac{1^2+2^2+3^2+4^2+5^2+6^2}{6} = \frac{91}{6} = 15.2$$

So the root-mean-square speed, $\sqrt{\overline{c^2}} = \sqrt{15.2} = 3.9$, which is about 12% different from the mean speed $\bar{c}$ in this simple case.

Introduction of Temperature in Kinetic Theory

Consider a volume V of gas, containing N molecules. The number of molecules per unit volume $n = N/V$. So the pressure of the gas, by equation (1) is

$$p = \tfrac{1}{3}nm\overline{c^2} = \tfrac{1}{3}\frac{N}{V}m\overline{c^2}$$

$$\therefore pV = \tfrac{1}{3}Nm\overline{c^2} \qquad . \qquad . \qquad . \qquad . \qquad (4)$$

But the ideal gas equation for 1 mole is

$$pV = RT$$

We can therefore make the kinetic theory consistent with the observed behaviour of a gas, if we write

$$\tfrac{1}{3}Nm\overline{c^2} = RT \qquad . \qquad . \qquad . \qquad . \qquad (5)$$

Essentially, we are here *assuming* that the mean square speed of the molecules, $\overline{c^2}$, is proportional to the absolute (kelvin) temperature of the gas. This is a reasonable assumption, because we have learnt that heat is a form of energy; and the translational kinetic energy of a molecule, due to its random motion within its container, is proportional to the square of its speed. When we heat a gas, we expect to speed-up its molecules. So we write

$$pV = \tfrac{1}{3}Nm\overline{c^2} = RT \qquad . \qquad . \qquad . \qquad . \qquad (6)$$

Variation of R.M.S. Speed

Since N molecules each have a mass m in the mole of gas considered, the molar mass M is Nm. Thus, from (5),

$$\tfrac{1}{3}M\overline{c^2} = RT$$

$$\therefore \text{ r.m.s. speed, } \sqrt{\overline{c^2}} = \sqrt{\frac{3RT}{M}}$$

So

1 the r.m.s. speed or velocity of the molecules of a *given gas* $\propto \sqrt{T}$, and
2 the r.m.s. speed of the molecules of different gases at the *same temperature* $\propto 1/\sqrt{M}$, so gases of higher molecular mass have smaller r.m.s. speeds.

To illustrate the numerical changes, hydrogen of relative molecular mass about 2 has a r.m.s. speed at s.t.p. (273 K and 1.013×10^5 Pa pressure) of roughly $1840 \, \text{m s}^{-1}$. At 100°C or 373 K and the same pressure, the r.m.s. speed c_r is given by

$$\frac{c_r}{1840} = \sqrt{\frac{373}{273}}$$

or $$c_r = 1840 \times \sqrt{\frac{373}{273}} = 1930 \, \text{m s}^{-1} \text{ (approx.)}$$

Oxygen has a relative molecular mass of about 32. From (2) above, $c_r \propto 1/\sqrt{M}$, it follows that at s.t.p. the r.m.s. speed of oxygen molecules is given by

$$\frac{c_r}{1840} = \sqrt{\frac{2}{32}}$$

or $$\text{or } c_r = 1840 \times \sqrt{\frac{2}{32}} = 460 \, \text{m s}^{-1}$$

Boltzmann Constant, Mean Energy of Molecule

The kinetic energy of a molecule moving at an instant with a speed c is $\tfrac{1}{2}mc^2$; the average kinetic energy of translation of the random motion of the molecule of a gas is therefore $\tfrac{1}{2}m\overline{c^2}$. To relate this to the temperature, we put equation (5) into the form

$$RT = \tfrac{1}{3}Nm\overline{c^2} = \tfrac{2}{3}N(\tfrac{1}{2}m\overline{c^2})$$

so $$\tfrac{1}{2}m\overline{c^2} = \tfrac{3}{2}\frac{R}{N}T \qquad . \qquad . \qquad . \qquad . \qquad . \qquad (7)$$

Thus, *the average kinetic energy of translation of a molecule is proportional to the absolute temperature of the gas.*

The ratio R/N in equation (7) is a universal constant, since R = molar gas constant, $8.31 \, \text{J mol}^{-1} \text{K}^{-1}$ for all gases, and $N = N_A$, the Avogadro constant, $6.02 \times 10^{23} \, \text{mol}^{-1}$ for all gases. Thus

$$\frac{R}{N_A} = k$$

The constant k, the gas constant per molecule, is called the *Boltzmann constant*. In terms of k equation (7) becomes

$$\tfrac{1}{2}m\overline{c^2} = \tfrac{3}{2}kT \qquad . \qquad . \qquad . \qquad . \qquad . \qquad (8)$$

The Boltzmann constant is usually given in joule per degree, since it relates energy to temperature: $k = \tfrac{1}{2}m\overline{c^2}/\tfrac{3}{2}T$. Its value is $k = 1\cdot38 \times 10^{-23}\,\mathrm{J\,K^{-1}}$.

Diffusion: Graham's Law

When a gas passes through a porous plug, a cotton-wool wad, for example, it is said to 'diffuse'. Diffusion differs from the flow of a gas through a wide tube because it is not a motion of the gas in bulk, but is a result of the motion of its individual molecules.

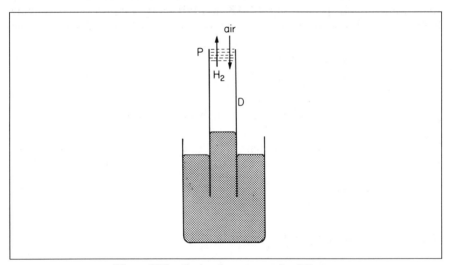

Figure 28.2 *Graham's apparatus for diffusion*

Figure 28.2 shows an apparatus devised by GRAHAM (1805–69) to compare the rates of diffusion of different gases. D is a glass tube, closed with a plug P of plater of Paris. It is first filled with mercury, and inverted over mercury in a bowl. Hydrogen is then passed into it until the mercury levels are the same on each side; the hydrogen is then at atmospheric pressure. The volume of hydrogen, V_H, is proportional to the length of the tube above the mercury. The apparatus is now left; hydrogen diffuses out through P, and air diffuses in. Ultimately no hydrogen remains in the tube D. The tube is then adjusted until the level of mercury is again the same on each side, so that the air within it is at atmospheric pressure. The volume of air, V_A, is proportional to the new length of the tube above the mercury.

The volumes V_A and V_H are, respectively, the volumes of air and hydrogen which diffused through the plug in the same time. Therefore the rates of diffusion of the gases air and hydrogen are proportional to the volumes V_A and V_H:

$$\frac{\text{rate of diffusion of air}}{\text{rate of diffusion of hydrogen}} = \frac{V_A}{V_H}$$

Graham found in his experiments that the volumes were inversely proportional to the square roots of the densities of the gases, ρ:

$$\frac{V_A}{V_H} = \sqrt{\frac{\rho_H}{\rho_A}}$$

thus

$$\frac{\text{rate of diffusion of air}}{\text{rate of diffusion of hydrogen}} = \sqrt{\frac{\rho_H}{\rho_A}}$$

In general:

$$\text{rate of diffusion} \propto \frac{1}{\sqrt{\rho}}$$

and in words

---

the rate of diffusion of a gas is inversely proportional to the square root of its density. **This is Graham's Law.**

Graham's law of diffusion is readily explained by the kinetic theory. At the same kelvin temperature T, the mean kinetic energies of the molecules of different gases are equal, since

$$\tfrac{1}{2}m\overline{c^2} = \tfrac{3}{2}kT$$

and k is the Boltzmann constant. So if the subscripts A and H denote air and hydrogen respectively,

$$\tfrac{1}{2}m_A\overline{c_A{}^2} = \tfrac{1}{2}m_H\overline{c_H{}^2}$$

and

$$\frac{\overline{c_A{}^2}}{\overline{c_H{}^2}} = \frac{m_H}{m_A}$$

At a given temperature and pressure, the density of a gas, ρ, is proportional to the mass of its molecule, m, since equal volumes contain equal numbers of molecules.

Therefore

$$\frac{m_H}{m_A} = \frac{\rho_H}{\rho_A}$$

so

$$\frac{\overline{c_A{}^2}}{\overline{c_H{}^2}} = \frac{\rho_H}{\rho_A}$$

$$\therefore \frac{\sqrt{\overline{c_A{}^2}}}{\sqrt{\overline{c_H{}^2}}} = \frac{\sqrt{\rho_H}}{\sqrt{\rho_A}} \qquad \cdot \quad \cdot \quad \cdot \quad \cdot \quad \cdot \quad (1)$$

The average speed of the molecules of a gas is roughly equal to—and strictly proportional to—the square root of its mean square speed. Equation (1) therefore shows that the average molecular speeds are inversely proportional to the square roots of the densities of the gases. And so it explains why the rates of diffusion—which depend on the molecular speeds—are also inversely proportional to the square roots of the densities.

Example on Kinetic Theory

Helium gas occupies a volume of $0.04\,\text{m}^3$ at a pressure of $2 \times 10^5\,\text{Pa}$ (N m^{-2}) and temperature $300\,\text{K}$.

Calculate (i) the mass of helium, (ii) the r.m.s. speed of its molecules, (iii) the r.m.s. speed at $432\,\text{K}$ when the gas is heated at constant pressure to this temperature, (iv) the r.m.s. speed of hydrogen molecules at 432 K. (Relative molecular mass of helium and hydrogen = 4 and 2 respectively, molar gas constant = $8 \cdot 3 \, \text{J} \, \text{mol}^{-1} \, \text{K}^{-1}$.)

(i) *Mass*   For n mols, $pV = nRT$

So
$$n = \frac{pV}{RT} = \frac{2 \times 10^5 \times 0 \cdot 04}{8 \cdot 3 \times 300} = 3 \cdot 2$$

Hence   mass of helium = $3 \cdot 2 \times 4 \, \text{g} = 12 \cdot 8 \, \text{g}$

(ii) *r.m.s. speed* Pressure $p = 2 \times 10^5 \, \text{Pa}$

$$\text{density } \rho = \frac{\text{mass}}{\text{volume}} = \frac{12 \cdot 8 \times 10^{-3} \, \text{kg}}{0 \cdot 04 \, \text{m}^3} = 0 \cdot 32 \, \text{kg} \, \text{m}^{-3}$$

So   r.m.s. speed $= \sqrt{\dfrac{3p}{\rho}}$

$$= \sqrt{\frac{3 \times 2 \times 10^5}{0 \cdot 32}} = 1369 \, \text{m} \, \text{s}^{-1}$$

(iii) *Temperature 432 K*   Since r.m.s. speed $\propto \sqrt{T}$, the new value c_r at 432 K is given by

$$\frac{c_r}{1369} = \sqrt{\frac{432}{300}} = \sqrt{1 \cdot 44} = 1 \cdot 2$$

So $c_r = 1 \cdot 2 \times 1369 = 1643 \, \text{m} \, \text{s}^{-1}$

(iv) *Hydrogen*   One mole of hydrogen has a mass of 2 g and one mole of helium has a mass of 4 g. So ratio of molar masses = $2:4 = 1:2$.

But r.m.s. speed at a given temperature $\propto 1/\sqrt{M}$, where M is the molar mass. So at 432 K,

$$\text{r.m.s. speed of hydrogen molecules} = \sqrt{2} \times 1643$$
$$= 2324 \, \text{m} \, \text{s}^{-1}$$

Distribution of Molecular Speeds

So far we have used the 'root-mean-square' speed and the 'mean' speed of the large number of molecules in a given mass of gas. The actual distribution of the speeds among the numerous molecules can be investigated by an apparatus whose principle is illustrated in Figure 28.3 (i) and from which all the air is evacuated.

A furnace F maintains a sample of molten metal at a constant high temperature T. Molecules from the vapour emerge from an opening O in the furnace and pass through a narrow collimator slit C which produces a parallel beam of molecules. This beam is incident on a wheel A, which rotates with the same angular speed as another wheel B on the same axle and is distant l from B. The wheel A has a narrow slit, A_1, and above it a wide slit YX through which all the vapour molecules may pass. The wheel B has a narrow slit B_1 in it which is displaced from the slit A_1 by an angle θ when both wheels are viewed end-on.

With B_1 originally displaced by an angle θ from A_1, both wheels are rotated at the same high angular velocity ω. Only molecules emerging from A_1, and which cross the distance l in the same time as the wheel B takes to turn through an

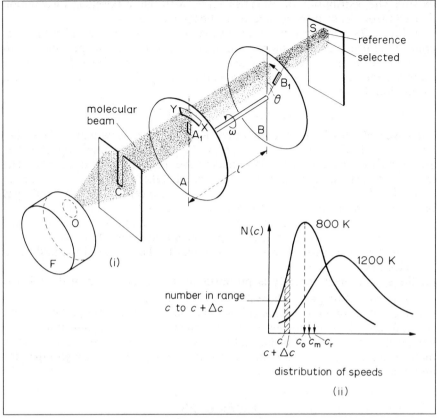

Figure 28.3 *Distribution of molecular speeds*

angle θ, will pass through the slit B_1. These molecules have a velocity v given by

$$\frac{l}{v} = \frac{\theta}{\omega}$$

so

$$v = \frac{\omega l}{\theta}$$

The molecules passing through A_1 and B_1 are thus a 'velocity selected' beam.

The slit YX, however, is wide enough to allow molecules of practically all speeds to pass through itself and through B_1.

These molecules form an 'unselected' or reference beam. The 'velocity selected' and reference beams are incident on a surface S cooled by liquid nitrogen and the ratio of the intensities of the two beams is a measure of the fraction of all the molecules which have velocities close to v in magnitude. Thus by rotating the wheels at different speeds, measurements can be made of the distribution of velocities among the molecules. Further, by varying the temperature T of the furnace, the molecular distribution can be found at different temperatures.

Maxwellian Distribution

The results are shown roughly in Figure 28.3 (ii). The quantity $N(c)$ plotted on the vertical axis represents the number of molecules ΔN in a small range of speeds c to $c + \Delta c$, so that $\Delta N = N(c) \cdot \Delta c =$ area of strip shaded in Figure

28.3 (ii). The distribution of velocities agrees with the Maxwellian distribution derived theoretically from advanced kinetic theory of gases.

In Figure 28.3 (ii), the value c_0 at the maximum of a curve is called the *most probable velocity*, because more molecules have velocities in the range c_0 to $c_0 + \Delta c$ than in any other similar range, Δc, of velocities. The value c_m is the mean velocity and c_r is the root-mean-square. For a Maxwellian distribution, calculation shows that

$$c_0 : c_m : c_r = 1 \cdot 00 : 1 \cdot 13 : 1 \cdot 23$$

The velocity distribution curves at 800 K and 1200 K show that at the higher temperature 1200 K, the distribution curve flattens more round its peak value. So at higher temperatures, more molecules have speeds near the peak value. The mean, root-mean-square and peak values are all higher at higher temperatures.

As already seen, the *mean-square velocity* is concerned in large scale gas properties such as 'pressure' and 'specific heat capacity'. This is because
(a) the pressure of a gas is proportional to the momentum change per molecule and to the number of molecules per second arriving at the walls of the container, which together are proportional to the square of the individual velocities,
(b) the specific heat capacity is proportional to the energy gained, which is proportional to $\frac{1}{2}M\overline{c^2}$ where M is the mass of gas.
On the other hand, the *mean velocity* is concerned in gas properties such as (1) 'diffusion' through porous partitions, since this rate of diffusion is proportional to the mean velocity, and (2) 'viscosity', since there is a transfer of momentum from fast to slow moving gas layers in gas flow through pipes, for example.

_____ Exercises 28 _____

1   Show that the relation $p = \frac{1}{3}\rho \overline{c^2}$ is dimensionally correct, where p is the pressure of a gas of density ρ and $\overline{c^2}$ is the mean square velocity of all its molecules.
    Write down
    (a) two assumptions made in deriving this relation from a simple kinetic theory,
    (b) the meaning of (i) mean velocity and (ii) mean square velocity.
2   Calculate the root-mean-square speed at 0°C of (i) hydrogen molecules and (ii) oxygen molecules, assuming 1 mole of a gas occupies a volume of $2 \times 10^{-2} \, m^3$ at 0°C and $10^5 \, N \, m^{-2}$ pressure. (Relative molecular masses of hydrogen and oxygen = 2 and 32 respectively.)
3   Assuming helium molecules have a root-mean-square speed of $900 \, m \, s^{-1}$ at 27°C and $10^5 \, N \, m^{-2}$ pressure, calculate the root-mean-square speed at (i) 127°C and $10^5 \, N \, m^{-2}$ pressure, (ii) 27°C and $2 \times 10^5 \, N \, m^{-2}$ pressure.
4   Using the kinetic theory, show that (i) the pressure of an ideal gas is doubled when its volume is halved at constant temperature, (ii) the pressure of an ideal gas decreases when it expands in a thermally insulated vessel.
5   Explain what is meant by the *root-mean-square velocity* of the molecules of a gas. Use the concepts of the elementary kinetic theory of gases to derive an expression for the root-mean-square velocity of the molecules in terms of the pressure and density of the gas.
    Assuming the density of nitrogen at s.t.p. to be $1 \cdot 251 \, kg \, m^{-3}$, find the root-mean-square velocity of nitrogen molecules at 127°C. (L.)
6   (a) (i) Explain how the molecules of a gas exert a pressure. (ii) Give two reasons why the pressure exerted by the molecules of a gas, maintained at constant volume, increases as the temperature increases. (iii) In an ideal gas pressure $p = \frac{1}{3}nm\overline{c^2}$ where n = number of molecules per unit volume, m = mass of one molecule, $\overline{c^2}$ = mean square speed of the molecules.

Show how this equation leads to the relationship between the pressure and volume of an ideal gas at constant temperature.
(b) (i) Calculate the root mean square speed of four molecules moving with speeds, in m s^{-1}, of 250, 500, 575 and 600 respectively.

(ii)

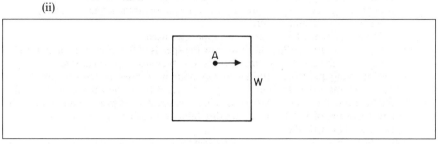

Figure 28A

Figure 28A shows a single molecule A of mass 4.6×10^{-26} kg moving with a speed of 500 m s^{-1} in a rigid cubical box. Calculate the change of momentum of the molecule when it strikes wall W elastically. If the box has side length 0·25 m, calculate the number of times the molecule strikes wall W each second, and deduce the average pressure exerted by molecule A on wall W.   (iii) The box in (ii) now contains 4.2×10^{23} such molecules. Assuming that all the molecules move with the same speed and in the same direction as A, calculate the pressure now exerted on wall W.   (iv) The actual pressure at this density and temperature is 1.03×10^{5} N m^{-2}. Show that this observation is consistent with your answer to (iii). (*AEB*, 1985.)

7   Calculate the pressure in mm of mercury exerted by hydrogen gas if the number of molecules per cm^3 is 6.80×10^{15} and the root-mean-square speed of the molecules is 1.90×10^{3} m s^{-1}. Comment on the effect of a pressure of this magnitude
(a) above the mercury in a barometer tube;
(b) in a cathode ray tube.
(Avogadro constant = 6.02×10^{23} mol^{-1}. Relative molecular mass of hydrogen = 2·02.) (*JMB*.)

8   (a) The kinetic theory of gases predicts that the root-mean-square (r.m.s.) speed of the molecules of an ideal gas is given by the expression $(3p/\rho)^{1/2}$, where p is the pressure and ρ is the density of the gas.

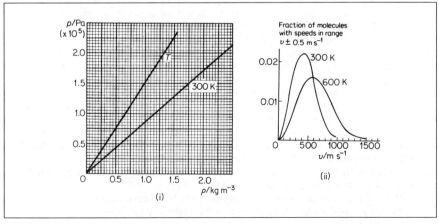

Figure 28B (i)   **Figure 28B (ii)**

The graphs in Figure 28B (i) show how the pressure of oxygen gas depends upon its density at two different constant temperatures, T and 300 K.   (i) Use

the graph to calculate a value for the r.m.s. speed of the oxygen molecules at 300 K. Explain your working. (ii) Is the temperature T higher or lower than 300 K? Explain your reasoning. (iii) The graphs above are based upon experimental results. What conclusion can you draw from them about the behaviour of oxygen? (iv) Outline a simple experimental procedure for investigating how the pressure of a known mass of air varies as its density changes at room temperature.
Include a labelled diagram of the apparatus used.

(b) The graphs in Figure 28B (ii) show how the speeds of the molecules in an ideal gas are distributed at two temperatures. Use them to help you answer the following questions. (i) In what two main ways does the temperature appear to affect the distribution of speeds? (ii) What is the value of v for which the fraction of molecules with speeds in the range $v \pm 0.5\,\text{m s}^{-1}$ is a maximum at a temperature of 300 K? How does this value compare with the r.m.s. speed calculated in (a)(i) above? (*L*.)

9   Define the term *mean square speed* as applied to the molecules of a gas. Explain why the pressure exerted by a gas is proportional to the mean square speed of its molecules.

Figure 28C shows apparatus designed to measure speeds of molecules. Atoms of vapour of a heavy metal emerge from oven O into an evacuated space. The atoms pass through fixed slit S′ in a well-defined beam and enter radially through slit S″ in the curved surface of a cylindrical drum D. When the drum is stationary the atoms strike the inner surface of the drum, giving a well-defined trace T. The drum is then set into rotation about its axis C and maintained at a high constant speed until a second trace has been produced. State and explain the ways in which this trace differs from the first trace.

Figure 28C

Oven O contains bismuth vapour (atomic weight 208) and is maintained at 1500°C. Calculate the root-mean-square speed of the atoms in the oven, assuming the vapour to be monatomic. (Take the gas constant R to be $8.314\,\text{J mol}^{-1}\,\text{K}$.)

At what angular speed must the drum rotate if the traces for atoms having speeds of 400 and 800 m s^{-1} are to be separated by a distance of 10 mm on the drum surface? The drum diameter is 0.5 m. (*O. & C.*)

10   Write down *four* assumptions about the properties and behaviour of molecules that are made in the kinetic theory in order to define an ideal gas. On the basis of this theory derive an expression for the pressure by an ideal gas.

Use the kinetic theory to explain why hydrogen molecules diffuse out of a porous container into the atmosphere even when the pressure of the hydrogen is equal to the atmospheric pressure outside the container.

Air at 273 K and $1.01 \times 10^5\,\text{N m}^{-2}$ pressure contains 2.70×10^{25} molecules per cubic metre. How many molecules per cubic metre will there be at a place where the temperature is 223 K and the pressure is $1.33 \times 10^{-4}\,\text{N m}^{-2}$? (*L*.)

11   Use the kinetic theory to derive an expression for the pressure exerted by an ideal gas, stating clearly the assumptions which you make.

What direct evidence is available to justify the belief that the molecules of a gas are in a continual state of random motion? Use the kinetic theory to explain qualitatively:
(a) Boyle's law,
(b) why the pressure of a gas increases if the temperature is increased at constant volume,

(c) why the temperature of a gas rises if it is compressed in a thermally insulated container. (*L.*)

12 The kinetic theory of gases leads to the equation $p = \frac{1}{3}\rho \overline{c^2}$, where p is the *pressure*, ρ is the *density* and $\overline{c^2}$ is the *mean square molecular speed*. Explain the meaning of the terms in italics and list the simplifying assumptions necessary to derive this result. Discuss how this equation is related to Boyle's Law.

Air may be taken to consist of 80% nitrogen molecules and 20% oxygen molecules of relative molecular masses 28 and 32 respectively. Calculate

(a) the ratio of the root-mean-square speed of nitrogen molecules to that of oxygen molecules in air,

(b) the ratio of the partial pressures of nitrogen and oxygen molecules in air, and

(c) the ratio of the root-mean-square speed of nitrogen molecules in air at 10°C to that at 100°C. (*O. & C.*)

13 Use a simple treatment of the kinetic theory of gases, stating any assumptions you make, to derive an expression for the pressure exerted by a gas on the walls of its container. Thence deduce a value for the root mean square speed of thermal agitation of the molecules of helium in a vessel at 0°C. (Density of helium at s.t.p. $= 0.1785 \, \text{kg m}^{-3}$; 1 atmosphere $= 1.013 \times 10^5 \, \text{N m}^{-2}$.)

If the total translational kinetic energy of all the molecules of helium in the vessel is 5×10^{-6} joule, what is the temperature in another vessel which contains twice the mass of helium and in which the total kinetic energy is 10^{-5} joule? (Assume that helium behaves as a perfect gas.) (*O. & C.*)

14 What do you understand by the term *ideal gas*? Describe a molecular model of an ideal gas and derive the expression $p = \frac{1}{3}\rho c^2$ for such a gas. What is the reasoning which leads to the assertion that the temperature of an ideal monatomic gas is proportional to the mean kinetic energy of its molecules?

The Doppler broadening of a spectral line is proportional to the r.m.s. speed of the atoms emitting light. Which source would have less Doppler broadening, a mercury lamp at 300 K or a krypton lamp at 77 K? (Take the mass numbers of Hg and Kr to be 200 and 84 respectively.)

What causes the behaviour of real gases to differ from that of an ideal gas? Explain qualitatively why the behaviour of all gases at very low pressures approximates to that of an ideal gas. (*O. & C.*)

15 Helium gas is contained in a cylinder by a gas-tight piston which can be assumed to move without friction. The gas occupies a volume of $1.0 \times 10^{-3} \, \text{m}^3$ at a temperature of 300 K and a pressure of 1.0×10^5 Pa.

(a) Calculate (i) the number of helium atoms in the container, (ii) the total kinetic energy of the helium atoms.

(b) Energy is now supplied to the gas in such a way that the gas expands and the temperature remains constant at 300 K.

State and explain what changes, if any, will have occurred in the following quantities: (i) the internal energy of the gas, (ii) the r.m.s. speed of the helium atoms, (iii) the density of the gas.

The Boltzmann constant $= 1.4 \times 10^{-23} \, \text{J K}^{-1}$. (*JMB.*)

16 (a) One mole of an ideal gas at pressure p and Celsius temperature θ occupies a volume V. Sketch a graph showing how the product pV varies with θ. What information can you obtain from the gradient of the graph and the intercept on the temperature axis?

(b) Some helium (molar mass $0.004 \, \text{kg mol}^{-1}$) is contained in a vessel of volume $8.0 \times 10^{-4} \, \text{m}^3$ at a temperature of 300 K. The pressure of the gas is 200 kPa. Calculate (i) the mass of helium present, (ii) the internal energy (the translational kinetic energy of the gas molecules). (*O. & C.*)

29

Transfer of Heat: Conduction and Radiation

Heat can be transferred from one place to another by conduction or convection or radiation. Transfer by convection was discussed in an earlier chapter (p. 641). In this chapter we deal with conduction and radiation and their laws. We shall see that the conduction of heat obeys similar laws to the conduction of electricity and we shall consider good and bad conductors and conductors in series.

Conduction

If we put a poker into the fire, and hold on to it, then heat reaches us along the metal. We say the heat is *conducted*. We find that some substances—metals—are good conductors, and others—such as wood or glass—are not. Good conductors such as a metal bar feel cold to the touch on a cold day, because they rapidly conduct away the body's heat.

Temperature Distribution along a Conductor

In order to study conduction in more detail consider Figure 29.1 (i), which shows a metal bar AB whose ends have been soldered into the walls of two metal

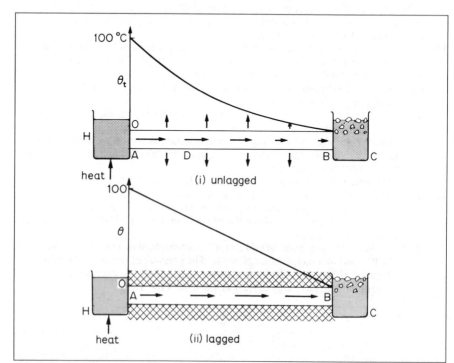

Figure 29.1 *Temperature fall along lagged and unlagged bars*

tanks, H, C. H contains boiling water, and C contains ice-water. Heat flows along the bar from A to B, and *when conditions are steady* the temperature θ of the bar is measured at points along its length. The measurements may be made with thermojunctions, not shown in the figure, which have been soldered to the rod. The curve in the upper part of the figure shows how the temperature falls along the bar, less and less steeply from the hot end to the cold.

The Figure 29.1 (ii) shows how the temperature varies along the bar, if the bar is *well lagged* with a bad conductor, such as asbestos wool. It now falls *uniformly* from the hot to the cold end.

The difference between the temperature distributions is due to the fact that, when the bar is unlagged, heat escapes from its sides, by convection in the surrounding air, Figure 29.1 (i). So the heat flowing past D per second is less than that entering the bar at A by the amount which escapes from the surface AD. The arrows in the figure represent the heat escaping per second from the surface of the bar, and the heat flowing per second along its length. The heat flowing per second along the length decreases from the hot end to the cold. But when the bar is lagged, the heat escaping from its sides is negligible, and the flow per second is now constant along the length of the bar, Figure 29.1 (ii).

We therefore see that the *temperature gradient* along a bar is greatest where the heat flow through it is greatest. We also see that the temperature gradient is uniform only when there is a negligible loss of heat from the sides of the bar.

Thermal Conductivity

Consider a very large thick bar, of which AB in Figure 29.2 (i) is a part, and along which heat is flowing steadily. We suppose that the loss of heat from the sides of the bar is made negligible by lagging. XY is a slice of the bar, of thickness

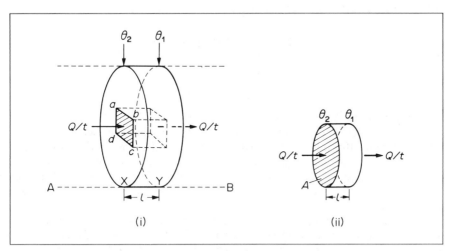

Figure 29.2 *Definition of thermal conductivity*

l, whose faces are at temperatures θ_2 and θ_1. Then the *temperature gradient* over the slice is

$$\frac{\theta_2 - \theta_1}{l}$$

We now consider an element *abcd* of the slice of unit cross-sectional area, and we denote by Q/t the heat flowing through it *per second*. The value of Q/t depends on the temperature gradient, and, since some substances are better conductors than others, it also depends on the material of the bar.

We therefore write

$$\frac{Q}{t} = k\frac{\theta_2 - \theta_1}{l}$$

where k is a factor depending on the material.

To a fair approximation the factor k is a constant for a given material; that is to say, it is independent of θ_2, θ_1, and l. It is called the *thermal conductivity* of the material concerned. From the above relation for Q/t, when the heat flow is *normal* to an area inside the material, k may be defined as *the heat flow per second per unit area per unit temperature gradient*.

This definition leads to a general equation for the flow of heat through any parallel-sided slab of the material, if no heat is lost from the sides of the slab. If the cross-sectional area of the slab is A in Figure 29.2 (ii), its thickness is l, and the temperature of its faces are θ_1 and θ_2, then the rate of flow of heat is

$$\frac{Q}{t} = \frac{kA(\theta_2 - \theta_1)}{l} \qquad . \quad . \quad . \quad . \quad (1)$$

So
$$\frac{Q}{At} = k\frac{\theta_2 - \theta_1}{l} \qquad . \quad . \quad . \quad . \quad (2)$$

heat flow per metre^2 per second = conductivity × temperature gradient . (2a)

The *U-value* or *heat transmittance* of a building material is the value *thermal conductivity(k)/thickness(l)*. So the U-value unit is $W\ m^{-2}K^{-1}$. A glass window of U-value $5·5\ W\ m^{-2}K^{-1}$, surface area $10\ m^2$ and a temperature difference of 12 K across it will lose heat per second $= 5·5 \times 10 \times 12 = 660$ W.

Lagged and Unlagged Bars

In terms of the calculus, (2) may be re-written

$$\frac{dQ}{dt} = -kA\frac{d\theta}{dl} \qquad . \quad . \quad . \quad . \quad (3)$$

the temperature gradient being negative since θ diminishes as l increases.

If a bar is *lagged* perfectly, as in Figure 29.1 (ii), then the heat per second, dQ/dt, flowing through every cross-section from the hot to the cold end is constant since no heat escapes through the sides. Hence, from (3), the temperature gradient, $d\theta/dl$, is constant along the bar. This is illustrated in Figure 29.1 (ii); the temperature variation with distance along the bar is a straight line.

If the bar is *unlagged*, as in Figure 29.1 (i), then heat is lost from the sides of the bar. In this case the heat per second, dQ/dt, flowing through each section decreases from the hot to the cold end. Hence, from (3), the temperature gradient, $d\theta/dl$, decreases with distance along the bar. This is shown by Figure 29.1 (i); the gradient at a point of the curve decreases with distance from the hot end of the bar.

$$Q/t = kA \times \textbf{temperature gradient}$$

Units and Magnitude of Conductivity

Equation (2) enables us to find the unit of thermal conductivity. We have

$$k = \frac{Q/At(\mathrm{J\,m^{-2}\,s^{-1}})}{(\theta_2 - \theta_1)/l(\mathrm{K\,m^{-1}})}$$

Thus the unit of thermal conductivity $= \mathrm{J\,s^{-1}\,m^{-1}\,K^{-1}}$, or since joule second^{-1} = watt (W), the unit of k is $\mathrm{W\,m^{-1}\,K^{-1}}$.

The thermal conductivity of copper, cardboard, water and air are roughly 380, 0·2, 0·6 and 0·03 $\mathrm{W\,m^{-1}\,K^{-1}}$ respectively. To a rough approximation we may say that the conductivities of metals are about 1000 times as great as those of other solids, and of liquids; and they are about 10 000 times as great as those of gases.

Examples on Conduction

1 Calculate the quantity of heat conducted through 2 m² of a brick wall 12 cm thick in 1 hour if the temperature on one side is 8°C and on the other side is 28°C. (Thermal conductivity of brick $= 0\cdot13\ \mathrm{W\,m^{-1}\,K^{-1}}$.)

$$\text{Temperature gradient} = \frac{28-8}{12 \times 10^{-2}}\,\mathrm{K\,m^{-1}} \text{ and } t = 3600\,\mathrm{s}$$

$$\therefore Q = kAt \times \text{temperature gradient}$$

$$= 0\cdot13 \times 2 \times 3600 \times \frac{28-8}{12 \times 10^{-2}}\,\mathrm{J}$$

$$= 156\,000\,\mathrm{J}$$

2 Figure 29.3 (i) shows a lagged bar XY of non-uniform cross-section. One end X is kept at 100°C and the other end at 0°C.

Describe and explain how the temperature varies from X to Y in the steady state.

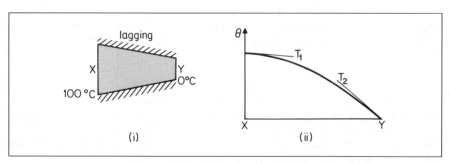

Figure 29.3 *Examples on conduction*

Figure 29.3 (i) shows how the temperature θ varies from the hot end X to the cold end Y. Since the bar is lagged, the heat per second, Q/t, through each section of the bar is the same. Now

$$\text{temperature gradient, } g, = \frac{1}{A} \cdot \frac{Q}{t}$$

So

$$g \propto \frac{1}{A}$$

So at X, where A is greatest, g is smallest, as shown by slope of the tangent T_1 to the temperature curve in Figure 29.3 (ii). At Y, where A is smallest, g is greatest as shown by the slope of the tangent T_2.

Thermal and Electrical Conductivity

We can make a useful analogy between thermal conductivity and electrical conductivity.

The electric current I flowing along a conductor $= Q/t$, where Q is the quantity of charge passing a given section in a time t. Also, $I = V/R$, where V is the potential difference between the ends of the conductor and R is its resistance (p. 244). Now $R = \rho l/A$, where ρ is the resistivity of the material, l is the length and A is its cross-sectional area (p. 254). So

$$I = \frac{Q}{t} = \frac{V}{\rho l/A}$$

Thus

$$\frac{Q}{t} = \frac{1}{\rho} A \frac{V}{l}$$

The quantity V/l is the *potential gradient* along the conductor. So

$$\frac{Q}{t} = \frac{1}{\rho} A \times \text{potential gradient} \qquad . \qquad . \qquad . \qquad . \qquad (1)$$

For heat conduction,

$$\frac{Q}{t} = kA \times \text{temperature gradient} . \qquad . \qquad . \qquad . \qquad (2)$$

Comparing (1) with (2), we see that $1/\rho$ is analogous to k. The inverse of resistivity is defined as the electrical *conductivity*, symbol σ. So thermal conductivity k is analogous to electrical conductivity σ.

Wiedemann and Franz discovered a law which states that, at a given temperature, the *ratio of the thermal to electrical conductivity is the same for all metals*. So a metal which is a good thermal conductor is also a good electrical conductor. This suggests that electrons are the carriers in both thermal and electrical conduction in metals. So on heating a metal bar the free electrons gain thermal energy and distribute this energy by collision with the fixed positive metal ions in the solid lattice.

Effect of Thin Layer of Bad Conductor

Figure 29.4 shows a lagged copper bar AB, whose ends are pressed against metal tanks at 100° and 0°C, but are separated from them by layers of dirt. The length of the bar is 10 cm or 0·1 m, and the dirt layers are 0·1 mm or $0·1 \times 10^{-3}$ m thick. Assuming that the conductivity of dirt is $1/1000$ that of copper, let us find the temperature of each end of the bar.

Suppose $\qquad\qquad\qquad k = $ conductivity of copper

$$A = \text{cross-section of copper}$$

$$\theta_2, \theta_1 = \text{temperature of hot and cold ends}$$

Since the bar is lagged, the heat flow per second Q/t is constant from end to end. Therefore,

$$\frac{Q}{t} = \frac{k}{1000} A \frac{100 - \theta_2}{0·1 \times 10^{-3}} = kA \frac{\theta_2 - \theta_1}{0·1} = \frac{k}{1000} A \frac{\theta_1 - 0}{0·1 \times 10^{-3}}$$

Dividing through by kA, these equations give

$$\frac{100 - \theta_2}{0·1} = \frac{\theta_2 - \theta_1}{0·1} = \frac{\theta_1}{0·1}$$

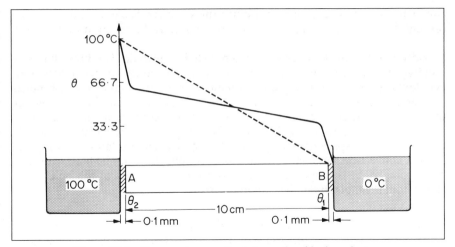

Figure 29.4 *Temperature gradients in good and bad conductors*

or
$$100 - \theta_2 = \theta_2 - \theta_1 = \theta_1$$

from which
$$\theta_2 = 66 \cdot 7°\text{C}, \theta_1 = 33 \cdot 3°\text{C}$$

So the total temperature drop, 100°C, is divided equally over the two thin layers of dirt and the long copper bar. The heavy lines in the figure show the temperature distribution; the broken line shows what it would be if there were no dirt.

Good and Bad Conductors

This numerical example shows what a great effect a thin layer of a bad conductor may have on thermal conditions; 0·1 mm of dirt causes as great a temperature fall as 10 cm of copper. We can generalise this result with the help of equation (2a):

$$\text{heat flow/m}^2\,\text{s} = \text{conductivity} \times \text{temperature gradient}$$

The equation shows that, if the heat flow is uniform, the temperature gradient is *inversely* proportional to the conductivity. So if the conductivity of dirt is 1/1000 that of copper, the temperature gradient in it is 1000 times that in copper; thus 1 mm of dirt sets up the same temperature fall as 1 m of copper. In general terms we express this result by saying that the dirt prevents a good thermal contact, or that it provides a bad one. The reader who has already studied electricity will see an obvious analogy here. We can say that a dirt layer has a high thermal resistance, and hence causes a great temperature drop.

Boiler plates are made of steel, not copper, although copper is about eight times as good a conductor of heat. The material of the plates makes no noticeable difference to the heat flow from the furnace outside the boiler to the water inside it, because there is always a layer of gas between the flame and the boiler plate. This layer may be very thin, but its conductivity is about 1/10 000 that of steel; if the plate is a centimetre thick, and the gas-film 1/1000 centimetre, then the temperature drop across the film is ten times that across the plate. So the rate at which heat flows into the boiler is determined mainly by the gas and not on the kind of metal used for the boiler plates.

If the water in the boiler deposits scale on the plates, the rate of heat flow is further reduced. For scale is a bad conductor, and, though it may not be as bad a conductor as gas, it can build up a much thicker layer. Scale must therefore be prevented from forming, if possible; and if not, it must be removed from time to time.

Badly conducting materials are often called *insulators*. The importance of building houses from insulating materials hardly needs to be pointed out. Window-glass is a ten-times better conductor than brick, and it is also much thinner. A room with large windows therefore requires more heating in winter than one with small windows. Wood is as bad a conductor (or as good an insulator) as brick, but it also is thinner. Wooden houses therefore have double walls, with an air-space between them. Air is an excellent insulator, and the walls prevent convection. In polar climates, wooden huts must not be built with steel bolts going right through them; otherwise the inside ends of the bolts grow icicles from the moisture in the explorer's breath.

Examples on Conduction

1 A cavity wall is made of bricks 0·1 m thick with an air space 0·1 m thick between them. (i) Assuming the thermal conductivity of brick is 20 times that of air, calculate the thickness of brick which conducts the same quantity of heat per second per unit area as 0·1 m of air, (ii) if the thermal conductivity of brick is 0·5 W m^{-1} K^{-1}, calculate the rate of heat conducted per unit area through the cavity wall when the outside surfaces of the brick walls are respectively 19°C and 4°C.

(i) Suppose θ_1, θ_2 are the respective temperatures at the ends of a brick of thickness l_B and thermal conductivity k_B, and at the ends of air of thickness l_A and thermal conductivity k_A. Then, with the usual notation,

$$\frac{Q}{t} = k_B A \frac{\theta_1 - \theta_2}{l_B} = k_A A \frac{\theta_1 - \theta_2}{l_A}$$

So

$$\frac{k_B}{l_B} = \frac{k_A}{l_A}$$

Then

$$l_B = l_A \times \frac{k_B}{k_A} = 0·1 \text{ m} \times 20 = 2 \text{ m}$$

(ii) Since the two bricks and the air are thermally in series, Figure 29.5 (i), we can replace the thickness of 0·1 m of air by 2 m of brick and add the three thicknesses of brick. So

$$\text{total brick thickness} = 0·1 + 2 + 0·1 = 2·2 \text{ m}$$

Then

$$\frac{1}{A} \cdot \frac{Q}{t} = k_B \frac{\theta_1 - \theta_2}{l_B} = 0·5 \frac{19 - 4}{2·2}$$

$$= 3·4 \text{ W m}^{-2}$$

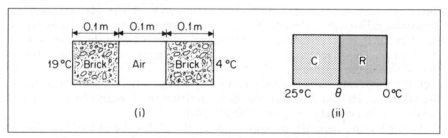

Figure 29.5 *Series conductors*

2 A sheet of rubber and a sheet of cardboard, each 2 mm thick, are pressed together and their outer faces are maintained respectively at 0°C and 25°C. If the thermal conductivities of rubber and cardboard are respectively 0·13 and 0·05 W m^{-1} K^{-1}, find the quantity of heat which flows in 1 hour across a piece of the composite sheet of area 100 cm^2.

We must first find the temperature, θ°C, of the junction of the rubber R and cardboard C, Figure 29.5 (ii). The temperature gradient across the rubber $= (\theta - 0)/2 \times 10^{-3}$; the temperature gradient across the cardboard $= (25 - \theta)/2 \times 10^{-3}$.

$$\therefore Q \text{ per second per m}^2 \text{ across rubber} = 0.13 \times (\theta - 0)/2 \times 10^{-3}$$

and Q per second per m^2 across cardboard $= 0.05 \times (25 - \theta)/2 \times 10^{-3}$

But in the steady state the quantities of heat above are the same.

$$\therefore \frac{0.13(\theta - 0)}{2 \times 10^{-3}} = \frac{0.05(25 - \theta)}{2 \times 10^{-3}}$$

$$\therefore 13\theta = 125 - 5\theta$$

$$\therefore \theta = \frac{125}{18} = 7°C$$

Now area $= 100 \text{ cm}^2 = 100 \times 10^{-4} \text{ m}^2$. So, using the rubber alone, Q through area in 1 hour (3600 seconds)

$$= \frac{0.13 \times 100 \times 10^{-4} \times 7 \times 3600}{2 \times 10^{-3}} = 16\,380 \text{ J}$$

Measurement of High Conductivity: Metals

When the thermal conductivity of a metal is to be measured, two conditions must usually be satisfied: (1) heat must flow through the specimen at a measurable rate, and (2) the temperature gradient along the specimen must be measurably steep. These conditions determine the form of the apparatus used.

When the conductor is a *metal*, it is easy to get a fast enough heat flow. The problem is to build up a temperature gradient. It is solved by having as the specimen a bar *long* compared with its diameter. Figure 29.6 shows the

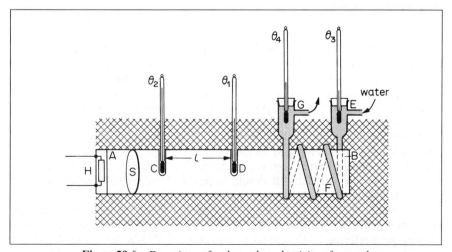

Figure 29.6 *Experiment for thermal conductivity of a metal*

apparatus, which is due to Searle. AB is the specimen, about 4 cm diameter and 20 cm long. In one form of apparatus it is heated by a coil at H, and cooled by circulating water at B. The whole apparatus is heavily lagged with felt. To measure the temperature gradient, thermometers are placed in the two mercury-filled cups C, D; the cups are made of copper, and are soldered to the specimen at a known distance apart. Alternatively, thermometers are placed in holes bored in the bar, which are filled with mercury. In this way errors due to bad thermal contact are avoided.

The cooling water flows in at E, round the copper coil F which is soldered to the specimen, and out at G. The water leaving at G is warmer than that coming in at E, so that the temperature falls continuously along the bar. If the water came in at G and out at E, it would tend to reverse the temperature gradient at the end of the bar, and might upset it as far back as D or C.

The whole apparatus is left running, with a steady flow of water, until all the temperatures have become *constant*: the temperature θ_2 and θ_1, at C and D in the bar, and θ_4 and θ_3 of the water leaving and entering. The steady rate of flow of the cooling water is measured with a measuring cylinder and a stop-clock.

Calculation. If A is the cross-sectional area of the bar and k is conductivity, then the heat flow per second through a section such as S is

$$\frac{Q}{t} = kA\frac{\theta_2-\theta_1}{l}$$

This heat is carried away by the cooling water; if a mass m of specific heat capacity c_w flows through F in 1 second, the heat carried away is $mc_w(\theta_4-\theta_3)$.

Therefore
$$kA\frac{\theta_2-\theta_1}{l} = mc_w(\theta_4-\theta_3)$$

With this apparatus we can show that the conductivity k is a constant over small ranges of temperature. To do so we increase the flow of cooling water, and thus lower the outflow temperature θ_4. The gradient in the bar then steepens, and $(\theta_2-\theta_1)$ increases. When the new steady state has been reached, the conductivity k is measured as before. Within the limits of experimental error, it is found to be unchanged.

To measure k for a good conductor, use a *long, thick* bar for measurable Q/t and temperature gradient.

Measurement of Low Conductivity: Non-metallic Solids

In measuring the conductivity of a bad conductor, the difficulty is to get an adequate heat flow. The specimen is therefore made in the form of a *thin* disc, D, about 10 cm in diameter and only a few millimetres thick (Figure 29.7 (i)). It is heated by a steam-chest C, whose bottom is thick enough to contain a hole for a thermometer.

The specimen rests on a thick brass slab B, also containing a thermometer. The whole apparatus is hung in mid air by three strings tied to B.

To ensure good thermal contact, the adjoining faces of C, D and B must be flat and clean; those of C and B should be polished. A trace of vaseline smeared over each face improves the contact.

When the temperatures have become *steady*, the heat passing from C through D escapes from B by radiation and convection. Its rate of escape from B is roughly proportional to the excess temperature of B over the room (Newton's law). Thus B takes up a steady temperature θ_1 such that its rate of loss of heat to the outside is just equal to its gain through D. The rate of loss of heat from the sides of D is negligible, because their surface area is small.

This apparatus is derived from one due to Lees, and simplified for elementary work. If we use glass or ebonite for the specimen, the temperature θ_1 is generally about 70°C; θ_2 is, of course, about 100°C. After these temperatures have become steady, and we have measured them, the problem is to find the rate of heat loss

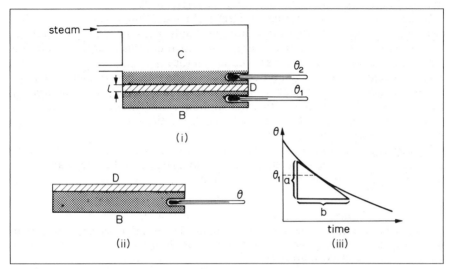

Figure 29.7   *Apparatus for thermal conductivity of a bad conductor*

from B. To do this, we take away the specimen D and heat B directly from C until its temperature has risen by about 10°C. We then remove C, and cover the top part of B again with the specimen D (Figure 29.7 (ii)). At intervals of a minute—or less—we measure the temperature of B, and afterwards plot it against the time (Figure 29.7 (iii)).

Calculation. (1) While the slab B is cooling it is losing heat by radiation and convection. It is doing so under the same conditions as in the first part of the experiment, because the felt prevents heat escaping from the top surface. Thus when the slab B passes through the temperature θ_1, it is losing heat at the same rate as in the first part of the experiment. The heat which it loses is now drawn from its own heat content, whereas before it was supplied from C via D; this is why the temperature of B is now falling, whereas before it was steady. The rate at which B loses heat at the temperature θ_1 is given by:

heat lost/second $= Mc \times$ temperature fall/second

where M, c are respectively the mass and specific heat capacity of the slab.

(2) To find the rate of fall of temperature at θ_1, we draw the tangent to the cooling curve at that point. If, as shown in Figure 29.7 (iii), its gradient at θ_1 would give a fall of a kelvin in b seconds, then the rate of temperature fall is $a/b\,\mathrm{K\,s^{-1}}$.

(3) We then have, if A is the cross-sectional area of the specimen, l its thickness, and k its conductivity,

$$kA\frac{\theta_2-\theta_1}{l} = Mc\frac{a}{b}$$

So k can be calculated.

To measure k **for a bad conductor, use a specimen of it which is** *wide* **(big area) and** *thin* **(high temperature gradient) for appreciable heat flow.**

Thermal Conduction in Solids

Metals. Metals are good thermal conductors and good electrical conductors. Wiedemann and Franz showed that, at a given temperature, *the ratio of thermal to electrical conductivity is the same for all metals* (p. 698).

Since electrons are the carriers in electrical conduction, it is considered that electrons transport thermal energy through metals. Thus on heating a metal bar the free electrons gain thermal energy and distribute this energy by collision with the fixed positive metal ions in the solid lattice.

Poor conductors. These have no free electrons. The transport of thermal energy through solids such as crystals is mainly due to waves. They are produced by lattice vibrations due to the thermal motion of the atoms. The waves are scattered by the atoms or by defects such as dislocations or impurity atoms and so distribute thermal energy to the solid.

The energy and momentum of the waves can also be considered carried by particles (p. 877). These particles are called *phonons*. Like the waves they represent, they travel with the speed of sound.

Examples on Tank Lagging and Double Glazing

1 Hot water in a metal tank is kept constant at 65°C by an immersion heater in the water. The tank has a lagging all round of thickness 20 mm and thermal conductivity $0.04 \text{ W m}^{-1}\text{K}^{-1}$ and its surface area is 0.5 m^2.

The heat lost per second by the lagging is 0.8 W per degree excess above the surroundings. Calculate the power of the immersion heater if the temperature of the surroundings is 15°C.

Let θ = surface temperature of lagging in °C

Then $(\theta - 15)$°C = excess temperature above the surroundings.

So   heat lost per second to surroundings = $0.8 \times (\theta - 15)$.   .   (1)

For conduction through the lagging,

$$\text{heat per second} = kA \times \text{temperature gradient}$$
$$= 0.04 \times 0.5 \times (65 - \theta)/(20 \times 10^{-3}) \quad . \quad . \quad (2)$$

In the steady state, (1) and (2) are equal

So $0.8(\theta - 15) = 0.04 \times 0.5 \times (65 - \theta)/(20 \times 10^{-3})$

Solving $\theta = 42.8$°C

Substituting in (1),

$$\text{power of heater} = 0.8 \times (42.8 - 15) = 22 \text{ W}$$

2 In order to minimise heat losses from a glass container, the walls of the container are made of two sheets of glass, each 2 mm thick, placed 3 mm apart, the intervening space being filled with a poorly conducting solid. Calculate the ratio of the rate of conduction of heat per unit area through this composite wall to that which would have occurred had a single sheet of the same glass been used under the same internal and external temperature conditions. (Assume that the thermal conductivity of glass and the poorly conducting solid = 0.63 and $0.049 \text{ W m}^{-1}\text{K}^{-1}$ respectively.)

We can replace the 3 mm thick solid ($k = 0.049$) by a thermally equivalent *greater* thickness of x mm of glass ($k_g = 0.63$). The value of x is given (p. 702) by

$$x = \frac{0.63}{0.049} \times 3\,\text{mm} = 38\tfrac{4}{7}\,\text{mm}$$

So   total equivalent glass thickness $= 2 + 2 + 38\tfrac{4}{7} = \dfrac{298}{7}\,\text{mm}$

If θ_2 and θ_1 are the outside temperatures of the respective glass sheets, then, with the usual notation,

$$\frac{1}{A}\frac{Q}{t} = k_\text{g}\frac{\theta_2 - \theta_1}{d} = k_\text{g}\frac{\theta_2 - \theta_1}{(298/7) \times 10^{-3}} \qquad . \qquad . \qquad . \qquad (1)$$

and for the single glass sheet of thickness 2 mm,

$$\frac{1}{A}\frac{Q}{t} = k_\text{g}\frac{\theta_2 - \theta_1}{d} = k_\text{g}\frac{\theta_2 - \theta_1}{2 \times 10^{-3}} . \qquad . \qquad . \qquad (2)$$

Dividing (1) by (2),

$$\text{ratio} = \frac{2 \times 7}{298} = 0.05\,(\text{approx.})$$

_____ **Exercises 29A** _____

Conduction

1   A closed metal vessel contains water   (i) at 30°C and then   (ii) at 75°C. The vessel has a surface area of $0.5\,\text{m}^2$ and a uniform thickness of 4 mm. If the outside temperature is 15°C, calculate the heat loss per minute by conduction in each case. (Thermal conductivity of metal $= 400\,\text{W}\,\text{m}^{-1}\,\text{K}^{-1}$.)

2   A uniform metal bar has one end kept at 100°C and the other at 0°C. Draw sketches showing how the temperature varies along the bar in the steady state
    (a) when its sides are well lagged,
    (b) when the sides are unlagged,
    (c) if the bar is not uniform but tapers or narrows from the hot to the cold end.
        Explain your answers in each case.

3   In measuring thermal conductivity of a metal, a long, thick, lagged bar is used. In measuring thermal conductivity of a bad conductor such as cardboard, a thin disc of large surface area is used which is not lagged.
        Explain the reasons for these practical arrangements.

4   A metal cylinder, containing water at 60°C, has a thickness of 4 mm and thermal conductivity $400\,\text{W}\,\text{m}^{-1}\,\text{K}^{-1}$. It is lagged by felt of thickness 2 mm and thermal conductivity $0.002\,\text{W}\,\text{m}^{-1}\,\text{K}^{-1}$. The room temperature is 10°C.
        Using the relation $Q/t = kAg$ to find the temperature gradient g for the metal and for the felt, show that
    (a) the temperature θ of the metal–felt interface is practically 60°C and
    (b) the rate of loss of heat by conduction is practically unaffected when the metal cylinder is replaced by a metal of smaller thermal conductivity $100\,\text{W}\,\text{m}^{-1}\,\text{K}^{-1}$ and the same thickness.

5   A copper hot water cylinder of length 1·0 m and radius 0·20 m is lagged by 2·0 cm of material of thermal conductivity $0.40\,\text{W}\,\text{m}^{-1}\,\text{K}^{-1}$. Estimate the temperature of the outer surface of the lagging, assuming heat loss is through the sides only, if heat has to be supplied at a rate of 0·25 kW to maintain the water at a steady temperature of 60°C.
        Assume that the temperature of the inside surface of the lagging is 60°C. (*L.*)

6   A double-glazed window consists of two panes of glass each 4 mm thick separated by a 10-mm layer of air. Assuming the thermal conductivity of glass to be 50 times greater than that of air calculate the ratios   (i) temperature gradient in the glass, to temperature gradient in the air gap,   (ii) temperature difference across one pane of the glass to temperature difference across the air gap.

Sketch a graph showing how the temperature changes between the surface of the glass in the room and the surface of the glass outside, i.e. across the double-glazed window, if there is a large temperature difference between the room and the outside. Explain why, in practice, the value of the ratio calculated in (i) is to high. (*L.*)

7 (a) The diagram (Figure 29A (i)) shows a section of a house wall one brick thick, the surfaces of the brick being at the temperatures shown. If the thermal conductivity of the brick is $0.6 \text{ W m}^{-1}\text{ K}^{-1}$, what is the rate of heat flow per unit area (W m^{-2}) through the bricks if steady state conditions apply?

Explain why, under these conditions, the temperature of the outer surface of the wall must be greater than the air temperature. At what rate must the outer surface be losing heat?

(b) The diagram (Figure 29A (ii)) shows a section of a cavity wall made up of brick, air and brick. The thermal conductivity of air is $0.02 \text{ W m}^{-1}\text{ K}^{-1}$. Explain why, when steady state conditions apply, the rate of heat flow across each layer is the same. Assuming this to be the case, draw a sketch graph to show how the temperature changes between the brick surface at 20°C and that of 5°C.

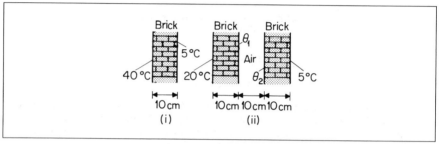

Figure 29A

Calculate, showing your working, the thickness of brick equivalent to 10 cm of air. Hence, or otherwise, calculate values for θ_1 and θ_2, the inner surfaces of the bricks, and compare the rate of heat loss through the cavity wall with that through the single brick described in (a).

Explain why, in practice, the introduction of a cavity does not produce the improvement suggested by the calculation unless, for example, the cavity is filled with plastic foam. (*L.*)

8 If the thermal conductivity of a gas is to be measured with any accuracy, immense care must be taken to minimise the amount of heat transferred through the gas in other ways. The diagram below shows apparatus used to measure the thermal conductivity of hydrogen, Figure 29B.

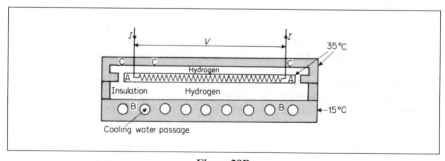

Figure 29B

A is a thin copper disc 15.00 cm in diameter. It is heated electrically to a constant temperature of 35.0°C by current I and potential difference V. B is a thick copper plate held at 15.0°C by cooling water which flows through it continuously. The lower face of A and the upper face of B are silver plated, highly polished, horizontal and 6·00 mm apart.

C is a thick copper lid and guard ring held at 35·0°C by electrical heaters (not shown). The temperatures of A, B and C are measured with thermocouples (not shown). The following measurements are recorded in steady state conditions and with the apparatus filled with hydrogen:

$$I_1 = 1·131 \text{ A} \qquad V_1 = 9·64 \text{ V}$$

With the apparatus evacuated and with steady state conditions once more,

$$I_2 = 0·230 \text{ A} \qquad V_2 = 1·96 \text{ V}$$

(a) Answer the following: (i) Why is there no net loss of heat from A upwards? (ii) Why are the opposing faces of A and B highly polished? (iii) Why must A and B be horizontal? (iv) Why is the diameter of A much larger than the distance from A to B?

(b) Explain why measurements are made with the space in the apparatus evacuated. Calculate values for (i) the energy per second transferred from A to B by radiation, (ii) the energy per second transferred from A to B by conduction, (iii) the thermal conductivity of hydrogen.

(c) How would you expect the values of I_1, I_2, V_1, V_2 to change if the opposing faces of A and B were painted a dull black colour? Explain your answer. (L.)

9 Define *thermal conductivity*.

Figure 29C represents, in outline, the apparatus used in Searle's bar method for determining the thermal conductivity of copper.

(a) Why is a thick bar used in this determination?

(b) Why must it be well insulated except at its two ends?

(c) Why does one wait for some time before taking readings?

(d) Does it matter where the thermometers T_1 and T_2 are placed along the bar? Explain.

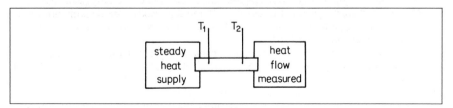

Figure 29C

(e) One end of the insulated copper bar, which is of length 0·2 m and cross-sectional area $1·2 \times 10^{-3} \text{ m}^2$, is maintained at a steady temperature by an electric heater which is supplying heat to the bar at the rate of 100 W. Thermometer T_1 is 0·06 m from the hot end and thermometer T_2 is 0·14 m from the hot end. At the cool end, water flows into a circulating coil at 15·3°C and leaves it at 16·7°C. Taking the thermal conductivity of copper to be 400 W m^{-1} K^{-1} and specific heat capacity of water 4200 J kg^{-1} K^{-1}, estimate the rate at which water is flowing through the circulating coil and also the reading of each of the thermometers T_1 and T_2. (O.)

10 The ends of a bar of uniform cross-section are maintained at steady different temperatures, both being above room temperature. Explain how the temperature varies along the bar if the bar is

(a) ideally lagged,

(b) unlagged.

A bar 0·20 m in length and of cross-sectional area $2·5 \times 10^{-4} \text{ m}^2$ (2·5 cm²) is ideally lagged. One end is maintained at 373 K (100°C) while the other is maintained at 273 K (0°C) by immersion in melting ice. Calculate the rate at which the ice melts owing to the flow of heat along the bar.

Thermal conductivity of the material of the bar $= 4.0 \times 10^2 \ \mathrm{W \ m^{-1} \ K^{-1}}$.
Specific latent heat of fusion of ice $= 3.4 \times 10^5 \ \mathrm{J \ kg^{-1}}$. (*JMB.*)

11 Give a critical account of an experiment to determine the thermal conductivity of a material of low thermal conductivity such as cork. Why it is that most cellular materials, such as cotton wool, felt, etc., all have approximately the same thermal conductivity?

One face of a sheet of cork, 3 mm thick, is placed in contact with one face of a sheet of glass 5 mm thick, both sheets being 20 cm square. The outer faces of this square composite sheet are maintained at 100°C and 20°C, the glass being at the higher mean temperature. Find
(a) the temperature of the glass-cork interface, and
(b) the rate at which heat is conducted across the sheet, neglecting edge effects.
(Thermal conductivity of cork $= 6.3 \times 10^{-2} \ \mathrm{W \ m^{-1} \ K^{-1}}$, thermal conductivity of glass $= 7.2 \times 10^{-1} \ \mathrm{W \ m^{-1} \ K^{-1}}$.) (*O. & C.*)

12 The thermal conductivity λ for a substance may be defined by the equation

$$\frac{dQ}{dt} = -\lambda A \frac{d\theta}{dx}$$

Identify briefly each term in this equation, and explain the minus sign.
Write down an analogous equation defining electrical conductivity σ, and identify briefly each term in your equation. Hence find the units of σ.

A cylindrical bar of metal has one end maintained at a steady high temperature and the other end is at room temperature. Sketch on the same axes, curves showing the variation of temperature along the cylinder when its curved surface
is (i) enclosed in a perfectly insulating jacket and   (ii) exposed to air. Explain briefly the difference between your curves.

A domestic hot water cylinder made of 3 mm copper sheet has a surface area of $2 \ \mathrm{m^2}$ and is fitted with a 50 mm thick insulating jacket. The water is maintained at 70°C and the temperature of the surrounding air is 20°C. (iii) Show from the data given that the copper makes a negligible contribution to conserving heat. (iv) Find the temperature gradient in the jacket. (v) Find the weekly cost of maintaining the tank at this temperature. It is not usual to insulate the bottom of a domestic hot water cylinder. Why is this so?
(Thermal conductivities of copper and insulating material are, respectively, $4 \times 10^2 \ \mathrm{W \ m^{-1} \ K^{-1}}$ and $8 \times 10^{-2} \ \mathrm{W \ m^{-1} \ K^{-1}}$; electricity cost is 5·1 p per kW-hour.)
(*W.*)

13 Define *thermal conductivity* and state a unit in which it is expressed.
Explain why, in an experiment to determine the thermal conductivity of copper using a Searle's arrangement, it is necessary
(a) that the bar should be thick, of uniform cross-section and have its sides well lagged,
(b) that the temperatures used in the calculation should be the steady values finally registered by the thermometers.
Straight metal bars X and Y of circular section and equal in length are joined end to end. The thermal conductivity of the material of X is twice that of the material of Y, and the uniform diameter of X is twice that of Y. The exposed ends of X and Y are maintained at 100°C and 0°C, respectively and the sides of the bars are ideally lagged. Ignoring the distortion of the heat flow at the junction, sketch a graph to illustrate how the temperature varies between the ends of the composite bar when conditions are steady. Explain the features of the graph and calculate the steady temperature of the junction. (*JMB.*)

14 Outline an experiment to measure the thermal conductivity of a solid which is a poor conductor, showing how the result is calculated from the measurements.
Calculate the theoretical percentage change in heat loss by conduction achieved by replacing a single glass window by a double window consisting of two sheets of glass separated by 10 mm of air. In each case the glass is 2 mm thick. (The ratio of the thermal conductivities of glass and air is 3:1.)
Suggest why, in practice, the change would be less than that calculated. (*L.*)

15 Outline, with the necessary theory, a simple method for finding the thermal
    conductivity of a bad conductor.
    Estimate the rate of heat loss from a room through a glass window of area $2\,m^2$
    and thickness 3 mm when the temperature of the room is 20°C and that of the air
    outside is 5°C.
    This estimate is too high because it is based on the assumption that the inner
    glass surface is at room temperature. A better approximation to actual conditions is
    to suppose that there is a thin uniform layer of still air in contact with the inner
    glass surface of the window. In practice, it is found that the rate of loss of heat
    through this window when the inside and outside temperatures have the above
    values is only 3 kW. Using the suggested approximation, find
    (a) the temperature difference between the faces of the glass,
    (b) the temperature difference across the air film,
    (c) the thickness of this film.
    (Take thermal conductivity of glass to be $1\cdot2\,W\,m^{-1}\,K^{-1}$ and that of air
    $2\cdot4 \times 10^{-2}\,W\,m^{-2}\,K^{-4}$.) (L.)

16 What is the relation between the *U-value* of glass and its *thermal conductivity*?
    An uninsulated roof has a U-value of 1·9. Its U-value when lagged is 0·4.
    Calculate the U-value of the lagging material and its thermal conductivity if it is
    2 cm thick.

17 The U-value of four construction components are given below.

Component	U-value W m^{-2}K^{-1}
Single-glazed window	5·6
Double-glazed window	3·2
Uninsulated roof	1·9
Well-insulated roof	0·4

    What do you understand by the *U-value* of a component?
    A house has windows of area $24\,m^2$ and roof of area $60\,m^2$. The occupier
    heats the house for 3000 hours per year to a temperature which on average is
    14K above that of the air outside. Calculate the energy lost per year through (i)
    the single-glazed windows, and (ii) the uninsulated roof, expressing your answers
    in kWh.
    If electricity costs 5·5p per unit, calculate the annual savings the occupier could
    make by (iii) installing double-glazing, and (iv) insulating the roof.
    If double-glazing costs £3000 and roof insulation costs £100, which if either, of
    the two energy-saving steps would you advise the occupier to take? (*L.*)

Radiation

All heat comes to us, directly or indirectly, from the sun. The heat which comes directly travels through 150 million km of space, mostly empty, and travels in straight lines, as does the light: the shade of a tree coincides with its shadow. Because heat and light travel with the same speed, they are both cut off at the same instant in an eclipse. Since light is propagated by waves of some kind we conclude that the heat from the sun is propagated by similar waves, and we say it is 'radiated'.

Measurements have been made which give the amount of radiant energy approaching the earth from the sun, called the *solar constant*. At the upper limit of our atmosphere, it is about $80\,000\,\text{J}\,\text{m}^{-2}\,\text{min}^{-1}$ or about $1340\,\text{W}\,\text{m}^{-2}$.

At the surface of the earth it is always less than this value because of absorption in the atmosphere. Even on a cloudless day it is less, because the ozone in the upper atmosphere absorbs much of the ultraviolet.

As we show later, more radiation is obtained from a dull black body than from a transparent or polished one. Black bodies are also better absorbers of radiation than polished or transparent ones, which either allow radiation to pass through themselves, or reflect it away from themselves. If we hold a piece of white card, with a patch of black drawing ink on it, in front of the fire, the black patch soon comes to feel warmer than its white surround.

Reflection and Refraction

If we focus the sun's light on our skin with a converging lens or a concave mirror, we feel heat at the focal spot. The heat from the sun has therefore been reflected or refracted in the same way as the light.

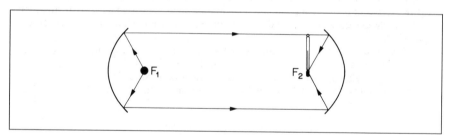

Figure 29.8 *Reflection of radiant heat*

If we wish to show the reflection of heat unaccompanied by light, we may use two searchlight (parabolic) mirrors, set up as in Figure 29.8. At the focus of one, F_1 we put an iron ball heated to just below redness. At the focus of the other, F_2, we put the bulb of a thermometer, which has been blackened with soot to make it a good absorber (p. 718). The mercury rises in the stem of the thermometer. If we move either the bulb or the ball away from the focus, the mercury falls back; the bulb has therefore been receiving heat from the ball, by reflection at the two mirrors. We can show that the foci of the mirrors are the same for heat as for light if we replace the ball and thermometer by a lamp and screen. (In practice we do this first, to save time in finding the foci for the main experiment.)

To show the refraction of heat apart from the refraction of light is more difficult. It was first done by the astronomer HERSCHEL in 1800. Herschell passed a beam of sunlight through a prism, as shown diagrammatically in Figure 29.9, and explored the spectrum with a sensitive thermometer, whose bulb he had blackened. He found that in the visible part of the ssectrum the mercury rose, showing that the light energy which it absorbed was converted into heat. But the mercury rose *more* when he carried the bulb into the darkened portion a little beyond the red of the visible spectrum. The sun's rays therefore carried energy which was not light.

Ultraviolet and Infrared

The radiant energy which Herschel found beyond the red is now called *infrared* radiation, because it is less refracted than the red. Radiant energy is also found beyond the violet and it is called *ultraviolet* radiation, because it is refracted more than the violet.

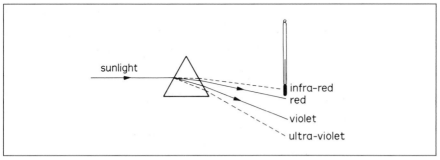

Figure 29.9   *Infrared and ultraviolet (diagrammatic)*

Ultraviolet radiation is absorbed by the human skin and causes sun-burn. More importantly, it stimulates the formation of vitamin D, which is necessary for the assimilation of calcium and the prevention of rickets. It is also absorbed by green plants; in them it enables water to combine with carbon dioxide to form carbohydrates. This process is called *photosynthesis*. Ultraviolet radiation causes the emission of electrons from metals, as in photoelectric cells; and it produces a latent image on a photographic emulsion. It is harmful to the eyes.

Ultraviolet radiation is strongly absorbed by glass—spectacle-wearers do not sunburn round the eyes—but enough of it gets through to affect a photographic film. It is transmitted with little absorption by quartz.

Infrared radiation is transmitted by rock-salt, but most of it is absorbed by glass. A little near the visible red passes easily through glass—if it did not, Herschel would not have discovered it. When infrared radiation falls on the skin, it gives the sensation of warmth. It is what we usually have in mind when we speak of heat radiation, and it is the main component of the radiation from a hot body; but it is in no essential way different from the other components, visible and ultraviolet radiation, as we shall now see.

Wavelengths of Radiation

In the section on Optics, we show how the wavelength of light can be measured with a *diffraction grating*—a series of fine close lines ruled on a glass. The wavelength ranges from $4 \cdot 0 \times 10^{-7}$ m for the violet, to $7 \cdot 5 \times 10^{-7}$ m for the red. The first accurate measurements of wavelength were published in 1868 by ANGSTROM, and in his honour a distance of 10^{-10} m is called an *Angstrom unit* (A.U.). The wavelengths of infrared radiation can be measured with a grating made from fine wires stretched between two screws of close pitch. They range from 7500 A.U. to about 1 000 000 A.U. Often they are expressed in a longer unit than the Angstrom, such as the micron (µm) or the nanometre (nm).

$$1\,\mu m = 10^{-6}\,m = 10^4\,\text{A.U.}$$
$$1\,nm = 10^{-9}\,m = 10\,\text{A.U.}$$

We denote wavelength by the symbol λ; its value for visible light ranges from 400 nm to 750 nm, or 0·4 μm to 0·75 μm, and for infrared radiation from 750 nm to about 10^5 nm, or 0·75 μm to about 100 μm.

We now consider that X-rays and radio waves also have the same nature as light, and so do the γ-rays from radioactive substances. For reasons which we cannot here discuss, we consider all these waves to be due to oscillating electric and magnetic fields. Figure 29.10 shows roughly the range of their wavelengths: it is a diagram of the *electromagnetic spectrum*.

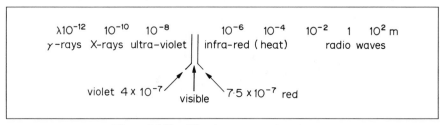

Figure 29.10   *The electromagnetic spectrum*

Figure 29A, B   *Infrared photography or thermal imaging. Photos A and B are of the same scene; a Boeing 747 aircraft being loaded at night. A is a normal (visible wavelengths) photo; it shows relatively little detail. B, however, is a clear photo. It uses only infrared wavelengths from the scene and is produced with infrared lenses and plates. In this photo, black is hot and white is cold. As can be seen, the aircraft has not been parked for very long as its tyres and engines are still warm. Note also the hot parts of the light beacon above the aircraft.*
(Courtesy of Barr and Stroud Limited)

Detection of Heat (Infrared) Radiation

A thermometer with a blackened bulb is a sluggish and insensitive detector of radiant heat. More satisfactory detectors, however, are electrical. One kind consists of a long thin strip of blackened platinum foil arranged in a compact zigzag on which the radiation falls (Figure 29.11).

The foil is connected in a Wheatstone bridge, to measure its electrical resistance. When the strip is heated by the radiation, its resistance increases, and the increase is measured on the bridge. The instrument was devised by LANGLEY in 1881; it is called a bolometer, *bole* being Greek for a ray.

Figure 29.11 *Bolometer strip*

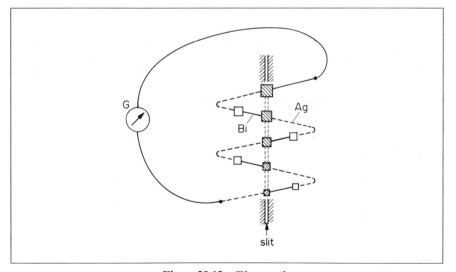

Figure 29.12 *Thermopile*

The other, commoner, type of radiation detector is called a *thermopile*. Its action depends on the electromotive force which appears between the junctions of two different metals, when one junction is hot and the other cold. The modern thermopile consists of many junctions between the fine wires, as shown diagrammatically in Figure 29.12; the wires are of silver and bismuth, 0·1 mm or less in diameter. Their junctions are attached to thin discs of tin, about 0·2 mm thick, and about 1 mm square. One set of discs is blackened and mounted behind a slit, through which radiation can fall on them; the junctions attached to them become the *hot* junctions of the thermopile. The other, cold, junctions are shielded from the radiation to be measured; the discs attached to them help to keep them cool, by increasing their surface area.

When radiation falls on the blackened discs of a thermopile, it warms the junctions attached to them, and sets up an e.m.f. This e.m.f. can be measured with a potentiometer, or, for less accurate work, it can be used to deflect a galvanometer, G, connected directly to the ends of the thermopile (Figure 29.12).

Reflection and Refraction: Inverse Square Law

With a thermopile and galvanometer, we can repeat Herschel's experiment more strikingly than with a thermometer. Using the simple apparatus of Figure 29.13 we can show that, when infra-red is reflected, the angle of reflection equals the angle of incidence. We can also show the first law of reflection; that the incident and reflected rays are in the same plane as the normal to the reflector at the point of incidence.

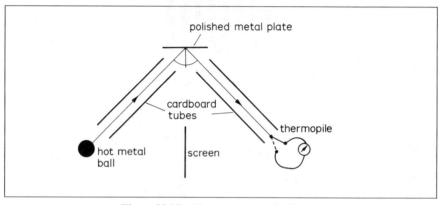

Figure 29.13 *Demonstration of reflection*

If heat is radiant energy, its intensity should fall off as the inverse square of the distance from a point source. We can check that it does so by setting up an electric lamp, with a compact filament, in a dark room preferably with black walls. When we put a thermopile at different distances from the lamp, the deflection of the galvanometer is found to be inversely proportional to the square of the distance.

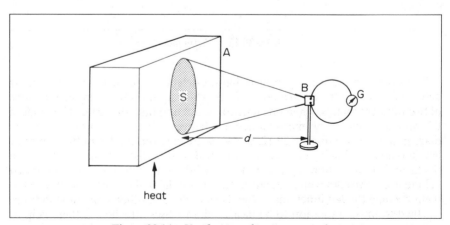

Figure 29.14 *Verification of inverse square law*

If we wish to do this experiment with radiation that includes no visible light, we must modify it. Instead of the lamp, we use a large blackened tank of boiling water, A, and we fit the thermopile, B, with a conical mouthpiece, blackened on the inside, Figure 29.14. The blackening prevents any radiation from reaching the pile by reflection at the walls of the mouthpiece. We now find that the deflection of the galvanometer, G does not vary with the distance of the pile from the tank, *provided that the tank occupies the whole field of view of the cone* (Figure 29.14). The area S of the tank from which radiation can reach the thermopile is then proportional to the square of the distance d. And since the deflection is unchanged when the distance is altered, the total radiation from each element of S must therefore fall off as the inverse square of the distance d.

The Infrared Spectrometer

Infrared spectra are important in the study of molecular structure. They are observed with an infrared spectrometer, whose principle is shown in Figure 29.15. Since glass is opaque to the infrared, the radiation is focused by concave mirrors instead of lenses; the mirrors are plated with copper or gold on their front surfaces. The sources of light is a Nernst filament, a metal filament coated with alkaline-earth oxides, and heated electrically. The radiation from such a filament is rich in infrared.

The slit S of the spectrometer is at the focus of one mirror which acts as a collimator. After passing through the rock-salt prism, A, the radiation is focused on to the thermopile P by the mirror M_2, which replaces the telescope of an optical spectrometer. Rotating the prism brings different wavelengths on to the slit of the thermopile; the position of the prism is calibrated in wavelengths with the help of a grating.

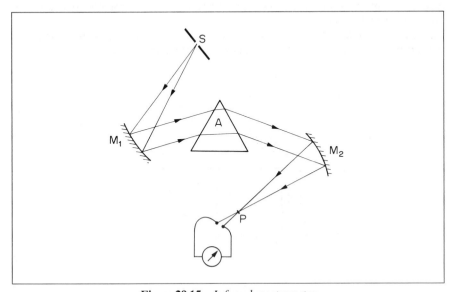

Figure 29.15 *Infrared spectrometer*

To a fair approximation, the deflection of the galvanometer is proportional to the radiant power carried in the narrow band of wavelengths which fall on the thermopile. If an absorbing body, such as a solution of an organic compound, is placed between the source and the slit, it weakens the radiation passing through the spectrometer, in the wavelengths which it absorbs. These wavelengths are therefore shown by a fall in the galvanometer deflection.

Radiation and Absorption

We have already pointed out that black surfaces are good absorbers and radiators of heat, and that polished surfaces are bad absorbers and radiators. This can be demonstrated by the apparatus in Figure 29.16, called a *Leslie cube*. It is a cubical metal tank whose sides have a variety of finishes: dull black, dull white, highly polished. It contains boiling water, and therefore has a constant temperature. Facing it is a thermopile, P, which is fitted with a blackened conical mouth-piece.

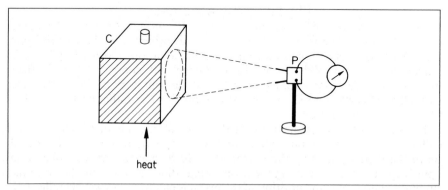

Figure 29.16 *Comparing radiators*

Provided the face of the cube occupies the whole field of view of the cone, its distance from the thermopile does not matter. The galvanometer deflection is greatest when the thermopile is facing the dull black surface of the cube, and least when it is facing the highly polished surface. The highly polished surface is therefore the worst radiator of all, and the dull black is the best.

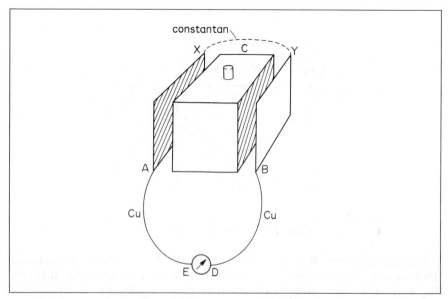

Figure 29.17 *Comparing absorbers*

Leslie's cube can also be used in an experiment to compare the absorbing properties of surfaces, Figure 29.17. The cube, C, full of boiling water, is placed between two copper plates, A, B, of which A is blackened and B is polished. The temperature differences between A and B is measured by making each of them one element in a thermojunction: they are joined by a constantan wire, XY, and connected to a galvanometer, by copper wires, AE, DB. If A is hotter than B, the junction, X, is hotter than the junction, Y, and a current flows through the galvanometer in one direction. If B is hotter than A, the current is reversed.

The most suitable type of Leslie's cube is one which has two opposite faces similar—say grey—and the other two opposite faces very dissimilar—one black, one polished. At first the plates A, B are set opposite similar faces. The blackened plate, A, then becomes the hotter, showing that it is the better absorber.

The cube is now turned so that the blackened plate, A, is opposite the polished face of the cube, while the polished plate, B, is opposite the blackened face of the cube. The galvanometer then shows no deflection; the plates thus reach the same temperature. It follows that the good radiating property of the blackened face of the cube, and the bad absorbing property of the polished plate, are just compensated by the good absorbing property of the blackened plate, and the bad radiating property of the polished face of the cube.

The Black Body

The experiments described before lead us to the idea of a *perfectly black body*; one which absorbs all the radiation that falls upon it, and reflects and transmits none. The experiments also lead us to suppose that such a body would be the best possible radiator.

A good black body can be made simply by punching a small hole in the lid of a closed empty tin. The hole looks almost black, although the shining tin is a good reflector. The hole looks black because the light which enters through it is reflected many times round the walls of the tin, before it passes through the hole again, Figure 29.18. At each reflection, about 80% of the light energy is reflected, and 20% is absorbed. After two reflections, 64% of the original light goes on to be reflected a third time; 36% has been absorbed. After ten reflections, the fraction of the original energy which has been absorbed is 0.8^{10}, or 0.1. So the closed tin with a hole in it is a very good absorber of radiation.

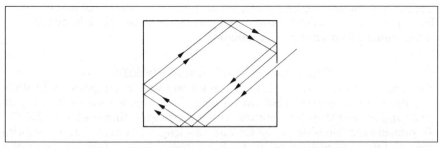

Figure 29.18 *Multiple reflections make a black body*

Any space which is almost wholly enclosed approximates to a black body. And, since a good absorber is also a good radiator, an almost closed space is the best radiator we can find.

A form of black body used in radiation measurements is shown in Figure 29.19. It consists of a porcelain sphere, S, with a small hole in it. The inside is blackened with soot to make it as good a radiator and as bad a reflector as possible. (The effect of multiple reflections is then to convert the body from nearly black to very nearly black indeed.) The sphere is surrounded by a high-temperature bath of, for example, molten salt (the melting-point of common salt is 800°C).

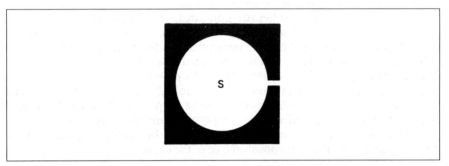

Figure 29.19   *A black body*

Quality of Radiation

The deepest parts of a coal or wood fire are black bodies. Anyone who has looked into a fire knows that the deepest parts of it look brightest—they are radiating most power. In the hottest part, no detail of the coals or wood can be seen. So the radiation from an almost enclosed space is uniform; its character does not vary with the nature of the surface of the space. This is so because the radiation coming out from any area is made up partly of the radiation emitted by that area, and partly of the radiation from other areas, reflected at the area in question. And if the hole in the body is small, the radiations from every area inside it are well mixed by reflection before they can escape. So the intensity and quality of the radiation escaping thus does not depend on the particular surface from which it escapes.

When we speak of the *quality* of radiation we mean the relative intensities of the different wavelengths in it; the proportion of red to blue, for example. The quality of the radiation from a perfectly black body depends *only on its temperature*. When the body is made hotter, its radiation becomes not only more intense, but also more nearly white; the proportion of blue to red in it increases. Because its quality is determined only by its temperature, black body radiation is sometimes called 'temperature radiation'.

Properties of Temperature Radiation

The quality of the radiation from a black body was investigated by LUMMER and PRINGSHEIM in 1899. They used a black body represented by B in Figure 29.20 and measured its temperature with a thermopile; they took it to 2000°C. To measure the intensities of the various wavelengths, Lummer and Pringsheim used an infrared spectrometer and a linear bolometer (p. 715) consisting of a single platinum strip.

The results of the experiments are shown in Figure 29.21. Each curve gives the relative intensities of the different wavelengths, for a given temperature of the body. The curves show that, as the temperature rises, the intensity of every wavelength increases, but the intensities of the shorter wavelengths increase

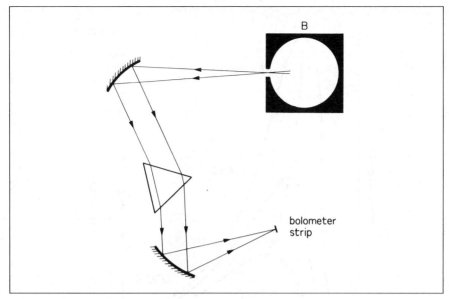

Figure 29.20 *Lummer and Pringsheim's apparatus for study of black body radiation (diagrammatic)*

more rapidly. Thus the radiation becomes, as we have already observed, less red, that is to say, more nearly white. The curve for sunlight has its peak at about 5×10^{-7} m in the visible green; from the position of this peak we conclude that the surface temperature of the sun is about 6000 K. Stars which are hotter than the sun, such as Sirius and Vega, look blue, not white, because the peaks of their radiation curves lie further towards the visible blue than does the peak of sunlight.

The actual intensities of the radiations are shown on the right of the graph in Figure 29.21. To speak of the intensity of a single wavelength is meaningless. The slit of the spectrometer always gathers a *band of wavelengths*—the narrower the slit the narrower the band—and we always speak of the intensity of a given band. We express it as follows ('s' represents 'second'):

$$\text{energy radiated m}^{-2}\text{s}^{-1}, \text{in band } \lambda \text{ to } \lambda + \Delta\lambda = E_\lambda \Delta\lambda . \qquad (1)$$

The quantity E_λ is called *emissive power* of a black body for the wavelength λ and at the given temperature; its definition follows from equation (1):

$$E_\lambda = \frac{\text{energy radiated m}^{-2}\text{s}^{-1}, \text{in band } \lambda \text{ to } \lambda + \Delta\lambda}{\text{bandwidth, } \Delta\lambda}$$

The expression 'energy per second' can be replaced by the word 'power', whose unit is the watt. Thus

$$E_\lambda = \frac{\textbf{power radiated m}^{-2}\textbf{ in band } \boldsymbol{\lambda} \textbf{ to } \boldsymbol{\lambda} + \boldsymbol{\Delta\lambda}}{\boldsymbol{\Delta\lambda}}$$

In the figure E_λ is expressed in watt per m^2 per nanometre (10^{-9} m).

The quantity $E_\lambda\Delta\lambda$ in equation (1) is the *area* beneath the radiation curve between the wavelengths λ and $\lambda + \Delta\lambda$ (Figure 29.22). Thus the energy radiated per metre^2 per second between those wavelengths is proportional to that area.

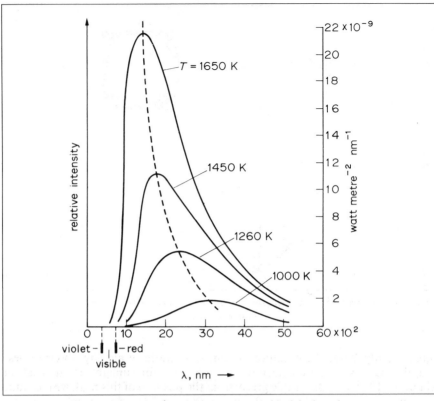

Figure 29.21 _Distribution of intensity in black body radiation_

Similarly, the _total radiation_ emitted per metre^2 per second over all wavelengths is proportional to _the area under the whole curve._

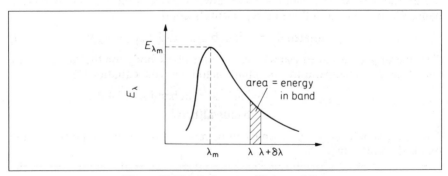

Figure 29.22 _Definition of E, λ_m and E_{λ_m}_

Laws of Black Body Radiation

The curves of Figure 29.21 can be explained only by Planck's quantum theory of radiation, whhch is outside our scope. Both theory and experiment lead to three generalisations, which together describe well the properties of black body radiation:

(i) If λ_m is the wavelength of the peak of the curve for T (in K), then

$$\lambda_m T = \text{constant} \qquad . \qquad . \qquad . \qquad . \qquad . \qquad (2)$$

The value of the constant is $2\cdot9 \times 10^{-3}$ m K. In Figure 29.21 the dotted line is the locus of the peaks of the curves for different temperatures.

The relationship in (2) is sometimes called *Wien's displacement law*.

(ii) If E_{λ_m} is the height of the peak of the curve for the temperature T, then

$$E_{\lambda_m} \propto T^5 \qquad . \qquad . \qquad . \qquad . \qquad . \qquad . \qquad (3)$$

(iii) The curve showing the variation of E_λ with λ at constant temperature T in Figure 29.21 obeys the *Planck formula*

$$E_\lambda = \frac{c_1}{\lambda^5(e^{c_2/\lambda T} - 1)}$$

where c_1 and c_2 are constants.

Stefan's Law

If E is the *total* energy radiated per metre^2 per second at a temperature T, which is represented by the *total area* under the particular $E_\lambda - \lambda$ curve, then

$$E = \sigma T^4$$

where σ is a constant. This result is called *Stefan's law*, and the constant σ is called the *Stefan constant*. Its value is

$$\sigma = 5\cdot7 \times 10^{-8} \, \text{W m}^{-2} \, \text{K}^{-4}$$

So in Figure 29.21, which shows four $E_\lambda - \lambda$ graphs at different temperatures T, the total area below the graphs should be proportional to the corresponding value of T^4.

The energy per second or *power P* radiated by an area A of a black body radiator is thus given by

$$P = A\sigma T^4$$

where P is in watts. This is illustrated in the examples which follow.

Examples on Stefan's Law of Radiation

1   The tungsten filament of an electric lamp has a length of $0\cdot5$ m and a diameter 6×10^{-5} m. The power rating of the lamp is 60 W. Assuming the radiation from the filament is equivalent to 80% that of a perfect black body radiator at the same temperature, estimate the steady temperature of the filament. (Stefan constant $= 5\cdot7 \times 10^{-8}$ W m^{-2} K^{-4}.)

When the temperature is steady,

$$\text{power radiated from filament} = \text{power received} = 60 \text{ W}$$

$$\therefore 0\cdot8 \times 5\cdot7 \times 10^{-8} \times 2\pi \times 3 \times 10^{-5} \times 0\cdot5 \times T^4 = 60$$

since surface area of cylindrical wire is $2\pi r h$ with the usual notation.

$$\therefore T = \left(\frac{60}{0\cdot4 \times 5\cdot7 \times 10^{-8} \times 2\pi \times 3 \times 10^{-5}} \right)^{1/4}$$

$$= 1933 \text{ K}$$

2 The solar constant, which is the energy arriving per second at the earth from the sun, is about $1400 \, \text{W m}^{-2}$. Estimate the surface temperature of the sun, given that the sun's radius $= 7 \times 10^5 \, \text{km}$, the distance of the sun from the earth $= 1\cdot5 \times 10^8 \, \text{km}$ and Stefan constant $= 5\cdot7 \times 10^{-8} \, \text{W m}^{-2} \, \text{K}^{-4}$.

Suppose T is the kelvin temperature of the sun's surface. Then

total energy per second radiated from sun's surface $= A\sigma T^4 = 4\pi r_s^2 \sigma T^4$

since the sun's surface is $4\pi r_s^2$ if r_s is its radius.

This energy falls all round a sphere of radius r_0 where r_0 is the radius of the earth's circular orbit round the sun. Since the area of the sphere is $4\pi r_0^2$, the energy per second falling on unit area

$$= \frac{1}{4\pi r_0^2} \times 4\pi r_s^2 \sigma T^4 = \left(\frac{r_s}{r_0}\right)^2 \sigma T^4$$

So $\qquad \left(\dfrac{r_s}{r_0}\right)^2 \sigma T^4 = 1400$

Hence $\qquad T^4 = \dfrac{1400}{5\cdot7 \times 10^{-8}} \times \left(\dfrac{1\cdot5 \times 10^8}{7 \times 10^5}\right)^2$

So $\qquad T = 5800 \, \text{K (approx.)}$

Hot Object in Enclosure

Consider a black body at a temperature of T_0 where T_0 is the temperature of the room or enclosure containing the body. Since the body is in temperature equilibrium, the energy per second it radiates must equal the energy per second it absorbs (see also Prevost's theory of exchanges, p. 726.) If A is the surface area of the body, then, from Stefan's law,

energy per second radiated $= \sigma A T_0^4$

So the *energy per second absorbed from the surroundings or enclosure* $= \sigma A T_0^4$.

Now suppose the black body X is heated electrically by a heater of power W watts and finally reaches a constant temperature T. In this case, from Prévost's theory,

energy per second from heater, $W =$ net energy per second radiated by X

The net energy per second radiated by X $= \sigma A T^4 - \sigma A T_0^4$, since $\sigma A T_0^4$ is the energy per second absorbed from the surroundings, as we showed before. So

$$W = \sigma A T^4 - \sigma A T_0^4 = \sigma A (T^4 - T_0^4)$$

Examples on Radiation

1 A metal sphere with a black surface and radius 30 mm, is cooled to $-73°\text{C}$ (200 K) and placed inside an enclosure at a temperature of $27°\text{C}$ (300 K). Calculate the initial rate of temperature rise of the sphere, assuming the sphere is a black body. (Assume density of metal $= 8000 \, \text{kg m}^{-3}$ specific heat capacity of metal $= 400 \, \text{J kg}^{-1} \, \text{K}^{-1}$, and Stefan constant $= 5\cdot7 \times 10^{-8} \, \text{W m}^{-2} \, \text{K}^{-4}$.)

Energy per second radiated by sphere $= \sigma A (T^4 - T_0^4)$

where A is the surface area $4\pi r^2$ of the sphere of radius r, $T = 200 \, \text{K}$ and $T_0 = 300 \, \text{K}$. Since the temperature of the surroundings is greater than that of the sphere, the energy per second, Q, *gained* from the surroundings is given by

$$Q = \sigma . 4\pi r^2 . (300^4 - 200^4)$$

The mass m of the sphere = volume × density = $\frac{4}{3}\pi r^3 \rho$, where ρ is the density. If c is the specific heat capacity of the metal, and θ is the initial rise per second of its temperature, then

$$Q = mc\theta = \frac{4}{3}\pi r^3 \rho c\theta = \sigma 4\pi r^2 (300^4 - 200^4)$$

Dividing by $4\pi r^2$, and then simplifying,

$$\theta = \frac{\sigma(300^4 - 200^4) \times 3}{r\rho c}$$

$$= \frac{5 \cdot 7 \times 10^{-8} \times (300^4 - 200^4) \times 3}{30 \times 10^{-3} \times 8000 \times 400}$$

$$= 0 \cdot 012 \text{ K s}^{-1} \text{ (approx.)}$$

2 Estimate the temperature T_e of the earth, assuming it is in radiative equilibrium with the sun. (Assume radius of sun, $r_s = 7 \times 10^8$ m, temperature of solar surface = 6000 K, distance of earth from sun = $1 \cdot 5 \times 10^{11}$ m.)

$$\text{Power radiated from sun} = \sigma \times \text{surface area} \times T^4$$

$$= \sigma \times 4\pi r_s^2 \times T_s^4$$

$$\text{Power received by earth} = \frac{\pi r_e^2}{4\pi R^2} \times \text{power radiated by sun}$$

since πr_e^2 is the effective area of the earth on which the sun's radiation is incident *normally*, Figure 29.23, and $4\pi R^2$ is the total area on which the sun's radiation falls at a distance R from the sun where the earth is situated.

$$\text{Now power radiated by earth} = \sigma . 4\pi r_e^2 . T_e^4$$

Assuming radiative equilibrium,

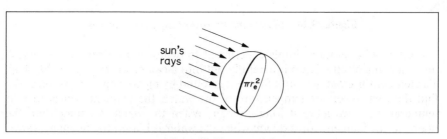

Figure 29.23 *Example*

$$\text{power radiated by earth} = \text{power received by earth}$$

$$\therefore \sigma . 4\pi r_e^2 . T_e^4 = \sigma . 4\pi r_s^2 . T_s^4 \times \frac{\pi r_e^2}{4\pi R^2}$$

Cancelling r_e^2 and simplifying, then

$$T_e^4 = T_s^4 \times \left(\frac{r_s^2}{4R^2}\right)$$

$$\therefore\ T_e = T_s \times \left(\frac{r_s}{2R}\right)^{1/2}$$

$$= 6000 \times \left(\frac{7 \times 10^8}{2 \times 1{\cdot}5 \times 10^{11}}\right)^{1/2}$$

$$= 290\ \text{K}$$

Note that the calculation is approximate, for example, the earth and the sun are not perfect black body radiators and the earth receives heat from its interior.

Prévost's Theory of Exchanges

In 1792 PRÉVOST applied the idea of dynamic equilibrium to radiation. He asserted that a body radiates heat at a rate which depends only on the nature of its surface and its temperature, and that it absorbs heat at a rate depending on the nature of its surface and the temperature of its surroundings. *When the temperature of a body is constant, the body is losing heat by radiation, and gaining it by absorption, at equal rates.*

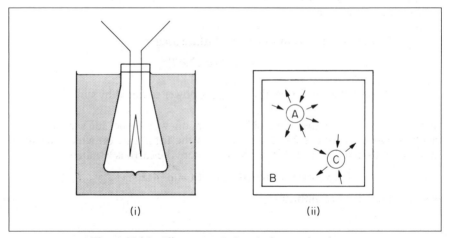

Figure 29.24 *Illustrating Prévost's theory of exchanges*

Figure 29.24 shows a simple experiment to demonstrate Prévost's theory. A high vacuum, electric lamp is carefully put into a can of water (Figure 29.24 (i)). The temperature of the lamp's filament is found by measuring its resistance. We find that, whatever the temperature of the water, the filament comes to that temperature, if we leave it long enough. When the water is cooler than the filament, the filament cools down; when the water is hotter, the filament warms up.

In the abstract language of theoretical physics, Prévost's theory is easy enough to discuss. If a hot body A (Figure 29.24 (ii)) is placed in an evacuated enclosure B, at a lower temperature than A, then A cools until it reaches the temperature of B. If a body C, cooler than B, is put in B, then C warms up to the temperature of B. We conclude that radiation from B falls on C, and therefore also on A, even though A is at a higher temperature. Thus A and C each come to equilibrium at the temperature of B when each is absorbing and emitting radiation at equal rates.

Now let us suppose that, after it has reached equilibrium with B, one of the bodies, say C, is transferred from B to a cooler evacuated enclosure D (Figure 29.25 (i)). It loses heat and cools to the temperature of D. Therefore it is radiating heat. But if C is transferred from B to a warmer enclosure F, then C gains heat and warms up to the temperature of F (Figure 29.25 (ii)). It seems unreasonable to suppose that C stops radiating when it is transferred to F. It is more reasonable to suppose that it goes on radiating but, while it is cooler than F, it absorbs more than it radiates.

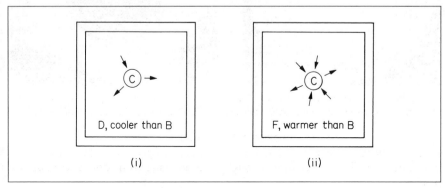

Figure 29.25   *Illustrating Prévost's theory*

Spectral Emissivity: Kirchhoff's Law

Most bodies are coloured; they transmit or reflect some wavelengths better than others. We can now see that, because they absorb them weakly, they must also radiate them weakly. To show this, consider a narrow band of wavelengths between λ and $\lambda + \Delta\lambda$. The energy falling per m^2 per second on the body, in this band, is $E_\lambda \Delta\lambda$ where E_λ is the emissive power of a black body in the neighbourhood of λ, at the temperature of the enclosure. If the body absorbs a fraction a_λ of this, we call a_λ the spectral absorption factor of the body, for the wavelength λ. In equilibrium, the body emits as much radiation in the neighbourhood of λ as it absorbs. So

$$\text{energy radiated} = a_\lambda E_\lambda \Delta\lambda \text{ watts per m}^2$$

We define the spectral emissivity of the body e_λ, by the equation

$$e_\lambda = \frac{\text{energy radiated by body in range } \lambda \text{ to } \lambda + \Delta\lambda}{\text{energy radiated in same range, by black body at same temperature}}$$

$$= \frac{\text{energy radiated by body in range } \lambda \text{ to } \lambda + \Delta\lambda}{E_\lambda \Delta\lambda} = \frac{a_\lambda E_\lambda \Delta\lambda}{E_\lambda \Delta\lambda}$$

Thus
$$e_\lambda = a_\lambda \quad . \qquad . \qquad . \qquad . \qquad . \qquad . \qquad (1)$$

Equation (1) expresses a law due to Kirchhoff:

> *The spectral emissivity of a body for a given wavelength is equal to its spectral absorption factor for the same wavelength.*

Kirchhoff's law is not easy to demonstrate by experiment. If a plate when cold shows a red pattern on a blue background, it glows blue on a red ground when heated in a furnace. Not all such plates do this, because the spectral emissivity of many coloured pigments vary with their temperature. However, Figure 29.26 shows two photographs of a decorated plate, one taken by reflected light at room temperature (left), the other by its own light when heated to a temperature of about 1100 K (right).

Figure 29.26 *Photographs showing how a decorated plate appears (left) by reflected light at room temperature and (right) by its own emitted light when incandescent at about 1100 K*

Absorption by Gases

An experiment which shows that, if a body radiates a given wavelength strongly, it also absorbs that wavelength strongly, can be made with sodium vapour (see p. 859).

The process of absorption by sodium vapour—or any other gas— is not, however, the same as the process of absorption by a solid. When a solid absorbs radiation, it turns it into heat—into the random kinetic energy of its molecules. It then re-radiates it in all wavelengths, but mostly in very long ones, because the solid is cool. When a vapour absorbs light of its characteristic wavelength, however, its atoms are excited; they then re-radiate the absorbed energy, in the same wavelength (589·3 nm for sodium). But they re-radiate it in all directions,

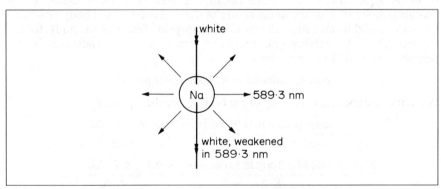

Figure 29.27 *Absorption by sodium vapour*

and therefore less of it passes on in the original direction than before (Figure 29.27). Thus the yellow component of the original beam is weakened, but the yellow light radiated sideways by the sodium is strengthened. The sideways strengthening is hard to detect, but it was shown by R. W. WOOD in 1906. He used mercury vapour instead of sodium. See *Fraunhofer lines*, p. 859.

Exercises 29B

Radiation

1 Explain why a body at 1000 K is 'red hot' whereas at 2000 K it is 'white hot'. (*C.*)

2 The silica cylinder of a radiant wall heater is 0·6 m long and has a radius of 5 mm. If it is rated at 1·5 kW estimate its temperature when operating. State *two* assumptions you have made in making your estimate. (The Stefan constant, $\sigma = 6 \times 10^{-8}\,\text{W m}^{-2}\,\text{K}^{-4}$.) (*L.*)

3 What is a *black body radiator*? Draw a diagram of a simple laboratory form of black body radiator.

Sketch roughly the energy distribution among the wavelengths of
(a) a black body radiator,
(b) a non-black body radiator and state the main differences.

4 Explain the difference between the *total radiation E* of a given black body and the *intensity of radiation* E_λ and state a law which each obeys.

Write down the units of E and E_λ.

5 What is the ratio of the energy per second radiated by the filament of a lamp at 2500 K to that radiated at 2000 K, assuming the filament is a black body radiator?

The filament of a particular electric lamp can be considered as a 90% black body radiator. Calculate the energy per second radiated when its temperature is 2000 K if its surface area is $10^{-6}\,\text{m}^2$. (Stefan constant $= 5·7 \times 10^{-8}\,\text{W m}^{-2}\,\text{K}^{-4}$.)

6 The sun is a black body of surface temperature about 6000 K. If the sun's radius is 7×10^8 m, calculate the energy per second radiated from its surface.

The earth is about $1·5 \times 10^{11}$ m from the sun. Assuming all the radiation from the sun falls on a sphere of this radius, estimate the energy per second per metre^2 received by the earth. (Stefan constant $= 5·7 \times 10^{-8}\,\text{W m}^{-2}\,\text{K}^{-4}$.)

7 Sketch graphs showing the distribution of energy in the spectrum of black body radiation at three temperatures, indicating which curve corresponds to the highest temperature. If such a set of graphs were obtained experimentally, how would you use information from them to attempt to illustrate Stefan's law? (*L.*)

8 Describe briefly how you would compare the thermal radiation emitted by a body at different temperatures.

How would you show that a matt black surface is a better radiator of heat than other surfaces? (*L.*)

9 An unlagged, thin walled copper pipe of diameter 2·0 cm carries water at a temperature of 40 K above that of the surrounding air. Estimate the power loss per unit length of the pipe if the temperature of the surroundings is 300 K and the Stefan constant, σ, is $5·67 \times 10^{-8}\,\text{W m}^{-2}\,\text{K}^{-4}$.

State *two* important assumptions you have made. (*L.*)

10 (a) Sketch graphs showing the distribution of the energy radiated by a black body with wavelength, for *three* temperatures. Label the axes of these graphs clearly. How is the total energy radiated by the body related to these graphs?

(b) A perfectly black sphere maintained at a temperature of 373 K in an enclosure at 0 K radiates heat at a rate of 200 W. At what rate will the black sphere radiate in the following cases? (i) Its radius is doubled; other factors remain unchanged. (ii) The temperature of the sphere is raised to 746 K; other factors remain unchanged. (iii) The temperature of the enclosure is raised to 300 K; other factors remain unchanged.

What would be the *net* rate of heat loss of the original sphere if the enclosure were also at a temperature of 373 K?

(Explain your answers in each of the above.)

(c) If the earth is assumed to be a perfectly black body radiating uniformly in all directions, a calculation of its equilibrium temperature gives a value of approximately 250 K. In fact the equilibrium temperature of the earth's surface is considerably higher than this. Explain why this is so. (*L.*)

11 (a) (i) What is meant by a *black body*? (ii) Sketch curves to show how the distribution of energy with wavelength in the radiation from a black body varies with temperature. (iii) With the aid of a diagram, describe a black-body source suitable for investigations over the temperature range 0°C to 1000°C.

(b) The solar power received at normal incidence on the surface of the Earth is $1·4 \times 10^3\,\text{W m}^{-2}$. Assume that the Earth and the Sun are uniform spheres and act as black bodies, and that atmospheric absorption can be neglected.

(Take the radius of the Sun to be $6·5 \times 10^8$ m, the radius of the Earth's orbit around the Sun to be 1.5×10^{11} m, the Stefan-Boltzmann constant σ to be 5.7×10^{-8} W m^{-2} K^{-4}, and the speed of light c in vacuum to be 3.0×10^8 m s^{-1}. The surface area of a sphere of radius r is $4\pi r^2$.) (i) Show that the power output per square metre of the Sun's surface is 75×10^6 W m^{-2}. (ii) Calculate the surface temperature of the Sun. (iii) Estimate the rate of loss of mass of the Sun in kg s^{-1} due to conversion of mass to energy. (iv) Estimate the mean surface temperature of the Earth, assuming that it intercepts solar energy as a disc and re-radiates it uniformly from the whole of its spherical surface. (*O.*)

12

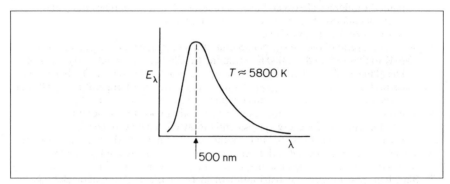

Figure 29D

Figure 29D shows how E_λ, the energy radiated per unit area per second per unit wavelength interval, varies with wavelength λ for radiation from the Sun's surface.

Calculate the wavelengths λ_{max} at which the corresponding curves peak for
(a) radiation in the Sun's core where the temperature is approximately 15×10^6 K, and
(b) radiation in interstellar space which corresponds to a temperature of approximately 2·7 K.
(You may use the relation $\lambda_{max} \times T = $ constant.)
Name the part of the electromagnetic spectrum to which the calculated wavelength belongs in each case. (*L.*)

13 Explain what is meant by the *Stefan constant*, defining any symbols used.
A sphere of radius 2·00 cm with a black surface is cooled and then suspended in a large evacuated enclosure the black walls of which are maintained at 27°C. If the rate of change of thermal energy of the sphere is 1.85 J s^{-1} when its temperature is -73°C, calculate a value for the *Stefan constant*. (*JMB.*)

14 Give an account of Stefan's law of radiation, explaining the character of the radiating body to which it applies and how such a body can be experimentally realised.
If each square cm of the sun's surface radiates energy at the rate of 6.3×10^3 J s^{-1} cm^{-2} and the Stefan constant is 5.7×10^{-8} W m^{-2} K^{-4}, calculate the temperature of the sun's surface in degrees centigrade, assuming Stefan's law applies to the radiation. (*L.*)

15 As the temperature of a black body rises, what changes take place in
(a) the total energy radiated from it and
(b) the energy distribution among the wavelengths radiated?
Illustrate (b) by suitable graphs and explain how the information required in (a) could be obtained from these graphs.
Use your graphs to explain how the appearance of the body changes as its temperature rises and discuss whether or not it is possible for a black body to radiate white light.
The element of an electric fire, with an output of 1.0 kW, is a cylinder 25 cm long and 1·5 cm in diameter. Calculate its temperature when in use, if it behaves as a black body. (The Stefan constant $= 5.7 \times 10^{-8}$ W m^{-2} K^{-4}.) (*L.*)

16 How can the temperature of a furnace be determined from observations on the radiation emitted?

Calculate the apparent temperature of the sun from the following information:

Sun's radius: 7.04×10^5 km. Distance from earth: 14.72×10^7 km.

Solar constant: 1400 W m^{-2}, Stefan constant: $5.7 \times 10^{-8} \text{ W m}^{-2} \text{K}^{-4}$. (*N.*)

17 Explain what is meant by
(a) a *black body*,
(b) *black body radiation*.

State *Stefan's law* and draw a diagram to show how the energy is distributed against wavelength in the spectrum of a black body for two different temperatures. Indicate which temperature is the higher.

A roof measures 20 m × 50 m and is blackened. If the temperature of the sun's surface is 6000 K, Stefan's constant $= 5.72 \times 10^{-8} \text{ W m}^{-2} \text{K}^{-4}$, the radius of the sun is 7.8×10^8 m and the distance of the sun from the earth is 1.5×10^{11} m, calculate how much solar energy is incident on the roof per minute, assuming that half is lost in passing through the earth's atmosphere, the roof being normal to the sun's rays. (*O. & C.*)

18 What is Prévost's Theory of Exchanges? Describe some phenomenon of theoretical or practical importance to which it applies.

A metal sphere of 1 cm diameter, whose surface acts as a black body, is placed at the focus of a concave mirror with aperture of diameter 60 cm directed towards the sun. If the solar radiation falling normally on the earth is at the rate of $0.14 \text{ watt cm}^{-2}$, Stefan's constant is taken as $6 \times 10^{-8} \text{ W m}^{-2} \text{K}^{-4}$ and the mean temperature of the surroundings is 27°C, calculate the maximum temperature which the sphere could theoretically attain, stating any assumptions you make. (*O. & C.*)

19 (a) State Prévost's theory of exchanges.

Discuss the observation that an object will not of its own accord become colder than its surroundings.

(b) The series of graphs given in the Figure 29E shows how the power radiated from unit surface area of a black body is distributed over the spectrum of its radiation at various temperatures.

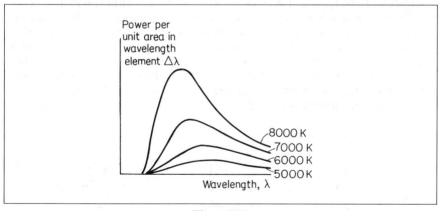

Figure 29E

Assuming that such graphs had been obtained experimentally how could they be used to verify the validity of Stefan's law of radiation?

(c) Explain, with an outline of experimental details, how the temperature of a bright star (other than the sun) could be determined.

(d) Compare the rate of fall of temperature of two hot, solid spheres made of the same material and with similar surfaces, where the radius of one sphere is four times the other and when the kelvin temperature of the large sphere is twice that of the small one. (Assume that the temperature of the spheres is so high that absorption from the surroundings may be ignored.) (*L.*)

30

Further Topics in Heat: Kinetic Theory, Real Gases, Thermodynamics

In this chapter we deal in more detail with (1) the internal energy of a gas and its mean free path on kinetic theory, (2) the laws about real gases and critical temperature, (3) topics in thermodynamics—work done by a gas in expansion, the adiabatic equation proof, the Carnot cycle, and the concept of entropy.

Kinetic Theory of Gases

Thermal Agitations and Internal Energy

The random motion of the molecules of a gas, whose kinetic energy depends upon the temperature, is often called the *thermal agitation* of the molecules. And the kinetic energy of the thermal agitation is called the *internal energy* of the gas, which we discussed earlier on p. 668.

The internal energy of a gas depends on the number of atoms in its molecules. A gas whose molecules consist of single atoms is said to be monatomic: for example, chemically inert gases and metallic vapours, Hg, Na, He, Ne, A. A gas with two atoms to the molecule is said to be diatomic: O_2, H_2, N_2, Cl_2, CO. And a gas with more than two atoms to the molecule is said to be polyatomic: H_2O, O_3, H_2S, CO_2, CH_4. The molecules of a monatomic gas we may regard as points, but those of a diatomic gas we must regard as 'dumb-bells', and those of a polyatomic gas as more complicated structures (Figure 30.1). A molecule which extends appreciably in space—a diatomic or polyatomic molecule—has an appreciable moment of inertia. It will therefore have kinetic energy of rotation, as well as of translation. A monatomic molecule, however, must have a much smaller moment of inertia than a diatomic or polyatomic; its kinetic energy of rotation can therefore be neglected.

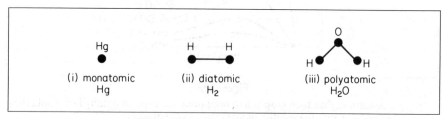

Figure 30.1 *Types of gas molecule*

Figure 30.2 shows a monatomic molecule whose velocity c has been resolved into its components u, v, w along the x, y, z axes:

$$c^2 = u^2 + v^2 + w^2$$

The x, y, z axes are called the molecules' degrees of freedom: they are three directions such that the motion of the molecule along any one is independent of its motion along the others.

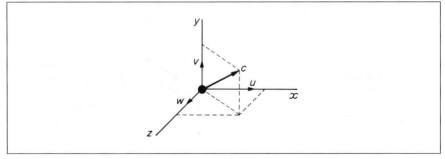

Figure 30.2 *Components of velocity*

If we average the speed c, and the components u, v, w, over all the molecules in a gas, we have

$$\overline{c^2} = \overline{u^2} + \overline{v^2} + \overline{w^2}$$

And since the molecules do not pile up in any corner of the vessel containing the gas, their average velocities in all directions must be the same. We may therefore write

$$\overline{u^2} = \overline{v^2} = \overline{w^2}$$

so

$$\overline{u^2} = \overline{v^2} = \overline{w^2} = \tfrac{1}{3}\overline{c^2}$$

The average kinetic energy of a molecule of the gas is given by (p. 686).

$$\tfrac{1}{2}m\overline{c^2} = \tfrac{3}{2}kT$$

Therefore the average kinetic energy of a monatomic molecule, in each degree of freedom, is

$$\tfrac{1}{2}m\overline{u^2} = \tfrac{1}{2}m\overline{v^2} = \tfrac{1}{2}m\overline{w^2} = \tfrac{1}{2}kT$$

Thus the molecule has kinetic energy $\tfrac{1}{2}kT$ per degree of freedom.

Rotational Energy

Let us now consider a diatomic or polyatomic gas. When two of its molecules collide, they will, in general, tend to rotate, as well as to deflect each other. In some collisions, energy will be transferred from the translations of the molecules to their rotations; in others, from the rotations to the translations. We may assume, then, that the internal energy of the gas is shared between the rotations and translations of its molecules.

To discuss the kinetic energy of rotation, we must first extend the idea of degrees of freedom to it. A diatomic molecule can have kinetic energy of rotation about any axis at right-angles to its own. Its motion about any such axis can be resolved into motions about two such axes at right-angles to each other, Figure 30.3 (i). Motions about these axes are independent of each other, and a diatomic molecule therefore has two degrees of rotational freedom. A polyatomic molecule, unless it happens to consist of molecules all in a straight line, has no axis about which its moment of inertia is negligible. It can therefore have kinetic energy of rotation about three mutually perpendicular axes, Figure 30.3 (ii). It has three degrees of rotational freedom.

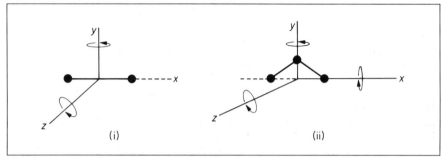

Figure 30.3 *Rotational energy of molecules*

We have seen that the internal energy of a gas is shared between the translations and rotations of its molecules. Maxwell assumed that the average kinetic energy of a molecule, in each degree of freedom, *rotational as well as translational*, was $\frac{1}{2}kT$. This assumption is called the *principle of equipartition of energy*; experiment shows, as we shall find, that it is true at room temperature and above. At very low temperatures, when the gas is near liquefaction, it fails. At ordinary temperatures, then, we have:

average k.e. of monatomic molecule $= \dfrac{3}{2}kT$ **(trans.);**

average k.e. of diatomic molecule $= \dfrac{3}{2}kT$ **(trans.)** $+ \dfrac{2}{2}kT$ **(rot.)** $= \dfrac{5}{2}kT$ **;**

average k.e. of polyatomic molecule $= \dfrac{3}{2}kT$ **(trans.)** $+ \dfrac{3}{2}kT$ **(rot.)** $= \dfrac{6}{2}kT.$

Internal Energy of a Gas

From the average kinetic energy of its molecules, we can find the internal energy of a mole of gas. Its internal energy, U, is the total kinetic energy of its molecules' random motions; so

$$U = N_A \times \text{average k.e. of molecule}$$

since there are N_A molecules per mole where N_A is the Avogadro constant. For a monatomic gas, therefore,

$$U = \frac{3}{2}N_A kT \text{ (monatomic)}$$

The constant k is the gas constant per molecule; the product $N_A k$ is therefore the gas constant R for one mole. So the internal energy per mole is

$$U = \frac{3}{2}RT \text{ (monatomic)} \qquad \cdot \qquad \cdot \qquad \cdot \qquad (1)$$

Similarly, for 1 mole of a diatomic gas,

$$U = \frac{5}{2}N_A kT = \frac{5}{2}RT \qquad . \qquad . \qquad . \qquad . \qquad (2)$$

Similar relations to (1) and (2) also apply to the internal energy of unit mass (1 kg) of a gas. In this case $U = 3rT/2$ (monatomic) and $U = 5rT/2$ (diatomic), where r is the gas constant per unit mass.

Molar Heat Capacities

We have seen that the internal energy U of a gas, at a given temperature, depends on the number of atoms in its molecule. For 1 mole of a monatomic gas its value is, from equation (1),

$$U = \frac{3}{2}RT . \qquad . \qquad . \qquad . \qquad . \qquad (i)$$

The molar heat capacity at constant volume is the heat required to increase the internal energy of 1 mole of the gas through 1 K. So, from (1),

$$C_V = \frac{3}{2}R$$

The molar heat capacity of a monatomic gas at constant pressure is therefore (p. 679)

$$C_p = C_V + R = \frac{3}{2}R + R = \frac{5}{2}R$$

Ratio of Molar Heat Capacities

Let us now divide C_p by C_V, their quotient is called the ratio of the molar heat capacities, and is denoted by γ (p. 672).
For a monatomic gas, its value is

$$\gamma = \frac{C_p}{C_V} = \frac{\frac{5}{2}R}{\frac{3}{2}R} = \frac{5}{3} = 1 \cdot 667$$

Similar relations apply to the principal specific heat capacities of a monatomic gas. In this case, we have, if r is the gas constant per kg,

$$c_V = \frac{3}{2}r, \; c_p = \frac{5}{2}r, \text{ and } \gamma = \frac{c_p}{c_V} = \frac{5}{3}$$

For a diatomic molecule, from equation (2),

$$U = \frac{5}{2}RT \qquad . \qquad . \qquad . \qquad . \qquad (ii)$$

So

$$C_V = \frac{5}{2}R$$

and

$$C_p = C_V + R = \frac{7}{2}R$$

Hence
$$\gamma = \frac{C_p}{C_V} = \frac{7}{5} = 1\cdot40$$

Similarly, we find that the principal specific heats of a diatomic gas are given by

$$c_V = \frac{5}{2}r \text{ and } c_p = \frac{7}{2}r, \text{ so } \gamma = \frac{c_p}{c_V} = 1\cdot40$$

In general, if the molecule of a gas has f degrees of freedom, the average kinetic energy of a molecule is $f \times \frac{1}{2}kT$ (p. 733). So

$$U = \frac{f}{2}RT$$

$$C_V = \frac{f}{2}R, \ C_p = C_V + R = \left(\frac{f}{2} + 1\right)R$$

and
$$\gamma = \frac{C_p}{C_V} = \frac{\frac{f}{2}+1}{\frac{f}{2}} = 1 + \frac{2}{f} \qquad . \qquad . \qquad . \qquad . \qquad \text{(iii)}$$

The ratio of the molar heat capacities of a gas thus gives us a measure of the number of atoms in its molecule, at least when the number is less than three. This ratio is easy to obtain by measuring the speed of sound in the gas. The poor agreement between the observed and theoretical values of γ for some of the polyatomic gases shows that, in its application to such gases, the theory is oversimple.

Mean Free Path

Although gas molecules move with high speeds such as several hundred metres per second, it still takes some time for two gases, initially separated, to diffuse into each other. This is due to numerous *collisions* which a molecule makes with other molecules surrounding it. Although the distance moved between successive collisions is not constant, we can consider the molecules of a gas under given conditions to have some average *mean free path, λ*.

Molecules have size. As an approximation, suppose a molecule in a gas is represented by a rigid sphere of diameter σ, and let us assume that a molecule moves a distance l while other molecules remain stationary. Then molecules such as A and B, situated at a distance σ from C, are those with which C *just* makes a collision, Figure 30.4. Consequently C makes a collision with all those molecules whose centres lie inside a volume $\pi\sigma^2 l$.

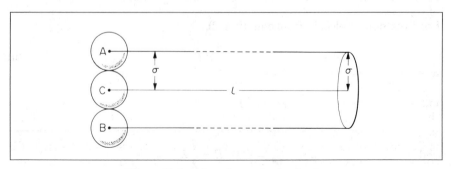

Figure 30.4 *Mean free path of molecule*

Suppose n is the number of molecules per unit volume in the gas. Then the number of collisions = the number of molecules in a volume $\pi\sigma^2 l$.

$$\therefore \lambda = \frac{\text{distance moved}}{\text{number of collisions}} = \frac{l}{n\pi\sigma^2 l}$$

$$\therefore \lambda = \frac{1}{n\pi\sigma^2} \qquad . \qquad . \qquad . \qquad . \qquad . \qquad (1)$$

Since the molecules are not stationary as we have assumed but moving about in all directions, the chance of a collision by a molecule is greater. Taking this into account, the mean free path can be shown to be $\sqrt{2}$ less than that given in (1). Thus

$$\lambda = \frac{1}{\sqrt{2}n\pi\sigma^2} \qquad . \qquad . \qquad . \qquad . \qquad (2)$$

The relation in (2) shows that the mean free path λ decreases in inverse proportion to the number of molecules per unit volume, n, or the pressure of the gas.

Calculation of Mean Free Path

Suppose that, for hydrogen, $\sigma = 3 \times 10^{-10}$ m approximately. At a standard pressure of 760 mmHg, a mole of the gas (2 g) has 6×10^{23} molecules approximately (the Avogadro constant) and occupies a volume of about 22·4 litres or $22\cdot4 \times 10^{-3}$ m^3. Then, from (2),

$$\lambda = \frac{22\cdot4 \times 10^{-3}}{\sqrt{2} \times 6 \times 10^{23} \times \pi \times 9 \times 10^{-20}}$$

$$= 9 \times 10^{-8} \text{ m (approx.)}$$

Thus the mean free path of hydrogen molecules at standard atmospheric pressure is about 10^{-7} m. Since the mean speed, the average distance travelled per second, of the molecule is about 2×10^3 m s^{-1}, we see that a molecule makes about 2×10^{10} collisions per second.

At extremely low pressures the mean free path becomes comparable with the dimensions of the container and no meaning is then attached to the term 'mean free path' of the molecules.

Viscosity of Gas

The viscosity of a gas is due to a transfer of momentum by molecules moving between fast and slow gas layers. On kinetic theory, the coefficient of viscosity η of a gas can be shown to be given by the formula

$$\eta = \frac{1}{3}mn\bar{c}\lambda \qquad . \qquad . \qquad . \qquad . \qquad (3)$$

where n is the number of molecules per unit volume, m is the mass of a molecule and $\bar{c}$ is the mean velocity and λ is the mean free path. But, from (2), $\lambda \propto 1/n$. It follows from (3) that η is *independent* of the pressure. This surprising result was shown to be true experimentally for normal pressures by Maxwell. He found that the damping of oscillations of a suspended horizontal disc was independent of the pressure of the gas when this was varied. The viscosity experiment is considered as one piece of sound evidence for the kinetic theory of gases, which leads to the relation in (3).

_____ **Exercises 30A** _____

Kinetic Theory

1 The kinetic energy of translation of 1 mole of an ideal monatomic gas is given by $3RT/2$. Show that the molar heat capacities of the gas at constant volume and constant pressure respectively are $3R/2$ and $5R/2$.

2 A diatomic gas has three translational degrees of freedom and two vibrational degrees of freedom. Show that the molar heat capacity at constant volume is $5R/2$ and that the ratio of the two molar heat capacities is 1·4.

3 What is an adiabatic change?
A cylinder contains compressed air at thirty times atmospheric pressure and at room temperature (300 K). The tap is opened so that air quickly escapes. Find the temperature of the air remaining in the cylinder when the pressure inside reaches atmospheric pressure. What assumptions have you made? (γ for air $= 1·4$) (W.)

4 2 moles of oxygen at 27°C are compressed isothermally from a volume of 0·40 m³ to 0·10 m³. Assume $R = 8·3\,\text{J mol}^{-1}\,\text{K}^{-1}$, calculate the work done on the gas. What amount of heat is given up by the gas in this change? Explain your answer.

5 An ideal gas (i) expands adiabatically and then (ii) expands isothermally.
   Explain what happens in each case to
   (a) the internal energy and
   (b) any heat exchanges.
   Write down gas equations which apply to the changes (i) and (ii), and apply the first law of thermodynamics to (a) and (b).

6 Argon has a molecular diameter about 3×10^{-10} m. The mean free path of the molecules is given by $\lambda = 1/(\sqrt{2}\pi\sigma^2 n)$, where λ is the mean free path, σ is the molecular diameter and n is the number of molecules per unit volume.
   Estimate the mean free path in one mole of argon at standard pressure when its volume is $22·4 \times 10^{-3}$ m³, assuming the Avogadro constant $= 6 \times 10^{23}\,\text{mol}^{-1}$.

7 List briefly the main assumptions on which the kinetic theory for an ideal gas is based.
   In what follows, the ideal gas equation $p = \frac{1}{3}nm\langle c^2 \rangle$ may be assumed. *Do not* deduce it.
   A gas consists of 5 molecules having speeds (in m s^{-1}) of 100, 600, 700, 800 and 1000. Find $\langle c^2 \rangle$.
   Write down an expression, containing only terms from the above equation, which is proportional to the temperature of an ideal gas. If the speed of each molecule is doubled, by what factor is the temperature changed?
   Show how Avogadro's law may be predicted from the above.
   Show that the molar heat capacity at constant volume, for an ideal gas, is $\frac{3}{2}R$.
   Deduce, with careful explanation, an expression for the mean free path λ between molecular collisions in a gas. How does λ vary with (i) pressure, when the temperature is kept constant, and (ii) temperature, then the pressure is kept constant? (W.)

8 (a) List the assumptions which are made in establishing the simple kinetic model of an ideal gas. Indicate which of these assumptions may not be justified for a real gas.
   (b) Van der Waals' equation of state for one mole of a real gas is written as

$$\left(p + \frac{a}{V^2}\right)(V - b) = RT$$

   (i) Explain the significance of the terms a/V^2 and b in this equation. (ii) with the aid of suitable sketch graphs relating p and V, compare the predictions of this equation with those of the ideal gas equation $pV = RT$. (*See next section*)
   (c) Show that the mean free path λ of a molecule in a gas is given approximately by the formula

$$\lambda = \frac{1}{n\pi\sigma^2}$$

where n is the number of molecules per unit volume and σ is the molecular diameter.

(d) Helium at 273 K (assumed an ideal gas) is introduced into a previously evacuated tube of length 0·40 m to such a pressure p that helium molecules, on average, traverse the length of the tube without colliding with each other.

(Take the volume of one mole of helium molecules at a temperature of 273 K and a pressure of 10^5 Pa to be 22×10^{-3} m^3, the molecular diameter σ of helium atoms to be $2·6 \times 10^{-10}$ m, and the Avogadro constant N_A to be $6·0 \times 10^{23}$ mol^{-1}.)

Calculate: (i) the number of helium molecules per unit volume under these conditions; (ii) the number of helium molecules per unit volume at 273 K temperature and 10^5 Pa pressure; (iii) the pressure p of the helium. (*O*.)

Real Gases, Critical Phenomena

An ideal gas is one which obeys Boyle's law (pV = constant at constant temperature) and whose internal energy is independent of its volume. No such gas exists but at room temperature and moderate pressures, many gases approach the ideal closely for most purposes.

We shall now consider the behaviour of *real gases*. We shall then appreciate better the relationship between liquid, vapour, and gas, and we shall see how gases such as air can be liquified.

Andrews' Experiments on Carbon Dioxide

In 1869 Andrews made experiments on carbon dioxide which have become classics. Figure 30.5 shows his apparatus. In the glass tube A he trapped carbon dioxide above the pellet of mercury X. To do this, he started with the tube open at both ends and passed dry gas through it for a long time. Then he sealed the end of the capillary. He introduced the mercury pellet by warming the tube, and allowing it to cool with the open end dipping into mercury. Similarly, he trapped nitrogen in the tube B.

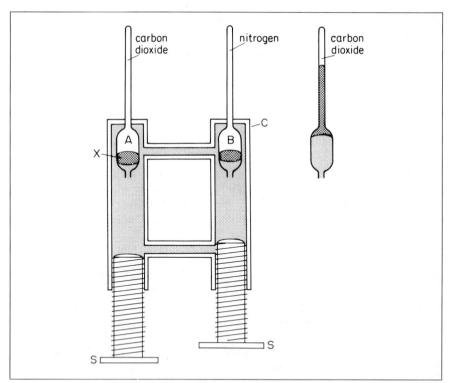

Figure 30.5 *Andrews' apparatus for isothermals of carbon dioxide at high pressures*

Andrews then fitted the tubes into the copper casing C, which contained water. By turning the screws S, he forced water into the lower parts of the tubes A and B, and drove the mercury upwards. The wide parts of the tubes were under the same pressure inside and out, and so were under no stress. The capillary extensions were strong enough to withstand hundreds of atmospheres. Andrews actually obtained 108 atmospheres.

When the screws S were turned far into the casing, the gases were forced into the capillaries, as shown on the right of the figure, and greatly compressed. From the known volumes of the wide parts of the tubes, and the calibrations of the capillaries, Andrews determined the volumes of the gases. He estimated the pressure from the compression of the nitrogen, assuming that it obeyed Boyle's law. Air was also used in place of nitrogen.

For work above and below room temperature, Andrews surrounded the capillary part of A with a water bath, which he kept at a constant temperature between about 10°C and 50°C.

p-V Curves of Carbon Dioxide, Critical Temperature

Andrews' results for the pressure (p)-volume (V) curves at various constant temperatures, called *isothermals*, are shown in Figure 30.6.

Let us consider the one for 21·5°C, ABCD. Andrews noticed that, when the pressure reached the value corresponding to B, a meniscus appeared above the mercury in the capillary containing the carbon dioxide. He concluded that the liquid had begun to form. From B to C, he found no change in pressure as the screws were turned, but simply a decrease in the volume of the carbon dioxide. At the same time the meniscus moved upwards, suggesting that the proportion of liquid was increasing. At C the meniscus disappeared at the top of the tube, suggesting that the carbon dioxide had become wholly liquid. Beyond C the pressure rose very rapidly; this confirmed that the carbon dioxide was wholly liquid, since liquids are almost incompressible.

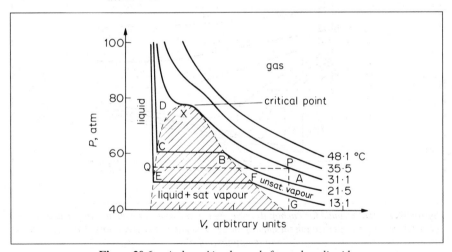

Figure 30.6 *Andrews' isothermals for carbon dioxide*

Thus the part CBA of the isothermal for 21·5°C is a curve of volume against pressure for a liquid and vapour, showing saturation at B. And the curve GFE is another such isothermal, for the lower temperature 13·1°C.

The isothermal for 31·1°C has no extended plateau; it merely shows a point of inflection at X. At that temperature, Andrews observed no meniscus; he concluded that it was the *critical temperature*. The isothermals for temperature above 31·3°C never become horizontal, and show no breaks such as B or F.

At temperatures above the critical, no change from gas to liquid can be seen.

The isothermal for 48·1°C conforms fairly well to Boyle's law; even when the gas is highly compressed its behaviour is not far from ideal.

The point X in Figure 30.6 is called the critical point. The pressure and volume (of unit mass) corresponding to it are called the *critical pressure* and *critical volume*; the reciprocal of the critical volume is the critical density.

Gases and Vapours, Continuity of State

We can now see the importance of Andrews' experiments. A gas *above its critical temperature* can not be liquefied by pressure. Early attempts to liquefy gases such as air, by compression without cooling, failed; and the gases were wrongly called 'permanent' gases. We still, for convenience, refer to a gas as a vapour when it is below its critical temperature, and as a gas when it is above it. The critical temperature of nitrogen is $-147°C$ (126 K). So nitrogen must be cooled below $-147°C$ to liquefy it by pressure. Air has a critical temperature of $-190°C$ (183 K) and oxygen a critical temperature of $-118°C$ (153 K).

Andrews' experiments also helped to illustrate the continuity between the liquid and gaseous states of a substance. Thus carbon dioxide is a gas above $31.1°C$. As shown in Figure 30.6, carbon dioxide can exist below $31.1°C$ as a liquid or as a vapour or as a mixture of liquid and vapour. One can also go directly from vapour to liquid along the line GP and then along PQ.

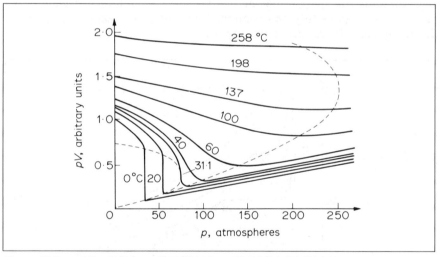

Figure 30.7 *Isothermals for carbon dioxide, as pV/p curves, at various temperatures. The small dotted loop passes through the ends of the vertical parts. The large dotted loop is the locus of the minima of pV*

Figure 30.7 shows some of the isothermals for carbon dioxide obtained by Andrews over a wide temperature range, this time with pV plotted against p. Corrections were made for the departure of nitrogen from Boyle's law. The vertical parts of the isothermals below the critical temperature of $31.1°C$ correspond to the change from the gaseous to the liquid state or to saturated vapour. If the gas obeyed Boyle's law, $pV =$ constant at constant temperature, the graph of pV against p would be a straight horizontal line. At the highest temperature shown, this is most nearly true.

Departures from Boyle's Law at High Pressure

In 1847 Regnault measured the volume of various gases at pressures of several atmospheres. He found that, to halve the volume of the gas, he did not have quite to double the pressure on it. The product pV, therefore, instead of being constant, decreased slightly with the pressure. He found one exception to this rule: hydrogen. By compressing the gases further, Regnault found the variation of pV with p at constant temperature, and obtained results which are represented by

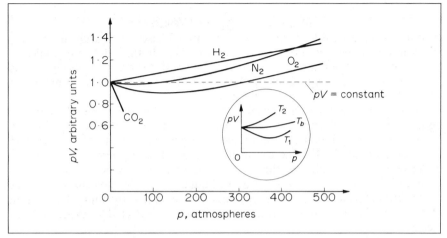

Figure 30.8 *Isothermals for various gases, at room temperature and high pressure*

the early parts of the curves in Figure 30.8. The complete curves in the figure show some of the results obtained by Amagat in 1892.

Real Gas Laws, Boyle Temperature

The results for real gases in Figure 30.7 can be expressed by the general relation

$$pV = A + Bp + Cp^2 + \ldots$$

where A, B, C are coefficients called 'virial coefficients' which are functions of temperature. At moderate pressures the gas obeys the relation

$$pV = A + Bp$$

to a good approximation. At very low pressures, $pV = A = RT$, that is, all real gases obey the same gas law. In these conditions the volume of the molecules themselves is negligible and the attractive forces between them are negligible, so the law, in fact, is that for an ideal gas. For this reason, gas thermometer measurements are extrapolated to zero pressure in accurate work (p. 625). From $pV = A + Bp + Cp^2 + \ldots$, it follows that the coefficient B is the *gradient* at $p = 0$ of the pV graph. Figure 30.8 shows that B is negative for some gases such as carbon dioxide or oxygen and positive for others such as hydrogen. From the gradient at $p = 0$ in Figure 30.7, we see that the value of B becomes less negative as the temperature rises.

At one particular temperature T_b called the *Boyle temperature*, $pV = $ constant for moderate pressures, to a good approximation, Figure 30.8, inset. For an ideal gas, the curve would be a straight line parallel to the axis of p, as shown in Figure 30.8. Below the Boyle temperature, the pV against p curves have a minimum, as at T_1. Above the Boyle temperature, the pV values increase, as at T_2. Hydrogen has a Boyle temperature of $-167°C$; for many other gases the Boyle temperature is above room temperature.

At high pressures and high temperatures, then, the laws for real gases depart considerably from those for real gases. At very low pressures, however, real gases obey the ideal gas laws.

Internal Energy and Volume

In our simple account of the kinetic theory of gases, we assumed that the molecules of a gas do not attract one another. If they did, any molecule approaching the boundary of the gas would be pulled towards the body of it, as is a molecule of water approaching the surface. The attractions of the molecules would thus reduce the pressure of the gas.

Since the molecules of a substance are presumably the same whether it is liquid or gas, the molecules of a gas must attract one another. But except for brief instants when they collide, the molecules of a gas are much further apart than those of a liquid. In 1 cubic centimetre of gas at s.t.p. there are $2·69 \times 10^{19}$ molecules, and in 1 cubic centimetre of water there are $3·33 \times 10^{22}$; there are a thousand times as many molecules in the liquid, and so the molecules in the gas are ten times further apart. We may therefore expect that the mutual attraction of the molecules of a gas, for most purposes, can be neglected, as experiment, in fact, shows.

The experiment consists in allowing a gas to expand without doing external work; that is, to expand into a vacuum. Then, if the molecules attract one another, work is done against their attractions as they move further apart. But if the molecular attractions are negligible, the work done is also negligible. If any work is done against the molecular attractions, it will be done at the expense of the molecular kinetic energies. So as the molecules move apart the internal energy of the gas, and therefore its temperature, will fall.

The expansion of a gas into a vacuum is called a 'free expansion'. If a gas does not cool when it makes a free expansion, then the mutual attractions of its molecules are negligible. Joule and Kelvin showed that most gases, in expanding from high pressure to low, do lose a little of their internal energy. The loss represents work done against the molecular attractions, which are therefore not quite negligible.

If the internal energy of a gas is independent of its volume, it is determined only by the temperature of the gas. The simple expression for the pressure, $p = \frac{1}{3}\rho c^2$, then holds and the gas obeys Boyle's law. Such a gas is called an *ideal* gas. All gases, when far from liquefaction, behave for most practical purposes as though they were ideal.

Van der Waals' Equation

In deriving the ideal gas equation $pV = RT$ from the kinetic theory of gases, a number of assumptions were made. These are listed on p. 683. Van der Waals modified the ideal gas equation to take account that two of these assumptions may not be valid:

---

1 *The volume of the molecules may not be negligible in relation to the volume V occupied by the gas.*
2 *The attractive forces between the molecules may not be negligible.*

In the bulk of the gas, the resultant force of attraction between a particular molecule and those all round it is zero when averaged over a period. Molecules which strike the wall of the containing vessel, however, are retarded by an unbalanced force due to molecules behind them. The observed pressure p of a gas is thus *less* than the pressure in the ideal case, when the attractive forces due to molecules is zero.

Van der Waals derived an expression for this pressure 'defect'. He considered that it was proportional to the product of the number of molecules per second striking unit area of the wall and the number per unit volume behind them, since this is a measure of the force of attraction. For a given volume of gas, both these numbers are proportional to the *density* of the gas. Consequently the pressure defect, p_1 say, is proportional to $\rho \times \rho$ or ρ^2. For a fixed mass of gas, $\rho \propto 1/V$, where V is the volume. Thus $p_1 = a/V^2$, where a is a constant for the particular gas. Taking into account the attractive forces between the molecules, it follows that, if p is the observed pressure,

$$\text{the gas pressure in the bulk of the gas} = p + a/V^2$$

The attraction of the walls on the molecules arriving there is to increase their velocity from v say to $v + \Delta v$. Immediately after rebounding from the walls, however, the force of attraction decreases the velocity to v again. Thus the attraction of the walls has no net effect on the momentum change due to collision. Likewise, the increase in momentum of the walls due to their attraction by the molecules arriving is lost after the molecules rebound.

Molecules have a particular diameter or volume because repulsive forces occur when they approach very closely and hence they cannot be compressed indefinitely. The volume of the space inside a container occupied by the molecules is thus not V but $(V - b)$, where b is a factor depending on the actual volume of the molecules. The magnitude of b is not the actual volume of the molecules, as if they were swept into one corner of the space, since they are in constant motion. b has been estimated to be about four times the actual volume.

Thus *van der Waals' equation* for real gases is:

$$\left(p + \frac{a}{V^2} \right)(V - b) = RT$$

At high pressures, when the molecules are relatively numerous and close together, the volume factor b and pressure 'defect' a/V^2 both become important. Conversely, at low pressures, where the molecules are relatively few and far apart on the average, a gas behaves like an ideal gas and obeys the equation $pV = RT$.

Real Gas and van der Waals' Equation

As stated on p. 743, the relation of the product pV to p for a real gas can be expressed by

$$pV = A + Bp + Cp^2 + \dots$$

where $A, B, C, \dots$ are coefficients decreasing in magnitude. The most important coefficients are A and B. So, at moderate pressures, $pV = A + Bp$. Further, when a real gas has a very low pressure, that is, p approaches zero, a real gas has the properties of an ideal gas. So $pV = A = RT$. At moderate pressures, then, we can write

$$pV = A + Bp = RT + Bp$$

or

$$p(V - B) = RT$$

From this relation we see that B represents a volume. Comparing this with the van der Waals equation, in which the term $(V-b)$ occurs, then B is roughly equal to the volume of the gas molecules themselves. Suppose 1 mole of a particular gas has a value B of 3×10^{-5} m^3. Since the Avogadro constant is about 6×10^{23} mol^{-1}, then roughly

$$\text{volume of 1 molecule} = \frac{3 \times 10^{-5}}{6 \times 10^{23}} = 5 \times 10^{-29} \text{ m}^3$$

The linear size of a molecule is of the order of the cube root of its volume. So an estimated size $= (5 \times 10^{-29})^{1/3} = 4 \times 10^{-10}$ m.

If required, we can expand van der Waals' equation to find a more exact relation for the coefficient B in terms of the constants a and b. Removing the brackets from the equation, then

$$pV - bp + \frac{a}{V} - \frac{ab}{V^2} = RT$$

or

$$pV = RT + bp - \frac{a}{V} + \frac{ab}{V^2}$$

Using the approximation $pV = RT$, then $1/V = p/RT$. Substituting for $1/V$ and $1/V^2$,

$$pV = RT + \left(b - \frac{a}{RT}\right)p + \frac{ab}{R^2T^2}p^2$$

Comparing this relation with $pV = A + Bp + Cp^2 + \ldots$, we see that

$$B = b - \frac{a}{RT} \quad \text{and} \quad C = \frac{ab}{R^2T^2}$$

The Boyle temperature T_b is the temperature when a real gas obeys the ideal gas laws (p. 743). From $pV = A + Bp$, this occurs when $B = 0$ since $A = RT$. So, from above, the Boyle temperature is given by $(b - a/RT_b) = 0$, or $T_b = a/bR$. So T_b can be calculated when the constants a, b and R are known.

Isothermals of Real Gas

A graph of pressure p against volume V at constant temperature is called an *isothermal* or *isotherm*. Figure 30.9 (i) shows some isothermals for an ideal gas,

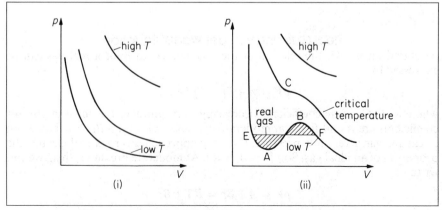

Figure 30.9 *Isothermals for ideal and van der Waals gases*

which obeys the perfect gas law $pV = RT$. Figure 30.9 (ii) shows a number of isothermals for a gas which obeys van der Waals' equation, $(p + a/V^2)(V - b) = RT$.

At high temperatures the isothermals are similar. As the temperature is lowered, however, the isothermals in Figure 30.9 (ii) change in shape. One curve has a point of inflexion at C, which corresponds to the *critical point* of a real gas. The isothermals thus approximate to those obtained by Andrews in his experiments on actual gases such as carbon dioxide, described on p. 740.

Below this temperature, however, isothermals such as EABF are obtained by using van der Waals' equation. These are unlike the isothermals obtained with real gases, because in the region AB the pressure increases with the volume, which is impossible. However, an actual isothermal in this region corresponds to a straight line EF, as shown. Here the liquid and vapour are in equilibrium (see p. 741) and the line EF is drawn to make the shaded areas above and below it equal. Thus van der Waals' equation roughly fits the isothermals of actual gases above the critical temperature but below the critical temperature it must be modified considerably. Many other gas equations have been suggested for real gases but quantitative agreement is generally poor.

_____ **Exercises 30B** _____

Real Gases, Critical Temperature

1. (a) Under what conditions would you expect the behaviour of a real gas to differ from that of an ideal gas? State and explain which assumptions of the kinetic theory are invalid under these conditions.

    (b) What is meant by the critical temperature of a gas? Sketch graphs, using the same axes, to show the variation of pressure with volume for a fixed mass of carbon dioxide under isothermal conditions: (i) well above the critical temperature, (ii) at the critical temperature, (iii) below the critical temperature.

    (c) For the graph of (b)(iii) describe the changes which occur as the pressure is increased for the range drawn. (*JMB*.)

2. Oxygen can be liquefied by first allowing the gas to expand very rapidly (to reduce its temperature to below 154 K) and then applying sufficient external pressure to it.

    (a) Explain why a sudden expansion can lead to cooling.

    (b) What is the significance of the temperature 154 K? (*L*.)

3. Define *critical temperature*.

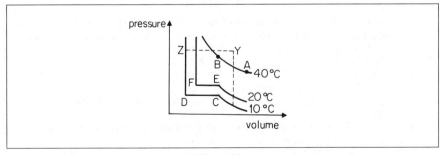

Figure 30A

Figure 30A shows the form of the relationship between the pressure and volume for isothermal changes in volume of carbon dioxide at three different temperatures.

Explain qualitatively, using kinetic theory,
(a) the rise in pressure from A to B,
(b) the constancy of the pressure from C to D, and
(c) why the pressure at E and F is greater than the pressure at C and D. Explain the energy interchanges which would take place in going from A to B and from C to D.

If the carbon dioxide were in a glass container, so that its condition could be seen, what would be observed as it was taken   (i) along the path CD, and   (ii) along the path indicated from 10°C to Y and then to Z? (*L*.)

4   What are the conditions under which the equation $pV = RT$ gives a reasonable description of the relationship between the pressure p, the volume V and the temperature T of a real gas?

Sketch $p–V$ isothermals for the gas-liquid states and indicate the region in which $pV = RT$ applies. Indicate the state of the substance in the various regions of the $p–V$ diagram. Mark and explain the significance of the critical isothermal.

Discuss a way in which the equation $pV = RT$ may be modified so that it can be applied more generally. Explain and justify on a molecular basis the additional terms introduced. Discuss the success of this modification. (*L*.)

5   (a) Draw a fully labelled diagram of the apparatus used in Andrew's experiments to investigate the relationship between the pressure p and the volume V for a fixed mass of a real substance over a range of pressures to produce a change of state, and for several different temperatures. Explain carefully how the pressure was varied and how corresponding values of p and V were determined.

(b) Sketch a graph of the results obtained with this apparatus with pV as ordinate and p as abscissa for three temperatures; one above, one below and one equal to the critical temperature. Indicate on your graph the corresponding isothermals you would expect for an ideal gas.

(c) (i) Which assumptions of the kinetic theory of an ideal gas need to be modified for a real gas?   (ii) Show how these modifications lead to an equation of state for a real gas different from that for an ideal gas. (*JMB*.)

6   (a) Explain what is meant by an ideal gas.

What properties are assumed for the model of an ideal gas molecule in deriving the expression

$$p = \tfrac{1}{3}\rho \overline{c^2} \tag{1}$$

where the symbols have their usual meaning?

(b) How is pressure explained in terms of the kinetic theory of gases?
Describe carefully, using diagrams where necessary, but without detailed mathematical analysis, the steps in the argument used to derive equation (1).

(c) Show that for a fixed mass of ideal gas at constant temperature, equation (1) can be written $pV = A$ where A is a constant.

For some real gases, the pressure can be described in terms of the equation

$$p(V + B) = A \tag{2}$$

where B is also a constant for a fixed mass of the gas at a particular temperature. Show that equation (2) implies a pressure less than the value predicted for an ideal gas. Suggest a reason for this in molecular terms. (*L*.)

7   Describe experiments in which the relation between the pressure and volume of a gas has been investigated at constant temperature over a wide range of pressure. Sketch the form of the isothermal curves obtained.

Explain briefly how far van der Waals' equation accounts for the form of these isothermals. (*L*.)

8   Describe, with a diagram, the essential features of an experiment to study the departure of a real gas from ideal gas behaviour.

Give freehand, labelled sketches of the graphs you would expect to obtain on plotting
(a) pressure P against volume V,
(b) PV against P for such a gas at its critical temperature and at one temperature above and one below the critical temperature.

Explain van der Waals' attempt to produce an equation of state which would describe the behaviour of real gases.

Show that van der Waals' equation is consistent with the statement that all gases approach ideal gas behaviour at low pressures. (*O. & C.*)

9 (a) With pressure p as ordinate and volume V as abscissa, draw sketch graphs of p–V isothermals for a real gas indicating (i) the region in which the gas approximately obeys Boyle's Law, (ii) the critical isothermal, (iii) the region in which the gas liquefies and gas and liquid are in equilibrium. Justify your answers in each case.

(b) At high pressure, the equation of state is modified from that of an ideal gas so that it becomes $pV = A + Bp$. Explain (i) how A depends on the temperature and (ii) why B is related to the size of the gas molecules.

In a particular experiment on 1 mole of oxygen, the value of B was found to be $20.2 \times 10^{-6} \, \text{m}^3$. Estimate the volume of the oxygen molecule. The Avogadro constant $N_A = 6.03 \times 10^{23} \, \text{mol}^{-1}$. (*JMB.*)

Thermodynamics, Entropy

Work Done in Reversible Isothermal Expansion

The heat taken in when a gas expands isothermally and reversibly is the heat equivalent of the mechanical work done. If the volume of the gas increases by a small amount ΔV, at the pressure p, then the work done is (p. 666)

$$\Delta W = p\Delta V$$

In an expansion from V_1 to V_2, therefore, the work done is

$$W = \int dW = \int_{V_1}^{V_2} p\,dV$$

Assuming we have 1 mole of gas, $p = \dfrac{RT}{V}$

so

$$W = \int_{V_1}^{V_2} p\,dV = \int_{V_1}^{V_2} RT\frac{dV}{V} = RT\ln\left(\frac{V_2}{V_1}\right)$$

The heat required, Q, is therefore

$$Q = W = RT\ln\left(\frac{V_2}{V_1}\right)$$

Equation of Reversible Adiabatic

Consider one mole of the gas, and suppose that its volume expands from V to $V+\Delta V$, and that an amount of heat ΔQ is supplied to it. In general, the internal energy of the gas will increase by an amount ΔU, and the gas will do an amount of external work equal to $p\Delta V$, where p is its pressure. The heat supplied is equal to the increase in internal energy, plus the external work done:

$$\Delta Q = \Delta U + p\Delta V \quad . \qquad . \qquad . \qquad . \qquad . \qquad (1)$$

The increase in internal energy represents a temperature rise, ΔT. We have seen already that the internal energy is independent of the volume, and is related to the temperature by the molar heat capacity at constant volume, C_V (p. 678). Therefore

$$\Delta U = C_V\Delta T$$

Equation (1) becomes

$$\Delta Q = C_V\Delta T + p\Delta V \quad . \qquad . \qquad . \qquad . \qquad . \qquad (2)$$

Equation (2) is the fundamental equation for any change in state of one mole of an ideal gas.

For a reversible isothermal change, $\Delta T = 0$, and $\Delta Q = p\Delta V$.
For a reversible adiabatic change, $\Delta Q = 0$ and therefore

$$C_V\Delta T + p\Delta V = 0 \quad . \qquad . \qquad . \qquad . \qquad . \qquad (3)$$

To eliminate ΔT we use the general equation, relating pressure, volume and temperature:

$$pV = RT$$

where R is the molar gas constant. Since both pressure and volume may change, when we differentiate this to find ΔT we must write

$$p\Delta V + V\Delta p = R\Delta T . \qquad . \qquad . \qquad . \qquad (4)$$

Substituting in (4) for $R(= C_p - C_V)$ and for $\Delta T(= -p.\Delta V/C_V$ from (3)), and then simplifying, we obtain

$$C_V V\Delta p + C_p p\Delta V = 0$$

or

$$V\Delta p + \gamma p\Delta V = 0 \left(\text{where } \gamma = \frac{C_p}{C_V} \right)$$

Integrating,

$$\int \frac{dp}{p} + \gamma \int \frac{dV}{V} = 0$$

or

$$\ln p + \gamma \ln V = A$$

where A is a constant.

Therefore,

$$\boldsymbol{pV^\gamma = \text{constant}} \qquad . \qquad . \qquad . \qquad . \qquad (5)$$

This is the equation of a reversible adiabatic; the value of the constant can be found from the initial pressure and volume of the gas.

Carnot Cycle Efficiency

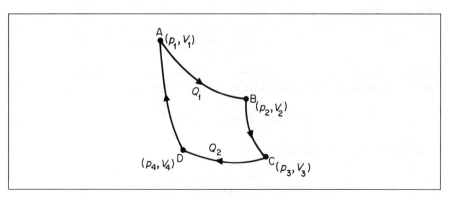

Figure 30.10 *Carnot cycle*

As we saw in an earlier chapter, the Carnot cycle ABCD consists of two reversible isothermal changes in the gas concerned, AB and CD in Figure 30.10, and two reversible adiabatic changes, BC and DA.

The heat Q_1 taken in along AB at a temperature T_1 = external work done.

So

$$Q_1 = RT_1 \ln (V_2/V_1) . \qquad . \qquad . \qquad . \qquad (1)$$

Similarly heat Q_2 given back at T_2 along CD is

$$Q_2 = RT_2 \ln (V_3/V_4) . \qquad . \qquad . \qquad . \qquad (2)$$

From the adiabatics for BC and AD, pV^γ = constant. So

$$p_1 V_1 . V_1^{\gamma-1} = p_4 V_4 . V_4^{\gamma-1}$$

and

$$p_2 V_2 . V_2^{\gamma-1} = p_3 V_3 . V_3^{\gamma-1}$$

Since $p_1 V_1 = p_2 V_2$ (isothermal AB) and $p_4 V_4 = p_3 V_3$ (isothermal CD), by division we have

$$\left(\frac{V_1}{V_2}\right)^{\gamma-1} = \left(\frac{V_4}{V_3}\right)^{\gamma-1}$$

So
$$V_1/V_2 = V_4/V_3 \qquad . \qquad . \qquad . \qquad . \qquad . \qquad (3)$$

Now
$$\text{efficiency} = \frac{\text{net work done}}{\text{heat taken in}}$$

$$= \frac{Q_1 - Q_2}{Q_1} = \frac{T_1 - T_2}{T_1} \qquad . \qquad . \qquad . \qquad (4)$$

using equations (1) and (2) for Q_1 and Q_2, and cancelling, by using equation (3), the ratio of the volumes.

From equation (4), we see that it is impossible to obtain 100% efficiency when a gas is taken through a cycle. The maximum efficiency is obtained for reversible changes, which is an ideal situation.

In a *refrigerator* or *heat pump*, heat is transferred from a cold body or 'sink' at T_2 to the hotter body or 'source' at T_1. In this case a motor does work by driving a reversible Carnot cycle 'backwards', taking in heat from the 'sink' and pumping it to the 'source'.

Entropy

If a system takes in an amount of heat ΔQ at a temperature T, its gain in *entropy* ΔS is defined as

$$\Delta S = \frac{\Delta Q}{T}$$

For the Carnot cycle in Figure 30.10, at the end of the cycle

$$\text{entropy change} = \frac{Q_1}{T_1} - \frac{Q_2}{T_2} \qquad . \qquad . \qquad . \qquad . \qquad (5)$$

From equation (4), we see that

$$1 - \frac{Q_2}{Q_1} = 1 - \frac{T_2}{T_1}$$

so
$$\frac{Q_1}{T_1} = \frac{Q_2}{T_2}$$

From equation (5), the entropy change $= 0$ in a Carnot cycle, which is reversible.

In practice, however, owing to friction and other irreversible processes, it can be shown that Q_1/T_1 is *less* that Q_2/T_2. So entropy *increases* in all operating machines. Irreversible processes are much more common or probable in everyday life than reversible processes. Natural processes appear to be irreversible. So though the total energy in the Universe is constant, its total entropy and disorder increases daily.

Distribution and Disorder

In any system, the more ways there are of achieving a particular distribution, the greater will be the probability of that distribution. For example, in tossing four coins, there are six ways of arranging two heads and two tails. Since this is greater than the number of ways of arranging any other distribution, tossing four coins produces this distribution a greater number of times than any other.

The number of ways W of achieving or getting a distribution is linked to the *disorder* of that system because the disorder increases as W increases. As an example, if some steam is released in the corner of a room, it is unlikely that all the steam molecules will remain in that corner. This would be low disorder for the system of steam and air molecules in the room. In practice, the steam molecules spread to other parts of the room and in time distribute themselves uniformly throughout the space available. The system of steam and air molecules thus increases in disorder as the molecules reach their eventual distribution.

Entropy and Disorder

We have seen before that the entropy S of a system tends to increase with irreversible processes or greater disorder. Boltzmann linked the entropy S of a system with the number of ways W of obtaining the distribution. From the way S and W change, Boltzmann stated that

$$S = k \ln W$$

where k is the Boltzmann constant we met in the kinetic theory of gases (p. 687).

The Boltzmann equation for S is widely used in a statistical approach to changes in entropy of a system. It is particularly useful in systems with a large number of particles or entities such as quanta. For example, it leads to the way energy is distributed among the atoms in a solid which is heated and to the distribution with temperature change of the energy among electrons in metals.

The link between entropy and disorder can be seen in several cases. At zero absolute temperature, a crystal is in equilibrium with all its atoms (ions) at fixed sites in the solid. Their energy is the lowest possible. This is therefore perfect order, or a disorder or entropy of zero. If the crystal is heated, the atoms now vibrate with increased energy and the system becomes more disordered. The entropy of the system has increased. If heat is supplied continuously, the solid melts at one stage and the ions in the liquid have more disorder than in the solid. So the entropy has increased further. When rubber is stretched the coiled molecules straighten (p. 169) but the rapid stretching increases the temperature of the rubber and produces a net increase in disorder or entropy in the rubber molecules.

Some Calculations of Entropy Change

1 *Isothermal expansion of ideal gas*

If an ideal gas expands isothermally at a kelvin temperature T from a volume V_1 to a volume V_2, then the heat Q absorbed $=$ external work done W, since there is no internal energy change of the gas. From our result on p. 750, $W = RT \ln(V_2/V_1) = Q$. So

$$\text{entropy increase } S = \frac{Q}{T} = \frac{RT \ln(V_2/V_1)}{T} = R \ln(V_2/V_1)$$

If p_1 is the initial gas pressure and p_2 the final pressure, then, since $p_1 V_1 = p_2 V_2$, $V_2/V_1 = p_1/p_2$. So

$$\text{entropy change } = R \ln(p_1/p_2)$$

At the greater volume V_2, the molecules of gas have greater disorder than at V_1.

2 *Heated substance*

Suppose a mass m of water of specific heat capacity c is heated from a kelvin temperature

T_1 to T_2. At a stage when the temperature reaches a value T, a small amount of heat δQ produces a small temperature rise δT. So

$$\text{entropy increase} = \int_{T_1}^{T_2} \frac{dQ}{T} = \int_{T_1}^{T_2} \frac{mc\,dT}{T}$$

$$= mc \ln(T_2/T_1)$$

At the higher temperature T_2, the molecules of the water have greater disorder than at T_1.

3 *Liquid to vapour*
Suppose a mass m of water at 100°C boils and forms steam at 100°C. If l is the specific latent heat of water at this temperature, then heat Q absorbed $= ml$. So

$$\text{entropy increase} = \frac{Q}{T} = \frac{ml}{373}$$

In steam, the molecules of water have a greater disorder than in water.

_____ Exercises 30C _____

Thermodynamics

1 1 mole of air expands isothermally at 27°C from a volume of $0.01 \, m^3$ to $0.04 \, m^3$. Calculate the work done, assuming $R = 8.3 \, J \, mol^{-1} \, K^{-1}$. What heat is taken in by the gas? Explain your answer.

2 Using 4 moles of air and assuming this is an ideal gas, an engine goes through the following reversible changes in one cycle: (i) Isothermal expansion from a volume of $0.02 \, m^3$ to $0.05 \, m^3$, at 100°C, (ii) at constant volume, cooling to 27°C, (iii) isothermal compression at 27°C to $0.02 \, m^3$ volume, (iv) at constant volume, compression to original pressure, volume and temperature in (i).
   Show these four changes in a p-V diagram. Calculate
   (a) the heat taken in at 100°C,
   (b) the work done by the engine during one cycle,
   (c) the efficiency of the engine,
   (d) the change in internal energy of the air at the end of the cycle. Assume $R = 8.3 \, J \, mol^{-1} \, K^{-1}$.

3 A Carnot cycle uses an ideal gas as a working substance in reversible changes. (i) What is meant by *reversible*? (ii) Draw a sketch of a Carnot cycle and state what happens along each of the four stages (iii) What is the work done in one cycle? (iv) Write down an expression for the efficiency of the cycle in terms of temperature. How can the efficiency be increased?

4 Working between temperatures of 27°C and 127°C, an engine does 40 000 J of work in a Carnot cycle. Find the amount of heat taken in at the higher temperature, assuming no heat losses.

5 In a refrigeration unit, a Carnot engine using an ideal gas is driven in reverse by a 1-kW electric motor of efficiency 80% to freeze water at 0°C. Assuming the working temperatures of the engine are 20°C and 0°C respectively, calculate the mass of water frozen in 5 min, assuming no heat losses in the unit and $l = 3.4 \times 10^5 \, J \, kg^{-1}$. (*Hint* $Q_2/W = Q_2/(Q_1 - Q_2) = T_2/(T_1 - T_2)$, where Q_2 is heat abstracted from water, Q_1 is the heat given up and W is work done by motor).

6 Define *entropy*. 0.2 kg of steam at 100°C condenses to water at 100°C. (i) Is there an increase or decrease in disorder of the water molecules? (ii) Calculate the entropy change, assuming l (steam) $= 2.26 \times 10^6 \, J \, kg^{-1}$.

Part 5
Electrons, Electronics, Atomic Physics

31
Electrons: Motion in Fields, The Cathode-Ray Oscilloscope

In this chapter, we first discuss how the charge and mass of the electron was measured and the properties of the electron. We then deal with the parabolic path of electrons in a uniform electric field and the circular and spiral path in uniform magnetic fields. The cathode-ray oscilloscope and its uses are described in the final section.

Basic Unit of Charge: The Electron

As we discussed earlier, matter has a particle nature—it consists of extremely small but separate particles which are atoms or molecules. Helmholtz, an eminent German scientist, stated about a century ago that charges were built up of basic units, that is, electricity is 'granular' and not continuous. He was led to this conclusion from studying the laws of electrolysis.

Particle Nature of Electricity

In electrolysis, we assume that the carriers of current through an acid or salt solution are ions, which may be positively or negatively charged.

Experiment shows that 96 500 coulombs (the Faraday constant) is the quantity of charge required to deposit one mole of a monovalent element. The number of ions carrying this charge is about $6 \cdot 02 \times 10^{23}$, the Avogadro constant, for a monatomic element. So the charge on each of the ions is given by $96\,500/6 \cdot 02 \times 10^{23}$ or $1 \cdot 6 \times 10^{-19}$ C. If $1 \cdot 6 \times 10^{-19}$ C is denoted by the symbol e, the charge on any ion is then e, $2e$, or $3e$, etc., depending on its valency. Thus e is a basic unit of charge.

All charges, whether produced in electrostatics, current electricity or any other method, are multiples of the basic unit e. Evidence that this is the case was obtained by MILLIKAN. In 1909, he designed an experiment to measure accurately the unit e. His method will be given later. Firstly, however, we describe a more direct experiment based on the Millikan method which can be carried out in the school laboratory.

Oil-drop Experiment for Electron Charge

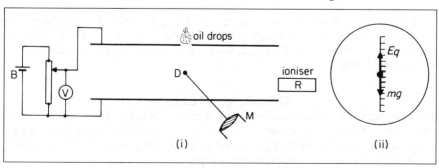

Figure 31.1 *Oil-drop experiment*

Figure 31.1 (i) shows the principle of the experiment, A variable p.d. V can be applied to two circular plates P and Q separated by a small distance d. The upper plate P has a small hole in the middle and by means of a spray above it, fine oil drops can pass through to the space between P and Q.

In a darkened room, the drops are illuminated by a light beam and are then seen by using a low-power microscope M to focus them. M has a vertical fine line in the eyepiece graduated in millimetres, so that the motion of the drop can be observed and measured, Figure 31.1 (ii).

A particular oil drop is then focused so that it is clearly seen. The spray produces a charge q on the drop by friction and by altering the p.d. V between P and Q, the weight mg of the drop can be exactly balanced by an upward force Eq, where E is the electric field strength between P and Q. Since $E = V/d$, then

$$Eq = \frac{Vq}{d} = mg$$

and so
$$q = \frac{mgd}{V} \qquad . \qquad . \qquad . \qquad . \qquad . \qquad . \qquad (1)$$

Strictly, mg is the weight of the drop less the upthrust on the drop due to the surrounding air. The upthrust is equal to the weight of air displaced by the drop and is taken into account in the accurate Millikan experiment described shortly. Since the air density is very small compared to the density of oil, we may neglect the upthrust in this experiment.

Results of Experiment

The charge q on the *same* oil drop can be increased by ionising the air between the plates, so that the drop picks up more charges from the ions. This is done by using an α-particle source from apparatus placed at R in Figure 31.1 (i). Each time a new charge is obtained by the drop, the p.d. V is altered until the drop becomes stationary.

When the charge q is varied, suppose the values for V are 150 V, 190 V, 250 V and 750 V. From equation (1), since mgd is constant, then $q \propto 1/V$. The values of $1/V$ are given roughly in the Table below:

V	150 V	190 V	250 V	750 V	
$1/V$	6·6	5·2	4·0	1·3	$\times 10^{-3}$
Ratios for $1/V$ or q	5	4	3	1	

Using the lowest value of $1/V$, which is $1/750$ or 1·3, as the 'unit' of charge, the different charges on the drop were practically whole numbers 5, 4 and 3 in terms of this unit 1. This leads to the conclusion that a charge is a *multiple* of a basic or fundamental charge.

Calculation of Charge

From equation (1), the actual value of a charge is given by $q = mgd/V$. The values of g, d and V are known. The unknown value of m, the mass of the oil drop, can be found by allowing the drop to fall under gravity and measuring its *terminal velocity* v, using the graduated eyepiece.

In this case, as shown on p. 110, Stokes' law for the upward frictional (viscosity) force on the sphere of radius a is $F = 6\pi\eta av$, where η is the coefficient of viscosity of air. So, at the terminal velocity (zero acceleration),

$$F = mg = 6\pi\eta av$$

Now $m =$ volume $V \times$ density ρ of oil $= 4\pi a^3\rho/3$, since the volume of a sphere is $4\pi a^3/3$. So

$$\frac{4\pi a^3\rho g}{3} = 6\pi\eta av$$

Simplifying,
$$a = \sqrt{\frac{9\eta v}{2\rho g}}$$

Knowing η, v, ρ and g, the radius a of the drop can be calculated from the formula. The mass m of the drop $= 4\pi a^3\rho/3$ and can now be found. So the charge q can then be calculated from $q = mgd/V$.

Millikan's accurate experiment, which we now describe, shows that the basic or smallest value of charge is $1\cdot6 \times 10^{-19}$C, which is the charge e on an electron. All charges are made up of units of e.

Theory of Millikan's Experiment

Millikan first measured the *terminal velocity* of an oil-drop through air. He then charged the oil-drop and applied an electric field to oppose gravity. The drop now moved with a different terminal velocity, which was again measured.

Suppose the radius of the oil-drop is a, the densities of oil and air are ρ and σ respectively, and the viscosity of air is η. When the drop, without a charge, falls steadily under gravity with a terminal velocity v_1, the upward viscous force $= 6\pi\eta av_1$, from Stokes' law $= kv_1$, where k is $6\pi\eta a$. This is equal to $m'g$, the weight of the drop less the upthrust due to the air. So

$$m'g = kv_1 \quad . \qquad . \qquad . \qquad . \qquad . \qquad \text{(i)}$$

Suppose the drop now gains a charge q and an electric field of intensity E is applied to oppose gravity and slow the falling drop. The drop then has a smaller terminal velocity v_2. Since the force on the drop due to E is Eq, then, if the mass of the drop has remained constant,

$$m'g - Eq = kv_2 \quad . \qquad . \qquad . \qquad . \qquad . \qquad \text{(ii)}$$
$$\therefore Eq = m'g - kv_2 = k(v_1 - v_2)$$

So
$$q = \frac{k}{E}(v_1 - v_2) \quad . \qquad . \qquad . \qquad . \qquad . \qquad \text{(ii)}$$

The weight of the oil-drop $=$ volume $\times$ density $\times g = \frac{4}{3}\pi a^3\rho g$; the upthrust on the drop due to the air $=$ weight of air displaced $= \frac{4}{3}\pi a^3\sigma g$. So

$$m'g = \text{weight} - \text{upthrust}$$

$$= \tfrac{4}{3}\pi a^3(\rho - \sigma)g = 6\pi\eta av_1, \text{ from (i)}$$

$$\therefore a = \left[\frac{9\eta v_1}{2(\rho - \sigma)g}\right]^{1/2}$$

Since $k = 6\pi\eta a$, it follows from (ii) that

$$q = \frac{6\pi\eta}{E}\left[\frac{9\eta v_1}{2(\rho-\sigma)g}\right]^{1/2}(v_1-v_2) . \qquad . \qquad . \qquad . \qquad \text{(iii)}$$

Experiment

In his experiments Millikan used two horizontal plates A, B about 20 cm in diameter and 1·5 cm apart, with a small hole H in the centre of the upper plate, Figure 31.2. He used a fine spray to 'atomise' the oil and create tiny drops above

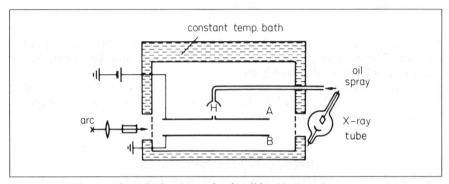

Figure 31.2 *Principle of Millikan's experiment*

H, and occasionally one would find its way through H, and would be observed in a low-power microscope by reflected light when the chamber was brightly illuminated. The drop was seen as a pin-point of light, and its downward velocity was measured by timing its fall through a known distance by means of a scale in the eyepiece. The field was applied by connecting a battery of several thousand volts across the plates A, B, and its intensity E was known, since $E = V/d$, where V is the p.d. between the plates and d is their distance apart. Millikan found that the friction between the drops when they were formed by the spray created electric charge, but to give a drop an increased charge an X-ray tube was operated near the chamber to ionise the air. In elementary laboratory experiments, increased charges may be obtained by using a radioactive source to ionise the air.

Results

In his experiment, Millikan found that, when an oil-drop gained charges, the *velocity change*, $v_1 - v_2$, varied by integral multiples of a basic unit. Now the charge q is proportional to $(v_1 - v_2)$, from (ii), for a particular drop. Consequently *the charge is a multiple of some basic unit*. So a particular charge consists of 'grains' or multiples of this basic unit.

From equation (iii) it follows that when v_1, v_2, E, ρ, σ and η are all known, the charge q on the drop can be calculated. Millikan found, working with hundreds of drops, that the charge q was always a simple multiple of a basic unit, about $1·6 \times 10^{-19}$ C, which was the *lowest* charge found. Earlier Sir J. J. Thomson had discovered the existence of the electron, a particle inside atoms which carried a negative charge $-e$ (p. 772). Millikan concluded that the charge e was $1·6 \times 10^{-19}$ C.

Example on Falling Oil Drop in Electric Field

Calculate the radius of a drop of oil, density $900\,\mathrm{kg\,m^{-3}}$, which falls with a terminal velocity of $2{\cdot}9 \times 10^{-4}\,\mathrm{m\,s^{-1}}$ through air of viscosity $1{\cdot}8 \times 10^{-5}\,\mathrm{N\,s\,m^{-2}}$. Ignore the density of the air.

If the charge on the drop is $-3e$, what p.d. must be applied between two plates $5\,\mathrm{mm}$ apart $(e = 1{\cdot}6 \times 10^{-19}\,\mathrm{C}.)$

When the drop falls with a terminal velocity, force due to viscous drag = weight of sphere. With the usual notation, if ρ is the oil density, we have

$$6\pi\eta a v = \text{volume} \times \text{density} \times g = \tfrac{4}{3}\pi a^3 \rho g$$

$$\therefore a = \sqrt{\frac{9\eta v}{2\rho g}} = \sqrt{\frac{9 \times 1{\cdot}8 \times 10^{-5} \times 2{\cdot}9 \times 10^{-4}}{2 \times 900 \times 9{\cdot}8}}$$

$$= 1{\cdot}6 \times 10^{-6}\,\mathrm{m} \qquad . \qquad . \qquad . \qquad . \qquad . \qquad . \qquad (1)$$

since $g = 9{\cdot}8\,\mathrm{m\,s^{-2}}$.

Suppose the upper plate is V volt higher than the lower plate when the drop is stationary, so that the electric field intensity E between the plates is V/d. Then upward force on drop $= E \times 3e =$ weight of drop.

$$\therefore E \times 3e = \tfrac{4}{3}\pi a^3 \rho g$$

$$\therefore E = \frac{4\pi a^3 \rho g}{9e} = \frac{V}{d}$$

$$\therefore V = \frac{4\pi a^3 \rho g d}{9e}$$

$$= \frac{4\pi \times (1{\cdot}6 \times 10^{-6})^3 \times 900 \times 5 \times 10^{-3} \times 9{\cdot}8}{9 \times 1{\cdot}6 \times 10^{-19}}$$

$$= 1576\,\mathrm{V}$$

Thermionic Emission of Electrons, Heated Cathodes

Nowadays a *hot cathode* is used to produce a supply of electrons. This may consist of a fine tungsten wire or filament, which is heated to a high temperature when a low voltage source of $6\,\mathrm{V}$ is connected to it. Metals contain free electrons, moving about rather like the molecules in a gas. If the temperature of

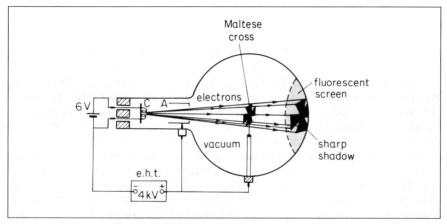

Figure 31.3 *Electrons travel in straight lines*

the metal is raised, the thermal velocities of the electrons will be increased. The chance of electrons escaping from the attraction of the positive ions, fixed in the lattice, will then also be raised. Thus by heating a metal such as tungsten to a high temperature, electrons can be 'boiled off'. This is called *thermionic emission*.

Figure 31.3 shows a tungsten filament C inside an evacuated tube. When heated by a low voltage supply, electrons are produced, and they are accelerated by a positive voltage of several thousand volts applied between C and a metal cylinder A. The electrons travel unimpeded across the vacuum past A, and produce a glow when they collide with a fluorescent screen and give up their energy.

Properties of Cathode Rays or Electrons

Before they were known to be tiny particles carrying a small charge e, beams of electrons were called *cathode rays* because they came from the cathode inside the tube producing them.

Fast-moving electrons emitted from a heated cathode C produce a sharp shadow of a Maltese cross on the fluorescent screen, as shown in Figure 31.3. Thus the electrons travel in straight lines. They also produce heat when incident on a metal—a fine piece of platinum glows, for example.

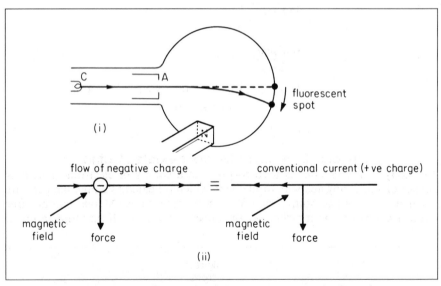

Figure 31.4 *Deflection shows electrons are negatively charged*

When a magnet is brought near to the electron beam, the glow on the fluorescent screen moves, Figure 31.4 (i). If Fleming's left hand rule is applied to the motion to deduce the conventional current (+ve charge) direction, the middle finger points in a direction *opposite* to the electron flow, Figure 31.4 (ii). This shows that electrons are particles which carry a *negative* charge.

This is confirmed by collecting electrons inside a *Perrin tube*, shown in Figure 31.5. The electrons are deflected by the magnet S until they pass into a metal cylinder called a 'Faraday cage' (see p. 194). The cylinder is connected to the plate of an electroscope, which has been negatively charged using an ebonite rod and fur. As soon as the electrons are deflected into the cage the leaf rises further, showing that an extra *negative* charge has been collected by the cage.

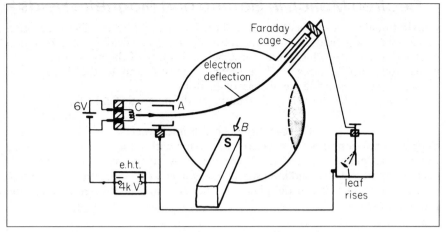

Figure 31.5 *Perrin tube. Direct method for testing electron charge*

Electron Motion in Electric and Magnetic Fields

Gravitational fields such as those round the Earth and the Sun affect the motion of masses in these fields. Similarly, electric and magnetic fields affect the motion of charged particles which enter these fields. A study of the motion has led to many useful applications, such as 'electron lenses', and to the discovery of unknown particles inside the nucleus of an atom.

First we consider the electric field and then the magnetic field.

Deflection in an Electric Field

Suppose a horizontal beam of electrons, moving with velocity v, passes between two parallel plates, Figure 31.6. If the p.d. between the plates is V and their distance apart is d, the field intensity $E = V/d$. Hence the force on an electron of charge e moving between the plates $= Ee = eV/d$ and is directed towards the positive plate.

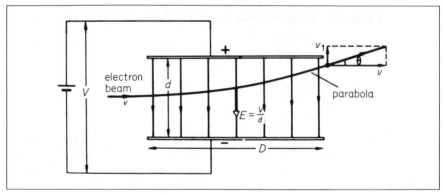

Figure 31.6 *Deflection in electric field*

Since the electricity intensity E is vertical, no horizontal force acts on the electron entering the plates. Thus the horizontal velocity, v, of the beam is unaffected. This is similar to the motion of a projectile projected horizontally under gravity. The vertical acceleration due to gravity does not affect the horizontal motion and the path of the projectile is a parabola (p. 15).

In a vertical direction the displacement, $y = \frac{1}{2}at^2$, where $a = $ acceleration $= $ force/mass $= Ee/m_e$ if m_e is the mass of an electron and t is the time.

$$\therefore y = \frac{1}{2}\frac{Ee}{m_e}t^2 \qquad . \qquad . \qquad . \qquad . \qquad \text{(i)}$$

In a horizontal direction, the displacement, $x = vt$ . . (ii)

Eliminating t between (i) and (ii), we obtain

$$y = \frac{1}{2}\left(\frac{Ee}{m_e}\right)\frac{x^2}{v^2} = \frac{Ee}{2m_e v^2}x^2$$

Between the plates, E and v stay constant. Since e and m_e are constant, we can write $y = kx^2$ for the motion, where k is a constant equal to $Ee/2m_e v^2$. This is the equation for a parabola. **So the path is a parabola between the plates**.

When the electron just passes the plates, $x = D$. The value of y is then $y = EeD^2/2m_e v^2$. Outside the plates or field, the beam then moves *in a straight line*, as shown in Figure 31.6. The time for which the electron is between the plates is D/v. Thus the component of the velocity v_1, gained in the direction of the field during this time, is given by

$$v_1 = \text{acceleration} \times \text{time} = \frac{Ee}{m_e} \times \frac{D}{v}$$

Hence the angle θ at which the beam comes out from the field is given by:

$$\tan \theta = \frac{v_1}{v} = \frac{EeD}{m_e v} \cdot \frac{1}{v} = \frac{EeD}{m_e v^2}$$

As the reader can verify, we can also write $\tan \theta = y/\frac{1}{2}D$, where y is the vertical displacement produced.

The energy of the electron is increased by an amount of $\frac{1}{2}mv_1^2$ as it passes through the plates, since the energy due to the horizontal motion is unaltered. This energy increase is equal to *force $\times$ distance*, where the force $= Ee$ and the distance in the direction of the force $=$ the vertical displacement of the electron in the field direction.

When an electron moves horizontally into a uniform vertical electric field it describes a parabolic path. The horizontal motion of the electron is not affected by the field. (Compare masses in gravitational field.) A charge gains energy when it moves in the direction of an electric field.

Example on Electron Motion in Electric Field

A beam of electrons, moving with a velocity of $1\cdot0 \times 10^7\ \mathrm{m\,s}^{-1}$, enters midway between two horizontal parallel plates P, Q in a direction parallel to the plates, Figure 31.7. P and Q are 5 cm long, 2 cm apart and have a p.d. V applied between them. Calculate V if the beam is deflected so that it just grazes the edge of the low plate Q. (Assume $e/m_e = 1\cdot8 \times 10^{11}\ \mathrm{C\,kg}^{-1}$.)

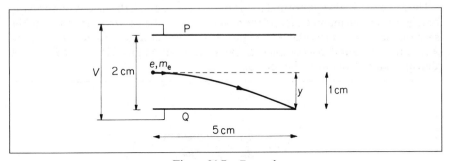

Figure 31.7 *Example*

Electric intensity between plates, $E = \dfrac{V}{d} = \dfrac{V}{2 \times 10^{-2}}$

Downward acceleration on electron, $a = \dfrac{\text{force}}{\text{mass}} = \dfrac{Ee}{m_e}$

So vertical distance, $y = \frac{1}{2}at^2 = \frac{1}{2}\dfrac{Ee}{m_e}t^2$

But $y = 1\,\mathrm{cm} = 10^{-2}\,\mathrm{m}$ and $t = \dfrac{5 \times 10^{-2}}{1 \times 10^7} = 5 \times 10^{-9}\,\mathrm{s}$, since the horizontal velocity is not affected by the vertical electric field and remains constant.

So
$$10^{-2} = \frac{1}{2} \times \frac{V}{2 \times 10^{-2}} \times 1 \cdot 8 \times 10^{11} \times (5 \times 10^{-9})^2$$

Simplifying,
$$V = 89 \text{ V (approx.)}$$

Deflection in a Magnetic Field

Consider an electron beam, moving with a speed v, which enters a *uniform magnetic field* of magnitude B acting perpendicular to the direction of motion, Figure 31.8. The force F on an electron is then Bev. The direction of the force is perpendicular to both B and v. Consequently, unlike the electric force, the magnetic force cannot change the *energy* of the electron. **It *deflects* the electron but does not change its speed or kinetic energy**.

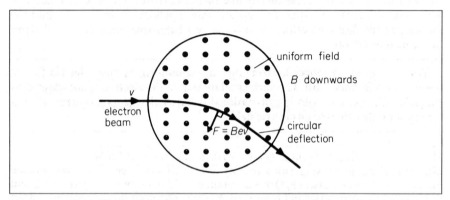

Figure 31.8 *Circular motion in uniform magnetic field*

The force Bev is always normal to the path of the beam. If the field is uniform the force is constant in magnitude and the beam then travels in a *circle* of radius r. If Bev is small, the path is a circular arc, Figure 31.8. If Bev is large enough, the electron beam can turn round in a complete circle, as shown in Figure 31B. Since Bev is the centripetal force (towards the centre),

$$Bev = \frac{m_e v^2}{r}$$

$$\therefore r = \frac{m_e v}{Be} = \frac{\text{momentum}}{Be}$$

If the velocity v of the electron decreases continuously due to collisions, for example, its momentum decreases. So, from the relation for r above, the radius of its path decreases and the electron will thus tend to spiral instead of moving in a circular path of constant radius.

Example on Magnetic Field Deflection

Protons, with a charge-mass-ratio of $1 \cdot 0 \times 10^8 \text{ C kg}^{-1}$, are rotated in a circular orbit of radius r when they enter a uniform magnetic field of $0 \cdot 5 \text{ T}$. Show that the number of revolutions per second, f, is independent of r and calculate f.

Suppose m_p is the proton mass and e is the charge. Then, with the usual notation,

$$Bev = \frac{m_p v^2}{r}$$

So
$$v = \frac{Ber}{m_p} = r\omega$$

where ω is the angular velocity. Since r cancels,

$$\omega = \frac{Be}{m_p} = 2\pi f$$

Then
$$f = \frac{Be}{2\pi m_p}$$

This result for f shows that it is independent of r. Also,

$$f = \frac{0.5 \times 1 \times 10^8}{2 \times 3.14} = 8 \times 10^6 \text{ rev s}^{-1}$$

Specific Charge of Electron by Magnetic Deflection

The 'charge per unit mass' or *specific charge* of an electron (e/m_e) can be measured in the school laboratory by a TELTRON tube designed for this purpose, Figure 31.9.

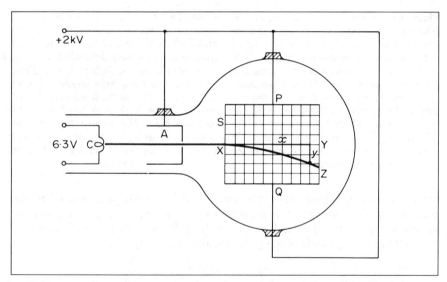

Figure 31.9 *Deflection of electron beam (Helmholtz coils not shown)*

The evacuated tube has (i) a hot cathode C which produces an electron beam, (ii) an accelerating anode A, (iii) a vertical screen S coated with luminous paint, between two plates P, Q not used in this experiment. The screen S has horizontal and vertical lines equally spaced and is slightly inclined to the electron beam incident on it. The beam then produces a luminous glow and its path can be seen.

Two Helmholtz coils H, H connected in series are on opposite sides of the tube. Plate 31A shows the coils and the tube in a photograph. The Helmholtz coils produce a *uniform* magnetic field B perpendicular to the beam over the region of S, so that the beam is deflected in a circular path. The circuit for the

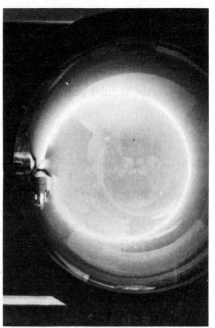

Plate 31A *Parabolic path of an electron beam after entering a uniform perpendicular electric field. The upper plate is +ve in potential relative to the lower plate. The screen, which is coated with a luminous paint, has squares marked on it for measuring the electron deflection in a charge/ mass ratio experiment. Helmholtz coils, the large circular coils in front and behind the tube, are used to apply a uniform magnetic field to the electron beam. (Courtesy of Teltron Limited)*

Plate 31B *Circular deflection of electron beam in a uniform magnetic field due to Helmholtz coils (not visible). Faster electrons produce the outer diffuse beam.*

Unlike the electric deflection tube (Plate 31A), this tube has a small amount of helium gas inside it. The gas molecules are ionised after collision with high-speed electrons and emit light. The electron track is thus made visible as a fine straight beam before the field is applied. (Courtesy of Teltron Limited)

coils is shown in Figure 31.10. The value of B is directly proportional to the current I in the coils, which is measured by the ammeter A, and is varied by means of a potential divider D with a battery B connected to it. Calculation shows that the magnitude of B is given by, approximately,

$$B = 0.72 \frac{\mu_0 N I}{R}$$

where N is the number of turns in each coil, I is the current in amperes and R is the radius in metres of the coils (p. 331).

Experiment. In an experiment, the cathode C is heated by a 6·3 V supply. The anode A has a high potential V relative to C of say 2 kV or 2000 V. The plates P, Q are joined together and kept at the same potential as A; this eliminates electric fields beyond A, which would alter the speed of the electron beam. After passing A, the beam produces a luminous horizontal line XY on S.

A current is now passed into the Helmholtz coils, H, H. The electron beam then deflected along a circular path XZ seen on S, Figure 31.9. By altering the current in the coils, the magnetic field B can be varied to produce a path XZ of suitable radius of curvature r. The horizontal and vertical distances x, y from X

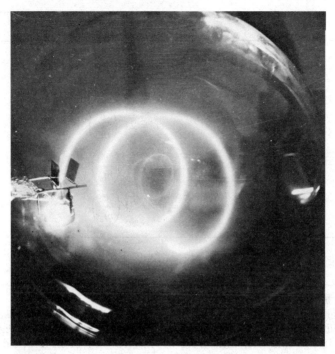

Plate 31C *Spiral electron path, produced when an electron beam enters a* magnetic field *at an angle to the field. The component of the velocity normal to the field produces the circular motion; the component parallel to the field produces the translational motion. Initially the beam passes through the two plates shown, which can be used to apply an electric field.* (Courtesy of Leybold-Heraeus GMB and Co)

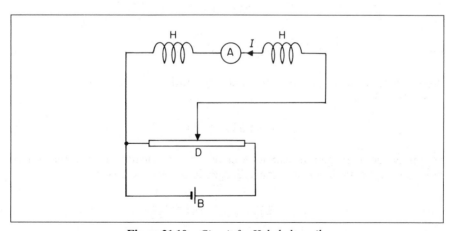

Figure 31.10 *Circuit for Helmholtz coils*

of a convenient point on XZ are then read from the graduations on S. As shown on p. 529, r is given by

$$r = \frac{x^2 + y^2}{2y}$$

Theory. If v is the velocity of the electrons at the anode A, then, assuming zero velocity at C, their gain in kinetic energy $= \frac{1}{2}m_e v^2$. So if V is the anode potential,

$$\tfrac{1}{2}m_e v^2 = eV \qquad \cdot \qquad \cdot \qquad \cdot \qquad \cdot \qquad \cdot \qquad (1)$$

The radius r of the circular path XZ of the beam is given by

$$Bev = \frac{m_e v^2}{r} \qquad \cdot \qquad \cdot \qquad \cdot \qquad \cdot \qquad (2)$$

From (1) $\qquad\qquad\qquad\qquad v^2 = 2V\left(\frac{e}{m_e}\right)$

From (2) $\qquad\qquad\qquad\qquad v = Br\left(\frac{e}{m_e}\right)$

So $\qquad\qquad B^2 r^2 \left(\frac{e}{m_e}\right)^2 = v^2 = 2V\left(\frac{e}{m_e}\right)$

Cancelling e/m_e, $\qquad\qquad\qquad \dfrac{e}{m_e} = \dfrac{2V}{B^2 r^2} \qquad \cdot \qquad \cdot \qquad \cdot \qquad \cdot \qquad (3)$

Knowing V, r and B, then e/m_e can be calculated.

The main errors in the experiment are: (i) the difficulty of measuring the radius r with accuracy, (ii) B may not be uniform over the whole region of S, (iii) error in the voltmeter measuring V. An error in r or in B would produce double the percentage error in e/m_e since r^2 and B^2 occur in (3) above.

Mass of Electron

Modern determinations show that

$$\frac{e}{m_e} = 1 \cdot 76 \times 10^{11}\,\mathrm{C\,kg^{-1}}$$

or $\qquad\qquad \dfrac{m_e}{e} = \dfrac{1}{1 \cdot 76} \times 10^{-11}\,\mathrm{kg\,C^{-1}} \qquad \cdot \qquad \cdot \qquad \cdot \qquad \cdot \qquad (ii)$

Now from electrolysis the mass-charge ratio for a hydrogen ion is $1 \cdot 04 \times 10^{-8}\,\mathrm{kg\,C^{-1}}$.

$$\therefore \frac{m_H}{e} = 1 \cdot 04 \times 10^{-8}\,\mathrm{kg\,C^{-1}}$$

assuming the hydrogen ion carries a charge e numerically equal to that on an electron, m_H being the mass of the hydrogen ion. Hence, with (ii),

$$\frac{m_e}{m_H} = \frac{1}{1 \cdot 76 \times 1 \cdot 04 \times 10^3} = \frac{1}{1830}$$

Thus the mass of an electron is about two-thousandths that of the hydrogen atom.

Helical Path of Electrons

Consider an electron P of charge e, mass m_e entering a uniform magnetic field B at a small angle θ with velocity v, Figure 31.11. The component $v \cos \theta$ parallel to B produces a *translational* (linear) motion, since no electromagnetic force acts in this case. The component $v \sin \theta$ normal to B, however, produces a *circular* motion. So the electron path is a helix (spiral). See Plate 31C, page 768.

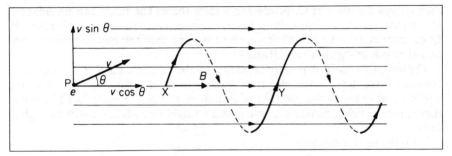

Figure 31.11 *Helical electron path*

The pitch XY of the helix, the distance between neighbouring turns, is given by $XY = v \cos\theta . T$, where T is the period of rotation of the electron. For circular motion (see p. 766),

$$Bev \sin\theta = \frac{m_e(v \sin\theta)^2}{r}$$

so

$$\frac{r}{v \sin\theta} = \frac{m_e}{Be}$$

Hence

$$T = \frac{2\pi r}{v \sin\theta} = \frac{2\pi m_e}{Be}$$

So

$$XY = v \cos\theta . T = \frac{2\pi m_e v \cos\theta}{Be}$$

If $v = 10^6 \, \mathrm{m\,s^{-1}}$, $e/m_e = 1.8 \times 10^{11} \, \mathrm{C\,kg^{-1}}$, $\theta = 10°$, $B = 2 \times 10^{-4} \, \mathrm{T}$,

then

$$XY = \frac{2\pi \times 10^6 \times \cos 10°}{1.8 \times 10^{11} \times 2 \times 10^{-4}} = 0.17 \, \mathrm{m}$$

Thomson's Experiment for e/m

In 1897, Sir J. J. THOMSON devised an experiment for measuring the ratio *charge/mass* or e/m_e for an electron, called its *specific charge*.

Thomson's apparatus is shown simplified in Figure 31.12. C and A are the cathode and anode respectively, and narrow slits are cut in opposite plates at A so that the electrons passing through are limited to a narrow beam. The beam

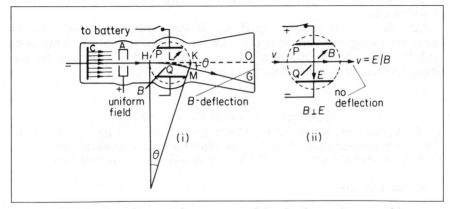

Figure 31.12 *Thomson's determination of* e/m_e *for electron* (not to scale)

then strikes the glass at O, producing a glow there. The beam can be deflected *electrostatically* by a high voltage battery connected to the horizontal plates P, Q, or magnetically by means of a current passing through two Helmholtz coils on either side of the tube near P and Q.

The *magnetic field B* is perpendicular to the paper, and if it is uniform a constant force acts on the electrons normal to its motion. With *B* alone, the particles move along the arc HM of a circle of radius *r*, Figure 31.12 (i). When they leave the field, the particles move in a straight line MG and strike the glass at G.

With the usual notation

$$\text{force } F = Bev = \frac{m_e v^2}{r}$$

where *e* is the charge on an electron and m_e is its mass.

$$\therefore \frac{e}{m_e} = \frac{v}{rB} \qquad . \qquad . \qquad . \qquad . \qquad . \qquad \text{(i)}$$

To find the radius *r*, from Figure 31.12 (i), $\tan \theta = OG/OL = HK/r$

$$\therefore r = \frac{HK \cdot OL}{OG}$$

L is about the middle of the magnetic field coils.

The *velocity v* of the electron beam was found by varying the p.d. or electric field between P and Q until the beam returned to O from its deflected position G, Figure 31.12 (ii). The beam now travels *undeflected* through both the electric field *E* and the magnetic field *B*. The two fields are sometimes described as 'crossed' fields, because *E* is vertical and *B* is horizontal and perpendicular to *E*.

Since the downward force *Bev* = upward force *Ee*,

$$\therefore v = \frac{E}{B}$$

If *E* is in $V m^{-1}$ and *B* is in T, then *v* is in $m s^{-1}$. Thomson found that *v* was considerably less than the velocity of light, $3 \times 10^8 \, m s^{-1}$, so that electrons are not electromagnetic waves.

On substituting for *v* and *r* in (i), the ratio charge/mass (e/m_e) for an electron was obtained.

Until Sir J. J. THOMSON's experiment, it was believed that the hydrogen atom was the lightest particle in existence. As explained on p. 770, comparison with the mass-charge ratio of a hydrogen ion showed that the electron was about 1/2000th of the mass of the hydrogen atom.

Examples on Electron Motion in Fields

1 An electron beam passes undeflected with uniform velocity *v* through two parallel plates, when a magnetic field of 0·01 T is applied perpendicular to an electric field between the plates produced by a p.d. of 100 V. The separation of the plates is 5 mm. Calculate *v*.

For no deflection,

$$Bev = Ee$$

So
$$v = \frac{E}{B} = \frac{V}{dB}$$

$$= \frac{100}{5 \times 10^{-3} \times 0{\cdot}01} = 2 \times 10^6 \, \text{m s}^{-1}$$

2 Describe and give the theory of a method to determine e the electronic charge. Why is it considered that all electric charges are multiples of e?

An electron having 450 eV of energy moves at right angles to a uniform magnetic field of flux density $1{\cdot}50 \times 10^{-3}$ T. Show that the path of the electron is a circle and find its radius. Assume that the specific charge of the electron is $1{\cdot}76 \times 10^{11} \, \text{C kg}^{-1}$. (L.)

With the usual notation, the velocity v of the electron is given by

$$\tfrac{1}{2}m_e v^2 = eV, \text{ where } V \text{ is } 450 \, \text{V}$$

$$\therefore v = \sqrt{\frac{2eV}{m_e}} \qquad . \qquad . \qquad . \qquad . \qquad (1)$$

The path of the electron is a circle because the force Bev is constant and always normal to the electron path. Its radius r is given by

$$Bev = \frac{m_e v^2}{r}$$

$$\therefore r = \frac{m_e}{e} \cdot \frac{v}{B} = \frac{m_e}{e} \cdot \frac{1}{B}\sqrt{\frac{2eV}{m_e}}, \text{ from (1)}$$

$$\therefore r = \frac{1}{B}\sqrt{2\frac{m_e V}{e}}$$

Now $e/m_e = 1{\cdot}76 \times 10^{11} \, \text{C kg}^{-1}$, $V = 450 \, \text{V}$, $B = 1{\cdot}5 \times 10^{-3}$ T

$$\therefore r = \frac{1}{1{\cdot}5 \times 10^{-3}}\sqrt{\frac{2 \times 450}{1{\cdot}76 \times 10^{11}}} \, \text{m}$$

$$= 4{\cdot}8 \times 10^{-2} \, \text{m}$$

3 A charged oil drop of mass 6×10^{-15} kg falls vertically in air with a steady velocity between two long parallel vertical plates 5 mm apart. When a p.d. of 3000 V is applied between the plates, the drop now falls with a steady velocity at an angle of 58° to the vertical. Calculate the charge Q on the drop. (Assume $g = 10 \, \text{m s}^{-2}$.)

The drop falls with steady velocity due to the viscosity of the air. Neglecting the upthrust, if v_1 is the vertical velocity

downward force on drop = upward viscous force

or $$mg = kv_1$$

Similarly, when the electric field of intensity E is applied,

horizontal force on drop = $EQ = kv_2$

where v_2 is the horizontal velocity.

By vector addition, if θ is the angle to the vertical,

$$\tan \theta = \frac{v_2}{v_1} = \frac{EQ}{mg}$$

Now
$$E = \frac{V}{d} = \frac{3000}{5 \times 10^{-3}} = 6 \times 10^5 \,\mathrm{V\,m^{-1}}$$

$$\therefore \tan 58° = \frac{EQ}{mg} = \frac{6 \times 10^5 \, Q}{6 \times 10^{-15} \times 10} = 10^{19} Q$$

$$\therefore Q = \frac{\tan 58°}{10^{19}} = 1.6 \times 10^{-19} \,\mathrm{C}$$

Exercises 31A

Electrons

1 Electrons are accelerated from rest by a p.d. of 100 V. What is their final velocity?
    The electron beam now enters normally a uniform electric field of intensity $10^5 \,\mathrm{V\,m^{-1}}$. Calculate the flux density B of a uniform magnetic field applied perpendicular to the electric field if the path of the beam is unchanged from its original direction. Draw a sketch showing the electron beam and the two fields. (Assume $e/m_e = 1.8 \times 10^{11} \,\mathrm{C\,kg^{-1}}$.)

2 A beam of protons is accelerated from rest through a potential difference of 2000 V and then enters a uniform magnetic field which is perpendicular to the direction of the proton beam. If the flux density is 0.2 T calculate the radius of the path which the beam describes. (Proton mass $= 1.7 \times 10^{-27}$ kg. Electronic charge $= -1.6 \times 10^{-19}$ C.) (*L*.)

3 (a) State what is meant by *thermionic emission*.
        Explain why, in early applications of thermionic emission, tungsten was used as the emitting material, while nowadays oxide-coated metals are preferred.
    (b) A potential difference of 2.0 kV is maintained between a heated thermionic cathode and a collector electrode in a vacuum, the latter being the more positive. Calculate the speed of electrons striking the collector.
        Determine the initial rate of rise of temperature of the collector when the potential difference is applied if the electron current is 0.105 mA and the thermal capacity of the collector is $2.10 \,\mathrm{J\,K^{-1}}$.
        (Specific charge of the electron, e/m, $= -1.76 \times 10^{11} \,\mathrm{C\,kg^{-1}}$.)
    (c) A beam of electrons moving at a speed v is allowed to enter a region where a uniform magnetic field of flux density B acts in a direction perpendicular to the beam's path. State the magnitude and direction of   (i) the force acting on one of the electrons   (ii) the acceleration of one of the electrons.
        Explain   (iii) the subsequent shape of the electron path,   (iv) whether the speed of the electrons would change,   (v) whether the velocity of the electrons would change. (*L*.)

4 (a) Describe an experiment to determine the ratio of the charge to the mass (e/m) for an electron. Show how the result is derived from the observations.
    (b) In an evacuated tube electrons are accelerated from rest through a potential difference of 3600 V and then travel in a narrow beam through a field free space before entering a uniform magnetic field the flux lines of which are perpendicular to the beam. In the magnetic field the electrons describe a circular arc of radius 0.10 m. Calculate   (i) the speed of the electrons entering the magnetic field,   (ii) the magnitude of the magnetic flux density.
    (c) If an electron described a complete revolution in a magnetic field how much energy would it acquire? ($e/m = 1.8 \times 10^{11} \,\mathrm{C\,kg^{-1}}$.) (*JMB*.)

5 Give an account of a method by which the charge associated with an electron has been measured.
    Taking this electronic charge to be -1.60×10^{-19} C, calculate the potential difference in volt necessary to be maintained between two horizontal conducting plates, one 5 mm above the other, so that a small oil drop, of mass 1.31×10^{-14} kg with two electrons attached to it, remains in equilibrium between them. Which plate would be at the positive potential? ($g = 9.8 \,\mathrm{m\,s^{-2}}$.) (*L*.)

6 Describe an experiment to determine the ratio of the charge to the mass of electrons. Draw labelled diagrams of
(a) the apparatus,
(b) any necessary electrical circuits,
and show how the result is calculated from the observations.

Two plane metal plates 4·0 cm long are held horizontally 3·0 cm apart in a vacuum, one being vertically above the other. The upper plate is at a potential of 300 V and the lower is earthed. Electrons having a velocity of $1·0 \times 10^7 \, \mathrm{m \, s^{-1}}$ are injected horizontally midway between the plates and in a direction parallel to the 4·0 cm edge. Calculate the vertical deflection of the electron beam as it emerges from the plates. (e/m for electron = $1·8 \times 10^{11} \, \mathrm{C \, kg^{-1}}$.) (*JMB*.)

7 (a) A charged oil drop falls at constant speed in the Millikan oil drop experiment when there is no p.d. between the plates. Explain this.
(b) Such an oil drop, of mass $4·0 \times 10^{-15}$ kg, is held stationary when an electric field is applied between the two horizontal plates. If the drop carries 6 electric charges each of value $1·6 \times 10^{-19}$ C, calculate the value of the electric field strength. (*L*.)

8 (a) Describe an experiment to determine the charge to mass ratio of electrons, indicating clearly the measurements made. How is the value of e/m calculated from them? What information from the experiment indicates that electrons are *negatively* charged?

(b)

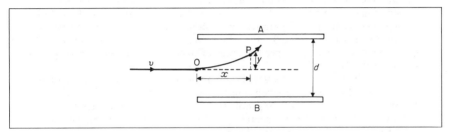

Figure 31A

A beam of electrons, travelling with a velocity v in the x-direction, enters at point O ($x = 0$, $y = 0$) a region of uniform electric field provided by applying a voltage V between plates A and B, separated by a distance d in the y-direction, Figure 31A. The electrons are deflected towards A as shown, the point P (x, y) being a point along the trajectory. (i) Is the potential of plate A positive or negative with respect to B? (ii) In terms of distance x, calculate the angle between the x-direction and the electron beam at P. (iii) Prove that the path is parabolic, namely $y = ax^2$, and find the value of a.

The position of the electron source is moved so that the direction of the incoming beam at point O is now at an angle θ to the x-direction towards plate B. The initial speed is unchanged. (iv) Find the distance L along the x-axis at which the beam again has $y = 0$. (v) Explain how this effect could be used as the basis for a velocity selector for electrons. (*O. & C.*)

9 (a) Explain how a beam of electrons may be produced in a vacuum tube, and describe an arrangement by which the beam may be deflected by a magnetic field. Draw a diagram showing clearly the directions of the beam, the field and the deflection.
(b) An electron moving at velocity v passes simultaneously through a magnetic field of uniform flux density B and an electric field of uniform intensity E. It emerges with direction and speed unaltered. (i) Explain how the fields are arranged to achieve this result. (ii) Derive the relationship between v, B and E.
(c) An electron is travelling at $2·0 \times 10^6 \, \mathrm{m \, s^{-1}}$ at right angles to a magnetic field of flux density $1·2 \times 10^{-5}$ T; its path is a circle. Uniform circular motion of an electron is accompanied by the emission of electromagnetic radiation of the same frequency as that of the circular motion.

(Take the specific charge of the electron e/m_e to be $1.8 \times 10^{11}\,C\,kg^{-1}$, and the speed of electromagnetic waves in air c to be $3.0 \times 10^8\,m\,s^{-1}$.) (i) Explain why the path of the electron is a circle. (ii) Calculate the radius of the circle. (iii) Calculate the frequency of the circular motion of the electron. (iv) Calculate the wavelength of the electromagnetic radiation emitted and identify in which part of the electromagnetic spectrum this radiation lies. (v) How would this wavelength be affected by a decrease in the speed of the electron? (*O.*)

10 Define *magnetic flux density* (*B*). Describe how the variation in *B* along the axis both inside and outside a long solenoid carrying a current may be investigated. Sketch a graph showing the results you would expect to obtain.

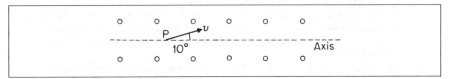

Figure 31B

Figure 31B represents a long solenoid in which a steady current is flowing. An electron is emitted at P with an initial velocity v in the direction shown. By considering the components of v (i) perpendicular to the axis, and (ii) parallel to the axis, deduce the path the electron will follow. If the value of v is $2.0 \times 10^6\,m\,s^{-1}$ and the flux density at P is $3.0 \times 10^{-4}\,T$, at what distance along the axis will the electron next cross it? (The specific charge of an electron, e/m_e, may be assumed to be $1.8 \times 10^{11}\,C\,kg^{-1}$.) (*L.*)

11 Show that if a free electron moves at right angles to a magnetic field the path is a circle. Show also that the electron suffers no force if it moves parallel to the field. Point out how the steps in your proof are related to fundamental definitions.

If the path of the electron is a circle, prove that the time for a complete revolution is independent of the speed of the electron.

In the ionosphere electrons execute 1.4×10^6 revolutions in a second. Find the strength of the magnetic flux density B in this region. (Mass of an electron $= 9.1 \times 10^{-31}\,kg$; electronic charge $= 1.6 \times 10^{-19}\,C$.) (*C.*)

12 The electron is stated to have a mass of approximately $10^{-30}\,kg$ and a negative charge of approximately $1.6 \times 10^{-19}\,C$. Outline the experimental evidence for this statement. Formulae may be quoted without proof. You are not required to justify the actual numerical values quoted.

An oil drop of mass $3.25 \times 10^{-15}\,kg$ falls vertically, with uniform velocity, through the air between *vertical* parallel plates which are 2 cm apart. When a p.d. of 1000 V is applied to the plates the drop moves towards the negatively charged plate, its path being inclined at 45° to the vertical. Explain why the vertical component of its velocity remains unchanged and calculate the charge on the drop.

If the path of the drop suddenly changes to one at 26° 30′ to the vertical, and subsequently to one at 37° to the vertical, what conclusions can be drawn? (*O. & C.*)

13 An electron with a velocity of $10^7\,m\,s^{-1}$ enters a region of uniform magnetic flux density of $0.10\,T$, the angle between the direction of the field and the initial path of the electron being 25°. By resolving the velocity of the electron find the axial distance between two turns of the helical path. Assume that the motion occurs in a vacuum and illustrate the path with a diagram. ($e/m = 1.8 \times 10^{11}\,C\,kg^{-1}$.) (*JMB.*)

14 Give an account of Millikan's experiment for determining the value of the electronic charge e.

In a Millikan-type apparatus the horizontal plates are 1.5 cm apart. With the electric field switched off an oil drop is observed to fall with the steady velocity $2.5 \times 10^{-2}\,cm\,s^{-1}$. When the field is switched on the upper plate being positive, the drop just remains stationary when the p.d. between the two plates is 1500 V.
(a) Calculate the radius of the drop.

(b) How many electronic charges does it carry?

(c) If the p.d. between the two plates remains unchanged, with what velocity will the drop move when it has collected two more electrons as a result of exposure to ionising radiation? (Oil density $= 900 \,\text{kg m}^{-3}$, viscosity of air $= 1\cdot8 \times 10^{-5}\,\text{N s m}^{-2}$.) (*O. & C.*)

15 A heated filament and an anode with a small hole in it are mounted in an evacuated glass tube so that a narrow beam of electrons emerges vertically upwards from the hole in the anode. A uniform magnetic field is applied so that the electrons describe a circular path in a vertical plane.

(a) Draw a diagram showing the path of the electrons and indicate the direction of the magnetic field which will cause the beam to curve in the direction you have shown. Explain why the path is circular.

(b) Derive an expression for the specific electronic charge (e/m) of the electrons in terms of the p.d. between the anode and filament, V, the radius of the circular path, r, and the magnetic flux density, B.

(c) What value of B would be required to give a radius of the electron path of $2\,r$, assuming that V remains constant? If B is now held constant as its new value, what value of V will restore the beam to its former radius?

(d) Describe and account for the changes in (i) kinetic energy, (ii) momentum which an electron undergoes from the instant it leaves the heated filament with negligible velocity until it has completed a full circle in the magnetic field. (*JMB.*)

16 An electron charge $-e$ and mass m is initially projected with speed v at right angles to a uniform field of flux density B. Show that the electron moves in a circular path and derive an expression for the radius of the circle. Show also that the time taken to describe one complete circle is independent of the speed of the electron.

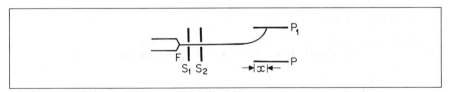

Figure 31C

Electrons are emitted with negligible speed, *in vacuo*, from a filament F. They are accelerated by a potential difference of 1200 V applied between the plates S_1 and F, as shown in Figure 31C. The electrons are collimated into a narrow horizontal beam by passing through holes in S_1 and in a second plate S_2 which is at the same potential as S_1. The electron beam subsequently enters the space between two large parallel horizontal plates P_1 and P_2 which are 0·02 m apart. The point of entry is midway between the plates. The mean potential of P_1 and P_2 is equal to that of S_1 but P_1 is at a positive potential of 150 V with respect to P_2. Neglecting the effect of gravity and of non-uniform fields near the plate boundaries, calculate the distance x travelled by the electrons between P_1 and P_2 before they strike P_1. (*O. & C.*)

17 Give an account of an experiment to obtain the value of the charge associated with the electron.

An electron beam after being accelerated from rest through a potential difference of 5000 V *in vacuo* is allowed to impinge normally on a fixed surface. If the incident current is 50 μA determine the force exerted on the surface assuming that it brings the electrons to rest. ($e = 1\cdot6 \times 10^{-19}\,\text{C}$.) (*L.*)

Cathode-Ray Oscilloscope (C.R.O.)

We now discuss an important type of electron tube, widely used, called a *cathode-ray oscilloscope*. It is called a cathode-ray oscilloscope because it traces a required wave-form with a beam of electrons, and beams of electrons were originally called cathode rays.

A cathode-ray oscilloscope is essentially an electrostatic instrument. It consists of a highly evacuated glass tube, T in Figure 31.13, one end of which opens out to form a screen S which is internally coated with zinc sulphide or other fluorescent material. A hot filament F, at the other end of the tube, emits electrons. These are then attracted by the cylinders A_1 and A_2, which have increasing positive potentials with respect to the filament. Many of the electrons, however, shoot through the cylinders and strike the screen; where they do so, the zinc sulphide fluoresces in a green spot. On their way to the screen, the electrons pass through two pairs of metal plates, XX and YY, called the *deflecting plates*.

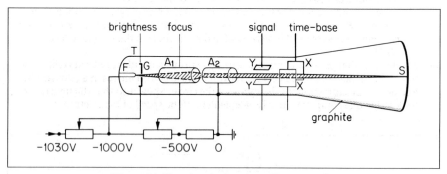

Figure 31.13 *A cathode-ray oscilloscope tube*

The inner walls of the tube are coated with graphite, which is connected to the final anode A_2. This makes the space between A_2 and S an equipotential volume so that the speed of the electrons is maintained from A_2 to S.

The *brightness* of the light on the screen is controlled by an electrode G in front of the filament F. If G is made more negative in potential relative to F, the increased repulsive force reduces the number of electrons per second passing G.

Electrode Potentials

In practice, the screen S, the tube and A_2 are all earthed to avoid danger due to high voltages. Further, touching the outside of S with earthed fingers does not then alter the electrostatic field inside the tube and affect the deflection of the beam. Thus, as indicated, the filament F is at a high negative potential relative to earth and is therefore dangerous to touch. The filament, electrode G and accelerating electrodes are often called the 'gun assembly' of a tube. Figure 31.13 illustrates a potential divider arrangement for the simple cathode-ray oscilloscope shown. The anode A_2, tube and screen are earthed; A_1 has a varying voltage for the focus control as explained soon; and the brightness control G has a varying negative potential relative to the potential of F, which is about -1000 V.

Deflection; Time-base

If a battery were connected between the Y-plates, so as to make the upper one positive, the electrons in the beam would be attracted towards that plate, and the beam would be deflected upwards. In the same way, the beam can be deflected horizontally by a potential difference applied between the X-plates.

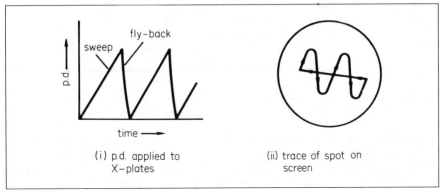

Figure 31.14 *Action of a C.R.O. time-base*

When the oscilloscope is in use, the varying voltage to be examined is applied between the Y-plates. If that were all, then the spot would be simply drawn out into a vertical line. To trace the *wave-form* of the varying voltage, the X-plates are used to provide a *time-axis*. A special type of circuit generates a potential difference which rises steadily to a certain value, as shown in Figure 31.14(i), and then falls rapidly to zero. It can be made to go through these changes tens, hundreds, thousands or millions of times per second. This *time-base voltage* is applied between the X-plates, so that the spot is swept steadily to the right, and then flies swiftly back and starts out again. The horizontal motion provides the *time-base* of the oscillograph. On it is superimposed the vertical motion produced by the Y-plates. Then as shown in Figure 31.14(ii), the wave-form of the voltage to be examined can be displayed on the screen.

In a *double beam oscilloscope*, the two Y-plates are joined to terminals labelled Y_1 and Y_2 respectively. An earthed plate between the plates splits the beam into two halves. One half can be deflected by an input voltage connected to Y_1 and the other half can be deflected by another input voltage connected to Y_2. With a common time-base applied to the X-plates, two traces can be obtained on the screen and two different waveforms, from Y_1 and Y_2 respectively, can thus be compared.

Focusing

To give a clear trace on the screen, the electron beam must be focused to a sharp spot. This is the purpose of the cylinders A_1 and A_2, called the first and second anodes.

Figure 31.15 shows the equipotentials of the field between them, when their difference of potential is 500 volts. Electrons entering the field from the filament experience forces from low potential to high at right angles to the equipotentials. They have, however, considerable momentum, because they have been accelerated by a potential difference of about 500 volts, and are travelling fast. Consequently the field merely deflects them, and, because of its cylindrical symmetry, it converges the beam towards the point P. Before they can reach this point, however, they enter the second cylinder. Here the potential rises from the axis, and the electrons are deflected outwards. However, they are now travelling

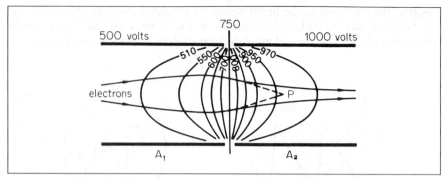

Figure 31.15 *Focusing in an oscilloscope tube by electron lens*

faster than when they were in the first cylinder, because the potential is everywhere higher. Consequently their momentum is greater, and they are less deflected than before. The second cylinder, therefore, diverges the beam less than the first cylinder converged it, and so the beam emerges from the second anode still somewhat convergent. By adjusting the potential of the first anode, the beam can be focused upon the screen, to give a spot a millimetre or less in diameter.

Electron-focusing devices are called *electron lenses*, or electron-optical systems. For example, the action of the anodes A_1 and A_2 is roughly analogous to that of a pair of glass lenses on a beam of light, the first glass lens being converging, and the second diverging, but weaker.

C.R.O. Supplies

The necessary voltage supplies or circuits for a cathode-ray oscillograph or oscilloscope are shown in block form in Figure 31.16.

1 The *power pack* supplies e.h.t.—very high d.c. voltage for the tube electrodes such as the electron lens; h.t.—rectified and smoothed d.c. voltage for the amplifiers, for example; and l.t.—low a.c. voltage such as 6·3 V for the valve heaters.

2 *Tube controls.* The brightness and focus, and correction of astigmatism, are controlled by varying the voltage of the appropriate tube electrode.

3 *Y- or Input-Amplifier.* This is a variable gain linear amplifier which amplifies the signal applied to the Y-plates.

4 *Y-shift.* This is in the Y-amplifier circuit and shifts the signal trace up or down on the screen.

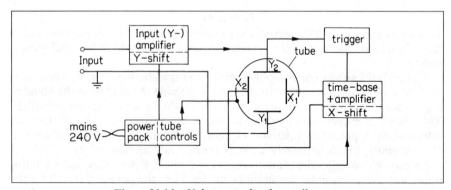

Figure 31.16 *Voltage supplies for oscilloscope*

5   *Time-base.* This is a 'sawtooth' oscillator (p. 779). In the oscillator circuit, the feedback can be adjusted with stability control until the oscillator is just not free running, and the trigger is then applied to 'lock' the trace.

6   *Trigger unit.* This applies a triggering pulse to the time-base oscillator from the Y- or input-voltage. The time-base frequency is then synchronised ('locked') with that of the input signal, so that a stationary trace is obtained on the screen.

7   *Time-base amplifier.* This amplifies the time-base voltage and applies it to the X-plates. The trace can be made to expand or contract horizontally by varying the amplification.

8   *X-shift.* This shifts the time-base horizontally to the left or right.

Oscilloscope controls are shown in Plate 31D. The *voltage sensitivity* of the Y-plates is the 'voltage per cm (or mm)' vertical deflection of the electron beam and the *time-base* is the 'time per cm' horizontal deflection of the electron beam due to the X-plates, so that voltage and time can be measured.

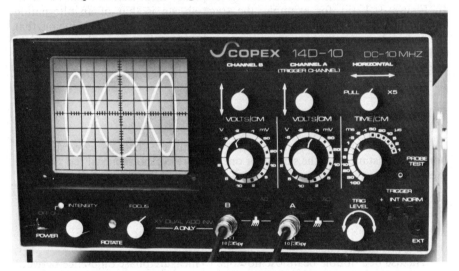

Plate 31D   *Oscilloscope display, showing the Lissajous figure obtained when the frequencies on the X and Y-plates are in the ratio $1:3$. In this modern double-beam instrument, the time-base (X-deflection) ranges from $1\,\mu s\,cm^{-1}$ to $100\,ms\,cm^{-1}$ and the voltage sensitivity (Y-deflection) from $2\,mV\,cm^{-1}$ to $10\,V\,cm^{-1}$.* (Courtesy of Scopex Instruments Ltd)

Uses of Oscilloscope

In addition to displaying waveforms, the oscilloscope can be used for measurement of voltage, frequency and phase, and act as a clock.

1   *A.C. voltage*

An unknown a.c. voltage, whose peak value is required, is connected to the Y-plates. With the time-base switched off, the vertical line on the screen is centred and its length then measured, Figure 31.17 (i). This is proportional to twice the amplitude or peak voltage, V_0. By measuring the length corresponding to a known a.c. voltage V, then V_0 can be found by proportion.

Alternatively, using the same gain, the waveforms of the unknown and known voltages, V_0 and V, can be displayed on the screen. The ratio V_0/V is then obtained from measurement of the respective peak-to-peak heights.

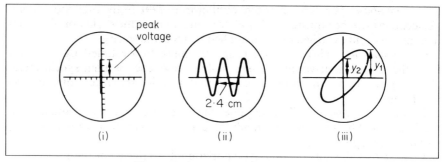

Figure 31.17 *Uses of oscilloscope*

2   Measurement and comparison of frequencies

If a calibrated time-base is available, frequency measurements can be made. In Figure 31.17 (ii), for example, the trace shown is that of an alternating waveform with the time-base switched to the '5 ms/cm' scale. This means that the time taken for the spot to move 1 cm horizontally across the screen is 5 milliseconds. The horizontal distance on the screen for one cycle is 2·4 cm. This corresponds to a time of $5 \times 2·4$ ms or $12·0$ ms $= 12 \times 10^{-3}$ seconds, which is the period T.

$$\therefore \text{frequency} = \frac{1}{T} = \frac{1}{12 \times 10^{-3}} = 83 \, \text{Hz}$$

If a comparison of frequencies f_1, f_2 is required, then the corresponding horizontal distances on the screen for one or more cycles are measured. Suppose these are d_1, d_2 respectively. Then, since $f \propto 1/T$,

$$\frac{f_1}{f_2} = \frac{T_2}{T_1} = \frac{d_2}{d_1}$$

3   Measurement of phase

The use of a double beam oscilloscope to measure phase difference between two a.c. voltages is given on p. 474. With the time-base switched off, in the single beam tube one input can be joined to the X-plates and the other to the Y-plates. We consider only the case when the frequencies of the two signals are the same. An ellipse will then be seen generally on the screen, as shown in Figure 31.17 (iii). This is a Lissajous figure (see p. 89).

The trace is centred, and peak vertical displacement y_2 at the middle O, and the peak vertical displacement y_1 of the ellipse, are then both measured. Suppose the x-displacement is given by $x = a \sin \omega t$, where a is the amplitude in the x-direction, and the y-displacement by $y = y_1 \sin(\omega t + \varphi)$, where y_1 is the amplitude in the y-direction and φ is the phase angle. When $x = 0$, $\sin \omega t = 0$, so that $\omega t = 0$. In this case, $y = y_2 = y_1 \sin \varphi$. Hence $\sin \varphi = y_2/y_1$, from which φ can be found. See also p. 89.

The frequencies f_x and f_y of two a.c. voltages can also be compared by connection to the X- and Y-inputs respectively of the oscilloscope. As shown in the oscilloscope display in Plate 31D, a Lissajous figure is obtained. Generally, if a horizontal line makes n_x intersections with the Lissajous figure and a vertical line makes n_y oscillations, then $f_x : f_y = n_y : n_x$. So if the frequency ratio $f_x : f_y = 2 : 1$, a figure of eight is obtained on the screen. In Plate 31D, the ratio $f_x : f_y = 1 : 3$, since $n_x = 6$ and $n_y = 2$. Here the beam moves horizontally through one cycle in the same time as it moves vertically through three cycles.

4 Use as clock

In measuring frequency, the time-base provides a value of a time interval. The time is needed in radar for calculating the distance of an aeroplane from a station or aerodrome.

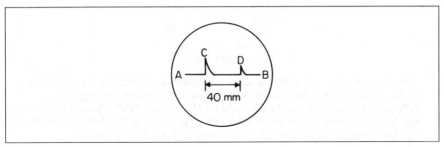

Figure 31.18 *C.R.O. as clock*

Figure 31.18 illustrates the principle. Using a time-base of 10 microseconds per millimetre ($10 \, \mu s \, mm^{-1}$), AB is the horizontal trace. A radar signal sent at C to a distant plane arrives back at D, 40 mm from C. So the time for the signal to travel to the plane and back $= 40 \times 10 \times 10^{-6} \, s = 4 \times 10^{-4} \, s$. Now a radio signal travels through space with a speed of $300\,000 \, km \, s^{-1}$ ($3 \times 10^{8} \, m \, s^{-1}$). So

$$\text{to-and-fro distance of plane} = 4 \times 10^{-4} \times 300\,000 = 120 \, km$$

and
$$\text{distance of plane} = \tfrac{1}{2} \times 120 = 60 \, km$$

_____ **Exercises 31B** _____

1 An oscilloscope is used to measure the time it takes to send a pulse of charge along a 200-m length of coaxial cable and back again. Figure 31D shows the appearance of the oscilloscope screen, A indicating the original pulse and B the same pulse after reflection.

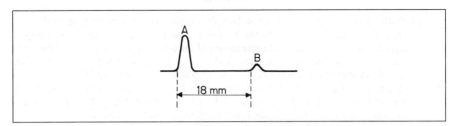

Figure 31D

   If the time base speed is set at $10 \, mm \, \mu s^{-1}$, calculate the speed of the pulse along the cable. (*L.*)

2 A cathode-ray oscilloscope has its Y-sensitivity set to $10 \, V \, cm^{-1}$. A sinusoidal input is suitably applied to give a steady trace with the time base set so that the electron beam takes 0·01 s to traverse the screen. If the trace seen has a total peak to peak height of 4·0 cm and contains 2 complete cycles, what is the r.m.s. voltage and frequency of the input signal? (*L.*)

3 When a sine-form voltage of frequency 1250 Hz is applied to the Y-plates of a cathode-ray oscilloscope the trace on the tube is as shown in Figure 31E (i).

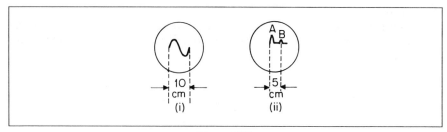

Figure 31E

If a radar transmitter sends out short pulses, and at the same time gives a voltage to the Y-plates of the oscilloscope, with the time-base setting unchanged, the deflection A is produced as shown in Figure 31E (ii). An object reflects the radar pulse which, when received at the transmitter and amplified, gives the deflection B. What is the distance of the object from the transmitter? (Speed of radar waves = $3 \times 10^8 \, \text{m s}^{-1}$.) (L.)

4 (a) Draw a clear labelled diagram showing the essential features of a single beam cathode-ray oscilloscope tube. Explain, without giving circuit details, how the brightness and focusing of the electron beam are controlled.

(b) What is the *time-base* in an oscilloscope? Sketch a graph showing the variation of time-base voltage with time.

(c) How would you use an oscilloscope, the time-base of which is not calibrated, to measure the frequency of a sinusoidal potential difference which is of the order of 50 Hz?

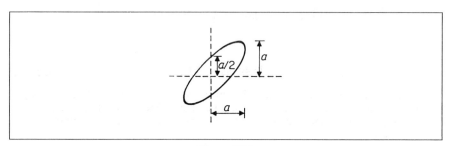

Figure 31F

(d) With the time-base disconnected, two alternating potential differences of the same frequency are applied to the X- and Y-plates respectively, the gains being equal. Figure 31F shows the appearance of the trace on the screen. The potentials may be represented by $x = a \sin \omega t$ and $y = a \sin (\omega t + \varphi)$. What is the value of φ, the phase angle between the potentials? Explain your reasoning. (L.)

5 (a) (i) Describe the layout of a cathode-ray tube that utilizes electrical deflection and focusing. Explain how the focusing and the brightness of the spot on the screen are controlled.   (ii) What is the function of the linear time-base in an oscilloscope?   (iii) Why is the tube of a cathode-ray oscilloscope always mounted inside a metal screen of very high magnetic permeability? Explain how the screen achieves its purpose.

(b) Explain in detail how you would use an oscilloscope: (i) to observe the waveform of the musical note emitted by a clarinet;   (ii) to check the frequency calibration of an audio sine-wave oscillator at any one chosen frequency (you may assume that the frequency of the mains supply is 50 Hz exactly);   (iii) to measure the r.m.s. value of a sinusoidal alternating current of about 0·2 A flowing in a 5·0 Ω resistor. (O.)

6 One sinusoidal voltage alternating at 50 Hz is connected across the X-plates of a cathode-ray oscilloscope and another 50 Hz sinusoidal alternating voltage of approximately the same amplitude is connected across the Y-plates. Sketch what

you would expect to observe on the screen if the phase difference between the voltages is

(a) zero,

(b) $\pi/2$,

(c) $\pi/4$.

If the voltage on the X-plates is replaced by a 100 Hz sinusoidal alternating voltage of similar amplitude sketch what you would observe on the screen.

Explain briefly why figures of this type are useful in the study of alternating voltages. (*JMB.*)

7 Explain what is meant by

(a) a linear time-base,

(b) a sinusoidal time-base, in a cathode-ray oscilloscope.

The X and Y deflection sensitivities of a cathode-ray oscilloscope are each 5 V cm^{-1}. A sinusoidal potential difference alternating at 50 Hz and of r.m.s. value 20 V is applied to the Y-plates of the instrument. A potential difference of the same form and frequency but of r.m.s. value 10 V is simultaneously applied to the X-plates. Sketch and explain the pattern seen on the oscilloscope when the potential differences are

(a) in phase,

(b) 90° out of phase.

Indicate the appropriate dimensions on your sketches. (*JMB.*)

32

Junction Diode. Transistor and Applications

In this chapter, we consider semiconductors and their applications in junction diodes and transistors. We begin with the different kinds of energy bands in solids with special reference to semiconductors such as silicon and then discuss the electron and hole carriers and their action in junction diodes and transistors. Rectifier circuits, amplifier circuits and switching circuits are then described.

Energy Bands in Solids

As we see later, the allowed energy levels in a single atom are discrete (separate) and spaced widely apart. In the solid state, however, as in a crystal, large numbers of atoms are packed closely together, and the electrons are influenced strongly by the assembly of nuclei. The allowable energy levels then broaden into *bands* of energy, Figure 32.1 (i). The bands contain allowable energy levels very close to each other, as at P. There may also be *forbidden bands* of energy, as at Q,

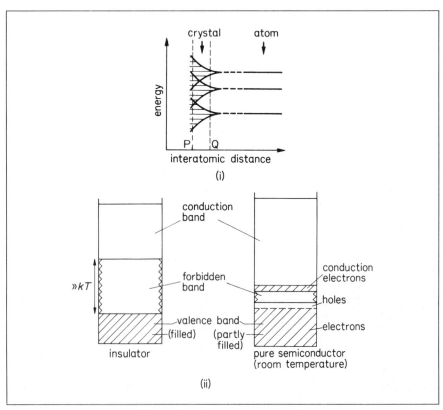

Figure 32.1 *Energy bands in solids*

which electrons cannot occupy. The lowest available energy band is called the *valence band*. The next available energy band is called the *conduction band*.

In an *insulator*, the valence band energy levels are completely filled by electrons. The conduction band is empty and the two bands are separated by a wide energy gap called the 'forbidden' band, Figure 32.1 (ii). This is much greater than kT in magnitude where k is the Boltzmann constant (p. 687). The electrons in the valence band, like molecules in a gas, have thermal energy of the order kT but at room temperature they cannot gain sufficient energy from an applied p.d. to move to higher unoccupied energy levels. So the material is an insulator.

Semiconductors are a class of materials with a *narrow* forbidden band between the valence and conduction bands; the energy gap is of the order kT. At 0 K, all the energy levels in the valence band are occupied and the material is then an insulator. At normal temperatures, however, the thermal energy of some valence electrons, of the order kT, is sufficient for them to reach the conduction band, where they may become conduction electrons. The gap left in the valence band of energies by the movement of an electron is called a *hole*, Figure 32.1 (ii). In semiconductor theory, both holes and conduction electrons play an active part, as we soon see.

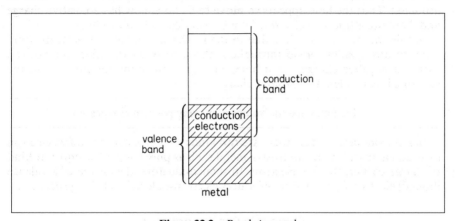

Figure 32.2 *Bands in metal*

In *metals*, however, the valence and conduction bands can overlap, as shown diagrammatically in Figure 32.2. The electrons in the overlapping region of energy are conduction electrons. Since there is a large number of conduction electrons, metals are good conductors.

Semiconductors, Electron and Hole Charge Carriers

Semiconductors have a resistivity about ten million times higher than that of a good conductor such as copper. Silicon and germanium are examples of semiconductor elements widely used in the electronics industry.

Silicon and germanium atoms are tetravalent. They have four electrons in their outermost shell, called *valence electrons*. One valence electron is shared with each of four surrounding atoms in a tetrahedral arrangement, forming 'covalent bonds' which maintain the crystalline solid structure. Figure 32.3 (i) is a two-dimensional representation of the structure, showing diagrammatically four silicon atoms, each having four valence electrons round them.

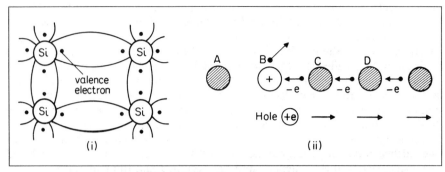

Figure 32.3 *Electrons and holes*

At 0 K, all the valence electrons are firmly bound to the nucleus of their particular atoms. At room temperature, however, the thermal energy of a valence electron may become greater than the energy binding to its nucleus. The covalent bond is then broken. The electron leaves the atom, B say, and becomes a *free electron*. This leaves B with a vacancy or *hole*, Figure 32.3 (ii). Since B now has a net positive charge, an electron in a neighbouring atom C may then be attracted. Thus the hole appears to move to C. So C now has a positive charge and therefore attracts an electron from D. This leaves D with a hole.

In this way we can see that, due to the movement of valence electrons from atom to atom. *holes* spread throughout the semiconductor. Since an electron carries a negative charge $-e$, a hole, moving in the opposite direction to the electron, is equivalent to a *positive* charge $+e$. So

---

moving holes are equivalent to moving positive charges $+e$

In semiconductors, then, there are *two* kinds of charge carriers: a free electron $(-e)$ and a hole $(+e)$. In contrast, a metal such as pure copper has only one kind of charge carrier, the free electron. In semiconductors, the escape of a valence (bound) electron from an atom produces electron-hole pairs of charge carriers.

Current in Semiconductor

The free electrons and holes move about randomly in different directions inside the semiconductor. But when a battery is connected to the semiconductor, the $+$ve charges or holes drift through the metal in opposite directions to the $-$ve charges or electrons. The small current I obtained is carried by both holes and electrons. Figure 32.4 (i).

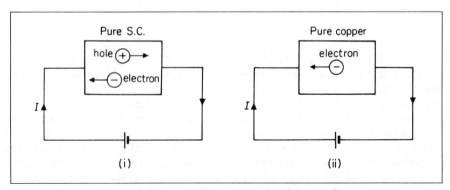

Figure 32.4 *Current in pure (intrinsic) semiconductor and pure copper*

In a pure metal such as copper, the current I is given by $I = nevA$, where n is the number of electrons per metre^3, A is the cross-sectional area in m^2 and v is the drift velocity of the electrons in $m\,s^{-1}$ (see p. 243), Figure 32.4 (ii). In a pure semiconductor, called an *intrinsic semiconductor*, there are equal numbers n of electrons and holes. The drift velocity v_n of the electrons or negative charges is actually greater than that v_p of the holes or positive charges. Taking both charges carriers into account, the current I is given by

$$I = nev_nA + nev_pA = neA(v_n + v_p)$$

In a pure (intrinsic) semiconductor:
1 there are two kinds of charge carriers, holes ($+e$) and electrons ($-e$)
2 the number of holes = the number of electrons
3 the drift velocity of the electrons is greater than that of the holes

Effect of Temperature Rise

As we have seen, the charge carriers in a metal such as copper are only free electrons. As the temperature of the metal rises, the amplitude of vibration of the atoms increases and more 'collisions' with atoms are then made by the drifting electrons. So the resistance of a pure metal increases with temperature rise (p. 243).

In the case of a semiconductor, however, the increase in thermal energy of the valence electrons due to temperature rise enables more of them to break the covalent bonds and become free electrons. Thus *more electron-hole pairs are produced* which can act as carriers of current. Hence, in contrast to a pure metal, the electrical resistance of a pure semiconductor *decreases* with temperature rise. This is one way of distinguishing between a pure metal and a pure semiconductor. Note that the pure or *intrinsic* semiconductor always has equal numbers of electrons and holes, whatever its temperature.

N-Semiconductors

A pure or intrinsic semiconductor has charge carriers which are thermally generated. These are relatively few in number. By 'doping' a semiconductor with a tiny amount of impurity such as one part in a million, forming a so-called *extrinsic* or impure semiconductor, an enormous increase can be made in the number of charge carriers.

Arsenic atoms, for example, have five electrons in their outermost or valence band. When an atom of arsenic is added to a silicon crystal, the atom settles in a lattice site with four of its electrons shared with neighbouring silicon atoms. The fifth electron may thus become free to wander through the crystal. Since an impurity atom may provide one free electron, an enormous increase occurs in the number of charge carriers. For example, 1 milligram of arsenic has about 8×10^{18} atoms and so provides this large number of free electrons in addition to the relatively small number of thermally-generated electrons.

Since there are a great number of negative (electron) charge carriers, the impure semiconductor is called an 'n-type semiconductor' or *n-semiconductor*, where 'n' represents the negative charge on an electron. Thus the *majority carriers* in an n-semiconductor are electrons. Positive charges or holes are also present in the n-semiconductor, Figure 32.5 (i). These are thermally generated, as previously explained, and since they are relatively few they are called the *minority carriers*. The impurity (arsenic) atoms are called *donors* because they donate electrons as carriers.

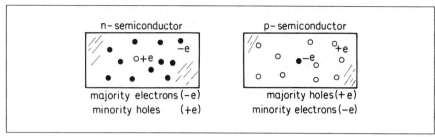

Figure 32.5   *N- and P-semiconductors*

The conductivity σ of an n-semiconductor increases with temperature from very low temperatures to about 250 K as more electron-hole pairs are formed. But from 250 K to about 650 K, σ decreases with temperature rise. In this temperature range the increase in electron-hole pairs is more than counteracted by the increase in the metal lattice vibration, which impedes the motion of the carriers.

P-Semiconductors

P-semiconductors are made by adding foreign atoms which are *trivalent* to pure germanium or silicon. Examples are boron or indium. In this case the reverse happens to that previously described. Each trivalent atom at a lattice site attracts an electron from a neighbouring atom, thereby completing the four valence bonds and *forming a hole* in the neighbouring atom. In this way an enormous increase occurs in the number of holes. Thus in a p-semiconductor, the majority carriers are holes or positive charges. The minority carriers are electrons, negative charges, which are thermally generated, Figure 32.5 (ii). The impurity atoms are called *acceptors* in this case because each 'accepts' an electron when the atom is introduced into the crystal.

Summarising: In an n-semiconductor, conduction is due mainly to negative charges or electrons, with positive charges (holes) as minority carriers. In a p-semiconductor, conduction is due mainly to positive charges or holes, with negative charges (electrons) as minority carriers.

P-N Junction

By a special manufacturing process, p- and n-semiconductors can be melted so that a boundary or *junction* is formed between them. This junction is extremely thin and of the order 10^{-3} mm. It is called a *p-n junction*, Figure 32.6 (i).

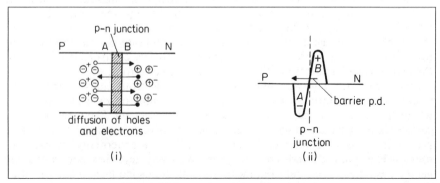

Figure 32.6   *n-p junction and barrier p.d.*

When a scent bottle is opened, the high concentration of scent molecules in the bottle causes the molecules to diffuse into the air. In the same way, the high concentration of holes (positive charges) on one side of a p-n junction, and the high concentration of electrons on the other side, causes the two carriers to diffuse respectively to the other side of the junction, as shown diagrammatically in Figure 32.6 (i). The electrons which move to the p-semiconductor side recombine with holes there. These holes therefore disappear and an excess negative charge A appears on this side, as shown in Figure 32.6 (ii).

In a similar way, an excess positive charge B builds up in the n-semiconductor when holes diffuse across the junction. Together with the negative charge A on the p-side, an e.m.f. or p.d. is produced which *opposes* more diffusion of charges across the junction. This is called a *barrier p.d.* and when the flow ceases it has a magnitude of a few tenths of a volt. The narrow region or layer at the p-n junction which contains the negative and positive charges is called the *depletion layer*. The width of the depletion layer is of the order 10^{-3} mm.

Junction Diode as Rectifier

When a battery B, with an e.m.f. greater than the barrier p.d., is joined with its positive pole to the p-semiconductor, P, and its negative pole to the n-semiconductor, N, p-charges (holes) are urged across the p-n junction from P to N and n-charges (electrons) from N to P, Figure 32.7 (i).

We can understand the movement if we consider the $+$ve pole of the battery to repel $+$ve charges (holes) in the p-semiconductor and the $-$ve pole to repel the $-$ve charges (electrons) in the n-semiconductor. Thus an appreciable current OA is obtained. The p-n junction is now said to be *forward-biased*, and when the applied p.d. is increased, the current increases along the curve OA.

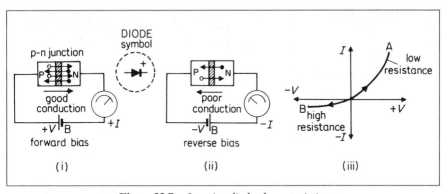

Figure 32.7 *Junction diode characteristics*

When the poles of the battery are reversed, only a very small current flows, Figure 32.7 (ii). In this case the p-n junction is said to be *reverse-biased*. This time only the *minority carriers*, negative charges in the p-semiconductor and positive charges in the n-semiconductor, are urged across the p-n junction by the battery. Since the minority carriers are thermally-generated, the magnitude of the reverse current OB depends only on the *temperature* of the semiconductors. It may also be noted that the reverse-bias p.d. increases the width of the depletion layer, since it urges more electrons in the p-semiconductor and holes in the n-semiconductor further away from the junction.

The characteristic **(I-V)** curve **AOB** in **Figure 32.7 (iii)** shows that the p-n junction acts as a *rectifier*. It has a low resistance for one direction of p.d., **+ V**, and a high resistance in the opposite p.d. direction, **− V.**

It is called a *junction diode*. In the diode symbol in Figure 32.7, the low resistance is from left to right (towards the triangle point) and the high resistance is in the opposite direction. The junction diode has several advantages, for example, it needs only a low voltage battery B to work; it does not need time to warm up; it is not bulky; and it is cheap to manufacture in large numbers.

Full Wave Rectification, Bridge Circuit

Figure 32.8 (i) shows how four diodes, two D1 and two D2, can be arranged in a so-called *bridge circuit* to measure alternating current using a moving-coil meter M.

On the *positive* half of the a.c. cycle of V, suppose the p.d. has its +ve side connected to R. The current then flows through the low-resistance path RSQP back to V. On the *negative* half of the same a.c. cycle, the +ve side of the p.d. is now joined to P. So the current now flows through the low-resistance path PSQR back to V.

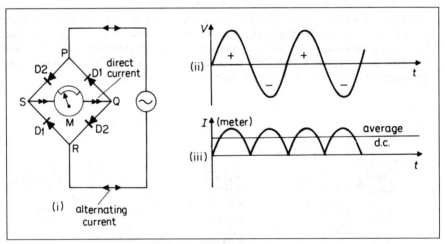

Figure 32.8 *Bridge rectifier circuit -a.c. measurement*

So on *both* halves of the cycle, the current flows the same way through the meter M. Figure 32.8 (ii) shows the alternating p.d. V and Figure 32.8 (iii) the varying current through M, which has an average positive or direct current (d.c.) value. So the meter registers a deflection, and the scale is calibrated in r.m.s. values of current or p.d. The r.m.s. value = $1 \cdot 1 \times$ average value in meter M.

Full Wave A.C. Voltage Rectification by Bridge Circuit

In a.c. mains transistor receivers, diodes are used to rectify the alternating mains voltage and to produce steady of d.c. voltage needed for the circuits in the receiver.

Figure 32.9 shows a *bridge circuit* which produces full-wave rectification four diodes are used, and the action is similar to that described for the measuring a.c.

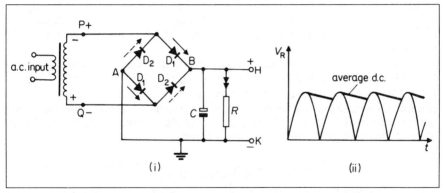

Figure 32.9 *Bridge rectifier circuit—d.c. voltage from a.c. mains*

with a moving-coil meter. Using a transformer, the mains a.c. voltage between P and Q is applied to the circuit.

On one half of a cycle, when P is +ve relative to Q, only the diodes D_1 conduct. On the other half of the same cycle, only the diodes D_2 conduct. In both cases the current goes through the resistor R in the *same* direction. Figure 32.9 (ii) shows the varying d.c. voltage across R. Small gaps may appear between half-waves as diodes need about 0·7 V before they conduct, so lowering the average voltage.

The large (electrolytic) capacitor C is used to produce a large and fairly steady d.c. voltage. C charges up to the *peak* value of the applied a.c. voltage and then discharges through R when this voltage begins to decrease. If the discharge takes place slowly (which depends on the time-constant CR) then C becomes re-charged to the peak value of the voltage again. The charge-discharge process takes place continuously and so the voltage across the C-R combination is a 'ripple' voltage as shown diagrammatically in Figure 32.9 (ii). With R alone, the average d.c. voltage would only be about two-thirds of the peak voltage. With the C-R arrangement, the average d.c. voltage is nearly equal to the peak voltage value, which is better, and so this removes (filters) the voltage variation.

Zener Diode

When the reverse bias or p.d. is increased across a p-n junction, a large increase in current is suddenly obtained at a voltage Z, Figure 32.10 (i). This is called the *Zener effect*, after the discoverer. It is partly due to the high electric field which

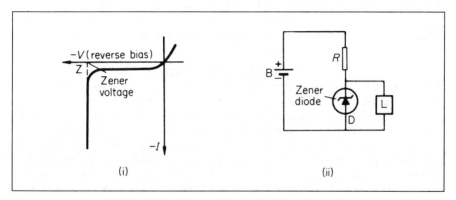

Figure 32.10 *Zener diode and voltage regulation*

exists across the narrow p-n junction at the breakdown or Zener voltage Z, which drags more electrons from their atoms and thus increases considerably the number of electron-hole pairs. Ionisation by collision also contributes to the increase in carriers.

Zener diodes are used as voltage regulators or stabilisers in circuits. In Figure 32.10(ii), a suitable diode D is placed across a circuit L. Although the battery supply B may fluctuate, and produce changes of current in L and D, if R is suitably chosen, the voltage across D remains practically constant over a reverse current range of tens of milliamperes at the Zener voltage shown in Figure 32.10(i). The voltage across L thus remains stable.

The Transistor

The junction diode is a component which can only rectify. The *transistor* is a more useful component; it is a *current amplifier*. A transistor is made from three layers of p- and n-semiconductors. They are called respectively the *emitter* (E), *base* (B) and *collector* (C). Figure 32.11(i) illustrates a *p-n-p transistor*, with electrodes connected to each of the three layers. In a *n-p-n transistor*, the emitter is n-type, the base is p-type and the collector is n-type, Figure 32.11(ii). The base is deliberately made very thin in manufacture. The transistor is thus a three-terminal device.

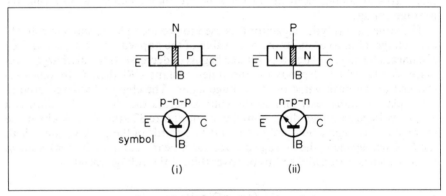

Figure 32.11   *Transistors and symbols*

Figure 32.11 shows the circuit symbols for n-p-n and p-n-p transistors. The arrows show the directions of conventional current (+ve charge or hole movement) between the emitter E and base B, so electrons would flow in the opposite direction. In an actual transistor, the collector terminal is displaced more than the others for recognition or has a dot near it.

Current Flow in Transistors

Figure 32.12(i) shows batteries correctly connected to a p-n-p transistor. The emitter-base is forward-biased; the collector-base is *reverse*-biased; and the base is common. This is called the *common-base* (C-B) mode of using a transistor. Note carefully the polarities of the two batteries. The positive pole of the supply voltage X is joined to the emitter E but the *negative* pole of the supply voltage Y is joined to the collector C. In the case of a n-p-n transistor, the negative pole of one battery is joined to the emitter and the positive pole of the other is joined to the collector, Figure 32.12(ii). In this way the emitter-base is forward-biased and the collector-base is reverse-biased.

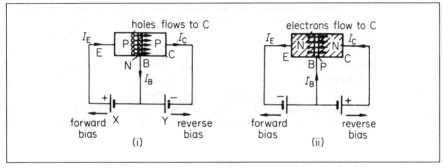

Figure 32.12   *Transistor action*

Consider Figure 32.12 (i). Here the emitter-base is forward biased by X, so that positive charges or holes flow across the junction from E to the base B. The base is so thin, however, that the great majority of the holes are urged across the base to the collector by the battery Y. Thus a current I_C flows in the collector circuit. The remainder of the holes combine with the electrons in the n-base, and this is balanced by electron flow in the base circuit, so a small current I_B is obtained here. From Kirchhoff's first law, it follows that always, if I_E is the emitter current,

$$I_E = I_C + I_B$$

Typical values for a.f. amplifier transistors are: $I_E = 1 \cdot 0 \, mA$, $I_C = 0 \cdot 98 \, mA$, $I_B = 0 \cdot 02 \, mA$.

Although the action of n-p-n transistors are similar in principle to p-n-p transistors the carriers of the current in the n-p-n transistor are mainly electrons but holes in the p-n-p transistor. Electrons are more speedy carriers than holes (p. 789). So n-p-n transistors are used in high-frequency and computer circuits, where the carriers are required to respond very quickly to signals.

Common-Emitter (C-E) Characteristics

In general, the transistor has an *input circuit*, for the voltage or current to be changed, and an *output circuit* for the amplified voltage or current, for example. In Figure 32.12, the emitter-base (EB) is the input circuit and the collector-base (CB) is the output circuit. This is called the *common-base* mode of using the transistor because the base is common to the input and output circuits. The transistor can be used in the *common-collector* or *common-emitter* mode, as we see later. Firstly, however, we need the characteristics of the common-emitter circuit, a circuit widely used for amplification. The characteristics help to estimate the performance of the common-emitter amplifier.

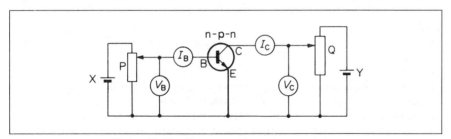

Figure 32.13   *Common-emitter characteristics*

Figure 32.13 shows a circuit for obtaining the characteristics of a n-p-n transistor in the common-emitter mode. X and Y may be batteries of 1·5 V and 4·5 V respectively, connected to rheostats P and Q of 1 kΩ and 5 kΩ which act as potential dividers. This enables the base-emitter p.d., V_{BE} or V_B, and the collector-emitter p.d., V_{CE} or V_C to be varied. The p.d. is measured by high resistance voltmeters, capable of measuring p.d. in steps such as 50 mV. The meter for base current, I_B, should be a microammeter and for the collector current, I_C, a milliammeter. Typical results are shown in Figure 32.14 (i), (ii) and (iii). The input circuit is the *base-emitter* circuit. The output circuit is the *collector-emitter* circuit.

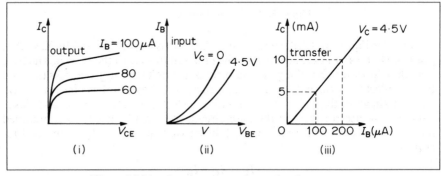

Figure 32.14 *Output, input and transfer characteristics*

Output characteristic ($I_C - V_C$, with I_B constant). The 'knee' of the curves shown in Figure 32.14 (i) correspond to a low p.d. of the order of 0·2 V. For higher p.d. the output current I_C varies *linearly* with V_{CE} or V_C for a given base current. The *linear* part of the characteristic is used in a.f. amplifier circuits, so that the output voltage variation is undistorted.

The *output resistance* r_o is defined as $\Delta V_C / \Delta I_C$, where the changes take place on the straight part of the characteristic. r_o is an a.c. resistance; it is the effective resistance in the output circuit for an a.c. signal input. It should be distinguished from the d.c. resistance, V_C / I_C, which is not required.

The small gradient of the straight part of the characteristic shows that r_o is high. For example, suppose $\Delta V_C = 2\,V$ and $\Delta I_C = 0·02\,mA = 2 \times 10^{-5}\,A$. Then $r_o = 2/(2 \times 10^{-5}) = 100\,000\,\Omega$. If a varying resistance load is used in the output or collector circuit, the high value of r_o relative to the load shows that the output current is fairly constant. So the output voltage is proportional to the load resistance.

Transfer characteristic ($I_C - I_B$, V_C constant). The output current I_C varies fairly linearly with the input current I_B, Figure 32.14 (iii). The current *transfer ratio* β, or *current gain*, is defined as the ratio $\Delta I_C / \Delta I_B$ under a.c. signal conditions. It should be distinguished from the d.c. current gain, I_C / I_B. From Figure 32.14 (ii),

$$\beta = \frac{(10-5)\,mA}{(200-100)\,\mu A} = 50$$

β, current gain, is also written h_{fe}, where '*fe*' is 'forward emitter'.

Input characteristic ($I_B - V_B$, V_C constant). The *input resistance* r_i is defined as the ratio $\Delta V_B / \Delta I_B$. As the input characteristic in Figure 32.14 (ii) is non-linear, then r_i varies. At any point of the curve, r_i is equal to the gradient of the tangent to the curve and is of the order of kilohms.

Current Amplification in C-E Mode

In general, the magnitude of β, $\Delta I_C/\Delta I_B$, for the common-emitter circuit is high, from about 20 to 500 for many transistors. So a small change in the base or input current can produce a large change in the collector or output current.

We can obtain a rough value for β by assuming that when electrons are emitted from the n-emitter towards the p-base, a constant fraction α reaches the n-collector where α is typically 0·98 (see p. 795). Thus $I_C = \alpha I_E$. Now from p. 795, we always have that

$$I_E = I_C + I_B$$

Substituting $I_E = I_C/\alpha$ and simplifying, we obtain

$$I_C = \left(\frac{\alpha}{1-\alpha}\right) I_B$$

$$\therefore \Delta I_C = \left(\frac{\alpha}{1-\alpha}\right) \Delta I_B$$

$$\therefore \frac{\Delta I_C}{\Delta I_B} = \beta = \frac{\alpha}{1-\alpha} = \frac{0·98}{0·02} = 49$$

In practice, the current gain β of a transistor may be much higher.

Voltage Amplification and Power Gain

As we have just seen, the transistor in the common-emitter mode is a current amplifier. To change the output (a.c.) current to a voltage V_o, a resistance load R must be used in the collector or output circuit. Figure 32.15 shows also diagrammatically the base bias (steady voltage) necessary for no distortion of V_o (p. 798).

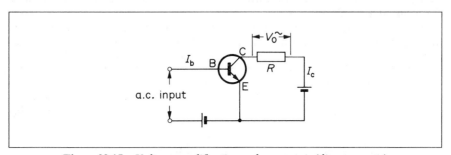

Figure 32.15 _Voltage amplification and power gain (diagrammatic)_

We can illustrate the voltage amplification in the output circuit by supposing that $R = 5\,\text{k}\Omega$, the input resistance $r_i = 2\,\text{k}\Omega$, the input (a.c.) voltage is 10 mV or 0·01 V peak value, and the current gain $\beta = 50$.

The peak a.c. current I_b flowing in the base circuit is, from $I = V/R$,

$$I_b = \frac{0·01\,\text{V}}{2000\,\Omega} = 5 \times 10^{-6}\,\text{A}$$

$$\therefore I_c = \beta I_b = 50 \times 5 \times 10^{-6} = 2·5 \times 10^{-4}\,\text{A}$$

$$\therefore V_o = I_c R = 2·5 \times 10^{-4} \times 5000$$

$$= 1·25\,\text{V peak}$$

$$\therefore \text{ voltage amplification } A_V = \frac{V_o}{V_i} = \frac{1 \cdot 25}{0 \cdot 01} = 125$$

Also, power gain = current gain × voltage gain = $50 \times 125 = 6250$

Note that the source of power gain is the battery or voltage supply.

C-E Amplifier Circuit

Figure 32.16 shows a n-p-n transistor in a simple C-E amplifier circuit. It uses one battery supply, V_{CC}. A load, R_L, is placed in the collector or output circuit. A resistor R provides the necessary bias, V_{BE}, for the base-emitter circuit. The base-emitter is then forward-biased but the collector-base is reverse-biased, that is, the potential of B is positive relative to E but negative relative to C.

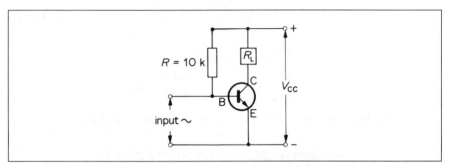

Figure 32.16 *Simple amplifier circuit*

As we show soon, the common-emitter circuit is very sensitive to temperature changes. So if no arrangement is made for *temperature stabilisation*, the output would become distorted when the temperature changed. Silicon transistors are much less sensitive to temperature change than germanium transistors and are hence used more widely.

In practice, Figure 32.16 is unsuitable as an amplifier circuit since there is no arrangement for temperature stabilisation (p. 799). A more reliable C-E a.f.

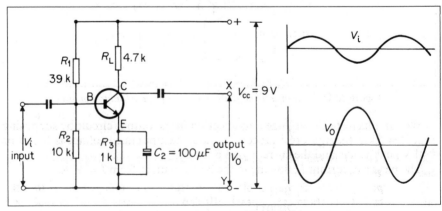

Figure 32.17 *Amplifier circuit*

amplifier circuit is shown in Figure 32.17. Its principal features are: (i) a potential divider arrangement, R_1, R_2, which provides the necessary base-bias; (ii) a load R_L which produces the output across X, Y; (iii) a capacitor C_1

which stops the d.c. component in the input signal entering the circuit; (iv) a large capacitor C_2 across a resistor R_3, which prevents undesirable feed-back of the amplified signal to the base-emitter circuit; (v) an emitter resistance R_3, which stabilises the circuit for excessive temperature rise. Thus if the collector current rises, the current through R_3 increases. This lowers the p.d. between E and B, so that the collector current is automatically lowered.

Effect of Temperature Rise on C-E Circuit

We now explain why the common-emitter circuit is sensitive to temperature change.

When the base current I_B is zero, some current still flows in the collector circuit in the common-emitter arrangement. This is due to the minority carriers present in the collector-base part of the transistor, which is reverse-biased. The collector current when I_B is zero is denoted by I_{CE0} and is called the *leakage current*. Since the minority carriers are thermally generated, the leakage current depends on the temperature of the transistor.

In the common-base arrangement in Figure 32.12, the leakage current obtained when I_E is zero is denoted by I_{CB0}. This is also due to minority carriers in the collector-base, which is reverse-biased. Thus, more accurately, we should write in place of $I_C = \alpha I_E$ (p. 797)

$$I_C = \alpha I_E + I_{CB0} \qquad \cdot \qquad \cdot \qquad \cdot \qquad \cdot \qquad \cdot \qquad (1)$$

Now from p. 797, $\alpha/(\alpha-1) = \beta$, so that $\alpha = \beta/(\beta+1)$. Further, from previous, $I_E = I_C + I_B$. Substituting in (1) for α and for I_E and simplifying, we obtain

$$I_C = \beta I_B + (\beta+1)I_{CB0} \qquad \cdot \qquad \cdot \qquad \cdot \qquad \cdot \qquad (2)$$

Thus the leakage current I_{CE0} in the common-emitter circuit $= (\beta+1)I_{CB0}$. The current I_{CE0} also flows in the base-emitter circuit when the transistor is operating. A temperature change from 25°C to 45°C, which would increase the current I_{CB0} by $10\,\mu A$ say, would then produce a current $I_{CE0} = (\beta+1)I_{CB0} = 50 \times 10\,\mu A = 0.5\,mA$, assuming $\beta = 49$. This is the increase in collector current if the temperature rose to 45°C from 25°C. The relatively large current change would have an appreciable effect on the output in the collector circuit; for example, it could lead to a distorted output in an a.f. amplifier circuit.

On this account the C-E circuit, which is very sensitive to temperature change, *must* be stabilised for excessive temperature rise.

C-E Amplifier, Phase Change

Figure 32.18 (i) shows a basic amplifier circuit with a n-p-n transistor, a base resistance R_B of $50\,k\Omega$ to limit the base current to a suitable value, a load resistance R_L of $4\,k\Omega$ in the collector circuit, and a supply voltage V_{CC} of 9 V.

The *input circuit* is the whole base-emitter circuit. So the input voltage V_i is connected to this circuit.

The *output circuit* is the whole collector-emitter circuit. Note carefully that the output voltage V_0 is always the voltage *between the collector C and the emitter E terminals* or earth. This is the voltage which is passed to another transistor circuit if this is required. So you should remember that $V_0 = V_{CE}$, as shown in Figure 32.18 (i).

The potential difference or 'potential drop' across the resistance $R_L = I_C R_L$.

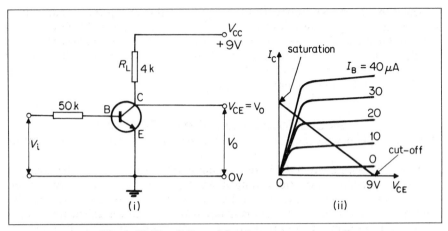

Figure 32.18   *C-E amplifier. Saturation and cut-off*

Since V_{CC} is the supply voltage, the output voltage V_o is

$$V_o = V_{CE} = V_{CC} - I_C R_L \qquad . \qquad . \qquad . \qquad . \qquad (1)$$

Suppose $I_C = 2\,\text{mA}$, $R_L = 4\,\text{k}\Omega = 4000\,\Omega$, then potential drop $= 2 \times 10^{-3} \times 4 \times 10^3 = 8\,\text{V}$. If $V_{CC} = 9\,\text{V}$, then $V_o = 9 - 8 = 1\,\text{V}$.

From equation (1), we also see that when I_C increases because the input voltage V_i increases, the output voltage V_o *decreases*. Similarly, if I_C decreases because V_i decreases, then V_o *increases*. So the input and output voltages are 180° out of phase or in antiphase.

D.C. Amplifier

When the transistor is used as a switch in computer circuits, the input voltage V_i is a d.c. voltage. Suppose V_i is $2 \cdot 0\,\text{V}$ in Figure 32.18(i) and the amplification factor or d.c. gain is 50 for the transistor. Neglecting the relatively small base-emitter resistance compared with the large base resistor R_B of 50 kΩ,

$$\text{base current, } I_B = V_i/R_B = 2\cdot0/(50 \times 10^3) = 4 \times 10^{-5}\,\text{A} \qquad . \qquad (1)$$

$$\text{So collector current, } I_C = 50 \times I_B = 50 \times 4 \times 10^{-5} = 2 \times 10^{-3}\,\text{A}. \qquad (2)$$

To find the output voltage V_o or V_{CE}, we next find the potential drop V across the load resistance R_L of 4 kΩ. Then

$$V = I_C R_L = 2 \times 10^{-3} \times 4000 = 8\,\text{V} \qquad . \qquad . \qquad (3)$$

Since the supply voltage V_{CC} is 9 V, $V_o = 9 - 8 = 1\,\text{V}$. So an input V_i from 0 to 2 V produces an output switch from 9 V (when $V_i = 0$) to 1 V. In practice, the base-emitter voltage is also taken into account.

Saturation and Cut-off

The *maximum* or *saturation current* I_S in the output or collector circuit will be obtained when the output voltage is zero.

A potential drop of about 9 V across the load resistance R_L of 4 kΩ will make V_o or V_{CE} zero because the supply voltage V_{CC} is 9 V. So

$$\text{saturation current } I_S = V/R = 9/4000 = 2\cdot25 \times 10^{-3}\,\text{A} = 2\cdot25\,\text{mA}$$

Although the base current I_B may rise, the collector current remains at its saturation value.

If no current flows in the collector circuit, that is, $I_C = 0$, then we have *cut-off*. In this case there is no potential drop across R_L. So $V_o = V_{CE} = 9\text{ V}$.

At cut-off, $I_B = 0$. Figure 32.18 (ii) illustrates saturation and cut-off values. The small collector current at $I_B = 0$ is due to minority carriers in the transistor.

Examples on D.C. and A.C. Transistor Amplification

1 In Figure 32.19, the transistor is just saturated with an input d.c. voltage V_i. If the gain is 100, calculate V_i. Neglect the base-emitter resistance and p.d.

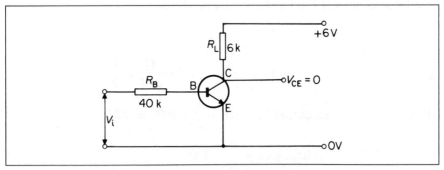

Figure 32.19 *Calculation on transistor*

Since the transistor is just saturated, the output voltage $V_{CE} = 0$. So potential drop across $R_L = 6\text{ V}$.

$$\therefore I_C = 6/6000 = 1/1000 = 1\text{ mA}$$

$$\therefore I_B = I_C/100 = 1\text{ mA}/100 = 1 \times 10^{-5}\text{ A}$$

Now
$$I_B = V_i/R_B$$

$$\therefore V_i = I_B \times R_B = 1 \times 10^{-5} \times 40\,000 = 0\cdot4\text{ V}$$

2 Figure 32.20 shows a simple form of silicon common-emitter amplifier. When the collector-emitter voltage V_{CE} is between $+1\text{ V}$ and $+9\text{ V}$, the collector current is about 100 times the base current and the base-emitter voltage V_{BE} is about $0\cdot7\text{ V}$.

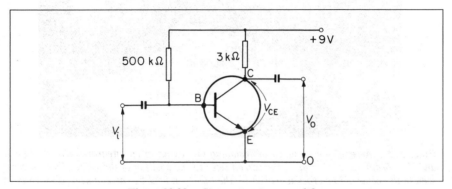

Figure 32.20 *Common-emitter amplifier*

Calculate (i) the base current I_B and the voltage V_{CE} in the circuit, (ii) the voltage gain if the input (base-emitter) a.c. resistance is $2000\,\Omega$, (iii) the largest (limiting) peak value of the input a.c. voltage if V_{CE} varies between $2\cdot2$ V and $8\cdot2$ V for linear amplification

(i) I_B. Since $V_{BE} = 0\cdot7$ V, p.d. across $500\,\mathrm{k}\Omega$ resistor $= 9 - 0\cdot7 = 8\cdot3$ V

So
$$I_B = \frac{8\cdot3}{5 \times 10^5} = 1\cdot66 \times 10^{-5}\,\mathrm{A} = 17\,\mu\mathrm{A}\ (\text{approx.})$$

V_{CE}. So $I_C = 100\,I_B = 100 \times 1\cdot66 \times 10^{-5}\,\mathrm{A} = 1\cdot66 \times 10^{-3}\,\mathrm{A}$

Thus
p.d. across $3\,\mathrm{k}\Omega = I_c R = 1\cdot66 \times 10^{-3} \times 3000 = 4\cdot98$ V

So
$$V_{CE} = 9\,\mathrm{V} - 4\cdot98\,\mathrm{V} = 4\,\mathrm{V}\ (\text{approx.})$$

(ii) *Voltage gain*. If V_i is input, a.c. base current is
$$I_b = \frac{V_i}{R_b} = \frac{V_i}{2000}$$

So
a.c. output $I_c = 100\,I_b = 100 \times \dfrac{V_i}{2000} = \dfrac{V_i}{20}$

Thus
a.c. output $V_o = I_c R = \dfrac{V_i}{20} \times 3000 = 150\,V_i$

So
voltage gain $= \dfrac{V_o}{V_i} = 150$

Plate 32A *An engineer engaged in designing the layout of an integrated circuit. The X and Y co-ordinates at the top of the screen refer to the position of the white tracking cross seen in the centre. The cross is moved by the light pen held by the engineer.* (Courtesy of Mullard Limited)

(iii) *Largest input voltage.* Since V_{CE} varies between 2·2 V and 8·2 V, peak value of voltage across collector load 3 kΩ is

$$V_o = \tfrac{1}{2}(8\cdot2 - 2\cdot2) = 3\text{ V}$$

Since gain = 150, peak value of largest input voltage V_i is given by

$$V_i = \frac{3\text{ V}}{150} = \frac{1}{50}\text{ V} = 20\text{ mV}$$

Transistor Switch

In addition to its use as a current amplifier, the transistor can be used as a *switch* in computer circuits. Millions of switching operations are needed daily in working computers, so swift switches are required. On this account n-p-n transistors are preferred. Here the charge carriers are mainly electrons, which have a much greater speed for a given voltage than holes or p-charges.

The basic circuit, shown in Figure 32.21 (i), consists of a n-p-n transistor connected in the common-emitter mode, with a resistance R in the output or collector circuit as we have already discussed.

A typical output voltage (V_o)-input voltage (V_i) characteristic of the circuit is shown in Figure 32.21 (ii). At very low input voltages the output voltage is practically +6 V, the supply voltage V_{CC} for the circuit. At input voltages of more than a fraction of a volt, however, the output voltage is nearly zero. This is explained shortly.

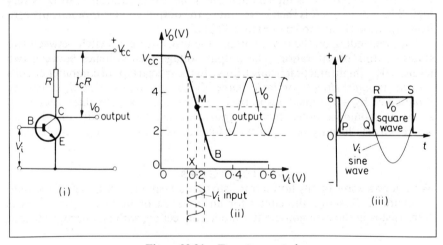

Figure 32.21   *Transistor switch*

Sine Wave Input, Amplifier Use

This type of transistor can be used
(a) to change a sine wave to a square wave and
(b) to amplify a.c. voltages.

Suppose the input V_i is a *sine wave* voltage of peak value 6 V, Figure 32.21 (iii). For a large part of the +ve half cycle, V_i will be greater than +0·4 V. So the output voltage V_o will be practically zero along PQ. When V_i is less than +0·4 V and negative on the −ve half of the cycle, V_o will be practically a constant high voltage RS as shown. So the output V_o is roughly a *square wave* voltage.

To use the transistor as an *amplifier*, the output voltage V_o must have the same waveform as the input a.c. voltage V_i. This time the straight inclined line AB of the characteristic in Figure 32.21 (ii) must be used. So, as shown,

(a) the base bias should correspond to the midpoint M of AB, a bias equal to OX or about 0·2 V, and

(b) the input voltage to be amplified must have a peak value not greater than XB (0·1 V), otherwise the output waveform is distorted during part of the input cycle.

In this and other a.c. amplification, it can be seen that the input voltage V_i is 180° out of phase, or antiphase, to the output voltage V_o.

States of Transistor

We can explain the characteristic curve in Figure 32.21 (ii) by noting that, if I_c is the collector current flowing for a particular input voltage, the output voltage V_o, or V_{CE}, is less than the supply voltage V_{CC} by the potential drop across R, which is $I_c R$. Thus

$$V_o = V_{CC} - I_c R \qquad . \qquad . \qquad . \qquad . \qquad (1)$$

In general, I_c depends on the base current I_b and this is governed by the base-emitter or input voltage V_i.

Suppose V_i is very low or practically zero. Then I_c is practically zero, and the transistor is said to be 'cutoff'. From equation (1), we can see that the output voltage V_o is then practically equal to V_{CC} or *high*, as shown in Figure 32.21 (ii). Conversely, suppose V_i is high so that the transistor is 'saturated', so I_c is very high. The p.d. across R is then large and so the output voltage is practically zero from equation (1), as shown in Figure 32.21 (ii).

Thus depending on the input voltage, the transistor can switch between two states—cutoff, or saturation. The output voltage then switches between two levels, $+ V_{CC}$ (high) and practically 0 (low). In the special type of computer circuits known as *logic circuits* or *logic gates*, the binary digits '1' and '0' can be represented by $+ V_{CC}$ and 0 respectively, or by 0 and $+ V_{CC}$, by this switching of states. It should be noted that the transistor acts 'non-linearly' in this case, whereas it acts 'linearly' in amplifiers as, for example, along AB in Figure 32.21 (ii).

Logic Gates

We can now show briefly how a transistor circuit can provide a useful *logic gate*.

Figure 32.22 shows the circuit for an INVERTER or NOT gate. It consists of a transistor in the common-emitter mode connection, with an appropriate load

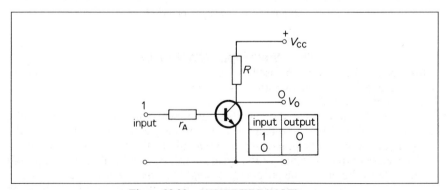

Figure 32.22 *INVERTER (NOT) gate*

resistance R and base resistance r_A. Suppose the input is a '1', for example, $+V_{CC}$ volt. A high base current then flows in r_A, and as explained before, the transistor becomes saturated and the output V_o is '0'. Conversely, if the input is '0' (zero volt), the transistor is cutoff and the output V_o is $+V_{CC}$ or '1'. Thus the output is always the *inverse* or opposite of the input. This is shown in a so-called 'truth table' in Figure 32.22.

Transistor circuits can provide a number of other useful logic gates, as we discuss more fully in the next chapter under Digital Electronics.

Exercises 32

1 (a) Why does the electrical conductivity of an intrinsic semiconductor increase as the temperature rises?
   (b) Introducing certain impurities into semiconducting material also increases its electrical conductivity. Describe briefly *either* an n-type *or* a p-type semiconductor and explain why this increase in conductivity occurs. (*L.*)

2 A semiconductor diode and a resistor of constant resistance are connected in some way inside a box having two external terminals, as shown in Figure 32A. When a potential difference V of $1·0$ V is applied across the terminals the ammeter reads 25 mA. If the same potential difference is applied in the reverse direction the ammeter reads 50 mA.
   What is the most likely arrangement of the diode and the resistor? Explain your deduction. Calculate the resistance of the resistor and the forward resistance of the diode. (*L.*)

3

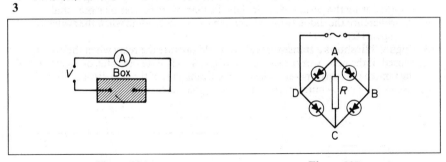

Figure 32A   **Figure 32B**

The circuit in Figure 32B shows four junction diodes and a resistor R connected to a sinusoidally alternating supply. Sketch graphs showing the variation with time over two cycles of the supply of
(a) the potential of C with respect to A,
(b) the potential of B with respect to A, and
(c) the potential of B with respect to D. (*L.*)

4 (a) (i) Explain the origin of holes in intrinsic semiconducting materials. What is the process by which holes participate in current flow? (ii) Explain how the presence of the donor impurities in an n-type semiconducting material raises the number of free electrons per unit volume without increasing the number of mobile holes per unit volume.
   (b)

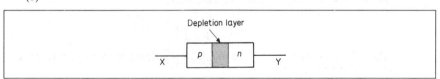

Figure 32C

Figure 32C shows two kinds of semiconducting material, *p*-type and *n*-type, in contact. What is the important characteristic which distinguishes the depletion layer from the rest of the assembly?

Explain the effect on the depletion layer of applying a *small* potential difference (about 0·1 V) across XY   (i) if X becomes negative with respect to Y, (ii) if X becomes positive with respect to Y. Hence explain the rectifying action of a *p-n* junction.

(c)

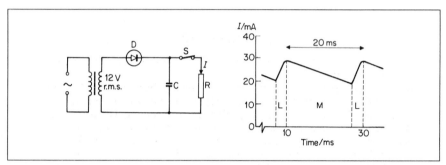

Figure 32D

Figure 32D shows a half-wave rectifying circuit connected to a resistor R through a switch S. The graph shows how the current *I* in resistor R varies with time. Write down the source of the current in R   (i) during the periods L, (ii) during the period M. Calculate the maximum reverse bias potential difference the diode D must be capable of withstanding when the switch S is open. (*L.*)

5   Figure 32E shows a transistor which should operate the relay when the switch is closed. If the relay works at a current of 8 mA, calculate the value of R_B which will just make the relay operate. The direct current gain of the transistor is 80, and assume the base-emitter resistance is negligible here.

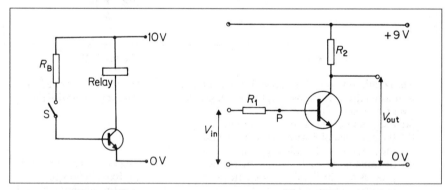

Figure 32E   **Figure 32F**

6   (a) Show that, in Figure 32F, a small change in the input p.d., δV_{in}, is accompanied by a change in the output p.d., δV_{out}, given by

$$\delta V_{out} = -\frac{\beta R_2}{R_1} \cdot \delta V_{in}$$

where β is the current gain of the transistor. Assume that the p.d. between the base and emitter is constant.

(b) With $V_{in} = 0$ and R_1 removed, a resistor, R_3, is connected between P and the positive (+9 V) rail. If $R_2 = 1 \text{ k}\Omega$ and $\beta = 150$, the output p.d., $V_{out} = 4.5 \text{ V}$.

Assuming the p.d. between base and emitter to be zero, calculate (i) the collector current, (ii) the base current, (iii) the resistance R_3. (*JMB*.)

7 Figure 32G shows an arrangement for investigating the characteristics of a transistor circuit. The input voltage V_i is varied using the potentiometer, P. The corresponding output voltage V_o is shown graphically in Figure 32H.

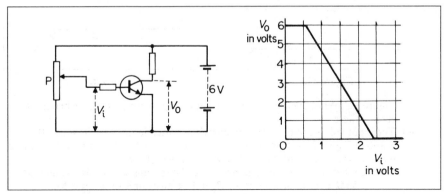

Figure 32G   **Figure 32H**

The circuit is to be used as an alternating voltage amplifier. The input voltage must first be fixed at a suitable value by adjusting P.

(a) Suggest the most suitable value for this fixed input voltage, explaining your answer.

(b) A sinusoidally alternating voltage of amplitude 0·5 V is superimposed on this fixed voltage. What will be the amplitude of the output voltage variations? Will the output variations be sinusoidal? Justify your answers.

(c) Sketch one complete cycle of the output voltage which would be obtained if the amplitude of the superimposed sinusoidal voltage were increased to 1·5 V. (*AEB*, 1983.)

8 In the transistor circuit shown in Figure 32I, R has a resistance of 150 kΩ, R_L has a resistance of 750 Ω, and the direct current gain of the transistor is 80.

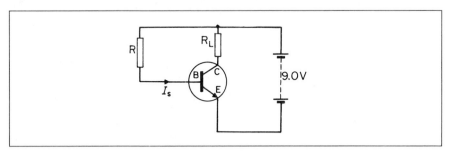

Figure 32I

Assuming that there is a negligible potential difference between B and E, calculate

(a) the base current I_B;

(b) the potential difference between the collector and emitter. (*AEB*, 1984.)

9 (a) Explain why, when a *p-n* junction diode is connected in a circuit and is reverse biased, there is a very small leakage current across the junction. How will the size of this current depend on the temperature of the diode?

(b) Sketch the output characteristic (I_C against V_{CE}) for a typical *n-p-n* transistor in *common-emitter* mode. Suggest typical values for I_C and V_{CE} on the axes and indicate how the base current varies over the family of curves.

Explain with the aid of the characteristic the meaning of the terms (i) saturation, and (ii) cut-off, applied to the silicon transistor in the circuit in

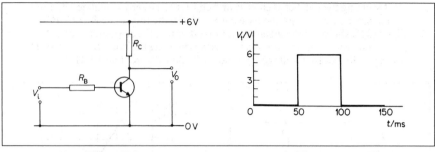

Figure 32J   **Figure 32K**

Figure 32J. What bias conditions exist at the transistor junctions when it is in each of these states?

(c) The input voltage to the circuit in part (b) is varied in the way shown on the graph in Figure 32K. Copy this graph and on a sketch graph drawn to the same scale show how the output voltage, V_o, varies with time.

Given that the d.c. current gain for this silicon transistor, h_{fe}, is 80 and that $R_C = 1.5 \text{ k}\Omega$, determine a value for R_B if the transistor is to saturate for a minimum input V_i equal to 3.0 V. (*L.*)

10 (a) (i) What are p-type and n-type semiconductor materials? (ii) Describe the charge flow that takes place when a junction is formed between p-type and n-type material, and hence explain how such a junction can act as a rectifier. (iii) With the help of a circuit diagram, describe how you would investigate the characteristic of a p-n junction diode. Sketch and explain the results that would be obtained for both forward and reverse bias voltages. (iv) Junction diodes made from certain materials will emit red light when forward biased above a threshold voltage. Suggest an explanation of this.

(b) Draw a circuit suitable for amplifying a small alternating input voltage without distortion which uses a n-p-n transistor in its common-emitter mode. Explain the functions of the individual components. (*O.*)

Analogue and Digital Electronics

Analogue Electronics

In analogue electronics we deal with circuits which process d.c. or a.c. signals, such as those in radio or television or in oscillators. In such circuits we have a continuously varying signal. The main circuit used is the operational amplifier (opamp) *and this is the 'building block' we use in this section. We shall see that the opamp can be used as an inverting, non-inverting and summing amplifier, in oscillators, in switching devices and as an integrator.*

Amplifiers, Voltage Gain

As we have seen, a single transistor can be used as an amplifier. Most practical amplifiers, however, use a combination of transistors. In this section the amplifiers contain many transistors and are called *operational amplifiers* or *opamps* for short.

We shall be concerned with the *voltage gain* of opamp circuits to be discussed later. If the input voltage to a circuit is V_{in} and the output voltage is V_{out}, the voltage gain is defined as:

$$Voltage\ Gain = \frac{V_{out}}{V_{in}}$$

The input and output voltages can be a.c. or d.c. and may be measured with a cathode-ray oscilloscope.

Opamps

Opamps are amplifiers built from many transistors and other circuit components. They can amplify a voltage from zero frequency (d.c.) up to very high frequencies. In practice the range of frequencies is often controlled by external components, especially capacitors.

The opamp is a *differential amplifier*, that is, the output is directly proportional to the *difference* in voltage between the two inputs of the opamp. The two inputs are the 'non-inverting input' (+) and the 'inverting input' (−). The function of the inputs will be explained later.

Figure 33.1 shows the symbol for the opamp:

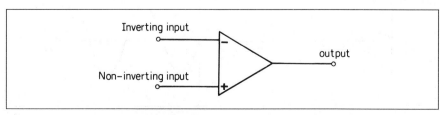

Figure 33.1 *Opamp symbol*

Function of Opamp

If an opamp is connected as in the simplified circuit of Figure 33.2, it acts as a *non-inverting amplifier*. This means that the output is the exact, amplified copy of

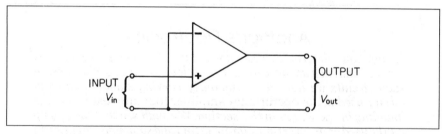

Figure 33.2 *Non-inverting amplifier*

the input as shown in Figure 33.3. So the output voltage V_{out} is *in phase* with the input voltage V_{in}.

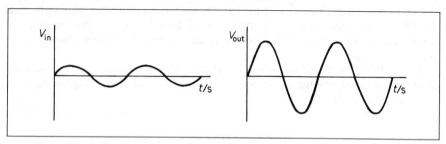

Figure 33.3 *Output voltage in phase with input voltage (non-inverting amplifier)*

If however the amplifier is connected as shown in Figure 33.4, it is said to be connected as an *inverting amplifier*. In this case the output is an exactly opposite, amplified copy of the input, as shown in Figure 33.5. So V_{out} is 180° out of phase or in antiphase with V_{in}.

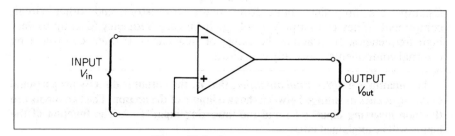

Figure 33.4 *Inverting amplifier*

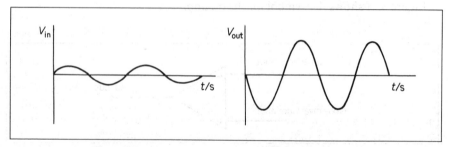

Figure 33.5 V_{out} *in antiphase to* V_{in} *(inverting amplifier)*

If an opamp were to be connected as shown in Figures 33.2 and 33.4, the ratio V_{out}/V_{in} would be very large. We call this the _open-loop gain_. Ideally it is infinite, for reasons we see later, although in practice the gain is about 10^5. In most practical uses of the opamp, the gain is limited by _negative feedback_.

Negative Feedback

'Negative feedback' is said to occur when a little of the output signal is fed back to the _inverting_ input. As we saw in Figure 33.5, the output is 180° out of phase with the input. So the feedback _reduces_ the signal that the amplifier has to amplify and therefore reduces the gain. The amount of the output fed back is controlled by a resistor R_2, as shown in Figure 33.6. Resistor R_1 serves a function that will be explained in the next section.

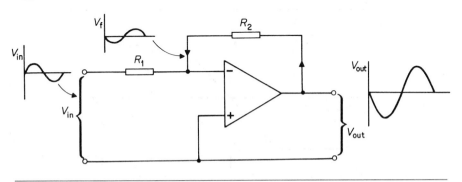

Figure 33.6 _Negative feedback—input and output voltages_

As we see later, there are several advantages in using negative feedback, for example:

1 Amplifiers with almost infinitely variable gain can be produced using one standard opamp circuit.
2 The use of negative feedback improves the range of frequencies that the amplifier will amplify and improves stability.

Some Opamp Characteristics

In order to calculate the gain of any given opamp circuit, it is necessary to know a little more about the characteristics of an opamp. These are:
1 There is a _very high impedance_ between the + or − input and ground. Ideally this impedance is infinite but in practice it is something like 2 MΩ. The high impedance means that no current (in practice very, very little current, called the _bias current_) will flow into or out of the inputs when a voltage is applied, so there is no effect on the input circuit when connected to the opamp.
2 Since the gain of the actual amplifier is so high, and remembering that the amplifier amplifies the _difference_ between the − and the + inputs, the slightest difference in voltage between the − and + inputs will make the output voltage go to its highest value, which is the voltage of the power supply. This is called the _saturation_ value (V_S) of the output as the output can go no higher. If the supply voltage is 15 V and the open-loop gain of the amplifier is 10^5, then a difference in voltage of 15 V/10^5, which is 150 μV, will produce saturation. We

see later that with such small voltage differences (V_{in}) between the non-inverting (V_+) and inverting (V_-) inputs, the output voltage (V_{out}) can swing either way between $+15$ V ($+V_s$) or -15 V ($-V_s$) (Figure 33.7). The output is $+V_s$ when V_+ is greater than V_-, and $-V_s$ when V_- is greater than V_+.

Thus for our amplifier to be of any use, the $-$ input must be at virtually the same voltage as the $+$ input. Now in the inverting amplifier circuit the $+$ input is connected to earth or ground which is also 0 V, so the $-$ input must be always virtually at the same voltage. The $-$ input is known as a *virtual earth*.

Saturation **occurs when output voltage reaches supply voltage, $+15$ V or -15 V. Very small voltage difference between inverting and non-inverting terminals produces saturation because gain is high.**

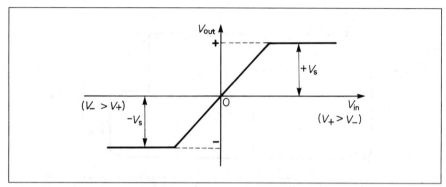

Figure 33.7 *Output voltage variation with input voltage.* $V_S =$ *saturation value* $V_+ =$ *non-inverting voltage,* $V_- =$ *inverting voltage*

Gain of Inverting Amplifier

We are now in a position to calculate the gain with negative feedback.

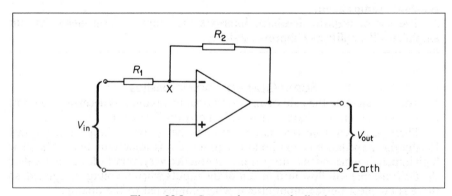

Figure 33.8 *Gain with negative feedback*

Referring to Figure 33.8, since the $-$ input is at earth potential (virtual earth) the current through R_1 will be V_{in}/R_1.

The current through R_2 will be V_{out}/R_2.

But since the input impedance of the amplifier is very high, no current can flow into the $-$ input. The sum of the currents at junction X must be equal to zero

(Kirchhoff's law). So whatever current goes into junction X through R_1 must leave the junction through R_2.

So
$$V_{in}/R_1 + V_{out}/R_2 = 0$$
$$\therefore V_{in}/R_1 = -V_{out}/R_2$$
$$\therefore V_{out}/V_{in} = -R_2/R_1 = \text{Gain}$$

So gain of this inverting amplifier $= -R_2/R_1$

The minus sign indicates that the output is *inverted*. Notice that the gain of the amplifier is independent of the open-loop gain of the opamp. Since the gain depends only on the *external* components R_1 and R_2, the gain is not affected by any changes that may take place inside the opamp, such as a change in gain due to temperature change. So the negative feedback provides stability.

Gain of Non-Inverting Amplifier

A similar argument can be applied to a circuit for a non-inverting amplifier. With a non-inverting amplifier, the input has to be applied to the + input but the feedback has to be applied to the − input. The circuit needed is shown in Figure 33.9. The fraction of the output signal to be fed back to the input is determined by the potential divider R_1 and R_2.

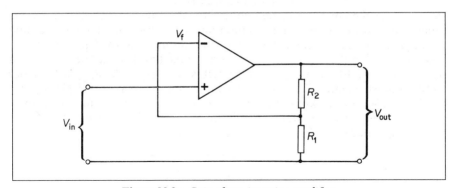

Figure 33.9 *Gain of non-inverting amplifier*

The fraction of V_{out} sent to the inverting (−) input is V_f, where
$$V_f = \frac{V_{out} \times R_1}{(R_1 + R_2)} \qquad \cdot \qquad \cdot \qquad \cdot \qquad \cdot \qquad \cdot \qquad (1)$$

The voltage *difference* between the two inputs $= V_T$, where
$$V_T = V_{in} - V_f \qquad \cdot \qquad \cdot \qquad \cdot \qquad \cdot \qquad (2)$$

V_T is the voltage actually amplified by the opamp.

Now
$$V_{out} = A_o \times V_T \qquad \cdot \qquad \cdot \qquad \cdot \qquad \cdot \qquad (3)$$

where A_o is the open-loop gain of the opamp (see p. 811). Substituting for V_T in (2) using (3), then
$$V_{in} - V_f = V_{out}/A_o$$
and
$$V_f = V_{in} - V_{out}/A_o \qquad \cdot \qquad \cdot \qquad \cdot \qquad \cdot \qquad (4)$$

Substituting for V_f in (1) using (4),

$$V_{in} - \frac{V_{out}}{A_o} = \frac{V_{out} \times R_1}{(R_1 + R_2)}$$

$$\therefore V_{in} = V_{out}\left(\frac{R_1}{R_1 + R_2} + \frac{1}{A_o}\right)$$

$$\therefore \text{Gain} = V_{out}/V_{in} = (R_1 + R_2)/R_1$$

since A_o is typically about 10^5 and $1/A_o$ is negligible.

So gain of this amplifier $= 1 + (R_2/R_1)$

As with the inverting amplifier, the gain depends only on the values of R_2 and R_1 and is independent of the open-loop gain of the opamp.
For example: If $R_1 = 1\,k\Omega$ and $R_2 = 100\,k\Omega$

$$\text{Gain} = 1 + 100/1 = 101$$

If $R_1 = 5\,k\Omega$ and $R_2 = 500\,k\Omega$, the gain would still be 101. But with large resistors $R_1 = 5\,M\Omega$, $R_2 = 500\,M\Omega$, the very small current drawn by the opamp at the input is now *not* negligible, as in the ideal opamp (p. 812). So the gain is not now given by $1 + (R_2/R_1)$.

Practical Opamps

Most practical opamps are integrated circuits. They are usually made as a 8-pin, dual-in-line package. The pin connections have become standard so that opamps can be readily interchanged with one another. The 741 type opamp, first manufactured in the late 1960's is still perhaps the best known, tried and tested opamp circuit. More modern replacements are available now with better specifications. Figure 33.10 shows the pin connections for the 8-pin, opamp package. The only variation between types is in pin 8. This pin is used for a variety of functions or if not required is not connected. The function of each pin will be explained in the following section.

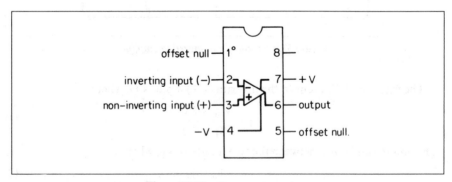

Figure 33.10 *Opamp connections*

An opamp is slightly unusual in its requirements for a power supply. It requires a *dual*, balanced supply, as shown diagrammatically below.

$+V(+15\text{ V})$	$+V$ is positive with respect to the centre 0 V line and has a maximum of about 15 V for most opamps.
$0\,(0\text{ V})$	
$-V(-15\text{ V})$	$-V$ is negative with respect to the centre 0 V line and must have the same magnitude as the $+V$.

0 V is connected to earth or ground. The potential difference between $-V$ and $+V$ will be $2V$. By convention, once the power supply has been established, the connections from the opamp to the power supply are not shown in the circuit diagram but are assumed to be made. Figure 33.12 shows an opamp connected as an inverting amplifier with power supply connections $+V_S$ and $-V_S$.

Offset Voltage

In practice, due to manufacturing tolerances, even when there is no signal applied to the input the internal components of the opamp may supply a small differential voltage to the inputs. Even if this voltage is small, the gain of the opamp results in a large offset voltage being present at the output, which should be at exactly 0 V under no signal conditions. In many circuits, this offset voltage at the output does not matter but for d.c. circuits it must be removed and this is achieved by using the two 'offset null' terminals of the opamp as shown in Figure 33.11. The $10\,k\Omega$ potentiometer is adjusted until the output is at exactly 0 V with no signal present at the inputs.

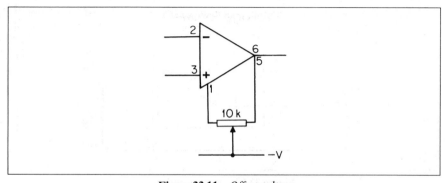

Figure 33.11   *Offset voltage*

A Practical Inverting Amplifier

The simplified circuit shown earlier in Figure 33.8 for an inverting amplifier needs some small modification. A resistor, R_3 needs to be connected between the $+$ input and 0 V, Figure 33.12. R_3 should be equal to the sum of R_1 and R_2 in

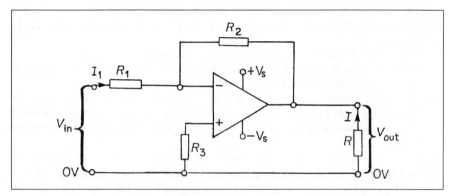

Figure 33.12   *Inverting amplifier*

parallel in order to keep the inputs balanced as far as possible. In this circuit, the power supply connections $+V_S$ and $-V_S$ are shown although these connections will be omitted in following circuits.

The input impedance of this circuit is determined by the value of R_1. Since the − input is virtual earth, the input is effectively connected across the parallel combination of R_1 and the (very large) input impedance of the opamp itself. The advantage of this circuit is that the input impedance can be varied to suit the preceeding stage.

Output of Opamp

The *output* is taken between the 0 V and the output terminal of the opamp circuit. Suppose R is the opamp resistance load (or the input resistance of the next circuit) and I is the current in R (Figure 33.12). Then $V_{out} = -IR$. So gain $= V_{out}/V_{in} = -IR/I_1R_1 = -R_2/R_1$ (p. 813), where I_1 is the current in R_1. So $I/I_1 = R_2/R$. The opamp can typically deliver a current of about 5 mA which means that the minimum load R is about 2 kΩ.

Since V_- is a virtual earth, *power gain* $= V_{out}^2/R \div V_{in}^2/R_1 =$ (voltage gain)^2 $\times (R_1/R) = (R_2/R_1)^2 \times (R_1/R)$.

Voltage Follower

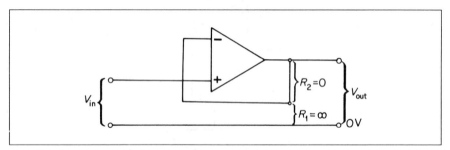

Figure 33.13 *Voltage follows*

The advantages of the non-inverting amplifier circuit discussed on page 813 are that
(a) the output is not an inverted copy of the input and
(b) since the + input is not a virtual earth, the input impedance of this circuit is high, typically 50 MΩ.

A further application of this circuit is shown in Figure 33.13. Here the output is connected directly to the inverting input so that the feedback is 100%. The gain is now 1 ($R_2 = 0$ and $R_1 =$ infinity in the expression gain $= 1 + (R_2/R_1)$, derived previously.) So V_{out} follows V_{in} exactly. This is why the circuit is called a *voltage follower*.

At first sight this does not appear to be useful. But
(a) the input impedance is very high so that practically no current is taken from the circuit connected to the input and
(b) the output impedance is low so that a current can be supplied to the following circuit or stage. So the voltage follower is used as a *buffer* between a high impedance (low current) circuit and a low impedance (high current) circuit.

Voltage follower: output directly connected to input so 100% feedback and V_{out} follows V_{in} exactly. Can act as a buffer connecting high and low impedance circuits.

Frequency Response of Opamp Circuit

The gain of an opamp depends on frequency. A graph of gain against frequency is shown in Figure 33.14 for a typical opamp. With no feedback, or _open-loop gain_ A_{OL}, the gain drops rapidly from 10^5 at low frequencies to 1 at very high frequencies. This situation is clearly not ideal since amplifiers generally require the gain to be constant at all frequencies. However, as the negative feedback is increased and the overall gain becomes lower, the gain remains constant over a wider range of frequencies and so the frequency response of the amplifier is improved. With feedback, the gain is called _closed-loop gain_ or A_{CL}.

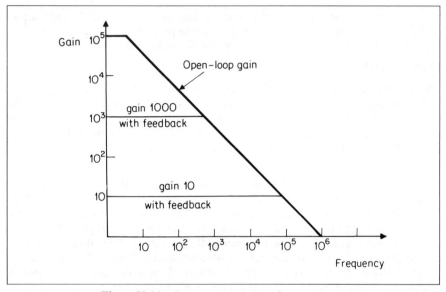

Figure 33.14 _Frequency response of opamp circuit_

Opamp as a Summing Amplifier

One of the most frequent uses of an opamp is in audio pre-amplifiers and mixers. An opamp can be used to add any number of signals or voltages and the circuit is called a summing amplifier. Figure 33.15 shows a simple _summing amplifier_ to add three voltages, V_1, V_2, V_3.

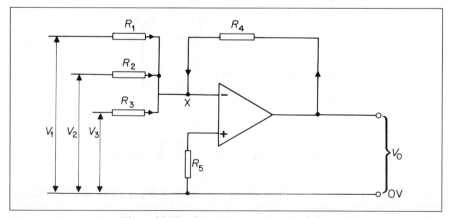

Figure 33.15 _Opamp as summing amplifier_

Point X is a virtual earth and therefore the input currents into point X are:

$$V_1/R_1, \; V_2/R_2, \; V_3/R_3$$

and the feedback current $= V_o/R_4$

So $$V_1/R_1 + V_2/R_2 + V_3/R_3 + V_o/R_4 = 0$$

since all the currents at a junction must add to zero.

$$\therefore \; -V_o = \frac{R_4}{R_1} \times V_1 + \frac{R_4}{R_2} \times V_2 + \frac{R_4}{R_3} \times V_3 \qquad . \qquad . \qquad (1)$$

So the output voltage, V_o, is the sum of the three inputs with each input multiplied by a factor R_4/R, where R is the corresponding input resistance.

Suppose $V_1 = 1 \, \text{V}$, $V_2 = 2 \, \text{V}$, $V_3 = 3 \, \text{V}$ and $R_1 = R_2 = R_3 = R_4 = 10 \, \text{k}\Omega$.

Then, from (1), Output $V_o = -6 \, \text{V}$

The minus sign for V_o indicates that V_o is antiphase to the inputs (p. 810).

This result applies to a.c. or d.c. voltages. It is especially useful for a microphone mixer since we do not want one microphone to affect another. It might also be necessary to change the gain of the amplifier for different inputs, in which case R_1, R_2 and R_3 can be varied.

Positive Feedback, Square Wave Oscillator or Astable Multivibrator

As we have seen, negative feedback *reduces* the differential voltage at the input. But if some of the output is taken back to the + (non-inverting) input, any change in the output will tend to *increase* the differential voltage at the input, since the output voltage will be in phase with the input voltage (page 810). The output voltage will then quickly reach the saturation voltage, $+V_S$ which is the supply voltage. If the opamp circuit is designed to make the output switch continually from $+V_S$ to $-V_S$, and from $-V_S$ to $+V_S$ an *oscillating* output voltage is obtained. Figure 33.16 shows a circuit with positive feedback which produces square wave oscillations and acts as an *astable multivibrator*.

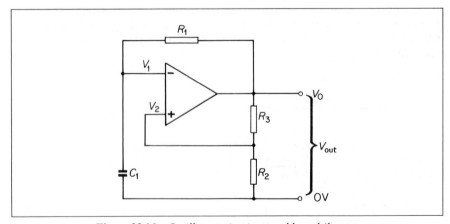

Figure 33.16 *Oscillatory circuit—astable multibrator*

To explain the action of this circuit, suppose the capacitor C_1 is initially uncharged and that the output voltage V_o has its maximum positive value $(+V)$ due to a small differential voltage at the inputs.

A fraction (V_2) of V_o is fed back to the non-inverting input:

$$V_2 = (V_o \times R_2)/(R_2 + R_3)$$

But V_o is also fed back into the inverting input through R_1. So C_1, which is initially uncharged, starts to charge up through R_1 towards $+V$ and the voltage V_1 then rises exponentially with time, as shown in Figure 33.17 (a). After a time which depends on the time constant $C_1 \times R_1$, V_1 reaches a value higher than V_2; whereupon the output of the opamp switches over so that the output is $-V$, as shown in Figure 33.17 (b). The positive feedback encourages the opamp to switch quickly since V_2 then drops, making V_2 much lower than V_1 and so forcing V_o to go negative even more quickly.

C_1 now discharges and starts to charge in the opposite direction until V_1 becomes lower than V_2. The opamp then switches back again so that V_o becomes positive ($+V$) again. The cycle is then repeated. Figure 33.17 (a) shows how the voltage V_1 varies with time and Figure 33.17 (b) shows how the output voltage V_o varies with time. The periodic time of the multivibrator is given by:

$$T = 2C_1R_1 \ln (1 + 2R_2/R_3)$$

the proof of which is beyond our scope.

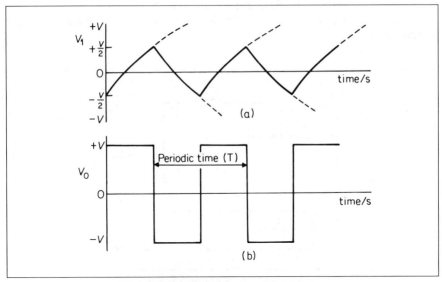

Figure 33.17(a) _Variation of_ V_1 _with time_
Figure 33.17(b) _Variation of_ V_o — _square wave_

A Practical Astable Multivibrator

The circuit in Figure 33.18 gives component values to construct an astable multivibrator from an opamp.

The periodic time $T =$

$$2 \times 0.1 \times 10^{-6} \times 100 \times 10^3 \ln (1 + 2 \times 10 \times 10^3/100 \times 10^3) = 0.00365 \text{ s}$$

Therefore frequency $= 1/T = 274$ Hz.

Oscillator, square-wave or astable multivibrator: (1) Negative feedback from $C_1 - R_1$ provides charge-discharge voltage. (2) Positive feedback from potential

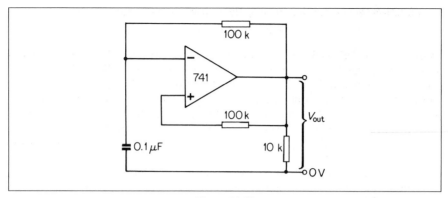

Figure 33.18

divider provides switching. (3) Period T depends on $C_1 R_1$ and feedback resistance ratio.

Sine Wave Oscillator

An oscillator based on a circuit originally due to Wien can provide a *sine wave* output.

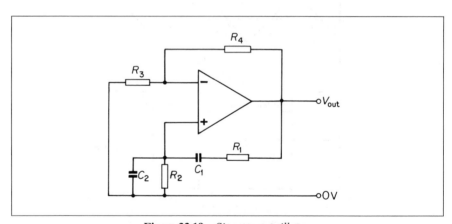

Figure 33.19 *Sine wave oscillator*

The necessary circuit is shown in Figure 33.19. The positive feedback is provided by the Wien bridge network, C_1, C_2, R_1, R_2. The particular property of the Wien bridge is that the voltage fed back to the $+$ input is only in phase with the output at one frequency (f) given by:

$$f = 1/(2\pi RC)$$

if $C_1 = C_2 = C$ and $R_1 = R_2 = R$. Thus positive feedback is only present at frequency f and the circuit will oscillate at this frequency provided that the overall gain of the amplifier is about 3. The overall gain is provided by resistors R_3 and R_4. To give a gain of about 3, $R_4 = 2R_3$.

In practice, in a simple circuit, R_3 is a 6 V–0·06 A lamp and R_4 is about 50 ohms. This arrangement prevents the output saturating at any stage, since if

V_{out} becomes too large, the current through the lamp increases. The lamp then heats up and its resistance increases, so decreasing the gain of the amplifier.

Opamp as Voltage Comparator, Switching Circuit

Without any feedback, an opamp can be used as a *voltage comparator*. Figure 33.20 illustrates the principle. R_1 and R_2 form a potential divider between $-V$ and $+V$. The relative values of R_1 and R_2 will determine the voltage at the non-inverting input (Y) of the opamp. The voltage at (Y) is a reference voltage.

If the inverting input $(-)$ of the opamp is at an even slightly higher voltage than the non-inverting input $(+)$, then the output will be at the maximum negative $(-V)$. As soon as the voltage of the inverting input $(-)$ falls below that of the non-inverting input $(+)$, or the voltage of the $+$ input rises above that of the $-$ input, the output immediately switches to the maximum positive $(+V)$.

A known reference voltage is applied to the $+$ input. If the voltage on the $-$ input is below the reference voltage, the output is $+V$. If the voltage on the $-$ input is above the reference, then the output is $-V$. This is especially useful in circuits that are required to detect a changing level of voltage and switch at a specific level.

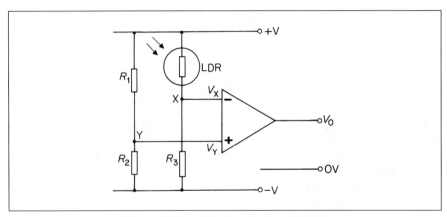

Figure 33.20 *Switching circuit with LDR*

In Figure 33.20 another potential divider is formed by the light dependent resistor (LDR) and R_3. As the light level becomes lower, the resistance of the LDR increases and the voltage on the inverting input falls until it is below the reference voltage at Y set by the potential divider R_1 and R_2. At this point the output, which had been at $-V$, switches rapidly to $+V$. This switching could be used to operate a relay. The whole circuit might then be used to switch on a street light.

The circuit in Figure 23.20 is called a 'voltage comparator' because it compares the voltage V_X at X with the voltage V_Y at Y. If V_X is higher than V_Y, then a high negative output is obtained. If V_X is less than V_Y, a high positive output is obtained.

The light level at which switching occurs can be altered by adjusting the reference voltage. The LDR can also be replaced by a *thermistor* to detect a pre-determined temperature and perhaps sound an alarm if a fire occurs and the temperature rises above a pre-set value. A thermistor is a resistance which varies with temperature; usually the resistance decreases with increasing temperature. So the voltage V_x relative to V_y will alter as the temperature rises or falls.

Switching circuit: V_X (– input) from potential divider with LDR (lighting control) or thermistor (temperature control). V_Y fixed by another potential divider. If LDR or thermistor resistance changes, output switches between $-V_S$ and $+V_S$, the supply voltage, and relay operates.

Opamp as Integrator

In certain circumstances, for example an analogue computer, a circuit is needed that will provide an output that is the integral of the input voltage. A basic integrator circuit is shown in Figure 33.21.

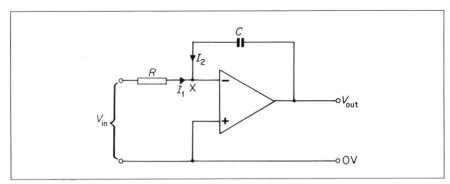

Figure 33.21 *Opamp as integrator*

Point X is a virtual earth as explained previously. The p.d. across R is V_{in} and the p.d. across C is V_{out}.

As before, the total current into point X must be equal to the total current leaving point X.

The current into point $X = I_1 = V_{in}/R$. I_2 is the current out of point X and is equal to the charging current for capacitor C.

So
$$I_2 = \frac{dQ}{dt} = \frac{C\,dV_{out}}{dt}$$

But since
$$I_1 + I_2 = 0$$

then
$$V_{in}/R = -C\,dV_{out}/dt \qquad . \qquad . \qquad . \qquad (1)$$

If we integrate this expression:
$$(1/RC)\int V_{in}\,dt = -\int dV_{out}$$

which gives:
$$V_{out} = -(1/RC)\int V_{in}\,dt \qquad . \qquad . \qquad (2)$$

Thus the output voltage is proportional to the *integral* of the input voltage.

If the input is a pulse as shown in Figure 33.22 (a), the output will be as shown in Figure 33.22 (b). We can see this from equation (1). Here V_{in} is constant, Figure 33.22 (a), and so

$$dV_{out}/dt = -V_{in}/CR = -k,$$

where k is a constant. So V_{out} varies *linearly* with time t and has a negative gradient as shown.

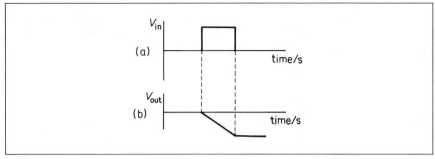

Figure 33.22 (a) *Input pulse* (b) *Integral of input*

From equation (2), one example of the use of the circuit would be to integrate an input proportional to the speed of a car, so as to give an output proportional to the distance covered.

When using an opamp as an integrator, any initial variation of the input from zero will cause an output which will be integrated. The output will thus tend to drift slowly towards saturation even before the signal to be integrated has been connected. It is therefore necessary to use the offset-null adjustment described earlier to remove any small output offset.

Examples on Opamp Circuits

1 An inverting opamp circuit has a voltage amplification of 100 and a supply of ±9V. If an input a.c. voltage of 50 Hz has a peak value of 0.6 V, sketch the output voltage waveform.

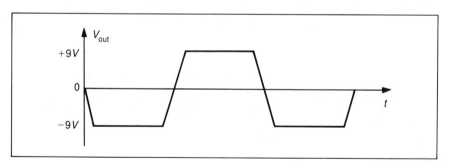

Figure 33.23 *Saturation of Opamp*

The inverted output voltage *saturates* at −9 V and +9 V, when the input voltage is respectively +0.09 V and −0.09 V. The 'clipped' waveform is shown roughly in Figure 33.23. The time t of saturation from zero first occurs whewn $0.09 = 0.6 \sin \omega t$, where $\omega = 2\pi f = 100\pi$, from which t can be calculated.

2 In the circuit shown in Figure 33.24, at time $t = 0$ the capacitor $C = 1\,\mu\text{F}$ is uncharged and the output voltage (V_{out}) is +14 V. At a later time the voltmeter connected to the output shows that V_{out} has changed to −14 V. Calculate the time at which V_{out} changes from +14 V to −14 V.

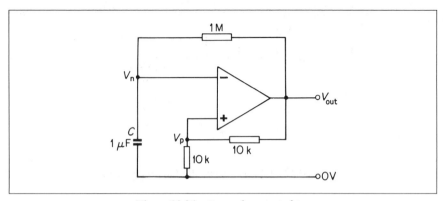

Figure 33.24 *Example on switching*

Initially, $\qquad V_{out} = +14\,V$

$$\therefore V_p = \frac{10 \times 10^3}{20 \times 10^3} \times 14\,V = 7\,V$$

The voltage V_n is the voltage across the capacitor C of $1\,\mu F$, which is in series with the $1\,M\Omega$ across the voltage V_{out} initially at $14\,V$. The time-constant of the circuit $= CR = 1 \times 10^{-6} \times 1 \times 10^6 = 1\,s$.

So $\qquad\qquad V_n = V_{out}(1 - e^{-t/CR}) = 14(1 - e^{-t})$

When V_n is just greater than V_p, or $7\,V$, V_{out} will switch to $-14\,V$. So time t when this happens is given by

$$7 = 14(1 - e^{-t})$$

from which $\qquad\qquad -7 = -14\,e^{-t}$

So $\qquad\qquad t = \ln(14/7) = \ln 2 = 0\cdot69\,s$

_____ **Exercises 33A** _____

1   In the opamp circuit shown in Figure 33A, R and S are resistances.
Write down (i) the gain, (ii) the phase difference between the input and output voltages. Why is X a 'virtual earth'?
Explain why the currents in R and S are equal.

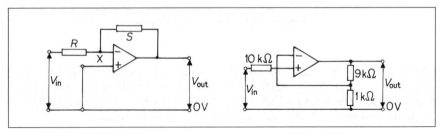

Figure 33A   **Figure 33B**

2   Calculate the gain of the circuit in Figure 33B and draw sketches showing roughly the input and output voltages if the input voltage is a square wave voltage varying from 0 to $0\cdot5\,V$.

Draw an inverting amplifier circuit with component values which give about the same gain as Figure 33B.

3   Figure 33C shows a summing amplifier. One of the input voltages V_1 is 0·55 V and the other is V_2. Calculate V_2 if the output voltage is 0·9 V.

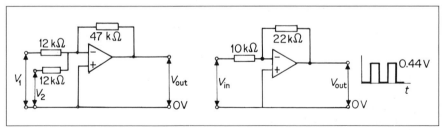

| **Figure 33C** | **Figure 33D** |

4   In Figure 33D, the output voltage V_{out} is a square-wave voltage of amplitude 0·44 V. Calculate the amplitude of the input voltage and draw a sketch of it.
5   Figure 33E shows an opamp circuit whose power supply is +15 V and −15 V. What is the gain of the circuit? Prove the formula used.
   Sine waves of frequency 1 kHz and amplitude
   (a) 100 mV and then
   (b) 1 V are applied to the input. In each case sketch the input and output waveforms, and comment on them.

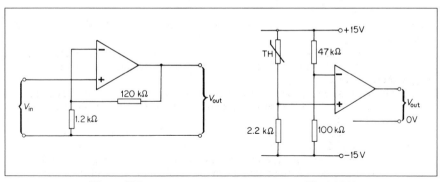

| **Figure 33E** | **Figure 33F** |

6   The circuit in Figure 33F shows the opamp used as a voltage comparator. The resistance of the thermistor shown decreases with increasing temperature. With the resistor cold, the output voltage is at maximum −15 V, the power supply being +15 V and −15 V.
   (a) What is the voltage at the inverting input of the opamp?
   (b) What voltage would the non-inverting input have to reach for the output to switch to maximum positive +15 V?
   (c) What would be the resistance of the thermistor at this point?
7   When an input voltage of constant value V_i is applied to the circuit in Figure 33G, the capacitor C becomes charged at a constant rate by current through R. If the output voltage is V_o, show that, if t is the time,

$$V_i/R = -C\,dV_o/dt$$

Using this relation, calculate the rate at which the output voltage changes when $V_i = 0·4$ V, $R = 50$ kΩ and $C = 0·2$ µF. Draw a sketch of V_o against t, and suggest a use of this circuit for one part of a cathode-ray oscilloscope.

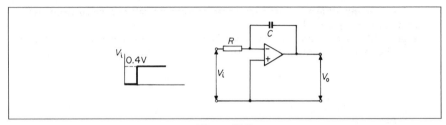

Figure 33G

8 Define the terms
  (a) open-loop gain,
  (b) closed-loop gain,
  (c) negative feedback,
  (d) off-set voltage.
    An amplifier with negative feedback has an improved frequency response compared with an amplifier without feedback. Draw sketches of frequency response curves which illustrate this improvement, labelling the axes.
9 Figure 33H shows a circuit for an electronic thermometer. The thermocouple generates an e.m.f. between the terminals X and Y of 50 μV per K temperature difference between its junctions. The thermocouple is connected to the non-inverting input of an operational amplifier and feedback is provided by the resistor of resistance R_f to the inverting input.
  (a) State what is meant by an *inverting input* and a *non-inverting input*. State the type of feedback used in this circuit and explain how it functions.
  (b) The voltage gain of this amplifier is given by

$$\frac{V_{out}}{V_{in}} = 1 + \frac{R_f}{R_a}$$

   If $R_f = 100\,k\Omega$ and $R_a = 1\,k\Omega$, calculate the reading on the voltmeter when there is a temperature difference of 100 K between the junctions of the thermocouple.
  (c) The thermometer is to be used to measure body temperatures over the range 35°C to 45°C, the voltmeter giving zero deflection at 35°C and full-scale deflection of 1·0 V at 45°C. State how the thermocouple can be set up so that the voltmeter reads zero when the temperature of the hot junction is 35°C.
    If $R_a = 1\,k\Omega$, calculate the value of R_f which will produce a full-scale deflection when measuring a temperature of 45°C. Explain why you would expect the value of R_f to be very much greater than the value of R_a in this application. (*JMB.*)

10

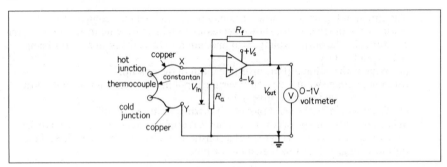

Figure 33H

In the above circuit P, Q, R and S form part of a Wheatstone bridge network in which an operational amplifier circuit is used to detect the difference between the

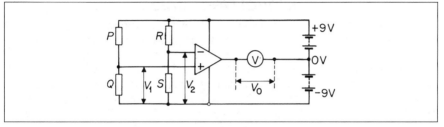

Figure 33I

voltages V_1 and V_2, Figure 33I. (i) The output voltage in this circuit, V_0, is given by $V_0 = A(V_1 - V_2)$ where A is the open-loop gain of the amplifier and V_0 must lie between 9 V and -9 V. Show that, if $A = 90\,000$, then V_0 should be either 9 V or -9 V if the difference between V_1 and V_2 exceeds 100 μV. (ii) If the resistances P and Q are each 10 kΩ and R is the resistance of a variable standard resistor, outline how you would use the arrangement to determine the resistance S, which is of the order of a few kΩ. (iii) A typical centre zero galvanometer has a resistance of 40 Ω and is graduated in divisions of 0·1 mA. If you were to choose between such a meter and the above operational amplifier as a null detector, which would you choose, and why? (*JMB*.)

11 Explain what is meant by *voltage gain* and *negative feedback* in relation to electronic circuits.

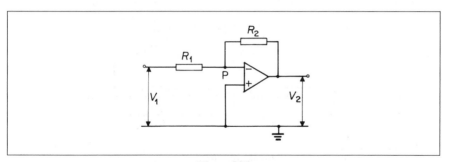

Figure 33J

Figure 33J shows a circuit containing an ideal operational amplifier where the point P is usually referred to as a *virtual earth*. Explain what you understand by *virtual earth* in this context and hence derive an expression for V_2 in terms of V_1 and the values of the circuit components.

The current, I, through a certain device varies with applied potential difference, V, according to the relation

$$I = I_0\, e^{kV}$$

where I_0 and k are constants. If R_2 is replaced by this device, write down an expression for the feedback current in terms of V_2. Hence show that V_2 is given by the expression

$$V_2 = 1/k \ln(V_1/R_1 I_0)$$

What is the possible advantage of this type of amplifier over a linear amplifier when a wide range of input signal amplitudes must be displayed? (*C.*)

Digital Electronics

Unlike analogue electronics, which deals with continuous signals, digital electronics is concerned with electrical signals that can be either ON or OFF. No other situation, for example 'HALF ON', can possibly be recognised.

In digital circuits, the 'ON' or '1' state can be represented as a high voltage (with respect to the grounded negative of the power supply). Thus in a typical circuit, '1' or 'ON' would be represented by a voltage of $+5$ V or 'HIGH' and a '0' or 'OFF' would be represented by a voltage of 0 V or 'LOW'. In this way, digital signals or numbers in binary can be recorded in a circuit without confusion.

Digital circuits are designed to process digital signals and there are a number of standard 'building blocks' available called *logic gates*, from which complex digital circuits can be built. For example, as shown later, they can be used in a simple burglar alarm. We shall show how logic gates are used in adding binary digits, in memory circuits and in the binary counter.

Logic Gates

Logic gates are generally made in the form of an integrated circuit. Several gates can then be made in one package, called a 'dual-in-line (dil)' package, Figure 33.25. Each gate will have one or more *inputs* and one *output*. A power supply is needed which is often 5 volts.

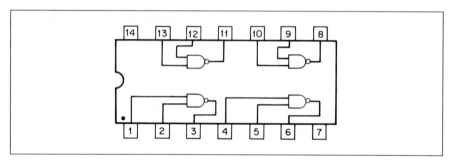

Figure 33.25 *Integrated circuit. 2-input NAND gate*

NOT Gate or INVERTER

The NOT gate is the simplest of all logic gates as it has only one input and one output. It changes the input A so that a '1' input becomes a '0' output at Q and a '0' input becomes a '1' output. In electronics, American standard symbols are commonly used, as shown in Figure 33.26.

The action of logic gates is most easily represented by means of a TRUTH TABLE.

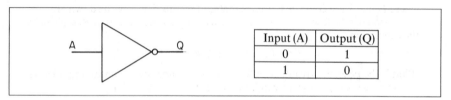

Input (A)	Output (Q)
0	1
1	0

Figure 33.26 *NOT gate*

The truth table for a NOT gate is shown in Figure 33.26. The function of the NOT (INVERTER) gate can be represented briefly by $\bar{A}$(A bar) $= Q$ or in words, 'not A equals Q'.

AND Gate

An AND gate can have any number of inputs but only one output. For simplicity, we deal with a two-input AND gate, Figure 33.27. An AND gate gives a high output (1) if input A AND input B are both high (1); otherwise the output is low (0). See Truth Table.

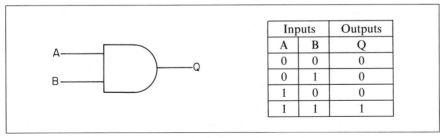

Inputs		Outputs
A	B	Q
0	0	0
0	1	0
1	0	0
1	1	1

Figure 33.27 _AND gate_

As we stated earlier, a burglar alarm might be required to sound if it is dark AND a door is opened. A simple electronic circuit could give a high voltage in the dark and a switch on the door could be arranged to give a high voltage when it is opened. These signals are then fed to an AND gate and the output used to sound an alarm, as shown in Figure 33.28. If either input A or B is 0, the output Q is low (0) and the alarm will not sound. If input A and input B are both 1, the output Q is high (1) and the alarm does sound.

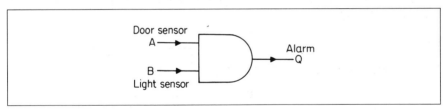

Figure 33.28 _Burglar alarm with AND gate_

NAND (NOT AND) Gate

Electronically, it is easy to make a gate which is equivalent to an AND gate followed by a NOT gate. The combination is given its own symbol and called a NAND (NOT AND) gate. Notice how the symbol is the same as for the AND gate but with a small circle at the output. This small circle is always taken to indicate a NOT or INVERTER operation. The truth table for the NAND gate is shown in Figure 33.29.

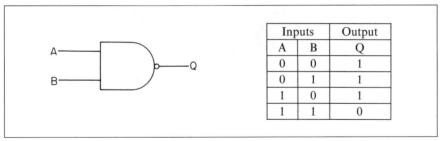

Inputs		Output
A	B	Q
0	0	1
0	1	1
1	0	1
1	1	0

Figure 33.29 _NAND gate_

As well as being a very useful gate in itself, the NAND gate is widely used because all other gates can be constructed using only NAND gates. For example, if a NOT gate is needed, it can be made from a NAND gate in two ways:

(a)   One input such as A is permanently tied (connected) high, so that A is 1 in the truth table, Figure 33.30 (i). In this case the only lines possible are lines 3 and 4, so if B is high, the output Q is low (0) and if B is low, Q is high (1).

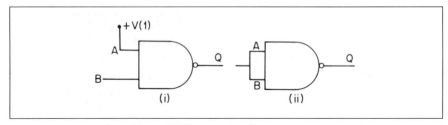

Figure 33.30   *NOT gate from* NAND *gate*

(b)   Both inputs may be tied (connected) together so that inputs A and B must both be high or low together, Figure 33.30 (ii). This means that only lines 1 and 4 of the truth table are possible and so, if A and B are both high, Q is low; and if A and B are both low, Q is high. However this method is not good electronic practice and should not be used.

OR Gate

An OR gate can have any number of inputs but for simplicity we shall consider a two-input OR gate. An OR gate gives a high output (1) if either input A or B, Figure 33.31, OR both are high (1), otherwise the output is low (0). The truth table shows the function of the OR gate.

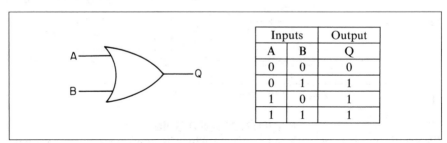

Inputs		Output
A	B	Q
0	0	0
0	1	1
1	0	1
1	1	1

Figure 33.31   OR *gate*

An OR gate could be used if a gardener wanted an alarm to sound when the air temperature in his greenhouse fell below 0°C OR rose above 30°C. Two sensors such as thermistors could be arranged so that either sensor gave a high input (1) to an OR gate, as, for example, in Figure 33.32. The output would then trigger an alarm.

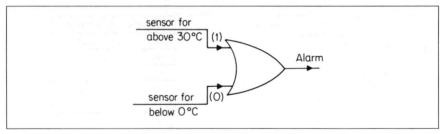

Figure 33.32   *Application of* OR *gate*

NOT OR or NOR Gate

A NOR gate is equivalent to an OR gate followed by a NOT gate (INVERTER), that is, all the outputs of the OR gate are inverted (changed from 0 to 1 or 1 to 0). The truth table for the NOR gate is therefore opposite to the OR gate as you should verify, Figure 33.33.

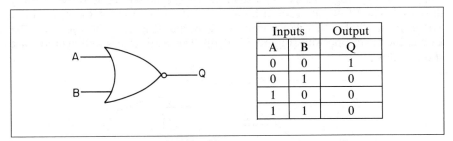

Inputs		Output
A	B	Q
0	0	1
0	1	0
1	0	0
1	1	0

Figure 33.33   *NOR gate*

Logic Gates Using NAND Gates

NAND gates can be used to construct all other logic gates such as OR, AND, NOR and NOT. Figure 33.34 shows three NAND gates that are equivalent to an OR gate. Figure 33.35 shows four NAND gates which are equivalent to a NOR gate. This can be seen from the truth tables, where Q_1, Q_2, Q are the outputs for the OR gate and Q_1, Q_2, Q_3, Q are the outputs for the NOR gate.

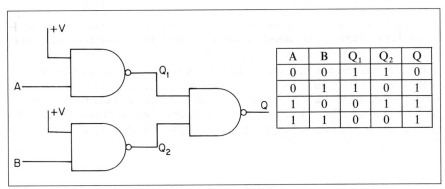

A	B	Q_1	Q_2	Q
0	0	1	1	0
0	1	1	0	1
1	0	0	1	1
1	1	0	0	1

Figure 33.34   *OR gate using NAND gates*

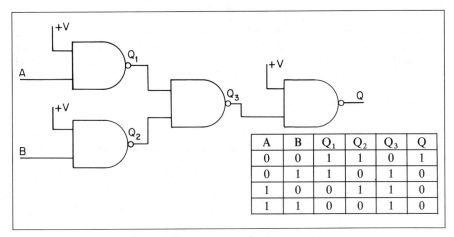

A	B	Q_1	Q_2	Q_3	Q
0	0	1	1	0	1
0	1	1	0	1	0
1	0	0	1	1	0
1	1	0	0	1	0

Figure 33.35   *NOR gate from NAND gates*

NOR gates can be used in a similar way to construct all other gates, AND, NAND, OR, NOT.

Exclusive OR (EOR) and Exclusive NOR (ENOR)

These last two logic gates are rather more specialised than those already discussed. The EOR gate gives a high output if EITHER input A OR input B but NOT BOTH are high, otherwise it gives a low output (Figure 33.36 (a)). The ENOR gate is the exact opposite (Figure 33.40 (b)). The truth tables and symbols are shown below.

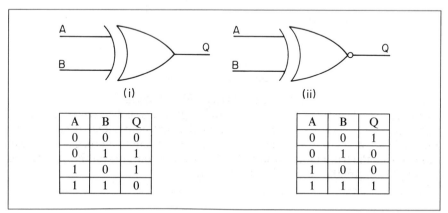

A	B	Q
0	0	0
0	1	1
1	0	1
1	1	0

A	B	Q
0	0	1
0	1	0
1	0	0
1	1	1

Figure 33.36 (*a*) *EXCLUSIVE OR* (*EOR*) *gate* (*b*) *EXCLUSIVE NOR* (*ENOR*) *gate*

It is also possible to construct an EOR gate using only NAND gates. Examination of the truth table for an EOR gate shows the output of an EOR to be the AND of the outputs of an OR and a NAND since only in lines 2 and 3 of their truth tables do both OR and NAND both give a high (1) output and the AND of two 1's is 1. This gives a clue to the final arrangement of NAND gates needed, which is shown in Figure 33.37 with a truth table.

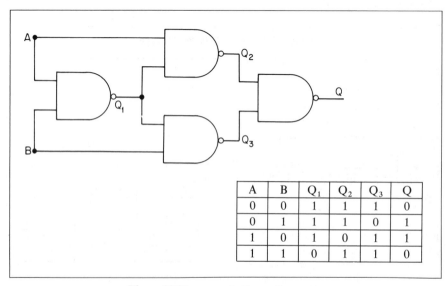

A	B	Q_1	Q_2	Q_3	Q
0	0	1	1	1	0
0	1	1	1	0	1
1	0	1	0	1	1
1	1	0	1	1	0

Figure 33.37 *EOR gate from NAND gates*

Central Heating Control Using Logic Gates

All the logic gates described can be connected together in various combinations to perform an endless variety of functions. These may range from simple control of traffic lights to arithmetic and data processing. One example is the use of logic gates in the control of a central heating system. The boiler might need to be on if: (i) The timeswitch is on AND the room thermostat is on OR (ii) There is a frost outside.

The logic combination for this would be as in Figure 33.38.

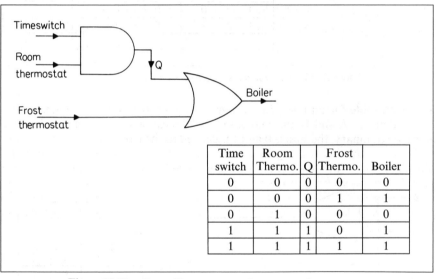

Time switch	Room Thermo.	Q	Frost Thermo.	Boiler
0	0	0	0	0
0	0	0	1	1
0	1	0	0	0
1	1	1	0	1
1	1	1	1	1

Figure 33.38 _Central heating control using_ AND _plus_ OR _gates_

The truth table shows how the output Q of the AND gate is combined with the input to the OR gate to make the boiler work correctly. As shown in Figure 33.39, page 834, this circuit could be redrawn using three NAND gates in place of the AND and OR gates. The truth table shows the outputs Q_1, Q_2 and the output Q to the boiler.

The truth tables in Figures 33.38 and 33.39 contain mainly those lines showing the conditions for the boiler to be 'ON'.

Half-Adder and Full-Adder

Another example of the use of logic gates is to add together two BInary digiTS (BITS). In binary addition, $0+0 = 0$, $0+1 = 1$, $1+0 = 1$ and $1+1 = 10$. In computers, binary addition has to be carried out very frequently and the circuit to do this addition with two bits (binary digits) is called a HALF-ADDER. A half-adder is shown in Figure 33.40. It consists of an EOR gate and an AND gate.

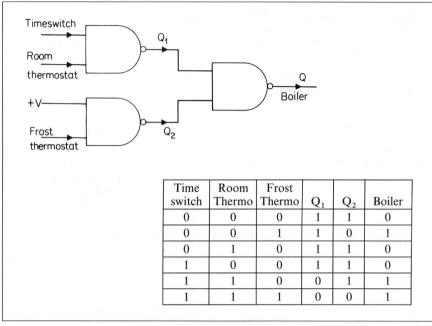

Figure 33.39 *Central heating control using only* NAND *gates*

The truth table for an EOR gate on page 832 gives the addition or *sum* of the binary digits at A and B; the AND gate gives the *carry* digit. For example, if we add $1 + 1$ in binary, the sum is 0 and 1 is carried to the next line.

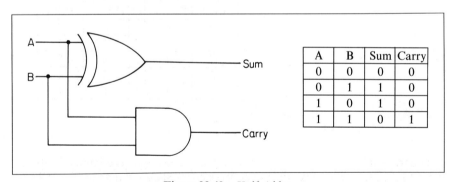

Figure 33.40 *Half-Adder*

In order to add other binary numbers, a circuit is needed that will add three bits together; two bits and the carry from the previous two bits. To add three bits at a time such as $1 + 1 + 1$ for example, we need a FULL-ADDER. A full-adder consists of two half-adders and an OR gate, as shown in Figure 33.41. The truth table shows the sum digit and the carry digit for the three inputs A, B, C.

In order to add two 4-bit numbers, four full-adders are needed.

Bistable Circuit, Flip-Flop

We now deal with circuitry in which the output is determined not only by the input at a particular time but also by previous inputs.

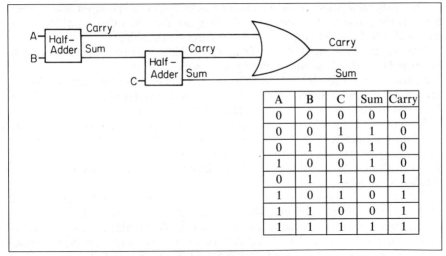

A	B	C	Sum	Carry
0	0	0	0	0
0	0	1	1	0
0	1	0	1	0
1	0	0	1	0
0	1	1	0	1
1	0	1	0	1
1	1	0	0	1
1	1	1	1	1

Figure 33.41 _Full-Adder_

The _bistable_ is one such circuit. As shown in Figure 33.42, it has two NAND gates N1 and N2, linked in a special way. The output of N_1 provides an input for N_2 and the output of N_2 provides an input for N_1. This circuit is known as a BISTABLE or a FLIP-FLOP because it has two stable states and it can be switched or 'flipped' between the two states, which are discussed shortly. Such circuits are the building blocks of computers.

The circuit has two inputs called SET (S) and RESET (R) and two outputs, (Q) and $(\overline{Q})$, called 'Q' and 'not Q'. $\overline{Q}$ is the complement of Q, so if Q = 1, $\overline{Q}$ = 0, and if Q = 0, $\overline{Q}$ = 1.

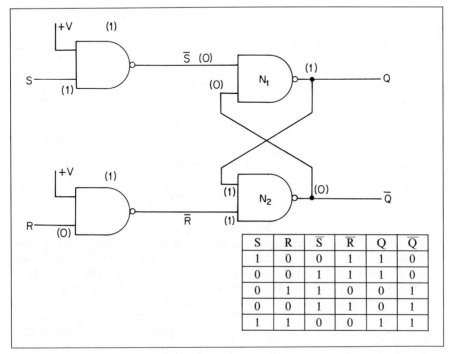

S	R	$\overline{S}$	$\overline{R}$	Q	$\overline{Q}$
1	0	0	1	1	0
0	0	1	1	1	0
0	1	1	0	0	1
0	0	1	1	0	1
1	1	0	0	1	1

Figure 33.42 _Bistable circuit_

The *truth table* shows how the circuit works. Suppose the input S = 1. Then
$\bar{S}$ = 0 is the input to the NAND gate N_1. Suppose that the other input to the N_1
gate is also 0. Then the output Q of N_2 = 1 (see truth table).

Now suppose the input R = 0. Then $\bar{R}$ = 1. Now the input for the NAND gate
N_2 from N_1 is also 1. So the output $\bar{Q}$ of N_2 = 0.

So the output 1 for N_1 provides the input 1 for N_2, and the output 0 for N_2
provides the input 0 for N_1. The circuit, then, is in a *stable state* with S = 1 and
R = 0. With the output Q = 1, the circuit is said to be SET.

By the same argument, the truth table shows that further changes to S have no
effect on the output Q (or $\bar{Q}$) until a change is made so that R = 1 and S = 0.
Then the Q output changes to 0 and $\bar{Q}$ to 1 and the circuit is said to be RESET.
This can also be shown to be stable state. Since it has *two* stable states, the circuit
is called a *bistable*.

The truth table shows that if R = S = 0, there is no change. If R = S = 1, both
Q and $\bar{Q}$ are 1, which is not desirable, and this situation is avoided.

The circuit is known as a SR FLIP-FLOP. The flip-flop is able to store one BIT of
information at Q such as 1 or 0. It therefore acts as a ONE-BIT MEMORY, since once
Q has been made 1, it will stay 1 until S and R are changed. A computer can
'read' the output Q at any time and find the output that has been stored. A
computer will have a large number of flip-flops as its memory.

Clocked SR FLIP-FLOP

A modification to the above circuit can provide a third input for a 'clock' circuit.
The clock (CK) sends out rectangular pulses which rise and fall from low (0) to high
(1) and back again at regular intervals. The low to high transitions occur at equal
time intervals. Figure 33.45 on page 837 shows clock pulses. Inputs to S and R
will only affect the outputs Q and $\bar{Q}$ when the clock input is high. This is known
as a CLOCKED SR FLIP-FLOP and is shown diagrammatically in Figure 33.43.

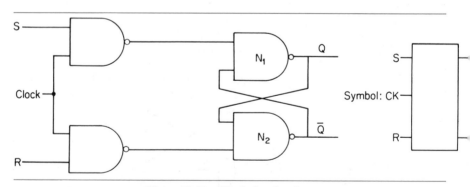

Figure 33.43 *Clocked SR flip-flop*

Binary Counter

A combination of these flip-flops can be used to make a *binary counter*. Two
modifications of the clocked SR flip flop are needed. Firstly, the output Q must
change state *only* as the clock pulse *changes* from low to high. If this is not done,
the binary counter will not work correctly. This so-called 'edge triggering' is
achieved by circuitry which we do not need to discuss.

Secondly, and importantly, the SR flip-flop must be made into a *T-type flip-
flop* or 'toggle' flip-flop, so called because the output 'toggles' or switches state
with each clock pulse, as we shall see later. The symbol for a T flip-flop is shown
in Figure 33.44. This has a chain of four flip-flops, each with input T and with

outputs Q_1, Q_2, Q_3, Q_4 respectively which are connected to light emitting diodes A, B, C, D. A LED will light or be ON when the output Q is HIGH or 1.

In the T flip-flop, there is no S input and the output Q changes state on each rising edge (0 to 1 transition) of the clock pulse. Some flip-flops change state on the falling edge of the clock pulse. There is a RESET (R) input and this causes Q to go to 0 irrespective of the state of the clock.

Four-Bit Binary Counter

We can now show how the four T flip-flops in Figure 33.44 act as a _four-bit binary counter_, that is, a counter with four binary digits which will record the number of incoming clock pulses. The $\overline{Q}$ output of the first flip-flop is connected to the toggle input T of the second flip-flop, the $\overline{Q}$ output of the second flip-flop is connected to the toggle input of the third flip-flop and so on. The outputs Q_1, Q_2, Q_3, Q_4 provide the binary digits A, B, C, D. The R connections on all four flip-flops are connected together so that a 1 on the combined reset inputs will reset the counter to 0000 regardless of the clock.

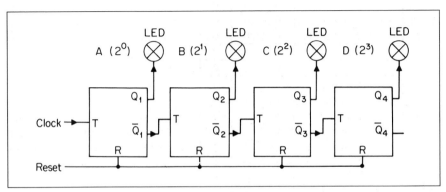

Figure 33.44   _Four-bit binary counter_

Figure 33.45 illustrates how the clock pulses or inputs affect the outputs Q_1, Q_2, Q_3, Q_4 of the flip-flop chain.

At the start, suppose that the outputs are all set to 0. As the first clock pulse rises from 0 to 1, Q_1 changes from 0 to 1 and so LED A is now ON. At this stage, $Q_1 = 1$, $Q_2 = 0$, $Q_3 = 0$, $Q_4 = 0$. So the binary digit recorded is 0001 reading from D (the Most Significant Bit) to A (the Least Significant Bit).

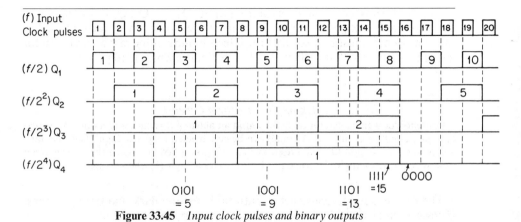

Figure 33.45   _Input clock pulses and binary outputs_

When the second clock pulse rises from 0 to 1 in a similar way, Q_1 changes from 1 to 0 and therefore Q_1 changes from 0 to 1. This provides the toggle for the second flip-flop and thus Q_2 changes from 0 to 1 as shown. LED B is now ON while LED A is OFF. The counter now records 0010 reading from D to A; which is 2 in decimal notation.

We can now see that if the frequency of the incoming clock pulses is f, the state of Q_1 changes at a frequency of $f/2$, Q_2 at a frequency of $f/2^2$, Q_3 at a frequency of $f/2^3$ and Q_4 at a frequency of $f/2^4$. This continual division by 2 shows that the chain of flip-flops can act as a binary counter. At the fourth incoming clock pulse, for example, $Q_1 = 0$, $Q_2 = 0$, $Q_3 = 1$, $Q_4 = 0$. The binary number recorded reading from D to A (most significant bit to least significant bit) is 0100.

Number of clock pulses	Outputs			
	Q_1	Q_2	Q_3	Q_4
0	0	0	0	0
1	1	0	0	0
2	0	1	0	0
3	1	1	0	0
4	0	0	1	0
5	1	0	1	0
6	0	1	1	0
7	1	1	1	0
8	0	0	0	1
9	1	0	0	1
10	0	1	0	1
11	1	1	0	1
12	0	0	1	1
13	1	0	1	1
14	0	1	1	1
15	1	1	1	1
16	0	0	0	0

Read binary from Q_4 to Q_1

The Table shows how the states of the outputs Q_1, Q_2, Q_3, Q_4 record in binary the incoming clock pulses. Figure 33.45 illustrates some binary recordings. The four-bit counter must be reset after the fifteenth pulse as the maximum count for a four bit counter is 0 to 15 (0000 to 1111 in binary), which is 16 pulses.

Example on Alarm Circuit

A householder wants a circuit that will sound an alarm if it is dark and a door is opened or a 'panic button' is pressed. A circuit can be constructed to give a 'high' (1) in the dark and a 'low' in the light (0) and a door switch will give a 'low' (0) when the door is open otherwise it gives a 'high' (1). The panic button gives a 'high' (1) when pressed.

The first stage is to construct a truth table for the dark and door switches. Using 1 for high and 0 for low gives the following truth table:

Door	Light	Output
0	0	0
0	1	1
1	0	0
1	1	0

This is the condition for the alarm.

The output looks a little like an AND gate output except that the door input would have to be a 1 for an AND gate to be possible. If we try *inverting* the door input we get the following truth table:

Inverted Door	Light	Output
1	0	0
1	1	1
0	0	0
0	1	0

This is the condition for the alarm.

The truth table above is the truth table for an AND gate (the lines are not in the usual order). The arrangement of gates would therefore be as in Figure 33.46.

The final condition can be met with an OR gate. The panic button gives a high when pressed, therefore the truth table would be:

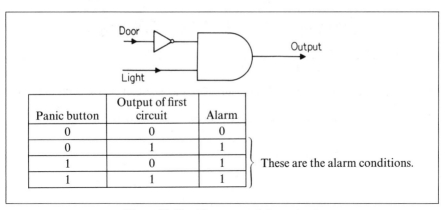

Panic button	Output of first circuit	Alarm
0	0	0
0	1	1
1	0	1
1	1	1

These are the alarm conditions.

Figure 33.46 *First stage of alarm circuit*

Therefore the complete circuit looks like Figure 33.47.

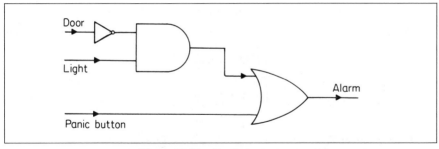

Figure 33.47 *Final alarm circuit*

We can see this is a solution because the alarm is sounded if it is dark AND a door is opened OR a 'panic button' is pressed.

_____ Exercises 33B _____

(The following symbols are used in the Exercise)

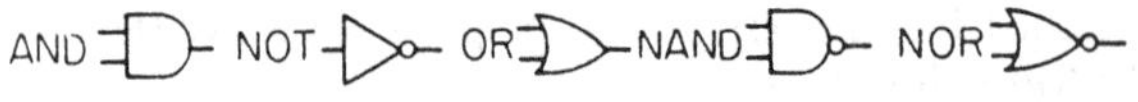

AND ⟩— NOT —▷∘— OR ⟩— NAND ⟩∘— NOR ⟩∘—

Figure 33K

1 Draw the circuit symbol for a two-input AND gate.
   An exclusive-OR gate is one where the output is high only when one or other but not both of the two inputs is high. Write out the truth table for this gate. (*C.*)

2 Figure 33L shows inputs X, Y, Z to a system of logic gates. Draw a truth table showing the outputs P and Q and the final output R.
   Show that a NAND gate system equivalent to all the NOR gates is:
   (i) NAND gate with X and Z inputs, (ii) NAND gate used as NOT gate with Y input, (iii) outputs P and Q to another NAND gate.

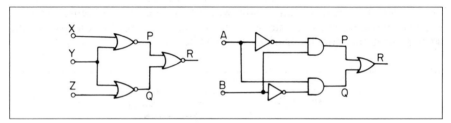

Figure 33L   **Figure 33M**

3 Figure 33M shows a system of logic gates. Draw a truth table showing the outputs P, Q and R.

4 Construct a truth table for the arrangement of NAND gates shown in Figure 33N.

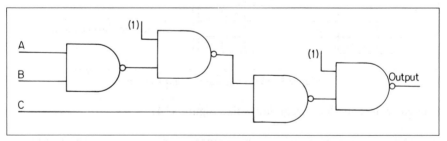

Figure 33N

5 Design a circuit that will give the truth table shown. (*Hint* Use EOR and NAND gates)

Inputs		Output
A	B	Q
0	0	1
0	1	0
1	0	1
1	1	1

6 Draw a system of
   (a) NOR gates, and then

(b) NAND gates, which can produce   (i) an AND gate and   (ii) an OR gate.
7   Figure 33O shows two NAND gates followed by a NOR gate. Construct a truth table showing the outputs P, Q and R.
    The system is equivalent to one gate G with inputs X, Y, Z and output R. What is G?

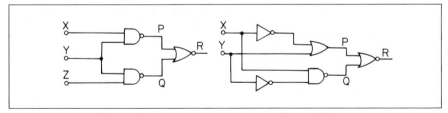

Figure 33O   **Figure 33P**

8   Figure 33P shows a system of logic gates. Construct a truth table showing the outputs P, Q and R.
    From the truth table, what inputs X and Y will produce a high output (1) at R?
9   The waveforms of inputs A and B to a logic gate are shown in Figure 33Q (a) and (b). In each case copy the inputs and show beneath them the output waveform at Q.

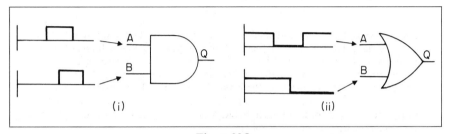

Figure 33Q

10   Figure 33R shows two NAND gates with feedback from each output. Construct a truth table starting with X = 1, Y = 1, P = 0 and show that the outputs P and Q have two stable states.

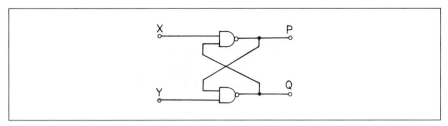

Figure 33R

11   Four NAND gates can produce an EOR (Exclusive-OR) gate. Draw a sketch of the system and construct a truth table to verify your answer.
12   Show how a full-adder is produced, starting with a half-adder. Explain your answer.
13   In a car, a red warning light comes on when the ignition is switched on if the door is not closed properly, or the seat belt is not fastened, or both door and seat belt are not properly secured.
    (a) Construct the truth table, showing the door and seat belt inputs to a logic gate system as high (1) if each is properly secured and low (0) if not properly secured, the ignition input as high (1) if switched on and low (0) if off, and the red warning light, the output, as high (1) if on and low (0) if off.

(b) Then draw a suitable logic gate system for operating the red warning light.

14 A big saw for cutting wood at a timber store is designed to operate by switching on an electric motor only when a safety guard is lowered completely round the saw. In this case a green light comes on when the motor is switched on. If the safety guard is not lowered, a red light comes on when the machine is switched on.

(a) Construct a truth table for a logic gate system showing the motor switched on as high (1) input and off as low (0) input, the safety guard input as high (1) when lowered and low (0) when not lowered, and the outputs high (1) for switching on the lights and low (0) for not switching.

(b) Draw a logic gate system showing how the green and red lights can be operated as required.

15

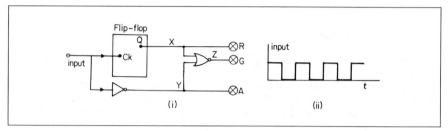

Figure 33S

Figure 33S (i) shows an input clock pulse to a toggle flip-flop and NOT gate, followed by a NOR gate. The output Q of the flip-flop changes from low to high *or* from high to low when the input pulse (Figure 33S (ii) goes only from low to high.

The circuit should operate the red (R), amber (A) and green (G), lights, which go on only with a high (1) input, in British traffic sequence. Draw sketches showing (i) the input clock pulse, (ii) below it the corresponding outputs X, Y and Z, (iii) the sequence or order of the lights R, A and G obtained from your X, Y and Z graphs (show R, A, G on the graphs).

16 (a) A light emitting diode (LED) which is connected to the output of a logic gate is required to light when the output is logic 1 (+5 V). For this to be achieved a forward current of 10 mA together with a forward p.d. of 2·0 V is required. Draw a circuit diagram to show how the LED is connected and calculate the value of any additional component required.

(b)

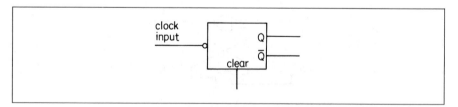

Figure 33T

The symbol shown in Figure 33T represents a T or toggle flip-flop circuit. The small circle on the clock input indicates that toggling occurs when an input pulse goes from logic 1 to logic 0. The output Q can be set to logic 0, i.e. cleared, by momentarily connecting the clear input to a zero voltage line. With the aid of a truth table for input and outputs, explain what is meant by the toggling action of this circuit.

The clock pulses shown in Figure 33U are applied to the flip-flop circuit. Using the same time scale as Figure 33U, sketch the outputs from Q and Q̄.

(c) Draw a diagram to show how four of the circuits shown in Figure 33T can connect together to make a four-bit binary counter to count upwards from zero.

Figure 33U

Show on your diagram where you would connect four LEDS to indicate the binary count, labelling them A to D with A being the least significant bit.

The clock pulses shown in Figure 33U are fed into the input of the four-bit counter after its outputs have been cleared. Using a common time scale, show by means of graphs how the output signal from each of the four flip-flops varies between logic 0 and logic 1 for nine input pulses. (*JMB.*)

Miscellaneous Electronics Questions

17 Draw an amplifying circuit for which the voltage gain is 8.
The output signal may carry more power than the input signal. What is the source of this power?
Explain, with reference to your circuit, the term 'negative feedback'. (*L.*)

18 (a) Construct a truth table for each part of the circuit outlined by a dotted box (as shown below). For values of the inputs at A and B give the corresponding values at each labelled point in the circuit and hence identify the logic gate equivalent of each box.
State the function of the complete circuit and distinguish between the two outputs.

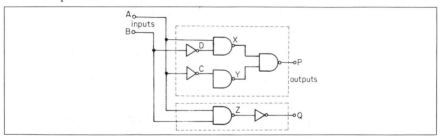

Figure 33V

(b) Draw a circuit diagram showing how two NAND gates can be used to construct a simple bistable latch or SR flip-flop.
Show the two inputs, S and R, connected to a two-way selector switch which allows either S or R to be grounded (connected to the negative terminal of the power supply). (*JMB. Part Qn*)

19 (a) Draw a circuit diagram of an operational amplifier used in the inverting mode as a summing amplifier with two inputs. Show that, if the input resistors of resistance R_1 and R_2 respectively are of equal magnitude and the feedback resistor has resistance R_f, the output voltage V_{out} is given by

$$V_{out} = -\frac{R_f}{R_1}(V_1 + V_2)$$

where V_1 and V_2 are the voltages applied to the two inputs.
Show how you would modify the summing amplifier so that it would accept positive input voltages but ignore negative input voltages. Explain your modification.

(b) Taking the values of R_1 and R_2 to be $5\,\text{k}\Omega$ draw another circuit diagram, giving component values which would *multiply* the sum of positive input voltages by -2 but would *divide* the sum of negative input voltages by -2. Explain your modification.
Calculate V_{out} when (i) $V_1 = +2\,\text{V}$ and $V_2 = +3\,\text{V}$, (ii) $V_1 = +2\,\text{V}$ and $V_2 = -8\,\text{V}$. Under what circumstances would the output voltage not have the expected value? (*JMB.*)

34

Photoelectricity, Energy Levels, X-Rays, Wave-Particle Duality

In this chapter, we first deal with photoelectricity and the Einstein photon theory. We then discuss the energy levels in the atom and their application to spectra, X-rays and their properties, the de Broglie formula and the wave-particle duality.

Photoelectricity: Particle Nature of Waves

Photoelectricity

In 1888 Hallwachs discovered that an insulated zinc plate, negatively charged, lost its charge if exposed to ultraviolet light. Later investigators such as Lenard and others showed that electrons were liberated from a zinc plate when exposed to ultraviolet light. Light thus gives energy to the electrons in the surface atoms of the metal, and enables them to break through the surface. This is called the *photoelectric effect*.

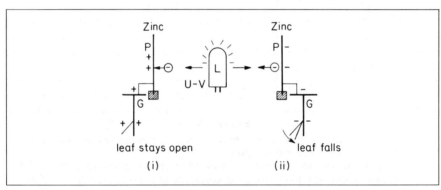

Figure 34.1 *Photoelectric demonstration*

Figure 34.1 (i) shows diagrammatically a simple demonstration of the photoelectric effect. The surface of a zinc plate P was rubbed with emery paper until the surface was clean and bright. P was then insulated and connected to the cap of a gold-leaf electroscope G, as shown, and given a *positive* charge by induction. Some of the charge spread to the leaf which then opened.

In a dark room, P was exposed to ultraviolet radiation from a small lamp L placed near it. The leaf stayed open. However, when the experiment was repeated with the plate P charged *negatively*, the leaf slowly collapsed, Figure 34.1 (ii).

The results are explained as follows. Electrons are usually emitted by the plate P when exposed to ultraviolet light. When P is positively-charged, any electrons (negative charges) would be attracted back to P. When P is negatively charged, however, the electrons emitted by P are now *repelled away from the plate*. So P loses negative charge and the leaf slowly falls.

Velocity or Kinetic Energy of Photoelectron

In 1902 Lenard found that the *velocity* or *kinetic energy* of the electron emitted from an illuminated metal was independent of the intensity of the particular incident monochromatic light. It appeared to vary only with the *wavelength* or *frequency* of the incident light.

Further, for a given metal, no electrons were emitted when it was illuminated by light of wavelength longer than a particular wavelength, called the *threshold wavelength*, no matter how great was the intensity of the light beam. But as soon as the metal was illuminated by light whose wavelength was *lower* than the threshold wavelength, electrons were emitted. Even though the light beam was made extremely weak in intensity, it was estimated that the electrons were emitted about 10^{-9} second after exposure to the light, that is, practically simultaneously with exposure to the weak light.

Classical or Wave Theory

On the wave theory of light, the so-called classical theory, these results are very surprising. If we assume light is sent out in waves from a source, the greater the intensity of the light the greater will be the energy per second reaching the illuminated plate. So the classical theory can explain why the number of electrons emitted increases as the light intensity increases.

But it can not explain the result that the velocity or kinetic energy of the emitted electrons is independent of the intensity of the incident light beam. On the classical theory, the greater the intensity of the beam, the greater should be the kinetic energy of the emitted electrons because the energy per second reaching the plate increases with the intensity of the light.

Further, on the classical theory electrons should always be emitted by light of *any* wavelength if the incident light beam is strong enough. Experiment, however, shows that however intense the light beam, *no* electrons are emitted if the wavelength is lower than the threshold values.

Experiments show that: (1) the maximum energy of the emitted electrons is *not* proportional to the light intensity, and that (2) electrons are *not* emitted for all wavelengths, which contradicts the classical (wave) theory, of light.

Quantum Theory of Radiation, Planck Constant

In 1902 Planck had shown that the experimental observations in black-body radiation could be explained on the basis that the energy from the body was emitted in separate packets of energy. Each packet was called a *quantum* of energy and the amount of energy E carried was equal to hf, where f is the *frequency* of the radiation and h was a constant called the *Planck constant*. So

---

$$E = hf \qquad . \qquad . \qquad . \qquad . \qquad . \qquad (1)$$

This is the *quantum theory of radiation*. Until Planck's quantum theory, it was considered that radiation was emitted continuously and not in separate packets of energy. Since $h = E/f$, the unit of h is joule second or J s. Measurements of radiation showed that, approximately, $h = 6·63 \times 10^{-34}$ J s.

The quantum of energy carried by radiation of wavelength 3×10^{-6} m can be calculated from $E = hf$. Since $f = c/\lambda$, where c is the speed of electromagnetic waves, about 3×10^8 m s^{-1},

$$f = \frac{3 \times 10^8}{3 \times 10^{-6}} = 10^{14}\,\text{Hz}$$

So $$E = hf = 6 \cdot 63 \times 10^{-34} \times 10^{14} = 6 \cdot 6 \times 10^{-20} \text{ J (approx.)}$$

Work Function

The minimum amount of work or energy necessary to take a free electron out of a metal against the attractive forces of surrounding positive ions is called the *work function* of the metal, symbol w_0. The work function is related to thermionic emission since this phenomenon is also concerned with electrons breaking away from the metal.

The work functions of caesium, sodium and beryllium are respectively about $1 \cdot 9$ eV, $2 \cdot 0$ eV and $3 \cdot 9$ eV. 1 eV (electronvolt) is a unit of energy equal to 1 electron charge $e \times 1$ volt, which is (p. 197)

$$1 \cdot 6 \times 10^{-19} \text{ C} \times 1 \text{ V} = 1 \cdot 6 \times 10^{-19} \text{ J}$$

So the work function of sodium, $w_0 = 2 \text{ eV} = 2 \times 1 \cdot 6 \times 10^{-19} = 3 \cdot 2 \times 10^{-19} \text{ J}$

Einstein's Particle (Photon) Theory

In 1905 Einstein suggested that the experimental results in photoelectricity could be explained by applying a quantum theory of light. He assumed that light of frequency f contains packets or quanta of energy hf. On this basis, light consists of *particles*, and these are called *photons*. The *number* of photons per unit area of cross-section of the beam of light per unit time is proportional to its intensity. But the *energy* of a photon is proportional to its frequency and is *independent* of the light intensity.

Let us apply the theory to the metal sodium. Its work function w_0 is $2 \cdot 0$ eV or $3 \cdot 2 \times 10^{-19}$ J. According to the theory, if the quantum energy in the incident light is $3 \cdot 2 \times 10^{-19}$ J, then electrons are just liberated from the metal. The particular frequency f_0 is called the *threshold frequency* of the metal. The *threshold wavelength* $\lambda_0 = c/f_0$.

We can now calculate f_0 and λ_0 for the metal sodium. Since $E = hf_0 = w_0$,

$$f_0 = \frac{w_0}{h} = \frac{3 \cdot 2 \times 10^{-19}}{6 \cdot 6 \times 10^{-34}} = 4 \cdot 8 \times 10^{14} \text{ Hz}$$

So $$\lambda_0 = \frac{c}{f} = \frac{3 \times 10^8}{4 \cdot 8 \times 10^{14}} = 6 \cdot 2 \times 10^{-7} \text{ m}$$

Thus electrons are not liberated from sodium if the incident light has a frequency *less* than $4 \cdot 8 \times 10^{14}$ Hz or a wavelength *longer* than $6 \cdot 2 \times 10^{-7}$ m. This agrees with experiment.

Einstein's Photoelectric Equation

Using Einstein's theory, a simple result is obtained for the maximum energy of the electrons liberated from an illuminated metal.

If the quantum of energy hf in the light incident on a metal is say $4 \cdot 2 \times 10^{-19}$ J, and the work function w_0 of the metal is $3 \cdot 2 \times 10^{-19}$ J, the *maximum* energy of the liberated electrons is $(4 \cdot 2 \times 10^{-19} - 3 \cdot 2 \times 10^{-19})$ or $1 \cdot 0 \times 10^{-19}$ J because w_0 is the minimum energy to liberate electrons from the metal. The electrons with maximum energy come from the metal surface, as we have already stated. Owing to collisions, others below the surface emerge with a smaller energy.

We now see that the maximum kinetic energy, E_{max} or $\frac{1}{2}m_e v_m{}^2$, of the emitted electrons (photoelectrons) is given generally by

$$E_{max} = hf - w_0 \quad . \quad . \quad . \quad . \quad . \quad \text{(i)}$$

So
$$hf = E_{max} + w_0 \quad . \quad . \quad . \quad . \quad . \quad \text{(ii)}$$

Equation (ii) or (i) is called *Einstein's photoelectric equation.*

Examples on Maximum Energy and Threshold Frequency

Sodium has a work function of 2·0 eV. Calculate the maximum energy and speed of the emitted electrons when sodium is illuminated by radiation of wavelength 150 nm.

What is the least frequency of radiation (threshold frequency) for which electrons are emitted? (Assume $h = 6·6 \times 10^{-34}$ J s, $e = -1·6 \times 10^{-19}$ C, $m_e = 9·1 \times 10^{-31}$ kg, $c = 3 \times 10^8$ m s^{-1}.)

(1) Incident photon energy $= hf = h\dfrac{c}{\lambda} = \dfrac{6·6 \times 10^{-34} \times 3 \times 10^8}{150 \times 10^{-9}}$

$$= 13·2 \times 10^{-19} \text{ J}$$

So   maximum kinetic energy $= hf - w_0 = 13·2 \times 10^{-19} \text{ J} - 2 \times 1·6 \times 10^{-19} \text{ J}$

$$= 10 \times 10^{-19} = 10^{-18} \text{ J} \quad . \quad . \quad . \quad \text{(i)}$$

Thus $\frac{1}{2}m_e v_m^2 = 10^{-18}$

So
$$v_m = \sqrt{\frac{2 \times 10^{-18}}{9·1 \times 10^{-31}}} = \sqrt{\frac{20}{9·1}} \times 10^6$$

$$= 1·5 \times 10^6 \text{ m s}^{-1} . \quad . \quad . \quad . \quad \text{(ii)}$$

(2) The threshold frequency is given by $hf_0 = w_0$. So

$$f_0 = \frac{w_0}{h} = \frac{2 \times 1·6 \times 10^{-19}}{6·6 \times 10^{-34}}$$

$$= 4·8 \times 10^{14} \text{ Hz} \quad . \quad . \quad . \quad . \quad \text{(iii)}$$

Measuring Maximum Kinetic Energy, Stopping Potential

The maximum kinetic energy of the liberated photoelectrons can be found by a method analogous to finding the kinetic energy of a ball moving horizontally. In Figure 34.2 (i), the moving ball is made to move up a smooth inclined plane PQ each time the slope of the plane is increased from zero. At one inclination the ball

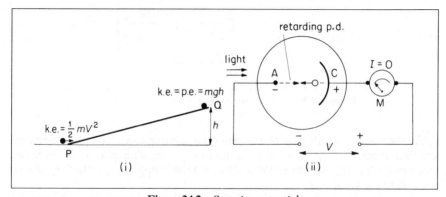

Figure 34.2   *Stopping potential*

will *just* reach the top of the plane. In this case the kinetic energy, $\frac{1}{2}mv^2$, at the bottom P is just equal to the potential energy *mgh* at Q. Knowing the height *h* of Q above P, the kinetic energy is equal to *mgh*.

Figure 34.2 (ii) shows an analogous electrical experiment using a potential 'hill' whose gradient can be varied. Here a varying p.d. *V* is applied between the plates A and C inside an evacuated glass tube, and the potential of A is *negative* relative to C. This means that when photoelectrons are liberated from C, they are acted on by a retarding force.

When C is illuminated by a suitable beam, the photoelectrons liberated have a varying kinetic energy as we have previously explained. If the negative p.d. *V* between A and C is very small, many electrons can reach A from C. As *V* is increased negatively, however, the current *I* recorded on the meter M decreases because fewer electrons have energies sufficient to overcome the retarding force.

At a particular negative value V_s the current *I* becomes zero. This is the value of the negative p.d. which just stops the electrons with maximum energy from reaching A. V_s is called the *stopping potential*. Since an electron would lose an amount of energy given by *charge × p.d.* in moving from C to A, we see that

$$E_{\text{max}} = e\,V_s$$

So $e V_s$ is a measure of the maximum energy of the emitted photoelectrons.

Verifying Einstein's Equation, Measurement of h

Figure 34.3 illustrates the basic features of a laboratory apparatus for investigating photoelectricity. It contains (1) a photoelectric cell X, which has inside it a photosensitive metal C of large area, the cathode, and a collector A of the electrons, in a vacuum, (2) a potential divider arrangement Y for varying the p.d. *V* between the anode A and cathode, C, and (3) a d.c. amplifier for measuring a small current.

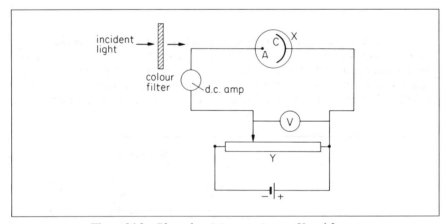

Figure 34.3 *Photoelectricity experiment:* $eV_s = hf - w_0$

As shown, A is made *negative* in potential relative to C. The photoelectrons emitted from C then experience a retarding p.d. The p.d. *V* is increased negatively until the current becomes zero and the 'stopping potential', V_s, is then read from the voltmeter.

Using pure filters, the frequency *f* of the incident light can be varied and the corresponding value of V_s obtained. Figure 34.4 (i) shows some typical results. For different wavelengths λ_1, λ_2, λ_3, the respective stopping potentials V_s (when $I = 0$) are V_1, V_2, V_3. The shortest wavelength λ_1 (greatest frequency) required the highest stopping potential V_1.

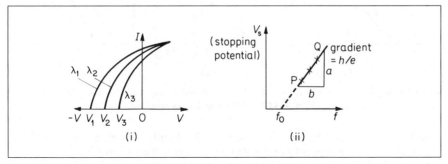

Figure 34.4 *Results of experiment*

The values of V_s are now plotted against the corresponding frequency f of the incident light. Figure 34.4 (ii) shows the result. A straight line graph PQ is obtained. Now from Einstein's photoelectric equation, $eV_s = hf - w_0$. So

$$V_s = \frac{h}{e}f - \frac{w_0}{e}$$

Thus from Einstein's theory, the gradient a/b of the line PQ is h/e, or

$$h = e \times \frac{a}{b}$$

Knowing e, the electron charge, h can be calculated. Careful measurements first carried out by Millikan gave a result for h of $6 \cdot 26 \times 10^{-34}$ J s, which was very close to the value of h found from experiments on black-body radiation. This confirmed Einstein's photoelectric theory that light can be considered to consist of particles (photons) with energy hf.

The line PQ in Figure 34.4 (ii) also enables the threshold frequency f_0 and the work function w_0 to be found. From $eV_s = hf - w_0$, we have $0 = hf - w_0$ when $V_s = 0$. So $f = w_0/h = f_0$. Thus the intercept of PQ with the frequency axis gives the value of f_0. The work function can then be calculated from $w_0 = hf_0$.

Experimental Results

Figure 34.5 (i) shows the results when a metal such as caesium is illuminated by monochromatic light of a given wavelength λ which is below the threshold value for the metal. When the retarding p.d. $-V$ is increased negatively, the stopping potential V_s is the *same* for a beam of low intensity and one of high intensity. This is because $eV_s = hf - w_0 = hc/\lambda - w_0$, and λ and w_0 are constant for a given metal and wavelength.

Figure 34.5 (i) also shows that when V is positive, (i) all the photoelectrons are now collected so that the current is constant, and (ii) a beam of high intensity produces more electrons than one of low intensity. If Q has twice the intensity of P, the current I is twice as much.

Figure 34.5 (ii) shows how the current I varies with intensity for a given

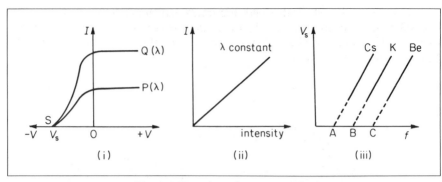

Figure 34.5 *Photoelectricity results*

wavelength λ below the threshold value. The number of electrons emitted is proportional to the intensity and a straight line graph is obtained.

Figure 34.5 (iii) shows the results when the stopping potential V_s is plotted against the frequency f for three different metals caesium (Cs), potassium (K) and beryllium (Be). The respective threshold frequencies A, B, C are different because the metals have different work functions; but the *slope of the three straight line graphs is the same* since the slope is given by h/e (see p. 849) and h and e are constants.

Example on Stopping Potential

Caesium has a work function of 1·9 electronvolts. Find (i) its threshold wavelength, (ii) the maximum energy of the liberated electrons when the metal is illuminated by light of wavelength $4·5 \times 10^{-7}$ m, (iii) the stopping p.d. (1 electronvolt = $1·6 \times 10^{-19}$ J, $h = 6·6 \times 10^{-34}$ J s, $c = 3·0 \times 10^8$ m s^{-1}).

(i) The threshold frequency f_0 is given by $hf_0 = w_0 = 1·9 \times 1·6 \times 10^{-19}$ J.

Now threshold wavelength, $\lambda_0 = c/f_0$

$$\therefore \lambda_0 = \frac{c}{w_0/h} = \frac{ch}{w_0}$$

$$= \frac{3 \times 10^8 \times 6·6 \times 10^{-34}}{1·9 \times 1·6 \times 10^{-19}}$$

$$= 6·5 \times 10^{-7} \text{ m}$$

(ii) Maximum energy of liberated electrons $= hf - w_0$, where f is the frequency of the incident light. But $f = c/\lambda$.

$$\therefore \text{ max. energy} = \frac{hc}{\lambda} - w_0$$

$$= \frac{6·6 \times 10^{-34} \times 3 \times 10^8}{4·5 \times 10^{-7}} - 1·9 \times 1·6 \times 10^{-19}$$

$$= 1·4 \times 10^{-19} \text{ J}$$

(iii) The stopping potential V_s is given by $eV_s = 1·4 \times 10^{-19}$ J. Since $e = 1·6 \times 10^{-19}$ C,

$$\therefore V = \frac{1·4 \times 10^{-19}}{1·6 \times 10^{-19}} = 0·9 \text{ V (approx.)}$$

Photo-emissive and Photo-voltaic Cells

Photoelectric cells are used in photometry, in industrial control and counting operations, in television, and in many other ways.

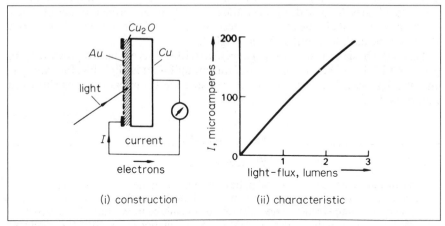

Figure 34.6 *A photo-voltaic cell*

Photoelectric cells of the kind we describe on p. 848 are called *photo-emissive* cells, because in them light causes electrons to be emitted. Another type of cell is *photo-voltaic*, because it generates an e.m.f. and can therefore provide a current without a battery. One form of such a cell consists of a copper disc, oxidised on one face (Cu_2O/Cu), as shown in Figure 34.6 (i). Over the exposed surface of the oxide a film of gold (Au) is deposited, by evaporation in a vacuum; the film is so thin that light can pass through it. When it does so it generates an e.m.f. in a way which we cannot describe here.

Photo-voltaic cells are sensitive to visible light. Figure 34.6 (ii) shows how the current from such a cell, through a galvanometer of resistance about $100\,\Omega$, varies with the light-flux falling upon it. The current is not quite proportional to the flux. Photo-voltaic cells are obviously convenient for photographic exposure meters, for measuring illumination in factories, and so on, but as measuring instruments they are less accurate than photo-emissive cells.

Solar batteries are photo-voltaic cells in series. Large panels of suitable semiconductor material are attached to satellites and form solar batteries which work when sunlight falls on them. Electrical apparatus inside the satellite is connected to the batteries.

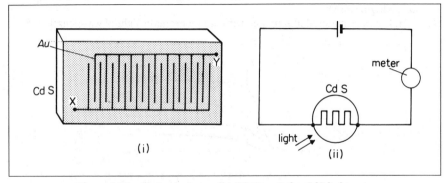

Figure 34.7 *(i) A selenium cell (ii) Circuit for CdS lightmeter*

Photo-conductive Cells

A photo-conductive cell is one whose resistance changes when it is illuminated. A common form consists of a pair of interlocking comb-like electrodes made of gold (Au) deposited on glass (Figure 34.7 (i)). Over these a thin film of cadmium sulphide (CdS) is deposited. In effect, the material forms a large number of strips, electrically in parallel. The resistance between the terminals, XY, falls from about 10 MΩ in the dark to about 100 Ω in bright light.

This is the basis of a *light dependent resistor* (LDR). As explained previously on p. 821, an LDR can be used in a switching circuit for street-lighting, for example, and as a light-sensitive cell for use in exposure meters of cameras, as shown in Figure 34.7 (ii).

_____ Exercises 34A _____

Photoelectricity

1 When light is incident on a metal plate electrons are emitted only when the frequency of the light exceeds a certain value. How has this been explained?
 The maximum kinetic energy of the electrons emitted from a metallic surface is 1.6×10^{-19} J when the frequency of the incident radiation is 7.5×10^{14} Hz. Calculate the minimum frequency of radiation for which electrons will be emitted. Assume that Planck constant $= 6.6 \times 10^{-34}$ J s. (*JMB.*)

2 When light of frequency 5.4×10^{14} Hz is shone on to a metal surface the maximum energy of the electrons emitted is 1.2×10^{-19} J. If the same surface is illuminated with light of frequency 6.6×10^{14} Hz the maximum energy of the electrons emitted is 2.0×10^{-19} J. Use this data to calculate a value for the Planck constant. (*L.*)

3 (a) State the main conclusions of Millikan's photoelectric experiments.
 (b) Light of wavelength 0.4 μm strikes the surface of potassium metal whose work function is 2.2 eV. What is the greatest kinetic energy that a photoelectron can have?
 (Planck constant $h = 6.6 \times 10^{-34}$ J s; Speed of light *in vacuo* $c = 3.0 \times 10^8$ m s^{-1}; $e = -1.6 \times 10^{-19}$ C.) (*W.*)

4 When a photon of energy E falls on a metal surface an electron may be emitted. Write down, and explain, an equation connecting photon energy, the work function ϕ of the metal and the maximum kinetic energy of the electron.
 Why, for photons of a given energy, are electrons emitted with a range of velocities? In what circumstances are no electrons emitted? (*AEB*, 1983.)

5 Explain the terms *photoelectric threshold frequency* and *work function* used in connection with the photoelectric effect.
 A photoelectric cell consists of a conducting plate coated with photo-emissive material and a metal ring as current collector, mounted *in vacuo* in a transparent envelope. Graphs are plotted showing the relation between the collector current I and the potential V of the collector with respect to the emitter, for the following cases (Figure 34A):
 A. Emissive material illuminated with monochromatic light of wavelength λ. *B.*

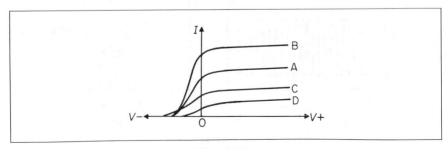

Figure 34A

As A, but the intensity of the incident light has been changed. C. As A, but the wavelength of the light has been changed. D. As A, but the emissive material is different.

Explain the general form of curve A. In what way do curves B,C,D (i) resemble A, (ii) differ from A? What explanation of the differences can be offered? (*O. & C.*)

6 Describe and explain one experiment in which light exhibits a wave-like character and one experiment which illustrates the existence of photons.

Light of frequency 5.0×10^{14} Hz liberates electrons with energy 2.31×10^{-19} J from a certain metallic surface. What is the wavelength of ultra-violet light which liberates electrons of energy 8.93×10^{-19} J from the same surface? (Take the velocity of light to be 3.0×10^8 m s^{-1}, and Planck constant (h) to be 6.62×10^{-34} J s.) (*L.*)

7 (a) A photocell has an anode and cathode made of the same material. When the cathode is illuminated with monochromatic light of constant intensity photoelectrons are released. Sketch a graph showing how the current in the cell varies as the potential of the anode is varied, the cathode being earthed. Include negative and positive values of anode potential on your graph.

(b) Explain why the graph has the form you sketch. (*JMB.*)

8 When a metallic surface is exposed to monochromatic electromagnetic radiation electrons may be emitted. Apparatus is arranged so that

(a) the intensity (energy per unit time per unit area) and

(b) the frequency of the radiation may be varied.

If each of these is varied in turn whilst the other is kept constant, what is the effect on (i) the number of electrons emitted per second, and (ii) their maximum speed? Explain how these results give support to the quantum theory of electromagnetic radiation.

The photoelectric work function of potassium is 2.0 eV. What potential difference would have to be applied between a potassium surface and the collecting electrode in order just to prevent the collection of electrons when the surface is illuminated with radiation of wavelength 350 nm? What would be (iii) the kinetic energy, and (iv) the speed, of the most energetic electrons emitted in this case? (Speed of electromagnetic radiation *in vacuo* $= 3.0 \times 10^8$ m s^{-1}, electronic charge $= -1.6 \times 10^{-19}$ C, mass of electrons $= 9.1 \times 10^{-31}$ kg, Planck constant $= 6.6 \times 10^{-34}$ J s.) (*L.*)

9 (a) The Einstein equation for the photoelectric effect can be written

$$hf = \tfrac{1}{2}mv^2 + \phi$$

(i) Explain the meaning of each term in the equation and how it relates to Einstein's explanation of the photoelectric effect. (ii) What does Einstein's explanation imply about the nature of light?

(b) Figure 34B shows the apparatus used in an experiment to investigate the photoelectric effect. Light falls on the photo-sensitive surface of a photocell, and the kinetic energy of the fastest moving emitted electrons can be measured by increasing the voltage provided by the battery until no current flows. The

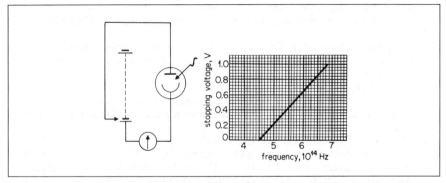

Figure 34B

results obtained for monochromatic light of various frequencies are shown graphically below.

(i) The graph indicates that there is a frequency below which no electrons are emitted. Why is this? (ii) When the frequency is 5.5×10^{14} Hz, the required stopping voltage is 0·43 V. Given that the electronic charge is -1.6×10^{-19} C, estimate the kinetic energy of the fastest moving electrons emitted by light of this frequency. Justify your answer. (ii) Use the graph to obtain a value for the Planck constant, explaining carefully how you obtain your result. (iv) State and explain the effect, if any, of increasing the intensity of the light on the experimental observations and on the graph. (*AEB*, 1984.)

10   Einstein's equation for the photoelectric emission of electrons from a metal surface can be written $hf = \frac{1}{2}mv^2 + \phi$, where ϕ is the work function of the metal, and consistent energy units are used for each term in the equation. Explain briefly the physical process that this equation represents. Outline an experiment by which you could determine the values of h (or of h/e) and ϕ.

For caesium the value of ϕ is 1·35 electronvolts.
(a) What is the longest wavelength that can cause photoelectric emission from a caesium surface?
(b) What is the maximum velocity with which photoelectrons will be emitted from a caesium surface illuminated with light of wavelength 4.0×10^{-7} m?
(c) What potential difference will just prevent a current passing through a caesium photocell illuminated with light of wavelength 4.0×10^{-7} m? ($e = 1.6 \times 10^{-19}$ C, $m = 9 \times 10^{-31}$ kg, $h = 6.6 \times 10^{-34}$ J s.) (*O.*)

11   Describe an experiment to measure the Planck constant using the photoelectric effect.

A strip of clean magnesium ribbon, surrounded by a cylinder of copper gauze maintained at a positive potential of 6·0 V with respect to the magnesium, is connected to the input of an amplifier which measures the p.d. across the resistor R as shown in the Figure 34C.

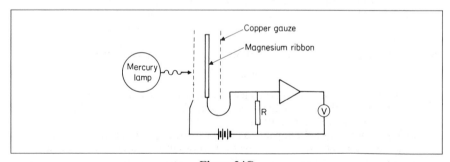

Figure 34C

When the magnesium is illuminated with mercury light of wavelength 254 nm the amplifier detects a current flowing in R. State the origin of this current.

Calculate the energies of photons of mercury light of wavelengths 254 nm and 546 nm.

Using these values and the data at the end of the question, describe and explain in detail what you would observe when each of the following experiments is carried out separately.
(a) The polarity of the 6·0 V battery is reversed.
(b) The mercury lamp is moved farther away from the magnesium ribbon.
(c) A filter selects mercury light of wavelength 546 nm in place of that of wavelength 254 nm.
(d) The e.m.f. of the battery is increased to 10 V.
(e) The magnesium ribbon is replaced by a strip of copper. (The photoelectric work function of magnesium $= 4.5 \times 10^{-19}$ J (2·8 eV) The photoelectric work function of copper $= 8.1 \times 10^{-19}$ J (5·05 eV) Planck constant $= 6.6 \times 10^{-34}$ J s.) (*O. & C.*)

Quantisation of Energy, Energy Levels

In 1911, Bohr proposed that an atom has a number of separated energy values or *energy levels*. These levels are characteristic of the particular atom so that a hydrogen atom, for example, has different energy levels to a nitrogen atom.

An important point in the theory, given later, is that an atom can not have any energies between these levels. We say that the energy is not continuous but 'quantised' in definite amounts. In an analogous way, a building has different floors from the ground floor upwards but no floors or levels between.

The energy of an atom is the energy of its electrons, which are moving round the positively-charged nucleus. These electrons therefore 'occupy' one of these separated energy levels.

The energy level values are expressed in terms of a number n, which can only be a whole number such as $n = 1, 2, 3$ and so on. n is called the *quantum number* for the particular energy level, as we now see for the hydrogen atom.

Hydrogen Energy Levels

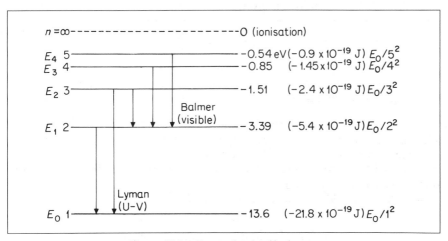

Figure 34.8 *Energy levels of hydrogen*

Figure 34·8 shows roughly some of the energy levels of the hydrogen atom in electronvolts (eV) and in joules (J). The quantum numbers $n = 1, 2, \ldots, \infty$ on the left correspond to values used to calculate the energy values. The *ground state*, $n = 1$, is $-13\cdot6\,\text{eV}$ $(-21\cdot8 \times 10^{-19}\,\text{J})$ and this is the lowest energy level, where the hydrogen atom is most stable. The higher energy levels are all calculated from the formula $-13\cdot6\,\text{eV}/n^2$, where $n = 1, 2, 3, 4, 5\ldots$ respectively, as illustrated in the diagram.

When an atom is raised from the ground state to any of its higher energy levels, it is said to be in an *excited state*. The *excitation energy* needed to raise a hydrogen atom from the ground state E_0 or $-13\cdot6\,\text{eV}$ to its energy level E_2 or $-1\cdot51\,\text{eV}$

$$= E_2 - E_0 = (-1\cdot51) - (-13\cdot6) = 12\cdot09\,\text{eV}$$

The energy levels *increase* to the value 0, which is the *ionisation level*. When the electron or atom is given an amount of energy equal to the *ionisation energy*, the electron is just able to become a 'free' electron. From the energy level values for hydrogen, the energy from the ground state $(-13\cdot6\,\text{eV})$ to the ionisation level (0) is $13\cdot6\,\text{eV}$, which is the ionisation energy. If the electron is given a greater amount of energy than the ionisation energy of $13.6\,\mathrm{eV}$ or $21.8 \times 10^{-19}\,\mathrm{J}$, such as $22.8 \times 10^{-19}\,\mathrm{J}$, the excess energy of $1.0 \times 10^{-19}\,\mathrm{J}$ is then the kinetic energy of the free electron outside the atom. In general, the free electron can have a continuous range of energies. Inside the atom, however, it can have only one of the energy level values characteristic of the atom.

Electromagnetic Radiation and Energy Change

When an atom is excited from the ground state to a higher energy level, it becomes unstable and falls back to one of the *lower* energy levels. This energy change occurs for any system. It becomes unstable at a higher energy than its normal energy value and tends to fall back to its stable state. So the excited atom may reach, for example, a level corresponding to $n = 2$ in Figure 34.8 and then fall back to the ground state where $n = 1$.

Bohr proposed that the decrease in energy when the fall occurs is released in the form of *electromagnetic radiation*. Now from quantum theory, the energy in radiation of frequency f if hf, where h is the Planck constant (p. 845). So if the atom falls from the first energy level E_1 to the ground energy level E_0 after it was excited, the radiation emitted is given by

$$E_1 - E_0 = hf = \frac{hc}{\lambda}$$

where c is the speed of electromagnetic waves in a vacuum or air and λ is the wavelength of the radiation.

Emission Spectra of Hydrogen

We can now calculate the wavelength of the emitted radiation when the hydrogen atom is excited from its ground state ($n = 1$) where its energy level E_0 is $-21.8 \times 10^{-19}\,\mathrm{J}$ to the higher level ($n = 2$) of energy E_1 $-5.4 \times 10^{-19}\,\mathrm{J}$, and then falls back to the ground state.

Since $E_1 - E_0 = hf = hc/\lambda$, then, using standard values,

$$\lambda = \frac{hc}{E_1 - E_0} = \frac{6.6 \times 10^{-34} \times 3 \times 10^8}{(-5.4 \times 10^{-19}) - (-21.8 \times 10^{-19})}$$

$$= \frac{6.6 \times 3}{16.4} \times 10^{-34+8+19}$$

$$= 1.2 \times 10^{-7}\,\mathrm{m}$$

This wavelength is in the *ultraviolet spectrum*.

Suppose we now calculate the wavelength of the radiation emitted from an energy level $-2.4 \times 10^{-19}\,\mathrm{J}$ ($n = 3$) to $-5.4 \times 10^{-19}\,\mathrm{J}$ ($n = 2$). This is an energy change E of $3.0 \times 10^{-19}\,\mathrm{J}$. Now from $\lambda = hc/E$ used above, we see that $\lambda \propto 1/E$ since h and c are constants. This means that the *smaller* the value of the energy change E, the *greater* is the wavelength. We have just shown that $\lambda = 1.2 \times 10^{-7}\,\mathrm{m}$ when $E = 16.4 \times 10^{-19}\,\mathrm{J}$. So, by proportion, for $E = 3.0 \times 10^{-19}\,\mathrm{J}$, λ is given by

$$\frac{\lambda}{1.2 \times 10^{-7}\,\mathrm{m}} = \frac{16.4 \times 10^{-19}}{3.0 \times 10^{-19}}$$

So
$$\lambda = \frac{1.2 \times 10^{-7} \times 16.4}{3.0} = 6.6 \times 10^{-7}\,\mathrm{m}$$

This wavelength is in the *visible spectrum*.

Excitation and Ionisation Potentials

The energy required to raise an atom from its ground state to an excited state is called the *excitation energy* of the atom. If the energy is eV, where e is the electron charge, V is known as the *excitation potential* of the atom (p. 856).

The ground state energy level $= -13.6\,e\,V$ and the energy level $E_4 = -0.85\,e\,V$ for hydrogen. So the excitation energy from the ground state to $E_4 = (-0.85) - (-13.6) = 12.75\,e\,V$. The excitation potential is then 12.75 V.

If the atom is in its ground state with energy E_0, and absorbs an amount of energy eV which *just* removes an electron completely from the atom, then V is said to be the *ionisation potential* of the atom. The potential energy of the atom is here denoted by E_∞, as the ejected electron is so far away from the attractive influence of the nucleus as to be, in effect, at infinity. E_∞ is taken as the 'zero' energy of the atom, and the lower energy levels are thus negative. The ionisation potential V is given by $E_\infty - E_0 = eV$, or by $-E_0 = eV$. In the case of the hydrogen atom, $E_0 = -13.6\,eV$. So the ionisation energy $= 13.6\,eV$ and the ionisation potential $V = 13.6$ V.

Emission Line Spectrum

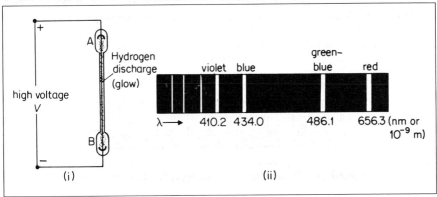

Figure 34.9 *Emission line spectrum*

Gases such as hydrogen, neon or carbon dioxide can be placed inside a narrow discharge tube at low pressure, as shown in Figure 34.9 (i). When the metal electrodes A and B at the ends of the tube are connected to a high voltage such as 1000 V, a discharge or light column is obtained between A and B. When the light from hydrogen or neon gas is examined using a diffraction grating, the *emission spectrum* is seen to consist of well-defined separated lines, such as those shown in Figure 34.9 (ii). This type of emission spectrum is called a *line spectrum*.

As they move through the discharge in the gas, some electrons obtain sufficient energy to excite atoms to a higher energy level. When the atoms fall to a lower energy level, the excess energy is emitted as electromagnetic radiation, as previously explained. The visible line spectrum of hydrogen shows the change in energy. Each line has a particular frequency or wavelength given by energy change $= hf$. **The fact that the lines are separated is *experimental evidence* for the existence of separate or 'quantised' energy levels in the atom.**

Types of Emission Spectra

Emission spectra are classified into *line*, *band* or *continuous* spectra. Line spectra are obtained from *atoms* in gases such as hydrogen or neon at low pressure in a discharge tube, Plate 34A shows the line spectrum of the gas krypton.

Gases such as carbon dioxide in a discharge tube produce a *band spectrum*. Each band consists of a series of lines very close together at the sharp edge or head of the band and farther apart at the other end or tail, Figure 34.10. Band spectra are essentially due to *molecules*. The different band heads in a band system are due to small allowed discrete energy changes in the *vibrational* energy of the molecule. The fine lines in a given band are due to still smaller allowed discrete energy changes in the *rotational* energy of the molecule.

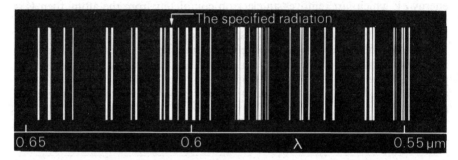

Plate 34A Line Spectrum (*atoms*). *Part of the spectrum of the gas krypton-86 in the range 0·65 μm (650 nm) to 0·55 μm (550 nm). The wavelength in a vacuum of the specified radiation shown is used in a definition of the metre.* (Crown copyright. Courtesy of National Physical Laboratory)

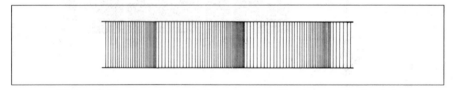

Figure 34.10 *Diagrammatic representation of band spectra (molecules)*

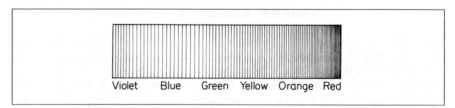

Figure 34.11 *Continuous spectrum from a solid or liquid glowing white hot*

The spectrum of the sun is an example of a *continuous spectrum*, and, in general, these spectra are obtained from *solids and liquids*. Figure 34.11. In these states of matter the atoms and molecules are close together, and the energy changes in a particular atom are influenced by neighbouring atoms to such an extent that radiations of all different wavelengths are emitted. In a gas the atoms are comparatively far apart, and each atom is uninfluenced by any other. The gas therefore emits radiations of wavelengths which result from energy changes in the atom due solely to the high temperature of the gas, and the line spectrum is obtained. When the temperature of a gas is decreased and pressure applied so that the liquid state is approached, the line spectrum of the gas is observed to broaden out considerably.

Absorption Spectra

The spectra just discussed are classified as *emission spectra*. There is another class of spectra known as *absorption spectra*, which we shall now briefly consider.

If light from a hot source having a continuous spectrum is examined after it has passed through a sodium flame at a lower temperature, the spectrum is found to be crossed by a dark line; this dark line is in the position corresponding to the bright line emission spectrum obtained with the sodium flame alone. The continuous spectrum with the dark line is naturally characteristic of the absorbing substance, in this case sodium, and it is known as an *absorption spectrum*. An absorption spectrum is obtained when red glass is placed in front of sunlight, as it allows only a narrow band of red rays to be transmitted.

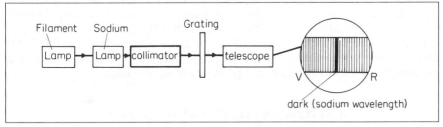

Figure 34.12 *Absorption spectrum of sodium (diagrammatic)*

Figure 34.12 shows how the absorption spectrum of sodium can be obtained. An over-run tungsten filament emits a continuous spectrum of light which is incident on a glowing sodium vapour lamp. The sodium lamp emits only visible yellow light. The light from the lamp passes through the hot sodium to a diffraction grating and the resulting spectrum is seen through a spectrometer telescope.

A darkened line is seen in the yellow part of the continuous spectrum. Also, when the sodium lamp is removed, the dark line becomes bright yellow. The dark line is the absorption spectrum of sodium. It has the same wavelength as the emission spectrum of sodium. Generally, an absorption spectrum is characteristic of the absorbing substance, so unknown elements can be identified from their absorption spectrum.

We now explain absorption spectra.

Absorption Spectra of Sun, Fraunhofer Lines

Atoms can absorb energy in a number of ways. In a flame, inelastic collisions with energetic molecules can raise atoms to higher energy levels. In a discharge tube, inelastic collisions with bombarding electrons can raise atoms to higher energy levels.

An atom can also absorb energy from a photon. If the photon energy $E = hf$ is just sufficient to excite an atom to one of its higher energy levels, the photon will be absorbed. When it returns to the ground state the excited atom emits the same wavelength as the photon but equally in all directions. So the intensity of the radiation in the direction of the incident photon is reduced. A *dark* line is thus seen whose wavelength is that of the absorbed photon.

Absorption of photons explain the dark lines in the sun's visible spectrum, first observed by Fraunhofer. The sun emits a continuous spectrum of photons. Vaporised elements in the outer or cooler parts of the sun's atmosphere absorb those photons which have the same frequency or energy to excite them to higher energy levels. The sun's spectrum is now darker at wavelengths *characteristic of the elements in the sun's atmosphere.* Since absorption spectra are always characteristic of the absorbing elements, these elements can be identified from their absorption spectra. In this way we know that elements such as iron exist in the sun.

In the sun's atmosphere, hot gases such as hydrogen are already excited to a higher energy level by collision with atoms. The absorbed photons are thus able to excite the atoms to still higher energy levels, which are in the sun's visible spectrum. So the Fraunhofer (dark) lines are present in the visible spectrum of the sun.

Normally, however, hydrogen is in the *ground* state, corresponding to the energy level $-13\cdot6\,eV$ (p. 855). The next higher energy level is about $-3\cdot4\,eV$. This is a relatively high energy 'jump' of $10\cdot2\,e\,V$ and corresponds to a photon in the ultraviolet. Visible light has photons of energy less than $3\cdot4\,eV$ (see p. 846). So photons in visible light are not able to be absorbed by hydrogen atoms in their normal or ground state. Consequently hydrogen is *transparent* to visible light. So if a wide band of wavelengths is passed through hydrogen, only those wavelengths in the ultraviolet would produce absorption lines.

Spontaneous and Stimulated Emission

As we have seen, an atom may undergo a transition between energy states if it emits or absorbs a photon of the appropriate energy. When an atom in the ground state, for example, absorbs a photon and makes a transition to a higher energy state or level, the photon is said to have 'stimulated' the absorption. The atom cannot increase its energy 'spontaneously', that is, in the absence of a photon.

We might expect that the atom in its higher energy state would emit a photon spontaneously, that is, the probability of emission would be independent of the number of photons present in the environment of the atom. In 1917 Einstein proved, however, that this is not the case. The probability per unit time that an atom will decay to a lower energy state and emit a photon is the sum of two terms. One is a spontaneous emission term. The other is a *'stimulated emission'* term, which is proportional to the number of photons of the relevant energy *already present in the environment of the atoms.* Further, the photon produced by stimulated emission is always *in phase* with the stimulating photon.

The Laser

This idea led to the invention of the *laser.* (The word 'laser' is short for *l*ight *a*mplification by *s*timulated *e*mission of *r*adiation.)

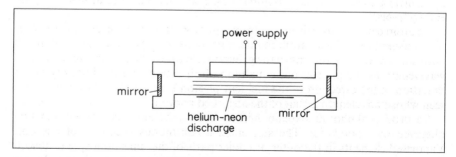

Figure 34.13 *Gas laser*

Figure 34.13 shows the basic features of a gas laser. A helium-neon gas mixture (carbon dioxide gas is also used) is contained inside a long quartz tube with optically plane mirrors at each end. A powerful radio-frequency generator is used to produce a discharge in the gas so that the helium atoms are excited or 'pumped up' to a higher energy level. By collision with these atoms, the neon atoms are excited to a higher energy level so that a 'population inversion' is obtained, that is, there are a much greater number of atoms in the higher energy level than in the ground state. Any stray photon, with energy equal to the difference in energies of the higher level and ground state, will stimulate many of the large population of excited atoms to fall from the higher energy level to the ground state. Repeated reflection at the silvered mirrors at the ends of the tube will multiply the number of photons rapidly. So strong stimulated emission is obtained and an enormously powerful burst of light or pulse of light is produced. The light pulse from a laser is:

(i) *monochromatic* light because all the photons have the same energy $(E - E_0)$ above the ground state and hence the same frequency $(E - E_0)/h$, where h is the Planck constant;

(ii) *coherent* light because all the photons or waves are in phase;

(iii) *intense* light because all the emitted waves are in phase or coherent. If the photons or waves were out of phase or incoherent, the resultant intensity would be the sum of the individual intensities and this is proportional to $n \times a^2$, where n is the number of waves and a is the amplitude of each wave. Since the waves are in phase, however, their total amplitude is na. Hence the resultant intensity is proportional to $(na)^2$ or n^2a^2. This intensity is greater by a factor n than that obtained from incoherent waves, such as those from an ordinary lamp filament. Since n is very large, this leads to an enormous increase in intensity.

Bohr's Theory of Hydrogen Atom

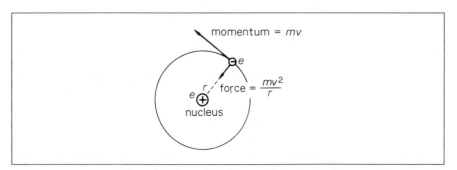

Figure 34.14 *Bohr's theory of hydrogen atom*

We conclude with a brief account of Bohr's theory of the hydrogen atom, which first gave important new ideas about atoms.

Bohr considered one electron of charge $-e$ and mass m, moving with speed v and acceleration v^2/r in an orbit round a central hydrogen nucleus of charge $+e$ (Figure 34.14). In classical physics, charges undergoing acceleration emit radiation and, therefore, they would lose energy. On this basis the electron would spiral towards the nucleus and the atom would collapse. Bohr, therefore, suggested that in those orbits where the angular momentum is a multiple of $h/2\pi$, the energy is constant. Twelve years later, de Broglie proposed that a particle such as the electron may be considered to behave as a *wave* of wavelength

$\lambda = h/p$, where h is the Planck constant and p is the momentum of the moving particle.

If the electron can behave as a *wave*, it must be possible to fit a whole number of wavelengths around the orbit. In this case a *standing wave* pattern is set up and the energy in the wave is confined to the atom. A progressive wave would imply that the electron is moving from the atom and is not in a stationary orbit.

If there are n waves in the circular orbit and λ is the wavelength,

$$n\lambda = 2\pi r \qquad . \qquad . \qquad . \qquad . \qquad . \qquad \text{(i)}$$

$$\therefore \lambda = \frac{2\pi r}{n} = \frac{h}{p} = \frac{h}{mv} \qquad . \qquad . \qquad . \qquad \text{(ii)}$$

So

$$mvr = \frac{nh}{2\pi} \qquad . \qquad . \qquad . \qquad . \qquad \text{(iii)}$$

Now $mv \times r$ is the moment of momentum or *angular momentum* of the electron about the nucleus. So equation (iii) states that the *angular momentum is a multiple of $h/2\pi$*. The quantisation of angular momentum is a key point in atomic theory.

In Figure 34.14, the electron moving round the nucleus has kinetic energy due to its motion and potential energy in the electrostatic field of the nuclear charge $+e$. Bohr calculated the total energy E of the electron in terms of its charge, mass, orbital radius and the number n which quantises the angular momentum. Thus for circular motion, (see p. 50)

$$mv^2/r = \text{force on electron} = e^2/4\pi\varepsilon_0 r^2 \qquad . \qquad . \qquad . \qquad \text{(1)}$$

Also

$$\text{potential energy of electron} = \frac{+e}{4\pi\varepsilon_0 r} \times -e = -\frac{e^2}{4\pi\varepsilon_0 r}$$

and

$$\text{kinetic energy of electron} = \tfrac{1}{2}mv^2 = \frac{e^2}{8\pi\varepsilon_0 r}, \text{from (1)}$$

So

$$\text{total energy of electron, } E = -\frac{e^2}{8\pi\varepsilon_0 r} \qquad . \qquad . \qquad . \qquad \text{(2)}$$

Using equations (iii) and (1), v can be eliminated to find r. From (2) we then obtain

$$E_n \text{ (quantum number } n\text{)} = -\frac{me^4}{8\varepsilon_0^2 h^2} \cdot \frac{1}{n^2} \qquad . \qquad . \qquad . \qquad \text{(3)}$$

The ground state, minimum energy E_0, corresponds to $n = 1$. If E is the energy value of a higher level corresponding to $n = n_1$, then

$$E - E_0 = \frac{me^4}{8\varepsilon_0^2 h^2}\left(1 - \frac{1}{n_1^2}\right) \qquad . \qquad . \qquad . \qquad \text{(4)}$$

Bohr assumed that if the atom moves from a higher energy level to a lower level, the difference in energy is released in the form of radiation of energy hf, where h is the Planck constant and f is the frequency of the radiation. So

$$E - E_0 = hf = hc/\lambda$$

where λ is the wavelength of the radiation, since $c = f\lambda$ for the electromagnetic wave of speed c. From the values of the energy, Bohr was able to calculate the wavelength emitted and this agreed with the experimental value obtained.

Spectral Series of Hydrogen

Before Bohr's theory of the hydrogen atom it had been found that the wavelengths of the hydrogen spectrum could be arranged in a formula or series named after its discoverer. The visible spectrum was the Balmer series, the ultraviolet was the Lyman series and the infrared was the Paschen series.

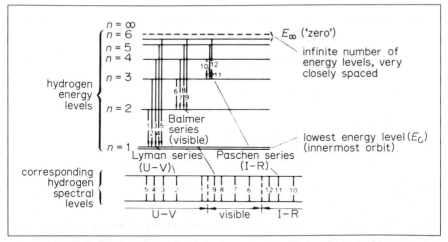

Figure 34.15 *Energy levels and spectra of hydrogen* (not to scale)

Bohr's theory of energy levels accounted for all the series. As shown in Figure 34.15, the hydrogen spectrum is obtained simply by using different numbers for n in calculating the energy levels. The ultraviolet series, for example, is obtained when the energy of the atom falls to the lowest energy level E_0 corresponding to $n = 1$; the visible spectrum is obtained for energy falls to a higher level corresponding to $n = 2$; and the infrared spectrum is obtained for energy falls to the higher level when $n = 3$. See also Figure 34.8.

From Figure 34.15, the energy change E for the ultraviolet series is greater than for the visible spectrum. Since $E = hf$, the frequency of ultraviolet radiation is greater than visible radiation. Wavelength, λ, is inversely-proportional to f, since $c = f\lambda$. Hence the ultraviolet wavelengths are *shorter* than those in the visible spectrum.

Bohr's theory of the hydrogen atom was unable to predict the energy levels in complex atoms, which had many electrons. Quantum or wave mechanics, beyond the scope of this book, is used to explain the spectral frequencies of these atoms. The fundamental ideas of Bohr's theory, however, are still retained, for example, the angular momentum of the electron has quantum values and the energy levels of the atom have only allowed separated values characteristic of the atom.

_____ **Exercises 34B** _____

Energy Levels of Atoms

1 (a) Explain briefly what is meant by an emission spectrum. Describe a suitable source for use in observing the emission spectrum of hydrogen.
   (b) Figure 34D, which is to scale, shows some of the possible energy levels of the hydrogen atom. (i) Explain briefly how such a diagram can be used to account

for the emission spectrum of hydrogen. (ii) When the hydrogen atom is in its ground state, 13·6 eV of energy are needed to ionise. it. Calculate the highest possible frequency in the line spectrum of hydrogen. In what region of the electromagnetic spectrum does it lie? (The frequency range of the visible spectrum is from 4×10^{14} Hz to 7.5×10^{14} Hz.) (iii) A number of transitions are marked on the energy level diagram. Identify which of these transitions corresponds to the lowest frequency that *would be visible*.

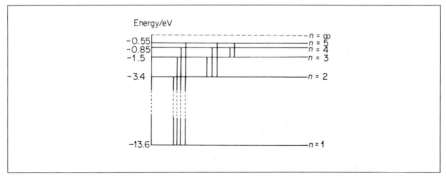

Figure 34D

(c) The wavelengths of the lines in the emission spectrum of hydrogen can be measured using a spectrometer and a diffraction grating. If the grating has 5000 lines per cm, through what angle would light of frequency 4.6×10^{14} Hz be diffracted in the first order spectrum?
(1 eV = 1.60×10^{-19} J. The Planck constant = 6.6×10^{-34} J s. Speed of light = 3.00×10^{8} m s^{-1}.) (*L.*)

2 Figure 34E shows three energy levels for the atoms of a particular substance, the energies being in electronvolts (eV). A beam of electrons passing through this substance may excite electrons from the various energy levels. What is the *minimum* potential difference through which the beam of electrons must be accelerated from rest to cause any excitation between two of these levels? At what speed would these electrons be travelling? (Mass of an electron = 9.0×10^{-31} kg. 1 eV = 1.6×10^{-19} J.) (*L.*)

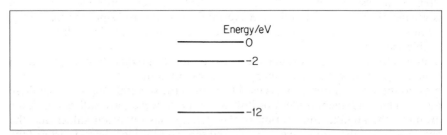

Figure 34E

3 Explain why a glowing gas, such as that in a neon tube, gives only certain wavelengths of light and why that gas is capable of absorbing the same wavelengths. (*L.*)

4 Describe briefly the Bohr model for the hydrogen atom.
What were the important assumptions that Bohr made which led to energy levels? Figure 34F shows some of the energy levels for the hydrogen atom.
(i) Define the electron volt. (ii) Why are the energies given negative values? (iii) In which of the levels is the electron nearest to the nucleus, and why? (iv) Which level corresponds to the ground state? (v) What is the ionisation energy for atomic hydrogen? (vi) How might the atom be changed from state e to state c?

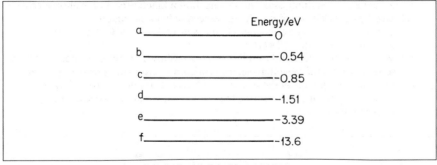

Figure 34F

(vii) Discuss whether, as the result of the transition in (vi), the speed of the electron goes up or down. (*W.*)

5   Describe an experiment which provides evidence for the belief that a quantity of electric charge is always a multiple of a fundamental unit of charge. Explain how, starting from the measurements made in this experiment, you would calculate the value of this unit.

The ground state of the electron in the hydrogen atom may be represented by the energy $-13{\cdot}6\,$eV and the first two excited states by $-3{\cdot}4\,$eV and $-1{\cdot}5\,$eV respectively, on a scale in which an electron completely free of the atom is at zero energy. Use this data to calculate the ionisation potential of the hydrogen atom and the wavelengths of three lines in the emission spectrum of hydrogen.

(Charge of the electron $= -1{\cdot}6 \times 10^{-19}\,$C. Speed of light in a vacuum $= 3{\cdot}0 \times 10^8\,$m s^{-1}. The Planck constant $= 6{\cdot}6 \times 10^{-34}\,$J s.) (*L.*)

6   What are the chief characteristics of a line spectrum? Explain how line spectra are used in analysis for the identification of elements.

Figure 34G, which represents the lowest energy levels of the electron in the hydrogen atom, specifies the value of the principal quantum number n associated

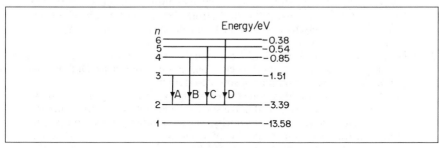

Figure 34G

with each state and the corresponding value of the energy of the level, measured in electronvolts. Work out the wavelengths of the lines associated with the transitions A, B, C, D marked in the figure. Show that the other transitions that can occur give rise to lines that are either in the ultra-violet or the infra-red regions of the spectrum. (Take 1 eV to be $1{\cdot}6 \times 10^{-19}\,$J; Planck constant h to be $6{\cdot}5 \times 10^{-34}\,$J s; and c, the velocity of light in *vacuo*, to be $3 \times 10^8\,$m s^{-1}.) (*O.*)

7   (a) The first excitation potential for sodium gas is $2{\cdot}11\,$V. Explain the term *excitation potential* and calculate the wavelength of the corresponding spectral line.

Describe and explain what happens when a parallel beam of white light passes through a glass bulb filled with hot sodium vapour.

(b) Describe briefly the phenomenon known as the photoelectric effect and explain the terms *work function* and *threshold wavelength*.

The threshold wavelength for a clean sodium surface is 680 nm. Calculate   (i) the work function for sodium, and (ii) the maximum kinetic energy for photoelectrons released by light of wavelength 390 nm. (Speed of light in vacuum, $c = 3.00 \times 10^8 \, \text{m s}^{-1}$. The electronic charge, $e = 1.60 \times 10^{-19}$ C. The Planck constant, $h = 6.63 \times 10^{-34}$ J s.) (*L.*)

8 (a) What do you understand by a *line spectrum*? Show how the existence of spectral series supports the view that electrons in atoms exist in discrete energy levels.
   (b) Outline the principle of the Franck and Hertz experiment. Give a diagram of a suitable apparatus, and describe the results that would be expected for a gas such as xenon.

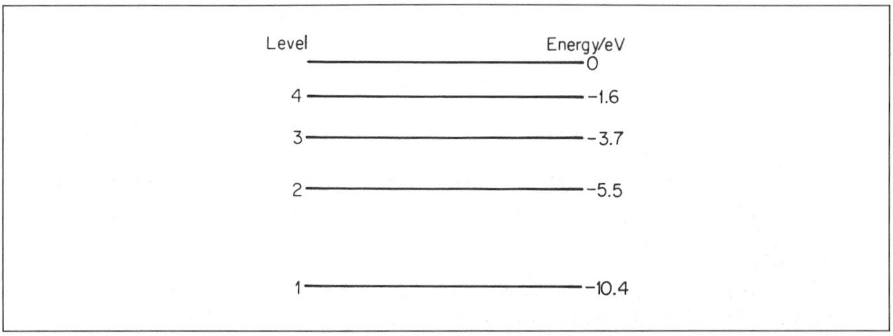

Level	Energy/eV
	——0
4	—-1.6
3	—-3.7
2	—-5.5
1	—-10.4

Figure 34H

   (c) Some of the energy levels of a mercury atom are shown in Figure 34H. Level 1 is the highest level occupied by electrons in an unexcited atom. (Take the electron charge e to be -1.6×10^{-19} C, the speed of light in vacuum c to be $3.0 \times 10^8 \, \text{m s}^{-1}$, and the Planck constant h to be 6.6×10^{-34} J s.) (i) Calculate the ionisation energy of a mercury atom in electronvolts (eV) and in joules (J). (ii) Calculate the wavelength of radiation emitted when an electron moves from level 4 to level 2. In what part of the electromagnetic spectrum does this wavelength lie? (iii) State what is likely to happen if a mercury atom in the unexcited state is bombarded with electrons with energies 4.0 eV, 6.7 eV, and 11.0 eV respectively. (*O.*) (*Note* Franck and Hertz experiment—see *Scholarship Physics* (Heinemann) Nelkon.)

9 Give an account of the Rutherford-Bohr model of the atom with special reference to the arrangement of the extra-nuclear electrons in orbits and shells. Outline some experimental evidence in support of your description.
   An electron of energy 20 eV comes into collision with a hydrogen atom in its ground state. The atom is excited into a state of higher internal energy and the electron is scattered with reduced velocity. The atom subsequently returns to its ground state with emission of a photon of wavelength 1.216×10^{-7} m. Determine the velocity of the scattered electron. ($e = 1.6 \times 10^{-19}$ C; $c = 3.0 \times 10^8 \, \text{m s}^{-1}$; $h = 6.625 \times 10^{-34}$ J s; m (electron) $= 9.1 \times 10^{-31}$ kg.) (*L.*)

X-Rays

In 1895, Roentgen found that some photographic plates, kept carefully wrapped in his laboratory, had become fogged. Instead of merely throwing them aside he set out to find the cause of the fogging. He traced it to a gas-discharge tube, which he was using with a high voltage. This tube appeared to emit a radiation that could penetrate paper, wood, glass, rubber, and even thick aluminium. Roentgen could not find out whether the radiation was a stream of particles or a train of waves—Newton had the same difficulty with light—and he decided to call it *X-rays*.

Nature and Production of X-rays

We now regard X-rays as waves, similar to light waves, but of much shorter wavelength, about 10^{-10} m or 0·1 nm. They are produced when fast electrons, or cathode rays, strike a target, such as the walls or anode of a low-pressure discharge tube.

In a modern X-ray tube there is no gas, or as little as high-vacuum technique can achieve; the pressure is about 10^{-5} mm Hg. The electrons are provided by thermionic emission from a white-hot tungsten filament. In Figure 34.16, F is the filament and T is the metal target embedded in a copper anode A. A cylinder C round F, at a negative potential relative to F, focuses the electron beam on T. Because there is so little gas, the electrons on their way to the anode do not lose

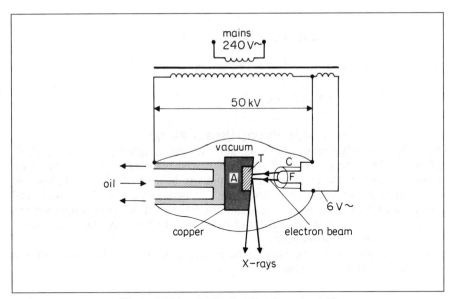

Figure 34.16   *An X-ray tube (diagrammatic)*

any appreciable amount of their energy in ionising atoms. From the a.c. mains, transformers provide about 6 volts for heating the filament, and about 50 000 volts for accelerating the electrons. On the half-cycles when the target is positive, the electrons bombard it, and generate X-rays. On the half-cycles when the target is negative, nothing happens at all—there is too little gas in the tube for it to break down. Thus the tube acts, in effect, as its own rectifier (p. 791), providing pulses of direct current between target and filament. The heat generated at the target by the electron bombardment is so great that the metal must be cooled. In large tubes this is done by circulating oil behind the anode, as shown. The target in an X-ray tube may be tungsten, for example, which has a high melting-point.

So X-rays (waves) are produced by bombarding matter with electrons (particles). The production of X-rays is therefore the inverse or opposite process to the photoelectric effect, where electrons (particles) are liberated from metals by incident light waves.

Example on Energy in X-ray Tube

An X-ray tube, operated at a d.c. potential difference of 40 kV, produces heat at the target at the rate of 720 W. Assuming 0·5% of the energy of the incident electrons is converted into X-radiation, calculate (i) the number of electrons per second striking the target, (ii) the velocity of the incident electrons. (Assume charge-mass ratio of electron, $e/m_e = 1·8 \times 10^{11}\,C\,kg^{-1}$.)

(i) Heat per second at target $= 99·5\% \times IV$, where I is the current flowing and V is the p.d. applied. So

$$0·995 \times I \times 40\,000 = 720$$

Then
$$I = \frac{720}{40\,000 \times 0·995} = 0·018\,A\ (approx.)$$

So number of electrons per second $= \dfrac{I}{e} = \dfrac{0·018}{1·6 \times 10^{-19}}$

$$= 1·1 \times 10^{17}$$

(ii) Energy of incident electrons $= \tfrac{1}{2}m_e v^2 = eV$

So
$$v = \sqrt{2V \times \frac{e}{m_e}} = \sqrt{2 \times 40\,000 \times 1·8 \times 10^{11}}$$

$$= 1·2 \times 10^8\,m\,s^{-1}$$

Effects and Uses of X-rays

When X-rays strike many minerals, such as zinc sulphide, they make them fluoresce. If a human—or other—body is placed between an X-ray tube and a fluorescent screen, the shadows of its bones can be seen on the screen, because they absorb X-rays more than flesh does. Unusual objects, such as swallowed safety-pins, if they are dense enough, can also be located. X-ray photographs can be taken with a metal plate in front of the screen. In this way cracks and flaws can be detected in metal castings.

X-rays are also used to investigate the structure of crystals. This important use of X-rays is discussed shortly.

Crystal Diffraction

The first proof of the wave-nature of X-rays was due to Laue in 1913, many years after X-rays were discovered. He suggested that the regular small spacing of atoms in crystals might provide a *natural diffraction grating* if the wavelengths of the rays were too short to be used with an optical line grating. Experiments by Friedrich and Knipping showed that X-rays were indeed diffracted by a thin crystal, and produced a pattern of intense spots round a central image on a photographic plate placed to receive them, Figure 34.17. The rays had thus been scattered by interaction with electrons in the atoms of the crystal. The diffraction pattern obtained gave information on the geometrical spacing of the atoms. An X-ray diffraction pattern produced by a crystal is shown on p. 875.

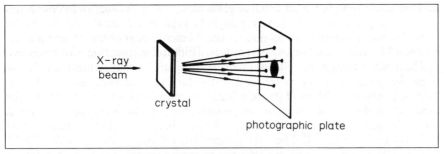

Figure 34.17   *Laue crystal diffraction*

Bragg Law

The study of the atomic structure of crystals by X-ray analysis was started in 1914 by Sir William Bragg and his son Sir Lawrence Bragg, with notable achievements. They soon found that a monochromatic beam of X-rays was reflected from a plane in the crystal rich in atoms, a so-called atomic plane, as if this acted like a mirror.

This important effect can be explained by Huygens' wave theory in the same way as the reflection of light by a plane surface. Suppose a monochromatic parallel X-ray beam is incident on a crystal and interacts with atoms such as A, B, C, D in an atomic plane P, Figure 34.18 (i). Each atom scatters the X-rays. Using Huygens' construction, wavelets can be drawn with the atoms as centres, which all lie on a plane wavefront reflected at an equal angle to the atomic plane P. When the X-ray beam penetrates the crystal to other atomic planes such as Q, R parallel to P, reflection occurs in a similar way, Figure 34.18 (ii). Usually, the beam or ray reflected from one plane is weak in intensity. If, however, the reflected beams or rays from all planes are *in phase* with each other, a strong reflected beam is produced by the crystal.

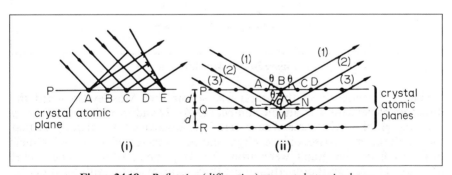

Figure 34.18   *Reflection (diffraction) at crystal atomic planes*

Suppose, then, that the glancing angle on an atomic plane ABCD in the crystal is θ, and d is the distance apart of consecutive parallel atomic planes, Figure 34.18 (ii). The path difference between the rays marked (1) and (2) = LM + MN = 2LM = $2d \sin \theta$. Thus a strong X-ray beam is reflected when

$$2d \sin \theta = n\lambda$$

where λ is the wavelength and n has integral values. This is known as *Bragg's law*. Hence, as the crystal is rotated so that the glancing angle is increased from zero, and the beam reflected at an equal angle is observed each time, an intense beam is suddenly produced for a glancing angle θ_1 such that $2d \sin \theta_1 = \lambda$. When the crystal is rotated further, an intense reflected beam is next obtained for an angle θ_2 when $2d \sin \theta_2 = 2\lambda$. Thus several orders of diffraction images may be observed.

The intense diffraction (reflection) images from an X-ray tube are due to X-ray lines *characteristic of the metal used* as the target, or 'anti-cathode' as it was originally known. The more penetrating or harder X-ray lines emitted from the metal are called the *K lines*; the less penetrating or softer lines are called the *L lines*. The wavelength of the K lines are shorter than those of the L lines, so that the frequencies of the K lines are greater than those of the L lines. The X-radiation from some metals thus consists of characteristic lines in the K series or L series or both.

Moseley's Law

In 1914 Moseley measured the frequency f of the characteristic X-rays from many metals, and found that, for a particular type of emitted X-ray such as K_a, whose origin is explained later, the frequency f varied in a regular way with the *atomic number Z* of the metal. When a graph of Z v. $f^{1/2}$ was plotted, an almost perfect straight line was obtained, Figure 34.19. Moseley therefore gave an empirical relation, known as *Moseley's law*, between f and Z as

$$f = a(Z-b)^2$$

where a, b are constants.

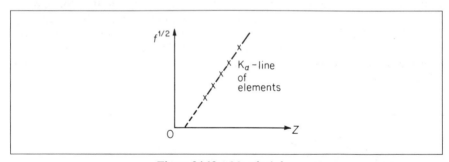

Figure 34.19 *Moseley's law*

Since the regularity of the graph was so marked. Moseley predicted the discovery of elements with atomic numbers 43, 61, 72 and 75, which were missing from the graph at that time. These were later discovered. He also found that though the atomic masses of iron, nickel and cobalt increased in this order, their positions from the graph were: iron $(Z = 26)$, cobalt $(Z = 27)$ and nickel $(Z = 28)$. The chemical properties of the three elements agree with the order by atomic number and not by atomic mass. Rutherford's experiments on the scattering of α-particles (p. 897) had shown that the atom contained a central nucleus of charge $+Ze$ where Z is the atomic number, and Moseley's experiments confirmed the importance of Z in atomic theory.

Explanation of Emission Spectra

Since the frequency of the X-ray spectra of elements is related to Ze, the charge on the nucleus, Moseley's results showed that the radiation was due to energy changes of the atom resulting from the movement of *electrons close to the nucleus*.

We now know, in fact, that the electrons in atoms are in groups which have various energy states. These groups are called 'shells'. Electrons nearest the nucleus are in the so-called *K-shell*. The next group, which are further away from the nucleus, are in the so-called *L shell*. The *M shell* is farther from the nucleus than the L shell. Electrons farther from the nucleus have greater energy than those nearer, since more work is needed to move them farther away. So electrons in the L shell have greater energy than those in the K shell; electrons in the M shell have greater energy than those in the L shell.

In the X-ray tube, energetic electrons bombard the metal target and may eject an electron from the innermost shell, the K shell. The atom is then raised to an excited state since its energy has increased, and it is unstable. An electron from the L shell may now move into the vacancy in the K shell, so decreasing the energy of the atom. Simultaneously, radiation is emitted of frequency f, where E = energy change of atom = hf. Thus $f = E/h$, and as E is very high for metals, the frequency f is very high and the wavelength is correspondingly short. It is commonly of the order of 10^{-9} m (1 nm) or less. The K series of X-ray lines is due to movement of electrons from the L (K_α line) or M (K_β line) shells to the K shell.

X-ray spectra are thus due to energy changes in electrons *close to the nucleus* of metals. In contrast, the optical spectra of the metals is due to energy changes in the outermost shell of the atom. Here the energy changes are about 1000 times smaller. So the frequencies of the optical lines are about 1000 times smaller, that is, their wavelengths are about 1000 times longer than those of X-rays.

Continuous X-ray Background Radiation, Minimum Wavelength

The characteristic X-ray spectrum from a metal is usually superimposed on a background of continuous, or so-called 'white', radiation of small intensity.

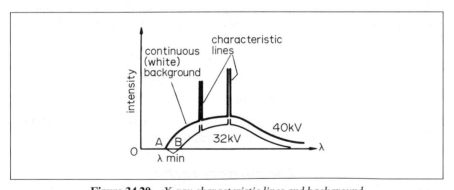

Figure 34.20 *X-ray characteristic lines and background*

Figure 34.20 illustrates the characteristic lines, K_α, K_β, of a metal and the continuous background of radiation for two values of p.d., 40 000 and 32 000 volts, across an X-ray tube. It should be noted that (i) the wavelengths of the characteristic lines are independent of the p.d.—they are characteristic of the metal, and (ii) the background of continuous radiation has increasing wavelengths which slowly diminish in intensity, but as the wavelengths diminish they are cut off *sharply*, as at A and B.

When the bombarding electrons collide with the metal atoms in the target, most of their energy is lost as heat. A little energy is also lost in the form of electromagnetic radiation. Here the frequencies are given $E = hf$ with the usual notation, and the numerous energy changes produce the background radiation in Figure 34.20.

The existence of a *sharp* minimum wavelength at A or B can be explained only by the quantum theory. The energy of an electron before striking the metal atoms of the target is eV, where V is the p.d. across the tube. If a direct collision is made with an atom and *all* the energy is absorbed, then, on quantum theory, the X-ray quantum produced has maximum energy.

$$\therefore hf_{max} = eV$$

$$\therefore f_{max} = \frac{eV}{h} \qquad \qquad \text{(i)}$$

$$\therefore \lambda_{min} = \frac{c}{f_{max}} = \frac{ch}{eV} \qquad \qquad \text{(ii)}$$

Verification of Quantum Theory

These conclusions are borne out by experiment. Thus for a particular metal target, experiment shows that the minimum wavelength is obtained for p.d.s of 40 kV and 32 kV at glancing angles of about 3·0° and 3·8° respectively. The ratio of the minimum wavelengths is hence, from Bragg's law,

$$\frac{\lambda_1}{\lambda_2} = \frac{\sin 3·0°}{\sin 3·8°} = 0·8 \text{ (approx.)}$$

From (ii), $\lambda_{min} \propto 1/V$ using quantum theory.

$$\therefore \frac{\lambda_1}{\lambda_2} = \frac{32}{40} = 0·8$$

With a tungsten target and a p.d. of 30 kV, experiment shows that a minimum wavelength of $0·42 \times 10^{-10}$ m is obtained, as calculated from values of d and θ. From (ii),

$$\therefore \lambda_{min} = \frac{ch}{eV} = \frac{3·0 \times 10^8 \times 6·6 \times 10^{-34}}{1·6 \times 10^{-19} \times 30\,000} \text{ m}$$

$$= 0·41 \times 10^{-10} \text{ m}$$

using $c = 3·0 \times 10^8 \text{ m s}^{-1}$, $h = 6·6 \times 10^{-34}$ J s, $e = 1·6 \times 10^{-19}$ C, $V = 30\,000$ V. This is in good agreement with the experimental result.

X-Ray Absorption Spectra

X-rays are absorbed by metals. The *coefficient of absorption* can be expressed by a relation of the form $I = I_0 e^{-\lambda x}$, where I_0 is the incident intensity of the X-ray beam, I is the transmitted intensity, x is the thickness of the metal sheet and λ is the linear coefficient of absorption. The absorption can also be expressed by the relation $I = I_0 e^{-\mu m}$, where μ, the mass absorption coefficient, $= \lambda/\rho$, ρ is the density of the metal, and m is its mass per unit area.

Figure 34.21 shows how the coefficient of absorption μ varies with the wavelength of the X-ray beam. Absorption occurs when the X-ray photons possess sufficient energy to knock out an electron from the K-shell. In this case the collisions are inelastic. At a particular wavelength λ_K, however, the collisions become elastic and in this case the X-ray photons lose no energy. So the coefficient of absorption now drops sharply, as shown. A similar explanation concerning the L, M and other shells accounts for the discontinuity at the wavelength λ_L and other wavelengths.

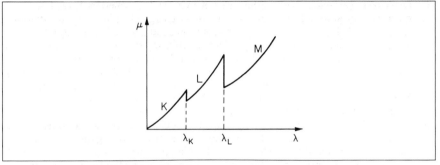

Figure 34.21 *X-ray absorption spectra*

_____ Exercises 34C _____

X-Rays

1 (a) Explain briefly why a modern X-ray tube can be operated directly from the output of a step-up transformer.
   (b) An X-ray tube works at a d.c. potential difference of 50 kV. Only 0·4% of the energy of the cathode rays is converted into X-radiation and heat is generated in the target at a rate of 600 W. Estimate   (i) the current passed into the tube, (ii) the velocity of the electrons striking the target. (Electron mass = $9\cdot00 \times 10^{-31}$ kg, electron charge = $-1\cdot60 \times 10^{-19}$ C) (N.)

2 In a modern X-ray tube electrons are accelerated through a large potential difference and the X-rays are produced when the electrons strike a tungsten target embedded in a large piece of copper.
   (a) What is the purpose of the large piece of copper?
   (b) Describe the energy changes taking place in the tube.
   (c) If the accelerating voltage is 40 kV, calculate the kinetic energy of the electrons arriving at the target, and determine the minimum wavelength of the emitted X-rays.
   (Electronic charge = $-1\cdot6 \times 10^{-19}$ C. Planck constant = $6\cdot6 \times 10^{-34}$ J s. Velocity of electromagnetic waves = $3\cdot0 \times 10^{8}$ m s^{-1}.) (*AEB*, 1984.)

3 Molybdenum K$_\alpha$ X-rays have wavelength 7×10^{-11} m. Find   (i) the minimum X-ray tube potential difference that can produce these X-rays, and   (ii) their photon energy in electronvolts.
   (The electronic charge, e = $-1\cdot6 \times 10^{-19}$ C; the Planck constant, $h = 6\cdot6 \times 10^{-34}$ J s; speed of light, $c = 3 \times 10^{8}$ m s^{-1}.) (*W.*)

4 Draw a labelled diagram of a modern X-ray tube. How may
   (a) the intensity,
   (b) the penetrating power, of the X-rays be controlled?
   An X-ray tube is operated with the anode potential of 10 kV and an anode current of 15·0 mA.   (i) Estimate the number of electrons hitting the anode per second.   (ii) Calculate the rate of production of heat at the anode, stating any assumptions made.   (iii) Describe the characteristics of the emitted X-ray spectrum and account for any special features. (Electron charge e = $1\cdot60 \times 10^{-19}$ C; Planck constant, $h = 6\cdot63 \times 10^{-34}$ J s; speed of light, $c = 3\cdot00 \times 10^{8}$ m s^{-1}.) (*C.*)

5 Describe the properties of X-rays and compare them with those of ultra-violet radiation. Outline the evidence for the wave nature of X-rays.
   The energy of an X-ray photon is hf joule where $h = 6\cdot63 \times 10^{-34}$ J s and f is the frequency in hertz (cycles per second). X-rays are emitted from a target bombarded by electrons which have been accelerated from rest through 10^{5} V. Calculate the minimum possible wavelength of the X-rays assuming that the corresponding energy is equal to the whole of the kinetic energy of one electron. (Charge of an electron = $1\cdot60 \times 10^{-19}$ C; velocity of electromagnetic waves *in vacuo* = $3\cdot00 \times 10^{8}$ m s^{-1}.) (*O. & C.*)

6 Describe the atomic process in the target of an X-ray tube whereby X-ray line spectra are produced. Determine the ratio of the energy of a photon of X-radiation of wavelength 0·1 nm to that of a photon of visible radiation of wavelength 500 nm. Why is the potential difference applied across an X-ray tube very much greater than that applied across a sodium lamp producing visible radiation? (*JMB.*)

7 The accelerating voltage across an X-ray tube is 33·0 kV. Explain why the frequency of the X-radiation cannot exceed a certain value and calculate this maximum frequency. (The Planck constant = $6·6 \times 10^{-34}$ J s; charge on an electron = $1·6 \times 10^{-19}$ C.) (*AEB*, 1981.)

8 In an X-ray tube the current through the tube is 1·0 mA and the accelerating potential is 15 kV. Calculate   (i) the number of electrons striking the anode per second,   (ii) the speed of the electrons on striking the anode assuming they leave the cathode with zero speed,   (iii) the rate at which cooling fluid, entering at 10°C, must circulate through the anode if the anode temperature is to be maintained at 35°C. Neglect any of the kinetic energy of the electrons which is converted to X-radiation. (Electronic charge = $1·6 \times 10^{-19}$ C; mass of electron = $9·1 \times 10^{-31}$ kg; specific heat capacity of liquid = $2·0 \times 10^3$ J kg^{-1} K^{-1}.) (*AEB*, 1982.)

9 State concisely what X-rays are.

Give a labelled diagram, of a modern X-ray tube. How is   (i) the predominant wavelength selected and   (ii) the intensity controlled, in practice? Sketch a typical X-ray spectrum and label its principal features. Explain briefly, at the atomic level, why there is a wavelength minimum.

List briefly **four** experimental properties of X-rays, mentioning in **each** case that character of X-rays which is responsible for the property concerned.

Outline **one** method of detecting X-rays. Comment on whether your method is able to distinguish between X-rays of different energies.

It is found that certain elements emit X-rays spontaneously. The X-ray wavelengths are those of the element's X-ray line spectrum, but it is found that in the process, the nucleus undergoes the transformation $^A_Z X \rightarrow {}^A_{Z-1} Y$. Suggest a possible explanation for this. (*W.*)

Wave-Particle Duality

Electron Diffraction

We have seen how the wave nature of X-rays has been established by X-ray diffraction experiments, as illustrated in Figure 34.22 (i). Similar experiments, first performed by Davisson and Germer, show that streams of *electrons* produce diffraction patterns and so these particles also show wave properties, see Figure 34.22 (ii). Electron diffraction is now as useful a research tool as X-ray diffraction.

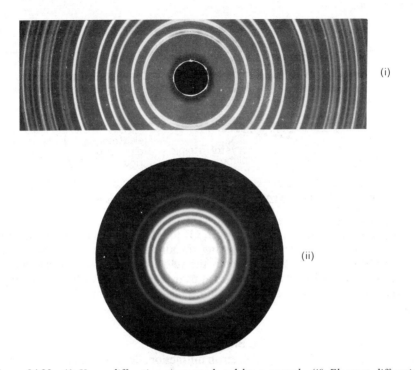

(i)

(ii)

Figure 34.22 (i) *X-ray diffraction rings produced by a crystal* (ii) *Electron diffraction rings produced by a thin gold film. Similarity of* (i) *wave and* (ii) *particle*

A TELTRON tube used for demonstrating electron diffraction is shown diagrammatically in Figure 34.23. A beam of electrons strikes a layer of graphite which is extremely thin, and a diffraction pattern, consisting of *rings*, is seen on the tube face. Sir George Thomson first obtained such a diffraction pattern using a very thin gold film, Figure 34.22 (ii). If the voltage *V* on the anode is increased,

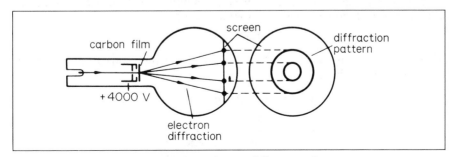

Figure 34.23 *Electron diffraction tube*

the velocity, v, of the electrons is increased. The rings are then seen to become narrow, showing that the wavelength λ of the electron waves decreases with increasing v or increasing voltage V.

If a particular ring of radius R is chosen, the angle of deviation ϕ of the incident beam is given by $\phi = 2\theta$, where θ is the angle between the incident beam and the crystal planes, Figure 34.24. Now $\tan \phi = R/D$, and if ϕ is small, $\phi = R/D$ to a good approximation. Hence $\theta = R/2D$. If the Bragg law is true for electron diffraction as well as X-ray diffraction, then, with the usual notation, $2d \sin \theta = n\lambda$.

$$\therefore \lambda \propto \sin \theta \propto \theta \propto R . \qquad . \qquad . \qquad . \qquad . \qquad (i)$$

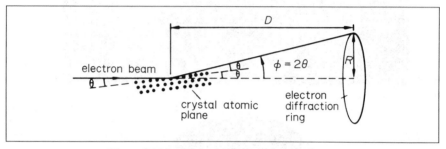

Figure 34.24 *Theory of diffraction experiment*

On plotting a graph of R against $1/\sqrt{V}$ for different values of accelerating voltage V, a straight line graph passing through the origin is obtained. Now $\frac{1}{2}m_e v^2 = eV$, or $1/\sqrt{V} \propto 1/v$, where v is the velocity of the electrons accelerated from rest. Hence, from (i), the electrons appear to act as *waves* whose wavelength is inversely-proportional to their velocity. This is in agreement with de Broglie's theory, now discussed.

De Broglie's Theory

In 1925, before the discovery of electron diffraction, de Broglie proposed that

$$\lambda = \frac{h}{p} \qquad . \qquad . \qquad . \qquad . \qquad . \qquad . \qquad (ii)$$

where λ is the wavelength of waves associated with particles of momentum p, and h is the *Planck constant*, $6\cdot63 \times 10^{-34}$ J s. The quantity h was first used by Planck in his theory of heat radiation and it is a constant which enters into all branches of atomic physics. It is easy to see that de Broglie's relation is consistent with the experimental result obtained with the electron diffraction tube. Here the gain in kinetic energy of the electrons is eV so that

$$\tfrac{1}{2}m_e v^2 = eV$$

where v is the velocity of the electrons, assuming they start from rest. Thus $v = \sqrt{2eV/m_e}$ and hence

$$p = m_e v = \sqrt{2eVm_e}$$

$$\therefore \lambda = \frac{h}{p} = \frac{h}{m_e v} = \frac{h}{\sqrt{2eVm_e}} \propto V^{-1/2}$$

We can now estimate the wavelength of an electron beam. Suppose

$V = 3600 \, \text{V}$. For an electron, $m = 9 \cdot 1 \times 10^{-31} \, \text{kg}$, $e = 1 \cdot 6 \times 10^{-19} \, \text{C}$, and $h = 6 \cdot 6 \times 10^{-34} \, \text{J s}$.

$$\therefore \lambda = \frac{h}{\sqrt{2eVm_e}} = \frac{6 \cdot 6 \times 10^{-34}}{\sqrt{2 \times 1 \cdot 6 \times 10^{-19} \times 3600 \times 9 \cdot 1 \times 10^{-31}}}$$

$$= 2 \times 10^{-11} \, \text{m}$$

This is about 30 000 times smaller than the wavelength of visible light. On this account electron beams are used in *electron microscopes*. These instruments, which use 'electron lenses', can produce resolving powers far greater than that of an optical microscope.

Wave Nature of Matter

Electrons are not the only particles which behave as waves. The effects are less noticeable with more massive particles because their momenta are generally much higher, and so the wavelength is correspondingly shorter. Since appreciable diffraction is observed only when the wavelength is of the same order as the grating spacing, the heavier particles, such as protons, are diffracted much less. Slow neutrons, however, are used in diffraction experiments, since the low velocity and high mass combine to give a momentum similar to that of electrons used in electron diffraction. The wave nature of α-particles is important in explaining α-decay.

Wave-Particle Duality

From what has been said, it is clear that particles can have wave properties, and that waves can sometimes behave as particles. As we saw in the photoelectric effect, electromagnetic waves appear to have a particle nature. Further, γ-rays behave as electromagnetic waves of very short wavelength but on detection by Geiger-Müller tubes, which count *individual* pulses, they behave as particles (see p. 884). It would appear, therefore, that a paradox exists since wave and particle structure appear different from each other.

Scientists gradually realised, however, that the dual aspect of wave-particle properties are completely general in nature. All physical quantities can be described either as waves or particles; the description to choose is entirely a matter of convenience. The two aspects, wave and particle, are linked through the two relations

$$E = hf; \qquad p = h/\lambda$$

On the left of each of these relations, E and p refer to a particle description. On the right, f and λ refer to a wave description. Note that the Planck constant is the constant of proportionality in *both* these equations, a fact which can be predicted by Einstein's Special Theory of Relativity. Further, from these equations, the frequency of the wave is proportional to the *energy* and the wavelength is inversely proportional to the *momentum*.

In the case of a photon of frequency f, we have $\lambda = c/f$. Thus $E = hf$ and $p = h/\lambda = hf/c = E/c$. Hence

$$E = pc = mc^2$$

So, as a particle, the mass of a photon is E/c^2 or hf/c^2. About 1923, Compton investigated the scattering of X-rays (high-frequency photons) by matter. He showed that the experimental results agreed with the assumption that a *particle* of mass hf/c^2 collided elastically with an electron in the atom of the material.

_____ Exercises 34D _____

Wave-Particle

(Assume $h = 6\cdot6 \times 10^{-34}$ J s, $e = -1\cdot6 \times 10^{-19}$ C, $m_e = 1 \times 10^{-30}$ kg, $c = 3 \times 10^8$ m s^{-1} unless other values are given)

1 Calculate the wavelength associated with the following objects: (i) electron moving with velocity of 10^6 m s^{-1}, (ii) bullet of mass 0·01 kg with velocity 400 m s^{-1}, (iii) sprinter of mass 60 kg with velocity 10 m s^{-1}.

2 The atomic spacing in a thin metal film is of the order 10^{-10} m. For diffraction of an electron beam, what order or velocity is required?
    Calculate the accelerating voltage to produce this velocity.

3 A parallel beam of electrons moving with velocity v is incident normally on a thin graphite film of atomic spacing $d = 1\cdot2 \times 10^{-10}$ m. The beam is diffracted through an angle θ of 11°, where $2d \sin \theta = \lambda$, the wavelength value.
    Calculate (i) the wavelength, (ii) the velocity v, (iii) the accelerating voltage needed to produce this velocity.

4 Electrons are accelerated by a p.d. of (i) 100 V, (ii) 400 V. Calculate the wavelength associated with the electrons in each case.

5 When an electron beam accelerated by a p.d. V is incident normally on a thin metal film, several diffraction rings are seen on a luminous screen. If V increases, explain why the radius of the rings decrease and the brightness of the rings decrease.

6 An X-ray photon has a wavelength of $3\cdot3 \times 10^{-10}$ m. Calculate the momentum, mass and energy of the particle associated with the photon, which moves with a velocity c.

7 Explain what is meant by
    (a) excitation by collision,
    (b) ionisation by collision.
    Light or X-radiation may be emitted when fast-moving electrons collide with atoms. Using atomic theory, explain in each case how the radiation is emitted.
    Write down two differences and one similarity between optical emission spectra and X-ray emission spectra produced in this way.
    Electrons are accelerated from rest through a p.d. of 5000 V. Calculate the wavelength of the associated electron waves. (Use $h = 6\cdot6 \times 10^{-34}$ J s, $e = 1\cdot6 \times 10^{-19}$ C, $m = 9\cdot1 \times 10^{-31}$ kg.)

8 (a) When atoms absorb energy by colliding with moving electrons, light or X-radiation may subsequently be emitted. For each type of radiation, state typical values of the energy per atom which must be absorbed and explain in atomic terms how each type of radiation is emitted.
    (b) State one similarity and two differences between optical emission spectra and X-ray emission spectra produced in this way.
    (c) Electrons are accelerated from rest through a potential difference of 10 000 V in an X-ray tube. Calculate (i) the resultant energy of the electrons in eV; (ii) the wavelength of the associated electron waves; (iii) the maximum energy and the minimum wavelength of the X-radiation generated. (Charge of electron = $1\cdot6 \times 10^{-19}$ C, mass of electron = $9\cdot11 \times 10^{-31}$ kg, Planck constant = $6\cdot62 \times 10^{-34}$ J s, speed of electromagnetic radiation *in vacuo* = $3\cdot00 \times 10^8$ m s^{-1}.)
    (*JMB*.)

35
Radioactivity, Nuclear Energy

In this chapter, we first discuss radioactive detectors and counters and the properties of α- and β-particles and γ-rays emitted by disintegration of the nucleus. We follow with half-life and its application in radioactive atoms. The discovery of the nucleus, proton and nucleon number, isotopes and the mass spectrometer are then discussed. The final part of the chapter describes nuclear energy and the Einstein mass-energy application, and the principles of nuclear fission and fusion and the nuclear reactor.

Radioactivity

In 1896 Becquerel found that a uranium compound affected a photographic plate wrapped in light-proof paper. He called the phenomenon *radioactivity*. We shall see later that natural radioactivity is due to one or more of three types of radiation emitted from heavy elements such as uranium whose nuclei are unstable. These were originally called α-, β- and γ-rays but α- and β-'rays' were soon shown to be actually particles.

α- and β-particles and γ-rays all produce ionisation as they move through a gas. On average, α-particles produce about 1000 times as many ions per unit length of their path as β-particles, which in turn produce about 1000 times as many ions as γ-rays. There are numerous detectors of ionising radiations such as α- and β-particles and γ-rays. We begin by describing two detectors used in laboratories.

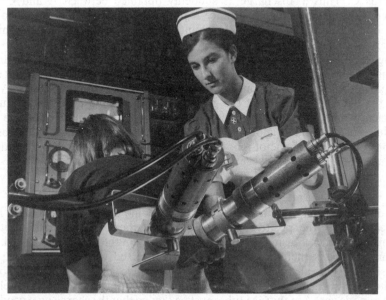

Plate 35A Radioisotopes in Medicine. *At Edinburgh Royal Infirmary, the patient has been given a dose of Iodine-131 which concentrates in her kidneys. The two counters shown measure the build-up of activity in the kidneys. Two graphs, one for each kidney, show on the recorder above the patient's head how the concentration of Iodine-131 varies.*
Copyright United Kingdom Atomic Energy Authority.

Geiger-Müller Tube

A Geiger-Müller (GM) tube is widely used for detecting ionising particles or radiation. In one form it contains a central thin wire A, the anode, insulated from a surrounding cylinder C, the cathode, which is metal or graphite-coated. The tube may have a very thin mica end window, Figure 35.1 (i). A is kept at a positive potential V such as $+400$ V relative to C, which may be earthed.

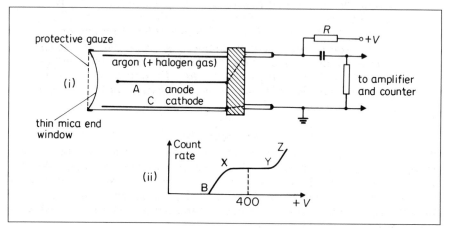

Figure 35.1 (i) *Principle of Geiger-Müller tube* (ii) *Count rate and voltage characteristic*

When a single ionising particle enters the tube, a few electrons and ions are produced in the gas. If V is above the so-called breakdown potential of the gas the number of electrons and ions are multiplied enormously, as explained shortly. The electrons are attracted by and move towards A, and the positive ions move towards C. Thus a 'discharge' is suddenly obtained between A and C. The current flowing in the high resistance R produces a p.d. which is amplified and passed to a counter, in this way the counter registers the passage of an ionising particle or radiation through the tube.

Figure 35.1 (ii) shows how the count rate of a GM tube varies with the anode voltage V. Along BX, after ionisation by collision starts, not all particles entering the tube produce sufficient ionisation for a pulse voltage to record them. Along XY, the flat part (plateau) of the graph, however, *all* the particles produce a sufficiently high pulse voltage for counting, although their initial ionisation may be different. Beyond Y along YZ a continuous discharge may occur as an uncontrollable avalanche of ions and electrons are obtained. So the middle of XY (about 400 V) is the most suitable anode voltage V.

Quenching, Ionisation

The discharge persists for a short time, as secondary electrons are emitted from the cathode by the positive ions which arrive there. This would upset the recording of other ionising particles following fast on the first one recorded. The air in the tube is therefore replaced by argon mixed with a halogen vapour, whose molecules absorb the energy of the positive ions on collision. So the discharge is quenched quickly. Electrical methods are also used for quenching.

The anode wire A in the GM tube must be *thin*, so that the charge on it produces an intense electric field E close to its surface (E is inversely-proportional to the radius of the wire). An electron-ion pair, produced by an ionising particle or radiation, is then accelerated to high energies near the wire, thus producing, by collision with gas molecules, an 'avalanche' of more electron-ion pairs.

Solid State Detector

A *solid state detector*, Figure 35.2, is made from semiconductors. Basically, it has a p-n junction which is given a small bias in the non-conduction direction (p. 791).

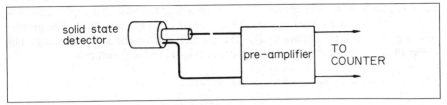

Figure 35.2 *Solid state detector*

When an energetic ionising particle such as an α-particle falls on the detector, more electron-hole pairs are created near the junction. These charge carriers move under the influence of the biasing potential and so a pulse of current is produced. The pulse is fed to an amplifier and the output passed to a counter.

The solid state detector is particularly useful for α-particle detection. If the amplifier is specially designed, β-particles and γ-rays of high energy may also be detected. This type of detector can thus be used for all three types of radiation.

Dekatron Counter, Ratemeter

As we have seen, each ionising particle or radiation produces a pulse voltage in the external circuit of a Geiger-Müller or solid state detector. In order to measure the number of pulses from the detectors, some form of counter must be used.

A *dekatron counter* consists of two or more dekatron tubes, each containing a glow or discharge which can move round a circular scale graduated in numbers 0–9, together with a mechanical counter, Figure 35.3. Each impulse causes the glow in the first tube, which counts units, to move one digit. The circuit is designed so that on the tenth pulse, which returns the first counter to zero, a

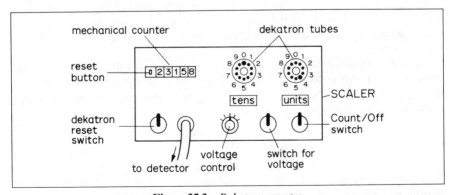

Figure 35.3 *Dekatron counter*

pulse is sent to the second tube. The glow here then moves on one place. The second tube thus counts the number of tens of pulses. After ten pulses are sent to the second tube, corresponding to a count of 100, the output pulse from the second tube is fed to the mechanical counter. This, therefore, registers the hundreds, thousands and so on. Dekatron tubes are used in radioactive experiments because they can respond to a rate of about 1000 counts per second. This is far above the count rate possible with a mechanical counter.

In contrast to a scaler, which counts the actual number of pulses, a *ratemeter* gives directly the average number of pulses per second or *count rate*. The principle is shown in Figure 35.4. The pulses received are passed to a capacitor C, which then stores the charge. C discharges slowly through a high resistor R and the average discharge current is recorded on a microammeter A. The greater the rate at which the pulses arrive, the greater will be the meter reading. The meter thus records a current which is proportional to the count rate.

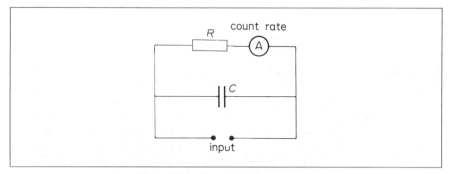

Figure 35.4   *Principle of a ratemeter*

A switch marked 'time constant' on most ratemeters allows the size of C to be chosen. If a large value of C is used, the capacitor will take a relatively long time to charge and correspondingly it will be a long time before the average count rate can be taken. The reading obtained, however, will be more accurate since the count rate is then averaged over a longer time (see below). For high accuracy, a small value of C may be used only if the count rate is very high.

Errors in Counting Experiments

Radioactive decay is random in nature (p. 887). If the count rate is high, it is not necessary to wait so long before readings are obtained which vary relatively slightly from each other. If the count rate is low, successive counts will have larger percentage differences from each other, unless a much longer counting time is used.

The accuracy of a count does not depend on the time involved but on *the total count obtained*. If N counts are received, the statistics of random processes show that this is subject to a statistical error of $\pm\sqrt{N}$. The proof is beyond the scope of this book. The percentage error is thus

$$\frac{\sqrt{N}}{N} \times 100 = \frac{100}{\sqrt{N}}\%$$

If 10% accuracy is required, $\sqrt{N} = 10$ and hence $N = 100$. Thus 100 counts must be obtained. If the counts are arriving at about 10 every second, it will be necessary to wait for 10 seconds to obtain a count of 100 and so achieve 10% accuracy. Thus a ratemeter circuit must be arranged with a time constant (CR) of 10 seconds, so that an average is obtained over this time. If, however, the counts are arriving at a rate of 1000 per second on average, it will be necessary to wait only 1/10th second to achieve 10% accuracy. Thus the 1–second time constant scale on the ratemeter will be more than adequate.

Existence of α-, β-particles and γ-rays

The existence of three different types of emission from radioactive substances can be shown by experiments such as those now outlined.

1. (a) When a radium-224 source S_1 is placed above a *spark counter*, which consists of a wire W close to an earthed metal base B, sparks are obtained between W and B, Figure 35.5 (i). If a sheet of thin paper is now introduced between S_1 and W, the sparks stop.

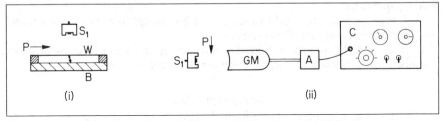

Figure 35.5 *Detection by (i) spark counter (ii) GM tube*

(b) The radium source S_1 is now placed in front of a GM tube connected to a counter C (or ratemeter) through a low-noise pre-amplifier A, Figure 35.5 (ii). The count rate is measured and the paper P is then introduced between S_1 and the GM tube. Practically no difference in the count rate is observed.

We thus conclude that there is a type of radiation from S_1 which is detected by a spark counter but is absorbed by a thin sheet of paper — these are α-particles.

2. There are other types of radiation which do not affect a spark counter but affect a GM tube. We can show this by placing a strontium-90 source S_2 so that its radiation is shielded from the GM tube by a lead plate about 1 cm thick, Figure 35.6. The count rate is then extremely low. A strong magnetic field B, in a direction perpendicular to the radiation, is now introduced beyond the lead as shown. When the field B is directed into the paper, the count rate increases. So the magnetic field has deflected the radiation towards the GM tube, which then detects it.

We conclude that there is a type of radiation which can be deflected by a strong magnetic field but is absorbed by lead 1 cm thick — these are β-particles.

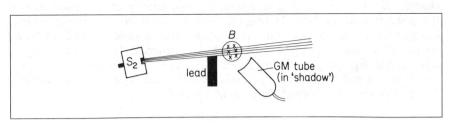

Figure 35.6 *Deflection of β-particles by magnetic field*

3. (a) We can repeat the experiment in Figure 35.6 with the radium source S_1 in place of S_2. This time the lead does not cut out all the radiation; some passes through it to the GM tube. But as before, there is an increase in the count rate when the magnetic field B is used as shown. So the radium source contains β-particles together with a third type of radiation which goes through 1 cm thickness of lead.

(b) Using a cobalt-60 source S directly in front of a GM tube as in Figure 35.5 (ii), the count rate is high. We can cut out any α-particles emitted by placing a thin sheet of paper between S and the GM tube. But using a fairly thick lead plate between S and the GM tube, the count rate continues to be high. Further, a magnetic field *B* between the lead and the GM tube makes no difference to the count rate. This type of radiation is called *γ-rays*

Conclusion. There are at least three different types of emission from radioactive sources: (1) A type recorded by a spark counter and cut off by thin paper—α-particles.

(2) A type recorded by a GM tube, cut out by lead, and deflected by a magnetic field—β-particles.

(3) A type recorded by a GM tube, not cut out except by very thick lead, and not affected by a magnetic field—γ-rays.

Further experiments show that the particles or radiation from most other radioactive sources fall into one of these three classes.

Alpha-particles

It is found that α-particles have a limited range in air at atmospheric pressure. This can be shown by slowly increasing the distance between a pure α-source and a detector. The count rate is observed to fall rapidly to zero at a separation greater than a particular value, which is called the 'range' of the α-particles. The range depends on the source and on the air pressure.

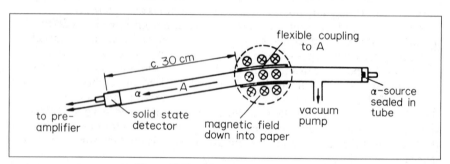

Figure 35.7 *Charge on an α-particle*

Using the apparatus of Figure 35.7, it can be shown α-particles have *positive* charges. When there is no magnetic field, the solid state detector is placed so that the tube A is horizontal in order to get the greatest count. When the magnetic field is applied, the detector has to be moved *downwards* in order to get the greatest count. This shows that the α-particles are deflected by a small amount downwards. By applying Fleming's left-hand rule, we find that particles are *positively* charged. The vacuum pump is needed in the experiment, as the range of α-particles in air at normal pressures is too small.

Nature of α-particle

Lord Rutherford and his collaborators found by deflection experiments that an α-particle had a mass about four times that of a hydrogen atom, and carried a charge $+2e$, where e was the numerical value of the charge on an electron. The relative atomic mass of helium is about four. It was thus fairly certain that an α-particle *was a helium nucleus*, that is, a helium atom which has lost two electrons.

In 1909 Rutherford and Royds showed conclusively that α-particles were helium nuclei. Radon, a gas given off by radium which emits α-particles, was collected above mercury in a thin-walled tube P, Figure 35.8. After several days some of the α-particles passed through P into a surrounding vacuum Q. After

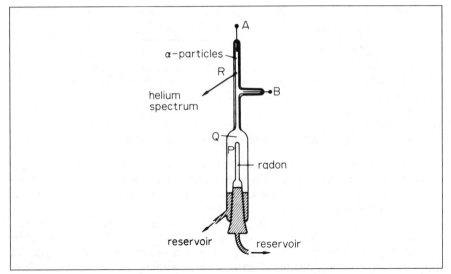

Figure 35.8 *Rutherford and Royd's experiment on α-particles*

about a week, the space in Q was reduced in volume by raising mercury reservoirs, and a gas was collected in a capillary tube R at the top of Q. A high voltage was then connected to electrodes at A and B. The spectrum of the discharge was observed to be exactly the same as the characteristic spectrum of helium.

Beta-particles and Gamma-rays

By deflecting β-particles with perpendicular magnetic and electric fields, their charge-mass ratio could be estimated. This is similar to Thomson's experiment, p. 771. These experiments showed that *β-particles are electrons moving at high speeds*. Generally, β-particles have a greater penetrating power of materials than α-particles. They also have a greater range in air than α-particles, since their ionisation of air is relatively smaller, but their path is not so well defined.

Using a strong bar magnet, it can be shown that β-particles are strongly deflected by a magnetic field. The direction of the deflection corresponds to a stream of *negatively*-charged particles, that is, opposite to the deflection of α-particles in the same field. This is consistent with the idea that β-particles are usually fast-moving electrons.

The nature of γ-rays was shown by experiments with crystals. Diffraction phenomena are obtained in this case, which suggest that *γ-rays are electromagnetic waves* (compare X-rays, p. 868). Measurement of their wavelengths, by special techniques with crystals, show they are shorter than the wavelengths of X-rays and of the order 10^{-11} m. γ-rays can penetrate large thicknesses of metals, but they have far less ionising power in gases than β-particles.

If a beam of γ-rays is allowed to pass through a very strong magnetic field no deflection is observed. This is consistent with the fact that γ-rays are electromagnetic waves and carry no charge.

Inverse-square Law for γ-rays

If γ-rays are a form of electromagnetic radiation and undergo negligible absorption in air, their intensity I should vary inversely as the square of the

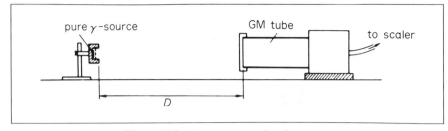

Figure 35.9 *Inverse square law for γ-rays*

distance between the source and the detector. The apparatus shown in Figure 35.9 can be used to investigate if this is the case. A suitable γ-source is placed at a suitable distance from a GM tube connected to a scaler, and the intensity inside the tube will then be proportional to the count rate C.

Suppose D is the measured distance from a fixed point on the γ-source support to the front of the GM tube. To obtain the true distance from the source to the region of gas inside the tube where ionisation occurs, we need to add an unknown but constant distance h to D. Then, assuming an inverse-square law, $I \propto 1/(D+h)^2$. Thus

$$D + h \propto \frac{1}{\sqrt{I}} \propto \frac{1}{\sqrt{C}}$$

A graph of $1/\sqrt{C}$ is therefore plotted against D for varying values of D. If the inverse-square law is true, a straight line graph is obtained which has an intercept on the D-axis of $-h$. Note that if I is plotted against $1/D^2$ and h is not zero, a straight line graph is *not* obtained from the relation $I \propto 1/(D+h)^2$. Consequently we need to plot $1/\sqrt{I}$ against D.

If a pure β-source is substituted for the γ-source and the experiment is repeated, a straight-line graph is not obtained. The absorption of β-particles in air is thus appreciable compared with γ-rays.

Absorption of Radiation by Metals

A GM tube and counter or ratemeter can be used to investigate the absorption of γ-rays or β-particles by metals, using equal thicknesses of thin sheets of lead for γ-rays and of aluminium sheets for β-particles.

Figure 35.10(i) shows roughly the results obtained in either case; the count rate C decreases with n, the number of sheets, along a curve. In the case of the β-particles, however, the count rate C reaches an alomost constant low value (not

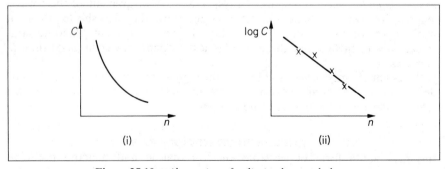

Figure 35.10 *Absorption of radiation by metal plates*

shown) as n becomes bigger. The log C result in Figure 35.10 (ii) shows that

$$I = I_0 e^{-\mu t}$$

where I is the intensity of the beam after passing through a thickness t of the metal, I_0 is the incident intensity, and μ is a constant which depends on the metal density.

Half-life and Decay Constant

We now consider the laws governing the rate of disintegration of radioactive atoms. These laws lead to the identification of atoms and to useful applications.

Radioactivity, or the emission of α- or β-particles and γ-rays, is due to disintegrating nuclei of atoms (p. 899). The disintegrations obey the statistical law of chance. Thus although we cannot tell which particular atom is likely to disintegrate next, the number of atoms disintegrating per second, dN/dt, is directly proportional to the number of atoms, N, present at that instant. Hence:

$$\frac{dN}{dt} = -\lambda N$$

where λ is a constant characteristic of the atom concerned called the *radioactivity decay constant*. The negative sign indicates that N becomes *smaller* when t increases. Thus, if N_0 is the number of radioactive atoms present at a time $t = 0$, and N is the number at the end of a time t, we have, by integration,

$$\int_{N_0}^{N} \frac{dN}{N} = -\lambda \int_{0}^{N} dt$$

$$\therefore \left[\ln N \right]_{N_0}^{N} = -\lambda t$$

$$\therefore N = N_0 e^{-\lambda t} \qquad . \qquad . \qquad . \qquad . \qquad . \qquad (i)$$

Thus the number N of undecayed atoms left decreases exponentially with the time t, and this is illustrated in Figure 35.11.

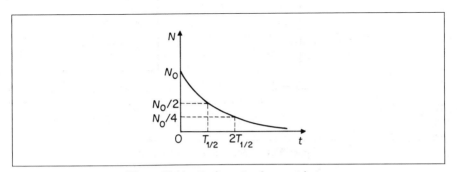

Figure 35.11   *Radioactive decay with time*

The *half-life* $T_{1/2}$ of a radioactive element is defined as the time taken for the atoms to disintegrate to half their initial number (see Figure 35.11), that is, in a time $T_{1/2}$ the radioactivity of the element diminishes to half its value. So if N_0 is the initial number of atoms, from (i),

$$\frac{N_0}{2} = N_0\,e^{-\lambda T_{1/2}}$$

$$\tfrac{1}{2} = 2^{-1} = e^{-\lambda T_{1/2}}$$

Taking logs to the base e on both sides of the equation and simplifying,

$$\therefore\ T_{1/2} = \frac{1}{\lambda}\ln 2 = \frac{0.693}{\lambda} \qquad\qquad . \qquad . \qquad . \qquad . \qquad \text{(ii)}$$

The half-life varies considerably for different radioactive atoms. For example, uranium has a half-life of the order of 4500 million years, radium has one about 1600 years, polonium about 138 days, radioactive lead about 27 minutes and radon about 1 minute.

The half-life is the same even if we start with different numbers of disintegrating atoms. So 8×10^6 atoms decay to 4×10^6 atoms in the half-life time $T_{1/2}$ and these decay to 2×10^6 atoms in the same time $T_{1/2}$. A decay from 8×10^6 to 1×10^6 atoms takes place in a time $3T_{1/2}$ (note that $8/1 = 2^3$). A decay from N_0 atoms to $N_0/32$ takes place in a time $5T_{1/2}$, since $1/32 = 1/2^5$.

The rate of nuclear disintegration or decay of an element, also known as its *activity*, is now measured in the SI unit of the *becquerel*, symbol Bq. A rate of nuclear disintegration of 5000 per second is 5 kBq.

Measurement of Short and Long Half-Life

The half-life of radon-220, a radioactive gas with a short half-life, can be measured by the apparatus shown in Figure 35.12 (i). C is a metal can or *ionisation chamber* containing a metal rod P insulated from C. B is a suitable d.c. supply such as 100 V connected between P and C, with a d.c. amplifier M joined in series as shown. This type of meter can measure very small currents.

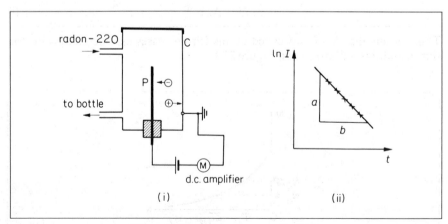

Figure 35.12 *Measurement of half-life*

A plastic bottle containing the gas radon-220 is connected to the chamber C. The radioactive gas is passed into C by squeezing the bottle and when the atoms decay the α-particles produced ionise the air in C. The negative and positive ions produced move between P and C due to the supply voltage B and so a small current or ionisation current is registered on the meter M.

Starting with an appreciable deflection in M, the falling current I is noted at equal intervals of time t such as 15 seconds. A graph of $\ln I$ against t is then plotted from the results and a straight line AB is drawn through the points. Figure 35.12 (ii). Now $I = I_0 e^{-\lambda t}$, since the number of disintegrations per second is proportional to the number of ions produced per second and hence to the current I. Taking logs to the base e,

$$\ln I = \ln I_0 - \lambda t$$

So the gradient a/b of the straight-line AB is λ numerically and this can now be found. The half-life $T_{1/2}$ is given by $T_{1/2} = 0.693/\lambda$ and so it can be calculated.

For a substance with a *long half-life* such as days, weeks or years, we can proceed as follows. Firstly, weigh a small mass of the specimen S, m say. If the relative molecular mass of S is M and S is monatomic, the number of atoms N in a mass m is given by

$$N = \frac{m}{M} \times 6.03 \times 10^{23}$$

using the Avogadro constant.

Secondly, determine the rate of emission, dN/dt, all round the specimen S by placing S at a distance r from the *end face*, area A, of a GM tube connected to a scaler. Suppose the measured count rate through the area A is dN'/dt. Then the count rate all round S, in a sphere of area $4\pi r^2$, is given by

$$\frac{dN}{dt} = \frac{4\pi r^2}{A} \times \frac{dN'}{dt}$$

Since $dN/dt = -\lambda N$, the decay constant λ can be calculated as N and dN/dt are now known. The half-life is then obtained from $T_{1/2} = 0.693/\lambda$.

Carbon Dating

Carbon has a radioactive isotope ^{14}C. It is formed when neutrons react with nitrogen in the air, as below. The neutrons are produced by cosmic rays which enter the upper atmosphere and interact with air molecules.

$$^{14}_{7}N + ^{1}_{0}n \rightarrow ^{14}_{6}C + ^{1}_{1}H$$

The radioactive isotope ^{14}C is absorbed by living material such as plants or vegetation in the form of carbon dioxide. The amount of the isotope absorbed is very small and it reaches a maximum concentration so long as the plants are living. Experiment shows that the activity of the radioactive isotope in living materials is about 19 counts per minute per gram. The half-life of the isotope is about 5600 years.

When a plant dies, no more of the isotope is absorbed. Wood formed from dead or decaying plants or vegetation, which were alive thousands of years ago, contain the radioactive isotope but its activity is much less owing to decay of the atoms. Measuring the activity of the isotope in ancient wood or similar carbon materials can provide information about its age, a method known as *carbon dating*. For example, suppose the measured activity I in a piece of ancient wood is 14 counts per minute per gram. Then, from $I = I_0 e^{-\lambda t}$,

$$\frac{I}{I_0} = \frac{14}{19} = e^{-\lambda t}$$

Taking logs to the base e on both sides we obtain

$$-0.305 = -\lambda t$$

So
$$t = \frac{0 \cdot 305}{\lambda}$$

But
$$\lambda = \frac{0 \cdot 693}{T_{1/2}} = \frac{0 \cdot 693}{5600 \, \text{y}}$$

Hence
$$t = \frac{0 \cdot 305 \times 5600 \, \text{y}}{0 \cdot 693} = 2465 \, \text{y}$$

So the ancient wood is about 2500 y old. By carbon dating the ancient material, Stonehenge in England was found to have a date of about 4000 years.

The following examples will show how to apply half-life values and to calculate the number of disintegrations per second or rate of decay (activity) after a given time. We shall need to use:

1 **Rate of decay (dN/dt) $= -\lambda N$ ($N =$ number of disintegrating atoms)**

2 $$N = N_0 \, e^{-\lambda t} \quad \text{or} \quad I = I_0 \, e^{-\lambda t}$$

3 $$T_{1/2} = \frac{0 \cdot 693}{\lambda}$$

Examples on Half-life

1 A point source of γ radiation has a half-life of 30 minutes. The initial count rate, recorded by a Geiger counter placed $2 \cdot 0$ m from the source, is $360 \, \text{s}^{-1}$. The distance between the counter and the source is altered. After $1 \cdot 5$ hours the count rate recorded is $5 \, \text{s}^{-1}$. What is the new distance between the counter and the source? (*L*.)

$1 \cdot 5 \, \text{h} = 3 \times 30 \, \text{min} = 3 \times$ half-life of source
So at *beginning* of $1 \cdot 5$ h, count rate $= 2 \times 2 \times 2 \times 5 \, \text{s}^{-1} = 40 \, \text{s}^{-1}$

$$= \frac{1}{9} \times 360 \, \text{s}^{-1} = \frac{1}{9} \times \text{initial rate}$$

$$\therefore \text{intensity of radiation} = \frac{1}{9} \times \text{initial intensity at } 2 \cdot 0 \, \text{m}$$

But
$$\text{intensity} \propto \frac{1}{d^2}, \text{ where } d \text{ is the distance}$$

$$\therefore \text{new distance} = 3 \times \text{initial distance} = 6 \cdot 0 \, \text{m}$$

2 Lanthanum has a stable isotope ^{139}La and radioactive isotope ^{138}La of half-life $1 \cdot 1 \times 10^{10}$ years whose atoms are $0 \cdot 1\%$ of those of the stable isotope. Estimate the rate of decay or activity of ^{138}La with 1 kg of ^{139}La. (Assume the Avogadro constant $= 6 \times 10^{23} \, \text{mol}^{-1}$.)

$$\text{Decay rate} \quad \frac{dN}{dt} = -\lambda N \qquad . \qquad . \qquad . \qquad . \qquad (1)$$

where λ is the decay constant and N is the number of atoms in ^{138}La. Now

number of atoms in 1 kg (1000 g) of $^{139}\text{La} = \dfrac{6 \times 10^{23} \times 1000}{139}$

Since $0 \cdot 1\% = 10^{-3}$, then

$$\text{number of atoms in } ^{138}\text{La}, \, N = \frac{10^{-3} \times 6 \times 10^{23} \times 1000}{139} = \frac{6 \times 10^{23}}{139}$$

Also,
$$\lambda = \frac{0.693}{T_{1/2}} = \frac{0.693}{1.1 \times 10^{10} \times 365 \times 24 \times 3600}$$

From (1),
$$\frac{dN}{dt} = \frac{0.693 \times 6 \times 10^{23}}{1.1 \times 10^{10} \times 365 \times 24 \times 3600 \times 139}$$
$$= 8600 \, s^{-1}$$

3 At a certain instant, a piece of radioactive material contains 10^{12} atoms. The half-life of the material is 30 days.

(1) Calculate the number of disintegrations in the first second. (2) How long will elapse before 10^4 atoms remain? (3) What is the count rate at this time?

(1) We have $N = N_0 \, e^{-\lambda t}$

$$\therefore \frac{dN}{dt} = -N_0 \lambda \, e^{-\lambda t} = -\lambda N$$

Hence, when $N = 10^{12}$,
$$\frac{dN}{dt} = -\lambda 10^{12}$$

Now
$$\lambda = \frac{0.693}{T} = \frac{0.693}{30 \times 24 \times 60 \times 60} \, s^{-1}$$

$\therefore$ number of disintegrations per second
$$= \frac{10^{12} \times 0.693}{30 \times 24 \times 60 \times 60} = 2.7 \times 10^5$$

(2) When $N = 10^4$, we have
$$10^4 = 10^{12} \, e^{-\lambda t}, \quad \text{so } 10^{-8} = e^{-\lambda t}$$

Taking logs to base 10,
$$\therefore -8 = -\lambda t \log e$$
$$\therefore t = \frac{8}{\lambda} \frac{1}{\log e} = \frac{8T}{0.693 \log e}$$
$$= 797 \text{ days (approx.)}$$

(3) Since
$$\frac{dN}{dt} = -\lambda N$$

$\therefore$ number of disintegrations per hour $= \dfrac{0.693}{30 \times 24} \times 10^4 = 9.6$

Cloud Chambers

We conclude with a brief account of the *cloud chamber*, which can be used for detecting or photographing ionising particles and radiation. C. T. R. Wilson's cloud chamber, invented in 1911, was one of the most useful early inventions for studying radioactivity. It enabled photographs to be taken of α- or β-particles or γ-rays.

Basically, Wilson's cloud chamber consists of a chamber containing saturated water or alcohol vapour. Figure 35.13 (i) illustrates the basic principle of a cloud chamber C which uses alcohol vapour. An excess amount of liquid alcohol is placed on a dark pad D on a piston P. When the piston is moved down quickly, the air in C undergoes an adiabatic expansion and cools. The dust nuclei are all carried away after a few expansions by drops forming on them, and then the dust-free air in C is subjected to a controlled adiabatic expansion of about 1·31 to 1·38 times its original volume. The air is now supersaturated, that is, the vapour pressure is greater than the saturation vapour pressure at the reduced temperature reached but no vapour condenses. Simultaneously, the air is exposed to α-particles from a radioactive source S, for example. Water droplets immediately collect round the ions produced, which act as centres of formation. The drops are illuminated, and photographed by light scattered from them.

α-particles produce short continuous straight trails, as shown at the top of Figure 35.13. β-particles, which have much less mass, produce longer but straggly paths owing to collisions with gas molecules. Wilson's cloud chamber has proved of great value in the study of radioactivity and nuclear structure, see Figure 35.18.

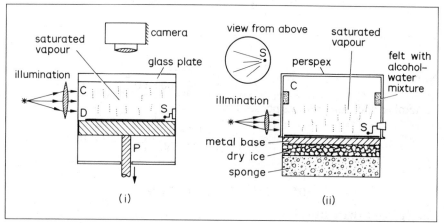

Figure 35.13 *Principle of* (i) *Wilson cloud chamber* (ii) *diffusion cloud chamber*

Diffusion cloud chamber. Figure 35.13 (ii) shows the principle of another type of cloud chamber. It has a perspex chamber C, with a strip of felt at the top containing excess of a mixture of water and alcohol. The dark metal base is kept at a low temperature of about − 50°C by dry ice (solid carbon dioxide) packed below it. Vapour thus diffuses continuously from the top to the bottom of the chamber.

Above the cold metal base there is supersaturated vapour. A radioactive source S near the base, emitting α-particles, for example, produces vapour trails which can be seen on looking down through the top of C. These trails show the straight line paths of the emitted α-particles, see Figure 35.13. Unlike the Wilson cloud chamber, this type does not require adiabatic expansion of the gas inside. In both cases, the ions formed can be cleared by applying a suitable p.d. between the top and bottom of the chamber.

The random nature of radioactive decay can be seen by using this cloud chamber. Particles emitted by a radioactive substance do not appear at equal intervals of time but are entirely random. The length of the track of an emitted particle is a measure of its initial energy. The tracks of α-particles are nearly all the same, showing that the α-particles were all emitted with the same energy. Sometimes two different lengths of tracks are obtained, showing that the α-particles may have one of two energies on emission.

_____ **Exercises 35A** _____

1 What is meant by the *half-life* of a radioactive element? Draw a labelled sketch of the relation $N = N_0 e^{-\lambda t}$ to illustrate your answer.

   The initial number of atoms in a radioactive element is 6.0×10^{20} and its half-life is 10 h. Calculate
   (a) the number of atoms which have decayed in 30 h,
   (b) the amount of energy liberated if the energy liberated per atom decay is 4.0×10^{-13} J.

2 A point source of alpha particles, a tiny mass of the nuclide $^{241}_{95}$Am, is mounted 7.0 cm in front of a GM tube whose mica window has a receiving area of 3.0 cm^2,

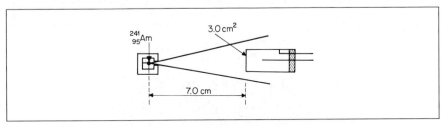

Figure 35A

Figure 35A. The counter linked to the GM tube records 5.4×10^4 counts per minute. Calculate
(a) the number of disintegrations per second within the source, and
(b) the number of $^{241}_{95}$Am atoms in the source.
(The decay constant, λ, for $^{241}_{95}$Am $= 4.80 \times 10^{-11}$ s^{-1}.) (L.)

3 $^{32}_{15}$P is a beta-emitter with a decay constant of 5.6×10^{-7} s^{-1}. For a particular application the initial rate of disintegration must yield 4.0×10^7 beta particles every second. What mass of pure $^{32}_{15}$P will give this decay rate? (C.)

4 A mixture of I^{131} (half-life 8 days) and I^{132} (half-life 2.3 hr) has an activity of 5 k Bq. After 12 days the activity is 1 k Bq. What proportion of the 5 k Bq was due to I^{132}? (The activity of the products of disintegration may be neglected.) (W.)

5 What do you understand by *radioactivity* and *half-life*? Plot an accurate graph to show how the number of radioactive atoms of a given element (expressed as a percentage of those initially present) varies with time. Use a time scale extending over five half-lives.

   The isotope $^{40}_{19}$K, with a half-life of 1.37×10^9 years, decays to $^{40}_{18}$Ar, which is stable. Moon rocks from the Sea of Tranquillity show that the ratio of these potassium atoms to argon atoms is 1/7. Estimate the age of these rocks, stating clearly any assumptions you make. Certain other rocks give a value of 1/4 for this ratio. By means of your graph, or otherwise, estimate their age. (C.)

6 Discuss the assumptions on which the law of radioactive decay is based.
   What is meant by the *half-life* of a radioactive substance?
   A small volume of a solution which contained a radioactive isotope of sodium had an activity of 12 000 disintegrations per minute when it was injected into the bloodstream of a patient. After 30 hours the activity of 1.0 cm^3 of the blood was found to be 0.50 disintegrations per minute. If the half-life of the sodium isotope is taken as 15 hours, estimate the volume of blood in the patient. (*JMB.*)

7 List the chief properties of α-radiation, β-radiation and γ-radiation. Describe a type of Geiger-Müller tube that can be used to detect all three types of radiation. If you were supplied with a radioactive preparation that was emitting all three types of radiation, describe and explain how you would use the tube to confirm that each of them was present.

   A radon ($^{222}_{86}$Rn) nucleus, of mass 3.6×10^{-25} kg, decays by the emission of an α-particle of mass 6.7×10^{-27} kg and energy 8.8×10^{-13} J.
   (a) Write down the values of the mass number A and the atomic number Z for the resulting nucleus.

(b) Calculate the momentum of the emitted α-particle.

(c) Find the velocity of recoil of the resulting nucleus. (*O.*)

8 What is meant by the *half-life period* (*half-life*) of a radioactive material? Describe how the nature of α-particles has been established experimentally.

The half-life period of the body polonium-210 is about 140 days. During this period the average number of α-emissions per day from a mass of polonium initially equal to 1 microgram is about 12×10^{12}. Assuming that one emission takes place per atom and that the approximate density of polonium is $10 \, \text{g cm}^{-3}$, estimate the number of atoms in $1 \, \text{cm}^3$ of polonium. (*N.*)

9 (a) A GM tube is exposed to a constant flux of alpha particles. The graph below shows how the recorded count rate depends on the potential difference across the tube. Figure 35B. Draw and label a diagram of a GM tube. Outline its

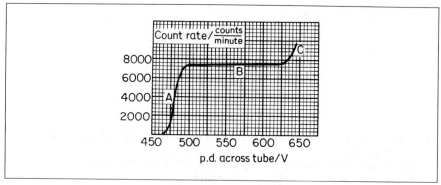

Figure 35B

working principle with reference to what happens when an alpha particle enters the tube. Explain why there is an upper limit to the rate at which a GM tube can detect α-particles.

How do you account for (i) the sharp rise in the recorded count rate at A, (ii) the 'plateau' at B, and (iii) the uncontrolled rise in the recorded count rate at C?

State what potential difference you would choose for the Geiger counter whose response is shown in the graph. Given one good reason for your choice.

(b) A small amount of ^{24}Na is smeared on to a card and its activity falls by 87·5% in 45 h. What is the half-life of ^{24}Na?

Describe how you would use a GM tube in conjunction with a suitable counter to measure the half-life of ^{24}Na. Explain carefully how the result is found from the measurements.

$$\left(\text{Decay constant} = \frac{0.693}{\text{half-life}} \right). (L.)$$

10 A radioactive isotope of thallium, $^{207}_{81}$Tl, emits beta particles (β^-) with an average energy of 1·5 MeV. The half-life of the isotope is 135 days, and it is also thought to emit gamma radiation.

(a) (i) Describe simple tests which could be used to confirm that beta particles are emitted, and to check for the presence of gamma radiation. (ii) What will be the atomic number and the atomic mass of the new isotope formed by the emission of a beta particle? What will happen to the nucleus of the new isotope if a gamma ray photon is emitted?

(b) (i) What is meant by an 'energy of 1·5 MeV', and what form does the energy take in this case? (ii) What is meant by a half-life of 135 days? (iii) Calculate the decay constant.

(c) Assuming that 207 g of thallium 207 contains 6×10^{23} atoms, calculate (i) the total energy, in joules, available from the beta particles emitted from 1 g of the isotope; (electronic charge $= -1·6 \times 10^{-19}$ C) (ii) the initial rate at which beta particles are emitted from 1 g of the freshly prepared isotope;   (iii) the initial power, in watts, available from the beta particles emitted at the rate calculated in (ii).

   (d) It has been suggested that thallium 207 could be used to power the amplifiers built into underwater telephone cables. Use the data and your answers to (c) to discuss whether the suggestion is worth pursuing. (*AEB*, 1984.)

11 (a) Radium has an isotope $^{226}_{88}$Ra of half-life approximately 1600 years. What is meant by the terms *isotope* and *half-life*?

   (b) A sample of $^{226}_{88}$Ra emits both α-particles and γ-rays. State, and account for, any change in   (i) nucleon number, A,   (ii) proton number, Z, which may occur as a result of the emission of these radiations.

   (c) A mass defect of $8{\cdot}8 \times 10^{-30}$ kg occurs in the decay of a $^{226}_{88}$Ra nucleus. Calculate the energy released.
      In a given sample it is found that most of the radium nuclei decay with the emission of an α-particle of energy 4·60 MeV and a γ-ray photon. What is the frequency of the γ-ray photon emitted? (Ignore the recoil energy of the decayed nucleus.)

   (d) Outline briefly how you could show experimentally that both α-particles and γ-rays are present in emissions from $^{226}_{88}$Ra.
      How is it possible that, with a half-life of 1600 years, $^{226}_{88}$Ra occurs in measurable quantities in minerals 10^9 years old?
      (Speed of light $= 3{\cdot}0 \times 10^8$ m s^{-1}. The Planck constant $= 6{\cdot}6 \times 10^{-34}$ J s. Electronic charge $= -1{\cdot}6 \times 10^{-19}$ C.) (*L.*)

12 (a) A sample initially consisting of N_0 radioactive atoms of a single isotope. After a time t the number N of radioactive atoms of the isotope is given by $N = N_0 e^{-\lambda t}$.   (i) Sketch a graph of this equation and show on the graph the time equal to the half-life of the sample, $T_{1/2}$.   (ii) Explain what is meant by the disintegration rate of the sample and represent this on the graph at zero time and at time $T_{1/2}$. State the ratio of these two disintegration rates.   (iii) Explain the physical significance of the constant λ in the equation above.

   (b) If you were provided with a small gamma ray source of very long half-life, describe the arrangement you would use, and the measurements you would make to investigate the inverse-square law for the gamma rays. Show how you would use your measurements to verify this law. (*JMB.*)

13 Describe how you would investigate the absorption of beta particles from a source of long half-life by different thicknesses of aluminium. Sketch a graph of the results you would expect to obtain and comment on any special features of the graph.
   A small source of beta particles is placed on the axis of a Geiger-Müller tube and a few centimetres from the window of the tube. State and explain three reasons why the observed count rate is less than the disintegration rate of the source.
   A source, of which the half-life is 130 days, contains initially $1{\cdot}0 \times 10^{20}$ radioactive atoms, and the energy released per disintegration is $8{\cdot}0 \times 10^{-13}$ J. Calculate
   (a) the activity of the source after 260 days have elapsed and
   (b) the total energy released during this period. (*JMB.*)

14 Describe the structure of a Geiger-Müller tube. Why are some tubes fitted with thin end windows? Why does the anode of a Geiger-Müller tube have to be made of a *thin* wire?
   Explain the principle of operation of a cloud chamber. Describe and explain the differences between the tracks formed in such a chamber by alpha and beta particles.
   A radioactive source has decayed to 1/128th of its initial activity after 50 days. What is its half-life? (*L.*)

15 Describe how you could detect and distinguish between beta-radiation and gamma-radiation from a radioactive source emitting both radiations.
   A source of radioactive potassium is known to contain two isotopes, $^{42}_{19}$K and $^{44}_{19}$K, both of which decay by emission of beta-radiation to stable isotopes of calcium. Write a nuclear transformation equation for one of these decays.
   The source is placed in front of a beta-radiation counter and the following count rates, corrected for background, are recorded.

time/hours	0	0·5	1·0	1·5	2·0	2·5	3·0
count rate/min^{-1}	10 000	3980	2125	1260	955	890	832

time/hours	4·0	5·0	6·0	7·0	8·0	9·0	10·0
count rate/min^{-1}	790	750	710	670	630	600	575

Plot the data on a graph of lg (count rate/min^{-1}) against time/hours.
From your graph estimate values for
(a) the half life of $^{42}_{19}K$ which is the longer lived isotope,
(b) the half life of $^{44}_{19}K$ which is the shorter lived isotope,
(c) the initial count rates due to $^{42}_{19}K$ and $^{44}_{19}K$,
(d) the ratio of the amounts of $^{42}_{19}K$ and $^{44}_{19}K$ present in the source at the start of the measurements. (*O. & C.*)

16 Living matter contains carbon. A tiny proportion of this carbon is the radioactive isotope ^{14}C. The average decay of the ^{14}C content per kilogram of carbon in living matter is 255 Bq (i.e. 255 disintegrations per second). The half-life of ^{14}C is $1·76 \times 10^{11}$ s.
(a) Calculate (i) the decay constant for ^{14}C, and (ii) the probable number of ^{14}C atoms per kilogram of carbon in living matter.
(b) Explain, with reference to the average decay of 255 Bq, (i) why radioactivity is described as a random phenomenon, and (ii) what is meant by *half-life*. (*L.*)

17 Nuclei of $^{238}_{94}Pu$ decay with a half-life of 90 years, emitting alpha particles with an energy of 5·1 MeV.
(a) Calculate (i) the decay constant for this disintegration, (ii) the number of disintegrations per second of 1·0 g of $^{238}_{94}Pu$.
(b) It is proposed to use the above isotope as an energy source for a heart pacemaker. (i) Estimate the minimum mass of plutonium which would give an initial power of 100 mW, stating any assumption you make. (ii) Suggest one reason why an alpha-emitter is preferable to a beta-emitter for this purpose.
1 MeV = $1·6 \times 10^{-13}$ J, the Avogadro constant = $6·0 \times 10^{23}$ mol^{-1} (*JMB.*)

The Nucleus, Nuclear Energy

Discovery of Nucleus, Geiger-Marsden Experiment

In 1909 Geiger and Marsden, at Lord Rutherford's suggestion, investigated the scattering of α-particles by thin films of metal of high atomic mass, such as gold foil. They used a radon tube S in a metal block as a source of α-particles, and limited the particles to a narrow pencil, Figure 35.14. The thin metal foil A was placed in the centre of an evacuated vessel, and the scattering of the particles after passing through A was observed on a fluorescent screen B, placed at the focal plane of a microscope M. Scintillations were seen on B whenever it is struck by α-particles.

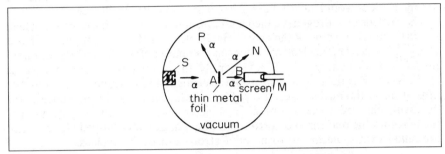

Figure 35.14 *Discovery of nucleus—Geiger and Marsden*

Geiger and Marsden found that α-particles struck B not only in the direction SA, but also when the microscope M was moved round to N and even to P. Thus though the majority of α-particles were scattered through small angles, some particles were scattered through very large angles. Rutherford found this very exciting news. It meant that some α-particles had come into the repulsive field of a highly concentrated positive charge at the heart or centre of the atom.

Paths of Scattered Particles

Rutherford assumed that an atom has a *nucleus*, in which all the positive charge and most of the mass is concentrated. The beam of α-particles incident on a thin metal foil are then scattered through various angles, as shown roughly in Figure 35.15. Those particles very close to the nucleus are deflected through a large angle, since the repulsive force is then very big.

Rutherford obtained a formula for the number N of α-particles scattered

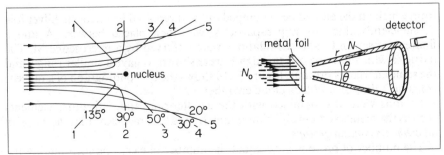

Figure 35.15 *Thin foil scattering of α-particles*   **Figure 35.16** *Rutherford scattering law*

through an angle θ by thin metal foil of thickness t and detected at a distance r. In a series of experiments using a detector as illustrated diagrammatically in Figure 35.16, Geiger and Marsden verified the Rutherford formula and thus confirmed the existence of the nucleus.

Atomic Nucleus and Atomic (Proton) Number

In 1911 Rutherford proposed the basic structure of the atom which is accepted today, and which subsequent experiments by Moseley and others have confirmed. A neutral atom consists of a very tiny nucleus of diameter about 10^{-15} m which contains practically the whole mass of the atom. The atom is largely empty. If a drop of water was magnified until it reached the size of the earth, the atoms inside would then be only a few metres in diameter and the atomic nucleus would have a diameter of only about 10^{-2} millimetre.

The nucleus of hydrogen is called a *proton*, and it carries a charge of $+e$, where e is the numerical value of the charge on an electron. The helium nucleus has a charge of $+2e$. The nucleus of copper has a charge of $+29e$, and the uranium nucleus carries a charge of $+92e$. Generally, the positive charge on a nucleus is $+Ze$, where Z is the *atomic number* of the element and is defined as the number of protons in the nucleus (see also p. 870). Under the attractive influence of the positively-charged nucleus, a number of electrons equal to the atomic number move round the nucleus and surround it like a negatively-charged cloud. These are called 'extra-nuclear' electrons, or electrons outside the nucleus.

Discovery of Protons in Nucleus, Mass (Nucleon) Number

In 1919 Rutherford found that energetic α-particles could penetrate nitrogen atoms and that protons were thrown out after the collision. The apparatus used is shown in Figure 35.17. A source of α-particles, A, was placed in a container D

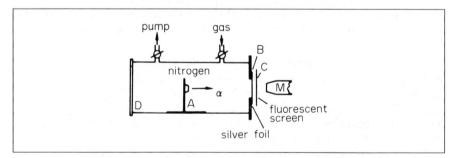

Figure 35.17 *Discovery of protons in the nucleus — Rutherford*

from which all the air had been pumped out and replaced by nitrogen. Silver foil, B, sufficiently thick to stop α-particles, was then placed between A and a fluorescent screen C, and scintillations were observed by a microscope M. The particles which have passed through B were shown to have a similar range, and the same charge, as protons. Figure 35.18 shows the first photograph of a nuclear collision, taken by a Wilson cloud chamber.

Protons were also obtained with the gas fluorine, and with other elements such as the metals sodium and aluminium. It thus becomes clear that *the nuclei of all elements contain protons.*

The number of protons must equal the number of electrons surrounding the nucleus, so that each is equal to the atomic number, Z, of the element. A proton

Figure 35.18 *Transmutation of nitrogen by collision with α-particle. An oxygen nucleus, short right-curved track, and a proton, left straight track are produced at the top centre*

is represented by the symbol, $_1^1H$; the top number denotes the *mass number* or *nucleon number A* (a nucleon is a particle in the nucleus) and the bottom number is the *atomic* or *proton number Z*. The helium nucleus such as an α-particle is represented by $_2^4He$; its mass or nucleon number A is 4 and its proton number Z is 2.

One of the heaviest nuclei, uranium, can be represented by $_{92}^{238}U$; it has a nucleon number A of 238 and a proton number Z of 92.

Discovery of Neutron in Nucleus

In 1932, Chadwick found a new particle inside a nucleus, in addition to the proton. It had about the same mass as the proton but carried no charge and so Chadwick called the new particle a *neutron*. It is now considered that *all nuclei contain protons and neutrons*. The neutron is represented by the symbol $_0^1n$ as it has a mass number of 1 and zero charge.

We can now see that a helium nucleus, $_2^4He$, has 2 protons and 2 neutrons, a total mass or nucleon number of 4. The sodium nucleus, $_{11}^{23}Na$, has 11 protons and 12 neutrons. The uranium nucleus, $_{92}^{238}U$, has 92 protons and 146 neutrons. Generally, a nucleus represented by $_Z^AX$ has Z protons and $(A-Z)$ neutrons.

Radioactive Disintegration, Uranium Series

Naturally occurring radioactive elements such as uranium, actinium and thorium disintegrate to form new elements, and these in turn are unstable and form other elements. Between 1902 and 1909 Rutherford and Soddy made a study of the elements formed from a particular 'parent' element. The *uranium series* is listed in the table on page 900.

Nuclear Reactions, Conservation of Mass and Charge

The new element formed after disintegration can be identified by considering the particles emitted from the nucleus of the parent atom. An α-particle, a helium nucleus, has a charge of $+2e$ and a mass number 4. Uranium I, of atomic number 92 and mass number 238, emits an α-particle from its nucleus of charge $+92e$, and so the new nucleus formed has an atomic number 90 and a mass

Element	Symbol	Atomic Number	Mass Number	Half-life Period (T)	Particle emitted
Uranium I	U	92	238	4500 million years	α
Uranium X_1	Th	90	234	24 days	β, γ
Uranium X_2	Pa	91	234	1·2 minutes	β, γ
Uranium II	U	92	234	250 000 years	α
Ionium	Th	90	230	80 000 years	α, γ
Radium	Ra	88	226	1600 years	α, γ
Radon	Rn	86	222	3·8 days	α
Radium A	Po	84	218	3 minutes	α
Radium B	Pb	82	214	27 minutes	β, γ
Radium C*	Bi	83	214	20 minutes	β or α
Radium C'	Po	84	214	$1 \cdot 6 \times 10^{-4}$ seconds	α
Radium C''	Tl	81	210	1·3 minutes	β
Radium D	Pb	82	210	19 years	β, γ
Radium E	Bi	83	210	5 days	β
Radium F	Po	84	210	138 days	α, γ
Lead	Pb	82	206	(stable)	

*Radium C exhibits branching; it produces Radium C' on β-emission or Radium C'' on α-emission. Radium D is then produced from Radium C' by α-emission or from Radium C'' by β-emission.

number 234. This was called uranium X_1, and since the element thorium (Th) has an atomic number 90, uranium X_1 is actually thorium.

We write this *nuclear reaction* as:

$$^{238}_{92}U \rightarrow {}^4_2He + {}^{234}_{90}Th$$

The top numbers of the nuclei are their mass or nucleon numbers. The bottom numbers are their atomic or proton numbers, which also represent the charges on the nuclei in terms of the numerical value e of the charge on an electron. In any nuclear reaction, we always apply the conservation of mass and of charge. So

1 Total mass (nucleon) number is constant before and after the reaction
2 Total charge (proton) number is constant before and after the reaction

We can apply these rules to β-emission from a radioactive nucleus. A β-particle is usually an electron of negligible mass and charge $-e$ or -1 in terms of e. On rare occasions, however, a β-particle is emitted which has a positive charge $+e$ or $+1$ in terms of e. This particle is called a *positron* and denoted by β^+.

Let us assume that a β-particle and a γ-ray are emitted by thorium Th which has a nucleon number 234 and a proton number 90. Then we write the reaction as:

$$^{234}_{90}Th \rightarrow {}^0_{-1}e + {}^{234}_{91}Pa$$

Note that the proton number has *increased* from 90(Th) to 91(Pa). As explained later, if a nucleus has too many neutrons for stability, a neutron changes to a proton inside the nucleus and a β-particle is then emitted.

The uranium series contains *isotopes* of uranium (U), lead (Pb), thorium (Th) and bismuth (Bi), that is, elements which have the same proton or atomic numbers but different nucleon or mass numbers. We now discuss how isotopes were found.

Thomson Mass Spectrometer, Isotopes

When a chemist measures the density of a gas such as hydrogen or chlorine, he or she finds the volume and then the masses of *all* the atoms present in the gas sample. Individual atoms can not be identified in this measurement.

In 1911, Sir J. J. Thomson measured the masses of individual atoms. The atoms were first ionised in a hot vapour or gas, that is, they were stripped of one or more electrons to form and *ion* with a positive charge and which had a mass practically equal to that of the atom since electrons are very light.

Thomson used parallel electric and magnetic fields to deflect the fast-moving ion in a parabolic path. The equation of the parabola is $y = kx^2$, where k is a constant which depends on the *charge-mass ratio* (Q/M) for that ion. For a given value of x and charge Q, the deflection y is greatest for a hydrogen ion, since this has the smallest mass, and least for the ion of heaviest mass.

Figure 35.19 *Mass spectrometer. Positive-ray parabolas due to mercury, carbon monoxide, oxygen and carbon ions*

Figure 35.19 shows a photograph of some of the parabolic paths due to different gas ions. By comparing the deflections obtained in the y-direction, the masses of individual ions or atoms can be compared.

With chlorine gas, two parabolas were obtained which gave atomic masses of 35 and 37 respectively. Thus the atoms of chlorine have different masses but the same chemical properties, and these atoms are said to be *isotopes* of chlorine. In chlorine, there are three times as many atoms of mass 35 as there are of mass 37, so that the average atomic mass is $(3 \times 35 + 1 \times 37)/4$, or 35·5. The element xenon has as many as nine isotopes. One part in 5000 of hydrogen consists of an isotope of mass 2 called deuterium, or heavy hydrogen. An unstable isotope of hydrogen of mass 3 is called tritium. Hydrogen isotopes are used in nuclear energy experiments (p. 913).

Bainbridge Mass Spectrometer

Thomson's earliest form of mass spectrometer was followed by more sensitive forms. Bainbridge devised a mass spectrometer in which the ions were incident on a photographic plate after being deflected by a magnetic field.

The principle of the spectrometer is shown in Figure 35.20. Positive ions were produced in a discharge tube (not shown) and admitted as a fine beam through slits S_1, S_2. The beam then passed between insulated plates P, Q, connected to a battery, which created an electric field of intensity E. A uniform magnetic field B_1, perpendicular to E, was also applied over the region of the plates.

Velocity selector   Suppose an ion of charge Q and mass M enters the region with a velocity v. If the magnetic force B_1Qv = the electric force EQ, the two

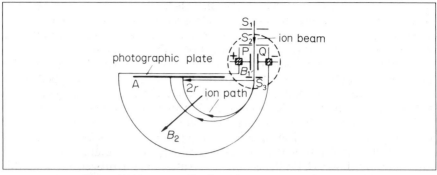

Figure 35.20 *Principle of Bainbridge mass spectrometer*

forces cancel each other and so the ion travels undeflected through PQ. The velocity v is given by $v = E/B_1$ from this equation. Since this value of v does not depend on the charge Q or the mass M of an ion, it follows that *all* ions, even though their masses are different, pass through the plates PQ undeflected and through a slit S_3. This arrangement of perpendicular electric and magnetic fields is therefore called a 'velocity selector'—it only allows ions through S_3 which have the same velocity v equal to E/B_1 (see page 772).

Magnetic Analyser   The selected ions were now deflected in a circular path of radius r by a uniform perpendicular magnetic field B_2, and an image was produced on a photographic plate A, as shown. In this case, if the mass of the ion is M,

$$\frac{Mv^2}{r} = B_2 Qv$$

$$\therefore \frac{M}{Q} = \frac{rB_2}{v}$$

But for the selected ions, $v = E/B_1$ from above

$$\therefore \frac{M}{Q} = \frac{rB_2 B_1}{E}$$

$$\therefore \frac{M}{Q} \propto r$$

for given magnetic and electric fields.

Since the ions strike the photographic plate at a distance $2r$ from the middle of the slit S_3, it follows that the separation of ions carrying the same charge is directly proportional to their mass. Thus a 'linear' mass scale is achieved. A resolution of 1 in 30 000 was obtained with a later type of spectrometer.

Example on Mass Spectrometer

In a mass spectrograph, an ion X of mass number 24 and charge $+e$ and an ion Y of mass 22 and charge $+2e$ both enter the magnetic field with the same velocity. The radius of the circular path of X is 0·25 m. Calculate the radius of the circular path of Y.

From above, $r \propto \dfrac{M}{Q}$

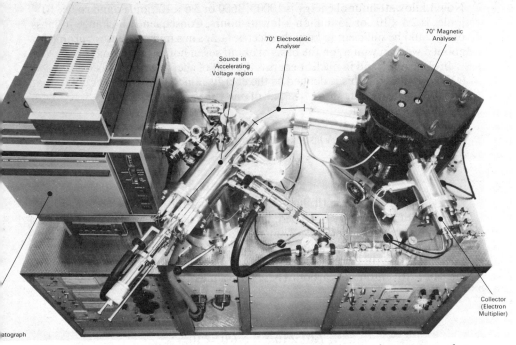

atograph

Plate 35B Industrial mass spectrometer. *Double-focusing by an electrostatic and a magnetic field is used, which produces mass and resolution superior to single focusing. The gas chromatograph (left) separates mixtures into their components. The ion sources are then passed into the front tube (left) to be accelerated by a high voltage. The ions are now deflected electrostatically by the analyser at the top and then deflected by the magnetic field of the second analyser. The separated ions are collected in the tube at the front. Here they are recorded electronically and their masses and relative abundance in the compound analysed are deduced from the readings. (Courtesy of VG Micromass Limited)*

For ion Y,
$$\frac{M}{Q} = \frac{22}{2e}$$

For ion X,
$$\frac{M}{Q} = \frac{24}{e}$$

So
$$\frac{r_Y}{0.25} = \frac{22/2e}{24/e} = \frac{11}{24}$$

$$\therefore r_Y = \frac{11}{24} \times 0.25 = 0.11 \text{ m (approx.)}$$

Einstein's Mass-Energy Relation

In 1905 Einstein showed from his Theory of Relativity that mass and energy can be changed from one form to the other. The energy E produced by a change of mass m is given by the relation:

$$E = mc^2$$

where c is the numerical value of the velocity of light. E is in joule when m is in kg and c has the numerical value 3×10^8, the speed of light in metres per second. So a change, in mass of 1 g could theoretically produce 9×10^{13} joules of energy.

Now 1 kilowatt-hour of energy is 1000×3600 or $3 \cdot 6 \times 10^6$ joules, and so 9×10^{13} joules is $2 \cdot 5 \times 10^7$ or 25 million kilowatt-hours. Consequently a change in mass of 1 g could be sufficient to keep the electric lamps in a million houses burning for about a week in winter, on the basis of about seven hours' use per day.

In electronics and in nuclear energy, the unit of energy called and *electron-volt* (eV) is often used. This is defined as the energy gained by a charge equal to that on an electron moving through a p.d. of one volt. So

$$1\,eV = 1 \cdot 6 \times 10^{-19}\,J$$

The *megelectronvolt* (MeV) is a larger energy unit, and is defined as 1 million eV.

So
$$1\,MeV = 1 \cdot 6 \times 10^{-13}\,J$$

Unified Atomic Mass Unit

If another unit of energy is needed, then one may use a unit of mass, since mass and energy are interchangeable. The *unified atomic mass unit* (u) is defined as 1/12th of the mass of the carbon atom $^{12}_{6}C$. Now the number of molecules in 1 mole of carbon is $6 \cdot 02 \times 10^{23}$, the Avogadro constant, and since carbon is monatomic, there are $6 \cdot 02 \times 10^{23}$ atoms of carbon. These have a mass 12 g.

$\therefore$ mass of 1 atom of carbon

$$= \frac{12}{6 \cdot 02 \times 10^{23}}\,g = \frac{12}{6 \cdot 02 \times 10^{26}}\,kg$$

$$= 12\,u$$

$$\therefore 1\,u = \frac{12}{12 \times 6 \cdot 02 \times 10^{26}}\,kg$$

$$= 1 \cdot 66 \times 10^{-27}\,kg\,(approx.)$$

From our previous calculation, 1 kg change in mass produces 9×10^{16} joules; and we have seen that $1\,MeV = 1 \cdot 6 \times 10^{-13}$ joule.

$$\therefore 1\,u = \frac{1 \cdot 66 \times 10^{-27} \times 9 \times 10^{16}}{1 \cdot 6 \times 10^{-13}}\,MeV$$

$$\therefore \mathbf{1\,u = 931\,MeV\,(approx.)} \qquad . \qquad . \qquad . \qquad . \qquad . \qquad (1)$$

This relation is used to change mass units to MeV, and vice-versa, as we shall see shortly. An electron mass, $9 \cdot 1 \times 10^{-31}$ kg, corresponds to about 0·5 MeV.

Binding Energy

The protons and neutrons in the nucleus of an atom are called *nucleons*. The work or energy needed to just take all the nucleons apart so that they are completely separated is called the *binding energy* of the nucleus. Hence, from Einstein's mass-energy relation, it follows that the total mass of all the separated nucleons is greater than that of the nucleus, in which they are together. *The difference in mass is a measure of the binding energy.*

As an example, consider a helium nucleus $^{4}_{2}He$. This has 4 nucleons, 2 protons and 2 neutrons. The mass of a proton is 1·0073 and the mass of a neutron is 1·0087 u.

$\therefore$ total mass of 2 protons plus 2 neutrons $= 2 \times 1 \cdot 0073 + 2 \times 1 \cdot 0087$

$$= 4 \cdot 0320\,u$$

But the helium nucleus has a mass of 4·0015 u.

$$\therefore \text{ binding energy} = \text{mass difference of nucleons and nucleus}$$

$$= 4 \cdot 0320 - 4 \cdot 0015 = 0 \cdot 0305 \text{ u}$$

$$= 0 \cdot 0305 \times 931 \text{ MeV} = 28 \cdot 4 \text{ MeV}$$

The *binding energy per nucleon* of a nucleus is binding energy divided by the total number of nucleons. In the case of the helium nucleus, since there are four nucleons (2 protons and 2 neutrons), the binding energy per nucleon is then 28·4/4 or 7·1 MeV.

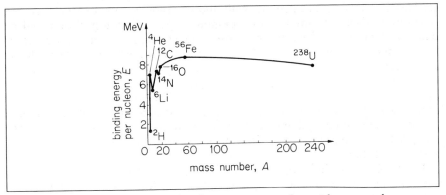

Figure 35.21   *Variation of binding energy per nucleon with mass number*

Figure 35.21 shows roughly the variation of the binding energy per nucleon among the elements. Excluding the nuclei lighter than ^{12}C, we can see from Figure 35.21 that the average binding energy per nucleon, E/A, is fairly constant for the great majority of nuclei. The average value is about 8 MeV per nucleon. The peak occurs at approximately the iron nucleus ^{56}Fe, which is therefore one of the most stable nuclei.

Later we shall use the curve in Figure 35.21 to show that energy is produced when heavy elements such as uranium undergo *fission* to form two lighter masses or where very light elements such as hydrogen undergo *fusion* to form a heavier element.

Nuclear Forces and Binding Energy

Inside the nucleus, the protons repel each other due to electrostatic repulsion of like charges. So for the nucleus to be stable there must exist other forces between the nucleons. These are called *nuclear forces*. They have short range, shorter than the interatomic distances, and are much stronger than electromagnetic interactions. They must provide a net attractive force greater than any repulsive electric forces.

A nucleus with a nucleon (mass) number $(N + Z)$ and proton (atomic) number Z has Z protons and N neutrons. When these particles come together in the nucleus there is an increase in potential energy due to the electrostatic forces of the protons but a *greater* decrease in potential energy due to the nuclear forces of the nucleons. There is therefore a net decrease in the potential energy of all the nucleons. This decrease per nucleon is the *binding energy per nucleon*.

So when the nucleons come together in the nucleus, there is a loss of energy equal to the binding energy. This results in a decrease in mass, from Einstein's mass-energy relation. As we have seen, the decrease in mass is the difference in the mass of the individual nucleons when they are completely separated and the mass of the nucleus when they are together; the so-called 'mass defect'. Expressed in symbols, the binding energy of the nucleus $^{N+Z}_{\ \ Z}X$ of an atom X is given by:

Binding energy

$$= \textbf{mass of } N \textbf{ neutrons} + \textbf{mass of } Z \textbf{ protons} - \textbf{mass of nucleus } ^{N+Z}_{\ \ Z}X$$

Energy of Disintegration

It is instructive to consider, from an energy point of view, whether a particular nucleus is likely to disintegrate with the emission of an α-particle. As an illustration, consider radium F or polonium, $^{210}_{84}Po$. If an α-particle could be emitted from the nucleus, the reaction products would be the α-particle or helium nucleus, $^{4}_{2}He$, and a lead nucleus, $^{206}_{82}Pb$, a reaction which could be represented by:

$$^{210}_{84}Po \rightarrow {}^{206}_{82}Pb + {}^{4}_{2}He \qquad . \qquad . \qquad . \qquad . \qquad \text{(i)}$$

Here the sum of the mass numbers, 210, and the sum of the nuclear charges, $+84e$, of the lead and helium nuclei are respectively equal to the mass number and nuclear charge of the polonium nucleus, from the law of conservation of mass and of charge.

If we require to find whether energy has been released or absorbed in the reaction, we should calculate the total mass of the lead and helium nuclei and compare this with the mass of the polonium nucleus. It is more convenient to use atomic masses rather than nuclear masses, and since the total number of electrons required on each side of equation (i) to convert the nuclei into atoms is the same, we may use atomic masses in the reaction. These are as follows:

$$\text{lead, } {}^{206}_{82}Pb, = 205 \!\cdot\! 969 \text{ u}$$

$$\alpha\text{-particle, } {}^{4}_{2}He, = \quad 4 \!\cdot\! 004 \text{ u}$$

$$\therefore \text{ total mass} = 209 \!\cdot\! 973 \text{ u}$$

Now $\qquad\qquad\qquad$ polonium, $^{210}_{84}Po, = 209 \!\cdot\! 982$ u

Thus the atomic masses of the products of the reaction are together *less* than the original polonium nucleus, that is,

$$^{210}_{84}Po \rightarrow {}^{206}_{82}Pb + {}^{4}_{2}He + Q$$

where Q is the energy released. It therefore follows that polonium can disintegrate with the emission of an α-particle and a release of energy (see *uranium series*, p. 900), that is, the polonium is unstable.

Suppose we now consider the possibility of a lead nucleus, $^{206}_{82}Pb$, disintegrating with the emission of an α-particle, $^{4}_{2}He$. If this were possible, a mercury nucleus, $^{202}_{80}Hg$, would be formed. The atomic masses are as follows:

$$\text{mercury, } {}^{202}_{80}Hg, = 201 \!\cdot\! 971 \text{ u}$$

$$\alpha\text{-particle, } {}^{4}_{2}He, = \quad 4 \!\cdot\! 004 \text{ u}$$

$$\therefore \text{ total mass} = 205 \!\cdot\! 975 \text{ u}$$

Now $\qquad\qquad\qquad\qquad$ lead, $^{206}_{82}Pb, = 205 \!\cdot\! 969$ u

Thus, unlike the case previously considered, the atomic masses of the mercury nucleus and α-particle are together *greater* than the lead nucleus, that is,

$$^{206}_{82}Pb + Q \rightarrow {}^{206}_{80}Hg + {}^4_2He$$

where Q is the energy which must be *given* to the lead nucleus to obtain the reaction products. It follows that the lead nucleus by itself is stable to α-decay.

Generally, then, a nucleus would tend to be unstable and emit an α-particle if the sum of the atomic masses of the products are together *less* than that of the nucleus; and it would be stable if the sum of the atomic masses of the possible reaction products are together *greater* than the atomic mass of the nucleus.

Stable and Unstable Nuclei

Many factors contribute to the binding energy, E, of nuclei and therefore to their stability. The α-particle, ^4_2He, appears to be particularly stable. Figure 35·22 (i)

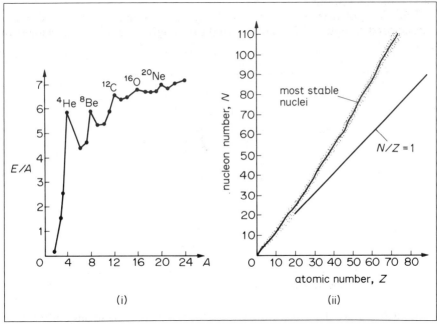

Figure 35.22 (*i*) *Variation of nuclear energy with atomic mass* (*ii*) *Most stable nuclei*

shows the binding energy per nucleon, E/A, for some light nuclei. Peaks occur for nuclei such as ^8_4Be, $^{12}_6C$, $^{16}_8O$, $^{20}_{10}Ne$. Each of these nuclei can be formed by adding an α-particle to the preceding nucleus.

Another factor affecting the stability of nuclei is the *neutron-proton ratio*. Figure 35.22 (ii) shows the number of neutrons, N, plotted against the number of protons (atomic number), Z, for all stable nuclei. It can be seen from the line of most stable nuclei that for light stable nuclei, such as $^{12}_6C$ and $^{16}_8O$, this ratio is 1. For nuclei heavier than $^{40}_{20}Ca$, the ratio N/Z increases slowly towards about 1·6. There are no stable nuclei above about $Z = 92$ (uranium).

Nuclear Emissions and Nuclear Stability

Unstable nuclei are radioactive. Their decay may occur in three main ways:

(1) *α-particle emission.* If the nucleus has excess protons, and α-particle

emission would reduce the protons by 2 and the neutrons by 2. So, generally, if A is the nucleon (mass) number and Z is the proton (atomic) number of the atom X concerned, then

$$\ce{^{A}_{Z}X} \rightarrow \ce{^{A-4}_{Z-2}Y} + \ce{^{4}_{2}He}$$

(2) β^- *particle (electron) emission.* If the nucleus has too many neutrons for stability, the neutron-proton ratio is reduced by β-particle (electron) emission. Here a neutron changes to a proton, so

$$\ce{^{1}_{0}n} \rightarrow \ce{^{1}_{1}H} + \ce{^{0}_{-1}e} \text{ (electron)}$$

Hence A remains unchanged but Z increases by 1.

(3) β^+ *particle (positron) emission.* If the nucleus is deficient in neutrons, a decay by β^+ (positron) emission may occur. A proton changes to a neutron:

$$\ce{^{1}_{1}H} \rightarrow \ce{^{1}_{0}n} + \ce{^{0}_{1}e} \text{ (positron)}$$

So A is unchanged but Z decreases by 1.

The effects of these three types of decays, (1), (2) and (3), on A and Z are summarised in Figure 35.23; the boxes indicate diagrammatically unit changes in A and Z.

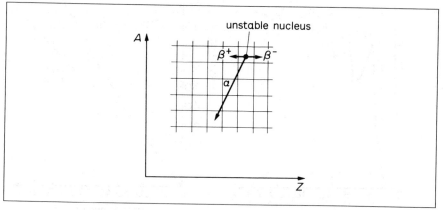

Figure 35.23 *Effect of particle emission on A and Z*

In nuclei lighter than the isotope ^{40}Co, the numbers of neutrons and protons are roughly equal. Heavier elements such as thorium or uranium have a greater number of protons which tend to repel each other and so make the nucleus less tightly bound. So most stable heavy nuclei have more neutrons than protons.

The nucleus formed by a decay may also be unstable. This gives rise to 'decay series or chains', which end when a stable nucleus, for example, ^{206}Pb (lead), is reached. A series consisting entirely of α-decays (α-particle emission) would increase the ratio of neutrons to protons as Z decreased. As an example, suppose $^{224}_{94}$Pu decays by α-emission to $^{240}_{92}$U. Now $^{240}_{92}$U has $N = 148$ and $Z = 92$, so the neutron-proton ratio, N/Z, is 1·609. *If* this nucleus were to decay by another α-emission to $^{236}_{90}$Th, the ratio would be increased to 146/90 or 1·622, which is too far from the line of stability shown in Figure 35.22 (ii). In fact, $^{240}_{92}$U decays to $^{240}_{93}$Np by β^- emission. The ratio N/Z is now 147/93 or 1·581, close to the line of the stability. In a natural decay series, then, there must always be some β^- decays in addition to α-decays.

β^+ (positron) decays are much rarer since they would result in an increase in N and a decrease in Z, thus increasing the ratio N/Z.

Artificial Disintegration, Nuclear Reactions

Uranium, thorium and actinium are elements which disintegrate naturally. The artificial disintegration of elements began in 1919, when Rutherford used α-particles to bombard nitrogen and found that protons were produced (p. 898). Some nuclei of nitrogen had changed into nuclei of oxygen, that is, transmutation had occurred,

---

From the laws of conservation of mass (nucleon number) and charge (proton number) the reaction can be represented by:

$$^{14}_{7}N + ^{4}_{2}He \rightarrow ^{17}_{8}O + ^{1}_{1}H$$

This reaction is often expressed briefly as

$$^{14}_{7}N(\alpha, p) \quad \text{or} \quad ^{14}_{7}N(\alpha, p)^{17}_{8}O$$

which means that the *incident* particle on the $^{14}_{7}N$ nucleus is an α-particle ($^{4}_{2}He$), the *emitted* particle is a proton $p(^{1}_{1}H)$ and the nucleus formed after the reaction is $^{17}_{8}O$. Similarly, the reaction $^{9}_{4}Be(p, \alpha)$ or $^{9}_{4}Be(p, \alpha)^{6}_{3}Li$ is

$$^{9}_{4}Be + ^{1}_{1}H \rightarrow ^{4}_{2}He + ^{6}_{3}Li$$

In 1932 Cockcroft and Walton produced nuclear disintegrations by accelerating protons with a high-voltage machine producing about half a million volts, and then bombarding elements with the high-speed protons. When the light element lithium was used, photographs of the reaction taken in the cloud chamber showed that two α-particles were produced. The particles shot out in opposite direction from the point of impact of the protons, and as their range in air was equal, the α-particles had initially equal energy. This is a consequence of the principle of the conservation of momentum. The nuclear reaction was:

$$^{7}_{3}Li + ^{1}_{1}H \rightarrow ^{4}_{2}He + ^{4}_{2}He + Q \qquad . \qquad . \qquad . \qquad . \qquad \text{(i)}$$

where Q is the energy released in the reaction.

To calculate Q, we should calculate the total mass of the lithium and hydrogen nuclei and subtract the total mass of the two helium nuclei. As already explained, however, the total number of electrons required to convert the nuclei to neutral atoms is the same on both sides of equation (i). So atomic masses can be used in the calculation in place of nuclear masses. The atomic masses of lithium and hydrogen are 7·016 and 1·008 u respectively, a total of 8·024 u. The atomic mass of the two α-particles is $2 \times 4·004$ u or 8·008 u. thus:

$$\text{energy released, } Q, = 8·024 - 8·008 = 0·016 \, u$$

$$= 0·016 \times 931 \, \text{MeV} = 14·9 \, \text{MeV}$$

Each α-particle has therefore an initial energy of 7·4 MeV, and this theoretical value agreed closely with the energy of each α-particle measured from its range in air. The experiment was the earliest verification of Einstein's mass-energy relation.

Cockcroft and Walton were the first scientists to use protons for disrupting atomic nuclei after accelerating them by high voltage. Today, giant high-voltage machines are built at Atomic Energy centres for accelerating protons to enormously high speeds. The products of the nuclear explosion with light atoms such as hydrogen yield valuable information on the structure of the nucleus. For example, the proton itself contains particles called *quarks*.

Energy from Radioactive Isotopes

Energy from the decay of unstable radioactive isotopes is sometimes used where a continuous, powerful and compact energy source is required. Such isotopes have been used to provide power for the batteries of heart 'pacemakers' and for scientific apparatus used in space vehicles.

As an example, consider the isotope Po 210. This has a half-life of about 140 days and emits α-particles each of energy 5·3 MeV.

Since the molar mass is 210 g and the Avogadro constant is about $6 \times 10^{23} \text{ mol}^{-1}$, each gram of the isotope contains $6 \times 10^{23}/210 = 2·9 \times 10^{21}$ atoms. Thus in one-half-life or 140 days, about $1·4 \times 10^{21}$ atoms decay.

These atoms release a total energy (using $1 \text{ MeV} = 1·6 \times 10^{-13} \text{ J}$)

$$= 1·4 \times 10^{21} \times 5·3 \times 1·6 \times 10^{-13} \text{ J}$$

$$= 1·2 \times 10^9 \text{ J (approx.)}$$

Thus the mean output power per gram

$$= \frac{1·2 \times 10^9}{140 \times 24 \times 3600} = 100 \text{ W (approx.)}$$

During the next half-life period, 140 days, the mean power output will only be 50 W, since half the remaining atoms decay in this time.

Energy Released in Fission, Chain Reaction

In 1934 Fermi began using neutrons to produce nuclear disintegration. These particles are generally more effective than α-particles or protons for this purpose, because they have no charge and are therefore able to penetrate more deeply into the positively-charged nucleus. Usually the atomic nucleus changes only slightly after disintegration, but in 1939 Frisch and Meitner showed that a uranium nucleus had disintegrated into two relatively-heavy nuclei. This is called *nuclear fission*, and as we shall now show, a large amount of energy is released in this case.

Natural uranium consists of about 1 part by mass of uranium atoms $^{235}_{92}\text{U}$ and 140 parts by mass of uranium atoms $^{238}_{92}\text{U}$. In a nuclear reaction with natural uranium and slow neutrons, it is usually the nucleus $^{235}_{92}\text{U}$ which is fissioned. If the resulting nuclei are lanthanum $^{148}_{57}\text{La}$ and bromine $^{85}_{35}\text{Br}$, together with several neutrons, then:

$$^{235}_{92}\text{U} + ^{1}_{0}\text{n} \rightarrow ^{148}_{57}\text{La} + ^{85}_{35}\text{Br} + 3^{1}_{0}\text{n} \qquad . \qquad . \qquad . \qquad \text{(i)}$$

Now $^{235}_{92}\text{U}$ and $^{1}_{0}\text{n}$ together have a mass of $(235·1 + 1·009)$ or $236·1$ u. The lanthanum, bromine and neutrons produced together have a mass

$$= 148·0 + 84·9 + 3 \times 1·009 = 235·9 \text{ u}$$

$\therefore$ energy released = mass difference

$$= 0·2 \text{ u} = 0·2 \times 931 \text{ MeV} = 186 \text{ MeV}$$

$$= 298 \times 10^{-13} \text{ J (approx.)}$$

This is the energy released per atom of uranium fissioned. In 1 kg of uranium there are about

$$\frac{1000}{235} \times 6 \times 10^{23} \quad \text{or} \quad 26 \times 10^{23} \text{ atoms}$$

since the Avogadro constant, the number of atoms in a mole of any element, is

$6 \cdot 02 \times 10^{23}$. Thus if all the atoms in 1 kg of uranium were fissioned, total energy released

$$= 26 \times 10^{23} \times 298 \times 10^{-13} \, J = 8 \times 10^{13} \, J \, (\text{approx.})$$

$$= 2 \times 10^{7} \, \text{kilowatt-hours}$$

which is the amount of energy given out by burning about 3 million tonnes of coal. The energy released per gram of uranium fissioned $= 8 \times 10^{10} \, J$ (approx.).

Fast neutrons are absorbed or captured by nuclei of U238 without producing fission. Slow or thermal neutrons in uranium, however, which have the same temperature as that of the uranium, produce fission when incident on nuclei of U235. About 2·5 neutrons per fission are released. When slowed, each of these neutrons produce fission in another U235 nucleus, and so on. Thus a rapid multiplying *chain reaction* can be obtained in the uranium mass, liberating swiftly an enormous amount of energy. This is the basic principle of the nuclear

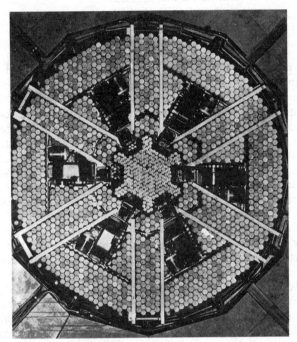

Figure 35.24 *Nuclear research reactor, ZEUS. This view shows the heart of the reactor, containing a highly enriched uranium central core currounded by a natural uranium blanket for breeding studies*

reactor or pile. As we now describe, graphite can be used to slow down the speed of the fast neutrons released in fission.

Nuclear Reactors

Figure 35.25 shows, in diagrammatic form, the principle of one type of nuclear reactor, used for commercial power generation.

Uranium fuel in the form of thick rods are encased in long aluminium tubes, which are air-tight to contain any gases released and prevent oxidation of the surrounding fuel. The tubes are lowered into hundreds of channels inside blocks.

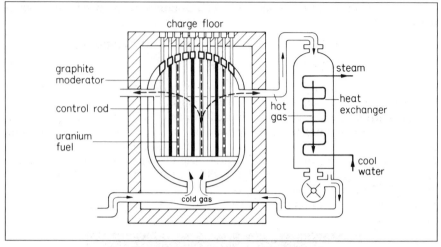

Figure 35.25 *Nuclear reactor principle*

As explained previously, the graphite is needed to moderate or reduce the speed of the neutrons released on fission until they become slow or thermal neutrons; they can then produce fission on collision with other U235 nuclei. The graphite *moderator* is in the form of blocks of pure carbon arranged in a stack.

Heavy water (deuterium oxide) has also been used as a moderator. It reduces the speed of a colliding neutron to about one-half, whereas graphite reduces the speed to about one-seventh. With a graphite moderator the separation of the uranium rods is about 20 cm. In this distance the fast neutron makes about 200 collisions with carbon nuclei and then becomes slow enough to produce fission on collision with a U235 nucleus. The separation of the uranium rods in a water moderator is less than in a graphite moderator as the speed of the colliding neutron is reduced much more with a deuterium nucleus.

The power level reached by a reactor is proportional to the *neutron flux density;* this is the number of neutrons per second crossing unit cross-sectional area of the reactor. If the neutrons are produced too fast, the reactor may disintegrate. Ideally, the neutron reproduction factor should be just greater than 1 to make the chain reaction self-sustaining. Boron-coated steel rods are used to control the rate of neutron production, Figure 35.25. The *control rods* are lowered or raised in channels inside the graphite block by electric motors operated from a control room until the neutron reproduction rate is just greater than 1, and the reactor is then said to go 'critical'. In the event of an electrical failure or other danger, the rods fall and automatically shut off the reactor. Boron atoms have a high absorption cross-section for neutrons and thus capture slow neutrons.

The energy produced by the nuclear reaction would make the reactor too hot. A *coolant* is therefore required. Water and gas such as carbon dioxide have been used as coolants. Molten sodium, which has a high value of (specific heat capacity $\times$ density) and a high thermal conductivity, has also been used as a coolant. In a *gas-cooled reactor* which produces power for the Grid system, the hot gas is led from the reactor into a heat exchanger, Figure 35.25. Here the heat is transferred to water circulating through pipes so that steam is produced and this is used to drive turbines for electrical power generation.

Finally, it should be noted that if the mass of the uranium is too small, the neutrons will escape and a chain reaction is not produced. The *critical mass* is the least mass to make a self-sustaining chain reaction. The critical mass for a reactor in the shape of a sphere is less than that in the shape of a cube since the surface (neutron escape) area is a minimum for a given volume or mass of material. Fermi, a distinguished pioneer in nuclear research, made one of the first pilot reactors in 1942 in roughly a spherical shape.

Further details of reactors must be obtained from the United Kingdom Atomic Energy Authority or from specialist books.

A nuclear reactor may contain.

1 **Uranium rods as** *fuel*
2 **Graphite as a** *moderator* **of the neutron speeds for the fission process**
3 **Boron-coated steel rods to** *control* **the neutron reproduction rate**
4 **A** *coolant* **to reduce the excessive heat produced in the reaction**
5 **A** *heat exchanger* **to remove the heat energy produced**

Energy Released in Fusion

In fission, energy is released when a heavy nucleus is spit into two lighter nuclei. Energy is also released if light nuclei are *fused* together to form heavier nuclei, and fusion reaction, as we shall see, is also a possible source of considerable energy. As an illustration, consider the fusion of the nuclei of deuterium, ^2_1H. Deuterium is an isotope of hydrogen known as 'heavy hydrogen', and its nucleus is called a 'deuteron'. The fusion of two deuterons can result in a helium nucleus, ^3_2He, as follows:

$$^2_1H + ^2_1H \rightarrow ^3_2He + ^1_0n$$

Now   mass of two deuterons

$$= 2 \times 2{\cdot}015 = 4{\cdot}030\,u$$

and   mass of helium plus neutron

$$= 3{\cdot}017 + 1{\cdot}009 = 4{\cdot}026\,u$$

$\therefore$ mass converted to energy by fusion

$$= 4{\cdot}030 - 4{\cdot}026 = 0{\cdot}004\,u$$

$$= 0{\cdot}004 \times 931\,MeV = 3{\cdot}7\,MeV$$

$$= 3{\cdot}7 \times 1{\cdot}6 \times 10^{-13}\,J = 6{\cdot}0 \times 10^{-13}\,J$$

$\therefore$ energy released per deuteron $= 3{\cdot}0 \times 10^{-13}\,J$

6×10^{26} is the number of atoms in a kilomole of deuterium, which has a mass of about 2 kg. Thus if all the atoms could undergo fusion,

energy released per kg

$$= 3{\cdot}0 \times 10^{-13} \times 3 \times 10^{26}\,J$$

$$= 9 \times 10^{13}\,J\,(approx.)$$

Other fusion reactions can release much more energy, for example, the fusion of the nuclei of deuterium, ^2_1H, and tritium, ^3_1H, isotopes of hydrogen, releases about 30×10^{13} joules of energy per kg according to the reaction:

$$^2_1H + ^3_1H \rightarrow ^4_2He + ^1_0n$$

In addition, the temperature required for this fusion reaction is less than that needed for the fusion reaction between two deuterons given above, which is an advantage. Hydrogen contains about 1/5000th by mass of deuterium or heavy hydrogen, needed in fusion reactions, and this can be obtained by electrolysis of sea-water, which is cheap and in plentiful supply.

Fusion and Binding Energy Curve, Thermonuclear Reaction

In Figure 35.21 (p. 905), the binding energy per nucleon is plotted against the mass number of the nucleons or nucleon number. Since the curve rises from hydrogen, the binding energy per nucleon of the helium nucleus, $^4_2$He, is greater than that for deuterium, $^2_1$H. Thus the binding energy of the helium nucleus, which has four nucleons, is *greater* than that of two deuterium nuclei, which also have four nucleons.

Now the binding energy of a nucleus is proportional to the difference between the mass of the individual nucleons and the mass of the nucleus (the so-called 'mass defect'), see p. 904. So the mass of the helium nucleus is *less* than that of the two deuterium nuclei. Hence if two deuterium nuclei can be fused together to form a helium nucleus, the mass lost will be released as energy.

The rising part of the binding energy curve in Figure 35.21 shows that elements with low mass number can produce energy by fusion. In contrast, the falling part of the curve shows that very heavy elements such as uranium can produce energy by *fission* of their nuclei to nuclei of *smaller* mass number (see p. 910).

For fusion to take place, the nuclei must at least overcome electrostatic repulsion when approaching each other. Consequently, for practical purposes, fusion reactions can best be achieved with the lightest elements such as hydrogen, whose nuclei carry the smallest charges and hence repel each other least.

In attempts to obtain fusion, isotopes of hydrogen such as deuterium, $^2_1$H, and tritium, $^3_1$H, are heated to tens of millions of degrees centigrade. The thermal energy of the nuclei at these high temperatures is sufficient for fusion to occur. One technique of promoting this *thermonuclear reaction* is to pass enormously high currents through the gas, which heat it. A very high percentage of the atoms are then ionised and the name *plasma* is given to the gas. The interstellar space of the *aurora borealis* contains a weak form of plasma, but the interior of stars contains a highly concentrated form of plasma. The gas discharge consists of parallel currents, carried by ions, and the powerful magnetic field round one current due to a neighbouring current draws the discharge together (see p. 327). This is the so-called 'pinch effect'. The plasma, however, wriggles and touches the sides of the containing vessel, thereby losing heat. The main difficulty in thermonuclear experiments in the laboratory is to retain the heat in the gas for a sufficiently long time for a fusion reaction to occur. The stability of plasma is now the subject of considerable research.

It is believed that the energy of the sun is produced by thermonuclear reactions in the heart of the sun, where the temperature is many millions of degrees centigrade. Bethe has proposed a cycle of nuclear reactions in which, basically, protons are converted to helium by fusion in the sun, with the liberation of a considerable amount of energy.

_____ **Exercises 35B** _____

1   Figure 35C shows a gold foil mounted across the path of a narrow, parallel beam of alpha particles. The fraction of incident alpha particles reflected back through more than 90° is very small.

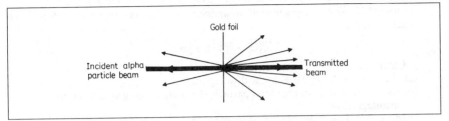

Figure 35C

How does this result lead to the idea that an atom has a nucleus
(a) whose diameter is small compared with the atomic diameter, and
(b) which contains most of the atom's mass? (*L.*)

2   When a nucleus of deuterium (hydrogen-2) fuses with a nucleus of tritium (hydrogen-3) to give a helium nucleus and a neutron, $2 \cdot 88 \times 10^{-12}$ J of energy are released.

The equation of the reaction is:

$$\,^2_1H + \,^3_1H \rightarrow \,^4_2He + \,^1_0n$$

Calculate the mass of the helium nucleus produced.
(Mass of deuterium nucleus = $3 \cdot 345 \times 10^{-27}$ kg. Mass of tritium nucleus = $5 \cdot 008 \times 10^{-27}$ kg. Mass of a neutron = $1 \cdot 675 \times 10^{-27}$ kg. Speed of light in a vacuum = $3 \cdot 00 \times 10^8$ m s^{-1}.) (*L.*)

3   (a) State the results of experiments on the scattering of alpha particles by a thin foil. Explain the importance of these results.
(b) In such experiments explain   (i) why the foil should be thin, and point out any differences in the results for foils of different materials;   (ii) why the alpha particles incident on the foil should be in a narrow parallel beam;   (iii) how the scattered alpha particles might be detected.
(c) An alpha particle travels directly towards a nucleus which may be assumed to remain stationary. Describe in qualitative terms how, during the motion, the kinetic energy and the potential energy of the alpha particle vary with the separation between the alpha particle and the nucleus. (*JMB.*)

4   Explain what is meant by
(a) the mass defect,
(b) the binding energy of an atomic nucleus.
The binding energy of the isotope of hydrogen ^3_1H is greater than that of the isotope of helium ^3_2He. Suggest a reason for this. (*JMB.*)

5   Define *nucleon number* (*mass number*) and *proton number* (*atomic number*) and explain the term *isotope*. Describe a simple form of mass spectrometer and indicate how it could be used to distinguish between isotopes.
In the naturally occurring radioactive decay series there are several examples in which a nucleus emits an α-particle followed by two β-particles. Show that the final nucleus is an isotope of the original one. What is the change in mass number between the original and final nuclei? (*L.*)

6   Explain the term *nuclear binding energy*. Sketch a graph showing the variation of binding energy per nucleon with nucleon number (mass number) and show how both nuclear fission and nuclear fusion can be explained from the shape of this curve.
Calculate in MeV the energy liberated when a helium nucleus (^4_2He) is produced
(a) by fusing two neutrons and two protons, and

(b) by fusing two deuterium nuclei (^2_1H). Why is the quantity of energy different in the two cases?

(The neutron mass is 1·008 98 u, the proton mass is 1·007 59 u, the nuclear masses of deuterium and helium are 2·014 19 u and 4·002 77 u respectively. 1 u is equivalent to 931 MeV.) (*L.*)

7   Using the information on atomic masses given below, show that a nucleus of uranium 238 can disintegrate with the emission of an alpha particle according to the reaction:

$$^{238}_{92}U \rightarrow \, ^{234}_{90}Th + \, ^4_2He$$

Calculate

(a) the total energy released in the disintegration,

(b) the kinetic energy of the alpha particle, the nucleus being at rest before disintegration.

   Mass of ^{238}U = 238·124 92 u. Mass of ^{234}Th = 234·116 50 u. Mass of ^4He = 4·003 87 u. 1 u is equivalent to 930 MeV. (*JMB.*)

8   Explain briefly what is meant by *nuclear fusion*.

   Write down the equation for a fusion reaction, making clear what the various symbols mean. What is the great difficulty in producing fusion in the laboratory? Where in nature does fusion occur continuously? (*W.*)

9   (a) Explain what is meant by the binding energy of a nucleus. For $^{56}_{26}Fe$, use data selected from the table below to calculate   (i) the nuclear mass in unified atomic mass units, u,   (ii) the binding energy per nucleon in MeV.

   (b) Sketch a graph of binding energy per nucleon against mass number for the naturally occurring isotopes, indicating approximate scales on the axes of your graph. Use your graph to explain why energy is released in nuclear fission.

   (c) A possible fusion reaction is represented by the equation

$$^3_1H + \, ^2_1H \rightarrow X + \, ^1_0n + 17 \cdot 6 \, MeV$$

(i) Identify the nuclide X.   (ii) Using data selected from the table below, calculate the atomic mass of X in u.

   In the following table the mass of an isotope is given for a naeutral atom of the substance and is quoted in unified atomic mass units, u. 1 u is equivalent to 931 MeV.

Name	electron	neutron	proton	deuterium	tritium	iron
Symbol	$^0_{-1}e$	^1_0n	^1_1p	^2_1H	^3_1H	$^{56}_{26}Fe$
Atomic mass	0·000 55	1·008 7	1·007 3	2·014 1	3·016 1	55·934 9

(*JMB.*)

10   In the fusion reaction $^2_1H + \, ^3_1H = \, ^4_2He + \, ^1_0n$, how much energy, in joules, is released? (Mass of ^2_1H = 3·345 × 10^{-27} kg, ^3_1H = 5·008 × 10^{-27} kg, ^4_2He = 6·647 × 10^{-27} kg, ^1_0n = 1·675 × 10^{-27} kg. Speed of light = 3·0 × 10^8 m s^{-1}.) (*L.*)

11   Given

$$^{235}_{92}U + \, ^1_0n \rightarrow \, ^x_{45}Rh + \, ^{113}_yAg + 2^1_0n$$

and

$$^2_1H + \, ^3_1H \rightarrow \, ^4_2He + A$$

(i) explain what is meant by the 235 and 92 in $^{235}_{92}U$;   (ii) determine x, y, and A; (iii) describe the importance of the reactions;   (iv) write down a similar equation for the fusion of two atoms of deuterium to form helium of atomic mass number 3.

   Given the mass of the deuterium nucleus is 2·015 u, that of one of the isotopes of helium is 3·017 u and that of the neutron is 1·009 u, calculate the energy released by the fusion of 1 kg of deuterium. If 50% of this energy were used to produce 1 MW of electricity continuously, for how many days would the station be able to function? (Speed of light c = 3·00 × 10^8 m s^{-1}.) (*W.*)

12 In the Rutherford α-particle scattering experiment, α-particles of mass 7×10^{-27} kg and speed 2×10^7 m s^{-1} were fired at a gold foil. What was the closest distance of approach between an α-particle and a gold nucleus? How could those α-particles which made such a closest approach be identified experimentally?
   (The atomic number for gold is 79; the electronic charge is -1.6×10^{-19} C; $1/4\pi\varepsilon_0 = 9 \times 10^9$ F^{-1} m.) (*W.*)

13 (a) Explain the meaning of the term *mass difference* and state the relationship between the mass difference and the *binding energy* of a nucleus.
   (b) Sketch a graph of nuclear binding energy per nucleon versus mass number for the naturally occurring isotopes and show how it may be used to account for the possibility of energy release by nuclear fission and nuclear fusion.
   (c) The sun obtains its radiant energy from a thermonuclear fusion process. The mass of the sun is 2×10^{30} kg and it radiates 4×10^{23} kW at a constant rate. Estimate the life time of the sun, in years, if 0.7% of its mass is converted into radiation during the fusion process and it loses energy only by radiation. (1 year may be taken as 3×10^7 s.) The speed of light, $c = 3 \times 10^8$ m s^{-1}. (*JMB.*)

14 (a) Z protons and N neutrons are combined to form a nucleus $^{Z+N}_{Z}$X. Describe the energy changes which occur as the $(Z + N)$ free particles are combined. Explain the concept of *binding energy per nucleon* which arises in this description. Sketch a graph of binding energy nucleon against nucleon number and use this graph to explain how the process of nuclear fission and nuclear fusion are possible, and the values of nucleon number at which they may occur.
   (b) A typical fission reaction is

$$^{235}_{92}U + {}^1_0n \rightarrow {}^{148}_{57}La + {}^{85}_{35}Br + \text{neutrons}$$

How many neutrons are released in this reaction? What is the importance of these neutrons in a nuclear reactor? Why are the products of lanthanum (La) and bromine (Br) likely to be radioactive and what type of radioactivity are they likely to exhibit? (*L.*)

15 (a) Describe in terms of nuclear structure the three different stable forms of the element neon, which have nucleon numbers 20, 21 and 22. Why is it impossible to distinguish between these forms chemically?
   A beam of singly ionised atoms is passed through a region in which there are an electric field of strength E and a magnetic field of flux density B at right angles to each other and to the path of the atoms. Ions moving at a certain speed v are found to be undeflected in traversing the field region. Show (i) that $v = E/B$ and (ii) that these ions can have any mass.
   If the emerging beam contained ions of neon, suggest how it might be possible to show that all three forms of the element were present.
   (b) Explain what is meant by the *binding energy* of the nucleus. Calculate the binding energy per nucleon for $^4_2$He and $^3_2$He. Comment on the difference in these binding energies and explain its significance in relation to the radioactive decay of heavy nuclei. (Mass of $^1_1$H = 1.007 83 u. Mass of $^1_0$n = 1.008 67 u. Mass of $^3_2$He = 3.016 64 u. Mass of $^4_2$He = 4.003 87 u. 1 u ≡ 931 MeV.) (*L.*)

16 (a) Sketch, on the same diagram, the paths of three alpha particles of the same energy which are directed towards a nucleus so that they are deflected through (i) about 10°, (ii) 90°, (iii) 180° respectively.
   (b) For the deflection of 180°, describe in qualitative terms how (i) the kinetic energy, (ii) the potential energy of the alpha particle varies during its path, assuming the nucleus remains stationary.
   (c) If, in (b) above, the alpha particle has an initial kinetic energy of 1.60×10^{-13} J and the nucleus has a charge of $+50\,e$, calculate the nearest distance of approach of the alpha particle to the nucleus. (Magnitude of electronic charge, $e = 1.60 \times 10^{-19}$ C, permittivity of free space, $\varepsilon_0 = 10^{-9}/36\pi$ or 8.85×10^{-12} F m^{-1}.) (*JMB.*)

36
Further Topics

In this section we discuss some basic topics in Energy and Tele-communications. Only the core principles are given. For further details reference must be made to specialist books.

Energy

There are many sources of energy in the World. The chemical energy from burning coal, oil and gas, called *fossil fuels*, is widely used. These fuels are non-renewable after they are burnt. Nuclear energy, wind power and wave power from the sea are other sources of energy. Wind and wave power are renewable sources.

All our energy comes primarily from solar energy. As stated on page 624, the Sun's ultraviolet rays are absorbed in the green matter of plants and make them grow. The plants and trees centuries ago are turned into coal and oil. Water power comes from the Sun. Water is evaporated by the Sun and this produces the rains which fill the lakes and reservoirs. Wind power also comes from the Sun. Unequal heating of air masses world-wide results in wind movement or kinetic energy.

In hot areas of the world, *solar energy* is collected by large concave mirrors and concentrated on water to produce steam to drive turbines, for example. Solar energy is also collected by solar panels on the roofs of houses for domestic heating purposes. As explained on page 914, the Sun's energy comes from nuclear fusion reactions of the Sun's elements.

Nuclear Energy, Fossil Fuels, Geothermal Energy

We have already described how *nuclear energy* is used in nuclear reactors for generating electrical power (p. 912). Heat exchangers pass the heat produced in fission to boilers, which then produce steam to drive the turbines in electrical power stations. In coal-fired or oil-fired power stations, these fossil fuels are burnt to produce the heat needed for the boilers.

Geothermal energy appears to come from nuclear energy changes deep in the Earth, which produces hot dry rock. In the United States in California, and in the Soviet Union in the Arctic Circle, deep holes are tunnelled into the Earth through hot rocks below. A depth of over 10 km has been reached. Brine or water is pumped under pressure through the holes and hot brine or steam at about 350°C can be obtained at the surface. About 24 MW (24×10^6 W) has been produced in this way for use in surrounding areas in California.

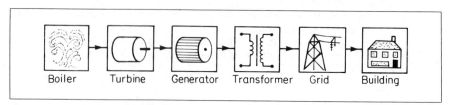

Figure 36.1 *Electrical power generation*

Figure 36.1 shows in block form the system needed for generating electrical power, which is widely used in industry and the home. The final power output depends on the *efficiencies* of all the machines used. For example:

(1) *Boiler.* Owing to unburnt fuel and the heat absorbed by the vessels used, only a small percentage of chemical energy of the fuel–air mixture provides useful heat for changing water to steam in the boiler.

(2) *Turbine.*

(a) Wasted energy is due to frictional forces at the moving parts.

(b) The Second Law of Thermodynamics limits the maximum energy obtainable. An ideal engine has a maximum efficiency of $(T_2 - T_1)/T_2$, where T_2 is the kelvin steam temperature and T_1 is that of the condenser for the steam condensed during the cycles. See p. 752. High pressure steam at 500°C or 773 K, and a condenser at 25°C or 228 K, produces a maximum efficiency = $(773 - 288)/773 = 0.61$ or 61%. With special design, 45% efficiency may be reached in turbines, taking losses into account.

(3) *Transformer.* Although losses occur, over 90% efficiency can be reached.

(4) *Grid system.* Heat losses due to current are produced in the cables (p. 259). These may be reduced in the future, following recent promising research for superconductors at normal temperatures using ceramics.

Wind Power

Blades on a horizontal or vertical axis can be rotated like windmills by wind power. By connecting the axle to turbines, generators can be driven to produce electrical power. High wind speeds near the coast round the British Isles produce sufficient power for local areas. In Scotland and the Isle of Man, wind turbines provide a back-up for electrical power supplies and saves fossil energy.

To estimate roughly the *available wind power*, suppose a fast-rotating vertical blade of 20 metres diameter or span is rotating about a horizontal axis O at its centre, and a horizontal wind of $13\,\mathrm{m\,s^{-1}}$ (30 mph) is blowing horizontally towards the blade. Figure 36.2.

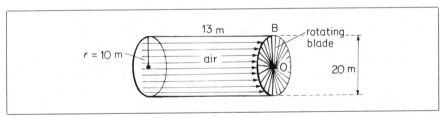

Figure 36.2 *Available wind power*

The cylindrical column of air moving to the circular area of the rotating blade in 1 second has a volume

$$= \pi \times 10^2 \times 13 = 4084\,\mathrm{m^3}$$

Assuming the density of air is $1.3\,\mathrm{kg\,m^{-3}}$, the mass per second reaching the blade = $4084 \times 1.3 = 5300\,\mathrm{kg}$ (approx).

The kinetic energy per second of the air = $\frac{1}{2}mv^2$ per second

$$= \frac{1}{2} \times 5300 \times 13^2 = 450\,000\,\mathrm{J\,s^{-1}} = 0.45\,\mathrm{MW}$$

Assuming the velocity of the air is reduced to zero at the blade, the available power would be $0.45\,\mathrm{MW}$.

If r is the radius of the rotating blade, v is the wind speed and ρ is the air density, then generally

$$\text{power available} = \pi r^2 \rho v^3$$

So the available power is proportional to the cube of the wind speed and the square of the blade diameter. A calculation involving the mass per second of moving air is given on page 26.

The *power extracted* by the rotating blade is much less than the available power of 0·45 MW. The velocity of the air is not reduced to zero at the blade, as we now explain.

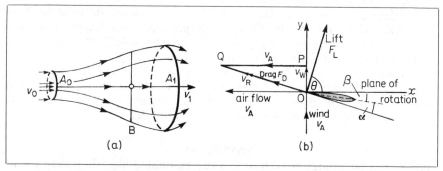

Figure 36.3   (*a*) *Air streamlines and rotor*   (*b*) *Forces on rotor*

Power Extracted, Forces on Rotor

Power extracted. Figure 36.3(a) shows an ideal situation. The air streamlines move with velocity v_0 through a cross-sectional area A_0 some distance from the rotor or blade and leave the blade with a final velocity v_1 through a greater cross-section A_1. As we now show, the broadening of the air stream passing through the blade is due to a decrease in energy.

Assuming a steady state and an incompressible fluid, then $v_1 A_1 = v_0 A_0$, since the volume per second through each area is the same. See *Bernoulli Principle*, p. 126. So v_1 is *less* than v_0. The power extracted by the blade is therefore $(\frac{1}{2}mv_0{}^2 - \frac{1}{2}mv_1{}^2)$, where m is the mass of air per second through the areas.

Betz showed that the power extracted has a maximum value of about 60% of the available power, discussed earlier. The electrical power output is also reduced by frictional forces at the turbine and alternator. In design, too, the alternator power must be matched to the mechanical power extracted.

Forces on rotor. Figure 36.3(b) shows a section of the rotating blade looking along it towards the axis O. The blade is at small angle β to the plane of rotation Ox and the wind velocity v_W at the blade is in the direction Oy of the axis. The rotating blade produces an air flow of velocity v_A in the plane of rotation which is many times greater than v_W. The tip-speed at the end of the rotor is taken as numerically equal to v_A.

From the triangle of velocities OPQ, the resultant air velocity or 'relative wind' velocity v_R is in a direction OQ at O. As in an aerofoil, the *lift* force F_L on the blade is 90° to v_R. See p. 128. There is also a *drag* force F_D on the blade in the direction v_R of the relative wind velocity. The net force or thrust on the blade in the plane of rotation is therefore $(F_L \cos \theta - F_D \sin \theta)$, where θ is the angle shown. In design, by varying the shape of the aerofoil and the angle α between the blade and the relative wind velocity, the ratio F_L/F_D is made as high as possible for maximum thrust and hence maximum power output, without the blade stalling.

A modern wind turbine has a high tip-speed to v_W ratio, typically 5, and a small pitch angle β. In practice, v_W is the wind velocity taking into account its slowing at the blade and the blade velocity v_A is slightly increased by an induced swirl behind the rotating blade.

Design of Wind Power Generators

Wind power turbines have long blades whose cross-sections may be shaped similar to an aerofoil. The wind is deflected by the blade and the sideways component of the reaction force causes the blade to rotate about its axis. See also page 920.

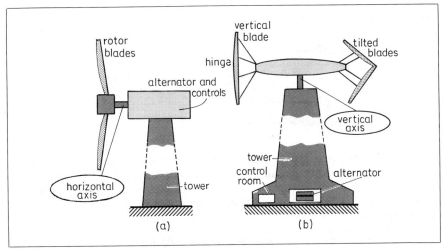

Figure 36.4

The two main types of wind turbines are horizontal-axis and vertical-axis types. The *horizontal-axis* turbine has two or more long vertical blades rotating about a horizontal axis. Figure 36.4(a). This machine needs turning into the wind to extract the wind energy effectively, which is a disadvantage in view of the cost of the device needed. The alternator is placed at the top of the supporting tower.

In the *vertical-axis* turbine, the blades are long and vertical and can accept wind from any direction. Figure 36.4(b). This is an advantage over the horizontal-axis type. The alternator can be placed on the ground at the base of the tower supporting the blades, which is another advantage over the horizontal-axis turbine. Due to centripetal forces, there are varying stresses on the blades as they rotate. Musgrove of Reading University has overcome the problem of limiting the power in very high winds by making the blade in two halves hinged at the middle. At high winds an operator tilts the two halves into an arrow-head shape, which reduces the stress.

The design of wind power systems is complex. For example, the generator power must be matched to the wind power extracted, and sensors and computers are needed in the control room for wind direction and speed. The world's most powerful wind turbine generator is installed on Orkney at one of the windiest places in the British Isles. Built by British Aerospace Wind Energy Group, the rotor has a span of 60 metres. It will turn at $34\,\text{rev}\,\text{min}^{-1}$ at wind speeds between $7\,\text{m}\,\text{s}^{-1}$ (15 mph) and $27\,\text{m}\,\text{s}^{-1}$ (60 mph) and will produce 3 MW of electrical power. Smaller models, with a 20 m span, are operating in North

Devon and in California, United States. An alternative to a large machine is a cluster of up to 100 medium machines (about 30 m blade diameter and 300 kW power) on a so-called 'wind farm'.

Tidal Power

Tides are due to the gravitational pull of the Moon on the waters surrounding the Earth. The pull varies during the monthly cycle of rotation of the Moon round the Earth and the tides vary from high to low tide twice per day.

For using tidal power, it is first necessary to build a dam across the tidal region of water. Sluice gates allow water to flow in at high tide. As the tide falls the gates are shut and water is allowed to run back through turbines to generate electricity.

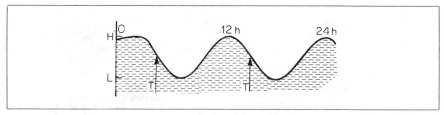

Figure 36.5 *Tidal power*

Figure 36.3 shows roughly the rise and fall of the trapped water at high (H) and low (L) tide during a 24 h period and the time of operation T of the turbines.

Suppose the water is trapped in a basin of area 40 km² or 40×10^6 m², and the maximum height of water is 10 m. Then, using water density = 1000 kg m⁻³ and $g = 10$ m s⁻²,

$$\text{weight of water, } mg = \text{volume} \times \text{density} \times g$$

$$= 40 \times 10^6 \times 1000 \times 10 = 4 \times 10^{12}\,\text{N}$$

If the maximum height of water above low tide is h, the centre of gravity is then at a height $h/2$ above low tide. So

$$\text{gravitational potential energy change from high to low tide}$$

$$= mg \times h/2 = 4 \times 10^{12} \times 10/2 = 2 \times 10^{13}\,\text{J}$$

From high to low tide, about 6 hours, ideally the average power obtained would be

$$\text{average power} = \frac{2 \times 10^{13}\,\text{J}}{6 \times 3600\,\text{s}} = 9 \times 10^8\,\text{W (approx)}$$

$$= 900\,\text{MW}$$

With system efficiencies taken into account, the available power is much less than that calculated.

In the tidal power system used in the Severn Estuary, the maximum height of the tide is about 10 m above low tide and the area of water is about 70 km² or 70×10^6 m². Ideally, this produces an energy change of 35×10^{12} J over a period of 6 hours, so about 1500 MW of power is obtained. Tidal power in the Bristol Channel between Cardiff and Weston-Super-Mare can produce about 6000 MW.

Wave Power

Waves in the sea have kinetic and gravitational potential energy as they rise and fall. Various systems have been designed to use wave energy and power.

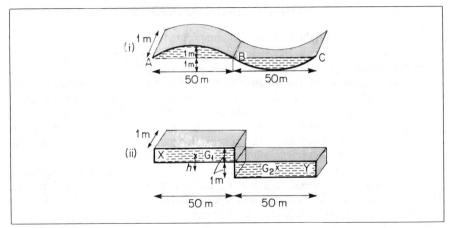

Figure 36.6 *Wave power*

To see what order of magnitude of energy and power may be obtained for a water wave, consider an ideal wave with straight wavefronts of width 1 m. Figure 36.6(i). Suppose it is a sine wave of amplitude 1 m and wavelength 100 m, which we approximate to the *rectangular wave* in Figure 36.6(ii) for a basic treatment.

When the wave crest in X falls into a trough at Y, the gravitational potential energy of the water changes by mgh, where m is the mass of water and h is the change in height of the centre of gravity G_1 to G_2. In this case $h = 1$ m. Since the density of water is 1000 kg m^{-3}, and assuming $g = 10 \text{ m s}^{-2}$,

$$mgh = \text{volume} \times \text{density} \times g \times h = (50 \times 1 \times 1) \times 1000 \times 10 \times 1$$

$$= 5 \times 10^5 \text{ J}$$

The wave has kinetic energy as it moves forward in addition to potential energy. It can be shown that the kinetic energy of a wave in deep water is equal to the potential energy change. So

$$\text{total energy} = 2 \times 5 \times 10^5 \text{ J} = 10^6 \text{ J}$$

Suppose the waves have a period of 5 s. In this time the water crest returns to its original height about the normal water level. So

$$\text{power} = \frac{\text{energy}}{\text{time}} = \frac{10^6 \text{ J}}{5 \text{ s}} = 200\,000 \text{ W} = 200 \text{ kW}$$

This is the power per metre wavefront. For a 1 km or 1000 m wavefront,

$$\text{power} = 1000 \times 200 \text{ kW} = 200 \text{ MW}$$

since 1 MW = 1000 kW.

The power from a sine wave is much less than this simplified rectangular wave. Further, although the wave travels with a constant speed, the water particles appear to have a circular motion which decreases rapidly with depth. More than 95% of the energy in a deep-water wave is contained in a depth $\lambda/4$, where λ is the wavelength. In winter, when waves are higher than in summer, more wave power is obtained.

Britain's first wave power scheme will be tested on Islay, an island in the Hebrides. The oscillating waves will flood into a special chamber and pump air to drive a turbine generator. The plant will produce about 200 kW of electrical power for a small village at very low cost.

Telecommunications

Optical Fibre Telecommunications

As we saw on p. 433, a laser is needed to provide a light signal from a pulsed electrical input in optical fibre telecommunications.

A *light emitting diode* (LED) can also provide a light signal, though it is not so intense as the laser. An LED consists of a forward-biased p-n diode made of gallium arsenide semiconductor material. As in normal diodes, electrons then drift into the p-region and holes into the n-region. The recombination of electron-hole pairs produces excess energy which is emitted as light. The intensity of the light is proportional to the input current.

LED's are available in a range of colour, such as red, orange, yellow, green. A series resistor R is needed to limit the current through an LED and prevent damage. Figure 36.5(i). For example, with colour red, an operating current of 20 mA, and a forward voltage of 2 V at 20 mA, a 400 Ω resistor would be needed with a 10 V supply, as the reader should check.

Basically, the *semiconductor laser* is an LED operating at high current. The p- and n-regions are made from gallium arsenide and gallium aluminium arsenide. Stimulated emission is produced by making the faces of the semiconductor optically flat so that some light is reflected back repeatedly as with the gas laser (p. 661).

The semiconductor laser is superior to the LED for telecommunications. The light is not only much greater in intensity but the spectral spread is typically 1 to 2 nm compared to a much greater spread using an LED. So low dispersion is produced in a fibre cable at high bit rates, which is an advantage.

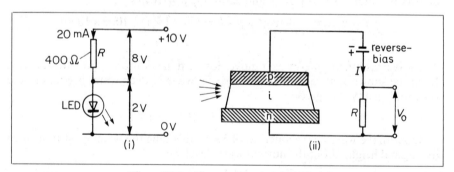

Figure 36.7 (i) LED (ii) *p-i-n photodiode*

Photodiode. At the output end of optical fibre communications, a photodiode is needed to change the arriving digital light pulses to an output voltage signal (p. 43).

The photodiode uses the principle that when light falls on a p-n junction, the photon energy hf can produce more electron-hole pairs in a crystal lattice. For example, visible light of wavelength 550 nm has photon energy of about 2·3 eV. This is greater than the ionisation energy of silicon, 1·1 eV.

Figure 36·5(ii) shows the principle of a *p-i-n photodiode*. It has an intrinsic (pure) silicon layer i between the p-layer and n-layer. Effectively this increases the size of the depletion layer. Many more electron-hole pairs are then generated compared to the normal p-n photodiode and so the current I, and the response to the illumination, are then greater.

As shown, the diode is *reverse-biased*. The output voltage V_0 across the resistor R increases with the intensity of illumination.

Radio Telecommunications, Aerials

We now turn to radio telecommunication and the topic of _aerials_. In a radio transmitter, the _aerial_ has the same function as the sounding board in a violin or as the resonance tube used with a tuning-fork. It helps the radio oscillator, which generates radio waves, to radiate its energy into space. If a tuning-fork is sounding when held in the hand, only a weak sound is heard. If, however, the sounding tuning-fork is moved over the top of a resonance tube of length $\lambda/4$, where λ is the wavelength of the sound in air, a loud sound is heard coming from the tube. Waves of large amplitude, stationary or standing waves, are now set up in the air in the tube (p. 601).

In the same way, the aerial of a transmitter is coupled to the oscillator or radio source, and adjusted so that standing electrical waves are set up along it. The radiofrequency waves from the oscillator are then most strongly radiated into space.

Quarter- and Half-Wave Aerials, Standing Waves

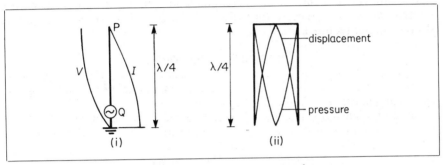

Figure 36.8   _Quarter-wave aerial_

Figure 36.8(i) shows a long vertical wire PQ with an r.f. oscillator Q near the bottom earthed end. A _current node_ is set up by Q at P, where electrons cannot move, and a _current antinode_ at Q where electrons are free to move. The electrons have greatest pressure at P, which is therefore a _voltage antinode_, and least pressure at Q which is a _voltage node_. Figure 36.8(ii) shows the analogy with air molecules in a closed pipe at resonance. The displacement curve is similar to the current $\lambda/4$ curve I and the pressure curve to the voltage curve V. The aerial and pipe have a length $\lambda/4$, where λ is the wavelength. So PQ is a _quarter-wave_ ($\lambda/4$) aerial. Medium-wave aerials are generally of the quarter-wave type. For 200 m wavelength, the aerial height is 50 m.

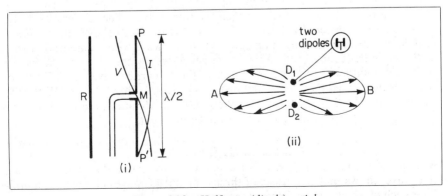

Figure 36.9   _Half-wave (dipole) aerials_

On short wavelengths such as television wavelengths, a *half-wave* or *dipole* aerial is used. Figure 36.9(i). Here the ends P, P' of the aerial are current nodes (voltage antinodes) while the middle M is a current antinode (voltage node). In Sound, this corresponds to a pipe closed at both ends and excited in the middle. A reflector R behind the dipole increases the radiation in a required direction.

Figure 36.9(ii) shows roughly how the radiation intensity varies round a two-dipole or H-aerial, D_1, D_2. In this so-called *polar diagram*, the lengths of the arrowed lines are proportional to the intensity in the direction concerned. So the radiation is mainly in the direction AB. More dipole arrays can provide greater intensity in a required direction.

Microwaves (very short wavelengths of the order of a few cm) can be transmitted from small dipoles placed at the focus of paraboloid dishes (see p. 414).

Receiving aerials are similar to transmitting aerials. Dipole aerials are used in medium-wave and television reception, and small dipoles for microwave reception with paraboloid dishes.

To avoid interference at receivers by neighbouring transmitters, the two radio signals are sent out polarised in different directions, horizontal and vertical. The receiving aerial must be correctly positioned, horizontally or vertically to receive the transmitter required.

Amplitude Modulation

The energy radiated from an aerial is practically zero when the frequencies are below about 15 kHz, which is in the audio-frequency (a.f.) or sound range. So speech and music cannot be radiated directly.

Aerials radiate most strongly only high frequencies, which are those known as *radio-frequency* (*r.f.*) waves. They may range from very high frequencies such as 100 MHz (10^8 Hz), 3 cm waves, to 200 kHz (2×10^5 Hz), 1500 m waves. For r.f. wave calculations, a middle value of 1 MHz (10^6) may often be taken.

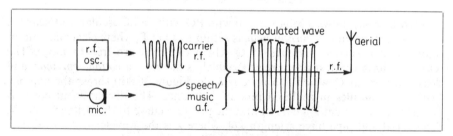

Figure 36.10 *Amplitude modulation*

Practical methods of radiating a.f. (sound) energy consists basically of carrying it on a r.f. wave, which is called the *carrier wave*. The carrier r.f. wave is then said to be *modulated* by the a.f. wave. Figure 36.10 shows an *amplitude modulated* wave radiated from an aerial. It is a r.f. wave whose amplitude varies at the same frequency as the a.f. voltage generated by speech or music by the studio microphone. If A is the amplitude of the r.f. carrier wave and B is that of the a.f. wave carried then B/A is defined as the 'depth of modulation'. With 50% modulation, $B = A/2$. With greater than 100% modulation, the amplitude no longer follows the a.f. wave and distortion would occur.

Sidebands, Bandwidth

Analysis of the modulated wave for a sine wave carrier, and a modulating a.f. sine wave, shows that it consists of three different r.f. waves of constant amplitude, each of which can be detected separately.

If f_c is the carrier r.f. and f_m is the much smaller modulating a.f., the three waves have respective frequencies of (i) f_c, (ii) (f_c+f_m), called the *upper sideband frequency*, and (iii) (f_c-f_m), called the *lower sideband frequency*. If $f_c = 1$ MHz and the highest modulating a.f. is 10 kHz or 0·01 MHz, then

$$\text{upper side frequency} = f_c+f_m = 1\cdot01 \text{ MHz}$$

and

$$\text{lower side frequency} = f_c-f_m = 0\cdot99 \text{ MHz}$$

In practice, the carrier is modulated by all frequencies in the a.f. range, from high to low. This is shown diagrammatically in Figure 36.11(i). The *bandwidth* of the sidebands transmitted is 20 kHz in this case, from 1·01 to 0·99 MHz.

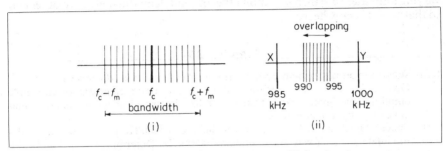

Figure 36.11 (i) *Sideband frequencies* (ii) *Overlapping*

Figure 36.11(ii) shows the frequencies X and Y of two transmitting stations which are less than 20 kHz apart. The carrier X has a frequency of 985 kHz (0·985 MHz) and the carrier Y has a frequency of 1000 kHz (1·0 MHz). In this case some of the upper sidebands of X have the same frequencies as some of the lower sidebands of Y, as shown, and overlapping occurs. So *interference* may occur from one transmitter when reception is required from the other. Since, in practice, two broadcasting stations may be separated by about 9 kHz, interference takes place between two powerful stations. It can be heard on an unselective radio receiver.

AM Radio Receiver (TRF)

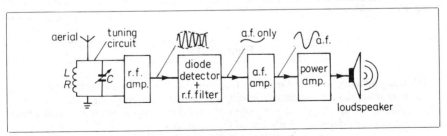

Figure 36.12 *Tuned radiofrequency (trf) receiver*

Figure 36.12 shows the basic system needed for a radio receiver when amplitude-modulated (AM) waves arrive at the aerial. The coil (L, R)-capacitor (C) circuit is a tuning circuit. See p. 396. Here a high voltage is obtained across C when it is tuned to the r.f. carrier wave.

The voltage is now passed to a r.f. amplifier for more amplification, and then to a suitable diode detector circuit. A diode conducts only one half, say the positive half, of the modulated wave. So the output voltage is a positive r.f. voltage with an average amplitude value which follows the a.f. variation. The diode thus 'detects' the a.f. voltage carried by the modulated wave.

An electrical filter circuit cuts off the r.f. voltage, leaving only the a.f. voltage to pass on to an a.f. amplifier. This is finally passed to an a.f. power amplifier, so that maximum sound energy is obtained from the loudspeaker.

Pulse amplitude modulation

To transmit many messages along a telephone line, the continuous or analogue sound signals are first modulated to digital (high and low) signals. *Pulses of varying amplitude are then sent along the line and demodulated at the other end so that sound is now heard.*

_____ **Exercise 36** _____

1 (a) What do you understand by the term *amplitude-modulated carrier wave*? Give a labelled block diagram showing the basic elements of a simple amplitude-modulated radio transmitter for the broadcasting of audio signals developed by a microphone.

(b) Speech signals in the frequency range 300 Hz to 3400 Hz are used to amplitude-modulate a carrier wave of frequency 200·0 kHz. Determine (i) the bandwidth of the resultant modulated signals, (ii) the frequency range of the lower sideband, (iii) the frequency range of the upper sideband.
The diagram below shows the range of speech signals and the carrier wave. Copy the diagram into your answer book and use it to represent your answers to (i), (ii) and (iii).

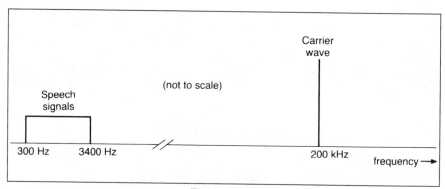

Figure 36.13

(c) Long distance intercontinental telecommunications may use free-space electro-magnetic wave propagation linking ground station to ground station. Use a diagram to show how this is achieved for (i) microwave signals, and (ii) radio signals. (Apparatus details are NOT required.)
Estimate the transit times for a U.K.–U.S.A. telephone signal link using (i) satellite communication, and (ii) surface wave propagation. Assume that the satellite is 36 000 km equidistant from each ground station and that the great circle distance between the two ground stations is 8400 km. (Speed of electromagnetic waves in free space $= 3.00 \times 10^8$ m s^{-1}.)(*L.*)

Answers to Exercises

Mechanics

Exercise 1A (p. 16)
1 (i) 5 s (ii) 62·5 m (iii) 17 m s^{-1}
2 (i) 4 s (ii) 20 m
4 3 s
5 (i) 3 s (ii) 30 m (iii) 72°
7 (i) 16 s (ii) 8868 m; 10 240 m
8 (i) 5 m s^{-2} (ii) 1·4 s (iii) 3 s (iv) 22·5 m
9 (i) 20 m s^{-1} (ii) 14·1 m s^{-1} at 45° to OA
10 (i) 10 km h^{-1} (ii) 2·4 km
11 21 650 m, 3125 m, 25 s, 465 m s^{-1}
12 B

Exercise 1B (p. 31)
1 2000 N, 3000 N
2 550 N, 500 N
3 (i) 600 N (ii) 400 N
4 (i) 10 N s (ii) 100 N
5 2·0 N
6 2·5 m s^{-2}
7 2 m s^{-1}, 4 m s^{-1}
8 10 m s^{-1}, 2·0 s
9 200 N
10 (a) (i) 0·38 m (ii) 3 m s^{-2}
   (b) (i) 5750 N (ii) 5000 N
11 0·9 v, 30°
12 (a) 4 m s^{-2} (b) 7·1 m s^{-1}
13 (a) (i) 250 m (ii) 4000 N (iii) 20 000 N,
   2000 N (b) 8 m s^{-1}
14 2 m s^{-2}
15 (i) 4·9 N (ii) 4·9 N (iii) 3·54 m s^{-1}
16 8/3 m s^{-1}, −4/3 m s^{-1}
17 254 m s^{-1}, 23° to orig. dir.
18 5·6 N s (i) 140 N (ii) 11·2 N s
   (iii) 35 m s^{-1}

Exercise 1C (p. 44)
1 180 J, 60 N s; 3 s, 9 m
2 600 J, 6 m
3 (i) 4 s (ii) 20 m (iii) 10 J, 10 J
4 (i) 5·2 m s^{-1}, 58 J (ii) 1·2 m s^{-1}, 314 J
5 20 J, 15 J
7 1000 J
8 1·25 m
9 4 kW, 14 kW
10 (i) B, C, E, G (ii) A, D, F, H
11 (i) 3 m s^{-2} (ii) 45 000 J (iii) 30 kW
12 $x = 4$

13 $x = \frac{3}{2}, y = \frac{1}{2}, z = -\frac{1}{2}; t = k\sqrt{a^3 \rho/\gamma}$
15 (a) 0·3 N (b) 0·6 W (c) 0·3 W
16 2/3
17 $x = 2, k = 2·2$
18 $m^2 Hg/(M + m)d$ $+ [(M + m)g]$
19 6×10^5 m s^{-1}, $2·12 \times 10^{-15}$ J
20 (a) 10 J (b) 5 J
21 (i) 400 kW (ii) 656 kW
22 2E (3E less E)
24 (a) 10 m s^{-1} (b) 24 J, 12·65 m s^{-1}
25 (a) 10/3 N (b) 5/9 W (c) 5/18 W
26 1667 N, 83 330 J

Exercise 2A (p. 56)
1 (i) 2 rad s^{-1} (ii) 96 N
2 (i) 118 N (ii) 32°
3 42°, 13 450 N
4 224 N, 64 N
5 15·8 N
6 7·3 × 10^{-5} rad s^{-1}, 0·068 N
7 0·5%
8 (i) 2·3 N (ii) 1·7 m s^{-1}
9 1·6 rev s^{-1}
10 (i) 4·2 m s^{-1} (ii) 11 N (iii) 7 N
11 90 N
12 0·675 rev s^{-1}
13 (a) mgl (b) $\sqrt{2}$ gl (c) 2 g up (d) 3mg
14 7·7 rad s^{-1}, 122 cm away
15 (a) 8·0 N (b) 8·0 N
16 (c) 11·5°
17 (i) 42·4 m s^{-1} (ii) 4500 N
   (iii) 4743 N at 18° to radius (v) 42°
18 (ii) 1·58 m s^{-1} (iii) 0·625 m

Exercise 2B (p. 72)
1 250 N
2 0·25 g
4 M^{-1}L^{3}T^{-2}
5 Y
6 9910 s
7 6 × 10^{24} kg
8 9·9 m s^{-2}
9 4·5 × 10^{-9} rad s^{-1}
10 (a) 7852 m s^{-1} (b) 5250 s
11 889 N
12 14·4 N, 24·5 h
13 (c) 0·012
14 24 h

15 (*c*) (i) 7·4 km s^{-1}, 100 min, 10·8 km s^{-1}
(*d*) 10^{-3}
16 3 m, 6:1
17 (*a*) 360 000 km from earth centre
(*b*) 60 MJ kg^{-1}
18 7 km s^{-1}, 1·46 × 10^{10} J, − 2·92 × 10^{10} J
19 42 600 km
20 7·1 × 10^{10} J

Exercise 2C (p. 88)
1 (i) 0·05 s (ii) 0, 3200 π^2 cm s^{-2}
(iii) 80 π cm s^{-1}
2 (*a*) 0·08 m s^{-1} (*b*) 1·57 s
3 (*a*) 2·5 m s^{-1} (*b*) 790 m s^{-2}
4 A, D, E, F
5 (i) 1·26 s (ii) 0·1 m s^{-2} (iii) 4 × 10^{-5} J
(iv) 4 × 10^{-5} J
6 10·0 m s^{-2}, 4·5 m; 2·45:1
7 (*a*) 0·25 m s^{-1} (*b*) 15·8 mJ
8 0·2 s
9 0·68
10 2·7 s; 0·24 J, 0·47 m s^{-1}
12 0·44 s; 3·7 cm, 47·6 cm s^{-1}, 740 cm s^{-2}
13 (*a*) (i) 20 mm (ii) 0·2 s (iii) 628 mm s^{-1}
(*b*) 7·9 × 10^{-2} J
14 (i) 0·4 s (ii) 0·5 N (iii) 0·005 J
15 0·2 J
16 6·3 cm
17 (*a*) 2·43 s (*b*) 0·150 J, 0·387 m s^{-1}
(*c*) 0·56 (*d*) 0·067 m
18 5%
19 1 s
20 8·75 cm, 3·2 rad s^{-1}, 90° 27·7 cm s^{-1},
87·5 cm s^{-2}, 2·015 N, 1·992 N,
5·4 × 10^{-5} J/cycle

Exercise 3 (p. 110)
1 E
2 (i) 0·8 N m (ii) 0·4 N m; 10·1 J
3 690 N, 1390 N
4 500 N m; 167 N
5 22·6 cm
7 41·7 N, 41·7 N
8 180 N, 159 N
9 0·0063 nm from N atom
10 1710 N
11 (i) 3:2 (ii) 40:9
12 1·62 N
13 5·1 cm
14 38 cm
15 773·8 mmHg
16 W ρ/σ (*a*) 0·95 (*b*) 1·19
17 (*a*) 750 kg (*b*) 270 kg (*c*) 0·42 m s^{-2}

Exercise 4 (p. 130)
1 (i) 2000 J (ii) 200 kg m^2 s^{-1}
(iii) 3·2 rev s^{-1}
2 (i) 8 rad s^{-1} (ii) 25 133 J
3 6·4 rev, 8 s
4 50 rad s^{-1}, 25 000 J
5 4 rev s^{-1}
6 (i) 2 rad s^{-1} (ii) 15 J
7 (i) 6 rad s^{-1} (ii) 12 N m (iii) 14·3 rev
8 (i) 2·2 rad s^{-1} (ii) 3·2 rad s^{-1}
9 1·3 × 10^{-4} kg m^2
10 18·3 rad s^{-1}
11 (*a*) 20 rad s^{-2} (*b*) 0·32 N m
12 7·3 × 10^{-4} kg m^2
13 6·3 × 10^{12} rad s^{-1}
14 (i) 2 × 10^7 J (ii) 10 km
15 (i) 2304 J, 38·4 kg m^2 s^{-1} (ii) 0·42 N m
(iii) 92·2 s
16 21·6 rad s^{-1}
17 1·5 kg s^{-1}
18 9 m s^{-1}
20 1 m s^{-1}, 9·25 × 10^4 Pa
21 4 m s^{-1}, 12 kg s^{-1}
22 (i) 2·8 m s^{-1} (ii) 5·7 × 10^{-3} m^3 s^{-1}

Elasticity, Solid Materials

Exercise 5A (p. 142)
3 (i) 1/2000 (ii) 10^{11} N m^{-2} (iii) 0·025 J
4 (i) Y
5 6·4 × 10^6 N m^{-2}, 6 × 10^{-5},
1·1 × 10^{11} N m^{-2}
6 67 N
7 0·08 mm
8 0·04 J, 0·08 J
9 1·2 × 10^8 N m^{-2}, 29·6 N
10 8·3 × 10^{-3} J
11 (*a*) C:1·25 × 10^{-3}, S:0·75 × 10^{-3};
(*b*) 471 N
12 (*a*) 2·0 × 10^{11} N m^{-2} (*b*) 4·8 × 10^{-2} J
13 (*a*) 1·5 (*b*) 6 mm, 4 mm (*c*) 780 N
14 20 m s^{-1}
15 (*a*) 2·3 × 10^{-6} m (*b*) 5·7 × 10^{-5} J
16 (*c*) (i) 4·02 × 10^{-3} (ii) 8·03 × 10^8 N m^{-2}
(iii) 0·21 J
17 40 N, 74 N
18 (*b*) 3·5 m (*c*) 8 kJ (approx.)
(*d*) 11 m (approx.)
19 (*a*) (i) 2 × 10^{11} Pa (*b*) (iii) 50 N, 10 J
20 (*c*) (i) 1·44 × 10^{-3} m^2 (ii) 0·098 m
(iii) 3890 J (iv) 6·4 × 10^5 W

Exercise 5B (p. 151)
7 2·8 × 10^{-10} m (*a*) 2 × 10^8 N m^{-2}
(*b*) 3 J (approx.) (*c*) 1·4 × 10^{11} . 3 mm

8 (i) 1000 N (ii) $8.7 \times 10^{-5}\,\text{m}^2$

Exercise 5C (p.172)
10 glass: $7.0 \times 10^{10}\,\text{Pa}$;
  wood: $1.8 \times 10^{10}\,\text{Pa}$
11 (i) A (ii) B (iii) B

Electricity

Exercise 6 (p. 208)
1 $2 \times 10^4\,\text{V m}^{-1}$, $1.6 \times 10^{-14}\,\text{N}$
3 $1.25 \times 10^{-18}\,\text{C}$
5 $1.6 \times 10^{-19}\,\text{C}$
7 $1.5 \times 10^6\,\text{V}$
9 3 electrons, $19.6\,\text{m s}^{-2}\,(2\,g)$
11 100 V
12 (a) $9 \times 10^4\,\text{N C}^{-1}\,(\text{V m}^{-1})$ (b) 9000 V
13 $41\,\text{kV m}^{-1}$
14 (i) $1.41 \times 10^3\,\text{N C}^{-1}$, $-3.38 \times 10^3\,\text{V}$
  (ii) 0.3 m from $-3Q$
17 (a) $1.45 \times 10^{-10}\,\text{N}$, $1.45 \times 10^{-11}\,\text{J}$
18 $1.33 \times 10^{-5}\,\text{C}$, $6 \times 10^5\,\text{V}$, yes

Exercise 7 (p. 236)
1 $0.2\,\mu\text{F}$, $2.5 \times 10^{-4}\,\text{J}$
2 (a) $9 \times 10^{-4}\,\text{C}$, $18 \times 10^{-4}\,\text{C}$; $0.135\,\text{J}$,
  $0.27\,\text{J}$ (b) $6 \times 10^{-4}\,\text{C}$, $6 \times 10^{-4}\,\text{C}$; $0.06\,\text{J}$,
  $0.03\,\text{J}$
3 (a) $6 \times 10^{-8}\,\text{C}$ (b) $3 \times 10^{-6}\,\text{J}$
  (c) $6 \times 10^{-6}\,\text{J}$; $1 \times 10^{-6}\,\text{J}$
4 $133.3\,\mu\text{C}$, $133.3\,\mu\text{C}$
5 (i) 160 V (ii) 0.14 J (iii) 0.128 J
6 $2/3\,\mu\text{C}$, $2\,\mu\text{C}$
7 (a) 4 V (b) $4\,\mu\text{C}$ (c) $8\,\mu\text{J}$
8 33.9 m
9 4
12 (a) 0.1 C (b) $10^{-2}\,\text{F}$ (c) $5\,\text{k}\Omega$
13 $171\,\text{k}\Omega$ (i) 9 mJ (ii) $500\,\mu\text{F}$
14 $1.35 \times 10^{-3}\,\text{C}$, 3 times, 84
15 (a) $2 \times 10^{-11}\,\text{F}$, (b) $1.2 \times 10^{-6}\,\text{J}$. 300 V
16 $1.13 \times 10^5\,\text{V m}^{-1}$ (a) 226 V
  (b) $2.2 \times 10^{-10}\,\text{F}$ (c) $5.65 \times 10^{-6}\,\text{J}$
17 $4.25 \times 10^{-5}\,\text{A}$
18 $200\,\mu\text{F}$, 43.2 s
19 (a) $10^{-3}\,\text{C}$ (b) 1.35 V (c) 6.9 s

Exercise 8 (p. 274)
1 $3.2\,\text{V}$, $10\,\Omega$
2 (i) 0.4 A (ii) $10\,\Omega$ (iii) 12 V
3 5 V, 0; 5.8 V, 4.8 V
4 14.4 V, 9.6 V; 8 V, 5.3 V
5 0.5 A, $6\,\Omega$
6 (a) $360\,\Omega$ (b) 0.96 V
7 (i) (a) 1.5 V (b) 9/16 V (c) 1.5 V (ii) 9.6 V
8 (i) $2 \times 10^{-4}\,\text{A}$ (ii) $4.99\,\text{M}\Omega$
9 (a) 6 V (b) 4 V

10 2 V, $4\,\Omega$ parallel
11 $5.2 \times 10^{-5}\,\text{m s}^{-1}$
12 $64\,n\,(n^2 + 16)$, 8 A, $3/8\,\Omega$
13 $X = 2000\,\Omega$, $1000\,\Omega$
14 (i) 0.6 A, 0.2 A, 0.4 A (ii) 6 V
  (iii) 5.8 W (iv) 7.2 W
15 1.54 V
16 0.01 A
17 2:1
18 4 m, 1 m
19 $0.16\,\Omega$
20 $R_1 = 16.7\,\Omega$, $R_2 = 12.5\,\Omega$
21 (a) $1\,\Omega$ (b) $20\,\Omega$ or $0.05\,\Omega$
22 $18\,\Omega$
23 (1) 1:5 (2) 1:7.5
24 $\frac{1}{2}$, 1, 2 kW
25 36, 39%
26 18.75 kW
27 7 V

Exercise 9 (p. 295)
1 1.53 V, 60.0 cm
2 0.12 A
3 $1197\,\Omega$, 3 mV
6 15.8 mV
7 0.825 m, 1.29 V
9 12, less
10 60 cm, 75 cm
11 9 mV, $101.6\,\Omega$
12 (a) 1.5 V (b) 1.6 V (c) 3 (d) 2 V
14 $5 \times 10^{-4}\,\text{A}$, $2.5 \times 10^{-3}\,\text{V}$, $1962\,\Omega$
15 $0.004\,\text{K}^{-1}$, 50°C
16 $14.9\,\Omega$
17 0.367 m
18 1.4:1, 1500 s
21 $0.0037\,\text{K}^{-1}$
22 $8.8 \times 10^{-5}\,\text{K}^{-1}$
23 $5.3 \times 10^{-4}\,\text{K}^{-1}$

Exercise 10 (p. 319)
1 (i) 0.2 N (ii) 60°
2 5 A
3 $5 \times 10^{-3}\,\text{N m}$, $2.5 \times 10^{-3}\,\text{N m}$
4 1.0 T, PQ
5 $1.6 \times 10^{-14}\,\text{N}$
6 $2.6 \times 10^{22}\,\text{m}^{-3}$
7 21.8°
8 $69\,\mu\text{A}$
9 (a) 10 (b) 1/4
10 $3.75 \times 10^{-24}\,\text{N}$
11 1.4 rad, 1.96 mm, $50\,\mu\text{A}$, $0.1\,\Omega$
12 $5.26\,\Omega$ parallel, $3900\,\Omega$ series
13 0.4 T
14 $3.5 \times 10^{-4}\,\text{m s}^{-1}$, BI/Net

15 (a) shunt 0.0125Ω (b) series 9995Ω
16 2.1×10^{22}

Exercise 11 (p. 338)
  1 (b) 1×10^{-5} T (c) 3×10^{-5} T, greater
  2 0.8 A
  3 (b) 0.04 m (c) 1.33×10^{-5} N m^{-1}
  4 6×10^{-6} N away from X
  6 3.7×10^{-3} V
  7 6 cm from 12 A wire
  8 3.64 A
  9 (i) 1.6×10^{-4} T (ii) 0
  11 1.6×10^{-6} N m^{-1}, 0.38 N
  12 39 mT, 7.9μN m

Exercise 12A (p.365)
  3 (a) 0 (b) 0.6 V . P
  4 (a) 0 (b) 0.1 V; 10^{-3} C
  5 3.1 V. (i) 3.1 V (ii) 1.6 V (iii) 0
  11 (i) Bvd (ii) 0 (iii) Bvd; (i) (iii) $I = Bvd/R$,
    $F = B^2vd^2/R$
  12 $\mu_0 nIAf$, $100 \mu_0 nAf$
  13 44.2 rev s^{-1}
  14 0.53 V
  15 1.05 V
  18 (i) $0.02 \pi f \sin 2\pi f t$ (ii) 67.5 Hz
  19 9.7×10^{-4} Wb, 3.9×10^{-5} Wb, 3.9 mV

Exercise 12B (p. 376)
  1 4 mH
  2 25μF
  4 (a) (i) 4.0 A (ii) 1.0 A s^{-1} (iii) 0.5 A s^{-1}
    (b) 26.8 div.
  5 4μH
  7 (a) 25 000 V (b) 2×10^{-2} A
  8 (a) 4 A (b) 2 A s^{-1}

Exercise 12C (p. 381)
  1 (a) 2.5×10^{-3} Wb (b) 2.5×10^{-5} C
  2 (ii) 0.1 T (iii) 1.25×10^{-9} J
  3 10^{-4} C

Exercise 13 (p. 402)
  1 (i) 2.1 A (ii) 2.1 A (iii) 3 A
  2 (i) 2 A (r.m.s.) (ii) 0.016 A (r.m.s.)
    (iii) 0.0031 A (r.m.s.)
  3 (i) R, 4 V; L, 3 V; C, 4 V (ii) 5 V
    (iii) 1 V (iv) 4.1 V; 0.8 W, 0, 0
  4 (a) 28.3 V (b) 64 Hz (c) 20 W
  6 0.5 A
  7 (b) 7.8 V, 6.2 V
  8 (b) 0.11 H (c) $2.56:1$
  10 3142Ω

11 0.33 H
12 (i) 0.4 mA, 3.2μF (ii) 40 mA
13 5 V, $53.1°$
14 greater, 111.8 V, $26.6°$, 11.2 mA, 1.25 W
15 37.5Ω
16 3.46 A
20 (a) 0.1 A (b) 3 W
21 60.6Ω
22 $R = 190.0 \Omega$, $C = 50.6 \mu$F
23 1990Ω
24 31.4Ω, 1592Ω, 356 Hz

Exercise 14A (p. 420)
  1 $35.3°$
  2 $41.8°$
  3 (i) $26.3°$ (ii) $56.4°$
  4 (i) $41.8°$ (ii) $48.8°$ (iii) $62.5°$
  6 1.60
  7 1.50
  8 (i) $60°04'$ (ii) $48°27'$

Exercise 14B (p. 428)
  1 $42°$
  2 $0 - 37.8°$
  3 $49°$
  5 $43°35'$
  6 (a) $53°$ (b) $46°$

Exercise 14C (p. 433)
  4 (a) $61°$ (b) $51°$
  5 (i) 0.4μs (iii) 2.5 MHz

Exercise 15A (p. 441)
  1 (i) 12 cm, $m = 1$ (ii) 12 cm, $m = 3$
  2 $6\frac{2}{3}$ cm, 0.6
  3 $13\frac{1}{3}$, 40 cm
  4 5 cm
  5 1.95 cm
  6 13.3 cm
  7 (i) 5.1 cm (ii) 22.5 cm
  8 1.82 mm, 20 cm from 2nd lens, virtual,
    9.1 mm
  9 4 cm
  10 -15 cm
  11 (a) 120 cm from conv. lens (b) 92.2 cm
    (c) 2.2

Exercise 15B (p. 452)
  1 (i) infinity (ii) 20
  3 11.25 cm
  4 0.02 m
  7 0.0091 rad
  8 7×10^{-6} rad

9 (*a*) 4 (*b*) 4·8
10 5°
12 f_e = 4 cm, dia. = 0·5 cm
13 (*a*) 22·4 (*b*) 4·9 cm dia.
14 $6·25 \times 10^{-3}$ rad
16 (i) 4·2 cm (ii) 6; 5
15 (*b*) 9·2 mrad, 2·76 mm, 11 mrad (*c*) 6
16 4·2 cm. (*a*) 6 (*b*) 5

Exercise 16 (p. 462)
1 20·2 cm sepn.
2 16 cm from second lens
3 3·2, 8·8 cm
4 8·7 cm, 46·7
5 16 mm
6 48 mm, 2 m, − 1 m
7 1/15 s
8 $6·8 \times 10^{-2}$ s
14 110 mm

Waves

Exercise 17 (p. 497)
1 $0·42 \, \text{m s}^{-1}$
3 (i) 1·33 m (ii) 400 Hz
4 (i) $5\pi/3$
    (ii) $y = 0·03 \sin 2\pi(250t − 25x/3)$
    (iii) 6 cm
7 3400 Hz
8 10^{10} Hz
9 4 max intensity per 3 s
11 3:1
12 (i) 1·73 A (ii) 1·5:1
13 (*a*) 1/9th (*b*) 2·9 cm, $1·03 \times 10^{10}$ Hz
14 (i) 100 Hz (ii) 1·7 m (iii) $170 \, \text{m s}^{-1}$
    (iv) π (v) $0·2 \sin(400\pi t + 20\pi x)/7$
15 (*a*) $330 \, \text{m s}^{-1}$ (*b*) $6·6 \times 10^{-4} \, \text{m s}^{-1}$
16 $330 \, \text{m s}^{-1}$, 579 mm
17 5·9 s
19 $349 \, \text{m s}^{-1}$
20 $332 \, \text{m s}^{-1}$
22 680 Hz

Wave Optics

Exercise 18 (p. 512)
3 18·6°
4 47°10′, 41°48′ to vertical at oil surface
7 34·8°, 34%
9 $4·0 \times 10^{-7}$ m
11 $1·9 \times 10^{-3}$ deg.
12 31·25 km, 300 rev s^{-1}
14 6250 m
15 250, 6×10^4 cycles
16 $2·23 \times 10^8 \, \text{m s}^{-1}$

Exercise 19 (p. 533)
2 $1·8 \times 10^{-3}$ m; $1·5 \times 10^{-3}$ rad
4 $1·8 \times 10^{-3}$ rad, 0·064 mm
6 $2·27 \times 10^{-5}$ m
7 643 nm
9 0·34 mm
13 (*b*) 9·375, 15 (*c*) 2·74 mm, 1536 nm,
    5·14 mm
14 1·5
15 $2·4 \times 10^{-3}$ rad
17 1·0 m
19 $1·5 \times 10^{-5}$ m
20 $7·1 \times 10^{-6}$ m, recede
21 2·11 mm, 4·33 mm

Exercise 20 (p. 557)
1 (*a*) $2·5 \times 10^{-3}$ mm (*b*) 13·9° (*c*) 4
4 2×10^{-6} m
6 $1376 \, \text{m s}^{-1}$
9 $97 000 \, \text{m}^{-1}$
10 1·2 mm, 2·4 mm from centre
11 $2·32 \times 10^{-7}$ m
12 2·35 mm, 0·0785 rad
13 2·7 mm
14 600 nm, $285 000 \, \text{m}^{-1}$, 43·2°
15 3; $5·895 \times 10^{-7}$ m
16 (i) $1·78 \times 10^{-6}$ m (ii) $7·52 \times 10^{-7}$ m
    (iii) 38·2°, 57·6°
17 7; 0°, ± 17·1°, ± 36·1°, ± 62·1°

Exercise 21 (p. 569)
3 67·5°
6 1·0 mm

Sound

Exercise 22 (p. 586)
5 (*a*) 4:1, 2500 Hz (*b*) 9:1 (*c*) 1:1
    (*e*) 12·4 cm
7 25 W
8 1·83 s
10 1091, 909 Hz
11 486 Hz
12 (*a*) away (*b*) $4 \times 10^5 \, \text{m s}^{-1}$
13 425 Hz
14 514, 545 Hz
15 12 Hz; 1007, 993 Hz
16 (i) 66 cm (ii) (1)550 (2)545 (3)600 Hz
17 $2·9 \times 10^{-6}$ rad s^{-1}
18 45 132 Hz; 45 265 Hz; 265 Hz;
    $0·02 \, \text{m s}^{-1}$

Exercise 23A (p. 605)
1 (i) $\lambda/2$ (ii) $\lambda/4$ (iii) $\lambda/2$, 567 Hz

2 20 cm

3 (i) 320 cm s^{-1} (ii) 10 cm (iii) 40 cm
(iv) 1200 Hz

4 (*a*) 0·322 m (*b*) 0·645 m

5 (i) max (ii) 0 (iii) max (iv) half-max

6 267 Hz

9 + 5·2°C

Exercise 23B (p. 615)

1 300 Hz

2 170 Hz

3 resonance at 115 Hz

4 $v = \sqrt{(F/\rho)}/r$

5 239 Hz

6 − 3·2%, + 6·8%

7 (*a*) 2 m (*b*) 100 Hz

8 pluck 1/6th from end

9 20·5 N

Heat

Exercise 25 (p. 632)

1 16·6°C, 17·0°C

3 1·43°C

5 12·5°C, 7·1°C

6 881°C

7 385°C

8 22–24°C; 0°C, 100°C

9 57·6°C

10 419·47 K

Exercise 26 (p. 647)

1 (i) 12 W (ii) 2 W (iii) 600 J kg^{-1} K^{-1}

2 (i) 100 s (ii) 0·2 kg

3 6000 J kg^{-1} K^{-1}, 10%

4 6·76 × 10^{-20} J molecule^{-1}

5 5·7 V

6 0·45 W

7 4200 J kg^{-1} K^{-1}

8 1·99 W, 1829 J kg^{-1} K^{-1}

9 (1) 1·3 W (2) 960 J kg^{-1} K^{-1}

10 2·6 × 10^3 J kg^{-1} K^{-1}, 1·5 W

11 3·78 × 10^5 J kg^{-1}

12 0·0935 kg

13 360 J K^{-1}, 0·035 kg

14 2·23 × 10^6 J kg^{-1}

15 1·6 × 10^5 J kg^{-1}

Exercise 27A (p. 664)

1 (i) 2·4 × 10^5 N m^{-2} (ii) 0·24 m^3
(iii) 0·13 m^3

3 1·1 kg

4 100°C

5 (i) 2086 J kg^{-1} K^{-1} (ii) 1·12 × 10^5 Pa

6 0·014 m^3

7 (c) 2·45 × 10^5 Pa (d) 2·5 × 10^5 Pa,
2·08 × 10^5 Pa. 31 strokes

Exercise 27B (p. 676)

1 1250 cm^3, 137·5 kPa

2 10·48 × 10^5 N m^{-2}, 666 K

3 (*a*) 586 K (*b*) 101·2 J (*c*) 355 J

4 222 K, 384 mmHg

6 4·40 × 10^5 Pa, 15°C; 7·66 × 10^5 Pa,
228°C

Exercise 27C (p. 681)

1 1·3 × 10^5, 2·3 · 10^5 N m^{-2}

2 (ii) 8·3 J mol^{-1} K^{-1}

3 (*b*) 1·06 M J

4 (i) 2000 J (ii) 450 K (iii) 60 J

6 1·44 × 10^4, 1·03 × 10^4 J kg^{-1} K^{-1}

7 8·3 J mol^{-1} K^{-1}, 25·9 J

8 164 J

9 (ii) 3·14 × 10^4 Pa, (iii) 188·6 K.
12·4 J K^{-1} mol^{-1}

10 1·67

Exercise 28 (p. 692)

2 (i) 1732 m s^{-1} (ii) 433 m s^{-1}

3 (i) 1039 m s^{-1} (ii) 900 m s^{-1}

5 597 m s^{-1}

6 (i) 501 m s^{-1} (ii) 4·6 × 10^{-23} N s,
1000 s^{-1}, 7·4 × 10^{-19} Pa
(iii) 3·1 × 10^5 Pa

7 0·21 mmHg

8 (i) 508 m s^{-1} (ii) 400–475 m s^{-1}

9 461 m s^{-1}, 64 rad s^{-1}

10 4·35 × 10^{16} m^{-3}

12 (*a*) 1·07 (*b*) 4/1 (*c*) 0·87

13 1305 m s^{-1}, 0°C

14 K r

15 (i) 2·4 × 10^{22} (ii) 150 J

16 (i) 2·6 × 10^{-4} kg (ii)240 J

Exercise 29A (p. 707)

1 (i) 4·5 × 10^7 J (ii) 1·8 × 10^8 J

5 51°C

7 (*a*) 90 W m^{-2} (*b*) 3 m, 19·5°C, 5·5°C,
1/32

8 (i) 0·45 W (ii) 10·5 W
(iii) 0·178 W m^{-1} K^{-1}

9 0·017 kg s^{-1}, 45·9°C, 29·2°C

10 1·47 × 10^{-4} kg s^{-1}

11 (*a*) 90°C (*b*) 57·6 W

12 1 K mm^{-1}, 137 p

13 89°C

14 94%

15 12 kW. (a) 3·75 K (b) 11·25 K
(c) 0·18 mm
16 0·5 W m^{-2}K^{-1}, 0·01 W m^{-1}K^{-1}
17 i 5645 kWh ii 4788 kWh iii £133
iv £207·90

Exercise 29B (p. 728)
2 1070 K
4 2·44, 0·82 W
5 4·55 × 10^{26} W m^{-2}, 1600 W m^{-2}
8 19·0 W m^{-1}
9 800 W, 3200 W, 200 W, 0
10 6023 K, 4·4 × 10^{-9} kg s^{-1}, 280 K
11 0·19 nm, 1·07 mm
12 5·7 × 10^{-8} W m^{-2} K^{-4}
13 5490°C
14 1105 K
15 5450°C
16 5·6 × 10^7 J
17 2140 K
18 4 : 1

Exercise 30A (p. 738)
3 113·5 K
4 6904 J, 6904 J
6 9 × 10^{-8} m
7 5 × 10^5 m^2 s^{-2}, 4 times
8 (i) 1·2 × 10^{19} m^{-3} (ii) 2·7 × 10^{25} m^{-3}
(iii) 0·043 Pa

Exercise 30B (p. 747)
9 3·35 × 10^{-29} m^3

Exercise 30C (p. 754)
1 3450 J, 3450 J
2 (a) 11 350 J (b) 2220 J (c) 20% (d) 0
4 160 000 J
5 9·6 kg
6 1212 J K^{-1}

Electrons, Electronics, Atomic Physics

Exercise 31A (p. 774)
1 6 × 10^6 m s^{-1}, 1·7 × 10^{-2} T
2 33 mm
3 2·65 × 10^7 m s^{-1}, 0·1 K s^{-1}
4 (b) (i) 3·6 × 10^7 m s^{-1} (ii) 2 × 10^{-3} T
(c) 0
5 2006 V
6 1·44 × 10^{-2} m
7 4·2 × 10^4 V m^{-1}
8 (ii) tan^{-1}(Vex/dmv^2) (iii) Ve/2dmv^2
(iv) L = dmv^2 sin 2θ/Ve
9 0·93 m, 3·4 × 10^5 Hz, 870 m
10 0·23 m
11 0·5 × 10^{-4} T

12 4e, changes to 2e and 3e
13 3·2 mm
14 (a) 1·5 × 10^{-6} m (b) 8
(c) 6·25 × 10^{-5} m s^{-1}
16 0·08 m
17 1·2 × 10^{-8} N

Exercise 31B (p. 783)
1 2·2 × 10^8 m s^{-1}
2 14·1 V, 200 Hz
3 60 km
4 (d) 30°

Exercise 32 (p. 805)
2 40 Ω, 40Ω
4 34 V
5 100 kΩ
6 (i) 4·5 mA (ii) 30 µA (iii) 300 kΩ
7 (a) 1·5 V (b) 1·5 V
8 (a) 6 × 10^{-5} A (b) 5·4 V
9 60 kΩ

Exercise 33A (p. 824)
1 (i) S/R (ii) 180°
2 10
3 −0·32 V
4 0·2 V
6 (a) +5·4 V (b) > 5·4 V (c) 1·03 kΩ
7 −40 V s^{-1}
9 0·505 V, 2 MΩ (about)

Exercise 33B (p. 840)
7 AND gate
8 X = 1, Y = 0
16 (a) 300 Ω
18 EOR, AND. Half-adder, P = sum, Q = carry
19 (i) −10 V (ii) +3 V

Exercise 34A (p. 852)
1 5·1 × 10^{14} Hz
2 6·7 × 10^{-34} J s
3 1·43 × 10^{-19} J
6 2·0 × 10^{-7} m
8 1·5 V; (iii) 2·5 × 10^{-19} J
(iv) 7·3 × 10^5 m s^{-1}
9 (ii) 6·9 × 10^{-20} J (iii) 6·9 × 10^{-34} J s
10 (a) 9 × 10^{-7} m (b) 7·9 × 10^5 m s^{-1}
(c) 1·7 V
11 7·8 × 10^{-19} J, 3·6 × 10^{-19} J

Exercise 34B (p. 863)
1 3·3 × 10^{15} Hz (c) 19°

2 $2\,\mathrm{V}, 8\cdot4 \times 10^5\,\mathrm{m\,s^{-1}}$
4 (v) $13\cdot6\,\mathrm{eV}$
5 $13\cdot6\,\mathrm{V}$
6 $A, B, C, D = 6\cdot5, 4\cdot8, 4\cdot3, 4\cdot0 \times 10^{-7}\,\mathrm{m}$
7 (a) $589\,\mathrm{nm}$ (b) $2\cdot93 \times 10^{-19}\,\mathrm{J}$,
$2\cdot17 \times 10^{-19}\,\mathrm{J}$
8 (i) $10\cdot4\,\mathrm{eV}, 1\cdot664 \times 10^{-18}\,\mathrm{J}$ (ii) $317\,\mathrm{nm}$,
UV
9 $1\cdot86 \times 10^6\,\mathrm{m\,s^{-1}}$

Exercise 34C (p. 873)

1 (i) $0\cdot012\,\mathrm{A}$ (ii) $1\cdot33 \times 10^8\,\mathrm{m\,s^{-1}}$
2 $6\cdot4 \times 10^{-15}\,\mathrm{J}, 3\cdot1 \times 10^{-11}\,\mathrm{m}$
3 $17\cdot7\,\mathrm{kV}, 17\cdot7\,\mathrm{keV}$
4 (i) $9\cdot375 \times 10^{16}$ (ii) $150\,\mathrm{W}$
5 $2\cdot0 \times 10^{-7}\,\mathrm{m}$
6 $5000:1$
7 $8 \times 10^{18}\,\mathrm{Hz}$
8 (i) $6\cdot25 \times 10^{15}$ (ii) $7\cdot3 \times 10^7\,\mathrm{m\,s^{-1}}$
(iii) $3 \times 10^4\,\mathrm{kg\,s^{-1}}$

Exercise 34D (p. 878)

1 (i) $6\cdot6 \times 10^{-10}\,\mathrm{m}$ (ii) $1\cdot7 \times 10^{-34}\,\mathrm{m}$
(iii) $1\cdot1 \times 10^{-36}\,\mathrm{m}$
2 $10^7\,\mathrm{m\,s^{-1}}, 136\,\mathrm{V}$
3 (i) $4\cdot6 \times 10^{-11}\,\mathrm{m}$ (ii) $1\cdot4 \times 10^7\,\mathrm{m\,s^{-1}}$
(iii) $450\,\mathrm{V}$
4 (i) $1\cdot2 \times 10^{-10}\,\mathrm{m}$ (ii) $5\cdot8 \times 10^{-11}\,\mathrm{m}$
6 $2\cdot2 \times 10^{-24}\,\mathrm{N\,s}, 6\cdot7 \times 10^{-33}\,\mathrm{kg}$,
$6 \times 10^{-16}\,\mathrm{J}$
7 $1\cdot7 \times 10^{-11}\,\mathrm{m}$
8 (i) $10^4\,\mathrm{eV}$ (ii) $1\cdot23 \times 10^{-11}\,\mathrm{m}$
(iii) $1\cdot6 \times 10^{-15}\,\mathrm{J}, 1\cdot24 \times 10^{-10}\,\mathrm{m}$

Exercise 35A (p. 893)

1 (a) $5\cdot25 \times 10^{20}$ (b) $2\cdot1 \times 10^8\,\mathrm{J}$
2 (a) $18\cdot5 \times 10^4\,\mathrm{Bq}$ (b) $3\cdot8 \times 10^{15}$

3 $3\cdot8 \times 10^{-9}\,\mathrm{g}$
4 $0\cdot43:1$
5 $4\cdot1 \times 10^9\,\mathrm{y}, 3\cdot2 \times 10^9\,\mathrm{y}$
6 $6000\,\mathrm{cm^3}$
7 (a) $218, 84$ (b) $1\cdot08 \times 10^{-19}\,\mathrm{N\,s}$
(c) $3\cdot1 \times 10^5\,\mathrm{m\,s^{-1}}$
8 $3\cdot36 \times 10^{22}$
9 $15\,\mathrm{h}$
10 (a) $207, 82$ (b) $5\cdot9 \times 10^{-8}\,\mathrm{s^{-1}}$
(c) $7 \times 10^8\,\mathrm{J}, 1\cdot7 \times 10^{14}\,\mathrm{s^{-1}}, 41\cdot3\,\mathrm{W}$
11 $7\cdot9 \times 10^{-13}\,\mathrm{J}, 8\cdot48 \times 10^{19}\,\mathrm{Hz}$
13 (a) $1\cdot54 \times 10^{12}\,\mathrm{s^{-1}}$ (b) $6 \times 10^7\,\mathrm{J}$
14 $7\cdot14\,\mathrm{days}$
15 (a) $13\cdot3\,\mathrm{h}$ (b) $0\cdot3\,\mathrm{h}$ (c) $960, 9040\,\mathrm{min^{-1}}$
(d) $4:1$
16 (i) $3\cdot94 \times 10^{-12}\,\mathrm{s^{-1}}$ (ii) $6\cdot5 \times 10^{13}\,\mathrm{kg^{-1}}$
17 (a) $2\cdot4 \times 10^{-10}\,\mathrm{s^{-1}}, 6\cdot2 \times 10^{11}\,\mathrm{Bq}$ (b) $0\cdot2\,\mathrm{g}$

Exercise 35B (p. 915)

2 $6\cdot646 \times 10^{-27}\,\mathrm{kg}$
6 (a) $28\cdot3\,\mathrm{MeV}$ (b) $23\cdot8\,\mathrm{MeV}$
7 (a) $4\cdot23\,\mathrm{MeV}$ (b) $4\cdot16\,\mathrm{MeV}$
9 (a) (i) $55\cdot9206\,\mathrm{u}$ (ii) $8\cdot8\,\mathrm{MeV/nucleon}$
(c) (i) $^4_2\mathrm{He}$ (ii) $4\cdot0026\,\mathrm{u}$
10 $2\cdot8 \times 10^{-12}\,\mathrm{J}$
11 $121, 47, ^1_0\mathrm{n}; 9 \times 10^{13}\,\mathrm{J}, 520\,\mathrm{days}$
12 $2\cdot6 \times 10^{-14}\,\mathrm{m}$
13 $1 \times 10^{11}\,\mathrm{y}$
14 (b) 3
15 $6\cdot78, 2\cdot39\,\mathrm{MeV/nucleon}$
16 $1\cdot44 \times 10^{-13}\,\mathrm{m}$

Exercise 36

1 (b)(i) $6800\,\mathrm{Hz}$ (ii) $199\cdot6{-}199\cdot7\,\mathrm{kHz}$
(iii) $200\cdot3{-}200\cdot4\,\mathrm{kHz}$
(c)(i) $240\,\mathrm{ms}$ (ii) $28\,\mathrm{ms}$

Index

Absolute temperature, 655
zero, 625, 655
Absorption of radiation, 718, 728
radioactivity, 886
spectra, 728, 859
A.C. circuits, 382
Acceleration, 5, 123
and force, 19–23
angular, 114
in circle, 49
of charge, 861
of gravity, 7, 20, 62
Accommodation (eye), 442
Achromatic doublet, 461
Adiabatic change, 671
, sound waves, 495, 497
curves, 671
equation, 671, 750
A.f. amplifier (transistor), 798
Aerial, 481, 925
Air breakdown, 194
Air wedge fringes, 523
Alpha-particle, 879, 883
, range of, 884
, scattering of, 897
Alternating current, 355, 382
Alternators, 357
Ammeter, a.c., 382
, d.c., 249, 251
Amorphous materials, 155
Ampere, 327
balance, 329
, circuit law of, 337
Amplification, transistor, 801
Amplifier, inverting, 812
, non-inverting, 813
, transistor, 798
Amplitude, 77, 465, 584
, a.c., 356, 383
Analogue electronics, 809
AND gate, 829
ANDREWS' experiments, 740
Angular acceleration, 114–20
magnification, 443, 451, 456
momentum, 114
speed, 48
Annealing, 162
Antinode, 479–83
Aperture, 459
ARCHIMEDES' principle, 107
Armature, 361–3
Artificial disintegration, 898, 909
Astable circuit, 818
Astronomical telescope, 443
Atmospheric pressure, 106
Atomic mass, 898
nucleus, 898
number, 898
structure, 898
unit, 904
Audio frequency, 572

AVOGADRO constant, 147, 659, 757

B (flux density), 307, 332, 345
, measurement of, 326, 379
Back-e.m.f. of induction, 370
of motor, 363
BAINBRIDGE spectrometer, 901
Ballistic galvanometer, 216, 378
Balmer series, 863
Band spectra, 375, 858
Bands, conduction, 786
, valence, 786
Banking of track, 52
Bar, the, 106
Barometer, 106
Barrier p.d., 791
Base (transistor), 794
Beats, 573
BECQUEREL, 879, 888
BERNOULLI's principle, 126
Beta-particles, 879
Bicycle rider, 54
Binary counter, 834
digits, 833, 837
Binding energy, 904
BIOT and SAVART law, 332
Bistable, 834
Black body, 719
radiation, 720
Blooming, lens, 530
BOHR's theory, 861
Boiling-point, 662
Bolometer, 715
BOLTZMANN constant, 687
equation, 753
Bonds, 151, 155
Boundary, crystal, 158
BOYLE's law, 651
BOYS, G, 61
BRAGG's law, 869
Breakdown potential, 194
Break strain, 150
stress, 135, 160
BREWSTER's law, 564
Bridge rectifier, 792
Brittle material, 135, 160
Brownian motion, 147
Bulk modulus, 494

Calibration of thermometer, 620
voltmeter, 285
Calorimeter, 636, 641
Camera lens, 458
Capacitance, 215
, measurement of large, 216
, measurement of small, 216
Capacitor, charging of, 212, 232
, discharging of, 213, 231

Capacitor, electrolytic, 218
mica, 212
paper, 212
parallel-plate, 219
variable, 218
Capacitors in parallel, 225
in series, 225
Carbon dating, 889
Carnot cycle, 675, 751
efficiency, 752
refrigeration, 752
CASSEGRAIN, 449
Cathode-ray oscilloscope, 778
Cathode rays, 762
Cells, series and parallel, 264
CELSIUS temperature scale, 621
Central forces, 119
Centre of gravity, 103
of mass, 102
Centripetal force, 51
CHADWICK, 899
Chain reaction, 910
Characteristics of sound, 572
Charge carriers, 318, 787
Charge on conductor, 184, 227
on electron, 757
CHARLES' law, 653
Chromatic aberration, 449, 461
Circle, motion in, 48
Circular coil, 303, 330
Cladding, fibre, 429
Clenched fist rule, 304
Closed pipe, 594
Cloud chamber, 891
COCKCROFT-WALTON, 909
Coefficient of viscosity, 109
Coherent sources, 516
Cold rolling, 162
Collector (transistor), 794
Collimator, 427
Colours of sunlight, 427
of thin films, 531
Common-base circuit, 794
-emitter amplifier, 797
characteristics, 795
power gain, 797
voltage gain, 797
Comparison of e.m.f.s., 280
of resistance, 286
Components of force, 14
of velocity, 15
resolved, 13
Composite materials, 164
Compound microscope, 456
Concave mirror, 413
Concentric spherical
capacitor, 220
Conductance, 244, 700
Conduction in metals, 242, 706

through gases (electrical), 253
Conduction, thermal 696
 of bad conductor, 704
 of good conductor, 703
Conductivity, electrical, 700
Conductors, 177
Conical pendulum, 53
Conservation of energy, 40, 622
 of momentum, 26, 117
Conservative forces, 39
Constant pressure experiment, 653
 volume gas scale, 625
 thermometer, 625
Constructive interference, 516
Continuous flow calorimeter, 637
 spectra, 427
Converging lens, 435
Cooling, correction for, 642
 NEWTON's law of, 644
Core, fibre, 429
Corkscrew rule, 303
Corona discharge, 194
Corpuscular theory, 505
COUDÉ, 449
Coulomb, the, 187
Counter, binary, 834
Couple on coil, 100, 309
Couples and work, 121
Covalent bond, 151
Cracks, 163
Creep, 164
Critical angle, 419, 430
 mass, 912
 temperature (gas), 741
Crystal diffraction, 868
Crystalline material, 155
Current balance, 329
 gain, 796
Current unit, 328
Current, potentiometer, 285
Curvature, 507
Curved mirror, 413
Cut-off, 800

DALTON's law of partial pressures, 662
Damped oscillation, 361, 468
Damping of galvanometer, 362
D.c. amplifier, 800
DE BROGLIE's law, 876
Decay constant, 887
 series, 900
Deceleration, 5
Deformation, elastic, 159
 , plastic, 160
Degrees of freedom, 733
Dekatron counter, 881
Density, 107
 of earth, 68
Depletion layer, 791
Depth of field, 460
Destructive interference, 516
Deuterium, 913

Deviations from gas laws, 742–7
Dielectric, 212, 222
 constant, 221
 polarization, 222
 strength, 221
Diffraction at lens, 543
 at slit, 486, 539–43
 , electron, 875
 , microwave, 490
 , X-ray, 869
Diffraction grating, 549–53
Diffusion, 688
 cloud chamber, 892
Digital electronics, 828
Dimensions, 41
 applications of, 42–4
Diode, junction, 791
Dioptre, 508
Dipole (electric), 222
Disc generator, 351
Discharge, capacitor, 213, 291
Disintegration, nuclear, 899, 908
Dislocation, 157
Dispersion, fibre, 431
 , lens, 461
 , prism, 426, 501
Displacement (waves), 465, 472
Distance-time graph, 8
Diverging lens, 435
Division of amplitude, 524
 wavefront, 519
DOPPLER effect, 575–82
Double refraction, 564
Drift velocity, 243
Ductile material, 135, 160
Dust-tube experiment, 603
Dynamo, 355

Earth, density of, 68
 , escape velocity, 70
 mass of, 68
 potential of, 68, 199
Earth's horizontal component, 326
 magnetic field, 350
 vertical component, 350
Eddy currents, 361
Efficiency (electrical), 268
EINSTEIN's mass law, 903
 photon theory, 846
Elastic collisions, 24, 30
 deformation, 159
 limit, 134
Elasticity, 133
 , modulus of, 136, 194
Electric field, 190, 764
 flux, 191
 potential, 196
 strength (intensity), 190
Electrical calorimetry, 636, 638
 symbols, 244
Electrolyte, 253
Electromagnetic induction, 342
 waves, 493

Electromagnetism, 302
Electron charge, 178, 757
 diffraction, 875
 , e/m_e, 767, 771
 lens, 780
 mass, 770
 motion, 242, 764
 orbit, 482, 861
 shells, 871
 -volt, 197
Electrons, 178, 757
Electroscope, 179
Electrostatic fields, 190
 intensity, 190–3
 shielding, 194
Emitter (transistor), 794
E.M.F., 261, 266
 , determination, 281
Emission spectra, 427, 857
End-correction, 598
Energy and matter, 896, 903
Energy bands in solids, 786
Energy, electrical, 258
 exchanges, 466
 gravitational, 71
 in capacitor, 228, 466
 in coil, 372
 levles (atom), 855
 light, 411
 mechanical, 35
 nuclear, 905
 shells, 871
 sound, 584
Energy in wire, 139
 , kinetic, 37
 , potential, 39
 rotational, 121, 733
 translational, 732
 vibrational, 81.
Enthalpy, 679
Entropy, 752–4
EOR gate, 832
Equation of state, 658
Equations of motion, 6
Equilibrium, conditions of, 94, 97, 101
Equipotential, 205
Errors, counting, 882
Escape velocity, 70
Evaporation, 663
Excitation potential, 857
Expansion, linear, 131
 of gas, 655
Explosive forces, 311
External work, 668, 674
Extraordinary ray, 565
Extrinsic semiconductor, 789
Eye, 442
Eye-ring, 446

Falling sphere, viscosity, 109
Far point of eye, 440
Farad, 215
Faraday constant, 757
FARADAY's ice-pail experiment, 183
 law of induction, 345
Feedback, energy, 811, 818
FERMAT principle, 412

Field of view of microscope, 457
of telescope, 446
Filter circuit, 793
pump, 127
Fine beam tube, 768
Fission, nuclear, 910
Fixed points, 625
FLEMING's left-hand rule, 305
right-hand rule, 348
Flip-flop, SR, 836
, T, 837
Flotation, 108
Fluid, 104
motion, 126
Flux, electric, 191
, magnetic, 345
Flux density, B, 307, 324, 332, 379
-linkages, 346, 378
f-number, 460
Focal length of lenses, 436
Focusing electron, 779
Force, 19–25, 94, 104
between currents, 327
constant, 83
due to expansion, 139
on charges, 187, 315
on conductor, 305
Forced oscillation, 470
Fossil fuels, 918
FRAUNHOFFER's lines, 859
Free fall, 7, 62
Free oscillation, 470
Frequency, 465
by beats, 573
by resonance tube, 602
by sonometer, 612
fundamental, 594, 596, 609
Friction, 21–2
Full-adder, 834
Fusion, nuclear, 913
(solid), 646

GABOR, 554
Galvanometer, 312
, sensitivity of, 313
Gamma-rays, 885
, inverse-square law, 885
, range, 885
Gas, 651
constant, 658
equation, 257
, ideal, 668, 673
laws, 651–63
, mole of, 659
, real, 740
thermometer, 625
velocity of sound in, 495
Gases, diffusion of, 668
Gaseous state, 651–63
GAUSS's theorem, 192
GEIGER and MARSDEN, 897
-MÜLLER (GM) tube, 880
Generators, 354, 359
Germanium, 787
Glassy material, 162, 169
Gold-leaf electroscope, 179
Graded index, 430

GRAHAM's law, 688
Grains, crystal, 158
Gravitation, law of, 60
Gravitational constant, 61
field strength (intensity), 20, 63
mass, 68
orbits, 59, 65–7
potential, 68, 71
Gravity, acceleration of, 7, 62
motion under, 7
Ground state, 855

HALE telescope, 447
Half-adder, 833
Half-life, 887
Hall probe, 324
voltage, 317–9
Hardness, 163
Harmonics of pipes, 595, 597
of strings, 609
Heat capacity, 635, 679
energy, 622
engine, 674, 751
losses, 638, 641
Heating effect of current, 257
Heavy hydrogen, 913
Helium, 884, 913
HELMHOLTZ coils, 331, 334, 769
Henry, the, 371
High tension transmission, 259
Holes, 788
Holography, 554
Hollow conductor, 183, 192
Homopolar generator, 351
HOOKE's law, 134
Hot-wire ammeter, 382
HUYGEN's construction, 502
Hydrogen atom, 855, 863
isotopes, 913
Hydrogen gas thermometer, 626
Hysteresis, rubber, 169

Ice pail experiment, 183
point, 620
Ideal gas, 657, 740
Images in lenses, 436
Impedance, 393
Imperfections, 187
Impulse, 23
Impurity, metal, 255
, semiconductor, 789
Induced charge, 180, 184
current, 342
e.m.f., 345, 370
Inductance, 371, 389
Induction, electrostatic, 180
Inelastic collisions, 30
Inertia, 18
, moments of, 114
Infrared radiation, 713
spectrometer, 717
Insulators, 177, 787
Integrator circuit, 822
Intensity, electric, 190
Intensity of sound, 584

Interaction, fields, 308
Interference (light), 515
in thin film, 531
in wedge films, 523
of waves, 486, 574
Intermolecular forces, 148, 745
energy, 148, 666
Internal energy of gas, 667, 732, 734
work, 667, 744
Internal resistance of cell, 261, 284
International temperature scale, 621
Inverse-square law (electrostatics), 187
(radiation), 716
Ionic bond, 151
Ionisation, current, 888
potential, 857
Ions, 855
Isothermal change, 669–70
curves, 670, 746
sound waves, 495
work, 750
Isotopes, 901

JOULE-KELVIN effect, 744
Joule, the, 35
JOULE heating, 243, 257
Junction diode, 791

KELVIN scale, 620, 655
KEPLER's laws, 59
Kilowatt, 36
Kilowatt-hour, 260
Kinetic energy, 37
theory of gases, 683, 732
KIRCHHOFF's laws (electricity), 270
(radiation), 727

Laminations, 361
LAPLACE's correction, 495
Laser, 554, 860
Latent heat, 644–6
of evaporation, 150, 644
of fusion, 646
Lateral magnification, 438
Laue diffraction, 868
Leakage current, 799
LEES' disc method, 704
Lens, 435
formula, 438
of eye, 442
LENZ's law, 343
Light dependent resistor, 852
emitting diode (LED), 924
Line spectra, 427, 857
Linear expansivity, 139
Lines of force, electric, 190–4
, magnetic, 302–5
Liquefaction of gases, 742
LISSAJOUS' figures, 89, 489, 782
LLOYD's mirror, 531
Logic gates, 828
Longitudinal waves, 472

LORENZ method, resistance, 353
Loudness, 581
Lyman series, 863

Magnetic fields, 301–5, 323, 766
  flux, 345
  flux-density, 307, 324
  circular coil, 330
  solenoid, 304, 323
  straight wire, 303, 336
  materials, 301
Magnification, angular, 443, 451
  lateral (linear), 438
  of microscope, 451
  of telescope, 444
Magnifying glass, 432
Mains frequency, 355
Majority carriers, 789
Mass, 19
Mass-energy relation, 903
  number, 898
  spectrometers, 901
Maximum power, 268
MAXWELL, 303, 691
  distribution law, 691
Mean free path, 736
Mean square velocity, 684
  velocity, 686, 692
Megelectron-volt, 904
Mer, 165
Mercury thermometer, 619
Metals, 133
Metal conductor, 214, 703
  fatigue, 164
Metallic bond, 151
Metre bridge, 292
Mica capacitor, 212
MICHELSON's method, 510
Microfarad, 215
Miscroscope, compound, 456
  , simple, 450
Microwaves, 487, 490
MILLIKAN's experiment, e, 759
  , photoelectric, 849
Minimum deviation by prism, 425
  distance, object-image, 437
Mirror galvonometer, 312
Modulation, amplitude, 926
Modulus of elasticity, bulk, 494
  YOUNG, 136–8
Molar gas constant, 658
  heat capacity, 678, 735
Mole, 659, 734
Molecular forces, 148–51
  potential energy, 148
  speeds, 685–7
Molecules, 147, 732
Moment of couple, 100
  of force, 98
Moment of inertia, 114
  of flywheels, 123
Momentum, angular, 116
  linear, 19, 22

conservation of angular, 117
  of linear, 27
Monomer, 165
Monomode fibre, 429
Moon, motion of, 60
MOSELEY's law, 870
Motion, NEWTON's laws of, 18
  in circle, 48
  in straight line, 4
  of projectile, 15
  simple harmonic, 75
  under gravity, 7
Motors, 362
Mount Palomar telescope, 447
Moving-coil ammeter, 311
  galvanometer, 312
  voltmeter, 251, 313
Multimeters, 250
Multimode fibre, 429
Multiplex system, 433
Multiplier, 249
Musical interval, 593
Mutual induction, 375

NAND gate, 829
Natural frequency, 470
Near point of eye, 442
Negative charge, 178
  feedback, 811
NEUMANN's law, 346
Neutral temperature, 274
Neutron, 899
  -proton ratio, 907
NEWTON's law of cooling, 641
  law of gravitation, 61
  laws of motion, 18
  rings, 527
  velocity (sound) formula, 495
Newton, the, 19
NICOL, prism, 565
Nodes, 478
  in air, 478
  in pipes, 494
  in strings, 609
Non-inductive resistance, 371
Non-ohmic conductors, 253, 791
NOR gate, 831
NOT gate, 828
Normal adjustment, 443
N-p-n transistor, 794
N-semiconductor, 789
Nuclear charge, 898
  forces, 905
  mass, 898
  reactions, 899, 906
  reactor, 911
  stability, 907
  structure, 899
Nucleon, 899
Nucleus, 897

Objective, microscope, 456
  , telescope, 443, 545
OERSTED, 302
Ohmic conductor, 253

OHM's law (electricity), 252
  electrolytes, 253
Oil-drop experiment, 757
Opamp, 809
  frequency-response, 817
  inverting, 810, 812
  non-inverting, 810, 813
  summing, 817
Open pipe, 596
Operational amplifer, 809
Optical fibre, 429–33, 449
  instruments, 442
  paths, 517
  spectra, 426, 857
OR gate, 830
Orbits, 70
Ordinary ray, 562
Oscillation, damped, 468
  , electric, 466
  , energy exchanges, 82
  of liquid, 88
  of spring, 811
  , square-wave, 88
  , undamped, 379
Overtones, 585
  of pipes, 595, 597
  of strings, 609

Paper capacitor, 212
Paraboloid reflector, 413
Parallel-plate capacitor, 217
Parallel a.c. circuits, 398
Parallel forces, 99
Parallelogram of forces, 95
  of velocities, 11
Parking orbit, 66
Particle, nature, 846
  -wave duality, 877
Pascal, the, 105
Peak value, 383
Pendulum, simple, 86
Period, 77
Permeability, 325
  relative, 325
  , vacuum, 324
Permittivity, 188, 223
  relative, 188
  , vacuum, 187
PERRIN tube, 762
Phase angle, 386–99, 467, 518
Phasor, 390, 467
Phonon, 706
Photon, 846
Photo-cell, 848, 851
Photodiode, 924
Photoelasticity, 568
Photoelectricity, 845
Pipe, wave in, 591
Pitch, 573
Pitot-static tube, 129
PLANCK constant, 845
Plane conductor, 193
  mirror, 412
  polarisation, 561
Plane-progressive wave, 473–7
Planetary motion, 599
Plasma, 582
Plastic deformation, 159
  materials, 160

Platinum resistance thermometer, 629
P-n junction, 790
P-n-p transistor, 795
Points, action, 181
Polar molecules, 223
Polarisation (e.m. wave), 492
  (light), 561
  and electric vector, 566
  by double refraction, 564
  by Polaroid, 561, 567
  by reflection, 563
Polarisation of dielectric, 222
Polarising angle, 564
Polycrystalline, 155
Polygon of forces, 97
Polymerisation, 168
Polymers, 165
  , molecules, 165
    branched, 166
    cross-linked, 166
    linear, 166
Polythene material, 165
Positive charge, 178
  feedback, 818
  ion, 901
Potential difference, 196, 258
  divider, 246
Potential, electric, 196–202
Potential energy (mechanical), 39
  (molecular), 148
  gradient, 203
  gravitational, 39, 68
Potentiometer, 280–90
Power (electrical), 258, 268, 919
  (a.c.), 399
  , mechanical, 36
  of lens, 507
P-semiconductor, 790
Pressure, atmospheric, 106
  curves, (sound), 472, 480
  , gas, 683
  liquid, 104–7
  standard, 106
PRÉVOST's theory, 724–7
Principal focus, 435
  section, 565
Principle of Superposition, 476
Prism, 422
  deviation by, 423–6
  minimum deviation by, 426
Probability and entropy, 753
Progressive wave, 473
Projectiles, 14
Proportional limit, 134
Proton, 898
  number, 898
  -neutron ratio, 907
P-semiconductor, 790
Pulse voltage, 233, 880, 928

Quality of sound, 585
Quantity of charge, 178, 213, 378
  of heat, 635
Quantisation of energy, 855

Quantum theory, 845
  of light, 845
Quark, 909
Quenching agent, 880

Radial field, 312
Radiation (thermal), 712, 718
  laws, 723
  wavelengths, 713
Radioactivity, 879
Radio telescope, 546
Radium, 900
Range (radioactivity), 885
Rarefaction, 472
Ratemeter, 882
Ratio of heat capacities, 735
Reactance of capacitor, 387
  inductor, 390
Reaction, 25
Reactor, 911
Real gases, 740, 743
  image, 412, 437
Rectification, 791–3
Rectilear propagation, 543
Red shift, 581
Reflection of light, 503
  of sound, 484
Reflector telescope, 447
Refraction at plane surface, 415
  through prism, 422
  wave theory of, 504
Refraction of sound, 484
Refractive index, 415
Refractor telescope, 443
Relative permeability, 325
  permittivity, 188, 221, 224
  velocity, 12
Resistance, 243
  absolute method, 353
  low, 287
Resistance thermometer, 629
Resistances in parallel, 245
  in series, 244
Resistivity, 254
Resolution of forces, 95
Resolving power, radio telescope, 544
  , telescope, 447, 544
Resonance (sound), 470, 599
  tube, 601
Resonance, series, 395
Resonant frequency, 396, 599
Retina, 442
Reversibility of light, 412
Reversible change, 671
Rigid body, motion of, 114–24
ROENTGEN, 867
Rolling object, 122
Root-mean-square current, 383
  speed, 685
Rotational dynamics, 114
Rotational energy, 121, 733
Rubber molecules, 160, 169
RUTHERFORD, 884, 897
  and ROYDS, 884

Saccharimetry, 568
Satellites, 59, 62, 71
Saturated vapours, 662
Saturation, 800
Scalars, 4
Scaler, 881
Scales of temperature, 619–21
Scattering law, 897
Screening, 194
Search coil, 326
SEARLE's method, 704
Secondary coil, 358
  electrons, 880
Selenium cell, 851
Self-induction, 370
Semiconductors, 787
Sensitivity of meter, 313
Series a.c. circuits, 393
Series-wound motors, 364
Shells, energy, 871
Shunt, 249
  -wound motors, 364
Sidebands, 926
Siemens, the, 244
Silicon, 787
Simple harmonic motion, 75, 465
  pendulum, 86
Slide-wire bridge, 292
Slip in metals, 161
Solar constant, 712
  energy, 712, 918
Solenoid, 304, 323
Solid state, 786
  detector, 881
Sonometer, 611
Sound, speed in, 474
Sound waves, 484–8
Spark counter, 883
Specific heat capacity, 635
  , constant pressure, 678
  , constant volume, 678
  of liquid, 637
  of solid, 636
  of water, 639
Spectra, 553, 858
  , hydrogen, 856, 863
Spectrometer, 427
Spectrum, hydrogen, 857, 863
  sunlight, 858
Speed, light, 509
  of sound, 474
  in circle, 48
Sphere capacitance, 219
  , field due to, 192
Spherical aberration, lenses, 462
  mirrors, 413
Spiral, electron, 768
Spontaneous emission, 860
Spring helical, 81, 83–6
Stability, nuclear, 907
Standard cell, 284
Starting resistance, motor, 363
Static bodies. 94
Stationary (standing) waves, 477
  aerial, 481

, electron orbit, 482
, light, 480
pipe, 591
sound, 592
string, 591, 608
Steam point, 619
Step-index fibre, 430
STEFAN constant, 723
law, 723
Stiffness, 160
Stimulated emission, 860
STOKES' law, 109
Stopping potential, 847
Straight conductor, 303, 325, 347
Strain, tensile, 136
Strength of materials, 160
Stress, tensile, 136, 163, 168
Stretched wire energy, 139
Strings, harmonics in, 611
resonance in, 609
, waves in, 608
Sun, mass of, 68
, radiation of, 712
Superposition principle, 476
Surface density charge, 185
Switching circuit, 803, 821

Telescope, astronomical, 443
, radio-, 449
, reflector, 447
Temperature, 619, 625
coefficient, 255, 293
gradient, 697-9
Tensile strain, 136
stress, 136
Terminal p.d., 261-4
velocity, 109
Terrestrial magnetism, 350
Tesla (T), the, 307
Thermal capacity, 635
conductivity, 697
expansion, 139
insulators, 701
Thermionic emission, 761
Thermistor, 294
Thermocouple, 273, 287, 630
Thermodynamic scale, 620
Thermodynamics, 666
first law, 667
second law, 675
Thermoelectric thermometer, 630
Thermoelectricity, 273
Thermometers, 626-31
Thermonuclear reaction, 914
Thermopile, 715
Thermoplastics, 167
Thermosets, 167
THOMSON experiment, e/m_c, 771
Threshold wavelength, 845
Tidal power, 922
Time constant, 232
Time-base (C.R.O.), 778

Toroid, 323
Torque, 62
on coil, 309
rotating body, 44
Torsion, 62, 312
Torsional oscillation, 62
Total internal reflection, 419
Toughness, 163
Transformers, 358
Transistor, 794
amplifier, 797-802
switch, 803
Translational energy, 686
Transmutation, 899
Tranverse waves, 471
Triangle of forces, 97
Triple point, 620
Tritium, 913
Tuning, 396

Ultraviolet rays, 713
Ultrasonics, 572
, medical use, 573
Uniform acceleration, 6
velocity, 4
Unsaturated vapour, 662
Upthrust, fluid, 107
Uranium, 879
fission, 910
U value, 698

Valence electron, 787
VAN DE GRAAF generator, 181
VAN DER WAALS' bond, 151
equation, 744
Vapour pressure, 662
Variable capacitor, 218
Variation of g, 62
Vectors, 4, 11
Velocities, addition of, 11
, subtraction of, 12
Velocity, angular, 48
relative, 12
selector, 772, 902
terminal, 109
-time graph, 9
, uniform, 4
Velocity of light, 509
FOUCAULT, 509
MICHELSON, 510
ROEMER, 509
Velocity of sound, 488, 494
in air, 489
in gas, 493
in pipe, 602
in rod, 493
in string, 608, 613
Venturi meter, 128
Vibrating reed switch, 214, 223
Vibrations, forced, 470, 599
in pipes, 601
in strings, 608
longitudinal, 472
resonant, 470, 599

transverse, 471, 591
Virtual image, 412
object, 413, 439
Viscosity, 109
falling sphere and, 109
, gas, 692, 737
Visual angle, 442
Volt, 244
Voltage amplification, 797, 809
comparator, 821
follower, 816
gain, 809, 817
Voltmeter, moving coil, 247, 251, 313
Volume expansivity, gas, 655

Watt, 36
Wattmeter, 314
Wave, 470
equation, 475
in pipes, 594-604
in strings, 608-14
nature (particle), 875-7
power, 922
properties, 471-9
theory of light, 501
velocity, 474
Wavefront, 501
and lens, 508
Wavelength of light, 521, 552
of sound, 471
Weber, the, 346
Weight, 19
Weightlessness, 64
WESTON cadmium cell, 284
WHEATSTONE bridge, 291
WIEDEMANN-FRANZ law, 706
WEIN's law of radiation, 723
WILSON cloud chamber, 892
Wind power, 919-22
Wood, 170
Work, 35
done by torque, 121
done by gas, 666, 669, 675, 750
hardening, 162
in stretching wire, 139
Work function, 846

X-rays, 867
, diffraction of, 868
spectra, 870, 872

Yield point, 135
YOUNG's fringes, 519-23
, modulus, 136-8
, determination of, 137

Zener diode, 793
Zero potential (atom), 855
(electrical), 199
(gravitational), 69
Zeroth law, 622